STANDARD HANDBOOK OF POWERPLANT ENGINEERING

Other McGraw-Hill Reference Books of Interest

Avalone and Baumeister • MARKS' STANDARD HANDBOOK FOR MECHANICAL ENGINEERS
Brady and Clauser • MATERIALS HANDBOOK
Callender • TIME-SAVER STANDARDS FOR ARCHITECTURAL DESIGN DATA
Chuse and Eber • PRESSURE VESSELS
Crocker and King • PIPING HANDBOOK
Croft and Summers • AMERICAN ELECTRICIAN'S HANDBOOK
Dudley • GEAR HANDBOOK
Fink • ELECTRONICS ENGINEER'S HANDBOOK
Fink and Beaty • STANDARD HANDBOOK FOR ELECTRICAL ENGINEERS
Harris • HANDBOOK OF NOISE CONTROL
Harris and Crede • SHOCK AND VIBRATION HANDBOOK
Hicks • STANDARD HANDBOOK OF ENGINEERING CALCULATIONS
Hicks and Mueller • STANDARD HANDBOOK OF PROFESSIONAL CONSULTING ENGINEERING
Higgins • MAINTENANCE ENGINEERING HANDBOOK
Juran • QUALITY CONTROL HANDBOOK
Karassik et al. • PUMP HANDBOOK
Kardestuncer • FINITE ELEMENT HANDBOOK
Kohan • PRESSURE VESSEL SYSTEMS
Koren • ROBOTICS FOR ENGINEERS
Kurtz • HANDBOOK OF ENGINEERING ECONOMICS
Maynard • INDUSTRIAL ENGINEERING HANDBOOK
Merritt • STANDARD HANDBOOK FOR CIVIL ENGINEERS
Parmley • MECHANICAL COMPONENTS HANDBOOK
Parmley • STANDARD HANDBOOK OF FASTENING AND JOINING
Perry and Green • CHEMICAL ENGINEERS' HANDBOOK
Rohsenow et al. • HANDBOOK OF HEAT TRANSFER APPLICATIONS
Rohsenow et al. • HANDBOOK OF HEAT TRANSFER FUNDAMENTALS
Rothbart • MECHANICAL DESIGN AND SYSTEMS HANDBOOK
Sandler and Luckiewicz • PRACTICAL PROCESS ENGINEERING
Schwartz • COMPOSITE MATERIALS HANDBOOK
Shigley and Mischke • STANDARD HANDBOOK OF MACHINE DESIGN
Smeaton • MOTOR APPLICATION AND MAINTENANCE HANDBOOK
Smeaton • SWITCHGEAR AND CONTROL HANDBOOK
Smith and Van Laan • PIPING AND PIPE SUPPORT SYSTEMS
Teicholz • CAD/CAM HANDBOOK
Teicholz and Orr • COMPUTER-INTEGRATED MANUFACTURING HANDBOOK
Tuma • ENGINEERING MATHEMATICS HANDBOOK
Woodruff, Lammers, and Lammers • STEAM-PLANT OPERATIONS

Encyclopedias and Dictionaries

CONCISE ENCYCLOPEDIA OF SCIENCE AND TECHNOLOGY
DICTIONARY OF MECHANICAL AND DESIGN ENGINEERING
DICTIONARY OF SCIENCE AND ENGINEERING

STANDARD HANDBOOK OF POWERPLANT ENGINEERING

Thomas C. Elliott

and the Editors of *Power* Magazine

McGRAW-HILL PUBLISHING COMPANY

New York St. Louis San Francisco Auckland Bogotá
Caracas Colorado Springs Hamburg Lisbon
London Madrid Mexico Milan Montreal
New Delhi Oklahoma City Panama Paris
San Juan São Paulo Singapore
Sydney Tokyo Toronto

Library of Congress Cataloging-in-Publication Data

Standard handbook of powerplant engineering.

Includes index.
1. Electric power plants. I. Elliott, Thomas C.
II. Power.
TK1191.S686 1989 621.31'21 88-8455
ISBN 0-07-019106-9

1234567890 DOC/DOC 8954321098

ISBN 0-07-019106-9

The editors for this book were Harold B. Crawford and Jim Halston and the production supervisor was Richard A. Ausburn. This book was set in Times Roman. It was composed by the McGraw-Hill Publishing Company Professional & Reference Division composition unit.

Printed and bound by R. R. Donnelley and Sons

CONTENTS

CONTRIBUTORS

(*Current situation is shown in parentheses, if changed from time of writing*)

James A. Baumbach *Nalco Chemical Co., Oak Brook, IL* [SEC. 4, PART 2, CHAP. 4]

Roger B. Bloomfield, P.E. *Bloomfield Associates, PC, Concord, NH* [SEC. 3, CHAP. 5]

J.D. Blue *Babcock & Wilcox Co., Barberton, OH* [SEC. 3, CHAP. 6]

Robert D. Bruce *Hoover Keith & Bruce Inc., Houston, TX* [SEC. 4, PART 3, CHAP. 1]

Peter M. Coates, President *CAM Industries, Inc., Kent, WA* [SEC. 1, CHAP. 5]

Albert F. Duzy, President *Duzy & Associates, Sun Lakes, AZ* [SEC. 3, CHAPS. 1 AND 2]

William Ellison, P.E., Director *Ellison Consultants, Monrovia, MD* [SEC. 4, PART 1, CHAP. 5]

Carlyle Esser *Harza Engineering Co., Chicago, IL (retired)* [SEC. 2, CHAP. 6]

David R. Gibbs *Babcock & Wilcox Co., Barberton, OH* [SEC. 3, CHAP. 6]

Dr. H. Haselbacher *Asea Brown Boveri Ltd., Baden, Switzerland (Univ. of Vienna)* [SEC. 2, CHAPS. 4 AND 5]

John C. Hensley, Marketing Services Manager *Marley Cooling Tower Co., Mission, KS* [SEC. 1, CHAP. 9]

Martin H. Hofheinz *H & G Engineering, Inc., Stockton, CA* [SEC. 5, CHAP. 5]

Robert M. Hoover *Hoover Keith & Bruce Inc., Houston, TX* [SEC. 4, PART 3, CHAP. 1]

Robert J. Jonardi *NUS Corp, Gaithersburg, MD (Dynamac Corp., Rockville, MD)* [SEC. 4, PART 1, CHAP. 1]

Neil H. Johnson *Detroit Stoker Co., Monroe, MI* [SEC. 1, CHAP. 8]

Roger M. Jorden, President *Water Management Associates, Steamboat Springs, CO* [SEC. 4, PART 2, CHAP. 5]

William M. Kauffmann, P.E., Consultant *Parma, OH (deceased)* [SEC. 2, CHAP. 3]

James D. Kilgroe *Environmental Protection Agency, Research Triangle Park, NC* [SEC. 4, PART 1, CHAP. 2]

Ronald Allen Knief, Staff Consultant *Nuclear Safety, GPU Nuclear Corp., Middletown, PA* [SEC. 1, CHAP. 6]

Harold L. Knox, President *Detroit Stoker Co., Monroe, MI (retired)* [SEC. 1, CHAP. 8]

Theodore C. Koss, Consulting Atmospheric Scientist *NUS Corp., Gaithersburg, MD* [SEC. 4, PART 1, CHAP. 1]

Leo A. Kriedler *Babcock & Wilcox Co., Barberton, OH* [SEC. 3, CHAP. 6]

R.W. Kronenberger *Babcock & Wilcox Co., Barberton, OH* [SEC. 3, CHAP. 6]

Ralph H. Lee, Consultant *Wilmington, DE (deceased)* [SEC. 5, CHAP. 3]

Leonard J. Lefevre *Larkin Laboratory, Dow Chemical Co., Midland, MI (retired/consultant)* [SEC. 4, PART 2, CHAP. 3]

Thomas W. Montgomery *Babcock & Wilcox Co., Barberton, OH* [SEC. 3, CHAP. 6]

J.A. Moore *Leeds & Northrup, a Unit of General Signal Corp., North Wales, PA (consultant)* [SEC. 6, CHAPS. 1, 2, 4, AND 5]

R.L. Nailen, Project Engineer *Wisconsin Electric Power Co., Milwaukee, WI* [SEC. 5, CHAP. 4]

Michael P. Polsky, President *Indeck Energy Services, Inc., Wheeling, IL* [SEC. 2, CHAP. 7]

James K. Rice, Consultant *Olney, MD* [SEC. 4, PART 2, CHAP. 1]

Douglas M. Rode *C-E Power Systems, Combustion Engineering, Inc., Windsor, CT* [SEC. 4, PART 1, CHAP. 6]

Earl H. Rothfuss *Baker Engineers, Beaver, PA (Remcor Inc., Pittsburgh, PA)* [SEC. 4, PART 2, CHAP. 6]

Kevin J. Shields *Sheppard T. Powell Associates, Baltimore MD* [SEC. 1, CHAP. 10]

John W. Siegmund *Sheppard T. Powell Associates, Baltimore, MD* [SEC. 1, CHAP. 10]

George J. Silvestri, Jr. *Power Generation Operations Div., Westinghouse Electric Corp., Orlando, FL* [SEC. 2, CHAP. 2]

Joseph G. Singer *Combustion Engineering Inc., Windsor, CT (consultant)* [SEC. 1, CHAPS. 1, 2, 4, AND 7]

Jerry Strauss *Environmental Protection Agency, Research Triangle Park, NC* [SEC. 4, PART 1, CHAP. 2]

J.R. Strempek *Babcock & Wilcox Co., Barberton, OH* [SEC. 3, CHAP. 6]

James H.T. Sun, Associate and Head *Hydraulic Machinery Section, Harza Engineering Co., Chicago, IL* [SEC. 2, CHAP. 6]

Russell A. Wills *Detroit Stoker Co., Monroe, MI* [SEC. 1, CHAP. 8]

PREFACE

The powerplant is largely a twentieth century phenomenon. Before 1900, power had to be used where it was made because there was no effective means of transmitting energy over long distances. Edison's pioneer central station, a late nineteenth century development, sparked the revolution (if you will) that has given us the incredibly versatile powerplant that we see today in countries around the world. The late twentieth century is the recipient of a century of power progress, a legacy that is distilled in these pages.

The *Standard Handbook of Powerplant Engineering* covers in considerable detail the systems and equipment comprising the modern industrial/utility powerplant. Although the emphasis throughout is on hardware, some theory and design information are included where it will help the reader to better understand how a powerplant works. Today's plant is a diverse mix of technologies, reflecting in no small measure the broad issues facing society worldwide. And so, pollution control, new schemes for producing energy and conserving water, and the "computer on a chip" adaptation to plant needs are covered integrally with such basic "power train" topics as steam generation, prime movers, and electrical interconnections.

It is the widening base of knowledge which those interested in powerplants must have that stamps this handbook with its special character. No other single reference work offers the in-depth coverage of such a broad field. Experts in steam generation may be woefully lacking in pollution control savvy, and vice versa, yet understanding of both areas is important. Although expertise in all facets of plant operation isn't necessary (or even possible), an appreciation of all the elements is highly desirable.

The handbook is designed to appeal to engineers in electric utilities, process and manufacturing industries, commercial and institutional buildings, plus consulting firms in power technology (those responsible for converting raw fuel into usable energy). These pages should also appeal to engineering students, engineers with vendors and service companies catering to the power field, and officials with energy and environmental agencies at all levels of government.

Handbook architecture starts with basics, *steam fundamentals,* and ends with the leading-edge *man/machine interface*. The six sections embrace steam generation, turbines and diesels, fuels and fuel handling, pollution control (air, water, noise), plant electric systems, and instrumentation and control. The 46 chapters within these sections describe the important equipment and systems found in the modern powerplant, with some chapters focusing on such signposts of the times as cogeneration, waste-to-energy fuels, coal cleaning, solid-waste disposal, zero-discharge systems, and pollution legislation.

My heartfelt thanks go to all the contributors—busy engineers, experts in their fields, willing communicators all—who took the time and energy to prepare chapters for the handbook. I am especially grateful to Joseph G. Singer, Combustion Engineering, and J.A. Moore, Leeds & Northrup, who each contributed four chapters;

and to Albert F. Duzy, Duzy & Associates, and Dr. H. Haselbacher, Asea Brown Boveri Ltd. (now a professor at the University of Vienna), who each contributed two chapters. Their multi-chapter efforts are appreciated, to put it mildly.

The editorial staff of *Power* magazine—Robert G. Schwieger, Sheldon D. Strauss, Jason Makansi, and John Reason—should also stand up and be recognized for helping with the organization of the handbook, providing candidate contributors, and preparing chapters based on special reports that appeared in the magazine earlier. These chapters, in my opinion, are equal in quality and readability to all other handbook chapters. I'd also like to thank my wife, Carol, for her understanding and support during this six-year odyssey.

The handbook has been metricated: in the text, SI (*Système International d'Unités*) appears in parentheses after English units; in the drawings, SI either appears after English units or is accounted for with conversion factors. As a convenience, the latter are listed on page xiii. A list of abbreviations of societies, associations, and terms common to the power field appears on p. xi.

THOMAS C. ELLIOTT
New York, NY

ABBREVIATIONS

[Associations, societies, institutes, standards, agencies, etc., referred to in the text. Abbreviations and definitions are also listed on pages 2.66 (Oil and Gas Engines), 4.5 (Clean Air Act Legislation and Regulated Pollutants), 4.32 (Coal Cleaning), 4.143 (Legislation and Water Pollutant Sources), 4.207 (Ion Exchangers), and 4.298 (Noise Control in the Powerplant). See Index for other abbreviations.]

ABMA	American Boiler Manufacturers Association
ACI	American Concrete Institute
ANS	American Nuclear Society
ANSI	American National Standards Institute
API	American Petroleum Institute
ASCII	American Standard Code for Information Interchange
ASME	American Society of Mechanical Engineers
ASTM	American Society for Testing and Materials
AWG	American Wire Gauge
DOE	Department of Energy
EEI	Edison Electric Institute
EIA	Electronics Industries Association
EPA	Environmental Protection Agency
EPRI	Electric Power Research Institute
FERC	Federal Energy Regulatory Commission
ICEA	Insulated Cable Engineers Association
IEEE	Institute of Electrical and Electronics Engineers
ISA	Instrument Society of America
ISO	International Standards Organization
JIC	Joint Industry Conference Electrical Standards for Industrial Equipment
MAP	Manufacturing Automation Protocol
NACA	National Advisory Committee for Aeronautics
NBS	National Bureau of Standards
NEC®	National Electrical Code®
NEMA	National Electrical Manufacturers Association
NFPA	National Fire Protection Association
NRC	Nuclear Regulatory Commission
NRDC	National Resources Defense Council
OSHA	Occupational Safety and Health Administration
PURPA	Public Utility Regulatory and Policies Act
RCRA	Resource Conservation and Recovery Act
SAMA	Scientific Apparatus Manufacturers Association
SCS	Soil Conservation Service
TEMA	Tubular Exchanger Manufacturers Association

TVA	Tennessee Valley Authority
USBR	U.S. Bureau of Reclamation
USGS	U.S. Geological Survey
UWAG	Utility Water Act Group

KEY CONVERSION FACTORS

acre × 0.405 = ha (hectare)
bar × 100 = kPa
Btu × 1.055 = kJ
Btu/ft^2 × 11.4 = kJ/m^2
Btu/ft^2 · h · °F × 5.68 = W/m^2 · °C
Btu/gal × 0.279 = kJ/L
Btu/h × 0.293 = W
Btu/h · ft^2 × 3.15 = W/m^2
Btu · in/ft^2 · h · °F × 144 = W · mm/m^2 · °C
Btu/lb × 2.33 = kJ/kg
Btu/lb · °F × 4.19 = kJ/kg · °C
(°F − 32)/1.8 = °C
ft × 0.3048 = m
ft/min × 0.00508 = m/s
ft of water × 2.99 = kPa
ft^2 × 0.0929 = m^2
ft^2 · h · °F/Btu × 0.176 = m^2 · °C/W
ft^3 × 28.3 = L
ft^3 × 0.0283 = m^3
ft^3/min (cfm) × 0.472 = L/s
ft-lb/min × 0.0226 = W
gallon × 3.79 = L
gal/min (gpm) × 0.0631 = L/s
hp × 0.746 = kW
in × 25.4 = mm
in Hg × 3.38 = kPa
in H_2O × 249 = Pa
in^2 × 645 = mm^2
in^3 × 16.4 = mL
kWh × 3.6 = MJ
mi (mile) × 1.61 = km
mm Hg × 0.133 = kPa
mm H_2O × 9.79 = Pa
oz (ounce) × 29.6 = mL
lb × 0.454 = kg
lb/ft × 1.49 = kg/m
lb/ft^3 × 16 = kg/m^3
lb/h × 0.126 = g/s
psi × 6.89 = kPa
qt (quart) × 0.946 = L

therm × 106 = MJ
ton × 0.907 = Mg
W/ft^2 × 10.8 = W/m^2
yd × 0.914 = m
yd^3 × 0.765 = m^3

SECTION 1

STEAM GENERATION

CHAPTER 1.1
STEAM FUNDAMENTALS

J. G. Singer
Combustion Engineering Inc.
Windsor, CT

INTRODUCTION

Before the equipment and systems that make up powerplants are discussed, certain basic relationships of mechanics, physics, and chemistry are introduced. From them come the engineer's initial understanding of how to design and operate a plant. Although the basics have been part of the engineer's lexicon for many years, the modern powerplant applies them in new and more efficient ways.

Covered briefly in this chapter are the fundamentals of steam and steam generators, combustion fundamentals and calculations, thermochemistry, and thermodynamic cycles. Later chapters draw on the concepts presented in this introduction of the basic principles. A Suggestions for Further Reading list at the end points the interested reader to leading texts in the field that discuss these fundamentals in much greater detail.

STEAM FUNDAMENTALS

The theoretical amount of work that can be obtained from steam used in a prime mover is equivalent to the change in its total heat content from its condition at the entering state to that at the exhaust state. The efficiency of the prime mover in converting the heat energy to mechanical effort governs the actual work obtained.

For economic studies involved in the selection and design of all steam and power generation equipment, it is necessary to understand thoroughly the properties of steam, the use of steam tables, and the application of superheat. A brief review of these fundamentals that apply to the generation of steam is helpful.

Properties of Steam

Steam results from adding sufficient heat to water to cause it to vaporize. This vaporization occurs in two steps: by adding heat to the water to raise it to the boiling temperature and by continuing the addition of heat to change the state from water to steam. When heated at average atmospheric pressure at sea level, each pound of water increases in temperature about 1°F for each Btu added until 212°F is reached. (One kilogram of water increases in temperature 1°C with an addition of approximately 4.19 kJ until 100°C is reached.) Adding heat does not cause the temperature to rise but, if continued, results in boiling, and the water changes its state from a liquid to a vapor.

When water is heated to the boiling point in a closed vessel, the vapor released causes the pressure in the vessel to rise. With the rise in pressure, the temperature at which the water boils also rises. It has been determined experimentally that, during the change in state of any substance from a liquid to a vapor at constant pressure, the vapor in contact with its liquid remains at constant temperature until the vaporization has been completed. Thus, the temperature at which boiling occurs for any given pressure is constant and is called the *saturation temperature*. It is the same for the water as it is for the vapor with which it is in contact.

Definitions. The *heat of the liquid* is the heat needed to raise a unit weight of water, normally 1 lb or 1 kg, from 32°F (0°C) to the saturation temperature corresponding to a given pressure. Also called the *enthalpy of the saturated liquid,* this is stated in Btus per pound or kilojoules per kilogram.

The latent heat of vaporization is the heat added to a unit weight of water at saturation temperature to vaporize it completely and produce dry saturated steam. This is the *enthalpy of evaporation or vaporization*. Dry saturated steam contains no moisture and is at a saturated temperature for the given pressure. Its total heat content, or enthalpy of the saturated vapor, is equal to the heat of the liquid plus the heat of vaporization.

Steam that contains water in any form—as minute droplets, mist, or fog—is called *wet steam,* which may result from the entrainment of water in boiling, from partial vaporization, or from partial condensation. In any case, the total heat content of the mixture is less than that of dry saturated steam, because vaporization is incomplete. The percentage of dry vapor by weight in the mixture is known as the *quality* of the steam. Thus, with 3 percent moisture in the steam, the quality is 97 percent. The total heat of wet steam is equal to the heat of the liquid plus the percentage of latent heat of vaporization represented by the steam quality. The temperature of wet steam is the same as that of dry steam at the corresponding pressure.

Steam with a temperature higher than that of saturated steam at the same pressure is called *superheated.* Superheating is accomplished by adding heat to the steam after its removal from contact with water, with pressure maintained (or held constant). Superheating results in an increase in temperature and volume. The total heat of superheated steam is equal to the total heat of dry saturated steam plus the heat of superheating. The term *total heat* is the general engineering expression applicable to any steam condition, whether wet, dry saturated, or superheated. Also known as the *enthalpy of steam,* it is the amount of heat that must be added to a unit weight of water at a given reference temperature to produce the end state under consideration.

The properties of steam have been the subject of considerable research in many countries of the world for many years. The Mollier diagram, constructed as an enthalpy-entropy chart to show the steam tables graphically, is particularly useful in the analysis of powerplant cycles. It visualizes the process of expansion of steam through the various sections of a steam turbine and helps in the quick solution of many other thermodynamic problems.

Superheated vs. Saturated Steam. The properties of superheated steam approximate those of a perfect gas. One important characteristic is the dependence of internal energy on temperature; thus, the closer steam approaches a perfect gas, the better it does its work. In addition, it contains no moisture, nor can it condense until its temperature has been lowered to that of saturated steam at the same pressure. This particular characteristic is valuable because, with the correct amount of superheat, it is possible to eliminate condensation in steam lines and to decrease the moisture in the steam turbine exhaust.

With saturated steam, the heat available depends entirely on pressure, while with superheated steam there is additional heat, proportionate to the degree of superheat. The added heat is obtained through an increased expenditure of fuel, but the economic benefits derived result in a net efficiency gain of considerable magnitude. By using a comparatively small amount of superheat, it is possible to reduce moisture at exhaust conditions and to effect an increase in the percentage of heat utilized.

Practical Limits

The materials of superheater construction govern the practical limits of steam temperature and pressure. Considerable development has occurred in the metallurgy of steel alloys and in the manufacturing of both the tubing and the finished sections of the superheater. These have made possible the design of superheaters and reheaters for high-temperature and high-pressure boiler installations, as have improvements in welding. While those installations which border on the limitations of available materials may be more interesting, and perhaps more sensational, the large majority are in the temperature range from 750 to 1050°F (approximately 400 to 566°C). Although commonplace, they represent the places where the greatest savings are made.

A study of steam temperatures accompanying installations using 2400-psig (16.5-MPa) throttle pressures discloses that 1050°F (566°C) was overwhelmingly selected as the primary temperature in the 1950s. The trend since that time has been in the direction of 1000°F (about 538°C) primary temperature, and is a well-established—but surely not exclusive—pattern. Again, experience has proved to be the arbiter. First, the initial cost of boilers and turbines for the higher temperature level proved to be less attractive as experience accumulated. Some owners therefore decided to reduce primary steam temperatures while continuing to lower capital cost through larger units. Other utilities chose to take advantage of the thermal gain from higher temperatures and maintained the established 1050°F (566°C) level.

Units using the nominal 3500-psig (24.1-MPa) cycle most commonly have 1000°F (538°C) superheat and 1000°F (538°C) reheat. About 30 percent of existing supercritical-pressure units have a second reheat stage at 1000°F (538°C), which is justified by high fuel cost and high anticipated load factor.

The curves in Fig. 1.1 show the variation in heat content of steam at different pressures. The values for plotting these curves were taken from ASME steam tables. When you are using these tables, or charts prepared from them, note that the pressure is stated as absolute (psia or MPa abs.). The steam gage on a boiler indicates the pressure in the vessel *above* that of the atmosphere. The absolute

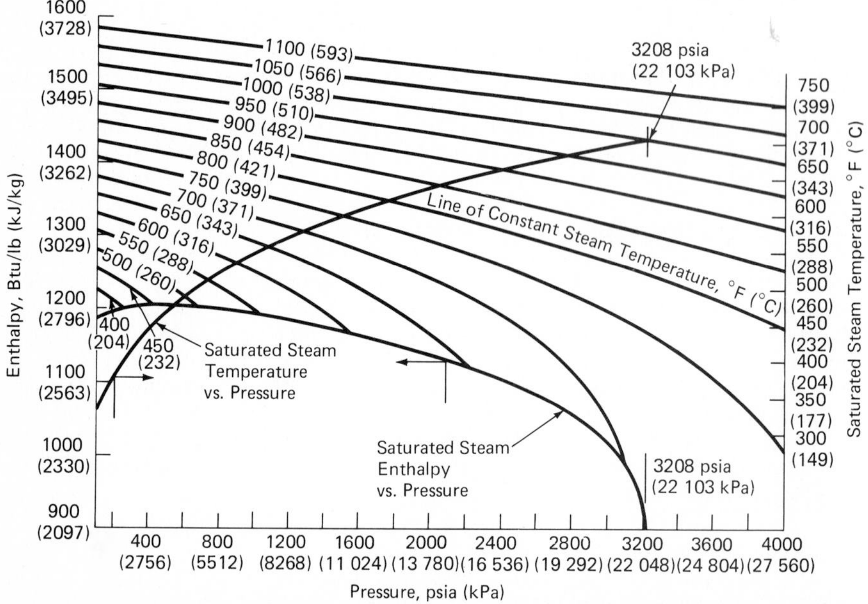

FIG. 1.1 Variation in heat content (enthalpy) of steam as a function of pressure and temperature.

pressure, then, is equal to the gage pressure plus atmospheric pressure. Condensers and similar equipment operate under vacuum, and their gages usually read in inches of mercury (in Hg or kPa). For these conditions, absolute pressure is equal to atmospheric pressure minus vacuum.

Superheating Advantages

Further discussion will show more clearly the manner in which superheating reduces steam consumption, particularly in regard to industrial boiler applications.

In the case of a steam turbine, high superheat is of utmost importance since the absence of moisture in steam decreases friction losses and erosion of turbine blades. If dry saturated steam enters a turbine, the condensation rate increases as the steam passes through the succeeding stages. Friction losses increase rapidly, because the condensate is actually an inert material that checks the speed of the turbine rotors. By using superheated steam, the condensation can be limited to a relatively few stages in the turbine discharge end, thus reducing windage loss and friction between rotor and vapor because of lower density and absence of moisture in the initial stages.

Pipelines carrying steam lose heat, principally by radiation. Thus, if steam entering a line is dry saturated, any loss in heat immediately results in some condensate, which is usually discharged from the line through traps and is frequently wasted. So, in addition to the heat loss from radiation, there is the loss of heat in the condensate. If the condensate is returned to the hot well, a portion of the heat in the liquid is recovered. By adding enough superheat to the steam, it can be piped without condensation loss.

Selecting saturated steam for industrial processes often results in minute quantities of moisture in the steam at the point of use. The use of dry saturated steam can translate to considerable savings in steam consumption and frequently increased process output. These advantages accrue from superheating the steam enough to overcome condensation in process-steam pipelines.

The installation of a superheater in a boiler has the effect of reducing the amount of work that must be done by the evaporative surfaces to produce the same power. In other words, installing a superheater has the effect of increasing the plant capacity for the same steam flow. Also, a properly designed superheater increases the thermal efficiency of the steam-generating cycle.

STEAM GENERATOR FUNDAMENTALS

Watertube boilers range in capacity from small low-pressure heating units generating small quantities of steam to large reheat steam generators operating in the supercritical-pressure region and serving turbine generators in the thousand-megawatt range. In slightly different terms, capacity may be magnified more than a thousand times from the smallest to the largest, pressure may extend from just above atmospheric to values as high as 5000 psig (35 MPa), and steam temperatures may vary from the boiling point to a highly superheated condition at 1050°F (566°C) or above.

Common Elements

What are the common elements in boilers having diverse design parameters? To answer this question, it is necessary to define the primary function of a boiler, which is simply to generate steam at pressures above atmospheric. Steam is generated by the absorption of heat produced in the combustion of fuel. In some instances, such as with waste-heat boilers, heated fluids (gases or liquids) serve as the heat source.

Generation of steam by heat absorption from products of combustion suggests that a boiler must have a pressure-parts system to convert incoming feedwater to steam; a structure within which the combustion reaction may take place, at the same time facilitating heat transfer and supporting boiler components; a means of introducing fuel and removing waste products; and controls and instruments to regulate and monitor operation.

A boiler designer has to work with such components as drums, headers, and tubing, which make up the pressure-parts system and enclose the furnace in which combustion takes place; burners and related fuel- and ash-handling equipment; and fans to supply combustion air and exhaust waste gases. Various types of instruments and controls link these elements in a physical and an operational sense.

Boiler Output

The output or capacity of a steam generator is either expressed in pounds per hour (kilograms per second) of steam or in the power output of a turbine generator in those cases where a single boiler provides the entire steam supply for an electric generating unit. Neither term is a true measure of the thermal energy supplied by the boiler.

The actual boiler output in terms of heat energy depends on several factors other than the quantity of steam. These include the temperature of the feedwater entering the economizer, the steam pressure and temperature at the superheater outlet, and the quantity, temperature, and pressure of steam entering and leaving the reheater. Similarly, because turbine and generator efficiencies affect boiler output, the generator output (in kilowatts) is not entirely a true measure of the energy output of the boiler alone.

These elements vary with the size and purpose of the powerplant in which the boiler is installed. A large central station in which high thermal efficiency is a primary requisite has many more refinements and auxiliaries than a small heating plant in which minimum capital investment may be an important criterion.

Primary Functions

In addition to its primary function of evaporating water to steam at high pressure, the modern boiler has to

- Produce that steam at exceptionally high purity, relying on stationary mechanical devices to remove impurities in the boiling water
- Superheat the steam generated in the unit to a specified temperature and maintain that temperature over a designated range of load
- Resuperheat (reheat) the steam that is returned to the boiler after expanding through the high-pressure stages of the turbine and keep that reheat temperature constant over a specified range of load
- Reduce the gas temperature leaving the unit to a level that satisfies the require-

ment for high thermal efficiency and at the same time is suitable for processing in the emissions control equipment downstream of the boiler

Boiler Efficiency

Figure 1.2 shows the distribution of heat energy in a reheat boiler for utility use. Clearly the primary source of heat is the fuel (supplemented by thermal energy from fans, pumps, and pulverizers), but the preheated air for combustion adds directly to the total heat in the furnace. The amount of heat in the preheated air corresponds to that extracted from the exhaust gases by the air heater.

Of the total heat entering the furnace, the major portion is absorbed as sensible and latent heat of vaporization in the heating surfaces of the economizer, furnace, superheater, and reheater. This absorbed heat represents the boiler output in the form of superheated and reheated steam. Losses that account for the remainder of the heat supplied to the furnace consist of the heat contained in the flue gas leaving the air heater (principally the sensible heat of the gas and latent heat in the moisture from the fuel), small losses from less-than-perfect combustion, and radiation from the boiler and its ancillary equipment.

Heat-Absorbing Surfaces

The boiler designer aims to arrange heat-transfer surface and fuel-burning equipment to optimize thermal efficiency and economic investment. Waterwalls, superheaters, and reheaters are selected, each of which absorbs heat from the furnace gas as it performs its respective function of heating water to the saturation point and of superheating and resuperheating steam. Also to be chosen are air heaters and economizers to recover heat from the furnace exit gases to preheat combustion air and increase the temperature of incoming feedwater.

The boiler designer must proportion heat-absorbing and heat-recovery surfaces to make the best use of the heat released by the fuel. Waterwalls, super-

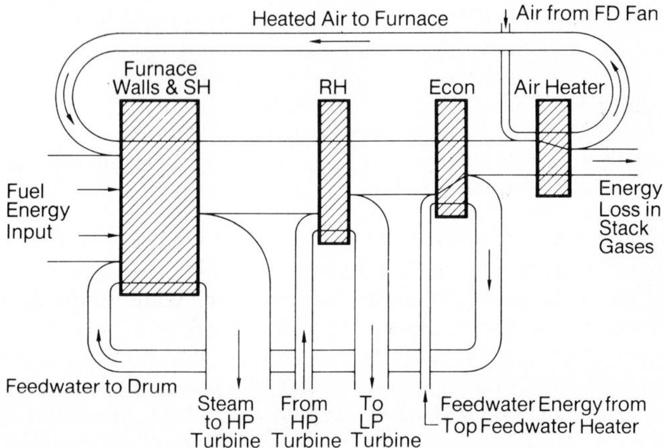

FIG. 1.2 Steam generator energy flow.

heaters, and reheaters are exposed to convection and radiant heat, whereas convection heat transfer predominates in air heaters and economizers. The relative amounts of such surfaces vary with the size and operating conditions of the boiler. A small low-pressure heating plant with no heat-recovery equipment has quite a different arrangement from a large high-pressure unit operating on a reheat regenerative cycle and incorporating heat-recovery equipment.

Figure 1.3 shows how the proportion of energy absorbed varies with different types of boilers. In a heating-plant boiler operating with a minimum of feedwater heating and no superheater, most of the heat absorbed is used in evaporating water to steam. In a large reheat unit with feedwater heaters and heat-recovery equipment, heat for evaporation is comparatively small, whereas heat for superheating and reheating accounts for more than one-half of the total input.

Figure 1.4 shows temperature versus heat absorption for four different high-pressure steam cycles: (1) the single-reheat 1800-psig (12.4-MPa) cycle at 5 percent overpressure; (2) the 2400-psig, 1000°F (16.5-MPa, 538°C) cycle with single reheat to 1000°F (538°C), again at 5 percent overpressure [2620 psig (18.1 MPa) at the superheater outlet]; and a (3) single-reheat and a (4) double-reheat supercritical cycle, both at 1000°F (538°C) at the turbine throttle. Such heat-absorption profiles will vary with the entering feedwater temperature, cold-reheat temperature, and relative size of the furnace and economizer.

Key Design Influences

Besides the basics of unit size, steam pressure, and steam temperature, the designer must consider other factors that influence the overall design of the steam generator.

Fuels. Although it is the most common fuel, coal is the most difficult to burn. The types of coal and their characteristics are covered thoroughly in Chap. 1 of Sec. 3. The ash in coal is composed of a number of objectionable chemical elements and compounds. Because of the high percentage of ash that coal can contain, it has a serious effect on furnace performance.

At the high temperatures resulting from the burning of fuel in the furnace, fractions of ash can become partially fused and sticky. Depending on the amount and the fusion temperature, the partially fused ash may adhere to surfaces contacted by the ash-containing combustion gases, causing objectionable buildup of slag on or bridging between tubes. Chemicals in the ash may attack materials such as the alloy steel in superheaters and reheaters.

In addition to the deposits in the high-temperature sections of the unit, the air heater (the coolest part) may be subject to corrosion and plugging of gas passages from sulfur compounds in the fuel acting in combination with moisture present in the flue gas.

Furnace. Heat generated in the combustion process appears as furnace radiation and sensible heat in the products of combustion. Water circulating through tubes that form the furnace wall lining absorbs as much as 50 percent of this heat which, in turn, generates steam by the evaporation of part of the circulated water.

Furnace design must consider water heating and steam generation in the wall tubes as well as the processes of combustion. Practically all large modern boilers have walls comprised of water-cooled tubes to form complete metal coverage of

Boiler Service	Fuel	Capacity, lb/h (g/s)	Outlet Pressure, psi (kPa)	Steam Temperature, °F (°C)	Reheat Temperature, °F (°C)	Feedwater Temperature, °F (°C)	Sensible Heat in Feedwater	Latent Heat of Evaporation	Superheat	Reheat
Heating	Oil	60,000 (7560)	125 (861)	Sat	—	212 (100)	14	86		
Small Industrial	Oil	100,000 (12 600)	650 (4479)	750 (399)	—	300 (149)	20	64	16	
Medium Industrial	Coal	250,000 (31 500)	650 (4479)	750 (399)	—	265 (129)	22	63	15	
Large Industrial	Coal	600,000 (75 600)	1500 (10 385)	880 (471)	—	275 (135)	32	46	22	
Large Utility	Coal	3,800,000 (478 800)	2620 (18 050)	1000 (538)	1000 (538)	480 (429)	26	23	35	16

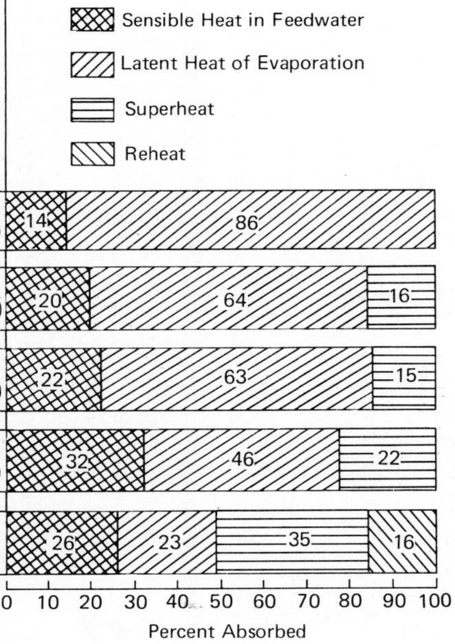

Percent Absorbed

FIG. 1.3 Heat absorption by various types of boilers.

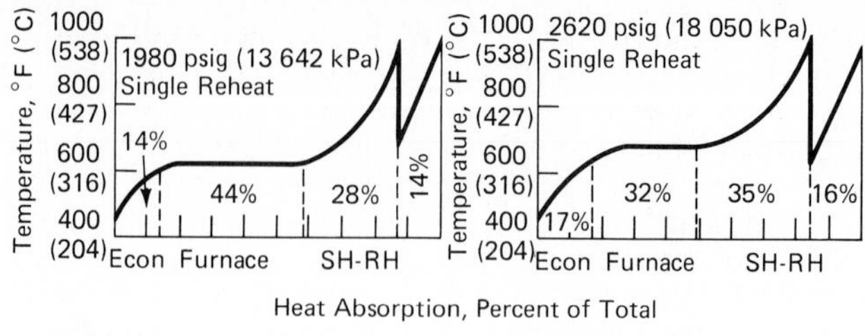

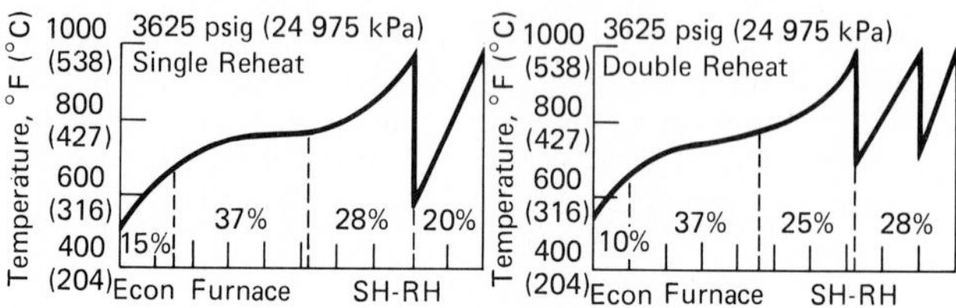

FIG. 1.4 Heat absorption variation with cycle pressure.

the furnace enclosure. Similarly, areas outside the furnace that form enclosures for sections of superheaters, reheaters, and economizers also use either water- or steam-cooled tube surfaces.

Waterwalls usually consist of vertical tubes arranged in tangent (or approximately so), connected at top and bottom to headers. These tubes receive their water from the boiler drum via downcomer tubes connected between the bottom of the drum and the lower headers. The steam, along with a substantial quantity of water, is discharged from the top of the waterwall tubes into the upper waterwall headers and then passes through riser tubes to the boiler drum. Here the steam is separated from the water, which together with the incoming feedwater is returned to the waterwalls through the downcomers.

Tube diameter and thickness are a concern from the standpoint of circulation and metal temperatures. Thermosyphonic (also called *thermal* or *natural*) circulation boilers generally have larger-diameter tubes than positive (pumped) circulation or once-through boilers. This practice is dictated largely by the need for more liberal flow area, to provide the lower velocities necessary with the limited head available.

The use of small-diameter tubes is an advantage in high-pressure boilers because the smaller tube thicknesses required result in lower outside tube-metal temperatures. Such tubes find application in recirculation boilers in which pumps provide an adequate head for circulation and maintain the desired velocities. The circulation of water and steam in both subcritical and supercritical boilers is discussed further in other sections of this chapter.

Superheaters and Reheaters. The function of a superheater is to raise the boiler steam temperature above the saturated temperature level. As steam enters the superheater in an essentially dry condition, further absorption of heat sensibly increases the steam temperature. The reheater receives superheated steam that has partly expanded through the turbine. The role of the reheater in the boiler is to resuperheat this steam to a desired temperature.

Superheater and reheater design depends on the specific duty to be performed. For relatively low final outlet temperatures, convection-only superheaters are generally selected. For higher final temperatures, surface requirements are larger, and, of necessity, superheater elements are located in very high gas-temperature zones. Wide-spaced platens or panels, or wall-type superheaters or reheaters of the radiant type, can then be selected. Figure 1.5 shows an arrangement of such platen and panel surfaces.

A relatively small number of panels are located on horizontal centers of 5 to 8 ft (1.5 to 2.5 m) to permit substantial radiant heat absorption. Platen sections, on 14- to 28-in (350- to 700-mm) centers, are placed downstream of the panel elements; such spacing provides high heat absorption by both radiation and convection. Convection sections are arranged for essentially pure counterflow of steam and gas, with steam entering at the bottom and leaving at the top of the pass, while gas flow is opposite. This arrangement allows a maximum mean temperature difference between the two media and minimizes the heating surface in the primary sections.

Economizers. Economizers help to improve boiler efficiency by extracting heat from flue gases discharged from the final superheater section of a radiant-reheat unit or the evaporative bank of an industrial boiler. In the economizer, heat is transferred to the feedwater, which enters at a temperature appreciably lower than that of saturated steam. Generally, economizers are arranged for downward flow of gas and upward flow of water.

Water enters from a lower header and flows through horizontal tubing that comprises the heating surface. Return bends at the ends of the tubing provide continuous tube elements, whose upper ends connect to an outlet header that is,

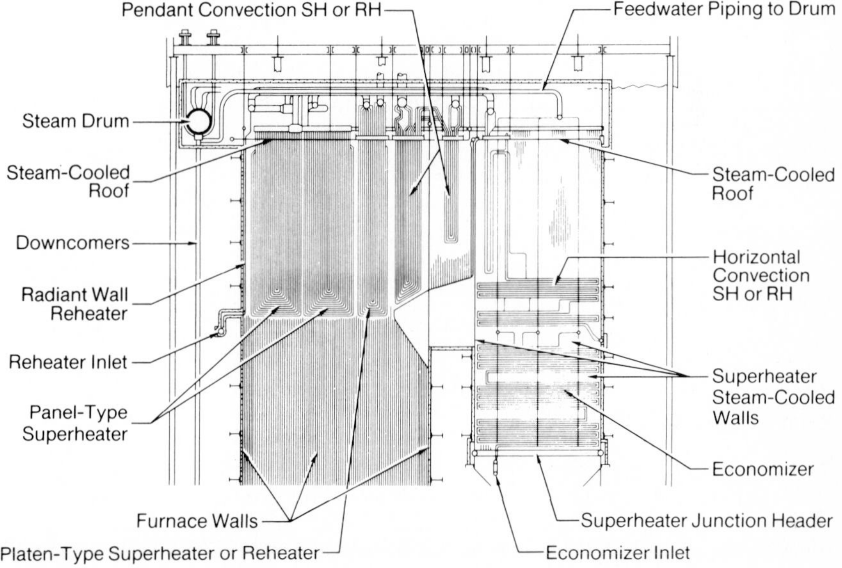

FIG. 1.5 Arrangement of superheater, reheater, and economizer of a large coal-fired steam generator.

in turn, connected to the boiler drum by tubes or large pipes. Tubes that form the heating surface may be bare or provided with extended surfaces such as fins.

Designing the economizer for counterflow of gas and water results in a maximum mean temperature difference for heat transfer. The upward flow of water helps avoid water hammer, which may occur under some operating conditions. To avoid generating steam in the economizer, the design ordinarily provides exiting water temperatures below that of saturated steam during normal operation.

As shown in Fig. 1.5, economizers of a typical utility-type boiler are located in the same pass as the primary or horizontal sections of the superheater, or superheater and reheater, depending on the arrangement of the surface.

Tubing forming the heating surface is generally low-carbon steel, so that it is necessary to provide water that is practically 100 percent oxygen-free to avoid corrosion. In central stations and other large plants, it is common practice to use deaerators for oxygen removal.

Small low-pressure boilers may have economizers made of cast iron, which is not subject to oxygen corrosion. However, the design pressure for this material is limited to approximately 250 psig (1.7 MPa). Although cast-iron tubes find little application today, cast-iron fins shrunk on steel tubes are practical and can withstand any boiler pressure.

Air Heaters. Steam generator air heaters have two important and concomitant functions: they cool the gases before passing to the atmosphere, thereby boosting fuel-firing efficiency, and they raise the temperature of the incoming combustion air. Depending on the pressure and temperature cycle, type of fuel, and type of boiler involved, one of the two functions will have prime importance.

For instance, in a low-pressure gas- or oil-fired industrial boiler, the combustion-gas temperature can be lowered in three ways—by a boiler bank, by an economizer, or by an air heater. Here, an air heater has principally a gas-cooling function, since no preheating is required to burn the oil or gas.

If the boiler is a high-pressure reheat unit burning a high-moisture subbituminous or lignitic coal, then high preheated-air temperatures are needed to evaporate the moisture in the coal before ignition can take place. Here, the air-heating function becomes primary. Without exception, then, large pulverized-coal boilers for either industry or electric power generation use air heaters to reduce the temperature of combustion products.

In theory, only the primary air must be heated, that is, air introduced to actually dry the coal in the pulverizers. Ignited fuel can burn without preheating the secondary and tertiary air. But a considerable advantage to the furnace heat-transfer process is realized from heating *all* the combustion air. This increases the rate of burning and helps raise the adiabatic flame temperature.

Control of Steam Temperature

To maintain turbine efficiency over a wide load range and to avoid fluctuations in turbine metal temperatures requires constant primary steam and reheat temperatures over the anticipated operating load range. To satisfy this need, a boiler must be equipped with the means for controlling and maintaining such steam temperatures over the desired range. If uncontrolled, steam temperatures will rise as the steam output increases.

To provide an economical installation that operates at minimum metal temperatures, superheaters and reheaters should be designed to give exactly the specified steam temperature at maximum output. An optimum design is one in which *all* the gas leaving the furnace passes over *all* the installed superheater and reheater surfaces at 100 percent boiler rating, without need of either superheater or reheater spray water. To satisfy this requirement, the means of control must maintain full steam and reheat temperatures over the total control range.

Steam-temperature control devices must be incorporated in the original design of the boiler firing system, in the superheater or reheater circuitry, or in arrangements of dampers for gas bypass.

Firing System Manipulation. Two common ways exist to vertically displace the zone of highest heat release in a furnace to achieve a change in the outlet gas temperature. The first, often used with front- or rear-wall fixed burners, is to insert or withdraw levels of burners as a function of the load; removing lower levels and firing through the remaining upper levels effectively moves the high-heat-release zone higher in the furnace. This method requires backup by spray desuperheating for vernier control.

Tilting fuel and air nozzles, found in corner (tangential) firing systems, are a positive means of controlling the outlet gas temperature smoothly without cycling equipment in and out of service. Superheater or reheater steam temperatures are regulated by changes in nozzle angle, above or below horizontal.

Desuperheating. Desuperheating is the reduction of temperature in superheated steam by spraying water into the piping either ahead of or behind a superheater or reheater section. To minimize the amount of water introduced, most large boiler plants pair desuperheating with one of the other temperature control methods. Other reasons include the water treatment and pumping costs.

Rising temperature with increasing output characterizes the performance of a superheater or reheater that receives its heat by convection from gas flowing over it. With desuperheating control only, the superheater must be designed for full temperature at some partial load. As a result, at higher loads the heat-transfer surface is excessive, with corresponding excessive temperatures, which the desuperheater can remove.

If located beyond the outlet of the superheater, a desuperheater will condition the steam before it is passed along to the turbine. Although this arrangement is practical for steam temperatures of 825°F (about 440°C) and lower, the preferred location of the desuperheater for higher temperatures is between sections of the superheater. In such interstage locations, the steam is first passed through one or more primary superheating sections, where it is raised to some intermediate temperature. It is then passed through the desuperheater, and its temperature is controlled so that, after the steam continues through the secondary or final stage of superheating, the required constant outlet temperature condition is maintained.

Desuperheaters are either the indirect or direct (mixing) type. The water available as the temperature-regulating medium governs the selection for any specific installation. This is not so important with the indirect noncontact type, where a tubular heat exchanger is used, since the steam to be desuperheated is separated from the cooling medium and the heat is transferred through the separating tube wall. In the direct type, however, the cooling medium is injected into, and mixed with, the superheated steam to reduce the temperature. To be used for this purpose, the cooling medium must be of condensate quality, containing very few solids.

Gas Recirculation. With this temperature control method, a portion of the flue gas is diverted from the main steam at a point following the superheater and reheater and is recirculated to the furnace. The gas passes through a recirculating fan and mixes with the gas in the furnace, causing a reduction in heat absorption. The heat available to the superheater and reheater increases, as does the quantity of gas passing over the surfaces. Both factors increase the steam temperature.

Bypass Damper. An arrangement of bypass dampers in a relatively cool gas zone downstream of superheater or reheater sections provides an acceptable means for maintaining a constant steam temperature; automatic controllers adjust the dampers to provide the required degree of temperature control. If load changes are abrupt, frequent, and of considerable magnitude, some hunting in positioning the damper is likely. Since there is also some lag in the response to temperature changes, the final temperature varies over a ±10°F (about ±5°C) range. This variation is characteristic where the regulation of steam temperature depends solely on the control of gas flow through the bypass-damper operation.

During the early damper opening periods, the gas flow rapidly increases and then falls off as the full-open position is approached. In one case, there is an increase of 15°F (about 8°C) in steam temperature for a change in damper from 100 to 40 percent open. However, the temperature increase from 40 percent open to fully closed damper is 50°F (about 30°C). Clearly sensitive responses, resulting in more uniform temperatures, are obtainable when regulator operation is confined to the early stages of damper opening.

Water-Steam Circulation

The term *circulation,* as applied to steam generators, means the movement of water, steam, or a mixture of the two through heated tubes. The tubes can be in furnace walls, boiler banks, economizers, superheaters, or reheaters. Adequate circulation results in the cooling fluid absorbing heat from the tube metal at a rate that maintains the tube temperature at or below design conditions.

Adequate circulation also keeps the tubes within the other physical and chemical limitations required by the inside and outside environment. In boilers, circulation through the varied systems of heated tubes can involve just the flow entering and leaving the system (called the *once-through flow*), a means of recirculating the fluid, or a combination of the two circulation concepts.

Steam generators of all manufacturers have similar pressure-part systems. For any given steam power cycle, the economizer contains fluid in the lowest temperature range, with its inlet temperature being that leaving the top feedwater heater. The superheater contains fluid in the highest temperature range, with its outlet temperature essentially fixed. The evaporator contains the middle range of fluid temperatures.

Circulation in economizers, superheaters, and reheaters is most commonly of the once-through type. The furnace-wall system of high-pressure units uses once-through or recirculation flow or some combination. By convention, these modes of circulation have become terms of reference for the complete steam-generating unit.

At pressures below the 3208-psia (22.1-MPa abs.) critical point, the major portion of the evaporator operates at a saturation temperature established by the pressure of the furnace-wall system. At supercritical pressures, the wall system has no fixed fluid temperature; a continuous temperature increase occurs in the furnace cooling fluid between the furnace-wall inlet and outlet.

The most critical circulating system in a large boiler is that of the furnace walls; they are at the same time the area of highest heat-absorption rates and a major structural component of the unit.

COMBUSTION FUNDAMENTALS

To the engineer concerned with boiler design and performance, combustion may be considered as the chemical union of the combustible part of a fuel and the oxygen of the air, controlled at such a rate as to produce useful heat energy. The principal combustible constituents are elemental carbon, hydrogen, and their compounds. In combustion, the compounds and elements are burned to carbon dioxide and water vapor.

Small quantities of sulfur are present in most fuels. Although sulfur is combustible and contributes slightly to the heating value of the fuel, its presence is generally detrimental because of the corrosive nature of its compounds.

Air, the usual source of oxygen for combustion in boilers, is a mixture of oxygen, nitrogen, and small amounts of water vapor, carbon dioxide, argon, and other elements. The composition of dry atmospheric air is given in Table 1.1.

In an ideal situation, the combustion process would occur with the exact pro-

TABLE 1.1 Composition of Combustion Air

Dry atmospheric air

The volumetric composition of dry atmospheric air given in NACA Report 1235 (Standard Atmosphere—Tables and Data for Altitudes to 65,800 ft or 20 055 m, November 20, 1952),[1] and the molecular weights of the gases constituting dry air are as follows:

	Volume, %	Mol. wgt.
Nitrogen	78.09	28.016
Oxygen	20.95	32.000
Argon	0.93	39.944
Carbon dioxide	0.03	44.010

(Neon, helium, krypton, hydrogen, xenon, ozone, and radon, combined, are less than 0.003%.)

Dry air with this composition has an apparent molecular weight of 28.97 lb/lb (kg/kg) mol and a density at 32°F and 14.7 psia (0°C and 101.325 kPa) of 28.97 ÷ 359 = 0.0807 lb/ft^3 (1.3 kg/m^3). The oxygen content is 23.14% by weight. The lb (kg) dry air/lb (kg) oxygen = 1 ÷ 0.2314 = 4.32 (2).

Wet atmospheric air

Wet atmospheric air is defined in this text as the above air plus 0.013 lb (0.006 kg) of water vapor/lb (kg) of dry air. [Air at 80°F (26.7°C), 60% relative humidity, and 14.7-psia (101.325-kPa) pressure contains 1.3% water vapor by weight. See Fig. 1.6.]

Wet air with this amount of water vapor has an apparent molecular weight of 28.74 lb/lb (kg/kg) mol and a density at 32°F and 14.7 psia (0°C and 101.325 kPa) of 0.0801 lb/ft^3 (1.3 kg/m^3). The oxygen content is 22.84% by weight. Then lb (kg) wet atmospheric air/lb (kg) oxygen = 1 ÷ 0.2284 = 4.38 (1.99).

The mass of nitrogen, argon, carbon dioxide, and water/lb (kg) oxygen = 77.16/22.84 = 3.38 lb (1.53 kg).

portions of oxygen and a combustible that are called for in theory (the stoichiometric quantities). But it is impractical to operate a boiler at the theoretical level of zero percent excess oxygen. In practice, this condition is approached by providing an excess of oxygen in the form of excess air from the atmosphere. The amount of excess air will vary with the fuel, boiler load, and type of firing equipment.

Combustion Equations

For combustion calculations, it is customary to write the combustion reaction equations on the basis of theoretical oxygen only, notwithstanding the presence of excess air and nitrogen. A partial list of these combustion equations and the approximate heat released in the reactions are given in Table 1.2. All combustion calculations are based on fundamental chemical reactions in the table. Not only do the equations indicate what substances are involved in the reaction, but they also show the molecular proportions in which the substances take part.

Each molecule has a numerical value that represents its relative weight or its molecular weight. This molecular weight is the sum of the atomic weights of the atoms composing the molecule. For example, carbon, C, has a molecular weight of 12; oxygen, O_2, has a molecular weight of $2 \times 16 = 32$; and carbon dioxide, CO_2, has a molecular weight of $12 + 2 \times 16 = 44$. These molecular weights are only relative values and may be expressed in any units. Note that the molecular weights in Table 1.2 are the whole-number values of the main isotopes of each substance.

A molecular weight of any substance in the gaseous state and under the same conditions of temperature and pressure will occupy the same volume. This rela-

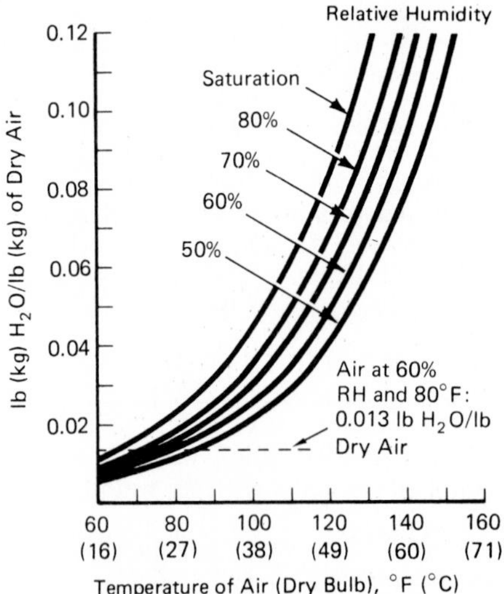

FIG. 1.6 Moisture content of dry air as a function of dry-bulb temperature and relative humidity.

TABLE 1.2 Combustion Equations

Combustible	Molecular weight	Reaction	Heat release* Btu/lb	Heat release* kJ/kg
Carbon	12	$C + O_2 \rightarrow CO_2$	14,100	32 853
Hydrogen	2	$H_2 + 0.5O_2 \rightarrow H_2O$	61,000	142 130
Sulfur	32	$S + O_2 \rightarrow SO_2$	4,000	9 320
Hydrogen sulfide	34	$H_2S + 1.5O_2 \rightarrow SO_2 + H_2O$	7,100	16 543
Methane	16	$CH_4 + 2O_2 \rightarrow CO_2 + 2H_2O$	23,900	55 687
Ethane	30	$C_2H_6 + 3.5O_2 \rightarrow 2CO_2 + 3H_2O$	22,300	51 959
Propane	44	$C_3H_8 + 5O_2 \rightarrow 3CO_2 + 4H_2O$	21,500	50 095
Butane	58	$C_4H_{10} + 6.5O_2 \rightarrow 4CO_2 + 5H_2O$	21,300	49 629
Pentane	72	$C_5H_{12} + 8O_2 \rightarrow 5CO_2 + 6H_2O$	22,000	51 260

*Higher heating value/lb (kg) of combustible.

tionship is very significant. The volume will, of course, vary numerically for different units of weight and for different conditions of temperature and pressure. For combustion calculations, the pound and the cubic foot are the units commonly used in the United States [in the International System of Units (SI), they are the kilogram and the cubic meter], and, unless otherwise stated, the temperature and pressure are understood to be, respectively, 32°F and 14.7 psia (0°C and 101.325 kPa). Thus, a molecular weight of 32 lb of oxygen at 32°F and atmospheric pressure will have the same volume as a molecular weight of 44 lb of carbon dioxide under the same conditions. This volume is 359 ft^3. (Similarly, 1 kmol of oxygen, with a molecular weight of 32 kg, will occupy 22.4 m^3 at 0°C and 101.325 kPa.)

Concept of the Mole

A molecular weight expressed in pounds is called a *pound-mole,* or simply a mole, and the volume it occupies is called a *molal volume*. [In SI, a molecular weight, expressed in kilograms, is called a *kilomole* (symbol: kmol).] The molal volume varies with changes in temperature and pressure according to Boyle's and Charles' laws and may be corrected to any desired conditions. The volume is directly proportional to the absolute temperature and inversely proportional to the absolute pressure. Because combustion processes in steam boiler furnaces usually take place at atmospheric pressure, or within 2 psi of 14.7 psia, pressure corrections are seldom necessary.

Returning to the combustion equation for carbon and oxygen and applying these concepts, we can now write this reaction in several ways. For the purposes of molal analysis, carbon may be treated as a gas.

$$1C + 1O_2 = 1CO_2 \tag{1.1}$$

$$1 \text{ mol C} + 1 \text{ mol } O_2 = 1 \text{ mol } CO_2 \tag{1.2}$$

$$12 \text{ lb C} + 32 \text{ lb } O_2 = 44 \text{ lb } CO_2 \tag{1.3}$$

$$(12 \text{ kg C} + 32 \text{ kg } O_2 = 44 \text{ kg } CO_2)$$

Dividing through by 12, we have

$$1 \text{ lb C} + 2.67 \text{ lb } O_2 = 3.67 \text{ lb } CO_2 \tag{1.4}$$

$$(1 \text{ kg C} + 2.67 \text{ kg } O_2 = 3.67 \text{ kg } CO_2)$$

$$1 \text{ volume C} + 1 \text{ volume } O_2 = 1 \text{ volume } CO_2 \tag{1.5}$$

Because there is 4.32 lb of dry air per pound of oxygen, the stoichiometric combustion of 1 lb of carbon requires 11.52 lb of dry air, or 11.68 lb of wet air (with 1.3 percent water vapor). (Similarly, the stoichiometric combustion of 1 kg of carbon requires 11.52 kg of dry air, or 11.68 kg of wet air.)

Each equation balances; there are the same number of atoms of each element and the same weight of reacting substances on each side of the arrow (or equal sign), but not necessarily the same number of molecules, moles, or volumes. Thus, one atom of carbon combined with one molecule of oxygen gives only one molecule of carbon dioxide; 2 mol of hydrogen plus 1 mol of oxygen yields 2 mol of water vapor. It is evident from a consideration of the mole-volume relationship that the percentage by volume is numerically the same as the mole percent.

Because a mole represents a definite weight as well as a definite volume, it is a means of converting analyses by weight to analyses by volume, and vice versa. Volumetric fractions of the several constituents of a gas can be multiplied by their respective molecular weights, with the sum of the products then being equal to the apparent molecular weight of 1 mol of gas. Then the percentage by weight of each component can be determined. Finally, the density of any gas at any temperature is found by dividing the molecular weight of the gas by the molal volume at that temperature (see Fig. 1.7).

COMBUSTION CALCULATIONS

Two methods of combustion calculations are presented here. The first is known as the *mole method* and is based on the chemical relationships previously explained. The second method uses the firing of 10^6 Btu as a basis for calculation. A similar method can be devised for the firing of a given amount of heat, using SI units.

The Mole Method

Table 1.2 gives the basic combustion reactions for the carbon, hydrogen, and sulfur in coal. Assume a high-volatile bituminous coal of the following analysis, burned at 23 percent excess air; perform calculations on the basis of 100 lb

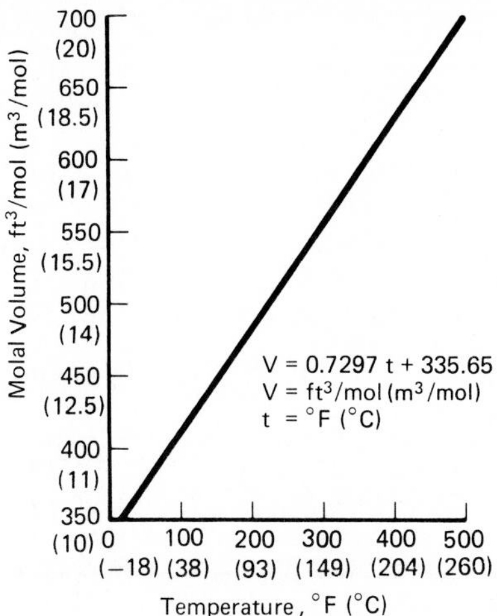

FIG. 1.7 Gas density determination.

(100 kg) of as-fired fuel. The fuel analysis as fired is as follows:

	% by weight	Mole weight
C	63.50	12
H_2	4.07	2
S	1.53	32
O_2	7.46	32
N_2	1.28	28
H_2O	15.00	18
Ash	7.16	
HHV	11,200 Btu/lb (26 100 kJ/kg)	

The calculation of air weight for combustion must be made on the basis of an oxygen balance, because oxygen is the only element common to all oxidizing reactions. The oxygen contained in the fuel must be deducted from the calculated quantity needed because it is already combined with carbon, hydrogen, or other combustible constituents of the coal.

The molar relations are as given in Table 1.2:

Mol	Mol O_2	Mol CO_2	Mol H_2O	Mol SO_2
C	1	1		
H	0.5		1	
S	1			1

Air for Combustion

$$O_2 \text{ for C} \quad \frac{63.5}{12} = 5.29 \text{ lb mol (5.29 kmol)}$$

$$O_2 \text{ for H} \quad \frac{4.07}{2 \times 2} = 1.02 \text{ lb mol (1.02 kmol)}$$

$$O_2 \text{ for S} \quad \frac{1.53}{32} = \underline{0.05 \text{ lb mol (0.05 kmol)}}$$

$$\text{Total} = 6.36 \text{ lb mol (6.36 kmol)}$$

$$\text{Less } O_2 \text{ in fuel} \quad \frac{7.46}{32} = \underline{-0.23 \text{ lb mol (0.23 kmol)}}$$

$$O_2 \text{ required} = 6.13 \text{ lb mol (6.13 kmol)}$$

$$O_2 \text{ in excess air} \quad 6.13 \times 0.23 = \underline{1.41 \text{ lb mol (1.41 kmol)}}$$

$$\text{Total } O_2 \text{ required} = 7.54 \text{ lb mol (7.54 kmol)}$$

$$\text{Dry air required} = 7.54 \text{ lb mol (kmol) } O_2 \times$$

$$\frac{100 \text{ lb mol air}}{20.95 \text{ lb mol } O_2} = 36.0 \text{ lb mol (36.0 kmol)}$$

$$36.0 \times 28.97 = 1043 \text{ lb (kg) dry air/100 lb (kg) fuel}$$

Dry Products of Combustion. The weight of gaseous products of combustion can be calculated from the volumetric analysis of the flue gas. Not only the weight of the flue gas per 100 lb (kg) of coal but also its analysis and volume can be calculated from the information given in the preceding example.

To obtain the wet products of combustion, or total wet gas when a fuel burns completely, the weight of the fuel is added to the weight of atmospheric air supplied for its combustion. If some of the fuel is ash or, because of incomplete combustion, does not leave the furnace with the gases, there will be less burned-out fuel in the products.

The wet products of combustion in the previous example, then, are the fuel (100 lb − 7.16 lb ash = 92.84 lb/100 lb) plus the air required for combustion, or (rounded) 93 + 1043 = 1136 lb/100 lb of fuel (1136 kg/100 kg of fuel).

The Million-Btu Method

This method for combustion calculations is based on the concept that the weight of air required in the combustion of a unit weight of any commercial fuel is more nearly proportional to the unit heat value than to the unit weight of the fuel. Consequently, the weights of air, dry gas, moisture, wet gas, and other quantities are expressed in pounds per million (10^6) Btu fired. (*Note:* SI conversion factors for the calculations with this method are kg/10^6 kJ = lb/10^6 Btu × 0.43 and kJ/kg = Btu/lb × 2.33.)

In connection with this calculation method, the following items are important:

Fuel in Products F. To define F, it is the portion of the fuel fired that appears in the gaseous products of combustion. Because all quantities are to be those required for, or resulting from, the firing of 10^6 Btu, F must be calculated on that basis. If a fuel contains no ash, F is obtained by dividing 10^6 by the as-fired heating value of the fuel. For solid fuels where ash and/or solid combustible loss must be considered;

$$F = \frac{10^4(100 - \%\ \text{ash} - \%\ \text{solid combustible loss})}{\text{fuel heat value}} \tag{1.6}$$

where
F = lb/10^6 Btu fired
% ash = % by weight in fuel as fired
% solid combustible loss = % by weight in fuel as fired
Fuel heat value = high heat value (HHV) as fired, Btu/lb

Atmospheric Air for Combustion A. In accordance with the molar method, the theoretical weight of dry air (zero excess) may be calculated from the fuel analysis and the formula

$$A_{\text{dry}} = \frac{11.52\ (\%\text{C}) + 34.57\ (\%\text{H} - \%\text{O}/8) + 4.32\ (\%\text{S})}{\text{HHV}} \times 10^4 \tag{1.7}$$

in which the numerator and denominator are on the same basis—as fired, moisture-free, or moisture-and-ash-free—and A_{dry} is in pounds per 10^6 Btu fired.

For air with 1.3 percent moisture by weight [80°F (26.7°C) and 60 percent relative humidity], the formula becomes

$$A_{\text{wet}} = \frac{11.68\ (\%\text{C}) + 35.03\ (\%\text{H} - \%\text{O}/8) + 4.38\ (\%\text{S})}{\text{HHV}} \times 10^4 \tag{1.8}$$

Values of A_{wet} range from 570 lb/10^6 Btu for pure hydrogen to above 800 lb/10^6 Btu for certain cokes and meta-anthracite coals, as shown in Fig. 1.8 for various

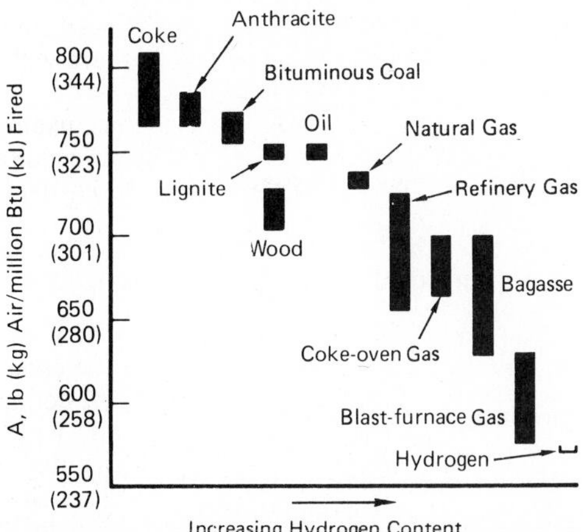

FIG. 1.8 Combustion-air requirements for various fuels at zero excess air.

fuels burned in steam generators. Any calculated values of wet air for combustion differing substantially from these values should lead to a cross verification of the ultimate analysis and the observed high heating value. The analysis and HHV of the fuel have to be from the same sample to avoid errors in air and gas weight determinations.

Effect of Unburned Combustible. In the combustion of solid fuels, even in pulverized form, it is not feasible to burn the available combustible completely. Thus, the air requirement per 10^6 Btu *fired* has to be reduced to the air required per 10^6 Btu *burned*. This is done by multiplying the combustion air A by the combustible-loss correction factor C.

The unburned combustible loss can be expressed either as the percentage of carbon heat loss or as the percentage of combustible weight loss. These are related by

$$\%\text{ Carbon heat loss} = \frac{14{,}600}{\text{HHV}} \times \%\text{ solid combustible weight loss} \tag{1.9}$$

in which 14,600 Btu/lb is the heat value for combustibles in refuse recommended by the ASME Performance Test Code and HHV is the high heating value of the as-fired fuel.

If a fuel has carbon as its only combustible constituent, then the factor C is

$$1 - \frac{\%\text{ solid combustible weight loss}}{100} \tag{1.10}$$

If, however, not all the heat in the fuel comes from the carbon alone (so that the air is not strictly proportional to the carbon burned), the factor C will not be exact. For high-carbon, low-volatile fuels, it will be nearly exact and will result in only a small error for fuels low in fixed carbon and high in hydrogen. The error involved by using Eq. (1.10) in all cases is quite within the limits of accuracy of all other combustion calculations.

Finally, C can be expressed as a function of the percentage of heat loss by combining the foregoing relationships:

$$C = 1 - \frac{\%\text{ carbon heat loss}}{100} \times \frac{\text{HHV}}{14{,}600} \tag{1.11}$$

Figure 1.9 is a graphical solution of this equation.

Products of Combustion P. The total gaseous products of combustion P become the sum of F and A (as corrected for combustible loss). Thus,

$$P = F + CA \tag{1.12}$$

where P = total gaseous products of combustion, lb/10^6 Btu fired
F = fuel fired exclusive of ash or solid carbon loss, lb/10^6 Btu fired
A = atmospheric air consumed, lb/10^6 Btu fired
C = combustible-loss correction factor

Moisture W_a and W_f. For heat-balance calculations, the moisture in air is 1.3 percent of the air weight per 10^6 Btu, or $W_a = 0.013A$, for ambient conditions of

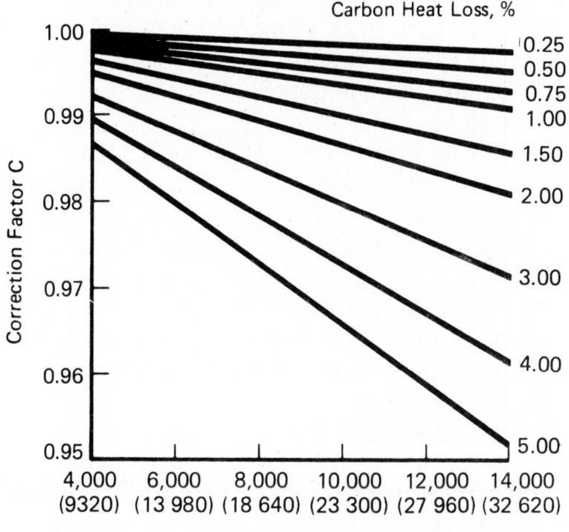

FIG. 1.9 Graphical solution of Eq. (1.11).

80°F (26.7°C) and 60 percent relative humidity. For air at a higher or lower temperature or relative humidity, the moisture content is as shown in Fig. 1.6; the air per 10^6 Btu A and W_a must be adjusted accordingly.

Moisture from the fuel is reported separately both in an ASME Performance Test Code heat balance and in a predicted heat balance. In the case of some fuels, the heat loss due to this moisture may be the largest single item in the heat balance: W_f includes the combined surface and inherent moisture W_c from a fuel plus the moisture formed by the combustion of hydrogen W_h; W_c will vary from zero, or a mere trace in fuel oil, to over 115 lb/10^6 Btu fired in the case of green wood; and W_h will vary from zero or a trace in lamp black to 100 lb/10^6 Btu fired in the case of some refinery gases. Thus

$$W_f = W_c + W_h \tag{1.13}$$

where

$$W_c = H_2O \times \frac{10^4}{HHV}$$

and

$$W_h = 9 \times H \times \frac{10^4}{HHV}$$

Here W_c and W_h are in pounds per 10^6 Btu fired, H_2O is the percentage of water by weight in the as-fired fuel, H is the hydrogen in the fuel (percentage by weight as fired), and HHV is the high heating value of the as-fired fuel in Btu per pound.

Dry Gas P_d. The dry-gas content of the combustion products is used in the calculation of the dry-gas loss item of a boiler heat balance. The dry gas may be determined by subtracting the water vapor from the total products:

$$P_d = P - (W_a + W_f)$$

where P_d = dry gas, lb/10^6 Btu fired
P = total products of combustion
W_a = moisture in air
W_f = moisture from fuel

THERMOCHEMISTRY

Energy is associated with the forces that bind atoms to form a molecule. A rearrangement of the atoms to form new molecules, such as occurs in a chemical reaction, entails the liberation or absorption of energy. The branch of thermodynamics that deals with this subject is referred to as *thermochemistry*. Unfortunately, the working definitions regarding this aspect of thermal analysis are often obscurely stated and contradictory between authors. The following definitions therefore are not universal or subscribed to by all authors in the field. In the main, the definitions rely on the authoritative treatises by Wark,[2] Reynolds and Perkins,[3] and Van Wylen and Sonntag.[4]

Standard Reference State

The reference datum for chemical thermodynamic tabulations is usually taken to be 1-atm pressure and 25°C (298 K, 77°F, or 537°R). This is called the *standard reference state,* and by convention the enthalpy of every elemental substance (such as H_2, O_2, or N_2) is defined to be zero at the standard reference state. Other definitions concerned with the emission or consumption of energy accompanying a thermochemical reaction follow.

Internal Energy of Reaction

If a chemical reaction occurs at constant volume and at 25°C initial temperature, and if the products of reaction are returned to 25°C, then the energy excess (liberated) or deficit (absorption) required to meet the conditions of this system is known as the *internal energy of reaction.* The symbol assigned to this energy is ΔU_R. The heat of reaction at constant volume is an alternate term for the internal energy of reaction. By first-law analysis, one can write for this item $\Delta U_R = Q - \Delta U$.

Enthalpy of Reaction and Combustion

If a chemical reaction is initiated at 25°C and the products are returned to 25°C and this process is carried out at constant pressure, then the energy supplied or disposed of is referred to as the *enthalpy of reaction*. The symbol assigned to this energy is ΔH_R. The heat of reaction at constant pressure is an alternate term for the enthalpy of reaction.

Although both ΔU_R and ΔH_R can be found listed in tables, ΔH_R is the more commonly encountered, because it is conveniently measured in a steady-flow device. The heat of reaction is a short form for the heat of reaction at constant pressure and is tabulated on a basis of Btu (kilojoules) per mole of fuel burned. Heating value is synonymous with the heat of reaction, except that it is always taken as positive.

The enthalpy of combustion is defined, sometimes, as the negative of the heating value. The symbol assigned to the enthalpy of combustion is ΔH_C.

Higher and Lower Heating Values

If water in the products of combustion of a fossil fuel is taken as being condensed and lowered in temperature to 77°F (25°C) at 1 atm, the condensation heat release h_{fg} is added to ΔH_C, and this is called the *higher heating value* (HHV). If the H_2O remains a vapor, then h_{fg} [1050.3 Btu/lb_m at 77°F (2443 kJ/kg at 25°C)] is not added to ΔH_C depending on whether the enthalpy of combustion is evaluated on a liquid or vapor (H_2O) basis.

Exothermic Heat

A chemical reaction that liberates energy (heat) is called *exothermic* and is denoted ΔH_R. The sign convention adopted for the liberated energy is negative, although not all textbooks do so.

Use of the negative convention invariably is justified by the argument that heat is removed from the system to return the system to 25°C. The negative convention for exothermic reactions is, in fact, physically correct and can be justified in thermochemical terms. Figure 1.10 demonstrates the principles on the basis of the formation of 1 lb mol or kmol of H_2O, where 1 mol of a substance whose molecular weight is M is defined as M unit masses of that material.[5]

Enthalpy of Formation

Although the internal energy and the enthalpy of reaction are basic quantities in the solution of thermochemical problems, the myriad chemical reactions effec-

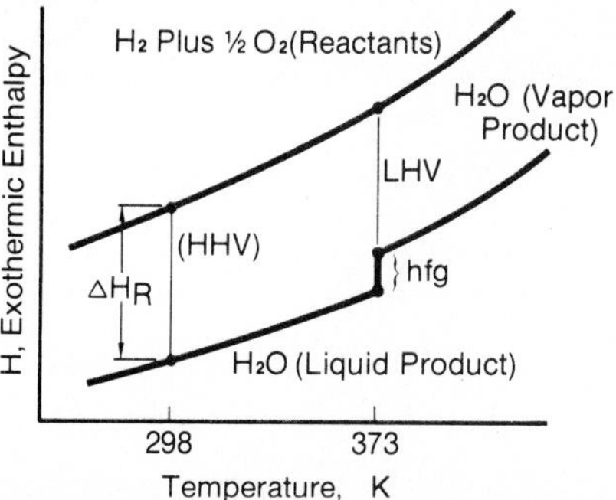

FIG. 1.10 Exothermic enthalpy of reaction for formation of H_2O at 1 atm, constant pressure.

tively prohibit the overwhelming task of tabulating ΔU_R and ΔH_R values. Therefore, the concept of the enthalpy of formation was introduced. Applying the first law to a chemical reaction in a closed system at constant pressure, one can write

$$\sum_{i \text{ reactants}}(\nu_i H_i) + \Delta H_R = \sum_{i \text{ products}}(\nu_i H_i)$$

where H_i is the molar enthalpy of any product or reactant at the pressure and temperature of the reaction and ν_i again is the stoichiometric coefficient for each, based on the balanced chemical equation. Clearly many sensible enthalpy tables for various gases account for changes in temperature at constant composition, whereas the thermochemical process being considered is defined in terms of constant temperature with variable composition. It follows that sensible enthalpy tables are not enough for evaluating energy transformations resulting from chemical reactions.

Definition. To circumvent this difficulty, the concept of the *enthalpy of formation* ΔH_f is introduced. It is defined as the energy released or absorbed when a specific chemical compound, such as CO_2 or H_2O, is formed from its elements. The reactions, again, are such that initial reactant temperature and final product temperature are the same, written symbolically as

$$Q = \Delta H_f = H_{\text{compound}} - \sum_i(\nu_i H_i)_{\text{elements}}$$

In this case, the reactants are taken arbitrarily as the stable form of the elements at the given initial state. Because a value of zero has been assigned to the enthalpy of all stable elements at 1 atm and 25°C, the enthalpy H_i of any compound at the same state is simply the enthalpy of formation of that substance, or

$$\Delta H_f = H_{\text{compound}} - 0 = H_{\text{compound}}$$

Absolute Enthalpy. Because the chemical enthalpy of a stable element is always zero, by definition, the enthalpy of any compound is composed of two parts: that associated with its formation from elements and the sensible enthalpy associated with a change in state at constant composition. The sum of these two parts is called the *absolute enthalpy* of a substance. The foregoing is illustrated by the reaction

$$C(s) + O_2(g) \rightarrow CO_2(g)$$

Since solid carbon and gaseous diatomic oxygen are the stable forms of these elements at 25°C and 1 atm, their chemical enthalpies are chosen to be zero. The enthalpy of reaction is easily measured experimentally for this reaction, and it is found to be −169,290 Btu/lb mol (−393 522 kJ/kmol). These values are also the enthalpy of formation of carbon dioxide. Thus ΔH_R, ΔH_C, and ΔH_f are related but not synonymous. This is shown somewhat more clearly by Table 1.3.

The usefulness of ΔH_f values is based on the assumption that the enthalpy of any compound is equal, arbitrarily, to its enthalpy of formation from stable elements. Therefore, substituting ΔH_f values for H_i values gives

TABLE 1.3 Enthalpy of Formation vs. Enthalpy of Combustion at 25°C and 1-Atm Pressure

Substance	Chemical formula	ΔH_f (25°C), kJ/kmol	ΔH_f (77°F) Btu/lb mol	ΔH_c (25°C), kJ/kmol	ΔH_c (77°F), Btu/lb mol
Carbon	C (solid)	0	0	−394 446	−169,290
Hydrogen	H_2 (gas)	0	0	−286 520	−122,970
Oxygen	O_2 (gas)	0	0	0	0
Carbon monoxide	CO (gas)	−110 768	−47,540	−283 678	−121,750
Carbon dioxide	CO_2 (gas)	−394 446	−169,290	0	0
Water	H_2O (vapor)	−242 413	−104,040	0	0
Water	H_2O (liquid)	−286 520	−122,970	0	0
Methane	CH_4 (gas)	−75 049	−32,210	−892 483	−383,040
Propane	C_3H_8 (gas)	−104 104	−44,680	−2 225 313	−955,070
Methyl alcohol	CH_3OH (gas)	−201 638	−86,540	−765 871	−328,700

For ΔH_c the water in products of combustion is assumed to be liquid. The ΔH_c values are therefore also HHV values (except that HHV values are always given as positive).

$$\Delta H_r = \sum_i (\nu_i \, \Delta H_f)_{\text{products}} - \sum_i (\nu_i \, \Delta H_f)_{\text{reactants}}$$

This equation is completely general for any reaction as long as the enthalpies of formation for all products and reactants are measured at the same temperature and pressure. This equation is widely used in thermochemistry.

Conservation of Mass. Several different definitions or labels have been presented here, but giving something a label does not necessarily constitute a satisfactory explanation. Suppose, in addition, that the words of Van Wylen[4] are literally interpreted:

> The justification of this procedure of arbitrarily assigning the value of zero to the enthalpy of elements at 25°C, 1 atm rests on the fact that, in the absence of nuclear reactions, the mass of each element is conserved in a chemical reaction. No conflicts or ambiguities arise with this choice of reference state, and it proves to be very convenient in studying chemical reactions from a thermodynamic point of view.

Then one is in a quandary regarding the origin of the energy being tabulated.

Obviously, no one is going to abandon the law of conservation of mass, because it is far too useful. However, energy cannot be created out of thin air. The problem is solved if the law of the conservation of mass is recognized as a near-exact approximation that has been replaced by the law of conservation of mass energy. There is nothing in Einstein's well-known equation that prohibits or invalidates it outside the world of nuclear transition. For the HHV to be extracted from the reaction and do work, something else must disappear—that something is a minute amount of mass.

Einstein's Equation. A conversion constant on a per-Btu or per-kilojoule basis can be readily calculated. The Einstein equation is

$$\delta m = \frac{E}{C^2}$$

where[5]

$$C = 2.997\ 930 \times 10^5 \text{ km/s}$$

$$= 1.86282 \times 10^5 \text{ mi/s}$$

For a ΔE of 1 Btu, C of 9.8357×10^8 ft/s, and gravitational constant of 32.174 ft • $\text{lb}_m/\text{lb}_f \cdot \text{s}^2$, the relationship for the mechanical equivalent of heat, 1 Btu = 778.16 ft • lb, is

$$1 \text{ Btu} \times \frac{778.16 \text{ ft} \cdot \text{lb}_f}{\text{Btu}} \times \frac{32.174 \text{ ft} \cdot \text{lb}_m}{\text{lb}_f \cdot \text{s}^2\ 96.741 \times 10^{16} \text{ ft}^2/\text{s}^2}$$

which yields 2.588×10^{-14} lb_m/Btu.

Using the more recent value of 777.649 for J (see Ref. 6), one gets 2.586×10^{-14} lb_m/Btu, or on a metric basis 1.112×10^{-14} kg/kJ. It follows that 2.586×10^{-14} lb_m/Btu and 1.112×10^{-14} kg/kJ are adopted as the mass equivalency constants. In the case of a powerplant that averages a heat fired of, say, 3000×10^6 Btu/h (0.88×10^6 kW), then over a 25-yr operating lifetime it is found that

$$3 \times 10^9 \times 24 \times 365 \times 25 \times 2.586 \times 10^{-14}\quad 17 \text{ lb}$$

So about 17 lb (7.7 kg) of matter has been converted to energy, of which roughly 6.8 lb (3.09 kg) became electric energy. This loss of mass is true only, of course, in the case where energy is transferred out of the system. In an adiabatic process, where the system is not cooled to the standard reference state, no mass is lost because no energy is transferred.

THERMODYNAMIC CYCLES

Historically, the roots of thermodynamics go back to attempts to quantify instinctive concepts of systems designed to produce work from various energy sources. For continuously converting heat to work, the first practical power systems used steam as a working fluid.

By definition, a *thermodynamic cycle* is a series of processes combined in such a way that the thermodynamic states at which the working fluid exists are repeated periodically. Customarily, in an electric generating station, the fluid is cycled through a sequence of processes in a closed loop designed to maximize generation of electric power from the fuel consumed consistent with plant economics.

The design of a specific powerplant represents an optimization of thermodynamic and economic considerations, the latter including initial, production, and distribution costs. In the following paragraphs, basic thermodynamics applicable to large utility and industrial powerplants is presented. This forms the basis for the selection of plant cycles and equipment.

Laws of Thermodynamics

The first and second laws of thermodynamics govern the thermodynamic analysis of fluid cycles. These laws are stated in equation form, in Eqs. (1.14) to (1.17): First law:

$$\Delta E = Q - W + \sum_i (\pm\ h_i \pm e_{xi})\ m_i \tag{1.14}$$

where ΔE is the change in energy content of the system, Q is the heat transferred to the system, W is the work transferred from the system, and $(h_i + e_{xi})m_i$ is the energy convected into or out of the system by mass m_i with enthalpy h_i and extrinsic energy e_{xi}.

The extrinsic energy is dependent on the frame of reference. For a fluid system, e_{xi} = kinetic energy + potential energy = $V_i^2/2_{gj} + z_i$, in which V_i is velocity and z is elevation above the datum. This equation applies equally to processes and cycles, steady- and transient-flow situations. For example, in a closed system where fluid streams do not cross the boundary, $m_i = 0$, and if the process is cyclic, then $\Delta E = 0$ and Eq. (1.14) becomes

$$\sum_{cycle} Q = \sum_{cycle} W \tag{1.15}$$

This implies 100 percent efficiency.

As another example, if the steady-state adiabatic expansion in a turbine is being analyzed, then $\Delta E = 0$, $Q = 0$, and Eq. (1.14) reduces to

$$W = -\sum_i (h_i + e_{xi}) m_i$$

$$= m\Big[(h + e_x)_{\text{in}} - (h + e_x)_{\text{out}}\Big]$$

$$W = m\left[\left(h + \frac{V^2}{2_{gj}} + z\right)_{\text{in}} - \left(h + \frac{V^2}{2_{gj}} + z\right)_{\text{out}}\right] \tag{1.16}$$

When changes in kinetic energy and elevation of the fluid stream can be neglected, Eq. (1.16) reduces to the familiar $W\ m\ (h_{\text{in}} - h_{\text{out}})$.

Second law:

$$\Delta S = \frac{Q}{T} + I + \sum_i S_i \tag{1.17}$$

where ΔS is change in entropy of the system, Q/T is the sum over the system boundaries of the heat transferred Q_i at a position on the boundary where the local temperature is T_i, and I is irreversibility. (For consistency with other second-law statements, $I \geqslant 0$. For reversible processes or cycles, $I = 0$; for irreversible processes or cycles, $I > 0$.) And $\Sigma_i S_i$ is the entropy flow into and out of the system associated with mass flow m_i into and out of the system. For a reversible cyclic process involving a closed system, $I = 0$, $\Delta S = 0$, $\Sigma_i S_i = 0$, and Eq. (1.17) reduces to

$$\frac{Q}{T} = 0 \tag{1.18}$$

The steady-flow, adiabatic expansion of a fluid through a turbine is governed by Eq. (1.19):

$$I = -\sum_i m_i s_i = -m\,(s_{\text{in}} - s_{\text{out}}) \tag{1.19}$$

$$= m(s_{\text{out}} - s_{\text{in}})$$

For reversible adiabatic expansion, $I = 0$ and the process is characterized by the familiar isentropic property $s_{\text{out}} = s_{\text{in}}$. Turbine expansion is not wholly isentropic. This nonisentropicity is taken into account by defining the turbine efficiency. The primary advantage of writing the second law in an equation, Eq. (1.17), rather than the usual inequality is its usefulness in analyzing processes and cycles in a direct, quantitative manner similar to the first-law analysis.[7]

Equations (1.14) and (1.17) thus provide quantitative means of examining all processes encountered in powerplant analysis regardless of the fluids used or the specific cycle applied.

Carnot Cycle

The Carnot cycle consists of several reversible isothermal and isentropic processes, which may be viewed as occurring in either the nonflow or flow device shown in Fig. 1.11*a* and 1.11*b*. In the first instance, the heat source and sink are placed in contact with the device to accomplish the required isothermal heat ad-

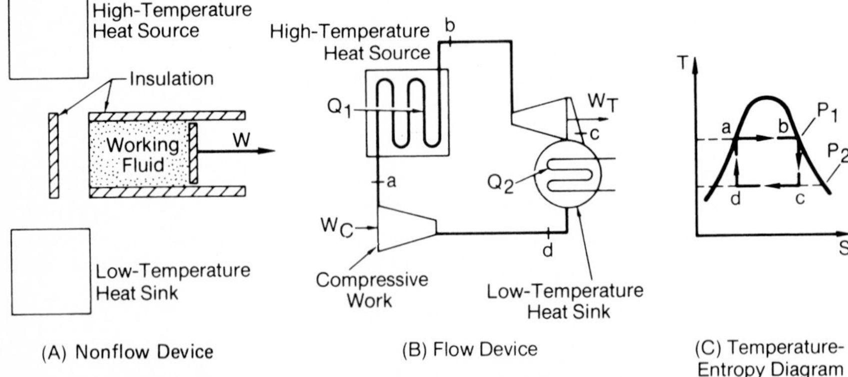

FIG. 1.11 Carnot cycle.

dition (*a-b*) and rejection (*c-d*). The insulation shown replaces the heat reservoirs for executing the reversible adiabatic processes involving expansion (*b-c*) and compression (*d-a*).

The process characteristics for good heat transfer and work transfer are not the same and are partially in conflict; Fig. 1.11*b*, therefore, shows a flow system for executing the Carnot cycle with the work- and heat-transfer processes assigned to separate devices. For both nonflow and flow systems, the state changes experienced by the working fluid are shown in the temperature-entropy diagram of Fig. 1.11*c*.

The classic Carnot cycle, then, is such that no other can have a better efficiency between the specified temperature limits than the Carnot value. Other cycles may equal it, but none can exceed it, so

$$\text{Carnot} = \frac{T_h - T_1}{T_h} = 1 - \frac{T_1}{T_h} \tag{1.20}$$

where T_h is the temperature of the heat source and T_1 is the temperature of the heat sink—all in terms of absolute temperature.

Practical attempts to attain the Carnot cycle encounter irreversibilities in the form of finite temperature differences during the heat-transfer processes and fluid friction during work-transfer processes. The compression process *d-a,* moreover, is difficult to perform on a two-phase mixture and requires an input of work ranging from one-fifth to one-third of the turbine output. When realistic irreversibilities are introduced, the Carnot cycle net work is reduced; the size and cost of equipment increase. Consequently, other cycles appear more attractive as practical models.[8]

In relation to the Carnot efficiency, the high-temperature heat source cannot be defined in terms of the maximum temperature. Instead, the weighted average of the temperature of the working fluid must be calculated, involving the heating of the feedwater as it leaves the last heater and the evaporation, superheating, and reheating processes.

Rankine Cycle

The elements comprising the Rankine cycle are the same as those appearing in Fig. 1.11*b* with one exception: because the condensation process accompanying the heat-rejection process continues until the liquid saturated state is reached, a simple liquid pump replaces the two-phase compressor. The temperature-entropy and enthalpy-entropy diagrams of Fig. 1.12*a* and *b* illustrate the state changes for the Rankine cycle. With the exception that compression terminates (state *a*) at the boiling pressure rather than at the boiling temperature (state *a'*), the cycle resembles a Carnot cycle. The triangle bounded by *a-a'* and the line connecting to the temperature-entropy curve in Fig. 1.12*a* signify the loss of cycle work because of the irreversible heating of the liquid from state *a* to a saturated liquid. The lower pressure at state *a,* compared to *a'*, makes possible a much smaller work of compression between *d* and *a*. For operating plants it amounts to 1 percent or less of the turbine output.

This modification eliminates the two-phase vapor compression process, reduces compression work to a negligible amount, and makes the Rankine cycle less sensitive than the Carnot cycle to the irreversibilities bound to occur in an actual plant. As a result, compared with a Carnot cycle operating between the

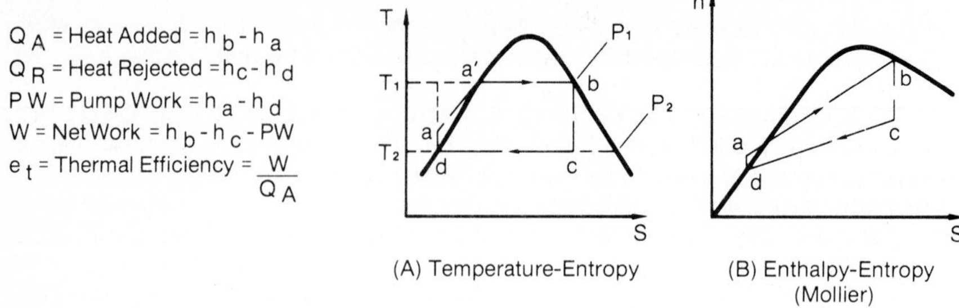

FIG. 1.12 Simple Rankine cycle (without superheat).

same temperature limits and with realistic component efficiencies, the Rankine cycle has a larger net work output per unit mass of fluid circulated, a smaller size, and a lower cost of equipment. Also, because of its relative insensitivity to irreversibilities, its operating plant thermal efficiencies will exceed those of the Carnot cycle.

Regenerative Design. Refinements in component design soon brought powerplants based on the Rankine cycle to their peak thermal efficiencies, with further increases realized by modifying the basic cycle. This occurred by increasing the temperature of saturated steam supplied to the turbine, by increasing the turbine inlet temperature through constant-pressure superheat, by reducing the sink temperature, and by reheating the working vapor after partial expansion followed by continued expansion to the final sink temperature. In practice, all these are brought into play with yet another important modification.

Earlier, the irreversibility associated with the heating of the compressed liquid to saturation by a finite temperature difference was cited as the primary thermodynamic cause of lower thermal efficiency for the Rankine cycle. The regenerative cycle attempts to eliminate this irreversibility by using as heat sources other parts of the cycle with temperatures slightly above that of the compressed liquid being heated. Figure 1.13 is an idealized form of such a procedure.

The condensed liquid at f is pumped to pressure P_1, passes through coils around the turbine, and receives heat from the fluid expanding in the turbine. The liquid and vapor flow counter each other, and by reversible heat transfer over the infinitesimal temperature difference d_T, the liquid is brought to the saturated state at T_1 (process b-c) and then rejects heat at constant temperature T_2 (process

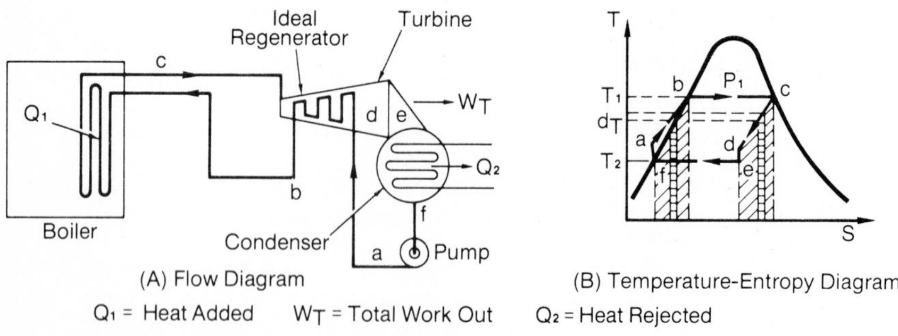

FIG. 1.13 Regenerative Rankine cycle with ideal regenerator.

e-f). Such a system, by the second law, will have a thermal efficiency equal to that of a Carnot cycle operating between the same temperatures.

Regenerative Heating. The procedure of transferring heat from one part of a cycle to another to eliminate or reduce external irreversibilities is called *regenerative heating,* which is basic to all regenerative cycles. Although it is thermodynamically desirable, the idealized regenerative cycle just described has several features that preclude its application in practice. Locating the heat exchanger around the turbine increases design difficulties and cost. Even if these problems were solved, heat transfer could not be accomplished reversibly in the time available; further, cooling as described causes the vapor to reach an excessive moisture content.

The scheme shown in Fig. 1.14 permits a practical approach to regeneration without encountering these problems. Extraction or "bleeding" of steam at state *c* for use in the "open" heater avoids excessive cooling of the vapor during turbine expansion; in the heater, liquid from the condenser increases in temperature by ΔT. (Regenerative cycle heaters are called *open* or *closed* depending on whether hot and cold fluids are mixed directly to share energy or kept separate with energy exchange occurring by the use of metal tubing.)

The extraction and heating substitute the finite temperature difference ΔT for the infinitesimal dT used in the theoretical regeneration process. This substitution, while failing to realize the full potential of regeneration, halves the temperature difference through which the condensate must be heated in the basic Rankine cycle. Additional extractions and heaters permit a closer approximation to the maximum efficiency of the idealized regenerative cycle,[9] with a further improvement over the simple Rankine cycle shown in Fig. 1.12.

Reducing the temperature difference between the liquid entering the boiler and that of the saturated fluid increases the cycle thermal efficiency. The price paid is a decrease in the net work produced per unit mass of vapor entering the turbine and an increase in the size, complexity, and initial cost of the plant.[10, 11]

Reheat Cycle

Superheat offers a simple way to improve the thermal efficiency of the basic Rankine cycle and to reduce the vapor moisture content to acceptable levels in the low-pressure stages of the turbine. But with the continued increase in higher

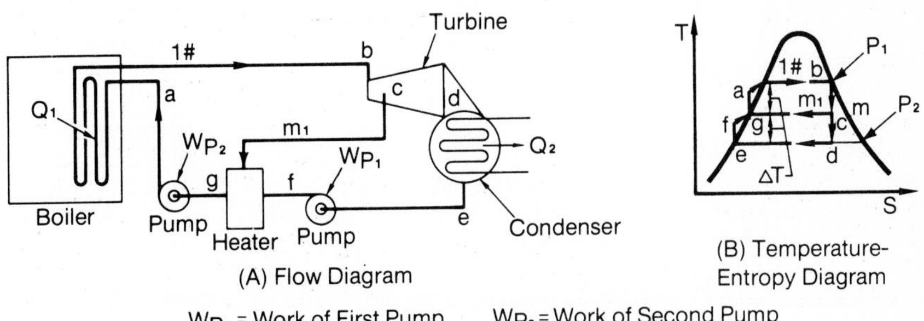

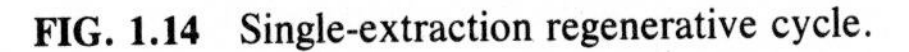

FIG. 1.14 Single-extraction regenerative cycle.

temperatures and pressures to achieve better cycle efficiency, in some situations available superheat temperatures are insufficient to prevent excessive moisture from forming in the low-pressure turbine stages.

The solution to this problem is to interrupt the expansion process, remove the vapor for reheat at constant pressure, and return the vapor to the turbine for continued expansion to condenser pressure. The thermodynamic cycle using this modification of the Rankine cycle is called the *reheat cycle*. Reheating may be carried out in a section of the boiler supplying primary steam, in a separately fired heat exchanger, or in a steam-to-steam heat exchanger. Most present-day utility units combine the superheater and reheater in the same boiler.

Usual central-station practice combines both regenerative and reheat modifications to the basic Rankine cycle. For large installations, reheat enables an improvement of approximately 5 percent in thermal efficiency and substantially reduces the heat rejected to the condenser cooling water.[12] The operating characteristics and economics of modern plants justify the installation of only one stage of reheat, except for units operating at supercritical pressure.

Supercritical-Pressure Cycle

A definite relationship exists between the operating temperature and the optimum pressure of a cycle. The supercritical-pressure cycle is used worldwide to obtain the highest possible thermodynamic efficiencies with fossil-fuel steam generation equipment.

A regenerative-reheat cycle may have six to eight stages of feedwater heating, and because of the high inlet temperature and pressure, two stages of reheat can be justified. Philadelphia Electric Company's Eddystone 1, a double-reheat supercritical unit with original outlet steam conditions of 5300 psig and 1210°F (36.5 MPa and 654°C) was the high-water mark of cycle efficiency for conventional steam plants. It was capable of producing electric output with an input of less than 8200 Btu (8650 kJ) per kilowatthour.

Process-Steam Cycles

In many cases, power and heating needs may be combined in a single powerplant that will operate at a high annual load factor and thermal efficiency. For heating service, the steam may be generated at a pressure and temperature high enough that exhaust from the turbine is at steam conditions suitable for delivery to the steam mains and distribution to users. The needs of industrial plants for process steam may be met by similar arrangements with either turbine exhaust steam or extracted steam from an appropriate turbine stage.

Cycle Requirements. The selection of optimum exhaust or extraction conditions will depend on the proportions of power and process heat required of the particular plant. Process-steam requirements are usually in the low-pressure range with modest superheat required (if at all); consequently, the initial steam temperature and pressure will generally be below the limits of current technology with a commensurate easing of boiler and turbine design requirements.

The great variety of power and process-steam requirements offers a continuing challenge to designers. Cycles have been developed and applied over a wide range of capacities, extending from small industrial powerplants to large central stations serving the heating needs of a metropolitan center or the power and process needs of a major petrochemical installation. The steel, chemical, and paper industries are three examples of the many important industrial users of process-steam cycles, also called cogeneration (see Sec. 2, Chap. 7).

Figure 1.15 is a schematic diagram of a power and process cycle paired with a backpressure turbine. After steam is generated at a suitable working pressure, it is admitted to the turbine and emerges usually in the superheated state *c*. After desuperheating, saturated steam *d* enters the heater and is entirely condensed. Because the steam required for power generation will not equal at all times that required for process, some means of controlling the exhaust steam pressure must be introduced to avoid variations in the pressure and, therefore, the steam saturation temperature.

Control Method. The method selected for controlling the exhaust pressure depends on the circumstances. An ordinary centrifugal governor fitted to the backpressure turbine will cause the quantity of available exhaust steam to be controlled by the load on the turbine. Should the available exhaust be too small, superheated steam may be passed through a reducing valve into the desuperheater. If the quantity of exhaust steam exceeds the requirements, then the excess steam may be blown to atmosphere, into an accumulator, or into a feed tank through the spill valve.

Another version of the process cycle uses high-temperature water circulated by pumps to supply energy for process and heating. The thermal head is provided by a water "boiler," the entire system being pressurized by steam or gas (compressed air or nitrogen) contained in an expansion drum.

Combined Cycles

A combined cycle is any one of a number of combinations of gas turbines, steam generators (or other heat-recovery equipment), and steam turbines assembled for the reduction in plant cost or improvement of cycle efficiency in the power generation process. This definition omits the wide variety of combined-cycle possibilities for industrial applications, which are limited in their variations only by the particular process requirement. Nevertheless, the underlying principle of all these arrangements is similar, in that they all depend on the energy contained in the gas-turbine exhaust and take advantage of the fact that the gas turbine is oth-

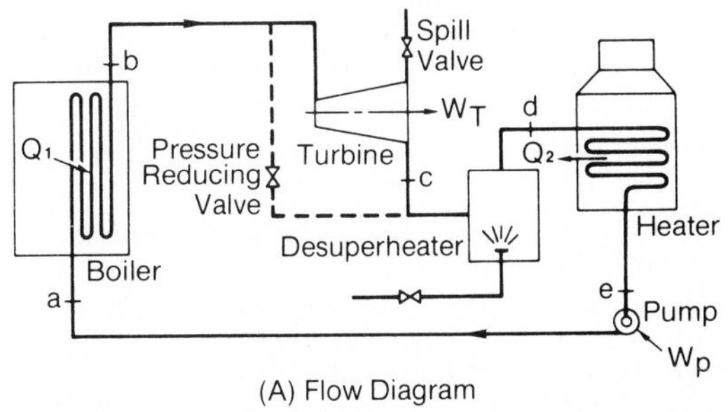

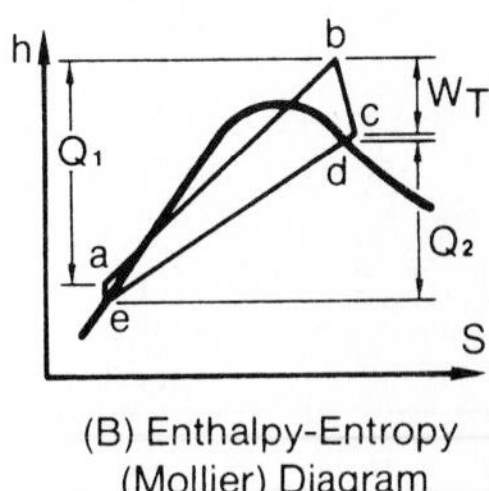

FIG. 1.15 Power process cycle using a backpressure turbine.

erwise subject to the losses associated with elevated stack-gas temperatures and the combustion of hydrogen-containing fuel.

Analogous to mercury-steam or steam-ammonia cycles, the combined gas-steam cycle is binary, although it is not normally thought of as such. The principle is to improve efficiency by increasing the working fluid temperature by tapping the high-temperature capabilities of the internal combustion turbine while utilizing the external combustion (steam) turbine to reduce the sink temperature. The penalty incurred is the required compressor work performed by the gas turbine.

Gas turbines fire natural gas, blast-furnace gas, jet fuel, light distillate oil, or, with proper prefiring treatment, a wide range of residual and crude oils. Open-cycle gas turbines satisfy both the peaking and reserve requirements of the utility industry because of their quick-starting capability and low capital cost. Their poor thermal efficiency and use of prime fuels, however, place them at an economic disadvantage when they are operated for long periods. Depending on the manufacturer and fuel selected, heat rates ranging from 12,000 to 16,000 Btu/kWh (12 700 to 16 900 kJ/kWh) (HHV) may be obtained.

A number of major design modifications may improve this heat rate, such as increased compressor pressure ratios and higher gas-turbine inlet-gas temperatures. In addition, the station heat rate associated with present gas-turbine designs may be improved further by capitalizing on the energy contained in the turbine exhaust gases. For example, a 50-MW gas turbine might discharge approximately 2.2×10^6 lb/h (277 kg/s) of exhaust gases at about 900°F (480°C), with an equivalent value of 450×10^6 Btu/h (132 MW) of energy or about 75 percent of the initial fuel input. But only a limited amount of this energy can be recovered in simple waste-heat equipment. .

ACKNOWLEDGMENT

Illustrations in this chapter are courtesy of *Combustion Engineering Inc.*

REFERENCES

1. "Standard Atmosphere Tables and Data for Altitudes to 65,800 Feet," National Advisory Committee for Aeronautics Report 1235 International Civil Aviation Organization, Montreal, Canada and Langley Aeronautical Laboratory, Langley Field, VA, 1952.
2. Wark, K., *Thermodynamics,* 4th ed., McGraw-Hill Book Co., New York, 1983.
3. Reynolds, W.C., and H.C. Perkins, *Engineering Thermodynamics,* McGraw-Hill Book, Co., New York, 1970.
4. Van Wylen, G.J., and R.E. Sonntag, *Fundamentals of Classical Thermodynamics,* 2d. ed., John Wiley & Sons, Inc., New York, 1973.
5. Condon, E.U., and Hugh Odishaw, *Handbook of Physics,* 2d. ed., McGraw-Hill Book Co., New York, 1967.
6. Bolz, R.E., and G.L. Tuve, eds., *Handbook (of Tables for) Applied Engineering Science,* Chemical Rubber Co., Cleveland, OH, 1970.
7. Bruges, E.A., *Available Energy and the Second Analysis,* Academic Press, New York, 1959.

8. Obert, E.F., and Richard A. Gaggioli, *Thermodynamics,* 2d. ed., McGraw-Hill Book Co., New York, 1963, pp. 384–416.
9. Weir, C.D., "Optimization of Heater Enthalpy Rises in Feed-Heating Trains," *Proceedings,* Institution of Mechanical Engineers, London, 174:769–796, 1960. Discussion by R.W. Haywood, pp. 784–787.
10. Chiantore, G., et al., "Optimizing a Regenerative Steam-Turbine Cycle, *Trans. ASME, Journal of Engineering for Power,* 83, Series A:433–443, October 1961.
11. Salisbury, J.F., *Steam Turbines and Their Cycles,* Part 3: Cycle Analysis, Robert E. Krieger Publishing Co., Huntington, NY, 1974.
12. "The Reheat Cycle—A Re-Evaluation," *Combustion,* 21 (12):38–40, June 1950. Papers given at the Symposium on the Reheat Cycle sponsored by the ASME and held in New York, November 29–December 3, 1948. *Trans. ASME,* 71:673–749, 1949

SUGGESTIONS FOR FURTHER READING

Clayton, W.H., and J.G. Singer, "Steam Generator Designs for Combined Cycle Applications," *Combustion,* 44(10):26–32, April 1973.

Gabrielli, F., and H.A. Grabowski, "Steam Purity at High Pressure," paper presented at ASME/IEEE/ASCE Joint Power Generation Conference, Charlotte, NC, Oct. 8–10, 1979.

Gaydon, A.G., and H.G. Wolfhard, *Flames: Their Structure, Radiation, and Temperature,* 4th ed. (revised), John Wiley & Sons, Inc., New York, 1979.

Goodstine, S.L., "Vaporous Carryover of Sodium Salts in High-Pressure Steam," *Proceedings of American Power Conference,* 36:784–789, 1974.

Habelt, W.W., "The Influence of Coal Oxygen to Coal Nitrogen Ratio on NO_x Formation," paper presented at 70th Annual AIChE Meeting, New York, Nov. 13–17, 1977.

Haywood, R.W., "Research into Fundamentals of Boiler Circulation Theory," *Proceedings of the General Discussion on Heat Transfer,* jointly sponsored by ASME and Institution of Mechanical Engineers, London, Sept. 11–13, 1951, and Atlantic City, Nov. 26–28, 1951.

Levy, J.M., et al., *Combustion Research on the Fate of Fuel-Nitrogen under Conditions of Pulverized Coal Combustion,* Final Task Report, No. PB-286 208 (EPA-600/7-78/165), National Technical Information Center, Springfield, VA, 1978.

MacKinnon, D.J., "Nitric Oxide Formation at High Temperature," *Air Pollution Control Association Journal,* 24(3):237–239, March 1974.

Powell, E.M., and H.A. Grabowski, "Drum Internals and High-Pressure Boiler Design," ASME Paper No. 54-A-242, ASME, New York, 1954.

Singer, J.G., ed., *Combustion—Fossil Power Systems,* 3d. ed., Combustion Engineering, Inc., Windsor, CT, 1981.

Straub, F.G., and H.A. Grabowski, "Silica Deposition in Steam Turbines," *Trans. ASME,* 67:309–316, May 1945.

Thermochemical Tables, 2d. ed., Joint Army, Navy, Air Force Project, NSRDS-NBS 37, U.S. Government Printing Office, Washington, D.C., 1971.

CHAPTER 1.2
BASIC POWERPLANT DESIGN

J. G. Singer
Combustion Engineering Inc.
Windsor, CT

INTRODUCTION

The design of equipment for a fossil-fuel steam supply system starts with the application of certain basics governing the relationship between properties of matter that define the conversion of energy from one form to another. Known as the first and second laws of thermodynamics (see Sec. 1, Chap. 1), these cornerstones of powerplant design provide a quantitative method of looking at the sequential processes of working fluids as a function of temperature, pressure, enthalpy, and entropy.

The design of a modern powerplant represents more than the application of thermodynamic data. It is a synthesis of economic considerations with thermal performance criteria that govern the selection of steam-generating equipment, whether the installation is for producing steam at an industrial site or at a utility powerplant.

The opening section describes certain general design approaches relating to industrial installations. The second section deals with the principles involved in the selection of steam-generating equipment for electric utilities. A third section covers combined cycles in some detail.

INDUSTRIAL POWERPLANT DESIGN

This section focuses on the overall design elements of the industrial or institutional powerplant. Many of these elements are involved also in the design of more complex utility plants (see the next section); other chapters of the book cover specifics of the design of steam generators, together with their various components and supporting systems.

The most elementary type of industrial powerplant incorporates a boiler as a heat source and a heating system as a load to dissipate the thermal energy released by fuel fired to the boiler. The thermal transport medium may be steam or hot water, but in either case it will be returned to the boiler at a lower temperature. Figures 2.1 and 2.2 show these elementary plants.

From this point onward, the industrial or institutional powerplant may encompass various degrees of complexity. Power may be generated in a condensing plant, thus giving the equivalent of a small central station. Steam may be supplied directly to the process and not be returned as condensed feedwater. However, it may be passed through a turbine acting as a reducing valve, as shown in Fig. 2.3.

Cycles may range from the simplicity and low level of thermal efficiency exemplified by the noncondensing steam locomotive to the most complex and efficient thermodynamic arrangements proposed for central stations, including combinations of steam and gas turbines. Boiler size may extend from the generation of a few thousand to several million pounds of steam per hour in large installations, and the same capacity spread is true of turbines for power generation. In short, the industrial powerplant illustrates virtually every aspect of the thermal engineering involved with a fossil-fuel plant.

Design Trends

Up to this point the term *industrial powerplant* has been used without definition. In engineering terms, no physical difference exists between industrial and utility types. At one time, power generation was a part of virtually every industrial

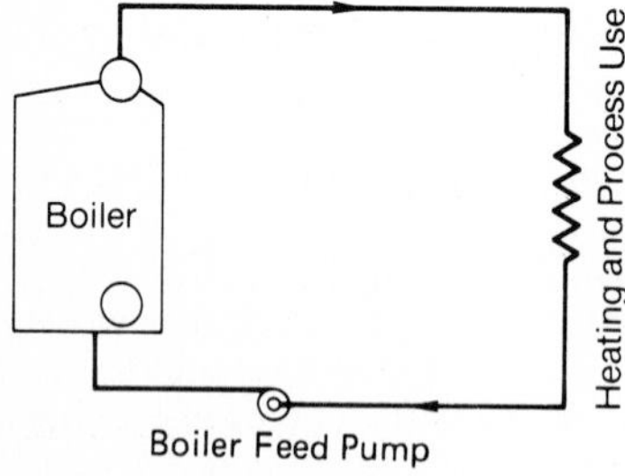

FIG. 2.1 Basic elements of a steam heating plant.

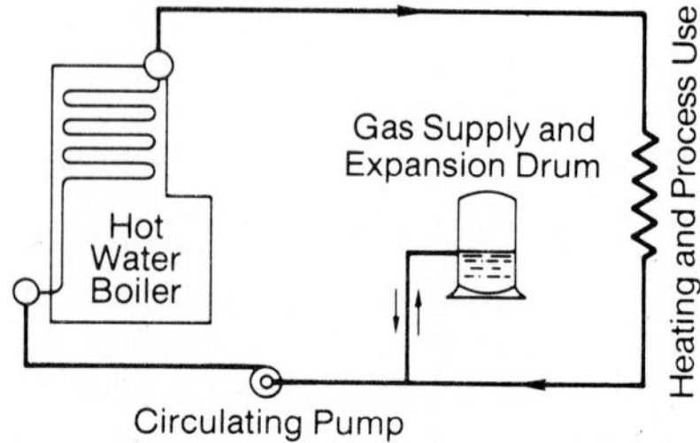

FIG. 2.2 Basic elements of a high-temperature-water heating plant.

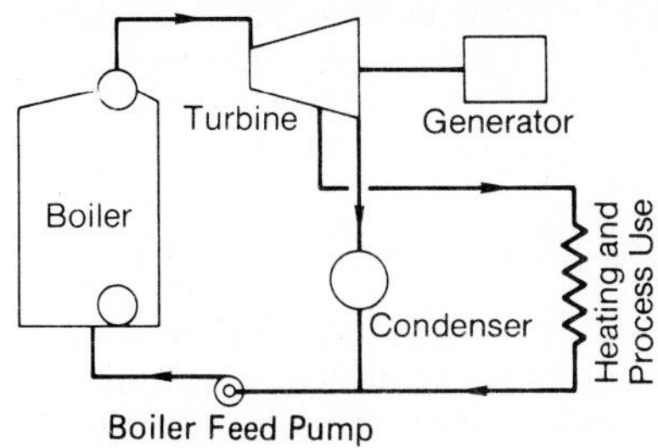

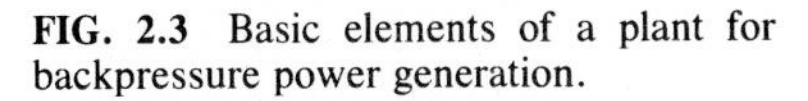

FIG. 2.3 Basic elements of a plant for backpressure power generation.

powerplant, but this is no longer the case. There are several reasons for this change, which accelerated during and after World War II:

- Public utility systems are more securely and effectively interconnected electrically, and service is very reliable.
- The cost of purchased electric power has resisted the effects of inflation better than almost any other commodity, even since 1973, the year of the Arab oil embargo.
- The demands for additional electric power have generally grown faster than the demands for heating, thereby exceeding the capabilities of backpressure generation.
- The same dollar investment may bring a greater return when applied to production equipment or facilities expansion than when spent in the powerplant.

The end result has been that the majority of industrial powerplants built since World War II have been for space heating and process steam. The exceptions have been based on situations such as these:

- Coordinated demands for steam and power, accompanied by the availability of waste fuels suitable for combustion in boilers. Examples are found in such segments of industry as pulp and paper, chemicals, petroleum, and steel, plus certain types of raw-material processing, such as bagasse and grain refuse.
- A balanced growth of electric and steam heating requirements, such as found in universities, hospitals, penal institutions, and some district-heating schemes in metropolitan areas. Here, backpressure power generation in a cogeneration cycle has marked economic advantages.

In recent years, because of energy uncertainties in many sections of the country, there has been a decided move back to on-site generation, which makes these exceptions more viable, more attractive today. A notable technology providing on-site power is cogeneration, which is covered in Sec. 2, Chap. 7.

Powerplant Studies

The starting point of an industrial powerplant is an engineering study. Power and steam loads must be ascertained and costs estimated before construction can be considered. In some cases, the organization for which the powerplant is being built may have an engineering staff of sufficient size and experience to make preliminary studies, develop the detailed design, evaluate bids, award contracts, and supervise construction. Most often, however, a consulting engineer is called in to perform one or more of these functions.

Preliminary Report. In any event, a preliminary report must be prepared to obtain authorization of capital funds. Not only does the report consider the total investment, but also it evaluates outlays for such items as operation, maintenance, depreciation, insurance, interest, and taxes. With its primary emphasis on economic factors, the preliminary report must be written so as to be intelligible to those whose background may be in finance or law rather than engineering. And whether the report deals with a new powerplant or an extension to an existing one, it always includes a tabulation of steam and power requirements.

A salient part of the preliminary report is the charting of anticipated loads for different conditions, such as daily load curves for winter and summer, weekdays, weekends, and holidays. Special consideration must be given to any unusual operating conditions and to the time of peak loads. If a manufacturing operation incorporates some process equipment with marked swings in demand, a detailed study should be made of the nature and frequency of the operation and its steam and power demands.

Load Curves. Typical load curves, as shown in Figs. 2.4, 2.5, and 2.6, help to determine the size of such equipment as boilers, turbines, and auxiliaries. Studies must then be made of the capability of the equipment to meet not only the conditions plotted on the basis of past experience, but also those forecast for a limited time in the future. Load curves should be made for both power and steam requirements, and due consideration should be given to the relative growth of power and steam in the future.

The engineer assigned to study the power and steam requirements must become thoroughly acquainted with the operating characteristics of the new plant or existing-plant extension. In the case of a hospital supplied by an isolated cogenerative power/heating plant, dependability of service is of paramount importance. The same is true of an industrial process where an emergency may be costly and possibly hazardous. In other situations, power and heating interruptions may be inconvenient but may not cause severe problems or losses.

As a final goal, the preliminary report should offer conclusions and recommendations, including estimates of the required capital investment along with operating costs and fixed charges. In most reports of this nature, a number of analyses of economic alternatives are offered to aid management in making decisions prior to authorization of construction.

Design Investigations

Detailed design follows acceptance of the preliminary engineering study and authorization to proceed. Some parts of the engineering study are investigated more

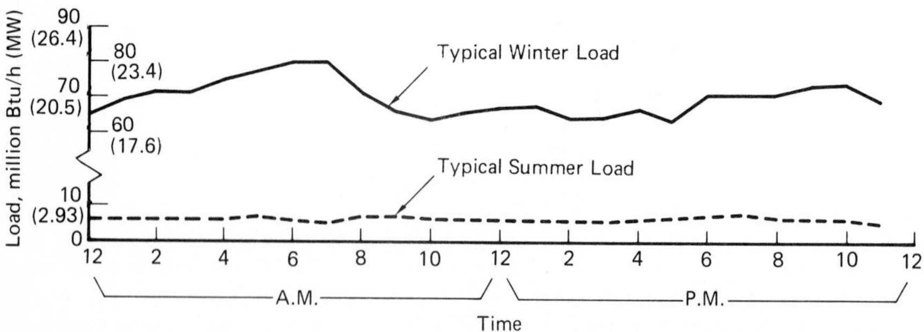

FIG. 2.4 Summer and winter heating loads in a high-temperature-water installation.

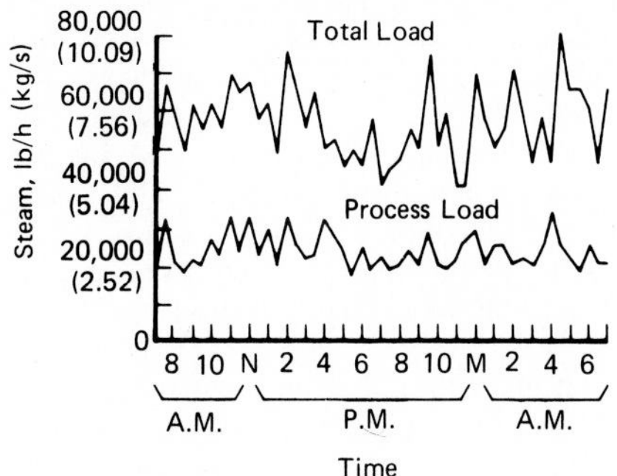

FIG. 2.5 Widely fluctuating process load curve.

thoroughly than others in developing the detailed design. The studies are coordinated with the purchase of materials and equipment and the creation of detailed construction drawings.

The number and scope of design investigations vary from plant to plant. As a minimum, decisions must be made as to choice of cycle, selection of fuel, number and types of auxiliaries, extent of instrumentation and automatic control, and plans for isolated or interconnected operation.

Choice of Cycle. In the design of a new plant, careful consideration must be given to the choice of thermal cycle. If an existing plant is to be enlarged, consideration might be given to "topping" the plant with a steam cycle operating at a higher pressure. Alternatively, the same cycle may be retained and expanded.

If electric power generation is a factor, the choice will be that of a superheated steam cycle. Although an almost infinite variety of steam conditions could be studied, designers usually limit themselves to choosing among a few widely used regenerative cycles for which standard equipment is available.

If space heating and process use constitute the entire load, then it is possible to make a choice between a steam and a high-temperature water cycle.

High-Temperature Water Systems. After World War II, high-temperature water systems became quite common. Conversely, the use of steam has been widespread since the early development of central heating. Much equipment is on

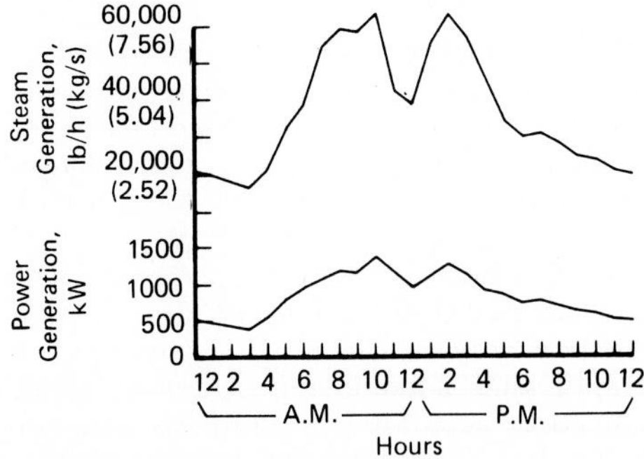

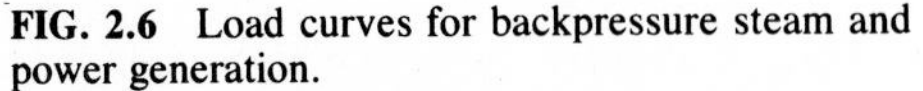

FIG. 2.6 Load curves for backpressure steam and power generation.

the market to tap steam as a source of heat, although some of this can be adapted to high-temperature water by the addition of heat exchangers.[1]

Because of its relatively large heat-storage capacity, a high-temperature water system provides a means of close temperature control. The heat-storage capacity also forms a reserve so that large-use demands may be met without affecting the temperature of individual heat-consuming units. The distribution lines for a high-temperature water system are much smaller than for a steam system of the same heating capacity, and no steam traps or condensate pumps are needed. Each of these factors must be evaluated in terms of the requirements of the proposed installation.

Powerplants serving institutional complexes may have to distribute steam or high-temperature water over a large area, with terrain varying sharply in elevation. Besides space heating, there may be limited requirements for low-pressure steam in kitchens and laundries. Distribution and return lines must be designed, and heat exchangers may be needed for local sources of steam or to reduce thermal losses.

Steam vs. High-Temperature Water. The choice of steam versus high-temperature water cycles must take all the previous factors into consideration as well as recognize that institutional powerplants such as prisons may have to function with somewhat less than the highest quality of operating and maintenance personnel.

Large industrial powerplants with substantial process and electric loads are likely to use more advanced cogeneration cycles. Some of these plants may be owned by an electric utility and operate under a sales contract in which the industrial plant purchases both steam and electric power. Conversely, sometimes the industrial or institutional powerplant sells its excess power to the utility.

Choice of Fuel. The selection of fuel is based on a combined investigation of availability, cost, and operating requirements. Most industrial powerplants use solid, liquid, or gaseous fuels, either singly or in combination. Although generally these are commercially available fuels, sometimes they are by-products of manufacturing and processing.

When comparing fuel costs, one must consider these expenses: the base price of fuel, fuel delivery, handling and reclaiming, labor (including firing and disposition of refuse), plant services, and fixed charges on fuel-handling and fuel-burning equipment. Careful study of these expenses, adjusted to an annual basis, will provide a foundation for fuel selection. Other less tangible factors should also be evaluated, such as reliability of supply, future availability, individual fuel cost trends, ease of conversion from one fuel to another, and extent of future plant expansion.

Charges for by-product fuels depend on the method of accounting for process and powerplant costs. In some industries, such as steel and petroleum refining, by-product fuels must be either burned in powerplant boilers as fast as they are produced or consumed as atmospheric flares. Other by-product fuels, such as some types of wood and plant refuse, may be stored for limited periods.

Choice of Auxiliaries. No steam powerplant is complete merely with a boiler and a turbine. Auxiliary equipment in the form of fans, pumps, heaters, tanks, and piping is essential. Frequently, heat-recovery equipment is added to boilers, and generally some form of water-conditioning equipment is required. Both fuel- and ash-handling systems are necessary if solid fuel is fired. If not all electrical requirements can be satisfied through backpressure operation, then a condenser is a must.

The designer of an industrial powerplant must investigate to some degree each of these auxiliaries. In the case of tanks and piping, the investigation may be to determine optimum sizes and selection from alternative piping and equipment options. In other cases, the designer may decide only whether or not to use the auxiliary equipment.

Instrumentation and Automatic Control. Industrial powerplants vary widely in their use of instrumentation and automatic control. Instrumentation may be installed as an aid to operation; as a means of keeping records of the use of fuel, steam, and electricity; or for both purposes. Automatic controls may be specified to reduce operating personnel to a minimum and to assist in maintaining the operation at a high level of efficiency. Microprocessor-based electronics are often featured in monitoring and control systems.

Instruments assist in the operation of a powerplant as well as in the collection of information about the cost of steam and power consumption. The designer must keep in mind the type of data that must be available at all times to the operator and must select the correct instruments to provide this information. Some of these instruments will be of the indicating type, while others will record data over prolonged periods.

Every industrial powerplant will incorporate instrumentation to indicate boiler and turbine loading. It is also common practice to provide instruments to show various steam and flue-gas temperatures, air and steam flow, feedwater and steam pressures, air and flue-gas pressures, and electrical outputs. The extent of this instrumentation will depend on the type of powerplant and the amount of information required for accounting purposes.

Drawings and Specifications. To build a powerplant, drawings and specifications must be prepared in order to obtain bids and exercise proper supervision over purchasing and construction. The objective of such drawings and specifications is to describe the work to be done, primarily from the point of view of the results to be achieved. The engineer must exercise extreme care that the drawings are clear, concise, and capable of but one interpretation. The drawings describe the work graphically and dimensionally, while the specifications represent verbal descriptions of important supplementary requirements.

Performance Specifications. In specifying powerplant equipment, it is sometimes expedient to write a performance specification, with only such descriptions or physical limitations as are necessary to ensure the desired quality of materials and work. Within this framework, the manufacturer is given as much freedom as possible to provide equipment that will best fulfill the functional requirements of the installation. By reason of specialized experience, frequently a manufacturer can best determine the detailed design of the particular equipment it supplies, given all the factors for performing economic analyses of available alternatives.

ELECTRIC-UTILITY POWERPLANT DESIGN

The objective here is to show the steps that must be taken and the decisions that must be made in the preliminary design of a central station for the generation of electric power. These steps and decisions must precede the detailed final design, which involves the efforts of large teams of engineers, designers, and drafters.

Designers of electric utilities are constantly thinking about the generating facilities that must be in operation 10 or more years in the future. The final design and construction of new capacity will typically require not less than 4 years. For these reasons, all basic decisions must be made at least that long before the capacity is needed.

Management must find the best practical answers to four fundamental questions about the provision of added capacity: When? How big? Where? What kind? In more formal language, the following steps must be taken:

Load Forecasts

The answers to When? and How big? are approached by forecasting future electric demands. Such activity is a continuous function in a major utility and necessarily involves judgment in interpreting company records, economic trends, population shifts, technological changes, and regulatory limits.

There are two significant characteristics of electric demand (usually called *load*) to watch for: the peak load, which is the maximum demand on the generating plant for a relatively short period, such as 15 min or 1 h, and the average demand over a longer period, usually a month or a year.

Load Factor. The ratio of minimum to average demand is called the *load factor* and typically is in the range of 40 to 60 percent.[2] A system with a relatively high load factor is fortunate in that it applies, on the average, more of its installed capacity profitably; it spreads its investment costs over a greater production than would be true of another system of the same size, having the same peak load but with a lower load factor.

The high-load-factor system can thus afford to spend more money for more efficient equipment that will reduce operating costs. The chief of these is the cost of fuel, expenditures for which vary directly with station output. Improvements in station thermal efficiency are one way to reduce the outlay for fuel.

The records of electric demands are usually kept in three major classifications: residential, commercial, and industrial. Special records applying to unusually large users of power are also available.

The forecaster studies the history of each type of load, but chiefly gathers all available data on the growth of the area being studied. The forecaster watches population shifts, growth of suburban shopping centers, trends toward air conditioning and electric heating, changing processes in industry, development programs to attract new employees, statistics on per-capita use of electricity—in short, every factor that bears on the future consumption of electric energy in the area under investigation.

Peak Load. The result of forecasting is a graph showing the expected peak loads and load factors for a period of years into the future. An example of a peak-load forecast is the line *X-X* in Fig. 2.7.

To be reasonably sure of carrying the peak load in any given year, the total generating capacity of the system must exceed the expected peak load by a margin for reserve. The necessary reserve is of two kinds:

- Spinning reserve is the excess generating capacity that is in operation and on the line at the time of peak load. It must be at least equal to the capacity of the largest single generating unit in service at that time on the system or its interconnections. Spinning reserve is necessary to guard against the possibility of a

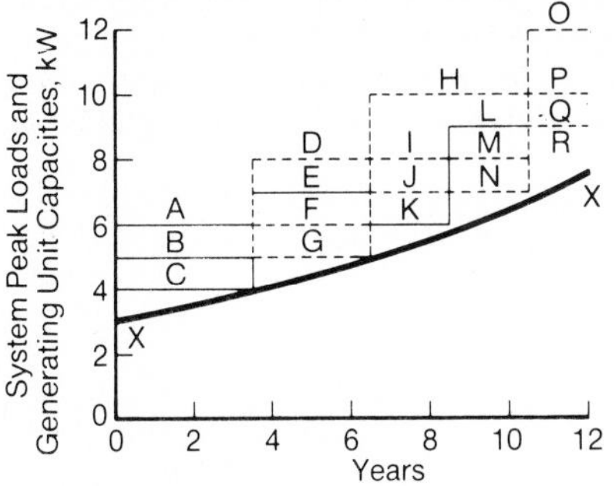

FIG. 2.7 Relationship between peak loads and installed capacity.

mishap to the largest unit causing loss of its generating capacity, even if only for a few minutes.

- Reserve for scheduled outages is generating capacity that is installed on the system but is not necessarily in operation at the time of peak load. Such reserve is needed because equipment must be inspected and overhauled at regular intervals. The amount of this reserve must not be less than the size of the largest unit that will be overhauled during the peak load season.

System Reserve Requirements

A simplified example will illustrate reserve requirements. Assume a system is composed of 12 identical generating units, each capable of generating a continuous output designated as *M,* kW, where *M* is any number. The total installed capacity is 12 *M,* kW. The spinning reserve is one unit, and the reserve for scheduled outages is another unit because an overhaul typically takes about a month. This system can be expected to carry a peak load of 10 *M,* kW, its firm capacity.

If the system consists of six units, each of 2-*M*-kW capacity, then the installed capacity is the same as before, but deducting the spinning reserve will bring the capacity down to 10 *M,* kW, and the further deduction of reserve for scheduled outages will leave a firm capacity of only 8 *M,* kW. However, it may be possible to schedule the overhauls of six units wholly outside the peak-load season, depending on the character of the system load variations through the year; this would make the firm capacity of the system 10 *M,* kW, as before.

Another Example. Another example is indicated in Fig. 2.7. The system is assumed to consist of six identical units each of *M*-kW capacity. Line *A* indicates present installed capacity. Deduction of spinning reserve gives line *B,* and further deduction of reserve for scheduled outages leaves the firm capacity, line *C.* Another unit must be in operation before the peak-load season of the fourth year. If this unit is of *M*-kW capacity, the new situation is indicated by lines *E, F,* and *G.* Similarly, other *M*-kW units must be added, one before the seventh year, one before the ninth year, and so on.

If the seventh unit is twice the size of the earlier ones, the installed capacity in the fourth year is shown by line *D.* Spinning reserve must now be 2 *M,* kW, the size of the largest unit. Reserve for scheduled outages, however, will be *M,* kW, making the firm capacity line *G,* the same as if unit 7 were of the smaller size. It follows that installing a large unit increases the firm capacity of a system only by an amount equal to the capacity of some smaller unit.

When the second larger unit is installed, as is indicated for years 7 to 10, the firm capacity will increase by 2 *M,* kW, if it is possible to schedule outages of the two larger units out of the peak-load season (lines *H, I,* and *J*). Otherwise, line *K* will give the firm capacity.

Further additions of capacity are also shown in Fig. 2.7. Solid lines apply to *M*-kW units and dashed lines to the larger size of 2 *M,* kW.

The foregoing is known as the *block system* of generating capacity addition. The objective is to install the equivalent of two blocks more than the peak load—one to provide spinning reserve, the other to permit scheduled outages. This method of capacity addition has the virtue of simplicity and has been commonly used since the early days of the utility industry.

System Capacity Additions

Several factors have led to modifications and deviations from the above method of capacity addition. Not only has load grown but so has available unit generating size. At the same time, knowledge of system operation has become more complete, and the reliability of components is better understood. Thus entire systems may be simulated by mathematical models, and unit outages may be predicted by probability methods.[3]

By using this approach, the necessary system reserve requirements may be less than under the block system, but the largest single unit should not exceed 5 to 10 percent of the total system capacity. As a matter of policy, many systems refrain from increasing unit size until the predicted load growth requires successive units of the larger size.

In practice, these considerations for system capacity additions are far from clear-cut, for a number of different reasons:

- The units will be of different types, sizes, ages, and vulnerability to accidental shutdown.
- The actual capacity of a plant varies with condensing water temperature and at times is subject to variations caused by changes in the fuel.
- A system usually includes transmission lines that are subject to interruption of service from bad weather and other such causes quite different from those that affect generating units. Outages from such interruptions may be greater than the loss of a single unit.
- There is the possibility that not only one unit, but two or more may suffer unexpected interruptions simultaneously or following closely on one another.

The graph of Fig. 2.7 plots predicted annual peak loads for certain future years. Another informative graph is constructed by getting from system records the numbers of hours per year in which the load equaled or exceeded a given amount. The system minimum load will correspond to 8760 h/yr, and the maximum load will appear, in all probability, for only 1 h. Such a graph is called a *load duration curve;* Fig. 2.8 represents an idealized example.

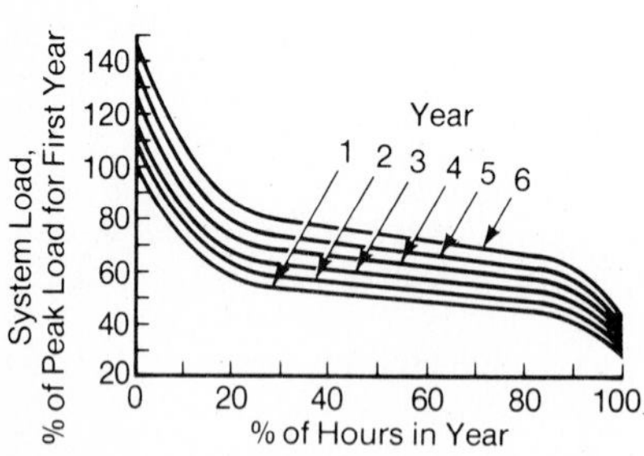

FIG. 2.8 Load-duration curves plot steam load against percent of hours in year.

The chief value of the load duration curve is to emphasize the brief period during which peak loads must be generated in the typical system. Even though the curve is drawn for a past year, its shape will not change markedly in the future unless the character of the load changes materially. Load-duration curves, therefore, can be drawn for future years, subject to the uncertainty of the predicted peak loads and load factors discussed earlier.

In Fig. 2.8, assume for the moment that all equipment in the system is in good condition and can be expected to operate successfully throughout the period under study. Given this assumption, the problem of meeting the future growth in peak load can be solved simply by providing the cheapest possible generating capacity large enough to take care of the growth in load. Fuel economy is not of prime importance because it will be used only a few hours per year, whereas investment costs go on continuously and must be kept to a minimum.[4]

Choice of Powerplant Additions

Generally, load distribution requirements for most utility systems follow the pattern of Fig. 2.8. A typical utility may have three distinct load requirements—base, intermediate, and peak. Prior to the mid-1970s, the traditional methods of meeting these load requirements were to have large nuclear and high-efficiency fossil-fuel steam plants designed for 6000 to 8000 h of operation per year supplying base load. Specially designed fossil-fuel steam plants and former base-loaded fossil plants would supply the intermediate load. Low-efficiency installations required to operate 500 to 2000 h/yr would supply the peak-load and reserve requirements of the system. Gas turbines, old steam plants, and hydro pumped-storage plants would supply this latter capacity.

Traditionally, utilities purchased new additions of the largest and most efficient steam cycle available. If the availability factor was judged to be decreased somewhat with such an optimal selection, then some compromise may have been made. These new units were assigned to base-load service and were expected to operate at a capacity factor equal to their availability for approximately the first 10 yr of commercial operation. This relegated previously base-loaded plants within the system to cyclic or lower-load-factor operation.

Currently, several factors are operating to change this pattern. Improvements in the heat rate have diminished. Utilities with a mix of nuclear and new fossil-fuel units may require the fossil-fuel plants to enter cycling service directly. Although the new fossil-fuel units may be designed for an efficient thermodynamic cycle, tail-end emissions control systems may have high parasitic power requirements. Similarly, the new plant may be fueled with low-sulfur coal transported great distances at considerable expense. Such factors may result in the new fossil-fuel units having higher generation costs than some of the older plants, with a corresponding influence on the selection and loading of equipment for the most economic dispatch. Finally, cogeneration powerplants at industrial and institutional sites are changing the load patterns facing many utilities, especially where buy-back provisions exist for excess power generated at the sites.

Final decisions as to size and timing may well be tied in with decisions concerning location and type of equipment, in addition to licensing and regulatory requirements that can result in exceedingly long construction lead times. However, the analysis and study described to this point will disclose a relatively small number of practical possibilities, each of which must be studied in some detail to set up comparative investment and operating costs over a period of years. At the same time, it is important not to neglect the effect of each possible alternative on the cost of operating the existing system.

Selection of Plant Location

The location of an electric generating station is determined by analysis of many factors that influence the selection in diverse ways. During the study and forecasting of future loads, it usually becomes apparent that only a few attractive sites are available. Each must be studied to determine the effect of individual factors on its desirability. The object is to make an engineering economic analysis that will disclose the best choice, which in general is the location that will result in the lowest total cost to the owner in the long run.

Condensing-Water Supply. As is well known, a steam turbine requires a substantial quantity of condensing water for its economical operation. Under average conditions estimates show that about 800 tons of water are needed for each ton (725 Mg for each megagram) of coal burned. The supply of water should not vary through the year, its temperature should be as low as possible, the water should not be corrosive to the usual materials, and the water should not contain suspended material that will interfere with the flow through pumps and tubes. These considerations suggest a river, lake, or ocean, and central stations are located whenever possible on the bank of a body of water like these.

The alternative to such a location is to provide cooling towers in which forced- or natural-draft circulation of air cools the condensing water nearly to the prevailing wet-bulb temperature. Although the investment in cooling towers is high compared to once-through cooling, the high cost is somewhat offset by the smaller waterfront construction and tunnels.

The extent to which condensing water is available limits the amount of power that can be generated at the site. Advanced steam conditions not only achieve greater cycle efficiency but also result in less rejection of heat to the condenser. For example, for the same initial steam conditions and a limited amount of condensing water, selection of a reheat regenerative cycle permits greater power output than a nonreheat cycle, everything else being equal. This selection can have considerable impact on central stations that are being redeveloped on sites having limited amounts of cooling water. Figure 2.9 shows the reduction of heat rejection to condenser cooling water as cycles of increased thermal efficiency and decreased net plant heat rate are incorporated in powerplants.[5]

Other Considerations. Additional concerns include the availability of municipal services and suitable labor, appropriate zoning, convenient access, climate, tax situation, and the like. Room for growth and freedom from undue risk of flood and earthquake are ordinarily essential. The nature of surrounding installations may play a part in the selection. For example, the height of the stacks will be subject to Federal Aeronautics Administration regulations and would be restrictive near an airport.

In many site selection problems, one possibility is to add to an existing station. To do so creates obvious economies in making use of existing facilities, no matter how much new construction may have to be provided. If the existing station site is large enough and has enough condensing water, and especially if it was designed for expansion, it is unlikely that a new site can compete successfully. Each case must be studied on its merits; the danger lies in assuming that the obvious answer is the best.

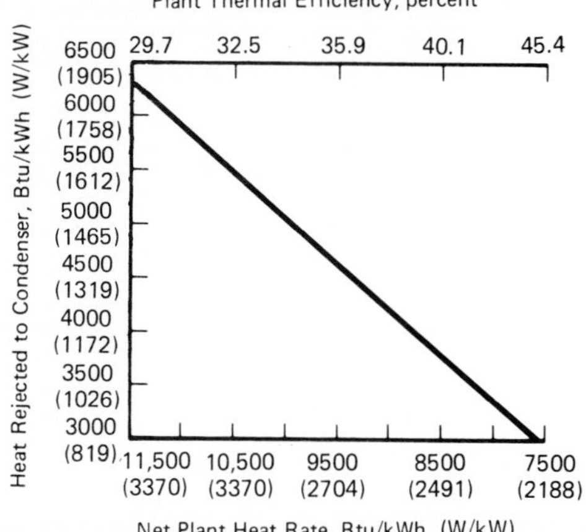

FIG. 2.9 Heat rejected versus plant heat rate based on 1.5 in Hg (5 kPa) backpressure.

Selection of Equipment Types

Once the size, timing, and location of the plant have been determined by the steps previously described, the question arises, What kind? In practice, the steps are not taken separately and in succession; they overlap considerably but may follow the sequence described below.

The primary choice is among the several plant fuel cycles: conventional fuel, nuclear energy, and internal combustion (diesel or gas turbine). The possible consideration of hydro or purchased power would occur before site selection and is outside the scope of this book. Strictly on the basis of total economy, steam with coal frequently is preferred in the larger sizes, with internal combustion having an advantage for smaller units.

The size range within which each type of plant fits best is reasonably well defined by experience. Capacity installed for peaking purposes may be subject to very different economic evaluation than is the case for base-load generating equipment. Heat rate economies dominate the latter, while availability of peak shaving capacity for short periods is most important in the former.

Cycle Characteristics. Once a conventional steam cycle has been selected, studies are made to find the best conditions with regard to steam and reheat pressures and temperatures, number and location of extraction stages, condenser pressure, and a host of similar cycle characteristics.[6] Only with experience and judgment as a guide is it possible to keep the combinations of variables within practical bounds. Even with the help of advanced methods of analysis, the basic method of solution is to assume a reasonable set of conditions and calculate the required investment in the central station and the resulting cost of operation under future load conditions.[7]

The sizes of the units will have been fairly well determined by the studies of load growth and estimated capital cost. The unit system, under which a single steam generator serves a single steam turbine, each with its own auxiliaries, is rather generally accepted. But there may well be exceptions, especially when a station has a process-steam load in addition to an electric load.

At this point, a number of subsidiary economic studies are required, for example, heater and condenser surface, turbine- versus motor-driven auxiliaries, voltage of electric auxiliary drives, and extent of building enclosure. The object is always to reduce the initial investment without increasing maintenance and operating costs unduly.

Combination Comparisons. The several assumed combinations are compared by standard methods of engineering economy to find the one that promises the lowest overall cost over the life of the station, giving due regard to the time value of money.

The result of this engineering process is to produce a set of preliminary arrangement drawings showing the station as it has been conceived; a set of abbreviated specifications covering the important equipment, structures, and systems; a reasonably reliable construction cost estimate; and a report covering the alternatives considered, advantages and disadvantages of each, comparative investment and operating costs, and other data specific to each case. Management will be able to base its decision on this material if it is complete and well presented. Then and only then can the detailed procurement and design process begin. The steam-turbine generator and the steam generator are purchased first, followed by other major components as dictated by their respective design and fabrication lead times.

As the equipment is purchased, hundreds of studies of the comparative economics of the vendor offerings are made, based on capital costs, energy costs, and probable costs of operation and maintenance.

TYPES OF COMBINED CYCLES

A *combined cycle* is a term that describes the thermal relationship among gas turbines, steam generators, and steam turbines in a powerplant, in a special configuration designed to boost plant efficiency. Combined-cycle installations fall into four broad classifications, each primarily dependent on how the steam generator is used in conjunction with the gas turbine:

- Gas turbine plus unfired steam generator
- Gas turbine plus supplementary-fired steam generator
- Gas turbine plus furnace-fired steam generator
- Supercharged furnace-fired steam generator plus gas turbine

Gas Turbine plus Unfired Steam Generator

In this concept, a steam generator is installed at the discharge of a gas turbine to recover the energy in the gas-turbine exhaust and to supply steam to a steam turbine (Fig. 2.10). All the fuel is fired in the gas turbine, and the steam generator depends entirely on the gas turbine for its heat. The steam generator is designed for low steam pressures and temperatures, with the available gas-turbine exhaust

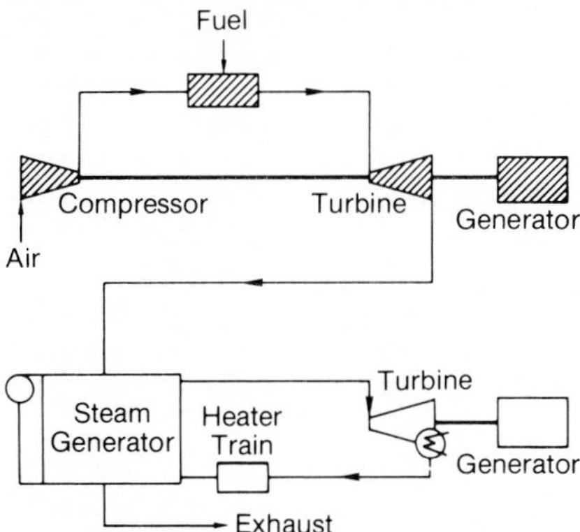

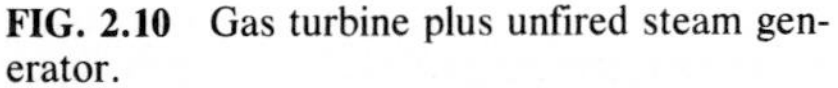

FIG. 2.10 Gas turbine plus unfired steam generator.

temperature limiting practical values to approximately 850 psi and 825°F (5860 kPa and 440°C).

In most applications, the steam turbine will produce about 30 to 35 percent of the total plant output, with the remaining 65 to 70 percent being supplied by the gas turbine(s). Because the power from the steam turbine will be produced without any additional fuel input and only a small decrease in gas-turbine efficiency occurs because of the backpressure of the steam generator, the overall plant thermal efficiency will be improved over that of open-cycle gas turbines. Depending on the gas-turbine exhaust temperature and the particular steam cycle and fuel, cycle efficiencies as high as 35 percent may be obtained.

Gas Turbine plus Supplementary-Fired Steam Generator

Gas-turbine exhaust contains from 16 to 18 percent oxygen and may be used as an oxygen source to support further combustion. Therefore, a modification of the simple waste-heat application is the addition of a supplementary firing system in the connecting duct between the gas turbine and the steam generator (Fig. 2.11). The firing system utilizes a portion of the oxygen contained in the gas-turbine exhaust and is selected to limit the maximum gas temperature entering the steam generator to approximately 1350°F (730°C).

With a given gas-turbine size and the above gas-temperature limit, the steam generation will double that of a simple waste-heat application, and the steam turbine will supply a greater proportion of the plant load—approximately 50 percent of the total electric output. The higher steam generator inlet-gas temperature will allow steam conditions to be increased to levels of 1500 psig and 950°F (10 300 kPa and 510°C). The steam-turbine designs are nonreheat and may be either condensing or noncondensing. For most arrangements, the final steam conditions are primarily based on steam-turbine economics. Cycle efficiencies of 34 to 38 percent are possible with this combination of equipment.

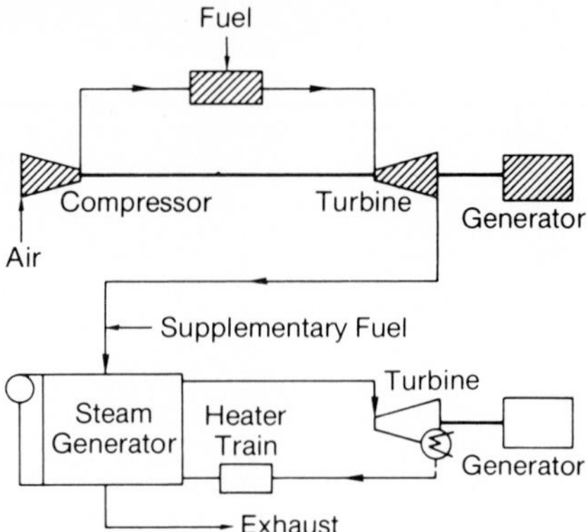

FIG. 2.11 Gas turbine plus supplementary-fired steam generator.

Gas Turbine plus Furnace-Fired Steam Generator

The previous cycle used only a small portion of the available oxygen in the gas-turbine exhaust. Another adaptation is the design of a plant which uses essentially all the oxygen in the turbine exhaust to support further combustion. Typical gas turbines operate with 300 to 400 percent excess air and thus can support the combustion of approximately 3 to 4 times as much fuel in a downstream boiler as was burned in the gas turbine. The majority of fuel can now be fired in the boiler, and 70 to 85 percent of the total plant power generation will be supplied by the steam turbine, with the remaining portion supplied by the gas turbine (Fig. 2.12).

In this cycle, the gas turbine may be considered as both an independent power supplier and a forced-draft fan for the boiler. Any of the high-pressure, high-temperature steam conditions found in modern steam turbines can be incorporated into this combined cycle. And although the gas-turbine fuel in the previous

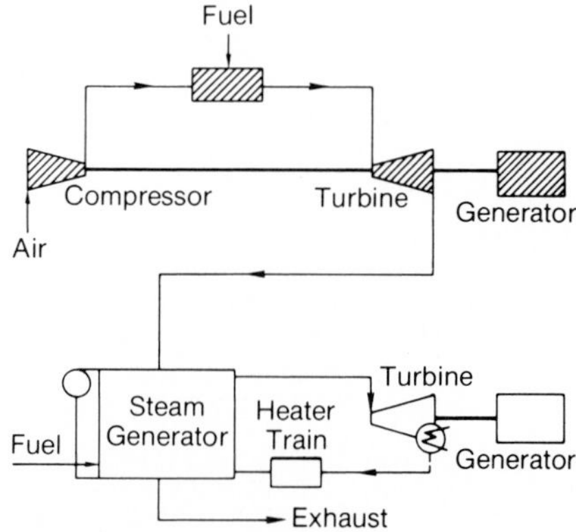

FIG. 2.12 Gas turbine plus furnace-fired steam generator.

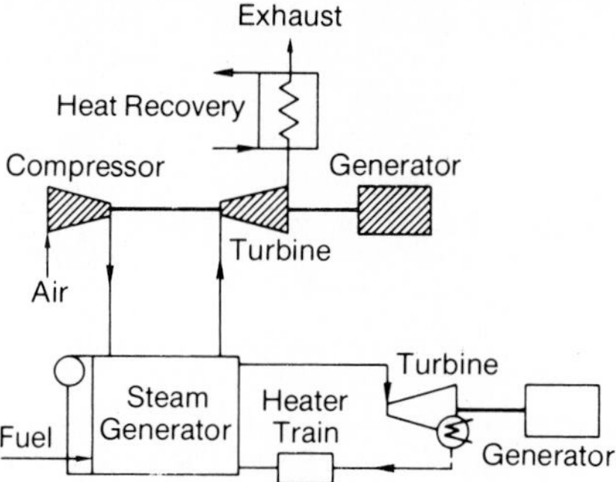

FIG. 2.13 Supercharged furnace-fired steam generator plus gas turbine.

arrangement is presently limited to gas and oil, this cycle allows the use of any fossil fuel in the steam generator. Cycle efficiencies of about 36 percent have been obtained.

Supercharged Furnace-Fired Steam Generator plus Gas Turbine

Another configuration for a combined cycle includes the introduction of a steam generator between the air compressor and the gas turbine (Fig. 2.13). The air compressor serves as a forced-draft fan and pressurizes the boiler, where all the fuel is fired. The products of combustion, having been partially cooled within the boiler complex, are then discharged through a gas turbine. Additional heat is recovered by heat exchangers installed at the exhaust of the gas turbine. They are used as economizers and/or feedwater heaters.

Although gas turbines require relatively high excess-air quantities, the firing system for the supercharged furnace is designed to operate with excess-air levels commensurate with conventional units; the air compressor is selected for this small air capacity. The steam turbine supplies the majority of the plant's electric power, with the gas turbine either selected to provide sufficient power to drive the air compressor (Velox cycle) or sized to supply additional electric power. Cycle efficiencies range from 34 to 38 percent.

ACKNOWLEDGMENT

Illustrations in this chapter are courtesy of *Combustion Engineering Inc.*

REFERENCES

1. Owen S. Lieberg, *High Temperature Water Systems,* Industrial Press, New York, 1958.
2. *Glossary of Electric Utility Terms,* Statistical Committee, Edison Electric Institute, Washington, D.C.
3. "Application of Probability Methods to Generating Capacity Problems," AIEE Paper

CP60-37, American Institute of Electrical Engineers, 1960. C. J. Baldwin, ''Modern Scientific Tools Used in the Power Industry for Tomorrow's Problems,'' *Proceedings of the American Power Conference,* **24**:94–105, 1962.

4. R. D. Brown and D. A. Harris, ''Large Coal-Fired Cycling Units,'' ASME-IEEE-ASCE Joint Power Generation Conference, Portland, Sept. 28–Oct. 2, 1975.
5. Stanley Moyer, ''Industry's Water Problems,'' ASME Paper No. 6 WA-141, American Society of Mechanical Engineers, 1961. Also in condensed form in *Mechanical Engineering,* **84**(3):46–49, March 1962.
6. James W. Lyons, ''Optimizing Designs of Fossil-Fired Generating Units,'' *Power Engineering,* **83**(2):50–56, February 1979.
7. W. A. Wilson, ''An Analytic Procedure for Optimizing the Selection of Power-Plant Components,'' *Transactions of the ASME,* **79**:1120–1128, July 1957.

CHAPTER 1.3
STEAM GENERATORS

Staff Editors
Power *Magazine*
New York, NY

INTRODUCTION

Steam generators are crucial to the modern powerplant. They convert the energy from fossil fuels to a usable form—steam—for powering the turbines which, in turn, drive the electric generators to produce electricity—the *raison d'être* of the powerplant. Indeed, most of the chapters in this handbook provide information that builds on this chapter, adding the details to complete the picture of our understanding of steam generator design and operation.

Basically, the steam generator has two separate but related loops—the combustion loop and the steam-water loop. Two common designs arising from this combination are the watertube boiler and the firetube boiler. The latter is essentially a water-filled cylinder with tubes running through it. Fuel is burned at one end of the cylinder, and the hot gas of combustion, the flue gas, passes through the tubes to the other end. In passing through the tubes, the hot gases heat the water to produce steam. The firetube boiler, once quite popular, still sees service in some small industrial applications, as covered later in the chapter.

Of primary interest here, however, is the watertube type of steam generator for large utility and industrial applications.

WATERTUBE DESIGN

In the watertube design, the boiler is basically a box whose walls are made up of tubes through which water flows, thus the expression *waterwall tubes*. The fuel is burned in the box, or *furnace*, as it is usually called. Heat is transferred to the water that flows through the tubes, and thus steam is generated. The flow of water is important not only to generate steam, but also to keep the tubes cool enough that they do not become overheated and fail. Figure 3.1 shows the general arrangement of a modern watertube boiler.

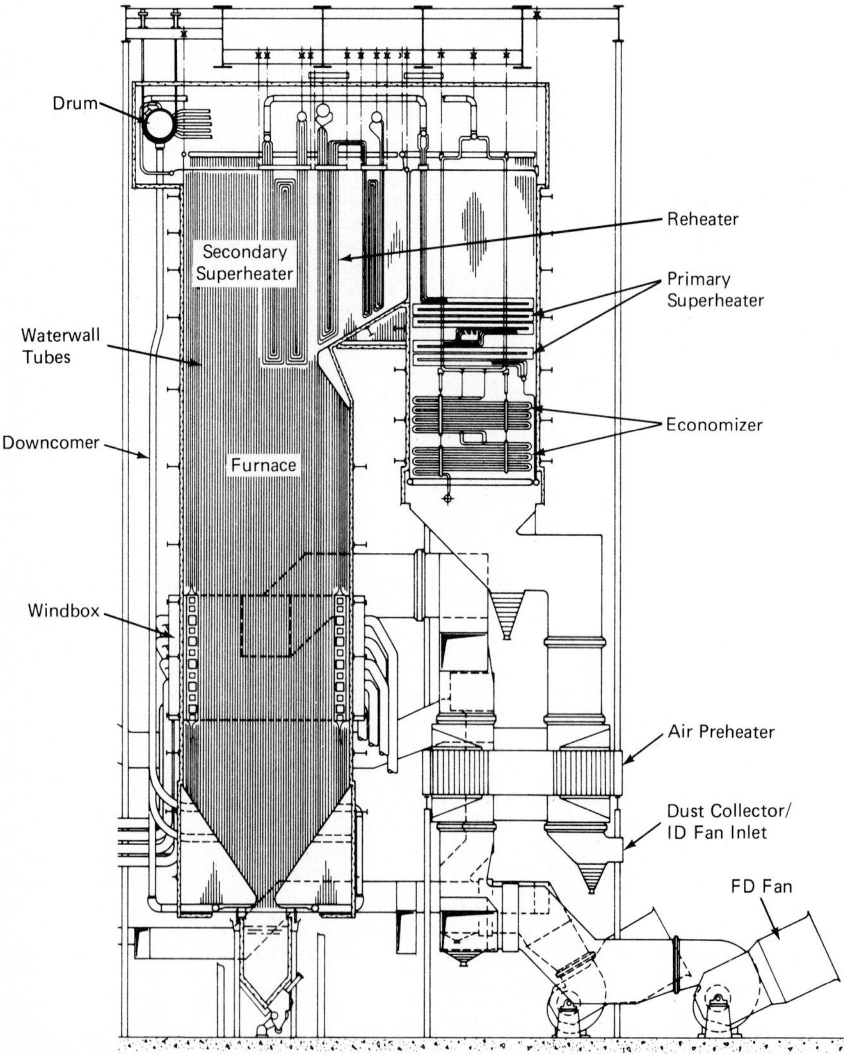

FIG. 3.1 General arrangement of watertube steam generator. *(Combustion Engineering Inc.)*

Steam-Water Loop

Flow through the waterwall tubes is provided by the steam-water loop. In most boilers, water circulates from a drum at the boiler top, through pipes called *downcomers* to the boiler bottom, and then up through the waterwall tubes to the drum again. For many boilers, the force of circulation from convection, or natural circulation, provides enough movement through waterwall tubes for both good heat transfer and adequate cooling of the tubes.

Natural Circulation. In a simplified steam generator with natural circulation (Fig. 3.2), heat is transferred from the burning fuel in the furnace to the water in the waterwall tubes. When the water is heated, its density decreases and it tends to rise in the tubes and flow toward the drum—the convection phenomenon. When the water becomes hot enough to reach the saturation temperature, steam bubbles form that also rise toward the drum because they are much less dense than the water around them. As the heated water and steam rise in the waterwall tubes, cooler water is drawn into the bottom of the tubes from the downcomers, which draw water from the boiler drum. The rate of circulation is much greater than the rate of steam generation in boilers of this type. The ratio of the number of pounds (kilograms) of water circulated to the number of pounds (kilograms) of steam generated may be as high as 8 or 10 to 1.

The rate of circulation because of convection depends on the difference in density between the heated water and steam mixture in the waterwall tubes and the water in the downcomers. That density difference is affected by two parameters: boiler height and boiler operating pressure. The greater the height of the boiler, the greater the difference in pressure (higher at the bottom due to the static head) and thus the greater the density between the bottom and top of the boiler. This is one reason that boilers are often as high as 150 to 200 ft (45 to 60 m).

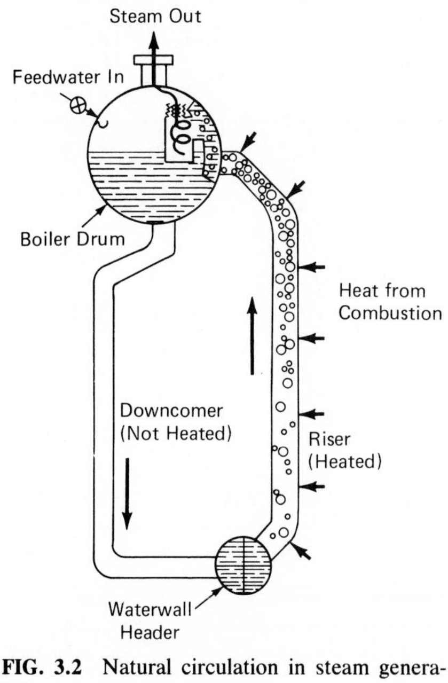

FIG. 3.2 Natural circulation in steam generator.

With boiler operating pressure, the difference in density between steam and water decreases as the pressure increases (Fig. 3.3). If the pressure in the boiler increases to what is known as the *critical pressure,* 3206.2 psia (22 090 kPa), no difference exists in density between steam and water, so there can be no natural circulation. The operating pressure need not be increased to the critical pressure before the point is reached where natural circulation no longer provides adequate flow through the waterwalls. Boilers with pressures as low as 1800 psia (12 400 kPa) may have inadequate natural circulation. Forced circulation is almost always necessary when operating pressures reach 2600 psia (17 900 kPa).

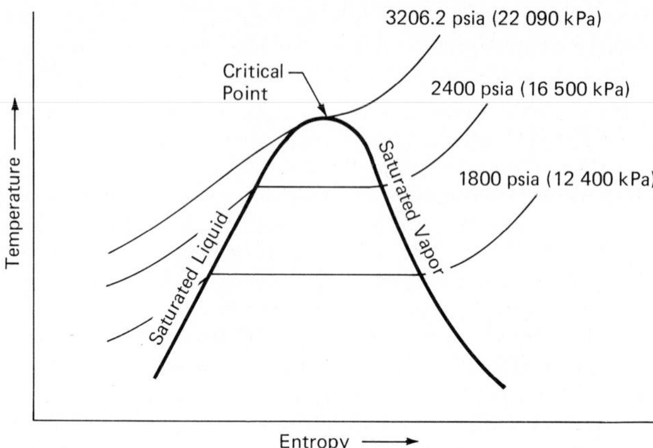

FIG. 3.3 Water phases in steam generator. As temperature rises, a critical point is reached at which density of water and steam is the same.

Forced Circulation. When natural circulation becomes inadequate, a boiler must use forced circulation. In this arrangement, pumps, usually located in the downcomer piping, aid in circulation, which permits higher operating pressures and better control of the circulation (Fig. 3.4).

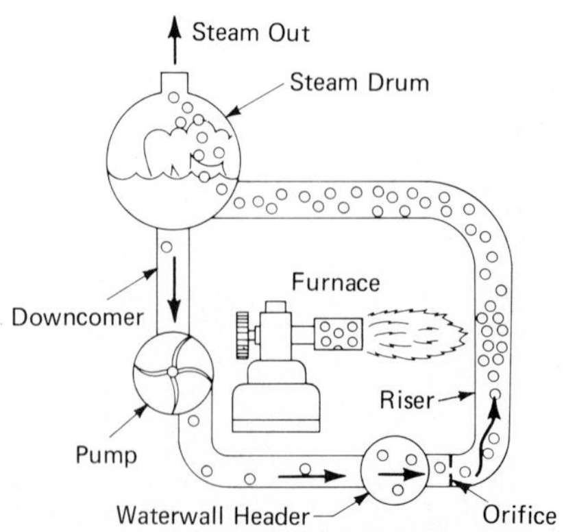

FIG. 3.4 Forced circulation in steam generator.

Circulation in the boiler can also be controlled by orifices, usually located in the bottom of the tubes. They ensure that the flow is evenly distributed through all the tubes in the waterwalls.

Recirculation: Drum vs. Once-through. As noted, boilers can be designed to operate at pressures above the critical pressure. No drum is necessary in such boilers since no separation of the water and steam occurs, the two being at the same density. Thus, there is no recirculation. The water enters the bottom of the tubes and is completely transformed to steam by the time it reaches the top, passing through the tubes only once. For this reason, it is known as a *once-through boiler* (Fig. 3.5).

Boilers can be designed to operate at subcritical pressures with a once-through arrangement, although they are not common. Almost all units that operate at subcritical pressures rely on a drum and recirculation, either natural or forced, and are commonly known as *drum boilers* (Fig. 3.6).

Many supercritical, once-through boilers use recirculation temporarily during start-up when the pressure is still below critical. Generally, a flash tank or mixing chamber and forced-circulation pumps are arranged to form a bypass

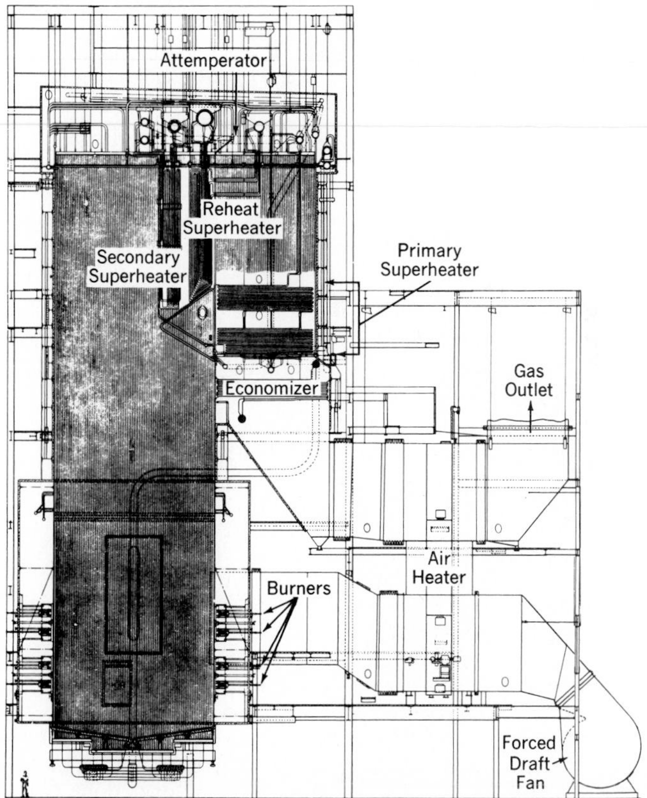

FIG. 3.5 Once-through supercritical boiler for natural gas firing has nominal operating pressure of 3850 psig (26 525 kPa). *(Babcock & Wilcox Co.)*

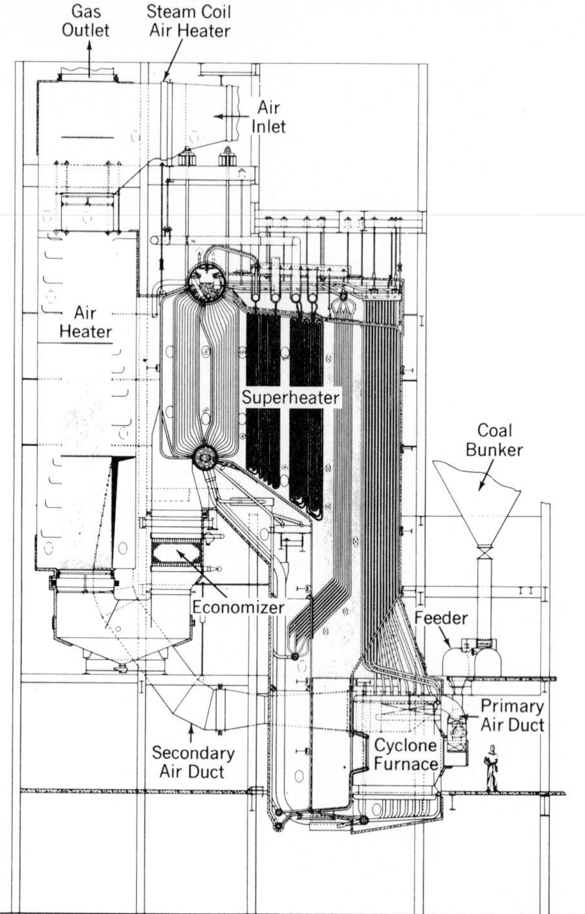

FIG. 3.6 Subcritical drum boiler has nominal operating pressure of either 1900 or 2600 psig (13 100 or 17 900 kPa). *(Babcock & Wilcox Co.)*

(Fig. 3.7). When the pressure and steam flow become large enough to cool the tubes adequately, the bypass is closed.

The steam in the drum of a drum boiler is at saturated conditions. If the operating pressure of the boiler is 2600 psig (17 900 kPa), the temperature is 669°F (354°C). Modern powerplants routinely go to temperatures of 950 to 1050°F (510 to 566°C) to increase efficiency. Thus, it is necessary to add more heat to the steam after it exits the drum and before it is piped to the turbine. A special section of the boiler called the *superheater* is used to raise the steam temperature.

Powerplant efficiency may be boosted by reheating the steam after it has passed through part of the turbine; reheating is sometimes done twice. Special sections of the boiler called *reheaters,* or reheat sections, are provided for this purpose. Superheating and reheating of steam are not confined to drum-type boil-

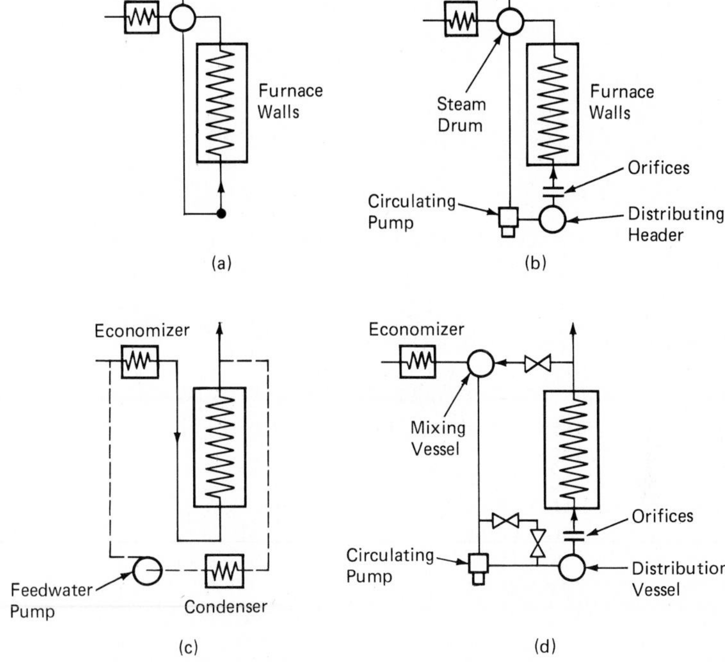

FIG. 3.7 Types of boiler circulation. (*a*) Natural; (*b*) controlled; (*c*) once-through; (*d*) once-through with recirculation (forced).

ers. The steam in a once-through boiler also relies on a superheater to heat steam to the required operating temperature. Moreover, reheat is applied as commonly in once-through as in drum boilers.

Another special heat-exchanger portion of the boiler is called the *economizer,* which draws on hot flue gas to heat the water that enters the boiler. This measure increases efficiency by recovering some of the heat in the flue gas. By transferring some heat over a lower temperature difference than in the furnace and waterwalls, the heat is transferred more efficiently. Superheaters, reheaters, and economizers are covered in more detail in the following chapter.

Heat transfer in the boiler waterwall tubes is complicated by the fact that it occurs as water changes phase to steam. The transformation occurs in two distinct ways: nucleate boiling and film boiling. In nucleate boiling, steam bubbles grow on the inside of the tube walls and are swept away from the walls by the water. In film boiling, a thin film of superheated steam covers the inside of the tube walls because the water flow through the tubes is not great enough, the tubes become too hot, or a combination of the two exists. Heat transfer through the film of steam is much less than that through water, which means that when film boiling occurs, the tubes are likely to overheat.

The point of transition from nucleate to film boiling is called *departure from nucleate boiling* (DNB). It is important in steam generator design and operation to limit heat release in the furnace and to provide enough flow that the point of

DNB is never reached. If DNB occurs, serious damage to the tubes is likely due to overheating.

Combustion Loop

The heating of water and the production of steam have been discussed to this point without identifying the mechanism of heat transfer from the burning fuel. On the furnace side of the boiler, two types of heat transfer predominate: radiation and convection. This situation exists because the fireball from the burning fuel is so hot that a tremendous amount of energy (infrared and other radiation) is generated. (Remember, radiation is proportional to the fourth power of the absolute temperature.)

Almost all the heat transferred to the waterwalls of the boiler is by radiation. After the fuel burns, the flue gas that remains is still very hot, but much less so than the fireball, thus there is negligible radiation. Heat is transferred from the flue gas by convection to other tubes in the boiler. Usually the superheater and reheater are in the convection section of the steam generator, and the economizer is always in this section.

Combustion Air and Flue Gas. The efficient combustion of fuel and transfer of heat in the boiler require that adequate combustion air be provided and that flue gas be removed from the area of combustion (Fig. 3.8). In some small or old boilers, removal is accompanied solely by natural convection, commonly called *natural draft*. The hot flue gas rises through the stack and draws in cool air for combustion.

As boilers become larger, however, natural draft becomes inadequate, and a fan must be added to blow enough air for combustion into the furnace. If the fan is big enough, it will pressurize the boiler furnace and aid in the removal of flue gas. Such a fan is called a *forced-draft* (FD) fan. Note that increasing the velocity of the flue gas through the convection portion of the steam generator increases the convective heat transfer. Figure 3.9 shows a typical pressurized system.

Boilers with FD fans and no *induced-draft* (ID) fans are called *pressurized-furnace boilers* because the boiler furnace is above atmospheric pressure. This situation can be troublesome because flue gas, which is toxic and corrosive, and flyash leak out of the smallest openings in the furnace, causing maintenance and personnel safety problems. Also, some large boilers with considerable convective heat-transfer area need more than just an FD fan to move the flue gas.

The solution to these problems is another fan that takes suction on the furnace at the flue gas exit—the ID fan. Boilers having both FD and ID fans include dampers or inlet vanes on the fans to balance them, and they operate at a slightly negative furnace pressure, −0.05 in H_2O (101 kPa). With this arrangement, the leakage of flue gas and flyash is eliminated. The configuration is called a *balanced-draft boiler* (Fig. 3.10).

Combustion Air Heater. Another feature seen in combustion air and flue gas systems of most large, modern utility boilers is the combustion air heater. The flue gas that exits from the steam generator is often as hot as 600°F (315°C) and represents a major loss of heat and a source of inefficiency in the powerplant. One way to reduce this inefficiency is to design a heat exchanger so that the flue gas can heat the incoming combustion air; in this fashion, it is possible to reduce the temperature of the exiting flue gas to 300°F (150°C) or less. A tubular heat ex-

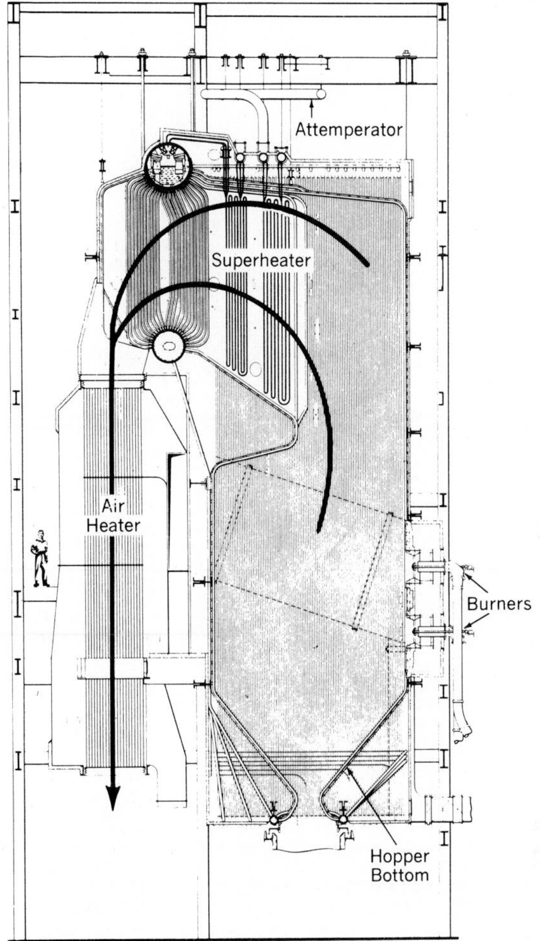

FIG. 3.8 Path of flue gas through boiler. *(Babcock & Wilcox Co.)*

changer, or more commonly a rotating drum type of air heater, is selected for this application. It is basically a large porous wheel that rotates from the flue gas duct, where it is heated, to the combustion air duct, where it gives up heat to the air. See the next chapter for more information on these air heaters.

Design and Construction

Steam generator design involves a series of compromises, both technical and economic. Some of the parameters that must be evaluated against each other to arrive at a final design include the heat-release rate, ash fusion temperature, percentage of excess air, production of NO_x, efficiency, fuel, and steam

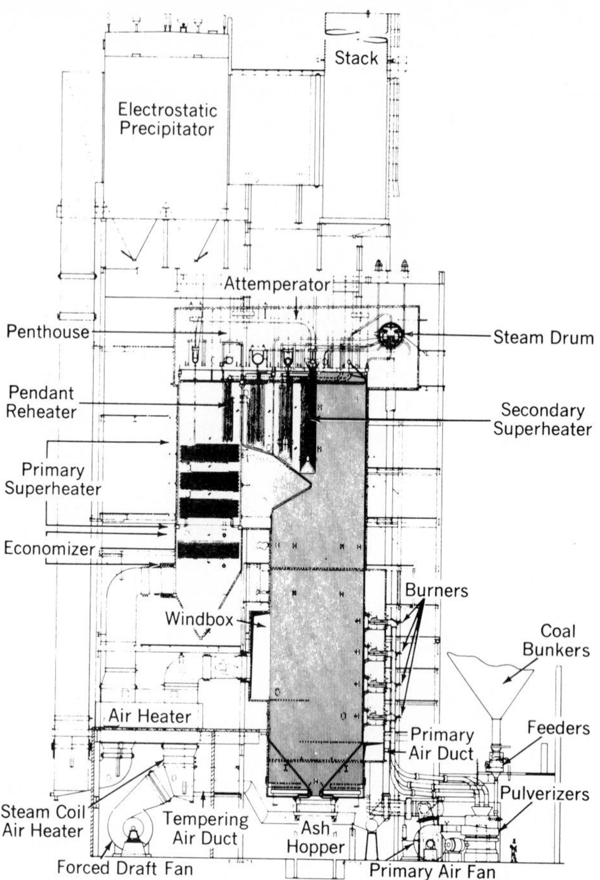

FIG. 3.9 Pressurized system contains forced-draft fans. *(Babcock & Wilcox Co.)*

temperature. Several of these parameters may act to oppose each other. Thus, low excess air will improve boiler efficiency, but may result in higher furnace temperatures that could cause slagging.

A furnace, which is basically a watertube-lined box with a tube-lined or refractory floor, is designed to take advantage of the high radiant heat flux near the burners. Normally, natural gas requires the smallest furnace because of its low emissivity, which results in a relatively even heat flux at the tube walls. Fuel oil, having a higher emissivity, requires that the walls of the boiler be farther from the burner to even out the peak heat flux.

Coal firing presents a different problem because the flue gas temperature at the entrance to the convective section must be at least 100°F (56°C) below the ash-softening point. Such a steep reduction in temperature can mean that a 15 percent or more increase in the amount of radiant surface is necessary.

Convective Section. The convective section of a boiler is designed to extract the maximum amount of heat from the partially cooled [2000°F (1093°C)] flue gas.

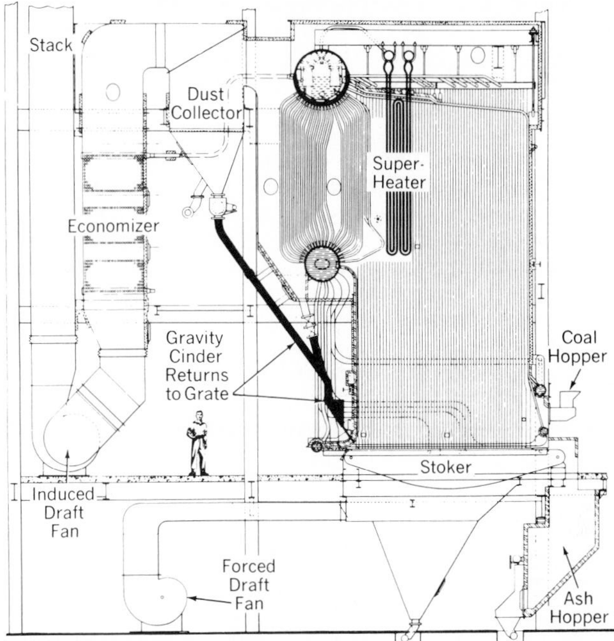

FIG. 3.10 Balanced-draft boiler has both forced- and induced-draft fans. *(Babcock & Wilcox Co.)*

The section should be as compact as possible to obtain the most efficient use of construction material and plant space. Design factors include flue gas velocities between tubes, tube spacing, and the use of finned tubes. Gas, oil, and coal fuels require different designs for optimum operation.

For coal, flue gas velocities should not exceed 60 ft/s (18.3 m/s) because of the highly erosive characteristics of coal ash, and tubes should be spaced much farther apart than for gas. Particulate matter is carried through the convective section by flue gas. Sootblowers are generally required for oil-fired steam generators and are always required for coal-fired units.

With any steam cycle, certain predominant factors must be considered if a satisfactory design is to result. Not only must the unit be capable of its designed output at full load, but also it must be able to maintain steam temperatures at partial loads—usually down to 60 percent to ensure that required steam quality at the low-pressure end of the turbine is provided.

Furnace size and shape should allow for adequate fuel residence time within the furnace to achieve complete combustion. The heat content of the gas leaving the furnace should be high enough to enable an efficient design of superheater, reheater, economizer, and air heater. The gas should yield a stack gas temperature low enough to minimize heat losses, but not so low that corrosion results in areas of low metal temperature. The design of firing equipment should minimize losses caused by unburned carbon, and boiler casings should be designed for a minimum infiltration of air.

Availability plays an important role in powerplant operation, and poor availability has often been associated with the blockage of gas passes by ash deposits.

Care must be taken to ensure that adequate clearance between tubes is provided, particularly in zones of high gas temperature and when inferior fuels are to be burned. The high failure rate of pressure parts, especially waterwall and superheater tubes, has emphasized the need for careful design of these parts and the wise selection of tube alloys.

Pressure Parts. Steam generators are designed in accordance with section I of the ASME Boiler and Pressure Vessel Code. Pressure part thicknesses are selected so that the primary stress (pressure plus dead load) falls within the allowable stress limit for the material at the design temperature. The manufacturer is responsible for determining the operating temperature and associated operating stresses (such as thermal and peak), as well as evaluating these stresses under different operating conditions and selecting an appropriate design approach.

Figure 3.11 summarizes the means of classification of stress types, allowable stresses, and loads currently included in section III of the ASME Code. These stresses must be analyzed for modern steam generators operating at high pressure and with relatively high-speed transient conditions. Also, consideration must be given to cyclic fatigue to avoid failure from stress rupture and creep.

Influence of Fuel on Design. The type and quality of fuel to be burned, especially coal, play a large part in the final boiler design. Coal quality varies widely, depending on the geographic locations of various coal seams. Table 7.3 in this section shows the large variations among classes and groups of coals for the selected constituents in the analyses.

In general, ignition stability in a pulverized-coal furnace varies directly with the ratio of volatile matter to fixed carbon. Thus, anthracite fuels cannot normally be burned in suspension without support ignition or blending with a coal containing a higher percentage of volatile matter. Similarly, coal with higher volatility can be more easily burned in suspension, which allows a lower furnace

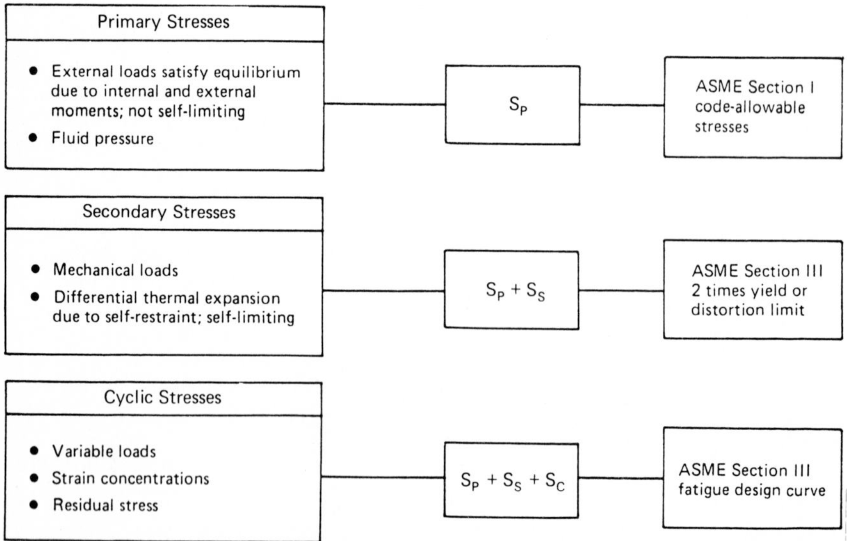

FIG. 3.11 Design basis for pressure parts.

temperature but requires a larger furnace heat-release area. Conversely, a lower furnace temperature is usually required when volatile coals are burned because the tendency toward lower ash-fusion temperatures (and thus higher slagging tendencies) increases as the volatile matter increases.

The type of coal burned will determine the furnace size (Fig. 3.12). Sufficient furnace volume must be available that the fuel can burn with the minimum amount of slag formation and very little unburned carbon in the ash. Figure 3.13 indicates the relationship between furnace exit gas temperature and heat-release rate for typical coal, gas, and oil fuels. For a coal having a relatively low ash-fusion temperature, a lower heat-release rate is required than for a coal having a higher fusion temperature, to avoid excessive slagging problems. The furnace exit gas temperature is primarily a function of the heat-release rate (available heat divided by equivalent water-cooled furnace surface), rather than the furnace liberation rate (available heat divided by furnace volume).

Typical analyses and properties are presented in Table 4.1 of Section 3 for various grades of fuel oils.

Obviously, fuel plays a major role in steam generator design. In recent years, serious operating problems have arisen in boilers large and small when fuels have been changed to satisfy environmental requirements without due regard for basic boiler parameters. In many cases, such changes have led to reduced steam capacity, increased slagging, and poorer ignition stability.

Furnace and Waterwalls. The watertube steam generator has been described as a tank with walls made up of water-filled tubes. The construction of these waterwalls is an important feature of the steam generator (Fig. 3.14). Originally,

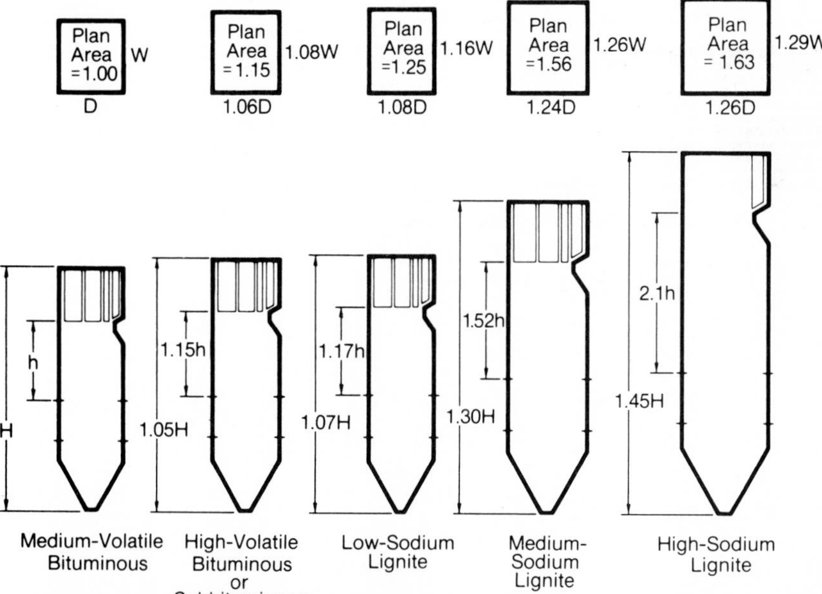

FIG. 3.12 Effect of coal type on furnace sizing, assuming constant fuel firing. H=distance between centerlines of lowest hopper headers and furnace roof tubes; h=distance between centerline of top burners and furnace nose apex; W=width; and D=depth. *(Combustion Engineering Inc.)*

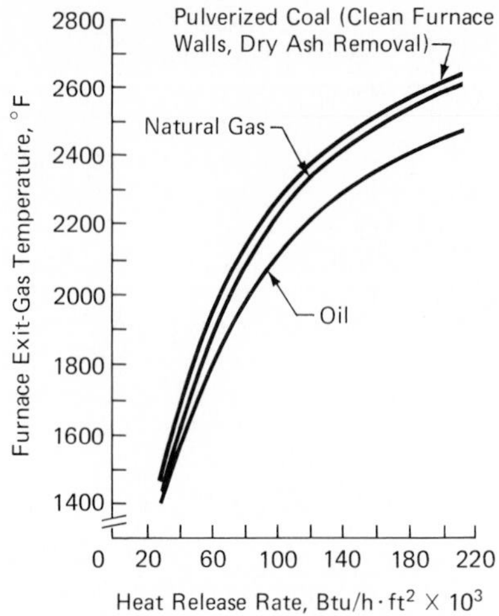

FIG. 3.13 Furnace exit-gas temperature versus heat release rate for three fuels. Note: °C=(F−32)/1.8; W/m^2=Btu/h • ft^2×3.15.

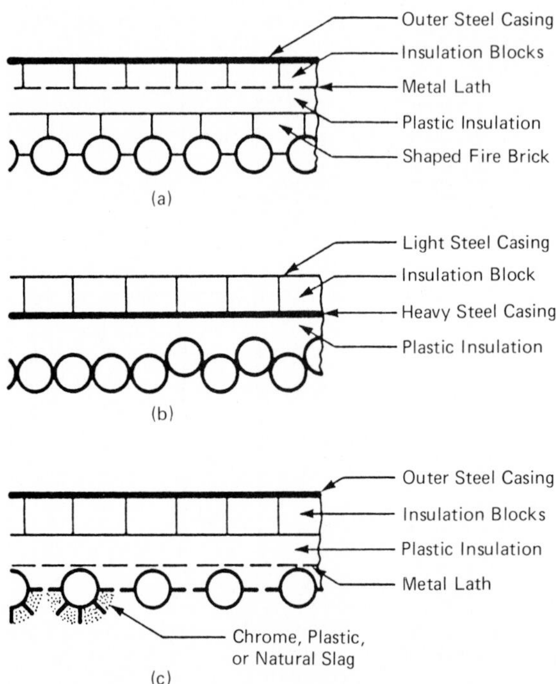

FIG. 3.14 Typical waterwall constructions. (*a*) Standard tube; (*b*) staggered tube; and (*c*) tangent tube waterwall construction. *(Combustion Engineering Inc.)*

tangent-tube construction was applied to obtain the greatest area of water-cooled, heat-absorbing surface compared to furnace surface area. When some of the tubes are removed and replaced by finned tubes, however, the circulation within the tubes is much more rapid; thus, greater heat absorption per tube is achieved without reaching extremely high metal temperatures. Fins do a better job of transferring heat to the tubes. Boilers with finned waterwall tubes must be designed with a fluid mass flow sufficient to hold the metal temperature within safe limits at every rating. Thus, the savings realized by eliminating tubes are partially offset if pumps are needed for more active circulation.

Tangent-tube construction has largely been replaced today with "membrane" walls, in which a steel bar or membrane is welded between adjacent tubes (Fig. 3.15). This type of construction, found in both natural- and forced-circulation units for all types of fuel firing, has flat wall sections composed of panels of single rows of tubes on centers wider than a tube diameter and connected by means of a membrane bar securely welded to each tube on its centerline. The design results in a continuous wall of rugged, pressure-tight construction. The individual panels are of a width and length suitable for economical manufacture and assembly, with bottom and top headers attached in the shop before shipment for field assembly. Membrane walls with refractory lining form the lower furnace walls of many cyclone-fired units.

Steam Drum. The steam drum serves two purposes in a boiler: It acts as a header for distributing water to the downcomers, and it physically separates the steam from the water. The steam leaves the drum and flows to the superheaters and the turbine. The steam drum runs the entire length of the boiler and has connections for the downcomers, waterwall risers, continuous blowdown, chemical addition, feedwater inlet, gage glass, pressure taps, and steam outlets (Fig. 3.16).

The inlets to the steam drum include feedwater, riser or generating tube, and chemical feed connections. The feedwater distribution manifold is at the bottom of the steam drum and admits the feedwater through holes or nozzles in the manifold, distributing the flow of water evenly along the whole length of the drum.

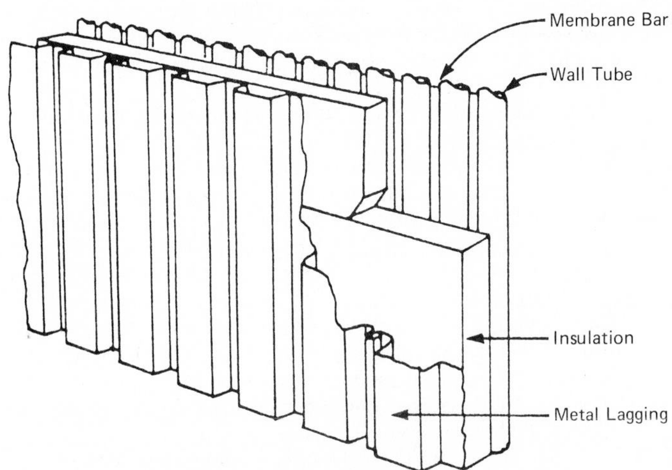

FIG. 3.15 Membrane waterwall construction. (Power *magazine*)

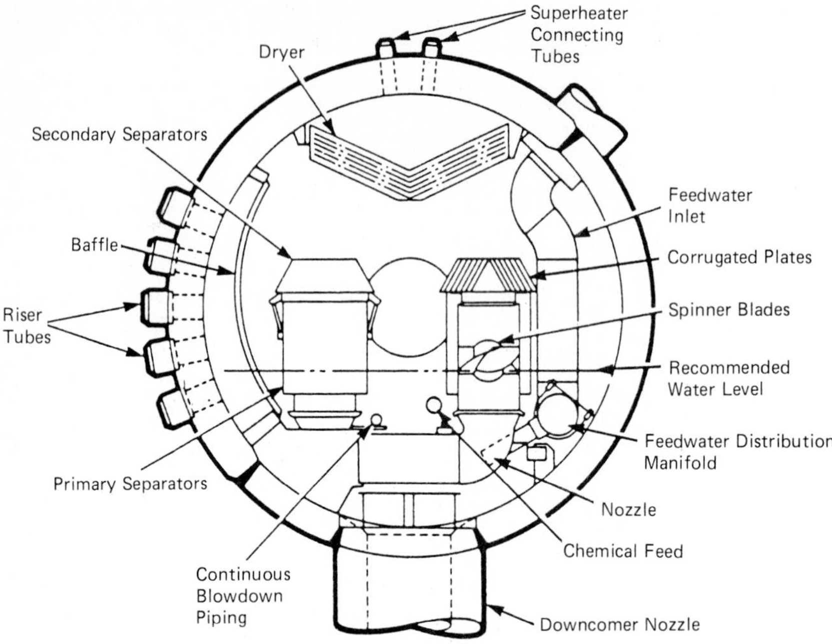

FIG. 3.16 Steam drum and internals. *(Combustion Engineering Inc.)*

The discharge of water is directed along the bottom of the drum and toward the downcomers. The baffle plates in the steam drum form a barrier between the feedwater injected and the steam and water rising out of the generating tubes. This barrier improves steam separation.

Chemicals are also admitted near the bottom of the drum to help ensure good mixing and distribution. As with the feedwater manifold, the chemical discharge is toward the downcomers. Injecting the chemicals into the flow causes them to be rapidly and evenly distributed throughout the boiler (also see the last chapter in this section).

In addition to the downcomer connections, the steam drum includes continuous blowdown and steam outlets.

Continuous Blowdown. Chemicals that are injected directly into the steam drum and other dissolved solids that may enter the boiler do not leave with the steam. As a result, the concentration of dissolved solids increases continuously. Continuous blowdown removes dissolved and suspended solids from the water by continually removing a portion of the water containing these solids. Feedwater replaces the water lost.

The continuous blowdown piping is placed in the drum at a point below the water level where the concentration of solids tends to be at its highest levels. Calibrated blowdown valves are often used to control the blowdown rate; the valves can indicate the percentage opening or number of turns. (Also see Section 4, Part 2, Chapter 5 on zero-discharge design.)

Steam Outlet. The steam outlet is normally a series of tubes at the top of the drum that carry steam to the superheater. In most utility boilers, a scrubber or steam dryer is introduced to eliminate any residual moisture in the steam leaving the drum. The scrubber or dryer consists of large, corrugated, closely fitted plates that cause the steam to flow sinuously between them. Some dryers use screens or mats of woven wire instead of plates. In either case, the objective is to have a large surface area on which moisture can collect.

Steam dryers operate on the principle of low-deposition velocity rather than on velocity separation. Water that deposits on the dryer's plates drains back to the boiler water by gravity. Because the steam velocity is low, the possibility of reentrainment is decreased.

Additional connections to the steam drum are primarily for control purposes. For example, a gage glass is mounted on the end of the drum to monitor the water level. Pressure gages and thermocouples are also mounted on the drum so that the pressure and drum metal temperature can be measured. Instrument lines provide inputs to the drum level control and, in some configurations, combustion control. The drum has several vents along its length. The vents are at the top and are operated during start-up to vent air from the drum. During shutdown, as the boiler cools, they are opened to prevent a vacuum from forming in the boiler.

Drum Internals. Besides baffles, the steam drum internals may consist of separators, which use centrifugal force to separate the water and steam, and scrubbers, which are corrugated sheets to deflect the steam and water.

Centrifugal Separators. Centrifugal separators, such as the vertical cyclone design (Fig. 3.17), cause the steam-water mixture to flow in a circular path. The resultant centrifugal forces set up in the mixture cause the heavier water droplets to be thrown to the separator's outside wall. Here, the water is collected and directed back to the water in the drum. The steam continues to rise to the scrubbers.

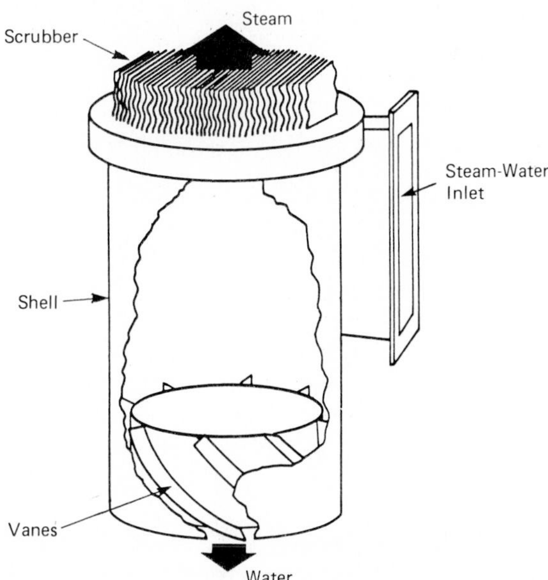

FIG. 3.17 Vertical cyclone steam separator.

Another common type of separator is the horizontal cyclone design. The principle is the same; only the arrangement is different. In the horizontal type, the water and steam flow along a horizontal axis and then change direction to pass into the scrubbers.

Scrubbers. In the scrubber, the steam-water mixture, which is mostly steam, flows through narrow corrugated slots. The water trapped in the steam adheres to the metal and drains back down to the water surface; from there it goes back into the cycle. Separators increase the quality of the steam, while scrubbers increase it even further.

The steam that rises from the scrubbers passes into the dryer, which is the final stage of separation, and out of the drum through the superheater supply tubes. The steam that leaves the drum is saturated because it is at the same temperature as the water. The quality of the steam (percentage of steam in the total mixture) is very high, normally over 99 percent.

Packaged Units

Industry's increasing reliance on solid fuels for new boilers, and on process by-products and wastes, has dramatically reduced sales of shop-assembled steam generators. These packaged units, almost always designed for oil and/or natural-gas firing, were extremely popular 25 years ago because of their relatively low cost and the short time required to fabricate and install them.

Packaged watertube boilers first came on the scene back in the late 1940s; the largest units then produced about 25,000 lb/h (3.15 kg/s) of steam. Capacities grew steadily, and boilers with ratings approaching 600,000 lb/h (75.6 kg/s) were fabricated in the early 1970s. The basic structural design of these units has changed little over the years. Most follow the A-, D-, and O-type configurations (Figs. 3.18, 3.19, and 3.20).

The majority of units built today are designed for operating pressures from 125 to about 1000 psig (860 to 6900 kPa), although pressures as high as 2000 psig (13 800 kPa) are feasible. Steam temperatures generally range from 353°F (saturation at 125 psig) to 950°F (178°C at 860 kPa to 510°C).

Main Components. Large packaged boilers, those rated more than 100,000 lb/h (12.6 kg/s), as well as many smaller ones have completely water-cooled furnaces. Advantages of this design include minimum weight and maintenance, plus maximum structural rigidity and safety. The use of welded rear and front waterwalls, in place of refractory walls, typically adds to boiler costs. But it increases availability, because the entire boiler is constructed of homogeneous material, and the unit expands and contracts as a cube.

Carry-over has been a recurring problem with packaged units over the years. As boiler capacities have increased, steam separator designs have become more sophisticated. Those installed in units producing relatively low-pressure saturated steam for process usually are the simplest in design, such as the baffle and chevron types. Boilers generating high-pressure superheated steam for turbines, or for specialized process use, may have centrifugal separators.

Superheaters are of the radiant or convective-radiant type and often have one or more of these features: (1) all-welded construction to eliminate handhole and tube-seat leakage, (2) optimum tube spacing to prevent slag bridging, and (3) location and orientation to keep tube supports out of the flue gas stream, for longer life and to permit full drainage.

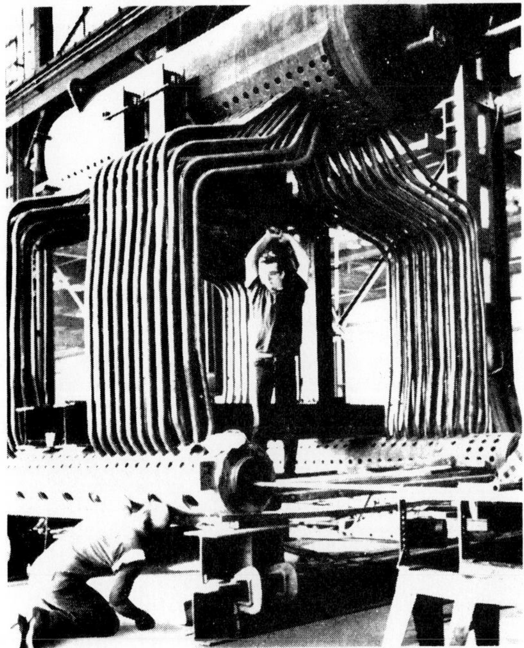

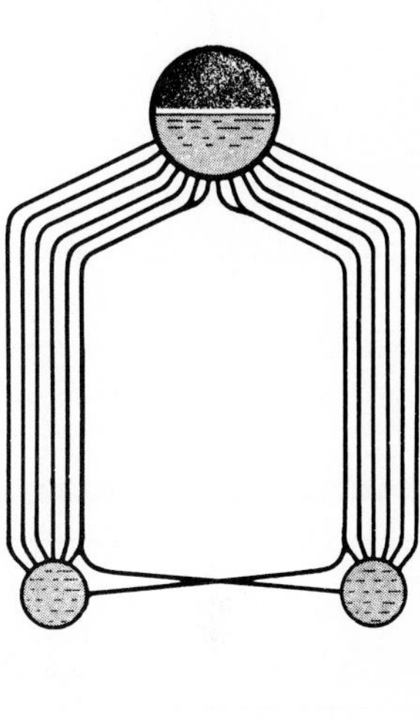

FIG. 3.18 A-type packaged watertube boiler has large upper drum and two small lower drums. *(Combustion Engineering Inc.)*

Recent Refinements. Improvements in boiler design that have helped boost availability include such refinements as (1) tube bank access ports for complete inspection and cleaning of convection surfaces, (2) water-wash troughs with drain connections in the furnace floor, and (3) seal air in soot blowers to prevent sulfur-laden flue gas from entering the blowing elements and corroding them.

Remember, packaged watertube boilers are not designed to burn just liquid and gaseous fuels. Some manufacturers offer units to fire coal, wood, and process wastes. The much larger furnace volumes required in units designed for solid fuels, however, restrict the capacity of these packaged units to about 80,000 lb/h (10.1 kg/s), or about one-third the rating of a comparable oil- or gas-fired unit that can be shipped by railroad.

Whether a sizable market for coal-fired shop-assembled boilers will develop remains to be seen. Industrial companies with substantial steam requirements do not necessarily obtain the same benefits from coal-fired packaged units as they do from oil- or gas-fired boilers.

Maximum unit ratings for coal-fired, rail-transportable models are relatively low; thus, the cost savings attributed to shop assembly are not so dramatic for the complete job, because the cost of fuel-handling and combustion equipment for coal firing is much greater than that for oil and gas.

Field-Erected Units

Field-erected boilers today are mostly coal-fired. Once a commitment to burn coal is made, the major decision facing designers is not whether the boiler should

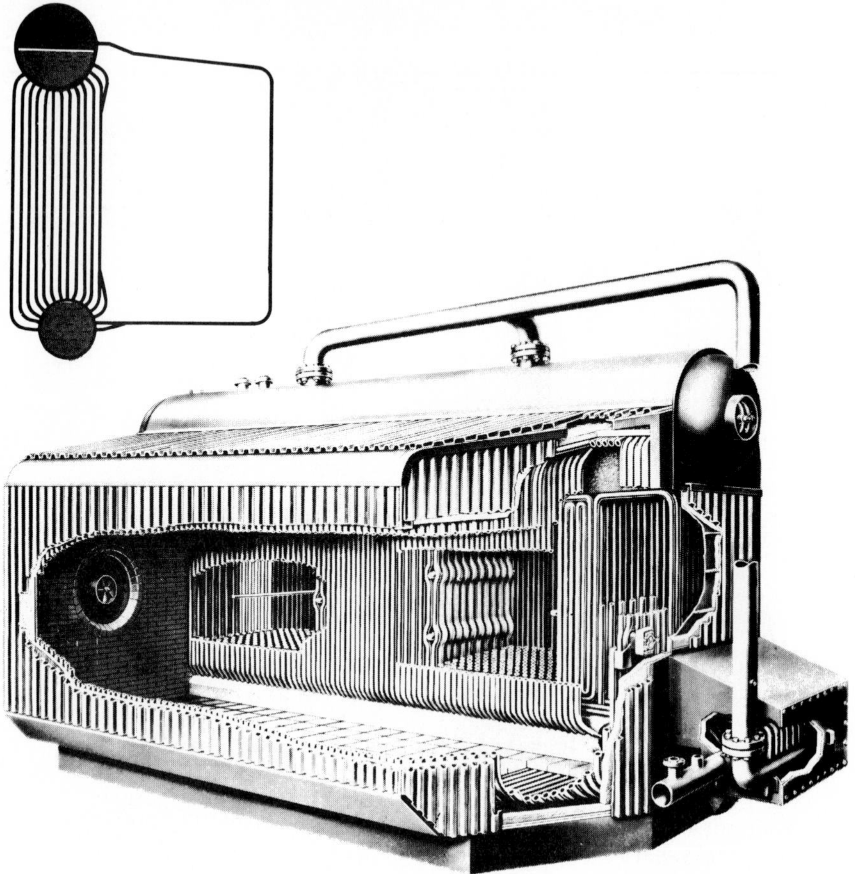

FIG. 3.19 D-type package has large drum over small drum, which allows more flexibility than A-type. *(Combustion Engineering Inc.)*

be shop-assembled or field-erected—required capacity almost always determines that—but whether pulverized-coal or stoker firing should be specified. In the past, most coal-fired boilers below 250,000-lb/h (31.5-kg/s) steam capacity were stoker-fired, while larger boilers almost always burned pulverized coal. As the price of coal rises, however, pulverized-coal-fired boilers with steam capacities as low as 100,000 lb/h (12.6 kg/s) often are economically attractive (Fig. 3.21).

Stoker vs. Pulverized-Coal Firing. Choosing between stoker and pulverized-coal firing involves a comprehensive investigation of economic and engineering factors, including capital and fuel costs, efficiency, auxiliary power and pollution control requirements, and maintenance and operating variables.

A pulverized-coal-fired boiler is generally more expensive than a spreader-stoker-fired unit of the same capacity, but the higher price tag sometimes can be justified. The controlling cost factors are usually the type of coal, steam conditions, and pulverizer requirements. Coal characteristics are particularly impor-

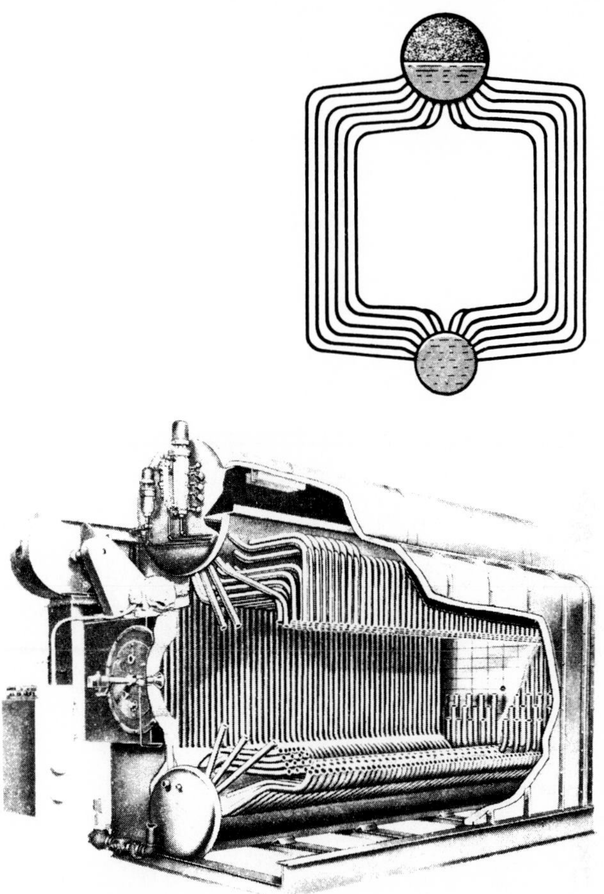

FIG. 3.20 O-type packaged boiler has symmetry of design that exposes the least tube surface to radiant heat.

tant. Furnace volume, for example, can vary by as much as 30 percent in a pulverized-coal-fired unit, depending on the slagging potential of the coal. Severely slagging coals require very large furnaces when pulverized-coal firing is specified. Stoker-fired boilers, by contrast, are affected less dramatically by slagging, although performance can be hindered by caking coals.

All field-erected coal-fired boilers permit the combustion of a variety of fuels because of their conservative design. The larger furnace associated with suspension firing may have an advantage over the stoker-fired boiler in that it can efficiently generate rated steam capacity, temperature, and pressure with auxiliary gaseous and liquid fuels.

When paired with stokers, auxiliary burners must be placed high enough above the grate to protect it from radiant heat. This precaution reduces effective furnace volume and may restrict steam production capability when all the energy comes from auxiliary fuels, that is, unless the unit is designed larger than normal.

But the spreader-stoker-fired boiler apparently has the edge in being able to burn a

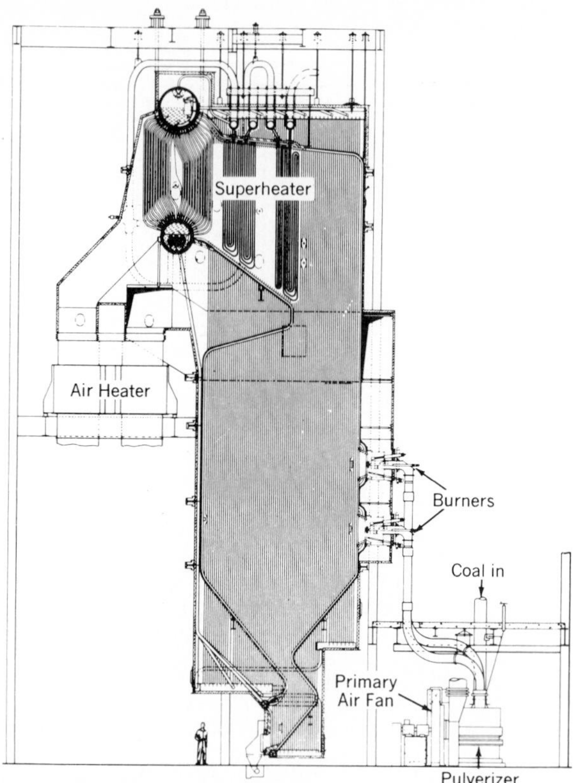

FIG. 3.21 Field-erected boiler with pulverized-coal firing. *(Babcock & Wilcox Co.)*

wide variety of solid by-product wastes, as well as residential refuse, with minimal preparation (Fig. 3.22). Many solid wastes burned in a pulverized-coal-fired unit require careful pretreatment, as well as the use of an auxiliary burnout grate in the furnace. (For more information about energy from waste, see Section 3, Chapter 6.)

Efficiency Comparisons. As fuel costs increase, the 3 to 5 percent higher efficiency possible with pulverized-coal firing looms as an important consideration, especially in large boilers. One reason that pulverized-coal firing is more efficient is that it offers lower carbon and excess air losses than stoker firing. A properly designed and operated unit can hold the efficiency loss caused by carbon to less than 0.5 percent, whereas a traveling-grate spreader stoker can do no better than about 4 percent—and even that only with 50 percent ash reinjection.

Although spreader-stoker firing demands about twice the excess air (on a percentage basis) that is needed for pulverized-coal firing, an effective overfire air system and good sealing between the grate and boiler can hold down air requirements.

One penalty for the higher efficiency of the pulverized-coal-fired boiler is the power required to operate the pulverizers, but this is small compared to the lower fuel costs associated with these units. Pollution control costs also may be higher

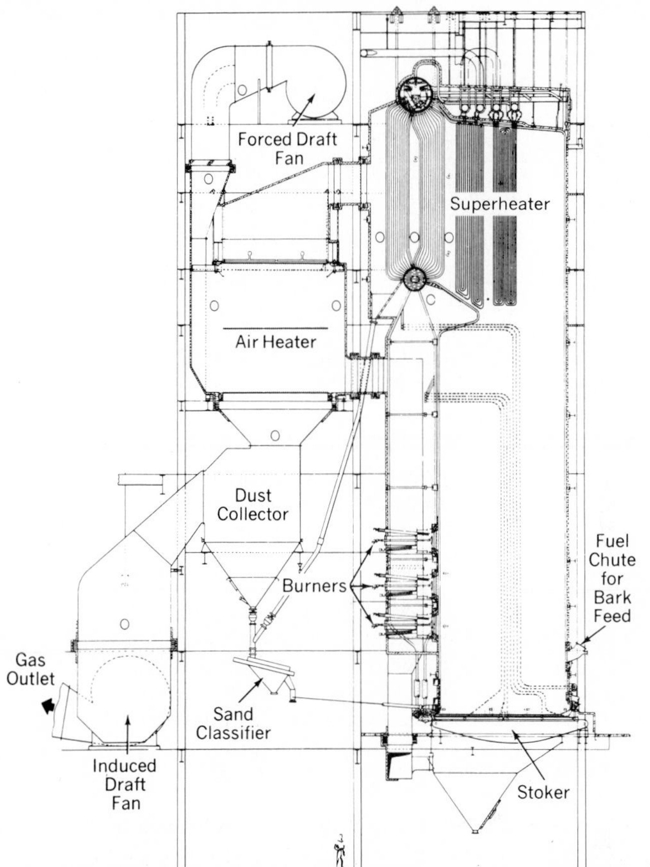

FIG. 3.22 Field-erected boiler with stoker firing, in which bark is the waste fuel used. *(Babcock & Wilcox Co.)*

with the pulverized-coal-fired boiler, because all fuel is burned in suspension. However, the stringent pollution control laws now in force throughout the country do not favor the installation of stoker-fired boilers as did yesterday's statutes.

Both stoker- and pulverized-coal-fired boilers are seeing increasing competition from fluidized-bed boilers, which are covered in Section 4, Part 1, Chapter 3.

FIRETUBE DESIGN

Firetube boilers of either the Scotch marine or firebox type serve in most industrial plants where saturated steam demand is less than about 50,000 lb/h (6.3 kg/s) and pressure requirements are less than 250 psig (1720 kPa). The general arrangement of these boilers has changed little in the past 25 years, although their size has increased (Fig. 3.23). Almost all firetube boilers made today are packaged designs.

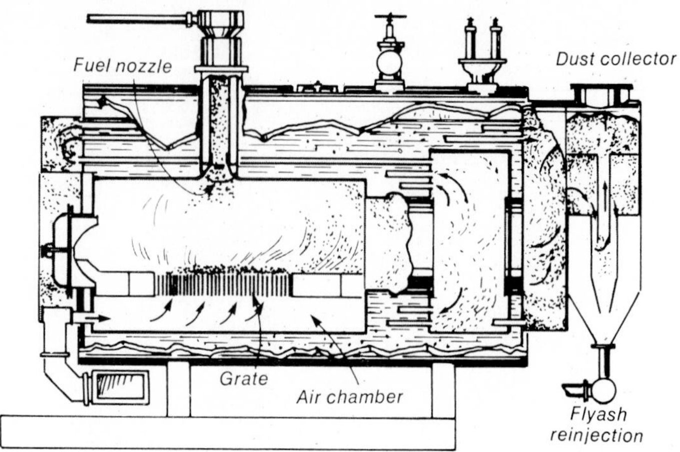

FIG. 3.23 Firetube boiler capable of burning solid fuels in small industrial and institutional plants has stationary grate. (Power *magazine*)

Scotch Marine Type

How They Work. In a Scotch marine boiler (Fig. 3.24), combustion takes place in a cylindrical furnace (usually corrugated to increase strength), which is located inside a cylindrical pressure vessel. Water in the welded steel boiler is heated by firetubes that carry the combustion gases. These tubes run the length of the shell at the sides, above, and sometimes below the furnace. Depending on the design, gases may make two, three, or four passes through the unit before being discharged (Fig. 3.25).

The largest Scotch marine boiler offered by a U.S. manufacturer is rated 2000 hp (19 620 kW). It has two combustion chambers in a single shell and measures about 13 ft (4 m) in diameter and more than 30 ft (9 m) long. Boilers with single combustion chambers come in sizes up to 1200 hp (11 772 kW). [The amount of energy in 1 *boiler* hp (=9.81 kW) is the same as that needed to produce 34.5 lb/h (4.35 g/s) of steam at 212°F (100°C) from 212°F (100°C) feedwater.]

Both these designs, like most other Scotch marine boilers, can be built to produce either hot water for low-temperature-water heating systems or saturated steam for process or space heating, at pressures up to 250 psig (1720 kPa). This pressure is near the practical operating limit for firetube designs. Since the shell strength of a cylinder to resist rupture is proportional to the pressure multiplied by the diameter, high pressures and large diameters lead to prohibitively thick shells. The thicker the shell, the higher the cost.

Advantages over Watertube Type. Today, Scotch marine boilers generally are cheaper than watertube boilers without economizers in sizes up to about 40,000 lb/h (5.04 kg/s). Other advantages of these boilers over watertubes include the following:

- An ability to respond to load swings more rapidly, because they have about 3½ times more water at or near the saturation temperature. Sometimes this can be a disadvantage since it takes firetube boilers longer to reach operating pressure after a surge in steam demand.

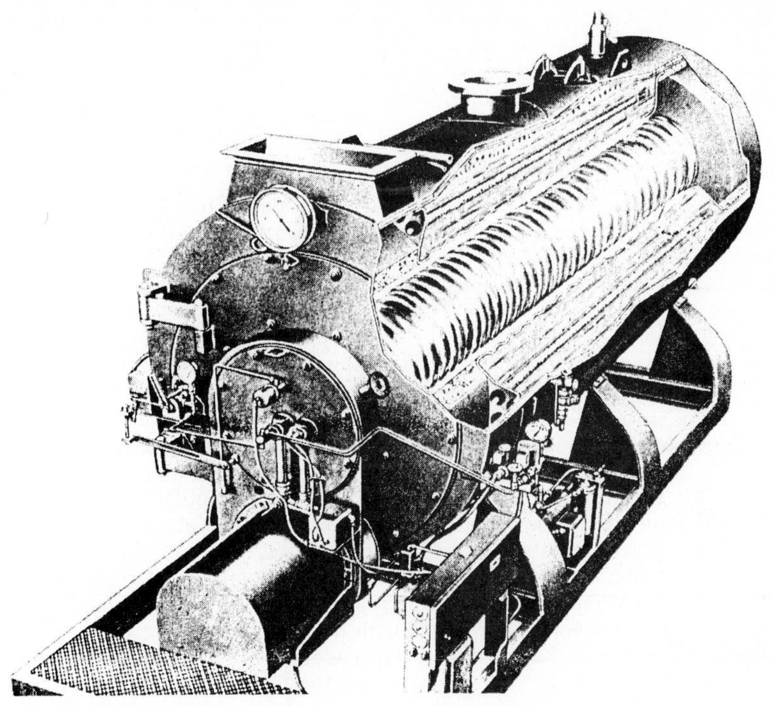

FIG. 3.24 Two-pass firetube boiler features a corrugated furnace to maximize radiant transfer from the turbulent flame.

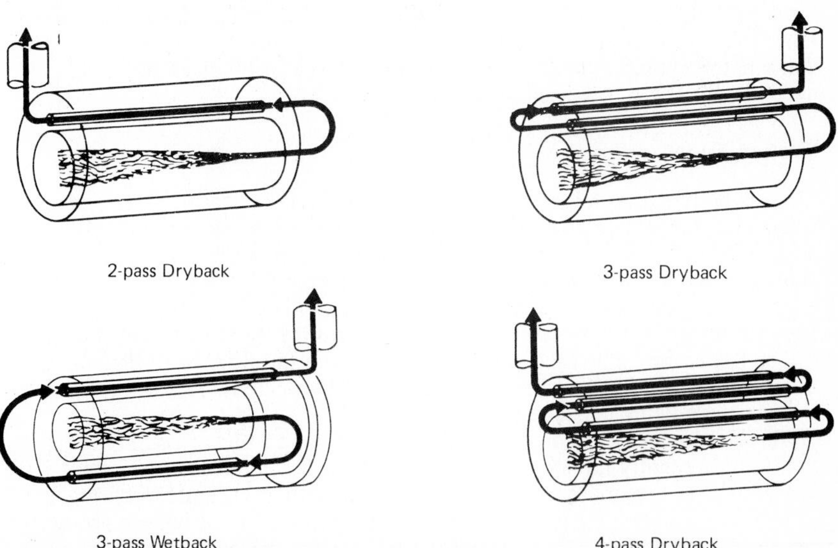

FIG. 3.25 Basic gas flow patterns used today in firetube boilers. All rely on internal furnace as first pass, then route gases into various tube layouts. The 3-pass wetback pattern is for the Scotch marine boiler. (Power *magazine*)

- Easier tube replacement. All firetubes are straight and are accessible from outside the boiler. By contrast, most watertubes are bent and can be reached only from inside the boiler. Also, tube orientation is such that good tubes sometimes must be removed to get at faulty ones.
- Installation in plants with low headroom. Watertube boilers are up to about 5 ft (1.5 m) taller than Scotch marine units of equal capacity.
- Low susceptibility to cold-end and casing corrosion. Critical parts of firetube boilers are maintained at or above the saturation temperature throughout the load range. Also, the high flue gas velocity, compared to that for a watertube unit, tends to carry any moisture formed out of the boiler.

Higher Efficiencies. Higher efficiencies may be still another advantage of Scotch marine boilers over watertube units. The major manufacturers of firetube boilers design their equipment with 5 ft^2/hp of heating surface (0.047 m^2/kW)—about 40 percent more surface than in watertube boilers of comparable capacity. The additional surface permits lower heat-transfer rates, which are conducive to higher efficiency. In fact, stack temperatures generally are within 100°F (56°C) of the saturation temperature at rated load. This allows some manufacturers to guarantee fuel-to-steam efficiencies greater than 81 percent when natural gas is burned and up to about 86 percent when distillate oil is fired by air-atomizing burners.

Even higher efficiencies may be possible if economizers are used and/or if combustion controls are outfitted with oxygen-trim capability. Cylindrical economizers, designed especially for plants with firetube boilers, are easy to install. The packaged units can be substituted for a section of flue piping of equivalent length. Control systems capable of correcting the fuel-air ratio are based on the flue gas–oxygen level.

Although distillates are preferred for Scotch marine boilers, heavy oil can be burned if it is properly strained and heated. However, solid fuels are difficult to burn efficiently in the small cylindrical furnaces provided with these units. Perhaps the most practical way to burn coal or wood is to use an external-combustion chamber, such as a refractory furnace or fluidized bed.

Firebox Type

Firebox boilers offer a larger ratio of furnace volume to heat-transfer surface than Scotch marine boilers, making them more suitable for burning solid fuels. Typical units have a combustion chamber—or *firebox*—with a flat bottom, and vertical sides going into an arched crown sheet that forms the top (Fig. 3.26). The furnace, which requires stays for structural support, is cooled at the sides and the back by water legs between it and the outer casing. Firetubes heat the water in the boiler as they convey combustion gases from the furnace to the stack, as in Scotch marine units. Combustion gases normally make three passes through the boiler before being discharged.

Firebox boilers were originally designed for hand firing of coal, but almost all those sold recently have been for gas or oil service in sizes up to about 700 hp (6870 kW). Today, many companies are buying either: (1) units to burn gas or oil as well as coal or a by-product fuel, or (2) gas- or oil-fired boilers arranged for future coal firing (Fig. 3.27). Since the basic firebox design has not been modified over the years, units bought for gas or oil can be converted to coal or by-product fuels if there is enough space underneath the boiler to add the firing equipment and furnace volume required.

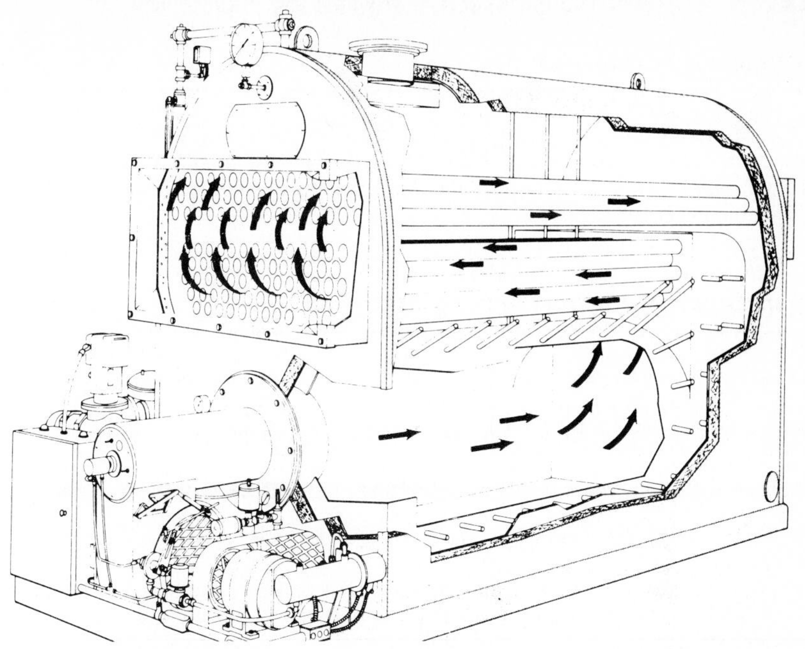

FIG. 3.26 Firebox design has arched crown sheet. The rear reversing chamber has stayed water legs at sides with a submerged top. (Power *magazine*)

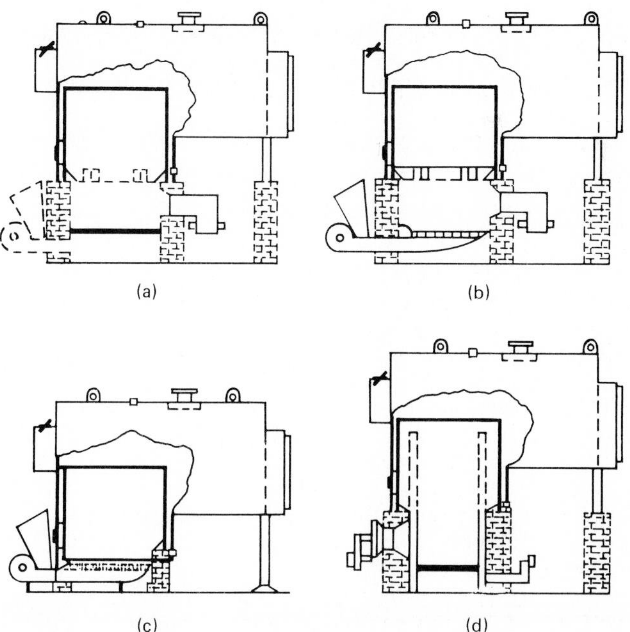

FIG. 3.27 Firebox boilers are designed to accommodate a variety of fuels, alone or in combination. (*a*) Gas/oil with future coal; (*b*) gas/oil and coal; (*c*) coal only; and (*d*) gas/oil and by-product fuels. (Power *magazine*)

ACKNOWLEDGEMENT

Illustrations in this chapter are courtesy of GP Publishing, except where noted otherwise. The section "WATERTUBE DESIGN" is adapted with permission from *Fundamentals of Power Plant Performance for Utility Engineers, Vol. 2*, GP Publishing, 1984.

SUGGESTIONS FOR FURTHER READING

ASME Boiler and Pressure Vessel Code—An American National Standard, ASME, New York, 1983.

Baumeister, T., et al., *Marks' Standard Handbook for Mechanical Engineers,* 8th ed., McGraw-Hill Book Co., New York, 1978.

Davis, R.F., "Expansion Theory of Circulation in Water Tube Boilers," *Engineering,* 163:145–148, 1947.

Kutateladze, S.S., *Fundamentals of Heat Transfer,* 2d. rev., Academic Press Inc., New York, 1963.

Lewis, W.Y., and S.A. Robertson, "The Circulation of Water and Steam in Water Tube Boilers and the Rational Simplification of Boiler Design," *Proceedings of Institution of Mechanical Engineers,* 143:147–178, 1940.

Powell, E.M., and H.A. Grabowski, "Drum Internals and High-Pressure Boiler Design," ASME Paper No. 54-A-242, ASME, New York, 1954.

Rohsenow, W.M., and H. Choi, *Heat, Mass and Momentum Transfer,* Prentice-Hall Inc., Englewood Cliffs, N.J., 1961.

Singer, J.G., ed., *Combustion—Fossil Power Systems,* 3d. ed., Combustion Engineering, Inc., Windsor, Conn., 1981.

Steam—Its Generation and Use, 39th ed., Babcock & Wilcox Co., Barberton, Ohio, 1978.

CHAPTER 1.4
KEY BOILER AUXILIARIES AND COMPONENTS

Joseph G. Singer
Combustion Engineering, Inc.
Windsor, CT

INTRODUCTION

Boiler auxiliaries and components are those pieces of equipment that allow a modern steam generator to function. They include burners (and ignitors), fans, air heaters, superheaters, reheaters, economizers, pulverizers (Chap. 7, Sec. 1), and stokers (Chap. 8, Sec. 1). Burners are covered first in this chapter. Combustion control and burner management systems are covered in Sec. 6, Chap. 3, Instrumentation and Control.

BURNERS

In the process of generating steam, burners provide controlled, efficient conversion of the chemical energy of fuel to heat energy which, in turn, is transferred to the heat-absorbing surfaces of the steam generator (see box). To do this, burners introduce the fuel and air for combustion, mix these reactants, ignite the combustible mixture, and distribute the flame envelope and the products of combustion.

Combustion Reaction Is Influenced by a Host of Factors

The rate and degree of completion of a chemical reaction, such as in the combustion process, are greatly influenced by the temperature, concentration, preparation, and distribution of the reactants; by catalysts; and by mechanical turbulence. All these factors have one effect in common: to increase contacts between molecules of the reactants.

Higher temperature, for example, increases the velocity of molecular movement, permitting harder and more frequent contact between molecules. A temperature rise of 200°F (110°C) at some stages can greatly increase the rate of reaction. At a given pressure, three factors limit the temperature that can be attained to provide the greatest intermolecular contact: the heat absorbed by the combustion chamber enclosure, the reactants in bringing them to ignition temperature, and the nitrogen in the air used as a source of oxygen.

The concentration and distribution of the reactants in a given volume are directly related to the opportunity for contact between interacting molecules. In an atmosphere containing 21 percent oxygen (the amount present in air), this rate is much less than it would be with 90 percent oxygen. As the reaction nears completion, the distribution and concentration of reactants assume even greater importance. Because of the dilution of reactants by the inert products of combustion, the relative distribution—and opportunity for contact—approaches zero.

Preparation of the reactants and mechanical turbulence greatly influence the reaction rate. These are the primary factors available to the burner system designer attempting to provide a desirable reaction rate.

The beneficial effect of mechanical turbulence on the combustion reaction becomes apparent when it is realized that agitation permits greater opportunity for molecular contact. Agitation improves both the relative distribution and the energy imparted. Agitation assumes greater significance if it is achieved later in the combustion process when the relative concentration of the reactants is approaching zero.

An ideal burner system fulfilling these functions would have the following characteristics:

- No excess oxygen or unburned combustibles in the end products of combustion
- Low rate of auxiliary ignition-energy input to start the combustion reaction
- An economic reaction rate between fuel and oxygen compatible with acceptable nitrogen oxide and sulfur oxide formation
- An effective method of handling and disposing of the solid impurities introduced with the fuel
- Uniform distribution of the product weight and temperature in relation to the parallel circuits of the heat-absorbing surface
- Wide and stable firing range

- Fast response to changes in firing rate
- High equipment availability with low maintenance

In practice, some of these characteristics must be compromised to achieve a reasonable balance between combustion efficiency and cost. For example, firing a fuel with the stoichiometric air quantity (no excess above the theoretical amount) would require an infinite residence time at temperatures above the ignition point at which complete burnout of the combustibles takes place. Thus, every firing system requires a quantity of air in excess of the stoichiometric amount to attain an acceptable level of unburned carbon in the products of combustion leaving the furnace. This amount of excess air is an indicator of the burning efficiency of the firing system. Similarly, if firing devices are used to control the formation of oxides of nitrogen, there is a reduction in the response time of the firing system to load variations.

Practical Burner Design

In the practical application of a burner system to a boiler, all fundamental factors influencing the rate and completeness of combustion must be considered with the degree of heat-transfer efficiency. Two methods of producing a total flow pattern in a combustion chamber are available to provide successful molecular contact of reactants through mechanical turbulence. One method is to divide and distribute the fuel and air into many similar streams and to treat each stream independently of all others. This provides multiple flame envelopes. By contrast, the second method provides interaction between all streams of air and fuel introduced into the combustion chamber to produce a single flame envelope.

The first concept requires that the total fuel and air supplied to a common combustion chamber be accurately subdivided. It also limits the opportunity for sustained mechanical mixing or turbulence—particularly in the early stages of combustion. The necessity of obtaining and sustaining good distribution of fuel and air is both a design and an operating problem. Sufficient opportunity must exist for contact of fuel and oxygen molecules as well as uniform distribution of product temperature and mass in relation to the combustion chamber.

Conversely, the single-flame-envelope technique provides interaction between all streams of fuel and air introduced into a common chamber. It allows more time for contact between all fuel and air molecules, and mechanical turbulence is sustained throughout the chamber. This technique avoids stringent requirements for accurate fuel and air distribution.

Firing systems representative of these concepts are the horizontally, tangentially, and vertically fired systems. (Boiler manufacturers usually have firing systems tailored to their own units.)

Horizontally Fired Systems. In horizontally fired systems, the fuel is mixed with combustion air in individual burner registers (Fig. 4.1). In this design, the coal and primary air are introduced tangentially to the coal nozzle, thus imparting strong rotation within the nozzle. Adjustable inlet vanes impart a rotation to the preheated secondary air from the windbox. The degree of air swirl, coupled with the flow-shaping contour of the burner throat, establishes a recirculation pattern extending several throat diameters into the furnace. Once the coal is ignited, the hot products of combustion are directed back toward the nozzle to provide the ignition energy necessary for stable combustion.

The burners are located in rows, either on the front wall only (Fig. 4.2) or on

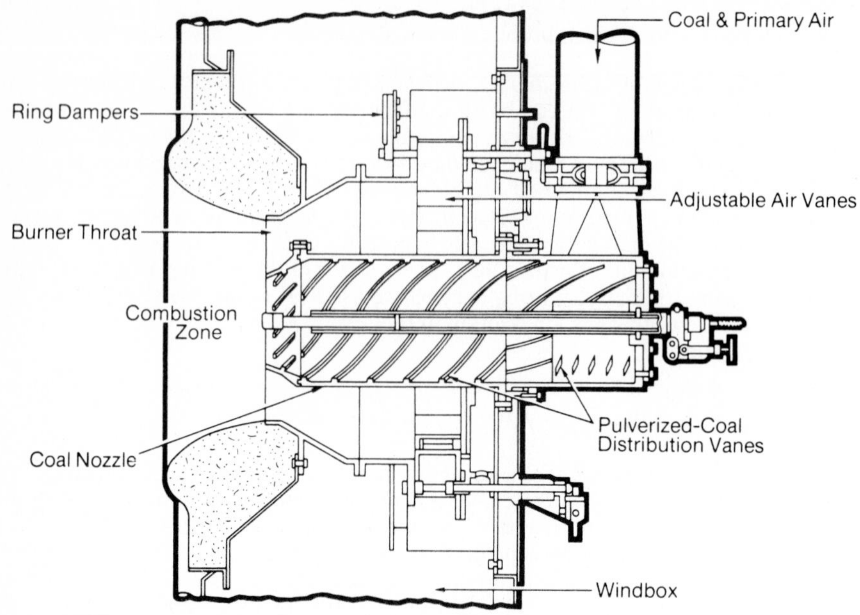

FIG. 4.1 Burner for horizontal firing of coal.

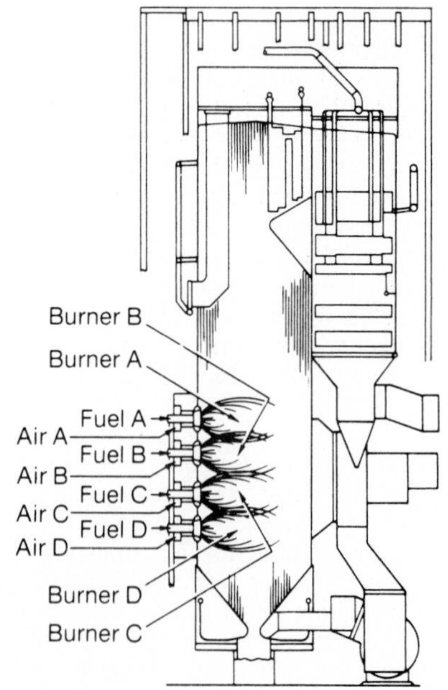

FIG. 4.2 Flow pattern of horizontal (wall) firing.

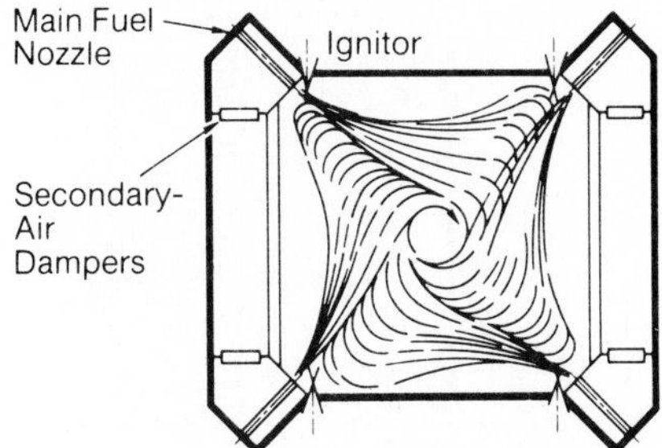

FIG. 4.3 Tangential firing pattern.

both front and rear walls. The latter is called *opposed firing*. Because the major portion of the combustion process must take place within the recirculation zone, it is imperative that the air-fuel ratio to each burner be within close tolerances. The rate of combustion drops off rapidly as the reactants leave the recirculation zone, and interaction between flames occurs only after that point. The degree of interaction depends on the burner and furnace configurations.

Tangentially Fired Systems. The tangentially (corner) fired system is based on the concept of a single flame envelope (Fig. 4.3). Both fuel and combustion air are projected from the corners of the furnace along lines tangent to a vertical cylinder at the center of the furnace. Intensive mixing occurs where these streams meet. A rotating motion, similar to that of a cyclone, is imparted to the flame body, which spreads out and fills the furnace area.

When a tangentially fired system projects a stream of pulverized coal and air into a furnace, the turbulence and mixing that take place along its path are low compared to horizontally fired systems. They are low because the turbulent zone does not continue for any great distance as the expanding gas soon forces a streamline flow. As one stream impinges on another in the center of the furnace, however, a high degree of turbulence is created, for effective mixing.

The fuel and air are admitted from the furnace corners in horizontal layers (Fig. 4.4). Dampers control the air to each compartment, making it possible to vary the distribution of air over the height of the windbox. It is also possible to vary the velocities of the airstreams, change the mixing rate of fuel and air, and control the distance from the nozzle at which the coal ignites.

The vertical arrangement of nozzles provides great flexibility for multiple-fuel firing. It is possible to provide for full-load capability with gas or oil by locating the additional nozzles for these fuels in the secondary-air compartments adjacent to the coal nozzles. Also, separate nozzles for injecting municipal refuse or other waste fuels are frequently provided in both utility and industrial boilers.

Fuel and air nozzles most commonly tilt in unison to raise and lower the flame in the furnace, to control furnace heat absorption and thus heat absorption in the superheater and reheater sections. In addition to controlling the furnace exit-gas temperature for variations in load, the tilts on coal-fired units automatically compensate for the effects of ash deposits on furnace-wall heat absorption.

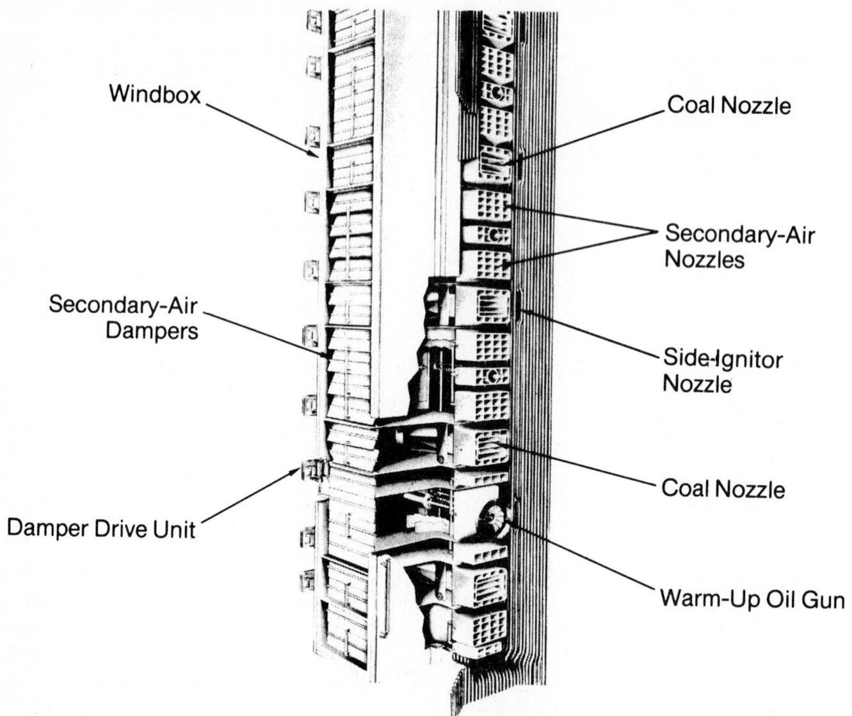

FIG. 4.4 Arrangement of corner windbox for tangential firing of coal.

Vertically Fired Systems. These systems are used today only to fire solid fuels that are difficult to ignite, such as coals with moisture- and-ash-free volatile matter of less than 13 percent. They require less supplementary fuel than the horizontally or tangentially fired systems, but have more complex firing equipment and so more complex operating characteristics.

In Fig. 4.5, pulverized coal is discharged through the nozzles. A portion of the heated combustion air is introduced around the fuel nozzles and through adjacent auxiliary ports. High-pressure jets help avoid short-circuiting the fuel-air streams to the furnace discharge. Tertiary air ports are in a row along the front and rear walls of the lower furnace.

The firing system produces a long, looping flame in the lower furnace, with the hot gases discharging up the center. A portion of the total combustion air is withheld from the fuel stream until it projects well down into the furnace. This arrangement has the advantage of heating the fuel stream separately from a significant portion of its combustion air, to provide good ignition stability. The delayed introduction of the tertiary air provides needed turbulence at a point in the flame where partial dilution from the products of combustion has occurred. The furnace flow pattern passes the hot product gases immediately in front of the fuel nozzles to provide a ready source of inherent ignition energy, which raises the primary fuel stream to the ignition temperature. The flow pattern also ensures that the

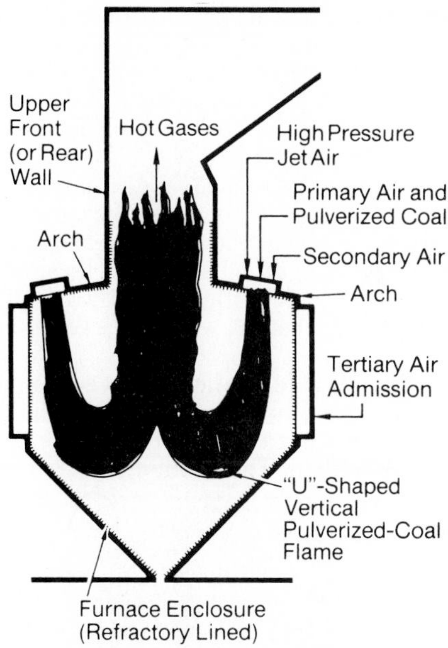

FIG. 4.5 Flow pattern of vertical firing.

largest entrained solid fuel particles, with the lowest surface-area-to-weight ratio, have the longest residence time in the combustion chamber.

Low-NO_x Burners

Prior to 1970, some steam generator firing systems produced great turbulence to achieve rapid mixing of the reactants and high energy release during the early stages of combustion. Surveys of these systems showed high levels of NO_x formation. With the advent of Environmental Protection Agency (EPA) restrictions in 1971, firing concepts underwent a change. Post-1971 firing systems are designed so that the fuel-bound nitrogen conversion is controlled by driving the major fraction of the fuel's nitrogen compounds into the gas phase under overall fuel-rich conditions.

In this atmosphere of oxygen deficiency, there occurs a maximum rate of decay of the evolved intermediate nitrogen compounds to molecular nitrogen. Following the admission of the remaining air, the slow burning rate reduces the peak flame temperature, thus curtailing the thermal NO_x production in the latter stages of combustion.

Tangential Firing. Early studies of NO_x emissions from all types of steam generators indicated that those from tangentially fired units were about one-half of those from horizontally fired systems. This difference was caused by the relatively low rate of mixing between the parallel streams of coal and secondary air emitted from the corner windboxes. Thus, ignition and partial devolatilization occur within an air-deficient primary combustion zone that exists from the fuel nozzles to a point within the furnace at which the stream is absorbed into the rotating mass of gases, termed the *fireball.*

The fireball itself is rich in oxygen because it contains all the air required for complete combustion of the fuel. Since the balance of the devolatilization occurs after the coal stream enters the fireball, the potential for fuel nitrogen oxide formation is limited.

Two significant modifications were made to the design and operation of the tangential firing system, to extend the oxygen-deficient combustion zone. The first added air compartments within the windbox above the uppermost coal nozzle (Fig. 4.6). Termed *overfire air* (OFA) *ports,* these compartments divert about 20 percent of the total combustion air to a burning zone above the windboxes. As a result, the fireball at windbox level is at or near stoichiometric air conditions.

The second modification was an operating procedure that moved the ignition point closer to the nozzle. This change extended the duration of the primary combustion zone so that a greater portion of the devolatilization took place before entry into the fireball. The ignition point is controlled by varying

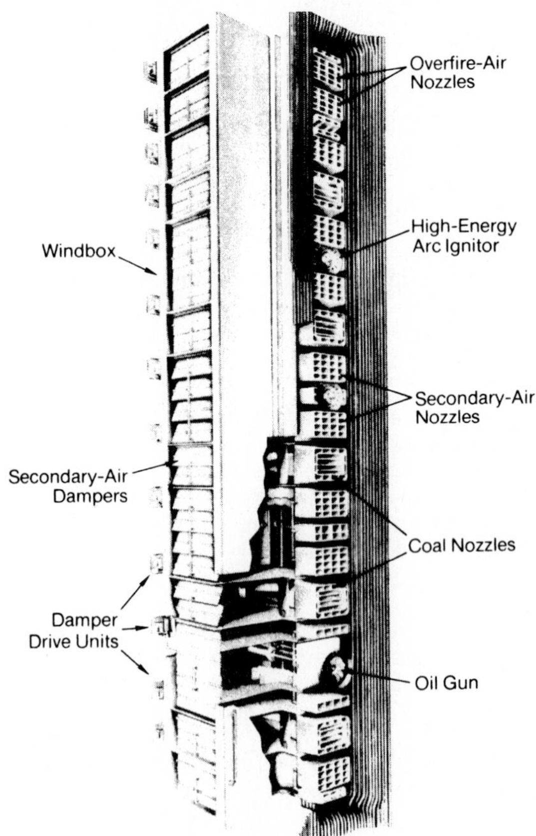

FIG. 4.6 Tangential-firing windbox with overfire air (OFA) ports and high-energy arc ignitors.

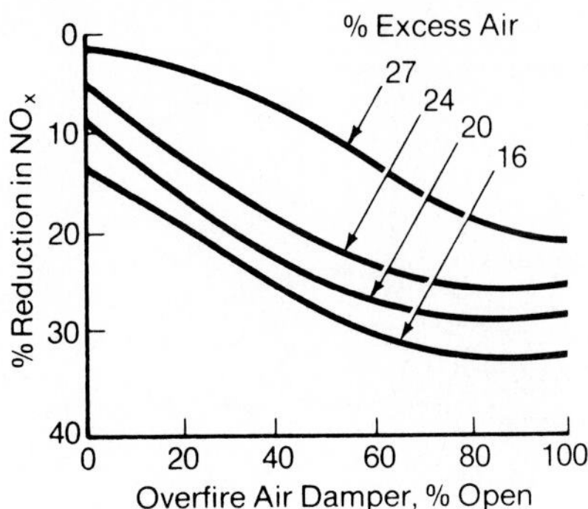

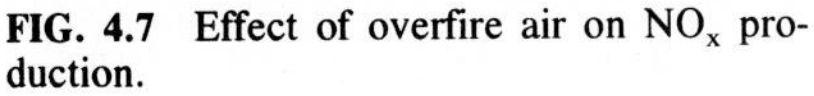

FIG. 4.7 Effect of overfire air on NO_x production.

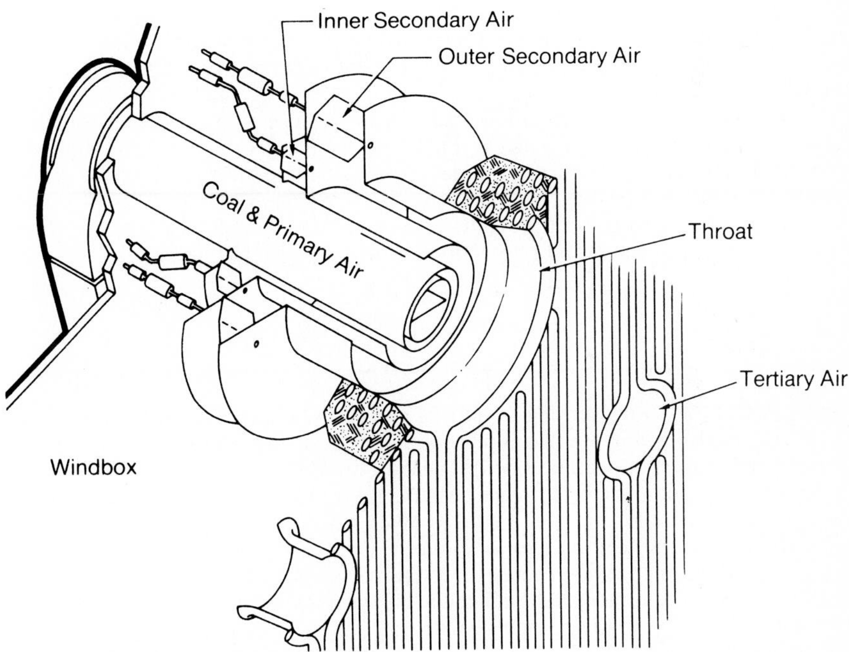

FIG. 4.8 Distributed mixing burner (DMB).

the quantity and velocity of the air through the annulus of the fuel compartment by changing the position of the damper. Figure 4.7 shows the effect on NO_x production of varying quantities of overfire air.

Horizontal Firing. A unique design developed for the EPA is the *distributed mixing burner* (DMB); see Fig. 4.8. Figure 4.9 shows how the design sequentially stages the fuel-air mixing in a distributed fashion. Conceptually, this is

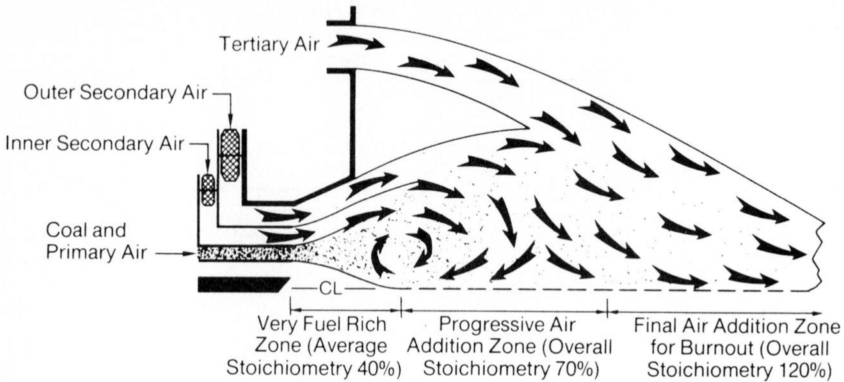

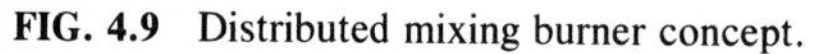

FIG. 4.9 Distributed mixing burner concept.

accomplished by dividing the total input to the furnace into three zones. In the first, the stoichiometry is maintained at a very rich condition during the start of the devolatilization process.

Supplementary secondary air is then added to the rich products to establish conditions that maximize the decay of the volatilized fuel-nitrogen compounds. The tertiary air provides the oxygen necessary for complete combustion and provides a complete oxidizing zone around the rich primary zone. This oxidizing zone serves two purposes: it prevents corrosion due to a reducing atmosphere in the furnace, and it limits the interactions between adjacent burners.

IGNITORS

The function of a fuel-burning system is to supply an uninterrupted flammable furnace input and to ignite that input continuously as fast as it is introduced and immediately on its appearance in the furnace. In this fashion, no explosive mixture can accumulate in the furnace because the furnace input is effectively consumed and rendered inert. Ignition takes place when the flammable furnace input is heated above the ignition temperature.

Supplying correct ignition energy for the furnace input is a substantial task. Many factors establish the combinations of ignition-energy quantity, quality, and location that provide a satisfactory furnace input ignition rate at any instant. Unfortunately, these factors change rapidly, so what constitutes sufficient ignition energy at one instant may be insufficient the next. From the rate-igniting standpoint, there is never excess ignition energy. The more stable a fire, the more likely that the ignition energy will substantially exceed the minimum necessary to maintain ignition.

Ignition Energy

As a safety factor, ignition energy for lightoff is not always supplied in great excess. It usually comes in the form of heat, large amounts of which may not always be desired, particularly at low loads. Also, the ignition fuel can be expen-

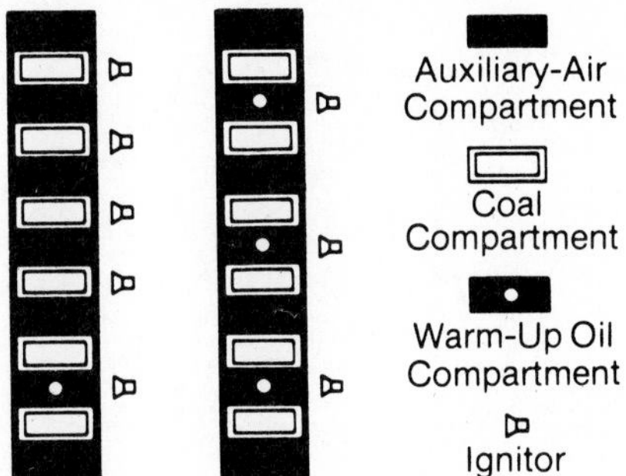

FIG. 4.10 Alternative step-ignition processes for tangential firing.

sive and, in some instances, difficult to obtain. Most stations use more costly fuel for auxiliary ignition than is supplied for main burner firing. If a large amount of premium lightoff fuel is consumed, the cost can be significant.

Igniting the input in prescribed steps also avoids undesirable furnace pressurizing during lightoff. A proportionately large increase in furnace input can be smoothly, adequately, and continuously ignited, but the sudden increase in average furnace temperature tends to produce a quick increase in specific volume of the reactants, which occurs because furnaces ordinarily cannot vent the excess material fast enough. The specific volume must not increase to the point that the furnace pressure increases.

Figure 4.10 shows two methods of achieving a step ignition for tangential firing. In the case on the left, the lowest elevation of ignitors is fired first, and after a suitable lapse of time, the elevation of oil warmup guns is ignited. As the furnace temperature stabilizes, the main coal nozzles adjacent to the oil guns are brought into operation. Then, additional ignitors can light off elevations of main coal nozzles directly. Experience indicates that firing by these step input methods will minimize furnace pressurization and unnecessary overpressure trips.

Systems for Tangential Firing

Ignition systems available for tangentially fired systems include:

- Spark ignition of distillate oils and high-calorific-value gases
- Spark ignition of residual oils
- Spark ignition of pulverized coal

Available ignitors can accommodate a variety of fuels, including natural gas, all grades of fuel oil, and most coals. The application of various ignitors is largely a function of ignitor fuel availability and economics.

IFM Ignitor

The *ionic flame monitoring* (IFM) side ignitor can spark-ignite high-Btu gases or distillate oils. The IFM design (Fig. 4.11) follows a traditional philosophy of pro-

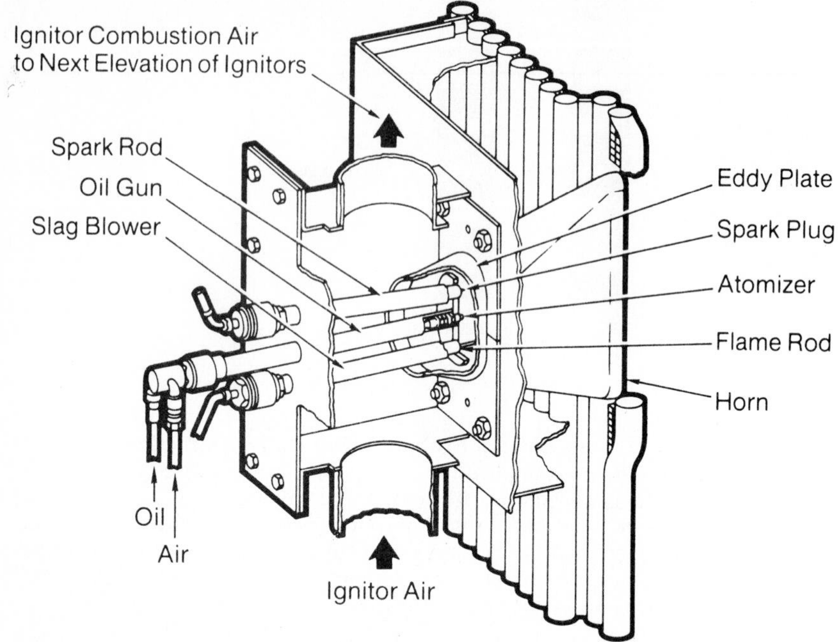

FIG. 4.11 Ionic flame monitoring (IFM) side ignitor.

viding an ignitor with both qualitative and quantitative indications of flame. The system incorporates the principle of flame ionization, which is present in all turbulent hydrocarbon flames, to detect the presence of combustion. In the burning process, energy is liberated by the combination of two or more reactants to form a product with a low energy level. During this burning process, many ions (charged particles) are liberated as electrons and charged nuclei. When a direct-current (dc) potential is placed across the flame, a varying current is generated because of the variable resistance that the flame presents to the rod.

The system operates by imposing a dc potential on the rod that is in contact with the flame. The dc voltage level is modulated plus or minus around the imposed level by the flame; then the imposed signal is filtered out. The variance is amplified, changed to a pulse shape, and used to drive a flame indication relay.

The circuit is designed to be fail-safe. If a component failure, a short circuit in the flame rod or lead wire, or a direct ac interference occurs, then a "no flame" indication is triggered. The common application of the ignitor is compatible with NFPA's class 2 classification. It is nominally sized for 4 percent of the heat input of the adjacent coal nozzle and is used for initial lightoff and low-load stabilization. The ignitor can also light off an oil-fired warmup gun, which then ignites the adjacent coal nozzle.

HEA Ignitor

The *high-energy-arc* (HEA) ignitor was developed to offset the decreasing availability and rising cost of ignition fuels such as natural gas and no. 2 fuel oil. The HEA ignitor effectively eliminates dependency on these fuels by igniting a no. 6

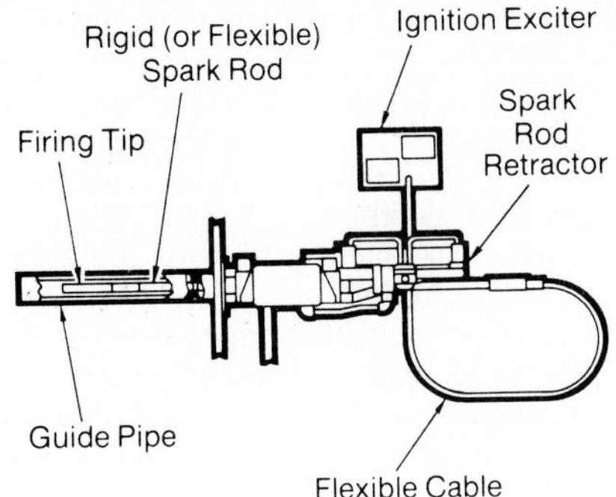

FIG. 4.12 High-energy-arc ignitor.

oil warmup gun which, in turn, ignites the main coal nozzles (Fig. 4.6). The HEA ignitor is paired with a discriminating scanner, which verifies the operation of the warmup gun. This gun is designed for a proven oil flow of about 10 percent of an adjacent coal nozzle, with the discriminating scanner proving the presence of a flame. The combination of an HEA ignitor and a no. 6 oil-fired warmup gun satisfies the NFPA definition of a class 1 or class 2 ignitor. The HEA ignitor alone would be a class 3 device.

The HEA ignition system consists of a

- High-energy-arc ignitor
- Warmup oil compartment capable of producing a stable flame at all loads
- Flame-detecting system sensitive only to its associated oil gun
- Control system to coordinate all the components and provide for unit safety

The HEA ignitor (Fig. 4.12) can ignite warmup fuel oils ranging from no. 2 to no. 6 and crude oils. The ignitor is a self-contained electric-discharge device that produces a high-intensity spark. A high-resistance transformer, which produces a full-wave charging circuit and controls the spark rate, enables the sealed power supply to store maximum energy and to deliver a greater percentage of this energy through insulated cables to the ignitor tip. A high spark energy also eliminates coking of the ignitor tip.

A key to the successful application of spark ignition is the presence of a strong recirculation pattern in the primary combustion zone. The recirculation provides the energy required to vaporize and heat the oil to its ignition point, thus maintaining stable ignition after the spark has been deactivated.

The discriminating scanners are ultraviolet (uv) devices with a reduced sensitivity, which permits each scanner to see only the flame from the associated warmup oil gun, and not the flames of the warmup oil guns in adjacent or opposite corners.

POWERPLANT FANS

Regardless of fuel and method of firing, all boilers for industrial and utility generation use mechanical-draft fans. They supply the primary air for the pulverization and transport of coal to the furnace. They also supply the secondary and tertiary air to the windboxes for completion of combustion. Fans remove the products of combustion from the furnace and move the gases through heat-transfer equipment and *flue-gas-desulfurization* (FGD) equipment. Sometimes, gas fans control the steam temperature. Numerous small fans are used for sealing and cooling of ignitors, scanners, and other equipment.

Power station fans are among the largest made; static pressures of 60 in H_2O (14.9 kPa) and individual fan volumes of 1.5×10^6 ft^3/min (0.71×10^6 L/s) are common. Usually they are custom-designed with the blade configuration and control systems highly sensitive to both owner preference and the evaluated cost of installed and operating power. Applications that require the largest fans on a boiler (and so use the greatest amount of plant power) fall into four categories—forced draft, primary air, induced draft, and gas recirculation—which combined account for more than one-half of total boiler auxiliary power requirements.

Forced-Draft Fans

Forced-draft (FD) fans supply the air necessary for fuel combustion, and they must be sized to handle the stoichiometric air plus the excess air needed for proper burning of the specific fuel for which they are designed. Also, they provide air to make up for air heater leakage and for some sealing air requirements. FD fans supply the total airflow except when an atmospheric-suction primary-air fan is used.

Radial airfoil (centrifugal) or variable-pitch (axial) fans are preferred for FD service. FD fans operate in the cleanest environment associated with a boiler and are generally the most efficient fans in the powerplant. They are particularly well suited for high-speed operation. Most FD fan installations have inlet silencers for noise reduction, with screens to protect the fans from entrained particles in the incoming air.

Both the air temperature at a powerplant and its elevation above sea level affect the air density, which in turn has a direct influence on FD fan capacity. If steam coils or other means of heating the air *ahead* of the FD fans are provided, the consequent air temperature to the fans must be taken into account. If hot-air recirculation gives air heater protection, then both the added volume of air and its higher temperature must be considered. If air moisture exceeds 1 or 2 percent by weight (as can happen in high-temperature tropical powerplants), the resulting greater volume proportionately increases all fan capacities on a unit.

Primary-Air Fans

Large high-pressure fans supply the air needed to dry and transport coal directly from the pulverizers either to the furnace or to an intermediate storage bunker. *Primary-air* (PA) fans may be located before or after the milling equipment. The most common applications are either as pulverizer exhauster fans or as cold (ambient-temperature) PA fans.

The mill *exhaust* fan draws hot air from the secondary-air duct and through the pulverizer. The coal-air mixture from the pulverizer then passes through the fan and discharges into the fuel pipes, which carry the mixture to the furnace for ignition. One fan is usually supplied for each pulverizer.

A materials-handling fan of the straight-blade type, the mill exhauster is sized for the maximum airflow needed by the pulverizer. It must develop sufficient pressure at maximum airflow to overcome the resistance of the air ducts, dampers, pulverizer, and fuel pipes to the furnace.

Located before the air heater, the *cold* PA fan draws air from the atmosphere and supplies the energy required to force the air through the ducts, air heater, pulverizer, and fuel piping. Usually two fans are supplied for each steam generator. Cold PA fans for ambient-air duty are usually of the centrifugal airfoil type and have silencers like FD fans. In situations involving severe particulates or high temperatures, straight radial or modified radial fans are recommended.

Induced-Draft Fans

Induced-draft (ID) fans exhaust combustion products from a boiler. In doing so, they create sufficient negative pressure to establish a slight suction in the furnace [usually from 0.2 to 0.5 in H_2O (50 to 125 Pa)]. This condition gives rise to the name "suction firing" or "balanced-draft" operation. These fans must have enough capacity to accommodate any infiltration caused by the negative pressure in the equipment downstream of the furnace and by any seal leakage in air preheaters.

Since ID fans are now typically located downstream of any particulate-removal system, they are a relatively clean service fan. In most instances, therefore, an airfoil centrifugal fan is selected. The airfoil designs can develop efficiencies of more than 88 percent and can have capacities greater than 1.5×10^6 ft^3/min (0.71×10^6 L/s). The airfoil blade shape minimizes turbulence and noise. The blades and center plates may also be fitted with wear pads and replaceable nose sections for greater wear protection. The structural strength, particularly important in larger sizes, is excellent with this design.

Where greater wear resistance is needed because of a dust burden or where a very conservative approach is desirable, a modified radial or forward-curved, backward-inclined design is selected. Without unduly sacrificing efficiency, these blade shapes minimize dust buildup and reduce downtime for cleaning. They have low noise characteristics, and their relatively simple design allows fabrication in special alloys, should they be required for service downstream of a scrubber discharging wet gas.

The ID fan is sometimes used as a booster fan with FGD scrubbers. In one such arrangement, ID fans follow the precipitator or fabric filter, and another set of fans—the booster fans—follows the scrubber. With pressurized scrubbers, the booster fans are placed directly behind the ID fans and *ahead* of the scrubbers, acting to "push" the gases through the FGD scrubbers.

Gas-Recirculation Fans

These fans draw gas from a point between the economizer outlet and the air preheater inlet and discharge it (for steam temperature control) into the bottom of the furnace. To control the steam temperature on coal-fired units, a high-efficiency, high-draft-loss mechanical dust collector must be installed ahead of the fan. If the recirculation fan on a coal-fired unit is only for standby or emergency oil firing, the dust collector is omitted.

Gas-recirculation duty provides the most severe test of a powerplant fan. The combination of heavy dust loads and rapid temperature changes demands the utmost in rugged, reliable fan design. Particularly important is how the fan hub is mated with the shaft; the conventional shrink fit may not be adequate. To cope with temperature excursions, fans with an integral hub are preferable. Straight or modified radials or forward-curved, backward-inclined centrifugal wheels meet these needs the best.

How Fans Work

A fan is a volumetric machine which, like a pump, moves quantities of air or gas from one place to another. In so doing, it overcomes resistance to flow by supplying the fluid with the energy necessary for continued motion. Physically, the essential elements of a fan are a bladed rotor and a housing to contain the incoming air or gas and to direct its flow. Because a fan does work, it needs energy to operate. The amount of energy depends on the volume of gas moved, the resistance against which the fan works, and the machine efficiency.

Fan-Pressure Relationships. The total pressure developed by a fan is the algebraic sum of the velocity pressure and static pressure. Specifically related to fans, total pressure is the air pressure that exists by virtue of the degree of compression and the rate of motion. When these definitions are applied to fan performance, distinct relationships exist between each variable.

Fan total pressure is the difference between the total pressure at the fan outlet and the total pressure at the fan inlet. *Fan velocity pressure* is that corresponding to the average velocity at the specified fan outlet area. *Fan static pressure* is the difference between the fan total pressure and the fan velocity pressure. Thus, the fan static pressure is the difference between the static pressure at the fan outlet and the total pressure at the fan inlet.

Static pressure rise, sometimes mistaken for fan static pressure, is the static pressure at the fan outlet minus the static pressure at the fan inlet. The difference between the fan static pressure and the static pressure rise is the inlet velocity pressure.

Power. With the equation for fan work and some basic physical constants, the equation that expresses air horsepower (air hp) can be developed:

$$\text{air hp} = \frac{VH}{6356}$$

where V is the volumetric flow through the fan in cubic feet per minute and H is the head or pressure difference in inches of H_2O across the fan. The air horsepower ($\times$ 0.746 = kW) may also be designated as either static or total. Because the resistance to be overcome in fan applications is primarily static pressure, the fan pressure developed is usually referred to in terms of static head. On this basis, the calculated fan power is known as the *static air horsepower* (air hp_s). When the power calculations are based on total head, fan power is referred to as the *total air horsepower* (air hp_t) and is equivalent to the power output. So

$$\text{Fan mechanical efficiency } \eta_t = \frac{\text{air hp}_t}{\text{power input}}$$

Transposing gives

$$\text{Power input (brake or shaft hp)} = \frac{\text{air hp}_t}{\eta_t} = \frac{VH_t}{6356\eta_t}$$

The power input formula assumes that air is an incompressible fluid. But the fact that air is compressible must be recognized in designing for high-pressure differentials. The fan power formula then becomes

$$\text{Power input, hp} = \frac{k_c VH_t}{6356\eta_t}$$

where k_c can be taken from Fig. 4.13, which is based on adiabatic compression.

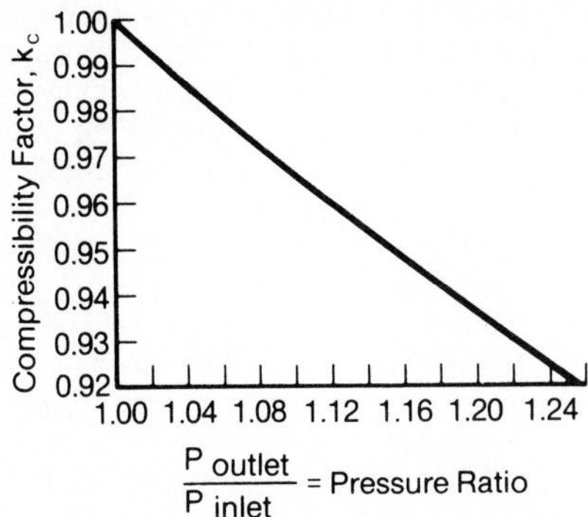

FIG. 4.13 Compressibility factor for use in calculating fan power consumption, assuming adiabatic compression.

Density. Because they affect gas density, the pressure and temperature of the air or gas also influence the power output and efficiency. A change in density changes the total and static pressure and their subsequent conversion to inches of H_2O (pascals) at standard conditions. Remember that head and horsepower vary inversely as the absolute fluid temperature and directly as the absolute fluid pressure (or directly with the fluid density), and adjustments often must be made for pressure and temperature variations in calculating performance or selecting a fan for a particular application.

Flow. The gas flow through the fan is usually expressed as a volumetric flow rate. It is necessary to determine the flow rate in actual cubic feet per minute (acfm), or liters per second, at the inlet to each fan from the density at the fan inlet and the mass flow rate, in pounds per hour (kilograms per second). Proper corrections for plant elevation and actual conditions of local pressure at the inlets of all fans must be made to the calculated air and gas volumes.

Types of Fans

From the point of view of fluid mechanics, fans represent a class of turbomachines designed to move fluids such as air, gases, and vapor against low pressures. From the point of view of mechanical design, fans have a very light casing because inlet pressures are atmospheric or lower. Simplified hydraulic forms and welded steel plate are generally encountered in fans.

Direct-connected drives most often are used in powerplant applications, with control obtained through variable-speed motors, hydraulic couplings, or variable-inlet vanes. Fans are broadly classed as either *centrifugal* or *axial,* according to the flow direction. The centrifugal (radial) fan moves air perpendicular to the rotational axis of the impeller; the axial-flow fan moves air parallel to the rotational axis of the impeller.

Centrifugal Fans. Centrifugal fans use blades mounted on an impeller (or rotor) that rotates within a spiral or volute housing. The blade design determines the fan

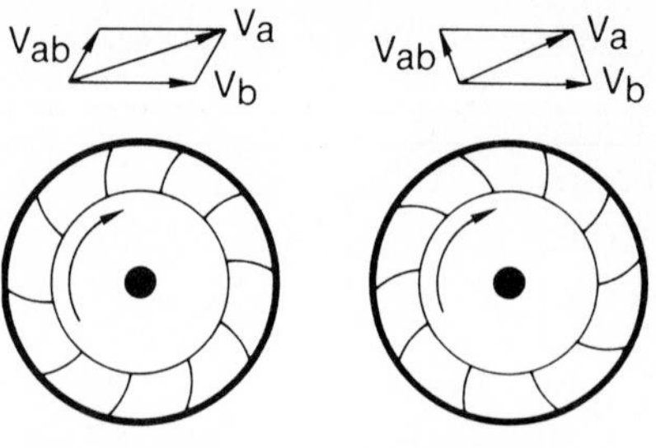

Forward Curved Backward Curved

FIG. 4.14 Velocity vector diagrams comparing forward-curved and backwardly curved centrifugal fan blades. At same tip velocity (V_b), each type blade produces different air velocity (V_a). Vector $\mathbf{V}_{ab}$ is air velocity relative to the blade.

characteristics, so by selecting blades of different shapes a fan engineer can develop an appropriate fan design. There are three basic blade types: forward-curved, straight (radial and radial tipped), and backward-curved (airfoil or flat).

A velocity vector diagram at the blade tip (Fig. 4.14) indicates that backward-curved blades produce low velocities for a given tip or peripheral speed and that forward-curved blades give high velocities. Radial and radial-tipped blades lie between these two extremes. The backward-curved type, therefore, operates at greater motor speeds than the other types for a given duty and is well adapted to direct drive with motors or steam turbines.

Blade Types and Shapes. Figure 4.15 shows some commonly used blade shapes. In general, the blade type limits the fan speed. Thus, the backward-curved blade machines can operate at a relatively higher speed than the forward-curved designs. Blade selection depends on speed limitations, allowable noise levels, efficiency demanded by specified load conditions, and desired fan performance characteristics in the most likely range of operation. Also, the maximum attainable mechanical efficiencies and tolerance of the blade to corrosion and erosion are selection factors (Table 4.1). Notice in the table that increasing the efficiency of a centrifugal fan sacrifices its erosion resistance.

Once the optimum fan has been determined, the aerodynamic selection process yields the width, diameter, and speed of the fan wheel. These parameters then become factors for subsequent elements of the fan design. In general, fan-wheel dimensions determine the basic dimensions of the fan housing and inlet boxes.

Backward Inclined Flat Blade

Backward Inclined Airfoil

Radial Tip

Radial

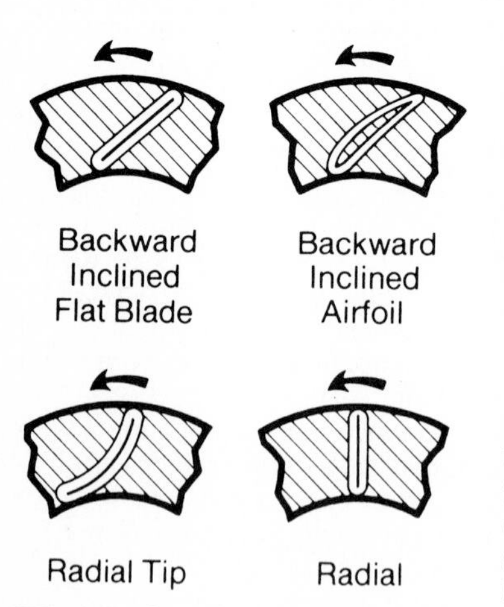

FIG. 4.15 Centrifugal (radial) fan blade types.

Fan Arrangements. A very popular centrifugal fan arrangement was a single-

TABLE 4.1 Effect of Blade Type on Erosion Resistance and Efficiency

Blade type	Typical maximum static efficiency, %	Tolerance to erosive environment
Straight	70	High
Forward-curved	80	Medium to high
Backward-curved	85	Medium
Airfoil	90	Low

speed motor drive that controlled flow by inlet vanes or inlet louvers (Fig. 4.16). The centrifugal fan in this instance is sized for specified conditions and is throttled by the inlet vanes to allow the fan to provide the flow and pressure required at lower operating loads.

As fuel costs and equipment efficiency have become increasingly important, new arrangements for centrifugal fans have developed. Because the welded blades of a centrifugal fan are not adjustable, the only other means of controlling the flow and pressure is to vary the speed of the fan. Two-speed motor drives, turbine drives, and fluid couplings have been installed with varying success on centrifugal fans and have helped to increase their efficiency at lower loads. These arrangements do add more moving parts to the fan system, however, with a potential negative effect on reliability.

The variable-speed and variable-frequency wound-rotor motors are fan drives that allow the speed to be varied from full to zero in infinite steps; they improve low-flow efficiencies dramatically. As discussed elsewhere, speed variation is the

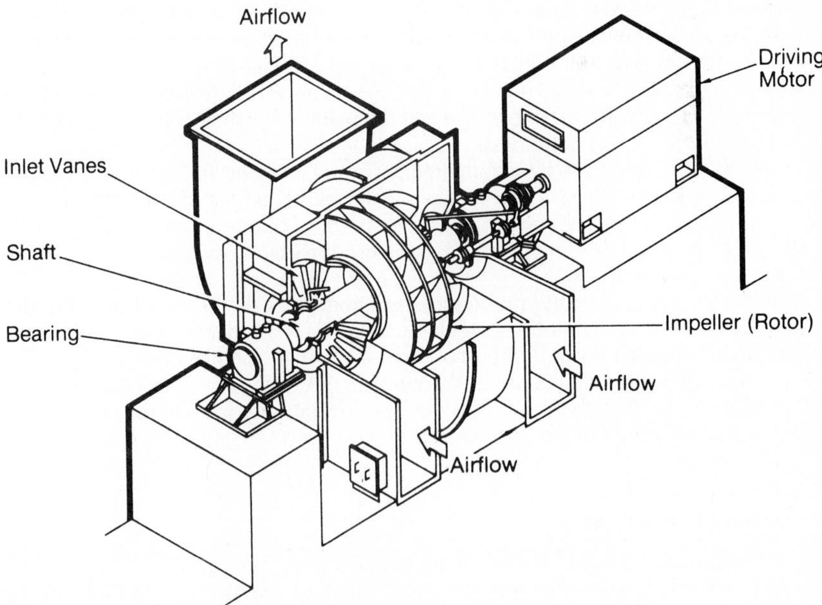

FIG. 4.16 Airfoil-blade centrifugal fan with inlet-vane control.

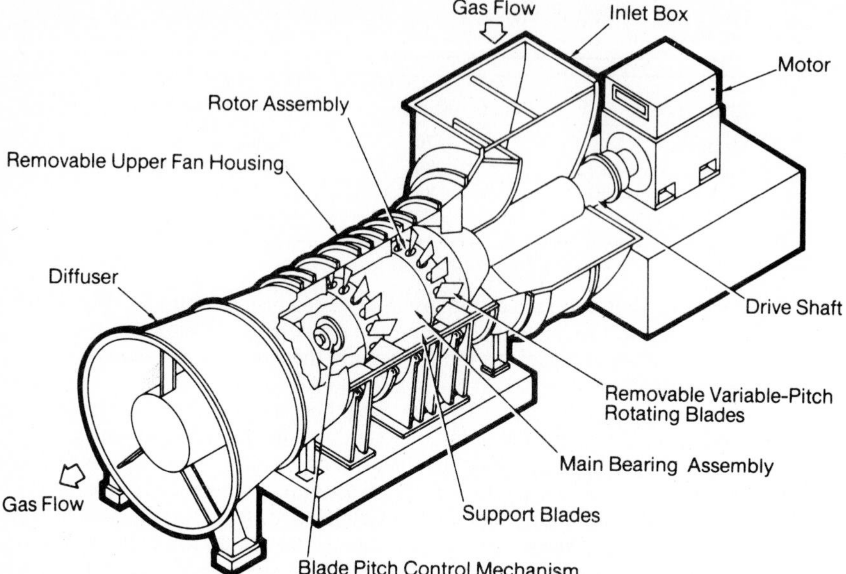

FIG. 4.17 Two-stage variable-pitch axial-flow fan for induced-draft service.

ideal way to change fan operating characteristics; thus, for centrifugal fans, these motor designs are of great interest.

Axial-Flow Fans. Axial fans with variable-pitch blading can maintain far higher efficiencies at various boiler loads than constant-speed centrifugal fans. As fuel costs continue to rise, the higher initial expense of axial-flow fans frequently is offset by operational cost savings over the life of the plant. A detailed evaluation is required, however, to show the advantages of using axials instead of centrifugals in any given situation.

In the most common axial arrangement (Fig. 4.17), the fan operates at constant speed, and the angle of the blades on the hub is adjusted to vary flow; no inlet vanes are required for control. For each point of operation, this arrangement enables the axial fan to develop a unique aerodynamic configuration that is as efficient as possible.

Either a mechanical or hydraulic mechanism adjusts the blade pitch while the fan operates at its design speed. Mechanical-type pitch-change mechanisms are usually insufficient to properly control fans of the size required for utility applications. Variable-pitch axials are best suited to PA, FD, ID, or booster fan applications on large industrial and utility boilers. Axial fans are not appropriate for small fan requirements or for gas recirculation.

Fan Characteristic Curves

Figure 4.18 shows a typical set of constant-speed characteristic curves for a centrifugal fan with backward-inclined blades. Although by applying the theory of similitude the performance of a single fan of a given type may determine the characteristics of a complete line of sizes, each separate type must be tested indepen-

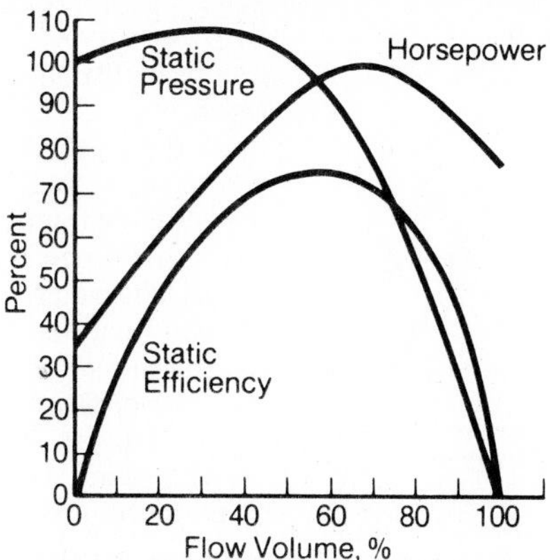

FIG. 4.18 Chart showing typical constant-speed characteristics for a fan with backwardly curved blades. Pressure decreases as fan capacity goes up, expanding the range of stable operation. Horsepower characteristic is nonoverloading.

dently. The curves are for one speed only; when a constant-speed motor drives the fan, the fan will operate somewhere on the characteristic curves, depending on the resistance imposed by the system through which the air passes. The system resistance, therefore, acts as an automatic control on the fan and limits the amount of overload that can be developed, despite the flatness of the pressure characteristic. The amount of overload depends also on the shape of the horsepower curve.

To fully appreciate the significance of fan characteristic curves, note that every fan is restricted to that performance defined by its curve. The fan *must* operate at a point that lies somewhere on the characteristic plot. For example, if the head required for a given volume is less than that specified by the curves, additional resistance must be placed in the system; otherwise, the fan will put out more capacity until it reaches a point on the characteristic curve at which the head matches the system resistance. The fan has no choice; it must operate at a point on the curve where the head, capacity, and system resistance balance.

Development. Fan performance curves are normally developed from base test curves which, in turn, are developed from test data recorded under controlled laboratory conditions. For continuity throughout industry, as well as for ease of understanding and ready comparison, test data are corrected or adjusted to what is known as standard conditions. These conditions for fan design work specify that all flow, pressure, and power values must be at 70°F (21°C) and sea level, with an air density of 0.075 lb/ft^3 (1.2 kg/m^3).

A normal curve format consists of a graph that has the air or gas volumetric flow numerically on the abscissa (horizontal axis), with both static pressure and brake horsepower numerically on the ordinate (vertical axis). Two separate curves—volume versus static pressure and volume versus power—are plotted on each performance curve.

Application. Figure 4.19 illustrates the application of typical centrifugal fan char-

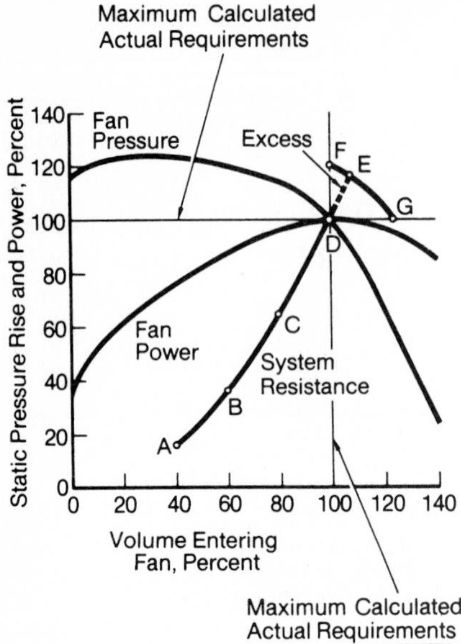

FIG. 4.19 Use of characteristic curves as applied to problems of centrifugal fan selection.

acteristic curves to a fan problem in which points *A, B,* and *C* are calculated requirements at four load points on a given boiler. The line through them defines the system resistance. The point where this line intersects the static-pressure characteristics of any fan, at any given speed, determines the point on the characteristic curve at which the fan will operate, if both curves are plotted for the same density. However, the fan can operate only on its characteristic curve. If any error has been made in calculating point *D*—in volume, pressure, or temperature—that point will not fall on the characteristic curve and the fan may not meet the requirements for operating at that particular speed.

To provide excess capacity, it is customary to specify the volume and pressure in excess of the actual calculated requirements and thereby obtain a larger fan. Suppose a portion of the pressure characteristic of this larger fan, operating at the same speed and density, is represented by line *FG* in Fig. 4.19. Then clearly this size fan would be selected by the manufacturer if the purchaser specified 24 percent excess volume. At the same time, the fan would satisfy the requirement of point *E,* which requires 8 percent excess volume and 17 percent excess pressure.

The only advantage in attempting to define point *E* on the extrapolated system resistance curve, instead of point *F* or *G,* is that the power requirement given by the manufacturer will then represent a closer estimate of the larger fan under actual operating conditions than if point *F* or *G* had been defined for fan selection. The fan finally chosen, however, can satisfy the requirements of all three points if sufficient power is available from the drive.

Fan Control

Very few boiler applications permit fans to operate continuously at the same pressure and volume. To meet the requirements of the system, therefore, some means of varying the fan output becomes necessary. Capacity control of a fan can be achieved in two ways: by controlling the aerodynamic flow into or within the fan and by controlling the fan speed.

The first method refers either to altering the flow of gas into the eye of a centrifugal-fan wheel, as with inlet vanes, or to changing the internal aerodynamics by altering the internal geometry, as with controllable-pitch axial fans. The second method refers to any speed-changing device such as a turbine, fluid drive, multiple-speed motor, or electronically adjustable motor drive connected to the fan.

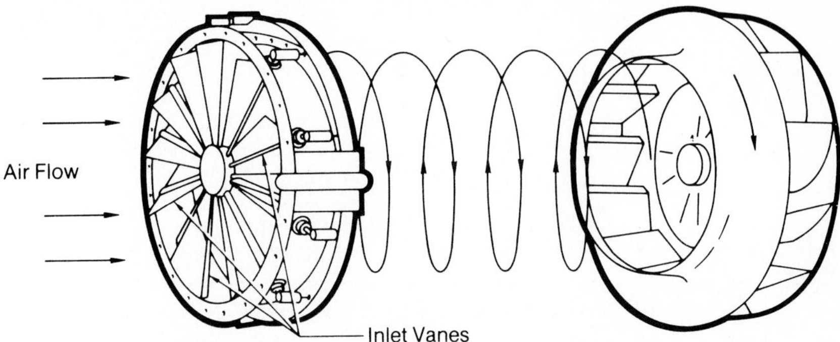

FIG. 4.20 Inlet vanes give an initial spin to air entering a centrifugal fan. By adjusting the angle of vanes, the degree of spin and volumetric output are regulated.

Variable-Inlet Vanes. Sometimes variable-inlet vanes are used to control fan performance by providing swirl to the fan impeller, saving significant power (Fig. 4.20). Variable-inlet vanes are more effective in saving power than parallel-blade inlet box dampers. When variable-inlet vanes are furnished for either centrifugal or fixed-pitch axial fans, fan manufacturers provide a complete performance envelope showing the effect of vane position on fan performance and power.

An inlet-vane-controlled centrifugal fan is selected to produce full specified flow and pressure with no inlet vanes present. Inlet vanes then throttle down this maximum performance capability so that the fan can operate over the range of normal boiler operating load points (Fig. 4.21). This type of control system is quite sensitive at lower load conditions. Extremely minor changes in inlet-vane openings have a dramatic effect on the flow produced by the fan, whereas at higher loads it requires increasingly larger movements of the inlet vanes to have any effect on the flow produced by a centrifugal fan.

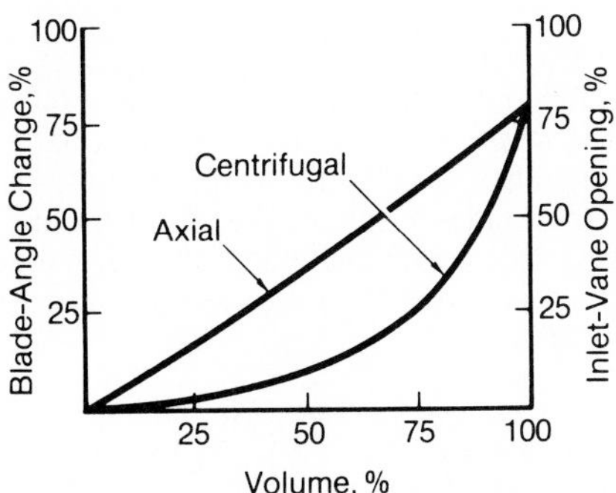

FIG. 4.21 Flow control for centrifugal and axial fans.

Axial-Fan Blade Pitch. An axial-flow fan can control flow at constant speed by varying its blade angle (Fig. 4.22). The effect is to create a unique aerodynamic configuration for the fan at each point of operation, so that the fan is operating at

FIG. 4.22 Closeup of adjustable airfoil blading of an axial-flow fan.

maximum possible efficiency. As the blade angle is adjusted from minimum to maximum position, the flow change is nearly linear (Fig. 4.21).

Another aspect of control rests in the response time of the fan. Most axial-flow fans can move the blades a full stroke, from the maximum open to fully closed position, in 30 s. This means that under the normal boiler operating range, an axial fan can respond or move from maximum continuous conditions to zero flow in approximately 20 s.

Fan Speed

Controlling fan speed is potentially the most efficient form of capacity control. The only significant inefficiency that a speed control system can introduce to a fan results from the inefficiencies of the system. All speed control systems yield certain operational and reliability improvements, especially to large fans in ID service. They include

- Reduction in erosion approximately proportional to the ratio of the squares of the impact velocities
- Reduction in mechanical shock at startup
- Adjustability of speed to any point within the operating speed range
- Reduction in potential fan-system-oriented problems (foundation, ductwork, noise) at reduced running speeds
- Reduction in electric power surge on motor startup

The primary disadvantage of adjustable-speed drives is their cost which, depending on the unit size, can be significantly more than a single-speed motor drive with inlet-vane control. Also, they add components to the drive train, which require more careful consideration of system dynamics to avoid torsional oscillations or other undesirable phenomena.

Fluid-Drive Control. For speed adjustment, fluid drives are either hydrokinetic or hydroviscous. In hydrokinetic operation, an impeller attached to the driving motor accelerates oil particles and impinges them against a runner attached to the driven fan. The speed is adjusted by changing the volume of oil in the system, and thus the transmittable torque. The hydroviscous method is similar in principle, but relies for the speed change on the increase in torque transmitted by the oil between alternate rows of driver-driven plates that are forced closer together.

Fluid drives have seen extensive service where speed changing was required on large centrifugal fans. They are more efficient than single-speed inlet-vane control as volume output drops below about 75 percent. They represent the least efficient adjustable-speed method of gas volume control, however, giving away roughly 1 percent of efficiency for each 1 percent of speed reduction. They also add another mechanical device in the drive train, which increases the cost of installation, requires cooling equipment, and possibly increases initial fan and drive-motor size because of slip at full speed.

Two-Speed Motor Drives. As a centrifugal fan is vaned or dampered down from the specified point, its efficiency falls off rather rapidly. If at the normal operating point, however, the fan speed is changed to a lower base speed and the vanes or dampers are reopened, then a system efficiency is achieved that matches the original full-load levels. This is the principle of application of two-speed motors to power fans. Many two-speed motors have been coupled to fans to gain the overall high efficiencies available from two running speeds.

Also, the two-speed motor drive is the least expensive and least complex speed control method. An objection frequently voiced is that operators hesitate to change to the lower speed until far below the design load limit, which of course negates the planned economy of the system.

Variable-Speed Drives. A variable-speed system significantly improves fan efficiency during periods when the boiler is operating at less than maximum load. A centrifugal fan can be equipped with an adjustable-speed drive and a control and monitoring system. The speed control range can be tailored to a particular installation, and drive system costs can be optimized. By equipping a fan with both adjustable-speed capability and inlet vanes, the fan output can be tailored to match all possible boiler operating conditions.

Slip-ring (or wound-rotor) motors provide step changes in speed by connecting the motor windings to an adjustable external resistance through slip rings and brushes. By changing the rotor circuit resistance, the motor speed-torque characteristic can be adjusted to fit changes in load conditions. With no resistance in the rotor circuit (rings-short-circuited condition), the wound-rotor motor will operate exactly the same as a squirrel-cage motor.

Another important difference is the starting characteristics of the motor. The wound-rotor motor inrush current can be as low as the magnetizing current—normally about 30 percent of the rated current. Most other induction motors draw a very high current (600 to 700 percent of rated) during startup. This lower-current characteristic can provide significant savings in the electric distribution system.

Synchronous motor adjustable-frequency systems use static electronic control equipment to adjust the frequency and voltage. The synchronous motor is an *alternating-current* (ac) device with two sets of windings—one fixed in the stator (frame), the other wound around field poles that are fixed to the rotor in pairs. For a given frequency, the number of poles determines the speed of the motor.

Fan Selection

Once the boiler or scrubber engineer determines the pressure and flow required by the system at the rated load, at the specified condition (with margin), and at various load points, this information can be plotted and the fan selected. A centrifugal fan will have its specified point fall within the area of maximum efficiency. At this point, the fan is operating wide open with no restrictions. To achieve the lower load points, inlet vanes are closed to throttle the maximum capability of the fan. As the fan performance moves down the system resistance line, it moves through areas of constant efficiency that fall perpendicular to the system resistance line.

Using the Characteristic Curves. The characteristic curve for an axial fan is considerably different from that of a centrifugal fan. Given exactly the same system resistance curve, an axial-fan designer will select a fan that operates at maximum efficiency at full boiler load. If normal fan performance tolerances have been used, the fan will usually have the capacity to achieve the specified requirements. If higher-than-normal tolerances have been applied, a less-than-optimum selection will have to be made, in which full boiler load does not fall inside the maximum efficiency area.

The oblong or egg-shaped areas on the axial-fan curve represent constant static efficiency. For the axial fan, these areas are oblong in a direction approximately parallel to the system resistance line, rather than perpendicular to it, as with the centrifugal fan. Changes in flow requirements along the system curve cause only slight changes in efficiency for constant-speed centrifugal fans. The lines perpendicular to the system line on the axial curve depict the angle of the blades at that point of operation.

Comparison of Fans and Fan Controls. Table 4.2 summarizes the available fans and operational control equipment; generally, with increasing efficiency of either

TABLE 4.2 Available Types of Fans and Fan Control Equipment

Fans, in order of increasing efficiency over normal operating range	Fan control equipment, in order of increasing efficiency at partial loads
Centrifugal	Single-speed motor
Radial	Outlet damper
Radial tip	Inlet louver
Backward-inclined solid	Inlet vane
Backward-inclined airfoil	
	Two-speed motor
Axial	Fluid coupling
Fixed pitch	Variable-speed turbine drive
Adjustable pitch	Variable-speed motor
Variable pitch	

draft equipment or controls, the cost increases. Figure 4.23 shows the power consumed by the various designs over the full range of capacity.

In comparing several types of fans, both capital and operating costs must be established over the anticipated load range. The costs of fans, controls, drives, silencers, foundations, switchgear, and other auxiliaries, plus any differences in ductwork, have to be obtained for equal specified capacities and sound-pressure levels. Based on the relative power-consumption curves, the operating hours at the individual load points for the entire life of the unit must be multiplied by the associated power consumptions of the fans being compared. Maintenance costs also must be taken into account by analyzing available historic data.

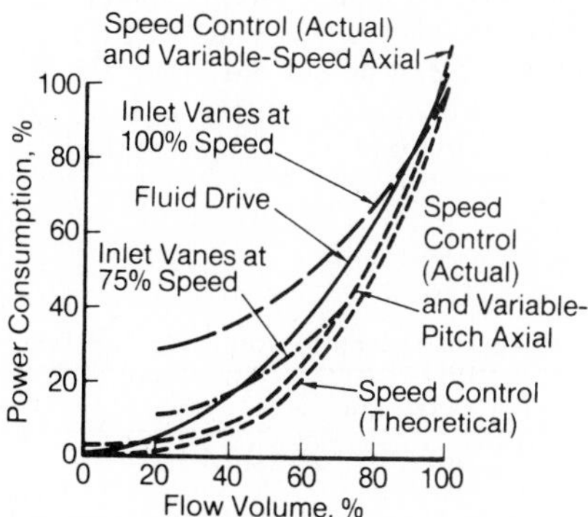

FIG. 4.23 Comparative fan power consumption versus volumetric flow with various types of control.

Fan Size Scale-up

Most fan designs are developed by using models of moderate size and input power. The performance obtained from the model provides an information base to calculate the performance of larger fans that are geometrically similar to the model. If only the basic fan laws are used to make these conversions (with no correction for the compressibility of air), the larger fan in many cases will perform better than predicted, providing that all geometric, kinematic, and dynamic similarity requirements are satisfied.

The three main performance factors of flow, speed, and head are linked in the concepts of specific speed and specific diameter. *Specific speed* is that revolutions per minute (rpm) at which a fan would operate if it were reduced proportionately in size so that it delivered 1 ft^3/min of air at standard conditions against a 1-in H_2O static pressure. *Specific diameter* is the fan diameter required to deliver 1 ft^3/min standard air against a 1-in H_2O static pressure at a given specific speed. From the fan laws, these equations are obtained:

$$\text{Specific speed } N_s = \frac{\text{rpm } (\text{ft}^3/\text{min})^{1/2}}{(\text{SP})^{3/4}}$$

$$\text{Specific diameter } D_s = \frac{D(\text{SP})^{1/4}}{(\text{ft}^3/\text{min})^{1/2}}$$

where flow in cubic feet per minute is at standard conditions, SP is static pressure in inches of water, and D is fan diameter in inches. (*Note:* For SI, $ft^3/min \times 0.472$ = L/s, in $H_2O \times 249$ = Pa, and in $\times$ 25.4 = mm.)

Table 4.3 shows the parameters for designing to specified pressure and volume. In the case of axial fans, maximum flow can be achieved by proper selection of the rotating-blade tip diameter, so an increase in tip size causes a shift of the entire fan curve along the flow (horizontal) axis to the right. Maximum pressure can be achieved by selection of the proper hub diameter, so an increase in hub size would result in a shift of the entire fan curve along the pressure (vertical) axis upward. The system resistance requirements given to the fan designer remain the same, so by selecting the proper hub and blade sizes, the fan characteristic can be shifted along the system resistance, such that boiler full-load conditions occur in the area of maximum efficiency. A centrifugal fan's characteristic can be shifted similarly along the system resistance line.

TABLE 4.3 Effect of Fan Parameters on Performance Capabilities

Capability	Axial fan	Centrifugal fan
To increase volume	Increase blade tip diameter Increase rotational speed	Increase wheel width Increase rotational speed
To increase pressure	Increase rotor diameter Increase number of stages Increase rotational speed	Increase wheel diameter Increase rotational speed

How a change in any operating condition affects a fan can be predicted by a set of rules known as the *fan laws*. These are summarized in the box and apply to fans of the same geometric shape and operating at the same point on the characteristic curve.

Fan Laws Reviewed for Key Parameters

1. For a given fan size, system resistance, and air density:
   a. When speed varies,
      (1) Capacity varies directly as the speed ratio
      (2) Pressure varies as the square of the speed ratio
      (3) Horsepower varies as the cube of the speed ratio
   b. When pressure varies,
      (1) Capacity and speed vary as the square root of the speed ratio
      (2) Horsepower varies as the 1.5 power of the pressure
2. *For constant pressure:* When density varies, the speed, capacity, and horsepower vary inversely as the square root of the density; that is, inversely as the square root of the barometric pressure and directly as the square root of the absolute temperature.
3. *For constant capacity and speed:* When the density of air varies, horsepower and pressure vary directly as the air density, that is, directly as the barometric pressure and inversely as the absolute temperature.
4. For constant mass flow:
   a. When the density of air varies,
      (1) Capacity, speed, and pressure vary inversely as the density, that is, inversely as the barometric pressure and directly as the absolute temperature
      (2) Horsepower varies inversely as the square of the density, that is, inversely as the square of the barometric pressure and directly as the square of the absolute temperature
   b. When both temperature and pressure vary,
      (1) Capacity and speed vary as the square root of pressure × absolute temperature
      (2) Horsepower varies as the square root of pressure cubed × absolute temperature

AIR HEATERS

Although justified by the increased efficiency resulting from lower exit-gas temperatures and the higher flame temperature in the boiler furnace, air heaters also make pulverized coal (PC) firing practical by providing the drying and transporting medium.

Two principal types of air heaters are in service: the tubular recuperative design and the regenerative design, typified by the Ljungstrom air preheater. The choice of the size and type of air heater depends on economic and engineering factors. Economic factors include the original cost of the air heater, maintenance costs, cost of the fuel, and cost of the fan power resulting from air heater draft losses. Engineering factors include the air temperature required for combustion and/or PC drying as well as unit reliability and installation space requirements. The heat-pipe air heater is finding increased use on power plant boilers.

Tubular Design

The air heater is arranged for vertical gas flow through tubes. Air flows horizontally across the tubes, usually 2 to 3 in (50 to 75 mm) in diameter, and in a staggered relationship for optimum heat transfer. The air passes over all the sections of the air heater in sequence, the effect of which is to provide counterflow heat transfer.

Tube sheets at top and bottom support and guide the tubes. Most frequently, the bottom tube sheet or sheets form the structural support; the upper tube sheet is welded to the outside casing. The tubes pass through slightly oversized holes in the upper sheet, which allows for expansion when the equipment is brought up to temperature. Many designs for use with sulfur-bearing fuels have separated cold-end sections to reduce the cost of tube replacement in the event of excessive corrosion.

In contrast to regenerative designs, tubular- or recuperative-type air heaters have more severe cold-end corrosion problems. With variations in cleanliness of the tube wall, entering air temperature, and flow intensities on gas and air sides, very low metal temperatures and correspondingly severe corrosion can occur. Although the tube-metal temperature may be considered to be the arithmetic average of the air temperature entering the air heater and the gas temperature leaving, actual field measurements have shown metal temperatures as low as 120°F (67°C) below the mean of the air and gas temperatures. Such conditions can result in deposits that reduce heat transfer and increase draft loss until they are removed, usually by water washing.

Regenerative Design

The Ljungstrom air preheater (Fig. 4.24) transfers sensible heat in the flue gas leaving the boiler to the combustion air through a regenerative heat-transfer surface in a rotor that turns continuously through the gas and airstreams from 1 to 3 rpm (depending on the diameter). The rotor, packed with Ljungstrom transfer surface, is supported through a lower bearing at the cold end of the preheater and guided through a guide-bearing assembly at the top, or hot, end of the air preheater.

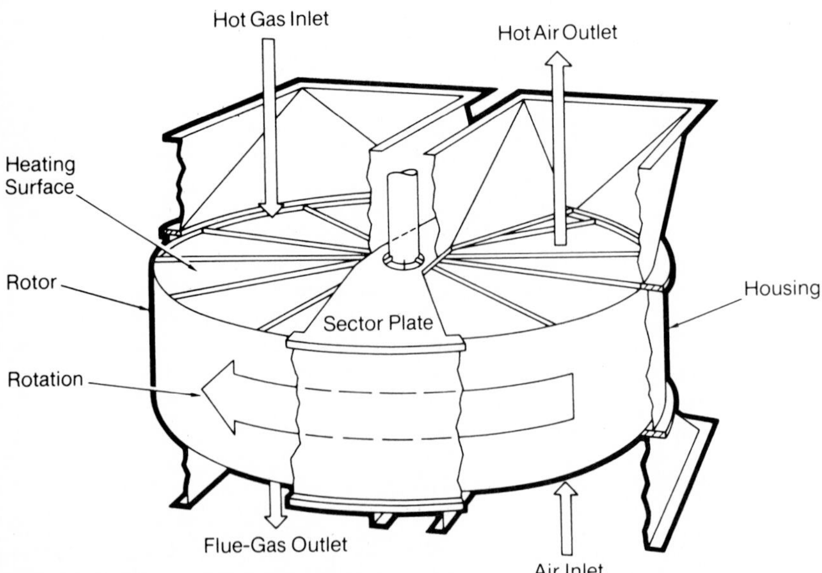

FIG. 4.24 Details of regenerative air preheater, bisector type.

Depending on its size, the rotor has either 12 or 24 radial members, which are attached to a center post; rotor compartments are closed with a shell plate. The rotor sealing system contains simple leaf-type labyrinth seals bolted to the rotor radial members at both the hot and cold ends. The radial seals compress against plates, again located at both the hot and cold ends of the rotor. This system effectively separates the airstream from the flue-gas stream.

Ljungstrom air heaters are designed with rotor diameters from 7 to 65 ft (roughly 2 to 20 m). The smaller units are completely shop-assembled, while the larger utility-size units are arranged for convenient field assembly.

Trisector Design

In the direct-fired PC system, air heating for coal pulverization is necessary. Ordinarily, hot air supplied to the pulverizers carries the heat to dry the coal and transports the pulverized fuel to the furnace. This portion of the total air for combustion is called the *primary air* (Fig. 4.25).

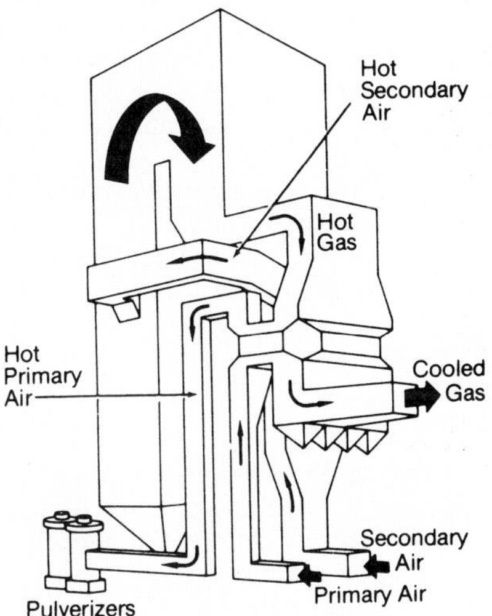

FIG. 4.25 Trisector air preheater arrangement.

The Trisector air preheater sees service on large, coal-fired boilers where a cold PA fan is desirable. The preheater is designed so that by dividing the air side of the unit into two sectors, the higher-pressure primary air and the lower-pressure secondary air may be heated by a single air preheater. These are some advantages of the Trisector over arrangements having separate primary- and secondary-air preheaters:

- No gas-biasing dampers and controls.
- Because average cold-end temperatures can be maintained with only secondary-air preheater coils, the requirement for PA steam coils is eliminated.
- Fewer sootblowers and water-washing devices are required with one heater in lieu of two.

- Even with a large variation in PA flow, relatively little impact on overall heat recovery occurs, because heat not recovered in the primary section will be picked up in the secondary section.

The Trisector is equipped with an additional radial sealing plate and an axial sealing plate separating the primary and secondary airstreams. The heater is readily adaptable to varying coal moisture content. The size of the PA opening depends on the coal being fired and its moisture content, with a 35° or 50° PA opening being normal.

Thermal Gradients

The transfer of heat from the hotter gas stream to the cooler airstream in air heaters creates temperature gradients that cause thermal distortions throughout the structural members. The relative distortion of the various components affects the clearances between the seals and sealing surfaces controlling leakage of air to the gas stream.

With fixed seals and a fixed sealing plate, rotor turndown would make a sizable opening between the leaf-type radial seal and the sealing plate. To close this opening during operation, the air heater includes a sealing-plate controller, which senses the position of the periphery of the rotor during operation and automatically positions the sealing plate to maintain an average clearance of $\frac{1}{16}$ in (1.6 mm).

Rotor turndown closes the seal opening between the cold-end radial seals and the cold-end sealing plate. In the same manner, the axial seals maintain minimum clearance during operation with the axial sealing plate. Therefore, all rotor seals are designed to maintain a close running clearance regardless of the mode of boiler operation.

Leakage Types

Direct leakage of rotary regenerative designs is that quantity of air that passes into the gas stream between the radial and circumferential seals and sealing surface as a result of the static pressure differential between the airstream and gas stream. The leakage across the sealing system is directly proportional to the square root of the pressure differential, but also it depends on the air and gas densities.

Entrained leakage is that quantity of air contained in the rotor as it passes from the air side to the gas side and from the gas side to the air side. The quantity of entrained leakage depends on the rotor depth, diameter, and speed.

Whether direct or entrained, leakage has no effect on the heat-transfer efficiency of the Ljungstrom air preheater. There is no difference in the heat transferred to the airstream from the gas stream because of leakage. However, the gas temperature leaving the preheater is diluted or decreased by 10 to 20°F (5.6 to 11.2°C) by the mixture of the cooler air with the hotter gas stream.

Low Exit-Gas Temperature

Although efficiencies can be improved by adding surface to reduce the air heater exit-gas temperature, this practice lowers the cold-end metal temperatures and the dewpoint of the flue gas. Consequently, steel construction materials are subject to corrosion from sulfuric acid as moisture condenses in the presence of sulfur in the gas.

A number of means have been developed to minimize the rate of corrosion as well as to provide for replacement of corroded surfaces. Because corrosion occurs on the lowest-temperature surface, air heater designs have been developed that incorporate replaceable cold-end sections. Other means to minimize corrosion aim at increasing the metal temperature. One such arrangement directs a portion of the preheated air to the inlet of the FD fan and recirculates it through the air heater. Thus, the temperature of air leaving the fan and entering the air heater is increased, correspondingly increasing the cold-end metal temperature.

Air bypass around the air heater is practiced to a certain extent. With reduced airflow, metal temperatures within the air heater are higher because of the influence of the higher gas-to-air ratio. Also, because the overall recovery is less as a result of the reduced airflow, the gas-outlet temperature rises, causing the cold-end metal temperature to rise.

The prevalent means of increasing the cold-end metal temperature is to introduce steam air heaters in the cold-air duct between the FD fan and the air heater. These increase the temperature of the air entering the heater, at the same time causing an increase in the metal temperature. Steam bled from the turbine serves as the heating medium in the steam air heater. In supplying heat to the cold air, steam is condensed and the condensate returned to the appropriate stage of the feedwater bleed heating system.

To assist specifiers and operators in arriving at reasonable cold-end temperatures, *average cold-end temperature* (ACET) guides are published. These guides take into account the variables of fuel type, sulfur content, and effect of excess air.

SUPERHEATERS AND REHEATERS

The function of a superheater is to raise the boiler steam temperature above the saturated temperature level. As steam enters the superheater in an essentially dry condition, further absorption of heat sensibly increases the steam temperature. The reheater receives superheated steam that has partly expanded through the turbine. Its role in the boiler is to resuperheat this steam to a desired temperature.

Design Elements

Superheater and reheater design depends on the specific duty to be performed. For relatively low final outlet temperatures, superheaters solely of the convection type are generally selected. For higher final temperatures, surface requirements are larger and, of necessity, superheater elements are located in very high gas-temperature zones. Wide-spaced platens or panels, or wall-type superheaters or reheaters of the radiant type, can then be used. Figure 4.26 shows an arrangement of such platen and panel surfaces. A relatively small number of panels are located on horizontal centers of 5 to 8 ft (1.5 to 2.4 m) to permit substantial radiant heat absorption. Platen sections, on 14- to 28-in (350- to 700-mm) centers, are placed downstream of the panel elements. Such spacing provides high heat

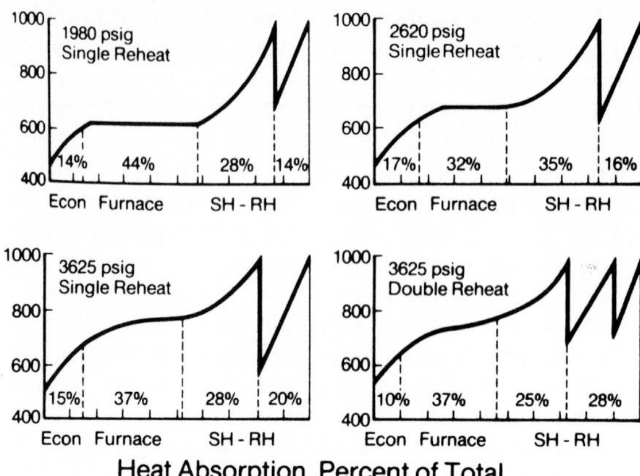

FIG. 4.26 Heat absorption variation with cycle pressure. Note: 1 psig = 6.89 kPa.

absorption by both radiation and convection.

Convection sections are arranged for essentially pure counterflow of steam and gas, with steam entering at the bottom and leaving at the top of the pass, while gas flow is in the opposite direction. The arrangement allows a maximum mean temperature difference between the two media and minimizes the heating surface in the primary sections.

Metallurgy

Superheaters and reheaters are designed for high pressures and temperatures above 1000°F (535°C) and thus require high-strength alloy tubing. Besides selection of materials for strength and oxidation resistance, high steam-pressure service requires greater tube thicknesses in all tubes subject to steam pressure. Furthermore, thicker tubes have higher outside metal temperatures, and because chemical action is accelerated at higher temperatures, the tube metal is more subject to external corrosion. This is of particular concern when fuels containing objectionable impurities are burned. The designer takes account of such conditions in choosing material and tube sizes.

ECONOMIZERS

Economizers help improve boiler efficiencies by extracting heat from flue gases discharged from the final superheater section of a radiant-reheat unit—or the evaporative bank of an industrial boiler. In the economizer, heat is transferred to the feedwater, which enters at a temperature appreciably lower than that of saturated steam. Generally, economizers are arranged for downward flow of gas and upward flow of water.

Water enters from a lower header and flows through horizontal tubing comprising the heating surface. Return bends at the ends of the tubing provide continuous tube elements, whose upper ends connect to an outlet header that is, in turn, connected to the boiler drum via tubes or large pipes. Tubes that form the heating surface may be plain or with extended surface, such as fins. Frequently, they are arranged in staggered relationship in the gas pass to obtain high heat transfer and to lessen the space requirements.

Designing the economizer for counterflow of gas and water results in a maximum mean temperature difference for heat transfer. Upward flow of water helps avoid water hammer, which may occur under some operating conditions. To avoid generating steam in the economizers, the design ordinarily provides exiting water temperatures below that of saturated steam during normal operation.

As shown in Fig. 4.26, economizers of a typical utility-type boiler are located in the same pass as the primary or horizontal sections of the superheater, or superheater and reheater, depending on the arrangement of the surface. Tubing forming the heating surface is generally made of low-carbon steel. Because steel is subject to corrosion in the presence of even extremely low concentrations of oxygen, it is necessary to provide water that is practically 100 percent oxygen-free. In central stations and other large plants, it is common practice to use deaerators for oxygen removal.

Small low-pressure boilers may have economizers made of cast iron, which is not as subject to oxygen corrosion. However, the design pressure for this material is limited to approximately 250 psig (1720 kPa). Whereas cast-iron tubes find little application today, cast-iron fins shrunk on steel tubes are practical and can be used at any boiler pressure.

SUGGESTIONS FOR FURTHER READING

1. Madison, R.D., ed., *Fan Engineering—An Engineer's Handbook,* 7th ed., Buffalo Forge Co., Buffalo, N.Y., 1970.

The latest issues of the publications referenced in numbers 2 through 6 are available from the Air Movement and Control Association, Arlington Heights, Ill.

2. *Fans and Systems,* AMCA Publication 201.
3. *Power Plant Fans—Specification Guidelines,* AMCA Publication 801.
4. "Test Code for Sound Rating Air Moving Devices," AMCA Standard 300.
5. "Methods of Calculating Fan Sound Ratings from Laboratory Test Data," AMCA Standard 301.
6. *Application of Sound Power Level Ratings for Ducted Air Moving Devices,* AMCA Publication 303.
7. "Method of Testing In-Duct Sound Power Measurement Procedure for Fans," ASHRAE Standard 68-78, American Society of Heating, Refrigerating and Air-Conditioning Engineers, Inc, Atlanta, Ga., 1978.
8. Rothemich, E.F., and G. Parmakian, "Tubular Air-Heater Problems," ASME Paper No. 52-A-124, ASME, New York, 1952.
9. Bagwell, F.A., et al., "Utility Boiler Operating Modes for Reduced Nitric Oxide Emissions," *Air Pollution Control Association Journal,* 21(11):702–708, November 1971.
10. Singer, J.G., ed., *Combustion—Fossil Power Systems,* 3d. ed., Combustion Engineering, Inc., Windsor, Conn., 1981.
11. *Steam—Its Generation and Use,* 39th ed., Babcock & Wilcox Co., Barberton, Ohio, 1978.
12. Potter, P.J., *Power Plant Theory and Design,* John Wiley & Sons, Inc., New York, 1959.

CHAPTER 1.5
ELECTRIC BOILERS AND HOT-WATER GENERATORS

Peter M. Coates
CAM Industries, Inc.
Kent, WA

INTRODUCTION

Electric boilers and hot-water generators were introduced in the United States more than 50 years ago. But it was not until the early 1970s, when oil and gas prices skyrocketed and pressure to clean up the environment gathered momentum, that they became economically attractive for producing steam and hot water in many industrial plants (see box). Today these boilers offer a means to utilize electric power available seasonally or at off-peak hours—or from hydroelectric generating plants, where water must be dumped if it is not put through the turbines.

Comparison of Electric Boilers and Fossil Fired Units

Knowing how high-voltage electrode boilers and hot-water generators compare to fired units on cost, space requirements, maintenance, and other key factors can help in deciding which design to select—assuming, of course, that fossil fuels offer no clear-cut cost advantage over electricity. Here are the items to consider:

- *Capital cost:* An electrode boiler can cost half as much as a comparable oil-fired unit, sometimes even less, and as little as one-tenth as much as a boiler that burns solid fuels.
- *Floor-space savings:* Savings in floor space with electrode boilers can be substantial. Whether inside or outside a building, the boiler location is not critical because no air must be supplied to the unit and no flue gas taken away. No special ventilation is needed, and at least some local codes simplify installation further by requiring only a 1-h fire wall. One potential problem for high-voltage boilers with vertically mounted electrodes, however, is that they may not fit in buildings with limited headroom.
- *Personnel:* Some states do not require full-time operators for electrode boilers; however, most demand that operators be in attendance whenever fired boilers produce high-pressure steam.
- *Boiler efficiency:* The efficiency of an electric boiler does not vary significantly with its load. A typical spread is 5 points in efficiency between 10 percent of rated capacity and full load (where efficiency is about 99.5 percent for a well-insulated unit). For a typical oil-fired boiler, the difference in efficiency from low fire to the point of best performance is about 15 percent.
- *Maintenance:* The highest temperature with which an electrode boiler's internal parts come in contact is that of the steam. Thus, their rate of deterioration is slower than that of some parts in a fired boiler, where much higher temperatures exist. Once drained, an electrode boiler can be entered almost immediately for inspection and repair. Also, all parts of an electrode boiler are accessible and easy to check.
- *Reliability:* Any well-operated, well-maintained boiler should perform reliably, but an electrode boiler may have the edge over a fired unit simply because there are fewer elements that can go wrong with it.
- *Operator performance:* An electrode boiler probably is less sensitive to serious damage from poor operation than a fossil-fuel-fired boiler, but it may be more prone to nuisance shutdowns. To illustrate, despite the use of redundant low-water protection systems on fired units, many cases of severe low-water damage are reported each year. By contrast, a low-water condition in an electrode boiler shuts down the unit automatically, because no current can flow. Operator inattention to proper water treatment is likely to be the biggest problem with electrode boilers.
- *Pollution control:* Electrode boilers emit no pollutants to the atmosphere and present no noise problem. Noise and air pollution controls may be needed on some fossil-fuel-fired boilers, depending on the type of fuel burned and the number and type of auxiliaries.

Basic Types

There are two basic types of electric boilers and generators: resistance and electrode. In resistance units, current is passed through a series of resistance-type heating elements—high-resistance wire encased in an insulated metal sheath—which are submerged in water. In electrode units, the current is passed through the water, not a wire, and the resistance of the water converts the electric energy directly to heat, that is, hot water or steam.

Applications

The first packaged electric boiler with replaceable heating elements began operating in the 1920s. Since then, many resistance-type boilers and hot-water generators have been installed in commercial and institutional buildings. Industrial application of these units has been limited, however, because they are economical only up to about 3600 kW [12,600 lb/h (1588 g/s) of saturated steam at 212°F (100°C) or to 12.2×10^6 Btu/h (3.6 MW)]. Myriad terminations, contactors, fuses, and other electric components required with large-capacity, resistance-type boilers make the electrode design cheaper and more practical at higher ratings.

ELECTRODE BOILERS

Electrode boilers are available for either low-voltage (600 V or less) or high-voltage (typically 2.3- to 25-kV) service in steam and hot-water systems. (Technically, electric equipment rated 1 to 72.5 kV is considered *medium* voltage by electrical engineers.) Steam pressures extend from about 12 to 500 psig (83 to 3445 kPa), water temperatures to 400°F (205°C). Higher pressures are possible.

Pressure vessel costs increase dramatically above 300 psig (2067 kPa). For example, a 300-psig (2067-kPa) vessel costs 30 percent more than one designed for one-half that pressure, and the rate of the price increase is greater at higher pressures. Note, too, that steam cannot be superheated without the addition of a separate resistance element or fired superheater.

Low-Voltage Models

Low-voltage electrode models compete primarily with resistance-type units. They offer a simpler electric control system than immersion-element boilers and hot-water generators, and they occupy less floor space. However, they demand close attention to water conductivity—a parameter that does not influence the performance of resistance heaters. Ratings for low-voltage electrode boilers extend up to about 2500 kW. High-voltage units are more economical in larger sizes.

High-Voltage Models

High-voltage electrode boilers and hot-water generators come in several different designs, which can be broadly classified as either submerged electrode or sprayed electrode. Although these units differ somewhat in operating characteristics, generally they are flexible enough to serve in a wide variety of applications.

Since the first packaged high-voltage steam boiler was built in the United

States in 1970—the first packaged high-voltage hot-water generator preceded it by two years—units have been installed in electric utility, pulp and paper, chemical, manufacturing, and district heating plants as well as in hospitals, schools, warehouses, mines, commercial buildings, etc.

CHEAP POWER

In most instances, high-voltage electrode boilers are being used in conjunction with oil- and gas-fired boilers, or hot-water generators, to take advantage of cheap electric power when it is available. In a typical arrangement, a demand controller allows the electric boiler all the power available (up to the unit's capacity) at favorable rates, and the steam produced is sent to a header at a slightly higher pressure than that from the fired boiler. When the header pressure drops below that required, the combustion-control system on the conventional steam generator responds and that unit picks up the slack.

The use of cheap electricity as an energy source can be even more effective in a hot-water system than in a steam system, because it allows for easier storage of heat. All the power available at off-peak rates can be consumed and stored as hot water in an accumulator for use during peak periods, when the electric hot-water generator is shut off. Any additional heat required can be supplied by burning oil or gas in a fired hot-water generator.

HIGH-VOLTAGE UNIT ADVANTAGES

High-voltage units offer advantages over low-voltage and resistance equipment at ratings above about 2500 kW. The most important are (1) the elimination of stepdown transformers and transformation energy losses, thereby reducing capital and operating costs; (2) the reduction of wiring—and possibly switchgear—costs, because less current is needed at higher voltage for the same power rating; and (3) the decrease in floor space and capital cost by using one large boiler instead of several small units.

The practical limit for high-voltage boilers today is around 170,000 lb/h (21.4 kg/s) at 13.8 kV and 260,000 lb/h (32.8 kg/s) at 25 kV. Such a unit requires a pressure vessel about 10 ft (3 m) in diameter—about the largest that can be shipped by conventional carriers. It is not practical to field-fabricate boilers of this type. Even if manufacturers could get greater capacity from their largest vessels, ratings of electric components might become the limiting factor. If so, it would be necessary to provide more than one three-phase, four-wire circuit—a change that would require a still larger shell to provide adequate phase-to-phase spacings.

IMMERSION-ELEMENT BOILERS

Pressure Vessels

Immersion-element boilers and hot-water generators consist of many fixed-resistance heating elements in a pressure vessel that is outfitted with operating and safety controls and standard boiler trim—safety valves, gage glasses, blowdown connection, etc., as shown in Fig. 5.1. Depending on the pressure and temperature requirements, the vessel is designed to specifications of either Section I or Section IV of the *ASME Boiler and Pressure Vessel Code*. Resistance-type boilers that operate above 15 psig (103.4 kPa) are built to Section I specifications; hot-water generators usually are designed for low-temperature heating systems [below 250°F (121°C)], allowing Section IV construction. High-temperature hot-water boilers are designed according to Section I.

Vessel Sizing. Engineers size the pressure vessel both to ensure adequate water velocity around the heating elements, so that there is no film boiling, and to minimize carryover. Proper selection of heating elements also helps eliminate these and other potential problems, such as scale formation. Leading manufacturers opt for replaceable low-watt-density elements [about 50 to 75 W/in^2 (0.078 to 0.116 W/mm^2)] with seamless copper sheaths.

Element Selection. Extruded copper tubing for hot-water generators and alloy sheaths for steam boilers fail sooner than other materials if the element remains energized during a low-water condition. This latter feature can prevent structural damage to the pressure vessel. Elements with higher heating densities, or alloy sheaths, are available and are specified when conditions warrant them.

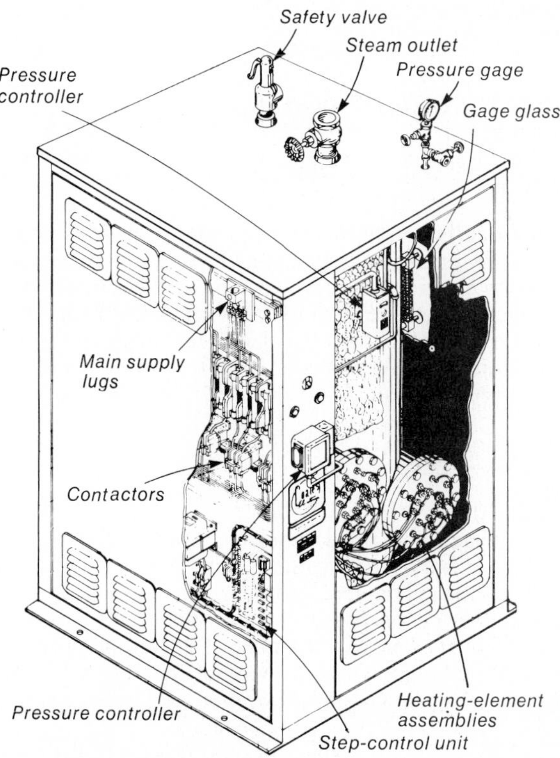

FIG. 5.1 Resistance-element boilers and hot-water generators. *(CAM Industries, Inc.)*

Automatic Controls

Controls regulate output by opening and closing heating-element supply circuits automatically. The simplest system, one for small heating loads, is of the on/off type. It consists of a single pressure-activated switch (for steam boilers) or thermostat (for hot-water generators), which energizes all circuits simultaneously.

Step Controls. Large commercial and small industrial systems require modulating step controls, which proportion input power to demand by selecting the number of heating-element groups needed to maintain the desired steam pressure or hot-water temperature. Both mechanical (cam-type) and solid-state controllers are available with automatic recycle capability at startup and restart—starting at zero load.

Sequencing Option. Solid-state controllers also come with a progressive sequencing option to equalize operating time among elements and to reduce failures. Standard safety devices, such as low-water and high-pressure cutoffs, can be connected to the controller to disconnect all power, should an unsafe condition arise.

Immersion-element boilers and hot-water generators can tolerate water of a poorer quality than electrode models can. Essentially all that has to be done is to control the hardness, so that scaling does not retard heat transfer, and to maintain dissolved solids at levels low enough to prevent foaming and carryover.

LOW-VOLTAGE ELECTRODE BOILERS

Method of Control

Low-voltage electrode boilers differ substantially from resistance-type units in several respects, a key one being the method of control. As just discussed, equipment with immersion elements turns one or more heater circuits on or off to meet demand. Low-voltage electrode steam boilers regulate steam output by varying the amount of water in contact with the electrodes; hot-water generators control the water temperature by using movable insulating shields to vary the exposure of electrodes to their neutral shields.

Boiler Operation

One type of low-voltage steam boiler works as shown in Fig. 5.2. At startup, water level is the same in both the generating and the balance chambers. As steam pressure in the generating chamber approaches the desired setting, however, the normally open regulator connecting both chambers begins closing, forcing water off the electrodes into the balance chamber. The rate of steam production decreases until the output matches the demand or the electrodes are fully uncovered; then it stops. As steam is consumed, pressure drops, the regulator opens, and water is readmitted to the generating chamber.

In the flooded high-voltage hot-water generator (Fig. 5.3), electrode insulating shields are mechanically adjusted by a solid-state load- and temperature-control

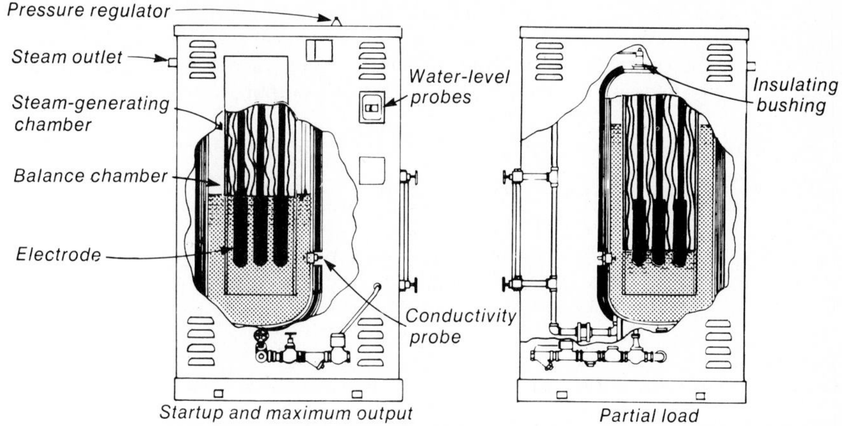

FIG. 5.2 Steam-generation control in low-voltage boiler.

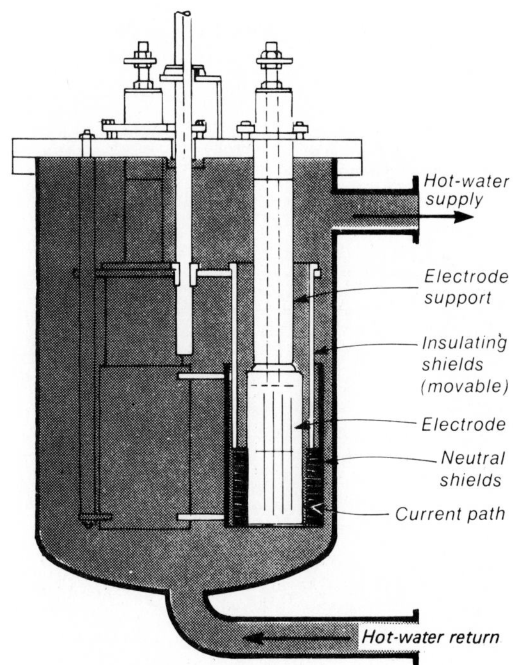

FIG. 5.3 Low-voltage hot-water generator runs flooded. (Power *magazine*)

system to regulate maximum power consumption and to provide close, stepless control of the water temperature from 5 to 100 percent of the rated unit capacity. As the insulating shields are interposed between their respective electrodes and neutral shields, the current path is lengthened and output is reduced.

HIGH-VOLTAGE ELECTRODE BOILERS

The control philosophy for high-voltage hot-water generators is similar to that for the low-voltage models. The vendor offering the low-voltage generator in Fig. 5.3, for example, has a high-voltage unit that operates in the same way.

Submerged-Electrode Design

As shown in Fig. 5.4, the high-voltage steam boiler with submerged electrodes has three generating compartments, one for each of the three electrodes. Its operation is similar to that of the low-voltage boiler in Fig. 5.2. Steam is generated as current flows from the electrodes to the neutral walls of the cylindrical compartments that contain them.

Another immersed-electrode design uses a movable insulating shield to con-

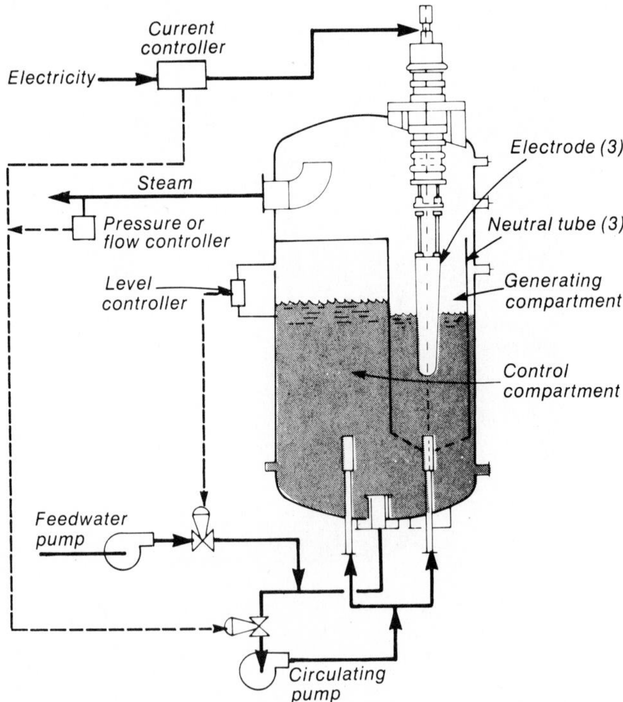

FIG. 5.4 High-voltage electrode boiler. (Power *magazine*)

trol steam production much as the low-voltage hot-water generator discussed above regulates temperature. The operating principles are identical, except that the boiler has a controller that maintains a fixed water level (the hot-water generator operates flooded). Insulating shields permit a 15:1 turndown; operation at loads less than 7 percent of the rated capacity requires interruption of electric service. Steam is produced in the space between the electrodes and neutral, and steam escapes from the surface at the waterline.

Sprayed-Electrode Design

Three other high-voltage boilers control the output by regulating the amount of water sprayed on electrodes suspended in the steam space, as shown in Figs. 5.5, 5.6, and 5.7. In these designs, a current path is created by the water as it flows to the electrodes through ports in a storage compartment.

Water that is not converted to steam runs down the electrodes and falls to a counterelectrode. Thus, a second current path is created for steam production—this one between the electrodes and the counterelectrode. The remaining water returns to the reservoir at the bottom of the vessel.

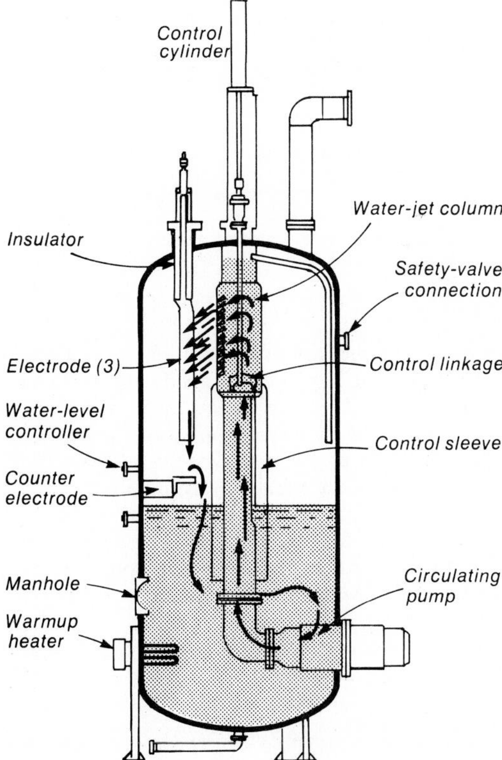

FIG. 5.5 Control sleeve covers or uncovers water jets. (Power *magazine*)

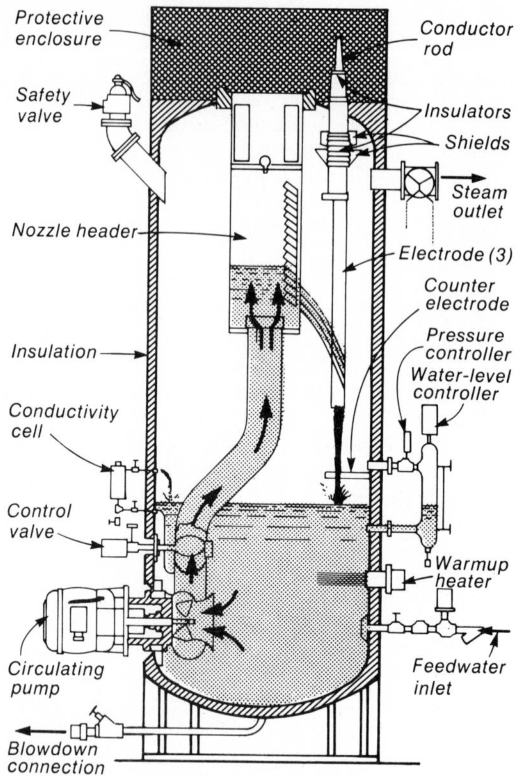

FIG. 5.6 Control valve regulates amount of water flow. *(CAM Industries, Inc.)*

Controlling Water Flow. The method of controlling water flow to the electrodes differs slightly among the three types of sprayed-electrode units. The boiler in Fig. 5.5 has a control sleeve that is raised and lowered hydraulically, which intercepts and diverts the streams of water from some or all of the jet ports, preventing them from striking the electrodes. The water level in the jet column is kept constant, so the flow from all nozzles is predictable.

The output of the unit shown in Fig. 5.6 is regulated by a butterfly valve, which controls the amount of water reaching the nozzle header. In both types, a signal proportional to the boiler pressure, plus one proportional to the power input, controls the water flow to keep the steam pressure constant or to restrict the power demand to the level desired.

A boiler designed for low-headroom applications (Fig. 5.7) uses a rotating distributor to regulate the water flow to the electrodes. Water falls through drilled passages in the distributor's nozzle plate, to and through the electrode assembly, and then to the counterelectrode.

Neutral Point. As in other types, the nozzle plates and counterelectrodes are

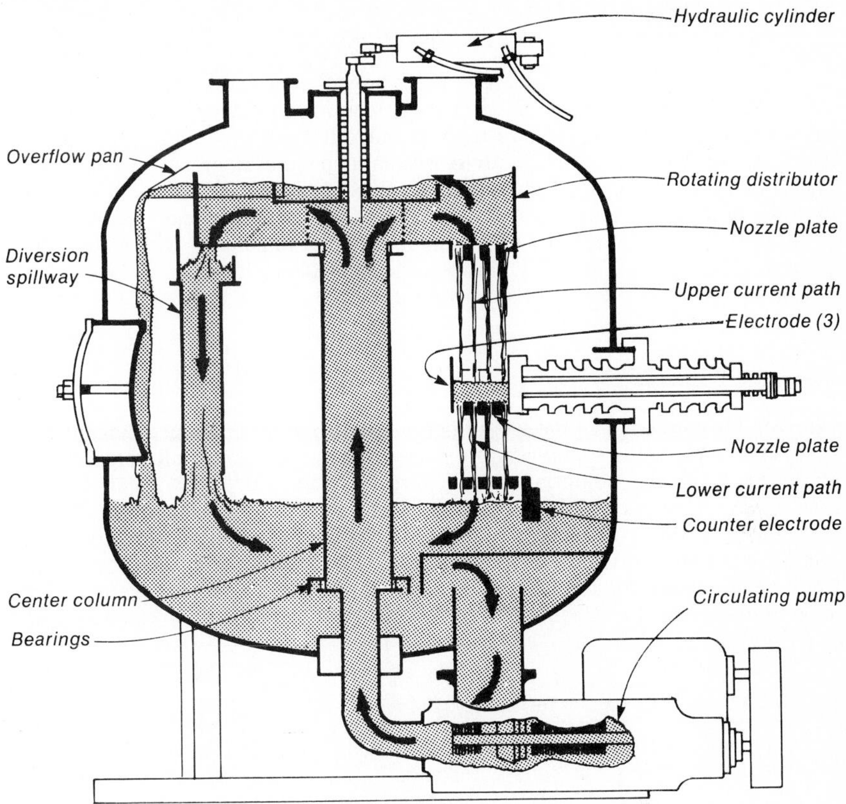

FIG. 5.7 Rotating distributor meters water flow. *(CAM Industries, Inc.)*

electrically common and form the neutral point of a wye-connected load. Thus, current flows (and steam is produced) in both the upper and lower paths shown in Fig. 5.7. Excess water in the distributor is returned to the bottom of the vessel via the diversion spillway. The overflow pan assembly keeps water in the distributor at a constant head.

UNATTENDED OPERATION

Electrode boilers and hot-water generators are usually arranged for unattended operation, that is, except for a visit each shift to check recorders and gages and to perform required blowoffs. Depending on the type of service, the units come equipped with safety devices designed to shut down the unit on high pressure, high temperature, short circuit, etc. Like resistance-type boilers, electrode units do not start up under load after a shutdown.

WATER CONDUCTIVITY

To choose between the submerged- and sprayed-electrode high-voltage boilers, their operating requirements must be examined in more detail. Perhaps the biggest difference between the two is the conductivity of the water demand. It is necessary to control the conductivity within a specific range because the water functions as a conductor. If the conductivity is too low, the boiler will not attain its rated output; if the conductivity is too high, short circuits can occur.

Most boilers have controllers to maintain the conductivity within the manufacturer's limits, as Table 5.1 shows. The controllers monitor the water, add chemicals, and blow down the unit when necessary.

Chemical Controls

One vendor recommends the use of sodium hydroxide to increase current flow; trisodium phosphate, sodium sulfite, and sodium sulfate are alternatives. The chemicals produce a solution with reasonably stable conductivity and minimum *total dissolved solids* (TDS) over the operating temperature range of the boiler.

Submerged-electrode boilers require extremely low conductivities when operating at the higher distribution voltages (12.5 to 15 kV). To illustrate, the conductivity specification for one unit is 50 μS • cm at 13.8 kV. In most applications, makeup water must be demineralized to keep the conductivity within acceptable

TABLE 5.1 Water Chemistry for Electric Boilers[a]

Boiler type	Hardness, ppm	pH	Total alkalinity, ppm	Iron, ppm	Oxygen, cm^3/L	Conductivity, μS • cm	TDS, ppm
Resistance							
Hot water	0.3–5	8.5–10.5	0–700	3	0.005	7000[b]	3500[b]
Steam	0.3–3	8.5–10.5	0–700	5	0.005	7000[b]	3500[b]
Low-voltage electrode							
Hot water	0.3–5	8.5–11	0–400	2.5	0.005	400–1400[c]	200–700
Steam	0	8.5–10.5	0–400	3	0.005	300–1500	150–750
High-voltage electrode							
Hot water	0–0.1	8.5–10	0–400	0.5	0.005	15 150[d]	
Steam[e]	0–0.1	8.5–12	0–400	2[f]	0.005	3500[b, g]	

[a]Boiler-water limits recommended by CAM Industries Inc. Other manufacturers of electric boilers specify many of the same limits, but there are some differences, particularly in the allowable conductivity. Hydro Steam Industries Inc., for example, recommends a maximum of 2100 μS•cm for its high-voltage sprayed-electrode boiler; General Electric Co., 25 to 300 μS•cm for its submerged-electrode unit, depending on the voltage.

[b]Upper limit.

[c]Varies with operating temperature and power requirements.

[d]15 μS•cm/MW at 13.8 kV and 150 μS•cm/MW at 4.16 kV.

[e]Both sprayed and submerged electrodes.

[f]Submerged electrode is 0.5 ppm.

[g]Submerged electrode is 50 μS•cm at 13.8 kV and 500 μS•cm at 4.16 kV.

limits without resorting to excessive blowdown. Since this degree of treatment is expensive, such boilers are seldom used in high-makeup applications.

High-Purity Feed

Where high-purity feedwater must be maintained and large quantities of steam are required, submerged-electrode boilers may be the only choice; sprayed-electrode units do not produce much steam at low conductivities. Startup boilers for certain nuclear powerplants—those where auxiliary- and primary-steam condensates mix—fall in this category. At central stations where deionized water is readily available, the less complex submerged-electrode design often is selected over the sprayed-electrode model and other boiler types on the basis of cost alone. Costs can mount in a multiunit powerplant, where auxiliary boilers may be used only once—at initial startup.

Submerged-electrode boilers also are a good choice where large amounts of steam must be produced with relatively low voltages. The reason is that conductivity can be increased appreciably to boost output as voltage decreases. This feature does not exist in a sprayed-electrode boiler because the conductivity is high already and short circuits will occur.

To put this relationship in perspective, a 30,000-kW sprayed-electrode boiler operating at 13.8 kV with a 3500-μS • cm conductivity (the upper limit for one vendor's equipment) is only capable of 7500 kW at 4.16 kV. The 50-μS • cm limit at 13.8 kV for one 30,000-kW submerged-electrode boiler can be increased 10-fold at 4.16 kV, permitting operation at 10,000 kW or more.

WATER TREATMENT

The key to successful operation of electrode boilers is proper water treatment. Although conductivity makes the boiler work, the unit health, and that of the steam or hot-water system, depends on such parameters as hardness, pH, alkalinity, dissolved oxygen, TDS, etc. (see Section 1, Chapter 10, Condensate and Feedwater Treatment).

These are the same water properties operators control now in their firetube and watertube boilers. But operator attention to water treatment can be more critical with electrode boilers because they are less forgiving.

Electrode Boiler Sensitivity

Changes in system chemistry influence unit performance much more rapidly than they would in other types of boilers. One example is failure to correct a high-alkalinity or other condition, which may lead to foaming, causing a short circuit and knocking the boiler out of service. A sporadic foaming condition in conventional boilers does not produce such severe consequences.

Too much chemical addition usually is more harmful than not enough. It can have an adverse effect on conductivity (thus on steam or hot-water production) and may promote foaming. Operators sometimes make the mistake of adding the same amount of chemicals to an electrode boiler as they would to a fired boiler of the same capacity. But electrode units usually demand and will tolerate far less. Since steam does not bubble up through the reservoir, carryover of water and chemicals from sprayed-electrode boilers is only about 10 percent of that experienced with most fired industrial boilers.

Preliminary Steps

To avoid potential water problems:

1. Consult the boiler manufacturer to determine the water quality and limits for the boiler.
2. Provide a complete analysis of feedwater and makeup water.
3. Ask for recommendations from the manufacturer, and attempt to do what is recommended.
4. Insist that the water treatment consultant or supplier discuss the proposed treatment plan with the boiler manufacturer.
5. Hire a qualified consultant to design a special treatment program for the plant's electrode boilers. She or he probably should not be affiliated with a firm that sells chemicals, because there is a tendency for such representatives to recommend more chemicals than are needed for best operation. Also, until more electrode boilers are installed, a consultant is more apt to be familiar with a user's requirements than a sales representative.

Treating Feedwater

Feedwater for electrode boilers should be softened to as near 0 ppm of total hardness as possible (Table 5.1 gives the maximum limits). Sodium zeolite softeners generally are recommended for makeup treatment. Depending on the boiler design and makeup characteristics, however, hydrogen zeolite softening or demineralization may be necessary. One manufacturer warns against the use of chemicals for scale control, because of potential side effects on conductivity levels or foaming tendency. Where silica scale is a problem, controlling alkalinity is recommended to hold the silica in solution. One user with high makeup requirements and a water source low in calcium and magnesium hardness suggests another approach: blow down to control silica and then increase the conductivity with commercial-grade sodium sulfite.

Scale Formation

Electrode boilers tend to scale less than other types of steam generators because of the moderate temperatures in the pressure vessel. The problem with scale in electrode boilers is not overheating, as it is in fossil-fuel-fired boilers (with tubes) and immersion-element units (with resistance elements), but rather electric insulation of the electrodes. Although the units cannot burn out, as scale forms, there is less electrode surface to carry the current required to meet a given steam demand. And if conductivity is increased to compensate for the insulating effect of the scale, this may eventually cause pitting corrosion of the electrodes and possibly of the boiler shell and erosion at the ends of the spray nozzles in sprayed-electrode boilers. The latter condition alters the spray pattern and causes premature shutdowns via phase-to-neutral arc-overs. Scale removal from electrodes is relatively simple once access to the pressure vessel is gained.

Highly alkaline water can cause both foaming and attack of the porcelain insulators used as lead-through bushings to bring power into the boiler. If the alkalinity must be kept above about 400 ppm to hold the silica in solution, or for some other reason, then specifying high-alumina insulators should be considered. One vendor claims it can handle up to 600-ppm alkalinity.

Treating Corrosion

Corrosion problems in the boiler, and in the condensate and feedwater systems, that are caused by high concentrations of oxygen also demand attention. More free oxygen is liberated in electrode boilers than in fired units; the actual amount is said to depend on the load, alkalinity level, presence of scale, and possibly other factors.

Deaeration and catalyzed hydrazine are recommended by some experts for oxygen control in systems where a large portion of the condensate is returned to the boiler. Filming amines also can be used to protect against the aggressive action of oxygen as well as of carbon dioxide; they form a nonwettable film on metal surfaces.

High iron concentrations—primarily in the form of rust—should be avoided. The use of catalyzed hydrazine for oxygen control helps retard the formation of iron oxide. Finally, to prevent attack on stainless-steel boiler internal parts, boiler-water chlorides must be held under 300 ppm. This limit may restrict the use of sodium-chloride-regenerated dealkalizers for makeup treatment.

ELECTRICAL CONSIDERATIONS

Where high-voltage electrode boilers and hot-water generators can be used, they require wye-connected, three-phase, four-wire service, and both the neutral and the boiler shell must be solidly grounded, as shown in Fig. 5.8. Manufacturers generally recommend that the neutral be the same size as the supply conductors. Consult the latest edition of the *National Electric Code* (NEC) before you install high-voltage service. The NEC provides essential guidelines on protective relaying, overcurrent detection, etc.

Available Voltage

A major factor in the selection of any electrode boiler is the voltage available from the utility. Generally, the higher the voltage, the smaller the pressure vessel for a given rating and the smaller the feeder needed to carry a specified amount of power. Where a suitable voltage is available directly from the utility's distribution lines, added benefits may be possible. For example, no stepdown trans-

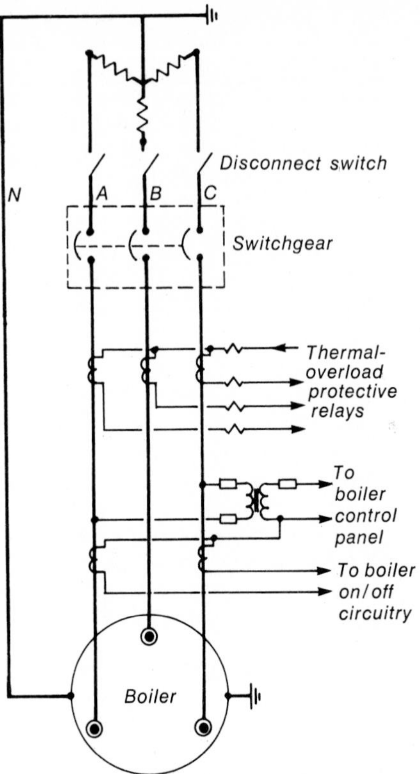

FIG. 5.8 Wye-connected, three-phase, four-wire service for high-voltage boilers. (Power *magazine*)

former is required, which saves a substantial amount of capital as well as transformer losses.

Switchgear Options

Plant management should think through its switchgear options thoroughly. Boilers and hot-water generators rated up to 720 A, for example, can use fuses or circuit breakers. Although the latter cost 2 to 3 times as much, many experts think circuit breakers are worth the premium. (At least one boiler manufacturer will not recommend fuses.) The reason: If close control of water treatment is not obtained (often difficult at startup), many fuses may blow. Large-size fuses are frequently hard to purchase on short notice, and are quite costly.

Fuses. A more difficult problem can arise if fuses are used that are slower to operate than the utility's circuit breaker. Any plans for a major installation should be handled by a qualified electrical engineer to avoid such trouble. Whether fuses or breakers are installed, they must be coordinated with upstream protective devices, so that the boiler protection always is triggered before the plant main or utility breakers.

When switchgear is chosen for boilers fed directly from the distribution system at high voltage, the impact of voltage fluctuations on unit operation should be studied. One user complains that transient voltage changes of perhaps 10 percent can shut down the boiler unexpectedly.

Contactors. If boilers and hot-water generators of the submerged-electrode type are to be operated at unusually low loads, contactors should be considered if the units must be cycled on and off. Maintenance considerations make it impractical to subject high-voltage breakers to frequent operation. During normal running, breaker wear from arcing can be minimized by first reducing the boiler load to a minimum with the circulating pump or electrode shields and then tripping the breaker.

Safety Measures

Here are three recommendations for promoting safety in high-voltage systems:

- Install an interlock system with a key that cannot be released to gain access to high-voltage areas until the main circuit is open.
- Specify a separate disconnect switch or breaker for the boiler room as added protection for operators.
- Provide the circuit breaker with a capacitor trip unit so that it will trip if the control power is lost.

CHAPTER 1.6
NUCLEAR STEAM-SUPPLY SYSTEMS

Ronald Allen Knief
Staff Consultant—Nuclear Safety
GPU Nuclear Corp.
Middletown, PA

INTRODUCTION

Nuclear power for peaceful uses began in 1956 with the start-up of the world's first full-scale generating plant, at Calder Hall in England, followed in this country a year later by the initial operation of a 60-MW unit at Shippingport, Pennsylvania. Since that auspicious beginning, a lot has happened—and much of it has not been promising for the continued expansion of nuclear power, especially in this country.

From a slow beginning, the middle 1960s saw a growing optimism on the part of utilities toward nuclear power as well as strenuous activity worldwide. It was becoming clear that electric generation via nuclear steam-supply systems could be as cheap as fossil-fuel firing. Adding to the optimism was the passage of a law in mid-1964 providing for private ownership of nuclear fuel. This action was hailed as paving the way for an independent nuclear power industry.

By the end of 1968, the combined capacity of U.S. nuclear generating plants in service, under construction, and on order added to a staggering 64,000 MW, more than twice the nuclear total for all other nations combined, and represented a capital investment of about $10 billion. Widespread enthusiasm for nuclear power had reached its high-water mark.

In the years since, as everyone knows, a continued erosion of confidence in nuclear power has taken hold in this country, although in western Europe and Japan the industry has grown despite strong opposition. Although the future is uncertain, to be sure, energy realists see that nuclear power is indispensable for electric generation well into the 21st century. It can fill a gap in the energy spectrum that is unfillable otherwise. Solar, geothermal, wind, and tidal power, while attractive, cannot be expected to supply energy in the vast amounts required in the immediate future. Oil and natural-gas supplies are shrinking and have become political toys. Although abundant, coal is difficult to recover, deliver, and burn. Even tight management of energy consumption is only part of the solution.

Current nuclear steam-supply systems use the fission process as their energy source. The characteristics that distinguish commercial nuclear power production from that of conventional fossil fuels are emphasized in the following treatment.

FISSION

Fission, or the splitting of a heavy nucleus such as ^{235}U (uranium 235), may occur when the nucleus is struck by a neutron, as shown in Fig. 6.1. Major advantages of the process are a large energy release (roughly 10^8 times that of burning one carbon atom in fossil fuel) and the production of extra neutrons, which allows the process to sustain a chain reaction for continued energy production. Materials that can support such a chain reaction by themselves are said to be *fissile*. A system in which the reaction is sustained is called a *nuclear reactor*—the heart of the *nuclear steam-supply system* (NSSS).

One major disadvantage of the process relates to the potential hazard from the radiation (including the neutrons) produced at the time of fission. Another is the presence of fission fragments which, being radioactive, give off radiation for

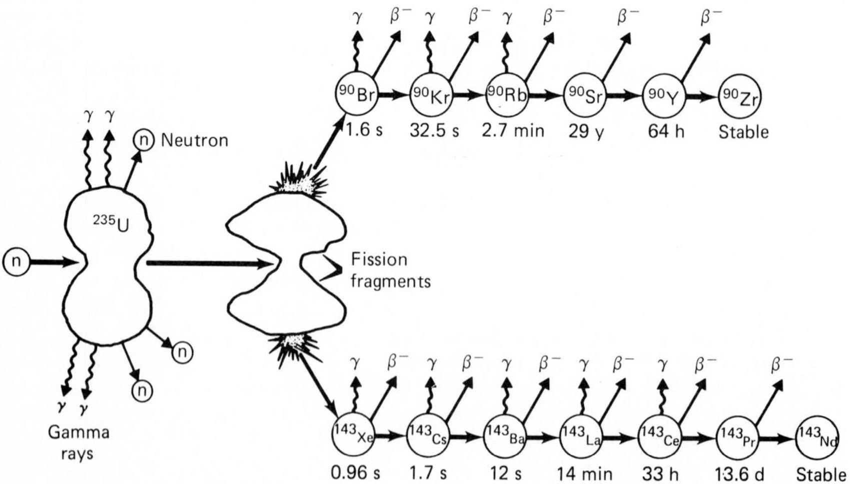

FIG. 6.1 Neutron fission of ^{235}U showing two representative fission product decay chains.[3]

varying periods after fission. Use of shielding and careful waste-handling procedures have been developed to address these concerns.

NUCLEAR FUEL CYCLE

Production of energy from most fuels depends on a fuel cycle consisting of at least the following elements: exploration, mining or drilling, processing or refining, consumption, waste disposal, and transportation. The nuclear fuel cycle is substantially more complicated for the following reasons:

- Uranium 235, the only naturally occurring fissile material, is less than 1 percent abundant in natural uranium.
- Two other fissile materials, ^{233}U and ^{239}Pu, are produced by neutron bombardment of ^{232}Th and ^{238}U, respectively. (For this reason, the latter two materials are said to be *fertile*.)
- All fuel-cycle materials contain small to large amounts of radioactive constituents.
- The neutron chain reaction that is used for power production could occur inadvertently outside a reactor.
- Some nuclear fuel materials have potential applications in nuclear explosive devices.

A schematic representation of the nuclear fuel cycle is shown in Fig. 6.2. The operations preceding reactor use, which contain materials with generally small amounts of radioactivity, are part of the *front end* of the cycle. Those following reactor use and characterized by high radiation levels are part of the *back end*. Other inherent steps not shown explicitly in Fig. 6.2 are recycle, transportation, nuclear safety, and material safeguards.

Front End

Uranium exploration, mining, and milling steps generally use procedures and processes adapted from the extractive metals industry. Exploration proceeds from general geologic evaluation to drilling and detailed ore-body mapping. The usually sparse, narrow ore seams are mined by open-pit or underground methods consistent with their locations. The milling operations separate the relatively small uranium content (often well under 1 percent) from the bulk of the ore to produce yellow cake (ammonium diuranate). Thorium is introduced separately if the fuel cycle is designed to produce ^{233}U.

The processing steps (conversion and enrichment) facilitate the usual solution to a common power-reactor requirement for a ^{235}U isotopic content in excess of the 0.71 percent (by weight) of naturally occurring uranium. In a multistep process, yellow cake is converted to UF_6 (uranium hexafluoride), which is a gas at atmospheric pressure and temperatures above 134°F (57°C). The small weight difference between these molecules of the scarce ^{235}U and the abundant ^{238}U provides the basis for physical separation and isotopic enrichment by gaseous diffusion, gas centrifuge, or gas-nozzle methods.

In the basic fabrication step, enriched UF_6 or yellow cake is processed to ce-

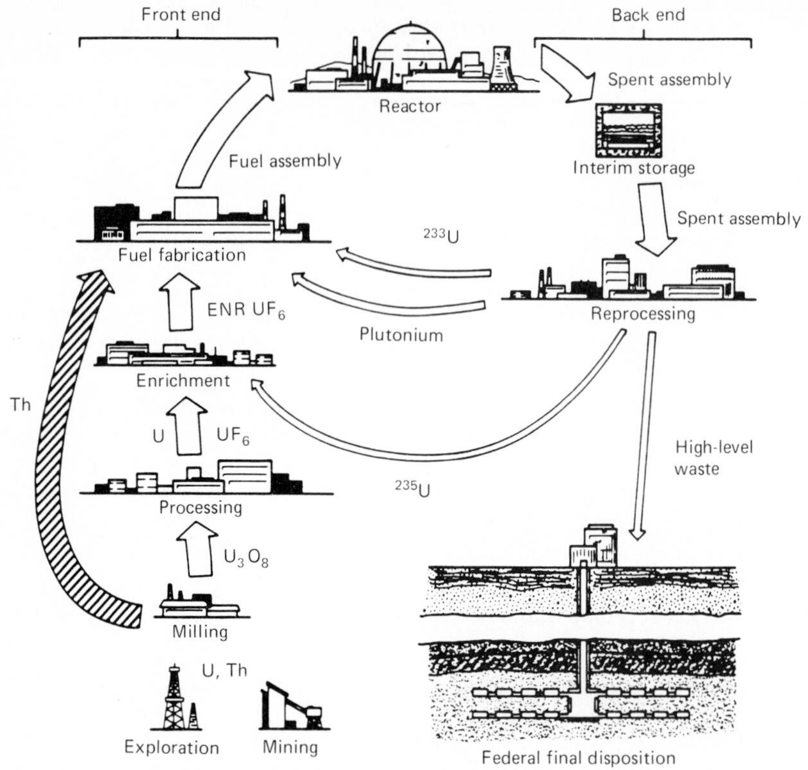

FIG. 6.2 Steps and material flow of the nuclear fuel cycle.[3]

ramic pellets of oxide (UO_2) or carbide (UC), which are then encased in metal or other cladding as fuel pins. Arrays of these pins form the fuel assemblies or bundles that are loaded into the reactor for power generation from fission. When plutonium or ^{233}U is recycled, it is also included in the fuel assemblies.

Back End

Each fuel assembly typically resides in the particular reactor for a time consistent with economical energy production. As ^{235}U is consumed, ^{239}Pu or ^{233}U is produced (and perhaps subsequently fissioned) if ^{238}U or ^{232}Th, respectively, is present. The buildup of fission products and their radioactive by-products tends to create a poisoning effect by absorbing neutrons that could otherwise participate in the chain reaction. When the loss of the initial fissile fuel and the buildup of poisons dominate, the effective lifetime of the fuel comes to an end.

Since fuel assemblies are very highly radioactive when discharged from the reactor, they are placed in interim storage in a water or other basin, to allow a period of "cooling." They may then be transferred to a reprocessing facility, where the fuel materials are separated first from the wastes and then from each other. Recycling of residual ^{235}U back to the enrichment step and/or of converted plutonium or ^{233}U directly to fabrication can reduce the need for new uranium ore by about 25 percent for light-water reactors up to no need for breeder reactors.

All steps of the nuclear fuel cycle produce some amounts of radioactive waste. Appropriate methods have been developed for their disposition, ranging from near-surface burial of "low-level" wastes to proposed salt-bed or granite geologic disposal of "high-level" wastes from reprocessing.

Since fuel-cycle operations tend to be distributed geographically, transportation is a key component. Containers are designed to accommodate radiological, chemical, criticality, economic, and theft concerns consistent with the specific fissile material and its form.

Nuclear safety is practiced in shielding and containing radioactive material and in preventing inadvertent criticality in fissile components. Material safeguards are designed to prevent theft or sabotage of materials or facilities in rough proportion to the perceived risks.

THEORETICAL CONSIDERATIONS

The fission process produces several hundred different fission fragments, each of which is radioactive and undergoes successive decays (Fig. 6.1) prior to reaching stability. Each species has a different characteristic lifetime, described in terms of a halflife—the average time required for one-half of a large neutron population to decay. The resulting overall radiation levels are substantial, having an energy equivalent to about 7.5 percent of the total fission output at the time a reactor is shut down and decreasing afterward only in rough proportion to the $-1/5$ power. Handling this lingering decay heat load is a principal consideration in the safe operation of nuclear reactors.

The neutron population participating in a fission chain reaction may be described in terms of the following balance equation:

$$\text{Rate of accumulation} = \text{rate of production} + \text{rate of absorption} + \text{rate of leakage}$$

The state of the chain reaction is generally described in terms of the neutron multiplication factor k, defined as

$$k = \frac{\text{rate of production}}{\text{rate of absorption} + \text{rate of leakage}}$$

or the reactivity ρ, defined as

$$\rho = (k - 1)/k$$

For $k = 1$ or $\rho = 0$, the system is said to be *critical;* for $k < 1$ or $\rho < 0$, *subcritical;* and for $k > 1$ or $\rho > 0$, *supercritical.*

The mathematical equations associated with the neutron balance are quite complex since the reaction rates are often a strong function of material (composition, location, and irradiation history) and neutron population (position, direction, and energy) characteristics. Neutron energy is a particularly important parameter, especially since fission produces high-energy neutrons, while those of lower energy (lower by 7 to 8 orders of magnitude) have the highest probability for causing fission. For this reason, a *moderator* material of low mass may be added to enhance the slowing down. (The closer the mass of the moderator nuclei to that of the neutron, the larger the potential energy loss in a single collision; thus, hydrogen with essentially the same mass is best in this regard, although it may not be selected for other reasons.)

OPERATIONAL CONSIDERATIONS

Specific characteristics of the fission process and the design of reactor fuel result in unique operational constraints relating to radiological control, reactivity control, and operating limits.

Radiological Controls

The potential hazards associated with radioactive fission and other products have led to the development of sophisticated control methods. In performing hands-on work, it is challenging to hold radiation exposures of personnel within specified limits on quarterly, annual, and lifetime doses. Overall, however, the requirement that all exposures be kept "as low as reasonably achievable" (Alara) has resulted in extensive preplanning of all radiological work activities.

Gaseous and liquid wastes are held for as long as practicable to allow maximum radioactive decay. Then they may be diluted and dispersed to the environment if the federal or international *maximum permissible concentrations* (MPCs) are not exceeded. Otherwise, solidification, containment, and disposal are required.

Reactivity Control

Although it is possible to control the neutron chain reaction by adjusting any or all of the production, leakage, or absorption terms in the neutron balance, control is generally achieved through absorption. Neutron poisons such as boron have been used in the form of both solid, movable control rods and in chemical solution. Other materials including silver, indium, cadmium, hafnium and gadolinium are also available as neutron poisons.

Several inherent mechanisms provide either negative or positive feedback to reactor power. One is based on expansion of the coolant or moderator (such as with increased temperature). Others relate to changes in fuel density and nuclear reaction properties. These essentially instantaneous contributions along with some long-term ones (such as from time-dependent changes in concentrations of xenon and samarium fission-product poisons) affect the stability of the chain reaction, may limit the ability to change the power level, and are a major consideration in overall reactor safety design.

Operating Limits

Reactor thermal operating limits are determined by local conditions. The maximum local linear heat rate (in kilowatts per foot or meter) is set to prevent fuel melting and to ensure minimal clad damage if coolant is lost during accident conditions. The other major limit focuses on the relationship between a fuel pin's heat flux and the coolant's flow and temperature characteristics, to avoid *departure from nucleate boiling* (DNB) and a resulting large temperature increase.

Both limits are typically established in terms of peaking factor ratios to core-averaged conditions. These factors may vary with power level, time in core life, and similar parameters.

NUCLEAR STEAM-SUPPLY SYSTEMS

Nuclear power reactors may be classified according to the following:

- *Coolant:* Primary heat-extraction medium.
- *Steam cycle:* The number of separate coolant loops, including secondary heat-transfer systems (if any).
- *Moderator:* Material (if any) used specifically to slow the neutrons produced by fission.
- *Neutron energy:* The general energy range for the neutrons that produce most of the fissions.
- *Fuel production:* The system is referred to as a *breeder* if it produces more fuel than it consumes; it is a *converter* otherwise.

The reactors associated with the world's five basic nuclear steam-supply systems are identified in these terms in Table 6.1 and shown conceptually in Fig. 6.3. Also see Figs. 6.4 through 6.10.

Light-Water Reactors

The world's two most popular reactor designs employ light water as both coolant and moderator. As Table 6.1 shows, these *light-water reactors* (LWRs) are the *boiling-water reactor* (BWR) and the *pressurized-water reactor* (PWR). The fuel for the two is very similar, consisting of long bundles of 2 to 4 percent (by weight) enriched uranium dioxide fuel pellets stacked in zirconium-alloy cladding tubes. The BWR fuel assembly, however, has a smaller number of fuel pins and is surrounded by a metal flow channel. The larger PWR assemblies are not enclosed.

The BWR design (Fig. 6.3*a*) consists of a single loop in which entering water is turned directly to steam for production of electricity. Since operating temperatures must remain below the critical temperature for water, steam separators and dryers are used with a "wet-steam" turbine.

The PWR (Fig. 6.3*b*) is a two-loop system that uses high pressure to maintain an all-liquid-water primary loop. Energy is transferred to the secondary steam loop through two to four steam generators. This LWR design also uses a wet-steam turbine.

TABLE 6.1 Typical Characteristics for Five Nuclear Steam-Supply Systems*

	Feature				
	BWR	PWR	PHWR	HTGR	LMFBR
	Reference design manufacturer				
	General Electric	Westinghouse	Atomic Energy of Canada, Ltd.	General Atomic	Novatome
	System (station)				
	BWR/6	Sequoyah/ SNUPPS	Candu-600	Fulton	Superpheni
General					
Steam cycle					
Loops	1	2	2	2	3
Primary coolant	H_2O	H_2O	D_2O	He	Liq. Na
Secondary coolant	—	H_2O	H_2O	H_2O	Liq. Na/H_2O
Moderator	H_2O	H_2O	D_2O	Graphite	—
Neutron energy	Thermal	Thermal	Thermal	Thermal	Fast
Fuel production	Converter	Converter	Converter	Converter	Breeder
Energy conversion					
Gross thermal power, MW(th)	3579	3411	2180	3000	3000
Net electric power, MW(e)	1178	1150	638	1160	1200
Efficiency, %	32.9	33.7	29.3	38.7	40
Heat transport					
Primary loops and pumps	2	4	2	6	4
Intermediate loops	—	—	—	—	8
Steam generators	—	4	4	6	8
Steam generator type	—	U-tube	U-tube	Helical coil	Helical coil
Fuel					
Particles	Short, cyl. pellets	Short, cyl. pellets	Short, cyl. pellets	Coated microspheres	Short, cyl. pellets
Chemical form	UO_2	UO_2	UO_2	UC/ThC	Mixed UO_2/Pu
Fissile (% by wt.)	2–4% ^{235}U	2–4% ^{235}U	Natural uranium	93% ^{235}U	10–20% Pu (co
Fertile	^{238}U	^{238}U	^{238}U	Thorium	^{238}U (core + blanket)
Pins	Pellet stacks in Zr-alloy tubes	Pellet stacks in Zr-alloy tubes	Pellet stacks in Zr-alloy tubes	Microspheres in graphite sticks	Pellet stacks in SS tubes
Assembly	8 × 8 sq. pin array	17 × 17 sq. pin array	37-pin cyl. array	Hex. graphite block with fuel sticks	271-pin hex. array
Core					
Axis	Vertical	Vertical	Horizontal	Vertical	Vertical
Assays on axis	1	1	12	8	1
Assays radially	748	193	380	493	364 (core), 233 (blanket)
Performance					
Equil. burnup, MWD/T	27,500	27,500	7500	95,000	100,000
Refueling sequence	¼ per yr	⅓ per yr	Continuous, on-line	¼ per yr	Variable

TABLE 6.1 Typical Characteristics for Five Nuclear Steam-Supply Systems* (*Continued*)

	Feature				
	BWR	PWR	PHWR	HTGR	LMFBR
	Reference design manufacturer				
	General Electric	Westinghouse	Atomic Energy of Canada, Ltd.	General Atomic	Novatome
	System (station)				
	BWR/6	Sequoyah/ SNUPPS	Candu-600	Fulton	Superphenix
Thermal hydraulics					
Primary system					
Pressure, MPa	7.17	15.5	10.0	4.90	~0.1
Inlet temp., °C	278	292	267	318	395
Avg. out. temp., °C	288	325	310	741	545
Core flow, Mg/s	13.1	18.0	7.6	1.42	16.4
Volume, L		3.36×10^5	1.20×10^5	(9550 kg)	(3200 Mg)
Secondary system					Na/H_2O
Pressure, MPa	—	6.89	4.7	17.2	~0.1/17.7
Inlet temp., °C	—	227	187	188	345/235
Outlet temp., °C	—	285	260	513	525/487
Power density					
Core avg., kW/L	54.1	105	12	8.4	280
Fuel avg., kW/L	54.1	105	60	44	280
Linear heat rate					
Core avg., kW/m	19.0	17.8	25.7	7.87	29
Core max., kW/m	44.0	42.7	44.1	23.0	45
Design peaking factors					
Radial	1.4		1.21		
(Total)		(2.5)		(2.9)	(1.55)
Axial	1.6		1.41		
Moderator					
Volume, L	(same as	(same as	2.17×10^6	(graphite blocks)	—
Inlet temp., °C	primary	primary	43	(graphite blocks)	—
Outlet temp., °C	coolant)	coolant)	71	(graphite blocks)	—
Reactivity control					
Control rods					
Geometry	Cruciform	Rod clusters	Rods	Rod pairs	Hex. pin bundles
Absorber matl.	B_4C	Ag-In-Cd	Various	B_4C/graphite	B_4C
Burnable poison	Gd. in fuel pellets	Borosilicate glass	—	B_4C/graphite	—
Other systems	Voids in coolant	Soluble boron	H_2O, various	Reserve shutdown	3-bundle secondary
Reactor vessel					
Inside dimen., m	6.05*D* × 21.6*H*	4.83*D* × 13.4*H*	7.6*D* × 4*L*	11.3*D* × 14.4*H*	21*D* × 19.5*H*
Wall thick., mm	152	224	28.6	(4.72 m min.)	25
Material	SS-clad carbon steel	SS-clad carbon steel	Stainless steel	Prestressed concrete	Stainless steel
Other features			Pressure tubes	Steel liners	Pool type

*Data summarized from R. A. Knief, *Nuclear Energy Technology*, Hemisphere Publishing, Washington, 1981, Tables 1-2 and IV-4.

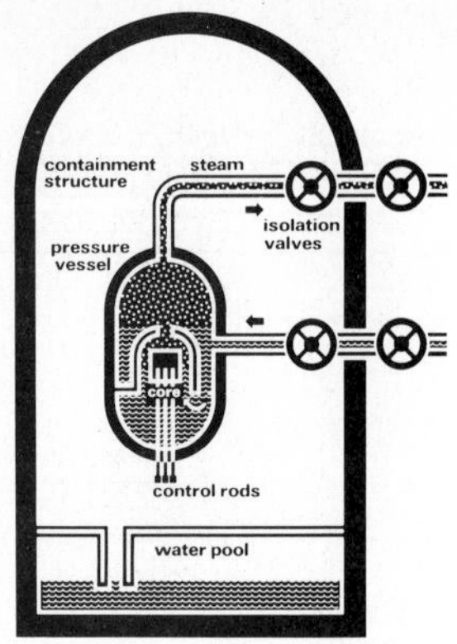

(a) Boiling-water reactor (BWR)

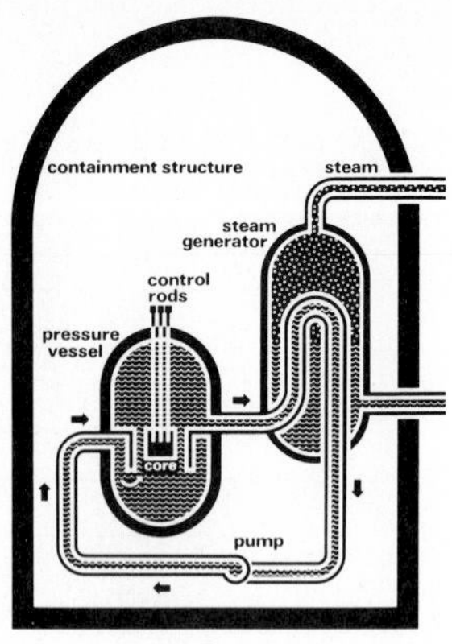

(b) Pressurized-water reactor (PWR)

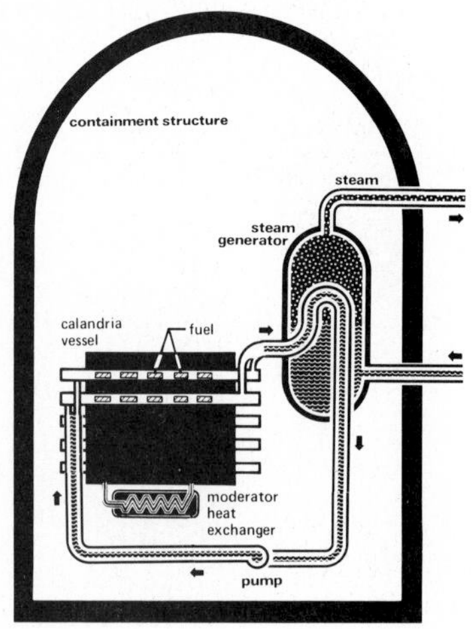

(c) Pressurized heavy-water reactor (PHWR)

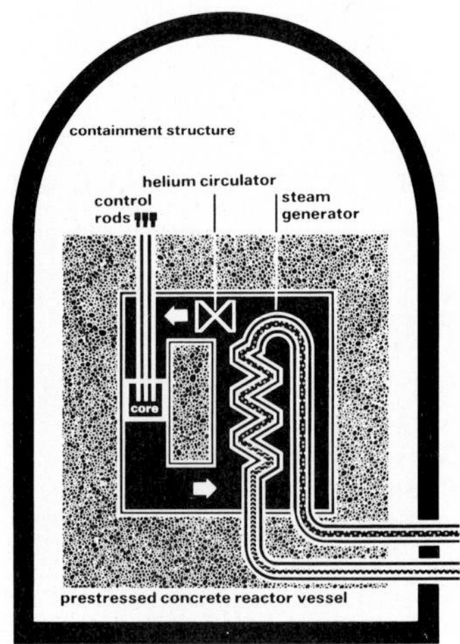

(d) High-temperature gas-cooled reactor (HTGR)

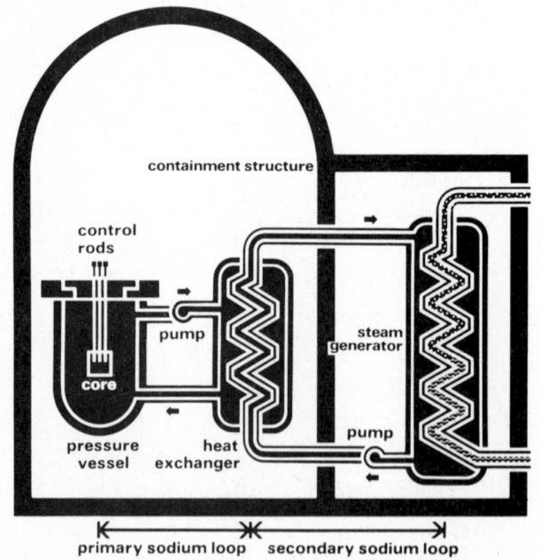

(e) Liquid-metal fast-breeder reactor (LMFBR)

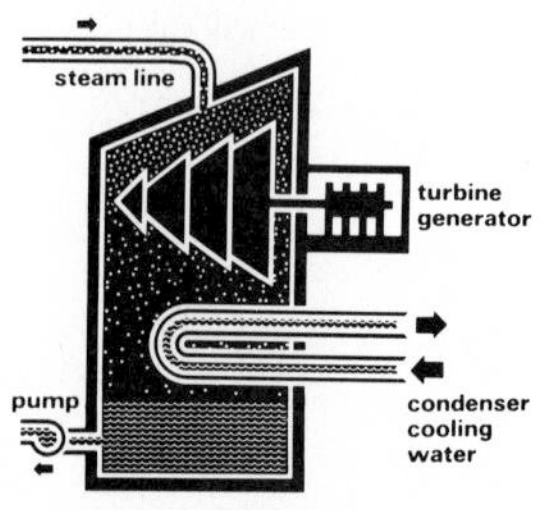

(f)

FIG. 6.3 Schematics for NSSS of five reactor types; (f) shows turbine/generator plant common to all types. *(Adapted courtesy of Atomic Industrial Forum Inc.)*

FIG. 6.4 Three light-water reactors of the PWR type built along the Ohio River at Shippingport, Pa. The 60-MW PWR located in the small rectangular building at far right was the first large-scale nuclear generating plant in the U.S.; it is now decommissioned. 833-MW Beaver Valley I and II reactors (domed buildings) began operation in 1976 and 1986, respectively. *(Courtesy Duquesne Light Co.)*

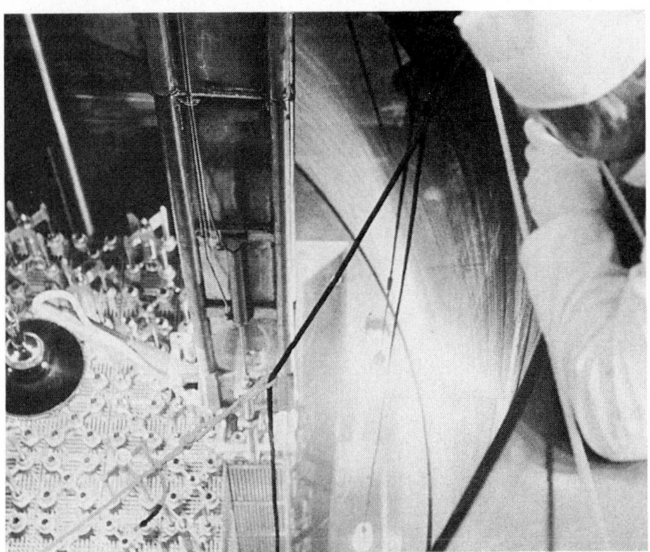

FIG. 6.5 PWR fuel assemblies being loaded into the core with the reactor vessel open. *(Courtesy Combustion Engineering Inc.)*

FIG. 6.6 The Pickering station in Ontario consists of eight 516-MW Candu PHWRs; the last went into service in 1985. *(Courtesy Atomic Energy of Canada Ltd.)*

FIG. 6.7 Unique pressure-tube and fueling-machine designs allow these PHWRs to change fuel while on-line. *(Courtesy Atomic Energy of Canada Ltd.)*

FIG. 6.8 Variation of the HTGR, 296-MW thorium high-temperature reactor (THTR) located at Hamm-Uenntrop in West Germany; it began operation in 1985. *(Courtesy Technische Vereinigung der Grosskraftwerksbetreiber EV.)*

FIG. 6.9 Spherical fuel elements of THTR in Fig. 6.8 give rise to the designation "pebble bed" reactor. *(Courtesy Technische Vereinigung der Grosskraftwerksbetreiber EV.)*

FIG. 6.10 1200-MW Superphenix LMFBR, located at Creys-Malville in France, began operation in 1985. The plutonium and uranium core, surrounded by a depleted uranium blanket, can produce more fuel than it consumes, thus leading to the designation "breeder" reactor. *(Courtesy Commissariat a l'Energie Atomique.)*

Pressurized Heavy-Water Reactors

The most common *pressurized heavy-water reactor* (PHWR) is the Canadian deuterium uranium (Candu) reactor (Table 6.1). Although its steam cycle (Fig. 6.3*c*) is similar to that of the PWR, this PHWR uses short, cyclindrical fuel bundles and a pressure-tube design to facilitate an on-line refueling method. This fueling mode coupled with the neutron economy of a heavy-water coolant and moderator allows the selection of natural-uranium fuel.

High-Temperature Gas-Cooled Reactors

The *high-temperature gas-cooled reactor* (HTGR) is the successor to a number of reactors that have used graphite moderator and gaseous coolant. It has a two-loop steam cycle (Fig. 6.3*d* and Table 6.1) like that of the PWR and PHWR, but uses helium coolant and steam generators contained in the concrete reactor vessel. Its high operating temperatures allow use of a "dry-steam" turbine.

The HTGR fuel is unique, consisting of tiny 93 percent (by weight) enriched UC microspheres with graphite coatings. These are formed directly into baseball-sized spheres coated with graphite or are fabricated into finger-sized sticks and loaded into the hexagonal graphite-block fuel assemblies. Mixing the UC with the ThC microspheres allows for substantial conversion and recycle of ^{233}U.

Liquid-Metal Fast-Breeder Reactors

The *liquid-metal fast-breeder reactor* (LMFBR) is designed to breed plutonium from ^{238}U by using fast neutrons (Table 6.1). Thus, no moderator is needed, and relatively heavy sodium coolant is employed.

The central core fuel consists of small-diameter PuO_2-$^{238}UO_2$ pellets in stainless-steel clad and short fuel bundles. The use of additional $^{238}UO_2$ in a blanket surrounding the core region can increase the production of plutonium to or past the point of breeding.

The LMFBR uses three heat-transport loops (Fig. 6.3*e*). The potential for an energetic chemical reaction between the liquid-sodium coolant and water led to the addition of an intermediate loop, so that core sodium with its complement of radioactivity will not have the opportunity to react with the water. High operating temperatures allow selection of a dry-steam turbine.

REACTOR SAFETY AND REGULATION

Commercial power reactors are characterized by large power densities and significant decay heat loads. Reactor safety depends on ensuring removal of the heat and containment of radioactive products from fission and by-product sources.

All reactors are designed for multiple-barrier containment of the radioactive products. In the five NSSS designs just described and in Table 6.1, ceramic fuel pellets and cladding provide the first two barriers; the vessel and primary system, a third; and a containment structure, the final barrier.

Engineered safety systems are designed and constructed to respond to accident conditions with these components:

- *Reactor trip (RT):* Insertion of solid or soluble neutron poisons to shut down the neutron chain reaction.
- *Emergency core cooling (ECC):* Coolant injection to maintain heat removal from the fuel, both short and long term.
- *Postaccident heat removal (PAHR):* Removal of heat from the containment and from the cooling water.
- *Postaccident radioactivity removal (PARR):* Spraying and filtering airborne radioactivity from the containment.
- *Containment integrity (CI):* The prevention of dispersal of radioactivity to the environment.

These capabilities are handled differently in each type of reactor consistent with its specific features.

The nuclear industry is among the most regulated in the world. Requirements for safety-system design and maintenance, quality assurance, facility and operator licensing, etc., have a major impact on the design and operation of nuclear powerplants.

SUGGESTIONS FOR FURTHER READING

The following books are recommended to the reader seeking additional information on commercial nuclear power in terms of general information,[3] reactor and operating theory,[2,5] reactor designs,[4] and nuclear fuel cycle.[1]

1. M. Benedict, T. H. Pigford, and H. W. Levi, *Nuclear Chemical Engineering,* 2d ed., McGraw-Hill, New York, 1981.
2. S. Glasstone and A. Sesonske, *Nuclear Reactor Engineering,* 3d ed., Van Nostrand Reinhold, New York, 1981.

3. R. A. Knief, *Nuclear Energy Technology,* Hemisphere Publishing, Washington, 1981.
4. A. V. Nero, *A Guidebook to Nuclear Reactors,* University of California Press, Berkeley, 1979.
5. F. J. Rahn, A. G. Adamantiades, J. E. Kenton, and C. Braun, *A Guide to Nuclear Power Technology: A Resource for Decision Making,* Wiley-Interscience, New York, 1984.

CHAPTER 1.7
PULVERIZERS

Joseph G. Singer
Combustion Engineering Inc.
Windsor, CT

INTRODUCTION

The main reason why *pulverized-coal* (PC) firing is currently the favored method of burning coal is because pulverized coal burns as a gas, and so fires are easily lighted and controlled. Almost any kind of coal can be reduced to powder and burned as a gas. PC burning has so dominated the utility market that power generation by stoker firing is no longer a consideration. A major reason for the success of PC burning is the ability to adapt operating conditions to all coal ranks from anthracite to lignite. Although certain accommodations must be made to handle fuels with such a wide range of properties, years of experience have simply served to prove the versatility and advantages of PC firing.

PULVERIZING PROPERTIES OF COAL

To predict pulverizer performance about a specific coal with some degree of accuracy, the ease with which the coal can be pulverized must be known, beginning with its grindability.

Grindability

A grindability index has been developed to measure the ease of pulverization. Unlike moisture, ash, or heating value, this index is not an inherent property of coal. Rather, it represents the relative ease of grinding coal when it is tested in a particular type of apparatus. The consistency of grindability test results permits the pulverizer manufacturer to apply the findings to a particular size and, to a lesser degree, type of pulverizer.

Grindability should not be confused with the hardness of coal. The same coal may have a range of grindabilities depending on other constituents in the coal. Figure 7.1 gives typical curves for North Dakota lignites and shows the variation in Hardgrove grindability as the moisture content changes. Typically, anthracites and some lignites have at least one point where their grindabilities are very close. Anthracite is a very hard coal, however, whereas lignite is soft—yet both are difficult to grind.

Pulverizing a small air-dried sample of properly sized coal in a miniature mill determines its grindability. Results may then be converted to a grindability index factor which, with appropriate correction curves, can help to interpret mill capacity.

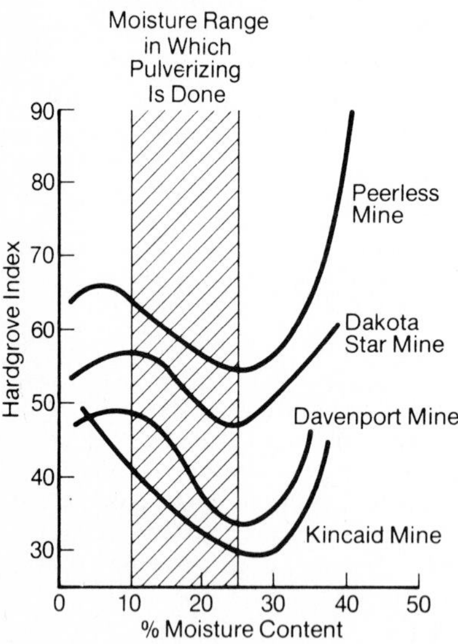

FIG. 7.1 Variation of grindability index with moisture content, North Dakota lignites.

The Hardgrove method was developed to measure the quantity of new material that will pass a 200-mesh sieve (Table 7.1, which also includes SI equivalent values). The apparatus for this method is quite simple. A 50-g sample of air-dried coal, sized to less than 16 mesh and greater than 30 mesh, is placed in the mortar of the test machine along with eight steel balls of 1-in (25-mm) diameter. A weighted upper race is placed on the ball and coal charge and is turned 60 revs. The sample is then removed and screened. The quantity passing the 200-mesh sieve is used in the preparation of a calibration chart, from which the grindability of the coal sample is determined in accordance with ASTM Standard D409, *Grindability of Coal by the Hardgrove-Machine Method.*

Frequently, too much emphasis is placed on grindability, while other factors affecting mill capacity (such as moisture) are almost entirely overlooked. The pulverizer capacity is pro-

TABLE 7.1 Comparison of Sieve Openings

U.S. standard sieve			W. S. Tyler sieve		
Mesh	in	mm	Mesh	in	mm
20	0.0331	0.841	20	0.0328	0.833
30	0.0234	0.595	28	0.0232	0.589
40	0.0165	0.420	35	0.0164	0.417
50*	0.0117	0.297	48*	0.0116	0.295
60	0.0098	0.250	60	0.0097	0.246
100*	0.0059	0.149	100*	0.0058	0.147
140	0.0041	0.104	150	0.0041	0.104
200*	0.0029	0.074	200*	0.0029	0.074
325	0.0017	0.043	325	0.0017	0.043
400	0.0015	0.037	400	0.0015	0.037

*Commonly used screens in PC practice for combustion purposes.

portional to the grindability index of the coal, but corrections must also be made for fineness of product and moisture of the raw feed.

Moisture

The total moisture content of coal consists of *equilibrium moisture* and *surface* or *free moisture*. Equilibrium moisture varies with coal type or rank and mine location, and would more accurately be called *bed* or *seam moisture*. In reality, surface moisture is the difference between the total moisture and the bed moisture.

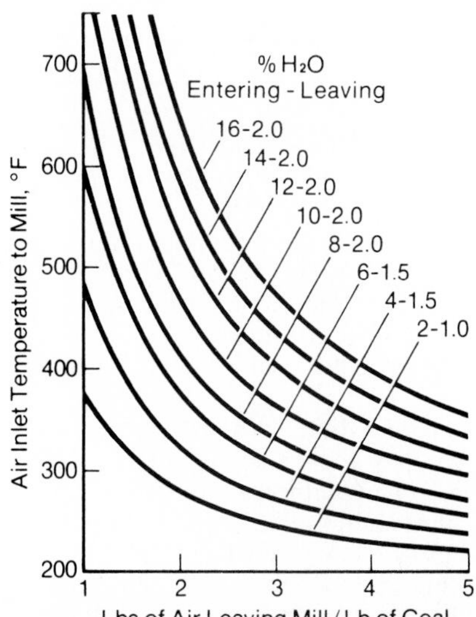

FIG. 7.2 Temperature of air to mill, eastern coals. Note: °C = (°F − 32)/1.8; kg = lb × 0.454.

Surface moisture adversely affects both pulverizer performance and the combustion process. It produces agglomeration of the fines in the pulverizing zone and reduces pulverizer drying capacity because of its inability to remove the fines as efficiently and as quickly as they are produced. Agglomeration of fines has the same effect as coarse coal during the combustion process, because the surface available for the chemical reaction is reduced. Since in-mill drying is the accepted method of preparing coal for PC burning, sufficient hot air at adequate temperature is necessary in the milling system. Curves such as those in Figs. 7.2 and 7.3 indicate the air temperature required to dry coal of varied total moistures and coal-air mixtures.

Achieving the rated pulverizer capacity depends on having sufficient heated air available to dry the coal. Without enough hot air, the mill output will be limited to the "drying capacity" and not the "grinding capacity."

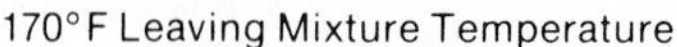

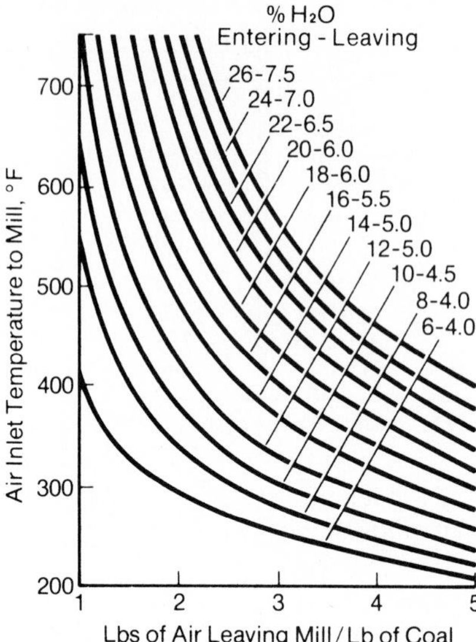

FIG. 7.3 Temperature of air to mill, midwestern coals. Note: °C = (°F − 32)/1.8; kg = lb × 0.454.

Thus, it may be possible to obtain more capacity with a relatively dry coal of lower grindability than with a high-moisture coal of higher grindability.

Pulverizer Capacity vs. Grindability

Mill capacity is not directly proportional to grindability. Thus, if the actual capacity of a pulverizer with 50 grindability is 10,000 lb/h (1260 g/s), then with 100 grindability it will be about 17,000 lb/h (2140 g/s), not 20,000 lb/h (2520 g/s). The nonlinearity is the result of differences between a commercial pulverizer and a grindability test machine which, with no provision for the continuous removal of fines, is of the batch type rather than the continuous type. The crushing pressure of the test equipment is also considerably less. As a result, some of its energy is dissipated in deforming the coal particles without breaking them.

Although the test equipment does not indicate a direct proportion of capability between hard and soft materials, the value of these tests is not reduced. Correction factors developed by pulverizer manufacturers on commercial equipment provide for overcoming these discrepancies (Fig. 7.4). As a rule of thumb, for every point the grindability index of a given coal changes, there will be a corresponding change of 1⅓ percent in pulverizer capacity. Similarly, for every percentage point of change in fineness from the basic design point of a pulverizer, there will be a corresponding change of 1½ percent in capacity.

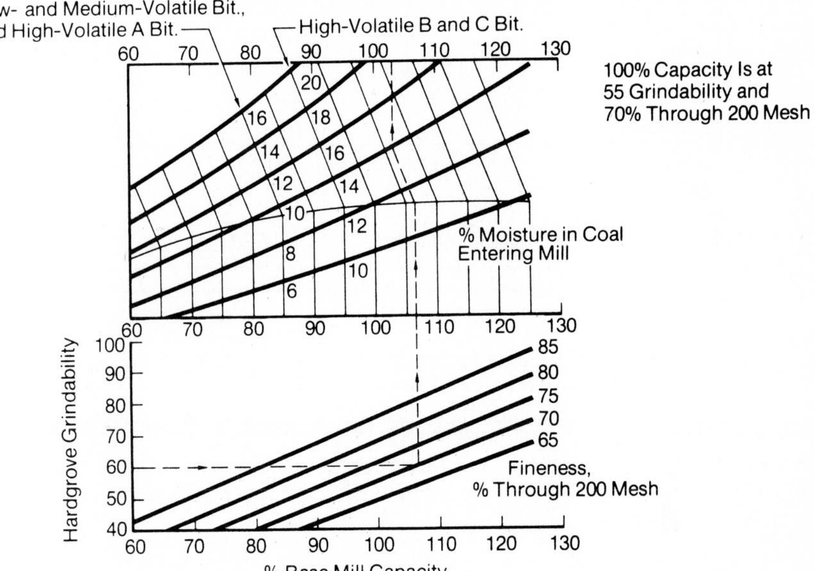

FIG. 7.4 Mill correction factors for grindability, fineness, coal type, and moisture—bituminous coals with preheated air.

Lignite Grindability vs. Moisture

The Bureau of Mines at Grand Forks, North Dakota, and others have reported the grindability of lignites at various moistures, and the results show a wide variation (Fig. 7.1). Some feel that such curves are of little value because it seems impossible to select the proper index from them; others feel that the curves do have significance. With the increased burning of lignites, solution of this problem is important.

In setting up the present ASTM code for grindability, the test specifies selection of an air-dried sample. In one manufacturer's mill, *all* the surface moisture and some of the equilibrium moisture are evaporated during pulverization with a hot-air sweep. The moisture content of the pulverized product leaving the mill is conceived to be the moisture level that exists in the grinding zone. Thus, grindability versus moisture indices above the equilibrium level are of little interest. Hardgrove indices therefore have meaning to the pulverizer designer only when they are below the equilibrium moisture level and in the general range of moisture content between 10 and 25 percent.

The actual choice of the grindability index for pulverizer design capacity requires consideration of total moisture, equilibrium (bed) moisture, and selected hot-air temperature.

Coal Rank vs. Required Fineness

Successful PC firing depends on recognizing differences in coals and on making whatever modifications are necessary to provide optimum conditions for efficient combustion. Experience has established that a relationship exists between the rank of bituminous coals and the degree of fineness required for successful operation.

TABLE 7.2 Range of Coals for PC Firing

Coal rank	Passing 200 mesh, wt. %	Retained on 50 mesh, wt. %
Subbituminous C coal and lignite	60–70	2.0
High-volatile bituminous C, subbituminous A, B	65–72	2.0
Low- and medium-volatile bituminous, high-volatile bituminous A and B	70–75	2.0

To ensure complete combustion within the furnace confines for minimal carbon loss, high-rank coals must be pulverized to a finer size than coals of lower rank. Although in part dependent on the type of firing system selected, the approximate limits within the ranges shown in Table 7.2 have been established. When certain coals in the low-volatile group are fired in small PC furnaces, the fineness percentage may be increased to as high as 80 percent to ensure adequate burnout of the carbon content.

Specifications for Classifying Coals. Table 7.3 lists the specifications set by ASTM for classifying coals by rank. Rank classifications are based on varying combinations of volatile-matter content, heating value, and agglomerating prop-

TABLE 7.3 Classification of Coals by Rank*

Class and group	Fixed carbon limits, % (dry, mineral-matter-free basis)		Volatile matter limits, % (dry, mineral-matter-free basis)		Calorific value limits, Btu/lb (kJ/kg) (moist,† mineral-matter-free basis)		Agglomerating character
	Equal or greater than	Less than	Equal or greater than	Less than	Equal or greater than	Less than	
I. Anthracitic							
1. Meta-anthracite	98	—	—	2	—	—	Nonagglomerating
2. Anthracite	92	98	2	8	—	—	
3. Semianthracite‡	86	92	8	14	—	—	
II. Bituminous							
1. Low-volatile bituminous coal	78	86	14	22	—	—	
2. Medium-volatile bituminous coal	69	78	22	31	—	—	
3. High-volatile A bituminous coal	—	69	31	—	14,000§ (32 620)§	—	Commonly agglomerating¶

TABLE 7.3 Classification of Coals by Rank* *(Continued)*

Class and group	Fixed carbon limits, % (dry, mineral-matter-free basis): Equal or greater than	Fixed carbon: Less than	Volatile matter limits, % (dry, mineral-matter-free basis): Equal or greater than	Volatile matter: Less than	Calorific value limits, Btu/lb (kJ/kg) (moist,† mineral-matter-free basis): Equal or greater than	Calorific value: Less than	Agglomerating character
4. High-volatile B bituminous coal	—	—	—	—	13,000§ (30 290)§	14,000 (32 620)	
5. High-volatile C bituminous coal	—	—	—	—	11,500 (26 795)	13,000 (30 290)	
					10,500 (24 465)	11,500 (26 795)	Agglomerating
III. Subbituminous							
1. Subbituminous A coal	—	—	—	—	10,500 (24 465)	11,500 26 795)	
2. Subbituminous B coal	—	—	—	—	9,500 (22 135)	10,500 (24 465)	
3. Subbituminous C coal	—	—	—	—	8,300 (19 340)	9,500 (22 135)	Nonagglomerating
IV. Lignitic							
1. Lignite A	—	—	—	—	6,300 (14 680)	8,300 (19 340)	
2. Lignite B	—	—	—	—	—	6,300 (14 680)	

*This classification does not include a few coals, principally nonbanded varieties, which have unusual physical and chemical properties and which come within the limits of fixed carbon or calorific value of the high-volatile bituminous and subbituminous ranks. All of these coals either contain less than 48% dry, mineral-matter-free fixed carbon or have more than 15,500 Btu/lb (36 115 kJ/kg) moist, mineral-matter-free.

†Moist refers to coal containing its natural inherent moisture but not including visible water on the surface of the coal.

‡If agglomerating, classify in low-volatile group of the bituminous class.

§Coals having 69% or more fixed carbon on the dry, mineral-matter-free basis shall be classified according to fixed carbon, regardless of calorific value.

¶It is recognized that there may be nonagglomerating varieties in these groups of the bituminous class, and there are notable exceptions in high-volatile C bituminous group.

Source: Reprinted from *ASTM Standards D388,* Classification of Coals by Rank.

erties. After the predrying step is accomplished in the pulverizing operation, the low-rank subbituminous coals and lignites manifest a higher degree of reactivity than do the higher-rank bituminous coals. Recent investigations confirm that this increased reactivity results primarily from the lack of agglomerating properties and increased O_2 content (Table 7.4).

TABLE 7.4 Key Properties of Three Ranks of Coals

	Bituminous					Subbituminous			Lignite
	Low volatile	Medium volatile	High volatile						
			A	B	C	A	B	C	A
Agglomerating character*	Agg.	Agg.	Agg.	Agg.	Agg.†	Non-agg.	Non-agg.	Non-agg.	Non-agg.
HHV, as-fired, Btu/lb (kJ/kg)	13,150 (30 640)	13,210 (30 780)	13,410 (31 245)	11,610 (27 050)	10,590 (24 675)	9,840 (22 930)	8,560 (19 945)	7,500 (17 475)	5,940 (13 840)
Flammability index, °F (°C)	1,010 (543)	1,030 (554)	950 (510)	1,030 (554)	990 (532)	970 (521)	970 (521)	990 (532)	890 (477)
Proximate, %									
Moisture (seam)	2.0	2.0	4.0	7.0	10.0	14.0	19.0	25.0	40.0
Volatile matter	21.1	32.3	38.4	33.8	35.9	35.3	34.5	25.8	25.9
Fixed carbon	68.6	55.8	51.5	47.3	43.3	41.2	37.5	40.9	27.4
Ash	8.3	9.9	6.1	11.9	10.8	9.5	9.0	8.3	6.7
Ultimate, %									
Hydrogen	5.0	5.5	5.6	4.6	5.5	5.4	5.1	5.6	4.3
Carbon	88.5	84.1	82.5	81.0	74.3	74.2	69.8	66.4	67.0
Sulfur	0.4	1.1	2.5	0.9	4.0	0.5	0.8	0.6	0.9
Nitrogen	1.3	1.7	1.5	1.3	1.4	1.2	1.1	1.3	1.2
Oxygen	4.8	7.6	7.9	12.2	14.8	18.7	23.2	26.1	26.6

*Agglomerating or nonagglomerating.
†Agglomerating but noncaking.

A comparison of flammability indices with volatile matter and heating value suggests that factors other than these have a great influence on ignition temperature and particle burnout. Because lower flammability temperatures are experienced primarily with lignitic and subbituminous coals, it is reasonable to ask what property is common to lower-rank coals. Exclusive of heating value, the most obvious difference between the two groups is that property known as *agglomerating character*—subbituminous coals and lignites do not agglomerate.

Agglomeration. As applied to coals, agglomeration is the property of particles fusing into a cokelike mass or bonding to a firm cake when the particles are heated to 1000°F (540°C) or above.

Although the determinations of both volatile matter and heating value are well defined, tests for establishing the agglomerating character of coals are less commonly known. ASTM Standard D388, *Specifications for Classification of Coals by Rank,* describes agglomerating character as

> The test carried out by the examination of the residue in the platinum crucible incidental to the volatile-matter determination. Coals which in the volatile-matter determination produce either an agglomerate button that will support a 500-gram weight without pulverizing, or a button showing swelling of cell structure, shall be considered agglomerating from the standpoint of classification.

Since the agglomerating property of coals is the result of particles transforming to a plastic or semiliquid state when heated, it reflects a change in the surface area of the particle. This surface change is manifested by a transformation of the particle from an angular, irregular shape to a spherical or spherelike particle. Also, the surface character of the particle changes from a porous, irregular, absorptive surface to a glasslike, nonporous surface. Thus, with the application of heat, agglomerating coals tend to develop a nonporous surface, while nonagglomerating coals become even more porous with pyrolysis. This explanation indicates why agglomerating coals require a correspondingly finer particle size to maintain an equivalent surface area for efficient, rapid ignition and burnout.

In addition to the correlation of agglomerating properties with coal reactivity, an equally strong correlation exists between the ultimate-analysis oxygen level of coals and their response to reactivity as reflected in the flammability temperatures. Data in Table 7.4 show ranges in oxygen from 4.8 to 14.8 percent for agglomerating coals, *moisture- and ash-free* (MAF), and from 18.7 to 26.6 percent MAF for the lower-rank coals that are nonagglomerating. Seemingly, the higher the inherent or organically bound oxygen content of the coal, the more reactive the coal.

Standards for Measuring Fineness. When solid fuels in suspension are burned, it is essential that the fuel-air mixture contain an appreciable quantity of extremely fine particles to ensure rapid ignition. Conversely, to obtain maximum combustion efficiency, a minimum amount of coarse particles in this same fuel-air mixture is desirable. The former condition is usually expressed as percentage through a 200-mesh screen, while the latter is designated as a percentage retained on a 50-mesh screen.

The number of openings per linear inch designates the mesh of a screen. Thus, a 200-mesh screen has 200 openings per inch, or 40,000 per square inch. The diameter of the wire forming the screen governs the size of the openings. The U.S. Standard and W. S. Tyler are the most common screen sieves (Table 7.1).

Classification and Size Consist. In some reactions, such as the setting of cement, the surface area is of extreme importance. In the combustion of pulverized coal, however, although it is important that a proper percentage of fine particles have a large surface area, it is equally necessary to eliminate the oversize on the coarser screen. Despite the percentage less than 200 mesh (−200 mesh), as little as 3 to 5 percent greater than 50 mesh (+50 mesh) may produce furnace slagging and increased combustible loss, even though combustion conditions are excellent for the finer coal. The small amount of oversize represents very little additional surface if it is pulverized to all −50 mesh and all +200 mesh.

As an illustration, assume a typical screen analysis of a high-volatile bituminous coal sample, pulverized to 80 percent −200 mesh:

- 99.5 percent: −50 mesh
- 96.5 percent: −100 mesh
- 80.0 percent: −200 mesh

This represents a surface area of roughly 6580 in^2/oz (150 000 mm^2/g), with over 97 percent of the surface in the −200-mesh portion. By overgrinding and poor classification, it would be possible on a commercial-size mill to have a sample of the following analysis:

- 95 percent: −50 mesh
- 90 percent: −100 mesh
- 80 percent: −200 mesh

This is not a satisfactory grind, because of the high percentage retained on the 50 mesh, even though the surface area remains the same.

In the pulverizing process, then, classification plays a major role in matching the particle size to the reactivity of the fuel. Both fine and coarse particles must be controlled within limits by mechanical classification techniques. Careful attention, therefore, must be paid to both the design and the operation of the classification system.

Sampling Pulverized Coal. Clearly product fineness has a considerable bearing on pulverizer performance. Thus, fineness samples should be analyzed periodically. In a storage system, this sample may be taken directly from the mill cyclone discharge. On a direct-fired system, obtaining the sample is more difficult because it must be taken from a flowing coal-air stream. A sampling device, consisting of a small cyclone collector, sample jar, and sampling nozzle with connecting hose, will do the job (Fig. 7.5).

The sample is obtained by traversing the pipe across its diameter from two points in the same plane and at 90° to each other. The entire pipe diameter must be traversed, and the rate of movement must be uniform. Samples must be taken in both directions for the same period. Although refinements of this technique are practiced in industry today, the principle is the same.

With collection completed, the PC samples from each mill are thoroughly mixed. The 50 g of the sample is placed in the top sieve of a nested stack of 50-, 100-, and 200-mesh sieves. The nest is then shaken by hand or mechanically until the procedure has separated the coal particles by size. The results of the percentages of coal passing through the individual screens are a straight line on a typical

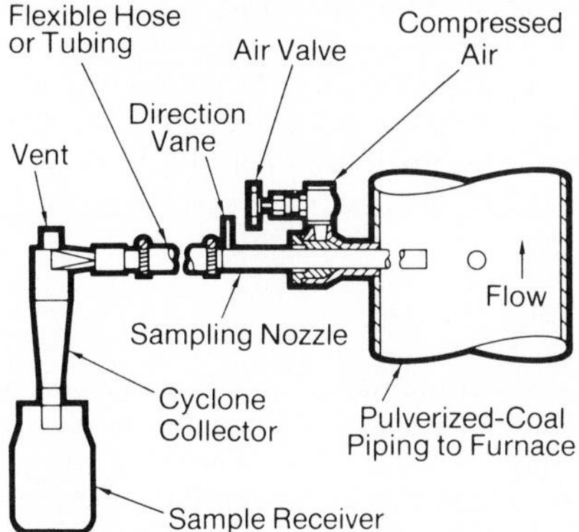

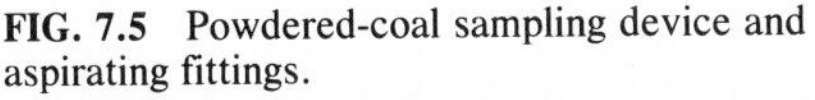

FIG. 7.5 Powdered-coal sampling device and aspirating fittings.

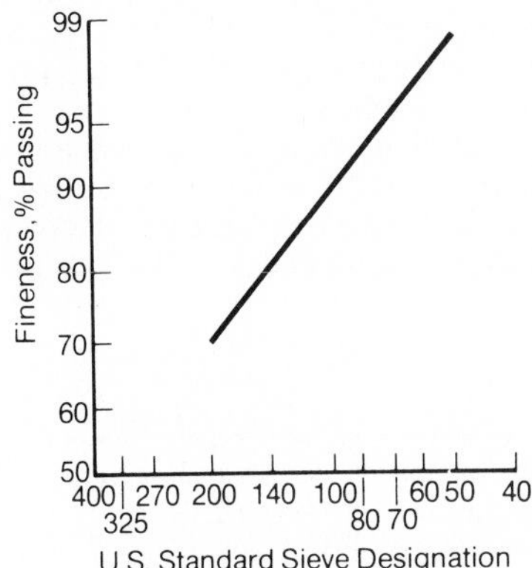

FIG. 7.6 Typical fineness sample results.

sieve distribution chart (Fig. 7.6). ASTM Standard D197 and ASME PTC 4.2 give additional information on recommended sampling techniques.

Closed-Circuit Grinding. When a large piece of coal is reduced to a number of smaller ones by any method, a larger number of finer particles will be produced simultaneously. Therefore, it is not possible for a pulverizer to yield a product that will pass a 50-mesh screen without also obtaining a large percentage of material finer than 200 mesh. Thus, if a quantity of coal at one stage of pulverization contains 50 percent material that will pass through a 50-mesh sieve, and if this −50-mesh material is removed from the grinding zone, then it will contain a smaller percentage of −200-mesh material than if it had been permitted to remain in the grinding zone until the total quantity had been reduced to pass a 50-mesh sieve.

Remember, however, in grinding finer than necessary, power is wasted and the pulverizing equipment must be larger than required. Removal of the fines from the pulverizing zone as rapidly as they are produced and return of the oversize for regrinding eliminate unnecessary production of fines and reduce energy requirements. Better product sizing and increased capacity result from the removal of the fines—a process called *closed-circuit grinding*. The pulverizing system component that accomplishes this size control is known as the *classifier*.

Abrasion

Pulverizing results in an eventual loss of grinding-element material. Balls, rolls, rings, races, and liners gradually erode and wear out as a result of abrasion and metal displacement in the grinding process. Thus, the power for grinding and the maintenance of the grinding elements make up the major costs of the pulverizing operation.

In itself, "pure coal" is relatively nonabrasive; however, such foreign materials as slate, sand, and pyrites, commonly found in coal as mined, are quite abrasive. These are the undesirable constituents that produce rapid and sometimes excessive wear in pulverizing apparatus. The economics of coal cleaning to remove such abrasive foreign materials depends on many variables and must be determined for each individual application.

The resistance of a smooth plane surface to abrasion is called its *hardness*. It is commonly recorded in terms of 10 minerals according to the Mohs scale of hardness (Table 7.5). No quantitative relation exists between these minerals, the diamond being much greater in hardness above sapphire than sapphire is above talc (Table 7.6).

TABLE 7.5 Mohs' Scale of Hardness

1 Talc	6 Feldspar
2 Gypsum	7 Quartz
3 Calcspar	8 Topaz
4 Fluorspar	9 Sapphire
5 Apatite	10 Diamond

TABLE 7.6 Common Materials and Their Mohs' Hardnesses

Coal	0.5–2.5
Slate	0.50–6.0
Mica	2.0–6.0
Pyrite	6.0–6.5
Granite	6.5
Marble	3.0
Soapstone	1.0–4.0
Kaolin clay	2.0–2.5
Iron ore	0.50–6.5
Carborundum	9.5

COAL PREPARATION

Coal should be prepared properly for its safe, economical, and efficient use in a pulverizing system. A controllable continuity of flow to the pulverizer must be maintained. Organic foreign materials such as wood, cloth, or straw should be removed—they may collect in the milling system and become a fire hazard, or they may impair material or airflow patterns in the mill. Although many mills are designed to reject small inorganic or metallic materials (or are not adversely affected by them), a magnetic separator should be installed in the raw-coal conveyor system to remove larger metallic objects. If this is not done, these objects may damage the pulverizer coal feeder or obstruct the coal flow.

The raw coal should be crushed to a size that will promote a uniform flow rate to the mill by the feeder. Favorable size consist will minimize segregation of coarse and fine fractions in the bunker and result in a more uniform rate of feed to various pulverizers being supplied from a given bunker. When it is mixed with relatively dry lump coal, fine coal with high surface moisture accentuates the segregation problem in bunkers. Crushing by size reduction of the dry lumps exposes additional dry surfaces for the adsorption of moisture from the wet fines, thereby producing a more uniform size and moisture distribution in the raw-coal mass.

Most commonly, direct firing of pulverized coal is the method used for steam generation. Here, an uninterrupted and uniformly controllable supply of pulverized coal to the furnace is an essential requisite. A steady and continuous flow of raw coal to the pulverizer will ensure this supply.

An ideal feed is one that is closely sized and double-screened [¾ in × ¼ in (roughly 19 mm × 6.35 mm)]. Coal of this size will permit excess water to drain off; it will flow freely from bunkers and can be fed uniformly. However, such favorable sizing can be obtained only at a considerable price premium, and this usually precludes its selection. In most cases, powerplants will receive coals classified as run-of-mine or screenings with lumps; therefore, crushing equipment must be installed to provide uniform raw-feed sizing. Generally, coal-feed sizing up to 1½ in (38 mm), as sieved through a round screen, is permissible with large pulverizers.

Coal Crushers

Although numerous types of crushers are commercially available, the one generally selected for smaller capacities is the swing-hammer type. This crusher has proved satisfactory for overall use and has demonstrated reliability and economy. The swing-hammer crusher consists of a casing enclosing a rotor to which are attached pivoted hammers or rings. Coal is fed through a suitable opening in the top of the casing, and crushing is effected by impact of the revolving hammers or rings directly on, or by throwing the coal against, the liners or spaced grate bars in the bottom of the casing. The degree of size reduction depends on hammer type, speed, wear, and bar spacing.

Roll crushers have been used but are not entirely satisfactory because of their inability to deliver a uniformly sized product. Probably the most satisfactory crusher for large capacities is the Bradford breaker. This design (see Fig. 7.7) consists of a large-diameter, slowly revolving (about 20-rpm) cylinder of perforated steel plates, the size of the perforations determining the final coal sizing.These openings are usually 1¼ to 1½ in (roughly 32 to 38 mm) in diameter.

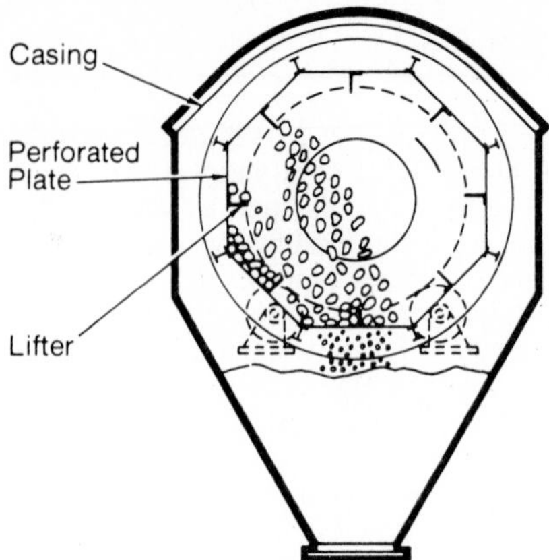

FIG. 7.7 Bradford breaker.

Coal Feeders

A coal feeder is a device that supplies the pulverizer with an uninterrupted flow of raw coal to meet system requirements. This is especially important in a direct-fired system. There are several types, including the belt feeder and the overshot roll feeder.

Belt Feeder. The belt feeder uses an endless belt running on two separated rollers, receiving coal from above at one end and discharging it at the other. Varying

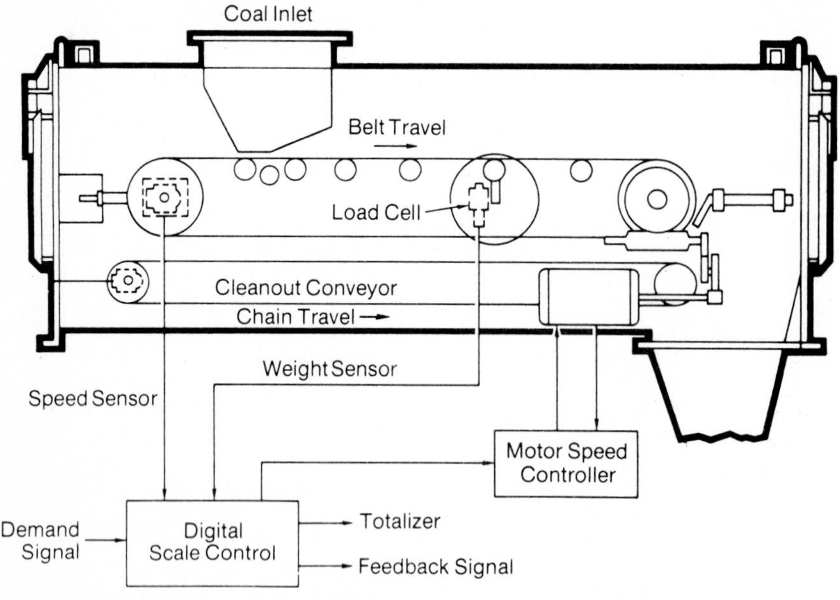

FIG. 7.8 Belt-type gravimetric coal feeder.

the speed of the driving roll controls the feed rate. A leveling plate fixes the depth of the coal bed on the belt.

The belt feeder sees service in either a volumetric or a gravimetric type of application (Fig. 7.8). The gravimetric type has gained wide popularity in the industry for accurately measuring the quantity of coal delivered to each individual pulverizer. Generally, they are applied to steam generators having combustion control systems requiring individual coal metering to the fuel burners.

Two accepted methods of continuously weighing the coal on the feeder belt involve (1) a series of levers and balance weights and (2) a solid-state load cell across a weigh span on the belt. Both are very accurate mechanisms, and both are well accepted by utilities. This same belt-feeder design can be used for volumetric measurement.

Overshot Feeder. The overshot roll feeder (Fig. 7.9) has a multiblade rotor that turns about a fixed, hollow, cylindrical core. This core has an opening to the feeder discharge and is provided with heated air, to minimize wet-coal accumulation on surfaces and to aid in coal drying. A hinged, spring-loaded leveling gate

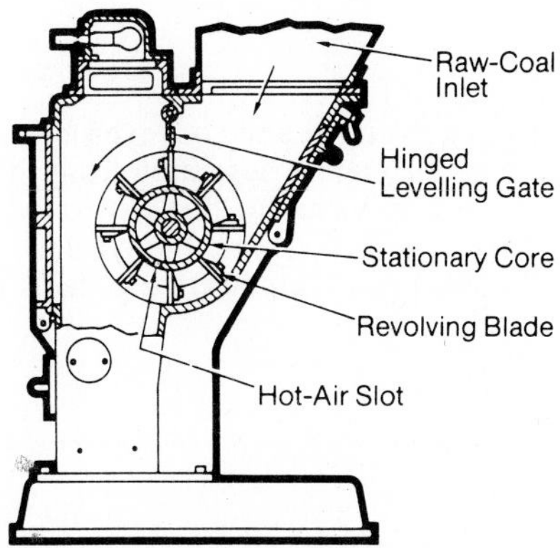

FIG. 7.9 Overshot roll feeder.

mounted over the rotor limits the discharge from the rotor pockets. This gate permits the passage of oversize foreign material.

Feeders of this type may be separately mounted, or they may be integrally attached to the side of a pulverizer. The roll feeder and the belt feeder, by virtue of their design, are considered highly efficient volumetric feeding devices.

PULVERIZING AND CONVEYING METHODS

Early coal-pulverizing installations received undried coal and relied on ambient air in the mill system. Because no heat was added to the system, the coal feed was limited to that of very low moisture content; therefore, maximum pulverizer capabilities were not realized. Subsequently, external coal dryers were added to the system. Because of the lack of cleanliness, high initial cost, fire hazard, and

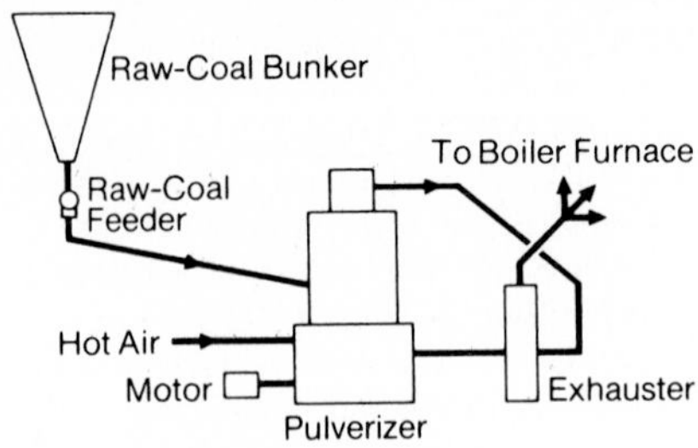

FIG. 7.10 Direct-fired system.

space requirements, these dryers were replaced with the now universally accepted mill drying.

Three methods of supplying and firing pulverized coal have been developed: the storage or bin-and-feeder system, the direct-fired system, and the semidirect system. These methods differ on the basis of their drying, feeding, and transport characteristics.

Storage System

In a storage system, coal is pulverized and conveyed by air or gas to a suitable collector where the carrying medium is separated from the coal, which is then transferred to a storage bin. The hot air or flue gas introduced to the mill inlet provides for system drying and is vented to rid the system of moisture evaporated from the fuel. From the storage bin, the pulverized coal is fed to the furnace, as required.

Direct and Semidirect Systems

In a direct-fired system (Fig. 7.10), coal is pulverized and transported with air, or air slightly diluted with gas, directly to the furnace where the fuel is consumed. Hot air or diluted furnace gas supplied to the pulverizer furnishes the heat for drying the coal and transporting the pulverized fuel to the furnace. Known as *primary air,* this air is a portion of the combustion air. As a reduction in oxygen concentration in this primary air affects the rapidity and stability of ignition, it is necessary, when hot gas is used, to draw the gas from a point of low CO_2 concentration and high temperature. This measure prevents the choice of flue gas for drying in direct firing.

In a semidirect system (Fig. 7.11), a cyclone collector between the pulverizer and furnace separates the conveying medium from the coal. The coal is fed directly from the cyclone to the furnace in a primary airstream, which is independent of the milling system. Thus, the drying medium can be the same as in a storage or bin system.

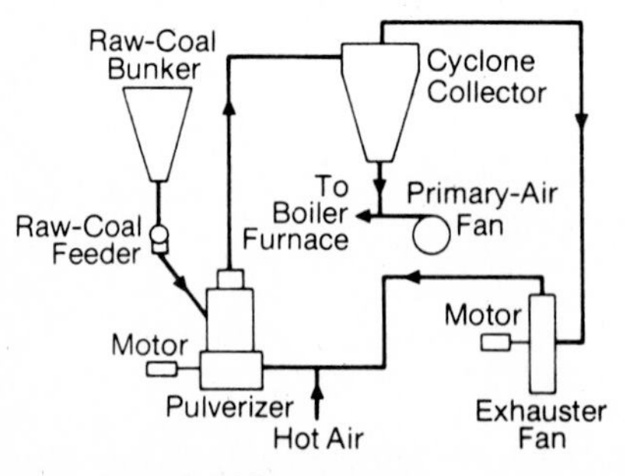

FIG. 7.11 Semidirect system.

Source of Heated Air

The best source of hot air for mill drying is either a regenerative or a recuperative air heater with combustion gas as the source of heat. Those used in connection with large boiler plants usually provide temperatures high enough for almost any fuel moisture condition. On small installations, where the moisture in the coal is not severe, steam air heaters may dry the coal. For higher-moisture conditions, a furnace-gas supplement may be necessary. Direct-fired air heaters, properly interlocked and protected, may also supplement air-heating requirements.

All the moisture contained in coal must be evaporated before ignition can take place. For rapid ignition, therefore, the surface moisture must be removed before the fuel is injected into the furnace. This same drying process facilitates pulverization. The type of fuel and its surface moisture govern mill-drying requirements.

The drying capability of a given pulverizer design depends on the extent of circulating load within the mill; the ability to mix the dry classifier returns rapidly with incoming raw, wet coal feed; and the air weight and air temperature that the design will tolerate. Some pulverizers are designed to operate satisfactorily with inlet air temperatures up to 825°F (440°C), while others are capable of using air at a maximum temperature of 600°F (315°C).

The type of fuel and the kind of system will determine the mill outlet temperature, which is lower for storage system mills than for direct or semidirect firing (Table 7.7), because most coals will not store safely at the temperatures created in direct firing. Storage-bin fires caused by spontaneous combustion of the fuel may result from inadequate mill outlet temperature control. These may be inhibited by maintaining an oxygen-deficient atmosphere, such as flue-gas inerting, over the bin coal level. Oxygen limits for various fuels are shown in NFPA 69, *Explosion Prevention Systems*.

TABLE 7.7 Allowable Mill-Outlet Temperatures

System	Storage		Direct		Semidirect	
	°F	°C	°F	°C	°F	°C
High-rank, high-volatile bituminous	130*	54*	170	77	170	77
Low-rank, high-volatile bituminous	130*	54*	160	71	160	71
High-rank, low-volatile bituminous	135*	57*	180	82	180	82
Lignite	110	43	110–140	43–60	120–140	49–60
Anthracite	200	93	—	—	—	—
Petroleum coke (delayed)	135	57	180–200	82–93	180–200	82–93
Petroleum coke (fluid)	200	93	200	93	200	93

*160°F (71°C) permissible with inert atmosphere blanketing of storage bin and low oxygen concentration conveying medium.

PULVERIZING AIR SYSTEMS

All coal-pulverizing systems utilize air or gas for drying, classification, and transport. Two methods accommodate the air requirements and overcome system resistance. In a direct-firing system, one method uses a fan behind the pulverizer, while the other has a fan ahead of it. The former is a suction system, and the fan handles coal-dust-laden air; the latter is a pressure system, and the fan handles relatively clean air.

In a suction system, the exhauster accounts for about 40 percent of the total power requirements (mill plus fan), while in a pressure system the blower is approximately 50 percent of the total.

With direct firing, the exhauster or blower volume requirement depends on the pulverizer size and is usually fixed by the base capacity of the pulverizer. The pressure or total head requirement is a function of the pulverizer and classifier resistance as well as the fuel distributing system and burner resistances. These resistances are, in turn, affected by the system design, required fuel-line velocities, and density of the mixture being conveyed.

In a storage system, the fan is located behind the dust collector and handles only a very small quantity of extremely fine (−200-mesh) dust at a relatively constant temperature [about 130°F (55°C)]. Thus, these fans are designed for high efficiencies and need not be designed for a head higher than the operating temperature requires.

Indirect Coal Storage Systems

Initial attempts at burning pulverized coal as a boiler fuel led to the development of the indirect-fired system. In this system, a cyclone collector separates the coal from the air used in the pulverizer for drying, classifying, and conveying. The pulverized coal is conveyed either mechanically or pneumatically to storage hoppers or bunkers. Mechanical, controllable feeders at the bunker outlets deliver the required quantity of coal to the fuel lines. At or near these feeders the coal is reentrained in air called *primary air* in proper proportions for transport to the burners.

In some installations, the primary air is taken from the room or a preheated air duct, or both. In other cases, the vented air from the pulverizer system is used for part or all of the primary-air source.

Primary Air. On installations of the first kind mentioned above, the *only* air needed is that to carry the coal to the furnace and to provide the required velocity of coal-air mixture at the burners. The quantity depends on the type of firing system, on the type of piping system selected (whether fuel lines are sloping, horizontal, or vertical), and on the burner and furnace arrangement.

PC feeder and fuel-piping arrangements determine the primary-air temperature, which may be as high as 600°F (315°C). Because any tempering air admitted to the system reduces the quantity of air passing through the air heater, which reduces the overall unit efficiency, a minimum amount of tempering air should be used in the primary-air system.

When full preheat cannot be justified, tempering is usually carried out at the suction side of the fan by dampering the preheated air and placing an adjustable cold-air opening between this damper and the fan inlet. If equipment arrangement permits, a dampered cold-air attachment from the *forced-draft* (FD) fan may be made at this location. The temperature of the air handled by the fan with such an arrangement should be the maximum usable with the particular feeder and piping arrangement.

If vented air is to be chosen when available, the temperature will depend on that of the vent, the position of the vent in the return air line, and whether preheated or cold air is added to the primary air at the fan inlet. The vented air usually will have a 70 to 90 percent relative humidity, will have a temperature of 110 to 160°F (43 to 71°C), and will contain coal fines vented from the collectors.

Vented Air. To pulverize and classify coal economically and properly, either it must be dry when it is delivered to the pulverizer, or it must be dried during the pulverizing operation. Drying is accomplished by supplying hot air or flue gas to the pulverizing system.

In a storage system, either a portion or all of the drying medium is discharged from the system by venting, which removes the moisture evaporated from the fuel. Aside from the adverse effects resulting from lack of drying in the pulverizer, a storage system requires the removal of moisture to eliminate or minimize difficulties in handling, storing, and feeding of the pulverized fuel. The drying medium supplies a major portion of the heat to evaporate the fuel moisture and, in so doing, approaches saturation.

Some of this almost-saturated mixture, about equal to the weight of hot drying medium supplied, must be removed or vented from the system. The amount vented will depend on pulverizer output, moisture removed from the fuel, initial temperature of the drying medium, temperature of the vented material, and efficiency of the drying system.

The vented mixture may be disposed of in two ways: (1) by venting to the atmosphere directly or through the boiler stack or (2) by using it as part of the air supply to the furnace. In the first, since the vent contains some extremely fine coal (up to 2 percent of the amount being pulverized), economically it cannot be vented directly without collecting the coal dust in cyclone collectors, fabric filters, air washers, or a combination of these. In the second method, the air need not be cleaned. In any case, sufficient resistance is in the venting system to require a separate fan (unless the PA fan is designed for the additional pressure requirements).

Direct-Firing Arrangements

Most direct-fired PC systems are for furnaces operated under suction (balanced-draft) conditions, as Fig. 7.12 shows. With a suction mill, the coal feeder discharges against a negative pressure; in the pressurized mill, the feeder discharges against a positive pressure of 18 to 21 in H_2O (4480 to 5230 Pa). No coal feeder can act as a seal; thus, the head of coal above the feeder inlet must be used to prevent backflow of the primary air. As a rule of thumb, the height of the coal column above the feeder should equal 1 ft (0.3 m) for every 3 in H_2O (750 Pa) of primary-air mill-inlet pressure. If the PA fan inlet pressure is 30 in H_2O (7500 Pa), then the coal column height will be 10 ft (3 m).

Suction System. The suction system has a number of advantages. It is quite easy to keep the area around the pulverizer clean. It is easy to control the airflow

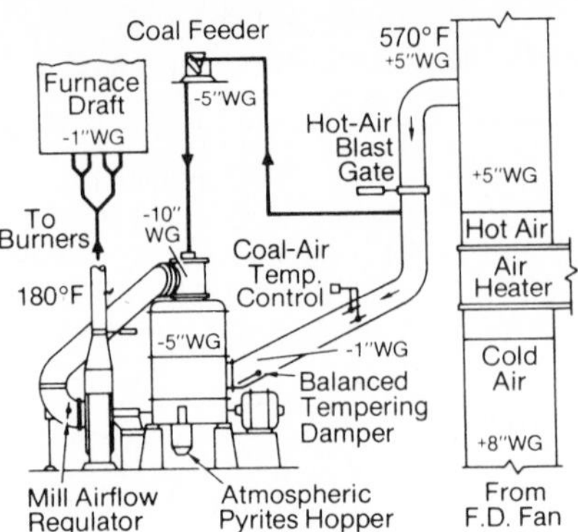

FIG. 7.12 Balanced-draft furnace with suction mill. Air pressures shown are illustrative only. Note: °C = (°F − 32)/1.8; Pa = in (H_2O) × 249.

through the pulverizer by dampering the constant-temperature coal-air mixture with a positioning device. Control of the coal-air-mixture temperature is achieved by a single hot-air damper and a barometric damper through which a flow of room air is induced by the suction in the pulverizer. With this control, the fan is designed for a constant, low-temperature mixture and has low power consumption, even though such material-handling fans have a relatively low efficiency of 55 to 60 percent.

The main disadvantage of the suction system is the maintenance required on the exhauster. Conversely, by using proper design techniques and wear-resistant materials, the maintenance on an exhauster can be minimized. Exhauster maintenance costs are more than offset by the power and capital savings of the system, which justifies the continued use of the suction system on smaller units.

Pressurized Exhauster System. To obtain sufficient pressure for firing a pressurized furnace, the pressurized exhauster system was developed; a number of these systems are in operation. The system retains the advantages of the suction system in the design of the fan for constant, low-temperature mixture and the relative ease of airflow control. Two dampers—one in the hot-air duct to the mill and the other in the cold-air duct to the mill—control the amount of pulverizer airflow. This flow varies with, but is not proportional to, the amount of fuel being fed to the pulverizer. Biasing the hot- and cold-air dampers controls the temperature of the mixture leaving the pulverizer.

One advantage of the pressurized exhauster system is that the low pressures in the pulverizers do not present as severe a problem in sealing the head of coal over the raw-fuel feeder as with pulverizers under direct blower pressure. The disadvantage of the system, as with the suction system, is exhauster maintenance.

Cold Primary-Air System. In this system, the PA fan handles only ambient air. The fan is located ahead of the air preheater, with a separate primary-air system through the air heater. Although it is not as simple as the other two systems, its chief advantage lies in its fan power and maintenance. As the fans handle cold air, they can be smaller, can run at higher speeds, and can use highly efficient airfoil blade shapes. Inlet vanes can control airflow and further increase fan efficiency.

Some other advantages of this system are as follows: With high-efficiency fans handling ambient air, design for higher pressure differentials is possible, and larger mills with longer fuel-pipe runs are practical. Thus, mills may be located farther from the boiler. Because individual fans for each pulverizer are not necessary, the space requirements for the pulverizer bays can be reduced. Moreover, experience has proved that metering the airflow on the inlet-air side of the pulverizer is most desirable. With the higher fan head available, the airflow can be measured quite easily by installing a venturi or other metering means.

Even with fewer pulverizer fans, controlling the airflow to the various pulverizers is still relatively simple. The total primary air required is a function of the number of pulverizers in operation. This fact permits the selection of a simple control for the airflow requirements. Because of possible variations of load and coal-moisture content between pulverizers, it is necessary to control not only the total airflow but also the temperature of the air to individual pulverizers. Temperature control is achieved with hot and cold primary-air ducts with a damper in each for each mill. The airflow requirement for a pulverizer is met by operating both dampers, while controlling temperature by properly proportioning the flow between the hot- and cold-air ducts.

As compared with other systems, total savings with this system—because of no exhauster maintenance and reduced fan power—may be 35 to 40 percent of the cost per unit of coal pulverized. This operating saving is partially offset by the capital charges for added ductwork, dampers, and controls. With larger units and pulverizers, the cold primary-air system becomes more favorable economically.

PRINCIPAL TYPES OF PULVERIZERS

To achieve the particle size reduction needed for proper combustion in PC firing, pulverizers or mills are used to grind or comminute the fuel. Grinding mills apply one, two, or all three of the basic principles of particle size reduction, namely, impact, attrition, and crushing. With respect to speed, these machines may be classified as low, medium, and high. The four most commonly selected pulverizers are the ball-tube or ball (low-speed), the ring-roll or ball-race (medium-speed), the impact or hammer mill, and the attrition (high-speed) type.

Ball-Tube Mills

A ball-tube mill (Fig. 7.13) is basically a hollow horizontal cylinder, rotated on its axis, whose length is slightly less than to somewhat greater than its diameter. Heavy cast, wear-resistant liners fit the inside of the cylindrical shell, which is filled to a little less than one-half with forged steel or cast alloy balls varying from 1 to 4 in (25 to 100 mm) in diameter. Rotating slowly at 18 to 35 rpm, the balls are carried about two-thirds of the way up the periphery and then continually cascaded toward the center of the cylinder.

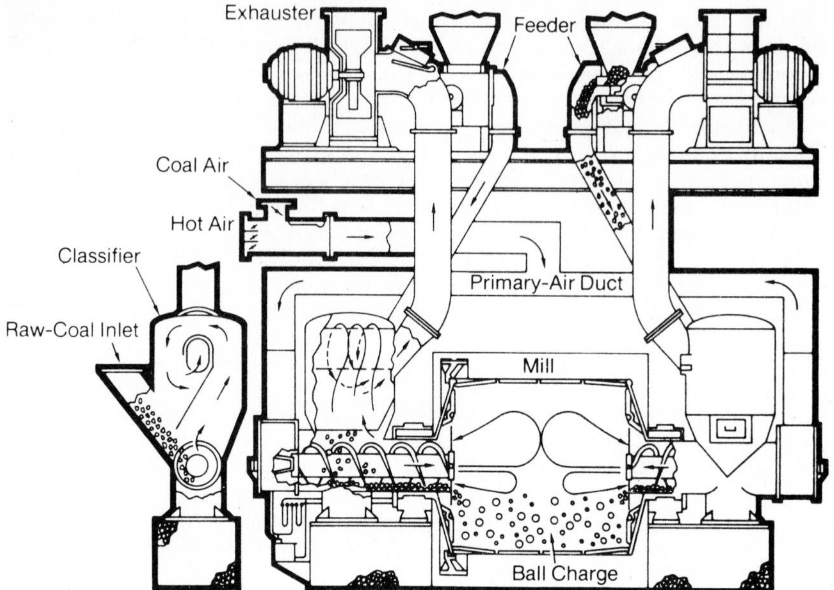

FIG. 7.13 Ball-tube mill.

Mill Operation. Coal is fed to the cylinder of the ball-tube mill through hollow trunnions and intermingles with the ball charge. Pulverization, which is accomplished through continual cascading of the mixtures, results from (1) impact of the falling balls on the coal, (2) attrition as particles slide over each other as well as over the liners, and (3) crushing as balls roll over each other and over the liners with coal particles between them. Larger pieces of coal are broken by impact, and the fine grinding is done by attrition and crushing as the balls roll and slide within the charge.

Hot airflow is passed through the mill to dry the coal and remove the fines from the pulverizing zone. In most designs for firing boilers or industrial furnaces, an external classifier regulates the size or degree of fineness of the finished product. The oversize or rejects from the classifier (sometimes called *tailings*), already dried, are returned to the grinding zone with the raw coal, which helps reduce the tendency for wet coal to plug the feed end in this type of pulverizer.

The power consumption of ball-tube mills, in kilowatts per unit of coal pulverized, is very high, particularly at partial loads. Relatively large phyically per unit of capacity, ball-tube mills require considerable floor space. Because of their size and weight, the initial cost is quite high.

Mill Limitations. Ball-tube mills are not well suited to intermittent operation, because the large amount of heat stored in the coal and ball charge may produce overheating and fires when the mill is idle. The mass of this mill type makes it necessary to select high-power, high-starting-torque motors. Also, these mills are very noisy. Most installations require that insulated "doghouses" be erected over each mill for noise attenuation.

Maintenance in the grinding zone of ball-tube mills is relatively easy to perform. Periodically, a ball charge is added to the mill to make up for metal lost in the grinding process. It may take years to wear out the cast liners, but considerable downtime is needed for their replacement. Over the long run, maintenance costs per unit of coal ground are about the same as for ring-roll-type pulverizers.

Impact Mills

An impact mill consists primarily of a series of hinged or fixed hammers revolving in an enclosed chamber lined with cast wear-resistant plates. Grinding results from a combination of hammer impact on the larger particles and attrition of the smaller particles on each other and across the grinding surfaces. An air system with the fan mounted either internally or externally on the main shaft induces a flow through the mill. An internal or external type of classifier may be used.

This class of mill is simple and compact, is low in cost, and may be built in very small sizes. Its ability to handle high inlet-air temperatures, plus the return of dried classified rejects to the incoming raw feed, makes it an excellent dryer. However, the high-speed design results in high maintenance and high power consumption for grinding fine.

Progressive wear on the grinding elements produces a rapid drop-off in product fineness, and it is difficult (if not impossible) to maintain fineness over the life of the wearing parts. Selecting an external classifier permits maintenance of fineness, but only at the expense of a considerable reduction in capacity as parts wear. The maximum capacity for which such mills can be built is lower than for most other types.

Attrition Mills

No true attrition mill is used for coal pulverizing because of the high rate of wear on parts. A high-speed mill that relies on considerable attrition grinding along with impact grinding is, however, used for direct firing of pulverized coal. In this type of mill, the grinding elements consist of pegs and lugs mounted on a disk rotating in a chamber. The periphery of the chamber is lined with wear-resistant plates, and its walls contain fixed rows of lugs within which the rotating lugs mesh.

The fan rotor is mounted on the pulverizer shaft. Instead of an external classifier, a simple shaft-mounted rejector is used. Wear-resistant alloy lug and peg facings and casing linings reduce the wear effect on fineness and extend the periods between parts replacement. This type of mill exhibits all the characteristics of the impact mills.

Ring-Roll and Ball-Race Mills

These mills comprise the largest number of pulverizers in service for coal grinding. They are of medium speed and use primarily crushing and attrition of particles, plus a small amount of impact to obtain size reduction of the coal. The grinding action takes place between two surfaces, one rolling over the other. The rolling element may be either a ball or a roll, while the member over which it rolls may be either a race or a ring. The ball diameter is 20 to 35 percent of the race diameter, which can be as large as 100 in (2500 mm). If the element is a roll, its

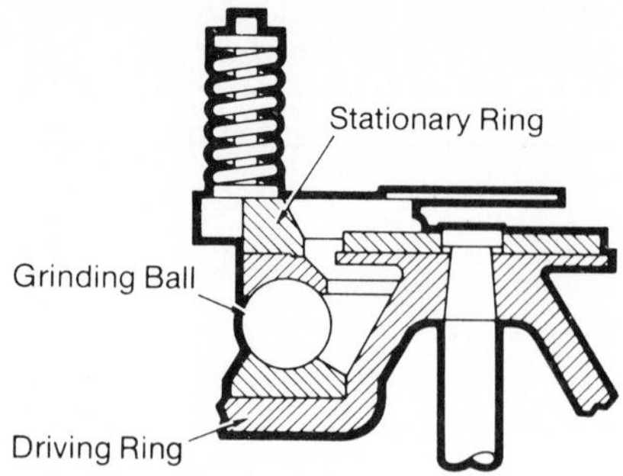

FIG. 7.14 Ball-race mill.

diameter may be 25 to 50 percent of the ring diameter, which can be 110 in (2800 mm) or greater. Its face width, depending on mill size, will vary from 15 to 20 percent of the ring diameter.

When the rolling elements are balls (Fig. 7.14), they are confined between races. In the majority of designs, the lower race is the driven rotary member, while the upper race is stationary. Some designs also utilize a rotating upper race. The required grinding pressure is obtained by forcing the races together with either heavy springs or pneumatic or hydraulic cylinders. Some additional grinding pressure is obtained from the centrifugal force of the rotating balls.

Mill Classes. Two general classes of mills rely on rollers as the rolling elements. In one (Fig. 7.15), the roller assemblies are driven, and the ring is stationary; in another (Fig. 7.16), the roller assembly is fixed, and the ring rotates.

The other classification of ring-roll mills, namely those in which the ring rotates, constitutes the largest number of mills selected for grinding coal; they are manufactured by the major boiler companies. The vertical driving shafts of these mills operate at 20 to 150 rpm, with larger mills running at slower speeds.

Generally, such mills are equipped with self-contained or integral classifiers to regulate the fineness of the finished product. In some cases, this device may be external to the mill itself and is then termed an *instream classifier*. Exhausters or PA fans create a flow of heated air through the mills. This heated air dries the coal, removes it from the grinding zone, carries it through the classifying zone, and conveys it to its point of use.

When this type of mill is provided with sufficient air at a temperature to produce a satisfactory mill-outlet temperature, it can handle very wet coals with only a small reduction in capacity. The high ratio of circulating load (rejects for further size reduction) to output, with the resulting rapid reduction of average moisture content, facilitates the grinding process. These mills require less power than any others.

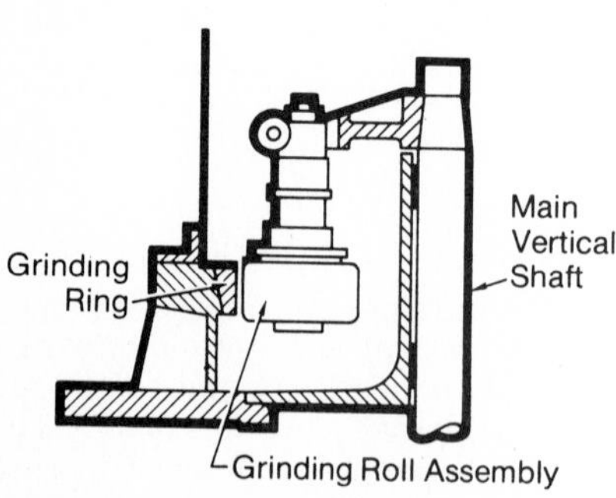

FIG. 7.15 Roller-type ring-roll-mill journal assembly.

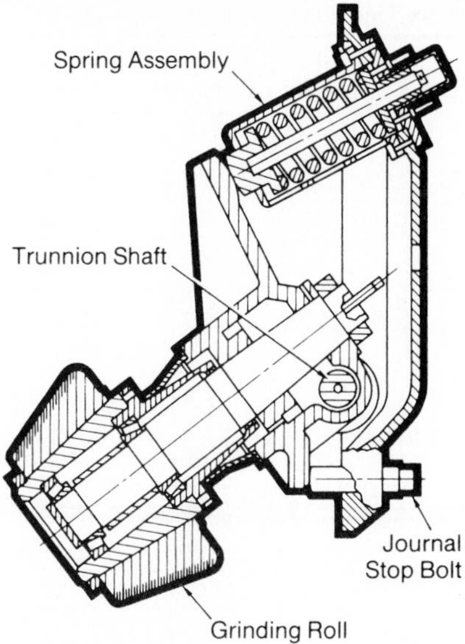

FIG. 7.16 Bowl-type ring-roll-mill journal assembly, spring-loaded.

Mill Materials. Grinding different coals economically and efficiently, as mentioned previously, is a reflection of grinding pressures available, method of application of this force, speed of moving elements, abrasion, and power and size limitations of the particular units. Although the materials capable of being pulverized by the mill types described here are quite extensive, let Table 7.8 cover those combustible materials of primary concern in utility and industrial powerplants.

TABLE 7.8 Types of Pulverizers for Various Materials

Type of material	Ball tube	Impact and attrition	Ball race	Ring roll
Low-volatile anthracite	x	—	—	—
High-volatile anthracite	x	—	x	x
Coke breeze	x	—	—	—
Petroleum coke (fluid)	x	—	x	x
Petroleum coke (delayed)	x	x	x	x
Graphite	x	—	x	x
Low-volatile bituminous coal	x	x	x	x
Medium-volatile bituminous coal	x	x	x	x
High-volatile A bituminous coal	x	x	x	x
High-volatile B bituminous coal	x	x	x	x
High-volatile C bituminous coal	x	—	x	x

TABLE 7.8 Types of Pulverizers for Various Materials *(Continued)*

Type of material	Ball tube	Impact and attrition	Ball race	Ring roll
Subbituminous A coal	x	—	x	x
Subbituminous B coal	x	—	x	x
Subbituminous C coal	—	—	x	x
Lignite	—	—	x	x
Lignite and coal char	x	—	x	x
Brown coal	—	x	—	—
Furfural residue	—	x	—	x
Sulfur	—	x	—	x
Gypsum	—	x	x	x
Phosphate rock	x	—	x	x
Limestone	x	—	—	x
Rice hulls	—	x	—	—
Grains	—	x	—	—
Ores—hard	x	—	—	—
Ores—soft	x	—	x	x

SUGGESTIONS FOR FURTHER READING

Benson, L.M., and C.A. Penterson, "A Comparison of Three Types of Coal Pulverizers," *Proceedings of American Power Conference,* 47:942–947, 1985.

Bogot, A., and T.B. Hamilton, "Maximizing Pulverizer Capabilities with Low Btu Western Coals," *Proceedings of American Power Conference,* 36:483–491, 1974.

Cross, D.A., "Pulverizer Maintenance and Operation," Proceedings of Kentucky Industrial Coal Conference, Lexington, Ky., 1987.

Hensel, R.P., "Coal Combustion," paper presented at Engineering Foundation Conference on Coal Preparation for Coal Conversion, Franklin Pierce College, Rindge, N.H., 1975; paper updated in 1978.

Hensel, R.P., "The Effects of Agglomerating Characteristics of Coals on Combustion in Pulverized Fuel Boilers," paper presented at Symposium on Coal Agglomeration and Conversion, Morgantown, W.Va., 1975.

Kmiotek, S.E., and R.F. Hickey, "Coal Pulverizer Inerting and Fire Fighting System," *Proceedings of the American Power Conference,* 49:158–164, 1987.

Parry, V.F., "Production, Classification and Utilization of Western United States Coals," *Economic Geology,* 45(6):515–532, Sept.–Oct., 1950.

CHAPTER 1.8
STOKERS

Russell A. Wills, Neil H. Johnson, and Harold L. Knox
Detroit Stoker Co.
Monroe, MI

INTRODUCTION

Stokers to burn coal were first used in the 1700s. Through the years stokers have evolved from crude machines of simple design to quite sophisticated devices designed to burn not only coal but many other solid fuels as well. Stokers can be divided into three general groups, depending on how the coal or other form of solid fuel reaches the grate of the stoker for burning—underfeed, overfeed, and spreader designs.

Underfeed Stokers

In this type, the coal comes from below the air-admitting surface of the grate and is fed by a screw or ram into a retort. Underfeed stokers use both stationary- and moving-grate tuyeres. Ash is discharged at either the side or rear end of the retorts, depending on the type of stoker.

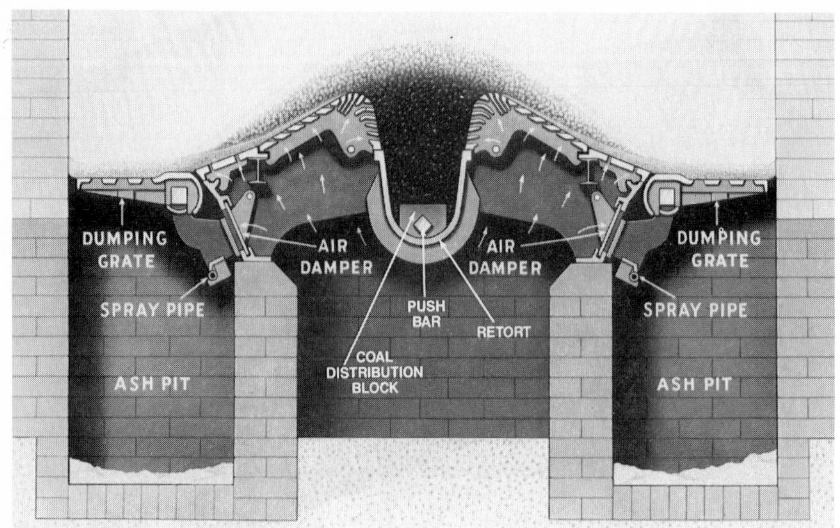

FIG. 8.6 Single-retort stoker showing contour of fuel bed and ash dumps.

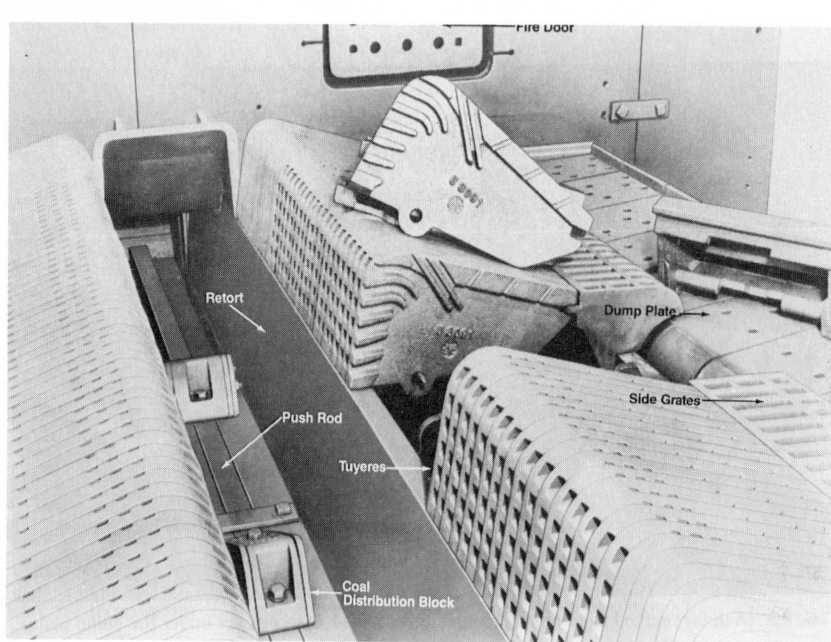

FIG. 8.7 Ram-fed stoker with fixed tuyeres.

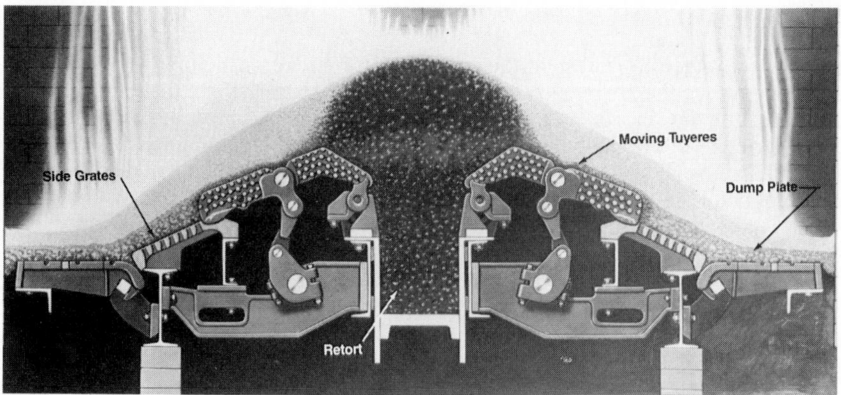

FIG. 8.8 Ram-fed stoker with moving tuyeres.

rod or ram plates are installed in the bottom of the retort to assist in distributing the coal over the length of the retort.

Ram-fed stokers with the moving-tuyere arrangement impart an undulating motion to the fuel bed, which tends to break up coke masses or agglomerated coal masses, thereby keeping the fuel bed porous and free-burning. Stokers of this type are more suited to handling coal having a higher free-swelling index.

Stoker Operation. On single-retort stokers, the majority of the air required to complete the combustion process is introduced to the fuel bed through the tuyeres. A small portion of the air is introduced through the side grates, when they are supplied by the manufacturer. Side dumps are normally furnished with all single-retort stokers. These stokers retain the ash so that complete burnout of the coal occurs. Prior to dumping the ash into the ash pit, air is introduced under the dump grates on some stoker models to cool the ash and to facilitate burnout of any remaining unburned carbon in the ash.

Total heat-release-rate capacities of single-retort stokers range from 900,000 Btu/h (0.26 MW) to 40 × 10^6 Btu/h (11.7 MW). The Btu input has been used for capacity rating of single-retort stokers in lieu of the steam capacity, since steam conditions and efficiency based on Btu input will determine steam output. Table 8.2 shows the capacity ranges of single-retort stokers.

TABLE 8.2 Capacity Ranges for Single-Retort Stokers

	Approximate steam		Total stroker heat input	
Stoker type	lb/h	g/s	10^6 × Btu/h	MW
Screw feed	600–8,000	76–1008	0.9–11.5	0.26–3.4
Ram feed with:				
Stationary tuyeres	2,000–25,000	252–3150	2.8–33.5	0.82–9.8
Side-moving tuyeres	9,000–30,000	1134–3780	13–40	3.8–11.7

Multiple-Retort Stokers

Unlike the single-retort design, the multiple-retort stoker is inclined at an angle of 20° to 25° above the horizontal, so that gravity assists in distributing the coal over the length of the retorts (Fig. 8.9). As the name implies, multiple-retort stokers use a series of retorts installed side by side and normally extending from front to rear of a boiler, with tuyere sections between the retorts and along the side walls of the boiler.

Round or square rams are installed to move the coal from the coal hopper into the upper or front end of the retorts below the point of combustion-air supply. Secondary rams or pusher plates move the coal along the retort and upward into the burning zone immediately above the retorts, where air for combustion is supplied to the fuel bed through tuyeres located between and above the retorts. An adjustable dump grate section is used at the rear end of the retorts to retain the ash and prevent an avalanching of unburned coal into the ash pit.

Coal Specifications

A relatively wide range of bituminous coal can be burned on single- and multiple-retort stokers; however, the top size is limited to a range of ¾ to 1½ in (19 to 38 mm), and typical specifications call for 1¼-in (32-mm) nut, pea, and slack, with equal portions of each. It is advisable to limit the amount of fines to 50 percent passing through a ¼-in (6-mm) round-hole screen.

The *free-swelling index* of the coal should be limited to 5 with single-retort stokers having stationary tuyeres. A coal having a free-swelling index up to 7 can be burned on single-retort stokers with moving tuyeres and on multiple-retort stokers. Underfeed stokers can burn coals at rated capacity with an ash fusion temperature above 2400°F (1316°C), based on $H = \frac{1}{2}W$, where H and W are stoker height and width, respectively. Coals having an ash fusion temperature as low as 2100°F (1149°C), based on $H = \frac{1}{2}W$, can be burned at reduced burning rates.

It is normally recommended that the iron content in the ash be no more than

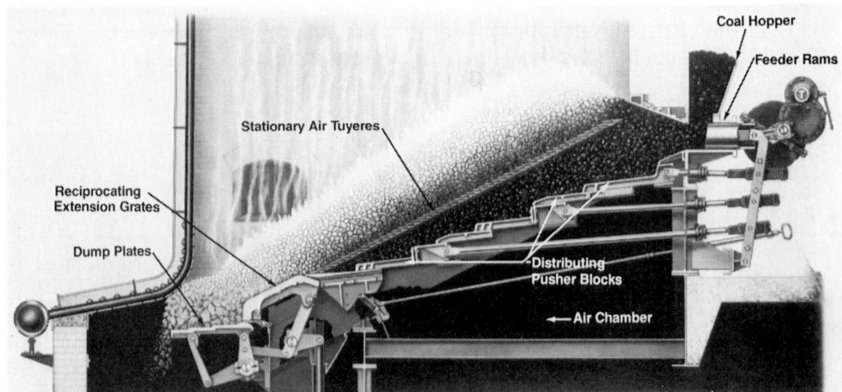

FIG. 8.9 Multiple-retort underfeed stoker with rear ash discharge.

20 percent as Fe_2O_3 with an ash fusion temperature above 2400°F (1316°C) and below 15 percent with coals having a lower ash fusion temperature.

Single-retort stokers have been used to burn coals ranging from lignite to anthracite; however, the vast majority of underfeed stokers sold are intended for free-burning midwestern, eastern caking, and mildly caking bituminous coals.

Boiler Furnaces

Water-cooled furnaces are preferred with underfeed stokers. On the smaller sizes of single-retort stokers, however, satisfactory performance can be obtained with refractory walls. Adequate furnace volume must be provided in accordance with the recommendations of the stoker manufacturer.

Overfire and Combustion Air

All underfeed stokers should be provided with an overfire-air system designed to provide adequate furnace turbulence to complete the burnout of the hydrocarbons. Depending on the width of the stoker, load conditions, and coal-burning characteristics, overfire air will represent 7½ to 12 percent of the theoretical air plus excess air.

A *forced-draft* (FD) fan furnished either as an integral part of the stoker or as a separate unit supplies the air required to complete the combustion process. When the forced-draft fan is sized, the test-block rating should be based on theoretical air plus excess air plus a margin of safety. Static-pressure capabilities of the fan should be sufficient to overcome damper losses, duct losses, and the resistance of the grate and fuel bed as specified by the stoker manufacturer, plus a margin of safety for fan test-block conditions.

OVERFEED STOKERS

The overfeed mass-burning stoker conveys coal or other types of solid fuel from the stoker hopper located at the front of the stoker. The depth of the fuel bed conveyed into the furnace is regulated by a vertically adjustable feed gate across the width of the unit. The fuel that is carried into and through the furnace passes over several air zones for stage burning. The ash is continuously discharged into an ash storage hopper at the rear end of the grate.

The three basic types of grates utilized with overfeed mass-burning stokers are the chain grate, traveling grate, and water-cooled vibrating grate.

Chain Grate

The grate surface of the chain-grate stoker consists of narrow grates which form links. The links are staggered and connected by rods extending across the stoker width to form a wide continuous-chain assembly, which is pulled or pushed through the furnace by an electric or a hydraulic drive. See Figs. 8.10 and 8.11.

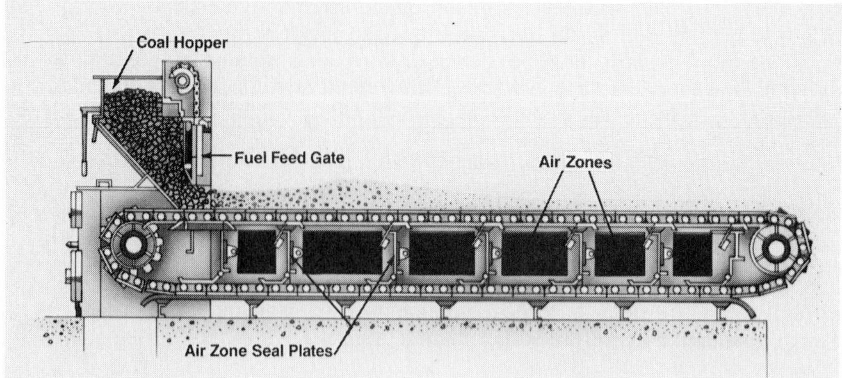

FIG. 8.10 Cross section of overfeed mass-burning chain-grate stoker.

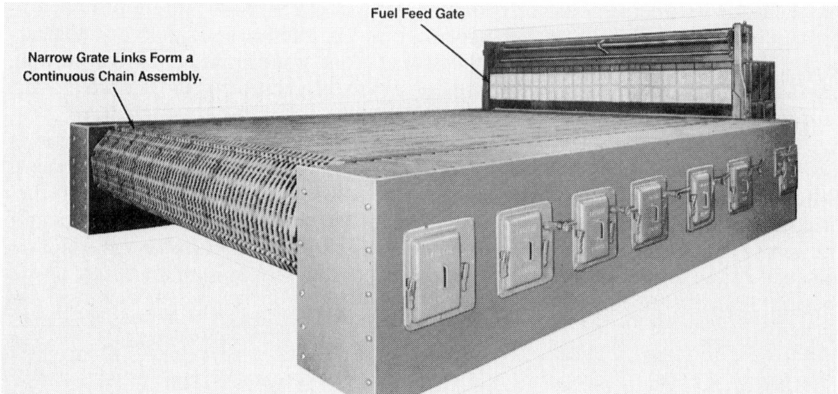

FIG. 8.11 Completely assembled overfeed mass-burning chain-grate stoker.

Traveling Grate

The grate surface of the traveling grate consists of narrow grate clips mounted on lateral carrier bars that are attached to the grate chains, which pull or push the grate through the furnace, using an electric or a hydraulic drive. See Figs. 8.12 and 8.13.

Water-Cooled Vibrating Grate

The grate surface of the water-cooled vibrating-grate stoker consists of tuyere-type grates mounted on and in close-fitting contact with a grid of water-cooled tubes, which are connected to the boiler circulation system for positive cooling. The entire structure is supported on steel flexing plates, permitting the entire water-cooled grid and grate surface to vibrate with a preset amplitude that moves the fuel bed through the furnace. Vibration of the grate is intermittent, and the

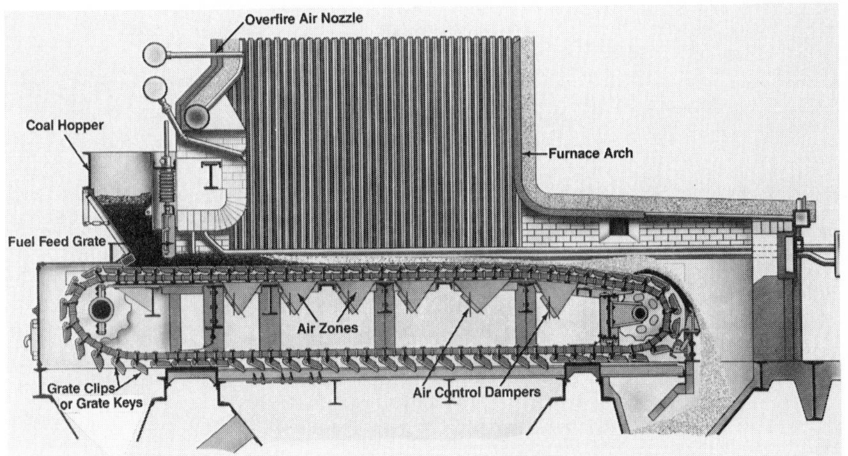

FIG. 8.12 Cross section of overfeed mass-burning traveling-grate stoker.

FIG. 8.13 Traveling-grate stoker showing carrier bars, grate keys and chain, which pulls or pushes grate through furnace.

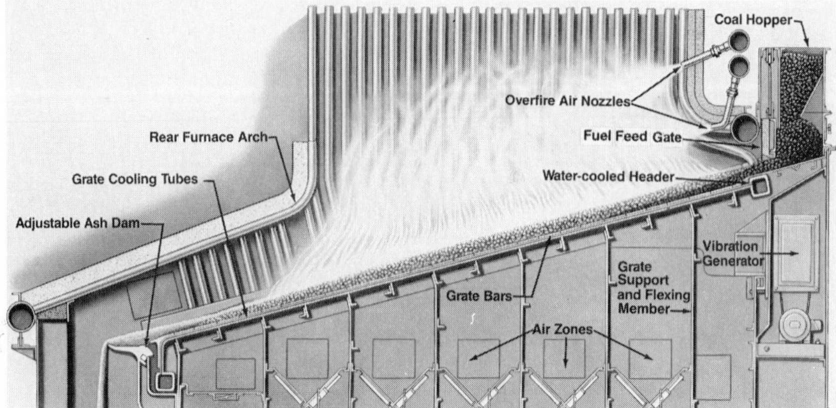

FIG. 8.14 Water-cooled, vibrating-grate stoker.

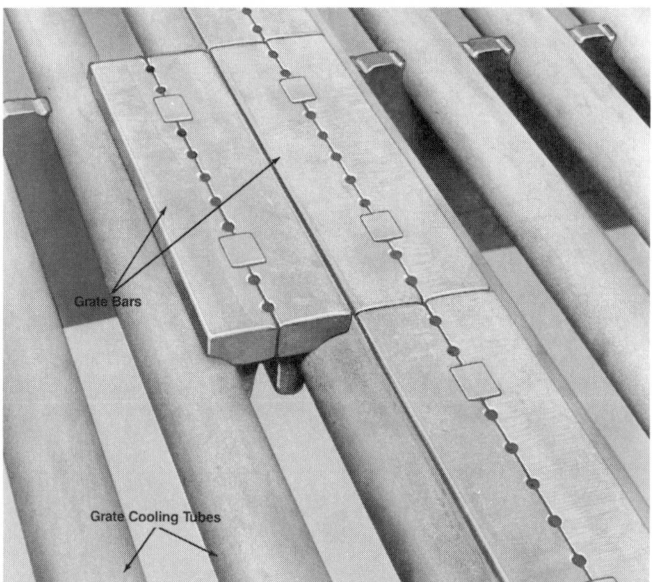

FIG. 8.15 Attachment of grate bars to cooling tubes of overfeed mass-burning, water-cooled vibrating-grate stoker.

vibration period and delay between vibrations are regulated by a timing device synchronizing the fuel feeding rate with steam demand. See Figs. 8.14 and 8.15.

Ignition

On all overfeed mass-burning stokers, the rate of ignition controls the grate speed and fuel-bed thickness. As the grate travels through the furnace, if the time that the fuel remains on the grate is too short for complete combustion, the burning fuel will be discharged into the ash hopper, resulting in high carbon loss.

It is mainly radiation that ignites the incoming raw fuel. Once the surface layer of the incoming fuel begins to burn, it warms the fuel beneath, causing rapid combustion of the fuel together with the hydrocarbon and other combustible gases driven off by distillation—*providing* the airflow rate through the forward air zones of the grate is correct. For proper combustion, the airflow rate must vary longitudinally along the grate. Forced-draft air beneath the grate is introduced through several air zones equipped with manually adjusted dampers, allowing for proper control of the airflow rate to each zone of the grate. The forced-draft fan supplying undergrate air should be sized to provide theoretical air plus excess air.

Overfire Air

Overfeed mass-burning stokers are provided with a high-pressure overfire-air system consisting of one or two rows of closely spaced air jets located above the fuel bed in the front wall of the furnace. The jets provide adequate furnace turbulence to complete the burnout of hydrocarbons. The high-pressure-air fan will be sized to provide approximately 10 percent of theoretical air plus excess air. The static pressure of the fan is selected to provide proper penetration of the air into the furnace.

Furnace Design

Water-cooled furnaces are preferred with all moving-grate stokers to prevent slag formation on the furnace walls. The tube spacing of the furnace walls should not exceed twice the tube diameter. Adequate furnace volume must be provided as recommended by the boiler and stoker manufacturers.

The furnace should include a rear arch, since a thin fire exists in the burnout zone of the stoker, causing high excess air in this area. The rear arch diverts the furnace gases forward into the rich zone of burning volatile gases released in the ignition zones. This arrangement increases the CO_2 at the furnace exit for optimum operating efficiency. A short front arch is adequate for most bituminous coals, since it serves to break up and mix rich streams of volatiles, which otherwise might travel through the furnace unburned.

Coal Specifications

A relatively wide range of bituminous coals can be burned on overfeed mass-burning stokers; however, the top size is limited to a range of ¾ to 1½ in (19 to 38 mm), and typical specifications call for 1¼-in (32-mm) nut, pea, and slack, with equal portions of each. The amounts of fines should be limited to 50 percent passing through a ¼-in (6-mm) round-hole screen.

Fuel Types. Almost any solid fuel of suitable size can be burned, including peat, lignite, subbituminous, free-burning bituminous, anthracite, or coke breeze. However, strongly caking bituminous coals may have a tendency to "coke over" and prevent proper passage of air through the fuel bed, thus causing high carbon loss to the ash hopper. The fuel bed can be made more porous by water or steam tempering of the fuel. This is done in the stoker coal hopper on chain- and traveling-grate units. Tempering is not required on water-cooled vibrating-grate stokers, since the vibrating action of the grate tends to keep the fuel bed uniform and porous.

The overfeed mass-burning stoker is quite sensitive to segregation or poor distribution of coal sizes. Unless the fuel-size consist is uniform across the width of the stoker, the fuel bed will not burn uniformly, resulting in unburned carbon being discharged to the ash hopper.

Grate Sizing. Overfeed mass-burning stokers are best suited for industrial and institutional powerplants having steady load demands. Since all the fuel is consumed on the grate, the stoker grate surface must be conservatively sized. The grate heat-release rate should be limited to a maximum of 425 Btu/(h • ft^2) (1340 kW/m^2) of active grate area. These stokers are normally designed for capacities up to 150,000 lb/h (18 900 g/ls) of steam. Larger units up to 200,000 lb/h (25 200 g/s) of steam have been installed but are not generally recommended. The chain-grate types are normally used for smaller capacities, while the traveling-grate and water-cooled vibrating-grate types are used for large capacities.

Since all the fuel is carried by and consumed on the grate surface, the mass-burning overfeed stoker is recognized for its inherently low dust-emission characteristics; therefore, less costly pollution control equipment is an advantage with these types of stokers. High operating efficiencies are maintained with proper fuel selection and conservatively sized grate surfaces, with total carbon loss approximately 1.5 percent without the use of fly-carbon recovery systems.

SPREADER STOKERS

Fuel, as the name implies with this type of stoker, is spread into the furnace over the grates from feeders located across the front of the unit. The purpose is to feed the fuel evenly over the grate surface in order to release an equal amount of energy from each square foot of active air-admitting grate surface. The air for combustion should then be admitted evenly through the grate to provide the oxygen for burning. Above the grates, an overfire-air system is provided for additional oxygen and turbulence in the lower-furnace zone of high temperature to ensure efficient performance. Since spreader stokers can burn a wide range of solid fuels and cover a wide range of boiler sizes, the arrangement and design of each of the three basic components—fuel feeders, type of grate, and overfire-air system—depend on the specific project.

Fuels and Fuel Sizing

The solid fuels burned on spreader stokers include bituminous coal, subbituminous coal, lignite, wood waste from the forest products industry, bagasse from the sugar cane industry, peat from peat bogs, peanut shells, furfural residue after extraction of the hydrocarbons from agricultural wastes, and the *refuse-derived fuel* (RDF) of municipal solid waste.

It is necessary to size the fuel properly for spreader stokers. Coals should have 95 percent less than 1¼ in (32 mm), with 2-in (51-mm) maximum size. Waste fuels can have top sizes up to 4-in (102-mm) cubes. Handling, feeding, and burning characteristics will preclude larger sizes. The amount of fine particles of fuel will vary widely, depending on a number of factors. Good practice would prescribe an even graduation of sizes between the top size and the fines. However, methods to adjust the spreading of fuel into the furnace for variations in fuel sizing can be incorporated into the feeding devices.

Spreader stokers can be applied to a wide range of boiler sizes. Inputs as low as 50×10^6 Btu/h (14.7 MW) are utilized up to inputs of 530×10^6 Btu/h (155.3 MW) for coal, and 900×10^6 Btu/h (263.7 MW) for certain waste fuels is possible. In another perspective, this might correspond to steaming rates of 40,000 lb/h (5040 g/s), 400,000 lb/h (50 400 g/s), and 600,000 lb/h (75 600 g/s) of steam, respectively.

Fuel Burning

As the fuel is fed to the furnace in a manner to spread it evenly over the complete grate surface, it burns both in suspension and on the grate. The amount burned in suspension depends on a number of factors. Fine fuel burns more in suspension; that is, with coal the size −16 mesh is significant. High-volatile fuels release more energy in suspension. High moisture can increase the energy released on the grate. The burning fuel bed on the grate may be only about 1 in (25 mm) thick. This depth depends on fuel characteristics and firing rate. Under the burning fuel an ash bed is formed, which should be 3 in (76 mm) to 4 in (102 mm) thick.

The ash-bed thickness also depends on fuel characteristics, but not on the firing rate since the rate of ash discharge is regulated to the firing rate. The ash bed provides an insulating barrier between the burning fuel and the metal-grate surface additionally cooled by the combustion airflow across the grate.

Since the fuel is burned in suspension and in a thin fuel bed on a spreader stoker, the inventory of energy in a furnace is small. If the fuel supply is interrupted, the fire will be completely out in a matter of 3 to 4 min. This burning characteristic allows the energy source level to be changed rapidly, and thus the level is able to follow rapid changes in steam demand. The ability to follow load swings makes the spreader stoker well suited to industrial applications where process loads fluctuate rapidly, that is, 10 percent of the rating per minute or more.

From an environmental standpoint, spreader stokers can meet 1984 federal standards. Particulate is controlled by electrostatic precipitators, fabric filters, or scrubbers. Nitrogen oxides are controlled by good operating practices, as is carbon monoxide. Sulfur oxides are a function of fuel sulfur content. Emission levels assume proper basic boiler design.

Fuel Feeders

Many types of feeders are manufactured today. They all strive to deliver fuel to the furnace in an even distribution over the grates. The two principal methods of fuel delivery are *revolving rotors,* which strike the fuel particles to impart velocity into the furnace, or *air jets* of high velocity to provide the energy to propel the fuel into the furnace.

Distribution Control. A combination of these methods can also be utilized. Lateral distribution can be controlled by the number of feeders, shape of the revolving rotors, air-jet direction, or a combination of all these. In every case, it is imperative that an adequate number of feeders be furnished to minimize a "V" at the front of the grate between the feeders. In general, the total feeder width should not be less than 40 percent of the grate width.

Longitudinal distribution is controlled by the speed of the revolving rotor, trajectory of the fuel (which in turn is controlled by the position of the device delivering the fuel to the rotor), velocity or mass of air-jet flow, or the size consist of the fuel being fed to the furnace.

Elevation of the feeder above the grate is also a factor. The feeder should be close to the grate for best control of distribution and to deliver fuel to the hottest part of the furnace for efficient combustion. A distance of 3 to 3½ ft (0.9 to 1.1 m) is typical.

Reciprocating Feeder. Coal feeders have the device which meters the coal from the combustion control system and delivers it to the built-in rotor. This device should have a bias means built in so that coal feed can be varied from one to another. This allows adjustment for segregation in fuel sizing from one feeder to another. The reciprocating feeder (Fig. 8.16) has one or more feed plates which travel back and forth on a spill plate. On the back stroke, coal drops from the hopper in front of the plate. On the forward stroke, coal is pushed off the spill plate onto the rotor. The amount of coal per stroke is adjusted by (1) the length of stroke of the feed plate through mechanical devices, (2) the number of strokes per minute through mechanical variable-speed devices, or (3) electronic devices such as variable-frequency power units to alternating-current (ac) motors or *silicon controlled rectifier* (SCR) control units to direct-current (dc) motors.

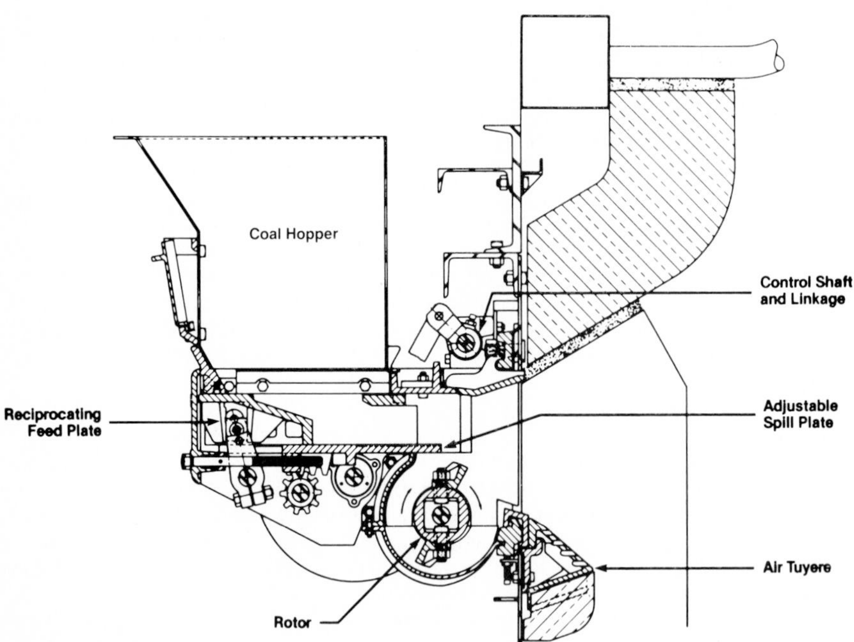

FIG. 8.16 Reciprocating coal feeder.

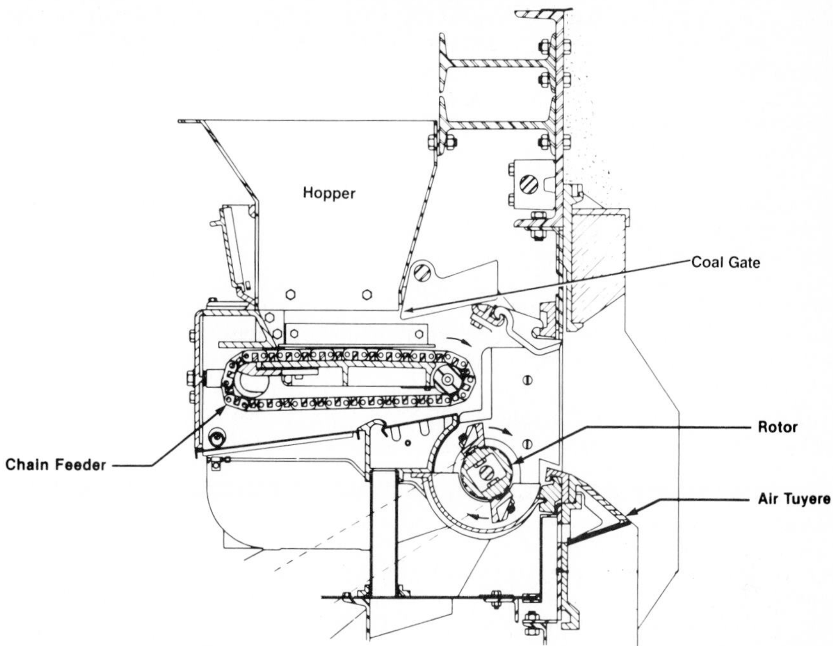

FIG. 8.17 Chain-type coal feeder.

Chain Feeder. A chain feeder (Fig. 8.17) has a drag conveyor deliver the coal out of a hopper, off the end of the chain, onto the rotor. The amount of coal is a function of the speed of the chain and the depth of the coal. The depth of coal is adjusted by the position of a gate above the chain. The speed can be regulated by an internal mechanical gearbox, external mechanical variable-speed device, or electronically, as described previously.

Drum Feeder. The drum feeder utilizes a revolving drum with semipockets to deliver the coal out of a hopper onto the rotor. As with the chain feeder, the amount of coal is regulated by the speed of the drum and the depth of coal coming out of the hopper on the drum. The vibrating feeder delivers the coal to the rotor out of the coal hopper on a vibrating conveyor. The amount of coal is adjusted by the frequency of vibration and the depth of coal on the conveyor.

Refuse Feeder. Refuse feeders differ from coal feeders in that the metering devices to control the rate of feed are not located at the feeder in the furnace wall, but upstream. Thus, refuse feeders are termed *distributors* since their sole purpose is to spread the fuel evenly over the grates. Refuse is a very difficult fuel to handle and keep flowing out of a hopper without bridging or pluggages. Therefore, special metering devices are installed upstream of the distributors to control the rate of fuel feed. Screw conveyors, drag conveyors, and drum feeders are most commonly used for this purpose, under hoppers very carefully designed with an ever-expanding cross section as they extend down to the conveyor.

Air-swept spouts are most commonly used as distributors of refuse into the furnace. Since the metering devices are upstream, the fuel falls downward to the distributors and must be fed to the furnace at the rate at which it is received. The air-swept spout picks up the refuse on an airstream, sweeping the floor of the spout and directing the refuse into the furnace. A rotating damper in the air supply to the spout can continuously vary the quantity of air to the spout and thus affect longitudinal distribution. Close-spaced air jets immediately under the spout can also aid in both lateral and longitudinal distribution.

Mechanical rotors much like those on coal feeders are also used to distribute refuse into the furnace, but without the built-in metering devices. High pressure, close-spaced air jets are also installed to aid distribution. The air-swept spout is much more popular because of its simplicity.

Combination Feeder. There are times when it is desirable to burn both coal and refuse fuels on the same unit either singly or in combination. To deliver either fuel to the furnace at the optimum location, a combination coal/refuse feeder would be applied (Fig. 8.18). The combination feeder has all the elements of the coal feeders and air-swept spouts described previously incorporated into a single unit.

Types of Grates

A number of grate types are utilized for spreader-stoker firing. They all serve the same purpose: they provide a floor on which the fuel can burn, a means of distributing air evenly through the grates, and a method of discharging the ashes that accumulate on the grate from the consumed fuel. The first purpose is self-

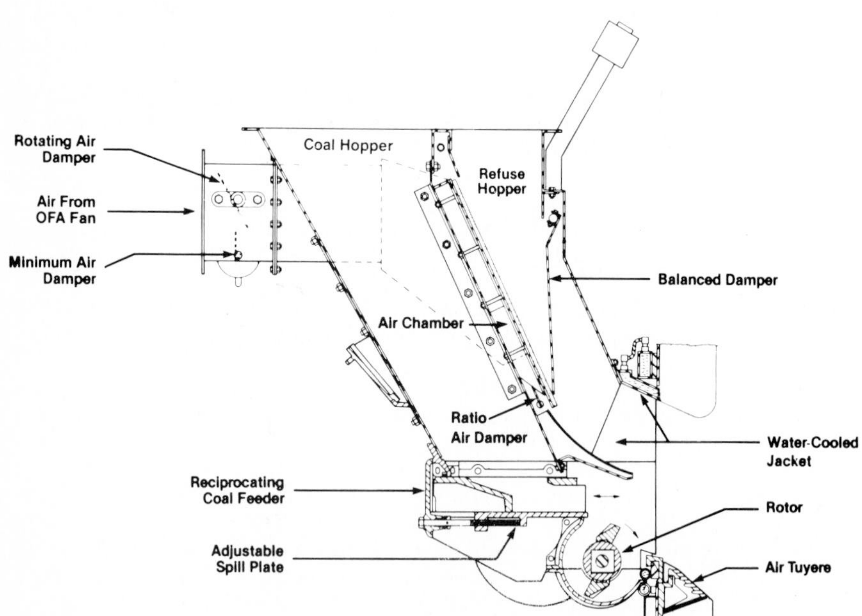

FIG. 8.18 Combination coal and refuse feeder.

evident, except to mention that the openings in the grate should be designed to minimize siftings of ash and fuel falling through the grate.

The second reason is most important to the efficient burning of the fuel. The fuel feeders are designed to spread fuel evenly over the grate surface. The air must also be distributed evenly to provide the correct amount of oxygen for the burning fuel. To accomplish this (and account for a changing ash bed depth, which has a low resistance), the grate should have an air-admitting-hole arrangement designed for high resistance across the air-admitting grate surface. A delta pressure range from 1.5 to 2.5 in H_2O (374 to 623 Pa) with ambient underfire air is recommended. With preheated underfire air, a delta pressure range from 2.0 to 3.5 in H_2O (498 to 872 Pa) is recommended. These pressures are at the maximum design rating of the boiler. The third item, ash discharge, is a function of the type of grate.

Stationary and Dumping. The intermittent cleaning types of grates are stationary and dumping (Fig. 8.19). The stationary grate is rarely used because of hazards to the operator when removing ashes through the open fire door and the resultant exposure directly to the furnace. It is also extremely difficult to clean fires without creating a smoky fire condition of high opacity.

The manual dumping type of grate has one section for each fuel feeder. To remove ashes, the fuel feed is stopped in front of one of the grate sections, to allow complete burnout of the combustible. The air supply to that section is then cut off with appropriate dampers, and the ashes can be dumped into a pit below.

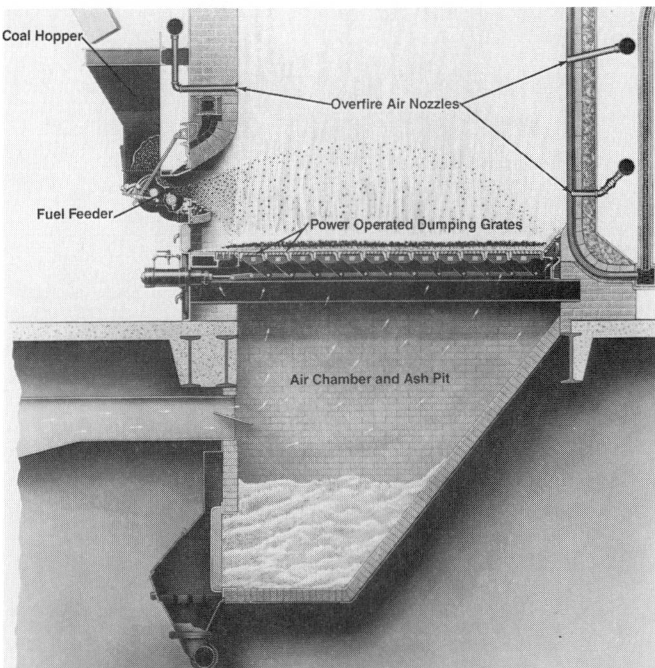

FIG. 8.19 Spreader stoker with dumping grates.

While this cleaning is done, it is necessary for the remaining sections to burn additional fuel to maintain the boiler rating. After the ashes are dumped, the fuel feeder is started again. When the fuel ignites, the undergrate damper to that section is reopened to resume normal combustion. It is very difficult to accomplish this cleaning process without high opacity. Consequently, the dumping grate is seldom used for coal burning, but is used for refuse burning having a high percentage of volatiles and low ash content.

Should a dump grate be used for coal, it would be for capacities of under 50,000 lb/h (6300 g/s) of steam and the heat release from the grate of no more than 450,000 Btu/h • ft^2 (1420 kW/m^2). When certain types of refuse are burned, the size of the unit fired might be as high as 150,000 lb/h (18 900 g/s) of steam, and the heat release from the grate (based on the amount of grate surface left when cleaning one section) might be a maximum of 1×10^6 Btu/h • ft^2 (3155 kW/m^2). The undergrate air temperature when coal is burned is normally ambient. Burning high-moisture refuse fuels (up to 50 percent by weight) requires preheated undergrate air temperatures up to 400°F (204°C).

Reciprocating. The reciprocating grate (Fig. 8.20) discharges ashes by a slow back-and-forth motion of moving grates alternating with stationary grates. Each row of grates rests on the row in front and has a raised nose which pushes the

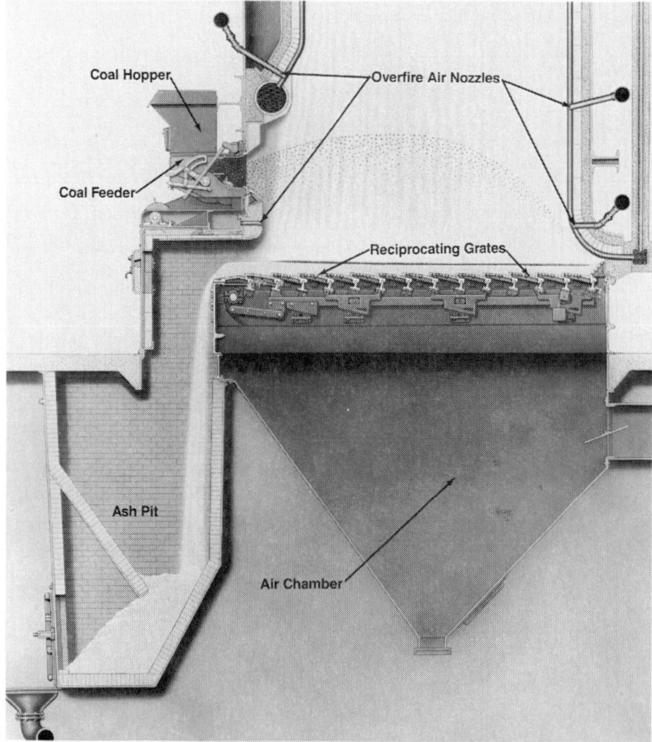

FIG. 8.20 Spreader stoker with reciprocating grates.

ashes forward and off the end of the grate into an ash hopper. It is far preferable to discharge the ashes at the front under the feeders. Since the fuel is thrown toward the rear, best control of ash discharge with minimum combustibles in the ashes can be achieved with front ash discharge from any type of conveying grate.

Because of the stepped nature of the reciprocating grate, it is used only for fuels with sufficient ash quantity to provide an adequate ash depth insulation on top of the grates. It is seldom used for refuse fuels, and undergrate air temperatures are ambient. It is generally used for boilers having steaming capacities up to about 75,000 lb/h (9450 g/s) of steam, with grate heat releases not over 650,000 Btu/h • ft^2) (2048 kW/m^2).

Vibrating. The vibrating or oscillating grate (Fig. 8.21) is suspended on flexing plates. Either an eccentric drive or rotating weights impart a vibrating action to the grate surface, which conveys ashes to the front and discharges them into an ash pit. The frequency of vibration is kept well below the natural frequency, to hold the forces on the support structure to a minimum. The rate of ash discharge is regulated by a timer system that controls the off time between vibrating cycles and the length of vibrating time.

Since the vibrating grate is flat, it is well suited to coal as well as refuse fuels. It is utilized on boilers up to about 100,000 lb/h (12 600 g/s) of steam on coal and 150,000 lb/h (18 900 g/s) of steam on refuse. The grate heat release should be limited to 650,000 Btu/h • ft^2 (2048 kW/m^2) on coal burning and 850,000 Btu/h • ft^2

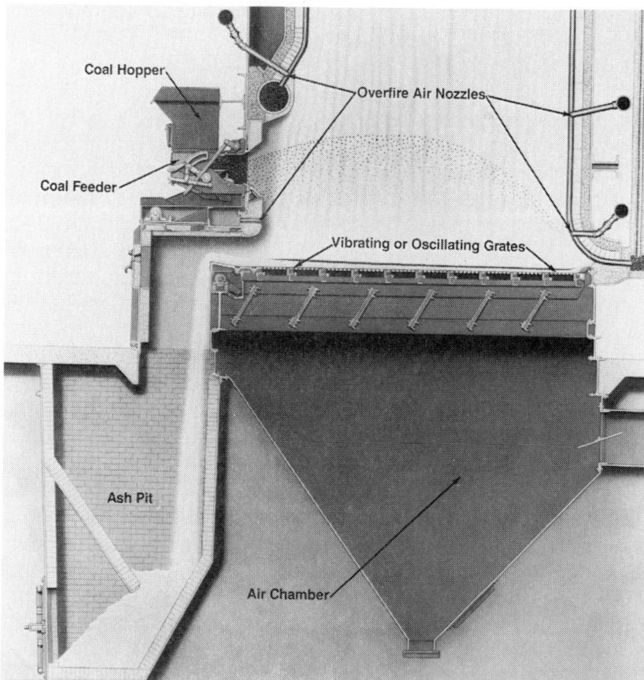

FIG. 8.21 Spreader stoker with vibrating or oscillating grates.

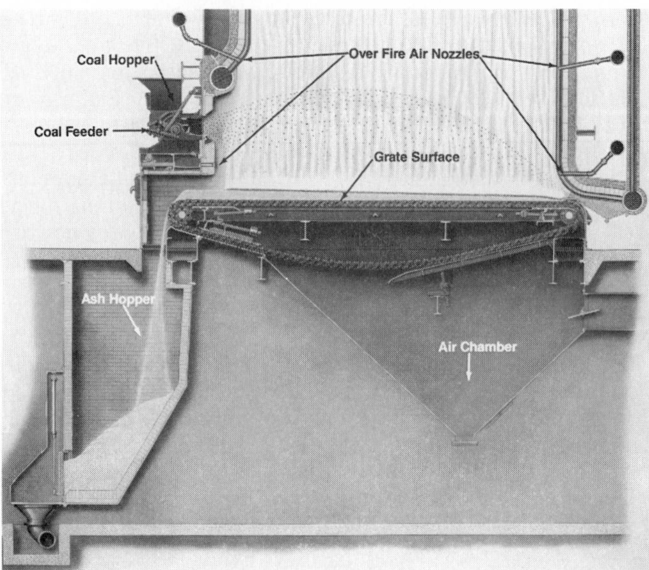

FIG. 8.22 Spreader stoker with traveling grates.

(2678 kW/m^2) on refuse burning. Preheated undergrate air can be used with temperatures up to 300°F (149°C) for burning coal and 400°F (204°C) for burning 50 percent moisture refuse.

Traveling. The traveling-grate spreader stoker (Fig. 8.22) is by far the most popular type, and there are numerous types of construction. The important consideration is that the design of the grate surface should provide a high resistance to airflow through the grate to ensure even distribution of air through the grate. The traveling grate moves forward to discharge the ashes at the front end under the fuel feeders. The return grate then passes underneath in the air chamber. The traveling grate is driven either mechanically or hydraulically through appropriate mechanical devices, electronic devices, or hydraulic flow-control valves.

When the ash content of the fuel is consistent, the speed of the grates can be regulated by the combustion controls to maintain an even ash-bed thickness. The grate bars can be made from cast iron with small amounts of alloy for heat and wear resistance, or from ductile iron for increased heat resistance and impact resistance. The traveling grate can be applied to boilers having ratings up to about 400,000 lb/h (50 400 g/s) of steam when burning coal, with grate heat releases up to 750,000 Btu/h • ft^2 (2366 kW/m^2). When refuse fuels are burned, the boiler size can approach 600,000 lb/h (75 600 g/s) of steam. The grate heat release on refuse fuels can be 1×10^6 Btu/h • ft^2 (3155 kW/m^2) with moisture content up to 55 percent and 1.25×10^6 Btu/h • ft^2 (3940 kW/m^2) with moisture content up to 30 percent. The undergrate air temperature can range from ambient up to 350°F (177°C) for most bituminous coals, 400°F (204°C) for lignite, and 550°F (288°C) for refuse with up to 55 percent moisture. Fuels with the higher moisture require the higher air temperatures.

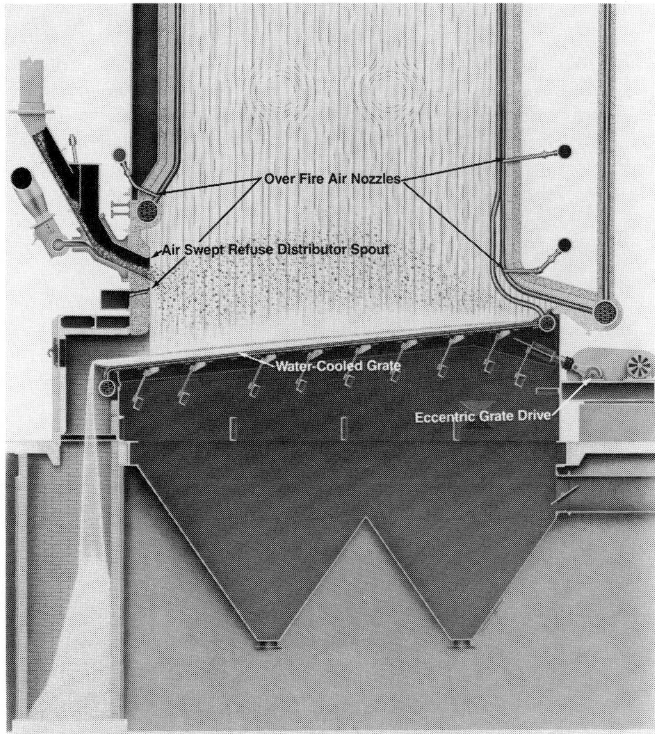

FIG. 8.23 Spreader stoker with water-cooled vibrating grates.

Vibrating, Water-Cooled. The water-cooled vibrating grate spreader stoker (Fig. 8.23) is used for refuse burning, but could conceivably be used for coal firing too. A water-cooled grid on which a grate surface is fastened is mounted on a frame. The whole assembly is supported on flexing plates on an angle of 6° down toward the ash discharge end. An eccentric drive vibrates the grate assembly through a timing circuit that controls the amount of off and on time. The vibrating action moves the ash to the front of the grate and into an ash pit. The vibration frequency is kept well below the natural frequency.

The water-cooled grid is cooled by tying into the boiler water-circulation circuit on a separate forced-cooled system. Which method to select is dictated by the needs of the individual project. This type of grate can be applied to refuse-fired boilers having capacities up to 600,000 lb/h (75 600 g/s) of steam. The grate heat-release rates burning refuse would be comparable to the traveling-grate spreader stoker. With this type of water-cooled grate, undergrate air temperatures up to 600°F (316°C) can be accommodated.

Overfire Air

The overfire-air system on spreader stokers has three functions: to provide oxygen and turbulence to mix the fuel and oxygen in the lower high-temperature zone of the furnace for complete burnout, to distribute the fuel and assist in distribution when the design dictates the use of air for this purpose, and to provide cooling air for mechanical fuel feeders. This air can come from a common fan that would have static pressure capabilities of 25 to 30 in H_2O (6.2 to 7.5 kPa) to provide the necessary energy for turbulence or fuel distribution.

Overfire-air-turbulence nozzles are placed in rows in the front and rear walls of the furnace. The design should place the air nozzles at elevations which will ensure mixing of the furnace gases in the high-temperature zones, so that the combustion process will be completed in the furnace. Air-nozzle spacing in each row should be on centers of 10 to 12 in (254 to 305 mm) for complete coverage of the furnace.

The size of the air nozzles in the various rows is a function of the depth of penetration desired and the amount of overfire air required. Bituminous-coal-fired boilers are designed with 15 to 25 percent overfire air. Lignite-fired boilers are designed with 20 to 25 percent overfire air. Refuse, which has the highest volatility, utilizes the highest amount of overfire air; 25 percent of total air would be a minimum amount with refuse firing, and some systems are designed with as much as 50 percent of the total air as overfire. The overfire-air temperature may be either ambient or the temperature of the undergrate air. The choice of temperature is generally a function of air heater and boiler system design.

Fly Carbon Reinjection

The particulate leaving the furnace of a spreader stoker can contain considerable combustible material and is thus termed *fly carbon*. It is beneficial to return some percentage of this fly carbon to the furnace for burning to improve the boiler efficiency. This is accomplished by the conveying system, which transfers the fly carbon from hoppers through a pipeline to the furnace using air as the energy source. By using a nozzle and venturi arrangement, air at up to 25 in H_2O (6.2 kPa) can create a suction to draw the fly carbon into the pipeline and pressure-convey it to the furnace.

The air supply is generally from the overfire-air fan; however, a separate fan can be used. The fly carbon can be collected in hoppers under the boiler's steam-generating section, economizer, air heater, or mechanical dust collector, although continuous evacuation of the hopper is required. The amount of reinjection sets the increase in boiler efficiency. The particulate collected in an electrostatic precipitator or fabric filter should not be returned to the furnace. The formulas and curves for carbon loss and reinjection most generally accepted are those proposed by the American Boiler Manufacturers Association.

ACKNOWLEDGMENT

Detroit Stoker Co. supplied all drawings and photos appearing in this chapter.

CHAPTER 1.9
COOLING TOWERS

John C. Hensley
Marketing Services Manager, Marley Cooling Tower Co.
Mission, KS

INTRODUCTION

In the power industry, energy in the form of heat is transformed to energy in the form of electricity. Unfortunately, this transformation is not accomplished on a one-to-one ratio. Although designers continuously seek newer and better ways to improve overall system efficiency, considerably more units of heat must be input than are realized as equivalent units of electric output. System equilibrium requires that this excess heat be dissipated—ultimately to the atmosphere.

The traditional vehicle used to transport the lion's share of this waste heat to atmosphere is water, using the principle of evaporation as the primary mechanism of heat transfer. Where available and ecologically acceptable, water from rivers, lakes, or oceans is often used on a once-through basis, schematically

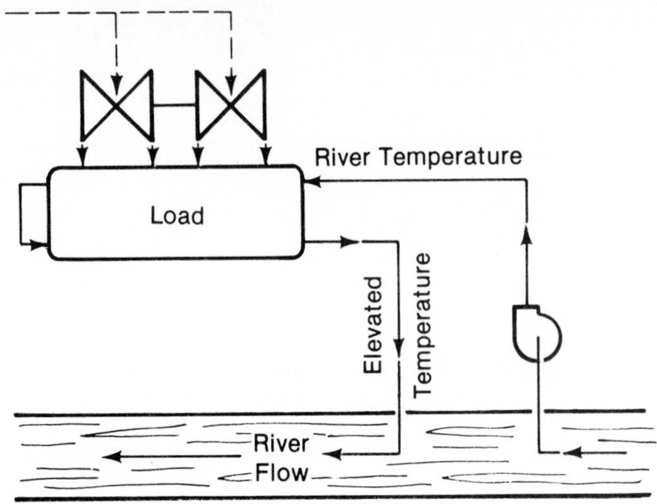

FIG. 9.1 Once-through system.

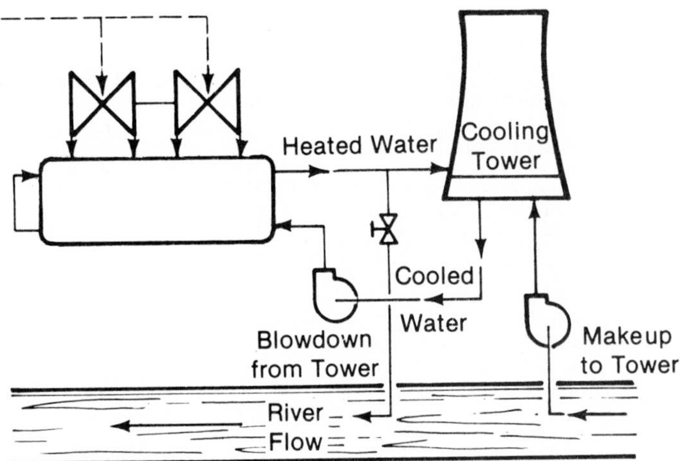

FIG. 9.2 Closed-circuit system.

shown in Fig. 9.1. Taken from the river, water absorbs heat from the steam condenser and returns to the river (at an elevated temperature) downstream from its point of extraction.

In an increasing number of cases, because of qualitative and quantitative restrictions regarding the use of natural waterways, plants make use of a closed-circuit system typical of that depicted in Fig. 9.2. As shown, the system's dependence on the river is limited to the requirement for a supply of makeup water and, perhaps, as a point of discharge of blowdown. In these cases, the entire heat loads are dissipated by cooling towers, the variety, application, and operation of which are the subject of this chapter.

the point of design, as evidenced by Eq. (9.1*a*) and (9.1*b*), the specifier must also establish the design wet-bulb temperature.

SELECTING THE DESIGN WET-BULB TEMPERATURE

The design wet-bulb temperature must be chosen on the basis of conditions existing at the site proposed for a cooling tower. Although performance analyses have indicated that most installations based on wet-bulb temperatures that are exceeded no more than 5 percent of the total hours during a normal summer give satisfactory results, the wet-bulb temperature ultimately chosen should permit the tower to produce the optimum cold-water temperature at or near the time of peak load demand.

Air temperatures, wet-bulb as well as coincident dry-bulb, are routinely measured and recorded by the U.S. Weather Bureau, worldwide U.S. military installations, airports, and various other organizations to whom anticipated weather patterns and specific air conditions are of vital concern. One compilation of such data is called *Engineering Weather Data,* available through the U.S. Printing Office, Washington, with satellite offices existing in many major cities.

However, the wet-bulb temperature determined from publications of this sort represents the ambient for a geographic area, and it does not take into account localized heat sources which may artificially elevate that temperature at a specific site. Before a final decision is made concerning the proper design wet-bulb temperature, it is good practice to take simultaneous wet-bulb readings at the proposed tower site and at other open, unaffected locations at the same plant. These readings should be compared to one recorded at the same time at the nearest source of weather data (airport, weather bureau, etc.), and the apparent design wet-bulb temperature should be adjusted accordingly.

The importance of selecting the correct design wet-bulb temperature is supported by Fig. 9.4, which shows the direct relationship between wet-bulb and

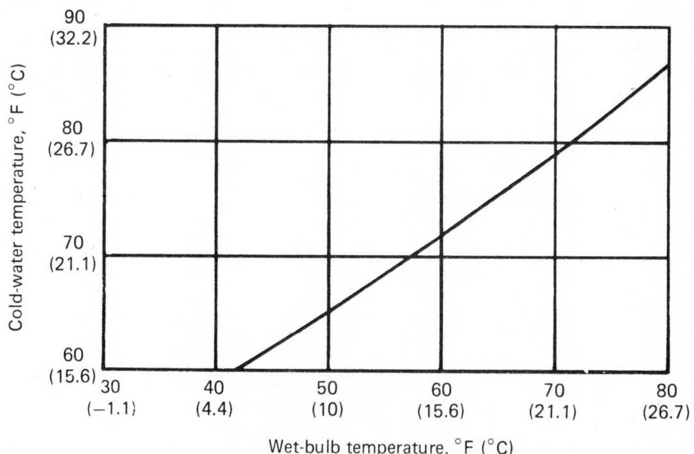

FIG. 9.4 Typical performance curve.

cold-water temperatures. If the actual wet-bulb temperature is higher than anticipated by design, then warmer cold-water temperatures will result. Conversely, if the actual wet-bulb temperature is lower than expected, then the user will probably have purchased a cooling tower larger than is needed. This and other considerations of heat-load configuration which affect tower size are discussed later in this chapter.

TYPES OF COOLING TOWERS

Although cooling towers are designed and manufactured in numerous types, sizes, and configurations, ranging from the relatively small factory-assembled types to the structurally imposing hyperbolic towers that dominate many plant sites, those capable of handling loads of the magnitude developed in the power generation industry are comparatively few. They are characterized by shape, flow relationship of air and water, manner of producing airflow, and heat-transfer methodology generally, as follows.

Induced Draft vs. Forced Draft

All the cooling towers pictured in this chapter are of the induced-draft type, in which the air-moving device is situated in the leaving airstream, thereby inducing air through the tower. In the forced-draft type, the air-moving device is in the entering airstream, forcing air through the tower. Except for those shown in Figs. 9.5 and 9.6, the induced-draft towers are of the mechanical-draft type, utilizing motor-driven fans to generate air movement.

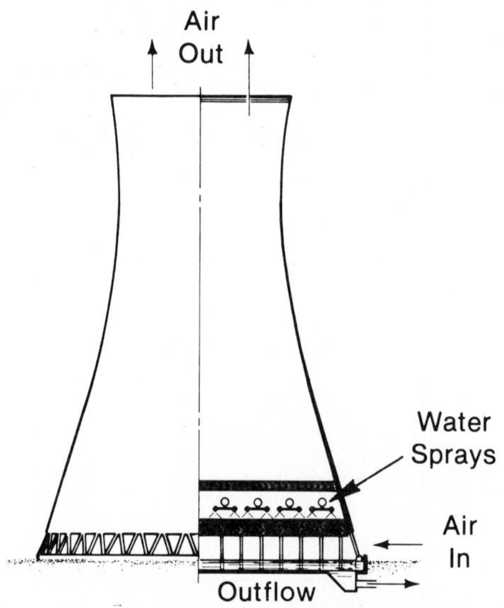

FIG. 9.5 Counterflow natural-draft tower.

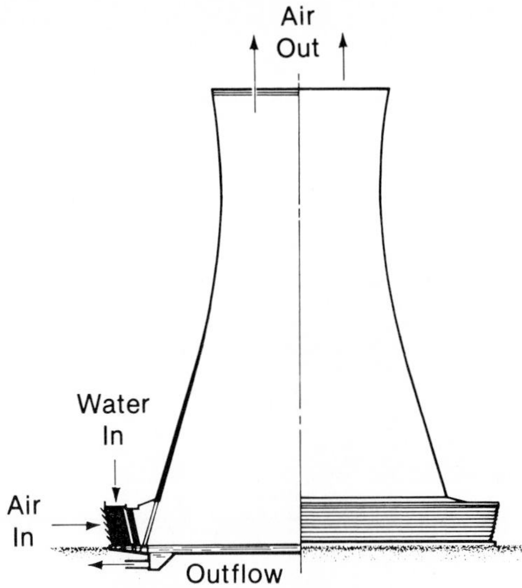

FIG. 9.6 Crossflow natural-draft tower.

Figures 9.5 and 9.6 are natural-draft types, sometimes referred to as *hyperbolic* because of their distinctive shape. Airflow through these towers is produced by the density differential that exists between the heated (less dense) air inside the stack and the relatively cool (more dense) ambient air outside the tower. Typically, these towers tend to be very large [250,000 gal/min (15 800 L/s) or more] and often in excess of 500 ft (150 m) in height.

Although natural-draft towers are typically more expensive than mechanical-draft types, they are used significantly in the field of power generation, where long amortization periods allow sufficient time for the absence of fan power to recoup the differential cost. However, because natural-draft towers perform most effectively in ambient air of consistently higher relative humidity (greater density), many plants located in arid and/or higher-altitude regions find mechanical-draft towers more applicable.

Crossflow vs. Counterflow

Crossflow towers (Fig. 9.7) have a fill configuration through which the air flows horizontally, across the downward fall of water. Water to be cooled flows into hot-water inlet basins located atop the fill areas and is distributed to the fill by gravity through metering orifices in the floor of those basins. Since the distribution basins are normally open to view, crossflow towers are considered to be somewhat easier to maintain than counterflow towers.

In *counterflow* towers (Fig. 9.8), water is delivered to an internal piping system at sufficient pressure to effect an even and constant spray pattern from the nozzles. Air movement is vertically upward through the fill, counter to the downward fall of water. Counterflow towers of current design give testimony to the technological advances which have occurred in the cooling-tower industry. Until the mid-1970s, the need for relatively high-pressure spray systems as well as greater static pressures in these towers caused them to require significantly more pumping head and fan power than their crossflow counterparts. However, devel-

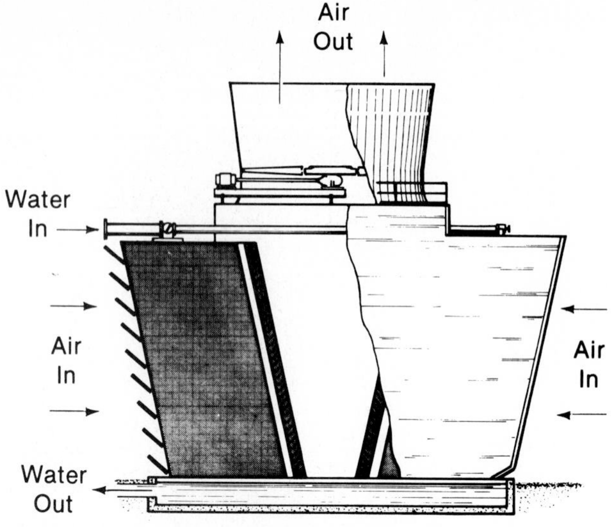

FIG. 9.7 Double-flow crossflow tower.

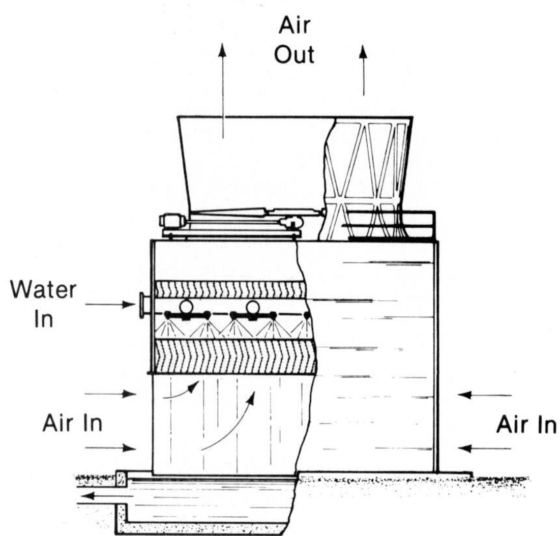

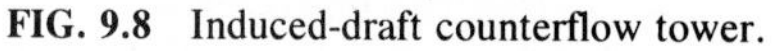

FIG. 9.8 Induced-draft counterflow tower.

opment of low-pressure, gravity-related distribution systems, plus refinements in fill and drift-eliminator techniques, have reversed this situation in many cases.

Accordingly, specifiers primarily concerned with the long-term cost of operation should avoid language that tends to exclude either type. Unless there is an overriding preference for one over the other in terms of maintenance procedures, manufacturer's reputation, or system peculiarities which dictate a particular choice, a better practice is to be guided by the "cost of ownership." In its simplest determination, this practice combines the initial cost of each bidder's offering with the long-term cost of energy reflected in the proposed pump head and fan power. If all else is considered to be equal, the lowest total indicates the bidder of choice.

Hybrid-Draft Towers

Hybrid-draft towers (Fig. 9.9) appear to be natural-draft towers with relatively short stacks. As seen in Fig. 9.10, however, they are also equipped with mechanical-draft fans to augment airflow. Because of this, they are often called *fan-assisted natural-draft towers*.

The primary purpose of the hybrid design is to minimize the horsepower required for air movement—with the least possible impact on stack cost. The fans, for example, may need to be operated only during periods of high loads and/or high ambient temperatures. In localities where a low-level discharge of the tower

FIG. 9.9 Fan-assisted natural-draft tower.

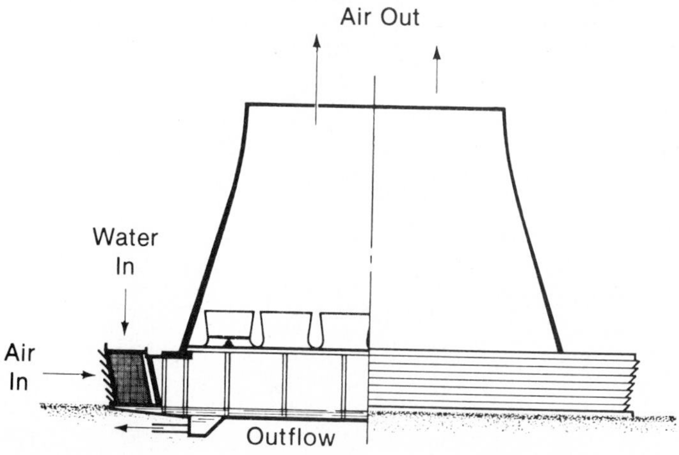

FIG. 9.10 Cutaway of Fig. 9.9 tower.

plume may prove to be unacceptable, however, the elevated discharge of a fan-assisted natural-draft tower can become sufficient justification for its use.

Rectilinear vs. Round

Rectilinear towers (Fig. 9.11) are cellular in design, increasing linearly to the length and number of cells necessary to accomplish a specific thermal perfor-

FIG. 9.11 Rectilinear counterflow tower.

FIG. 9.12 Round mechanical-draft towers.

mance. In contrast, Fig. 9.12 shows two towers of the round mechanical-draft type located at one plant, each serving units of approximately 550-MW capacity. The foreground tower is counterflow; the background tower is crossflow. As the name implies, these towers are essentially round in plan configuration, with the fans clustered symmetrically around the center point of the tower. Such towers can handle enormous heat loads with relatively minimal site-area impact—as evidenced by their proximity to each other and to the plant.

PURPOSE AND TYPES OF FILL

Although cooling-tower fill is often referred to as a *heat-transfer surface,* such terminology is not true in its strictest sense. The heat-transfer surface in the classic cooling tower is actually the exposed surface of the water itself. Fill is merely a medium by which more water surface is caused to be exposed to the air for a longer time. Hence its use increases both the rate and the amount of heat transfer far beyond the capability of the spray-filled tower shown in Fig. 9.3.

Splash Type

The two basic types of fill used are the splash type (Fig. 9.13) and the film type (Fig. 9.14). Splash-type fill causes the flowing water to cascade through succes-

FIG. 9.13 Splash-type fill.

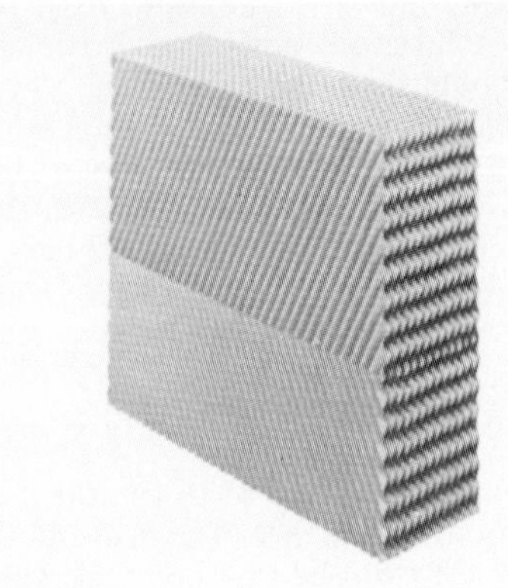

FIG. 9.14 Film-type fill.

sive elevations of parallel splash bars, continuously breaking up the flowing stream of water into relatively small droplets. Of equal importance is the increased time of air-water contact brought about by repeated interruption of the water's flow progress.

Although either type of fill may be installed in towers of either crossflow or counterflow configuration (positioned as shown in Figs. 9.7 and 9.8, respectively), splash-type fill lends itself most usefully to towers of crossflow design. Since the movement of water within a cooling tower is essentially vertical, splash-type fill must be arranged with the wide dimension of the splash bars situated in a horizontal plane, to achieve maximum dispersal and retardation of the water. The resultant small vertical dimension, plus relatively wide vertical spacing of the splash bars, therefore provides the least opposition to airflow in the horizontal direction typical of crossflow. Its use in counterflow towers, where airflow is essentially vertical, requires an increase in fan power, which is normally considered unacceptable in today's market.

Film Type

Film-type fill has gained prominence in the cooling-tower industry because of its ability to expose greater water surface within a given packed volume. As can be seen in Fig. 9.15, water flows in a thin "film" over vertically oriented sheets of fill which are spaced appropriately for either vertical or horizontal air passage. Therefore, film-type fill is equally effective in either type of cooling tower.

For purposes of clarity, the fill sheets in Fig. 9.15 are shown to be flat. In practice, these sheets [predominantly of polyvinyl chloride (PVC) material] are molded into unique patterns to create a certain amount of turbulence within the airstream and to further extend the exposed water surface. The fill pack indicated in Fig. 9.14, for example, is manufactured in a cross-corrugated configuration, with the contact points of the corrugations providing the proper spacing. Other shapes may include regular protrusions to maintain spacing.

Although film-type fill currently predominates in the cooling-tower industry, there is no present indication that splash-type fill will become obsolete in the foreseeable future. The narrow passages afforded by close spacing of fill sheets

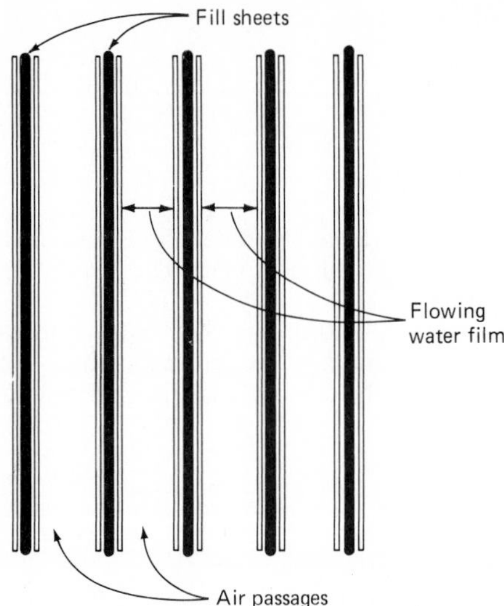

FIG. 9.15 Film-fill concept.

makes film fill very sensitive to water quality. High turbidity, leaves, debris, or the presence of algae, slime, or coagulants can diminish passage size or, in extreme cases, lead to complete plugging. In geothermal applications, therefore, as well as installations tapping turbid water as a source of makeup, splash fill continues to be the fill of choice.

FACTORS AFFECTING TOWER SIZE AND PERFORMANCE

As one might gather from a study of Fig. 9.16, several parameters of choice affect the size of a cooling tower. Among these are the system-related aspects of heat load, range, and approach; and wet-bulb temperature, which is site-related. In

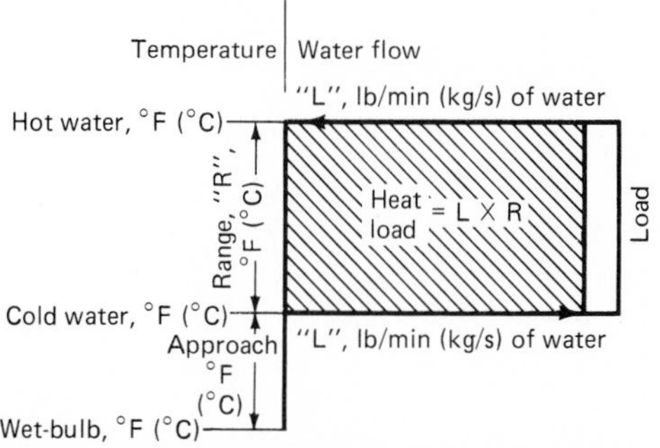

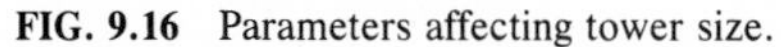

FIG. 9.16 Parameters affecting tower size.

addition to these parameters, site-related interference must be considered, as well as recirculation, which is influenced by tower design and orientation. The effect of all these factors on tower size is discussed here.

Heat Load, Range, and Approach

As seen in Fig. 9.17, at a given range, approach, and wet-bulb temperature, tower size varies directly and linearly with heat load. The larger the heat load, the bigger the required tower will be.

Tower size varies inversely with range (Fig. 9.18). Increasing the range increases the temperature differential between the incoming hot water and the incoming air, thereby increasing the driving force for enthalpy exchange and reducing the size of the tower.

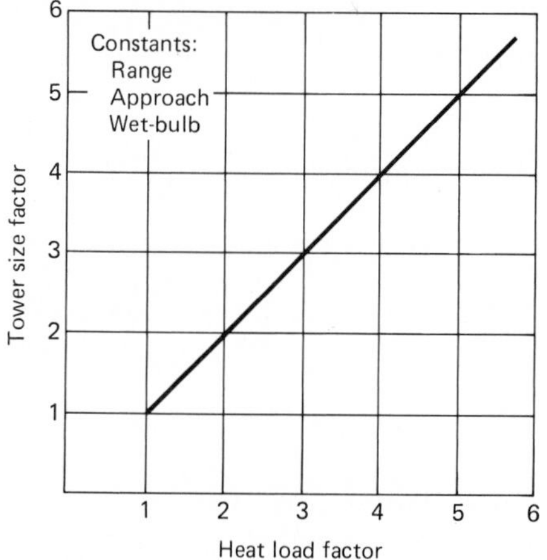

FIG. 9.17 Effect of heat load on tower size.

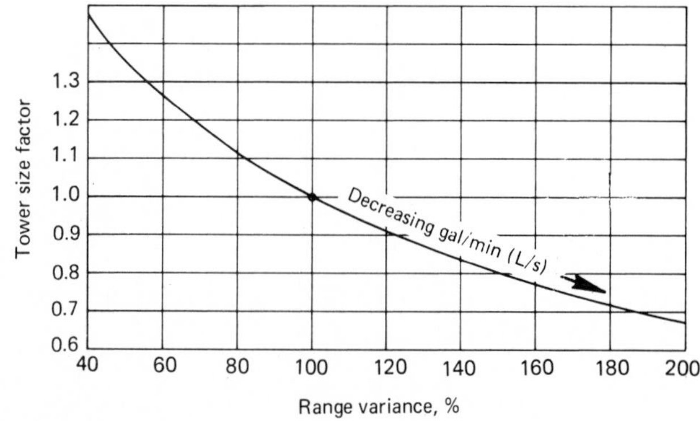

FIG. 9.18 Effect of varying range on tower size.

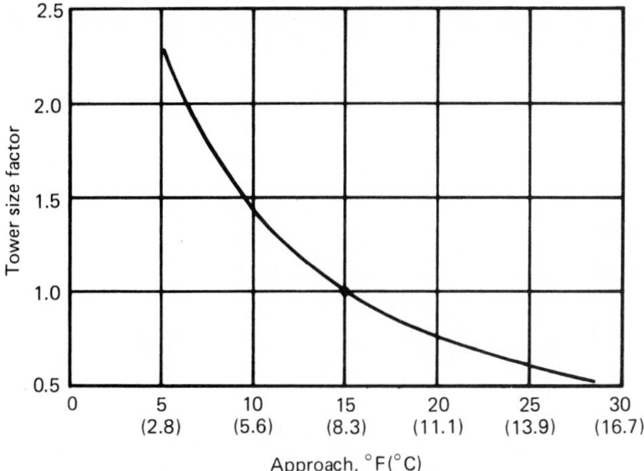

FIG. 9.19 Effect of chosen approach on tower size.

Tower size varies inversely with approach. A longer approach requires a smaller tower (Fig. 9.19). Conversely, a smaller approach requires an increasingly large tower, and at an approach of 5°F (2.8°C), the effect on tower size begins to become asymptotic. For that reason, it is not customary in the cooling-tower industry to guarantee approaches of less than 5°F (2.8°C).

Wet-Bulb Temperature

Tower size varies inversely with the design wet-bulb temperature. When heat load, range, and approach values are fixed, increasing the design wet-bulb temperature reduces the size of the tower (Fig. 9.20). Since the primary exchange of heat in a cooling tower is through evaporation, the fact that air's ability to absorb moisture increases with higher temperatures accounts for this size reduction.

In attempting to understand these effects, readers must not allow themselves to become confused regarding the relative relationship of cold-water tempera-

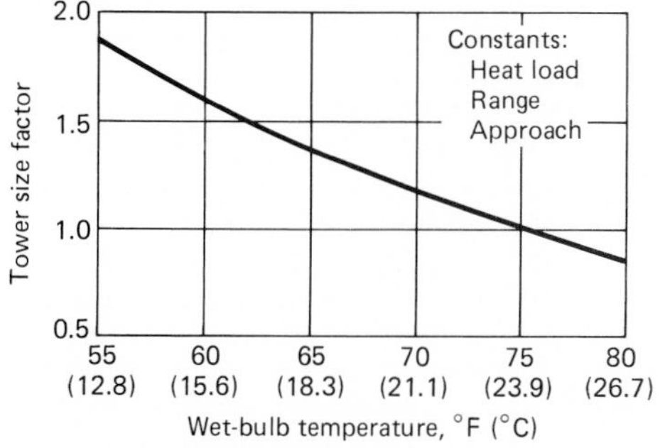

FIG. 9.20 Effect of wet-bulb temperature on tower size.

ture, wet-bulb temperature, and approach. The inverse effect of design wet-bulb temperature on tower size applies *only* when approach values are held constant, that is, when the cold-water temperature is altered in the same direction as the changed wet-bulb temperature and by the same number of degrees.

Working an Example

To establish a base example, assume a fixed heat load, range, and water flow rate, with a cooling tower required to cool the water to 90°F (32.2°C) when the incoming air wet-bulb temperature is 75°F (23.9°C) [15°F (8.3°C) approach]. This condition essentially represents unity, as defined by the curves in Figs. 9.19 and 9.20. Also assume that, for some reason, it is determined that a more appropriate design wet-bulb temperature would be 80°F (26.7°C).

If the approach is held constant, Fig. 9.20 reveals that the elevation in wet-bulb temperature will have afforded an approximate 20 percent reduction in tower size—but the specifier must also realize that *the design cold-water temperature will have also risen to 95°F (35°C)*. If the cold-water temperature were held constant, as is usually the case, this 20 percent reduction in tower size would be more than offset by the 45 percent increase in size caused by reducing the approach (Fig. 9.19) from 15°F (8.3°C) to 10°F (5.6°C). As a result, tower size would be approximately 25 percent larger than the base case.

There are two principal reasons why the design wet-bulb temperature in the previous example might have been reconsidered. They are both wind-related in their effect and so require an understanding of the flow pattern of winds as they encounter structures such as cooling towers. In these cases, the normal path of the wind is disrupted, and a reduced-pressure zone forms on the lee side (downwind) of the structure. Obviously, the wind will try to fill this "void" by the shortest possible route, including flow around the ends and over the top of the tower.

Recirculation and Interference

To the extent that wind flow is over the top, some of the warm, saturated effluent leaving the tower is trapped in this downflow, and recirculation occurs (Fig. 9.21). Since all the heat removed from the water is contained in this effluent plume, the net effect of recirculation is an artificial elevation of the wet-bulb temperature entering the tower and a commensurate elevation in cold-water temperature.

The amount of wind forced to take the "over-the-top" route depends on the shape of the tower. If the projection of the tower opposing the wind is relatively tall and narrow (that is, the end of a long tower), then the easiest flow path for the wind will be around the sides of the tower to fill a relatively small void at the far end. In this case, what little recirculation is induced will affect an almost negligible portion of the tower as a whole. However, when the long dimension of the tower is broadside to the wind, the results are quite different.

Figure 9.22 represents a measured example of what can happen in this situation. Note that the air inlets on the north face of the tower are experiencing wet-bulb temperatures up to 8°F (4.4°C) higher than those seen by the south face. The resultant increase in enthalpy of the entering air has, of course, degraded thermal performance significantly.

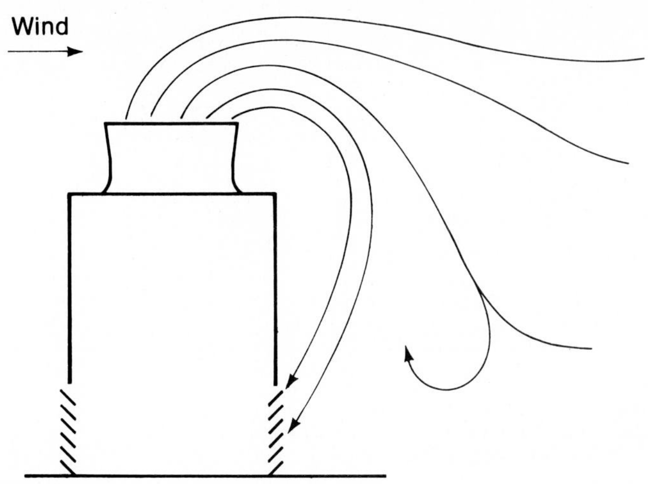

FIG. 9.21 Tower recirculation due to wind.

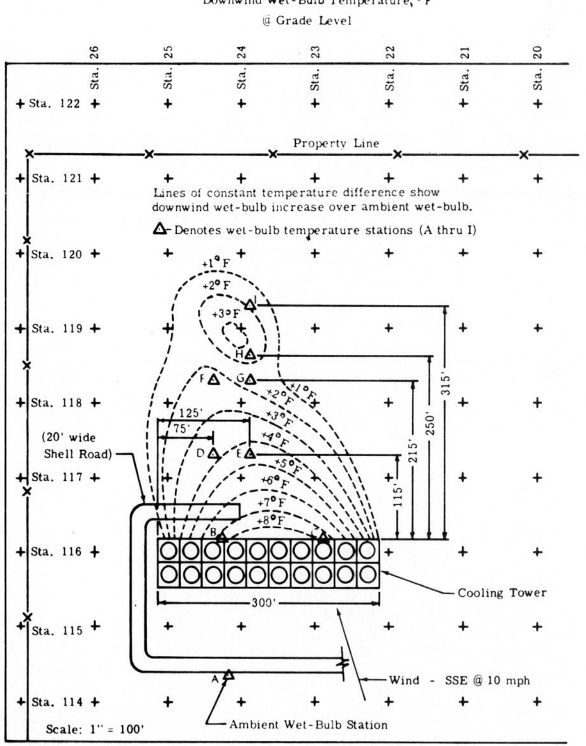

FIG. 9.22 Downwind wet-bulb contour of large existing tower. Note: °C = (°F − 32)/1.8; m = ft × 0.3048; mm = in × 25.4; km/h = mph × 1.61.

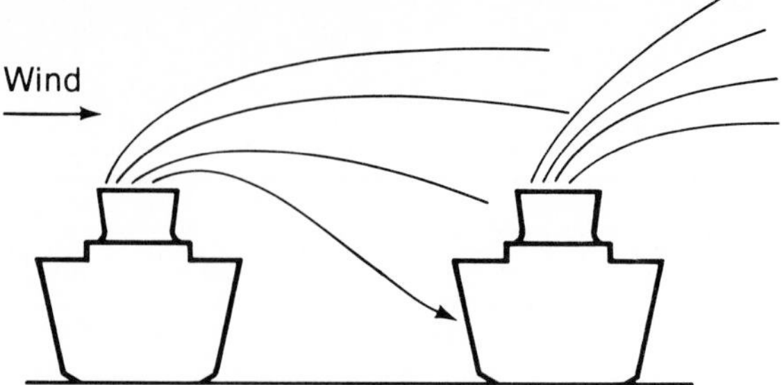

FIG. 9.23 Interference caused by upwind tower.

The concept of interference (Fig. 9.23) is made obvious when one considers the probable effect of siting a proposed tower about 300 ft (91 m) downwind of the tower in Fig. 9.22. No doubt the new tower would experience wet-bulb temperatures some 2 to 3°F (1.1 to 1.7°C) higher than local weather bureau data would lead the designer to expect and, unless anticipated in the design wet-bulb chosen for the new tower, would result in higher-than-desired cold-water temperatures.

IMPORTANCE OF PROPER SITING

The performance of a cooling tower is thus dependent on the quantity and thermal quality of the entering air. External influences that raise the entering wet-bulb temperature or restrict airflow to the tower reduce its effective capacity. It is the responsibility of the specifier to situate the tower such that these performance-influencing effects will be minimized. Recirculation is minimized by orienting rectilinear towers such that the primary-air inlet faces are parallel to the prevailing wind direction coincident with the design wet-bulb condition (Fig. 9.24).

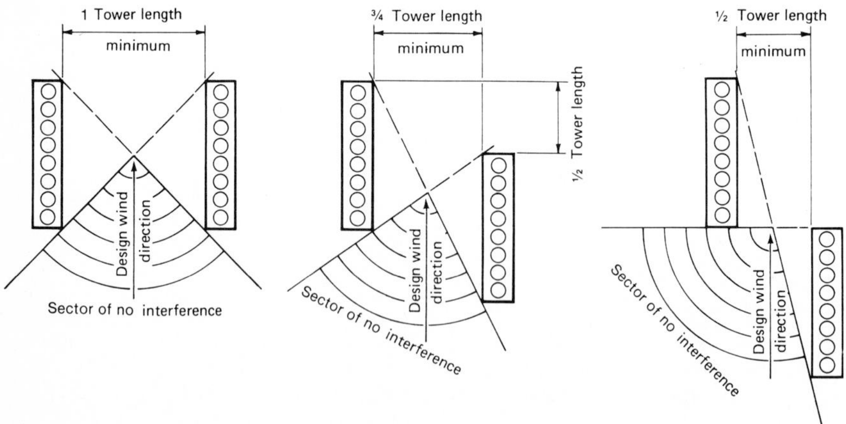

FIG. 9.24 Tower orientation in a prevailing longitudinal wind.

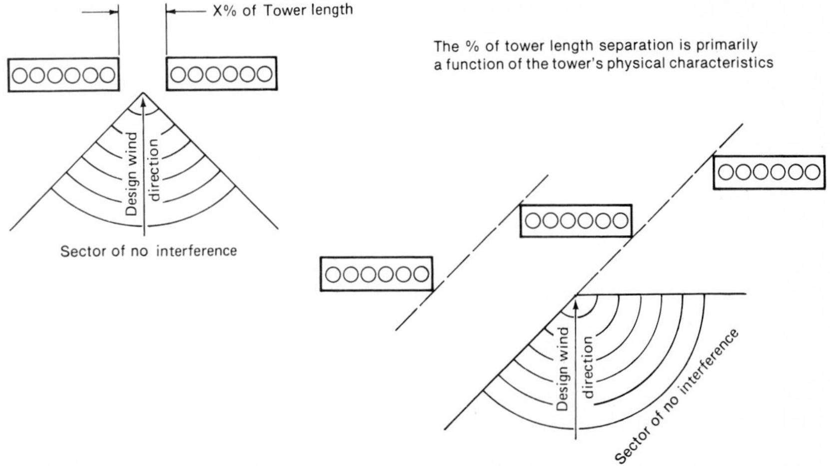

FIG. 9.25 Tower orientation in a prevailing broadside wind.

Because of the restricted siting areas available in some plant locations, however, users may have no choice but to orient towers broadside to a prevailing wind and to adjust the design wet-bulb temperature accordingly (Fig. 9.25). The amount of adjustment necessary may be reduced by recognizing that the recirculation potential increases with tower length, and by splitting the tower into multiple units of lesser length, with a significant ventilating airspace in between. For example, if the tower in Fig. 9.22 had been installed as two 150-ft (46-m) towers in line, with a 50-ft (15-m) space between the ends of the towers, the net amount of recirculation might well have been halved.

Because wind flow is far less disrupted by a circular or cylindrical shape, round towers of the type depicted in Fig. 9.12 experience very little recirculation, and because of their extreme height hyperbolics (Figs. 9.5 and 9.6) suffer none at all. Although round towers are relatively immune to recirculation, they are *not* unaffected by interference. Care still has to be taken to ensure that they are not situated downwind of a heat source, such as another tower.

MATERIALS OF CONSTRUCTION

To establish a basis for the selection of standard construction materials, the following "normal" water conditions have been arbitrarily defined:

- A circulating water with a pH between 6 and 8; a chloride content (as NaCl) below 750 ppm; a sulfate content (SO_4) below 1200 ppm; a sodium bicarbonate ($NaHCO_3$) content below 200 ppm; a maximum temperature of 120°F (48.9°C); no significant contamination with unusual chemicals or foreign substances; and adequate water treatment to minimize corrosion and scaling.
- Chlorine, if used, to be added intermittently, with a free residual not to exceed 1 ppm and maintained for relatively short periods.
- An atmosphere surrounding the tower no worse than "moderate industrial," where rainfall and fog are only slightly acid and they do not contain significant chlorides or hydrogen sulfide (H_2S).

Water conditions falling outside these limits would warrant an investigation of the combined effect of all the constituents on each component material. In many cases, very few components require a change in materials. Wood and plastic components, for example, are very tolerant of chemical excursions far beyond these limits. Conversely, carbon-steel items are relatively unforgiving of all but the most limited variations.

Wood Components

Because of its availability, relative low cost, and durability under the severe operating conditions encountered in cooling towers, wood continues to be one of the most utilized materials. Although California redwood and coastal region Douglas fir predominate, various other wood species may be used in cooling towers, provided they exhibit proven durability in severe exposures and meet the structural requirements of the installation.

Regardless of species, however, any wood used in cooling-tower construction should be treated with a reliable preservative to prevent decay. Currently, the preferred treatment has been acid-copper-chromate (ACC), containing salts of chromium, chromic acid, and copper. Such preservatives are diffused into the wood by total immersion in a pressure vessel, with pressure being maintained until a prescribed amount of the preservative is retained by the wood or until the wood refuses to accept further treatment.

Metal Components

Metals, predominantly carbon steel, are selected for many components of the cooling tower where high strength is required. These would include fan hubs for larger-diameter fans, unitized supports for stabilization of the mechanical equipment, driveshaft guards, bolts, nuts, and washers. (Because of their sensitivity to the imbalance that corrosion can bring about, drive shafts normally are made of stainless steel.)

In the majority of cases, circulating-water conditions will be considered sufficiently benign to warrant galvanizing as the coating of choice for steel items. For severe water service, such as saltwater cooling, those components whose requirements are not economically met by the use of alternative materials are usually coated with epoxy coal tar.

Cast iron is used for gear cases, anchor castings, valve bodies, and fan hubs for intermediate-size fans. Although cast iron tends to build a self-protecting patina of iron oxide, it is usually either galvanized or coated with a high-grade enamel for cooling-tower use. For severe water service, it is normally sandblasted and coated with epoxy coal tar. Valve bodies are often porcelainized to guard against erosion.

Bolts, nuts, and washers do not lend themselves to a precoat, other than galvanizing or cadmium plating. Severe water conditions normally dictate a change in materials, and an appropriate grade of stainless steel is the popular choice because of its excellent corrosion resistance in the aerated conditions existing in cooling towers.

Copper alloys are sometimes used to resist special conditions, such as saltwater or brackish water service. Silicon bronze fasteners are suitable for service in saltwater, but must be protected against erosion in flowing water locations. The use of Naval Brass is usually discouraged because of its tendency toward stress-corrosion cracking. The use of more sophisticated metals, such as monel and titanium, is usually too expensive.

Plastic Components

The capability of plastics to be molded into complex shapes is an advantage, particularly for such close-tolerance components as fan blades and fan cylinders. Their many desirable characteristics, combined with the advancements being made in both plastic materials and their production, ensure that their rate of use in cooling towers will increase.

Plastics are currently used in such components as structural connectors, fan blades, fan cylinders, fill, fill supports, drift eliminators, piping, nozzles, casing, louvers, and louver supports. Most commonly used are *fiber-reinforced polyester* (FRP), fiber-reinforced epoxy, PVC, polypropylene, fiber-reinforced nylon, and fiber-reinforced polyphenylene oxide.

As a general rule, plastic components are of a dimension, location, or formulation that makes them least susceptible to abnormal water conditions. However, the relatively thin cross section of PVC film-fill sheets makes them somewhat sensitive to temperature, requiring something more than routine thought in fill support design. Some plastics, such as PVC, are inherently fire-resistant. Where considered necessary, others may be formulated for fire retardancy, such as fiber-reinforced polyester for casing and louvers.

Concrete Components

Concrete has been used for many years in Europe and other regions of the world, and its use is increasing in the United States. Following the first concrete hyperbolic cooling tower installed in the United States in 1961, numerous others have been installed, and the technology has expanded to include large mechanical-draft towers. Those shown in Fig. 9.12, for example, are constructed principally of concrete. In many cases, the higher first cost of concrete construction is justified by decreased fire risk as well as higher load-carrying capacity for larger structures.

Essentially, design philosophies for concrete cooling-tower construction coincide with those developed by the *American Concrete Institute* (ACI), except denser mixes and lower water/cement ratios are used than would normally be expected in more commercial construction. Typically, type I cement is selected, except where the presence of above-normal sulfate concentrations dictates the use of type II in water-washed areas.

Circulating water in want of calcium (quantified by a negative saturation index) can be "corrosive" to concrete components, in which case the concrete gives up a portion of its calcium content in an effort to neutralize the water. Chemical treatment of the circulating water should be aimed at maintaining a slightly positive saturation (Langelier) index.

FIRE PROTECTION, PREVENTION, AND CONTROL

Although concrete or steel cooling-tower structures will not burn, they can be rendered useless by significant exposure to a very hot fire. If their contents (fill, drift eliminators, etc.) are combustible, therefore, they can be placed in a high-risk category by fire insurance underwriters. However, with the advent of PVC fill and eliminators, insurers have begun to reassess the risk factors and, in many such cases, have permitted installation without a fire protection sprinkler system and with no increase in premium.

Obviously, wood towers are most susceptible to fire, particularly after they have remained inoperative for a period sufficient to allow them to dry out. The use of PVC (or other materials formulated for fire-retardant characteristics) for fill and eliminators in wood-framed cooling towers also has a risk-reducing effect, although not usually to the extent recognized in concrete- or steel-framed towers. Depending on the criticality of the system served by the tower, or the severity of local fire codes, insurance carriers may insist on a simple wet-down system, at least, or may require a full-fledged fire protection sprinkler system.

Wet-Down Systems

Wet-down systems are simple piping and nozzle arrangements designed to deliver a relatively small continuous flow to the top regions of an inoperative tower by means of a pump that takes suction from the tower's cold-water basin. Flow in such a system needs to be no more than that required to maintain dampness in the primary wood components; and system sophistication is usually limited to an interlock that starts the wet-down system pump on main-pump shutdown, as well as a sensor that will prevent operation of the wet-down system pump at air temperatures below approximately 40°F (4.4°C). It is important that the flow through such a system be continuous, because alternate wetting and drying of the wood can degrade its service life.

Periodic operation of the main pumps will keep specific areas of the tower wet, but may leave critical areas unaffected. Circulating water over a crossflow tower, for example, will moisten the distribution basins, fill, major outboard structure, and louver areas, but without airflow will accomplish no appreciable moistening effect on the fan deck, drift eliminator, and plenum areas. Similarly, intermittent pump operation on a counterflow tower will moisten the fill and lower structure, but will leave the drift eliminators and entire upper areas relatively dry.

Fire Protection Sprinkler Systems

Fire protection sprinkler systems for cooling towers are defined and governed by *National Fire Protection Association* (NFPA) Bulletin 214 (latest revision). The system normally consists of a rather intricate arrangement of piping, nozzles, valves, and sensors or fusible heads that cause the tower to be automatically deluged with water soon after the start of a fire. Piping within the tower is usually free of water, to prevent freezing. Water at a prescribed residual pressure is required to be available at an automatic valve, located within a nearby heated space. Operation of the valve is initiated either by thermostatic sensors, which react to an abnormal rate of temperature rise, or by fusible heads, which cause pressure loss within a pneumatic control system.

In many cases, insurance underwriters for a plant will alter premium values in recognition of thoughtful modifications made to the cooling tower, whether or not it is equipped with a fire protection sprinkler system. Among those modifications are the following:

- Where plastic items of significant scope are utilized in the tower, they may be formulated to retard or resist fire. Primary areas of concern would be casings, louvers, fan cylinders, fill, and drift eliminators.
- Selected top areas of the tower (notably fan decks) may be covered by a specified thickness of fireproof material.
- Partition walls between cells of a rectilinear tower may be designed to act as "fire walls" to prevent or delay the spread of fire. Typically, a ½-in (13-mm) thickness of either treated Douglas fir plywood or fireproof material on both sides of the transverse column line that constitutes a partition bent is recognized as a 20-min barrier to the spread of fire. Fire walls of increased rating are accomplished by increasing the thickness of the material used.

Depending on the scope of required modifications, their cost should be evaluated against the cost of a fire protection sprinkler system and/or the benefit of reduced insurance premiums.

WATER TEMPERATURE CONTROL AND ENERGY MANAGEMENT

Given a specific heat load and water flow rate, a cooling tower is designed to produce a certain cold-water temperature at a particular entering air wet-bulb temperature. Since the chosen wet-bulb temperature is usually one that will seldom be exceeded, the cooling tower becomes a device which may be considered to have "excess capacity" during a period approaching 95 percent of its operating life. This is because the cold-water temperature that it produces reduces with any declination in wet-bulb temperature or heat load, assuming continued full-fan operation and constant pumping rate (Fig. 9.4).

Most steam-driven systems generating electricity derive a net benefit from the cooling tower's ability to produce colder and colder water as the season progresses into fall and winter. Within limits, the resultant decrease in turbine backpressure can measurably increase system operating efficiency. Usually, the limit is reached when the system can no longer extract noncondensable gases economically. This is particularly true in older plants having condenser leakage problems.

For whatever reason, however, there invariably comes a time when cold-water temperatures below a certain level are either nonrewarding or actually detrimental to system efficiency. Those situations require means of controlling the cold-water temperature and can offer the opportunity for significant energy savings brought about by proper operation of the tower.

Modulation of the water flow to the tower should never be employed as a means to control the water temperature. This is because cooling-tower water-distribution systems, whether spray or gravity type, are calculated to produce maximum efficiency within a relatively narrow range of flow rates. Increasing flow can produce either overpressuring of spray systems or overflowing of hot-water basins. Conversely, flow reductions can result in inadequate water coverage on the fill, which not only disrupts heat-transfer efficiency but also increases the work load of the fan(s).

Influencing Factors

The effectiveness of air-side control depends primarily on the number of fan cells comprising the tower as well as the speed-change characteristics of the motors

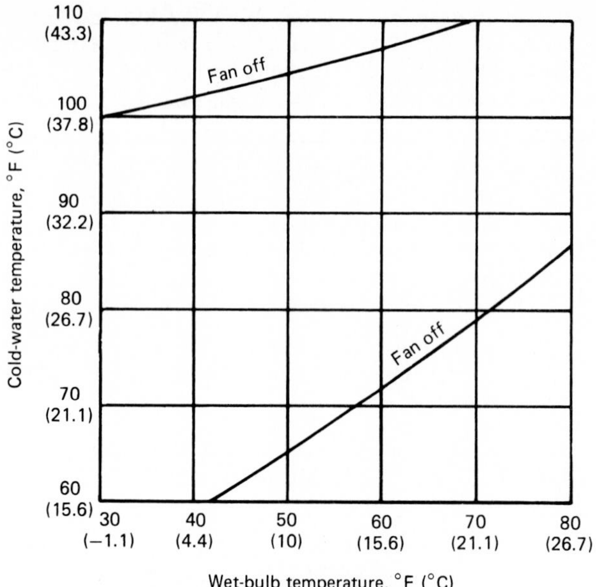

FIG. 9.26 Single-cell tower with single-speed motor.

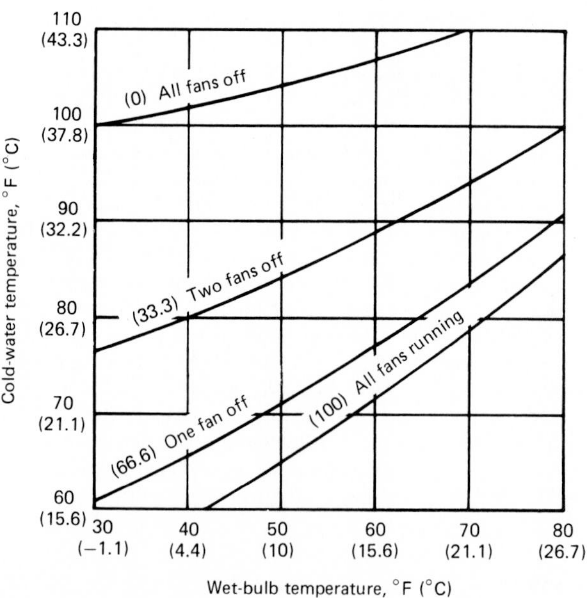

FIG. 9.27 Three-cell tower with single-speed motors.

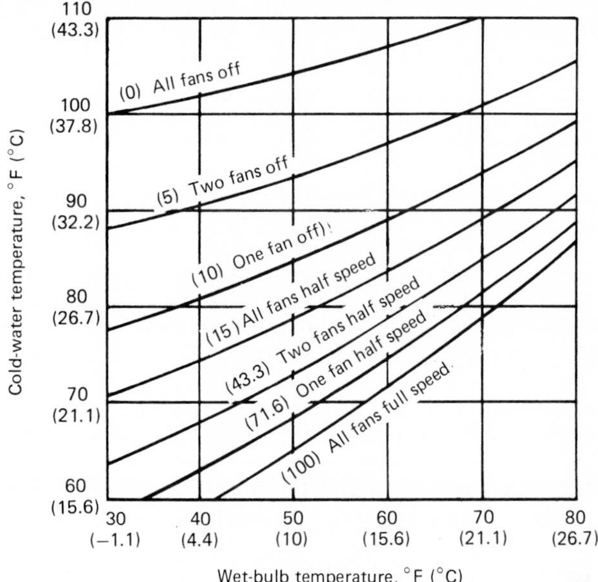

FIG. 9.28 Three-cell tower with two-speed motors.

driving the fans. Figure 9.26 defines the operating modes available with a single-cell tower equipped with a single-speed motor. In this most rudimentary case, the fan motor can only be cycled on and off for erratic control of the cold-water temperature, and great care must be taken to prevent an excessive number of starts from burning out the motor.

To illustrate the increased control capability afforded by a multiplicity of cells and fan speed options, the operating characteristics of a three-cell tower (selected for the same performance as the aforementioned single-cell tower) are indicated in Figs. 9.27 and 9.28, equipped with single- and two-speed motors, respectively. The numbers in parentheses represent the approximate percentage of total fan power consumed in each operating mode. Note that the opportunity for both temperature control and energy management is considerably enhanced by the use of two-speed motors.

At any selected cold-water temperature, an increase in the number of fan/speed combinations causes the operational mode lines to become closer together. It follows, therefore, that the capability to modulate a fan's capacity or speed (within the range from 0 to 100 percent) would represent the ultimate in temperature control and energy management. Two technologies currently exist by which to achieve this ideal situation: *automatic variable-pitch* (AVP) fans change air-movement capacity at constant speed and frequency-modulating control devices change air-movement capacity by changing the speed.

Basic Fan Laws

The energy management principles on which these technologies work are related to the following two basic fan laws:

- The capacity (ft^3/min or L/s) of a fan operating at constant pitch varies directly as the speed ratio; that is, a fan turning at 70 percent of its design speed moves 70 percent of its design airflow.
- The horsepower (or kW) of a fan varies as the cube of its capacity ratio; that is, a fan moving at 70 percent of its design airflow develops 34.3 percent (0.7^3) of its design horsepower (kW).

The result of applying these principles is shown in Fig. 9.29, where the cold-water temperature (dashed line) is assumed to be controlled such that it will not go below 75°F (23.9°C). When the temperature reaches that level, the fan power (solid line) begins to decline rapidly and the cold-water temperature remains constant.

Although both technologies represent effective methods of capacity control, designers must be made aware of some physical limitations that may be encountered in the use of frequency-modulating (that is, variable-speed) devices. First, overspeeding the fan would place it in structural jeopardy; therefore, frequencies representing fan speeds greater than 100 percent of design must not be allowed. Second, most fans have at least one critical speed that occurs between 0 and 100 percent of design rpm, and some have several. Therefore, it becomes necessary to determine the critical speed(s) and to prevent corresponding frequencies from being used. Finally, most gear-reduction (power transmission) devices in the cooling-tower industry begin to suffer lubrication problems at operating speeds below approximately 30 to 40 percent of design. Therefore, frequencies below the appropriate level must be blocked out.

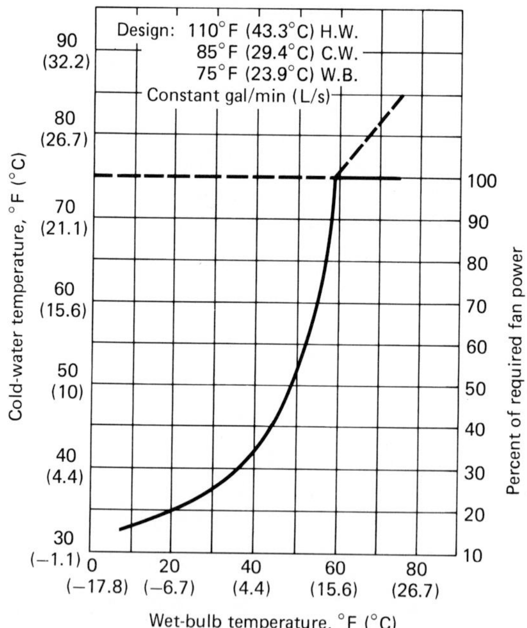

FIG. 9.29 Fan power versus wet-bulb temperature.

TOWER OPERATION IN FREEZING WEATHER

In geographic areas where winter-time air temperatures drop well below freezing, cooling-tower operators find themselves significantly concerned with the buildup of ice on the tower. Of particular concern is the formation of ice in such critical areas as fill and air intakes, where an appreciable buildup can jeopardize thermal performance either temporarily (as in the case of blocked airflow) or permanently (as in the case of collapsed fill).

Air-Side Control

Operators will find that the principles applied in the previous segments of this chapter also serve to good advantage for operation in freezing weather. Manipulation of the airflow is an invaluable tool, not only in the retardation of ice formation but also in the reduction or elimination of ice already formed. In addition to bringing less cold air into contact with the circulating water, reducing the entering air velocity alters the path of the falling water, allowing it to impinge upon—and melt—ice previously formed by random droplets which wind gusts or normal splashing may have caused to escape the protection of the relatively warm mainstream of water.

Single-speed fans afford the least opportunity for airflow variation, and towers so equipped require maximum vigilance on the part of the operator to determine the proper cyclic operation of the fans that will result in the best ice control. Two-speed fan motors offer significantly greater operating flexibility and should be given maximum consideration in the purchase of towers for use in cold climates. Fans may be individually cycled back and forth between full speed and half speed as required to achieve a balance between the cooling effect and ice control, limited only by the maximum allowable motor insulation temperature, which an abnormal number of speed changes per hour may cause to be exceeded. In many cases, all the fans operating at half speed produce the best combination of cooling effect and ice control.

On towers having two or more fans evacuating a common plenum (such as the round towers depicted in Fig. 9.12), those fans should be brought to the off position in unison to prevent a down draft of cold, moisture-laden air from icing up the mechanical equipment of an inoperative fan.

On multicell towers of rectilinear configuration (Fig. 9.11) equipped with a separate plenum for each fan, individual fans may be cycled as necessary to control ice. However, it must be understood that cycling the fan on a particular cell accomplishes nothing with respect to deicing of adjacent cells.

Ultimately, severe ice formations may require that the fan(s) be reversed for a time. This causes the falling water pattern to be shifted outward, bringing a deluge of relatively warm water in contact with ice formations for rapid melting. The warmed air exiting the air inlets also promotes melting of ice formations not reached by the falling water. This mode of operation should be utilized only for short periods because of the possibility of ice forming on the fan cylinders, fan blades, and mechanical equipment. The allowable length of time will be a function of atmospheric conditions and should be established, and monitored, by the operator. On multifan towers, individual fan reversal should be avoided; otherwise, the discharge vapors from adjacent fans may contribute to severe icing of a reversed fan.

Water-Side Control

Understanding how to anticipate and control ice requires some knowledge of the water-temperature gradients that occur in a cooling tower. Without such knowledge, operators often assume that controls which will automatically cycle fans to maintain a leaving cold-water temperature well above freezing are sufficient insurance against the formation of ice. Occasionally, operators are bewildered to find ice beginning to form even before the leaving water temperature has depressed to that presumably "safe" level.

This problem is caused by the previously mentioned temperature gradients which occur transversely in all towers, as well as longitudinally in multicell towers where fans are cycled progressively. For example, Fig. 9.30 defines transverse temperature gradients in a bank of crossflow fill designed for a specific performance. In this case, water is entering the tower at 64.5°F (18.0°C) and leaving at 44.5°F (6.9°C), which are temperatures seemingly indicative to an operator that a 12.5°F (7°C) "safe" zone exists between the 44.5°F (7.0°C) control point and freezing.

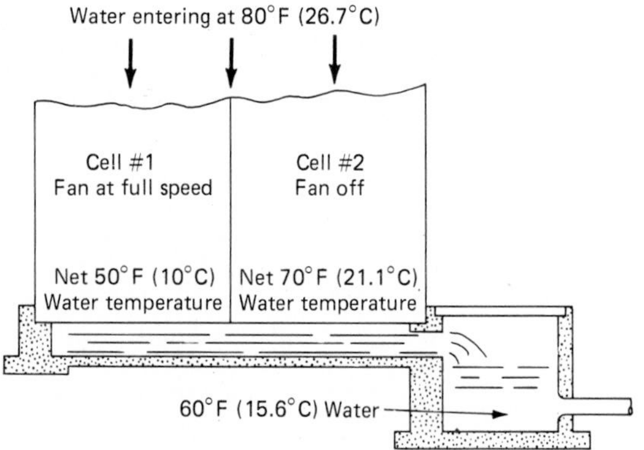

FIG. 9.30 Fill transverse-temperature gradients.

Obviously, such a zone is not the case. As can be seen, the net outlet temperature of 44.5°F (6.9°C) results from a mixture of water temperatures varying from about 53°F (11.7°C) at the inboard edge of the fill to about 33°F (0.6°C) at its coldest point. Consequently, the real margin of safety is virtually nil.

Water temperatures at the coldest point of the fill are very sensitive to the tower's operating range. At a given cold-water-temperature control point, reduced ranges (that is, reductions in heat load at a constant water flow rate) will cause the water temperature at the coldest point of the fill to *rise*. Conversely, increased ranges (that is, reductions in water flow at a fixed heat load) will cause the water temperature at the coldest point of the fill to depress.

For example, if the tower in which the Fig. 9.30 fill is installed were operating at a 10°F (5.6°C) range [cooling the water from 54.5°F (12.5°C) to 44.5°F (6.9°C)], the entering wet-bulb would be 29°F (−1.7°C), and the water temperature at the coldest point of the fill would be approximately 38.5°F (3.6°C). As the wet-bulb temperature reduces further, measures could be taken to diminish airflow through the fill (by fan manipulation), and the coldest water in the fill would reduce only negligibly below that level.

Thermal Gradients

There is also a longitudinal temperature gradient that develops as individual fans are manipulated in a multicell tower. This is because cells with fans operating at full speed contribute much more to the tower's overall cooling effect than do cells with fans operating at reduced speed or turned off. For example, with one fan running and one fan off, if water were entering the two-cell tower in Fig. 9.31 at 80°F (26.7°C) and leaving at a net 60°F (15.6°C), the actual water temperature produced by cell 1 would be 50°F (10°C) and water at the coldest point of its fill would be at or near freezing.

Figures 9.32 and 9.33 indicate net performance—and thermal gradients—of a two-cell tower cooling through a 20°F (11.1°C) range, equipped with single-speed and two-speed fans, respectively. These curves are drawn on the premise that the operator will manipulate the fans to prevent the net cold-water temperature from dropping below 60°F (15.6°C). The solid line indicates the net water temperature sensed by the operator's thermometers or control devices, the dashed line indicates the net water temperature from the cell operating at the greatest fan speed, and the dotted line indicates the coldest water temperature in the fill. Figure 9.32

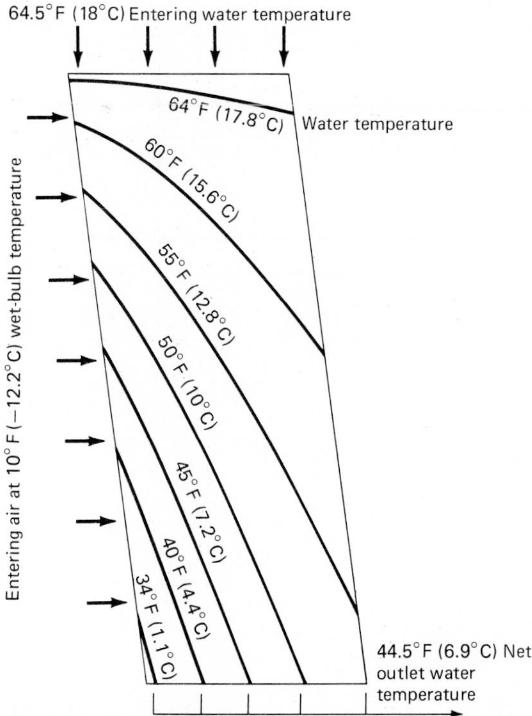

FIG. 9.31 Longitudinal-temperature gradient.

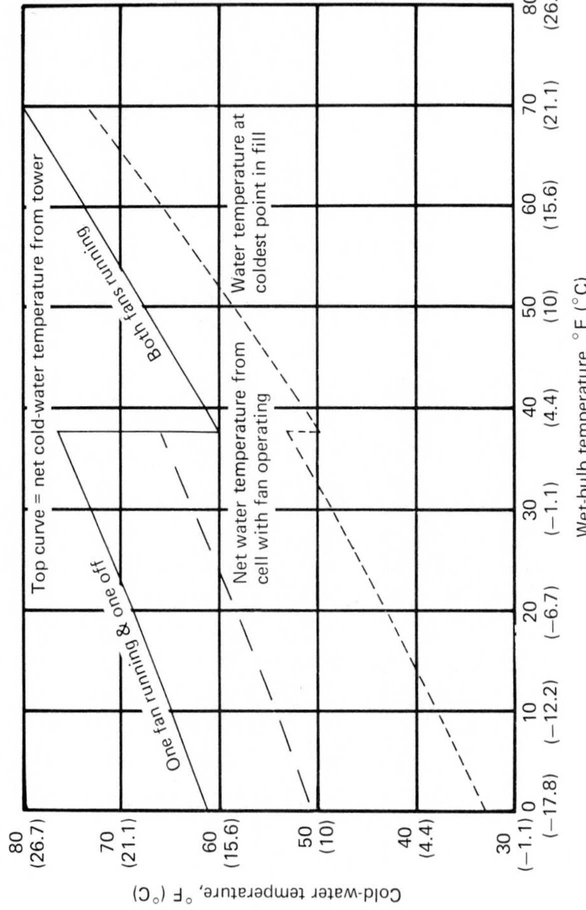

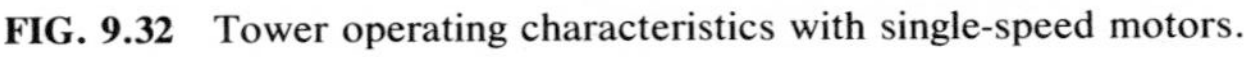

FIG. 9.32 Tower operating characteristics with single-speed motors.

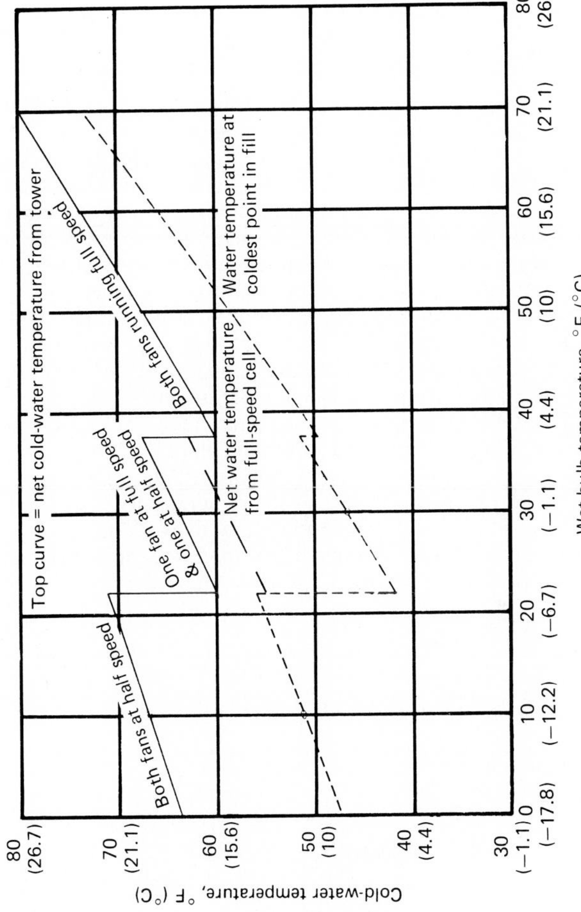

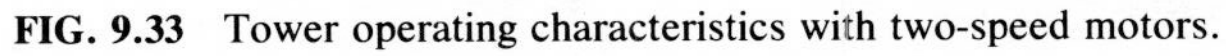

FIG. 9.33 Tower operating characteristics with two-speed motors.

shows that the situation depicted in Fig. 9.31 would occur at about a −2°F (−18.9°C) wet-bulb temperature for that particular tower.

Comparing Figs. 9.32 and 9.33, one can also see the advantage afforded by two-speed motors, both in operating flexibility and in the reduction of fill temperature gradients. By logical extrapolation, clearly the civilized behavior of a cooling tower under potential icing conditions improves as the ability to modulate airflow increases. Therefore, the ultimate in ice management, under current technology, would be achieved by applying either the AVP fans or the frequency-modulation speed controllers, as discussed previously.

Water Distribution

Larger towers designed for operation in freezing weather should be equipped with a water distribution system that can be manipulated to place the greatest concentration of flowing water nearest the air intakes of the tower. This applies particularly to natural-draft (hyperbolic) towers, where no means of air-side control is available. Not only does this give the most difficult cooling job to the coldest air, but it also ensures a rapid rise in air temperature to preclude freezing on the fill. Most importantly, it places the maximum amount of relatively warm flowing water in close proximity to the areas of greatest ice concern.

To provide for start-up and operating flexibility, provision for total water bypass directly into the cold-water basin (Fig. 9.34) is highly advisable on mechanical-draft towers and must be considered mandatory on natural-draft towers. During cold-weather start-up, the basin water inventory may be at a temperature very near freezing, at which time the total water flow should be directed back to the cold-water basin on its return from the plant load—*without going over the fill.* This bypass mode should be continued until the total water inventory reaches an acceptable temperature level [usually about 80°F (26.7°C)], at which time the bypass may be closed to cause total flow over the fill.

Even during operation, combinations of low load, low ambient temperatures, and high winds can promote ice formations which may defy normal air-side and

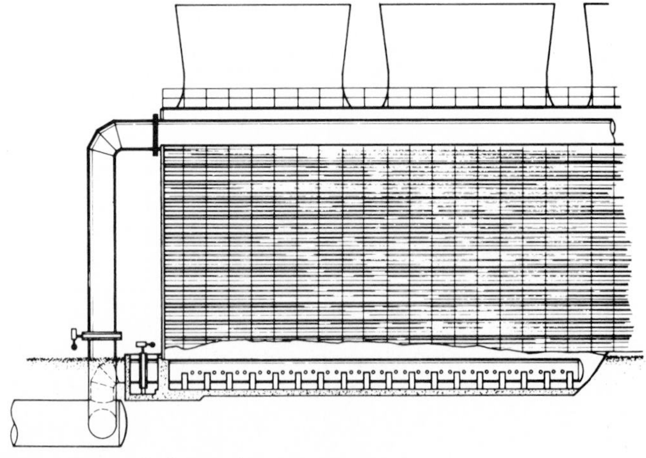

FIG. 9.34 Typical water-bypass arrangement.

water-side control procedures. In those cases, it may become necessary intermittently to revert to total bypass flow in order to build a heat content in the circulating water. The importance of *total* bypass flow is emphasized because the opportunity for fill icing increases as water flow over the fill decreases. Modulation of the bypass (whereby a portion of the water flow is allowed to continue over the fill) *must not be permitted under freezing conditions!*

ACKNOWLEDGMENT

The contents of this chapter were adapted by the author from *Cooling Tower Fundamentals* and *Cooling Tower Information Index,* published by Marley Cooling Tower Co., P.O. Box 2912, Mission, Kansas 66201. The author also wrote these publications. Drawings, charts, and photographs are from these publications.

REFERENCES

1. F. Merkel, "Verdunstungskuehlung", VDI Forschungsarbeiten No. 275, Berlin, 1925.
2. D. R. Baker and H. A. Shryock, "A Comprehensive Approach to the Analysis of Cooling Tower Performance," *Journal of Heat Transfer,* Aug. 1961.

CHAPTER 1.10
CONDENSATE AND FEEDWATER TREATMENT

John W. Siegmund and Kevin J. Shields
Sheppard T. Powell Associates
Baltimore, MD

INTRODUCTION

Numerous schemes are available for production of steam and electric power. In simple terms, all consist of the introduction of feedwater to a boiler or steam generator, generation of steam, condensation of steam in turbine-generator, heat transfer in condenser, and return of condensed steam to feedwater system. Since the condensate and feedwater systems are directly connected, an arbitrary definition of where one ends and the other begins is necessary. Generally, the condensate system is considered to end at the point where makeup water is introduced to the cycle. In many power generation systems this point is the deaerator. Steam may also be extracted from the turbine for process applications.

Regardless of the definitions applied to a given system, the treatment objectives for condensate and feedwater systems are the same: to prevent or control contamination of the water by species that may corrode cycle metals or form scale on heat-transfer surfaces and to provide a chemical treatment program compatible with system materials. Both equipment and chemical approaches are available for condensate and feedwater treatment. Selection of a suitable treatment approach requires consideration of boiler water and steam purity requirements. Boiler operating pressure, steam quality (percentage of moisture in the steam), furnace-wall heat input in Btu/h • ft^2 (or W/m^2), and where and how the steam is used must be included in these considerations.

CONDENSATE TREATMENT

Under design operating conditions, condensed steam should exhibit purity as good as at any other point in the cycle. However, the condensate is subject to several sources of contamination. Two major sources are air in-leakage and cooling-water leakage.

The condenser, some turbine parts, and low-pressure heaters normally operate under vacuum, an ideal condition for air in-leakage. Air brings with it oxygen (O_2), carbon dioxide (CO_2), and other gases which can dissolve in the condensed steam. Dissolved oxygen (DO) can be removed more readily under condenser vacuum than more soluble gases such as CO_2 and ammonia (NH_3). Devices such as steam-jet ejectors and vacuum pumps remove air and noncondensable gases from the condenser steam space.

Condenser tube integrity is essential to prevent contamination of the condensate by cooling water. Corrosion of waterside surfaces of the tubes can be particularly troublesome in systems circulating seawater or brackish water for cooling or in freshwater-cooled systems that operate with cooling towers at high cycles of concentration.

Other common sources of contamination can be salts, organic acids, and regenerants in the makeup water; corrosion products (metal oxides) from the condensate and feedwater systems; and oil from lubrication and fuel systems. Condensate can also be contaminated by plant service water used for pump seal water or other applications. However, contamination of this sort is minor compared to that resulting from condenser tube leakage.

To minimize adverse impacts resulting from contamination of the condensate, a variety of equipment and chemical treatment alternatives are available. Note that the treatment measures described are useful in reducing the effects of contamination, but do not prevent it. Only an effective maintenance program can ensure that equipment remains a low failure risk, thus slashing condensate system contamination. The treatment measures described in the following paragraphs are intended to back up the maintenance program, not to be a substitute for regular inspections and maintenance.

Polishing Equipment

A variety of equipment alternatives are available for removal of contaminants from condensate. The type of polisher selected for a given system depends on the character and quantity of contaminants and the water chemistry requirements of the boiler or steam generator. Most polishers rely on ion-exchange technology to remove contaminants and exchange them for less objectionable species. Most polishers also serve as filters for suspended metal oxides, or they may be preceded by septum-type filters for this purpose.

Polishers can be specified as an original component during the design of new systems or retrofitted to existing systems. In recent years, retrofit installation of magnetic filters to remove corrosion products from condensate systems has become another polishing approach of interest where condensate temperatures exceed 120°F (48.9°C).

Typical polisher locations in condensate and feedwater systems are shown in Figs. 10.1 and 10.2 for utility and industrial powerplants.

Sodium-Cycle Polishers. Sodium-cycle polishers function on the principle of ion exchange. Cations such as dissolved calcium, magnesium, ammonia, amines, and soluble species of metal ions are removed and replaced by sodium. Because the sodium-cycle polisher can contribute sodium to the condensate (and subsequently the feedwater and boiler water), its application is limited to low-pressure cycles.

Iron oxide is found primarily in insoluble form in condensate, and it is removed in the polisher principally by filtration. Filtration efficiency varies substantially with pressure drop and iron oxide form, but suspended-iron removal on the order of 40 to 70 percent or more should be achievable.

Should tests of the polisher effluent show hardness or should the pressure drop through the polisher exceed predetermined limits, the polisher vessel is removed from service for backwashing and regeneration with a sodium chloride–brine solution. Even with proper regeneration, the effective exchange capacity of the resin will gradually decrease as the resin service life increases. Normal practice is to include two or more polisher vessels in a system to keep one or more vessels in service while others are being regenerated or are on standby status.

Pressure drop across the bed increases during the service run as the bed compacts. Other factors can result in a more rapid increase in pressure differential. The resin beads function fairly effectively as filter media to remove suspended matter from the condensate. In sodium-cycle polishers, suspended iron oxide is by far the most significant material removed by physical filtration. Regular surveillance of suspended iron across the polisher bed is desirable.

Gradual increases in suspended iron oxides in condensate streams may be indicative of deficiencies in chemical treatment and/or control of air in-leakage. Rapid increases in levels of suspended iron oxides reaching the polishers are often related to system operations such as unit start-up, shutdown, or trip. Mechanical breakdown of the resins occurs during service and regeneration. Whole beads are fractured into small pieces known as *fines,* which also increase the pressure drop through a polisher. Proper backwashing of the bed will remove these fines.

During service operation of sodium-cycle polishers, a small amount of sodium is present in the polisher effluent water, even when there is negligible contamination of the inlet condensate. This phenomenon is commonly referred to as *sodium leakage*. Coordinated phosphate and other solids-based boiler-water-treatment approaches that provide pH buffering capacity are of some benefit in considering installation of this polisher type.

Deep-Bed Demineralizers. Deep-bed demineralizers use a mixed bed of cation and anion exchange resins. Deep beds see service in many high-pressure systems where a high degree of water purity is desired and concern over condensate contamination exists. Deep-bed demineralizers are also of benefit in removing suspended metal oxides from condensate. Removal efficiency is similar to that of sodium-cycle polishers. It is customary to install multiple units in parallel to permit continuous polishing of the condensate. Removal of sodium, ammonia, and

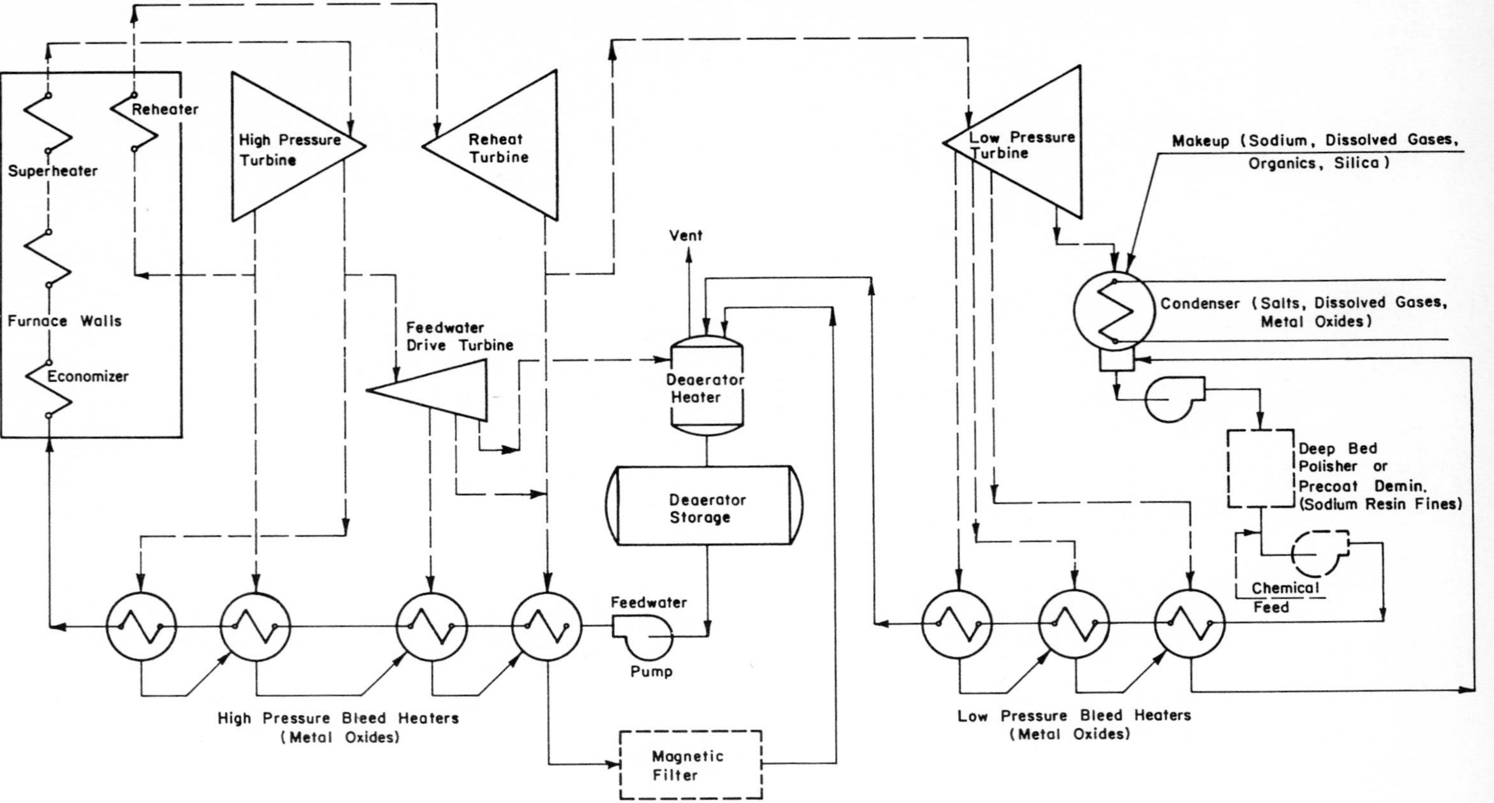

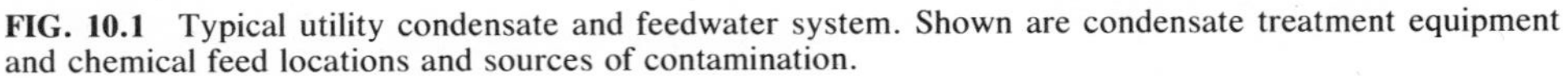

FIG. 10.1 Typical utility condensate and feedwater system. Shown are condensate treatment equipment and chemical feed locations and sources of contamination.

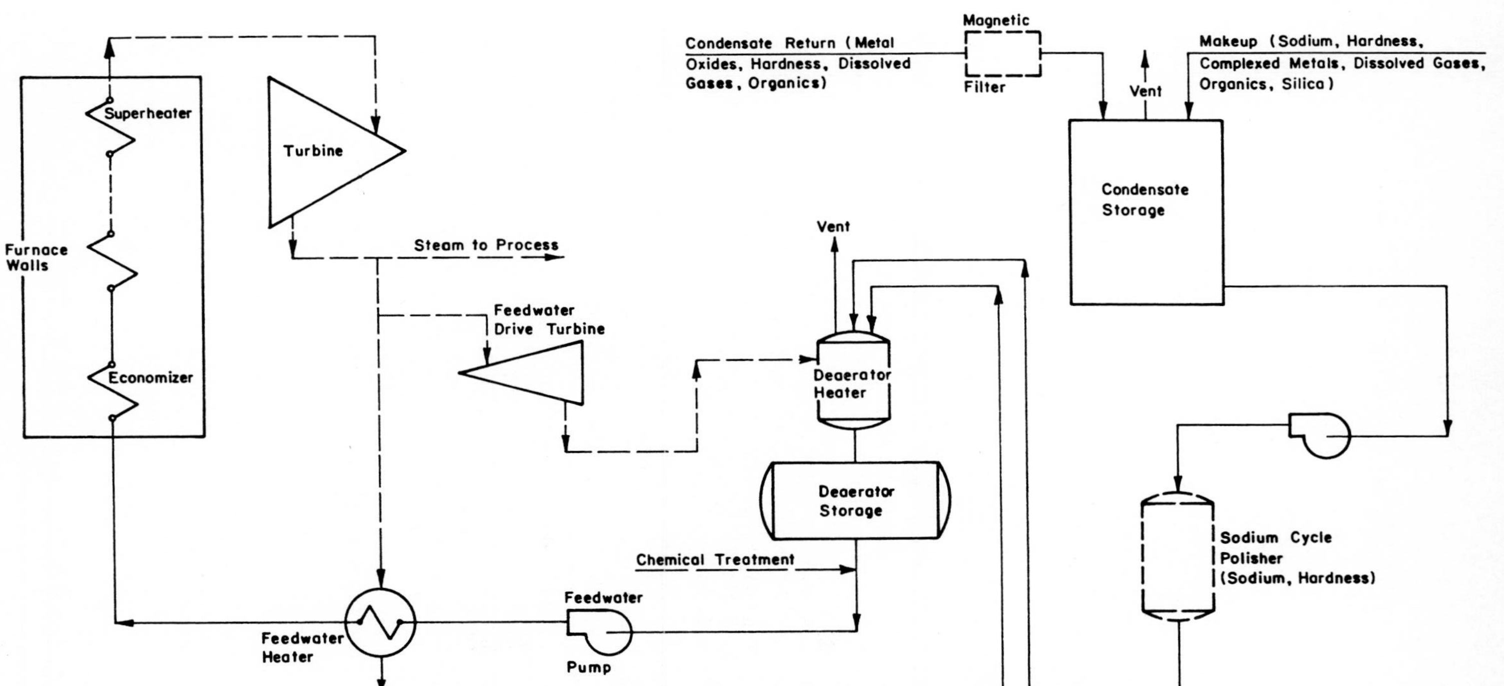

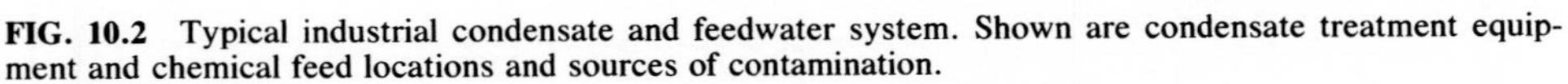

FIG. 10.2 Typical industrial condensate and feedwater system. Shown are condensate treatment equipment and chemical feed locations and sources of contamination.

other cations is accomplished by the cation resin. The usual location of this equipment is shown in Fig. 10.1.

Regeneration of deep-bed demineralizers is more difficult than for sodium-cycle polishers. The cationic and anionic portions of the mixed bed must be separated prior to regeneration, and the degree of separation influences the purity of the polisher effluent. Regeneration is performed either in place in the polisher vessel or in separate cation and anion regeneration vessels after separation. Sulfuric acid is commonly selected to regenerate cation resin, while caustic soda is the choice for anion resin regeneration. Provisions must be made to neutralize regenerant wastes before disposal.

Deep-bed polishing demineralizers find application in many once-through fossil-fuel boilers and also extensively in nuclear and drum-type fossil-fuel condensate systems on both original equipment and retrofits. Many of these systems use ammonia for pH control. In systems without copper alloys, relatively high concentrations of these chemicals permit operation at the optimum pH for protection of steel surfaces.

Ammonia can rapidly consume available exchange sites on the cation resin, after which, if the resin is not regenerated, ammonia passes through to the polisher outlet. Sodium leakage may also increase slightly if operation is past the ammonia "breakpoint." Provided that adequate measures are taken to monitor polisher outlet water chemistry, deep-bed units may be run successfully beyond the ammonia breakpoint. Under normal operating conditions of no condenser leakage or other significant condensate contamination, deep-bed polishers are regenerated on the basis of time or volume of water treated. Pressure drop is usually a secondary criterion for regeneration.

A major difference between makeup and polishing demineralizers is the service flow rates. In makeup demineralizers, flow rates are typically 4 to 6 gal/min • ft^2 (2.72 to 4.08 L/s • m^2) of resin surface, while rates of 40 to 50 gal/min • ft^2 (27.2 to 34.0 L/s • m^2) are necessary in deep-bed polishers, depending on the bed depth. Operation at higher service flow rates in a polisher is required for suspended-metal-oxide removal. In some polishing systems, cartridge filters have been installed ahead of and behind the deep-bed demineralizers. Placement of a filter ahead of the polisher simplifies regeneration of the resin by removing suspended metal oxide from the condensate before it reaches the polisher. Post filters are of benefit in capturing resin fines that can pass through demineralizer underdrain systems and resin traps.

All ion-exchange resins tend to lose exchange capacity with increased service life until replacement is ultimately required. Resin losses often occur during the operating life of a resin bed, thus addition of replacement resins is necessary. The importance of a regular resin sampling and analysis program is clear.

Precoat Demineralizers. Powdered ion-exchange resin is applied as a precoat material to septum-type filters. Available resin types include cation (hydrogen or ammonia form), anion (hydroxide form), and inert (no available ion-exchange sites). The resin precoat mixture applied may be varied if desired to suit specific needs, for example, during start-up and normal operation.

Precoat demineralizers are superior to deep-bed units with respect to suspended-metal-oxide removal efficiency. The fine size of the resin used (325 mesh) makes this possible, permitting low service flow rates—4 gal/min • ft^2 (2.72 L/s • m^2) of filter surface area—relative to the flow rates for a deep-bed polishing demineralizer. Since a thin precoat of powdered resin is selected, the available ion-exchange capacity is much less than for deep beds. For this reason, the ability of precoat demineralizers to remove dissolved-solids contamination is limited in comparison to deep-bed units sized to handle the same throughput.

The use of powdered resin, however, offers some advantages over deep-bed resins. When a precoat is exhausted (normally from increased pressure drop caused by suspended-solids accumulation), the vessel is backwashed and recoated with virgin resin. This process requires perhaps one hour while regeneration of a deep-bed unit takes several hours. As with other polishing demineralizers, a battery of vessels is normally specified in the system design. Since the powdered resin is not regenerated, problems related to improper separation and regeneration of deep beds are avoided. High air in-leakage rates can quickly exhaust the anion portion of the precoat because of CO_2 that dissolves in condensate from the air.

The precoat demineralization technique is applicable to virtually all steam-water cycles. It is not commonly found in condenser cooling systems that circulate seawater, brackish water, or other water high in total dissolved solids. Under these conditions, deep-bed demineralizers are usually specified, since they provide greater ion-exchange capacity to maintain condensate purity longer than precoat units during episodes of condenser leakage. Another approach is to place precoat demineralizers ahead of the deep-bed units. This condensate polishing scheme has not been applied widely because of associated capital and operating expenses.

Precoat demineralizers are normally placed in the condensate system in the same location as deep-bed polishers (Fig. 10.1).

Magnetic Filters. Magnetic filters operate by applying an electromagnetic field to a steel matrix. Condensate or feedwater is directed through the matrix, where magnetic iron oxide is removed. The filters are capable of high removal efficiencies of suspended iron in magnetic form. A chemical feed, acting as an oxygen scavenger and a metal passivator ahead of the magnetic filters, reduces nonmagnetic iron to the magnetic form, thereby enhancing filter efficiency. It has been observed during operation of magnetic filters that some limited capture of nonmagnetic suspended solids does occur, including nonmagnetic iron oxides, copper oxides, and even hardness. However, typical performance specifications for these filters require control of air in-leakage to maximize the percentage of suspended iron present in magnetic form—thus no efficiency claims for nonmagnetic species are made.

Removal of filtered-out solids is typically required once every 8 to 12 h, during which time the matrix is demagnetized and accumulated solids are washed to waste. Normally, the entire cleanup cycle requires only 30 s. Because of the very brief time the filters are not in service, installation of redundant units is not usually specified.

Although magnetic filters are well suited for removal of suspended solids from condensate, they have no ability to remove dissolved-solids contamination. Magnetic filters are applied most extensively in industrial power generation systems, in some cases ahead of sodium-cycle polishers. They have also seen service in a number of utility power cycles. Because they are not affected by the temperature of the water being filtered, the filters are useful for cleaning up high-temperature water sources (Fig. 10.1).

Chemical Treatment

In spite of the best efforts of engineers to prevent contamination of the condensate, some minor contamination is inevitable, particularly as the result of air in-leakage. Dissolved gases present in the condensate can increase the corrosion of base metals in the cycle. DO and CO_2 are normally of greatest concern, but others such as hydrogen sulfide (H_2S) and sulfur dioxide (SO_2) may sometimes be present in small amounts. Dissolved gases, particularly CO_2, depress condensate pH, which renders system metals more vulnerable to corrosion.

Chemical treatment is available to ameliorate the undesirable effects of incidental air in-leakage, but is not intended to be a substitute for preventive maintenance. The chemicals selected generally may be divided into three generic classes: neutralizing amines and ammonia, filming amines, and oxygen scavengers and metal passivators. Each class is discussed in the following paragraphs.

Neutralizing Amines and Ammonia. One approach to minimizing pH depression and corrosion of condensate system metals is to add an alkaline chemical. When air in-leakage is significant, it reacts with carbonic acid formed from dissolved CO_2. Volatile chemicals such as ammonia and ammonia derivatives are well suited for this purpose and are added at the beginning of the condensate system or directly into the steam system when needed. Because of their volatile nature, amines are partially removed from cycle water at points where gases are vented. The neutralizing amines most frequently used in condensate treatment include cyclohexylamine, morpholine, 2-amino-2-methyl-1-propanol, methoxypropylamine and diethylaminoethanol, dimethylaminoethanol, diethylamine, and dimethylamine.

Neutralizing amines in solution exhibit a tendency to corrode copper and copper alloys and to dissolve copper oxides. Carbon and stainless steel are not affected by neutralizing amines. Higher dosage levels are therefore permissible in systems that contain no copper materials. Condensate pH may be controlled to 9.2 to 9.4, the place at which steel corrosion rates are lowest. In systems containing copper, lower levels of amine are desirable. In most cases, a pH range of 8.7 to 8.9 is desirable to achieve acceptable corrosion control for both copper and steel alloys in the cycle.

On the basis of equal concentrations in water, ammonia is more aggressive to copper than other amines such as morpholine and cyclohexylamine. However, these chemicals should be applied only at the concentration needed to maintain condensate pH in the desired range. Ammonia, by virture of its superior ability to neutralize carbonic acid, can be introduced to control condensate pH at concentrations lower than those necessary with amines. When used at the concentration required for control of condensate pH, even with minimal air in-leakage, ammonia is less aggressive to copper than other amines.

Adding other neutralizing amines may be desirable in systems where steam is used for purposes besides power generation. The lower distribution ratios associated with other neutralizing amines may be beneficial in systems where steam condenses at various temperatures. Care should be taken in applying all neutralizing amines, however, since the amount of chemical required to maintain condensate pH increases as air in-leakage and the presence of CO_2 increase, which in turn results in higher corrosion rates for copper alloys in the cycle.

Filming Amines. Another approach to controlling waterside corrosion in condensate systems is to create a barrier between the metal and the water. The barrier must be sufficiently thin that the heat-transfer efficiency is not significantly reduced, yet durable enough to prevent localized penetration of the film, which could result in pitting corrosion. A class of compounds collectively referred to as *filming amines* has been used in condensate treatment to meet these requirements.

Amine molecules applied in this manner generally consist of a long hydrocarbon chain, which is hydrophobic in nature, with a hydrophylic amine group at one end. In the cycle, the hydrophylic amine groups tend to attach to metal surfaces in a monomolecular film. Once a complete film is formed, the hydrophobic portion of the molecules serves as a water barrier, protecting the metal surfaces from dissolved corrosive species. Octadecylamine has been most widely applied as a filming amine, although other amine types are also available to minimize corrosion through film formation.

When condensate is treated with filming amines, proper control of amine residual is essential. An insufficient feed rate will lead to localized depletion of the protective film, permitting accelerated corrosion in these areas. Overfeed of filming amine can result in fouling of various cycle equipment. Filming amines may also form boiler deposits or contaminate the steam if overfed.

Oxygen Scavengers and Metal Passivators. Control of air in-leakage is often difficult, especially in industrial steam and power cycles where steam may be applied in a variety of process or heating operations. In some cases, the work force required to maintain the system properly becomes cost-prohibitive. In other instances, large portions of the condensate return system are inaccessible.

Many times, volatile chemicals are applied to condensate systems to control the effects of oxygen dissolved in the condensate. Some chemicals such as hydroquinone function as oxygen scavengers at typical condensate temperatures, reacting with DO before it can corrode metals in the cycle. Hydrazine and other chemicals may be added to function as metal passivators in the condensate cycle. These reducing agents do not react directly with DO to a significant extent, but instead react with iron oxide to restore it to its reduced form.

Additional chemicals are available as oxygen scavengers and metal passivators, but these are typically applied as feedwater treatments.

Combination Treatments. In well-maintained cycles, a single chemical treatment, such as ammonia or another neutralizing amine, is usually satisfactory to control pH and minimize corrosion rates. In cycles that include condensate returns from a number of sources, two neutralizing amines are sometimes applied in an effort to control corrosion better. Applying the two amines is based on the variance in vapor/liquid distribution ratios for the individual amines. The preferred condensate treatment for a particular application is specific to the system in question.

Air In-leakage Control

Control of air in-leakage into the condensate system of any type of steam generator is crucial, since the air contains O_2, CO_2, and occasionally SO_2 and other corrosive gases. These gases consume treatment chemicals, and corrosion products created as the result of air in-leakage are transported to the boiler. These products may deposit in the boiler, where directly or indirectly they can cause additional corrosion of boiler metal. To control corrosion in the condensate system, sources of air in-leakage must be minimized through proper fabrication, assembly, and sealing techniques. Principal sources of air in-leakage are found in equipment under vacuum, such as turbines and condensers.

FEEDWATER TREATMENT

Feedwater frequently consists of condensate returns combined with cycle makeup water ahead of the deaerating heater, which is the usual case in industrial preboiler systems (Fig. 10.2). This concept differs from a utility system's, where the makeup is most often added to the condensate in the condenser hot well (Fig. 10.1). In either case, that portion of the cycle starting with the deaerating heater may be considered the beginning of the feedwater system.

Treated makeup, as well as contaminated condensate, represents potential sources of cycle contamination, although makeup-water-treatment technology—if it is selected, operated, and monitored properly—will not contribute to cycle contamination. Approaches to control condensate purity have been discussed. Under all circumstances, careful surveillance of the feedwater as well as condensate chemistries is crucial to steam generation operations. As a minimum, surveillance of pH, specific and cation conductivities, and DO is required for high-purity feedwater maintenance. Indications of feedwater contamination suggest a need for corrective action in either the condensate or the makeup treatment system. In cases of significant contamination, it is necessary to remove the boiler and turbine from service to protect them from damage.

It is important to recognize that corrosion rates increase with temperature. As condensate enters the high-pressure heaters of the feedwater system, increasing temperatures render the water more corrosive to system metals, particularly steel. If contaminants are present in the water, its corrosivity can be increased.

Deaeration

In most fossil-fuel systems, a deaerating heater separates the condensate and feedwater systems, as described earlier. These devices remove dissolved gases from the water. Removal efficiency is highest for DO. Carbon dioxide and other volatile substances such as ammonia and hydrazine, if present in the feedwater, will be only partially removed by the deaerator because of their solubility in water.

Besides deaerating heaters, deaerating condensers also see service in utility condensate and feedwater systems. Although deaerating condensers protect mainly the condensate portion of these systems, they offer additional protection against DO corrosion in the feedwater system.

Deaerating heaters are direct-contact devices using steam as a stripping agent to remove noncondensable gases from feedwater. During unit start-up these heaters do not function until turbine extraction steam is available, unless steam can be tapped from another unit or an auxiliary boiler. Manufacturers generally guarantee deaerator effluent water to have a maximum DO content of seven parts per billion (7 ppb).

Storage of hot deaerated feedwater is provided either in a section of the deaerator proper or in an attached storage tank, most often beneath the deaerator. Water from the makeup treatment system is normally introduced to the heater section of the deaerator.

Chemical Treatment

Under normal operating conditions, minimal chemical treatment of feedwater is required. This statement assumes good control of condensate purity, whether or not the system includes a polisher. The principal chemicals applied to feedwater are an oxygen scavenger and ammonia or an amine for pH control. These and other boiler-water-treatment chemicals, such as chelants and iron dispersants, are preferably introduced to the feedwater via condensate entering the deaerating heater.

Oxygen Scavengers and Metal Passivators. The selection of certain oxygen scavengers and metal passivators has been discussed briefly with reference to the condensate system. Control of corrosion by DO becomes even more critical in the feedwater system since corrosion rates increase with temperature. A variety of chemicals are available to minimize this problem by either direct reaction with dissolved oxygen or passivation of metal surfaces in the feedwater system.

Sodium sulfite and sodium sulfite catalyzed with cobalt have been applied for many years in low-pressure cycles. Sulfite, a reducing agent, functions strictly through reaction with DO. Hydrazine has been widely used in intermediate- and high-pressure cycles. As a volatile compound, hydrazine treatment adds no dissolved solids to the feedwater, an advantage over sodium sulfite. Hydrazine functions as both an oxygen scavenger and a metal passivator. However, relatively high temperatures such as those present in high-pressure feedwater heaters are necessary before the rate of direct reaction with oxygen becomes significant. Also, it has been identified as a suspected carcinogenic material, and proper care should be taken near it.

A number of hydrazine substitutes have been developed and are in use in some cycles. Hydroquinone is an oxygen scavenger that has been applied in blends with hydrazine to "catalyze" its reaction with DO, but it may also be used alone as a hydrazine substitute. Other chemicals such as carbodihydrazide decompose at feedwater temperatures and form hydrazine as a by-product. Organic oxygen scavengers, such as erythorbic acid and diethylhydroxylamine, are also available.

Amine Additions. Both neutralizing amines and filming amines were discussed in detail as part of condensate treatment. However, volatile neutralizing amines, particularly ammonia, are frequently introduced to the feedwater. This practice is required in cycles that contain condensate polishing demineralizers, since ammonia is removed by cation resin prior to the ammonia breakpoint. In boilers treated with volatile chemicals, the ammonia feed rate determines the pH.

All-Volatile Treatment. *All-volatile treatment* (AVT) is based on treatment with hydrazine and ammonia or volatile amines. No other materials that add dissolved solids to the boiler water are used.

Solids-Based Additions. Solids such as sodium phosphates and caustic soda are most often added directly to the boiler water in the steam drum. This treatment maintains boiler-water pH in the alkaline range and provides buffering capacity against boiler-water pH excursions as the result of cycle contamination. Solids-based additions also react with scale-forming ions such as calcium, magnesium, and silica to form relatively soft, nonadherent sludges. These sludges tend to be less resistant to heat transfer and more easily removed by chemical cleaning than the mineral scales that would form in the absence of these treatments. Sludge formation in the feedwater system would be undesirable, however, and for this reason these chemicals are normally added directly to the boiler.

Organic chelating agents (chelants) are used widely in low-pressure drum boilers, particularly in industrial plants. Chelants such as ethylenediaminetetraacetate (EDTA) salts function by forming complexes with divalent cations including calcium, magnesium, iron, and copper. Formation of a stable complex prevents formation of deposits in the boiler, and boiler blowdown removes the complexed cations from the boiler.

Iron dispersants are also recommended to help remove suspended iron from the boiler. These and other nonvolatile, organic treatment chemicals such as chelants are useful in maintaining boiler cleanliness where tube-metal temperatures are low enough to permit their selection.

TREATMENT APPLICATIONS

Selection of the best equipment and chemical treatment approaches to maintain desired feedwater and condensate purity is dependent on several site- and plant-specific factors. These include (but are not limited to) makeup water characteristics, operating pressure, system metallurgy, and any applications for steam other than for power generation. The following paragraphs address some of the considerations in applying condensate and feedwater treatment to some types of power generation systems.

Drum-Type Boilers

This classification encompasses a wide range of industrial and utility power generation systems. Operating pressures of under 1000 to 2800 psig (6890 to 19 290 kPa) are encountered with drum-type, fossil-fuel boiler systems. As such, any of the equipment and chemicals discussed may be selected.

Condensate polishers have long been viewed as optional equipment for drum-type boilers, but new plant designs usually include some form of condensate purification equipment. Magnetic filters and/or sodium-cycle polishing demineralizers are most commonly included in industrial power generation systems. Deep-bed and precoat demineralizers are usually specified for higher-pressure, utility drum-type boilers.

Numerous condensate chemical treatments, including combinations of neutralizing amines, filming amines, and oxygen scavengers and metal passivators, have been applied with success in low- to intermediate-pressure industrial and utility boiler systems. Ammonia and hydrazine are frequently the only treatment chemicals present in the condensate of utility systems operating at pressures above 2000 psig (13 780 kPa).

Once-Through Boilers

These boiler systems operate at both subcritical and supercritical pressures and require very pure feedwater. Solids-based boiler-water treatment such as coordinated phosphate cannot be used since chemicals like these will carry over with the steam. AVT, consisting of ammonia and hydrazine, is usually applied in these systems, since whatever is introduced in the feedwater can readily form deposits to restrict boiler tube circuits or carry through with the steam into the turbine.

Once-through boilers are vulnerable to contamination, so full-flow deep-bed condensate polishers are required. Makeup water must also be of the highest quality. Control of air in-leakage in these systems is important, to protect the treatment chemicals and thus control corrosion.

SUGGESTIONS FOR FURTHER READING

Abrams, I.M., "Developments in Deep-Bed Condensate Polishing," 5th Mexican Water Treatment Congress, Mexico City, Mexico, March 1985.

Crits, G.J., "Co-countercurrent Ion-Exchange Regeneration," 5th Mexican Water Treatment Congress, Mexico City, Mexico, March 1985.

Emmett, J.R., and P.M. Grainger, "Ion-Exchange Method in Condensate Polishing," International Water Conference, Pittsburgh, Pa., October 1979.

Gabrielli, F., and H.A. Grabowski, "Steam Purity at High Pressure—System Considerations," Joint Power Generation Conference, Charlotte, N.C., 1979.

Gabrielli, F., and W.R. Sylvester," Water Treatment Practice for Cyclic Operation of Utility Boilers," International Water Conference, Pittsburgh, Pa., October 1979.

Gabrielli, F., et al., "Prevent Corrosion and Deposition Problems in High-Pressure Boilers," *Power*, July 1978.

Handbook of Industrial Water Conditioning, 8th ed., Betz Laboratories Inc., Trevose, Pa., 1980.

Kemmer, F.J., *The Nalco Water Handbook*, McGraw-Hill Book Co., New York, 1979.

Nitti, N.J., "A New Condensate-Polishing Technique for Medium-Pressure Boilers," International Water Conference, Pittsburgh, Pa., October 1979.

Strauss, S.D., "Boiler, Turbine Reliability Continue to Dictate Water and Steam Purity," *Power*, November 1982.

Strauss, S.D., "Water Treatment—A Special Report," *Power*, June 1974.

Yost, W.H., "Chemical-Treatment Options for Steam Generation," *Power*, February 1982.

SECTION 2

TURBINES AND DIESELS

CHAPTER 2.1

STEAM-TURBINE FUNDAMENTALS

Staff Editors
Power *Magazine*
New York, NY

INTRODUCTION

In powerplants, steam turbines convert the thermal energy developed by steam generators and boilers into mechanical energy, or shaft torque. This turbine output drives electric generators, which produce the electricity that is subsequently transformed and transmitted for use by consumers large and small. Smaller turbines, sometimes called *mechanical-drive turbines,* provide rotational power for feedpumps, motors, and other equipment in the powerplant. Obviously, the steam turbine is a vital link in the production of electric and mechanical power in the modern utility or industrial plant. How it works, key turbine components and design elements, and basic turbine types are the concerns of this chapter. The next chapter builds on these fundamentals to discuss steam turbines in practice.

HOW TURBINES USE STEAM'S ENERGY

In a reciprocating engine, steam presses equally on cylinder walls and piston (Fig. 1.1). Since the piston moves, the steam does work, using some of its internal energy to do this. The steam cools as its pressure drops. Similarly, steam in a nozzle box presses equally on all walls, but it escapes through the nozzle to form a high-speed jet (Fig. 1.2). Reaction pressure P_R on the wall area opposite the

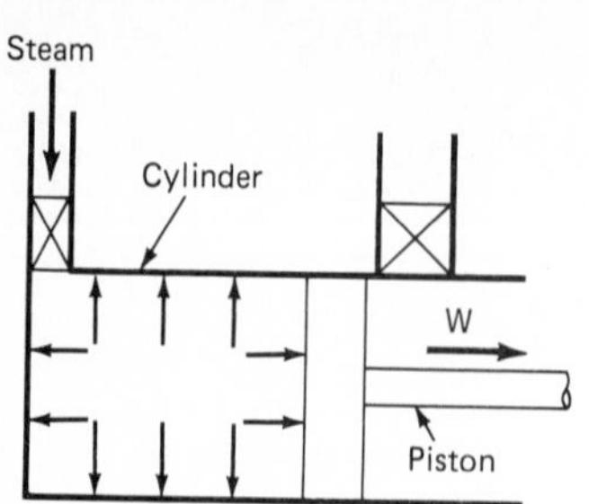

FIG. 1.1 Expanding steam in reciprocating engine moves the piston and converts internal energy to work.

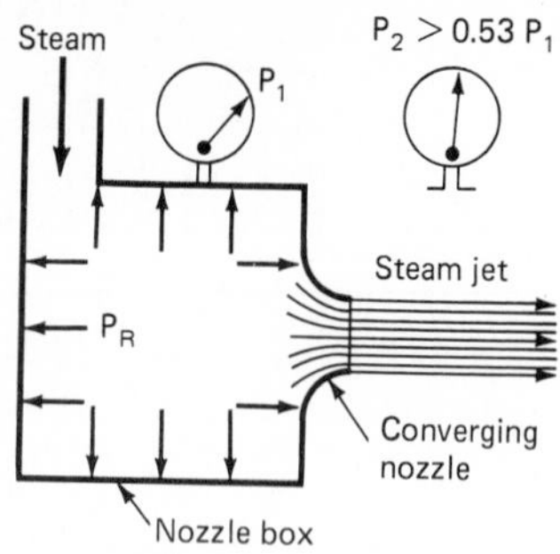

FIG. 1.2 Converging nozzle expands steam, creating an unbalanced reaction force on the opposite wall.

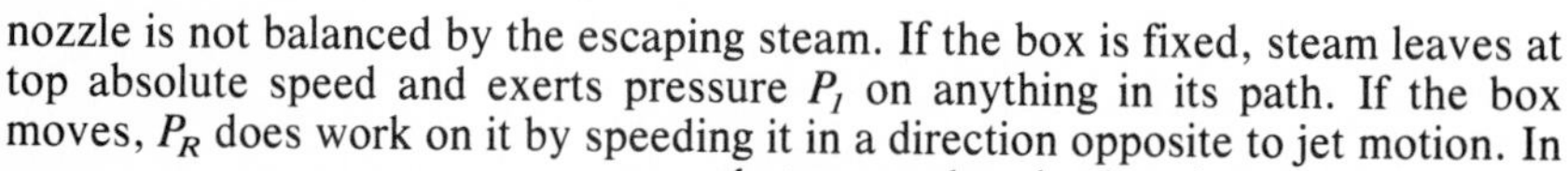

nozzle is not balanced by the escaping steam. If the box is fixed, steam leaves at top absolute speed and exerts pressure P_1 on anything in its path. If the box moves, P_R does work on it by speeding it in a direction opposite to jet motion. In that case, the absolute jet speed is correspondingly slower.

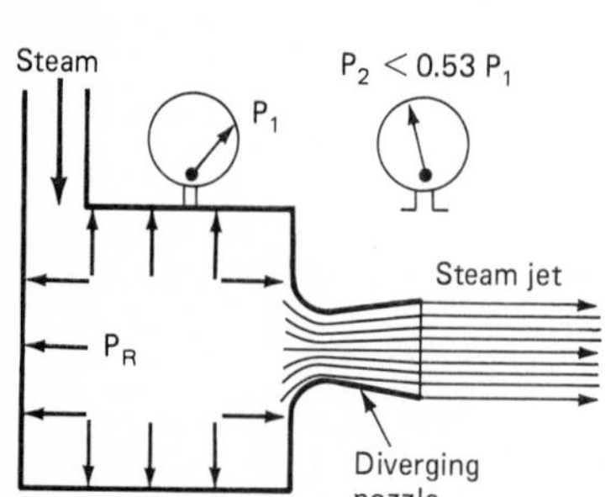

FIG. 1.3 Diverging nozzle is required if the discharge pressure is under 53 percent of initial pressure.

When the steam discharge pressure is 53 percent or more of the box pressure, the nozzle needs only a converging cross section. When the discharge pressure is much less than 53 percent, the converging section should be followed by a diverging section (Fig. 1.3). The area of exit cross section depends on the pressure ratio.

Impulse-turbine nozzles organize the steam so that it flows in well-formed high-speed jets. Moving buckets absorb the jet's kinetic energy and convert it to mechanical work in a rotating shaft (Fig. 1.4). When the bucket is locked, the jet enters and leaves with equal speed and develops maximum force F, but no mechanical work is done. As the bucket is allowed to speed up, the jet moves more slowly and force F shrinks. Figure 1.5 shows how both force and work done vary with the blade speed. The steam jet does maximum work when the bucket speed is just one-half of the steam speed. In this condition, the moving bucket leaves behind it a trail of inert steam, since all kinetic energy is converted to work. The starting force or torque of this ideal turbine is double the torque at its most efficient speed.

For practical reasons, most impulse turbines mount their buckets on the rims of disks, and nozzles feed steam from one side (Fig. 1.6). Pressurized steam from the nozzle box flows through parallel converging nozzles formed by vanes or foils. Steam leaves as a broad high-speed jet to flow through the slower-moving-bucket passages, which turn the steam flow to an axial direction as they absorb its kinetic energy. The steam leaves with lower internal energy and speed.

Steam pressure and speed vary through the true impulse stage. When the impulse stages are pressure-compounded, exhaust steam from one stage flows through following similar impulse stages, where it expands to a lower pressure. If

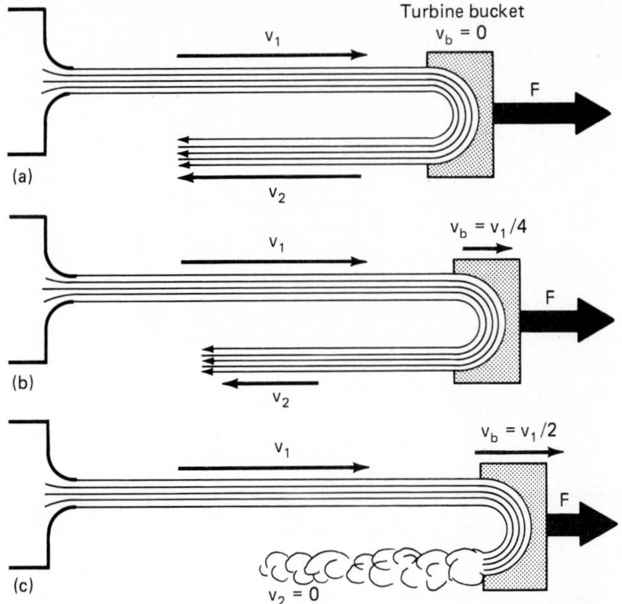

FIG. 1.4 In a steam turbine: (*a*) The steam jet exerts maximum force on the locked impulse bucket, but no work is done since the bucket does not move; (*b*) when the bucket moves with one-quarter speed of the steam jet, the force diminishes and steam leaves the bucket at lower speed but first does work by moving the bucket; and (*c*) when bucket speed equals one-half the steam-jet speed, the force drops to half the locked condition. Steam leaves the bucket with zero speed and does maximum work.

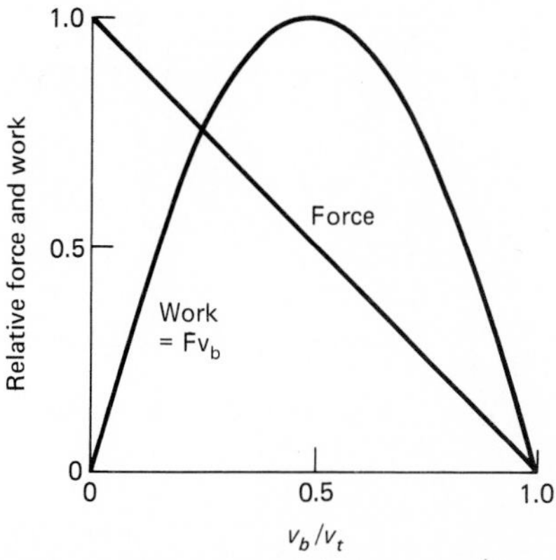

FIG. 1.5 Curves show how the reactive force and work vary with bucket speed.

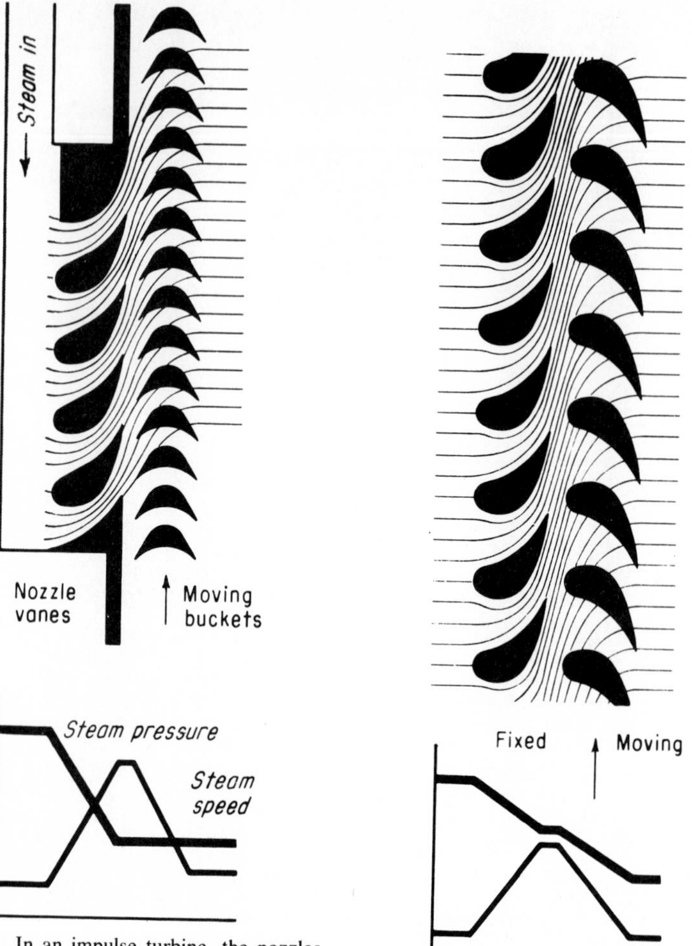

FIG. 1.6 In an impulse turbine, the nozzles direct steam into buckets mounted on the rim of a rotating disk. The steam flow changes to an axial direction as it goes through the moving passages. The pressure drops only across the nozzle.

FIG. 1.7 In the reaction turbine, stationary blades direct steam into passages between moving blades; pressure drops across both the fixed and moving blades.

the impulse stages are velocity-compounded, steam velocity is absorbed in a series of constant-pressure steps.

In the reaction stage (Fig. 1.7), steam enters the fixed-blade passages; it leaves as a steam jet that fills the entire rotor periphery. Steam flows between moving blades that form moving nozzles. There it drops in pressure, and its speed rises relative to the blades, which creates the reactive force that does work. Despite the rising relative speed, the overall effect reduces the absolute steam speed through one stage. When the enthalpy drop is about equal in moving and stationary blades, it is called a *50 percent reaction stage.*

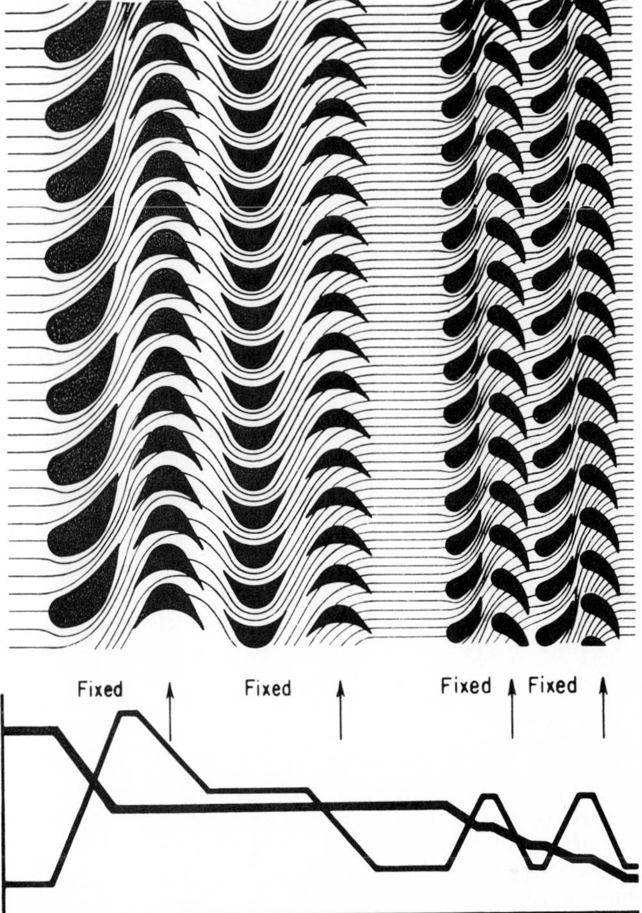

FIG. 1.8 In the velocity-compounded turbine (above), steam is discharged into two reaction stages. The velocity stage uses a large pressure drop to develop the high-speed jet. Part of the kinetic energy is absorbed in the first moving bucket row. Fixed buckets turn partly slowed steam before it enters a second row of moving buckets, where most of the remaining energy is absorbed. In pressure compounding, there are multiple nozzles, each followed by one row of moving buckets. Each stage is essentially designed as a single-stage impulse turbine.

A velocity-compounded control stage is followed by two reaction stages (Fig. 1.8). The high-speed steam jet gives up only part of its kinetic energy in the first row of moving buckets. Then come reversing blades that redirect the slowed-up steam into the second row of moving buckets, where most of its remaining kinetic energy is absorbed. Steam then enters the series of reaction stages.

In practice, steam turbines combine these impulse and reaction stages, although manufacturers usually label their products as either reaction or impulse turbines. The latter is often compared to a waterwheel (driven by steam), while the reaction turbine can be compared to a rotary lawn sprinkler.

KEY TURBINE COMPONENTS

The previous paragraphs covered the basic methods of converting the internal energy of pressurized steam to mechanical shaft work. By allowing the steam to drop in pressure while it flows through a nozzle, a high-speed jet is formed. A large amount of leeway exists as to where the pressure drop takes place in the turbine.

In the impulse design—single, pressure-compounded, or velocity-compounded—the pressure drops are mainly in the stationary nozzles; the pressure level across the moving buckets drops slightly. In the reaction design, the stages lose pressure in both the stationary and moving passages. The ratios of the two drops fall in a wide range. One that is often used is a 50 percent reaction, giving a pressure ratio nearly equal across stationary and moving passages. The 50 percent reaction produces the most efficient energy conversion when the speed of the moving blades about equals the steam-jet speed as it leaves the stationary blade passages.

Many of today's impulse turbines use about 5 to 10 percent reaction in their design. Thus there is a small steam pressure drop through the moving-blade passages. Instead of taking a symmetric shape, these buckets have a longer tail to form a slightly converging passage at the exits.

Blade and Bucket Sections

Multistage steam turbines have large cross-sectional areas of steam flow, so the choice of blade and bucket sections becomes complicated. A stage may begin as pure impulse at the root of the bucket and nozzle and then increase the percentage of reaction along the length of bucket and blade toward the outer tips. Other parts must guide the steam flow and carry the mechanical shaft work out of the unit, if nozzles and buckets are to do their job. The buckets and moving blades ride on rotating parts, disks carried on either a shaft or a rotating drum (Fig. 1.9). Where the percentages of reaction are higher, more stages will be used over a given pressure range to limit steam leakage past blade tips and seals. Because of the larger number of stages, drum construction is more practical.

Shaft Assembly

A shaft carries the drum and is supported in turn by journal bearings at each end. A thrust bearing controls the rotor's axial position to hold clearances between stationary and moving elements of the stages. Couplings mounted on shaft ends transmit the mechanical energy from the buckets to the driven equipment (generators, pumps, motors, etc.).

Motor-driven turning gears on the shaft rotate the drum when the turbine is warming up prior to loading and while the unit is cooling after it has been unloaded and the steam has been shut off. This rotation prevents unequal expansion or contraction that might kink the rotor assembly, creating a serious imbalance.

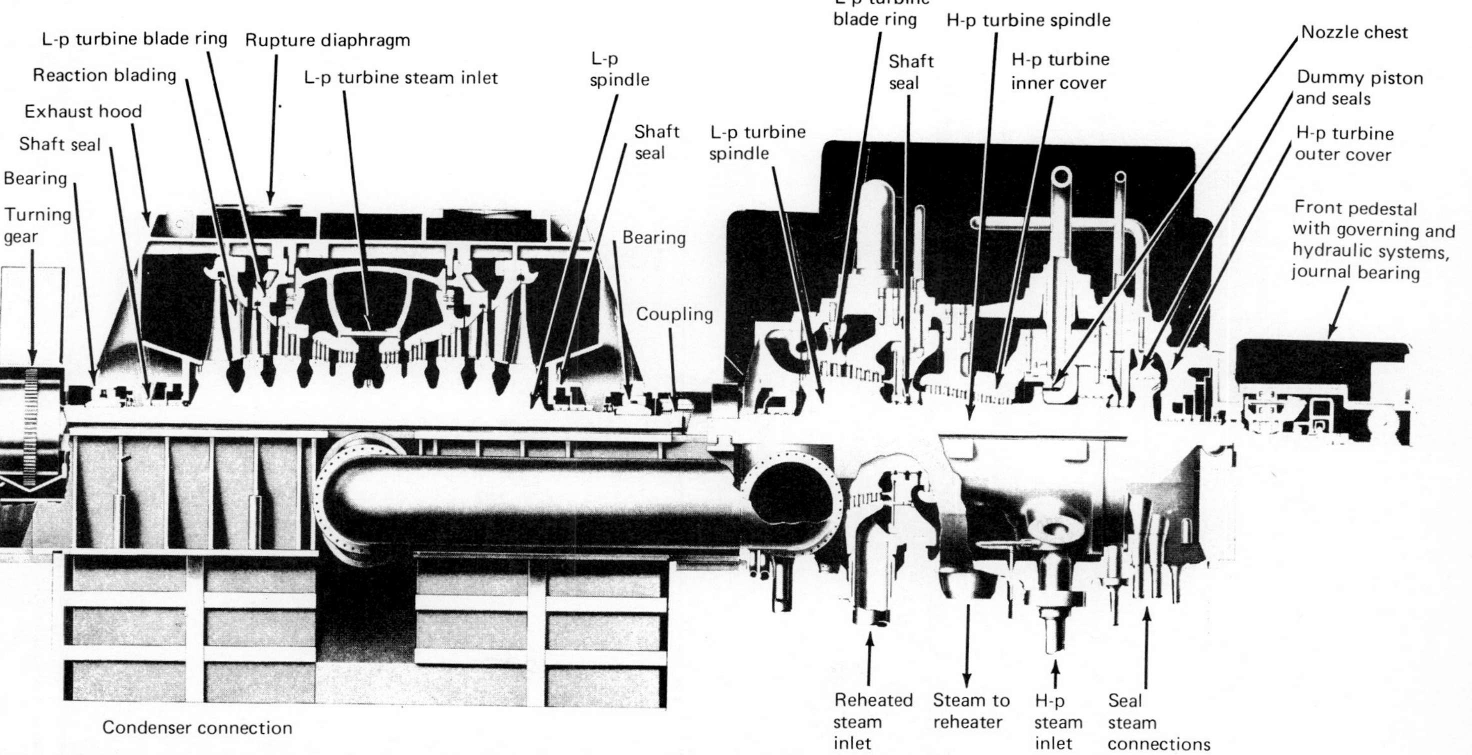

FIG. 1.9 This drum-type turbine has two main casings. The buckets and moving blades ride on the rotating drum.

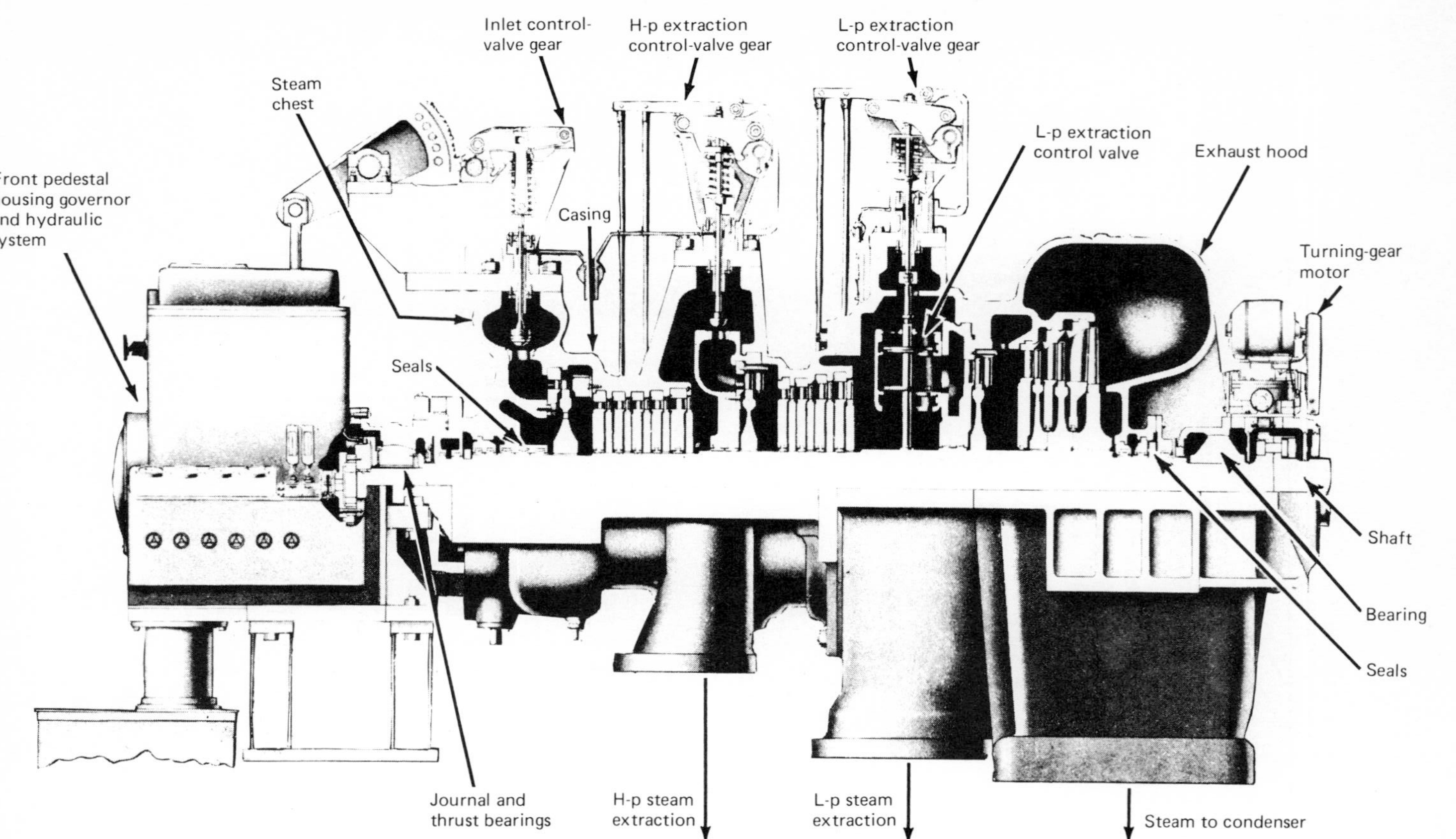

FIG. 1.10 A double-automatic-extraction condensing turbine receives boiler steam in its steam chest; governor-controlled valves admit steam to nozzle chests. Extraction control valves hold the steam pressure by regulating its flow to the stages following.

The turning-gear motor automatically declutches when the rotor accelerates above a given speed; some engage automatically at this speed limit when the rotor is decelerating during shutdown.

The casing carries bearings for the rotor and holds the structure needed to guide steam flow through nozzles and buckets. Steam enters a turbine through a stop valve and steam chest. In high-temperature units, these elements are separate from the main turbine structure. In smaller units (Figs. 1.10 and 1.11), the steam chest usually mounts directly on the casing, either as an integral part of it or as a separate casing.

Governing valves in the steam chest admit steam to the nozzle bowls, chests, chambers, or boxes of the first stage. The first stages in Fig. 1.9 are velocity-compounded impulse or control stages.

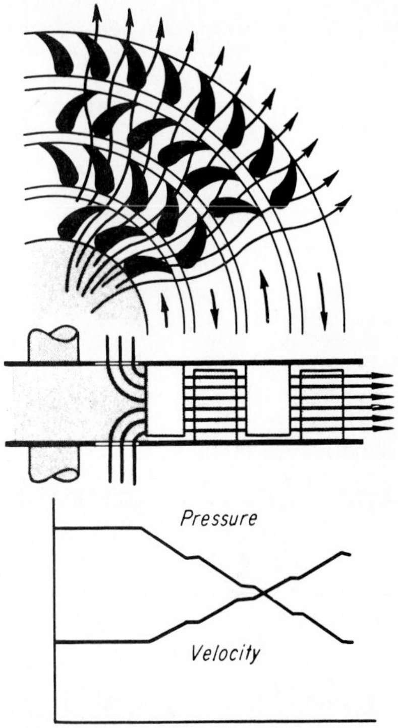

FIG. 1.11 A radial-flow, double-rotation reaction turbine has only moving blades. It can handle high-speed steam jets because there are no stationary blades; steam can be extracted.

Casing and Exhaust Hood

The casing supports nozzles and reversing blades of all stages held in the blade rings and covers. It also holds the seal strips, which limit steam leakage at all points where the shaft passes. Seals must also be provided at diaphragm openings and reaction blade tips.

Since a reaction force acts on stationary nozzles and blading, the casing must be securely anchored to resist these forces. Conversely, the steam temperature swings widely between shutdown and operating conditions, which means that the rotor and casing expand and contract. The front, high-pressure end must be allowed to move axially, so that it can expand and contract as needed. At the same time, the casing must stay axially aligned to maintain clearances from the rotating parts.

An exhaust hood guides the flowing steam from the last stage of the turbine to the point of disposal—a pipeline, condenser, etc. This hood must be designed to minimize pressure loss, which otherwise would reduce the turbine's thermal efficiency. In highly refined central station units, the hood may act as a diffuser to develop pressure at the outlet of the last-stage buckets lower than at the hood outlet. It does this by converting leaving kinetic energy of the last stage to internal energy.

Rupture diaphragms in the exhaust hood prevent excessive pressure build-up, should the condenser lose its vacuum. A governor geared to the shaft in the front bearing pedestal operates the governor valves through a hydraulic servomotor system. The main governor can be mechanical or hydraulic; it opens and closes the valves in response to the rise and fall of shaft rpm. An overspeed governor is usually located in the shaft at the front bearing pedestal. Its function is to shut a main stop valve ahead of the governing valves when shaft rpm exceeds the rating by 10 percent (15 percent on small turbines). In recent years, solid-state technology has invaded the realm of governor functions on steam turbines; however, the aims and end results are the same as for mechanical and hydraulic devices.

Disk-and-Diaphragm Turbine

The disk-and-diaphragm type of turbine sees service where impulse staging predominates with higher pressure drops per stage, and the percentage of reaction is relatively low (Fig. 1.10). It features two points of automatic extraction in which parts of the steam may be withdrawn from the turbine at constant pressure; the steam chest is integral with the casing. Control and extraction valves respond to both shaft speed and steam pressure at the extraction points.

In this turbine, the moving buckets ride on wheels or disks that are shrunk and keyed to the shaft. The nozzles are carried in diaphragms that have centered holes through which the shaft passes. Seals at these openings set a limit to steam leakage bypassing the nozzles; construction of the diaphragm limits the cross-sectional area available for leakage through the seal.

All turbines need a lubricating system for their bearings. Usually it also acts as the hydraulic system for actuating valves and servomotors. The oil reservoir is often remote and at a lower level to promote drainage.

TURBINE DESIGN ELEMENTS

Every stage of a turbine has two basic design elements: the stationary nozzle and the moving bucket or blade. The design of these parts depends on factors such as entering steam conditions, exhaust steam pressure, shaft speed, rated capacity, and steam flow.

Nozzles and Blades

As mentioned earlier, nozzles may be either converging or converging-diverging; selection depends on the pressure ratio of steam across the nozzle. Nozzles are grouped so they can be controlled by individual valves. In the high-pressure end of the turbine, nozzle vanes are usually welded into the nozzle diaphragm; in the low-pressure end, nozzle vanes may be cast as an integral part of a diaphragm.

In a small turbine, converging-diverging nozzles are drilled into the block, and a set of stationary reversing buckets is attached for the two-row control or velocity-compounded stage. Steam passages through the nozzles have a circular cross section. A single converging-diverging nozzle usually has attached reversing blading and fits a small turbine with a two-row single wheel. A typical nozzle diaphragm for the lower stage of a disk-and-diaphragm turbine is shown in Fig. 1.12 (top). In a nozzle row for a drum-type block (Fig. 1.12, bottom), no inner diaphragm is needed because the drum occupies this space. Figure 1.13*a* shows nozzle, wheel, and stationary blading for the two-row single-wheel unit.

Velocity compounding can be done with a single row of moving buckets.

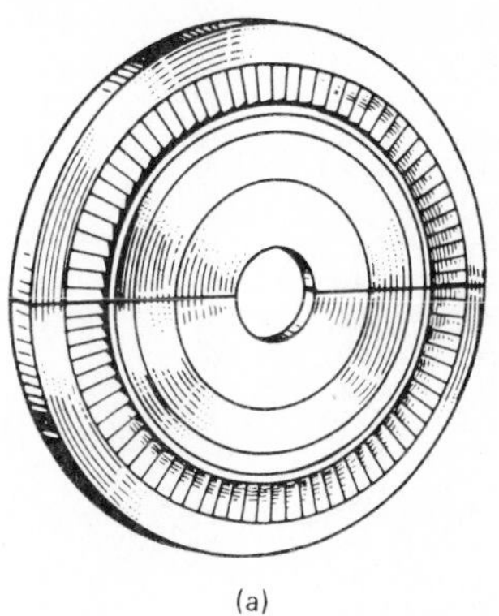

(a)

FIG. 1.12 Typical nozzle setups are for disk-and-diaphragm turbine (top), and drum-type unit (bottom).

Steam may pass just once through the blades (Fig. 1.13*a*), or a stationary reversing nozzle (Fig. 1.13*b*) may guide high-speed steam into a second pass through the moving buckets. In another method of velocity compounding, steam flowing from a converging-diverging nozzle makes three passes through buckets milled into the edge of a solid wheel. Stationary reversing chambers return the high-

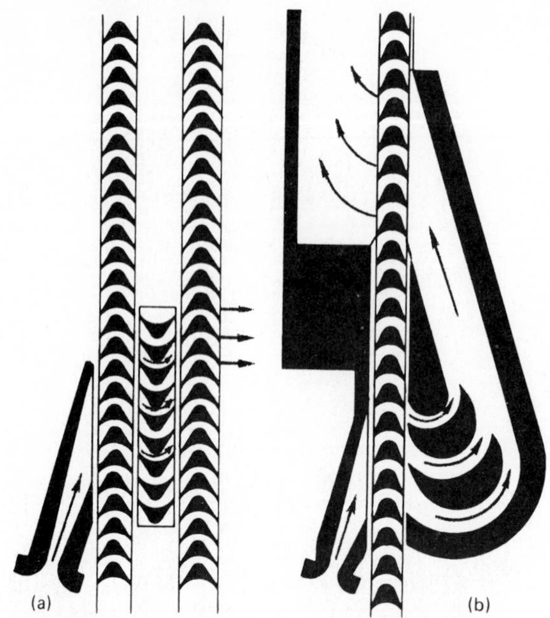

FIG. 1.13 At (*a*) a velocity-compounded stage passes steam once through moving blades; at (*b*), steam flows twice through the blades.

FIG. 1.14 The control stage for a large turbine has low-percentage reaction and steam pressure drop.

FIG. 1.15 Blade height and stage diameter increase in succeeding stages to accommodate the volume of steam.

speed steam to the wheel buckets along a helical path. The control stage for a large turbine (Fig. 1.14) differs slightly from that of a small turbine, since vanes have a small percentage of reaction built in.

Blades (or buckets) take many forms. They receive the working force of the steam and transmit it as a moving torque to the wheels and disks to which the blades are mounted. The blade height and stage diameter increase in succeeding stages to accommodate the volume of steam that expands as its pressure drops (Fig. 1.15). Shroud bands cover the blade tips to keep steam from spilling out radially.

In high-efficiency turbines, every effort is made to confine steam flow to the working passages through nozzles and buckets. Sealing strips between bucket shrouds and casing diaphragms minimize steam leaking past the bucket tip and into the following nozzle. Some turbines use pure impulse buckets without any pressure drop across them. In other designs (Fig. 1.14), the stage may have some degree of reaction—10 percent or less—to produce a small pressure drop across the bucket. This small drop can be very useful in keeping all buckets running full of steam, with the shaft thrust positive in one direction.

In many turbines that rely on velocity-compounded control stages, steam flows over only a part of the total periphery of the stage. In the "idle section," buckets churn the stagnant steam. This windage loss reduces turbine efficiency; to minimize the loss, shields fit closely over the idle part of the bucket travel. The

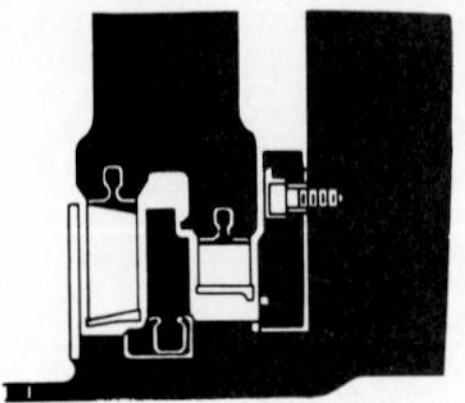

FIG. 1.16 Shields cut windage loss on idle buckets.

shields limit the amount of steam whipped about by the nonworking buckets (Fig. 1.16).

Blade Geometry

When blade height becomes a significant part of the total stage diameter, the ratio of steam to bucket speed changes over the length of the bucket. To counteract this change, warped buckets are installed (Fig. 1.17). An exploded view (Fig. 1.18) shows a nozzle diaphragm, steam vortex flow, a bladed or bucketed wheel, and the leaving exhaust steam. Ideally, steam enters the nozzles in an axial direction and leaves in a circumferential direction, forming a vortex flow or eddy that is contained by the turbine casing before steam enters the moving buckets. To avoid cross currents in the vortex flow, the product of linear velocity of steam and the radius of the circle it travels in must be constant, that is, $v_1 r_1 = v_2 r_2$. The steam pressure must be higher at the outer rather than the inner radius.

Steam leaves nozzles at the inner radius with higher linear speed than at the outer radius. But the bucket's linear speed increases with radius. So a steadily growing ratio of blade to steam speed occurs as one moves from root to tip. Figure 1.17 shows velocity diagrams at the root and tip of a blade that receives a

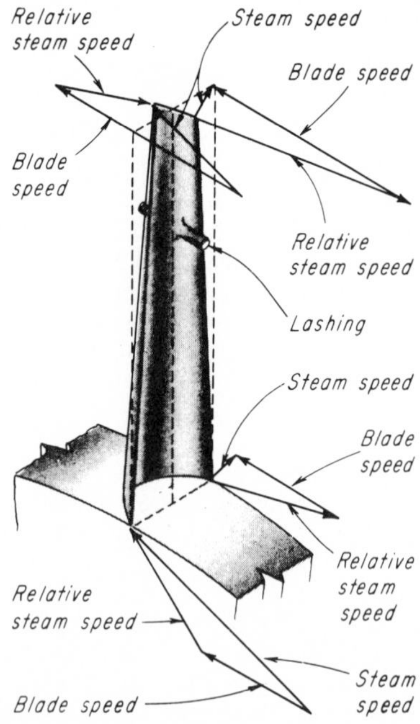

FIG. 1.17 This twisted blade works with impulse flow at root and reaction at tip.

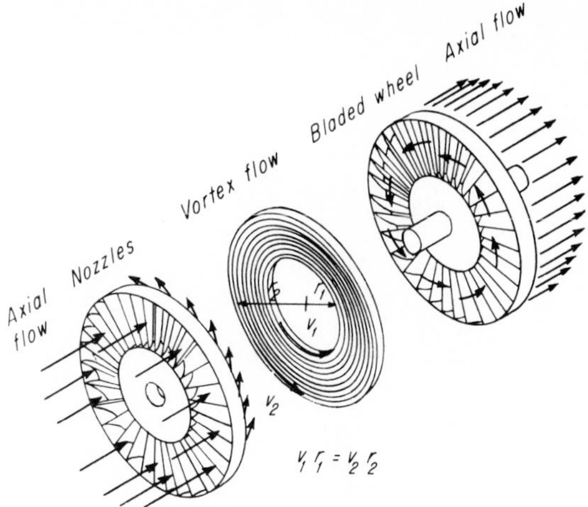

FIG. 1.18 Steam flows axially in entering-stage nozzles, leaves as vortex flow. A twisted-blade design returns it to axial.

steam jet moving in vortex flow. The blade root has been designed for impulse flow, which is equivalent to 0 percent reaction and no pressure drop.

The blade's entrance angle is fixed by the angle of approach of the steam's relative speed, so steam slides smoothly over the blade. In the ideal situation, the absolute steam speed should just about double the blade speed. At the blade's exit edge, the vector difference of relative steam and blade speeds shows that steam has a residual absolute speed in an axial direction.

Since the blade speed at its tip is about double the absolute steam speed, the steam must approach the blade from a direction almost opposite to its motion. The blade section must be twisted to receive the steam smoothly all the way up the blade. But since entering steam pressure is higher at the tip than at the root, there will be a pressure drop through the blade. Thus a reaction blade section must be used with the relative steam speed higher at the blade exit. A pure reaction force acts at the blade tip, a pure impulse force at the root. At the tip exit, the vector difference of relative steam speed and blade speed (Fig. 1.17) indicates that steam leaves with low velocity in an axial direction, just as at the root.

Rotors and Casings

Rotors for disk-and-diaphragm turbines may be machined from a single forging for smaller-capacity units or built up of separate disks shrunk and keyed on a forged and machined shaft. Edges of the disk are milled to hold the buckets. The shaft usually has integral journals for the bearings and also carries a thrust collar. Drum-type turbine rotors increase their diameter from the front to the rear end of the unit. They are usually made as one forging and then milled to receive the blade roots. Figure 1.19 shows cross sections of the solid or monoblock (top), the shrunk-on disk (middle), and welded (bottom) rotors. Steelmakers worldwide now have the capability to produce very large rotors from one forging, called the *monoblock*.

Casings hold the blade rings, bearings, and other stationary parts. A small unit's casing may split vertically, but all large ones split horizontally at the shaft

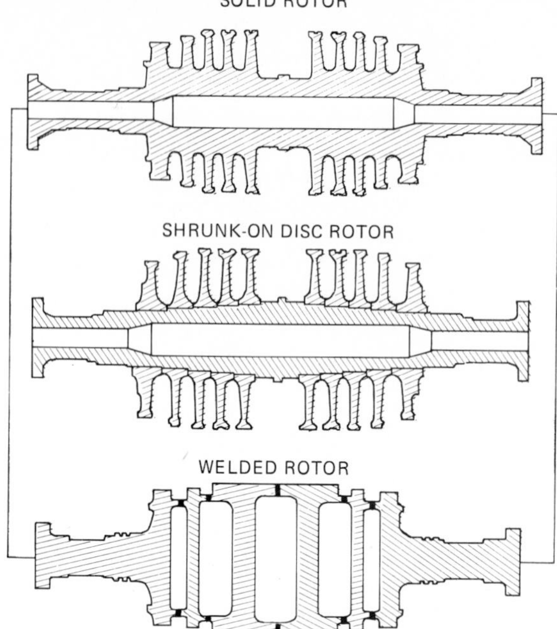

FIG. 1.19 Key types of rotors for large and small turbines.

centerline. For high-pressure, high-temperature service, front elements have been built with two casings; steam flows between the inner and outer one. This design allows thinner shells, so expansion and contraction are not as much of a problem. Casings may be cast or built up of welded plate.

Valves

Valves control the flow of steam through a turbine. A combined stop-throttle valve regulates steam flow during warmup by opening the inner valve first. When the turbine is ready to take the load, the main valve is opened wide and nozzle-chest valves pick up the control function. In an emergency, the stop valve slams shut when oil pressure is released from the operating cylinder below.

Reheat turbines need both intercept and stop valves between the boiler reheater and turbine-reheat inlet. Steam in the reheater and its connecting lines can seriously overspeed the unit during a large load drop or loss. The intercept valve closes part way during rapid load changes; both valves shut on overspeed.

Changes in steam flow to meet varying loads are controlled by valves in a steam chest. These admit steam to groups of nozzle chests, an arrangement giving high part-load efficiency. Valves are typically opened by cams on a governor-controlled shaft. A balanced governor-controlled single-admission valve (Fig. 1.20) handles the flow of small steam turbines. It may feed several paralleled nozzle chests on the unit; intervening hand valves can cut them in and out as desired.

In automatic-extraction turbines, three different types of valves may control

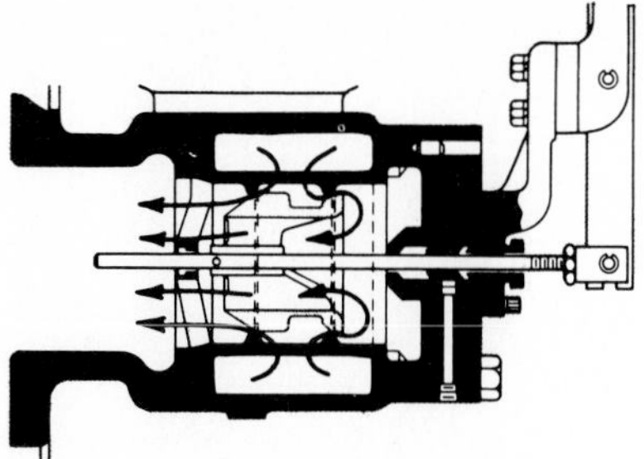

FIG. 1.20 Balanced throttle valve controls steam flow to smaller turbines.

steam flow through the downstream sections. In Fig. 1.10, cam-operated poppet valves control the first opening, with cam-operated spool valves for the second. A piston-operated grid valve may be used for low pressure drop across the valve—about 50 psig (345 kPa) or less.

Bearings and Seals

Bearings range from pressure-lubricated journal types for large turbines (Fig. 1.21) through ball bearings for small turbines. Larger bearings are almost universally designed with oil grooves in their top halves to build an oil wedge that presses down on the journal.

The steam pressure differential across most turbine stages creates a net thrust along the shaft. This must be counterbalanced to keep the rotor in proper position. Figure 1.22 shows a Kingsbury thrust bearing, in which individual movable thrust shoes bear on leveling plates (left). A thrust collar, fixed to the shaft, pushes on the shoes from the right and holds the clearance between moving and fixed components.

Steam leakage cannot be avoided entirely. Some steam will leak out of turbine casings where the shaft must pass through. In condensing turbines, air tries to

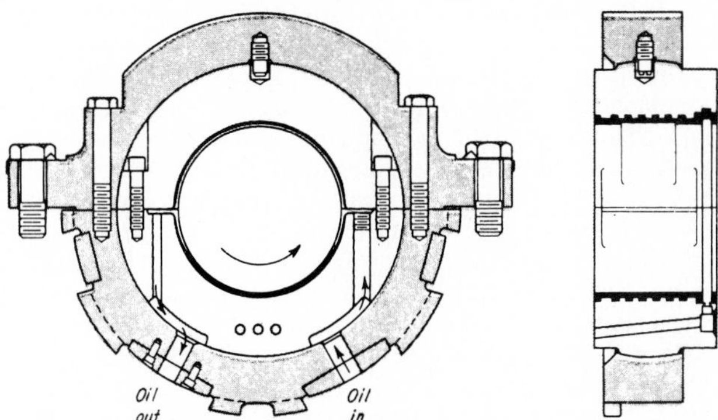

FIG. 1.21 Journal bearing has a spherical seat for self-alignment. The groove in top builds an oil wedge and stops the shaft from whipping.

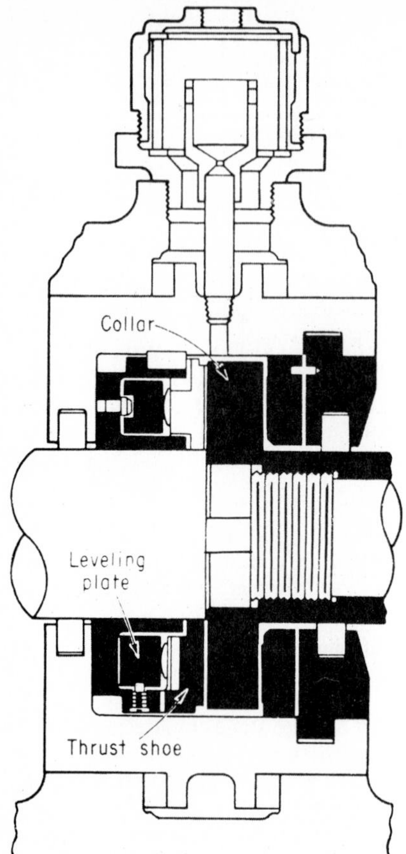

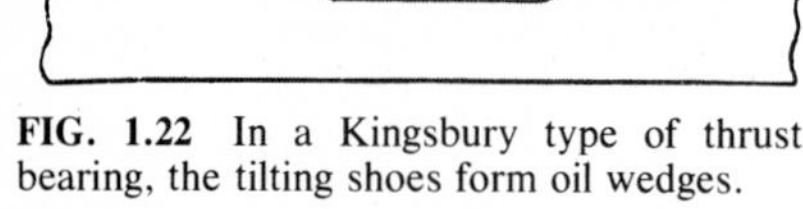

FIG. 1.22 In a Kingsbury type of thrust bearing, the tilting shoes form oil wedges.

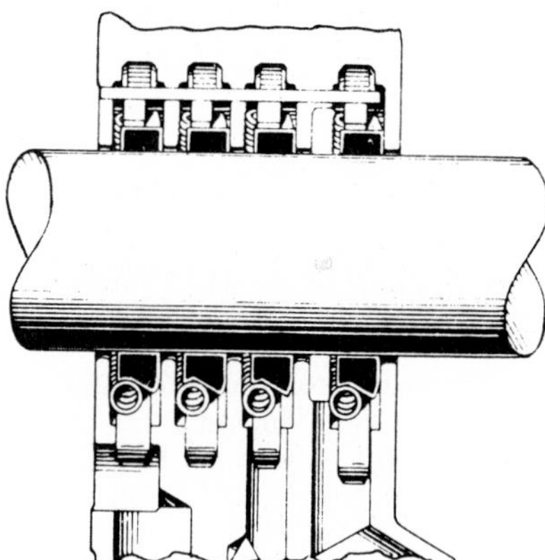

FIG. 1.23 Carbon packing rings form a simple seal against shaft steam leakage.

leak along the shaft into the low-pressure condensing space. Seals are installed to stop leakage. In a slant-top packing-ring seal (Fig. 1.23), carbon (or comparable material) packing rides directly on the shaft. Springs anchored at the top hold packing segments in place and against the sealing surfaces. Steam that does manage to leak past the seals may be channeled to a lower-pressure stage of the turbine, channeled to a heater, or vented to atmosphere. Condensed steam usually accumulates in the last section of the seal and then drains to waste.

Stepped labyrinth-gland seals also control shaft leakage. Intermediate leakoffs direct the steam to lower turbine stages or heaters. The large intermediate chamber may connect to the suction of a blower that holds a vacuum lower than the turbine's last stage. This would draw in low-pressure steam from one side and air from the blower side. The blower discharges the mixture to a condenser, where the steam is recovered.

Seal strips have tapered edges, so any accidental rubbing will wear them down quickly without overheating the shaft. Steps milled in the shaft match longer seal strips, forming a long, tortuous path with high flow resistance. Enough axial dis-

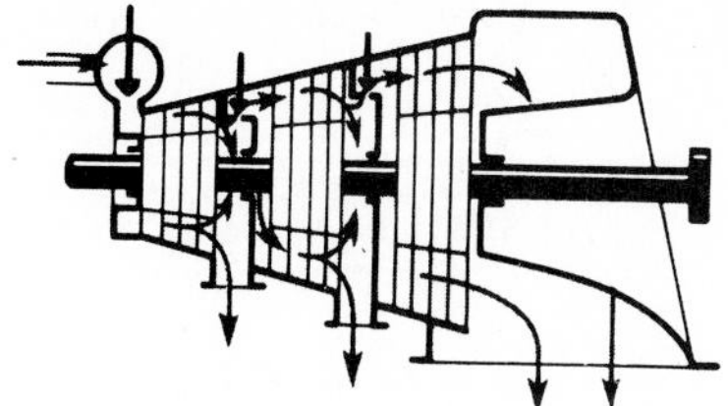

FIG. 1.30 Double-automatic-extraction unit, condensing type.

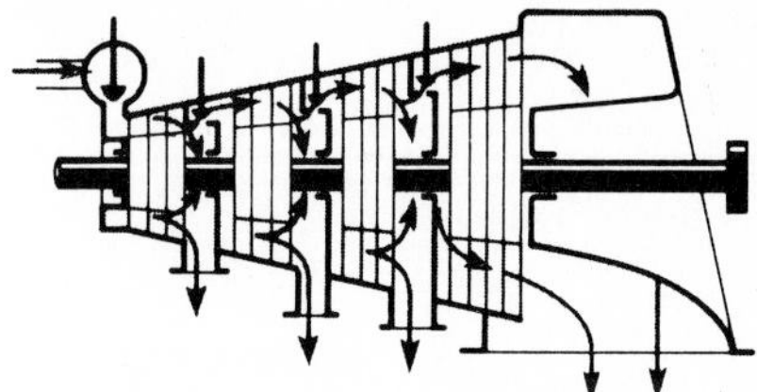

FIG. 1.31 Triple-automatic-extraction unit, condensing type.

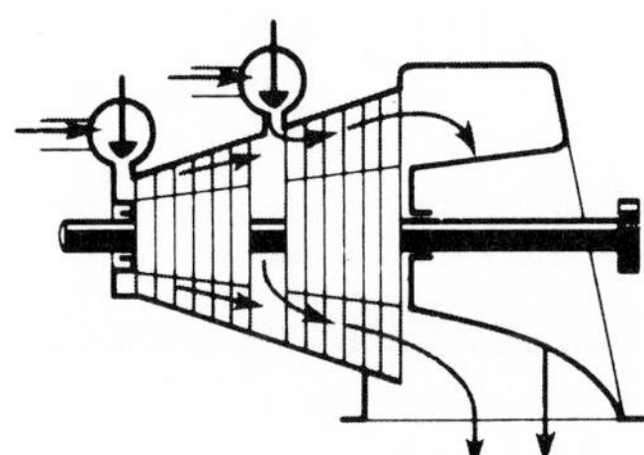

FIG. 1.32 Induction turbine, condensing type.

such variations can seldom be tolerated for process work, they are acceptable for feedwater heating service.

Induction turbines work as extraction turbines—except in reverse (Fig. 1.32). That is, steam at lower than throttle pressure is injected into the unit downstream of the throttle valve to produce a portion of the total output. Combination units incorporate both functions: extraction and induction. Induction machines generally are found in process plants, although at least one manufacturer also specifies them for utility combined-cycle service.

Most importantly, although the simplicity of straight-through machines keeps first cost low, operating flexibility and efficient use of steam dictate the selection of extraction or induction units in many process applications, and reheat, nonautomatic-extraction turbines in central station service.

Tandem Compounding vs. Cross Compounding

Simple, small multistage turbines are generally built with all the stages on a single shaft that is housed in one casing. As turbine sizes increase beyond 40 MW, however, single casings become impractical. To avoid long casing runs, the different stages may be split among two or more casings on separate shafts; as many as six different casings may see service today.

If all the shafts of the different casings are bolted together in line to drive the same generator, it is called a *tandem-compound turbine* (Fig. 1.33). In other installations, the sections may be arranged with two shafts, or groups of shafts, side by side driving two separate generators. This is called a *cross-compound turbine* (Fig. 1.34). Cross compounding for large units has the advantage of it being

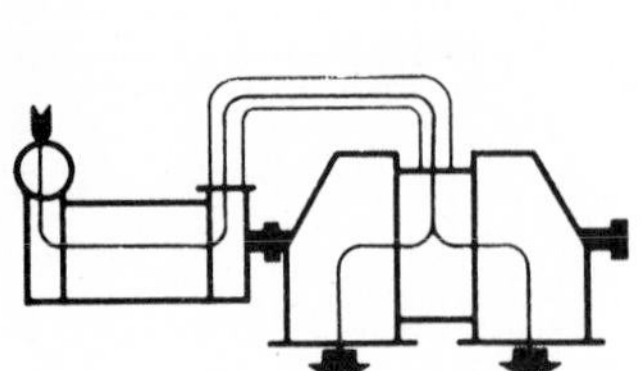

FIG. 1.33 Tandem-compound unit is a two-casing, double-flow design.

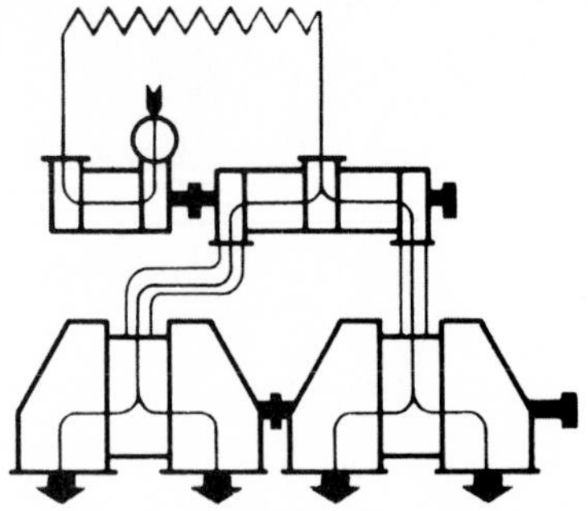

FIG. 1.34 Cross-compound unit. It is a four-casing, quadruple-flow, reheat design.

easier to construct two half-sized generators than one large unit.

Nearly all larger turbines built in the last 10 years are condensing, single or double reheat, with simple multiple extraction for regenerative feedwater heating, and nearly all are compound units, with most being tandem-compound rather than cross-compound machines. Steam conditions for modern turbines have also largely been standardized at 1800, 2400, and 3600 psia (12 400, 18 600, and 24 800 kPa). Initial and reheat steam temperatures range between 950 and 1050°F (510 and 566°C). Exhaust pressures, while limited by the temperature of the cooling water available, range from about 1 to 3½ inHg (3.38 to 11.83 kPa). Recently, there has been considerable interest in high-backpressure condensing turbines for pairing with dry cooling towers and condensers. It requires a special design of the last-stage blading.

ACKNOWLEDGMENT

Illustrations in this chapter are courtesy of GP Publishing, and *Power* magazine.

SUGGESTIONS FOR FURTHER READING

Akiba, M., et al., "The Design of a 700 MW Steam Turbine with Advanced Steam Conditions," *Proceedings of American Power Conference,* 47:190–195, 1985.

Baumeister, T., et al., *Marks' Standard Handbook for Mechanical Engineers,* 8th ed., McGraw-Hill Book Co., New York, 1978.

Faires, V.M., *Applied Thermodynamics,* The Macmillan Co., New York, 1949.

Nock, H.T., and K.T. Sullivan, "Upgrading Older Turbines for Today's Needs," *Proceedings of American Power Conference,* 46:158–164, 1984.

Ortolano, R.J., and M.L. Franz, "Improving the Availability of Electric Utility Steam Turbines," *Proceedings of American Power Conference,* 46:141–149, 1984.

Potter, P.J., *Power Plant Theory and Design,* John Wiley & Sons, Inc., New York, 1959.

Salisbury, J.K., *Steam Turbines and Their Cycles,* Robert E. Krieger Publishing Co., Huntington, NY, 1974.

Skrotzki, B.G.A., and W.A. Vopat, *Power Station Engineering and Economy,* McGraw-Hill Book Co., New York, 1960.

CHAPTER 2.2
STEAM TURBINES

George J. Silvestri, Jr.
Power Generation Operations Division
Westinghouse Electric Corp.
Orlando, FL

INTRODUCTION

The function of a steam turbine is the conversion of stored thermal energy to mechanical work by controlled expansion of the steam through stationary nozzles and rotating blades. Steam turbines can be classified in a number of ways. They can be classified by cycle and steam conditions such as Rankine cycle, Rankine regenerative cycle, reheat cycle (both single and double reheat), condensing or noncondensing. The turbine can be classified according to the number and arrangement of the turbine shafts and casings. When there are two or more casings, the designs are designated as *tandem compound* (all casings on the same shaft) or *cross compound* (casings on two or more shafts). The smaller-size units using a nonreheat cycle accomplish the expansion from the initial steam (throttle) conditions to the condenser in a single casing, as shown in Fig. 2.1.

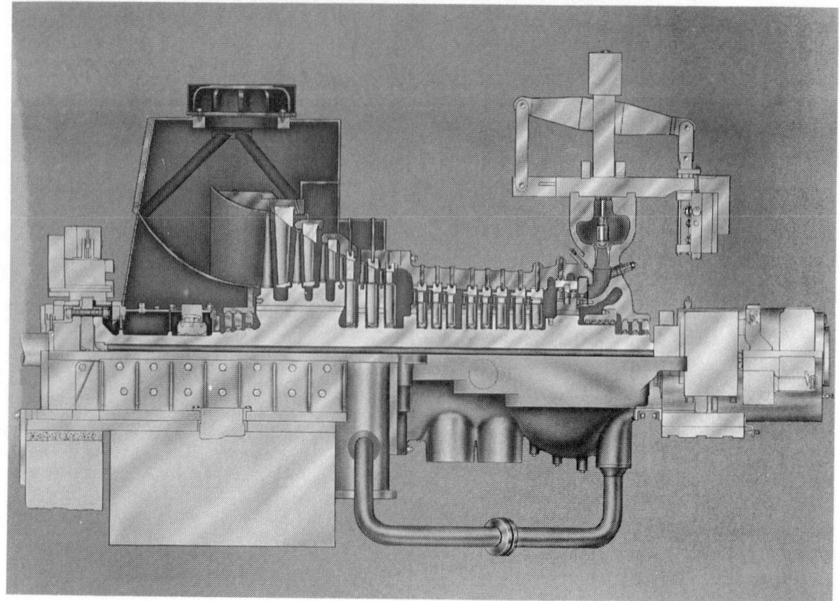

FIG. 2.1 Single-case turbine. *(Westinghouse Electric Corp.)*

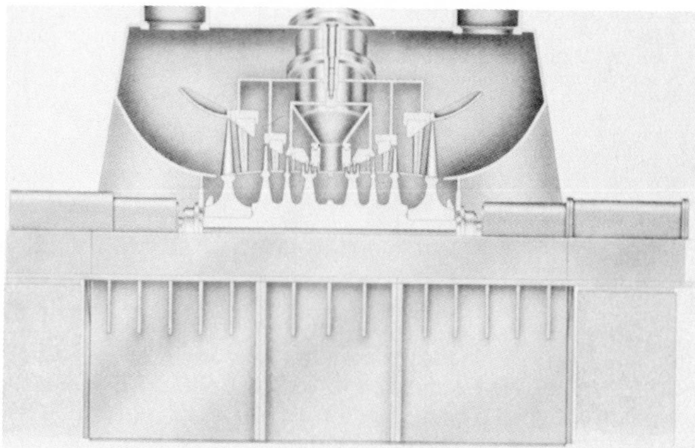

FIG. 2.2 Double-flow LP element. *(Westinghouse Electric Corp.)*

As the unit size increases, the last-stage annulus area requires that there be two parallel-flow paths, or double flow, resulting in a separate low-pressure (LP) casing (Fig. 2.2) and a separate high-pressure (HP) casing. For reheat turbines the expansion of the steam occurs in three stages: HP (Fig. 2.3), intermediate pressure (IP) (Fig. 2.4), and LP with the IP element expanding the steam from the hot reheat to the LP element inlet. Moreover, for unit ratings up to about 600 MW,

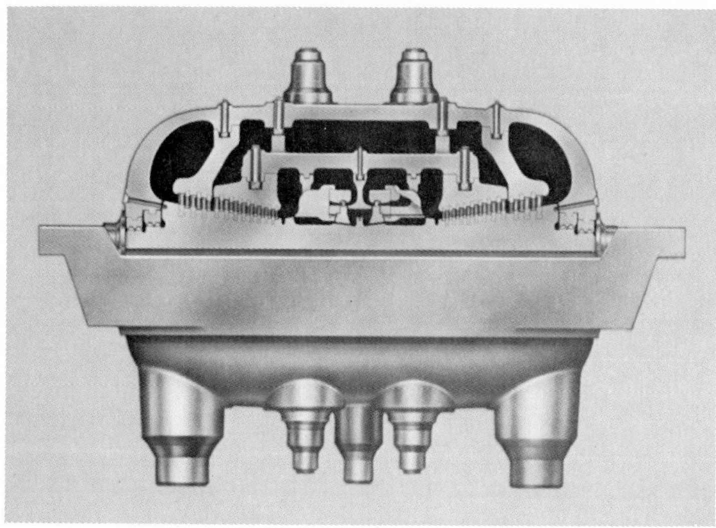

FIG. 2.3 Reheat HP element. *(Westinghouse Electric Corp.)*

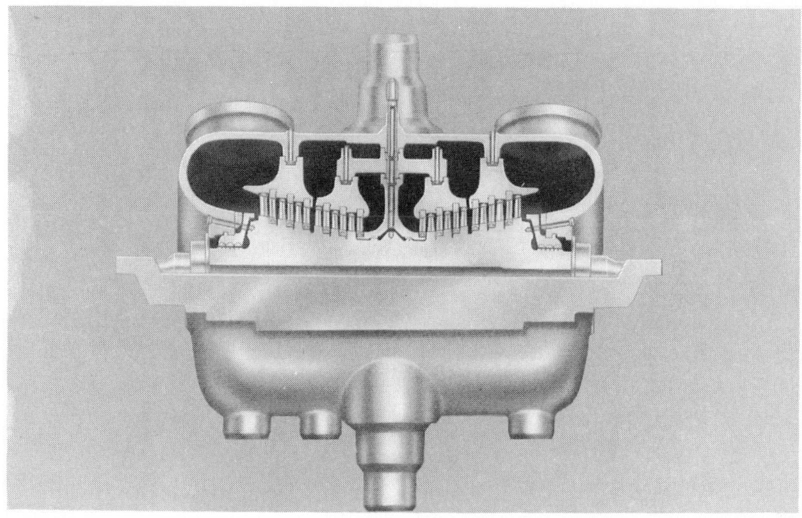

FIG. 2.4 Reheat IP element. *(Westinghouse Electric Corp.)*

the HP and IP expansions may be accomplished in a combined HP-IP element (Fig. 2.5), with a common outer shell or casing with appropriate subdivisions in the inner shell.

Turbines can also be characterized by the level of throttle steam temperature and pressure. Fossil steam conditions are typified by high pressure and high temperature (high superheat). Nuclear turbine steam conditions, typically light-water

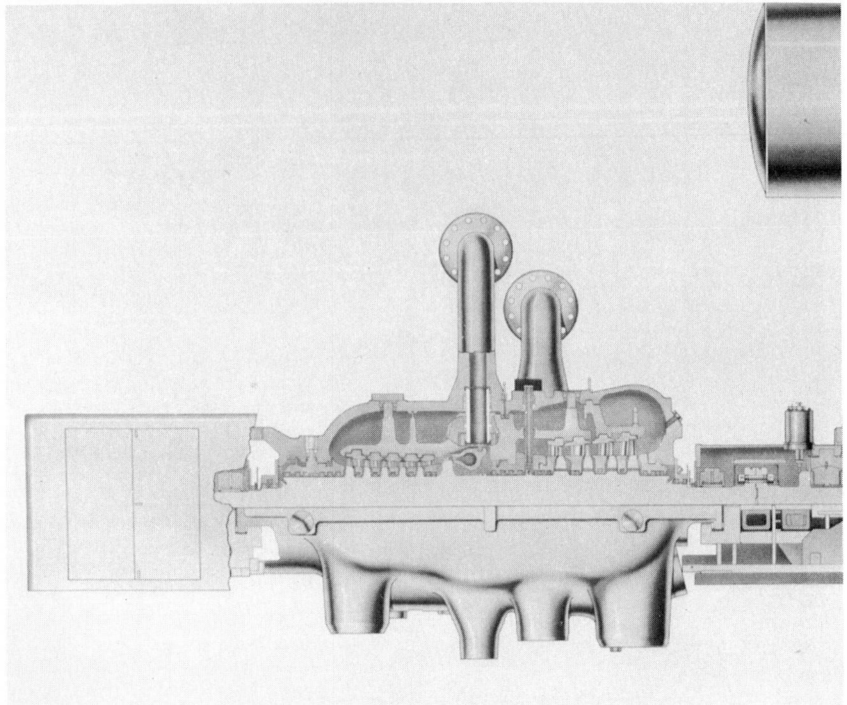

FIG. 2.5 Combined HP-IP element. *(General Electric Co.)*

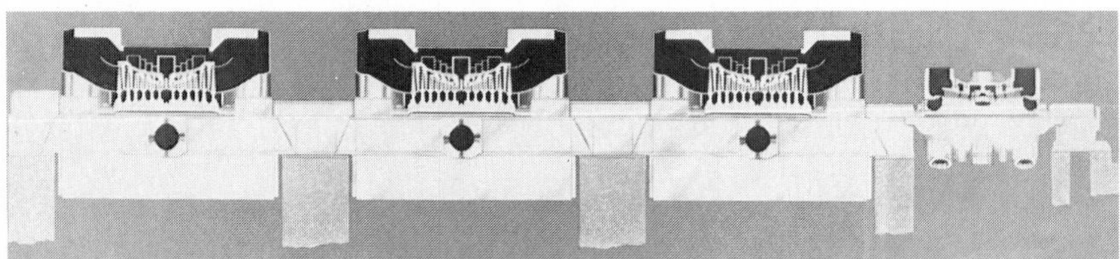

FIG. 2.6 Nuclear turbine configuration. (*Westinghouse Electric Corp.*)

reactors, are essentially dry and saturated steam or modest superheat with pressures in the range from 700 to 1100 psia (4823 to 7579 kPa). For nuclear turbines rated at 800 MW and above, the turbine configuration consists of a double-flow HP element and two or more double-flow LP elements, as illustrated in Fig. 2.6.

Finally, there are the classical descriptions: impulse and reaction turbines. All the pressure drop occurs in the stationary nozzle or vane of a classical impulse turbine, while 50 percent of the pressure drop occurs in the rotating blade of a classical reaction turbine. The combination of a stationary vane and a rotating blade is called a *turbine stage* except for one type of impulse stage. Impulse stages are further classified as velocity-compounded or Curtis stages and

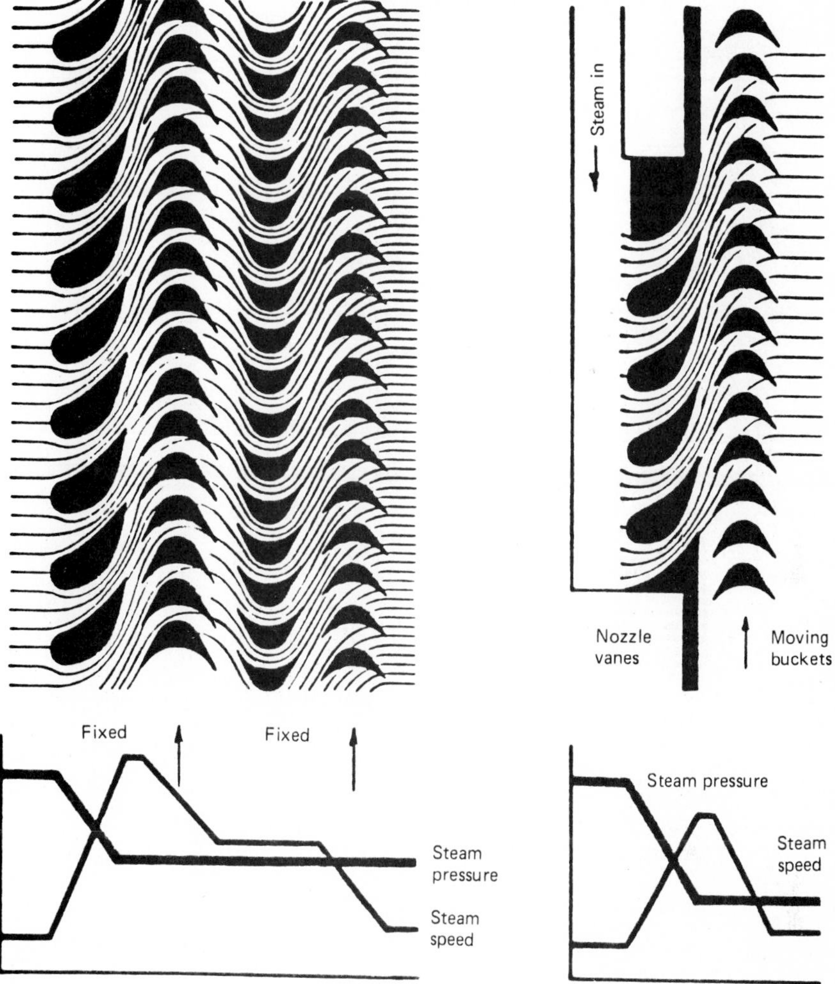

FIG. 2.7 Velocity-compounded (Curtis) stage. (Power *magazine)*

FIG. 2.8 Rateau stage. (Power *magazine)*

pressure-compounded or Rateau stages, named after their inventors. In a velocity-compound stage, the velocity resulting from the pressure drop in the nozzle is so high that velocity is converted to work in two or more rotating rows with reversing vanes interspersed between the rotating rows, as shown in Fig. 2.7. In the classical Curtis stage, no pressure drop occurs in the rotating blades and reversing vanes.

With pressure compounding, the pressure drop is divided between two or more Rateau stages, each consisting of a row of stationary nozzles and a row of rotating blades, as shown in Fig. 2.8. Because one-half of the stage pressure drop of a classical or 50 percent reaction stage occurs in the rotating blade, the profiles

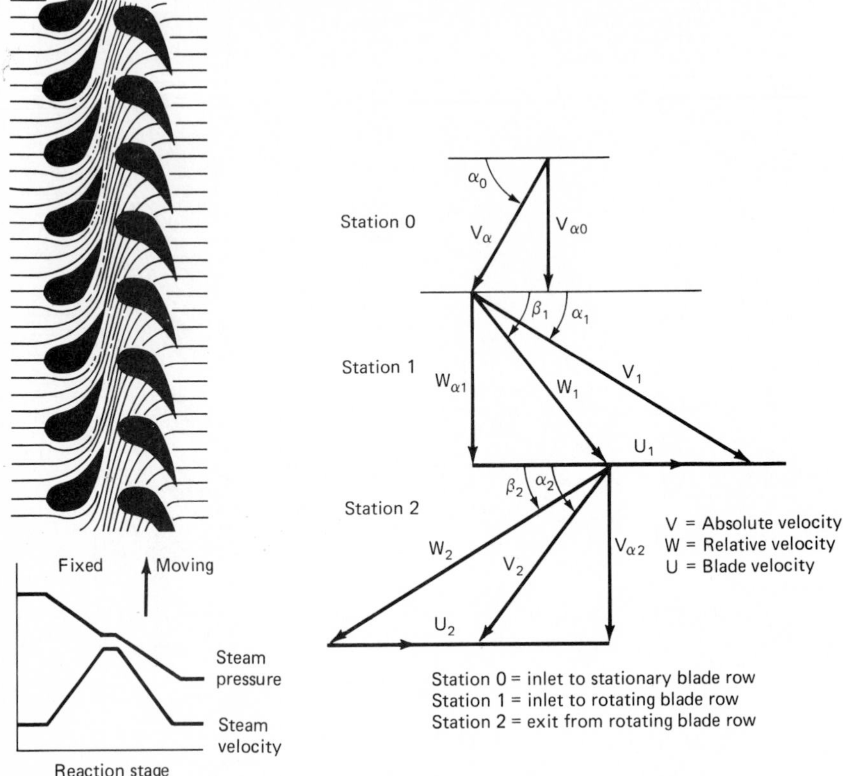

FIG. 2.9 Reaction stage. *(Westinghouse Electric Corp. and* Power *magazine)*

of the rotating and stationary blades are similar, and the variations in steam pressure and absolute steam velocity are similar. The stage velocity diagram is shown in Fig. 2.9.

Modern turbine stages are more accurately classified by the level of reaction or portion of the stage pressure drop that occurs in the rotating row. Consequently, impulse designs are more properly designated as low-reaction stages and the conventional reaction designs as high-reaction stages. Moreover, part of the stage pressure drop occurs in the rotating row of most modern impulse stages. In addition, the level of reaction of an impulse stage is important to ensure against excessive thrust-bearing load, to minimize blade leakage losses, and to guard against negative reaction at the base of the rotating blade, which increases blade losses.

Radial equilibrium results in variations of turbine intrastage pressure from base (hub) to tip, as illustrated in Fig. 2.10, with a corresponding increase in reaction. The longer the blade, the larger the variation in reaction from base to tip.

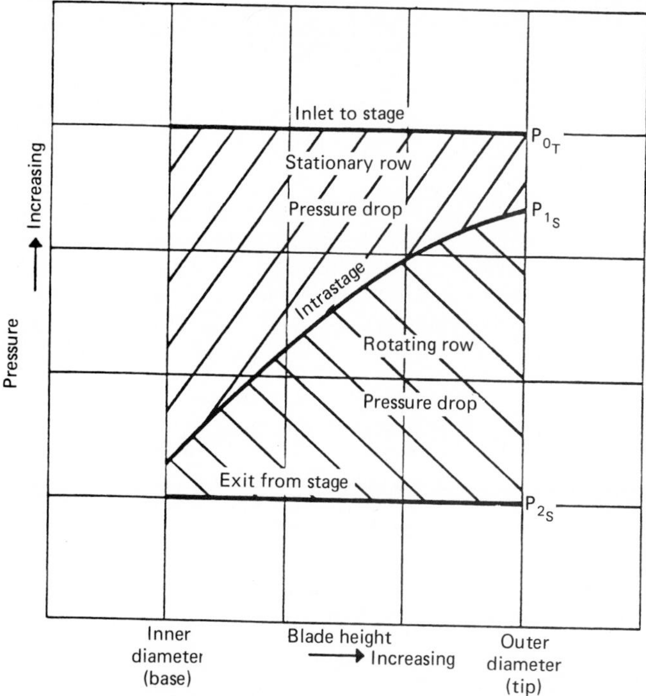

FIG. 2.10 Radial equilibrium. *(Westinghouse Electric Corp.)*

Consequently, with long blades having low base-tip ratios, the differences between impulse and reaction blades are academic.

CYCLE DESIGN PARAMETERS

Since the vast majority of central station fossil-fueled units installed since 1950 and those contemplated in the future use the reheat cycle, the discussion of steam conditions relates to those designs. Currently, typical throttle conditions for reheat turbines are 2400 psig and 1000°F with reheat to 1000°F [2400 psig, 1000°F/1000°F (16 536 kPa, 538°C/538°C)].

Throttle Pressures

Single-reheat turbines have been built for throttle pressures of 1450, 1800, 2000, 2400, and 3500 psig (9990, 12 402, 13 780, 16 536, and 24 115 kPa) supercritical pressure. The temperature combinations have been 1000°F/1000°F, 1050°F/1000°F, 1050°F/1050°F, and 1100°F/1050°F (538°C/538°C, 566°C/538°C, 566°C/566°C, and 593°C/566°C). All the double-reheat turbines had throttle pressures of 3500 psig (24 115 kPa) with the exception of the pioneering Philo Unit 6,

American Electric Power Corp, at 4500 psig, 1150°F/1050°F/1000°F (31 000 kPa, 621°C/566°C/538°C) and the Eddystone Unit 1, Philadelphia Electric Co., at 5000 psig, 1200°F/1050°F/1050°F (34 450 kPa, 649°C/566°C/566°C).[1, 2] The most prevalent double-reheat temperature combinations have been 1000°F/1000°F/1000°F, 1000°F/1025°F/1050°F, and 1050°F/1050°F/1050°F (538°C/538°C/538°C, 538°C/552°C/566°C, and 566°C/566°C/566°C).

The optimum first-reheat pressure for plants with practical final-feedwater temperatures (highest temperature heat supplied with cold reheat steam) is 20 to 25 percent of the throttle pressure. The incorporation of reheat in a regenerative-reheat cycle improves the heat rate efficiency by 4 to 5 percent over a regenerative Rankine cycle, the higher values corresponding to higher throttle pressure.[3, 4] The optimum second-reheat pressure of a double-reheat cycle is 20 to 25 percent of the first-reheat pressure. The heat rate improvement is 2 to 2.5 percent as compared to a single-reheat unit.

Higher throttle pressures are more practical with reheat cycles than with Rankine regenerative cycles because of the reduction in the level of moisture present during the turbine expansion. For example, the highest practical throttle pressure on nonreheat cycles has been 1800 psig (12 402 kPa) at a throttle temperature of 1050°F (566°C). In contrast, throttle pressures up to 3500 psig (24 115 kPa) have been adopted with 1000°F (538°C) throttle and reheat steam temperatures. The exhaust moisture of the 3500-psig (24 115-kPa) reheat cycle is lower than on the 1800-psig (12 402-kPa) nonreheat cycle.

Exhaust Pressures

The choice of turbine exhaust pressure is determined primarily by the type of heat-rejection system and the temperature of the cooling medium. With once-through cooling systems (rivers, lakes or ponds, and oceans), the condenser cooling-water inlet temperature, the amount of cooling water, and the allowable cooling-water exit temperature (which may be imposed by environmental regulations) are of major importance. With wet-cooling towers, the ambient wet-bulb temperature is of prime importance, while the ambient dry-bulb temperature is the major factor with dry-cooling systems.

Choke Point. Since the size of the turbine exhaust annulus is a major cost factor, the annulus-area selection for a given turbine rating is justified by the fuel savings achievable at a given exhaust temperature. For a given size exhaust and exhaust flow, a minimum exhaust pressure exists below which there is no increase in last-stage work when the exhaust pressure is reduced. This is often termed the *choke point*. On modern turbines with diffusing exhaust hoods, the choke point corresponds to an exhaust-annulus axial Mach number of 0.9 ($M =$ the 0.9 line of Fig. 2.11). The hood is the structure that connects the space immediately after the last stage to the condenser.

As the turbine-exhaust mass flow is reduced, the choking pressure is reduced. The exhaust mass flow is often normalized by dividing the mass flow per unit time by the exhaust-annulus area, a term referred to as *end loading,* in lb/h • ft^2 or g/s • m^2.[5,6] The higher the exhaust pressure, the higher will be the end loading corresponding to choking, as shown in Fig. 2.11.

For many years, maximum allowable turbine-exhaust pressures ranged from 5 to 5.5 inHg abs. (17 to 18.6 kPa), with the exhaust end loadings up to 15,000 lb/h • ft^2 (20 340 g/s • m^2). This corresponds to the 0.39 axial Mach number line of Fig. 2.11. For higher allowable exhaust pressures, it was necessary to increase the exhaust end loading.[7] There is an accompanying increase in the choking exhaust pressure.

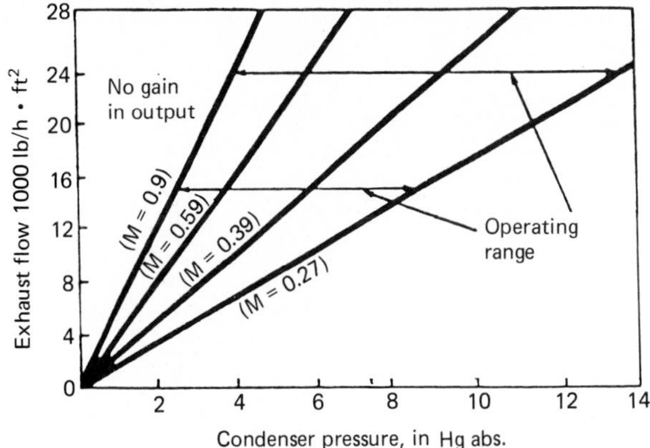

FIG. 2.11 Exhaust pressure versus end loading. Note: in Hg abs. × 3.38 = kPa; lb/h • ft^2 × 1.356 = g/s • m^2. *(Westinghouse Electric Corp.)*

Mach Number Ranges. With designs limited to the 0.39 to 0.90 axial Mach number, an end loading of 24,000 lb/h • ft^2 (32 544 g/s • m^2) would be required for an exhaust pressure of about 9.0 in Hg abs. (30.4 kPa). However, if the allowable Mach number were increased to the 0.27 to 0.90 range, it would allow the maximum exhaust pressure to be increased to 8.7 in Hg abs. (29.4 kPa) without increasing the choking exhaust pressure at an end loading of 15,000 lb/h • ft^2 (20 340 g/s • m^2). The 0.27 Mach number line corresponds to the condition at which the last-stage net work is zero.

Designs with this wider allowable Mach number range are now available.[6] They provide economic benefits not only for units with dry-cooling systems (high exhaust pressures) but also for units with conventional cooling systems and conventional exhaust pressures.[6]

Specific turbine ratings have evolved in recent years and have clustered about certain size ranges. In many cases, the nominal sizes have been related to specific levels of turbine exhaust flow, reflecting the low energy prices that prevailed prior to the oil embargo of 1974. Table 2.1 lists the specific rating ranges and the corresponding range of exhaust-annulus areas for high-pressure, high-temper-

TABLE 2.1 Ratings vs. Exhaust Flow for High-Pressure, High-Temperature Turbines

	Total exhaust-annulus area	
Unit rating, MW	ft^2	m^2
180–200	66–83	6.13–7.71
225–250	83–111	7.71–10.30
300–350	111–222	10.30–20.62
400–450	132–222	12.26–20.62
500–560	164–264	15.24–24.53
650–700	222–381	20.62–35.39
800–850	264–397	24.53–36.88
1050–1300	508–690	47.19–64.10

TABLE 2.2 Ratings vs. Exhaust Flow for Nuclear-Fueled Turbines

	Total exhaust-annulus area	
Unit rating, MW	ft^2	m^2
500–650	423–508	39.30–47.19
800–1000	508–764	47.19–70.98
1150–1350	634–1058	58.90–98.29

ature turbines, while Table 2.2 relates to nuclear-fueled turbines. The lower end of the exhaust-area range relates to units with low fuel costs and/or high site exhaust pressures.

OPERATION

Main steam flow and load can be controlled by varying the active nozzle area of the first turbine stage (nozzle control) or by varying the steam pressure at the inlet to the first-stage nozzles (pressure control). The first method is called *partial-arc admission* or *multivalve point operation,* and the second is called *full-arc admission* or *single-valve point operation*. In the latter instance, the steam flow is controlled either by throttling on the governor valves (constant-throttle pressure operation) or by varying the throttle pressure while holding the governor valves in a fixed position.

With full-arc admission, the first-stage, turbine-cycle efficiency is higher with sliding-throttle pressure operation than with constant-throttle pressure operation because the valve throttling is eliminated and the boiler feed pump power is reduced. Sliding-throttle pressure operation can also improve low-load efficiency of partial-arc admission designs. In this instance, valves are successively closed as the load is initially reduced from the maximum value. When a specific valve point is reached, the valve position is held constant and further load reductions are achieved by reducing the throttle pressure. This is sometimes called the *hybrid mode* of operation. Published studies have shown that the optimum transfer point from constant- to sliding-throttle pressure operation occurs when one-half the governor valves are wide open and one-half are fully closed, about 70 percent load and corresponding to 50 percent active-arc admission on the first stage.

Operation in the hybrid mode results in higher efficiency for practically the entire load range, except above 95 percent load, as compared to full-arc admission, sliding-throttle pressure designs, as shown in Fig. 2.12.[8] Moreover, operation in the hybrid mode also reduces low-cycle thermal fatigue as compared to constant-pressure, partial-arc admission operation because of the smaller change in the first-stage exit temperature; see Fig. 2.13. In addition, the load response approaches that of constant-pressure operation during sliding-throttle pressure operation and is much faster than that of sliding-throttle pressure, full-arc admission operation.

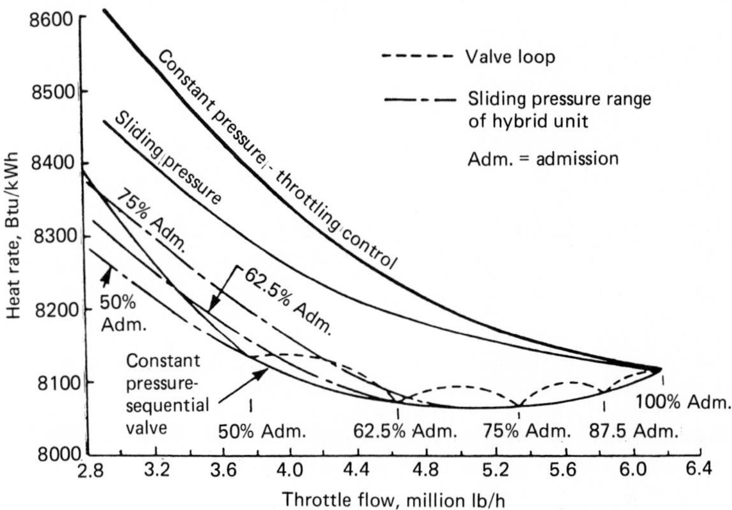

FIG. 2.12 Throttle-pressure performance comparison. Note: Btu/kWh × 0.293 = W/kW; lb/h × 0.126 = g/s; psig × 6.89 = kPa; (°F − 32)/1.8 = °C. *(Westinghouse Electric Corp.)*

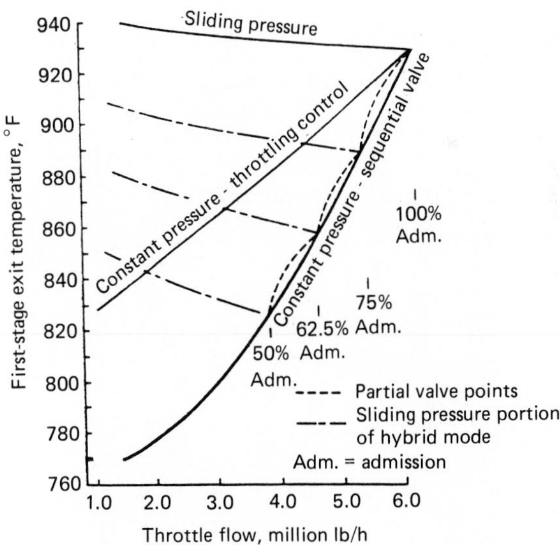

FIG. 2.13 First-stage exit-temperature variation. Note: (°F − 32)/1.8 = °C; lb/h × 0.126 = g/s. *(Westinghouse Electric Corp.)*

STRUCTURAL ELEMENTS

A turbine casing is comprised of blading, a shell or cylinder, and a rotor. In addition, there are a number of auxiliary systems or equipment such as valves and controls, bearings, lubrication system, and shaft seals.

Blading

The blades convert the pressure and temperature energy of the steam to velocity and then to rotary motion (torque) of the rotor. One-half of the blades are attached to the rotor, and one-half to the casing or shell. Generally, the stationary blades are shrouded at the base while the rotating blades are shrouded at the tip, as illustrated in Fig. 2.14. As the blades get longer, centrifugal force may make it undesirable to have a shroud. The shroud has a dual function: improved sealing with reduced leakage losses and frequency tuning of the blades. Also the shroud limits the vibration amplitude of the individual blades, thereby reducing the stress level. The shrouds may be integral to the blades, may be welded to them, or may be mechanically fastened to them. Lashing wires are another means of tuning the

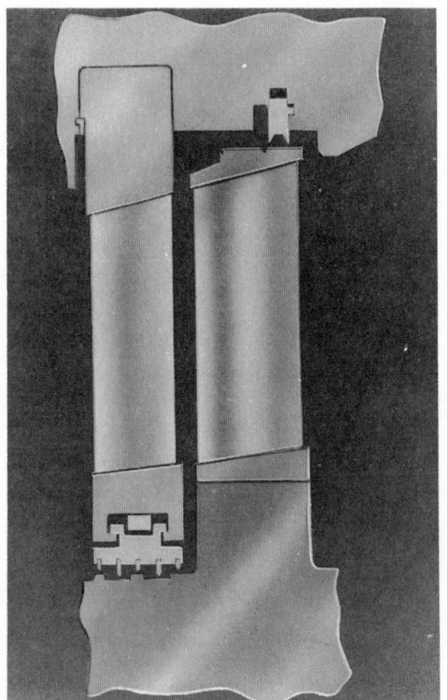

FIG. 2.14 Shrouded blades. *(Westinghouse Electric Corp.)*

blades by joining them at an intermediate location along the blade length. Finally, some designs use free-standing blades without lashing wires or shrouds.

Rotors

The rotor absorbs the torque produced in the rotating blades from the expansion of the steam and transmits it through the coupling at the shaft ends to the following shaft and eventually to the generator. The rotor material will vary depending on the operating temperature. Likewise, there will be a variation in rotor diameter, with the smallest diameters occurring at the HP inlet because of the high temperature and small blade heights. The rotor diameter may increase as the steam expands through a casing and is invariably larger as the steam passes from one casing into the following one.

Rotor speeds for the HP and IP elements of high-pressure, high-temperature designs are typically 3600 rpm, 60 Hz. The LP elements generally operate at 3600 rpm although in some instances, to obtain additional exhaust-annulus area, the LP elements are on a separate shaft and rotate at 1800 rpm. The HP and LP elements for light-water-reactor (LWR) applications operate at 1800 rpm because of the large volumetric flows at both the HP-element inlet and the LP-element exhaust.

Shells and Blade Rings

The turbine pressure vessels are subjected to pressure loads, which produce the primary stresses, and thermal loads, which create bending moments, that are resisted by the horizontal joint bolting. The designer's chief concern is protection against tensile or creep-rupture failure. Also, transient temperature changes can cause repeated plastic strain, leading to fatigue damage and cracking. There are widely varying temperatures and therefore metal temperature gradients between the zones of the turbine shells. There may also be a radial temperature gradient across the shell wall. These temperature gradients can often be reduced by judicious location of the zones and/or flow passage of the steam.

Rather than cutting grooves in the turbine casing or shell wall and making it undesirably thick, thus creating a potential source of high thermal stress, designers often mount the blades in a structure separate from the shell—a blade ring. The blade ring material is often different from that of the shell. The blade rings are supported and guided by keys and dowels so as to allow unobstructed thermal expansion. Figures 2.3, 2.4, and 2.5 show the use of blade rings in the HP and IP sections of various turbine elements.

To maintain a permanently tight horizontal joint and to avoid gasket yielding and creep, the cylinder joints are designed with metal-to-metal contact. To seal the horizontal joint surface and to maintain a clamping force, the bolts are prestressed when cold. The bolt materials are selected primarily for their relaxation strength.

Auxiliary Equipment and Systems

The turbine controls can be classified as protective or emergency systems and regulating systems that control flow or load or speed either singly or in combina-

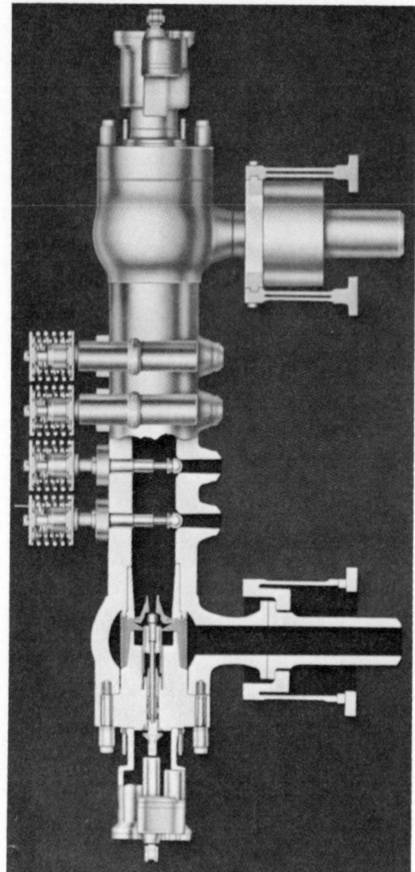

FIG. 2.15 Combined throttle-valve and governor-valve body (steam chest). *(Westinghouse Electric Corp.)*

tion. Dedicated instrumentation or sensors provide the basis for control actuation, equipment shutdown, or activation of alarms. Some control systems perform both the protective and the regulating function.

Turbine Valving. While turbine valving is the most readily identifiable protective system, there are a number of other systems that activate alarms or trips, including low condenser-vacuum trip, low bearing-oil and main-oil pressure trips, rotor-position alarm or trip, rotor and differential-expansion trip, and overspeed trip.

The turbine has throttle and governor valves at the main-steam inlet and reheat stop and interceptor valves at the hot reheat inlet. The throttle valve normally functions as a protective device when the overspeed trip is activated. However, this valve is often used during start-up for the initial turbine roll. The governor valves are used for load and flow control during normal on-line opera-

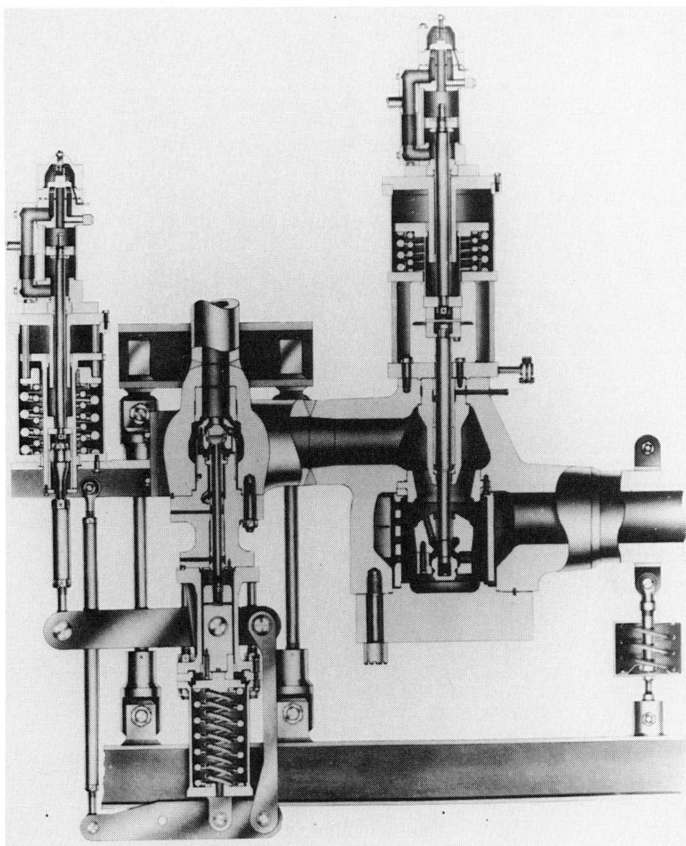

FIG. 2.16 Separate throttle-valve and control-valve design. *(General Electric Co.)*

tion. The throttle and governor valves may share a common valve body (steam chest), as shown in Fig. 2.15, or may have separate valve bodies, as in Fig. 2.16.

Because of the large stored steam volume in the reheater and associated piping, additional protective valves are located at the turbine reheat inlet. The interceptor valve looks similar to a throttle valve, being wide open during normal operation. During overspeed, this valve closes and then modulates to control speed unless a trip occurs. The reheat stop valve, a clapper design, provides redundancy in case the interceptor valve malfunctions. The combined interceptor-reheat stop-valve assembly is shown in Fig. 2.17. Some LWR turbines employ interceptor valves at the LP inlet that are similar to fossil-fuel designs, while other designs use butterfly valves as both reheat stop and interceptor valves.

Bearings and Seals. The turbine employs thrust and journal bearings. The former positions the turbine rotor axially in relation to the stationary parts while absorbing any steam-thrust loads on the rotor. The journal bearings position the rotor radially in the turbine shell, support the weight of the rotor, and absorb vertical and transverse loads on the rotor. The journal bearings may be the sleeve or

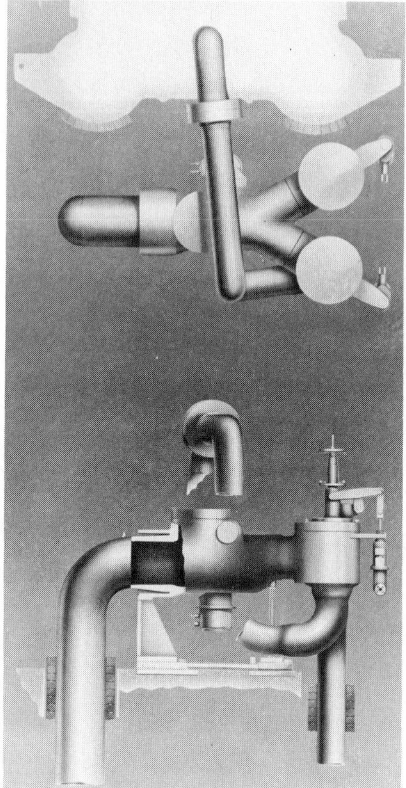

FIG. 2.17 Combined interceptor-reheat stop-valve assembly. *(Westinghouse Electric Corp.)*

tilting-pad types and must be capable of running without oil whip. Tilting-pad bearings have greater resistance to oil whip and can reduce the magnitude of steam whirl.

The oil system consists of a number of pumps, shaft- and motor-driven, that supply oil for controls and lubrication of bearings. Many oil systems use a low-capacity, high-pressure pump for control requirements and to supply an ejector which delivers a relatively large quantity of low-pressure lubricating oil. The combined power consumption of the oil pump, thrust bearing, and journal bearings is about 1 percent of the rated turbine output.

Wherever the turbine shafts pass through the shell wall, seals are used to reduce the leakage. The leakage zones may occur at large diameters (dummy or balance pistons that offset the thrust produced in the rotating blades) or modest diameters such as glands. To reduce the amount of leakage, a multiplicity of seals is installed to dissipate the pressure and velocity of the leakage steam. Modern

FIG. 2.18 Radial shaft seals. *(Westinghouse Electric Corp.)*

units use axially insensitive, radial seals (Fig. 2.18) because of the large differential expansions and shaft lengths.

DESIGN PROCESS

The design of a steam turbine involves aerodynamics, thermodynamics, mechanics, and material selection. The aerodynamic and thermodynamic factors are more closely interwoven as they relate to efficiency. However, the thermodynamic aspects are broader because they relate to not only the blading efficiency and flow losses but also cycle variables, such as regenerative feedwater heating, initial steam conditions, reheat- and exhaust-steam conditions.

Aerodynamic Design

Several types of efficiency losses occur in a turbine stage with the magnitude of loss of its occurrence depending on the level of reaction in the stage, the blade length, and the fluid state. The losses include profile losses, end-wall losses, leakage losses, secondary-flow losses, overexpansion and underexpansion losses, incidence losses, windage losses, disk and shroud friction losses, partial-arc admission losses, moisture and supersaturation losses, and hood and leaving losses.

Profile Losses. Profile losses are related to the shape or profile of the blade foil and include the losses associated with the trailing-edge wakes. High-work, low-reaction stages have higher profile losses than nominal 50 percent reaction

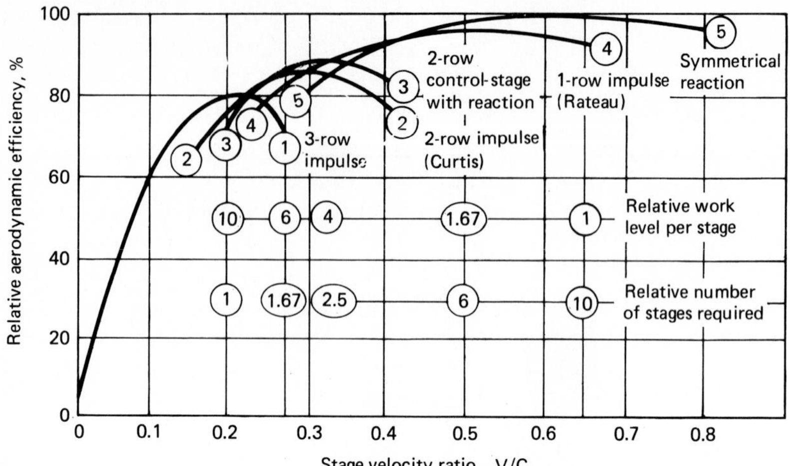

FIG. 2.19 Aerodynamic efficiency versus stage type. *(Westinghouse Electric Corp.)*

stages. The Curtis velocity-compounded stage has the lowest aerodynamic efficiency of the three types of blading described earlier. The high heat drop and high turning of the blade passages result in high friction and high secondary-flow losses. The lower heat drop and lower velocities of the Rateau stage result in higher efficiency than the Curtis stage.

For a given wheel velocity V, as the stage heat drop and associated ideal steam velocity C_o are decreased, the stage aerodynamic efficiency improves. However, this requires an increase in the number of stages to achieve the same level of work. Each type of stage has a unique relationship between the aerodynamic efficiency and the stage isentropic velocity ratio V/C_o. The variation of aerodynamic efficiency and the stage velocity ratio is illustrated in Fig. 2.19.

End-Wall, Leakage, and Secondary-Flow Losses. In addition to the profile losses, there are losses at the ends of the blade passage, top and bottom. Because of friction at these end walls, the centrifugal force is reduced, resulting in a local secondary flow. Radial (secondary) flow then occurs along the blade foil, both inward and outward, with the formation of vortices and distortion of flow.

Leakage losses result from flow that bypasses the blade passages and escapes through the clearance space between rotating and stationary parts. The leakage loss occurs because not only is no work done by the leakage flow, but also additional turbulence losses are incurred when this flow reenters the blade path and mixes with the blade passage flow.

High-reaction stages have greater tip (rotating-blade) leakage than low-reaction designs because of the large pressure drops across the rotating blades. Moreover, the stationary-blade leakage is lower on low-reaction stages because the seals are located at a smaller diameter and a greater number of seals are used. However, the reentry loss associated with the leakage stream is less severe on high-reaction stages. Since high-reaction stages have lower profile losses and low-reaction stages have lower leakage losses, the comparative performance of the two types on stages with short blades depends on the relative importance of the two losses.

Secondary-flow losses result from flow that is at right angles to the main flow, resulting in a partial loss of fluid contact with the blade foil and an accompanying loss of useful energy. This loss affects the movement of moisture in the boundary layer; it has been observed that stationary vanes with less severe secondary flow reduced the erosion associated with moisture.[9]

Overexpansion and Underexpansion Losses. Overexpansion and underexpansion losses occur when the blade passage areas do not match the area required to expand the steam most efficiently. Convergent or nonexpanding blade passages match the area requirement at exhaust pressures at or above the critical exit pressure (related to the passage-inlet pressure and temperature), which is identified as point *D* in Fig. 2.20. At a lower exhaust pressure, between points *B* and *D*, the passage exit or throat pressure remains at the critical value, and the expansion from the throat to the exhaust conditions is accompanied by additional losses and a deflection of the steam jet toward the axial direction. Since the expansion is not completed within the confines of the blade passage, this condition is known as *underexpansion*.

With a convergent-divergent (C-D) or expanding passage, the blade passage initially converges until the throat is reached and then increases in area. With a C-D passage the steam can expand efficiently at exhaust pressures below the critical. The C-D passage also reaches a condition, depending on the ratio of mouth area to throat area, where further decreases in exhaust pressure are not accompanied by decreases in mouth pressure, for example, point *E* of Fig. 2.20. Decreases in exhaust pressure below point *E* result in an expansion from the mouth to the exit pressure accompanied by additional losses and jet deflection. When the exhaust pressure of the C-D passage is high at point *G*, the velocity and pressure at the throat never reach the critical value, and the steam diffuses up to the exhaust pressures with attendant losses in the expanding or divergent portion of the passage. When the exhaust pressure is low enough and critical pressure occurs at the throat, the steam continues to expand in the divergent section to point *I*, for example, which is below the exhaust pressure, point *H*. A normal shock, point *J*, occurs in the passage, and the steam then diffuses and achieves the exhaust pressure, a process known as *overexpansion*.

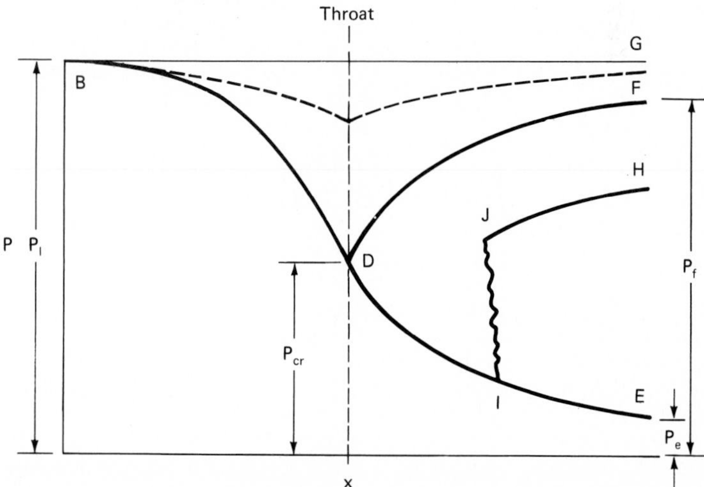

FIG. 2.20 Pressure distribution across a nozzle. *(Westinghouse Electric Corp.)*

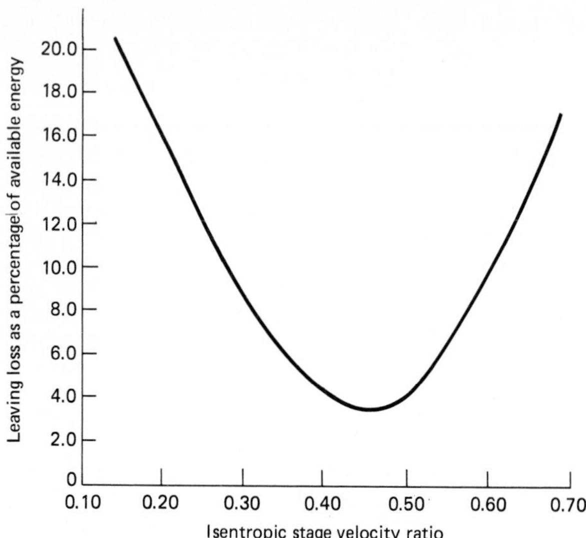

FIG. 2.21 Stage leaving loss. *(Westinghouse Electric Corp.)*

Overexpansion losses are much greater than underexpansion losses. As a result, designers use converging passages wherever possible unless the passages operate for long periods at exhaust pressures considerably below the critical pressure.

Incidence, Windage, Friction, and Moisture Losses. When the inlet steam-flow angles are not at the optimum blade-inlet angles, additional losses are incurred. The magnitude of this incidence loss is greater on blades with sharp or selective inlet. Large incidence angles can result in stall flutter (flow-induced vibration) or buffeting of the blade. This condition is a concern at the tip section of last-row LP blades operating with low mass flow and high exhaust pressure.

On partial-arc admission stages there is a windage loss associated with the pumping action of the blades on the stagnant steam in the unadmitted or inactive areas of admission. In addition, there is a displacement loss in the partial-arc stage when the rotating blading enters and leaves the steam jet of the active nozzle groups. Because low-reaction stages have small pressure drops in the rotating blades, there are friction losses on both sides of the blade disk and the outer surface of the blade shroud. This loss and the windage and displacement losses are not present on high-reaction designs with drum construction.

The presence of moisture in the steam results in additional losses that are related to nonequilibrium or supersaturation effects, momentum interchange and friction between the moisture and the steam, and retardation or braking losses from moisture droplets impacting on the backs of the rotating blades. This loss varies between 1.0 and 1.2 percent for each percentage of average moisture in the stage. The moisture-droplet impacts also erode the blade inlet edges.

Exhaust Losses. When the absolute velocity leaving a turbine stage is partially or totally dissipated, the inefficiency is called *leaving loss*. In well-designed stages, a sizable fraction of the velocity leaving a stage is available for conversion to work in the next stage. The leaving velocity is a minimum at a specific value of the stage-velocity ratio or exit-specific volume with a characteristic shape, as illustrated by Fig. 2.21.

When the turbine stage is the last one in a casing, especially the stage discharging to the condenser, there may also be a loss in static pressure in the exhaust chamber or hood joining the blade exhaust annulus and the exhaust opening in the casing. The term *hood loss* has been applied to this pressure loss. However, in high-efficiency turbine elements, diffusing exhaust hoods have been developed which convert some of the exhaust velocity to static pressure, as illustrated in Fig. 2.22. As a result, the blade-exit static pressure is lower than the static pressure at the exhaust opening, increasing the work of the last stage.[10] For simplicity, the hood and leaving losses are combined into one quantity, *exhaust loss*, which represents the energy dissipation between the blade exit and the casing exhaust, as shown in Fig. 2.23.[11]

The passage of steam into and out of the turbine shells is accompanied by pressure losses in the associated valves and piping. The pressure loss in the valves (steam chest) and piping at the HP inlet is typically about 4 percent, while that associated with the interceptor valves and IP inlet piping ranges between 2.0 and 2.5 percent. At the shell penetrations for extracting steam for regenerative feedwater heating, there may be not only a loss in total pressure but a loss in static pressure as well.

Thermodynamic Design

All thermodynamic cycles consist of a compression process, a heat-addition process, an expansion process, and a heat-rejection process.

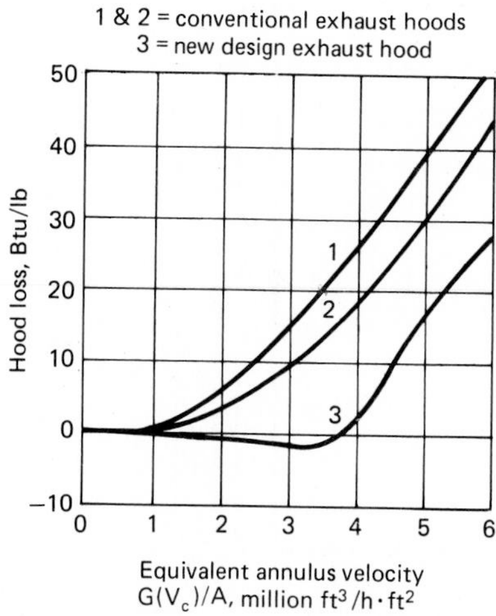

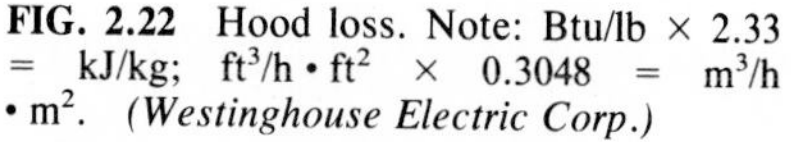

FIG. 2.22 Hood loss. Note: Btu/lb × 2.33 = kJ/kg; $ft^3/h \cdot ft^2 \times 0.3048 = m^3/h \cdot m^2$. *(Westinghouse Electric Corp.)*

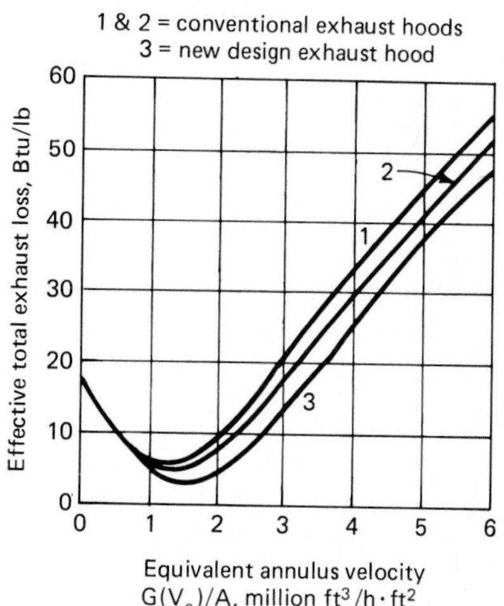

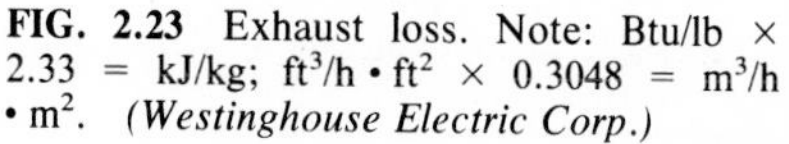

FIG. 2.23 Exhaust loss. Note: Btu/lb × 2.33 = kJ/kg; $ft^3/h \cdot ft^2 \times 0.3048 = m^3/h \cdot m^2$. *(Westinghouse Electric Corp.)*

Rankine Cycle. The simplest of steam cycles, the Rankine cycle involves a straight expansion of the steam with no internal heat transfer and a single stage of external heat addition. Although the major portion of the heat added to the working fluid of a saturated-steam Rankine cycle is used for evaporation, a sizable fraction is needed to heat the water to its boiling or saturation point. This lowers the effective heat-addition temperature of the cycle. The heavy solid line in Fig. 2.24 is an idealized 400-psia (2756-kPa) saturated-steam cycle plotted on a temperature-entropy (*T-s*) diagram. Line *AB* corresponds to the liquid heating from 70 to 444.6°F (21 to 229°C). Line *BC* represents the heat of vaporization at 444.6°F (229°C), while *CD* represents the turbine expansion from 400 to 0.363 psia (2756 to 2.5 kPa), and *DA* represents the heat rejection at 70°F and 0.363 psia (21°C and 2.5 kPa).

The area below line *ABC* corresponds to the total heat added Q_A to the cycle, while the area below line *DA* corresponds to the heat rejected Q_R from the cycle.

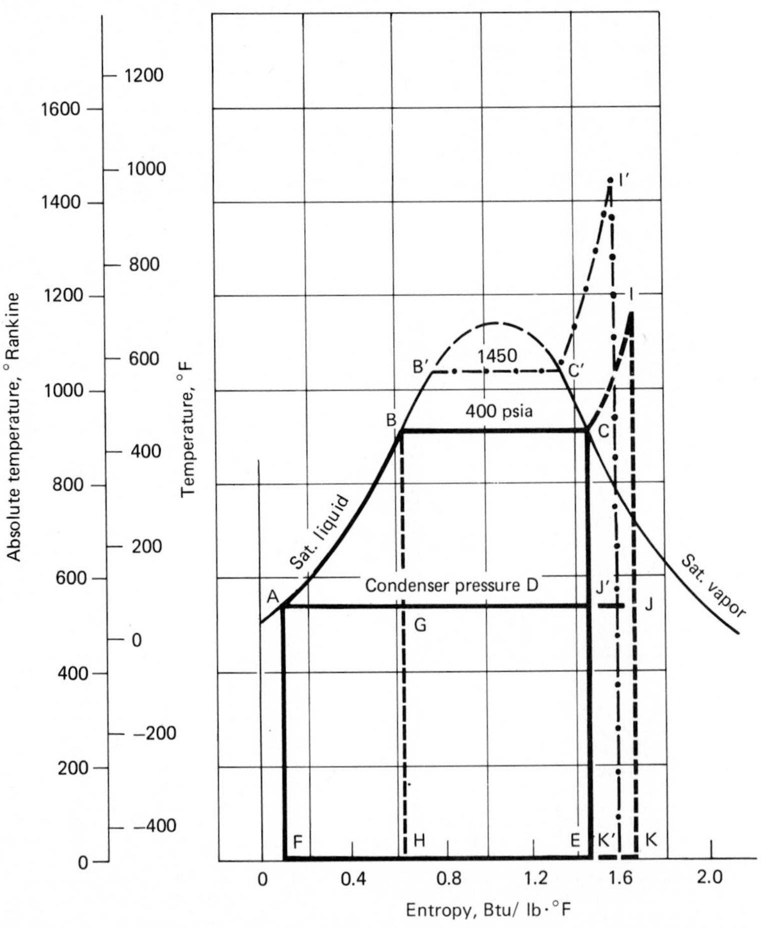

FIG. 2.24 *T-s* diagram for nonreheat cycle. Note: (°F − 32)/1.8 = °C; (R − 32)/1.8 = K; psia × 6.89 = kPa; Btu/lb • °F × 4.19 = kJ/kg • °C. *(Westinghouse Electric Corp.)*

Since work W equals heat added Q_A minus heat rejected Q_R, the area between line ABC and line DA represents the work.

Regenerative Rankine Cycle. If the liquid heating (line AB) could be eliminated, the ratio of the area above heavy solid line DG to the area below line BC would be greater than the ratio of the area above line AG to the area below line AB. This would improve the cycle thermal efficiency e. The thermal efficiency is defined as

$$e = \frac{W}{Q_A} = \frac{Q_A - Q_R}{Q_A} \tag{2.1}$$

The cycle in which part of or all the external heating of the liquid is eliminated is called the *regenerative Rankine cycle*. The liquid is heated from point A to point B by internal heat exchange in which partially expanded steam is extracted from a series of pressure zones during the expansion process and heats the liquid in a series of heat exchangers. The net effect is a reduction in both the heat addition and the heat rejection of the cycle. The cycle work is reduced because flow is continuously removed during the expansion to heat the liquid. However, the ratio of work to heat added, the thermal efficiency, is improved.

While a finite number of stages of feedwater heating are practical, the optimum final feedwater temperature increases as both the main steam pressure and the number of feedwater heaters increase. Figure 2.25 illustrates the effect of the

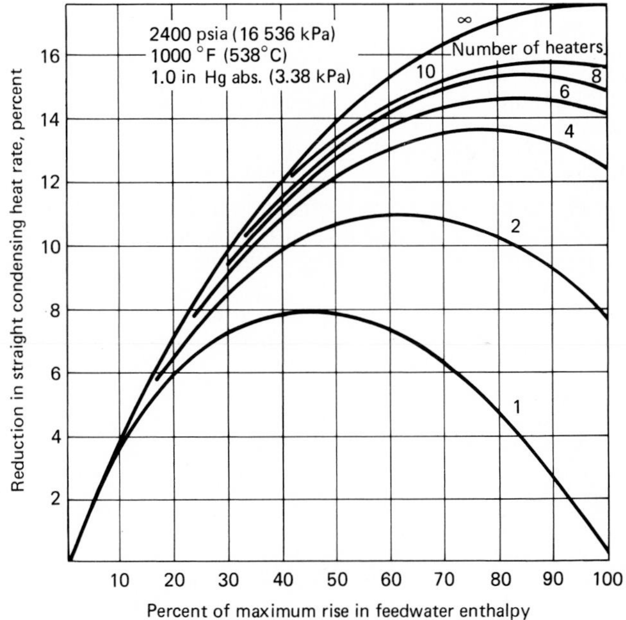

FIG. 2.25 Regenerative-cycle heat-rate effect. *(Westinghouse Electric Corp.)*

TABLE 2.3 Pressure and Temperatures for Regenerative Rankine Cycle

Pressure		Saturation temperature		Practical final-feedwater temperatures	
psia	kPa	°F	°C	°F	°C
400	2756	444.6	229.2	300	148.9
1000	6890	544.6	284.8	400–450	204.4–232.2
1465	10 094	593.0	311.7	450	232.2
1815	12 505	622.1	327.8	460–500	237.8–260.0
2015	13 883	636.9	336.1	470–520	243.3–271.1
2415	16 639	663.0	350.6	480–530	248.9–276.7
3000	20 670	695.3	368.5	500–540	260.0–282.2
3515	24 218	—*	—*	510–550	265.6–287.8

*Above the critical pressure.

number of feedwater heaters. Just as the saturation temperature increases with increasing pressure and represents the maximum feedwater temperature of an ideal regenerative Rankine cycle, the optimum final-feedwater temperature increases with increasing pressure for actual cycles with a finite number of feedwater heaters. Table 2.3 illustrates the increase in saturation temperature and practical final-feedwater temperatures with pressure.

Superheat Cycles. When steam is heated above its saturation temperature—that is, when it is superheated—the thermal efficiency of both ideal and real cycles is improved. This additional heating (heavy dashed line *CI*) improves the ratio of the work to the heat added, as shown in Fig. 2.24. Besides improving the ideal cycle efficiency, superheating improves the turbine efficiency by reducing the moisture level (and the associated efficiency loss) in the steam during the expansion cycle of real turbines. This, in turn, improves the cycle efficiency.

Although increases in initial (throttle) temperature increase the thermal efficiency of ideal and real cycles, increases in both initial pressure and temperature can also increase thermal efficiency, as indicated by the dot-dashed line of Fig. 2.24. The potential for higher cycle efficiency from increased throttle pressure can be realized only if the turbine efficiency is not degraded excessively. For example, with a given turbine power rating, an increase in throttle pressure will reduce the blade heights and flow areas of the initial turbine stages, thereby increasing the blade leakage and other blading losses.

An increase in power rating may be required to maintain the turbine efficiency or to minimize the decrease in turbine efficiency. So throttle-pressure increases must be judicious in order to preserve the theoretical increases in thermal efficiency. In addition, increases in throttle pressure without an accompanying increase in the throttle temperature of a Rankine cycle would increase the moisture in the latter stages of the turbine. The increased moisture would decrease turbine efficiency and offset some of the theoretical gain resulting from increased throttle pressure.

As was noted earlier, regenerative feedwater heating reduces output for a given throttle steam flow. To maintain the cycle power rating, the throttle flow must be increased. This results in increased blade heights in the initial turbine stages while the exhaust stages, which are already quite long, receive less flow.

As a result, the leakage losses of the initial stages are reduced, the leaving loss of the last stage is reduced, and the overall turbine efficiency improves.

Reheat Cycle. In the reheat cycle, after the steam has partially expanded through the turbine, it is returned to the reheater section of the boiler where more heat is added to the steam to increase its temperature. Then the steam completes its expansion in the turbine. The cycle events of a theoretical reheat cycle are illustrated in the *T-s* diagram of Fig. 2.26. When the throttle steam at point *D* expands only slightly to point *E* and is reheated to point *F*, the effective heat-addition temperature is very high. However, only a small amount of heat is added, so it has a small effect on the combined heat-addition temperature of the total heat added, lines *ABCD* and *EF*. In contrast, a larger initial expansion, from

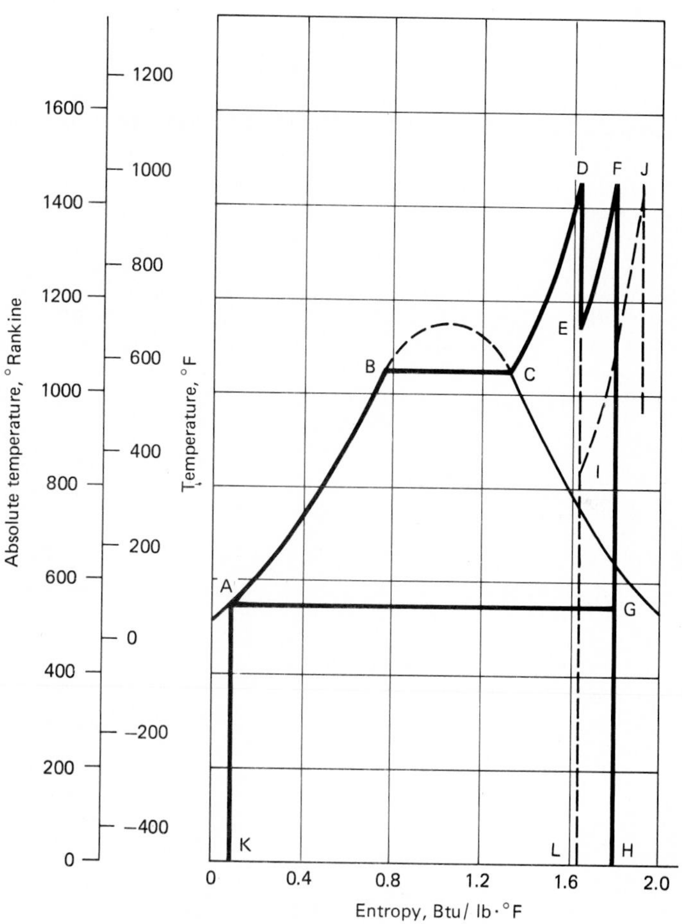

FIG. 2.26 Theoretical heat cycle. Note: (°F − 32)/1.8 = °C; (R − 32)/1.8 = K; Btu/lb • °F × 4.19 = kJ/kg • °C. *(Westinghouse Electric Corp.)*

point *D* to point *I* (the expansion indicated by the dashed line), has a lower effective heat-addition temperature (line *IJ*) as compared to *EF*. Because more heat is added during *IJ*, it has a greater weight in the total heat added to the cycle.

With practical reheat-regenerative cycles, the optimum reheat pressure is 20 to 25 percent of the throttle pressure. In addition, steam for the highest-pressure (and highest-temperature) feedwater heater is generally extracted from the cold reheat point.

The incorporation of reheat in a regenerative Rankine cycle improves the heat rate or thermal efficiency by 4.0 to 5.0 percent, depending on the throttle pressure. Adoption of reheat in the early 1950s resulted in rapid increases in throttle pressure from 1450 to 1800 to 2000 and finally 2400 psig (9990 to 12 402 to 13 780 to 16 536 kPa). Throttle temperatures ranged from 1000 to 1100°F (538 to 593°C), and reheat temperatures ranged from 1000 to 1050°F (538 to 566°C).

A second stage of reheat improves the thermal efficiency by 2.0 to 2.5 percent with optimum first- and second-reheat pressures. Double or two-stage reheat has been practical only on plants with supercritical throttle pressures (above the critical pressure of 3208 psia or 22 103 kPa). Moreover, in all the double-reheat plants, the highest-pressure feedheater extracts steam from the cold reheat point of the first reheat.

Moisture Separation and Steam-to-Steam Reheat Cycles. The application of nuclear energy produced a major change in the steam-turbine cycle and related hardware as compared to fossil-fueled technology. Although the steam conditions of nuclear plants were similar to those of the early days of power generation, the large equipment sizes and plant complexity resulted in cycle innovations. These cycle innovations, although similar to fossil-fuel units, were specifically adapted to meet the reliability and efficiency concerns related to the high moisture levels present in nuclear cycles.

Because of the high moisture levels, 20 to 25 percent, that would be present during an expansion from the throttle to condenser pressure, the first-generation plants used an external moisture separator at an intermediate location in the expansion of the steam. The separator received the partially expanded stream at a pressure ranging from 10 to 20 percent of the throttle pressure, returned it to an essentially dry condition, and discharged it to the succeeding turbine stages, reducing the turbine exhaust moisture to a level of 13 percent. This resulted in an acceptable level of blade erosion and a 1.5 percent improvement in thermal efficiency as compared to a simple regenerative cycle.

The expansion process is illustrated on the Mollier or enthalpy-entropy (*h-s*) diagram in Fig. 2.27. The HP expansion ends at point *A*. The external moisture removal is indicated by the dashed line between points *A* and *B* after which the LP expansion occurs (dot-dashed line between points *B* and *H*). The discontinuities at points *C, D, E, F,* and *G* represent internal (blade path) moisture-removal devices.

Because there are no losses in an ideal engine, the moisture removal would be of no benefit and would actually produce a loss in cycle efficiency. In the case of actual engines, however, the reduction in moisture loss results in a net improvement in cycle efficiency. The cycle just described is typically referred to as a *nonreheat cycle,* although the reduction in moisture between points *A* and *B* is thermodynamically similar to reheating.

In more modern plants, the external moisture-removal process, *A* to *B* in Fig. 2.27, is followed by steam-to-steam reheating. Throttle steam (single-stage reheat) or a combination of throttle and partially expanded steam (two-stage reheat)

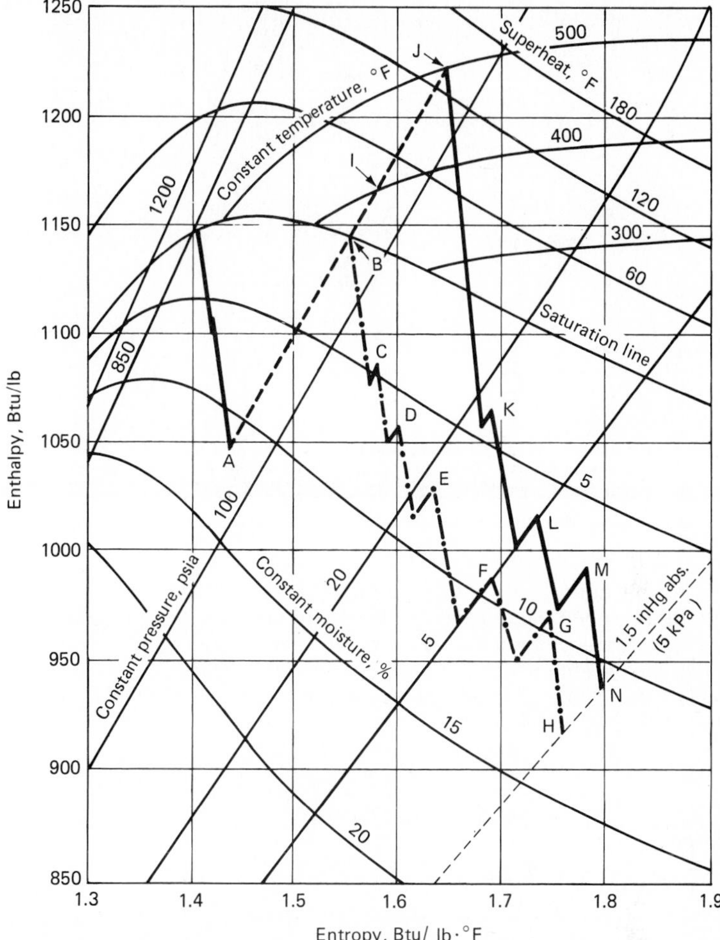

FIG. 2.27 Nuclear-turbine expansion lines. Note: Btu/lb × 2.33 = kJ/kg; psia × 6.89 = kPa; (°F − 32)/1.8 = °C; Btu/lb • °F × 4.19 = kJ/kg • °C. *(Westinghouse Electric Corp.)*

is used to reheat the separator exhaust steam to a temperature that is about 25°F (14°C) below the saturation temperature of the heating steam. In an ideal engine, steam-to-steam reheating would produce a loss in cycle efficiency because of the degradation of the available energy in the heating steam. Again, as was the case for moisture separation in actual engines, the reduction in moisture losses results in a net cycle improvement.

With single-stage reheat, the steam is heated from point *B* to point *J* (Fig. 2.27) on the dashed line with throttle steam; with two-stage reheat, the heating from point *B* to *I* is accomplished with partially expanded steam and from *I* to *J* with throttle steam. Single-stage reheat improves the cycle efficiency by 1.5 percent over a cycle with external moisture separation. There is an additional improvement of 0.3 to 0.5 percent from two-stage reheat. The LP expansion of the reheat

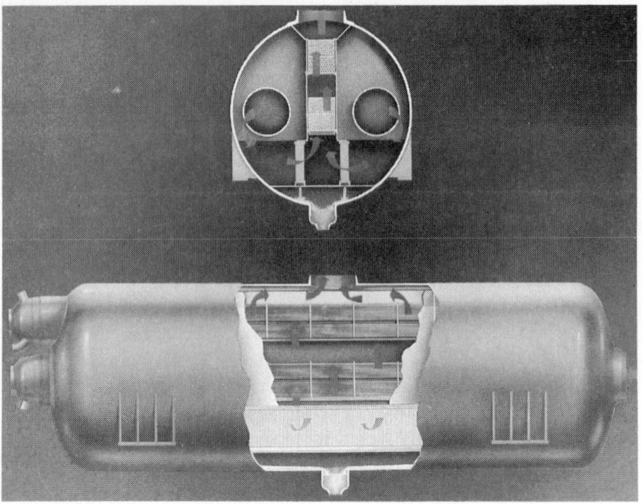

FIG. 2.28 Moisture separator reheater. *(Westinghouse Electric Corp.)*

cycle (solid-line connection points *J* and *N* of Fig. 2.27) also uses internal moisture separation (points *K, L,* and *M*) when appropriate. The moisture separation and reheating are accomplished in a combined moisture separator reheater (Fig. 2.28). The steam enters through two distribution manifolds, passes downward and toward the center through the separator section, and then passes upward through the two-stage reheater.

MECHANICAL DESIGN AND MATERIAL SELECTION

The turbine components are subjected to both steady forces and unsteady forces. The forces result from pressure loads, thermal loads, centrifugal loads, vibratory loads, and other time-varying loads.

Blading

The turbine blades are subjected to centrifugal loads related to the rotational speed and mass of the blades. Bending stresses are also present because of the pressure drops across the blades and the change in axial velocity across the blade rows. In addition to the steady forces, centrifugal and bending, there are vibratory forces associated with blade wakes and other sources of exertion. In addition, partial-arc admission stages experience shock loading as the rotating blades successively pass through active and inactive arcs of steam, as illustrated in Figs. 2.29 and 2.30.

In addition to partial-arc shocks, there are other high-frequency alternating forces which result from nonuniformity in the steam velocity leaving the nozzles

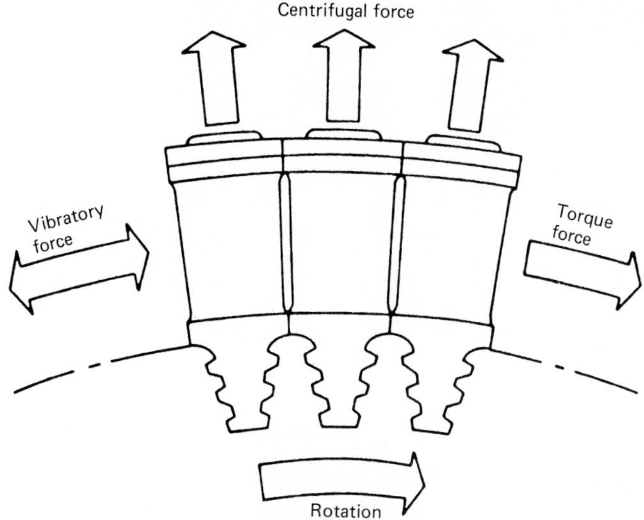

FIG. 2.29 Forces acting on partial-arc admission blades. *(Westinghouse Electric Corp.)*

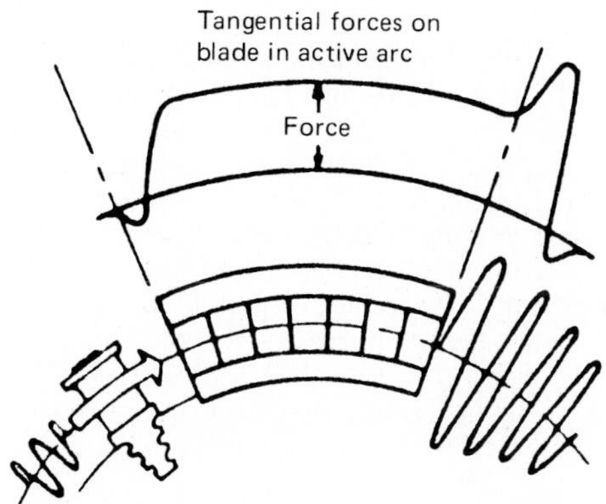

FIG. 2.30 Partial-arc admission shock. *(Westinghouse Electric Corp.)*

and from the finite thickness of the nozzle vanes (Fig. 2.31) Resonance occurs when the nozzle wake frequency coincides with the natural frequency of one of the blade group modes.

Blades may be tuned or untuned in regard to vibration. Untuned blades—which comprise all the full-arc admission HP blading, the IP blading, and the initial LP blading—utilize the blade-material damping to limit the amount of vibratory buildup due to damping. By using a combination of vibration theory and experience, the blades are designed to withstand operation under resonant con-

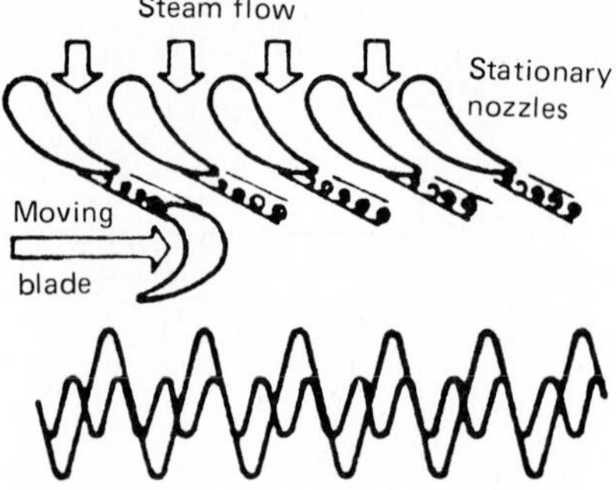

FIG. 2.31 Nozzle wake resonance. *(Westinghouse Electric Corp.)*

ditions in the fundamental modes of vibration. As noted earlier, the blade shrouds reduce the vibratory stress levels by limiting the vibration amplitude of the individual blades.

Tuned blades have low fundamental frequencies (being long, tapered, and twisted) and are generally used in the last three stages of the LP element. Sufficient strength to withstand resonance makes it difficult to achieve high aerodynamic efficiency. So the large vibratory stresses are eliminated by tuning the lower natural frequencies away from resonance at integral multiples of running speed.

Type 403 stainless steel (12 percent Cr) is the most widely applied blading material and is used in both the high- and low-temperature sections. The highest-temperature blades, especially partial-arc admission first stages, may use a modified 12 percent Cr, type 422. On the longer last-row blades, 17-4 PH (precipitation hardened) steel may be selected, with titanium sometimes being used on the next-to-last rotating blades to resist corrosion resulting from contaminants in the steam. Ferritic steels have seen service for blading at steam temperatures up to 1050°F (566°C). Austenitic superalloys such as K42B and Discaloy have been used for higher temperatures such as the throttle steam of Eddystone 1 at 1200°F (649°C).

Among the blade root fastenings that have been used are single-tee roots, double-tee roots, side-entry roots, and a variety of pinned types, as shown in Figs. 2.32 and 2.33.

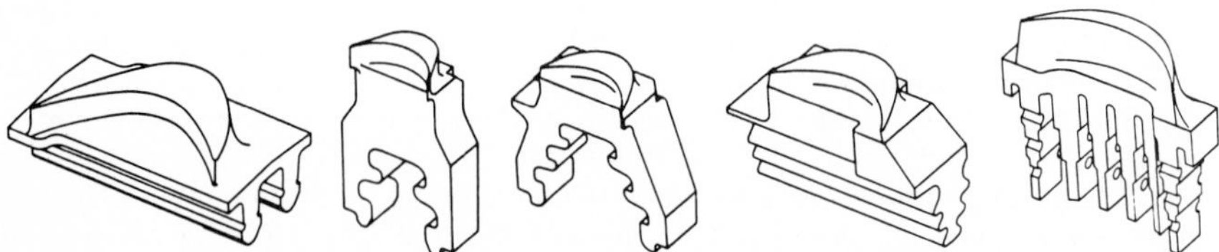

FIG. 2.32 Blade-root designs. *(General Electric Co.)*

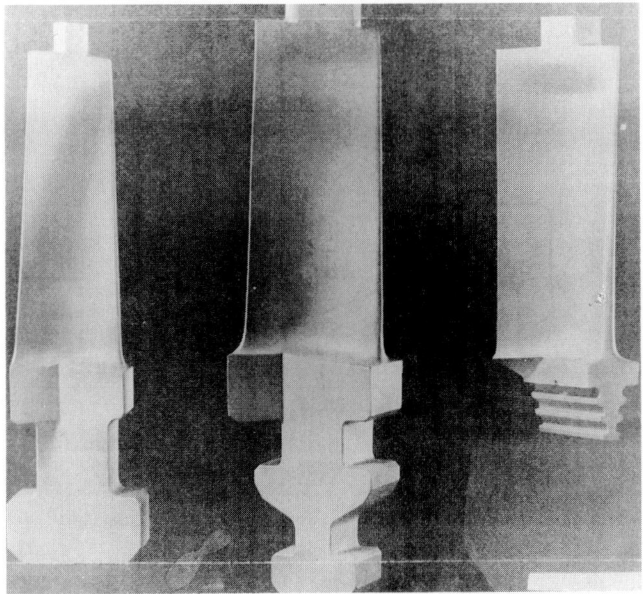

FIG. 2.33 High-temperature blade fastenings. *(Westinghouse Electric Corp.)*

Rotors

High-temperature rotors, especially those exposed to steam temperatures above 900°F (482°C), generally use the familiar Cr-Mo-V composition, although a number use a 12 percent Cr alloy at 1000°F (538°C). While the 12 percent Cr alloy has somewhat higher high-temperature strength than Cr-Mo-V, it has significantly higher oxidation resistance at 1000°F (538°C) and above. However, the alloying elements used in the high-temperature rotors increase their *fracture appearance transition temperature* (FATT). The rotor materials are notch-sensitive below FATT, and brittle fracture is a concern.

Consequently, LP rotors use an alloy composition Ni-Cr-Mo-V that has a low transition temperature. This material becomes increasingly sensitive to temper embrittlement as steam temperatures increase beyond 700°F (371°C).[12] This condition relates to the segregation of tramp elements, such as phosphorus, tin, arsenic, and antimony, at the grain boundaries and subsequent embrittlement. In the case of LWR turbines with modest steam temperatures, below 650°F (343°C), the HP rotors can use LP rotor materials as well as carbon steel. The primary selection criteria for high-temperature rotor materials are resistance to rupture and elongation (creep-rupture) and high-temperature notch sensitivity (toughness). Toughness is enhanced in modern forgings by improved melting techniques such as vacuum-carbon deoxidation, vacuum-arc remelting, electroslag remelt, and low-sulfur melting.[13]

Cooling methods are often employed at the HP- and IP-inlet stages to increase the rotating-blade groove wall strength. Cooler steam from the HP exhaust is directed under the inlet flow guide of the IP element and at the base of the blade roots to reduce the rotor temperature, as shown in Fig. 2.34. Another technique,

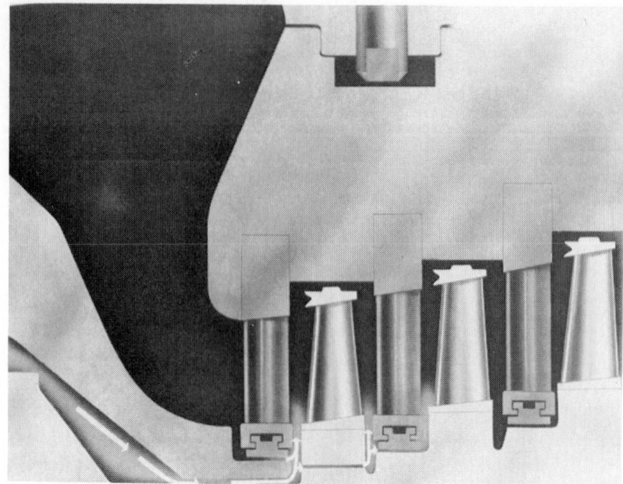

FIG. 2.34 IP rotor-cooling arrangement. *(Westinghouse Electric Corp.)*

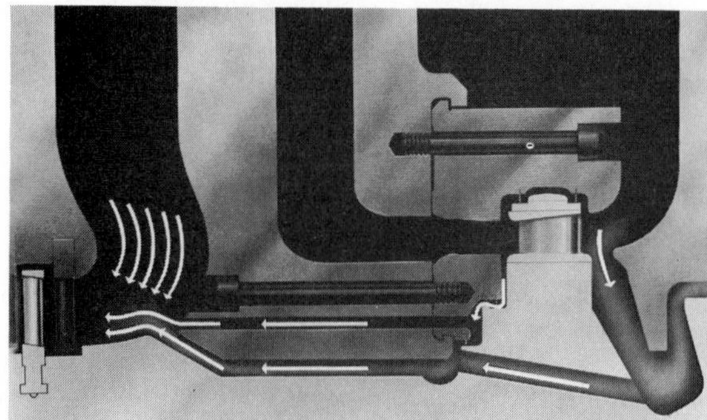

FIG. 2.35 HP rotor-cooling arrangement. *(Westinghouse Electric Corp.)*

employed on HP rotors, is to pass steam through holes in the first-stage disk to cool the bore and the rotor ahead of the following stage, as shown in Fig. 2.35, thereby eliminating bore creep-to-rupture cracks.

In addition to centrifugal stresses, the rotor is subjected to (1) alternating bending stresses caused by the weight of the rotor and blades, (2) transverse vibratory forces resulting from unbalance, (3) steady torsional stresses imparted by the blading and upstream rotors, (4) torsional oscillations resulting from electrical interactions with the transmission system, and (5) rotor misalignment stresses. Finally, there are three self-excited rotor whirl categories: *oil whip,* an instability inherent in certain bearing configurations; *steam whirl,* the result of aerodynamic forces of the steam on the blades and seals; and *friction whirl,* the result of the internal hysteresis energy added to the rotor due to shrink fits, material damping, etc.[14]

Shells and Blade Rings

For steam temperatures above 800°F (427°C), creep becomes increasingly important, and Cr-Mo steels of various compositions are used up to 1050°F (566°C). The Mo improves the high-temperature strength, and the Cr stabilizes the carbon and enhances the oxidation resistance. These shell materials are ferritic steels. Austenitic steels have higher creep strength than the ferritic steels but are susceptible to thermal distortion and cracking because of their low thermal conductivity and higher thermal expansion. For this reason, ferritic steels are preferred as pressure-vessel materials for steam temperatures up to 1050°F (566°C). In contrast, steam turbines for LWRs can use carbon steel for the HP shells because of the modest steam temperatures.

Although partial-arc admission is used to improve turbine efficiency, it results in large temperature differences between the nozzles of the active and inactive arcs of admission. When the active and inactive nozzles share a common chamber, thermal stresses are produced in the walls separating the two groups of nozzles and in the adjoining structure. When the active and inactive nozzles are in separate structures (nozzle chambers or nozzle boxes) and are separate from the shells, the thermal stress is reduced considerably. An example of separate nozzle-chamber construction is shown in Fig. 2.36.

Large-temperature mismatches are imposed on the steam chest and/or control-valve bodies at the main steam inlet during unit start-up. Large-pressure

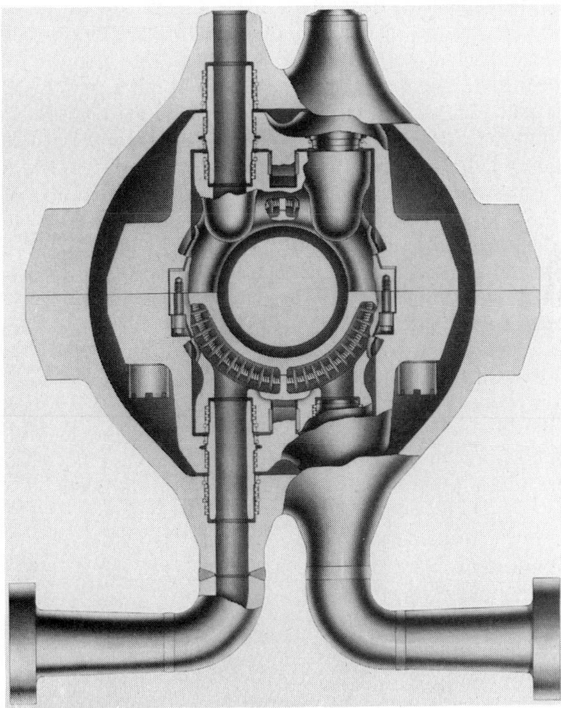

FIG. 2.36 Separate nozzle chambers. *(General Electric Co.)*

differentials also exist. Although large-temperature mismatches occur on the valve bodies at the hot reheat inlet, the pressure stresses are low at start-up, increasing as the load increases.

CURRENT AND FUTURE DEVELOPMENTS

In the late 1950s the rising costs of high-temperature material, especially austenitic steels, trends in fuel costs, and the lower availability (reliability) of powerplants with steam temperatures above 1000°F (538°C) led to the adoption of 1000°F (538°C) as a standard steam temperature. To offset the loss of efficiency, many utilities resorted to 3500-psig (24 115-kPa) supercritical throttle pressure with an overall improvement in plant efficiency. The advantages of 3500 psig (24 115 kPa) were documented in a study by Public Service Electric and Gas (PSE&G) of New Jersey.[15]

PSE&G found that the cost of austenitic materials was making their 2400-psig, 1100°F/1050°F (16 536-kPa, 593°C/566°C) plants uneconomical. It was reported that for a 342-MW plant, the efficiency of a 3500-psig, 1000°F/1000°F (24 115-kPa, 538°C/538°C) unit was practically the same as for the lower-pressure unit. In addition, the higher-pressure unit had a capitalized operating and equipment cost advantage of $2.65 per kilowatt of capacity. Increasing the unit size from 342 to 400 MW resulted in a comparatively low incremental increase in capital cost. Moreover, the adoption of double reheat with steam temperatures of 1000°F/1025°F/1050°F (538°C/552°C/566°C) [ascending reheat] reduced the capitalized operating and equipment cost by $1.31 per kilowatt of capacity as compared to the 3500-psig, 1000°F/1000°F (24 115-kPa, 538°C/538°C) unit. The double-reheat unit had a 309-Btu/kWh lower heat rate than the 3500-psig (24 115-kPa) single-reheat unit. The *heat rate,* the inverse of efficiency, is defined as Q_A/W, where the heat added Q_A is in Btu per hour (watts) and the work W is in kilowatts.

Over one hundred fifty 3500-psig (24 115-kPa) single-reheat units have been purchased in the United States as well as twenty-six 3500-psig (24 115-kPa) double-reheat units. However, the subsequent poorer reliability of the 3500-psig (24 115-kPa) units and the adoption of nuclear-fueled units for base-load operation led to a decline in the application of 3500-psig (24 115-kPa) units. A major factor in the poorer reliability was the rapid increase in unit size, particularly in the 3500-psig (24 115-kPa) designs. Moreover, boiler reliability was poorer because of the need for once-through designs which required more restrictive water chemistry and greater control complexity. Finally, the 3500-psig (24 115-kPa) units were poorly suited to cycling.

In recent years, most units have been designed for steam conditions of 2400 psig, 1000°F/1000°F (16 536 kPa, 538°C/538°C) and more modest ratings. More recently, research efforts have been directed at increasing plant efficiency through higher pressure and temperatures while maintaining unit reliability to offset the rapid increases in fuel costs that followed the 1974 oil embargo. The development efforts to improve plant operation have been related to design, materials, efficiency, and operation.

Design

One facet of current and future design efforts, particularly for steam turbines, relates to what have been termed *ruggedized* designs in which the stress margins have been increased. For turbines, in many instances the blade proportions are more robust. Some new LP-turbine designs will use freestanding blades in the last few stages, eliminating the need for lashing wires. Integrally shrouded blades or freestanding blades have eliminated the crud traps where steam contaminants collected in blades with mechanically fastened shrouds. The accummulation of contaminants often led to component distress.

To achieve improved reliability at higher pressures and temperatures, more modest steam conditions were proposed in two recent Electric Power Research Institute (EPRI) funded studies—4500 psig, 1100°F/1050°F/1050°F (31 000 kPa, 593°C/566°C/566°C) and 4500 psig, 1050°F/1075°F/1100°F (31 000 kPa, 566°C/579°C/593°C) as compared to the pioneering Eddystone and Philo units.[16, 17] EPRI is currently soliciting proposals to implement the research programs recommended in the two studies.

The ability to repair weld rotors where damage has occurred and to achieve material properties equal or superior to the parent material indicates that there may be more widespread application of welded rotors fabricated from a series of disks.[18, 19] In addition, monoblock LP rotors, as opposed to designs with separate last-row disks or entirely built-up constructions of shrunk-on disks and a central shaft, will predominate, reducing the incidence of stress corrosion cracking. The built-up rotor construction is illustrated in Fig. 2.37.

To improve load-cycling capability while reducing low-cycle thermal fatigue and improving part-load efficiency, sliding-throttle pressure operation is applied on partial-arc admission turbines (hybrid operation) coupled with supercritical-pressure boilers. The boiler manufacturers have recently developed supercritical designs in which conventional sliding-throttle pressure operation is possible—as is the case for subcritical boilers.

Improved blading, high- and low-pressure, is being developed for both new turbine designs and retrofit applications that can improve the turbine heat rate by a total of 1.5 to 2.0 percent, 100 to 150 Btu/kWh (29.3 to 44 W/kW).[20, 21] In addition, many of the improved LP last-row blade designs will have increased exhaust-pressure capability, benefiting both units with more conventional ex-

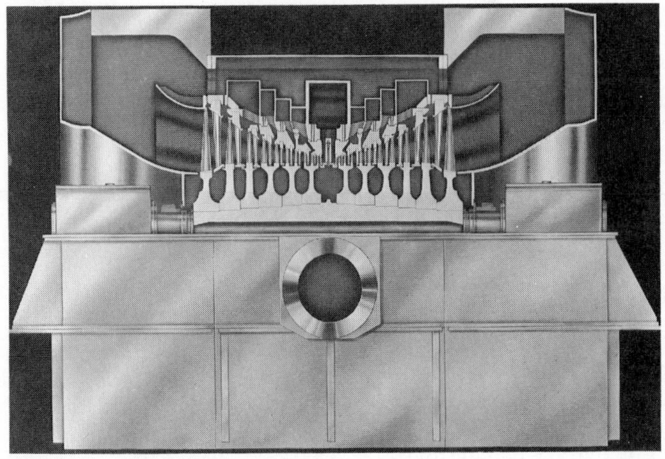

FIG. 2.37 Builtup LP rotor construction. *(Westinghouse Electric Corp.)*

haust (condenser) pressures as well as those with considerably higher exhaust pressures.[6] Operation at low load and high exhaust pressure can result in buffeting or stall flutter, which is a flow-induced vibration that occurs when there is appreciable flow incidence on the turbine blade. Some currently operating designs have demonstrated that last-row blades can be built that will avoid stall flutter under low-volumetric-flow conditions.

Materials

The advanced melting techniques for high-temperature Cr-Mo-V rotors—identified in a previous section and in references 12, 13, and 22—improved toughness and lowered FATT, which would improve the reliability of rotors with 1050°F (566°C) steam temperatures as proposed in the EPRI-funded studies. Materials research on LP rotor components to eliminate or reduce the amount of tramp elements will not only reduce temper embrittlement on Ni-Cr-Mo-V rotors and allow an increase in steam-inlet temperature, but also will increase material toughness. Moreover, heat treatments are being employed which result in lower yield strengths than in the past but which have improved toughness.

For the 1100°F (593°C) steam temperatures proposed in the EPRI studies, ferritic rather than austenitic steels are used primarily. Type 422 (12 percent Cr) steel rotor forgings would be used, coupled with some advanced melting techniques such as electroslag remelt, low-sulfur melting, and vacuum-arc remelting, similar to the processes used for Cr-Mo-V rotors.[22]

The inner shells and nozzle chambers which are exposed to 1100°F (593°C) steam would use a cast 12 percent Cr alloy steel with composition refinements. The lower-temperature shells would probably use 2¼ percent Cr castings. In contrast, type 316 austenitic steel would be used for the forged valve bodies and turbine-inlet steam piping that is exposed to 1100°F (593°C) steam. This material has given good service at 1100 and 1200°F (593 and 649°C) on units that have been operating for over 20 years. The material would have a lower carbon content and somewhat higher nitrogen content than past applications, particularly to improve creep strength. The bolting on the 12 percent chrome shells might be a 12 percent Cr alloy to obtain the same thermal-expansion coefficient and thereby alleviate bolt relaxation. The high-C_o materials that have been used are susceptible to stress-corrosion cracking and are austenitic steels that have higher thermal-expansion coefficients.

To achieve partial-arc admission at 1100°F (593°C), a new blade material would be required for the first stage. The selection involves trade-offs between materials with internal damping and materials with stress-corrosion-cracking resistance and high fatigue strength.

Efficiency

We noted earlier that expected blading improvements will reduce heat rate by 100 to 150 Btu/kWh (29.3 to 44 W/kW). The increase in steam pressure and temperature and other cycle and hardware improvements increased the efficiency of the 4500-psig, 1150°F/1025°F/1050°F (31 000-kPa, 621°C/552°C/566°C) advanced plant[16] by about 9.5 percent over the 3500-psig, 1000°F/1025°F/1050°F (24 115-kPa, 538°C/552°C/566°C) base plant. The wider-range exhaust-pressure turbines described in reference 6 had higher efficiencies than their conventional exhaust-pressure predecessors in spite of earlier predictions that their performance would be poorer.[23] Increasing the final-feedwater temperature by adding a heater that extracts above the cold reheat pressure can improve the heat rate by 0.5 to 0.7 percent. This was fairly common practice in the early 1950s and only recently has been given serious consideration again.[24, 25]

Operation

The recent development of supercritical boilers that can slide throttle pressure will considerably enhance the cyclic capability of high-pressure, high-temperature units in addition to improving part-load efficiency and reducing low-cycle thermal fatigue on the turbine. Faster loading rates can be achieved without sacrificing turbine life. The availability of control systems that allow full-arc admission start-up, coupled with sliding-throttle pressure capability, reduces hard-particle erosion on the initial turbine stage. These improved control systems are also capable of on-line transfer from full-arc to partial-arc operation and vice versa, improving efficiency and operational flexibility. The use of turbine bypass systems can improve steam-metal temperature matching at start-up, thereby reducing thermal stresses and reducing start-up times. Smaller bypass systems are practical when they are employed with partial-arc admission turbines that have control systems that can switch from full-arc to partial-arc operation.

REFERENCES

1. S. N. Fiala, "First Commercial Supercritical-Pressure Steam-Electric Generating Unit for Philo Plant," ASME Annual Winter Meeting, Chicago, Nov. 1955.
2. J. H. Harlow, "Engineering the Eddystone Plant for 5000 Lb and 1200 Deg Steam," *Transactions of the ASME,* Vol. 79, American Society of Mechanical Engineers, New York, 1957.
3. R. L. Reynolds, "Recent Development of the Reheat Steam Turbine," *Mechanical Engineering,* New York, Jan. 1952.
4. C. Schabtach and R. Sheppard, "Modern Reheat Turbine," *Mechanical Engineering,* New York, Feb. 1952.
5. G. J. Silvestri, Jr., and J. Davids, "Effects of High Condenser Pressure on Steam Turbine Design," *Proceedings of the American Power Conference,* Vol. 33, Illinois Institute of Technology, Chicago, 1971.
6. G. J. Silvestri, Jr., and J. Davids, "Laboratory and Field Tests on Low Pressure Steam Turbines at High Exhaust Pressure," *Proceedings of an EPRI Workshop on Water Conserving Cooling Systems,* Palo Alto, Calif., May 1982.
7. R. J. Gray, W. C. Brauer, and S. C. Leland, "Design and Initial Operation of the Wyodak Plant," *Proceedings of the American Power Conference,* Vol. 41, Illinois Institute of Technology, Chicago, 1979.
8. G. J. Silvestri, Jr., O. J. Aanstad, and J. T. Ballantyne, "A Review of Sliding Throttle Pressure for Fossil Fueled Steam Turbine Generators," *Proceedings of the American Power Conference,* Vol. 34, Illinois Institute of Technology, Chicago, 1972.
9. T. Vuksta, Jr., "Tangential Blade Velocity and Secondary Flow Field Effect on Steam-

Turbine Exhaust-Blade Erosion,'' ASME Paper 63-WA-238, ASME Winter Annual Meeting, Philadelphia, Nov. 17–23, 1963.

10. R. O. Brown, F. J. Heinze, and J. Davids, ''High Performance–Low Pressure Turbine Elements,'' ASME Paper 63-Pwr-15, 1963 ASME-IEEE National Power Conference, Cincinnati, Sept. 22–25, 1963.
11. C. E. Siglem and R. O. Brown, ''Turbine Exhaust Losses,'' ASME Paper 60-Pwr-7, 1960 ASME-AIEE Power Conference, Philadelphia, Sept. 21–23, 1960.
12. R. Viswanathan and R. I. Jaffee, ''Toughness of Cr-Mo-V Steels for Steam Turbine Rotors,'' ASME Paper 82-JPGC-Power-35, presented at Joint Power Generating Conference, Denver, Oct. 1982.
13. R. M. Curran, D. L. Newhouse, and J. C. Newman, ''The Development of Improved Rotor Forgings for Modern Large Steam Turbines,'' ASME Paper 82-JPGC-Power-25, presented at Joint Power Generating Conference, Denver, Oct. 1982.
14. H. Gunter, ''Dynamic Stability of Rotor Bearing Systems,'' NASA SP-113, 1966.
15. R. A. Baker, ''Mercer Generating Plant,'' *Proceedings of the American Power Conference,* Vol. 24, Illinois Institute of Technology, Chicago, 1962.
16. S. B. Bennett, R. L. Bannister, D. V. Giovanni, and A. F. Armor, ''The Design and Performance of the Next Generation Low-Heat-Rate Pulverized-Coal Power Plant,'' *Proceedings of the American Power Conference,* Vol. 43, Illinois Institute of Technology, Chicago, 1981.
17. A. F. Armor, D. V. Giovanni, R. H. Ladino, E. H. Miller, G. M. Yasenchak, and R. J. Waltz, ''The Next Generation of Pulverized-Coal Power Plants,'' *Proceedings of the American Power Conference,* Vol. 43, Illinois Institute of Technology, Chicago, 1981.
18. J. Conway, G. Kuhnen, and D. G. Graves, ''Salvaging Condemned Turbine Rotors by Welding Technique,'' *Proceedings of the American Power Conference,* Vol. 46, Illinois Institute of Technology, Chicago, 1984.
19. L. D. Kramer, R. E. Clark, and D. Amos, ''Welding Repair of Low Pressure Rotors for Increased Availability,'' ASME Paper 84-JPGC-Pwr-26, presented at the Joint Power Generating Conference, Toronto, Canada, Oct. 1984.
20. M. F. O'Connor, K. E. Robbins, and J. C. Williams, ''Redesigned 26-Inch Last Stage for Improved Turbine Reliability and Efficiency,'' ASME Paper 84-JPGC-Pwr-17, presented at Joint Power Generation Conference, Toronto, Canada, Oct. 1984.
21. E. J. Barsness, F. R. Vaccaro, and J. P. Kessinger, ''Turbine-Generator Technology Advancements Enhance Reliability and Life Cycle Extension,'' *Proceedings of the American Power Conference,* Vol 46, Illinois Institute of Technology, Chicago, 1984.
22. V. P. Swaminathan, J. E. Steiner, and R. I. Jaffee, ''High Temperature Turbine Rotor Forgings by Advanced Steel Melting Technology,'' ASME Paper 82-JPGC-Power-24, presented at the Joint Power Generating Conference, Denver, Oct. 1982.
23. R. D. Mitchell and J. P. Rossil, ''Economics of High Back Pressure Turbines with Dry Cooling Systems,'' *Proceedings of the American Power Conference,* Vol. 39, Illinois Institute of Technology, Chicago, 1977.
24. H. H. Audi and E. A. Jodidio, ''Optimization of Coal Fired Reference Plant Feed Cycle for Plant Sizes 400–600 and 800 MW,'' Paper presented at Joint Power Generation Conference, Indianapolis, Sept. 1983.
25. P. J. Adam and P. G. Davidson, ''Design Coal Units for Best Heat Rate to Reduce Cost per Kilowatt,'' *Proceedings of the American Power Conference,* Vol. 46, Illinois Institute of Technology, Chicago, 1984.

CHAPTER 2.3
OIL AND GAS ENGINES

William M. Kauffmann, P.E.
Consultant
Parma, OH

INTRODUCTION

Internal combustion engines for powerplants are built to operate on one of two basic principles, namely, four- and two-stroke cycles. To be competitive and profitable in the heavy-duty industrial, generating, chemical, or pipeline fields, large commercial engines are normally turbocharged to obtain the maximum output; like engines, turbocharger systems vary widely in design and construction. Current heavy-duty diesel engines operate on fuel oil, a combination of gas and fuel oil, or gas only. Each mode of operation requires specific control systems and equipment.

The wide variety of designs has led to numerous patents being granted to engine manufacturers. Equally important as engine design is the installation of the units. Correct sizing of heat exchangers, air filters, exhaust silencers, radiators, cooling towers, and piping will ensure proper operating conditions. These components are generally selected by the engine builder. Important, too, are foundation design and soil analysis, for they ensure adequate engine support.

Four-Cycle Engine

The four-cycle engine completes its sequence in two engine revolutions, or four strokes of the piston. As shown in Fig. 3.1, the events are suction (or admission), compression, expansion, and exhaust. The pressure-volume changes are indicated in Table 3.1.

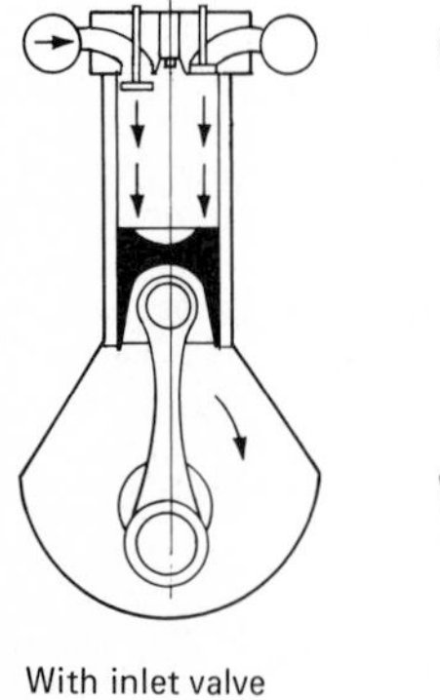

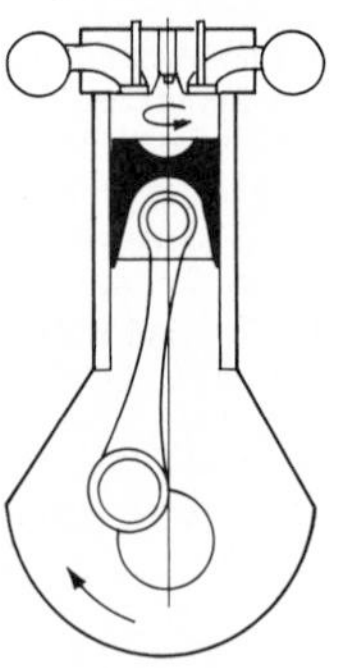

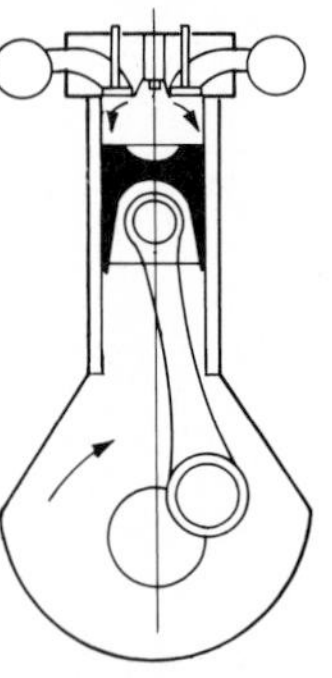

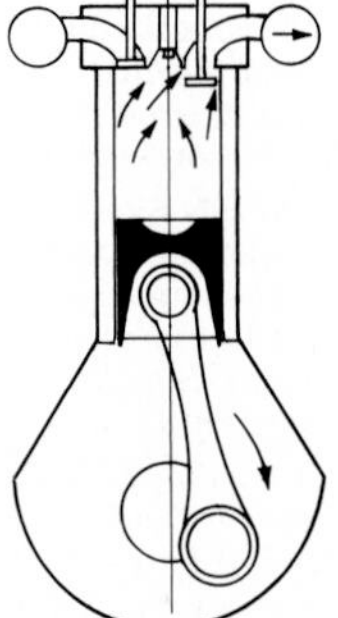

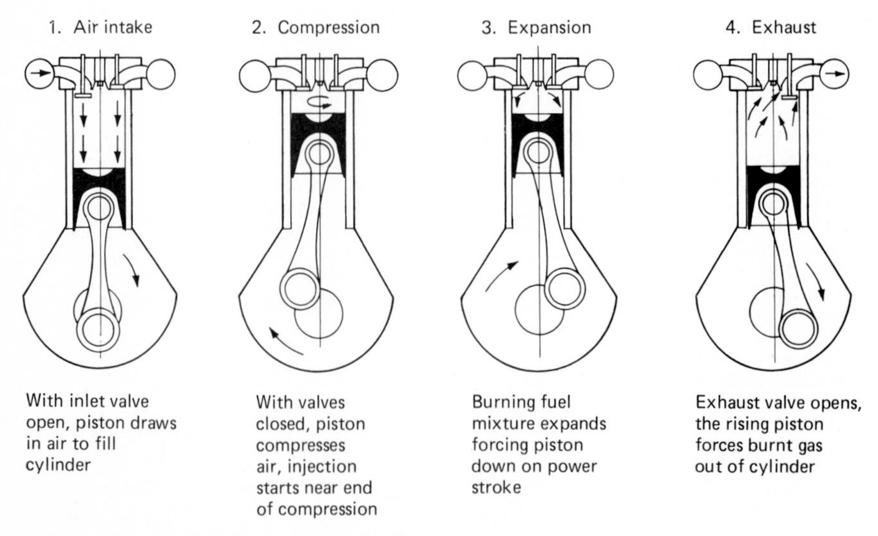

FIG. 3.1 Four-cycle engine. *(W.M. Kauffmann.)*

TABLE 3.1 Engine Pressure-Volume Changes

Stroke	Volume in cylinder	Pressure
Suction	Increases	Little change
Compression	Decreases	Increases
Expansion	Increases	Decreases
Exhaust	Decreases	Little change

Two-Cycle Engine

As Fig. 3.2 shows, two-cycle engines have inlet ports in the lower part of the cylinder, through which pressurized air for scavenging enters the cylinder. Exhaust gases exit through two or four poppet valves in the cylinder head. In the fully ported design, a row of exhaust ports is located on one side and a row of inlet ports on the other, forming a loop flow of gases. Exhaust-poppet-valve, two-cycle diesels are referred to as *uniflow engines.*

Another type of uniflow two-cycle unit is the opposed-piston engine. Two crankshafts are used, one at the top of the cylinder and the other at the bottom. Each has a piston and rod that reciprocate toward the center of the cylinder. The bottom piston controls exhaust ports; the top piston, scavenging ports. Burnt gases and scavenging air flow in the same direction.

Definitions of terms related to engine performance assist in evaluating power data. Key terms are defined in the box.

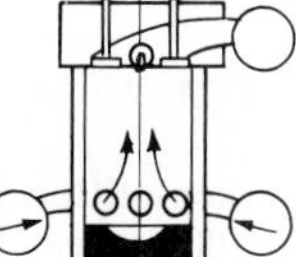
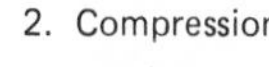
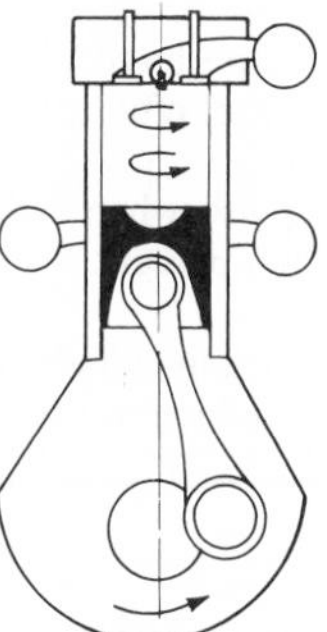
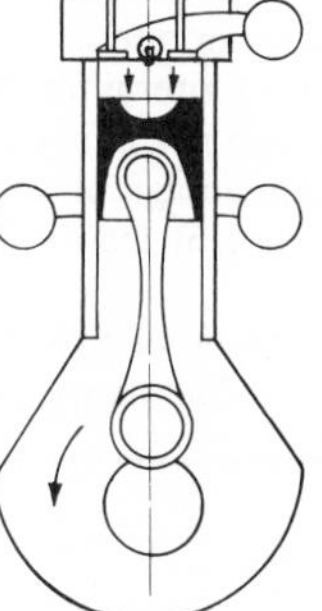

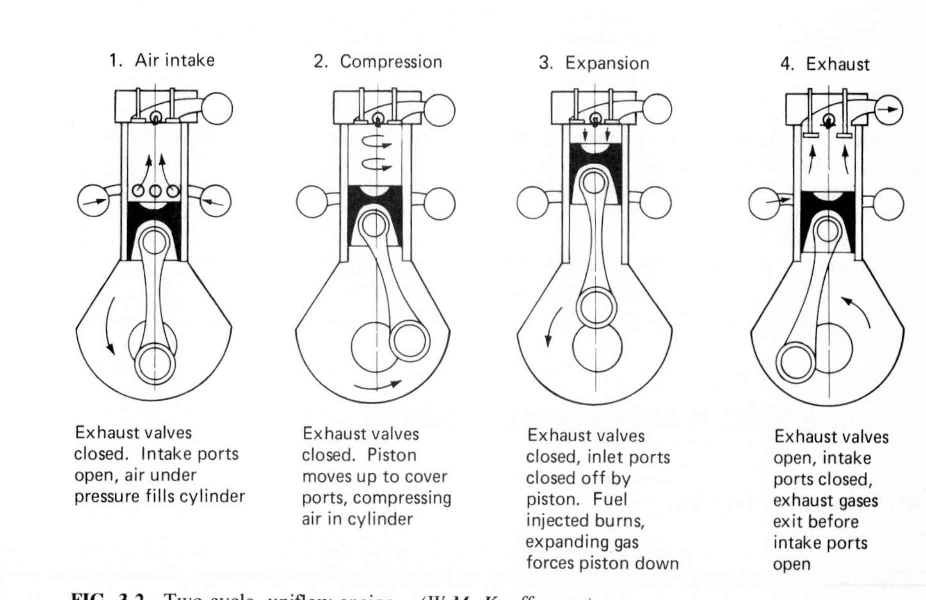

FIG. 3.2 Two-cycle, uniflow engine. *(W.M. Kauffmann.)*

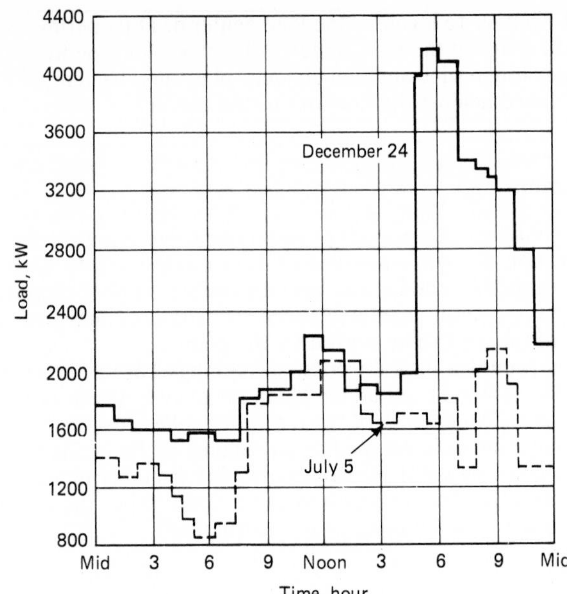

FIG. 3.4 Daily load curves. *(W.M. Kauffmann.)*

curves (Fig. 3.4). Daily load curves show changes in load during a 24-h period and shifts in load during normal seasons and help operators to anticipate maximum load periods. The load duration curve, derived from daily log curve data, shows how load in kilowatts varies in percentage of total hours in a year. For a 2200-kW peak load, for example, the load exceeded 600 kW 20 percent of the total hours; it exceeded 900 kW 50 percent of the total hours. In anticipating future load parameters, possible duration curves may be developed to assist in projecting plant growth.

Defining the High-Output Engine

High-output turbocharged engines are currently operating successfully and economically in continuous heavy-duty service. There are a number of reasons for this.

Research and Development. Research and development of high-pressure-ratio turbocharging engines require high combined efficiencies of the turbocharger-blower and gas-turbine sections. Proper matching of turbocharger and engine has yielded bmep ratings above 200. Improved air-intake and exhaust manifolding, through-flow scavenging, four valve heads in both the four- and two-cycle uniflow engines, and valve-port streamlining contributed to improved flow coefficients and higher volumetric efficiencies.

Valve timing and inlet/exhaust overlap are designed for better scavenging and increased trapped air volume. Aftercooling lowers the blower discharge-air temperature, increasing the charge air density to the cylinder as well as lowering the initial cycle temperature and thermal load on critical parts. A greater mass of air allows more fuel to be burned, increasing the power output.

Flood-cooled pistons reduce the ring belt temperature, improving piston ring life and lube oil consumption. High-velocity jacket-water cooling reduces liner-wall thermal stresses as well as the piston and cylinder-head deck. An insulated exhaust manifold contributes maximum energy to the engine turbocharger. Elimination of water-cooled exhaust piping reduces the thermal load of the radiator and heat exchanger.

Four Main Types. Based on fuels burned, the four main types of high-output engines are diesel, dual fuel, high-compression spark-ignition gas, and the so-called tripower. The diesel principle, patented in 1893 by Rudolph Diesel, a German engineer, consists of compressing air in the engine cylinder to a temperature above the ignition point of the injected fuel. The high compression ratio, above 12:1, gave a lower specific fuel consumption than any other prime mover operating at the time.

Dual-fuel engines operate on the diesel cycle, except that instead of introducing air only in the cylinder, a mixture of gas and air is used. Most modern units use natural gas, although other gases, particularly those of the paraffin series, have been chosen.

High-compression spark-ignition gas engines operate on the principle that the gas-air mixtures will not ignite at diesel compression ratios provided that a very lean air/fuel ratio is ensured. An ignition source is needed to fire the charge, such as an electric spark from a special-design high-voltage spark plug.

Tripower engines are basically dual-fuel but capable of conversion to spark-ignition gas operation. The fuel-injection nozzle is replaced by an integral spark plug and casing; a magneto drive is built in. Fuel pumps are deactivated, or removed, and the fuel-oil system is valved off. Governor controls selected are similar to those for dual-fuel operation.

Turbocharging

As Fig. 3.5 shows, turbocharging provides increased air density to the working cylinder. Two basic systems are popular, pulse charging and constant-pressure, for both two- and four-cycle engines. Turbochargers consist of a blower, or compressor section, and an exhaust-gas turbine section. The compressor impeller is mounted on a common shaft with the turbine wheel. In large engines, the turbine casing is water-cooled. Smaller units (and automotive types) are noncooled.

Pulse Charging. Turbochargers for pulse charging have two or more gas entry passages to a bladed nozzle ring that directs gas against the turbine wheel. (Constant-pressure charging uses a single entry passage.) Shown in Fig. 3.6 is a typical cross section of a conventional turbocharger. The pulse-charging system shown in Fig. 3.7 may be applied to both two- and four-cycle engines, and it consists of one to four pipes, with diameters equal to the exhaust passage of the cylinder head. These are combined to two, three, or four cylinders per pipe. In V-12 engines, three pipes enter each of four manifolds; V-16 engines may have four cylinders entering four manifolds. Here the option for using constant pressure becomes attractive (see Fig. 3.7).

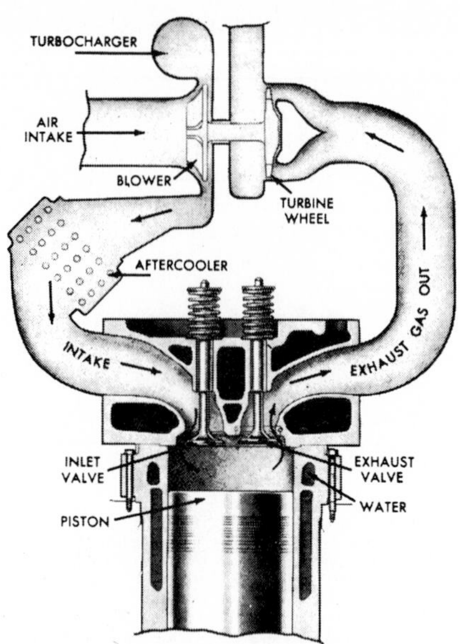

FIG. 3.5 Turbocharging system on high-output engine. *(Worthington Corp.)*

FIG. 3.6 Conventional turbocharger cross section. *(Elliott Co.)*

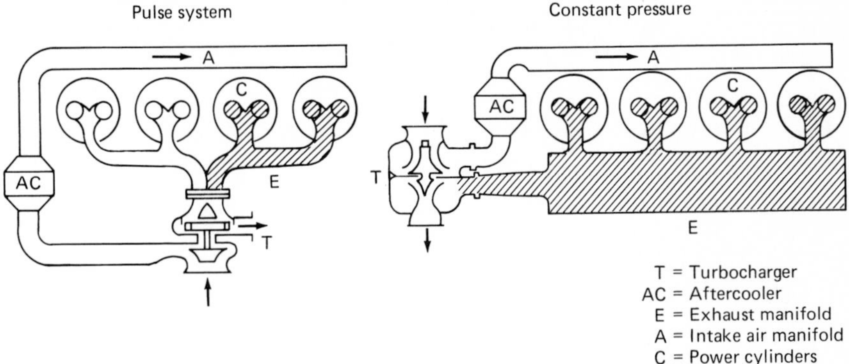

FIG. 3.7 Pulse versus constant-pressure turbocharging. *(W.M. Kauffmann.)*

Constant-Pressure Charging. For constant-pressure charging, an air jet assist against the turbine wheel permits the engine to be supplied with pressurized air for starting. The inherent simplicity of the single-insulated exhaust manifold is shown in Fig. 3.8, a powerplant diesel with two-cycle uniflow design. In the constant-pressure system, the cylinder exhaust pulses are eliminated by the relatively large manifold to the turbine.

Engine Heat Balance

The engine heat balance is a breakdown of the total heat available in the fuel consumed by the engine. Included in the breakdown are power delivered, jacket-water cooling, turbocharger cooling, air aftercooling, lube oil cooling, exhaust,

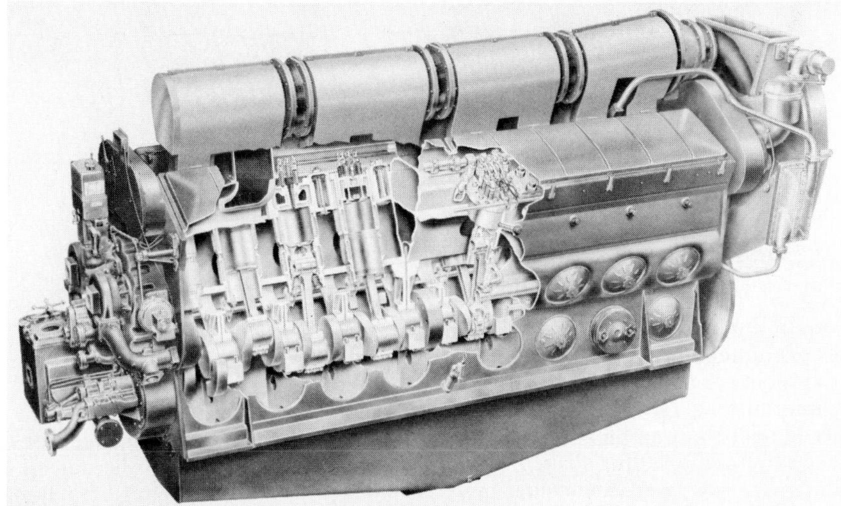

FIG. 3.8 Single-insulated exhaust manifold. *(Electromotive Corp.)*

and radiation. Each is expressed in Btu/bhp • h (W/kW) or as a percentage of the total heat. The procedure for obtaining these data consists of measuring coolant flow and temperature differentials over a given test period. The product of pounds (kilograms) per hour and the degrees differential gives the rate of heat rejection to the coolant for both water and lubricant flowing through the engine.

Table 3.2 shows a typical heat balance for full rated load and is based on the *specific fuel consumption* (SFC) of the engine. For full diesel-oil operation, SFC is about 0.36 lb/bhp • h (219 g/kWh). Based on 19,350 Btu/lb (44 965 kJ/kg) of fuel, the total heat available is 6967 Btu/bhp • h (2736 W/kW). For dual-fuel engines using 6 percent pilot fuel, about 6400 Btu/bhp • h (2514 W/kW) is an average value. For high-compression spark-ignition gas operation, about 6150 Btu/bhp • h (2415 W/kW) of fuel economy is obtained. This would indicate the latter type of engine provides the most efficient power mode.

In calculating heat balance, the first step is to obtain full-load SFC by test. The data will assist in sizing heat exchangers circulating oil and cooling water. For example, find the heat balance for a 1000-bhp (746-kW) diesel with a rated full-load consumption of 0.36 lb/bhp • h (219 g/kWh). From Table 3.2, the heat rejected to the jacket water is 800 Btu/bhp • h (314 W/kW), to the turbocharger is 139 Btu/bhp • h (55 W/kW)—or 939,000 Btu/h (275 kW) for the water system and 264,000 Btu/h (57 kW) to the lube oil heat exchanger.

TABLE 3.2 High-Output Heat Balance

Engine system	Btu/bhp • h	W/kW	Percent
Jacket water	800	314	11.5
Turbocharger	139	55	2.0
Lubricating oil	264	104	3.8
Aftercooling	278	109	4.0
Exhaust	2420	951	34.7
Radiation	521	205	7.5
Power	2545	1000	36.5
Total	6967	2738	100.0

ENGINE CONSTRUCTION

Current engine designs cover a wide range of construction features. Engine designers must account for type of service, whether the unit will be portable or stationary, bmep rating, speed, and whether operation will be continuous or intermittent. Given approval by management, a prototype unit is built and turned over to research and development for evaluation and performance, followed by field tests for shakedown and reliability. Only then will the unit be offered for sale. Let us consider how these designs vary by diagnosing major parts and materials.

Base and Frame

There are two types of base-and-frame construction—bedded crankshaft and hung shaft. Bedded-shaft bases are generally semisteel castings, with deep heavily ribbed main-bearing supports and an oil sump. The dry-sump type has an external oil reservoir. The wet-sump type has a deep base to hold all the oil circulated through the engine. A screen is placed above the sump to collect any metal particles. Base-bearing saddles are line-bored for true alignment. Foundation-bolt bosses are rugged to avoid distortion during tightening.

The hung-shaft design (Fig. 3.9) has a crankshaft hung in the frame with a steel-fabricated oil sump attached at the frame support level. The cylinder block and water jackets are integral with the frame structure. The frame design for the bedded-shaft engine either may be with an integral cylinder block or a separate block. Frames are either semisteel or welded fabrications, with bearing caps inverted and bolted to the frame. Generally, through-bolts are inserted.

In some bedded-shaft designs, through-bolts are used for fastening the frame to the base and to the main bearing caps. Frame cover doors are fitted with explosion covers, or spring-loaded plates. In some cases, through-bolts extend from

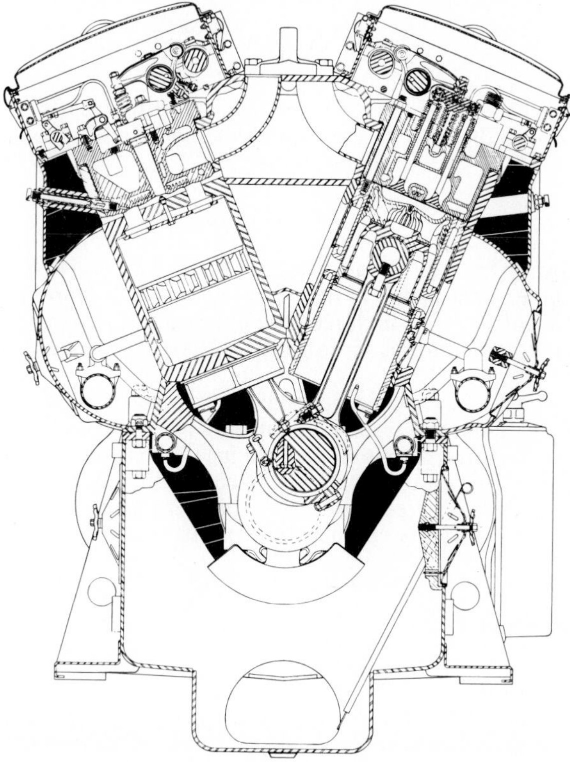

FIG. 3.9 Hung-shaft design for two-cycle uniflow engine. *(Electromotive Corp.)*

FIG. 3.10 Integral-liner design. *(DeLaval Engine Co.)*

the bottom of the base to the top of the frame, placing these parts in compression while the steel takes the load, resulting in lightweight construction.

Cylinder Liners

Cylinder liners may be dry, wet sleeve, or integral with the water jacket. Wet sleeves fit into the water jacket of the block, with two or more neoprene sealing rings at the bottom. For high-temperature cooling, sealing rings must be able to withstand temperatures up to 250°F (121°C), so a special ring material is needed.

Liner seals should be checked periodically to avoid lube oil contamination caused by leakage. Figure 3.10 shows an integral liner design. In this case, nodular iron is used, requiring porous chrome plating of the cylinder bores to prevent piston scuffing. Dry liners are normally specified for automotive-type engines and are not recommended for large diesels. The problem here is the contact of the thin sleeve and the cylinder for proper heat transfer.

Cylinder Heads and Pistons

For high-output engines, cylinder heads have four valves (Fig. 3.11)—two intake and two exhaust. Uniflow two-cycle engines have four valve heads for improved cooling and minimum distortion. Nodular iron is the material selected. It is a magnesium-inoculated iron casting having high tensile strength and a microscopic structure that shows nodules in ferrite formations—versus stringlike structures in ordinary iron.

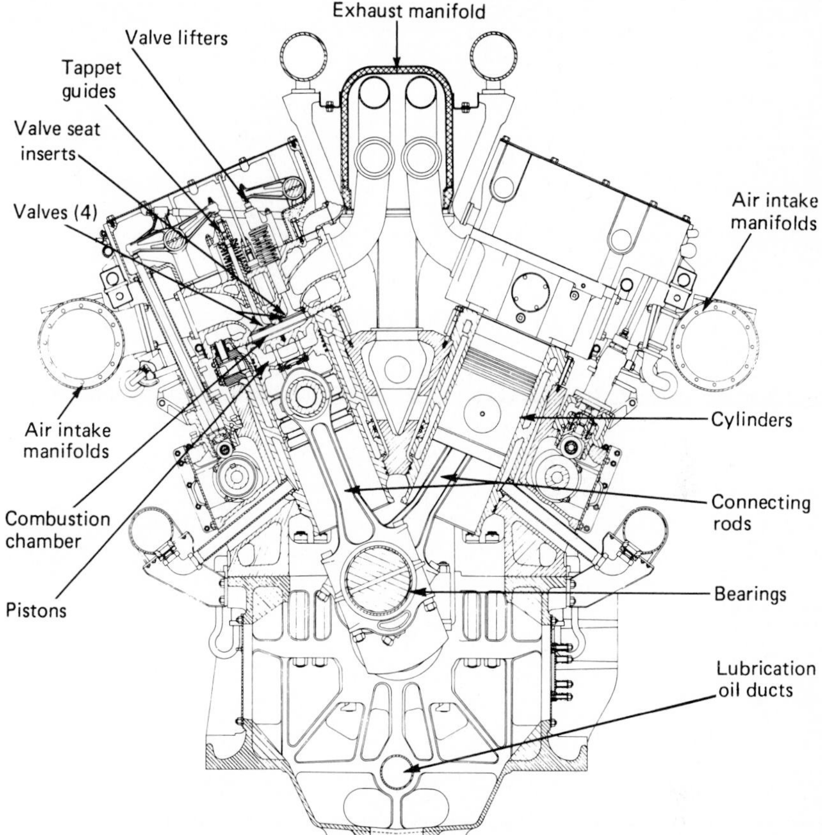

FIG. 3.11 Cylinder heads and pistons. *(Worthington Corp.)*

Pistons transmit combustion pressure to the reciprocating connecting rod and rotating crankshaft. Piston rings seal the gases and funnel heat to the cylinder wall and water jacket. Large, high-output engines have oil-cooled pistons of the flood-cooled design, in which lube oil passes up through the rod and piston pin to a closed cavity under the piston crown. Oil is returned to the crankcase from a cast-in standpipe or drilled passage.

Spray Cooling. For spray cooling in older engines, a nozzle at the top of the rod sprayed oil to the underside of the crown. This practice has been discontinued in large engines because of clogging of the nozzle with babbit when bearing failures occurred. In engines with a 6- to 8-in. (152- to 203-mm) bore, a stationary pipe sprays piston cooling oil from an oil gallery in the frame.

Currently, pistons are of two-piece construction, made of forged- or cast-aluminum alloy material, with a thin-wall steel crown bolted to a nodular iron

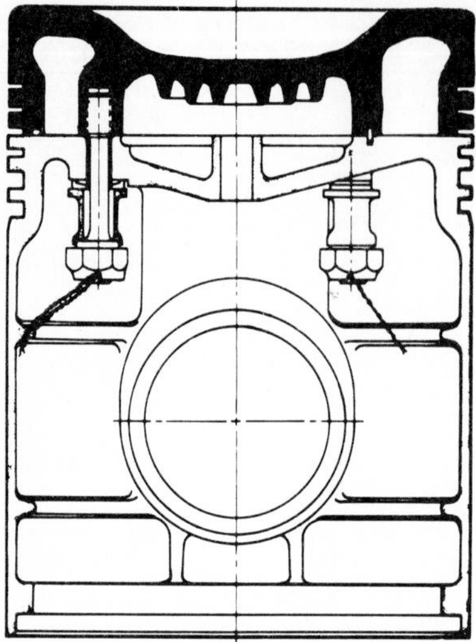

FIG. 3.12 Two-piece steel crown piston. *(DeLaval Engine Co.)*

skirt (Fig. 3.12). This design promotes improved heat transfer and more lightweight pistons and offers less heat load on the rings. For this reason, the trend is to reduce the number of power rings from five to three, depending on the cylinder-bore diameter. In some engines, the top rings are made of porous chrome-plated steel. Where porous chrome liners are the choice, tinnized iron rings are generally used.

Drain Rings. Conformable oil drain rings control lubrication of the cylinder walls. These are machined with a hollowed contour inside, which houses a coiled spring. The outer scraper edges draw in oil and discharge it to the crankcase. Two-cycle engines may have two oil rings, one above the piston pin and one below. Four-cycle engines may have all rings above the piston pin to improve skirt lubrication. Refer to Fig. 3.11.

Crankshafts

Crankshafts are the most expensive part of an engine, amounting to about 7 to 10 percent of the total engine cost; therefore, extreme care in alignment and lubrication is essential. Normally, the shaft material is a heat-treated 0.40 carbon-steel forging. Nodular iron castings have been used successfully in 9-in (229-mm) bore, two-cycle, opposed-piston engines to reduce both weight and cost.

Some builders hollow out crankpins to reduce main-bearing inertia loads. Engine crankshafts are subject to torsional vibrations that excite the shaft at certain critical speeds and cause eventual failure. Vibration dampers, with a silicone-absorbing medium, are placed at the free end of the shaft to reduce amplitudes to a safe stress level. Main and crankpin bearings are either bronze-back with thin babbit lining or a steel-back trimetal design.

Connecting Rods

All connecting rods are heat-treated forgings, fitted with four bolts to retain the rod cap and bearing. To reduce engine length, some builders construct articulated rods for V-engine configurations. Several types are in production. In one, rods are bolted to a knuckle pin that is part of the master rod. In another, the master rod is split vertically to permit removal through the cylinder bore. Here, a link rod is bolted to the knuckle pin opposite the master rod.

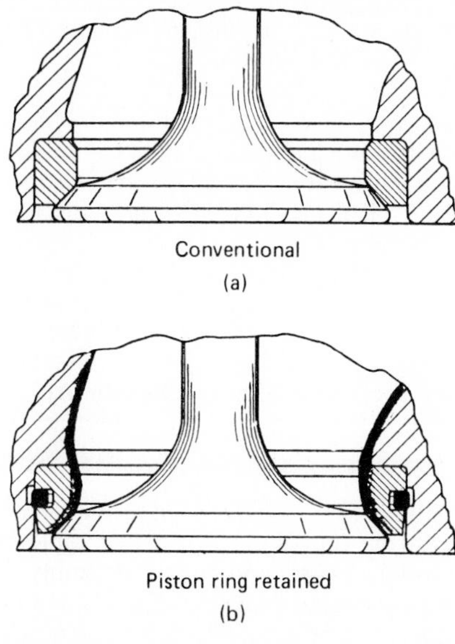

FIG. 3.13 Valve-seat inserts: (*a*) conventional; (*b*) piston ring retained. *(W.M. Kauffmann.)*

Valves

Inlet and exhaust functions are controlled by poppet valves, which are actuated by cams of the engine camshaft for four-cycle units; exhaust valves for two-cycle uniflow diesels are also cam-operated. The design of these elements requires extensive engineering analysis. Valve timing, lift, spring design, and cooling all require careful study.

Valve-seat inserts minimize valve-seat wear, as Fig. 3.13 shows. The seats are either shrunk in or held by a snap ring; materials are generally a high-nickel-alloy steel. Inlet valves are of one-piece construction, whereas exhaust valves are a two-piece welded design with a forged-steel stem and iron-alloy head. In two-cycle engines, an Inconel steel is used.

The valve gears for both four- and two-cycle uniflow engines are similar in construction. The two-cycle camshaft rotates at engine speed, whereas the four-cycle camshaft rotates at half engine speed. Two valve levers operate two valves each, either by a T piece or crossover pushrod, as used in four-cycle engines with inlet and exhaust manifolds at opposite sides of the cylinder head.

Fuel Injection

Key fuel-injection systems include the rail, plunger pump, unit injector, and distributor types.

Rail. Common rail injection, one of the oldest systems, has been used for over 70 years. Since the introduction of the plunger pump in the early 1920s, however, it has gradually disappeared from medium-speed engines. Recently, it has become the basis for new types of electronically controlled fuel injection. The system comprises an engine-driven fuel pump that maintains a line pressure of about 4000 psi (27 560 kPa) with a spring-loaded regulator valve. Injection nozzles are cam-operated in the conventional system and solenoid-operated in the electronic system. A governor-controlled mechanism adjusts the quantity of fuel injected into the engine cylinder.

Plunger Pump. A plunger pump consists of a cast housing, ported sleeve, and plunger with a machined scroll. The plunger is actuated by a cam of the engine camshaft and roller guide. The plunger operates at constant stroke. Fuel delivery terminates when the scroll edge reaches an orifice in the sleeve. Rotation of the scroll is controlled by a rack and pinion operated from the engine governor.

Unit Injector. The unit injector system is used primarily in the two-cycle uniflow diesel because of its low inertia (Fig. 3.14). It is completely self-contained with plunger pump, needle stem, spring, and nozzle assembly. The plunger section is operated by a lever off the engine camshaft. Control is effected by the rack and pinion, which rotate the plunger scroll from a governor linkage. High-pressure piping is eliminated, reducing both mechanical and hydraulic lag in fuel injection (Fig. 3.15).

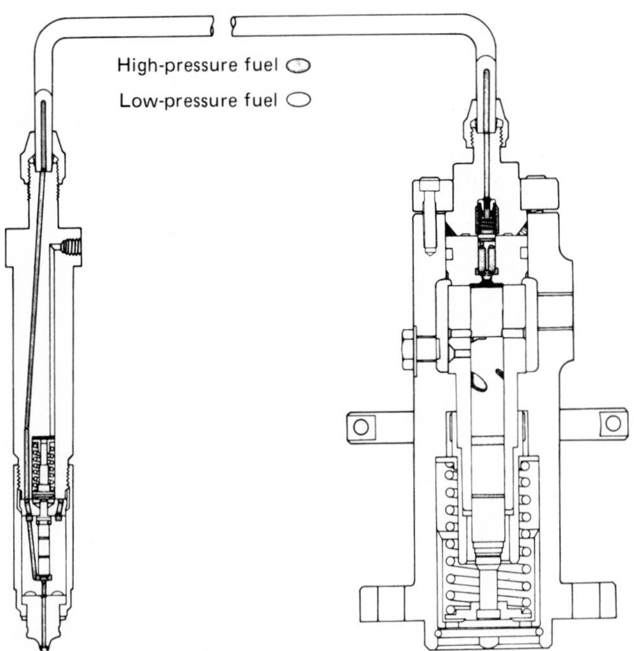

FIG. 3.14 Diesel-fuel injection equipment. *(Bendix Scintilla Corp.)*

For dual-fuel engines, a double-plunger pump was developed to maintain a constant and minimum pilot oil quantity to the cylinder during gas operation. The smaller plunger is on top of the main plunger. The main plunger is similar in operation to the single-plunger pump, with scroll and orifice control from the governor. Pilot-plunger delivery is determined by a land and orifice. By changing the land length, the pilot quantity can be varied from 3 to 6 percent of the total value of the fuel burned at full load.

Distributor. The distributor fuel-injection system is essentially for automotive and small industrial engines. A rotary horizontal cam plate actuates the pump plungers; delivery is controlled by a governor-operated sleeve or collar.

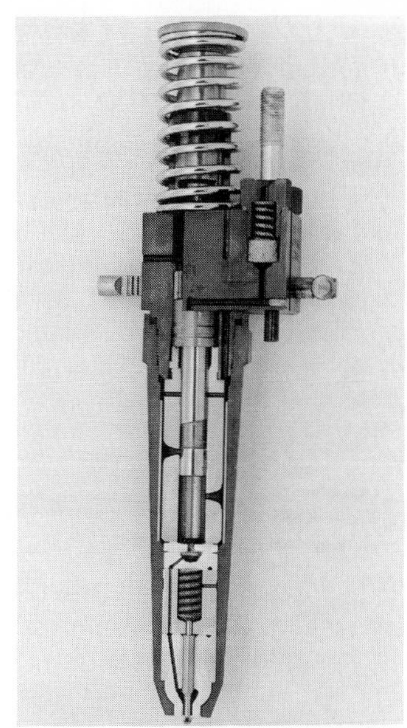

FIG. 3.15 Unit fuel injector. *(Electromotive Corp.)*

Diesel Combustion Chamber

Criteria for diesel combustion chambers include the fuel-spray dispersion, air swirl or quiescent flow, crown shape, and squish. Large high-output diesels use a simple flat-dished crown, which results in low thermal stresses in the material. The "Mexican hat" design conforms to the spray envelope. Both contours will burn the fuel efficiently; however, too strong an air swirl will cause jets from the nozzle to collide, upsetting combustion.

A quiescent airflow is favored by some, particularly when 10 to 12 nozzle holes are in operation, since the nominal included angle of a jet spray in air compressed to 450 psi (3100 kPa) is about 18° between jets. The spray envelope will contain a central core of liquid droplets and an outer layer of finely vaporized fuel. Combustion begins at random in this outer layer, increasing the temperature and igniting other jets. Some air movement essential for fuel/air mixing comes from the squish of air forced from the crown perimeter inward at top dead center of the piston.

IGNITION SYSTEMS

In theory, increasing the compression ratio in high-compression spark-ignition gas engines improves thermal efficiency, as shown by

$$E = 1 - \left(\frac{1}{r}\right)^{n-1} \tag{3.1}$$

where E = indicated thermal efficiency, %

r = compression ratio

n = 1.21 = adiabatic coefficient

Thus, increasing *r* from 6 to 12 will theoretically improve the efficiency from 32 to 39 percent, which is a 22 percent gain. In other words, 22 percent less fuel is needed to develop each brake horsepower per hour. The limit is detonation, or combustion knock. The knock limit may be raised via turbocharging, charge-air aftercooling, humidification under low-humidity conditions, and a proper air/fuel ratio.

Ignition starts at the spark plug, and the flame front advances to the opposite end of the combustion chamber, or toward the periphery for a centrally located plug. Gases compress unburned mixtures of fuel and air ahead of the front. Under adverse conditions of overload, cylinder combustion unbalance, faulty air/fuel ratio, and inadequate cooling, the compressed mixture reaches the critical temperature and ignites prematurely, causing sharp knock or detonation. This auto ignition is virtually at constant volume and is accompanied by severe pressure waves in the end zone.

Preignition

Other cylinder combustion sounds, mistaken for detonation, are caused by preignition. In high-compression gas engines, preignition may be caused by improper spark plug heat range, oil droplets in the gas, or condensed gaseous fuel that will ignite at the normal compression temperature of 450°F (232°C) before the spark plug fires. Damage to piston rings and cylinder heads will result.

Backfires occur when raw gas enters the exhaust manifold because of misfiring spark plugs. Subsequent opening of an exhaust valve will cause the unburned gas to ignite and flash back into the intake manifold during the scavenging period. For this reason, spring-loaded safety covers are usually provided on the air-intake manifold of high-compression gas engines.

Combustion Knock

The knock characteristics of hydrocarbon gases are related to their molecular structure. Figure 3.16 shows the critical compression ratio versus the number of carbon atoms in the molecular chain. The longer the chain, the greater the knock tendency down to *n*-heptane, which has zero octane number. Chemical prereactions, or cool flame, precede combustion, breaking down gases into proknock formaldehyde and hydrogen peroxides. Reducing compression and temperature by cooling will inhibit these prereactions.

Flame velocity is another factor in the combustion of various gases. The velocity reaches its maximum value at the stoichiometric mixture, or zero-air excess. The optimum operating range is on the lean side of the stoichiometric mixture, resulting in a lower exhaust temperature and reduced fuel consumption. The spark advance for high-compression gas engines is set about 10° before top center versus 30° for low-compression engines. Precise timing is essential for proper operation.

High Compression

New developments in electric ignition for high-compression gas engines have produced compact and lightweight units. Pulsetronic ignition has these principal components: magnetic timer generator, transistor, control unit for each engine

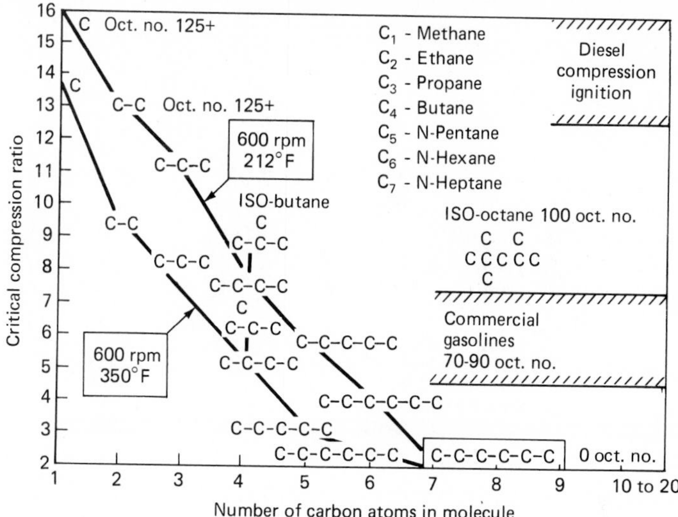

FIG. 3.16 Critical compression ratio versus molecular structure. Note: (°F − 32)/1.8 = °C. *(W.M. Kauffmann.)*

cylinder, mounting console, ignition transformer for each power cylinder, and some form of external power supply.

Figure 3.17 shows a conventional low-tension magneto ignition system. Here, a four-pole motor, rotating between two-pole shoes, generates four flux reversals. These flux changes build up current in the windings. When the main breaker closes, the circuit is completed and current passes to the transformer coils back to ground, collapsing and building up as each pole piece passes through the neutral position. At about 15° after the distributor contacts close, the main breakers are separated by a cam. The sudden interruption induces current in the secondary winding of the transformer coil, causing a high-voltage discharge at the spark

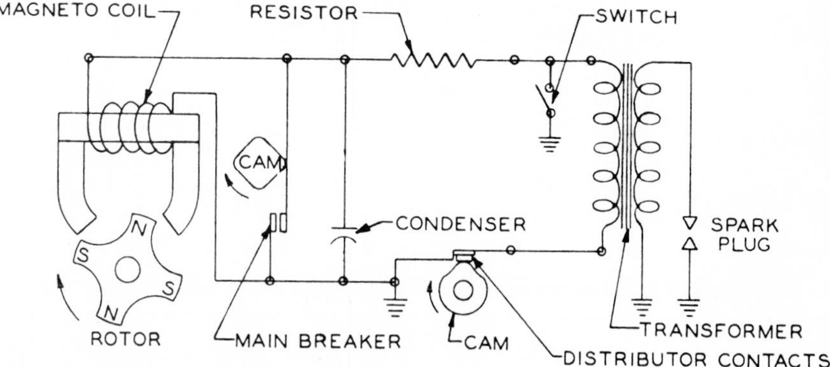

FIG. 3.17 Low-tension magneto ignition system. *(W.M. Kauffmann.)*

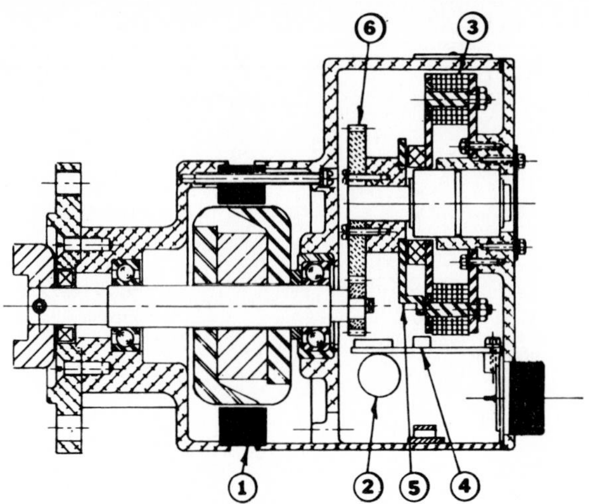

FIG. 3.18 Altronic ignition system. *(American Bosch Corp.)*

plug. A capacitor placed across the main breaker serves to reduce arcing by acting as a current reservoir. A resistor also eliminates arcing at distributor contacts. The transformer coils, which provide stepup in voltage, have windings sealed in a thermosetting potting compound to prevent any movement due to engine vibration. Breakerless magnetos are available which, with transformer coils, provide 32-kV open-circuit voltage.

Altronic System

The Altronic system uses a complete electronic circuit (Fig. 3.18). The alternator (1) provides power to a charge-energy stored capacitor (2) and *silicon controlled rectifier* (SCR) solid-state switch (4). A rotating timer arm (5) driven by speed-reducing gears (6) passes over the pickup coils to trigger the SCR switches in sequence. This releases the capacitor's stored energy to ignition coils that step up the voltage to fire the spark plug.

Spark Plug

A key to proper combustion at high compression is the spark plug. Figure 3.19 shows a section of a tripower spark plug installation. In this novel design, the plug has a threaded upper body that is screwed into a cylindrical casing. The lower part is a smooth cylinder with a collar for the gasket. The cage assembly is held in place by the fuel-nozzle clamp device, making these interchangeable. The spark plug center electrode is generally 96 percent nickel, and the insulator body is high-alumina ceramic material. Its bottom internal gasket is generally silver-copper alloy, and the hermetic seal is copper-glass.

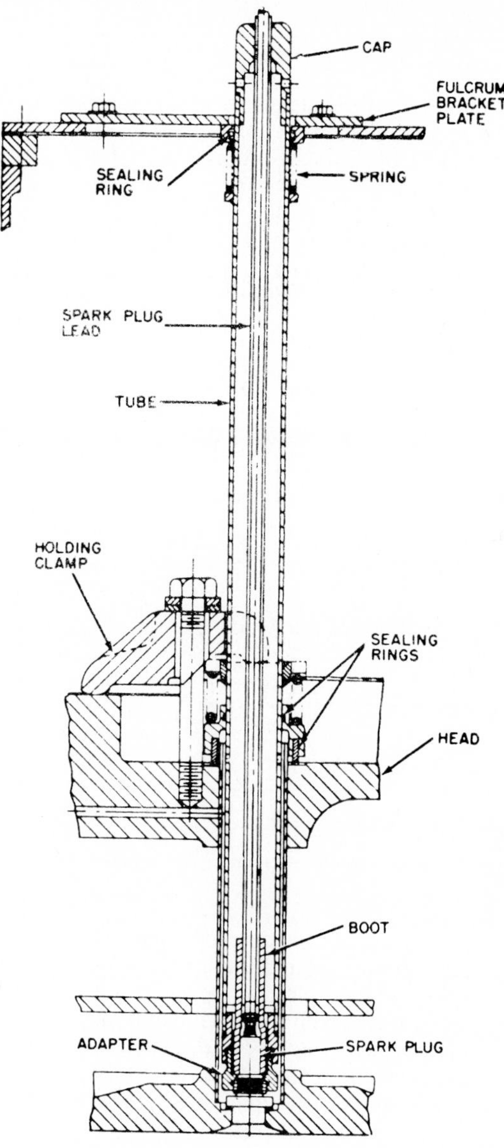

FIG. 3.19 Tripower spark plug. *(Worthington Corp.)*

COOLING SYSTEMS

The cooling system design depends on the plant environment, which affects the selection of such system elements as the radiator, evaporative cooler, cooling tower, high-temperature or ebullient cooling, and total-energy heat recovery.

Radiator Cooling

Where the water supply is a problem, radiator cooling is attractive (Fig. 3.20). Simplicity is a factor in its favor for portable powerplants in which water is often

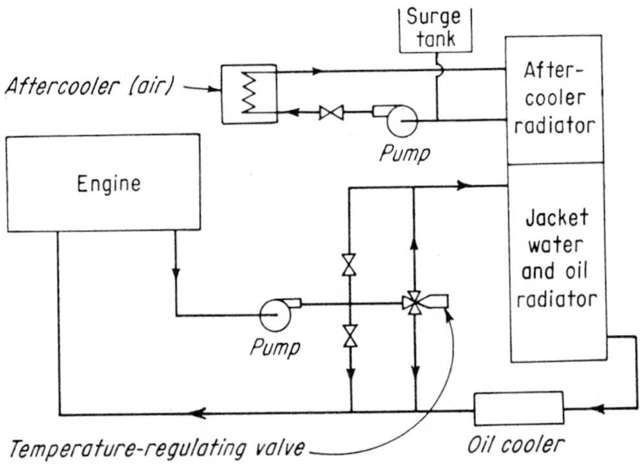

FIG. 3.20 Radiator cooling system. (Power *magazine*)

scarce or of poor quality. The turbocharged engine (Fig. 3.5) uses a two-section radiator—one for the aftercooler and the other for the jacket-water and lube oil cooler. Each section has a circulating pump. The engine circuit has a temperature-regulating valve, with a manual bypass in the piping to and from the engine. The aftercooler radiator cores are designed to provide 100°F (38°C) coolant with 90°F (32.2°C) ambient temperature. With a 20°F (11°C) rise in aftercooler temperature, a possible air temperature of 120°F (49°C) to the engine is obtained.

Cooling Tower

A cooling-tower system (Fig. 3.21) is selected where the ambient dry-bulb temperature exceeds 100°F (38°C) and the relative humidity averages 50 percent or less during the year. The coolant temperature can be maintained about 10°F (5.6°C) above wet bulb. Separate pumps are used for soft or treated water and raw-water circuits. An oil cooler is placed in the return line from the heat exchanger to the engine. Large engine-plant radiators are of the horizontal type with an electric or hydraulic motor drive for the fan. In some diesel installations, the aftercooler may be placed in the soft-water circuit to simplify piping.

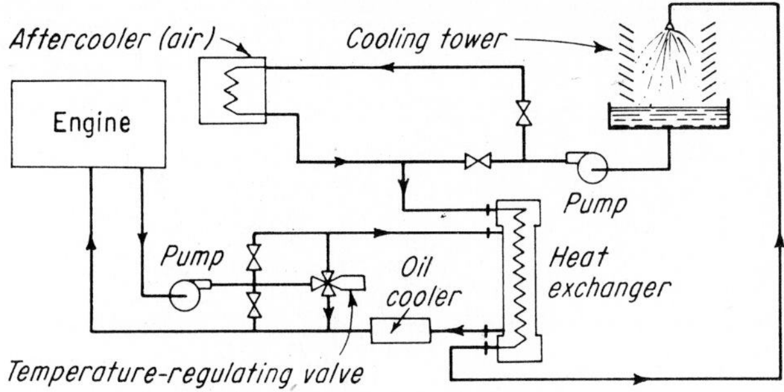

FIG. 3.21 Cooling-tower system. (Power *magazine*)

Evaporative Cooling. An evaporative cooling-tower arrangement (Fig. 3.22) for the air aftercooler and for the radiator for jacket-water and lube oil cooling is popular in such dry areas as the southwest. It is essential that raw water be chemically treated to prevent scale and corrosion problems.

Ebullient Cooling. High-temperature, or ebullient, cooling has had wide acceptance in high-output engine installations (Fig. 3.23). Centrifugal pumps are deleted from the engine water circuit. Aftercoolers use an evaporative cooling tower to reduce the air-manifold temperature as much as possible. Steam from a flash tank is piped to process, where it condenses and is returned to the flash-tank standpipe. The temperature rise is only 3 to 5°F (1.7 to 2.8°C) across the engine.

The water temperature in ebullient cooling will depend on the system pressure. Normally this is about 15 psi (103.4 kPa) when a nominal jacket temperature of 250°F (121°C) is maintained. The system frequently sees service where steam is needed for plant operations, where residual fuel is burned for diesel operations

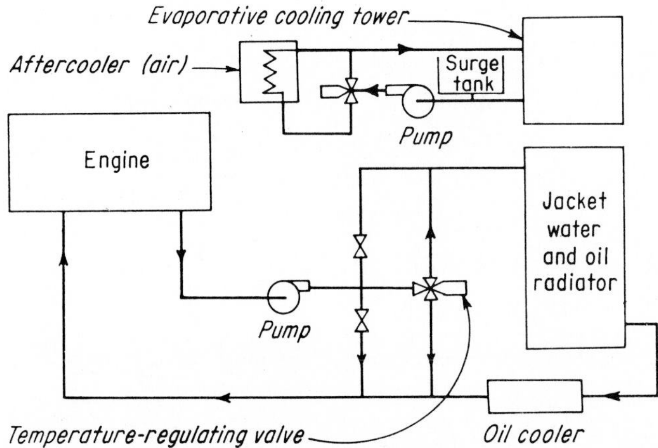

FIG. 3.22 Evaporative cooling-tower arrangement. (Power *magazine*)

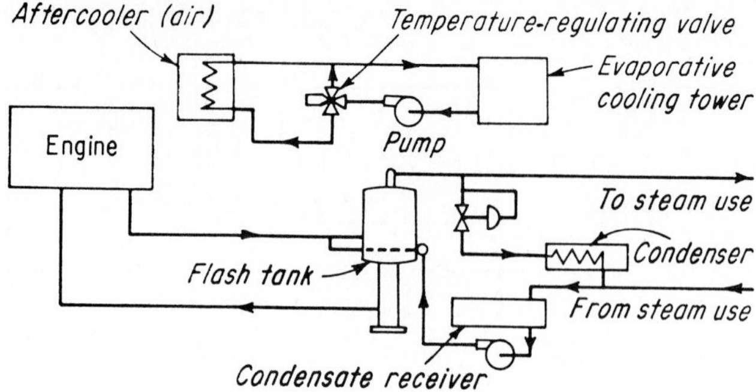

FIG. 3.23 Ebullient-cooling system. *(Power magazine)*

for heating fuel, and where corrosion caused by condensation of contaminants in residuals can be a problem.

Total-Energy Cooling

The total-energy system (Fig. 3.24) reclaims practically all the heat from the engine jackets and exhaust, within certain limits. A waste-heat boiler is placed in series with the engine jackets. The steam pressure for engine wet-seal liners is limited to 15 psi (103.4 kPa). Where liners are cast integral with cylinder jackets, a pressure of 35 psi (248 kPa) can be tolerated. If higher pressures are needed, the exhaust waste-heat boiler can be provided with a separate circuit set to maintain pressures as high as 150 psi (1034 kPa).

For both ebullient and total-energy cooling, it is recommended that the coolant flow velocity be limited to about 4 ft/s (1.22 m/s). Piping sizes should be designed on this basis, to obtain the most efficient performance. Cooling equipment

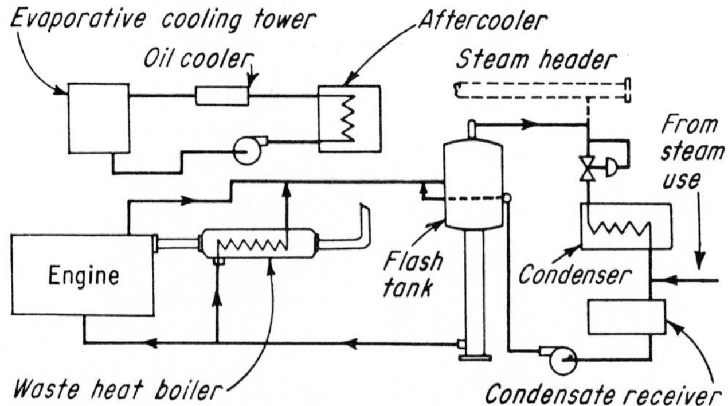

FIG. 3.24 Total-energy system. *(Power magazine)*

should be sized with contingency factors based on supplier experience, plus a 10 percent overload capability.

Slant-Diagram Calculations

For proper powerplant performance, cooling systems require careful analysis. The author has developed slant-diagram calculations that facilitate (1) sizing of heat-exchanger equipment, pump capacities, and piping sizes and (2) establishing of temperature gradients in the cooling system (Fig. 3.25). By referring to Table 3.2, the jacket-water flow to maintain a 10°F (5.6°C) differential across the engine is derived from

$$t_2 - t_1 = \frac{H}{KQ} \tag{3.2}$$

where H = heat rejection, Btu/h (kW)
t_2 = jacket-water outlet temperature, °F (°C)
t_1 = jacket-water inlet temperature, °F (°C)
Q = flow rate, gpm (L/s)
K = constant = 500 (4.23)

Heat rejection of 939,000 Btu/h (275 kW) plus a 10 percent contingency factor is 1.03×10^6 Btu/h (303 kW). Substituting for H, we see that the required flow is 207 gpm (13.06 L/s). Pumps are sized for this capacity at specified head conditions. Radiators operate at a 35°F (19.4°C) differential for both jacket-water and lube oil coolers. Heat rejection for the oil cooler is 290,000 Btu/h (85 kW). The radiator should handle 1.033×10^6 Btu/h (303 kW) plus 290,000 Btu/h (85 kW). From Eq. 3.2, the flow is calculated to be 75 gpm (4.73 L/s).

To determine the temperature drop of lube oil through the oil cooler, substitute 207 (1.75) for K in Eq. 3.2. Thus, H = heat rejection to the lube

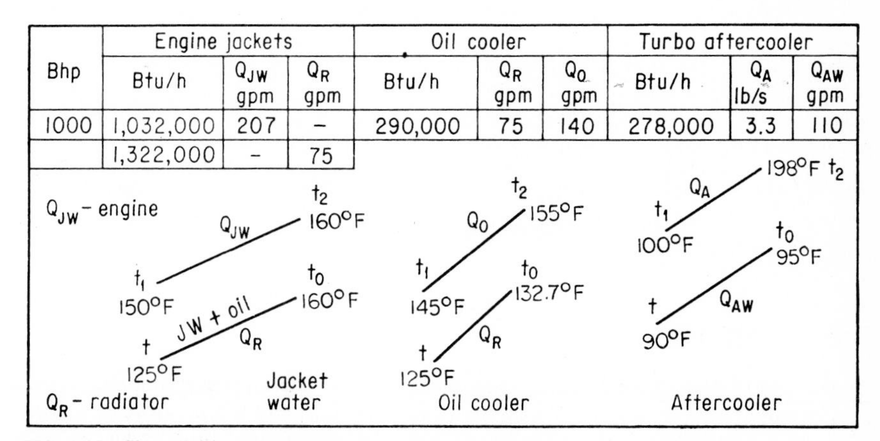

Bhp	Engine jackets			Oil cooler			Turbo aftercooler		
	Btu/h	Q_{JW} gpm	Q_R gpm	Btu/h	Q_R gpm	Q_O gpm	Btu/h	Q_A lb/s	Q_{AW} gpm
1000	1,032,000	207	–	290,000	75	140	278,000	3.3	110
	1,322,000	–	75						

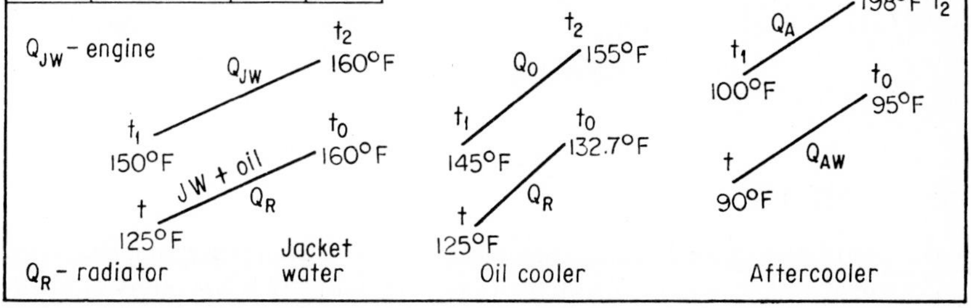

FIG. 3.25 Slant diagrams for heat exchangers. Note: Btu/h × 0.293 = W; gpm × 0.0631 = L/s; lb/s × 0.4535 = kg/s; (°F − 32)/1.8 = °C. *(W.M. Kauffmann.)*

oil = 290,000 Btu/h (85 kW); Q = lube oil flow = or 177 gpm (11 L/s); and $t_2 - t_1$, the temperature drop across the lube-oil cooler, is 7.9°F (4.4°C).

When ethylene gylcol antifreeze is selected, the change in specific gravity and specific heat of solution requires more coolant flow. For a 50% glycol and 50% water solution, substitute 436 (3.7). About 15 percent greater coolant flow is needed.

Aftercooler Heat Exchangers

Placed in the turboblower discharge piping of the turbocharger, these heat exchangers are the most effective means for reducing the air-manifold temperature for diesels and for increasing the knock limit for dual-fuel and high-compression spark-ignition gas engines. The exchanger consists of an outer shell and inner core, in either a single- or double-pass configuration. Another means, used almost exclusively in hot, dry climates, is an evaporative cooler placed directly before the intake air filter, which acts to cool and humidify air to the engine. Such units can reduce the air temperature to within 5°F (2.8°C) of the wet-bulb reading.

The correct piping size should be determined for the air intake, exhaust, cooling water, starting air, fuel, and lube oil. Locate piping in amply sized trenches with provision for support and expansion. For estimating the cooling-water pipe size, use 5 gpm (0.316 L/s) per brake horsepower, with pressure loss not to exceed 1 psi/100 ft (22.6 kPa/100 m). Rule of thumb indicates that the fluid velocity equals the nominal pipe size, that is, 2 ft/s (0.61 m/s) for 2-in (50.8-mm) pipe, up to 10 ft/s (3.05 m/s) for 10-in (254-mm) pipe. For pipe sizes larger than 10 in (254 mm), the velocity should not exceed 10 ft/s (3.05 m/s).

The actual pipe size is calculated from

$$D = \sqrt{\frac{Q}{2.5\,V}} \tag{3.3}$$

where D = pipe diameter, in (mm)
Q = pipe flow, gpm (L/s)
V = mean velocity through pipe, ft/s (m/s)

For example, assume that the flow is 20 gpm (1.26 L/s) and the mean velocity is 2 ft/s (0.61 m/s). Thus, $D = \sqrt{20 / (2.5)(2)} = 2$ in (50.8 mm). Refer to handbooks for pressure loss per pipe length and include losses in fittings, valves, etc., to check this selection.

Recoverable Heat

Recovery of heat rejected to exhaust increases the overall plant efficiency. The recoverable heat is determined by the steam pressure and temperature specified for process, specific fuel consumption, and inlet temperature to the waste-heat boiler. Refer to Fig. 3.24. The available exhaust-gas energy H_e is calculated from

$$H_e = WC\,(t_{e1} - t_{e2}) \tag{3.4}$$

where W = exhaust gas flow, lb/h (g/s)
t_{e1} = exhaust-gas temperature to boiler, °F (°C)
t_{e2} = exhaust-gas temperature from boiler, °F (°C)
C = specific heat of exhaust gas

To avoid condensation, t_{e2} is kept at least 75°F (41.7°C) above the steam temperature.

ENGINE CONTROL

Engine-governing systems are responsible for maintaining the desired engine speed regardless of the load. In full diesels, the governor is connected to the fuel-pump control shaft. In dual-fuel and spark-ignition engines, the governor is connected to the dual-fuel control shaft and gas-regulating valves via a mechanical linkage. Two types of governors are currently in use, the hydraulic and the electrohydraulic.

Hydraulic Governor

This type, for constant-speed heavy-duty service, is mounted on the engine and generally driven off the engine camshaft. Governors are self-contained and have their own lubrication system. Figure 3.26 shows a modern hydraulic governor. A gerotor pump, driven off the engine, pumps oil to an accumulator, which is fitted with a piston and spring. A power piston is attached to the output shaft. As the differential piston pressure increases and moves the piston up, it causes the output shaft to rotate in the "increase" direction. A pilot-valve system and ball-head assembly function to change the speed setting of the governor. A compensation mechanism stabilizes piston movement, allowing the engine speed to return to normal. The pressure generated by the accumulator determines the work capacity of the governor.

Electrohydraulic Governor

Figure 3.27 shows an electrohydraulic governor, which is used for high-performance service in which fast-acting engine response is needed with low work output. Here the actuator pilot-valve plunger controls the flow of oil to and from the power piston. The plunger is connected to an armature magnet suspended in the field of two-coil polarized solenoids. The electric-control output signal is applied to the polarized coil, producing a force in proportion to the signal. Under steady-state operation, the restoring spring force, spring compression force, and magnetic forces are in balance. Any change in speed causes the actuating pistons to move, rotating the terminal shaft to increase or decrease fuel, as required.

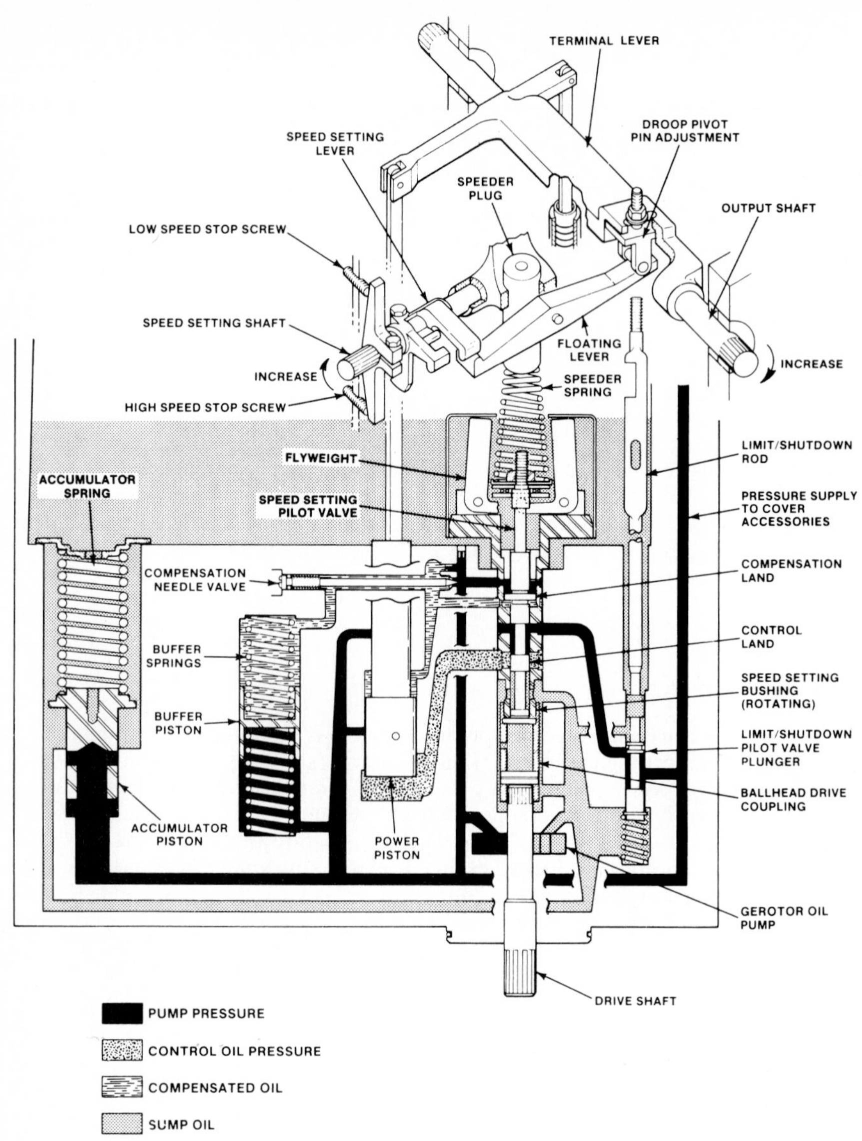

FIG. 3.26 Modern hydraulic governor. *(Woodward Governor Co.)*

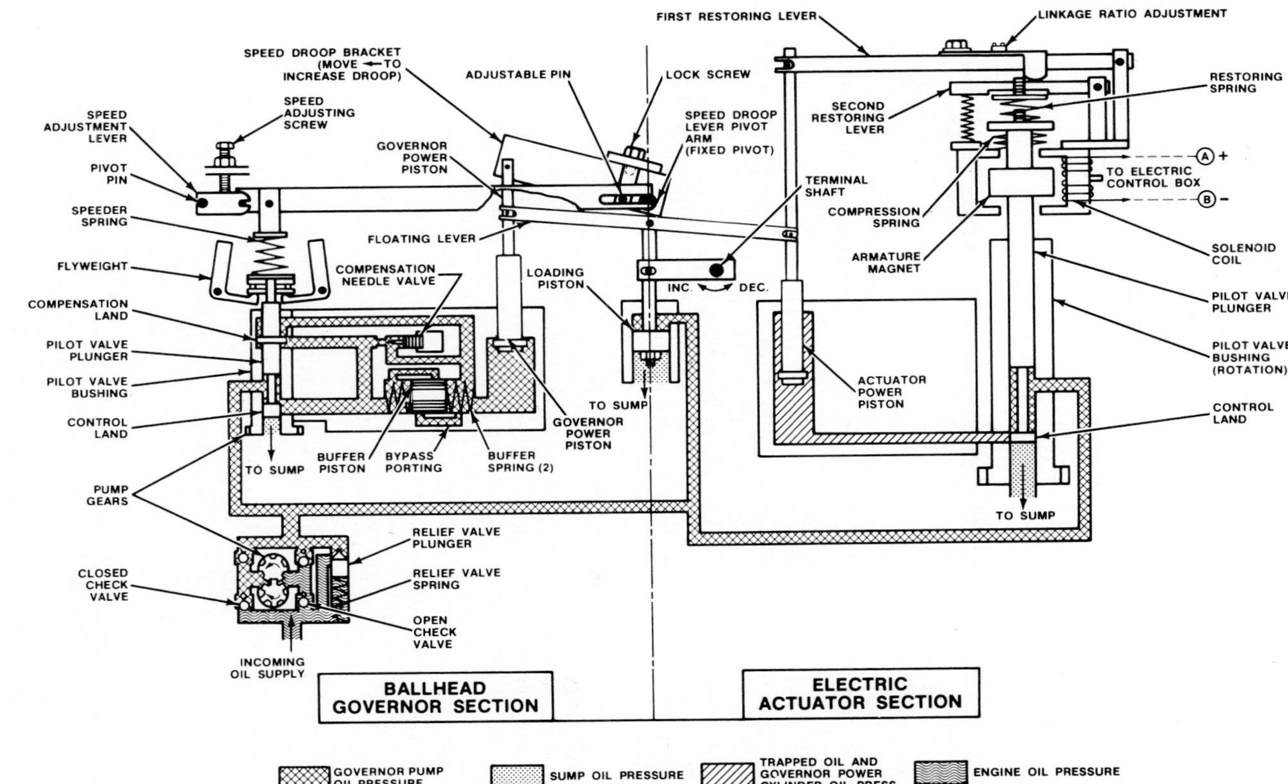

FIG. 3.27 Electrohydraulic governor. *(Woodward Governor Co.)*

INTAKE AND EXHAUST SYSTEMS

Proper performance of an engine powerplant, whether industrial or utility, is determined by not only the engine but also its installation. Intake and exhaust systems need careful analysis, especially the filters, exhaust and tail pipes, and mufflers.

Air Filters

Air to the engine must be filtered to remove pollution and abrasives. The air filter size must be adequate to limit the pressure drop across elements, provide accessibility for cleaning, and ensure easy maintenance.

Filter types may be dry, viscous impingement, oil bath, or electrostatic precipitation. The dry type is built of fabric or spun-glass fiber elements. The viscous impingement type uses an oil-saturated wire-strand basket, grouped in a common housing. The oil-bath type has air passing through an oil spray, thus coating dust particles, which are trapped in the filtering elements. The heavier particles remain in the bath and do not get to the filter. The precipitator has the disadvantage of requiring high voltage and is expensive compared to other types, although its high efficiency and low pressure drop are recognized.

Air filters have some silencing; however, in residential areas a separate silencer may be necessary, or filters can be placed in a plenum chamber. Filters should be located above ground level, preferably on a concrete slab. For roof mounting, provide steel girders for reinforcement, and keep filters well above the roof level to avoid vibration. The pressure drop across air filters is normally limited to 3 in H_2O (747 Pa) for turbocharged engines.

Exhaust Elements

Exhaust systems consist of an exhaust pipe, silencer (or muffler), and tail pipe. The function of the silencer is to reduce pulsations generated in the engine to an acceptable level with minimum backpressure. The piping size should be adequate to reduce the pressure drop of the system to a maximum of 12 in H_2O (3 kPa). *Recommendations:* Provide a flexible joint in the piping at the turbocharger exhaust outlet, avoid bends where possible, and extend the tail pipe above the powerplant roof. In any case, the exhaust pipe should not exceed 80 ft (24.4 m) in length, with not more than three 90° bends.

In sizing pipes, use 3.5 ft^3/min (1.65 L/s) per brake horsepower for four-cycle turbocharged engines and 4 to 5 ft^3/min (1.89 to 2.36 L/s) per brake horsepower for two-cycle engines. An air pressure drop of 6 in H_2O (1.5 kPa) at the blower inlet requires that the filter design pressure drop preferably not exceed 2 in H_2O (498 Pa). The piping velocity should not exceed 3300 ft/min (17 m/s). For exhaust systems, the gas velocity should be limited to 6000 ft/min (30 m/s).

LUBRICATION AND LUBRICANTS

Lubrication systems for powerplant diesels vary with the type of engine and service conditions. Figure 3.28 shows a piping diagram for a 4000-kW, 16-cylinder V

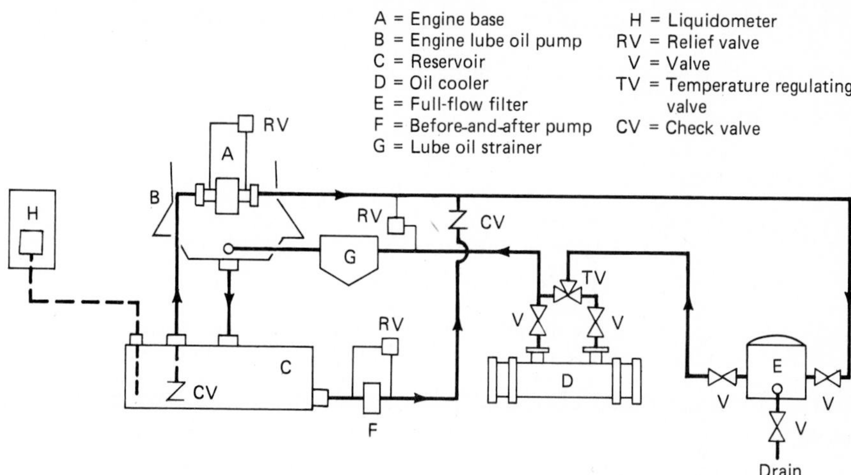

FIG. 3.28 Engine lube oil piping. *(W.M. Kauffmann.)*

engine. The dry sump is used where oil in the base is drained to a reservoir. In the wet-sump type, the base is deep enough to act as a reservoir. In either case, an engine-driven lube oil pump circulates oil through a full-flow filter (E) and oil cooler (D) to a through-strainer (G) to the oil header in the engine base.

When the engine is started with cold oil, a thermostatic temperature-regulating valve bypasses oil around the cooler to permit the oil to reach its operating temperature quickly. When this temperature is reached, the valve's bypass port closes and the outlet port opens, passing oil through the cooler. The TV valve's ability to pass some or all of the oil permits the proper temperature to be maintained at all loads. A built-in pressure-relief valve (RV) in the engine pump returns excess oil to the pump suction if the pressure exceeds the normal setting.

Before-and-After Pump

A before-and-after pump (F) may be added to allow prelubrication before start-up. The RV valve serves as a bypass in the event of a clogged full-flow filter. Strainer (G) is placed directly ahead of the entry to the oil header and is generally of the scraper-edge type. The before-and-after pump is started about 3 min before engine start-up and is allowed to run for 5 min after shutdown to cool the bearings.

Oil in the header flows to the main bearings through drilled passages, then to connecting-rod bearings through holes drilled in the crankshaft, and on up to piston pins through rifle-drilled holes in the connecting rods. Here, a large part is diverted to the piston cooling cavity under the crown, where it returns to the crankcase from a standpipe in the cavity.

Cylinder walls are lubricated by splash thrown off the connecting-rod bearing ends. A small oil header supplies lube oil to the camshaft bearings and overhead valve gear, governor drive, and camshaft drive gears or chains. Turbocharger lubrication must be careful and meticulous. In some cases, a separate pump, filter, oil cooler, and reservoir are used, particularly for large V-type engines. In smaller units, lubricant is taken from the engine system. Before-and-after lubrication may be provided with a separate air-operated drive from the compressed-air starting tank, by taking oil from either the turbo oil reservoir or engine system.

Lube Oil Selection

Selection of an engine lubricating oil is based on cycle type (two or four), engine rating, number of cylinders, oil filter system, fuel specifications, and operating load factor. Engine builders will not recommend any particular brand of oil, but will specify viscosity and detergent level. Knowledgeable operators will submit their selection to the engine manufacturer for approval. Recommended for heavy-duty, high-output engines is a viscosity of 1000 SSU at 100°F (38°C), or 75 SSU to 80 SSU at 210°F (99°C). Oil should contain oxidation and corrosion inhibitors.

Dual-fuel engines that operate on diesel oil for 50 days/yr or more should use such diesel specifications as MIL-L-2104 B ordnance type. The performance of an oil is based on the detergency level, dispersancy, base oil, and oxidation resistance. The oil base will determine the oil's inherent resistance to oxidation and whether hard or flocculent piston-ring zone deposits are formed. The oxidation-resistant additives inhibit chemical reaction with oxygen that is normally accelerated at high temperatures. Oxidized oil clogs piston rings, hydraulic lifters, wear cams, and gear train.

FUELS AND FUEL SYSTEMS

Distillate Fuel Oils

A typical distillate fuel oil system is shown in Fig. 3.29. Fuel is supplied from the tank car or truck to a fuel reservoir tank, from which it is pumped to the engine

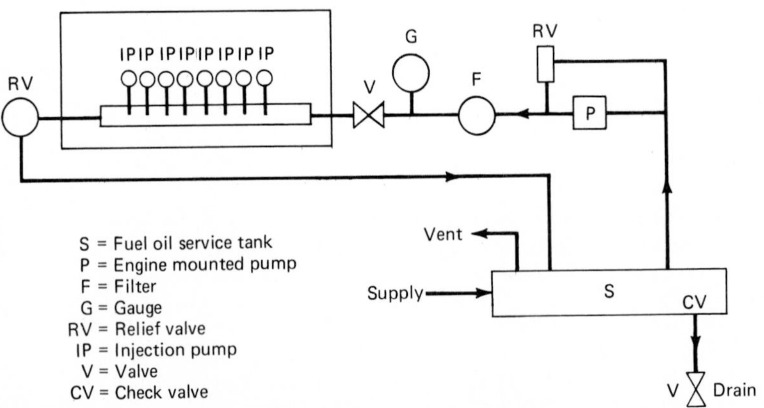

FIG. 3.29 Typical distillate fuel oil system. *(W.M. Kauffmann.)*

fuel header through a filter. A pressure-relief valve maintains a predetermined pressure in the fuel inlet header, or manifold, and returns excess fuel to the vented tank. Specifications for recommended distillate fuels are listed in Table 3.3.

TABLE 3.3 Specifications for Distillate Fuels

Parameter	Recommended	Satisfactory
Cetane number	45	40
Gravity (API)	28–32	24–34
Viscosity, SSU @ 100°F (38°C)	30–50	30–150
Sulfur not over, % wt	1.5	2
Conradson carbon, max., % wt	1.0	2.5

Gas Fuels

Figure 3.30 shows the piping for a gas fuel system. The gas supplied is brought in to scrubbers (when specified) and then to one or two pressure regulators on a gas receiver near the engine. The fuel then goes to gas-metering valves on the engine cylinder heads, which are timed by camshaft to inject the correct amount of gas into the inlet passage or directly into the cylinder—depending on the builder's design.

Gas fuel for high-compression spark-ignition and dual-fuel engines should be dry and clean, or provided with scrubbers if this is not possible. Gases containing more than 60 gr/100 ft^3 (137 g/100 m^3) of hydrogen sulfide should be avoided, to limit corrosion. Hydrogen should be limited to 20 percent of the total gas composition. For operation at the diesel compression ratio, natural gas has the highest knock resistance. Typical composition of natural gas is 87% methane, 7% nitrogen, 3% ethane, 1% propane, 0.05% butane, with trace amounts of isobutane, pentane, carbon dioxide, and helium. The high heat value of natural gas is 1050 to 1150 Btu/ft^3 (39 165 to 42 895 kJ/m^3), and the density is 0.05 lb/ft^3 (0.8 kg/m^3).

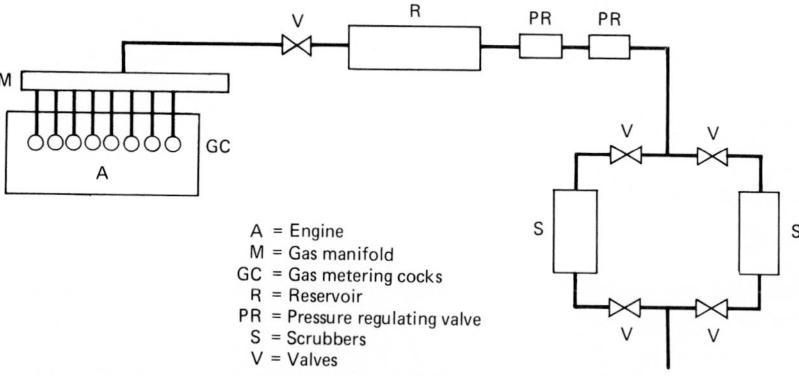

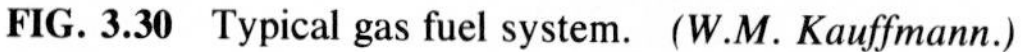

FIG. 3.30 Typical gas fuel system. *(W.M. Kauffmann.)*

TABLE 3.4 Specifications for Residual and Distillate Fuels

	Kero-sene no. 1 fuel oil	Distillates			Heavy fuels		
		Diesel oil	No. 2 fuel	No. 4 fuel	Navy Special	No. 5 fuel	No. 6 fuel
Price average factor	1.00	0.92	0.87	0.65	—	0.55	0.45
Gravity (API)	43	38	34	21	11.5	18	11.5
Viscosity, SSU @ 100°F (38°C)	33	35	40	140	450	530	—
Viscosity, SSF @ 122°F (50°C)	—	—	—	—	—	—	150
Flash point, °F (°C) (min.)	115 (46)	150 (66)	130 (54)	150 (66)	150 (66)	150 (66)	150 (66)
Conradson carbon, % wt	0	0.10	0.25	0.02–0.06	0.15	0.07–0.16	10–20
Sulfur, % wt (max.)	0.15	0.5	0.01	1.9	1.9	—	—
Total ash, % wt (max.)	0	0.01	0.01	0.02	0.10	0.09	0.12
Cetane number	50–55	49–59	30–40	—	—	—	—
Diesel index	60–65	50–65	30–40	—	—	—	—
End boiling point, °F (°C)	560 (293)	675 (357)	680 (360)	—	—	—	—

Alternative fuels are considered where natural gas or distillate fuels are in short supply. Table 3.4 shows specifications for various distillate and residual fuels, with the relative cost based on 1.00 for no. 1 fuel. Prime engine fuels are diesel oil and no. 2 distillate; some plants have burned no. 4 successfully. Navy Special is a controlled-specification heavy fuel for marine and military plants. Number 5 is preferred to no. 6 residual because of its lower metallic content. Number 6 fuel needs careful examination for composition and special treatment prior to its selection.

Residual Fuel Oils

Figure 3.31 shows a typical residual fuel oil system. Since vanadium causes severe burning of exhaust valves, water washing and inhibitor addition are essential. Figure 3.32 is a temperature/viscosity chart for various fuels, showing ranges for pumping, centrifuging, and injection in the engine. The main system elements are the storage tank and heater, demulsifier tank, flow meter, wash solution tank, pumps, and strainer. Fuel treatment is tailored to the particular fuel.

Economics of Fuel Selection

Fuel selection is determined largely by availability. Where natural gas is a reliable source, dual-fuel or spark-ignition engines are attractive. Market requirements control the processing of crudes to produce gasoline, kerosene, light and heavy distillates, residual fuel, and gas. Heavy distillates are about 3 percent of Gulf Coast crude, when processed. Catalytic cracking of heavier cuts increases the yield of light fractions to meet seasonal demand, which will vary from 17 to 27% gasoline, 10 to 20% distillate, and 1 to 2% residual.

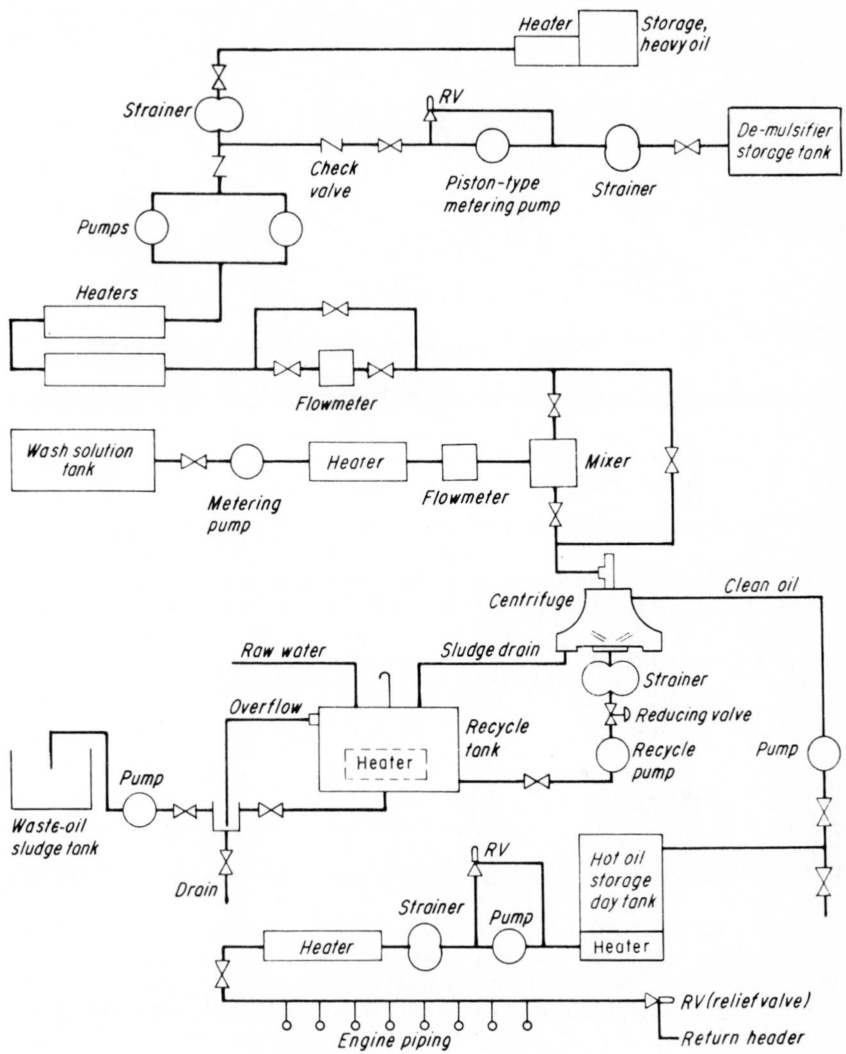

FIG. 3.31 Typical residual fuel oil system. (Power *magazine*)

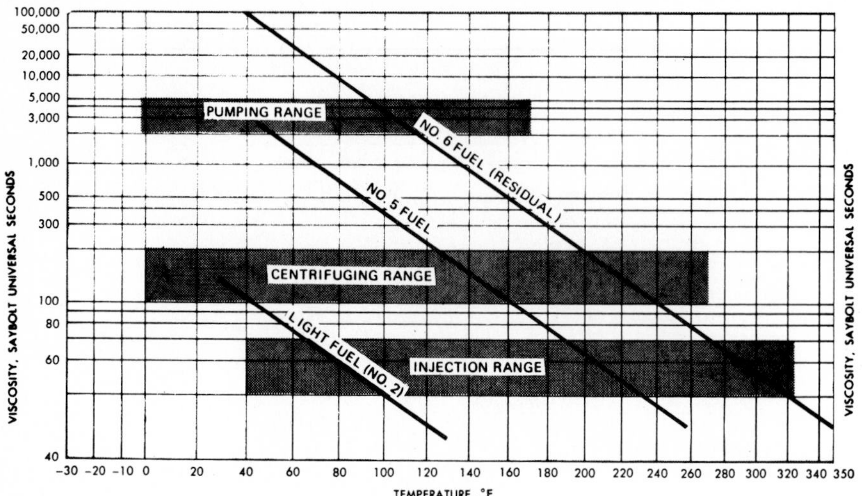

FIG. 3.32 Temperature/viscosity chart for diesel fuels. Note: (°F − 32)/1.8 = °C. *(ASTM.)*

Residual fuel is analogous to scrap and is thus sold at lower cost. Plants operating on residual fuel report a 35 to 40 percent savings in fuel cost, but show an increase in maintenance costs of 25 to 30 percent. Thus, only savings of 5 to 15 percent overall are possible. To run on residuals, diesels require extensive design changes, such as valve rotators and preinjection pumps, which add to the basic engine cost.

ENGINE FOUNDATIONS

Engine foundations serve to maintain a correct initial elevation and rotating shafting in permanent alignment. Also they should reduce the transmission of vibration from unbalanced forces in the engine. The nature of supporting soil—its physical properties, density, moisture content, cohesion character, and groundwater level—is critical. Figure 3.33 is a typical foundation design for an eight-cylinder engine, generator, and belt-driven exciter. The foundation width should not be less than the distance from the crankshaft to the bottom of the block.

For stability, the center of gravity of the combined engine and foundation should be below the foundation top. The generator pit depth should be at least 60 percent of the stator outside diameter. Plants adjacent to dwellings, and generating sets for office buildings, are flexibly mounted on springs, neoprene disks, or other means of suspension (Fig. 3.34). Here a fabricated steel chassis is cast into the concrete-reinforced block with projecting H beams that rest on spring isolators. These, in turn, are supported by a subfoundation of suitable design for the soil conditions at the site.

Soil Conditions

Where soil conditions may lead to settlement, piling is recommended. The following guidelines are suggested. The excavation should be braced at the sides,

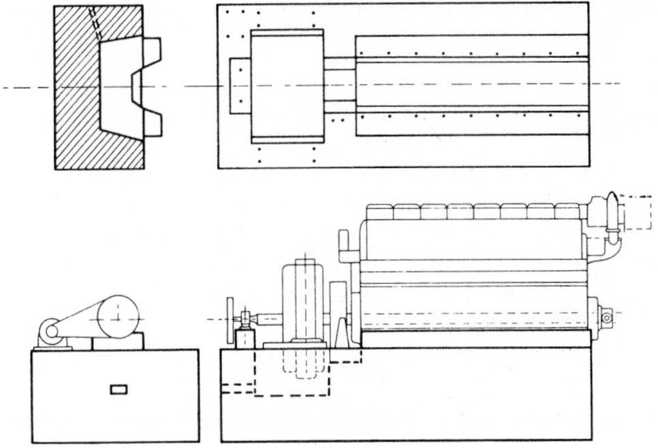

FIG. 3.33 Foundation for engine/generator. (Power *magazine*)

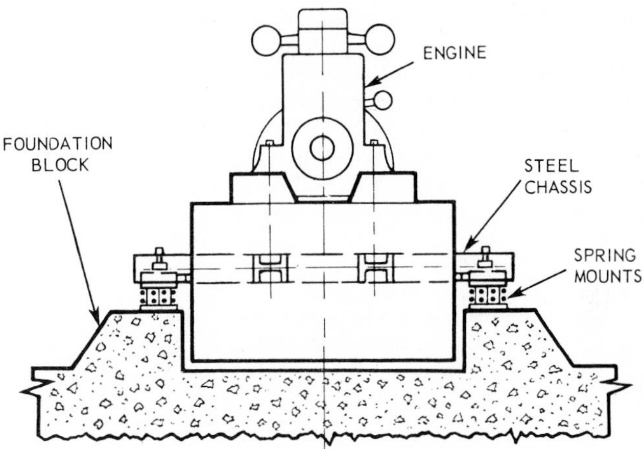

FIG. 3.34 Spring-supported foundation. (Power *magazine*)

cut 6 in (152 mm) below the foundation bottom, and hand-cut to finish so as not to exceed a tolerance of 0.5 in in 10 ft (4.16 mm in 10 m). A bed of well-graded sand should bring the level up to that of the concrete; then it should be compacted to give correct bearing capacity and density.

Various methods are available for reinforcement of foundation blocks. Rebars will extend vertically and horizontally at spacings determined by the block size, with 3-in (76.2-mm) cover over the outside diameter of the bars. The bar size may vary from no. 4 to no. 7 [numbers are in ⅛-in (2-mm) segments]. For engines in the 3000- to 6000-kW range, use no. 7 bars for the sides and bottom, with no. 6 at the top, for 12-in (305-mm) spacing. In cross section, total steel to concrete areas may average 0.13 percent. Rebars should conform to ASTM A 305 and A 15 intermediate-grade steel. Concrete should meet ASTM C 150 or C 175. Water for concrete should be free of harmful solids, alkali, or organic material (never use seawater). Aggregate should comply with ASTM C 33.

Foundation Analysis

The analysis of engine foundations depends on supporting soil frequency. A system can be modified by changing the block mass, increasing the area of support to reduce the unit soil load pressure, and eliminating unbalanced forces, if possible. The soil natural frequency is calculated from soil deflection, impact tests, or mechanical oscillators. The basic equation is

$$f_n = \frac{35{,}400}{d} \qquad (3.5)$$

where f_n = soil natural frequency and d = soil deflection, in. The deflection d is determined from the allowable dynamic bearing pressure for the soil at the site.

The dynamic bearing pressure is the total weight of an engine/generator set and foundation divided by the total area of support at the bottom of the block. The safe allowable static soil load should be 4 times the dynamic bearing pressure. Table 3.5 lists allowable static and dynamic soil loads for various soils. For example, soil at the support level is dry gravel, whose safe static load is 6000 lb/ft^2 (29 300 kg/m^2). At 1:4, the dynamic bearing pressure is 1500 lb/ft^2 (7320 kg/m^2). The deflection for a 1500 lb/ft^2 (7320 kg/m^2) dynamic soil load is 0.15 in (0.381 mm), so $f_n = \sqrt{35{,}400/0.015}$ = 1520 cycles/min. For estimating purposes, figure the foundation weight from 0.03 to 0.05 yd^3/hp (0.03 to 0.05 m^3/kW), or a foundation weight 3 to 4 times the engine weight.

TABLE 3.5 Static and Dynamic Soil Load Data

	Safe static load		Safe dynamic load	
Type of soil	lb/ft^2	kg/m^2	lb/ft^2	kg/m^2
Rock	8000	39 000	2000	9760
Dry gravel	6000	29 300	1500	7320
Dry, coarse sand	4000	19 500	1000	4880
Dry, fine sand	3000	14 650	750	3660
Firm clay	2500	12 200	625	3050
Soft clay	2000	9760	500	2440
Wet clay	1000	4880	250	1220
Wet sand and clay*	880	4300	220	1075
Alluvial soil†	800	3900	200	975

*Consider piling to support block and mat. Piling to be driven down to refusal. Concrete-filled steel pipe preferred.

†High water table present.

Foundation Growth

A problem in some installations has been foundation growth, which results in crankshaft failures. Fretting of main bearings, bearing wear, broken main-bearing studs, and cracking of the engine base have resulted from foundation distortion.

Failures may occur at 2000 or 50,000 h of operation. One cause is alkali aggregate deterioration, because of the reaction between finely divided amorphous silica and free alkali in the concrete cement. This reaction creates high osmotic pressure, causing disruption of the foundation block, and shows up in random cracks in the block. Growth can amount to 2 in in 40 ft (167 mm in 40 m) in severe cases and 0.025-in (0.635-mm) distortion of the top surface, leading to a 0.005-in (0.127-mm) crankweb deflection [maximum allowable is normally 0.001 in (0.025 mm)].

Silicious rocks, once thought to be safe, have shown reactive aggregate composition. The reactivity may be controlled by limiting the alkali content of cement to 0.6 percent following an ASTM C 227 test procedure. The temperature effect due to the differential between the hot engine base and the cold block is another contributing factor. Placing the engine on steel rails allows air to circulate under the engine base and permits correction of any deviations in crankshaft alignment.

PLANT AUXILIARIES

Heat Exchangers

The type and size of heat exchangers are based on heat-balance data supplied by the engine builder, who generally includes such equipment in the purchase contract. Normally, heat exchangers such as lube oil and jacket-water coolers are designed with a fouling factor for clean treated water. The sizing of oil coolers is based on both the heat-rejection rate to the lube oil and the coolant flow, with a suitable fouling factor.

Air-Starting Auxiliaries

As Fig. 3.35 shows, these auxiliaries consist of a motor-driven compressor (*B*), a gasoline-engine-driven compressor (*A*), and vertical air tanks (*C*) with drain

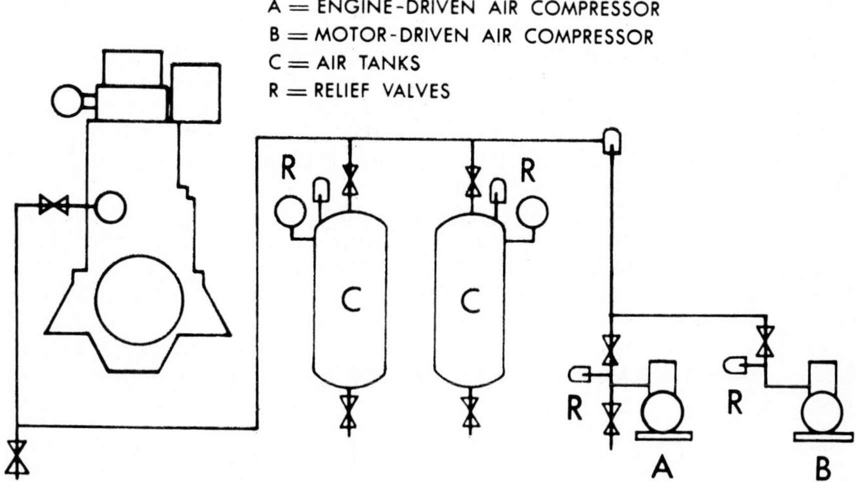

FIG. 3.35 Air-starting system. *(Worthington Corp.)*

valves, pressure gages, and relief valves (*R*). The engines are two-stage, air- or water-cooled, and are suitable for 250-psi (1723-kPa) pressure.

Engine Control Panels

These panels contain meters and gages to record pressures and temperatures of the engine system. Temperatures recorded continuously are for the cooling-water inlet and outlet, lube oil to and from the engine, and (in some cases) the inlet and outlet of the oil-cooler coolant. Pyrometers record exhaust temperatures from cylinders, turbocharger gas inlet and outlet temperatures, and stack temperatures.

The fuel oil level in storage tanks is indicated by a liquidometer placed on the control panel. Warning lights alert operators of any malfunction or deviation from preset readings for minimum or maximum values. Alarms on the panel activate on low lube oil temperature, high jacket-water temperature, and high main-bearing temperature when this shutdown is furnished.

Engine Generators

Electric equipment is a critical part of the generating plant, since the generator and switchgear produce and control electric energy to consumers. Selecting the generator requires study of such service conditions as rating, speed, voltage, temperature rise, frequency supplied, excitation, and cost. Alternating-current (ac) generators furnish the greater part of electrical service. Direct-current (dc) units are used in melting operations and for mills.

Direct-current generators are shunt or compound-wound. Because of its flexibility, the compound-wound generator is used extensively, since the voltage can drop, stay constant, or rise with a load increase. Voltages normally specified are 125, 250, and 600 V. Alternating-current generators (Fig. 3.36) are built for frequencies of 25, 50, or 60 Hz, with voltages of 240, 480, 600, and 2400 V. Municipal plants will generally specify 2400 V.

The generator efficiency is the ratio of power produced to power required to drive it, or the power input. Efficiencies of large diesel/generators average about 95 percent. The capacity is limited by stator heating and the field-winding temperature rise. Standard ratings are based on 50°C at full load, assuming an ambient temperature of 40°C and a 0.80 power factor. Units at altitudes above 3300 ft (1000 m) will require design modifications.

Excitation for a generator's alternator field is supplied by a dc exciter, direct- or belt-driven from the engine extension shaft. Generators for dual-fuel and spark-ignition gas engines should specify closed-damper windings. The extension-shaft outboard bearing is insulated to prevent stray currents from forming a path through the bearing, which would cause arcing and pitting of bearing rollers and eventual failure. Fig. 3.37 shows a fully insulated outboard bearing. Holding studs, as well as the base, are insulated with fiber tubing and sheet.

FIG. 3.36 Ac generator. *(Worthington Corp.)*

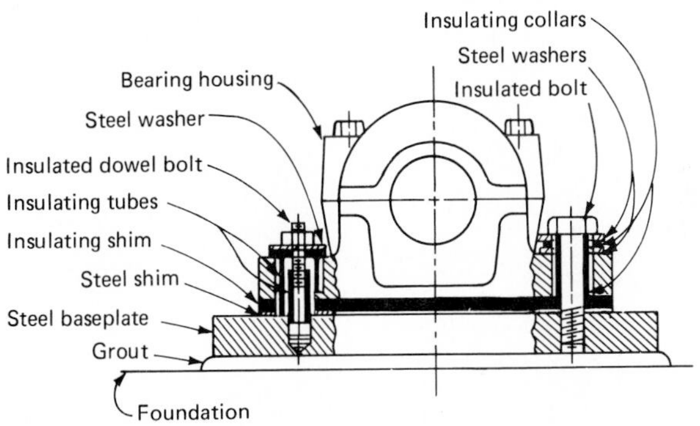

FIG. 3.37 Fully-insulated outboard bearing. (Power *magazine)*

Precise Power

Rigid electrical requirements of computer system controls have required extremely close regulation of diesel/generator sets. Figure 3.38 shows two charts which describe the steady-state speed band and underspeed from a speed-recording instrument. Table 3.6 gives governing limits for general-purpose, indus-

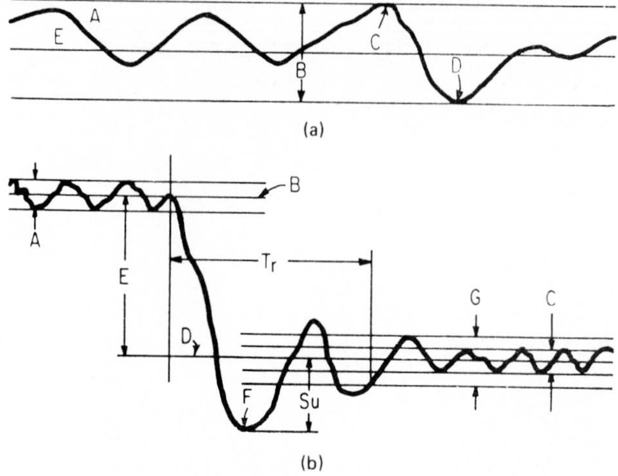

FIG. 3.38 Precise-power charts: (*a*) steady-state governing; (*b*) underspeed chart. *(W.M. Kauffmann.)*

TABLE 3.6 Speed-Governing Limits

Parameter	Key application: General-purpose, industrial*	Utilities; parallel operation*	Precise power
Steady-state regulation, %	±1.5 to ±3.0	±0.33	±0.25
Steady-state governing speed band, %	±0.50	±0.33	±0.25
Momentary underspeed, %	3.00	2.00	1.00
Underspeed recovery time, s†	4.00	3.00	1.50

*IEEE-ASME Speed Governing Code 606.
†For quarter-load transient.

trial, and utility controls, plus precise power. Note that precise power is the most demanding, limiting frequency of the steady-state speed band to ±0.25 percent, with recovery time limited to 1.5 s. Currently, governors are available with load-sensing features that reduce the transit time to 0.02 s, to obtain nearly isochronous automatic load sharing of units operating in parallel. Generators for precise power are heavier in construction; excitation is through silicon or selenium rectifiers.

AUTOMATION IN PLANT OPERATION

Automation, or control sequencing, has become standard for many powerplants as well as many smaller unattended stations. The advantages of automation are

- Savings in cost due to increased plant efficiency
- Improvement in performance from integration of starting procedures and modulation of fuel admission to supercede manual control
- Increased availability because of minimum downtime, with safety devices overriding sequence controls
- More incentive for operating personnel to upgrade engine inspection and improve overall maintenance
- Ease of training personnel to maintain automated equipment

Typical Sequences

A typical control sequence for dual-fuel engine operation consists of

- Prestart condition, purging
- Start operation and sequence
- Idling and warm-up, if required
- Engine loading
- Safety alarm system and annunciators activated
- Safety shutdown system and annunciators activated
- Incomplete sequence indicator and control activated

Prestart includes activating the prelube pump, radiator fans, water pumps, air-filter screens (if motor-operated), and positioning controls on driven equipment (other than the electric generator).

The sequence for spark-fired engines is to

- Apply starting air
- Rotate the engine at least 10 revolutions to purge manifolds and cylinders
- Apply ignition and starting fuel
- Introduce running fuel

Automatic operation in diesel plants consists of pushing a button on the control panel, which activates the auxiliaries, sets the fuel-rack position for start, and sequences the rated-speed position after a short warm-up. The engine is synchronized and put on line automatically, and the load is applied by devices sensitive

ENGINE SHUTDOWN FOR ABNORMAL OPERATION

Shutdowns protect the engine against abnormal operating conditions that would damage it. High-output engines are supplied with shutdown controls for low lube oil pressure, high jacket-water temperature, and engine overspeed. Gas engines are provided with a diaphragm-operated gas shutoff valve. Also builders have added (generally at extra cost) shutdowns for high main and connecting-rod bearing temperatures, excessive engine vibration, turbocharger low lube oil pressure, turbocharger thrust-bearing abnormal wear, and low fuel pressure. Shutdowns are connected in series with an electric circuit to a master solenoid valve. When any shutdown is activated, the master solenoid vents the air pressure from control devices to cut off fuel and, in some cases, combustion air to the engine.

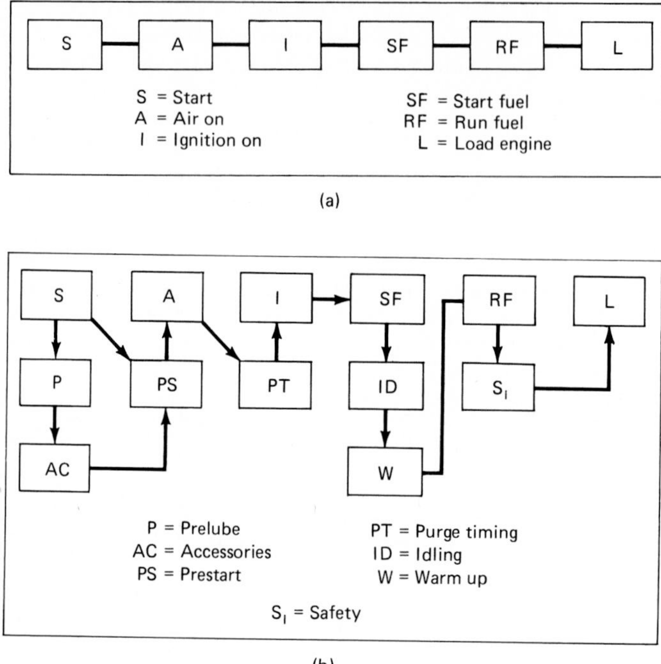

FIG. 3.39 Control sequence systems: (*a*) simple sequence system; (*b*) extended sequence system. *(W.M. Kauffmann.)*

to plant demand, with additional units added or dropped off the line as the load changes. When the unit shuts down, the load is automatically removed and the circuit breaker trips. After a short cooldown period, the controls are set for the next start (also see box on page 2.105).

Sequence Systems

Figure 3.39 is a typical block diagram for simple-sequence and extended-sequence systems. A speed switch, sensing the rate of speed change, checks the starting speed, acknowledges the onset of combustion, and initiates the shutoff of starting air. Fuel controls set the correct air/fuel ratio and limit the combustion pressure and rate of acceleration. Pressure- and temperature-sensing devices set correct operating conditions. The "control sequence" pertains to application or rejection of control power. Actuation of a momentary-contact pushbutton will trigger a relay, causing pumps to start. Time-delay relays count the number of seconds, or minutes, for the next function; thus, the sequence is fixed and cannot deviate from a planned control program.

ENGINE PLANT MAINTENANCE

Proper engine plant maintenance requires daily records diligently kept up, systematic inspection of all equipment, a routine plant cleanliness program, and periodic overhaul of parts subject to wear. Engine builders will supply a form for recording temperatures and pressures and lube oil use. Engine instruction books give essential data on construction, inspection, and overhaul periods.

Systematic Inspection

Inspection of jacket-water pumps, fuel system, lube oil filters, and intake and air filters should be scheduled for not less than 30-day intervals. To ensure correct readings, instruments, gages, and thermometers are a vital part of maintenance; they should be checked periodically and calibrated, if necessary. To ensure that they can be activated when needed, regularly check safety devices, such as high-bearing-temperature shutdowns, high-water-temperature alarm, and low-pressure shutdown of the lube oil system. Plant structural parts need regular checking and repair to avoid the expense of replacement or failure.

Most engine builders retain a staff of qualified service personnel, or a service facility, to assist plant operators. Currently, independent service organizations are geared to perform many functions of engine repair and overhaul.

Other Checks

The importance of periodic checks of crankshaft alignment cannot be overemphasized. Crankweb deflection readings are taken with a micrometer fitted with an extension piece and are taken at the heel of the web. The deviation in one complete revolution of the crank should not exceed ± 0.001 in (0.025 4 mm) when the engine is hot. (Checking alignment of a cold engine will give erroneous results.) Important, too, is reestablishing correct alignment with shims or rails, or regrouting, as is done in older engines.

Another vital check involves the bolt torques of main bearings and connecting rods. Where through-bolts are selected, some builders prefer measuring bolt stretch or extension. This should be done about every 8000 h of operation, or as suggested by the builder.

Lube Oil Analysis

An important part of plant maintenance is a monthly lube oil analysis by a reputable laboratory to test for oxidation, dilution, and metal content. An oil change should be considered when the viscosity drops 15 percent, or increases 15 percent due to dilution or soot; a 15 percent drop represents 3 percent dilution. The water content should not exceed 0.5 percent, nor should benzene and pentane insolubles exceed 0.6 percent. Total insolubles should be limited to 3 percent.

SUGGESTIONS FOR FURTHER READING

Lilly, L.R.C., ed., *Diesel Engine Reference Book,* Butterworth Publishing, Ltd., London, England, 1981.

O'Connor, J.J., and John Boyd, eds., *Standard Handbook of Lubrication Engineering,* "Diesel and Gas Engines," McGraw-Hill Book Co., New York, 1968.

The following articles by the author relate to oil and gas engines; all appeared in *Power* magazine:

"Cooling System Design," September 1964.
"Know BMEP," November 1969.
"Foundation Growth," January 1973.
"Alternative Engine Fuels," May 1973.
"Total Energy Guidelines," April 1974.
"Analyze Engine Foundations," May 1976.
"Standby Diesel Power," December 1979.
"Parallel Operation Guidelines," May 1980.
"Prevent Crankcase Failure," September 1980.
"Build in Fuel Flexibility," February 1981.
"Prevent Crankcase Explosions," March 1982.
"Large Engine Piling Foundations," July 1982.
"Heavy Fuel Reduces Cost," September 1983.

CHAPTER 2.4
GAS-TURBINE FUNDAMENTALS

Dr. H. Haselbacher
Asea Brown Boveri Ltd.
Baden, Switzerland

Glossary

η_{th} = thermal efficiency
q_r = amount of heat removed
q_s = amount of heat supplied
h = enthalpy
T_m = average temperature
$T_{m,r}$ = average temperature, removed
$T_{m,s}$ = average temperature, supplied
P_2, P_1 = inlet, outlet compressor pressures
π = compressor pressure ratio = P_2/P_1
γ = ratio of inlet turbine T_3 and compressor T_1 temperatures
W_t = specific useful work
ω_t = W_t/h_1

INTRODUCTION

In the utility industry today, gas turbines are second only to steam turbines in importance as power generation machines. The thermodynamic cycles available for converting the chemical energy to useful mechanical work are customarily classified as either open or closed cycles. In the closed cycle, the working-medium circuit is separated materially from the environment. As a result, the supply and removal of heat by heat transfer in heaters or coolers are an additional important feature of this cycle.

In the open cycle, the working-medium circuit is connected materially with the environment. For this reason, air is the only medium with which the open cycle can be operated. The energy required can be supplied either wholly, as in the closed cycle, or in part by means of heat transfer in heaters. In practice today, however, the increase in temperature necessary for operation is attained only by the direct burning of oil or gas in the airstream. It is possible, nevertheless, to do calculations as if an equivalent heat flow had been supplied indirectly to the cycle. (In other words, the closed cycle is, by definition, a true thermodynamic cycle, while the open cycle with internal combustion is not or is only conditionally so.)

The historical perspective described next will show how these major lines of development emerged and why today only the simple combustion gas turbine is of importance. That type of turbine is the primary subject of this chapter, which is limited mostly to machines with a medium to high power rating (greater than 30 MW), reflecting their industrial and economic roles at present. Whether designs deviating from present standards will become established in the future depends on many factors, the development of which are difficult to predict. Refer to the Glossary for a handy explanation of equation terms.

HISTORICAL PERSPECTIVE

After the crankshaft and the planetary gear were introduced in 1780, the steam engine, brought to commercial reality by James Watt a few years earlier, found rapid acceptance in industries that needed rotational movement for the transmission of power. Ironically, the striving for simplification and improvement of steam engine plants soon led to the development of the gas-turbine concept. As early as 1791, J. Barber was granted a patent for what was a forerunner to the gas turbines of today. This machine consisted of a reciprocating compressor, a heating system, and a single-stage impulse turbine—only partially fulfilling the requirement of rotational movement, however.

The Barber patent provided for the simplest open cycle imaginable, with constant-pressure combustion. In operation, the compressor takes in air from the atmosphere, compresses it, and drives the air to the combustion chamber, where its temperature is raised by burning fuel. Then the hot gas is expanded in the turbine to drive the compressor and provide useful power. The exhaust gas is released directly to the atmosphere. This process was the only one among many possibilities that later assumed a broad economic significance.

Many detours and wrong turns were taken, however, before a technological breakthrough occurred. The date worth noting is the year 1872, when Franz Stolze is said to have designed the first constant-pressure gas turbine. (It was tested from 1900 to 1904.) His design appears quite modern, consisting of a multistage turbine and a multistage axial compressor on a drum rotor suspended between two journal bearings (Fig. 4.1).

While use of the steam turbine, which was introduced at about the same time, spread quickly, gas turbines met with no success for a long time because knowledge of aerodynamics was inadequate for building high-efficiency axial or radial

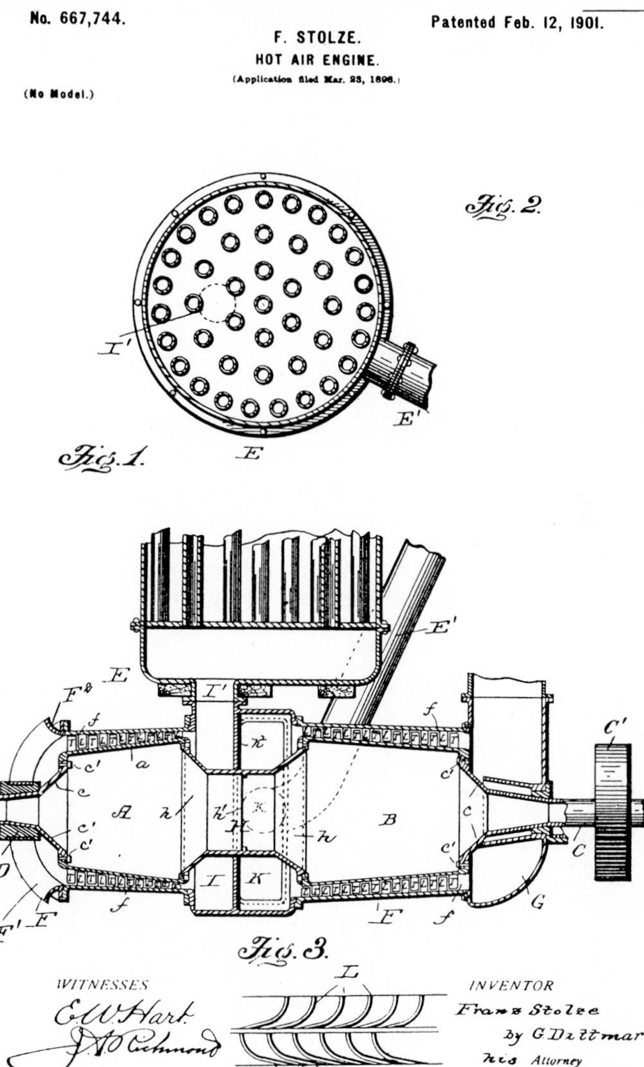

FIG. 4.1 F. Stolze's gas turbine, patented in 1901. Note: A = compressor, B = turbine.

compressors. Also, no materials were available with adequate high-temperature capability.

In 1906, H. Holzwarth tried to bypass these difficulties with the explosion, or constant-volume, gas turbine. In this type of machine, the fuel was burned at a constant volume in a valve-controlled combustion chamber. As a result, the pressure rose to a level several times its initial value (the compressor exit pressure). The combustor and other hot parts were water-cooled so that the amount of excess air could be relatively small. For this reason, these gas turbines were not as sensitive to the poor compressor efficiencies of those days. They did not find acceptance, however, because they were quite complicated and expensive.

Major progress in the development of axial compressors did not occur until the 1930s—advances closely linked with the name of C. Seippel. Axial compressors were first used successfully by Brown Boveri, Cie (BBC) in the turbochargers for Velox boilers, which have now, in a transmuted form, regained importance as *pressurized fluidized-bed combustion* (PFBC) boilers (see Sec. 4, Part 1, Chap. 3).

Early Commercial Installations

By the 1930s it was just a short step to using the gas turbine directly—that is, without the detour via steam—for the production of electricity. The first installation of this sort anywhere in the world went into operation in Neuchatel, Switzerland, in 1939 (Fig. 4.2). The turbine was equipped with one combustion chamber, operated on heavy-grade oil, and produced 4 MW. The predecessor of the combustion chamber was the Velox boiler, from which the fuel supply system and burner design were adapted.

It was clear that such small, simple installations with turbine inlet temperatures of 1022 to 1110°F (550 to 600°C) could not compete with the steam turbines of that time with respect to efficiency and output. Already at that time the main

FIG. 4.2 Neuchatel gas turbine, an early installation.

application of gas turbines was seen as standby and peaking. The desire grew to make gas turbines more attractive economically in countries where fuel costs were high, by increasing their efficiencies and power capacities. Thus, installations having intermediate cooling, air preheating, and reheating were introduced.

Multishaft machines were developed to improve part-load efficiencies. With efficiencies of approximately 31 percent attainable with such equipment, together with the burning of relatively inexpensive heavy-grade oils, it was possible in the 1940s to compete with steam turbines. The picture changed again when clear improvements were made in steam-turbine cycles over the ensuing 10 to 15 years. During these years, however, there was also an increase in the temperature potential of gas-turbine materials, and fuels became less expensive. These changes brought about a return to the simple concept employed in the first industrial gas turbines.

One important parallel line of development may well assume great significance in the future in the production of energy in conjunction with coal gasification plants. This is the blast-furnace gas turbine, which was integrated into steel mill processes and burned blast-furnace gas—a low-Btu fuel containing large amounts of dust and having a heating value of 1070 to 1290 Btu/lb (2500 to 3000 kJ/kg).

The first turbojet engines for use in aircraft were developed shortly before World War II in England and Germany. Because minimum weight and low drag were important criteria, lightweight construction prevailed throughout. To minimize the outside diameter, a fairly large number of small combustors were arranged around the engine. As everyone knows, aircraft engine technology developed rapidly and started to exert a continuing influence on the design of industrial gas turbines as well.

This influence can be seen, for example, in the industrial gas turbines built by General Electric Co. (GE), production of which started on a large scale a few years after World War II. At first, the primary emphasis was on drives for pipeline compressors. These units show design features typical of aircraft engines, but their most striking feature is their compactness. For this reason, it was possible to preassemble them on frames at the factory together with their auxiliaries and to ship them to their erection sites largely ready to operate. The concept of packaging installations at the factory into as few blocks as possible can likewise be applied to gas-turbine powerplants, and has also been adopted by manufacturers of other types of big equipment.

Closed and Semiclosed Cycles

At the same time as the open-cycle design was evolving, the closed-cycle gas turbine was also being developed in Switzerland, specifically by Escher Wyss. During the 1930s, Escher Wyss saw the main application for the gas turbine as being not in peaking plants but in base-load energy production. The power capacity of a gas turbine was therefore to be as large as possible; with respect to efficiency, the machine was to compete with steam turbines. It was to have the capability of burning inexpensive fuels as well, mainly coal, and good efficiencies at part load were demanded. These criteria led Escher Wyss to turn in 1935 to the closed-cycle gas turbine.

When the circuit is closed, it can be supercharged and its power output can be raised directly proportional to the increased density of the medium within the circuit. The inventory in the circuit can also be varied during operation, making it possible to run at part load with high efficiency. J. Ackeret and C. Keller were

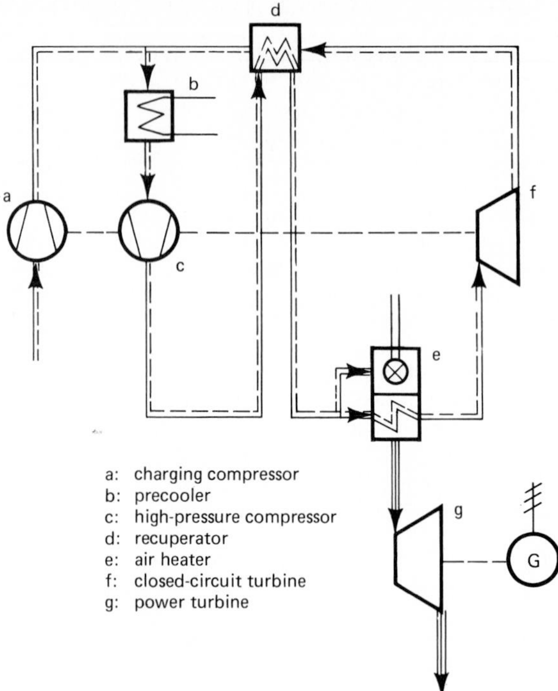

FIG. 4.3 Semi-closed gas-turbine cycle.

granted their well-known patent for this method of load control in 1935. Construction of the first installation was completed in 1939.

The only major new component to be found in the closed-cycle design is the gas heater, which replaces the combustion chamber in the open-cycle plant. Because of its limited temperature potential and high cost, on one hand, and of the rapid development of open-cycle gas and steam turbines, on the other, the application of closed-cycle gas turbines using conventional fuels to generate electricity alone soon became outdated. However, the design allows convenient use of a portion of the heat removed in the coolers for heating purposes. This increases utilization of fuel energy from roughly 30 to 85 percent. Because of this, most of the closed-cycle plants built were cogeneration types (see Chap. 7 in this section).

In 1942, shortly after the first CO_2-cooled reactor was put into operation, Keller suggested that a reactor of this type be directly connected to a closed-cycle gas turbine. This led in 1946 to a U.S. proposal that helium be the cooling and working medium for high-temperature gas-cooled reactors (HTGRs) and gas turbines in a closed cycle (direct-cycle HTGR). This combination is ideal because helium cannot become activated and possesses very good heat-transfer properties and a high sonic speed. Moreover, it is possible to attain far higher hot-gas temperatures with this reactor than with conventional gas heaters. For these reasons, the direct-cycle HTGR has met with great interest. The future of this concept will depend on the further development of the reactor.

The disadvantages of the large air heater in conventional closed-cycle gas-turbine powerplants can be avoided by charging it. In 1938, Sulzer was the first to follow this path, using oil or gas as the fuel in a semiclosed cycle. In principle, the closed and open cycles are superimposed on each other (Fig. 4.3). At maximum pressure a precise amount of air is tapped from the "closed" circuit that is needed in the heater for combustion. A compressor continually returns the same amount of air to the "closed" circuit at its lowest pressure. Thus, the compressor and the combustion gas turbine after the heater have to process only relatively small volume flows. Because the same heater pressure prevails on the air and flue-gas sides, not only are dimensions relatively small, but also stress on the piping is less than in the closed cycle. In a commercial installation built in 1947, difficulties were encountered in operation on heavy oil. These would be manageable today, but they probably contributed to the fact that no further semiclosed-cycle plants have been built since then.

In brief, the foregoing represents the major lines of development of powerplant gas turbines. This retrospective view, however, is incomplete to the extent that it does not address the developmental work done by all firms whose efforts also contributed to the field.

GAS-TURBINE CYCLES

Thermodynamic cycles are made up of sequences of thermodynamic processes. The two parameters most frequently used for cycle evaluation are the specific useful work

$$W_t = q_s - q_r = q_s\left(1 - \frac{q_r}{q_s}\right)$$

and the thermal efficiency

$$\eta_{th} = \frac{W_t}{q_s} = 1 - \frac{q_r}{q_s}$$

where q_s and q_r represent the amount of heat supplied and removed respectively, per unit of mass of the working medium. They depend on the processes making up the cycle and on the quality of the components involved in the processes. Normally, W_t and η_{th} should be as large as possible.

In the ideal case, all cycle components operate loss-free, and heat transfer takes place without temperature differences. Another parameter is the reversible cycle. Its thermal efficiency represents the theoretical maximum value if the characteristic process temperature and pressure ratios are otherwise constant. Although these limit cases cannot actually be realized, reversible cycles are important because useful conclusions for practical applications can be drawn from them and because the real processes can be measured against them. For reversible cycles,

$$\oint \frac{dq}{T} = 0$$

where dq is the amount of heat transferred at temperature T. The average temperature of the heat transferred in a process is defined by

$$\int \frac{dq}{T} = \frac{q}{T_m}$$

From these two equations, we obtain

$$\frac{q_s}{T_{m,s}} = \frac{q_r}{T_{m,s}}$$

and

$$W_t = q_s \left(1 - \frac{T_{m,r}}{T_{m,s}}\right) \qquad \eta_{\text{th}} = 1 - \frac{T_{m,r}}{T_{m,s}}$$

The values of T_m depend on the changes in state that have occurred. If an ideal gas with constant specific heat is assumed, for example, we obtain for the isobar between points A and E

$$\frac{T_m}{T_A} = \frac{T_E/T_A - 1}{\ln(T_E/T_A)} \approx \frac{T_E/T_A + 1}{2}$$

The important general conclusion to draw from these relationships is this: To achieve high thermal efficiencies, the heat must be supplied at as high an average temperature as possible and must be removed at as low an average temperature as possible.

Cycle Design Points

Figure 4.4 shows the schematic flow diagrams of the simplest open and closed cycles. If, at the start, pressure losses are neglected, the closed cycle is made up of four processes:

- Polytropic compression
- Isobaric addition of heat
- Polytropic expansion
- Isobaric removal of heat

The closed cycle appears in the enthalpy-entropy diagram in the form of familiar trapezoidlike lines (Fig. 4.5). The diagram for the open cycle is similar, with these exceptions:

- The addition of heat, as explained in the Introduction.
- The removal of heat to the atmosphere takes place through a cooler.

Specific useful work and thermal efficiencies can now be expressed in terms of the enthalpies of the working medium as

$$W_t = q_s - q_r = h_3 - h_2 - (h_4 - h_1)$$

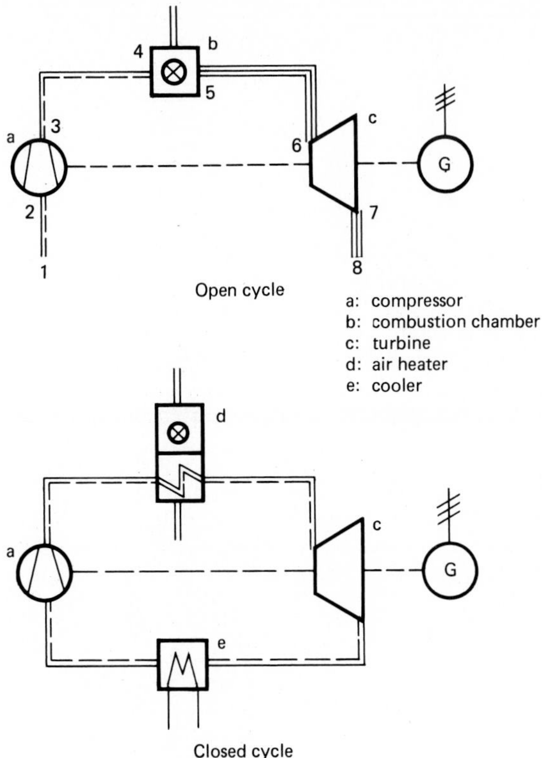

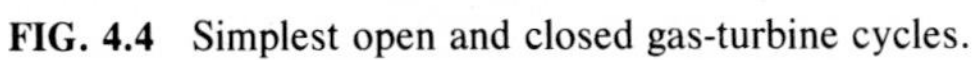

FIG. 4.4 Simplest open and closed gas-turbine cycles.

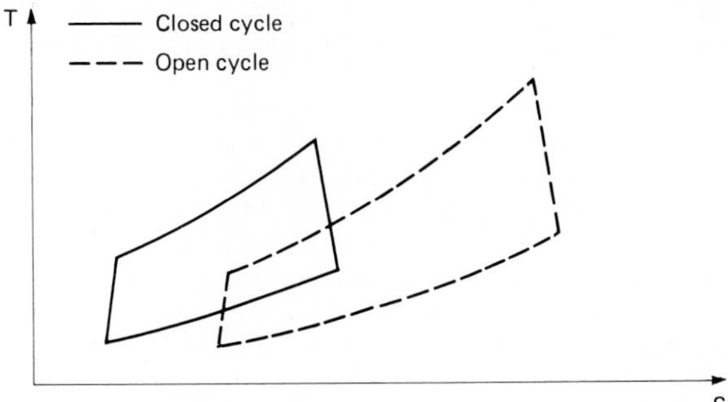

FIG. 4.5 Enthalpy/entropy diagrams for the Fig. 4.4 cycles.

or in nondimensional terms as

$$\omega_t = \frac{W_t}{h_1} = \frac{h_3 - h_2 - (h_4 - h_1)}{h_1}$$

$$\eta_{th} = \frac{W_t}{q_s} = 1 - \frac{h_4 - h_1}{h_3 - h_2}$$

In dealing, as here, with working media in a continuous flow, the corresponding total enthalpies must be substituted.

In design, it is important to select characteristic pressure and temperature ratios with which certain values for ω_t and η_{th} can be attained. Suitable parameters, and those most easily understood, are the compressor pressure ratio $\pi = P_2/P_1$ and the ratio of the inlet temperatures to the turbine and compressor, $\gamma = T_3/T_1$.

Ideal Cycles. The shift mentioned earlier from a real to an ideal cycle produces very clear relationships, particularly in the case of a simple-cycle gas turbine, as shown in Fig. 4.4. The cycle in question, also known as the *Joule* or *Brayton cycle,* consists of two isentropes and two isobars. Normally, calculations are also made by using ideal gases with a constant specific heat. One thus obtains

$$\omega_t = \left(\gamma - \pi^{(k-1)/k}\right)\left(1 - \frac{1}{\pi^{(k-1)/k}}\right)$$

and

$$\eta_{th} = 1 - \frac{1}{\pi^{(k-1)/k}}$$

where k is the ratio of the specific heats (see Fig. 4.6). In present-day gas-turbine design, π is approximately 10 to 15, and γ is approximately 4 to 5. Higher values can also be found in individual cases. In the range of $\omega_{t(max)}$, then, η_{th} is approximately 0.5.

With the same temperature ratio, the Carnot cycle yields a thermal efficiency of about 0.8. With the temperatures customary in steam-turbine design, the thermal efficiency obtained is still about 0.65. The poorer performance of the Joule cycle results mainly from the higher average temperature level during heat removal. Significant improvements can be realized by using waste-heat recuperation, intercooling during compression, and reheating during expansion.

Waste-Heat Recuperation. The equations that apply for a gas turbine with a recuperator are

$$\omega_t = (\gamma - \pi^{(k-1)/k)}\left(1 - \frac{1}{\pi^{(k-1)/k}}\right)$$

and

$$\eta_{th} = 1 - \frac{\pi^{(k-1)/k}}{\gamma}$$

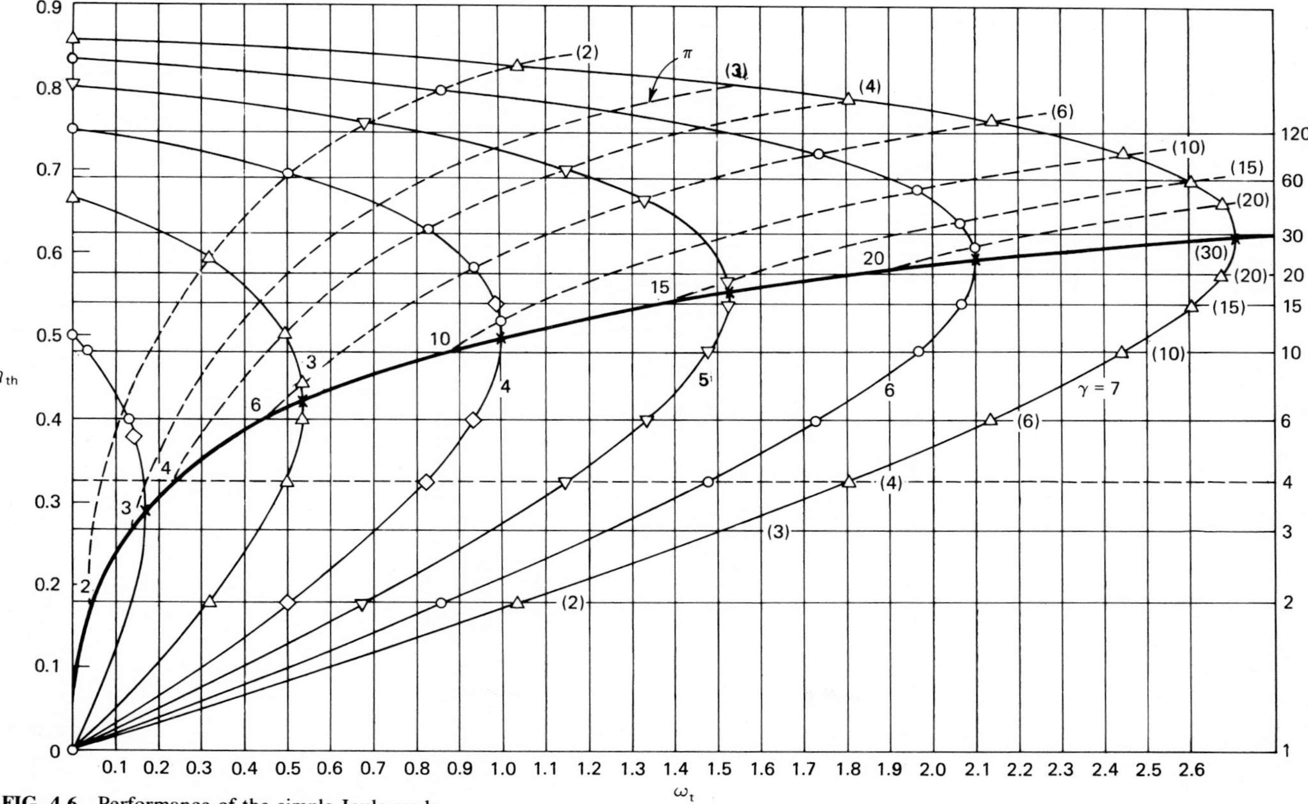

FIG. 4.6 Performance of the simple Joule cycle.

If π and γ remain constant, the thermal efficiency is higher than previously, but the useful work remains unchanged. See Fig. 4.7, in which the lines for π = constant and γ = constant for the simple cycle (that is, without recuperation) have been drawn once again in broken lines. Clearly, with a recuperator the thermal efficiency increases as π drops and γ rises. In practice, however, limits are soon reached for this gain because the turbine outlet temperature also increases rapidly and so can become problematic for the design of the recuperator. Furthermore, the mere addition of a recuperator will not make possible any further improvement in a cycle that is already optimum without it. *Optimum* here means that for a given γ, there is a π for which ω_t reaches a maximum.

The introduction of intercooling and reheating increases not only the efficiency but also the useful work. As the number of stages of reheating and intercooling increases, one comes closer and closer to isothermal addition and removal of the heat. This limiting case is referred to as the *Ackeret-Keller* or the *Ericsson cycle*. For this,

$$\omega_t = \frac{k-1}{k}(\gamma - 1)\ln \pi$$

$$\eta_{\text{th}} = 1 - \frac{1}{\gamma}$$

Thus, the thermal efficiency is as high as that of the Carnot cycle with the same temperature ratio. Not only is there a clear-cut increase in thermal efficiency, but also the useful work produced increases sharply, specifically, with the pressure and temperature ratios customary for the simple Joule cycle, by a factor of approximately 2.

Combined-Cycle Plants. All the foregoing improvements require more complicated—and therefore more expensive—cycle arrangements, but only gas-turbine installations are involved. A practical alternative is the combination of a gas-turbine cycle with a steam-turbine cycle (also see Sec. 1, Chap. 2). One thereby combines the advantages of high-temperature heat addition in the gas-turbine cycle and low-temperature heat removal in the steam-turbine cycle. For the ideal case, ω_t and η_{th} are determined from

$$\omega_t = \gamma - \pi^{(k-1)/k} - \ln \frac{\gamma}{\pi^{(k-1)/k}}$$

and

$$\eta_{\text{th}} = 1 - \frac{\ln(\gamma/\pi^{(k-1)/k})}{\gamma - \pi^{(k-1)/k}}$$

Figure 4.8 offers convincing proof that the efficiencies are indeed relatively high. They do not, however, reach the Carnot levels because the heat is supplied at an average temperature between the compressor outlet and the turbine inlet.

Even though the useful work and the thermal efficiency of the combined cycles do not, in all cases, come up to those of the improved cycles using gas-turbines alone, combined cycles still have advantages because it is possible to make direct use of the ordinary simple-cycle gas turbine; thus, any improvements in these turbines also benefit the combined cycle as well. These advantages have contributed to the relatively rapid propagation of combined-cycle plants. With re-

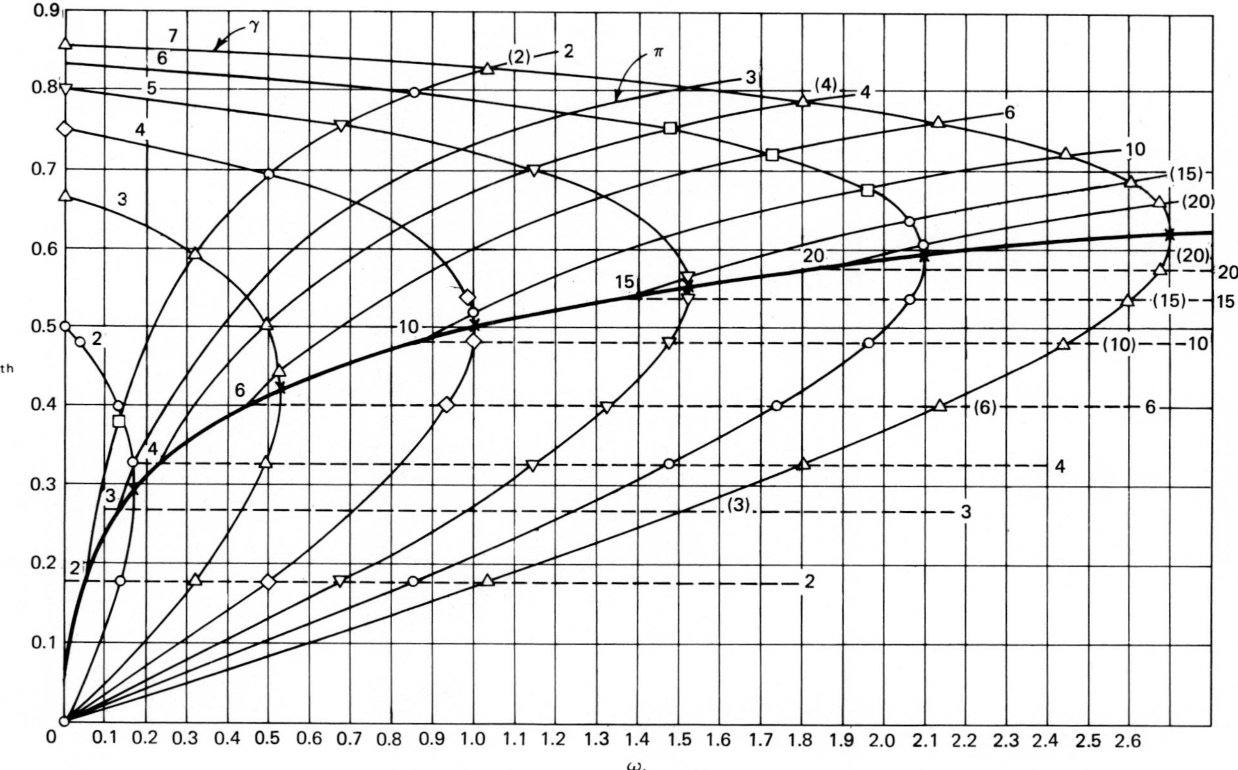

FIG. 4.7 Performance of the recuperated Joule cycle.

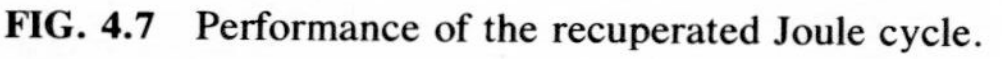

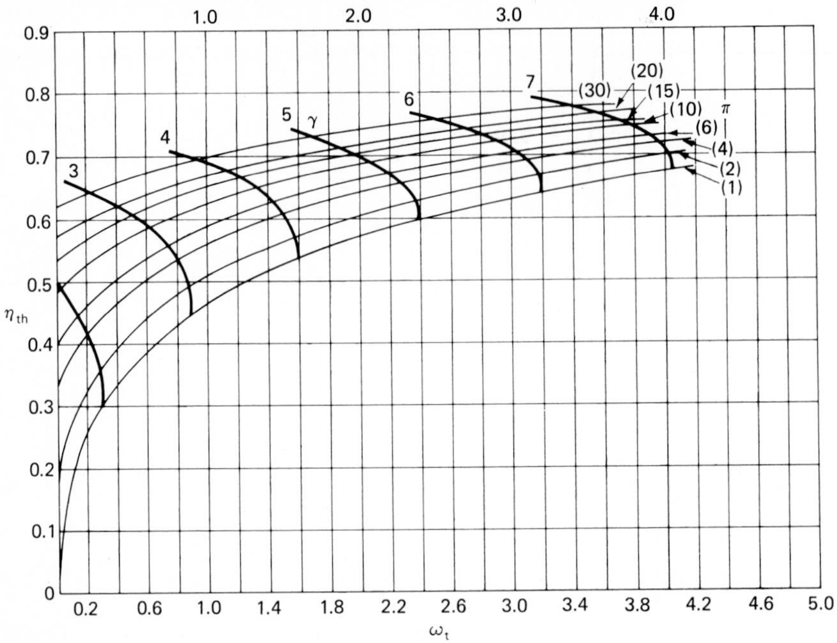

FIG. 4.8 Performance of the ideal combined cycle.

gard to the temperature after the gas turbine, however, if it is too high, expensive materials are necessary for the exhaust-gas duct and parts of the waste-heat boiler.

Real Cycles. The following applies mostly to the simple-cycle gas turbine—the only cycle that plays a significant role in practice today. The difference between a real cycle and the corresponding ideal cycle is due to

- Aerodynamic losses—losses in compression and expansion (polytropic instead of isentropic) and pressure losses outside the compressor and turbine
- Cooling losses
- Differences between the properties of the medium in the cycle and those of the ideal gas
- Incomplete combustion

With the technical documentation and computer programs available today, it is not difficult to take into account all these factors in designing the cycle. Normally, aerodynamic and cooling losses predominate; for this reason, they are considered in more detail. It is still assumed that the cycle medium is an ideal gas.

Aerodynamic Losses. The change of state of an ideal gas in a multistage turbomachine is, by definition, polytropic if all the stages are working with the same efficiency. Because this is largely true in real machines, the following calculations have also been made on that basis.

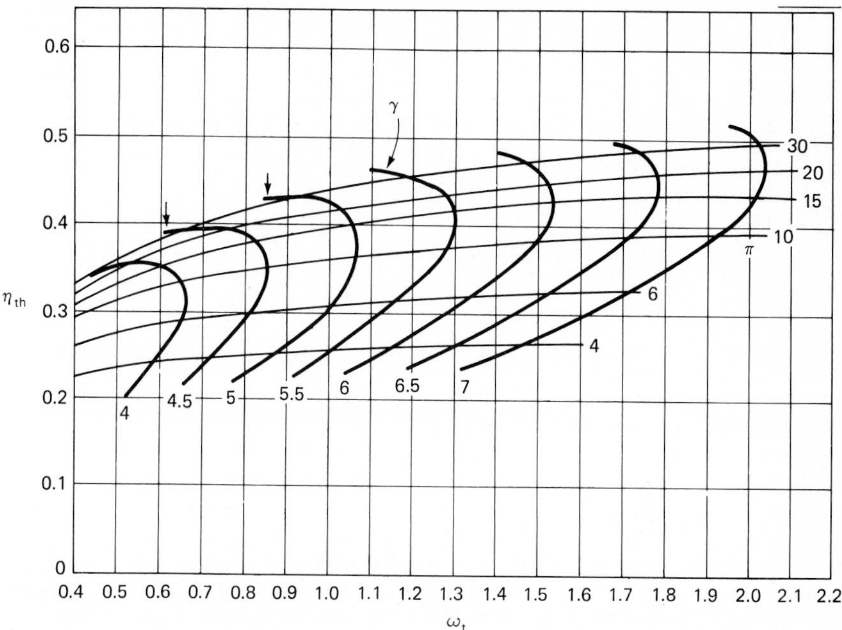

FIG. 4.9 Performance of the simple cycle, taking pressure losses into consideration.

If η_T and η_C are the polytropic efficiencies of the turbine and compressor, respectively, then for the turbine

$$\frac{T}{T_r} = \left(\frac{P}{P_r}\right)^{[(k-1)/k]\eta_T}$$

and for the compressor

$$\frac{T}{T_r} = \left(\frac{P}{P_r}\right)^{[(k-1)/k]\eta_C}$$

The index r refers to an arbitrary reference point along the expansion and compression curves.

In cycle studies, the relative pressure losses occurring along the circuit can be considered as gathered together at one point (Fig. 4.9). Their sum $\Sigma\ \Delta P/P$ or the pressure loss ratio π_p is related to the pressure ratios of the turbine π_T and the compressor π_C as

$$\Sigma \frac{\Delta P}{P} = \ln \pi_P = \ln \pi_C - \ln \pi_T$$

Another measure for the pressure losses is

$$\eta_P = \frac{\ln \pi_T}{\ln \pi_C} = 1 - \frac{\Sigma(\Delta P/P)}{\ln \pi_C}$$

For ideal gases with a constant specific heat, the product $\eta' = \eta_C \eta_T \eta_P$ reduces the nondimensional specific useful work and the thermal efficiency of the simple-cycle gas turbine to functions of γ, π_C, η_C, and η', namely,

$$\omega_t = \gamma \left(1 - \pi_C^{[-(k-1)/k]\eta'/\eta_C}\right) - \left(\pi_C^{[(k-1)/k]\,1/\eta_C} - 1\right)$$

and

$$\eta_{\text{th}} = \frac{\gamma \left(1 - \pi_C^{[-(k-1)/k](\eta'/\eta_C)}\right) - \left(\pi_C^{[(k-1)/k]1/\eta_C} - 1\right)}{\gamma - \pi_C^{[(k-1)/k]1/\eta_C}}$$

Typical values for η', which is a good measure of the aerodynamic quality of the cycle, are on the order of 0.75 (Fig. 4.9).

Cooling Losses. With the turbine inlet temperatures specified today, it is necessary to cool the turbine blades, rotors, and housing to obtain economically justifiable life expectancies for these components. With the current state of the art, this means cooling with air withdrawn from the compressor at as low a pressure as possible. (Cooling with an outside medium—for example, water—would be more effective but is more difficult to carry out than air cooling.)

The cooling air can be withdrawn from the compressor at several points and, if need be, recooled to attain an optimum effect. After it has fulfilled its cooling function, the cooling air is mixed with the hot working gas. As a rule, leakage to the outside is negligible. Obviously, cooling entails losses, that is, the increased power output and efficiency *theoretically* possible from raising the turbine inlet temperature are not fully realized. Part of the power output deficit is due to reducing the turbine mass flow by the amount of the cooling airflow. But additional losses also occur from the increases in entropy within the turbine. These are the consequence of impulse and mixing losses when the cooling air rejoins the mainstream, and they can be determined by means of mass, energy, and entropy balances for the mixing zones in question.

In practice, the calculations for the turbine take into account all these effects in detail, row by row, and determine the average status at the exit from each row. No procedures are available that are simple and still accurate enough to describe the effect of the cooling on the process as a whole. For this reason, detailed analysis is often used to determine a resultant turbine efficiency and a defined turbine inlet temperature for cycle calculations. Frequently, this is a mixed temperature, which is obtained from the heat balance:

$$\dot{m}_4 h_4 + Q_s = \dot{m}_6 \bar{h}_6 = \dot{m}_8 h_8 + \dot{m}_4 (h_4 - h_1) + P_o$$

where Q_s is the heat supplied and P_o is the power output. Variable $\bar{T}_6$ can be determined from $\bar{h}_6$, by using the equation of state; $\dot{m}_8 = \dot{m}_1 + \dot{m}_F$, where $\dot{m}_F$ is the mass flow of fuel (Fig. 4.4).

SELECTION OF DESIGN PARAMETERS

Gas turbines are built for many applications. The power generation industry uses them in peaking, medium-, and base-load operations. The type of operation depends on overall economics. In industrialized nations, gas turbines selected nowadays are intended only as peaking and medium-load machines. With the development of new technologies (for example, integrated coal gasification), gas turbines can also be reconsidered as base-load machines—even in industrialized nations. At the same time, the problem of unit power capacity can once again take on new significance.

For the sake of low plant and operating costs, the market generally demands low specific plant costs and high thermal efficiencies. Low specific plant costs require primarily a high specific power capacity, but they also need a high air mass flow unless the power output itself were to be limited, for whatever reason. Thus, the parameters that must be discussed are as follows:

Turbine Inlet Temperature

The turbine inlet temperature selected will be as high as possible to attain a high specific power capacity and a high thermal efficiency. Customary values today are from 1725 to 2100°F (940 to 1150°C). Because the maximum temperature allowable for turbine materials is considerably lower than that, cooling becomes unavoidable. As the gas temperature rises, the amount of cooling air required increases so sharply that with the cooling systems and materials selected today the corresponding increase in thermal efficiency is slight and would ultimately disappear altogether. Very careful attention to the cooling air path thus becomes indispensable, and unneeded bypasses, excessive cooling-air pressures, etc., must be avoided.

Compressor Pressure Ratio

After the turbine inlet temperature has been specified, a decision must be made as to the compressor pressure ratio—one optimum value with respect to the specific power capacity and another with respect to the thermal efficiency. As the temperature increases, however, these values differ more and more. Thus, a compromise must be made, based on considerations of overall economy. The selection is simplified somewhat because, near the maximum point, the curve for efficiency as a function of the pressure ratio is quite flat. Values customary today for the pressure ratio are between 9 and 18; that is, at the higher pressure ratios, the compressor discharge temperatures are such that, for the sake of the rotor, either a partial recooling of the cooling air or the use of austenitic materials becomes necessary.

Air Mass Flow Rate

In the past, economic strictures have always led to the demand for machines with a maximum power output. One result has been that the mass flow capacity of the gas turbine, in the course of development, has also risen steadily. (The flow does, in fact, affect thermal efficiency as well, but there it plays a subordinate role.) The critical point in this case is the state of development of the first compressor and last turbine stages, for these have to handle the greatest air volume flows. Today, the power capacity stands at more than 150 MW, which means that the unit power capacities that can be built at present are greater than those demanded by the market. When large coal gasification plants are built, however, another shift is conceivable.

Rotational Speed and Number of Shafts

Assume that all the machines in a family of gas turbines are based on the same process and the same type of blading. The velocity triangles for the gas turbines are then identical, and the units themselves are geometrically similar. Thus,

$$\frac{P}{P_1} = \frac{\dot{m}}{\dot{m}_1} = \left(\frac{D}{D_1}\right)^2 = \frac{1}{(n/n_1)^2}$$

where the index 1 refers to the variant with the rotational speed $n = 3600$ rpm (synchronous speed). For power outputs P less than P_1, then, there is a correspondingly higher turbine rotational speed, and a reduction gear must be put between the gas turbine and the generator.

Gears are being built today for power throughput capacities of up to 80 MW. Losses in the gears are approximately 1 to 2 percent. Whenever the rotational speed of the gas turbine is clearly greater than the synchronous speed (greater than 4500 rpm), a single-shaft installation of this sort is still an economical solution despite the costs and losses for the gear. This is true because the dimensions of the gas turbine are relatively small, and so the turbine is relatively inexpensive.

As an alternative for small and medium power capacities, it is also possible to avoid the gear by building machines with two shafts. The first (high-speed) rotor accommodates the compressor and the stages of the turbine required to drive it. The second rotor, running at the same speed as the generator, accommodates only the power turbine. With designs of this type, relatively high efficiencies can be attained because there are no gear losses and the power turbine is, as a rule, a multistage machine (up to five stages). However, a prerequisite for success here is that the flow between the compressor turbine and the power turbine be handled very carefully.

With regard to efficiencies at part load and at start-up, gas turbines with two shafts are also superior to those with a single shaft. With single-shaft machines, it is easier to deal with overspeeds because the compressor acts as a brake.

ACKNOWLEDGMENT

Illustrations courtesy of Asea Brown Boveri Ltd.

SUGGESTIONS FOR FURTHER READING

Ball, R. W., "Alternative Fuels for Gas Turbines That Only Burn Oil or Gas," ASME Paper No. 85-GT-228, ASME, 1985.

Cleveland, A., "Energy Recovery with Turbo Expanders," ASME Paper No. 86-GT-66, ASME, 1986.

DelBueno, R., "Gas Turbines Fire Heavy Oils," *Power,* May 1972.

Javetski, John, "The Changing World of Gas Turbines—A Special Report," *Power,* September 1978.

Johnson, R. H., and C. Wilkes, "Environmental Performance of Industrial Gas Turbines," ASME Paper No. 74-GT-23, ASME, 1974.

"Pretreat Liquid Fuels for Gas Turbines," *Power,* July 1974.

Puntel, W., "Utility System Economic Benefits of Improving Gas Turbine Reliability," *Proceedings of American Power Conference,* 47:525–531, 1985.

Schemenau, W., "Extension of Gas Turbine Powerplants to Combined-Cycle Power Generation," ASME Paper No. 86-GT-64, ASME, 1986.

Skrotzki, B. G. A., "Gas Turbines—A Special Report," *Power,* December 1963.

Low Rotational Speeds. At low rotational speeds (and low pressure ratios)—for example, during start-up—flow conditions in the front stages of the compressor may be so unfavorable that stall occurs on the blade profiles. This does not happen simultaneously on all the profiles, but instead is restricted to certain zones that are not stationary but rotate at a constant speed—approximately one-half that of the compressor. This phenomenon is referred to as *rotating stall.* Within the stall zones, there are localized backward flows. These blockages cause the net compressor mass flow to drop, but that flow does not fluctuate with time. Such cases are called *local instabilities.*

If the pressure ratio increases while the rotational speed remains constant (which reduces the mass flow), these stall zones spread. Ultimately, they can block the entire ring cross section to such an extent that the mass flow suddenly begins to vary between values at which the compressor is free of stall and values at which rotating stall or even flow reversal (negative flow) occurs. This *global instability* is referred to as a *compressor surge.*

High Rotational Speeds. If a drop in the mass flow or an increase in the pressure ratio occurs at high rotational speeds, flow conditions are created that can also produce a rotating stall. This stall takes place only in the last stage at first, while there is little change in conditions for the front stages. Because of the relatively low aspect ratio of the blading, stall sets in along the entire blade height, so that rotating stall and surge coincide.

The curves for the rotating stall and surge lines have also been shown in Fig. 5.4. The lower one goes, the more difficult they become to follow, particularly surge, since the flow generally becomes very rough.

It is not permissible to operate compressors in the rotating stall or surge zones because the blading would be subjected to unacceptably high transient stresses. For this reason, the nominal operating point is set at a safe margin from the surge limit. As a rule, the margin is approximately 15 to 20 percent, based on the pressure. This is sufficient; even operating conditions deviating from the nominal operating point can normally be handled without the risk of compressor surge. Because fouling of the compressor blading lowers the surge limit, it is important that the blading always be clean.

If no special precautions were taken, the compressor would cross through the rotating stall zone during start-up and shutdown. A common precaution is the addition of compressor blow-off valves that are opened up during start-up and shutdown, which increases the axial velocity of the air up to the blow-off slots and prevents the flow separation from the blade profiles. The same effect can be realized by using variable guide vanes. Compared to blowing off air, this method has the advantage that less starting power is required, although it is more expensive and more sensitive to fouling on the vane stem. The blow-off valves are closed successively, while the variable guide vanes are opened wider as the rotational speed increases.

COMBUSTORS

The air leaving the compressor is raised to the preset turbine inlet temperature by the burning of fuel in a combustor. The demands imposed on the combustor can vary in several respects, depending on

- Uniformity of temperature at the combustor exit or turbine inlet
- Pressure losses
- Pressure pulsations
- Cooling-air requirements
- Completeness of combustion
- Emissions of nitrogen oxides
- Fuel flexibility

Because these parameters not only interact with one another but also affect the power output and efficiency of the machine, a careful balance among them must be attained. The processes that take place within the combustor are so complex that they cannot be adequately described in a purely mechanical treatment. In any event, many different designs have been developed for combustors over the years, and it still remains necessary to back up each one with tests and extrapolation of data from existing models.

Combustors fall into three design groups, for which certain basic principles can be stated. The combustion chambers for these key groups are single, multiple, or annular (Fig. 5.5).

Principle of Operation

Generally speaking, most combustors in use today operate on this principle (also see Fig. 5.6): The fuel and a portion of the air coming from the compressor (the primary air) are brought into the combustion zone through the swirl basket and/or fuel nozzle. The swirl stabilizes the flow and the flame, and the fuel continually being supplied is ignited by gases circulating in the zone's center. Because the combustor wall in this zone is exposed to intensive radiation, it generally is provided with cooling on both the inside and the outside with air taken from the compressor. On the inside, this compressor air is usually sent as a film along the wall.

Air entering through secondary air nozzles helps to stabilize the flow and lowers the mixed-gas temperature to the preset turbine inlet temperature. The deviations in temperature must not exceed a given level, since an increase in overheating makes a greater amount of cooling air necessary and/or reduces the life expectancy of the turbine blades. Table 5.1 lists state-of-the-art values for several characteristic parameters.

Combustor Design Details

The ratios of length to diameter in typical single combustors range from 3 to 6. Because the length of the combustion chamber (essentially the length of the flame plus that of the mixing section) is determined by the flow processes, it can be reduced by dividing the supply of fuel among several nozzles that operate in parallel, thus producing a corresponding number of shorter flames.

The same effect is also obtained with multiple or annular combustors. In a multiple combustor, each element can in turn be equipped with several fuel nozzles. The annular combustor, which is most commonly selected for jet drives,

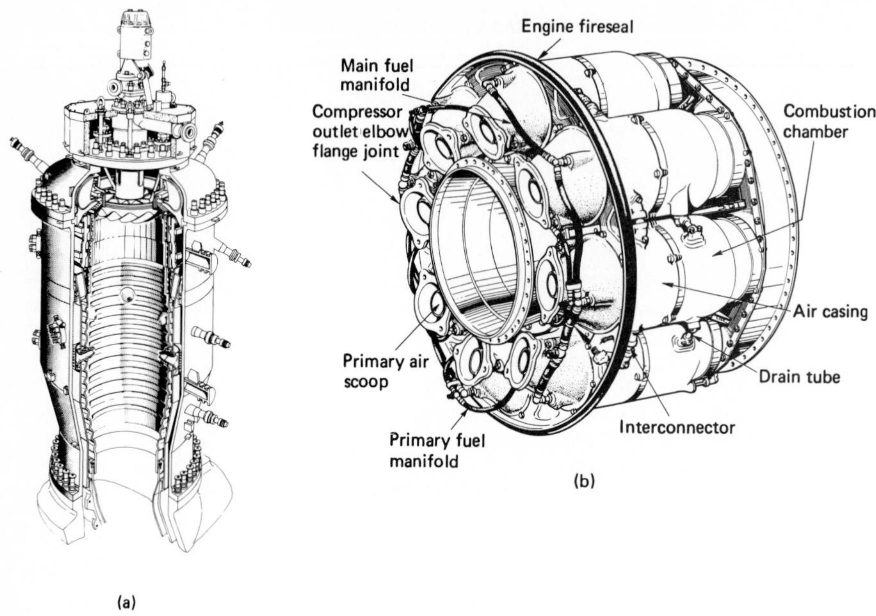

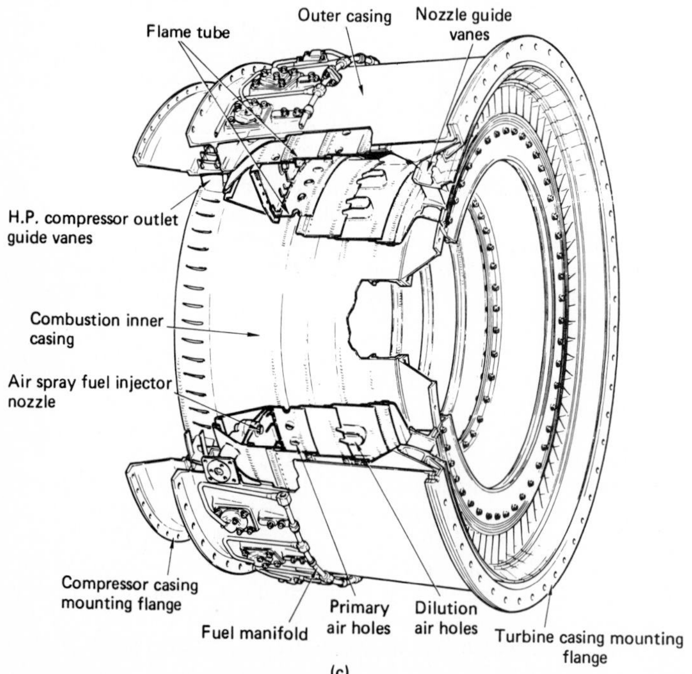

FIG. 5.5 Three types of straight-through combustors: (*a*) single; (*b*) multiple; and (*c*) annular.

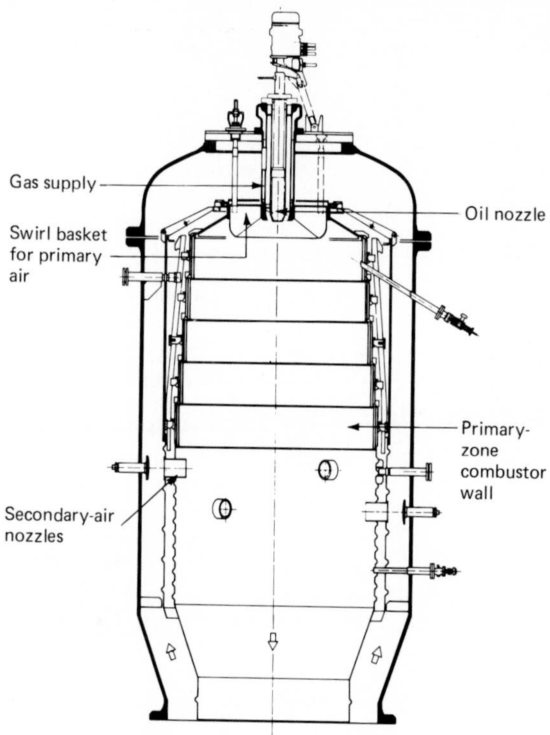

FIG. 5.6 Combustor operating principle (see page 2.135).

TABLE 5.1 State-of-the-Art Combustor Values

Parameter	Equation*	Range of values
Thermal loading, B	$= \frac{Q}{pA}$	110,000–330,000 Btu/h • psi • ft^2 (5000–15 000 kW/bar • m^2)
Primary air reference velocity, v	$= \frac{m}{A}$	50–150 ft/s (15–45 m/s)
Residence time, t	—	0.005–0.1 s
Maximum temperature nonuniformity, ϵ	$= \frac{T_{\max} - T_5}{T_5 - T_4}$	0.1–0.2

*Q = heat input, Btu/h (kW)
p = pressure, psi (bar)
A = area of combustor cross section, ft^2 (m^2)
m' = mass flow rate, primary air, lb/s (kg/s)
$T_{\max}$ = peak temperature in front of turbine inlet guide vane, °F (°C)
T_5 = mean temperature in front of turbine inlet guide vane, °F (°C)
T_4 = compressor exit temperature, °F (°C)

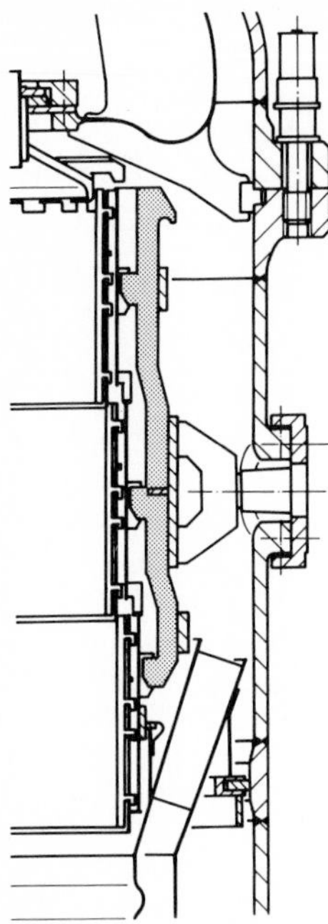

FIG. 5.7 Tile elements hung on a support frame.

offers several advantages because of its compact dimensions. The advantages of a large single combustor with only one nozzle are its greater fuel flexibility and the simplicity of the system for supplying fuel to it. A single combustor is also easier to monitor.

With regard to the fuel nozzles, the injection characteristic must be such that the fuel and the primary air become well mixed. Liquid fuels must be atomized into droplets as small as possible so that they vaporize quickly. Fuel oil is atomized either by hydraulic pressure or by compressed air, the latter method frequently being chosen for heavy oils.

Cooling Provisions. The combustor walls and the connecting housing between combustor and turbine inlet are subjected to great thermal stress and must be cooled adequately, as mentioned. At the same time, expansion must not be impeded. One design for combustor walls that has proved itself over the years in large single combustors uses tilelike metal elements hung on a support frame (Fig. 5.7). Weld material resistant to heat and corrosion is applied on the flame side. The backs of the tiles are finned, with cooling air flowing between the fins. The cooling air leaves the end of each tile to form a cooling film on the inside surface of the next row. Ceramic coatings can also be sprayed on the inside of the tiles as an insulating barrier and a protection against corrosion. Another alternative, ceramic tiles (Fig. 5.8), makes possible a significant reduction in the amount of cooling air required.

Smaller combustors can be made up of several cylindrical tube segments, staggered in diameter and overlapped, so that the cooling air can enter the combustor at the overlap points. In other designs, the cylindrical wall of the combustor is perforated with several holes to provide correct dosing of the cooling air. In principle, annular combustors are cooled in exactly this way.

Transition Housing and Ignition. The connection between the annular combustor and the turbine inlet is the simplest conceivable, but for the other two types a relatively complicated transition housing is required, which presents a considerable manufacturing challenge. Despite the relatively small gas-pressure forces, in operation the high temperatures can cause the structure to bulge. For this reason, special attention must be paid to providing a bulge-resistant shape. A well-designed housing can also counter the thermal fatigue and cracking resulting from high temperatures.

Ignition is generally provided by high-voltage ignition sparks, often with a gas torch that makes it possible to avoid overheating of the spark plug during normal operation. Another alternative is retraction of the spark plug, but that requires appropriate mechanical linkages. In multiple combustors, only two chambers positioned across from each other are equipped and lighted with spark plugs; in the others, the flame propagates instantaneously through ignition tubes that provide a cross-connection. These cross-tubes must be sufficiently cooled so that they do not burn during normal operation.

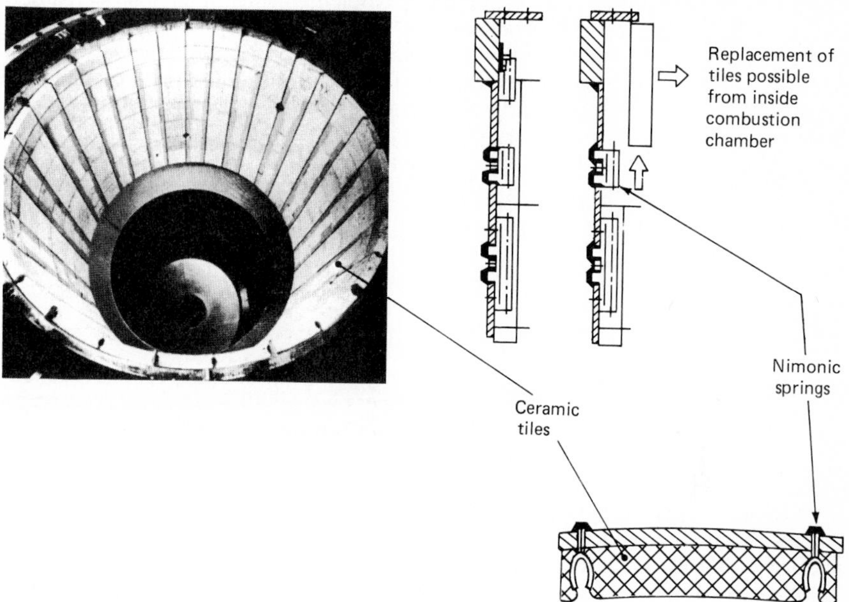

FIG. 5.8 Ceramic tiles reduce amount of cooling air required.

NO_x Reduction

The tightening of regulations concerning NO_x emissions is making intensive developmental work in this field necessary. Nitrogen oxides form in the hottest portion of the flame during the combustion process. The critical factor is the adiabatic flame temperature in the vicinity of the fuel droplet and/or in the mixing zone between air and gaseous fuel. At the start of the reaction, the conditions that prevail there are stoichiometric to a large extent. Inert gases, such as water vapor brought into the flame either as water or steam, lower the adiabatic flame temperature and thus reduce the formation of NO_x. This method represents the state of the art. When it is evaluated, both the costs involved (equipment plus water) and the changes in efficiency must be taken into account. Because gas turbines otherwise function without requiring water, some objections have been raised to this method of NO_x reduction, and "dry" processes have been developed as a substitute. These dry approaches are possible:

- Diffusion combustion, with short reaction zones and an early intensive admixture of quenching air with as much excess air as possible
- Two-stage combustion, with the first stage fuel-rich and the second fuel-lean
- Mixing of the fuel with combustion air, with as much excess air as possible

The last has the greatest potential for NO_x reduction. It can be implemented by retrofitting existing single or multiple combustors. The area of the combustion chamber is broken down into several combustion elements, in each of which occurs a partial premixing of the fuel with the combustion air. One reason for splitting up the air is that flame stability is guaranteed in a premixed flame only within narrow limits for the amount of excess air. To remain within the required range at every possible load, the burners are switched on or off in groups. Starting from approximately 1500 service hours per year, this NO_x reduction method is more economical than the one using water injection.

A further reduction in NO_x would require the introduction of ammonia for catalytic reduction of the NO.

TURBINES

The hot gases coming from the combustor expand in the turbine. A large portion of the mechanical work produced in this process is used to drive the air compressor—roughly two-thirds in the simple-cycle gas turbine, the type that prevails today. The remaining one-third of mechanical work is available for useful power output.

Turbines differ from compressors in having a relatively small number of stages, about three to five in the case of today's single-shaft machines. Another feature is that the elevated gas temperatures require cooling of the first few turbine stages to attain an economically justifiable life expectancy for the turbine blading. The rotor itself and the stator must also be protected from the high temperatures of the hot gases by so-called cooled heat shields.

Turbine Operation

In the *h-s* diagram of the turbine process (Fig. 5.9), the jags in the expansion curve reflect the mixing of the cooling gas into the mainstream of combustion gases, which occurs at several distinct points. When they leave the final stage, as a rule the gases still contain a great deal of kinetic energy, which is recovered by partial conversion to pressure in a diffuser. While the flow passes through the sound dampers and stacks following the diffuser, pressure losses are incurred, and the kinetic energy at the stack outlet is lost.

Although the main criteria determining the flow dynamics and mechanical design of the first stages are the turbine inlet temperature and pressure, the main criterion for the final stage is the mass flow. The designer is compelled to work with high axial velocities. As a result, the quality of the diffuser and a good interaction between turbine and diffuser take on special significance.

For the sake of cooling-air economy, the tendency is generally toward selection of turbines with few stages. For the same reason, very careful attention must be paid to the design of the cooling system itself, which consists of channels within the blading and the heat shields. The main point is that no cooling air is wasted and no hot gas enters the cooling system.

If the cooling airflows are not taken into consideration, the computing tools available for designing the turbine blading are basically the same as those for development of the compressor. The cooling-air exit from blade and vane rows and its mixing with the mainstream complicate the calculations. Of course, allowance must also be made for air leakage from the rotor and stator. The basic procedure to follow was outlined earlier.

Once again, as with compressors, extensive tests on models are required for

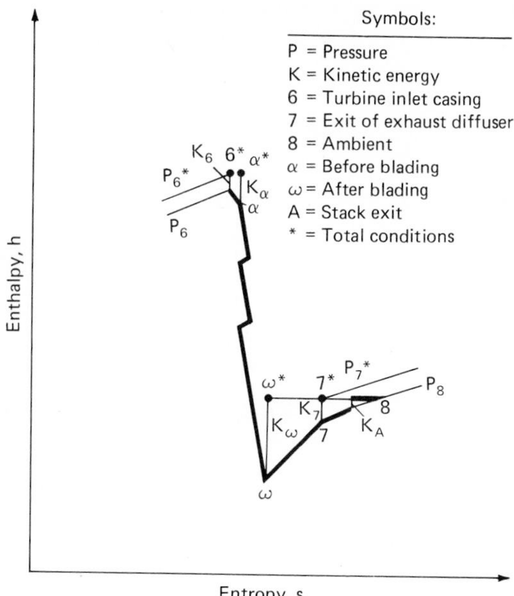

FIG. 5.9 A *h-s* diagram for the turbine process.

turbines whenever completely new blading is being developed. For process calculations, it is customary to use one-dimensional characteristics. The mass flows of cooling air and their admixture are included in the considerations.

Blade Cooling

The steady increase in the turbine inlet temperature over the years has been made possible by improved high-temperature materials and by advances in cooling techniques. Because modern cooled blades are quite complicated in form, improvements in manufacturing processes—in casting technology, in particular—have also played an important role.

Types of Cooling. Two broad types of blade cooling techniques—*convection* and *film*—are found in turbines today. With convection cooling, the heat is transmitted from the blade foil to the cooling air by means of convective heat transfer, the effect of which can be enhanced by including longitudinal and transverse ribs and other turbulence generators. Another possibility is to allow the cooling air to impinge on the inside surface of the blade. For this, an insert is required to guide and distribute the cooling air; the insert has small holes through which the cooling air flows toward the surface. Known as *impingement cooling,* this technique is quite effective, especially on the nose of the blade profile.

Film cooling occurs whenever the outside surface of the blade is covered with a protective film of cooling air. To accomplish this, air is taken from the inside through holes and slits and is directed along the blade's surface. In real machines, film cooling and impingement cooling occur not by themselves, but only in combination with convective cooling.

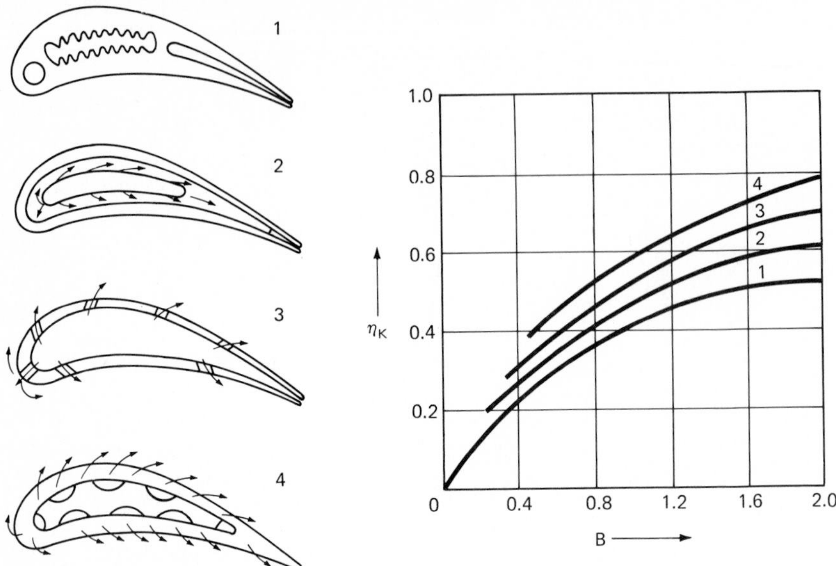

FIG. 5.10 Cooling efficiency η_K of (1) convection cooling, (2) impingement cooling, (3) film cooling, and (4) transpiration cooling.

Transpiration cooling is an extreme case of film cooling. Here, the entire surface of the blade consists of porous material, and the cooling air flows out through it everywhere. It is difficult to achieve this principle in practice, since the pressure in the stream flowing past on the outside, against which the cooling air must discharge, varies along the blade. Moreover, the channels are very fine and thus sensitive to fouling. For these reasons, this method of cooling, which in itself is quite effective, has not been used in industrial gas turbines.

Comparing Cooling Methods. Figure 5.10 makes it possible to compare the effectiveness of the cooling obtained from the various methods described above. The cooling effectiveness has been plotted as

$$B = \frac{\dot{m}_K C_p}{\bar{\alpha}_G A_G}$$

where $\dot{m}_K$ = mass flow, cooling air
C_p = specific heat, cooling air
$\bar{\alpha}_G$ = average coefficient of heat transfer
A_G = area of blade surface

Here $m_K C_p$ and $\alpha_G A_G$ can be interpreted as indicators for the cooling effort and the difficulty of the cooling problem, respectively.

On the whole, the stress condition in a cooled blade is better than that in a solid blade without cooling. For a cooled blade in steady-state operation, compression stresses corresponding to the temperature distribution on the outer surface are superimposed, so the effects of the alternating fatigue stresses occurring during operation are tempered. Thermal stresses are relatively small during start-up as well because (1) the wall thicknesses are less than those of solid blades and (2) the heating takes place simultaneously from outside and inside.

It is important that the mechanical stresses and vibration for the turbine blading be carefully calculated. High-performance computer codes are available for this purpose. In many cases, the only problem is ensuring that the input data for blade clamping conditions in the rotor accurately reflect real conditions. Because the results obtained from the computation entail uncertainties, they are checked experimentally in several testing steps (on models, prototypes) unless the deviations from proven design solutions are minor.

Materials

Nickel-base superalloys such as IN738 and IN739 are frequently specified today for the first stages. Their relatively high chrome content makes them fairly resistant to corrosion. Their strength and ductility can be further improved by a high-temperature heat treatment called *hot isostatic pressing* (HIP).

Further steps in the development of alloys include *directionally solidified* (DS) materials, *single-crystal* (SC) superalloys, and *oxide-dispersion-strengthened* (ODS) superalloys. DS-material blades are currently being produced for larger industrial gas turbines, but the SC and ODS superalloy technologies have not yet reached that stage.

Another possible way for allowing an increase in turbine inlet temperatures is to spray the blades with thermal-barrier coatings. The problems encountered include thermal stresses and chemical attack originating from fuel contaminants. Protective coatings prevent corrosion attack and oxidation. They are produced by galvanic methods or by plasma spraying.

SHAFT AND BEARINGS

The shaft supports the turbine and compressor blades and transmits their output differential to drive the generator. The turbine portion of the rotor is protected from the hot gases by air withdrawn from the compressor. Thermal stresses during start-up are caused by the rotor being heated by the compressor air and not (or only to a lesser degree) by the hot gases. Besides the practice of directing compressor air under the heat shields, there are also some machines in which the rotor is bathed in the cooling airflow. Uneven thermal expansion, particularly during quick starts, can cause problems with unbalances. Thus, the uniformity of cooling around the entire rotor circumference is important.

With the exception of smaller gas turbines whose rotors can be manufactured from a single piece, industrial gas-turbine rotors are made of several disks either welded together or held together by gear teeth and one or more bolts (Fig. 5.11). The welded construction, which is also specified in modern jet drives, produces a monolithic structure (Fig. 5.12). Thus, no relative movement between the individual elements is possible, preventing the vibrations that originate there. Moreover, because the disks do not have a central bore, they are free of the stress concentrations produced thereby.

Industrial gas turbines are frequently single-shaft machines with rotors supported in two bearings only. The thrust bearing is, as a rule, on the compressor

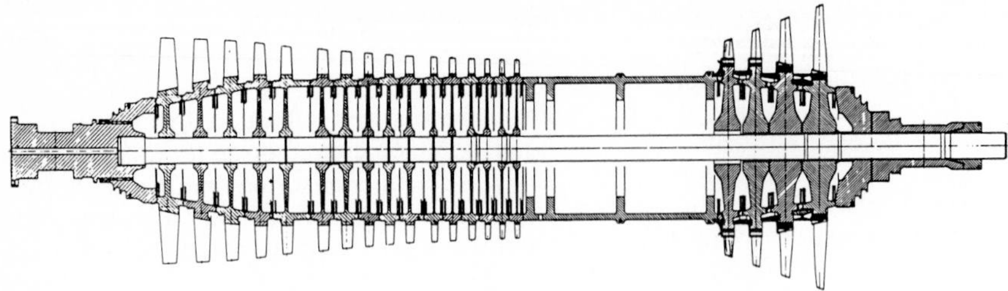

FIG. 5.11 Mechanically joined rotor. *(Kraftwerk Union AG.)*

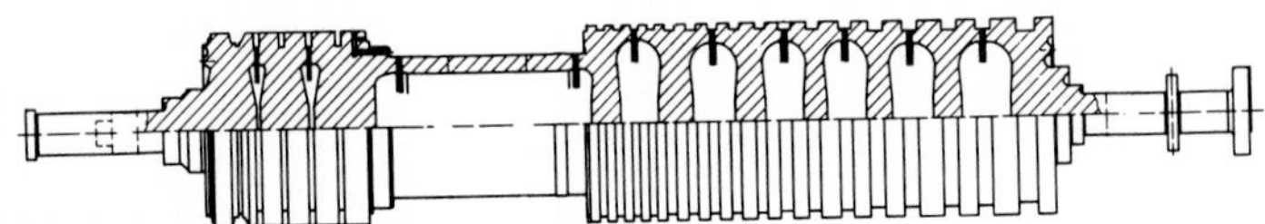

FIG. 5.12 Welded rotor construction.

end. In two-shaft machines with a separate power turbine, one bearing is placed between the compressor and the compressor turbine. The two turbines are therefore overhung, which makes an aerodynamically proper interface between them possible.

Generally, the journal bearings selected are fully enclosed friction devices. Sometimes tilting pad segmental bearings find application in special cases where the stability limit is approached. However, thrust bearings are practically always tilting pad designs; no major problems exist in supplying oil to the bearings.

AUXILIARIES AND ACCESSORIES

The auxiliaries and accessories for a gas turbine include filters, sound dampers, lube oil system, fuel systems for oil and gas, and open- and closed-loop control systems. Even a concise description of these systems would be quite lengthy. For that reason, the remarks that follow are confined to a few major components and to the control principle employed in modern gas turbines.

Lube Oil Pumps

Generally, the pumps selected for the lube oil system are positive-displacement designs—for example, gear pumps—and are driven directly by the gas-turbine shaft across gears. This makes it possible for the gas turbine to run down even when there is a power failure without the oil supply to the bearings being fully interrupted.

Power failures have caused fairly severe damage when motor-driven pumps were installed in the lube oil system. Sufficient redundancy must therefore be provided to keep the risk of damage within bounds. The use of elevated oil tanks cannot be recommended because of the fire hazard and other problems in fire protection.

Fuel Pumps

Either volumetric or centrifugal pumps can be used for liquid fuels. Which one to select depends on the properties of the fuel, with viscosity being one of the most important criteria. For higher viscosities, volumetric pumps are generally a better choice.

In machines that have several combustors, the dosing of fuel is not left to the fuel nozzles, but is ensured by using so-called flow dividers. These are constructed from an appropriate number of gear pump elements rigidly coupled in parallel so that an equal amount of fuel flows to each nozzle—even, for example, if one nozzle has eroded.

When the fuel being burned is gas, sometimes it may not be supplied at high enough pressure. Sliding-vane rotary compressors or Roots blowers will correct this problem.

Temperature Control

The gas-turbine process is controlled by adjusting the fuel flow and, with that, the turbine inlet temperature. In a modern gas turbine, however, that temperature is so high that it cannot be measured directly. It is calculated from the turbine discharge temperature and the compressor discharge pressure, by using a linearized formula. This value is used as a signal both for the temperature controller and for protection against excess temperature.

OPERATION AND MAINTENANCE

Compared with other types of powerplants, the operation of gas turbines is simple and not labor-intensive. Many gas-turbine powerplants, particularly peaking stations, are operated by the load dispatcher—a fully automatic, remote control operation. Providing maintenance is adequate, gas-turbine plants attain high reliability and availability levels. Typical availabilities of base-load machines range from 85 to 95 percent.

Fouling

Severe fouling of the compressor can occur, depending on the site, type of filter, and effort expended on filter maintenance. Fouling often entails a loss in power output and, in extreme cases, can even cause surge. It thus presents a hazard for the machine.

Cleaning of the compressor at reduced rotational speed by injecting water containing cleaning solvents, followed by running the machine until dry, is a useful corrective measure. In many cases, the full power output can be regained in this way. Cleaning the turbine at full speed with abrasive materials such as nutshells is not effective. Wet cleaning at full speed can indeed be effective in the first stages, but as a result of evaporation of the drops during compression, it can cause deposits to be left on the stages further back. All cleaning procedures done at full speed have the disadvantage that the pressure side of the blade profiles is not reached and remains fouled.

When the fuels contain a large amount of ash, fouling of the turbine can be severe, particularly with crude and heavy oils. Because ash almost always contains corrosive components such as alkalines and heavy metals, additives such as magnesium and silicon must be added to the fuel to keep corrosion under control. However, these additives increase the total amount of ash and the buildup of ash coatings on the blades. The coatings can be very hard. The changes in the blade profiles reduce the turbine's efficiency and flow capacity. Cleaning can be carried out during operation by injecting fine powdered quartz or by washing at reduced speed—provided the coatings are water-soluble or at least hygroscopic.

Aging

The power output and efficiency of a gas turbine change during its service life and even during a prolonged period at standstill due not only to fouling but also to aging. The following changes should be watched for:

- Increasing roughness of the blading caused by erosion
- Increasing roughness in the turbine from oxidation
- Changes in shape of the vane carriers, sealing cases, housings, etc., resulting in increased aerodynamic losses in the blading and changes in the flows of cooling and sealing air
- Scraping of the vanes, blades, and labyrinth seals
- Rusting on rotors, housings, and piping

After 1000 service hours, these aging factors can lead to a drop of approximately 0.5 percent in power output and efficiency.

Key Precautions

Because the hot-gas parts in particular are relatively expensive and their replacement can be time-consuming, it is important that inspections and preventive repair or replacement be made regularly, if more expensive interruptions in service later are to be avoided.

The maintenance intervals vary depending on the mode of operation. The instructions from the manufacturer with regard to the timing of the inspections and the activities to be completed should be followed closely. Special attention should be paid to signs of corrosion. Moreover, regular checking of the open- and closed-loop control and monitoring systems is also very important.

Most important of all is assurance that the pertinent regulations limiting the emissions of noxious materials produced in the combustion process are met—and that noise levels are not exceeded. Heat removal, however, is not a problem.

Regulations for SO_2 emissions levels can be met only by selecting a sufficiently low-sulfur fuel. Restriction of the formation of NO_x has been discussed; unburned hydrocarbons and CO can be held at low levels, at least in operation at full load.

ACKNOWLEDGMENT

Illustrations courtesy of Asea Brown Boveri Ltd. unless otherwise noted.

CHAPTER 2.6
HYDRAULIC TURBINES

Carlyle Esser and James H. T. Sun
Harza Engineering Co.
Chicago, IL

INTRODUCTION

Hydraulic turbines convert potential energy to mechanical energy, which can be used as the prime mover for various types of machines and generating equipment.

Note: Nomenclature and notations in this chapter conform to modern standard terminology commonly used in the hydroelectric field, including the parameters introduced recently for computer-aided design and analysis. Notations are defined wherever cited in the chapter.

A generator converts the mechanical energy to electric energy. The amount of energy which can be converted depends on the available head, flow, and efficiency of the hydraulic turbine. Mechanical energy can be calculated by the following equations:

English units:

$$P = \frac{HQw\eta}{550} = \frac{HQ\eta}{8.81}$$

where P = turbine output, hp [the trend is to adopt kilowatts for turbine output (1 hp = 0.7457 kW)]
H = net head, ft
Q = turbine discharge, ft^3/s
w = weight of water (standard conditions) = 62.4 lb/ft^3
η = turbine efficiency

Metric units:

$$\text{kW} = \frac{HQw\eta}{102} = 9.8HQ\eta$$

where kW = turbine output, kW
H = net head, m
Q = turbine discharge, m^3/s
w = weight of water (standard conditions), 1000 kg/m^3
η = turbine efficiency

In the above equations, the following efficiencies at rated output are representative conservative values which may be used for planning and estimating in the absence of information for the selected turbine:

0.93 for Francis and propeller turbines

0.90 for Pelton and tubular turbines

Proportionality laws for homologous hydraulic turbines are shown in Table 6.1.

TABLE 6.1 Proportionality Laws*

Constant head	Constant runner, diameter	Variable runner, diameter and head
$P \sim D^2$	$P \sim H^{3/2}$	$P \sim D^2H^{3/2}$
$n \sim \frac{1}{D}$	$n \sim h^{1/2}$	$n \sim \frac{h^{1/2}}{D}$
$Q \sim D^2$	$Q \sim H^{1/2}$	$Q \sim D^2H^{1/2}$

*P = turbine output; hp (kW)
D = runner discharge diameter, ft (m)
n = turbine rotating speed, rpm
Q = turbine discharge, ft^3/s (m^3/s)
H = net head, ft (m)

TYPES OF HYDRAULIC TURBINES

Modern hydraulic turbines evolved from crude waterwheels to today's sophisticated modern designs which were developed to accommodate a wide range of heads, flows, and variance in operating conditions. The Francis turbine was developed by James B. Francis in the mid-1800s. The Pelton turbine was first produced in the late 1800s and is named for its developer, Lester Pelton. The fixed-blade propeller turbine was introduced in the early 1900s, and later came the adjustable-blade "Kaplan," named after Victor Kaplan, the developer of both versions. The Deriaz or diagonal turbine was developed by Paul Deriaz in the 1900s. The straight-flow-type turbine and rim generator concept was patented by Leroy F. Harza in 1919 (the modern version of this concept is manufactured under the tradename *Straflo*). Tubular-, bulb-, and pit-type turbines were developed in the mid-1900s. The determination of the turbine type is made in consideration of expected unit-load regulation, head and flow variations, powerhouse arrangement, headwater source and power conduit, geological conditions, necessity for a surge chamber or pressure-relief device, plant capability needed, and comparison of construction costs.

The typical range of application for the various turbine types in relation to head, flow, and output is represented in Fig. 6.1. The applications which each type of turbine can accommodate overlap. Consequently, for conditions in this

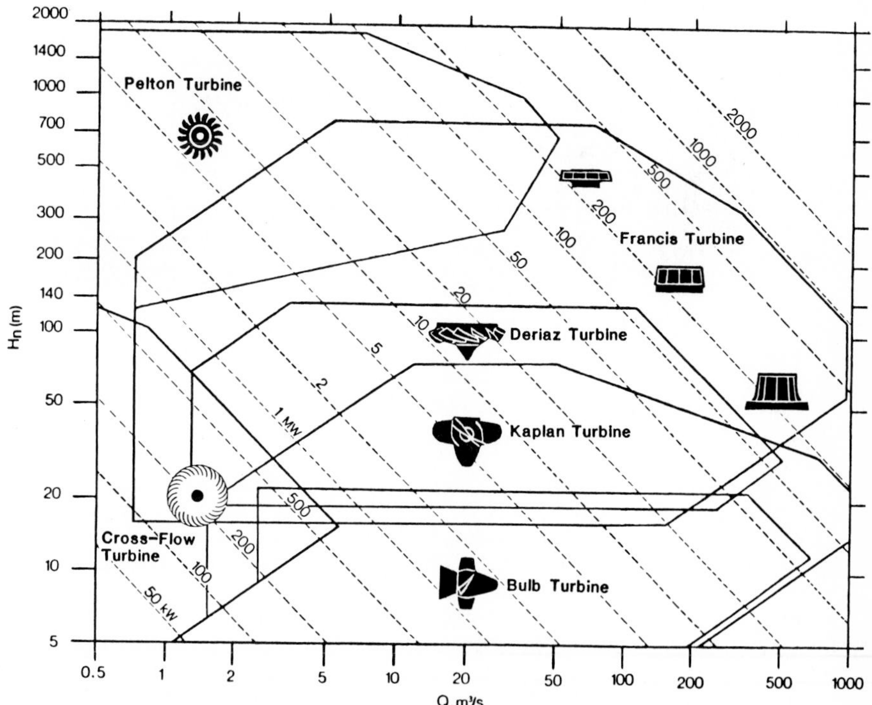

FIG. 6.1 Typical range of applications for the various turbine types in relation to head, flow, and output. Note: m × 3.28 = ft; $m^3/s \times 35.3 = ft^3/s$. *(Sulzer-Escher Wyss, Ltd.)*

FIG. 6.2 Francis turbine runner rated at 200 MW. *(Toshiba Corp.)*

overlap range, more than one type of turbine may be suitable. The actual operating ranges are continually being extended with the advancement in turbine design technology.

Hydraulic turbines are classified into two broad categories as either reaction (pressure type) with continuous water column from headwater to tailwater or impulse (pressureless type). Reaction turbines are driven by the difference in pressure between the pressure side and suction side of the runner blades. Impulse turbines are driven by one or more water jets directed tangentially into the buckets of a disk-shaped runner rotating in air. Hydraulic turbines are also classified by head: low head, below 100 ft (30.5 m); medium head, 100 to 1000 ft (30.5 to 305 m); and high head, 1000 ft (305 m) and higher. The most discriminating classification is by the type of runner and configuration of the turbine water passage. The major types of turbines are described now along with their unique features, advantages, and normal application.

Reaction Turbines

Reaction turbines are provided with a concrete semispiral case, or steel spiral case or water distributor housing, a mechanism for controlling the rate of flow and for distributing the flow equally to the runner inlet, a runner and shaft, and a draft tube for regaining part of the kinetic energy.

Francis Turbine. The Francis type of turbine (mixed flow) was initially developed for medium-head applications; however, the head range has continually been extended with the advancement in technology (Fig. 6.2). The runner has multiple fixed scoop-shaped blades attached at the top to the runner crown and at the bottom to the runner band. The runner geometry and dimensional proportions vary with the runner's specific speed, as shown in Fig. 6.3. The efficiency of a Francis turbine remains relatively good down to approximately 40 percent of rated load (see typical performance curves in Fig. 6.28).

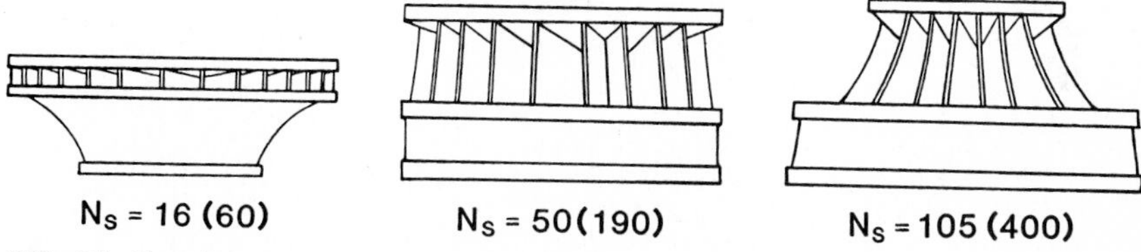

FIG. 6.3 Francis runner geometry and dimensional proportions with respect to changing specific speed.

Propeller Turbines. Propeller-type turbines (axial flow) were developed for relatively low-head applications. The propeller runner theoretically may have from 3 up to 10 radial blades that may be fixed or movable; however, the majority of modern propeller-type turbines have runners with three to six blades, which experience has proved to be practicable and the most favorable design range for current practice (Fig. 6.4).

The fixed-blade propeller turbine has a slightly higher (0.2 to 0.5 percent) peak efficiency than the adjustable-blade Kaplan turbine, because of the reduced blade-tip clearance and smaller runner hub size. However, its efficiency peak is narrower than that of the Kaplan turbine, as can be seen in Fig. 6.28. The Kaplan turbine usually has the mechanism in the hub, which allows the blade tilt to be changed automatically with respect to the wicket gate opening while the unit is operating. A cam in the governor (mechanical or electronic) correlates the blade tilt with the gate position. The adjustable-blade runner optimizes the gate and blade position, which enables operation in its high efficiency zone over a broad range of operating load and head conditions.

Tubular Unit. The turbine runner for the tubular unit is of the propeller type (Fig. 6.5). The tubular or S-type turbine is classified by its water passage configuration from the intake to the draft tube, which is S-shaped and allows the generator to be located outside the water passage. The generator may be oriented vertically or horizontally or may be inclined, and it may be direct-driven or driven at a higher speed by means of a speed-increasing gear drive. The speed increaser allows a higher operating speed and thus a smaller (and usually less expensive) generator. The tubular turbine water passage excavation is shallow and therefore much less than that required for the vertical-type turbine.

FIG. 6.4 Kaplan adjustable-blade runner rated at 26.8 MW. *(Fuji Electric Co., Ltd.)*

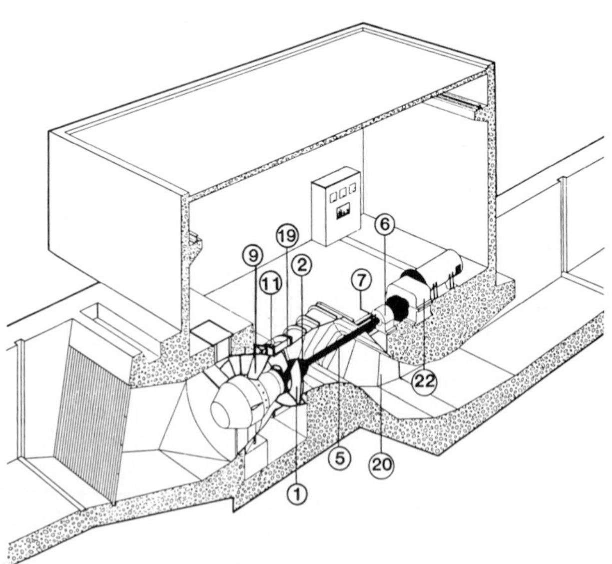

FIG. 6.5 Typical tubular-type turbine arrangement. Note: see pp. 2.159–2.165 for identification of numbered components. *(Voith Hydro, Inc.)*

FIG. 6.6 Bulb-type turbine. *(Hitachi, Ltd.)*

Bulb Unit. The bulb unit also utilizes a propeller-type turbine runner (Fig. 6.6). The bulb-type turbine configuration and arrangement locate the generator inside a watertight housing (or bulb) in the water passage. The powerhouse for a bulb unit is usually integral with the dam with a straight draft tube. The submerged and enclosed configuration makes access to the turbine and generator less inviting—with the exception of the turbine runner. The degree that accessibility to the internal parts is affected by the bulb configuration is related to physical size and whether the bulb is of the pressurized design. Inherent in the bulb unit configuration is that the mechanical inertia of the generator rotating parts is much lower than that of a vertical unit, even with the maximum additional mechanical inertia that can be built into the machine. Consequently, the bulb unit has an inherent disadvantage for isolated operation or where a high degree of unit stability is required for speed regulation. The low inertia of the bulb unit can be compensated to some degree by the use of modern hydraulic turbine governors and a *power system stabilizer* (PSS).

Straflo Turbine. The Straflo turbine is designed so that the rim of the generator rotor is attached directly to the periphery of the blades of a propeller turbine runner (Fig. 6.7). A dependable water seal must be provided at the edge of the rotor rim, to separate the rotor poles and stator from the water passage. The water passage configuration is essentially the same as for a bulb turbine, except for external mounting of the generator. Unlike in the bulb unit generator, the mechanical inertia in the generator of the Straflo unit is equivalent to a conventional type of turbine generator. Consequently, the Straflo can overcome the disadvantages of the bulb unit with respect to stability and regulation.

FIG. 6.7 Straflo turbine. *(Sulzer-Escher Wyss, Ltd.)*

FIG. 6.8 Deriaz pump-turbine runner rated at 41.7 MW. *(Voest-Alpine International Corp.)*

Deriaz Turbine. This turbine (mixed axial flow) is to the Francis turbine what the Kaplan is to the fixed-blade propeller; both the Deriaz and the Kaplan configurations are capable of having either adjustable or fixed blades (Fig. 6.8). The Deriaz adjustable-blade turbine can provide high efficiency over a broad load and head range, as can the Kaplan turbine; however, its peak efficiency may be slightly lower (1 percent) than for a Kaplan or Francis turbine. The Deriaz turbine design has a lower specific speed, lower runaway speed, lower hydraulic thrust, higher plant sigma (cavitation coefficient), and higher operating head range than does a Kaplan turbine. Due to the hub-blade configuration, a Deriaz turbine has a higher efficiency at part-load operation than a Kaplan. Another characteristic of the Deriaz turbine is the reduced discharge at overspeed (or choking of the flow), as is characteristic of a low-specific-speed Francis runner. Deriaz-type turbines can also be designed for reversible pump-turbine service.

Impulse Turbines

Impulse turbines are provided with a water distributor pipe and nozzles (jets), a mechanism for controlling the water jets to the runner, and a runner rotating in air.

Pelton Turbine. The Pelton turbine is designed for high-head application and has one or more nozzles, up to a maximum of six (Fig. 6.9). It can be arranged for operation with a single nozzle or multiple nozzles. The nozzles of a Pelton turbine are directed tangentially into the center of the runner buckets. Horizontal arrangements are typical for small units with one or two nozzles, while vertical arrangements are used for large, modern high-power units. Water is conducted and distributed to the nozzles by a spiral distributor pipe. The nozzles are fitted with adjusted needles to control flow; in addition, each nozzle can be provided with a fast-acting jet deflector to deflect the jet from the runner in the event of large load changes or emergency conditions. The deflector control feature is es-

FIG. 6.9 Pelton turbine runner rated at 34.2 MW. *(Sulzer-Escher Wyss, Ltd.)*

pecially useful on long high-head penstock arrangements, which allows large changes in load without creating pressure changes in the penstock. Water discharges from the Pelton runner directly into the discharge pit and flows to the tailrace. Pelton impulse turbines discharge to the atmosphere and therefore do not regain any kinetic energy from the flow downstream of the runner.

Cross-Flow Turbine. The cross-flow turbine is a variation of the impulse turbine, which is sometimes described as a radial, impulse-type turbine with partial flow admission (Fig. 6.10). In the cross-flow turbine the water flow path makes two passes through the runner. Flow is regulated by guide vanes, allowing the turbine to operate over a wide range of flows. Cross-flow units up to 1000 kW and heads up to 600 ft (183 m) are in service.

Turgo Turbine. The Turgo is an impulse-type turbine similar to the Pelton wheel, except that the jet enters the runner bucket and discharges free from ob-

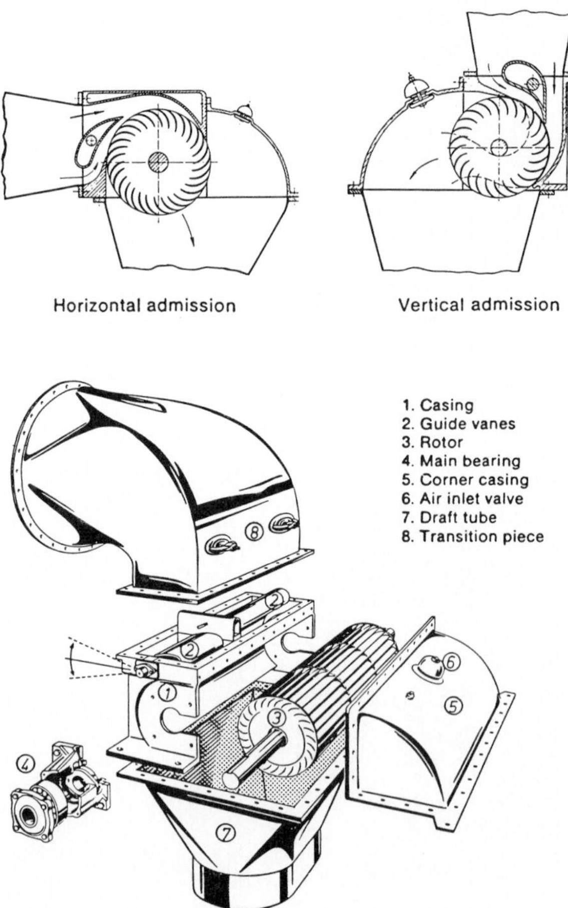

FIG. 6.10 Cross-flow turbine with horizontal and vertical admission arrangement. *(Ossberger Turbines, Inc.)*

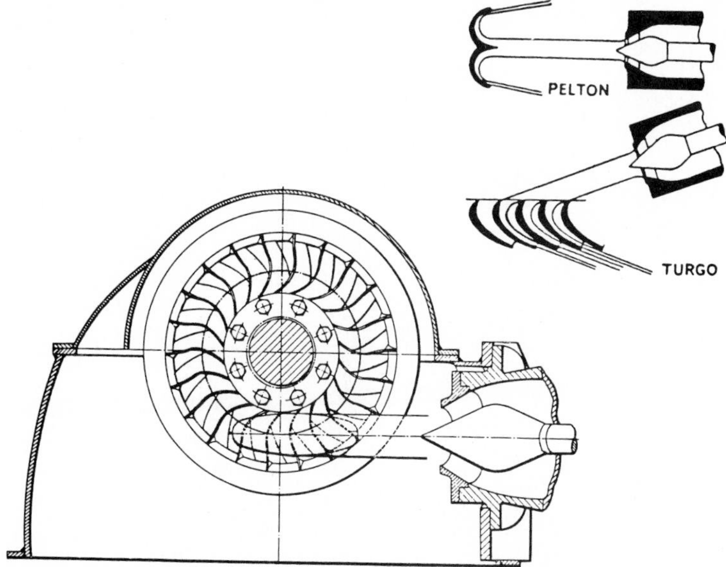

FIG. 6.11 Comparison of Turgo and Pelton impulse turbines. *(Gilbert Gilkes & Gordon, LTD.)*

struction of the buckets (Fig. 6.11). The Turgo impulse turbine has a higher specific speed than the Pelton unit. Turgo turbines with outputs up to 7500 kW and heads up to 770 ft (235 m) are in service.

MAJOR TURBINE COMPONENTS

Component Definitions

The principal components of hydraulic turbines are defined in the following nomenclature. Each definition is preceded by an identification number that also appears on one or more of the accompanying drawings to assist in identifying the component (Figs. 6.12 to 6.18).

Runner Components.

1. *Runner:* The rotating element of the turbine that converts hydraulic energy to mechanical energy. Francis-type runners consist of multiple contoured blades connected on one end by the runner crown and on the other end by the runner band. Pelton runners consist of buckets connected radially to the periphery of a disk, and propeller runners consist of blades attached to a hub.
2. *Runner cone:* The extension of the runner crown or hub that guides the water as it leaves the runner.
3. *Runner seals:* Close running clearances located between the rotating crown and the stationary head cover, and the rotating band and the stationary bottom ring. These close clearances restrict leakage from the high-

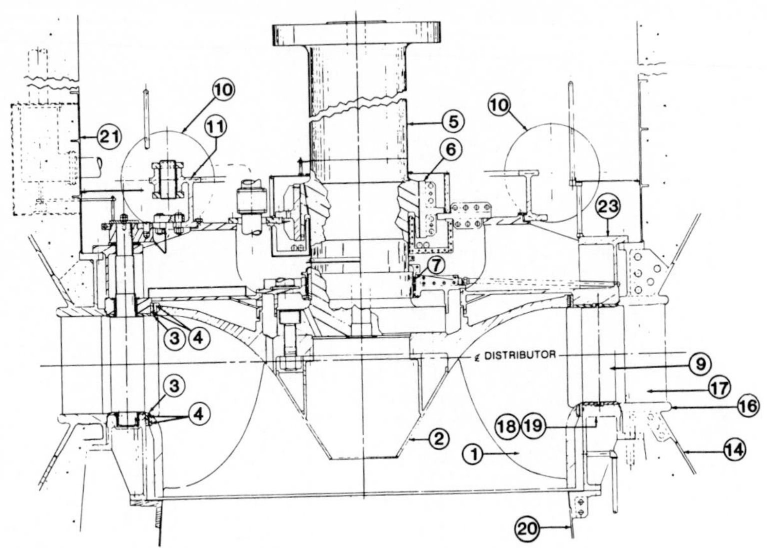

FIG. 6.12 Vertical Francis turbine. *(Voith Hydro, Inc.)*

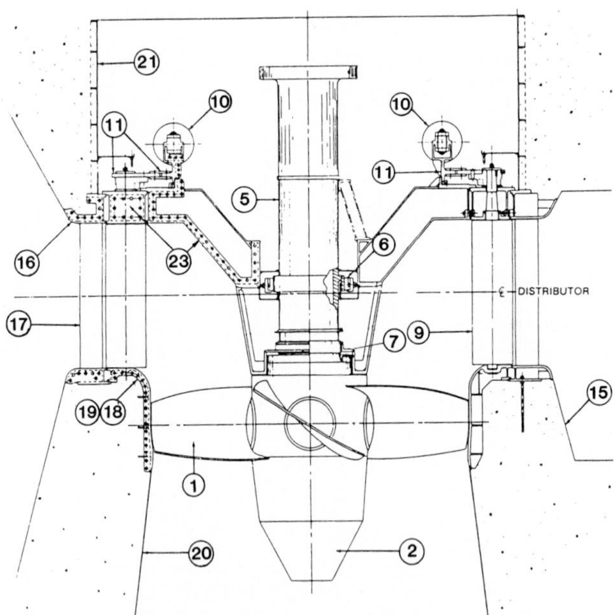

FIG. 6.13 Vertical Kaplan turbine. *(Voith Hydro, Inc.)*

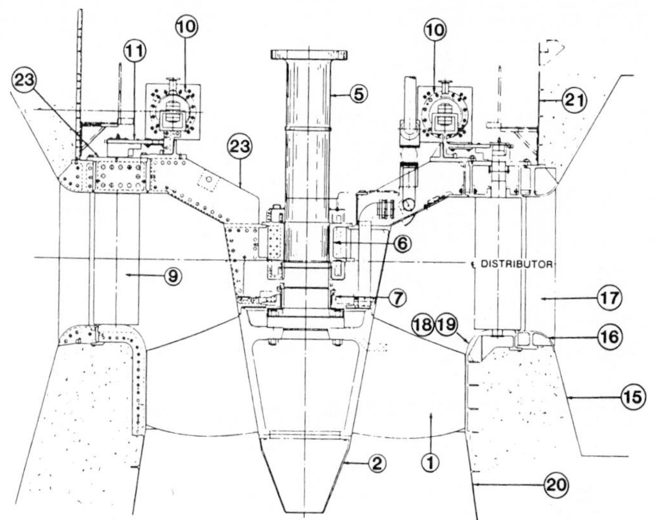

FIG. 6.14 Vertical fixed-blade propeller turbine. *(Voith Hydro, Inc.)*

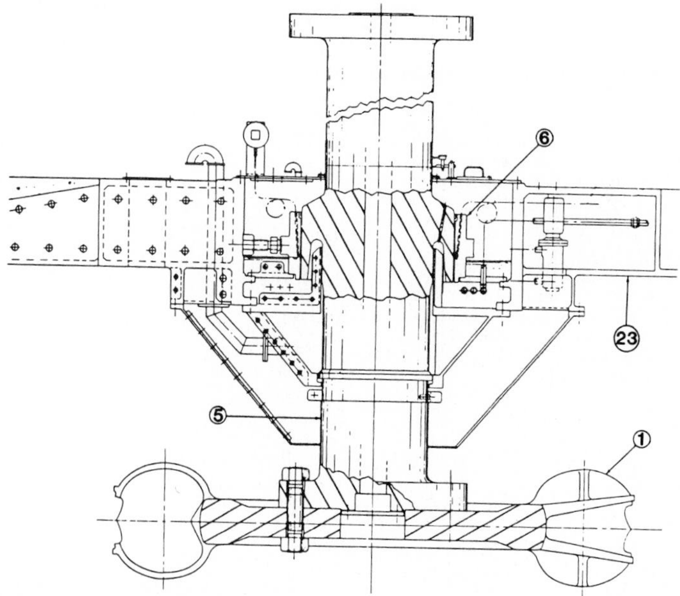

FIG. 6.15 Pelton impulse turbine. *(Voith Hydro, Inc.)*

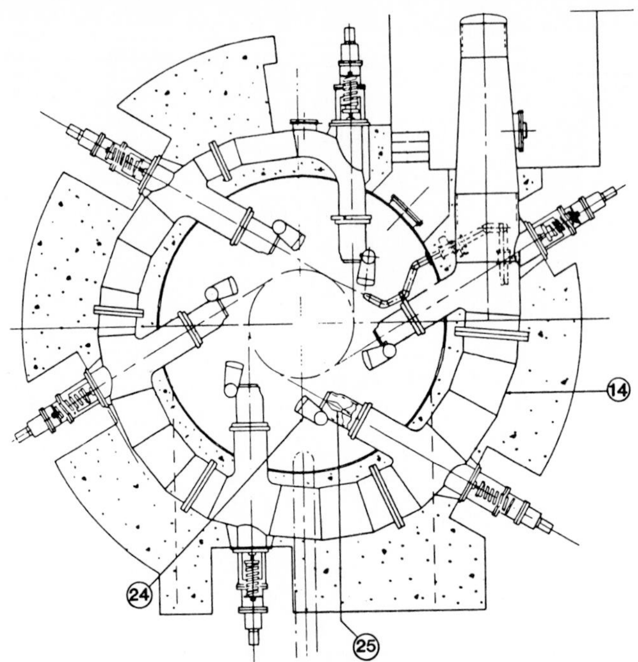

FIG. 6.16 Plan view of Pelton impulse turbine. *(Voith Hydro, Inc.)*

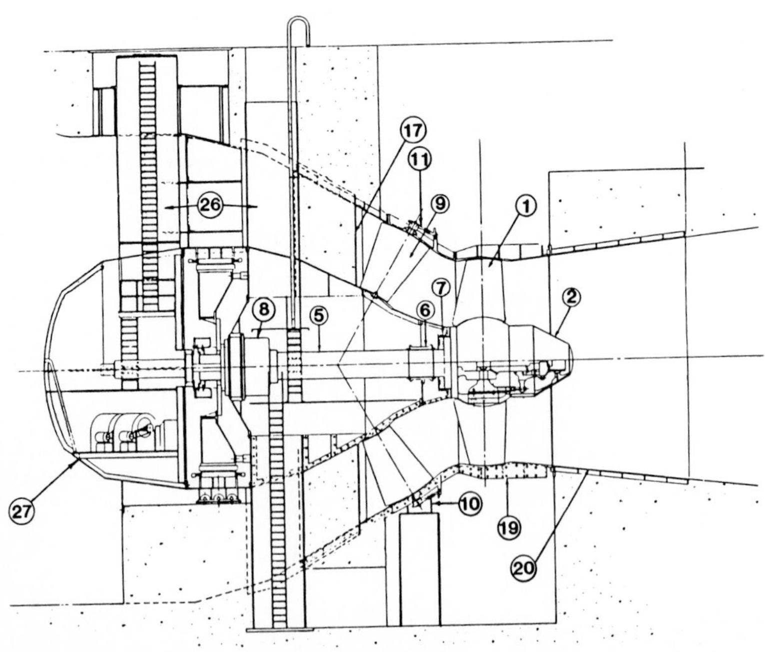

FIG. 6.17 Bulb-type turbine. *(Voith Hydro, Inc.)*

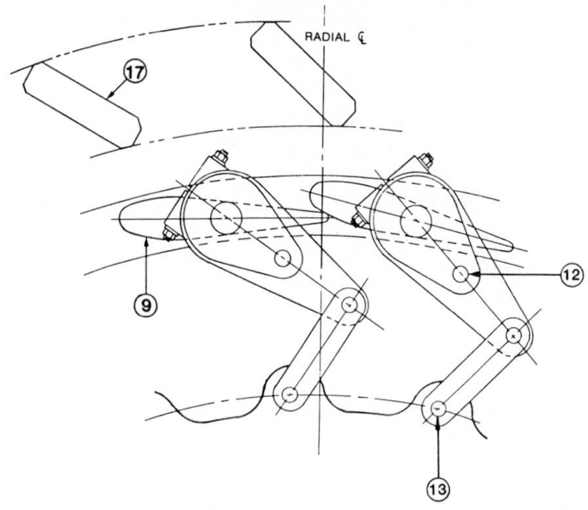

FIG. 6.18 Wicket gate linkage. *(Voith Hydro, Inc.)*

pressure zone to the lower-pressure zones, which affects efficiency and hydraulic thrust.

4. *Runner wearing rings:* Rotating and stationary rings that form the runner seals, which are usually designed for either one or both to be removable and replaceable.

Shafts and Bearings.

5. *Shaft:* Rotating structural element that transfers the torque developed by the runner to the generator rotor or speed increaser. The shaft is typically provided with a shaft sleeve and a bearing journal.
6. *Main guide bearing:* The bearing located closest to the turbine runner that radially supports the rotating runner and shaft assembly.
7. *Shaft seal:* The seal that is located where the shaft penetrates the turbine housing or head cover, to minimize leakage past the shaft. The seal can be a packing box, mechanical ring, or labyrinth (noncontact) type. The type of seal selected is influenced by the head, water quality, and type of service.
8. *Thrust bearing:* The bearing which carries the weight of the turbine and generator rotating parts and the hydraulic down thrust for vertical arrangements. It can be located directly above or beneath the generator rotor or mounted on the turbine head cover. Horizontal arrangements may have two thrust bearings, or a double-acting thrust bearing to carry the axial thrust in both directions.

Gate Components.

9. *Wicket gates:* Angularly adjustable, streamlined elements that direct and control (throttle) water flow to the runner.

10. *Wicket gate servomotors:* Hydraulic cylinders actuated by oil pressure to operate the wicket gates.
11. *Gate mechanism:* The components that actuate the wicket gates consisting of the gate operating ring, linkages, and servomotors (Fig. 6.18).
12. *Shear pin:* Replaceable protective device which is designed to fail by shearing when obstructions prevent the wicket gate from closing. The shear pin connects two linkages in the gate mechanism and transmits the force to move the wicket gate.
13. *Eccentric pin:* An eccentric pin that connects the gate linkage and gate operating ring, designed for adjusting and calibrating the wicket gate positions.

Embedded Components.

14. *Spiral case:* The spiral-shaped water passage which completely surrounds the turbine distributor, providing a uniform distribution of water flow to the runner.
15. *Semispiral case:* A concrete water passageway which opens directly to the upstream of the turbine, with a spiral case around the downstream of the turbine providing a uniform distribution of water flow to the runner. Large turbines are usually provided with piers to reduce the span at the semispiral case inlet.
16. *Stay ring:* An annular member with upper and lower rings connected by a number of fixed guide vanes in the water passage, which guide water from the spiral or semispiral case to the wicket gates. The stay ring is designed to support the weight of the superimposed powerhouse structure and machinery.
17. *Guide vanes:* Streamlined structural members that connect the two annular rings of the stay ring and direct the flow from the spiral or semispiral case to the wicket gates.
18. *Bottom ring:* The stationary ring that contains the lower wicket gate bushings and sometimes is combined with the discharge ring.
19. *Discharge ring:* The structural member that surrounds the runner band and forms the lower portion of the runner chamber on a Francis turbine. The discharge ring of a propeller turbine surrounds the runner blades, forming the water passage in the area and leading to the draft tube. The downstream end of the discharge ring is attached to the top of the upper draft tube liner.
20. *Draft tube:* The turbine water passage (diffuser) that guides the water leaving the runner to the tailrace. The draft tube is designed to regain a portion of the kinetic energy losses. It can be conical, elbow-shaped, square, or tubular. Large turbines are usually provided with piers in the horizontal section of the elbow-type draft tube, to reduce the span.

Other Components.

21. *Turbine pit liner:* The steel liner surrounding the open space on a vertical unit between the turbine head cover and the generator which provides access to the gate mechanism, shaft seals and bearings, and turbine auxiliary systems. The pit liner protects the surrounding concrete and serves as a concrete form during construction.

22. *Speed increaser:* Geared drive unit that increases turbine shaft speed to an optimum generator speed. It contains thrust and guide bearings, which are sometimes used for carrying the load of turbine and generator rotating parts.
23. *Head cover:* Structural member that spans the upstream end of the runner chamber, to enclose the water passage and provide support for the shaft seal and bearings. It carries the upper wicket gate bushings and provides structural support for the gate operating ring. On larger turbines the head cover is sometimes divided into inner and outer annular sections, to permit the removal of the turbine components without disturbing the wicket gates.
24. *Deflector:* The device on Pelton impulse turbines used to deflect part or all of the water jet from the runner buckets to regulate power output without changing penstock flow.
25. *Needle:* The moving element which controls the size of the water jet impinging on the runner buckets of a Pelton or Turgo impulse turbine.
26. *Access shaft:* The watertight passage of a bulb unit providing access between the generator compartment and the powerhouse and between the turbine compartment and the powerhouse.
27. *Bulb:* The streamlined watertight housing for the generator of a bulb unit.

Design Considerations

With respect to hydraulic turbines, design considerations involve both embedded and removable components.

Embedded Components.

1. *Spiral case:* Steel spiral cases are used for heads greater than 100 ft (30.5 m). Concrete semispiral cases are limited to heads below 100 ft (30.5 m), because of strength limitations of concrete to high-pressure loading conditions. The spiral case (Fig. 6.19) is fabricated from formed plate steel in radial sections and welded. The steel case should be designed for the maximum static head plus the pressure rise caused by water hammer, disregarding any support from the concrete embedment. The spiral cases of large-capacity and high-head units are usually embedded in concrete, to provide additional support and minimize vibrations.

   Embedded spiral cases should ideally be hydrostatically tested and then pressurized at the normal maximum static pressure during embedment, to reduce the tensile load transferred to the surrounding concrete. A permanent access mandoor should be provided for entering the interior of the spiral case; the temporary test head should be provided with handholes to ease preparations for pressure testing and embedment.
2. *Stay and discharge rings:* The stay ring for large vertical units must be designed to transmit the superimposed loads from the equipment and powerhouse superstructure through the stay vanes to the substructure. The stay ring is made of steel—integrally cast, fabricated of plate, or cast-welded and fabricated.

   The discharge ring should be made of stainless steel, be stainless-steel-clad, or have a stainless-steel overlay for a distance of not less than 10 percent of the runner diameter, downstream of the runner blades.

FIG. 6.19 Spiral-case shop assembly for 730-MW Francis turbine, Guri Project Powerhouse II in Venezuela. *(Hitachi, Ltd.)*

3. *Draft tube liner:* The draft tube should be steel-lined from the discharge ring to the point where the water velocity reduces to about 20 ft/s (6.1 m/s), which is considered below concrete scouring velocity. Draft tube piers should be provided with steel nose liners. The draft tube liner should be designed for the external hydrostatic pressure from seepage equivalent to the maximum design flood tailwater elevation, with an absolute pressure of 0.5 atm inside the liner.

   The draft tube mean exit velocity under rated conditions should be limited to 6 to 8 ft/s (1.8 to 2.4 m/s), representing less than 1-ft (0.3-m) velocity head loss. Piping connections for turbine aeration and tailwater depression, when required, are provided on the liner immediately below the runner. Smaller pipe connections for water-level controls and pressure and vacuum instruments may also be needed on the liner. Vertical fins are sometimes mounted on the liner wall below the runner to alleviate swirl and pressure pulsations.

Removable Components.

1. *Runner:* The runner must be designed to structurally withstand the maximum hydraulic thrust, forces due to maximum runaway speed, and pressures from transient hydraulic conditions. It must also be able to support its own weight plus the weight of the turbine shaft when it is resting on the runner band (Francis turbine) or on the runner support (propeller turbine). The runner should be balanced statically and, when practicable, balanced dynamically in the manufacturers' shops. Static balancing for large slow-speed conventional runners is considered adequate.

   Runners are made of steel or stainless-steel one-piece castings or are fabricated from weldments of component castings and formed-steel plates.

Large-diameter Francis runners, because of shipping limitations, may be of a split design arranged for flanged and bolted connections, or very large runners such as Grand Coulee Powerhouse III units may be fabricated at the site.

Francis runners are provided with replaceable wearing rings at the runner seals on the periphery of either the crown, the band, or both. The proper design seal clearance between the rotating and stationary parts is essential to the turbine efficiency and hydraulic thrust and is usually on the order of 0.04 to 0.05 percent of the runner nominal diameter. The thrust bearing design should consider the hydraulic thrust load with the runner seals worn to at least twice the design clearance.

The hub of a Kaplan and Deriaz turbine contains the blade servomotor and operating linkages for adjusting the runner blade position. The servomotor is located either above or below the blades, and the hub is maintained full of oil at a positive pressure, to preclude leakage of water into the hub. In very early designs, the blade servomotor was often located inside the turbine shaft instead of the runner hub.

2. *Shaft:* The shaft can be either a one-piece solid forging of vacuum-degassed steel or a fabricated weldment of forged, cast, or rolled plate steel. Very large shafts that exceed foundry capabilities are weld-fabricated. Generally, it is also more economical and practical to weld-fabricate large-diameter shafts. Solid forged shafts should be provided with a concentric inspection hole at least 6 in (150 mm) in diameter through the entire length. The shafts for Kaplan turbines must be designed large enough to accommodate the concentric pipes for supplying pressure oil to the runner blade servomotor.

   The shaft should be designed both to operate safely at any speed up to maximum runaway speed without detrimental vibration or objectionable distortion and to withstand the working stresses due to the maximum-overload operating output. The shaft's first critical speed should be at least 25 percent above the maximum runaway speed, to avoid resonance. The individual turbine shaft runout and combined turbine and generator shafts should be checked for runout in the manufacturers' shops. The combined shafts, when practicable, should be aligned and checked for runout in the shop.

3. *Bearings:* The guide bearing should be designed for radial support of the shaft (vertical arrangement) or to support the weight of the rotating parts plus any unbalanced loads from the runner or generator (horizontal arrangement), without exceeding the allowable operating temperature (usually 60°C) under continuous normal operation. The guide bearing design clearance is normally on the order of 0.006 to 0.012 in (0.15 to 0.30 mm). The most common types of guide bearings used for hydraulic turbines are the sleeve and the multiple-shoe babbitted types, either immersed in an oil sump or lubricated by oil supplied from an external pump. The multiple-shoe type of bearing has the advantage of being adjustable and is more easily replaced.

   Small units sometimes utilize water-lubricated rubber bearings and, in the past, wood (lignum vitae) bearings, but these bearings are limited to applications where the bearing loads and speeds are low. The modern trend is to use self-lubricated antifriction-type bearings for small units.

   The types of thrust bearings commonly used in hydroelectric units are (1) the adjustable, pivoted-shoe type (either Mitchell or Kingsbury); (2) the pressure-equalizing spring; or (3) the hydropneumatic type. Thrust bearings should be designed to carry the weight of the rotating parts plus the hydraulic thrust (vertical arrangement) and the axial thrust in both directions (horizontal arrangement).

4. *Wicket gates:* Wicket gates should be designed to withstand the maximum static head plus water hammer on the upstream side when the gates are in the closed position. Torque on adjacent gates of a malpositioned gate can be up to five times the normal. This condition should be considered in the gate design. They should also resist the edge loading and the torsional and bending loads caused by the operation of the gate linkage.

   Wicket gates can be one-piece steel or stainless-steel castings or fabricated of welded components. The critical seal areas of the gates should be accurately machined and finished to provide a tight seal when the gates are closed.

   The number of wicket gates, stay vanes, and runner vanes should be carefully coordinated so as not to cause objectionable flow-induced vibrations during turbine operation. The wicket gates should be designed with a hydraulic closing tendency from the full-gate-output position to approximately the speed-no-load position. They should also be designed so as to not contact the runner in the event of gate rotation due to shear pin and gate stop failure.

5. *Gate mechanism:* The gate operating mechanism should be designed to withstand the maximum load that could be imposed by the most severe operating conditions. The servomotor(s) should have sufficient capacity to operate the wicket gates under all conditions of head and load and to move the gates through the full opening or closing stroke in the specified time. The servomotors should be provided with a control device to retard the closure rate at the end of the closing stroke, to preclude shock loading in the gate mechanism and to avoid pressure spikes in the water conduits.

   A shear-pin-failure indication device and wicket gate friction-restraining device may be provided to annunciate a broken pin and to prevent wicket gate flailing following shear pin failure. Parts with relative motion and sliding contact are provided with self-lubricated or grease-lubricated bushings.

6. *Head cover:* The portions of the head cover outside of the wicket gate circle should be designed to withstand the pressure due to the maximum static head, including water hammer. The head cover should also be designed to take the loads imposed from the guide bearing, shaft seal, gate mechanism, and other auxiliary components. The head cover is made of steel, either integrally cast or fabricated of plate and cast components welded together. It contains the upper wicket gate stem bearings and seals, the upper wicket gate seals and facing plates, and the upper runner seal stationary wearing ring.

7. *Shaft seal:* The shaft seal should be located and designed to facilitate access for maintenance and replacement of sealing elements without disturbing or dismantling the other turbine components. An inflatable auxiliary maintenance seal is sometimes provided, to enable the seal components to be replaced without dewatering the turbine. Filtered water is normally used for lubrication and for cooling the seal.

AUXILIARIES

This section is conveniently divided into turbine auxiliaries (Fig. 6.20) and powerhouse auxiliaries.

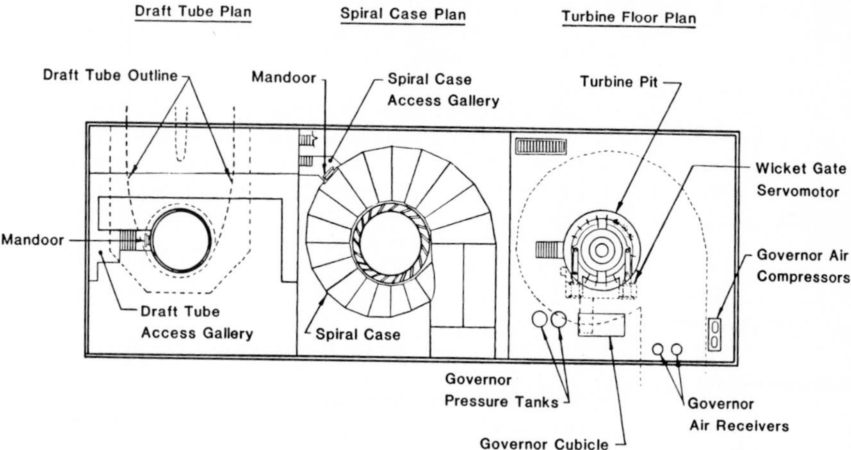

FIG. 6.20 Typical turbine equipment arrangement.

Turbine Auxiliaries

Governing System. While the development of a variable-speed hydroelectric unit continues to hold promise for the future, almost all the hydroelectric units today operate at constant speed regardless of the loading requirements. However, owing to advancement in variable-speed motors and generators, future hydroelectric units may utilize this feature when it is desirable or beneficial. A small pump turbine unit with variable-speed capability is being manufactured in Japan. The speed-controlling mechanism that automatically monitors and maintains the speed of the rotating parts of the hydroelectric unit is the governing system. The governing system is also used to control starting of the turbine, synchronize the generator to the system frequency for connection to the electric grid, and to shut down the unit in an emergency because of mechanical or electrical faults. The mechanical hydraulic governor for the control of hydraulic turbines has proved to be a simple and reliable device for speed and output control. A compensating dashpot is used to stabilize the unit. The dead time of a typical mechanical hydraulic governor is 0.25 s, which is the elapsed time from the initial speed change of the unit to the first movement of the wicket gate servomotor for a sudden change of more than 10 percent rated load.

The electric hydraulic governor has the additional capability to accept and react to the demands of external signals from head-level sensing device, pressure-relief valves, and other devices. The modern three-term PID solid-state electric hydraulic governor provides individual *proportional, integral,* and *derivative* channels for wide-range, continuous-gain adjustments for unit stability. Electric hydraulic governors should be selected for hydroelectric units connected to relatively long penstocks and especially for multiple units served by a common penstock. Although the speed-droop characteristic, which requires a decrease in speed to produce an increase in the wicket gate opening, is commonly used for governor regulation, the speed regulation characteristic has been developed for the electric hydraulic governor. By utilizing a solid-state watt transducer, the generator output is used to develop the speed regulation characteristic. The linear relation between unit speed and generator output is manually adjustable over a range of 0 to 10 percent, as compared to 0 to 5 percent for the speed-droop adjustments. The dead time is less than 0.2 s for a typical electric hydraulic governor, as compared to 0.25 s for a mechanical hydraulic governor.

A more recently developed digital governor is the electric-position operator type. The governor consists of a programmable controller and a hydraulic control unit. The programmable controller is composed of a microprocessor designed for controlling and operating the turbine. The hydraulic control unit converts the electric signal from the programmable controller to a proportional displacement of the wicket gate servomotor. The industry trend is toward digital governors.

Cooling-Water System. Cooling water for the turbine shaft seal, turbine and generator bearings, generator air coolers, and other powerhouse equipment typically is supplied from headwater or is pumped from the tailrace. Gravity-fed systems are sometimes used for plants with heads up to approximately 300 ft (91 m), where the water is supplied directly from headwater, or from the penstock or spiral case, and by reducing the water pressure when necessary. A disadvantage to gravity-type systems can be the added equipment for maintenance and adjustment of the pressure-reducing equipment.

Pumped cooling-water systems are generally more economical for high-head plants, since the energy that could be produced through the turbine from the high head is more than that required for pumping from the tailrace. Pumped systems are more easily matched to the cooling-water flow and pressure requirements of the equipment than gravity-fed systems, and backup systems can easily be provided. Pumping schemes for multiple-unit plants can be an individual pumping system for each unit or per plant; or per pairs of units or selected number of sets of units. Either scheme should have standby backup pumping capability, to avoid the generating unit being shut down for loss of pumping capability.

Where a gravity or combination pumped and gravity system is selected, the water supply can be taken from the spiral case or penstock. Connection points for the cooling-water piping on the spiral case should be approximately 45° to 60° from the horizontal. Top and bottom connections should be avoided, to reduce the possibility of entrainment of air or debris into the system. In addition to the gravity system, when necessary, a supply system can be provided from the headwater reservoir, independent of the units, which can be used for unit filling.

Consideration should be given in the design of pumped systems to avoid operation of pumps at run-out conditions. Often these systems are designed to operate as a closed loop without the benefit of static head. In such cases, artificial head can be introduced by venting the high point of the discharge line or installing an orifice plate.

Compressed Air System. A low-pressure (100-psi, or 689-kPa) compressed air system is usually provided for the generator brakes, utility hose connections, and miscellaneous station services. A higher-pressure system (up to a maximum of approximately 1000 psi, or 6890 kPa) supplies air to the governor pressure tanks. Even higher-pressure systems are provided for large valves such as inlet valves, 1500 psi (10 335 kPa) and higher. When the unit is to be used for synchronous condenser operation, a compressed air system is provided to depress the water level in the draft tube below the runner. Draft tube depression systems are used in pumped storage units to reduce the starting load when the unit is started as a pump. Another compressed air system sometimes provided is a turbine aeration system.

Common speeds are 4 to 5 ft/min (1.2 to 1.5 m/min) for hoisting and 10 to 15 ft/min (3.0 to 4.6 m/min) for traveling. For modern large hydroelectric units, overhead cranes having a capacity of 1100 tons (1000 Mg) have been built.

Unit Unwatering and Filling System. The unwatering system consists of the piping, valves, and pump(s) necessary for unwatering the turbine water passages between the intake gate or turbine inlet valve and the draft tube gate at the outlet of the draft tube. When the unwatering operation is complete, the normally submerged turbine parts, spiral case, and draft tube may be inspected and repaired. A closed system in which the unwatering pumps are directly connected to the unwatering drain lines is preferred because it reduces the risk of accidental flooding of the powerhouse.

Unwatering pumps are sized to permit unwatering within the desired time, usually 1 to 4 h. A separately valved drain line is provided in the unwatering manifold, to permit water which leaks past the intake and draft tube gates to drain to the station sump. This line is routed directly to the station sump for removal by the relatively smaller station sump pumps which operate under automatic control.

A filling system, which usually is connected to tailwater, is also required for filling the draft tube to tailwater level after inspection of the turbine has been completed. Then the draft tube gate may be removed under balanced head conditions. A draft tube air vent is required to allow removal of the air being displaced by the water filling the draft tube. This vent also allows air to enter the draft tube during unwatering operations.

TURBINE SELECTION

The various types and designs of hydraulic turbines can be compared by their specific speeds. The *specific speed* is the speed, in revolutions per minute, at which a turbine of homologous design would rotate with the runner reduced to the size required to develop one unit of power under one unit of head. The specific speed can be calculated as follows:

English units (ft • hp)

$$N_s = \frac{N\,(\mathrm{HP})^{0.5}}{H^{1.25}}$$

Metric units (m • kW)

$$N_s = \frac{N\,(\mathrm{kW})^{0.5}}{H^{1.25}}$$

where N_s = speed, rpm
H = net head, ft (m)
HP = turbine output, hp
kW = turbine output, kW

Economic Considerations

Economic considerations influence and often are the deciding factor in the selection of the turbine specific speed as well as the number and type of units. For a given output and head, the equipment's physical size is reduced with an increase in specific speed. The smaller the turbine and generator, the smaller the size and cost of the powerhouse. There is a continuous effort to increase the specific speed as rapidly as improvements in design and materials will permit. However, high-specific-speed units are more susceptible to problems with vibration and rough operation, cavitation pitting, and other less serious problems. Increased specific speeds require a deeper setting of the unit, to maintain the cavitation coefficient (plant sigma) within a reasonable margin. The additional excavation required for deeper submergence adds to the powerhouse construction cost.

The optimum specific speed for a given type of turbine has been refined and proved over the years by design, research, and unit operating experience. However, because of the uniqueness of each hydroelectric site (geographic conditions, power grid, etc.), the selection of the optimum unit speed, type, and size is influenced by the civil works cost, anticipated power demands and load requirements, and available head and flow conditions.

It might be more economical to build a powerhouse with one large unit, but the economies of operation could dictate that a powerhouse with three small units would better utilize the available flow. Even though economic considerations play an important part in the selection process, the expected operating needs and requirements of the plant, characteristics of the available resource (head and flow), current technical limitations in manufacturing and construction, and, above all, experience need to be considered.

Preliminary Specific Speed

The selection of preliminary specific speed has been historically based on experience and uses the following equation: $N_s = K/(H)^{0.5}$, where the K factor is based on the K factors of previous units in satisfactory operating condition. Current values (1987) used by the authors' company for the preliminary selection of specific speed are shown in Table 6.2.

The preliminary specific speed can be substituted into the specific-speed for-

TABLE 6.2 *K* Factors for Primary Types of Hydraulic Turbines

Turbine type	*K* factor	
	ft • hp	m • kW
Kaplan and bulb	1250	2625
Propeller and tubular	1150	2415
Francis	1050	2205
Pelton (per jet)	200	420

mula to determine a trial rotating speed. This procedure is described in a following section.

Turbine Runner Setting

A simplified description of the cavitation phenomenon is that it occurs when the absolute-pressure head of the water in a closed conduit drops to the vapor pressure of the water. At this point the water vaporizes, and vapor cavities appear in the water passage and within the flowing stream. The formation of the vapor-filled cavities in the stream is called *cavitation*. When these vapor cavities reach a point downstream where the absolute pressure again exceeds the vapor pressure, under the higher pressure the vapor cavity is suddenly condensed and returns to a liquid, simultaneously producing a collapse of the cavity and an implosion. These implosions or cavity collapses are accompanied by localized water hammer action with extremely high pressure. Continuous repetition of these minute implosions acting on the surface of the water passage causes a fatiguelike destruction of the surface and eventual failure of the material. This destruction, caused by cavitation, is *cavitation pitting*. Cavitation pitting occurs in hydraulic prime movers such as turbines and pumps as well as in low-pressure areas of water conduits and passageways, such as at valves, sluice gates, spillways, etc.

The principal flow conditions associated with cavitation are high-velocity flow, low pressures, and abrupt changes in the direction of flow. Theoretically, if the pressure of the fluid can be controlled, then the occurrence of cavitation can also be regulated. Selection of the turbine runner setting should be such that the critical pressure is not reached, which should minimize cavitation damage to the turbine components. This parameter, known as the *Thoma cavitation coefficient*, can be expressed as follows:

$$\sigma_{\mathrm{p}} = \frac{H_a - H_v - H_s}{H}$$

where σ_{p} = cavitation coefficient (plant sigma)
H_a = atmospheric pressure, psi (mm Hg) (see Table 6.3)
H_v = vapor pressure of water, ft (m) (see Table 6.3)
H_s = distance from runner throat or blade centerline to minimum tailwater (see Fig. 6.23), ft (m)
H = maximum net operating head, ft (m)

The turbine sigma is the cavitation coefficient determined by model tests and is the critical value below which the turbine performance parameters of efficiency, power, and discharge cease to remain substantially constant.

The plant sigma should be greater than the turbine sigma to avoid cavitation. If the turbine is set too high in relation to the minimum tailwater level, causing the plant sigma to be smaller than the turbine sigma, then cavitation may be excessive. Figure 6.24 shows typical experience curves used for recommending the minimum cavitation coefficient (plant sigma) as a function of the turbine specific speed.

The following selections and remedies should be considered when one is dealing with or trying to avoid damaging cavitation:

TABLE 6.3 Atmospheric Pressure and Vapor Pressure of Water

Atmospheric pressure					
Altitude, ft	H_a, psi	H_a, ft H_2O	Altitude, m	H_a, mm Hg	H_a, m H_2O
0	14.696	33.959	0	760.00	10.351
1,000	14.17	32.75	500	715.99	9.751
2,000	13.66	31.57	1000	674.07	9.180
3,000	13.17	30.43	1500	634.16	8.637
4,000	12.69	29.33	2000	596.18	8.120
5,000	12.23	28.25	2500	560.07	7.628
6,000	11.78	27.21	3000	525.75	7.160
7,000	11.34	26.20	3500	493.15	6.716
8,000	10.91	25.22	4000	462.21	6.295
9,000	10.50	24.27			
10,000	10.10	23.35			

Vapor pressure of water			
Temperature, °F	H_v, ft	Temperature, °C	H_v, m
40	0.28	5	0.089
50	0.41	10	0.125
60	0.59	15	0.174
70	0.84	20	0.239
80	1.17	25	0.324

Source:
U.S. Bureau of Reclamation (USBR).

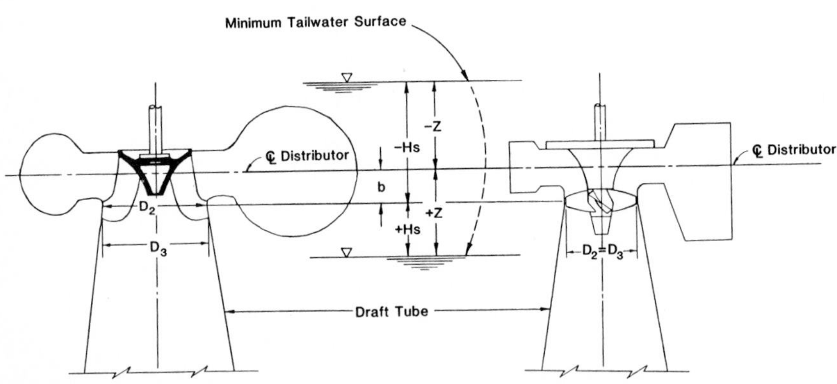

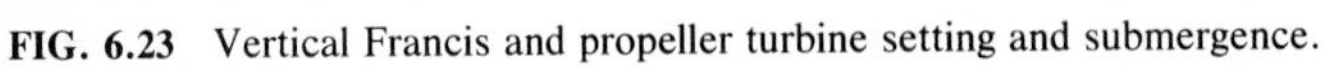

FIG. 6.23 Vertical Francis and propeller turbine setting and submergence.

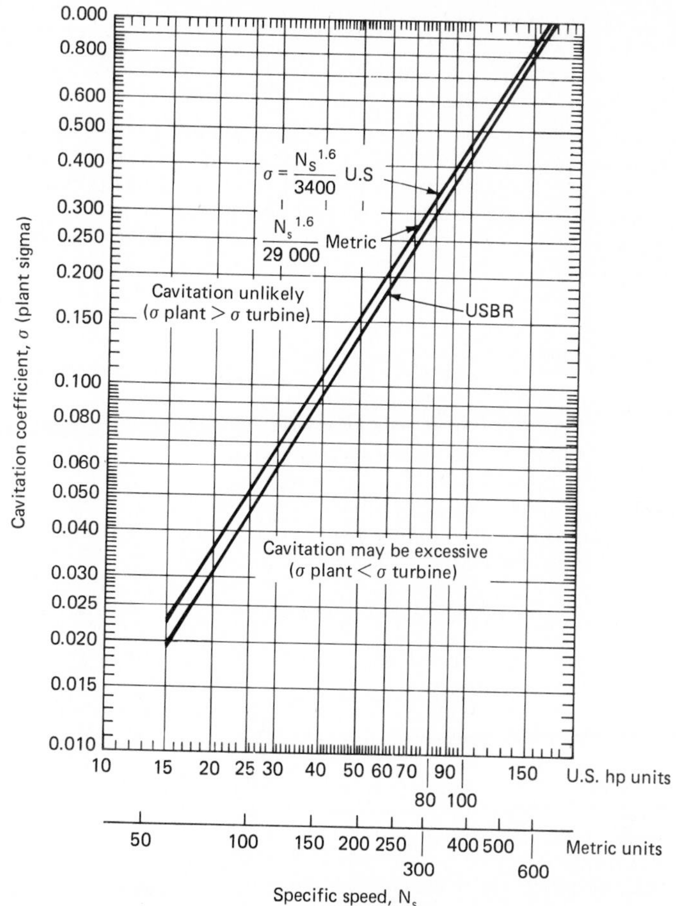

FIG. 6.24 USBR experience curve with modified version used by authors' firm for planning.

- Set the turbine conservatively with respect to tailwater, to provide sufficient backpressure below the runner to maintain high absolute water conduit pressure.
- Ensure that the water flow path is smooth and continuous and without abrupt changes in direction.
- Select suitable materials for turbine components susceptible to cavitation. Stainless-steel overlay should be provided on the vulnerable areas on the low-pressure side of carbon-steel runner blades, or preferably the runner should be made of all stainless steel of high cavitation resistance. Discharge rings should be provided with protective stainless steel.
- There are cases where flow splitters have been mounted on the inlet of the tur-

bine runner blades to correct the flow streamlines where a flow separation exists, causing cavitation damage.

- Compressed air introduced to the runner chamber has been used under certain conditions to reduce the noise and damage from cavitation pitting.
- Impulse turbines do not experience the same magnitude of cavitation problems as reaction turbines. However, cavitation pitting can occur on the runner buckets due to improper velocity triangles, jet disturbances, blade profile problems, or local vortices caused by surface erosion. The most common cavitation problem with the Pelton runner is on the back side of the runner buckets, caused primarily by jet interference.

Selection Procedure

The data required for the selection of a hydraulic turbine include the available head, flow, tailwater rating curve, site elevation and water temperature, expected load demands, regulation and operation requirements, and water quality. After preliminary studies have shown the project to be feasible, project planning studies are needed to optimize the project ratings. The settings in relation to tailwater elevations, design head, operating ranges, number of units, rotational speed, and output ratings are established.

Operating Data. The operating data required for selection include

- Headwater elevations (headwater duration curve determined from planning studies)
- Tailwater elevations (tailwater rating curve)
- Gross heads
- Net heads on the turbine, including
  Maximum (top of spillway gate, one unit operating)
  Average weighted or design (expected value determined from reservoir operation studies)
  Minimum (for dependable capacity)
  Extreme minimum (infrequent drawdown); the net head is the difference between the headwater and the tailwater elevations less the hydraulic losses in the penstock and intake
- Design head (net head at which maximum hydraulic efficiency of turbine is reached)
- Rated head (the net head at which turbine full-gate guaranteed power output corresponds)
- Rated output (the turbine full-gate guaranteed output at rated head)
- General project arrangement (location of powerhouse with respect to headwater and approximate penstock size, length, and profile)

Range of Heads. The operating head range is the first guide to the selection of the type of turbine. The approximate range of heads for which each type of turbine is best suited is shown in Fig. 6.1 and is as follows:

Turbine type	Approximate range of heads	
	ft	m
Impulse	500–5500	150–1700
Francis	50–850	15–260
Propeller	10–250	3–75

The head ranges shown are not rigid and often overlap. The head ranges are exceeded when special conditions exist. In the overlapping head ranges, the advantages, disadvantages, and physical size of each type of turbine, particularly their part-load efficiencies, should be considered.

Making the Selection.

- Select rated net head and output.
- Select type of turbine.
- Calculate trial specific speed with K factors from Table 6.2:

$$N_s = \frac{K}{H^{0.5}}$$

- Calculate the trial rotational speed:

$$N = \frac{N_s H^{1.25}}{P^{0.5}}$$

where N = rotational speed, rpm
N_s = specific speed, ft • hp (m • kW)
H = rated net head, ft (m)
P = rated turbine output, hp (kW)

- Adjust the unit rotational speed to the nearest standard synchronous speed. Select the unit speed nearest to the synchronous speed:

$$N = \frac{120f}{\text{no. poles}}$$

where f = line frequency (Hz).

$$N = \frac{7200}{\text{no. poles}} \qquad \text{for 60 Hz}$$

$$N = \frac{6000}{\text{no. poles}} \qquad \text{for 50 Hz}$$

The number of poles should be divisible by 2, but preferably by 4.

- Calculate the specific speed for selected synchronous speeds:

$$N_s = \frac{NP^{0.5}}{H^{1.25}}$$

where N_s = turbine specific speed, ft • hp (m • kW).

- Calculate the runner discharge diameter D of Francis, propeller, or pitch diameter of impulse (Pelton) runner:

$$D = \frac{153\ \phi H^{0.5}}{N} \qquad \text{English}$$

$$= \frac{84.47\ \phi\ H^{0.5}}{N} \qquad \text{metric}$$

where D = discharge diameter, ft (m)
ϕ = velocity ratio at runner discharge diameter
H = rated net head, ft (m)
N = rotational speed, rpm

The velocity ratio at the runner discharge diameter is calculated from the following equations:

Francis:

$$\phi = \begin{cases} 0.057N_s^{2/3} & \text{English} \\ 0.0234N_s^{2/3} & \text{metric} \end{cases}$$

Propeller type:

$$\phi = \begin{cases} 0.063N_s^{2/3} & \text{English} \\ 0.0258N_s^{2/3} & \text{metric} \end{cases}$$

Impulse (Pelton):

$$\phi = 0.46 \qquad \text{English and metric}$$

- Calculate the outside diameter of a Pelton runner:

$$D_0 = \begin{cases} (1.028 + 0.0522n_{sj})D & \text{English} \\ (1.028 + 0.0137n_{sj})D & \text{metric} \end{cases}$$

where D_0 = runner outside diameter, ft (m), and

$$n_{sj} = \left(\frac{P}{i}\right)^{0.5} \frac{N}{H^{1.25}}$$

where n_{sj} = specific speed per jet
P = turbine output, hp (kW)
i = number of jets
N = rotational speed, rpm
H = rated net head, ft (m)

- Calculate σ, the Thoma cavitation coefficient for Francis and propeller-type turbines:

$$\sigma = \frac{N_s^{1.6}}{3400} \qquad \text{English}$$

$$= \frac{N_s^{1.6}}{29\,000} \qquad \text{metric}$$

- Compute the setting and submergence:

$$H_s = H_a - H_v - \sigma H$$

where H_s = distance from D_2 to minimum tailwater level, ft (m)
H_a = atmospheric pressure for minimum operating tailwater elevation, psi (mm Hg) (see Table 6.3)
H_v = vapor pressure of water [use maximum expected water temperature, °F(°C)], ft (m) (see Table 6.3)
H = maximum operating head, ft (m)
σ = Thoma cavitation coefficient
b = distance from D_2 to centerline of distributor, ft (m) (see Fig. 6.25)
D_2 = minimum diameter through turbine runner band, ft (m)
D_3 = runner discharge diameter, ft (m)
Z = distance from centerline of distributor to minimum tailwater = $H_s + b$, ft (m) (see Fig. 6.23)

Note: For horizontal orientations (both Francis and propeller), H_s is measured from the minimum tailwater elevation to the top of the runner outlet diameter, since the minimum pressure will be at the upper side of the water passage.

Turbine Main Dimensions. The turbine main water passage dimensions can be approximately scaled up by the ratio of the runner discharge diameter of a recent ex-

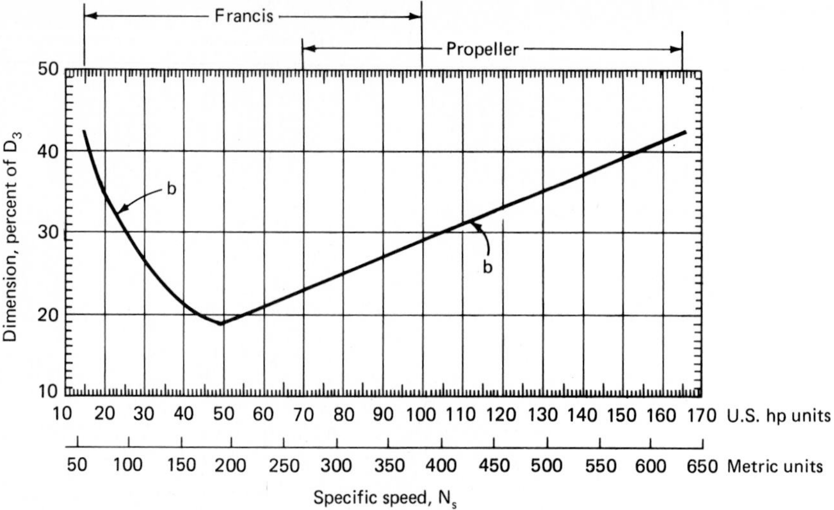

FIG. 6.25 Distance from D_2 to distributor centerline. *(USBR)*

isting prototype turbine having a similar specific speed or can be determined from the various sources of published experience data.

In selecting a turbine, in addition to the speed and setting, many other parameters need to be analyzed. Since it is beyond the scope of this section to treat the following items in depth, they are mentioned to alert the reader that they need to be considered. Further information regarding these subjects can be found in the references listed in Suggestions for Further Reading. Unit stability for speed regulation and transient behavior (water-hammer pressure rise and pressure drop) are critical parameters that determine the effectiveness and safety of the unit operation. If the unit is located far from the headwater, and the penstock is long, then the need for a surge chamber or a pressure-relief valve has to be analyzed. The draft tube pressure pulsation frequencies and the generator natural frequency should be adequately separated to ensure that the generator and draft tube do not operate in resonance.

Additionally, the number of runner vanes, wicket gates, and stay vanes needs to be coordinated to avoid resonance. Large-capacity units require special consideration during the selection and design process to preclude fatigue cracking, objectionable rough operation, and damage to the turbine components and powerhouse structure due to the extremely high amount of energy dissipated through losses during unit operation. The magnitude of this energy and the accompanying loading conditions need to be considered and compensated for.

TURBINE PERFORMANCE

Model Tests and Prototype Performance

Testing of homologous scale models under controlled conditions in a hydraulic laboratory historically has been accepted as the most scientific and accurate method of predicting the characteristics of hydraulic turbine design. Through model testing, the designer can predict the basic turbine performance characteristics with high accuracy, namely, the unit output, discharge, and efficiency. In addition, cavitation characteristics, runaway speed, hydraulic thrust, wicket gate torque, draft tube pressure pulsations, and flow path characteristics can also be predicted with reasonable accuracy. Until the recent development of the sophistication in computer-aided design of turbine runners, considerable time and effort were required in the design of a new model runner (1 month to prepare drawings and 10 months to manufacture and test a new model runner). Today, aided by the use of computers to develop new runner designs, the designers can concentrate on refining the hydraulic and mechanical design details and complete the design in a much shorter time.

Recent sophistication with the computer enables the designer to model the turbine flows by using three-dimensional flow programs for analyzing flow velocities and pressures; two- and three-dimensional boundary layer analyses are used to predict losses in the flow field. Coupled with the flow analysis, the designer can now use three-dimensional finite-element stress analysis, fracture mechanics, and fatigue analysis programs to design the mechanical structure of the turbine. These analyses are linked through interactive graphics design programs which allow the designer flexibility in creating a three-dimensional runner geometry.

Computer design allows all drawings and analyses to be generated automatically and ensures that the analysis, model, and prototype manufacturing are performed by using identical geometry. The model and prototype runners are manufactured by using the computer to drive multiple-axis milling machines, ensuring uniform and accurate blade profiles, shapes, and surface finish.

Model testing is essential; it is needed to confirm the computer-generated designs and to refine the analysis program's modeling capabilities. Transient behavior and the prediction of unsteady flows are still done by model testing, because the sophisticated programs to analyze these phenomena analytically have not been fully developed. Model testing must be closely controlled and performed under strict compliance with recognized and accepted codes. Where performance guarantees are to be verified by model tests, it is customary and recommended that the tests be witnessed by a third-party qualified expert.

Model Hill Chart and Prototype Performance Curves

The model hill chart is a convenient and useful way to define the model performance characteristics. The hill chart (similar to a topographical map) is made up of a family of performance curves. It is especially useful in the turbine selection and application process and in predicting turbine prototype performance. A representative turbine model hill chart is depicted in Fig. 6.26 with constant wicket gate openings and constant efficiency parameters plotted with respect to varying unit speed and unit discharge.

Model test data are customarily represented in unit values of speed, output, and discharge corresponding to a homologous turbine runner having a unit diameter (1 ft or 1 m) and operating under a unit net head (1 ft or 1 m).

The unit values from the hill chart can be stepped up for an entire family of homologous prototype turbines of various sizes and heads from the same model hill chart. Different types of model hill charts can be produced. Some designers prefer to use unit speed and discharge as the variable parameters, others select unit speed and output, while still others favor the unit peripheral speed coeffi-

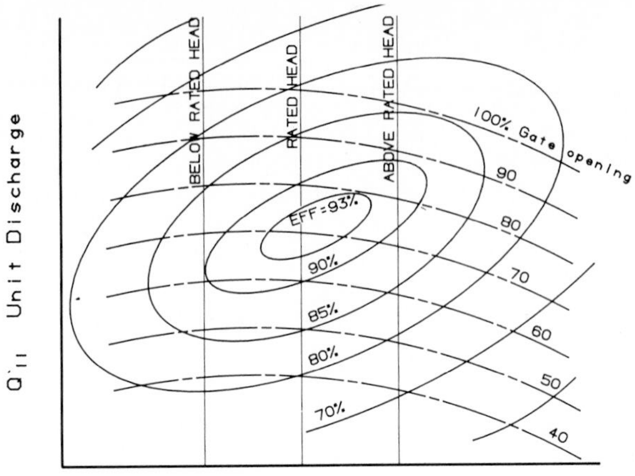

FIG. 6.26 Typical model hill chart.

cient ϕ and unit output. Various constant values of net head, specific speed, output, turbine sigma, etc., can also be identified on the model hill charts, which help ensure that the optimum operating ranges and conditions can be selected for the prototype turbine. However, there is a trend with modern analytical procedure to adopt the use of dimensionless parameters and coefficients, which greatly enhance the capability to analyze, evaluate, and predict performance (Fig. 6.27).

The prototype performance characteristics are stepped up from the model test results. The prototype turbine power is calculated from the model power by using the proportionality laws and correcting for efficiency. The prototype efficiency for reaction turbines is determined from the corresponding model efficiency, corrected by an experience step-up formula. The efficiency for impulse turbines is not significantly affected by scale effects, and customarily the model efficiency is not stepped up for the prototype.

There are two well-known step-up formulas, the Moody and Hutton formulas, which are listed below. In using these formulas, the peak or maximum model efficiency is stepped up to become the maximum prototype efficiency, and the remaining prototype efficiency points are determined by adding the same increment $\Delta\eta$ as for the point of maximum efficiency. The Moody formula tends to allow a larger efficiency step-up than the Hutton formula for various combinations of parameters, such as (1) if the prototype-to-model scale ratio is more than 10 with the head ratio less than 15 or (2) if the scale ratio is more than 8 with the head ratio less than 8. The authors' firm, through experience, has favored using two-thirds of the Moody step-up value, especially in stepping up propeller turbines and large, low- to medium-head Francis turbines. The two-thirds Moody formula is more conservative than the Hutton formula.

- Model to prototype step-up formulas:

Power step-up:

$$P_p = \left(\frac{D_p}{D_m}\right)^2 \left(\frac{H_p}{H_m}\right)^{3/2} \left(\frac{\eta_p}{\eta_m}\right) P_m$$

PARAMETER	DESIGNATION	DEFINITION	SYMBOL	FUNDAMENTAL QUANTITY	UNIT
Unit Speed	ω_{ed}	$\frac{\omega D}{(gH)^{0.5}}$	D	Reference diameter	ft (m)
Unit Discharge	Q_{ed}	$\frac{Q}{D^2 (gH)^{0.5}}$	g	Acceleration due to gravity	ft/s^2 (m/s^2)
Unit Torque	T_{ed}	$\frac{T}{\rho D^3 gH}$	h_b	Barometric pressure head	ft (m)
Unit Power	P_{ed}	$\frac{P}{\rho D^2 (gH)^{1.5}}$	h_s	Geodetic suction head	ft (m)
Energy Coefficient	$E_{\omega d}$	$\frac{gH}{(\omega D)^2}$	h_{va}	Vapor pressure head	ft (m)
Discharge Coefficient	$Q_{\omega d}$	$\frac{Q}{\omega D^3}$	H	Head difference between inlet and outlet sections	ft (m)
Torque Coefficient	$T_{\omega d}$	$\frac{T}{\rho \omega^2 D^5}$	P	Shaft power ($T \bullet \omega$)	Hp (W)
Power Coefficient	$P_{\omega d}$	$\frac{P}{\rho \omega^3 D^5}$	Q	Flow or discharge	ft^3/s (m^3/s)
Specific Speed	ω_s	$\frac{\omega Q^{0.5}}{(gH)^{0.75}}$	T	Torque	lb_f-ft (Nm)
Cavitation Coefficient	σ	$\frac{h_b - h_{va} - h_s}{H}$	ρ	Mass density	lb_m/ft^3 (kg/m^3)
			ω	Angular velocity	rad/s

FIG. 6.27 Dimensionless parameters and definitions.

Discharge step-up:

$$Q_P = \left(\frac{H_p}{H_m}\right)^{1/2} \left(\frac{D_p}{D_m}\right)^2 Q_m$$

- Moody formula:

$$\Delta\eta = (1 - \eta_m)\left[1 - \left(\frac{D_m}{D_p}\right)^{1/5}\right]$$

- Hutton formula:

$$\Delta\eta = (1 - \eta_m)1 - \left\{0.3 + 0.7\left[\left(\frac{D_m}{D_p}\right)\left(\frac{\nu_p}{\nu_m}\right)\left(\frac{H_m}{H_p}\right)^{1/2}\right]^{1/5}\right\}$$

where
p = prototype values
m = model values
P = turbine output, hp (kW)
Q = turbine discharge, ft^3/s (m^3/s)
η = turbine efficiency, %
D = runner discharge diameter, ft (m)
H = rated net head, ft (m)
ν = kinematic viscosity of the fluid, ft^2/s (m^2/s)

From comparison of the equations, the Hutton formula introduces two parameters not included in the Moody formula, namely, the fluid viscosity and head.

Verification of the guaranteed prototype performance values, based on the model test results, can be done with performance tests on the prototype turbine. Prototype testing to determine absolute values of the turbine efficiency requires the accurate measurement of large quantities of flow, which is very difficult and expensive to perform to get a flow accuracy of ± 2 percent.

Unless there is some indication that the prototype performance is suspect, the plant owners will usually opt for supplementing the model test with index and capacity tests, where the turbine relative efficiencies and the shape of the power-versus-efficiency curve is determined. Index and capacity tests are relatively simple to perform and inexpensive compared to absolute efficiency tests.

Performance of Different Turbine Types

Each turbine type is designed to operate within a specific range of heads and flows. Discussion of prototype turbine performance here is limited to the Francis, propeller (fixed-blade), Kaplan, and Pelton types of turbines (Fig. 6.28).

Specific speed is one method of categorizing the various turbine types in relation to their performance characteristics. Pelton turbines, with very low specific speeds due to high heads, have high part-load and overload efficiencies and suffer only a small impairment of efficiency over a wide range of nozzle openings and outputs. They are characterized by a flat-topped efficiency curve. Low-specific-speed Francis turbines have similar characteristics. These two turbine types are suitable for taking load variations, for medium- to high-head plants.

High-specific-speed Francis turbines and fixed-blade propeller turbines have

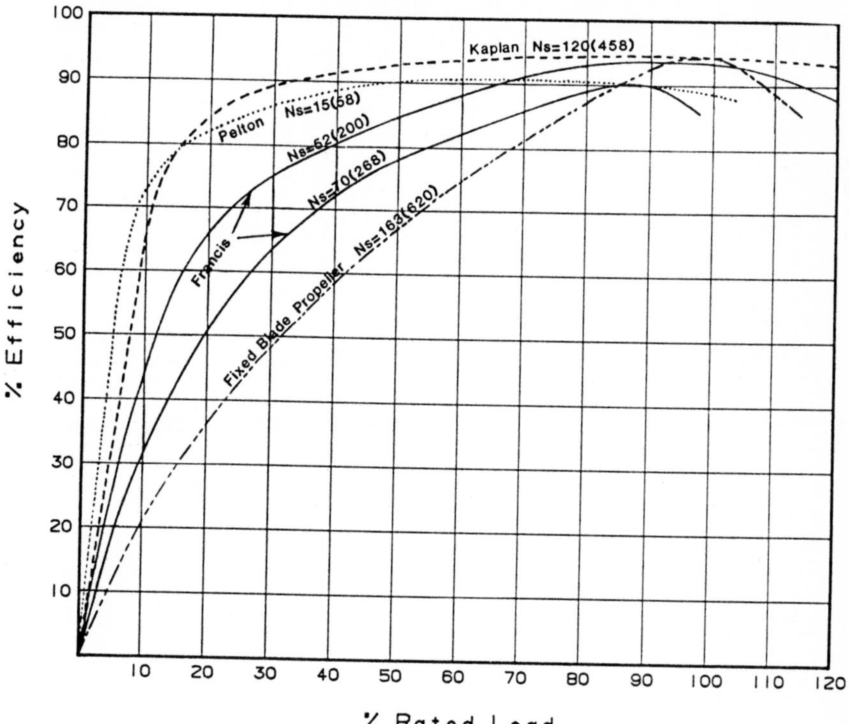

FIG. 6.28 Turbine performance chart.

poor part-load and overload efficiencies and are best suited for operation under block load conditions or within a narrow range of load variations.

Kaplan turbines (high specific speed) have an even more favorable power-efficiency curve than Francis or Pelton turbines. Kaplan turbines are suitable where a high demand of load variations is required in low-head plants or in an operation where there are wide variations in head.

Figure 6.29 shows (to scale) the variance in size and shape of different specific-speed runners operating under the same constant head and output. A low-specific-speed Francis runner is much larger and slower in operating speed than a Kaplan runner for the same output with high specific speed. Consequently, much larger water passages and generator are required for the Francis turbine, which in turn would require a larger powerhouse.

Unit size, speed, and type play an important role in the vulnerability of cavitation, vibration, draft tube pulsations, and regulating stability of the generating equipment. The trend to large-capacity and higher-specific-speed machines has increased the vibration and pressure pulsation problems inherent with fixed-blade turbine runners at part-load operation. Francis turbines with specific speed N_s smaller than 70 (267) have a tendency toward choking flow (throttling) by the runner with a resulting pressure rise during transient operation. However, the discharge of high-specific-speed Francis and propeller-type turbines increases through the runner at the runaway speed. This characteristic is often overlooked in selection of the turbine. Both condi-

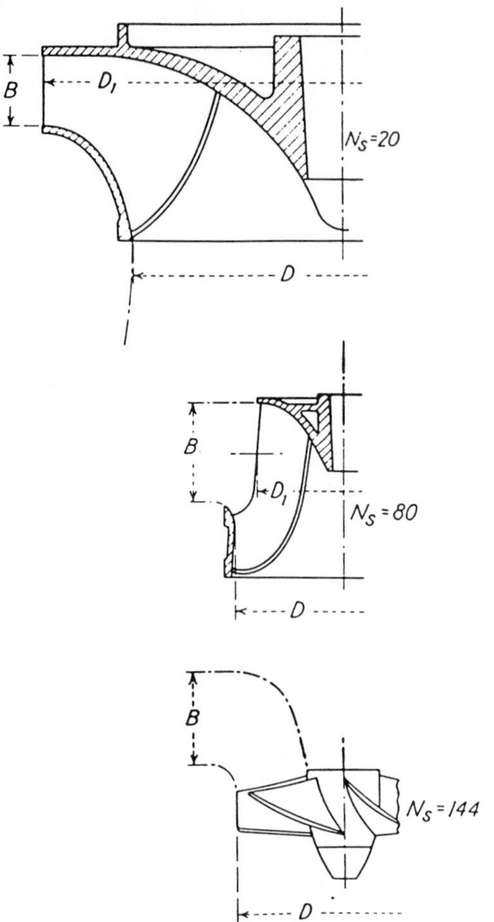

FIG. 6.29 Relative runner dimensions with respect to specific speed.

tions should be reviewed during turbine selection and design, and one should confirm that they do not present problems with the penstock, draft tube, and other components in the power station. Figure 6.30 shows the change in percentage of rated discharge of Francis turbines as a function of specific speed.

The expected operating conditions of the turbine determine the type and speed of the turbine selected. High head, poor water quality (suspended solids, sand, etc.), long penstock, no surge chamber, and large variation in load are conditions which would best be met by an impulse (Pelton) type of turbine. Medium to high head, large capacity, and relatively constant load are conditions that generally favor the use of a Francis-type turbine. U.S. Bureau of Reclamation (USBR) guidelines give the operating range of a Francis turbine as 65 to 125 percent of the design head and 45 to 140 percent of the design output.

Low head, large flow, and wide variations in operating heads and outputs are

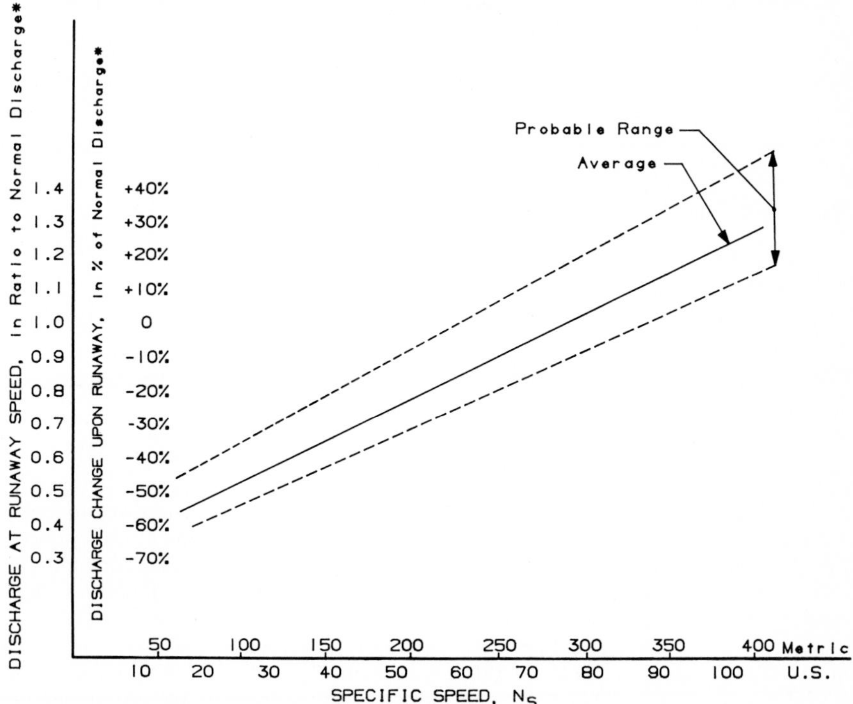

FIG. 6.30 Discharge of Francis turbine at runaway speed with respect to specific speed.

conditions generally favored by the use of a Kaplan type of turbine. USBR guidelines indicate the operating range of a Kaplan turbine to be from 65 to 125 percent of the design head and from 60 to 130 percent of the design output, while the fixed-blade propeller turbine operating range is from 90 to 110 percent of the design head and from 90 to 110 percent of the design output. A multiple-unit powerplant can be used to accommodate wide variations in flow and power requirements.

PROJECT TYPES

Here is a brief review of the main types of hydroelectric power projects being built today.

Run-of-River Operation

The output of a run-of-river project (without flow regulation) is limited by the day-to-day flow of the river, that is, the natural flow of the river. The operation utilizes the river flow without being able to pond or store. Generally, power generation is an incidental product of this type of project. Typically in a run-of-river power development, a short penstock is required to direct water to the turbine(s), using the natural flow of the river with little or no alteration of the stream channel and little impoundment of the water. Adjustable-blade propeller-type turbines are frequently used for run-of-river conditions to best take advantage of the wide variations in heads and flows.

Reservoir Storage

Flow regulation is achieved by providing storage which requires building a dam upstream of the powerhouse. The reservoir is sized to provide the energy storage required by the project, which may be a combination of hourly, weekly, monthly, and seasonal storage.

Pumped Storage

Pumped-storage systems are the most widely used type of energy storage system in use today. The need for peaking capacity to smooth load distribution, improve system reliability, reduce cycling duties of base-load powerplants, and provide low-cost on-peak energy (by pumping in off-peak hours with must-run thermal and nuclear power stations) has greatly sparked the need for pumped storage. Modern pumped-storage plants have the ability to provide peaking power, spinning reserve, synchronous condenser operation, etc., with rapid response. It is possible to bring the turbine generating unit on line from standstill within 90 s and to full power within 120 s, while changeover from pumping to generating can be done, under ideal conditions, in 180 to 240 s.

Pumped-storage site selection is more complex than for conventional hydroelectric plants, because of the need for an upper and lower reservoir (one of which is usually natural) and a site location having suitable geographical characteristics (elevation differential, distance proximity, etc.) to situate the water passages and powerhouse properly. The equipment required for pumped storage is also more complex, because the flow passes through the system in both directions. Commonly, the water in the system is recirculated between the reservoirs.

Even though the overall efficiency of a pumped-storage plant is approximately 10 percent lower than that of a conventional hydroelectric plant, the ability to utilize lower-cost off-peak energy for pumping and to produce on-peak energy by generating makes pumped-storage plants very attractive in the electric utility industry. Underground pumped-storage plants are getting more consideration in areas where environmental, geological, and economic issues are major factors in the selection and design. The application of pumped-storage powerplants is expected to show a marked increase in the United States during the next decade.

Small Hydroelectric Projects

Small hydroelectric projects have been defined in the United States by the *Department of Energy* (DOE) as those that have a capacity of 15,000 kW or less.

Low-head hydroelectric projects are defined by DOE as those with heads of less than 65 ft (20 m); however, many small hydroelectric installations fall within the category of low-head hydro plants.

Renovation of Hydroelectric Powerplants

Hydroelectric power structures are usually designed with the assumption of a useful life of somewhere on the order of 50 to 100 yr. Many units installed in the early 1900s are still operational, although others have been shut down due to lack of readily available spare parts, excessive maintenance requirements, and expense, among other reasons.

Recently, a marked increase of interest in the rehabilitation of hydroelectric powerplants has been apparent, primarily because of the economic benefits afforded by the existing structures of the dam, powerhouse, and water passageways—many of which can be reused at a significantly lower cost than that for new construction. Also, capitalizing on existing structures can (1) save on feasibility siting studies, (2) simplify plant licensing procedures (government license amendments in lieu of new licensing), and (3) take advantage of existing power transmission systems.

Hydraulic machinery, generators, and electric equipment in general have a useful life of approximately 30 to 50 yr, subject to the way that they have been maintained and operated. Often by performing a major overhaul of the equipment and rehabilitating the powerhouse and dam, the life of a hydroelectric powerplant can be extended for another 30 yr or more.

Upgrading and uprating of the powerhouse equipment are also done to increase the unit output, efficiency, automation, reliability, and safety of the powerplant, thus improving its operating capabilities and reducing outage and downtime. The usual equipment components that need to be analyzed in the renovation of the hydroelectric generating equipment include the following:

- *Turbine:* Runner, bearings, shaft seal, bushings, wicket gates, discharge ring, draft tube
- *Generator:* Stator windings, excitation system, main leads (busbars), neutral grounding, current transformers
- *Governor:* Oil pumps, distributing valves, piping
- *Electrical:* Power transformers, switchgear, control switchboards, buses, motors

If upgrading is performed, all the equipment parts need to be analyzed to ensure that they will safely withstand the increased load conditions.

ACKNOWLEDGMENTS

The authors wish to acknowledge R.E. Israelsen, P.E., and J.H. Mellado, Harza Engineering Co., for their contributions to this chapter; and R.W. Fazalare, Harza Engineering Co., for reviewing this chapter.

SUGGESTIONS FOR FURTHER READING

Baumeister, T., et al., *Marks' Standard Handbook for Mechanical Engineers,* 8th ed., McGraw-Hill Book Co., New York, 1978.

Carson, J.L., "The Selection of Hydroturbines Considering the Effect of Type and Size of Energy Production," *Proceedings of American Power Conference,* 47:1144–1149, 1985.

Considine, D.M., *Energy Technology Handbook,* McGraw-Hill Book Co., New York, 1977.

Fazalare, R.W., "Bulb Turbine Selection for the Main Canal Project," *Water Power,* October 1985.

Fazalare, R.W., "Trends in Selecting and Procuring Hydro Turbines," *Water Power,* December 1986.

Pruce, L.M., "Power from Water," *Power,* April 1980.

Raabe, J., *Hydro Power,* VDI-Verlag GmbH, Dusseldorf, West Germany, 1985.

"Selecting Hydraulic Reaction Turbines," U.S. Bureau of Reclamation Engineering Nomograph No. 20, 1976.

Wislicenus, G.F., *Fluid Mechanics of Turbomachinery,* 2d ed., Dover Publications, Inc., New York, 1965.

CHAPTER 2.7
COGENERATION

Michael P. Polsky
President
Indeck Energy Services, Inc.
Wheeling, IL

INTRODUCTION

Cogeneration is the simultaneous production of useful thermal energy and electric power from a fuel source or some variant thereof. It is more efficient to produce electric power and steam or hot water together than electric power alone, as utilities do, or thermal energy alone, which is common in industrial, commercial, and institutional plants (see box).

Cogeneration systems of 1 to 10 MW are the most popular of the alternative-energy sources available in total megawatts planned or installed nationwide. Unlike small cogeneration systems that primarily use reciprocating engines firing natural gas or large systems that rely on combustion turbines or coal-fired boilers, these medium-size cogeneration systems are based mainly on characteristics of the specific application, rather than their technical limitations—often the case with large and small cogeneration systems.

THE BUSINESS SIDE OF COGENERATION

Cogeneration is a business originally driven by rising fuel costs and the watershed *Public Utility Regulatory and Policies Act* (Purpa) of 1978. Essentially, the act gave license to anyone—someone who used steam or hot water and/or electric power or a third party—to build powerplants and sell the electric power back to the utility at its avoided cost, without reprisals, such as taking away standby power. Utilities, unable to own more than 49 percent of a cogeneration plant, had to negotiate in good faith.

Prior to Purpa, cogeneration was economical in only a few industries because many power systems produce more electricity per thermal requirement than can be used internally, and the excess usually could not be sold to a utility at attractive rates.

Now firms whose primary line of business is not power production have become small utilities. Equipment suppliers, architect-engineers, entrepreneurs, and others can band together and finance a project, or convince a bank to lend the money, based on the revenue streams from electric power and steam or hot-water sales. In recent years, these third parties have come to dominate the cogeneration market. Many have little equity and rely on debt financing to carry the project through.

Because suppliers and/or banks and entrepreneurs are putting up the capital, the revenue streams have to be guaranteed. For this reason, there is little incentive to try anything new. Components must offer absolute reliability and availability so investors can recover their money as quickly as possible. Some entrepreneurs have the capability both to raise the capital and to operate the system after it has been installed.

The major and most widely used medium-size cogeneration systems are focused on below. Later in the chapter other technologies are covered, along with regulations and projections about the future of cogeneration.

KEY OPERATING CYCLES

Topping, combined, and bottoming cycles are the main types covered. Prime movers are steam turbines, gas turbines, and reciprocating engines.

Steam-Turbine Topping Cycle

Because of its high initial cost, the steam-turbine topping cycle is probably selected least often for medium-size cogeneration. The main advantage of this cycle is its ability to capitalize on most types of fuels, including such solid fuels as wood, coal, petroleum, coke, refuse, etc., which no other cycle can do. Capital costs of the cycle burning solid fuel could range from \$1100 to \$1500 per kilowatt for 8- to 10-MW plants and \$2000 per kilowatt and higher for plants less than 5 MW. The costs are a function of the type of fuel selected, emissions control technology, and design and construction philosophy. Figures 7.1 and 7.2 show two steam-turbine topping cycles.

The cycles have inherent disadvantages that significantly limit their application, the most common of which is a relatively low power-to-heat ratio that becomes even smaller at high process-steam pressures. Also, for most industrial applications, they provide a poor match between process steam and electrical requirements, thus making them less attractive economically.

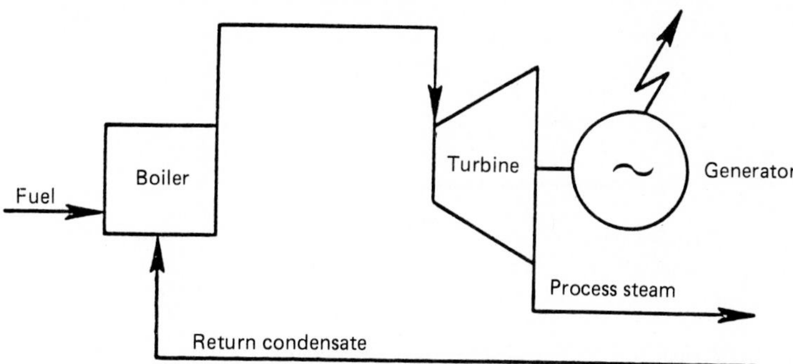

FIG. 7.1 Cogeneration plant with backpressure steam turbine.

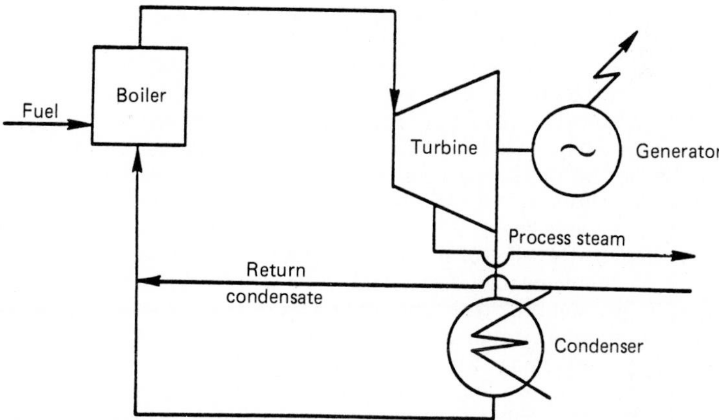

FIG. 7.2 Cogeneration plant with extraction-condensing turbine.

The greatest application of steam-turbine topping cycles is expected in larger sizes (8 MW and above) and in applications where low-quality refuse and by-product fuels are readily available. With very few exceptions, there would be practically no attraction for this type of cycle where natural-gas or other high-quality gaseous or liquid fuels are to be used.

Gas-Turbine Topping Cycles

Thanks to cogeneration, the popularity of gas-turbine prime movers has increased dramatically in the past decade. The working fluid in a gas turbine is the hot gases at temperatures near 2000°F (1093°C). Although the overall turbine thermal efficiency could exceed 30 percent, the exhaust gases at temperatures up to 1000°F (538°C) represent more than 68 percent of the total fuel energy input. Exhaust-gas thermal energy can be recovered in different ways. It can be consumed directly for air preheating or drying, for example, or indirectly by passing the gases through a heat exchanger, producing hot water or steam for process needs.

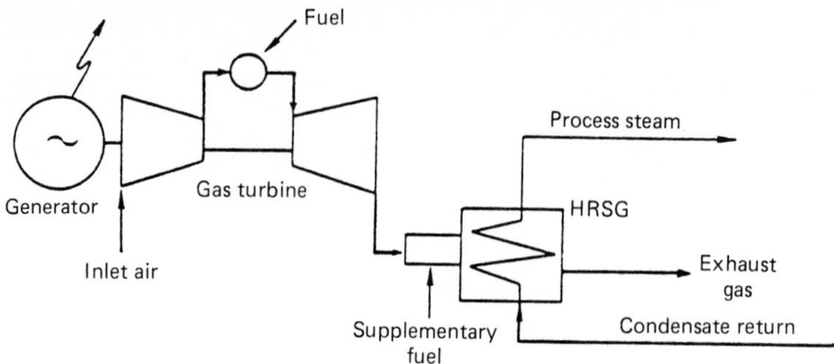

FIG. 7.3 Gas turbine with heat recovery cogeneration system.

Most Common System. The most commonly used gas-turbine topping cycle is one that has an unfired or supplementary fired *heat-recovery steam generator* (HRSG) to produce steam, later used for process (Fig. 7.3). Steam can be generated at any pressure level (saturated or superheated) or at several pressure levels, to match process requirements. If the amount of steam required by the process exceeds the amount that can be produced by simply recovering the thermal energy in the gas-turbine exhaust, then supplementary fuel can be burned in the ducting between the turbine and HRSG, to increase the gas energy to the required level. A limiting factor on the amount of supplementary firing is the gas temperature after the burners. It is good design practice to maintain this temperature below 1700 to 1800°F (927 to 982°C), to prevent failure of uncooled ducting materials. With supplementary firing held to a safe amount, it is still possible to increase steam generation 2 to 3 times over what would be available without it.

Another important aspect of supplementary firing is that the thermal efficiency of converting supplementary fired fuel energy into thermal energy is almost 100 percent, since no additional loss occurs with the exhaust gases. The efficiency of a conventional boiler is only 80 to 85 percent; thus supplementary firing can lead to additional energy saving.

Gas-turbine cogeneration plants have a number of distinct advantages: (1) high power-to-heat ratios, which nicely fit many industrial applications; (2) low capital costs, normally 50 to 70 percent of steam-turbine plants; (3) low operating and maintenance costs; (4) short construction periods (usually less than 12 months); and (5) no cooling water requirements. Because of these advantages, gas-turbine systems are, and will remain, the most popular selection for medium-size cogeneration. Figure 7.4 shows the exhaust from a gas turbine serving as preheat air for an incinerator.

Selection Factors. Besides the specific experience and track record of gas-turbine manufacturers (only three in the United States build machines between 1 and 10 MW), the selection of turbine type and size is heavily dependent on the unit's electrical efficiency. Higher electrical efficiency could be a major and sometimes determining factor in the selection.

In general, smaller gas turbines are not as efficient as larger units and are significantly less efficient than reciprocating gas engines. As a result, few installations have combustion turbines rated below 2.5 MW. However, most gas-fired cogeneration plants above 2.5 MW use combustion turbines. The capital cost of

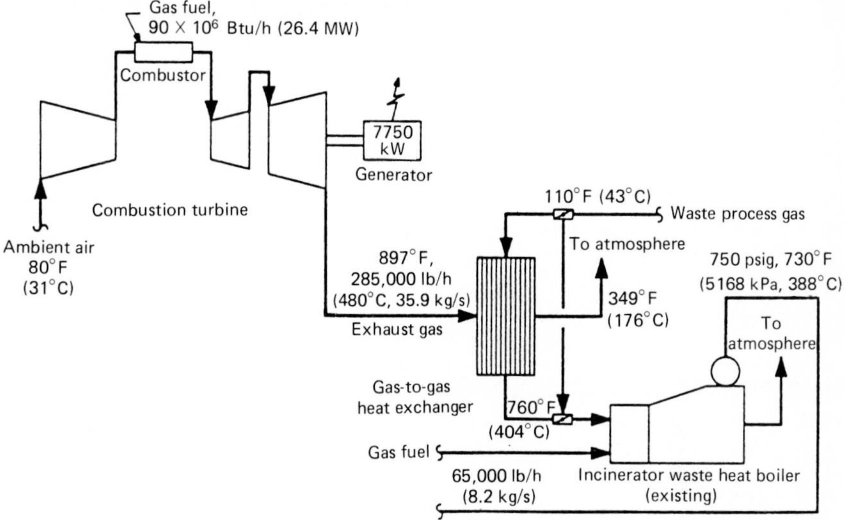

FIG. 7.4 Gas turbine cogeneration plant in which exhaust from the turbine serves as preheat air for an incinerator.

these systems ranges between $400 and $700 per kilowatt, depending on the complexity of site-specific conditions.

Optimum reliability of most modern medium-size combustion turbines is achieved when they are operating continuously or almost continuously on natural gas. Even the use of no. 2 oil shortens turbine life expectancy considerably, and burning heavier grades of fuel oil (such as no. 6) is totally prohibitive. As a result, applications where frequent starting and stopping or continuous use of nongaseous fuel is necessary must be carefully evaluated. An alternative prime mover may be the best choice.

Another area in which gas turbines have not yet reached their full potential involves the direct utilization of exhaust gas in process furnaces for air or oil preheating and for boiler repowering (Fig. 7.5). Since the selection of the gas turbine for these applications is normally associated with substantial modification of the existing processes, expect more engineering and development work.

Combined Cycles

A shortcoming of the gas-turbine cycle is that the electric energy and thermal energy produced are in a fixed relationship. Unfortunately, many industrial processes have wide fluctuations in the amount of energy required. The combined cycle solves this problem. Here is how it works: Thermal energy from the gas-turbine cycle is recovered in HRSGs, which generate relatively high-pressure steam that is expanded through a steam turbine driving an electric generator (Fig. 7.6). Since process thermal energy in the combined cycle is produced by extracting steam from the steam turbine (eliminating loss in the condenser), both thermal and electric energy are available from the steam-turbine cycle.

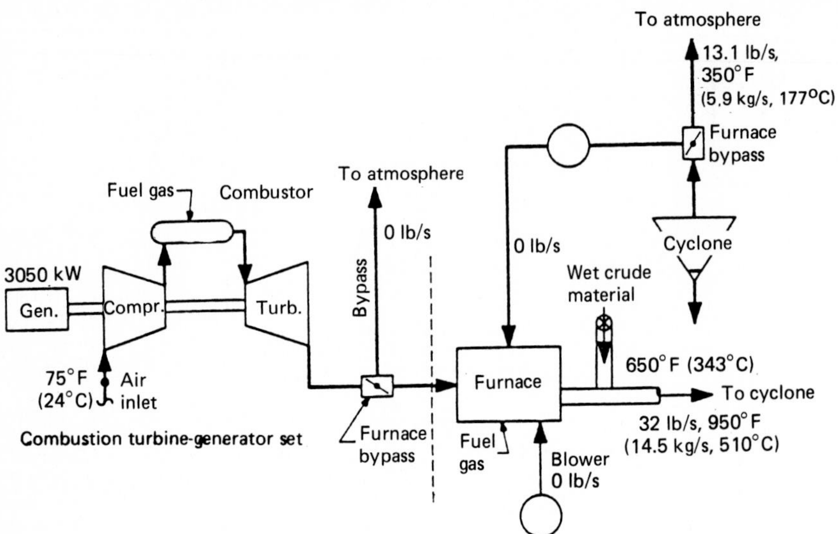

FIG. 7.5 Repowering of drying furnace using gas turbine exhaust.

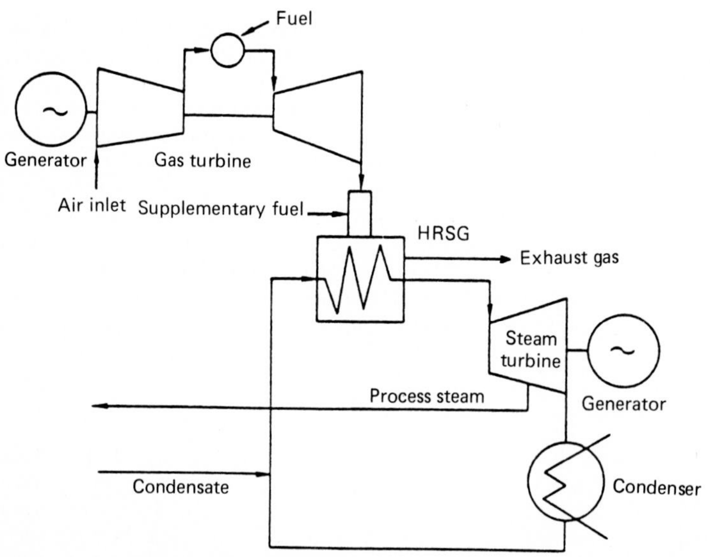

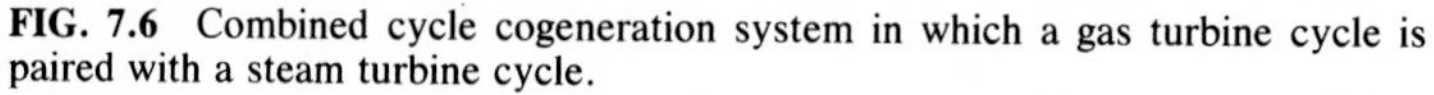

FIG. 7.6 Combined cycle cogeneration system in which a gas turbine cycle is paired with a steam turbine cycle.

The combined cycle is the most efficient, commercially available power-generating cycle. The thermal efficiency of even a small combined cycle is well above 40 percent and in some designs could reach 46 percent. Steam-turbine electricity production varies between 10 and 40 percent of gas-turbine output, depending on process-steam requirements.

The steam turbine can be either extraction-condensing or backpressure. Since steam is produced by recovering waste heat, however, the selection of the condensing turbine for combined-cycle operation is almost always justified economically, especially if process-steam requirements are below the steam production capability of the gas-turbine cycle.

HRSGs in the combined cycle are similar to those used in heat-recovery applications, except that steam is produced at pressures and temperatures [600 to 1400 psig (4130 to 9650 kPa), 700 to 950°F (371 to 510°C)] that make its expansion through the turbine economical. As with the gas-turbine cycle, supplementary firing could be implemented to achieve greater steam cycle flexibility. Since the combined cycle is a combination of the gas and steam cycles, it also combines all their advantages while eliminating most of their disadvantages. The capital cost of the medium-size combined cycle is higher than for the gas-turbine cycle and may vary from $600 to $1000 per kilowatt, depending on specific design conditions.

Because production of electric and thermal energy is independent and can be controlled efficiently over wide ranges by using the combined cycle, it is being favored in many medium-size cogeneration applications.

Reciprocating-Engine Cycles

Reciprocating engines can be spark-ignited, gas-fired, dual-fuel (gas and oil), or diesel fuel, depending on fuel-firing capability. Dual-fuel and diesel engines are ignited by compression of the fuel.

Engine Particulars. The main thermodynamic difference between gas-turbine and reciprocating-engine cycles is that almost two-thirds of the waste heat from reciprocating engines is rejected in the form of low-temperature [up to 250°F (121°C)] energy rather than as a high-temperature exhaust (Fig. 7.7). As a result, re-

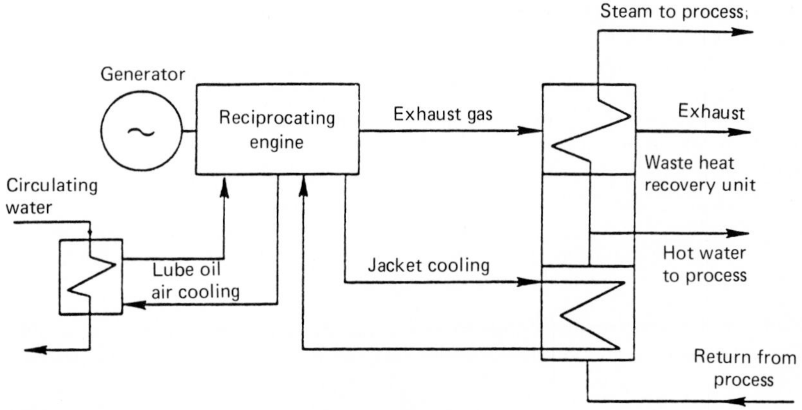

FIG. 7.7 Reciprocating engine cogeneration system in which waste heat from the exhaust, and jacket and lube oil cooling, are recovered.

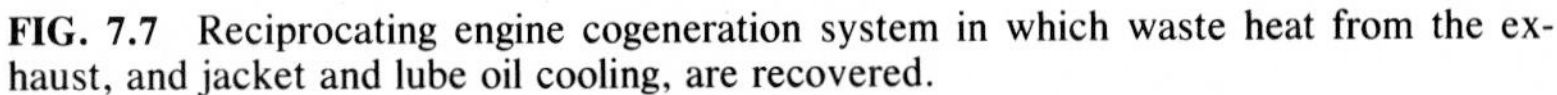

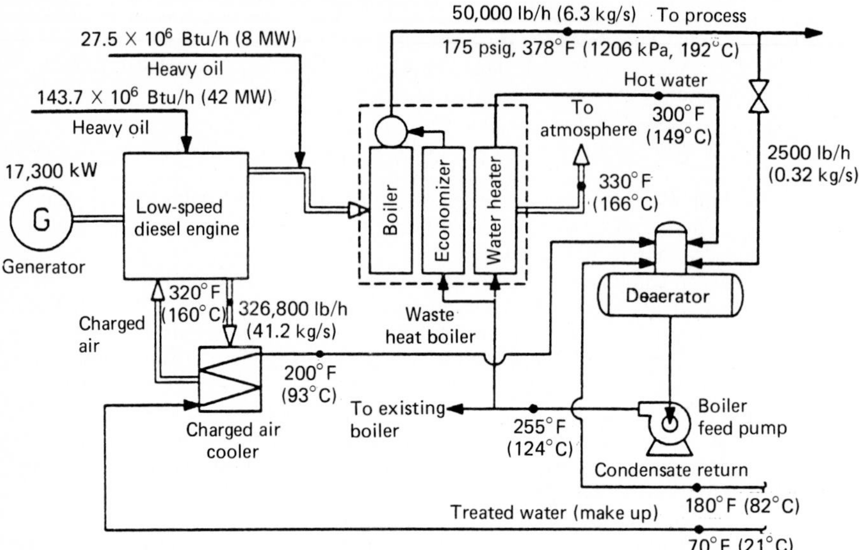

FIG. 7.8 Low-speed diesel engine cogeneration.

ciprocating engines can produce only one-half to one-third the amount of high-pressure [above 30-psig (207-kPa)] steam produced by a gas turbine of equivalent capacity. If the process can accept hot-water or low-pressure [less than 25-psig (172-kPa)] steam, an ebullient type of heat recovery can be applied to boost the amount of useful waste heat recovered from the engines.

Diesel engines have seen service for base-load and power generation for decades (Fig. 7.8). Contemporary diesels have an electrical efficiency of 42 to 45 percent, while gas engines reach efficiencies of 35 percent, making diesels the most efficient type of prime mover for cogeneration.

Unlike the gas turbine, the reciprocating engine is available in practically any size up to 2000 kW. Units above 2000 kW are available in discrete sizes. Large sizes of 5 to 10 MW are primarily liquid-fuel diesels. Reciprocating engines are designed for high speed (1200 rpm), medium speed (400 to 900 rpm), and low speed (below 400 rpm). Most of the small reciprocating engines (less than 2000 kW) are high speed, while large diesels are medium or low speed. Modern diesels can accommodate practically any type of gaseous or liquid fuel, including the heaviest grades of no. 6 oil.

Comparison with Gas Turbines. Comparison between gas turbines and reciprocating gas engines is the subject of many cogeneration studies. Each type has different applications and needs and may prove to be more or less economical for a specific application. Some basic comparisons are made:

- Medium-size reciprocating engines have substantially higher electrical efficiencies than equivalent combustion turbines (34 to 40 versus 26 to 30 percent). Better efficiencies may lead to much lower fuel consumption, creating a decreased dependency on fuel cost.

- Reciprocating engines require 20 to 40 psig (138 to 276 kPa) gas fuel pressure versus 180 to 400 psig (1240 to 2756 kPa) for combustion turbines. This difference results in additional cost savings, reduced capital costs, and increased operating reliability.
- The electrical output of reciprocating engines is much less sensitive to ambient air temperature. For example, the electrical output of a 3000-kW gas turbine drops approximately 125 kW for each 10°F (5.6°C) increase in ambient air temperature, with a respective decrease in fuel efficiency. This decrease will also lead to an additional electrical demand and a higher cost of electricity from the utility.
- Because of a higher part-load efficiency, reciprocating engines provide better load dispatch and load following than gas turbines. Most of the engines are also designed for frequent start-ups and shutdowns; thus, it is easier to start and stop them in accordance with the plant's electrical loads and with utility power purchase schedules.
- The capital cost of reciprocating plants is normally 10 to 25 percent higher than that of same-size gas-turbine plants. The operating and maintenance costs (excluding fuel) are also higher (1 to 1.2 ¢/kWh versus 0.4 to 0.5 ¢/kWh).

For the reasons outlined above, a sharp distinction exists between reciprocating engines and gas turbines. In applications calling for high power-to-heat ratios, hot water or low-pressure steam, frequent start-ups and shutdowns, reciprocating engines clearly offer better economic opportunities than combustion turbines.

Bottoming Cycles

In the topping cycles previously described, fuel energy is first used to produce electricity. The heat rejected from these cycles can be characterized as useful thermal energy. In the bottoming cycle, almost the exact opposite sequence of events takes place (Fig. 7.9). First, fuel energy is consumed in the process; then the waste heat discharged from the process is recovered and converted to electricity..

Most bottoming cycles recover waste heat and convert it to steam, which is then used to produce electricity in the steam turbine-generator. To capitalize on steam as a working medium, the waste energy should have a relatively high temperature, 600°F (316°C) or above. However, working fluids other than steam, primarily hydrocarbon liquids, could be selected if the exhaust energy is below 600°F (316°C).

The thermal efficiency of the bottoming cycle is quite low—10 percent for low-temperature and up to 20 percent for high-temperature applications. Given that the waste heat is free, however, cycle economics are often favorable. The water-steam bottoming cycle relies on conventional technology; in fact, most of the equipment is identical to equipment in the steam portion of the combined cycle. The technology of low-temperature bottoming cycles (also called *organic Rankine cycles*) is not fully proved and normally requires a capital investment up to 2 times greater than for steam cycles.

The main application areas of bottoming cycles are in the exothermic processes of the chemical industry and in the material processing industries, such as cement, limestone, and steel. The economics of these cycles are directly related

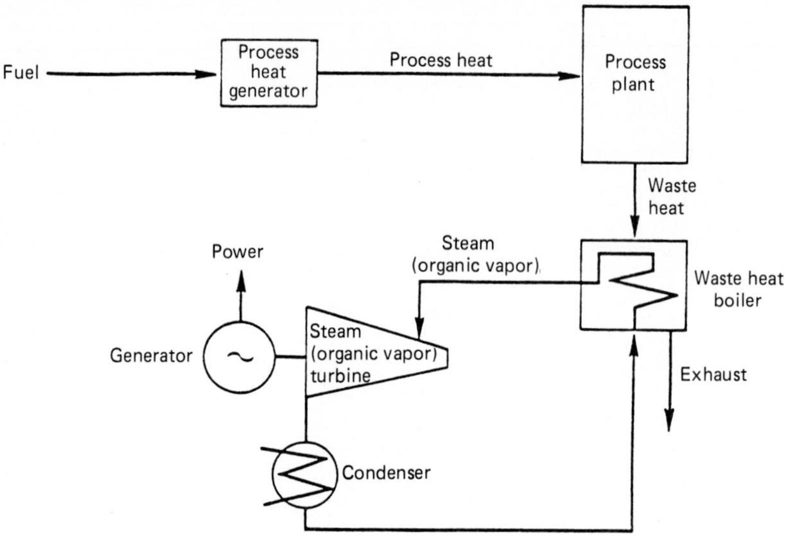

FIG. 7.9 Cogeneration plant using a steam or organic bottoming cycle.

to the cost of electricity. Higher electrical costs in the future could boost the attraction for this cogeneration system.

OTHER TECHNOLOGIES

Other technologies can be applied to cogeneration. Some are a combination or modification of previously described systems, others are new and innovative. Consider these systems:

Direct Mixing Cycle

This system is presently being offered by a California company. Called a *Cheng cycle,* it incorporates a conventional 3500-kW combustion turbine with a high-pressure, waste-heat-recovery steam generator. The steam produced by the HRSG that is not used by process could be reinjected into the gas turbine to produce additional electric energy. The electrical output of the Cheng cycle ranges from 3500 to 6000 kW.

The electrical efficiency of the cycle is less than that of the condensing combined cycle. However, the absence of a steam turbine, condenser, and other auxiliaries may lower the capital cost and simplify operation. Because of its inherent design, applications of the Cheng cycle are mainly limited to processes with wide and quick variations of thermal load.

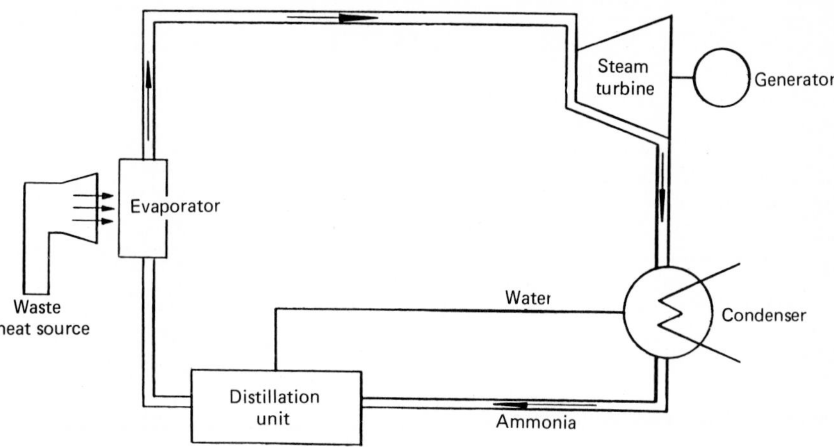

FIG. 7.10 Cogeneration plant using the Kalina cycle.

Wood and Coal Gasification

Gasification systems convert solid organic fuels to low- or medium-Btu gases, which are used as fuel by the boilers or various types of prime movers. Although they have been technically proved, the economic viability of these systems in the near future is questionable because of the soft price for high-grade fuels.

Kalina Cycle

The Kalina cycle can be described as a bottoming cogeneration cycle (Fig. 7.10). It pairs a low-temperature, waste-heat-to-power Rankine cycle with a variable-ratio ammonia-water mixture. Although the cycle resembles the widely publicized organic Rankine cycle to some degree, the principal difference is that the Kalina cycle can achieve an incredible electrical efficiency of 50 percent, versus only 10 to 14 percent for the organic Rankine cycle.

At this time, the Kalina cycle exists only on paper, but widespread implementation of this technology may surface, especially in bottoming cycle applications—if the prototype installation confirms the theory. The real advantage of the Kalina cycle is that it draws on mostly conventional technology. Components and equipment in the system may be applied immediately without sizable developmental costs.

GOVERNMENT REGULATIONS

Energy Regulations

A great deal of speculation exists over how future government regulations could affect the cogeneration market. Issues such as Purpa (see the box at the beginning of the chapter) efficiency standards, the wheeling of power, and utility standby power are among those widely discussed and publicized. The National Energy Act of 1978 has aimed at conserving energy, which is what the majority of medium-size cogeneration systems do.

Unlike the large "Purpa machine" type of cogeneration systems, medium-size plants are mostly designed and sized to supply internal energy requirements and to maximize energy savings with minimal or no electricity sales to the local utility. Also, most third-party developers find that it is much less risky to size these systems to supply only the customer's basic electric load rather than to be exposed to the many uncertainties associated with power sales. If the power sales agreement with the utility is a worthwhile effort for large plants, it is only an unnecessary complication for medium-size systems. Because of this design restriction, these systems are much less sensitive to any future changes in energy regulations.

Environmental Regulations

By its very nature, the market for medium-size congeneration systems is quite sensitive to changes in environmental regulations. The cost of pollution control equipment can play a critical role in the economic justification of these systems. The only possible impact is in the area of emissions regulations. As of now, practically all medium-size, gas-fired cogeneration systems meet federal, state, and local air pollution regulations without any special control technology.

Only regulations enforced by the state of California require special equipment to limit emissions from gas-fired systems. Recently, several California air quality districts have tightened these regulations substantially; expensive controls are now required to limit emissions by combustion turbines and reciprocating engines. The effect of the regulations may be devastating to medium-size gas-turbine cogeneration development in that part of the country. However, there is hope that the cost of control equipment will decrease because of new developments and further manufacturer experience.

It is expected that other states, such as New Jersey, may also tighten their regulations, which may mean additional emissions controls. However, it is unlikely that these regulations would be as severe as those imposed by the South Coast Air Quality District in California. Gas turbines may require injection of steam or water, and it may be necessary to adjust the timing in reciprocating engines. No changes in federal air pollution regulations that would affect gas-fired installations are expected in the near future.

There is significantly more uncertainty about air emissions regulations for solid-fuel sources. Many states have already promulgated regulations that are more stringent than the federal standards. Anticipated acid-rain legislation may also impose much tougher federal standards for sulfur dioxide emissions. Fluidized-bed combustion technology, and perhaps sorbent injection, is the safest approach to burning coal and other sulfur-containing solid fuels.

TECHNOLOGY PROJECTIONS

Since its rebirth in 1978, several shake-outs have affected the development of the cogeneration market. They were caused by regulatory, fuel, economic, environmental, and geographic considerations. Most of these shake-outs have affected very large, third-party-owned cogeneration systems. Fortunately, medium-size cogeneration systems were largely immune to these past shake-outs because of their conservative design and economic factors that favored them. Fortunately also, corporate acceptance of internal generation by many industries has removed institutional and bureaucratic barriers.

What is cogeneration's future over the next several years? As it exists today, major growth will most likely be in medium-size systems, divided among the operating cycles as follows:

Future for Steam-Turbine Topping Cycles

Although there will be many new plants using this type of technology, they will not occupy a major part of the medium-size cogeneration market. Steam-turbine topping cycles will burn mostly solid and waste fuels that cannot be used by any other technology.

The majority of the systems will be 5 MW or larger. Since the cost of fuel determines their economics, a technology capable of accepting various kinds of very low cost fuels could attract the most attention. Recent price developments in oil and natural gas have, and will continue to, put pressure on all alternative-energy systems to become more competitive.

Future for Gas-Turbine and Combined Cycles

Lower gas prices will be a major influence contributing to the dominance of these systems in the upper range of the medium-size market. The spread between electric rates (retail) and natural-gas prices becomes quite favorable for gas turbine and combined cycles in many parts of the country. The possibility of obtaining medium-range, low-cost gas supply contracts made many previously marginal and risky projects a reality.

Recent experience has indicated that gas turbines can successfully burn waste gaseous fuels, such as landfill and digester gases. Gas-turbine projects in which these waste fuels are specified are being planned and constructed.

As with small systems, a number of attempts have been made to standardize medium-size cogeneration packages. The fact is that almost all medium-size systems have different design criteria and require various degrees of customization to achieve maximum economic effectiveness. Thus, a trend toward standard packages for medium-size gas-turbine cogeneration is not likely.

Future for Reciprocating Engines

Falling gas and oil prices are like a strong fertilizer for this type of technology. Sharp competition will be evident for reciprocating engines vis-à-vis gas turbines in the range of 1 to 3 MW. Reciprocating engines will be especially strong in areas where electric prices vary significantly with the time of day (on peak, off peak) or for industries and institutions operating noncontinuously. In these applications, engines can be started and stopped more effectively, to ensure the most efficient cogeneration system.

A dramatic fall in oil prices may stimulate the market for large oil-fired diesel engines, which are often less expensive than natural-gas machines—another stimulus. Up to this point, the technology has not received significant attention in the United States because of the low price and availability of natural gas; also, the uncertainty of oil prices has been a concern.

Future for Bottoming Cycles

As was indicated previously, the attraction of the bottoming cycle is heavily dependent on the price of electricity. Escalating electric rates will uncover new applications for bottoming cycles. Because electric prices have generally increased more slowly than expected, however, there probably will not be many applications for new types of bottoming cycles, even the organic Rankine cycle.

SECTION 3

FUELS AND FUEL HANDLING

CHAPTER 3.1
COAL CHARACTERISTICS

Albert F. Duzy
President
Duzy & Associates
Sun Lakes, AZ

INTRODUCTION

Coal may be broadly grouped into two major categories—steam coal and metallurgical coal. This discussion is concerned mostly with the former, or that used mainly for the production of steam in boilers. Practically all the coals used in the production of metallurgical coke are of the bituminous rank. The bituminous coals have caking or coking (agglomerating) characteristics; that is, they become plastic or semiplastic when heated in the absence of air, which makes them amenable for use in coal blends for coke. Bituminous coals that have excessive amounts of sulfur or ash for coke production are classified as steam coals.

ELEMENTS OF ASTM STANDARDS

Coals are sampled, prepared, and analyzed by methods and procedures in accordance with ASTM Standards. Standard analyses may be reported as shown in Table 1.1. The *free-swelling index* (FSI) is another standard, and it indicates the plastic properties of bituminous coals. It ranges from 0 (nonbituminous) to about 10.

Still another standard is the *hardgrove grindability index* (HGI). It is a mea-

TABLE 1.1 Standard ASTM Analyses

Proximate	Short proximate	Ultimate*
Moisture	Moisture	Moisture
Volatile matter	Volatile matter	Carbon
Fixed carbon†	Fixed carbon†	Hydrogen
Ash	Ash	Nitrogen
		Sulfur
	Sulfur	Chlorine
	GCV‡	Oxygen†

*By definition, the moisture content is not part of the ultimate analysis, but it is commonly used. Occasionally, the moisture content may be included as hydrogen and oxygen; if so, approximately one-ninth is hydrogen and eight-ninths is oxygen. Chlorine is now also commonly used in the ultimate analysis.

†Obtained by difference.

‡Gross calorific value, Btu/lb (kJ/kg), or higher heating value (HHV).

sure of the work input to obtain a size reduction of the coal. The range will be from about 25 to 110, with the higher indices indicating ease of grinding.

Ash Fusibility

When coal is oxidized (burned), a remaining residue called *ash* is formed from the mineral matter in the coal. This ash may be tested to indicate its fusibility in reducing or oxidizing conditions. Table 1.2 lists the tests. The fusibility of ash is a rough indicator of its behavior when heated in the range of 1900°F (1038°C) to 2730°F (1500°C).

Oxide Constituents

The constituents in the ash may also be determined and reported as oxides. Based on mineral analyses of ash, the usual oxides are SiO_2, Al_2O_3, TiO_2, Fe_2O_3, CaO, MgO, Na_2O, K_2O, P_2O_5, and SO_3.

As a matter of routine, some laboratories also present three other oxides—SrO, BaO, and Mn_3O_4—and they will also show the undetermined amount. Obviously, the total should be 100 percent. Also, some trace-element data in the

TABLE 1.2 Ash-Fusibility Tests

Type of test	ASTM Standard nomenclature
Initial deformation temperature (IDT)	IT
Spherical softening temperature (SST), where the height of an ash cone is equal to its width ($h = w$)	ST
Hemispherical softening temperature (HST), where the height of an ash cone is equal to one-half its width ($h = w/2$)	HT
Fluid temperature (FT), where the ash cone is melted down to a 1/16-in (1.6 mm) height	FT

coal may be determined. Generally, these trace elements should have concentrations expressed as a *parts-per-million* (ppm) weight of the dry coal. Trace-element analysis involves the elements in Table 1.3. Analysis data are generally developed for environmental use.

Coal Rank

ASTM classifies coal according to its rank, that is, according to the degree of metamorphism in a natural series of coals from the low-rank brown or lignitic coals to the high-rank anthracitic coals, as shown in Table 1.4, which is ASTM number D 388. Classification is according to *fixed carbon* (FC) and *gross calorific value* (GCV) calculated on the mineral-matter-free basis. The higher-rank coals are classified according to fixed carbon on a dry basis; the lower-rank coals, gross calorific value on a moist basis.

Agglomeration. The agglomerating type of coal is used to differentiate between adjacent groups, such as semianthracite and low-volatile bituminous coals, and between high-volatile C bituminous and subbituminous A coals. If the coal is agglomerating, it is of the bituminous type; if nonagglomerating, of the anthracitic, subbituminous, or lignitic type. Generally, if the carbon residue from the determination of the *volatile matter* (VM) is noncohesive (discrete particles), the coal is of the nonagglomerating type. One can request this fact to be reported by the testing laboratory. Calculation on the mineral-matter-free basis should be done in accordance with Eqs. 1.1, 1.2, and 1.3 or the approximation formulas in Eqs. 1.4, 1.5, and 1.6.

Parr Formulas. For normal steam-coal use, it is recommended that the approximation formulas be used; in the case of litigation, however, use the Parr formulas:

$$\text{Dry, Mm-free FC} = (\text{FC} - 0.15S)/[100 - (M + 1.08A + 0.55S)] \times 100 \quad (1.1)$$

$$\text{Dry, Mm-free VM} = 100 - \text{Dry, Mm-free FC} \quad (1.2)$$

$$\text{Moist, Mm-free Btu} = (\text{Btu} - 50S)/[100 - (1.08A + 0.55S)] \times 100 \quad (1.3)$$

Note: The above formula for fixed carbon is derived from the Parr formula for volatile matter. The approximation formulas are as follows:

TABLE 1.3 Trace Elements in Coal

Uranium	Lanthanum	Strontium	Nickel
Thorium	Barium	Rubidium	Cobalt
Lead	Cesium	Bromine	Chromium
Mercury	Antimony	Selenium	Vanadium
Europium	Tin	Arsenic	Scandium
Samarium	Molybdenum	Germanium	Fluorine
Neodymium	Niobium	Gallium	Boron
Praseodymium	Zirconium	Zinc	Beryllium
Serium	Yttrium	Copper	Lithium

TABLE 1.4 Classification of Coals by Rank*

Class	Group	Fixed-carbon limits (dry, mineral-matter-free basis), %: Equal to or greater than	Fixed-carbon limits: Less than	Volatile-matter limits (dry, mineral-matter-free basis), %: Greater than	Volatile-matter limits: Equal to or less than	Calorific value limits (moist,† mineral-matter-free basis), Btu/lb: Equal to or greater than	Calorific value limits: Less than	Agglomerating character
I. Anthracitic	1. Meta-anthracite	98	—	—	2	—	—	Nonagglomerating
	2. Anthracite	92	98	2	8	—	—	Nonagglomerating
	3. Semianthracite‡	86	92	8	14	—	—	Nonagglomerating
II. Bituminous	1. Low-volatile bituminous coal	78	86	14	22	—	—	Commonly agglomerating¶
	2. Medium-volatile bituminous coal	69	78	22	31	—	—	Commonly agglomerating¶
	3. High-volatile A bituminous coal	—	69	31	—	14,000§	—	Commonly agglomerating¶
	4. High-volatile B bituminous coal	—	—	—	—	13,000§	14,000	Commonly agglomerating¶
	5. High-volatile C bituminous coal	—	—	—	—	11,500	13,000	Commonly agglomerating¶
						10,500	11,500	Agglomerating
III. Subbituminous	1. Subbituminous A coal	—	—	—	—	10,500	11,500	Nonagglomerating
	2. Subbituminous B coal	—	—	—	—	9,500	10,500	Nonagglomerating
	3. Subbituminous C coal	—	—	—	—	8,300	9,500	Nonagglomerating
IV. Lignitic	1. Lignite A	—	—	—	—	6,300	8,300	Nonagglomerating
	2. Lignite B	—	—	—	—	—	6,300	Nonagglomerating

*This classification does not include a few coals, principally nonbanded varieties, which have unusual physical and chemical properties and which come within the limits of fixed carbon or calorific value of the high-volatile bituminous and subbituminous ranks. All these coals either contain less than 48 percent dry, mineral-matter-free fixed carbon or have more than 15,500 moist, mineral-matter-free Btu/lb. To obtain kilojoules per kilogram, multiply by 2.33.

†*Moist* refers to coal containing its natural inherent moisture but not including visible water on the surface of the coal.

‡If agglomerating, classify in low-volatile group of the bituminous class.

§Coals having 69 percent or more fixed carbon on the dry, mineral-matter-free basis shall be classified according to fixed carbon, regardless of calorific value.

¶It is recognized that there may be nonagglomerating varieties in these groups of the bituminous class and that there are notable exceptions in high-volatile C bituminous group.

$$\text{Dry, Mm-free FC} = \text{FC}/\ [100 - (M + 1.1A + 0.1S)] \times 100 \quad (1.4)$$
$$\text{Dry, Mm-free VM} = 100 - \text{Dry, Mm-free FC} \quad (1.5)$$
$$\text{Moist, Mm-free Btu} = \text{Btu}/\ [100 - (1.1A + 0.1S)] \times 100 \quad (1.6)$$

where Mm = mineral matter
Btu = British thermal units per pound GCV (kJ/kg)
FC = fixed carbon, %
VM = volatile matter, %
M = moisture, %
A = ash, %
S = sulfur, %

These quantities are all on the inherent (equilibrium) moisture basis. This basis refers to coal containing its natural inherent or bed moisture, but not including water adhering to the surface of the coal.

Carbon dioxide in coal is another standard, and it is required in estimating the mineral matter content of high-carbonate coals. Its use is primarily for determination of coal rank. U.S. coals arranged in order of ASTM classification by rank are shown in Table 1.5.

Other Physical Characteristics

Bulk Density. Bulk density of coal [cubic foot (cubic meter) weight of crushed bituminous coal] is covered by ASTM D 291. There are two procedures for determining an uncompacted weight and a compacted weight, with both on an as-received basis. The Standard indicates the main use of such weight is primarily for that as crushed and charged into coke ovens, and it states that the moisture content and size distribution of the coal should also be reported with the bulk density, since these are two main factors that affect the cubic-foot (cubic-meter) weight. The Standard excludes powdered (pulverized) coal and the bulk densities of coals in storage piles.

From a practical standpoint, Standard D 291 is not of much use for application to steam-coal utilization problems, since many nonbituminous coals are fired as steam coals. Furthermore, the bulk density of coal is also affected by its ash content; that is, greater ash content equals greater bulk density. Commonly burned bulk densities of coal for use in determination of bunker and coal-feeder sizes are listed in Table 1.6.

For coal-bunker structural design, some engineers use the density of water, that is, 62.4 lb/ft^3 (1000 kg/m^3). For some high-ash coals, the greater bulk density value should be determined and used.

For safe storage of compacted coals having relatively low-ash contents (less than 25 percent), a safe bulk density of crushed coal is in the range of 68 to 72 lb/ft^3 (1089 to 1153 kg/m^3). Some high-ash coals of about 30 to 45 percent ash may have much higher bulk densities.

Size Stability. The relative size stability (and friability) of sized coal is covered by ASTM Standard D 440, Drop Shatter Test for Coal. For steam-coal use, this standard has found very limited application.

Friability. The relative friability of a particular size of sized coal is covered by ASTM Standard D 441, Tumbler Test for Coal. It indicates the liability of coal to break into smaller pieces when it is subjected to repeated handling or to indicate

TABLE 1.5 ASTM Classification for U.S. Coals

Coal rank			Coal analysis, inherent equilibrium basis						Rank
Class	Group	Designation*	M, %	VM, %	FC, %	A, %	S, %	GCV, Btu/lb†	(FC) GCV, Btu/lb†
I	1	ma	4.5	1.7	84.1	9.7	0.77	12,745	(99.2)
	2	an	2.5	6.2	79.4	11.9	0.60	12,925	(94.1)
	3	sa	2.0	10.6	67.2	20.2	0.62	11,925	(88.7)
II	1	lvb	1.0	16.6	77.3	5.1	0.74	14,715	(82.8)
	2	mvb	1.5	20.8	67.5	10.2	1.68	13,720	(77.5)
	3	hvAb	1.5	30.7	56.6	11.2	1.82	13,325	(65.8)/15,230
	4	hvBb	5.8			11.7	2.70	11,910	13,710
	5	hvCb	12.2			9.0	3.20	11,340	12,630
III	1	subA	14.1			7.0	0.43	11,140	12,075
	2	subB	25.0			3.7	0.30	9,345	9,745
	3	subC	31.0			4.8	0.55	8,320	8,790
IV	1	ligA	37.0			4.2	0.40	7,255	7,610

*ma = meta-anthracite; an = anthracite; sa = semianthracite; lvb = low-volatile bituminous coal; mvb = medium-volatile bituminous coal; hvAb = high-volatile A bituminous coal; hvBb = high-volatile B bituminous coal; hvCb = high-volatile C bituminous coal; subA = subbituminous A coal; subB = subbituminous B coal; subC = subbituminous C coal; ligA = lignite A.

†kJ/kg = Btu/lb × 2.33.

TABLE 1.6 Coal Densities for Finding Bunker and Feeder Size

Coal size	lb/ft^3(kg/m^3)	ft^3/ton (m^3/Mg)
	Anthracite	
Stoker (or larger)	54 (865)	37 (1.15)
Cyclone*	54 (865)	37 (1.15)
Pulverized	50 (801)	40 (1.25)
	Bituminous and subbituminous	
Stoker (or larger)	50 (801)	40 (1.25)
Cyclone*	50 (801)	40 (1.25)
Pulverized	45 (721)	44.5 (1.39)
	Lignite	
Stoker (or larger)	45 (721)	44.5 (1.39)
Cyclone*	45 (721)	44.5 (1.39)
Pulverized	42 (673)	47.5 (1.48)

*¼ in × 0 in (6.4 mm × 0 mm).

the relative extent to which sized coals will suffer size degradation in certain mechanical-feed devices.

***Dustiness*.** The method covering the relative index of the dust produced in handling coal is described in ASTM Standard D 547. It may be used to determine the amount of coarse dust and float dust for coal having moisture as sampled or on an air-dried moisture basis. The Standard does not specify showing the moisture content for the dustiness index. Moreover, reproducibility of results may be extremely poor. This standard is scheduled to be withdrawn. As for the drop-shatter and tumbler tests, they have found limited application for steam coal in recent years, but more frequent use may be possible in the future.

EVALUATION OF ASTM STANDARDS

Moisture

The moisture content of coal is extremely difficult to measure. By definition, the *moisture content* is that water which exists as H_2O and can be driven off at 219°F (104°C) to 230°F (110°C) in an atmosphere of dried air for a period of 1 h. Water in coal may be obtained from decomposition of organic molecules, which is sometimes called *combined water,* surface-absorbed water, capillary-condensed water, dissolved water, and water of hydration of inorganic constituents.

The water of decomposition may not be driven off until a temperature of 525°F (274°C) has been reached, whereas carbon dioxide, methane, and nitrogen may be driven from the coal at temperatures up to 230°F (110°C). Moreover, the coal itself may gain weight by absorption of oxygen. Reproducibility of results of moisture content may vary up to 0.5 moisture point.

Different bases are used to express coal properties: as-determined (ad), as-received (ar), equilibrium-moisture (eqm), air-dried [from *air-dry loss* (ADL)], dry (d), dry ash-free (daf), and as-fired (af).

TABLE 1.7 Conversion of Coal to Different Bases

Given	Wanted			
	As determined	As received	Dry	Dry, ash-free
As determined (ad)	—	$\frac{100 - M_{ar}}{100 - M_{ad}}$	$\frac{100}{100 - M_{ad}}$	$\frac{100}{100 - M_{ad} - A_{ad}}$
As received (ar)	$\frac{100 - M_{ad}}{100 - M_{ar}}$	—	$\frac{100}{100 - M_{ar}}$	$\frac{100}{100 - M_{ar} - A_{ar}}$
Dry (d)	$\frac{100 - M_{ad}}{100}$	$\frac{100 - M_{ar}}{100}$	—	$\frac{100}{100 - A_{d}}$
Dry ash-free (daf)	$\frac{100 - M_{ad} - A_{ad}}{100}$	$\frac{100 - M_{ar} - A_{ar}}{100}$	$\frac{100 - A_{ad}}{100}$	—

TABLE 1.8 Conversion Results of Coal Data to Different Bases

Item	Basis				
	As received	Equilibrium moisture	As determined	Dry	Dry, ash-free
Moisture (ad), %	—	—	9.00	—	—
Moisture (eqm), %	—	27.00	—	—	—
Moisture, total (ar), %	29.02	—	—	—	—
VM, %	31.78	32.68	40.74	44.77	49.0
FC, %	33.07	34.01	42.40	46.59	51.0
Ash, %	6.13	6.31	7.86	8.64	—
Carbon, %	46.86	48.19	60.08	66.02	72.26
Hydrogen, %	3.46	3.56	4.43	4.87	5.33
Nitrogen, %	0.69	0.71	0.88	0.97	1.06
Chlorine, %	0.00	0.00	0.00	0.00	0.00
Sulfur, %	0.57	0.58	0.73	0.80	0.88
Oxygen, %	13.27	13.65	17.02	18.70	20.47
GCV, Btu/lb	8,365	8,603	10,724	11,785	12,900
GCV, kJ/kg	19 457	20 011	24 944	27 412	30 005
lb ash/MBtu	7.33	7.33	7.33	7.33	—
kg ash/MJ	3.15	3.15	3.15	3.15	—
lb/SO_2/MBtu	1.36	1.36	1.36	1.36	1.36
kg SO_2/MJ	0.585	0.585	0.585	0.585	0.585

The as-fired basis is not defined in ASTM, but it is used by the *American Boiler Manufacturers Association* (ABMA), where the definition is "fuel in the condition as fed to the fuel-burning equipment." In reality, this is the as-received total moisture basis. Conversion of coal data to different bases appears in Table 1.7 while Table 1.8 shows the calculated results of coal data to these bases. Reference is to ASTM Standard D 3180, Calculation of Coal and Coke Analysis from As-determined to Different Bases.

The total moisture content of the coal reflects the ease of handling the fuel and how that will affect the operating efficiency of the boiler. If one cannot calculate the new boiler efficiency, assume a 1.25 percent efficiency change for a 10-moisture-point change.

Volatile Matter

The volatile matter represents the percentage of gaseous products in the coal, exclusive of moisture vapor. Normally, for nonsparking-type coals, the procedure is to heat the coal to 1742°F ± 36°F (950°C ± 20°C) for 7 min. For sparking-type coals (some low- and high-rank varieties), an alternative procedure is used in which a temperature of 1112°F ± 90°F (600°C ± 50°C) is reached in 6 min. After the preliminary heating, the sample is heated for exactly 6 min at 1742°F ± 36°F (950°C ± 20°C).

The volatile content is the weight loss of the coal sample on a dry basis. Not all the gaseous products from heating may be limited to combustible gases, but may include carbon dioxide, tightly bound moisture, etc. As with the moisture content, this is an empirical test. Reproducibility of results expressed as percentage points are bituminous, 1.0; subbituminous, 1.4; and lignite, 2.0.

For steam-coal use, the volatile matter may indicate its relative ease of ignition, or burnability. Some engineers have established a relationship among the volatile-matter content, dry basis, and theoretical air required for combustion. This practice is inaccurate and should not be followed. For low-rank coals having high-carbonate content, it may be advisable to deduct the carbon dioxide content from the volatile-matter content to determine the true content of the latter.

Ash Content

The ash content is found by weighing the residue remaining after oxidizing the coal under controlled temperature, time, and atmosphere. The coal sample is gradually heated to redness, precluding any mechanical loss because of rapid expulsion of volatile matter. The oxidation is finished at a temperature between 1292°F (700°C) and 1382°F (750°C) until a constant weight is obtained. Some problems may occur for duplication of results for coals having a high-calcite content and pyrites. For this case an alternative method is used, in which the dried coal sample is gradually heated with ample air supply to 932°F (500°C) in 1 h and 1382°F (750°C) in 2 h, and oxidation is continued until a constant weight is obtained.

Two-Step Heating. The oxidation of the coal at high temperature causes an expulsion of water from the clays and calcium sulfate, an expulsion of carbon dioxide from the carbonates, and the conversion of pyrites to ferric oxide, which entails a loss of weight in the original inorganic material. With the two-step heating method, the pyrites are oxidized and expelled before the calcite is decomposed.

Reproducibility of results yields the following: no carbonates present, 0.3; carbonates present, 0.5; and coal with more than 12 percent ash containing carbonates and pyrites, 1.0.

Coals containing large amounts of carbonates with some pyrites, such as those low-rank coals of the western United States, will tend to have less ash than the determined ash content unless these values are corrected. Thus, with some U.S. coals, the accurate determination of the ash content is difficult.

Effects on Coal Components. For steam-coal utilization, the ash content may affect the following:

- Coal storage, cubic-foot (cubic-meter) weight
- Coal handling, if much of the ash is obtained from clayey materials
- Possibility of erosion of air-conveyor coal systems such as pulverizers, classifiers, coal-air pipes, and burners
- Possible erosion of boiler pressure parts
- Dust loading to bottom-ash equipment and dust-collecting equipment

Sulfur Content

The total sulfur content is part of the ultimate analysis of coal; it is one of three elements that contribute to coal's heating value. Three alternative methods are available, but methods A and B (Eschka and bomb washing) are more commonly selected, although their precision for reproducibility is less than that determined by method C (high-temperature combustion). Reproducibilities are thus: method A or B, less than 2% sulfur, 0.10; 2% sulfur or more, 0.20. And method C, less than 2% sulfur, 0.15; 2% sulfur or more, 0.25.

For steam-coal use, the total sulfur content of a coal is a very important parameter. Most commonly, the sulfur content may be a maximum limit, as determined by the various government agencies. It is used to determine the temperature of the exit flue gases from combustion equipment, to preclude the formation of sulfurous acids that are corrosive. Sulfur content is a key parameter for the design of electrostatic precipitators. It may also have other uses, such as an indicator of potential high-temperature slagging of furnaces, high-temperature corrosion, etc.

Another standard pertains to the test method for the determination of three commonly recognized forms of sulfur: sulfate sulfur (which is usually less than 0.10 percent in coal), pyritic sulfur (disulfides), and organic sulfur (determined by difference).

The sulfur content may be used to indicate the degree of high-temperature furnace slagging, of pulverizer grinding element wear rates, and of erosion of some pulverizer components, including classifiers, coal-air pipes, and burners. It can also help evaluate coal preparation and processing operations, to reduce coal sulfur levels.

Ultimate Analysis

This analysis involves the determination of carbon and hydrogen in coal as found in the gaseous products of its complete combustion; the determination of sulfur, nitrogen, and ash in coal as a whole; and the calculation of oxygen by difference. The carbon, hydrogen, and nitrogen have no reproducibility limits. The reproducibilities are summarized in Table 1.9.

For steam-coal use, the ultimate analysis determines the amount of theoretical air required for combustion. As a convenient working tool, the dry theoretical air required per 10,000-Btu (10 550-kJ) heating value of the coal is

$$144 \times \frac{8C + 24\,(H_2 - O_2/8) + 3S}{GCV}$$

where C = carbon, % by wt
O_2 = oxygen, % by wt
H_2 = hydrogen, % by wt
S = sulfur, % by wt
GCV = gross calorific value, Btu/lb (kJ/kg)

As an alternate method of determining the theoretical air required per 10,000 Btu (10 550 kJ) of the GCV of the coal, one may simply select 7.56 lb (3.43 kg) per 10,000 Btu (10 550 kJ) of coal.

Analysis of Coal Ash

Standard methods are used for finding the major elements of coal ash by rapid and relatively inexpensive analysis. The coal is ashed under standard conditions and ignited to standard weight to isolate SiO_2, Al_2O_3, TiO_2, Fe_2O_3, CaO, MgO, Na_2O, and K_2O.

Phosphorus pentoxide (P_2O_5) is also determined; although these values may be in small amounts, they are important in metallurgical coke use. A separate standard method is needed to determine the sulfur trioxide (SO_3) portion of the major-element ash analysis.

Reproducibility Limits. Reproducibility limits from a major-element ash analysis are SiO_2, 2.0; Al_2O_3, 2.0; TiO_2, 0.25; P_2O_5, 0.15; Fe_2O_3, 0.7; CaO, 0.4; MgO, 0.5; Na_2O, 0.3; and K_2O, 0.3.

The SO_3 is determined, since it may be significantly large, in an attempt to obtain an accounting of 100 percent of the constituents. At the time of this writing, it is known that the ignition to constant-weight (oxidation) process volatilizes

TABLE 1.9 Sulfur and Ash Reproducibilities

Sulfur	Points
Coal, under 2%	0.10
Coal, over 2%	0.20
Ash	**Points**
No carbonates present	0.30
Carbonates present	0.50
Coal with more than 12% ash containing carbonates and pyrites	1.00

TABLE 1.10 Reproducibility Limits

Call point*	Temperature, °F (°C)
IT	125 (70)
ST	100 (55)
HT	100 (55)
FT	150 (85)

*See Table 1.2

some of the sodium and potassium in the coal. As a result, those values of reported Na_2O and K_2O in ash will be biased low.

Ash Constituents. For steam-coal use, the constituents in the ash are very important, since they may indicate the behavior of the ash at elevated temperatures and under different conditions. As an example, the slag viscosity of different ash types may be found by the base-acid ratio, where the basic constituents are Fe_2O_3, CaO, MgO, Na_2O, and K_2O; and the acidic constituents are SiO_2, Al_2O_3, and TiO_2. Other use may indicate the presence of abrasive material such as rutile or brookite, based on an abnormally high TiO_2 content, or the presence of quartz by an abnormally high silica-alumina ratio.

Use of the SO_3 retention in ash by standard methods should never be adapted for the prediction of sulfur retention of coal ash in combustion systems. Nor should the sulfur content of the ash be added to that of the coal, since it is already accounted for.

Until a new procedure is adopted, some coals containing a relatively large amount of NaCl, or a relatively large amount of Na_2O, may contain much more sodium oxide than is reported through the current method.

Fusibility of Coal Ash

The fusibility of coal ash is covered by an empirical method. Reproducibility limits for reducing atmospheres are shown in Table 1.10. Note that there are relatively large potential inherent errors associated with fusibilities. Thus, they should never be used by themselves for evaluation of the performance of coals in utility and industrial combustion systems.

ACKNOWLEDGMENT

American Society for Testing and Materials as source for standards and tabular information.

CHAPTER 3.2
COAL ANALYSIS AND COAL SIZING

Albert F. Duzy
President
Duzy & Associates
Sun Lakes, AZ

INTRODUCTION

When an engineer reads a coal analysis sheet containing data that was created by using ASTM standards, she or he should realize that three distinct procedures were performed first: procurement of the gross sample, preparation of the coal sample for laboratory analysis, and the laboratory analysis itself. All three procedures must be accomplished in accordance with current ASTM standards.

With respect to the coal analysis, a review and study of data indicate the following probable sources of error:

- Collection of gross sample, 65 percent
- Preparation of sample, 20 percent
- Laboratory analysis, 15 percent

Thus, one can readily see the importance of performing each of these three procedures as meticulously as possible, especially sample collection, and adhering strictly to the appropriate ASTM method.

COAL ANALYSIS

In performing the first procedure, ASTM Standard D 2234 should be followed. This standard method gives the overall requirements for the collection of coal samples. The wide variety of coal-handling facilities precludes the publication of detailed procedures for every sampling situation.

Collection of Gross Samples

The proper collection of the sample involves the understanding and consideration of the physical character of the coal, the number and weight of increments, and the overall precision required. The collection of a gross sample of coal is a difficult task if the sample is to be truly representative. Here is an approach that is recommended.

The sample increment collection classification, in regard to increment types, conditions, and spacing, appears in Table 2.1. The number and weight of increments for the general sampling procedure are shown in Table 2.2. Furthermore, one gross sample can represent the total tonnage, provided the number of increments per Table 2.1 are increased as follows:

$$N_2 = N_1 \sqrt{L/1000 \text{ tons (908 Mg)}}$$

where N_2 = number of increments required
N_1 = number of increments specified in Table 2.1
L = total lot size, tons (Mg)

For example, for a unit train holding approximately 10,000 tons (9080 Mg) of crushed −2-in (−50-mm) raw coal, the number of increments required, after it is determined that N_1 = 35 from Table 2.1, would be

$$N_2 = 35 \sqrt{10{,}000/1000} = 110.6 \text{ increments}$$

The minimum total gross sample weight from Table 2.2 would be 111 × 6 lb = 666 lb (333 kg).

Preparing Coal Samples

In performing this procedure, ASTM Standard D 2013 should be followed. Sample preparation is a process that may include air drying, crushing, division, and mixing of a gross sample for the purpose of obtaining an unbiased analysis sample.

TABLE 2.1 Increment Types, Conditions, and Spacing

Condition of increment collection from the main body of coal	Spacing of increments			
	Type I increment*		Type II increment†	
	1. Systematic	2. Random	1. Systematic	2. Random
A, stopped belt cut	I-A-1	I-A-2	II-A-1	II-A-2
B, full stream cut	I-B-1	I-B-2	II-B-1	II-B-2
C, part stream cut	I-C-1	I-C-2	II-C-1	II-C-2
D, stationary sampling	I-D-1	I-D-2	II-D-1	II-D-2

*No human discretion is used.
†Human discretion is used.

TABLE 2.2 Number and Weight of Increments for General-Purpose Sampling

	Top size		
	⅝ in (16 mm)	2 in (50 mm)	6 in (150 mm)
Mechanically cleaned coal			
Minimum number of increments	15	15	15
Minimum weight of increments, lb (kg)	2 (1)	6 (3)	15 (7)
Raw (uncleaned) coal			
Minimum number of increments	35	35	35
Minimum weight of increments, lb (kg)	2 (1)	6 (3)	15 (7)

Sample reduction is the process whereby a sample is reduced in particle size by crushing or grinding without a change in weight. Sample division is the process whereby a sample is reduced in weight without changing the particle size.

Standard D 2013 includes lists of definitions, apparatus, and various procedures to follow to obtain unbiased analysis samples from the gross sample of coal.

Laboratory Analysis of Sample

In performing this final procedure, these ASTM Standards of laboratory test methods for analysis should be followed:

Moisture

- *D 1414:* Equilibrium moisture of coal at 96 to 97 percent relative humidity and 30°C
- *D 3302:* Moisture in coal, total
- *D 2691:* Moisture in coal, total reduced to no. 8 top sieve size
- *D 3173:* Moisture in the analysis sample of coal and coke

Proximate

- *D 3172:* Proximate analysis of coal and coke
- *D 3175:* Volatile matter in the analysis sample of coal and coke
- *D 3174:* Ash in the analysis sample of coal and coke

Ultimate

- *D 3176:* Ulimate analysis of coal and coke
- *D 3178:* Carbon and hydrogen in the analysis sample of coal and coke
- *D 3179:* Nitrogen in the analysis sample of coal and coke
- *D 3177:* Sulfur in the analysis sample of coal and coke
- *D 4239:* Sulfur in the analysis sample of coal and coke using the high-temperature tube furnace combustion method

- *D 2361:* Chlorine in the analysis sample of coal and coke
- *D 4208:* Total chlorine in coal by the oxygen bomb combustion/ion selective electrode method

Gross Calorific Value

- *D 2015:* Gross calorific value of solid fuel by the adiabatic bomb calorimeter
- *D 3286:* Gross calorific value of solid fuel by the isothermal-jacket bomb calorimeter

Mineral Matter in Coal

- *D 2795:* Analysis of coal and coke ash
- *D 3682:* Major and minor elements in coal and coke ash by the atomic absorption method
- *D 4326:* Major and minor elements in coal and coke ash by X-ray fluorescence
- *D 1857:* Fusibility of coal and coke ash
- *D 3683:* Trace elements in coal and coke ash by the atomic absorption method
- *D 3684:* Total mercury in coal by the oxygen-bomb combustion/atomic absorption method
- *D 3761:* Total fluorine in coal by the oxygen-bomb combustion/ion selective electrode method
- *D 4606:* Arsenic and selenium in coal by the hydride generation/atomic absorption method
- *D 1757:* Sulfur in ash from coal and coke

Other Test Methods

- *D 409:* Grindability of coal by the hardgrove machine method
- *D 720:* Free-swelling index of coal
- *D 1756:* Carbon dioxide in coal
- *D 2492:* Forms of sulfur in coal

Petrographic Analysis

- *D 2796:* Megascopic description of coal and coal seams and microscopic description and analysis of coal
- *D 2797:* Preparing coal samples for microscopic analysis by reflected light
- *D 2798:* Microscopic determination of the reflectance of the organic components in a polished specimen of coal
- *D 2799:* Microscopic determination of the volume percentage of the physical components of coal

ASTM Standards for coal and coke are reissued annually, so the year of original adoption, the last revision, or the last reapproval have not been included.

The latest version of these Standards should be obtained from ASTM before the coal analysis is begun. Also note that ANSI numbers have not been used.

COAL SIZING

In performing this procedure, ASTM Standard D 310 for anthracitic coals and ASTM Standard D 431 for other ranks of coal should be followed.

Anthracitic Coals

Standard anthracitic sizing specifications are listed in Table 2.3 for screen openings specified by the Commonwealth of Pennsylvania, Specifications for Coal, Anthracite.

Other Ranks of Coal

Standard D 431 covers coal sizing from sieve analysis tests of samples taken to represent the condition of the coal as sold. The sieve defining the upper limit shall be the smallest sieve upon which is retained a total of less than 5 percent, by weight, of the sample. The sieve defining the lower limit shall be the largest sieve through which passes a total of less than 15 percent, by weight, of the sample. The screen and sieve sizes in Table 2.4 should be used.

For wire-cloth specifications, refer to ASTM Standard E 11. For perforated-plate specifications, refer to ASTM Standard E 323. For a method to cover sieve analysis of crushed bituminous coal, such as is charged into coke ovens, use ASTM Standard E 311.

ASTM Standard D 197, Sampling and Fineness Test of Pulverized Coal, is a key standard. The method covers collection of the gross sample, preparation of the laboratory sample, and determination of the fineness of pulverized coal by sieve analysis from no. 16 mesh (1.18 mm) to no. 200 mesh (75 μm).

TABLE 2.3 Standard Anthracite Sizing Specifications

	Size of round-hole openings in test screens, in (mm)	
Item	Passing	Retained on
Broken	$4\frac{3}{8}$ (111)	$3\frac{1}{4}$ to 3 (82.6 to 75)
Egg	$3\frac{1}{4}$ to 3 (82.6 to 75)	$2\frac{7}{16}$ (61.9)
Stove	$2\frac{7}{16}$ (61.9)	$1\frac{5}{8}$ (41.3)
Nut	$1\frac{5}{8}$ (41.3)	$1\frac{3}{16}$ (30.2)
Pea	$1\frac{3}{16}$ (30.2)	$\frac{9}{16}$ (14.3)
Buckwheat	$\frac{9}{16}$ (14.3)	$\frac{5}{16}$ (8)
Rice	$\frac{5}{16}$ (8)	$\frac{3}{16}$ (4.75)

TABLE 2.4 Screen and Sieve Sizes for Other Coals

Round-hole screens, in (mm)		Wire-cloth sieves with square openings
8 —	1½ (37.5)	No. 4 (4.75 mm)
6 —	1¼ (31.5)	No. 8 (2.36 mm)
5 (125)	1 (25)	No. 16 (1.18 mm)
4 (100)	¾ (19.0)	No. 30 (600 μm)
3 (75)	½ (12.5)	No. 50 (300 μm)
2½ (63)	⅜ (9.5)	No. 100 (150 μm)
2 (50)		No. 200 (75 μm)

FINAL COMMENTS

To illustrate the importance of sample preparation, refer to the commonly used term in coal contracts (sales agreements) whereby it is specified that the gross sample shall be divided into three equal parts. The implication is that the gross sample will then be in three equal parts as to quantity and quality—a virtual impossibility. In accordance with ASTM methods, four subsamples should be prepared from the gross sample—a more realistic approach.

Petrographic methods are listed, although from a practical standpoint they are used primarily in metallurgical coal applications. Still, they may find some limited use in steam-coal applications. For larger coal sizing, square-hole plates may be selected; it is most important that the coal sizing be specified as square- or round-hole openings.

It is impossible to explain in detail all the problems associated with obtaining a gross sample, preparing it, and making the laboratory analysis. A thorough and careful reading of the appropriate ASTM Standards, however, should permit the engineer to avoid most of, if not all, these problems.

ACKNOWLEDGMENT

American Society for Testing and Materials as source for standards and tabular information.

CHAPTER 3.3
COAL HANDLING

Robert G. Schwieger
Power *Magazine*
New York, NY

INTRODUCTION

Coal usually arrives at steam-generating plants by rail, barge, or truck. (In special cases, ships, conveyor belts, or pipelines may be preferable.) Rail and truck transportation are the most common methods; however, the movement of coal by barge has increased substantially in recent years. Then the coal is unloaded, prepared, and transferred to either storage or the plant for firing.

BARGE TRANSPORTATION

Practically all river coal travels in barges that are either 175 ft × 26 ft or 195 ft × 35 ft (53 m × 8 m or 59 m × 11 m). The smaller size, sometimes called the *standard* or *Pittsburgh standard* barge, carries about 1000 tons (about 900 Mg); the larger, or *jumbo* barge, holds 1500 tons (1360 Mg). Most barges in service today have watertight cargo spaces; also, most have open tops. Covered barges are not used, except when required by a backhaul cargo, such as grain, because they are harder to load and unload and many docks cannot accept them.

Barge coal usually contains a high proportion of fines. In addition, its moisture content can increase significantly during transport if heavy precipitation is encountered. For example, 1 in (1 mm) of rain falling on an open jumbo barge will increase its payload by 12 tons (0.44 Mg), the water generally being absorbed by the upper few feet (meters) of coal.

Such conditions require that coal sampling and weighing be done at the same time and place. To illustrate, if sampling is done at the origin and weighing at the destination, the consumer will pay for the weight of rainwater and not receive credit for the corresponding increase in moisture.

RAIL TRANSPORTATION

Three basic ways to ship coal by rail are bulk rate and in unit and integral trains. The bulk-rate train generally consists of shipments from one or more points of origin, with a definite minimum tonnage per train to qualify for special rates. Unit trains move coal between one point of origin and one destination, with the possibility of altering trips to other destinations to obtain better utilization of equipment. The integral train has cars and motive power coupled from one point of origin to one destination and return, exclusively, that is, except when maintenance is required.

Operation of the bulk-rate train is much the same as that with standard individual-car shipments. Motive power is usually dispatched from the regular railroad pool; so are hopper cars, which may vary in capacity or be limited in certain minimum capacities. Cars may be loaded and unloaded singly or in small groups (as is done in normal rail shipment), but time limitations are imposed for these operations.

Unit trains generally are considered only by industrial plants consuming at least 150,000 tons/yr (136 000 Mg/yr) of coal. However, there are exceptions. The unit train is but one element of a total coal transportation and coal-handling system, which requires careful overall analysis before equipment is committed. Variables such as minimum train capacity, distance between mine and plant, car type and capacity, motive power requirements, loading and unloading facilities, and storage must be evaluated because they all affect scheduling and economics.

Integral trains have virtually no application in the industrial sector. They handle only the largest shipments of coal; consequently, integral trains are confined primarily to electric utility and export trade.

UNLOADING COAL

The first step in the coal-handling process at any plant is unloading the incoming fuel. This phase of the operation should be accomplished as quickly as possible, to minimize labor costs and avoid costly demurrage charges. A rule of thumb commonly applied suggests that enough equipment capacity be on hand to unload at least a 3-day supply in an 8-h shift.

Just what kind of equipment will do the best job basically depends on the transport vehicle and the quantity of coal received in a typical shipment. If the coal is delivered by truck and the plant arrangement is favorable, then equipment specifically intended for unloading may not be required. For example, all that may be necessary is for a truck to dump to a small outdoor storage pile or into a basement bin.

If coal is received by rail, some means of bringing the hopper cars to the right spot for rapid unloading probably will be needed. In small plants, where only a few cars are handled at a time, this often will be a capstan (Fig. 3.1), sometimes called a *car puller* or *car spotter*. If many cars must be unloaded in rapid sequence, such as with unit-train deliveries, a far more complex piece of equipment, called a *positioner,* will be needed (Fig. 3.2), which permits sequenced, automatic, centralized control of the positioning-dumping process.

Hopper Cars

The unloading system also should be designed to accommodate the specific type of hopper cars that will be delivering the coal. They come in three basic designs: the conventional, manually locked-door, sawtooth hopper; a bottom-dump type, where virtually the entire coal-supporting area is opened for quick unloading; and a top-dump car, which is unloaded by rotary-car dumpers (Figs. 3.3 to 3.6).

With the sawtooth design typically selected for small shipments and bulk-rate trains, some type of shaker or vibrator is normally required for fast, complete discharge, especially when the coal is wet or frozen. These units range from the relatively inexpensive lightweight shaker (Fig. 3.7) to the more complex hoist-mounted shaker (Fig. 3.8). Under normal operating conditions, a good hoist-mounted shaker can handle from six to eight cars per hour.

The majority of unit trains also use bottom-dumping cars, but most of these are of a more modern type that is designed for rapid unloading [100 tons (91 Mg) in 20 s, for example] without the aid of shakers. These cars may have steeper interior slopes and special hopper doors that can be opened remotely.

Some unit trains and most integral trains have top-dump cars. They require

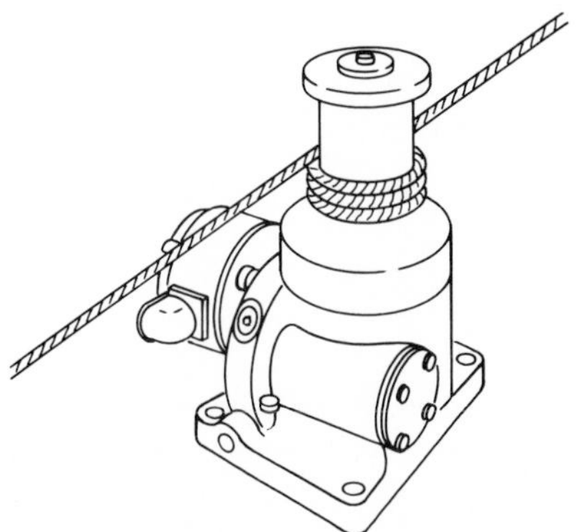

FIG. 3.1 Winch helps pull rail cars into position for unloading.

FIG. 3.2 Positioner permits rapid, automatic unloading of coal cars.

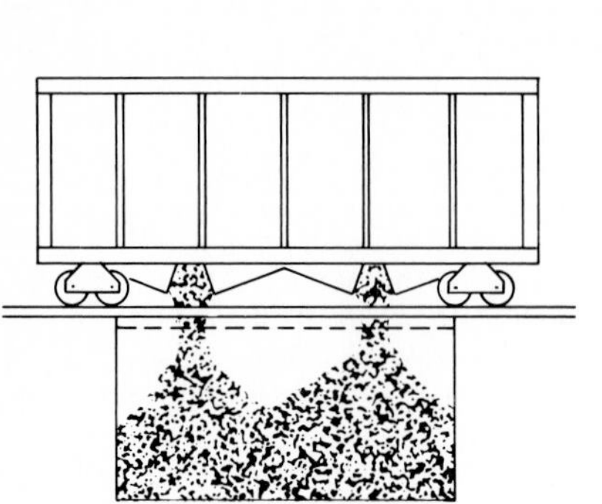

FIG. 3.3 Hopper-bottom cars for small-capacity unloading.

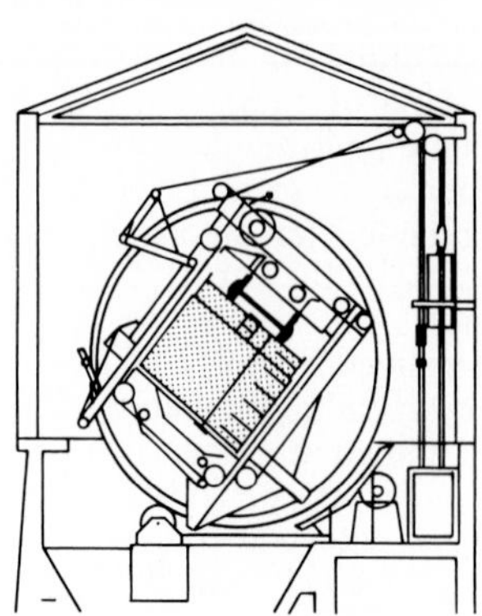

FIG. 3.4 Rotary dumper for high-capacity unloading.

rotating dumpers at the plant (Fig. 3.4). This machine is automatic and so simple to operate (just one button must be pressed) that unskilled labor can be employed. The entire dump-and-return cycle takes about 30 s.

In cold weather, coal in the hopper cars frequently arrives frozen, and a thawing system is needed to facilitate unloading. Most systems operating today use either electric or gas-fired radiant heaters. For both types, heaters are installed in a thawing shed with a section of track running through it. When a hopper car is in position, heaters located beneath it (between the rails) and alongside it are turned on.

Receiving Hoppers

An important feature in the design of an undertrack hopper is the slope of its sides. A minimum slope of 50° from the horizontal is needed; this angle should be increased to 60° or more where wet coals are handled.

Some key points worth remembering about hoppers include these:

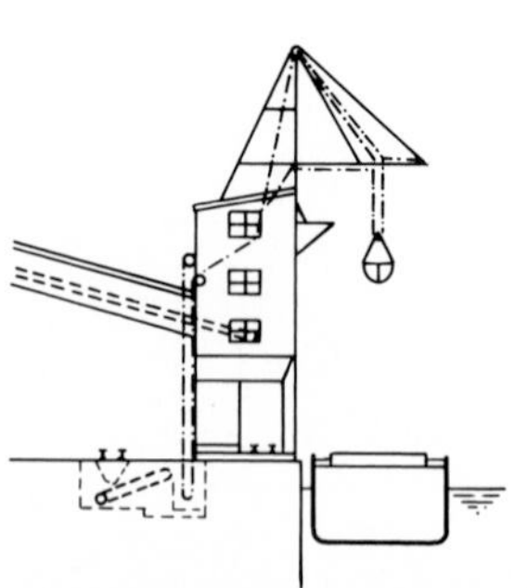

FIG. 3.5 Coal tower with clamshell bucket.

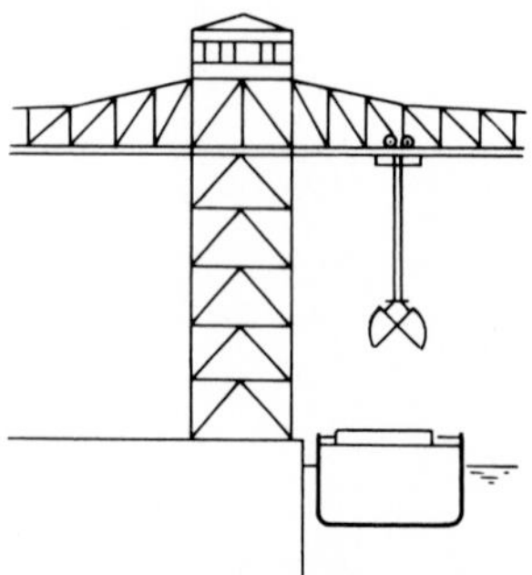

FIG. 3.6 Unloading bridge with clamshell bucket.

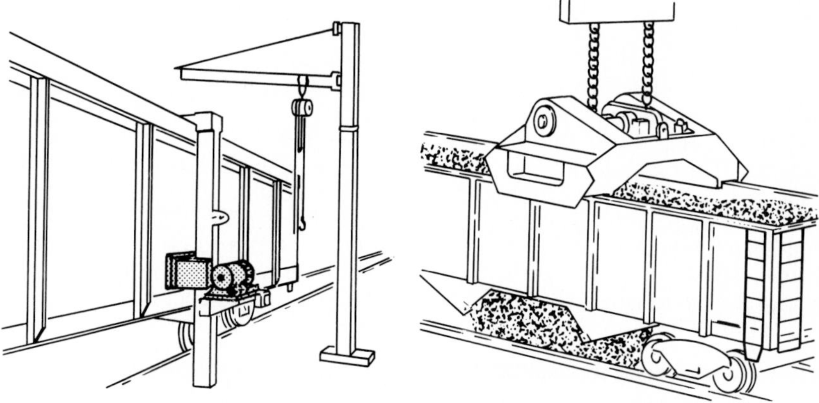

FIG. 3.7 Lightweight shaker.

FIG. 3.8 Heavy-duty shaker.

- Stainless-steel linings enhance hopper performance significantly and minimize maintenance.
- Waterproofing and flood protection should be considered when the hopper is near water.
- Hoppers must be deep enough to hold more than a carload of coal for each car length.

Waterborne Carriers

The general run of barges and ships that deliver coal cannot self-unload, so an external system is needed to discharge cargo. For large operations, specially designed towers and bridges are selected (Figs. 3.5 and 3.6). A typical tower has a steel structure that houses hoisting machinery and a trolley boom for swinging the bucket out over the barge. In the in-board position, the bucket dumps into the hoppers feeding a conveyor. Rail-mounted movable towers also see service.

Coal bridges take many specialized forms, but an essential feature is the ability to deliver coal a short distance inshore for further handling or stockpiling. Bucket elevators also are used for unloading barges and boats.

DELIVERING COAL

Most unloading methods already discussed deliver fairly large slugs of coal at intervals. Track hoppers and other temporary-storage piles absorb these variations in delivery rate, but not all forms of conveying equipment can reclaim directly from them. Devices that deliver coal at a controlled rate, from a storage area to a point where a conveyor can handle it conveniently for transport, are called *feeders*.

Apron Feeders

Heavy and rugged in construction, apron feeders consist of overlapping steel pans, 18 in (450 mm) wide and up, mounted on double rolls of steel rolling chain (Fig. 3.9). They differ from the familiar flight conveyor in that the pans carry the coal, rather than drag or scrape it along.

Apron feeders travel at low speeds, usually less than 25 to 30 ft/min (0.13 to 0.15 m/s). They commonly have fixed side plates or skirt boards to protect the chains and to permit coal to be carried at a considerable depth before it spills over. As a rule, apron feeders do not operate well at inclines greater than 25° from horizontal, that is, unless they are fitted with cleats to prevent coal from flowing backward. Note that the pit depth can be reduced with an apron feeder having an upward slope, to control coal flow from a track hopper.

Bar-Flight Feeders

Where the pit depth must be held to a minimum, the bar-flight feeder serves well. The feeder in Fig. 3.10 operates with about 2 ft (0.6 m) of headroom and delivers 20 tons/h at 20 ft/min (18 Mg/h at 0.1 m/s). It has a simple design and is very effective in handling small capacities. The feeder consists of bars or flights attached to two strands of chain in a manner that permits the bars to slide along the flat bottom of a trough, dragging the coal.

Reciprocating Feeders

These feeders, which agitate incoming coal and thereby work against hangups in the hopper, do a good job with lump material. One design for feeding coal from hopper to crusher or conveyor consists of a steel plate—6 to 9 ft long and 15 to 30 in wide (1.8 to 2.7 m long and 380 to 760 mm wide)—mounted on tracked wheels between guiding skirt plates (Fig. 3.11). Rods connected to adjustable cranks or eccentrics drive the pan, from a motor. As the eccentrics revolve, they push the pan forward with its load of coal. Additional coal then falls from the hopper to the back end of the pan, filling space left by that carried forward. On the return stroke, coal does not move back with the pan, and some falls off the end.

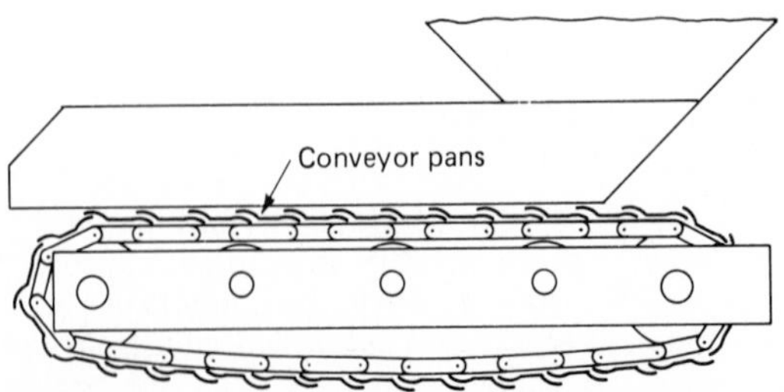

FIG. 3.9 Apron feeder.

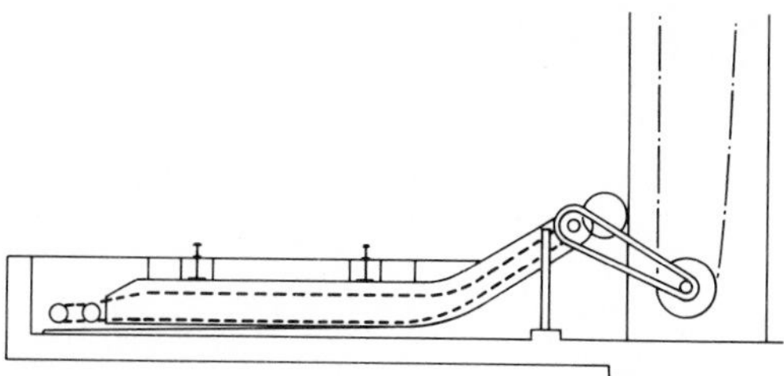

FIG. 3.10 Bar-flight feeder.

Vibrating Feeders

In one type of vibrating feeder (Fig. 3.12), a feeder pan is attached to vibrator arms or bars. A laminated armature is connected to the vibrator bars, and an electromagnet is anchored to the mainframe. When energized, the electromagnet causes the vibrators to deflect toward the magnet as current builds up in its coil and to release when the current passes through its zero point. This causes the bars and feeder pan to vibrate back and forth.

In a more modern design, a rotary vibrator drive creates a sinusoidal excitation force, which is transmitted and magnified through a spring system that holds the feeder pan. The result is a constant, straight-line vibration of the trough system. The trough movement and material delivery remain stable regardless of changes in material headloads or material damping properties.

Although the vibration amplitude in the systems described is so small that the feeder pan does not appear to be in motion, coal flows uphill or downhill in a smooth stream. The feed rate is changed by varying the intensity of vibration. Since vibrating feeders have few, if any, parts subject to friction wear, maintenance (like power demand) is light.

Screw and Belt Conveyors

Screw conveyors (Fig. 3.13) serve well in small plants, that is, plants that receive not more than two hopper cars of coal per week and those that get their coal by truck. They see service where coal comes in small lumps and is free of tramp iron, wood, and other foreign objects. Simple and low in cost, screw conveyors can be built into restricted spaces and can easily be made dusttight.

A screw conveyor usually consists of a long-pitch steel-plate helix on a spindle, carried by bearings within a U-shaped trough. As the spindle rotates, coal moves forward under the thrust of the lower part of the helix and discharges through openings in the bottom or open end of the trough. Capacity suffers when the helix is inclined. For steep inclines, it may be given a short pitch and the trough made tubular to reduce capacity loss.

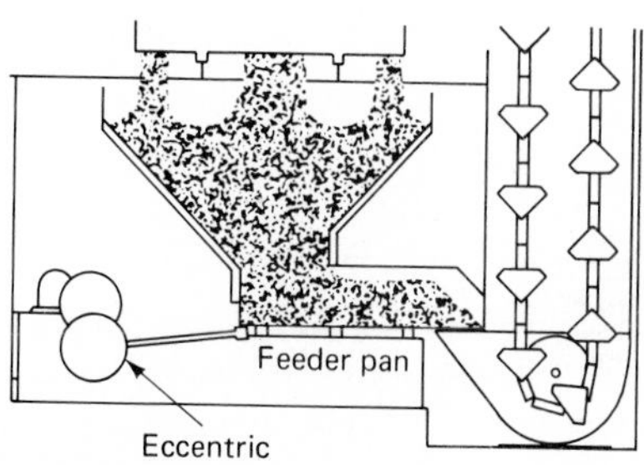

FIG. 3.11 Reciprocating feeder.

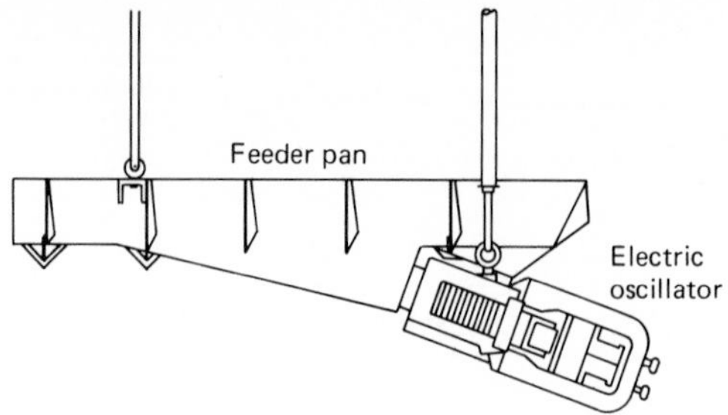

FIG. 3.12 Vibrating feeder.

Simplicity, smooth operation, and uniform discharge make belt conveyors attractive as feeders (Fig. 3.14). When so used, they are short and have closely spaced idlers for support against the impact of hopper-fed coal.

WEIGHING COAL

Since fuel is a major cost item in plant operation, it is important to weigh coal when it is delivered and again just before it is burned. The first measurement ensures that only the energy received is paid for; the second provides input for plant performance calculations. The amount of coal delivered can be weighed either at the unloading point or somewhere in the handling system. Coal flowing to individual boilers is measured during final handling.

Basic Scale Types

The basic types of scales for weighing coal as received are the truck, rail, combination truck and rail, and conveyor-belt units. Weight indicators for these devices include dials, with or without an automatic printout; registering beams; and digital indicators (Figs. 3.15, 3.16, and 3.17).

The weight of a given load is transmitted from a scale to its weight indicator mechanically, through a system of levers, or electrically through load cells. A combination system extends the capabilities of the mechanical-lever scale. It takes output from the lever system and converts it to an electric signal via a load cell. The advantage of this combination is that the weight indicator may be lo-

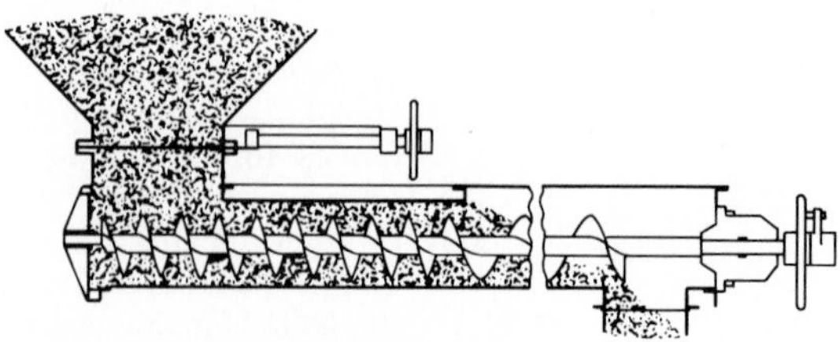

FIG. 3.13 Screw conveyor.

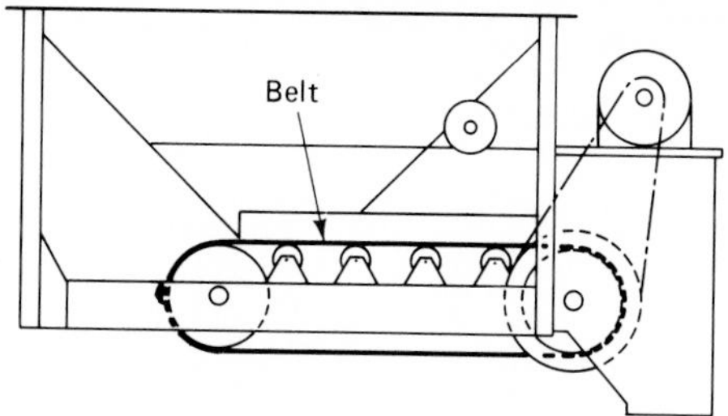

FIG. 3.14 Belt conveyor.

cated away from the scale, just as it can be with load-cell weight sensing. Lever scales require that the indicator be connected mechanically to the scale at the weigh station.

In planning a coal-receiving system, remember that truck and track scales commonly are supplied in the following standard sizes:

- Mechanical-lever type with or without load-cell output. Truck lengths from 8 to 70 ft (2.4 to 21 m) and capacities from 10 to 100 tons (9 to 90 Mg); track lengths from 50 to 75 ft (15 to 23 m) and capacities from 60 to 200 tons (54 to 180 Mg) per section.
- Load-cell weight-sensing type with truck lengths from 8 to 70 ft (2.4 to 21 m) and capacities from 20 to 75 tons (18 to 68 Mg); track lengths from 60 to 80 ft (18 to 24 m) and capacities from 75 to 200 tons (68 to 180 Mg) per section.

Load-cell scales generally are more expensive than mechanical-lever scales in the low- to middle-capacity ranges. Full load-cell scales, however, sometimes offer a shallower, less costly pit and a lower total installed cost. Also, they have fewer moving parts than the mechanical-lever type, so maintenance costs may be slightly lower.

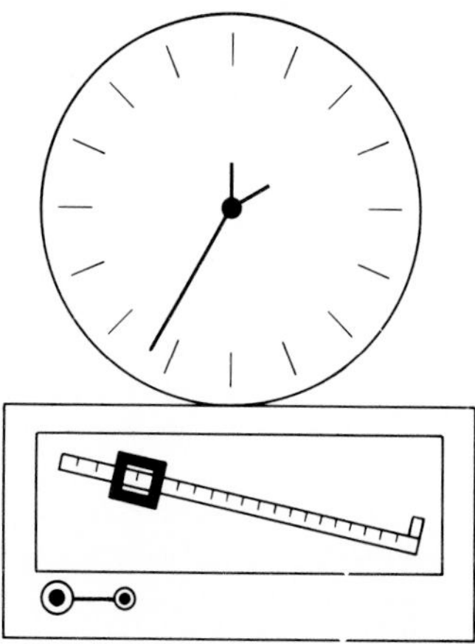

FIG. 3.15 Dial-type weight indicator.

Conveyor-belt scales also come in both mechanical and electrical types. Both designs continuously weigh moving coal by multiplying the varying load on the conveyor belt by the belt speed. The total flow over the conveyor can be calculated at any time by a totalizer, or integrator, and displayed.

Accuracy attained by belt scales is within the industry standard of ± 0.25 percent at full load. Standard scales accommodate belt widths from 18 to 84 in (450 to 2100 mm).

Automatic Coal Scales

Coal is weighed between the bunker and boiler by a scale or feeder. In stoker firing, scales are used exclusively; in pulverized-coal firing, ei-

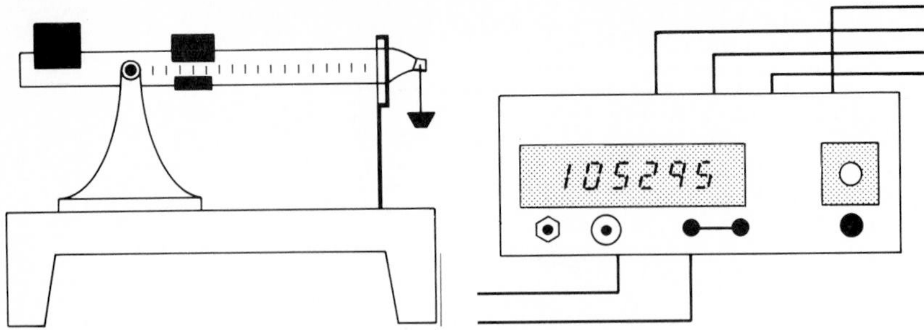

FIG. 3.16 Registering-beam indicator.

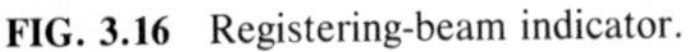

FIG. 3.17 Digital indicator.

ther may be selected. Typically, coal scales handle between 10 and 30 tons/h (about 9 and 27 Mg/h), while feeders process up to about 100 tons/h (90 Mg/h).

An automatic coal scale essentially consists of three major assemblies: a weigh-lever system; a weigh hopper, which has a bottom dump gate and closing mechanism; and a belt feeder, arranged with skirt bars, to transfer coal from the bunker or silo to the weigh hopper.

The scale operates like this: When the weigh-hopper gate closes after dumping a load of coal, the unit's control system locks the hopper gate shut and starts the belt feeder. Coal is fed into the weigh hopper until its weight overcomes the scale's balancing force—typically, 200 to 500 lb (91 to 227 kg). As the weigh lever raises, it triggers action by the control system to stop the feeder and release the weighed coal to the hopper below.

Feeder Weighing

In small industrial plants that burn pulverized coal, economic considerations generally dictate the use of table, pocket, drag, or apron feeders for controlling the fuel flow from the bunker to the pulverizer.

The first unit operates on a stop-start basis and is controlled by a pulverizer level indicator. It delivers quantities of coal that vary widely, even though the device runs at constant speed. The other three feeders are more sophisticated and can be considered volumetric devices, with the rate of coal flow *by volume* being directly proportional to the motor and belt speed. But for one reason or another, none of these is accurate enough to provide weight data for combustion or air pollution calculations. Therefore, plants using this equipment normally install a coal scale between the bunker and the feeder.

Perhaps the most accurate volumetric feeder is the belt type. Normally finding application in moderately large industrial plants, the belt feeder has a fixed leveling bar to maintain the load of coal at a constant height and width. As with the other volumetric feeders, however, it does not compensate for changes in fuel density. Thus it, too, may be used in conjunction with a coal scale.

Where the cost of fuel is high and precise control of stack emissions is critical, the price of a gravimetric feeder usually can be justified. Such a feeder is extremely accurate. Its scale controls a leveling bar that maintains a constant weight of coal per foot of belt. Therefore, the rate of coal flow *by weight* is directly proportional to motor and belt speed.

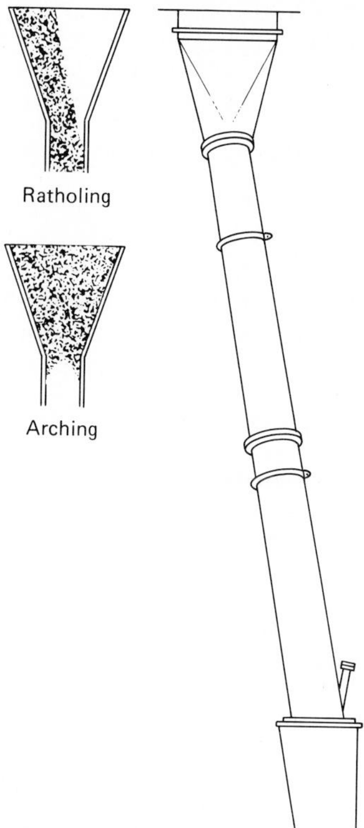

FIG. 3.19 Ratholing and arching can be troublesome at the final handling step if bunkers, chutes, and hoppers are not designed properly. Chutes, such as one at right, should be inclined steeply, preferably be round or oval, and flare in the direction of flow.

the paddle and gamma-ray types. The first has a paddle that is deflected by coal flow. Deflection turns a shaft, which rotates cams to operate limit switches that stop or start coal flow to the bunker. A radium source in the second type of monitor sends gamma rays across the bunker to a Geiger counter. The instrument works on the principle that more gamma rays reach the tube if there is no coal in the bunker to absorb them. Bunkers may also be mounted on load cells that indicate, by weight, the height of coal in the container.

Monitors using gamma rays also find application in downspouts, to detect ratholing and arching. They often are used to start a vibrator on the bunker when a void is detected and to stop it when coal is restored in the downspout.

MOVING COAL INTO PLANT

Transferring coal from the unloading point or storage to the bunkers may be extremely simple, as in a plant where trucks dump coal in a yard storage pile and a lift truck with a scoop takes it directly to stoker hoppers. Generally, the system is far more complex, because both reserve and live-storage inventories must be manipulated. In such cases, a conveying system is required to route coal to either or both of the storage piles. Also, coal must be moved from live storage to bunkers in the plant.

In the transfer operation, coal normally is lifted as well as moved. Most conveying equipment does both, but some designs serve best for vertical or nearly vertical lifts whereas others are better for horizontal runs or slight inclines.

Bucket Elevators

Bucket elevators find extensive application for vertical lifts (Fig. 3.20). Key designs today are the centrifugal-discharge and continuous-bucket types. The former has buckets bolted at regularly spaced intervals along a single strand of endless chain that travels at relatively high speed. It is a medium-capacity device, capable of handling small to medium-size lumps. Buckets dig the material from the elevator's boot and discharge it by centrifugal force.

The continuous design has an overlapping arrangement of buckets that travel at much slower speed. Of larger capacity than the centrifugal types, it handles a wider range of coal sizes, from fines to large lumps. Buckets are fed directly from a loading leg or chute and are emptied by gravity at the discharge point. Thus, continuous elevators require greater headroom and a deeper boot. For large-capacity handling of run-of-mine coals, there is a supercapacity continuous elevator. It feeds at inclines of from 35° to 70° and works at slow speeds, usually under 150 rpm.

Perfect-discharge elevators, which have their buckets between two strands of chain and use snubbing sprockets under the headwheel to bring buckets into the right position above the discharge chute, are most often specified when coal is wet and sticky.

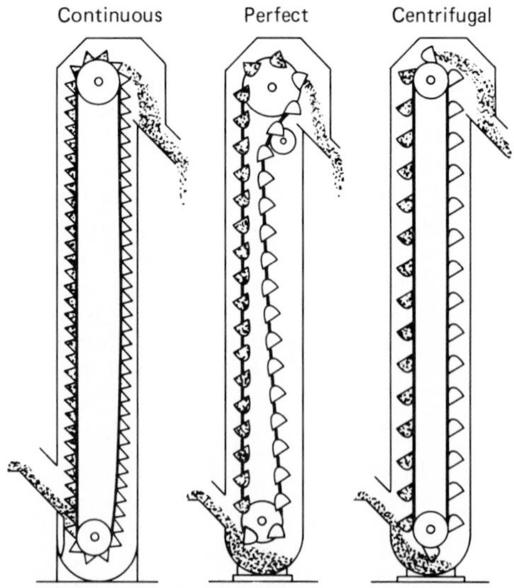

FIG. 3.20 Bucket elevators are centrifugal, perfect-discharge, and continuous types.

Flight conveyors serve well on horizontal and slightly inclined runs where there is little headroom. Discharge can be at the end or at an intermediate point through a gate in the trough.

Continuous-Flow Conveyors

The continuous-flow conveyor (often called a *bulk-flow,* or *Redler, conveyor;* see Fig. 3.21) essentially is a duct within which closely spaced skeleton flights act as the conveying element. The conveyor feeds itself at any point with a uniform load that fills the duct and entirely surrounds the conveying element. When the flights move, they cause the material to flow upward or forward with them in a solid column. Coal can be discharged at any opening where it can fall away from the flights.

Continuous conveyors require no deep loading pit, only enough room for gravity feed. Connecting joints are flexible, permitting easy movement around corners. Further, casings are dusttight, and connections can also be made dusttight. This design serves well for horizontal, vertical, or inclined runs.

Belt Conveyors

Belt conveyors offer relative simplicity, few moving parts, and low power requirements (Fig. 3.22). They serve most effectively on horizontal runs and on inclines up to 15° or 20°. The simplest discharge setup is from the end. When distribution along the belt's length is needed, trippers are used.

Another distribution scheme centers on the shuttle conveyor (Fig. 3.23). It has a reversible belt mounted on a frame for longitudinal travel. Coal can be discharged from either end of the conveyor.

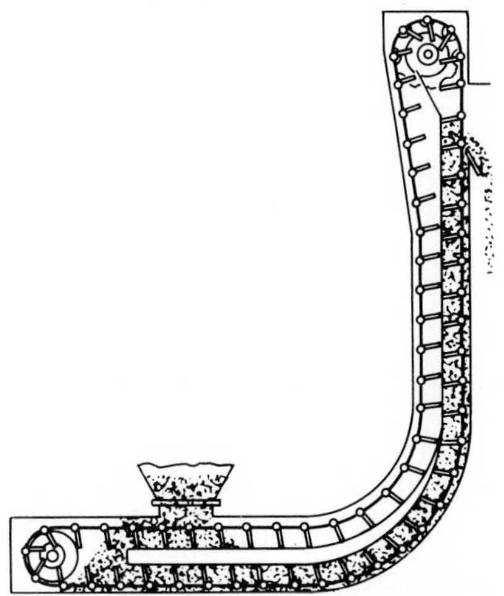

FIG. 3.21 Continuous-flow conveyor (Redler type) lifts vertically, horizontally, and on inclines.

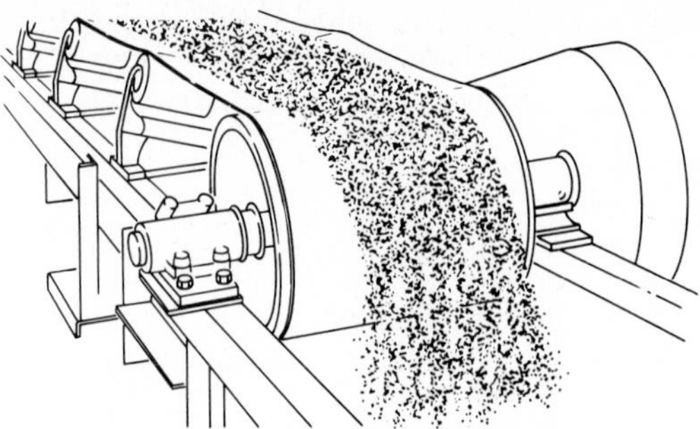

FIG. 3.22 Belt conveyor with fixed-end discharge.

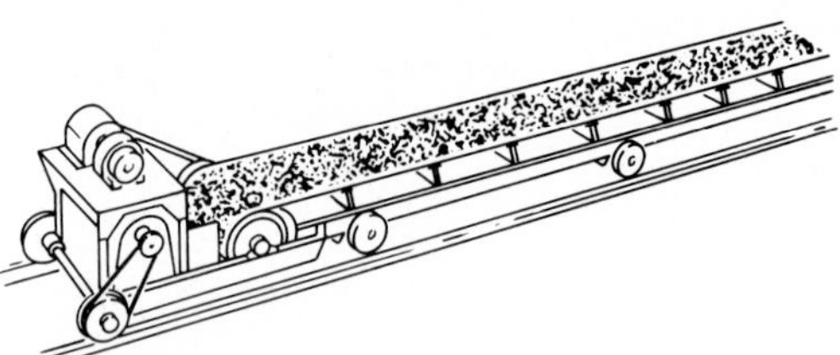

FIG. 3.23 Shuttle conveyor with coal discharge possible from either end.

Most belt conveyors come equipped with magnetic protection equipment. One way to remove tramp iron is with a magnetic head pulley. The magnetic field keeps iron from passing into the coal spout, but allows it to drop into a collection bin. Fine coal carried over with the tramp iron is recovered on a vibrating screen installed in the bin and is returned to the system. Another design features a self-cleaning magnetic separator over the head pulley.

ACKNOWLEDGMENT

Illustrations courtesy of *Power* magazine.

SUGGESTIONS FOR FURTHER READING

Carpentier, P.T., "Weigh-Bin for High Accuracy Conveyor Belt Scale Calibration," *Proceedings of Coal Technology Conference,* Houston, 1985.

Duffy, T.J., et al., "How to Correct Flow-Related Problems in Large Coal Silos," *Power,* May 1986.

Engstrom, S.P., and P.S. Hunt, "Practical Design for Enhanced Coal Flow," *Proceedings of American Power Conference,* 47:367–372, 1985.

"Extend Life of Conveyor Belts by Proper Slicing Methods," *Power,* February 1984.

"Freeze Conditioning Agents Prevent Coal Freezeup," *Power,* December 1982.

Goldberg, H.J., "Chemical Technology for Coal Handling—A Plan for All Seasons," *Proceedings of Coal Technology Conference,* Houston, 1986.

Kelley, M.E., "Selecting Mobile Equipment for Different Yard Layouts," *Proceedings of Coal Technology Conference,* Houston, 1985.

Mattison, P., "Organize Unit-Train Maintenance to Minimize Cost," *Power,* May 1985.

"Power from Coal—A Special Report," *Power*, April 1974.

"Upgrading and Refurbishing Coal-Handling Systems," *Power,* March 1983.

Zbasnik, G., et al., "Coal Dust Control and Explosion Mitigation," *Proceedings of American Power Conference,* 47:373–380, 1985.

CHAPTER 3.4
OIL FUEL AND OIL HANDLING

Staff Editors
Power *Magazine*
New York, NY

INTRODUCTION

Liquid fuels are an ideal energy source for industry and utilities, although their availability in recent years has worsened and their cost has soared. Nevertheless, they have relatively constant heating values, are generally easy to handle and store, and are easy to burn. Efficient combustion of fuel oil, however, is not quite so simple. It requires the optimization of many variables, including the fuel's physical and chemical characteristics, the type of combustion equipment, and powerplant operating practice.

With the goal of increasing a plant's combustion efficiency in mind, a look at the properties of fuel oils that influence handling, storage, and burning is instructive. Once these properties are understood, the unloading, storage, transfer, and combustion of fuel oil can be investigated.

FUEL OIL PROPERTIES

Grades of Oil

The best starting point in learning how fuel oil behaves is to identify the various grades of oil available: no. 1, no. 2, no. 4, no. 5 (light), no. 5 (heavy), and no. 6. The first two are distillates; the last three, residual oils. Number 4 can be a distillate or a mixture of refinery products. All these oils are classified according to their physical characteristics by the specifications set forth in ASTM Standard D 396, Detailed Requirements for Fuel Oils (Table 4.1).

Distillate Oils. Distillate oils can be divided into two classes: straight-run and cracked. A straight-run oil is refined directly from crude oil by heating it and then condensing the vapors at various temperatures and at atmospheric pressure. Cracking processes depend on higher temperatures and pressures, or a catalyst,

TABLE 4.1 ASTM Specifications Classify Oils According to Grade

						Saybolt viscosity, s				Kinematic viscosity, cSt					
						Universal at 100 °F		Furol at 122 °F		at 100 °F		at 122 °F			
Grade	Flash point, °F	Pour point, °F	Water and sediment, vol %	Max ash, wt %	Carbon residue on 10% bottoms, %	Min	Max	Min	Max	Min	Max	Min	Max	Minimum gravity, deg API	Sulfur, %
1	100[a]	0[c]	Trace	—	0.15	—	—	—	—	1.4	2.2	—	—	35	0.5[a,e]
2	100[a]	20[c]	0.05	—	0.35	(32.6)[b]	(37.9)	—	—	2.0[c]	3.6	—	—	30	0.5[a,e]
4	130[a]	20[c]	0.5	0.1	—	45	125	—	—	(5.8)	(26.4)[f]	—	—	—	Legal
5 light	130[a]	—	1	0.1	—	150	300	—	—	(32)	(65)[f]	—	—	—	Legal
5 heavy	130[a]	—	1	0.1	—	350	750	(23)	(40)	(75)	(162)[f]	(42)	(81)	—	Legal
6	150	—[d]	2[g]	—	—	(900)	(9000)	45	300	—	—	(92)	(638)[f]	—	Legal

[a]Or legal.

[b]Viscosity values in parentheses are for information only and not necessarily limiting.

[c]Lower or higher pour points may be specified whenever required by conditions of storage or use. When pour point less than 0°F is specified, the minimum viscosity for grade no. 2 shall be 1.8 cSt (32.0 s Saybolt Universal), and the minimum 90% point shall be waived.

[d]Where low-sulfur fuel oil is required, grade no. 6 fuel oil will be classified as low-pour (60°F max) or high-pour (no max). Low-pour fuel oil should be used unless all tanks and lines are heated.

[e]In countries outside the United States, other sulfur limits may apply.

[f]Where low-sulfur fuel oil is required, fuel oil falling in the viscosity range of a lower-numbered grade, down to and including no. 4, may be supplied by agreement between purchaser and supplier. The viscosity range of the initial shipment shall be identified, and advance notice shall be required when changing from one viscosity range to another. This notice shall be in sufficient time to permit the user to make the necessary adjustments.

[g]The amount of water by distillation plus the sediment by extraction shall not exceed 2%. The amount of sediment by extraction shall not exceed 0.5%. A deduction in quantity shall be made for all water and sediment in excess of 1%. Note: Data outlined in this table come from the American Society for Testing & Materials' Standard D 396, Standard Specification for Fuel Oils. When using this table, bear in mind that the failure to meet any requirement of a given grade does not automatically place an oil in the next lower grade unless it meets all requirements of the lower grade.

°C = (°F − 32)/1.8.

to produce distillate from heavier fractions. The difference between the two types of oil is that cracked distillate contains a substantial amount of olefinic and aromatic hydrocarbons, which are more difficult to burn than the paraffinic and naphthenic hydrocarbons produced in the straight-run process.

Straight-run grades nos. 1 and 2 are distilled or vaporized from crude at relatively low temperatures [300 to 650°F (149 to 343°C)]. Number 1 oil is used almost exclusively for home heating and as a light grade of diesel fuel. It is a little heavier than kerosene, but in many locations today both are in the same class. Number 2 comes from the refinery fractionating tower after no. 1 oil. It frequently is called a *gas oil,* reminiscent of the days when it was used mainly in gas manufacture. Number 2 fuel oil is suitable for both domestic heating and industrial applications.

Distillates are very fluid and run like water at room temperature. The availability of these fuels is related directly to the type of crude oil refined and the demand for gasoline, home-heating oil, diesel fuel, and residual. When straight-run supplies are low, some domestic heating oils are blended with a small percentage of cracked distillates. Sometimes called *industrial no. 2,* cracked distillate often is used in medium-size fuel-burning installations—such as small smelting furnaces, ceramic kilns, small packaged boilers, etc.—where the combustion equipment can accommodate this type of fuel.

Residual Oils. Number 6 fuel oil—sometimes called *residual, resid, Bunker C, vacuum bottoms,* and *reduced crude*—is produced by many methods, but basically it is the residue left after most of the light volatile products have been distilled from the crude. It is a very heavy oil, with a viscosity ranging from 900 to 9000 Saybolt Universal Seconds (SUS) at 100°F (38°C). Thus it can be used only in installations with heated storage tanks and sufficient heat available at the burner for correct atomization.

Number 6 oil is essentially a refinery by-product. Its production rate historically has been determined by many complex variables, but seldom by the market demand for this type of oil. The supply picture is changing today, however; the high market price for no. 6 has encouraged some refiners to increase the production of resid at the expense of light fractions.

Number 4 and no. 5 oil can be produced by blending from 20 percent to 85 percent of no. 2 with no. 6 to meet ASTM specifications. Blending, however, is not the only way to make these oils. The composition of no. 4 can vary widely within the established ASTM viscosity range of 45 to 125 SUS at 100°F (38°C). It can be a straight-run distillate, a cracked distillate, a mixture of refinery by-products, a contaminated oil, or an off-specification refinery product. Number 5 also can be produced in the refinery fractionating tower. It is that part of the crude barrel removed from the tower just ahead of no. 6 oil.

Number 4 oil generally is used in small installations, such as schools, apartment houses, and brick kilns. Before selecting it for one's plant, be sure a supply contract for the type of oil the particular burner requires can be obtained. To illustrate: Some installations can use a no. 4 distillate, but combustion problems result with a no. 4 that is a blend of distillate and resid. Often, these problems can be traced to the use of distillate in the blending process. Distillate can produce an unstable fuel, and stratification or separation of the two blending components occurs in the storage tank.

Number 5 fuel oil comes in two classes: light (or cold) with a viscosity range of

150 to 300 SUS at 100°F (38°C), and heavy (or hot), with a viscosity range of 350 to 750 SUS at 100°F (38°C). The reason for the two classes is that the light oil should be capable of atomization without preheating; the heavy oil needs preheating.

Testing

ASTM has developed several standardized test methods and procedures for determining the quality and grade of fuel oils (Table 4.2). Most testing is done to ascertain an oil's physical characteristics—gravity, heating value, viscosity, water and sediment, and pour, flash, and fire points—which are important to powerplants, being related directly to handling and combustion. An ultimate or chemical-element analysis normally is not included in fuel oil specifications—except for sulfur. Complete data from an ultimate analysis are used for more detailed combustion and flue-gas analyses and for experimental work.

The gravity of a fuel oil, which may be indicative of its grade and heating value, is found by using a standard hydrometer and correcting the reading at the observed temperature to the gravity at 60°F (16°C). The gravity scale used in the

TABLE 4.2 ASTM Tests for Fuel Oil Analysis

Standard no. 1*	Method of test for
D 56	Flash point by tag-closed tester
D 88	Viscosity, Saybolt
D 92	Flash and fire points by Cleveland open cup
D 93	Flash point by Pensky-Martens closed tester
D 95	Water in petroleum products and bituminous materials by distillation
D 97	Pour point
D 129	Sulfur in petroleum products (bomb method)
D 189	Carbon residue, Conradson, of petroleum products
D 240	Heat of combustion of liquid hydrocarbon fuels by bomb calorimeter
D 287	API gravity of crude petroleum and petroleum products (hydrometer method)
D 341	Viscosity/temperature charts, standards for liquid petroleum products
D 445	Viscosity of transparent and opaque liquids (kinematic and dynamic)
D 473	Sediment in crude and fuel oils by extraction
D 482	Ash from petroleum products
D 524	Carbon residue, Ramsbottom, of petroleum products
D 1266	Sulfur in petroleum products (lamp method)
D 1298	Density, specific gravity, or API gravity of crude petroleum and liquid petroleum products (hydrometer method)
D 1322	Sampling petroleum and petroleum products
D 1551	Sulfur in petroleum oils (quartz-tube method)
D 1552	Sulfur in petroleum products (high-temperature method)
D 1796	Water and sediment in distillate fuels by centrifuge
D 2161	Viscosity, conversion of kinematic to Saybolt Universal or to Saybolt Furol
D 2622	Sulfur in petroleum products (X-ray spectrographic)
D 2709	Water and sediment in distillate fuels by centrifuge

*All listed standards are also approved ANSI standards.

petroleum industry is the American Petroleum Institute (API) scale, *not* specific gravity (Table 4.3).

Recall that (1) the unit weight of a liquid, in pounds per gallon (kilograms per liter), is the density of that liquid and (2) the ratio of the density of any liquid to the density of water is the specific gravity of that liquid. Therefore, if the weight of a gallon (liter) of water at 60°F (16°C) is 8.328 lb (3.781 kg) an oil that weighs 8.099 lb/gal at 60°F (3.677 kg/L at 16°C) would have a specific gravity of 0.973 (8.099 divided by 8.328) (or 3.677 divided by 3.781).

The relationship between API gravity (expressed in degrees) and specific gravity is defined by this formula: API gravity = (141.5 ÷ sp gr at 60°F or 16°C) − 131.5. Thus, an oil with a specific gravity of 1 would have an API gravity of 10 deg. Since specific gravity appears in the denominator, the heavier the oil, the lower the API gravity, and vice versa. For example, no. 1 oil has an approximate API gravity range of 46 to 41 deg; no. 2, 39 to 30; no. 4, 28 to 24; no. 5, 22 to 18; no. 6, 17 to 9.

All petroleum products are sold as net gallons or barrels (liters) at 60°F (16°C). Since loading temperatures are rarely 60°F (16°C), expansion factors based on the oil's gravity are used to convert from the gross volume received to net volume at 60°F (16°C). These expansion factors are obtained from ASTM tables nos. 6 and 7, Reduction of Volume to 60°F against API Gravity at 60°F, in ASTM D 396.

Table 7 is a simplified version of Table 6. In Table 7, one expansion factor is used for a range of gravities, although it is based on only one gravity. Table 6 gives a different expansion factor for each gravity reading. Although it is more accurate, Table 6 requires an exact gravity reading, which is subject to error in determination, frequently causing controversy over differences between the supplier's and the consumer's test results. Table 6 has not been used to any extent in the past; however, there is a trend by industry and government to refer to it.

Heating Value

There are two heating values for any oil, the gross (or high) heating value and the net (or low) heating value. The difference is that the gross heating value includes the latent heat of evaporation of the water vapor formed during combustion; the net heating value does not. Gross heating values typically are higher by 8400 to

TABLE 4.3 API Gravity, Specific Gravity, and Density at 60°F (15.6°C)

Deg API	Specific gravity	Density, lb/gal*
10	1.0000	8.328
15	0.9659	8.044
20	0.9340	7.778
25	0.9042	7.529
30	0.8762	7.273
35	0.8498	7.076
40	0.8251	6.870
45	0.8017	6.675

*kg/m^3 = lb/gal × 120.

8500 Btu/gal or 1000 Btu/lb (2344 to 2372 kJ/L or 2330 kJ/kg) than net heating values.

The reason for the large difference is that when 1 lb (2.2 kg) of water vapor is condensed at 60°F (16°C) it will release 1058 Btu (1116 kJ). Depending on the amount of hydrogen in the oil, 7 to 9 lb (3.2 to 4.1 kg) of water is produced for each gallon (liter) of oil burned.

Gross heating value generally is used in buying and selling fuel oil. Net heating value, indicative of the heat produced during combustion, is seldom used in this country, except for exacting combustion calculations.

The method for finding the heating value of an oil is a long and complicated process, requiring expensive equipment. Where a high degree of accuracy is not needed, as for most industrial use, one can use tables that correlate heating value with gravity. Several different tables are in use today, but they are not in complete agreement with one another.

The first table correlating heating values to gravity was developed over 70 years ago by the National Bureau of Standards (Publication no. 97). This table was the standard for many years, but recently more accurate data have replaced it. Some oil suppliers have prepared their own tables from tests within their refineries, others from data accumulated from various sources.

A recent trend has been to reduce the heating value by an amount based on sulfur content. Reductions of 20 to 120 Btu/lb (47 to 280 kJ/kg) per 1 percent of sulfur are typical. The reduction often is justified on the basis that the heat of combustion of 1 lb (2.2 kg) of sulfur is only 4000 Btu (4220 kJ), compared to 14,600 Btu (15 400 kJ) for carbon and 62,000 Btu (65 410 kJ) for hydrogen. But an increase in the sulfur content of an oil does not necessarily mean the carbon or hydrogen content will be decreased proportionately. There are indications that the hydrogen actually increases in some high-sulfur oils, offsetting the decrease attributed to sulfur content. Table 4.4 lists the heating values for gravities from 10 to 45 deg API. Accuracy is close to laboratory tests and is satisfactory for practical purposes.

Since fuel oils are purchased on a volume basis, products with a low-API gravity will give the consumer more heat than will oils with a high gravity. An oil of

TABLE 4.4 Heating Values of Fuel Oils vs. API Gravity at 60°F (15.6°C)

	Heating value			
	Gross		Net	
Deg API	Btu/lb*	Btu/gal†	Btu/lb*	Btu/gal†
5	17,980	155,470	16,990	146,860
10	18,260	152,280	17,270	144,000
15	18,500	149,030	17,480	140,750
20	18,740	145,880	17,660	137,510
25	18,940	142,820	17,830	134,350
30	19,130	139,660	17,980	131,300
35	19,300	136,720	18,110	128,350
40	19,450	133,760	18,230	125,390
45	19,590	130,910	18,340	122,530

*kJ/kg = Btu/lb × 2.33.
†kJ/L = Btu/gal × 0.279.

15-deg API gravity, for example, can have 240 Btu/lb (560 kJ/kg) less than a 20-deg gravity oil, but it also will provide an additional 3150 Btu/gal (879 kJ/L).

Viscosity

Viscosity is the relative ease or difficulty with which an oil flows or is pumped. Quantitatively, it is the time in seconds that it takes 60 cm^3 of oil to flow through a standard-size orifice at standard temperature. The Saybolt viscosimeter is generally used for determining the viscosity of fuel oils in the U.S. petroleum industry. But the use of kinematic viscosity is becoming more pronounced here, especially for the lighter distillates. In Britain, Redwood no. 1 is the accepted viscosimeter; the rest of Europe uses Engler.

The Saybolt viscosimeter has two variations, Universal and Furol. The only differences between them are in the size of the orifice and the sample temperature required. The Universal instrument has the smaller opening and is used mainly for lighter oils than the Furol. Sample temperatures are 100°F (38°C) for the Universal and 122°F (50°C) for the Furol; higher temperatures can be used for extremely high-viscosity oils. When one is quoting a viscosity, the type of instrument and temperature should be included; otherwise the reading is meaningless. Frequently, the viscosities of no. 4, no. 5, and no. 6 oils are listed by one instrument, and it is necessary to convert to another. This is simplified by the use of conversion tables (Tables 4.5, 4.6, and 4.7).

Since viscosity is a measure of flow characteristics, it is an important factor in the design and operation of oil handling and burning equipment, the efficiency and sizing of pumps, pumping temperatures, and pipeline sizing. Distillate oils have low viscosities and present no problems in handling and burning. By contrast, the range of viscosities specified in Table 4.1 for the heavy oils is quite broad: No. 5 extends from 150 to 750 SUS at 100°F (38°C); no. 6, from 900 to 9000 SUS at 100°F (38°C).

Pour Point

The *pour point* of an oil is the lowest temperature at which it flows under standard conditions. In a controlled laboratory environment, the pour point is 5°F (2.8°C) above the oil's solidification temperature. The pour point is influenced substantially by the oil's wax content: the more wax, the higher the pour point. The standard pour point test, however, is only indicative of what can be expected in actual service, because of the many variables that affect the handling, storage,

TABLE 4.5 Saybolt Furol vs. Saybolt Universal

Saybolt Universal, s* at 100°F (37°C)	Saybolt Furol, s at 122°F (50°C)	Ratio
300	21	14:1
1,000	50	20:1
3,000	135	22:1
5,000	220	27:1
10,000	342	29:1

*Sometimes abbreviated SUS, or Saybolt Universal Seconds.

TABLE 4.6 Kinematic Viscosity vs. Saybolt Universal

Kinematic, cSt at 100°F (37.8°C)	Saybolt Universal, s at 100°F (37.8°C)
2.68	35
4.27	40
7.37	50
14.37	75
20.60	100
32.00	150
53.90	250
64.30	300
75.50	350
86.30	400
97.20	450
107.90	500
118.70	550
129.50	600
140.30	650
151.00	700
161.80	750
172.60	800
183.40	850
194.20	900
205.00	950
215.80	1000
258.90	1200
323.80	1500
366.80	1700
431.50	2000
474.80	2200

and use of fuel oils, such as tank size, pipeline size, oil pressure, and the structure of the wax crystals when solidifying.

Modern refining processes usually produce distillate oils that will pour freely above 10°F (−12.2°C). Between 0°F (−17.8°C) and 10°F (−12.2°C), some wax may congeal, but it does not adversely affect system operation unless the oil has remained stationary for a long period. At temperatures below 0°F (−17.8°C), consideration should be given to the use of a pour point depressant.

Pour points for domestic heavy oils range from 25 to 65°F (−3.9 to 18.3°C). But normal heating of storage tanks, pipelines, and pumping equipment to improve viscosity keeps the oil well above this range of temperatures.

Some foreign oils, or residual oils refined in this country from foreign crudes, have a high wax or paraffin content. Thus, pour points can be as high as 125°F (52°C), and these oils present a delicate handling problem. Here's why: For normal domestic residuals, as the oil cools, it increases in viscosity or thickens before solidifying. This congealing process occurs over a relatively wide temperature range. High-wax oils, however, can change from a fluid to a solid, greaselike consistency with a temperature drop of only a few degrees, plugging handling and storage equipment.

High-pour-point residuals also do not have predictable physical characteristics. Some will follow the normal viscosity/temperature curve once they are

TABLE 4.7 Saybolt Universal, Redwood No. 1, and Engler Viscosity Equivalents at Same Temperature

Saybolt Universal, s	Redwood no. 1, s	Engler, deg
35	34	1.16
40	36	1.32
45	40	1.45
50	45	1.59
60	53	1.88
75	66	2.31
100	87	2.98
125	111	3.63
150	130	4.40
200	173	5.82
250	221	7.28
300	259	8.65
350	303	10.10
400	346	11.54
450	388	12.90
500	432	14.10
750	647	21.40
1,000	863	28.70
1,200	1,042	34.00
1,500	1,356	42.80
1,800	1,560	50.40
2,000	1,727	57.50
2,200	1,903	63.10
2,500	2,158	71.60
3,000	2,591	86.00
3,500	3,022	100.00
4,000	3,500	113.00
4,500	4,000	127.00
5,000	4,450	143.00
5,500	4,900	157.00
6,000	5,325	172.00
6,500	5,765	186.00
7,000	6,210	201.00
7,500	6,680	215.00
8,000	7,100	229.00
8,500	7,520	244.00
9,000	8,010	258.00
9,500	8,450	272.00
10,000	8,950	287.00

heated above the pour point; others become highly fluid only a few degrees above the pour point. This property creates the need for greater attention to burner preheat temperature, to prevent viscosity from dropping too low at the nozzle for efficient combustion. Further, under certain storage conditions the pour point of waxy oils actually can increase. This can be caused by the raising and lowering of storage tank temperatures.

To avoid problems with high-pour-point oils:

- Keep storage tank temperatures higher than normal.
- Eliminate cold spots in the fuel handling system with proper tracing and adequate insulation.
- Purge all flow lines and fluid handling equipment with a light oil before shutting down the system.
- Heat tank vents to prevent plugging by wax condensation.
- Make sure all flow lines and fluid handling equipment are above the pour point before starting the fuel oil system. Although high-pour oils may be completely fluid in the storage tank, if they impinge on a cold surface—such as pipe walls, filters, pumps, etc.—wax particles precipitate and can build up quickly to block flow. This problem often occurs with high-wax oils that have been blended or cut with distillate oils.

Flash Point

The *flash point* is the temperature at which heated oil vapors flash when ignited by an external flame. If the heating continues, sufficient vapors are driven off to produce continuous burning. The temperature at which this latter phenomenon occurs is called the *fire point*. The flash point is important to operating engineers because it is a measure of the oil's volatility and indicates the maximum temperature for safe handling. Results of a flash-point test depend on the apparatus, so this is specified as well as the temperature.

The most popular instrument for determining the flash point is the closed-cup tester. It differs from open-cup apparatus in that oil vapors produced in heating the sample are contained in the tester and are not permitted to escape. The result is a flash point lower than that obtained with the open cup, but one more representative of actual conditions.

Distillate oils normally have flash points from 145 to 200°F (63 to 93°C), while the flash points for heavy oils can be as high as 250°F (121°C). Thus, fuel oils do not present a fire hazard at ambient temperatures unless they are contaminated with a volatile product.

Sulfur Content

The *sulfur content* of fuel oil has a definite relationship to the sulfur content of the crude oil from which it was refined. Some 70 to 80 percent of the sulfur in the crude, which is bound up in a variety of complex compounds, is concentrated in the distillate and residual fractions. The different grades of fuel oils have varying percentages of sulfur, ranging from about 0.3 to 3 percent. Distillates usually are at the lower end of the range, and heavy fuels generally at the upper end. Depending on crude composition, however, it is possible for a distillate to contain more sulfur than some resids.

Until recently, maximum sulfur limits were set by ASTM and customer specifications. But, as shown in Table 4.1, the limit now is quoted as "legal." This means the maximum limit is dictated by Environmental Protection Agency regulations, which vary considerably in different localities. More stringent limits may be set by industry to meet special process requirements, for example, where the combustion gases come in contact with the manufactured product.

The need for low-sulfur heavy oils has increased the use of desulfurization processes at the refinery. Most desulfurized products are light distillates, which can be blended with high-sulfur resid to reduce the latter's sulfur content. Blending also influences other properties of oils. For example, API gravity increases, resulting in a lower heating value per gallon of oil; viscosity is reduced, facilitating pumping and handling; and the ash content drops slightly, reducing boiler deposits and emissions.

Ash

Fuel oil is composed of organic compounds of carbon and hydrogen, with small amounts of oxygen, nitrogen, sulfur, and chlorine, plus traces of impurities classified as organometallic compounds. During combustion, these impurities produce a metallic-oxide ash in the furnace.

Ash-producing compounds that are present in crude end up concentrated in fuel oils, particularly resid. Since they are chemically bound in the oil, physical methods of separation (such as centrifuging and filtering) cannot reduce their volume. The ash content of fuel oils varies substantially. Distillates have about 0 to 0.1 percent ash, and the heavier grades from 0.2 to 1.5 percent. Although the percentages are quite small, a considerable amount of ash can accumulate on the fireside of a boiler if firing rates are high.

Some 25 different metals can be found in oil ash. Iron, nickel, calcium, aluminum, and sodium are the most prevalent. Vanadium, present in small amounts in most domestic crudes, may appear in substantial quantities in Venezuelan and Arabian crudes. Any ash deposit can present problems, but those with sodium, nickel, and vanadium are especially troublesome. They cause accelerated corrosion of boiler tubes, fouling of superheaters and gas-turbine blades, reduced heat transfer in boilers, and deterioration of refractories.

Contaminants

Water, sediment, and sludge are contaminants found in all grades of fuel oil. The total amount of water and sediment usually is determined by a centrifuge test. An extraction method is used to determine sediment by itself; distillation to find the amount of water. The clean fuels, such as distillate oils, have only from a trace to 0.2 percent of water and sediment; the heavier grades have from 0.1 to 2 percent. Bear in mind that combustion units have varying tolerances for sediment and water, so make sure the oil one contracts for is compatible with one's equipment.

Water and Sediment. Water in fuel oil usually comes from condensation, leaking tanks, leaking manholes, or, with residual oils, leaking heating coils. Sediment is caused by dirt carried along with the crude oil through the refining process, plus impurities picked up in transportation and in storage. Although the normal amount of sediment and water found in oil may not be high enough to cause problems in a system, pickup of accumulated tank bottoms can produce large amounts of these contaminants. Headaches that sediment can cause include line plugging, and strainer, control-equipment, and burner-nozzle blockage. Water can lead to corrosion, erratic and inefficient combustion, and even flame failure.

Sludge. Sludge found in some heavy fuels often is classified incorrectly as sediment. Actually, it is a mixture of organic compounds that have precipitated after different oils have been blended. The most notable is the asphaltene group, consisting of heavy hydrocarbons. After precipitation, they disperse in the oil and contribute to plugging. Another heavy-oil contaminant is wax.

Unfortunately, asphaltenes and wax are not detected with normal test methods, because the solvents used, benzol and toluene, dissolve them. The presence of these compounds usually is not detected until they cause problems. Heat can eliminate wax, but asphaltenes require a solvent for dissolution, and this generally is impractical in a fuel oil system.

Taking Samples

Any fuel oil received by an industrial plant should be tested by the consumer to make sure it meets specifications and to supply the data needed for proper handling and combustion. Test results are meaningless, however, unless the tests are performed with a representative sample. Classification of samples falls into two categories: line and delivery carrier.

From Unloading Line. During the unloading of any oil carrier, samples can be taken from the unloading line by either the drip or the slug method. A drip sample is the accumulated composite obtained by the continuous dripping through a small nipple and valve on the unloading line. The volume of the composite sample should range from 1 qt to 1 gal (0.946 to 3.79 L) depending on the size of the carrier and on test requirements.

Slug samples are small quantities [½ to 1 pt (0.237 to 0.473 L)] taken periodically from the unloading line. The number and frequency of these samples depend on the unloading time. With a barge or tanker, hundreds of thousands of gallons are unloaded over several hours, so a slug sample should be taken about every 30 min; for a truck, sample every 5 min. The separate samples are then mixed to form a composite.

From Storage Tank. Samples from a storage tank are taken through the manhole in the tank roof. The number of samples taken to form a composite depends on the depth of oil in the tank, but never take fewer than three—top, center, and bottom. Top samples should be taken at least 1 ft (0.3 m) below the surface; the bottom sample, about 1 ft (0.3 m) from the tank bottom. This way, one will not pick up light or stratified oils floating at the surface or sludge lying on the bottom.

Direct sampling of oil barges and tankers requires sampling of each compartment at the top, center, and bottom and then mixing the samples to form a composite. The procedure for direct sampling of trucks and tank cars is the same as for storage tanks. Here, three samples should be sufficient, because of the limited volume of oil.

OIL UNLOADING AND STORAGE

The design of a fuel oil system for utility central stations and industrial powerplants can be divided into two work packages: equipment needed to transfer oil from the transport vehicle to a storage tank or the day tank and equipment

length and elevation, pumping rate, type and viscosity of oil, pumping temperature, insulation, tracing, and method of operation.

If fuel oil is delivered to an industrial plant or central station in barges and lake tankers, discharge rates will vary from about 80,000 to 400,000 gal/h (84 000 to 420 000 mL/s), at oil temperatures ranging from about 110 to 130°F (43 to 54°C). For these rates, the diameter of the unloading pipeline normally ranges from 10 to 14 in (250 to 350 mm), depending on length and elevation. Tank car and truck discharge rates typically are 125 to 250 gal/min at 140 to 200°F (8 to 16 L/s at 60 to 93°C), requiring a pipeline diameter between 3 and 6 in (75 and 150 mm). Before settling on a size, however, be sure it is known how many trucks or tank cars are to be unloaded at one time.

Viscosity

Viscosity is, perhaps, the most important factor in line sizing. Since viscosity can vary from delivery to delivery and with any change in source of supply, the pipe size should be based on the maximum viscosity of the grade of oil expected to be handled—for no. 6, for instance, this is 9000 s on the Saybolt Universal scale at 100°F (38°C) (Table 4.8). Estimates of pipeline friction losses are given in Table 4.9. Remember that good design should avoid reliance on a high pumping temperature, to reduce viscosity and minimize pipeline size. Reasons: High temperature may not always be available, and some oils, particularly the blended ones, can gasify at higher-than-normal pumping temperatures.

Pipelines may be aboveground or below ground, depending on terrain and site congestion. Aboveground lines are easier to install, are less expensive to heat-trace, provide a visual observation of their condition, and are easy to maintain. They can be at ground level, supported by piers, or suspended. Overhead lines, however, are far more costly.

Aboveground Pipelines. All aboveground pipelines carrying heated heavy oil should be insulated. No insulation is necessary on distillate pipelines unless operation requires the oil to remain stationary for long periods or the plant is lo-

TABLE 4.8 Viscosity vs. Temperature*

Saybolt Universal, s at 100°F	Saybolt Furol, s at 122°F	Viscosity, SSU† at		
		120° F	130°F	140°F
1,000	50	540	400	300
2,000	100	1,020	700	575
3,000	135	1,400	1,000	750
4,000	160	1,800	1,250	900
5,000	190	2,200	1,500	1,100
6,000	220	2,600	1,800	1,250
7,000	260	2,900	2,000	1,400
8,000	285	3,200	2,200	1,550
9,000	320	3,600	2,400	1,700
10,000	344	4,000	2,600	1,900

*(°F − 32)/1.8 = °C.
†Conversion factors: SSU-centistokes × 4.635, Centistokes-centipoises ÷ specific gravity.

TABLE 4.9 Estimates of Pressure Loss in psi/100 ft (× 22.6 = kPa/100 m) of Pipe

Flow rate, gal/min*	Pipe size, in†	Viscosity, SSU at pumping temperature			
		1000	2000	3000	4000
25	3	1.7	3.3	5.0	6.6
30		2.0	4.0	6.0	8.0
40		2.6	5.3	8.0	10.3
50		3.3	6.6	10.0	14.0
100		6.5	12.0	20.0	26.0
30	4	0.7	1.3	2.0	2.7
40		0.9	1.8	2.7	3.5
50		1.2	2.2	3.3	4.5
100		2.3	4.5	6.6	9.0
150		3.5	6.6	10.0	14.0
100	6	0.5	0.9	1.3	1.8
150		0.7	1.3	2.0	2.6
200		0.9	1.8	2.6	3.5
250		1.1	2.0	3.3	4.3
300		1.4	2.7	4.0	5.2
200	8	0.3	0.6	0.9	1.2
300		0.5	0.9	1.3	1.8
500		0.8	1.5	2.2	3.0
600		0.9	1.8	2.6	3.5
800		1.2	2.3	3.5	4.5
1000		1.5	3.0	4.5	6.0

*L/s = gal/min × 0.0631.
†mm = in × 25.

cated in an area that has abnormally low temperatures and high winds. Remember that although the viscosity of distillate oils is low even at cold temperatures, there is a tendency in some grades for wax to drop out of solution and plug strainers. If heat is not available, movement of the oil by continual pumping can help retard wax formation.

The insulation should be of good quality and sized for optimum heat retention. Pipes frequently are insulated with molded insulation or fiberglass blankets or panels and are covered with aluminum jacketing or heavy tar paper. The external covering is needed to protect against bad weather and plant wear and abuse.

Underground Pipelines. Underground pipelines generally are buried in the soil. But in some cases, such as when line inspection and maintenance, tracing, or valving make it necessary, they are placed in a culvert or trench with other lines.

Direct burial is simplest. The insulated pipe is placed underground without additional construction or protection, other than perhaps loose stone under the pipe for drainage. One approach calls for insulation by covering with felts, with asphaltic sealing. Generally accepted practice is to place the pipe below the frost line.

Cellular-glass insulation is also used for direct burial. Fiberglass mesh between asphalt layers, to a total thickness of ⅛ in (3 mm), is applied with an aluminum coating and another asphalt layer. Joint sealing is important.

Insulation for direct burial must support soil loading without deformation. This means resisting both earth loads and soil movement. Live loading can occur under roads, too. The jacketed insulation must be able to support all this by itself, if rigid protective conduit is lacking.

Loose-fill-type insulation in a trench is another direct-burial approach. The piping is laid on supports at suitable depth, and a drain is run at the trench bottom. The insulating fill can be either processed gilsonite or insulating concrete. Gravel around the gilsonite helps entering water to reach either the bottom drain or the top vent. The soil above the insulating fill must be tamped. Factory-insulated pipe is still another method for direct burial.

Cathodic protection of buried piping is a big help in preventing external corrosion. Coatings can protect where they are unbroken, but damage to the coating leaves bare metal susceptible to pitting. Cathodic protection can come from graphite or magnesium anodes that direct current to base metal, making that metal a cathodic area and preventing it from corroding. Magnesium anodes (giving galvanic protection) need no external source of current but are slowly consumed while they protect. A wire connects each anode to the pipe. Graphite anodes (electrolytic protection) are not consumed, but they must be energized by an external source of current, rectified to direct current.

Heat Tracing. Pipelines carrying heavy residual oil should be heat-traced. The type of tracing—steam or electrical—depends on the location of the line (buried or aboveground), the type of oil, and the heating medium available.

The carrier for the steam is either copper tubing or steel pipe. The simplest method of tracing for short- or medium-length lines is to wrap the tubing around the pipeline, or to secure it to the pipeline in a wave form pattern. For long and large-diameter pipelines, steel pipe usually is used to carry the heating medium, because of the large volume of steam required. The heating pipe is attached to the product pipeline. When designing a steam-tracing system, provide a significant number of drains to prevent freeze-ups, particularly in low-lying areas. Also, be sure to allow for heat expansion of tracing lines.

Under certain conditions, tracing may not be necessary if the volume of hot oil is great enough to keep the line warm by continuous movement. But some method must be available for start-up after a prolonged shutdown. For example, purge the pipeline with a light distillate oil or compressed air just before system shutdown, or use a very small circulating pump to keep the oil flowing continuously in the pipeline by recirculating it back to the storage tank.

Transfer Pumps

Transfer pumps are used to move oil from storage to the day tank or directly into the plant. They also may be used to unload oil from the carrier. While tankers and the larger, modern barges have their own discharge pumps, some of the older barges may not. Trucks have their own pumps, but usually there is a usage charge—typically \$6 to \$8 per load. If the volume of receipts is large, this fee could, within a short time, finance the purchase of a pump for the plant.

Parameters to consider in pump selection include the type of oil, viscosity, pump use factor, pumping temperature, pumping rate, discharge pressure, suction lift (positive or negative pressure), specific gravity, and type of drive.

Distillate oils, with their low viscosities, are relatively easy to pump, but the high viscosities of the heavier grades can present problems. To illustrate: When one is handling heated resid, there is a tendency, under the high vacuum sometimes created by pumping from underground tanks, for the light fractions to gasify, causing pump cavitation.

Classes of Pumps. The three classes of pumps used for handling oil are rotary, centrifugal, and steam-reciprocating. The three types of rotary pumps in general use are the gear, screw, and sliding vane (Fig. 4.2). The rotary types generally are preferred—especially for high-viscosity fuel oils. They are self-priming, and their close clearances can produce a high vacuum on the suction side. Rotary pumps are designed to handle fixed volumes at constant speed; pressure limits are determined by the amount of motor horsepower available.

Centrifugal pumps typically are less expensive than rotary pumps and require a smaller drive. They are not self-priming, however, and have little lift. Centrifugal pumps are sometimes used for pumping light-viscosity petroleum distillates from aboveground tanks. Steam-reciprocating pumps are seldom used on new installations.

Positive-displacement pumps should be specified for truck and tank-car unloading, because they maintain their prime when air is drawn into the suction line along with oil. This condition can occur when a particular compartment of a multicompartment carrier empties before another.

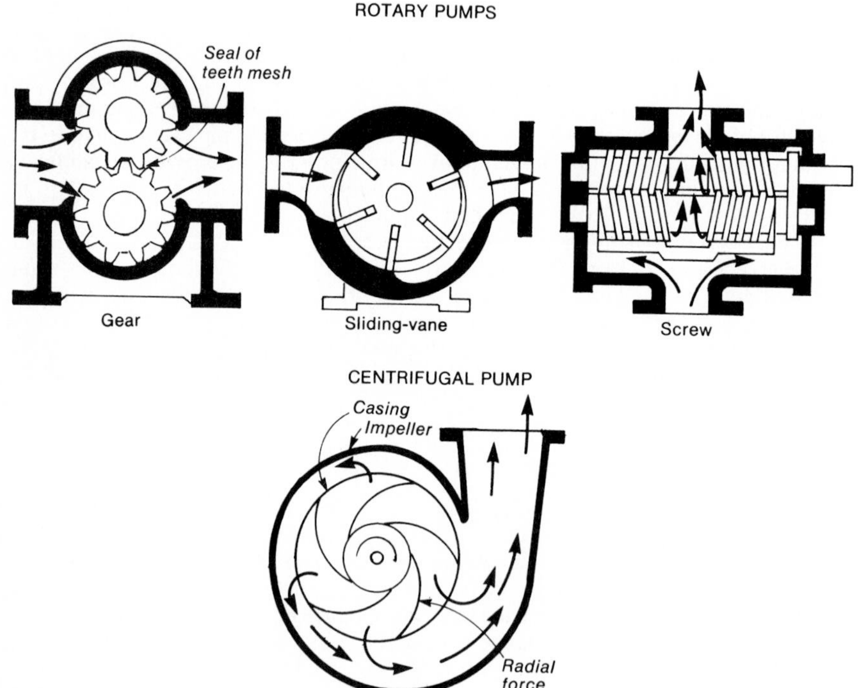

FIG. 4.2 Pumps used in fuel-oil transfer service.

Pump Design. When sizing a pump to transfer oil from storage to the plant, remember that at least 25 percent more oil is needed at the burners than is consumed; the balance is returned to storage. This factor, plus the need to compensate for pump wear and inefficiencies, means that the transfer pump should have a capacity of about 50 percent more than the consumption rate.

All heavy-oil transfer pumps that are installed outside should be traced and well insulated. Avoid pump start-ups with cold, congealed residual oil; they may overstress critical parts. Pumps also should have strainer protection on the suction side; duplex strainers are most practical. Perforated metal baskets for these should have about ⅛-in (3-mm) openings for viscous oils; $\frac{1}{16}$-in (1.5-mm) for distillates. Finally, do not forget a pressure gage on the suction side of the pump. It can reveal leaks, cold oil, a plugged line, or a dirty strainer.

OIL TRANSFER AND PREPARATION

Clean oil, supplied to the burners at the required volume, pressure, and viscosity, is essential for efficient combustion. Oil is prepared for atomization by a *pumping, heating, and straining* (PHS) set—an engineered system of piping, valves, pumps, heat exchangers, and strainers.

If the PHS set is to function efficiently and accommodate variations in fuel oil characteristics and in boiler-load swings, the selection and assembly of these components require the same quality of engineering thought necessary as that for the boiler, its auxiliaries, and the other critical parts of the steam generation system.

Experience indicates that it is possible to design and build a PHS set to meet nearly all operating conditions at a particular plant. This demands, however, that all components, including the base on which they are mounted, be matched carefully.

System Design

PHS sets for heavy-oil service can be designed with one pump and one heater (a so-called simplex system) or with duplicate, equal-capacity pumps and heaters (duplex). For some installations, it is desirable to install three 50 percent capacity pumps—usually because of total-capacity and turn-down considerations.

In this last type of arrangement, referred to as *triplex,* two pumps operating in parallel are required for full capacity; the third is a standby unit. Two or three heaters may be used on triplex sets, depending on personal preference. Obviously, economic considerations favor a system with two heaters. Note that oil piping on duplex or triplex sets is arranged to permit operation of any pump with any heater.

A typical duplex PHS set, designed to serve one or more boilers requiring heavy oil at moderate pressures and moderate flow rates, is shown in Fig. 4.3. Oil enters the unit through a duplex basket-type strainer, flows through either pump and through either heater, and exits through a duplex basket-type strainer. Accessories include:

- Pump-suction-header thermometer and compound gage with diaphragm seal

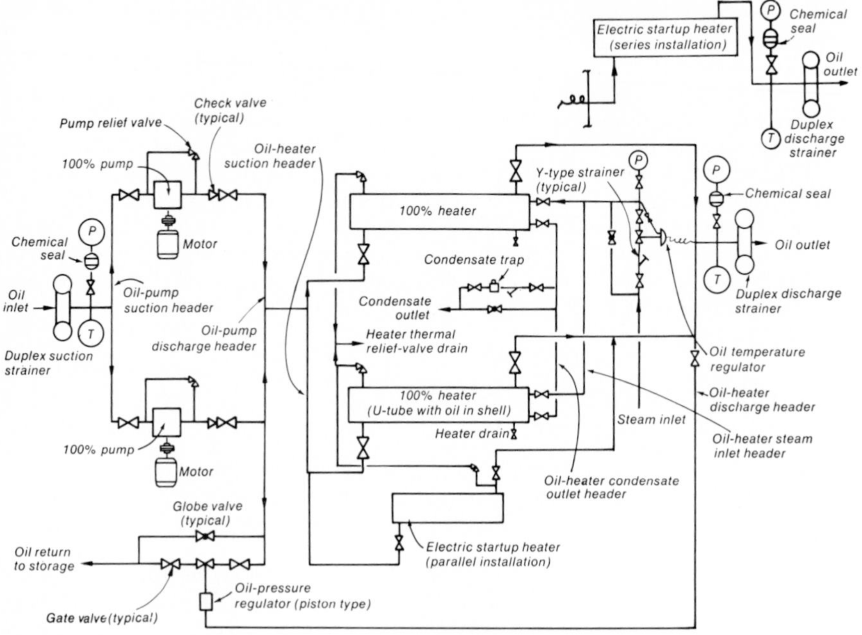

FIG. 4.3 Duplex pumping, heating, and straining (PHS) set.

- Pump-discharge relief valve
- Oil-pressure regulator with bypass
- Oil-heater thermal relief valve
- Oil-discharge-header pressure gage with diaphragm seal
- Oil-discharge-header thermometer
- Oil-temperature controller with strainer, bypass, and thermal system
- Steam-pressure gage with siphon
- Condensate trap with strainer and bypass
- Pipe, valves, and fittings

An electric heater, with a built-in adjustable thermostat, is provided for plant start-up when steam is not available for the main heaters.

All this equipment—except for the suction and discharge strainers—normally is shop-mounted and piped as a completely assembled unit on a structural-steel base. Strainers are shipped loose for field installation. After strainers are in place, the skid-mounted system is installed by making several connections to plant systems, including oil suction, discharge, and return lines; steam inlet; condensate outlet; electric wiring; and an air connection for pneumatic controls (if required). Designers also recommend that the drain in the base be connected to a disposal system, so any oil collecting in the base can be removed easily.

Key Components

Now, let's discuss in detail each component in the PHS set.

Suction Strainers. The suction strainer protects fuel oil pumps, which have close clearances, against damage from foreign material in the oil. Flow is through one basket of the duplex strainer. When cleaning is required, oil is diverted to the other basket by a plug valve, which is arranged so oil begins flowing through the idle basket before the operating basket is closed off. The valve handle extends over the basket in use, to prevent operators from opening that chamber.

On sets with large strainers [8-in (200-mm) diameter and larger], sliding-gate or globe valves sometimes are used on the inlet and outlet sides of the chambers. A chain ties the valve handwheels together, ensuring that the inlet and outlet ports for a given chamber are opened or closed simultaneously.

Three parameters to consider in strainer applications: (1) the ratio of open-basket area to the equivalent line size—4:1 or higher is recommended; (2) the size of the perforations required in the baskets—1⁄16-in (1.5-mm)-diameter holes usually provide adequate protection; and (3) the maximum desirable pressure drop—2 to 3 psi (14 to 21 kPa) is typical.

Pumps. Pump selection is possibly the most important factor in the design of a fuel oil set. Considerations include capacity, suction lift or pressure, discharge pressure, speed, and the type, viscosity, and temperature of the oil.

Rotary positive-displacement pumps are best-suited for oil burner service. One reason is that at constant speed, and with oil of uniform viscosity, the rate of flow through a rotary pump, on a volume basis, is relatively constant—regardless of the pressure level. When designing systems with rotary pumps, be sure to remember that the pressure developed by the pump is that required to overcome system resistance, and is not a characteristic of the pump itself, as in the case of centrifugal pumps.

All rotary pumps have close internal clearances between gears or screws. At constant pressure and speed, the slip of oil back from the pump discharge to its suction is inversely proportional to the viscosity; therefore, the capacity of a given pump is greater at high viscosities than at low. If operating conditions require a rotary pump to handle oils with different viscosities, the pump capacity should be based on the minimum viscosity expected, and provisions should be made to recirculate the excess oil flow at higher viscosities back to storage. Further, since horsepower requirements at constant pressure and speed increase with viscosity, the pump drive should be selected for the maximum viscosity.

Centrifugal pumps seldom are used in PHS sets. An exception is booster service, where unusually high-pressure operation [1000 to 1500 psig (6890 to 10 335 kPa)] is required with a low-viscosity oil. When they are needed, arrange centrifugal pumps for constant differential-pressure operation.

Reciprocating-steam pumps generally are not desirable either, because of their inherent pressure pulsations, which are not compatible with burner, boiler-safety, and combustion-control equipment. If reciprocating pumps are selected, however, install a generously sized air-charged bottle, or other type of damping device, in the pump-discharge piping, to minimize pulsations.

Pump Drives. Pump drives normally are motors or turbines. Select motors with enclosures suitable for job site conditions. Open, dripproof enclosures may be satisfactory for indoor or outdoor installations, depending on the type and amount of contamination to which they may be subjected. By contrast, motors installed in hazardous indoor or outdoor locations, as defined by Section 500 of the National Electrical Code, may require explosionproof enclosures.

Duplex PHS sets often are supplied with one motor-driven and one turbine-driven pump. Either can be selected as the preferred drive, with the other as the standby. A turbine normally is arranged for direct connection (no reduction gear) to the pump. Thus, its steam rate is high, but the total required horsepower is low enough so that the total steam consumption is not excessive. When connecting pumps and drives, use couplings that are protected by guards conforming to ANSI B 15.1, Section 8, requirements.

Oil Heaters. Oil heaters used in fuel oil service normally are designed in accordance with Section VIII, Division 1, of the ASME Boiler and Pressure Vessel Code, as well as with the requirements of the Tubular Exchanger Manufacturers Association (TEMA). If any state or other regulations also are applicable, be sure to identify them in the specification.

Heaters meeting TEMA C requirements generally are acceptable and are the least expensive. Occasionally, however, TEMA R or B standards are required. The ASME code stamp and the manufacturer's data report for pressure vessels (Form U-1) are not mandatory for all heaters, but they are desirable.

The most common fuel-oil-heater designs are the U-tube, baffled, bundle type with oil in the shell; the straight-tube, multipass type with oil in the tubes; and the hairpin-section type with single or multitube surface, with oil in the shell (Fig. 4.4). All units come with bare or finned tubes.

The U-tube heater is the most compact and economical of the three types. Oil flow is directed over the heating surface by segmented baffles, the location and cut of which can be quite critical. Long, small-diameter heaters offer the advantage of low first cost, but they require a substantial amount of space for installation and tube removal.

Straight-tube heaters have a lower heat-transfer coefficient than the U-tube designs. Typical coefficients for straight-tube designs range from 10 to 28 Btu/h•ft^2•°F (56.8 to 159 W/m^2•°C); for U-tube heaters, 30 to 65 Btu/h•ft^2•°F (170 to 369 W/m^2•°C). The most economical design for a straight-tube heater is one with fixed tubesheets and removable heads.

While straight-tube designs are more expensive than U-tube heaters, the additional cost may be justified if frequent cleaning is required, or if plant layout permits easier cleaning of the fixed-tube bundle than of the U-tube bundle (which should be removed from the shell for proper cleaning). Straight tubes can be rodded out or mechanically cleaned without disturbing the tube bundle.

Discharge Strainers. Discharge strainers are required only if oil heating is necessary. The earlier discussion on suction strainers is generally applicable to discharge strainers also. But since heating of the oil may result in some coking, baskets with $\frac{1}{32}$-in (0.8-mm)-diameter perforations are recommended on the discharge side. In addition, pressure drops of from 5 to 10 psi (34.5 to 68.9 kPa) across the discharge strainers normally do not affect pump selection adversely, because they represent only a small percentage of the required system resistance.

When high system resistances are involved, the pressure drop through the discharge strainer may be even greater, but will still represent only a small percentage of the total system resistance.

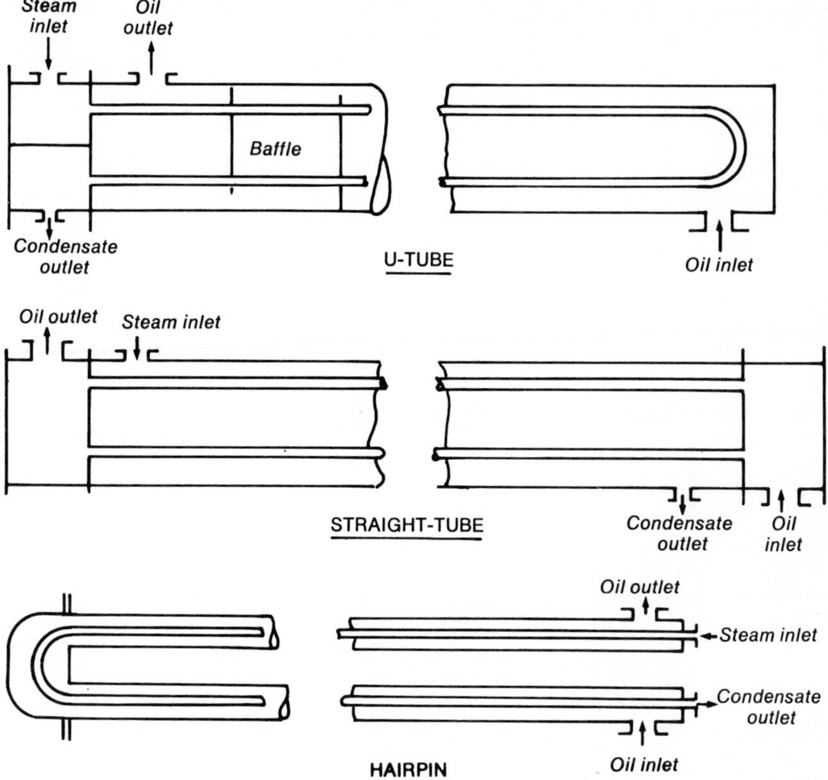

FIG. 4.4 Fuel-oil heater designs: U-tube with oil in shell (top), straight-tube with oil in tubes (center), and hairpin section with oil in shell (bottom).

Key Accessories

Key accessories for PHS sets must be matched carefully to design requirements. These accessories include compound gages, thermometers, oil-pressure regulators, temperature-regulating systems, valves, and fittings.

Compound Gages. Compound gages are recommended for installation in the pump suction header and pressure gages in the oil-heater discharge header. They should be graduated to encompass the range of pressures anticipated. In general, select gages that are graduated so that the pointer is approximately vertical under normal operating conditions. The total range then is approximately double the normal operating pressure. On PHS sets where two or more pumps are operating in parallel, it is desirable to have individual pump discharge-pressure gages.

Install a steam-pressure gage, with syphon, in the oil-heater steam-inlet line. It, too, should be graduated so that the pointer is approximately vertical under normal operating conditions. And to reflect the conditions actually existing in the heater, locate the gage downstream of the steam-flow regulating station.

Thermometers. Any style of thermometer may be used in oil piping, as long as the scale is compatible with anticipated temperatures. Suggestion: Install all thermometers in wells, and pack the wells with a suitable heat-transfer compound. Use of wells allows the instrument to be removed for any purpose without shutting down the set.

Relief Valves. Relief valves located immediately downstream from each pump discharge protect the system against high pressures. Size these valves to relieve the maximum pump capacity. Set them to open at a pressure 25 psig (172 kPa) or more (but not more than 10 percent) above the pump-discharge pressure. Remember that one should never install shutoff valves in relief-valve piping—that is, unless they are of the lock-open type, with possession of the key limited to supervisory personnel.

Thermal relief valves are installed on each steam/oil heater and on the electric start-up heater—when it is installed in parallel with the steam/oil heaters. They provide thermal overpressure protection if steam is admitted accidentally to a heater through which there is no oil flow.

Oil-Pressure Regulator. An oil-pressure regulator is essential. The pressure drop in the piping system varies directly as the square of the flow through the system. Therefore, if the flow doubles, the pressure drop quadruples. To ensure sufficient pressure at the discharge of the PHS set, select a pump with a capacity 10 to 20 percent greater than the system requires. During normal operation, the excess oil is bypassed back to storage, as shown in Fig. 4.3. The volume of oil bypassed to storage is controlled by the pressure regulator, which is actuated by the oil pressure at the pump discharge or at the outlet of the set. As system flow requirements decrease, the regulator allows more oil to return to storage, resulting in a fairly uniform pressure at the outlet.

Two more facts to remember: Size the oil-return line to ensure sufficient free-flow area to handle the maximum volume of oil that can be returned to storage; and regulator valves with tight shutoff characteristics are not required, since there is always flow through the pressure regulator.

For any type of oil-pressure regulator, size the control valve to handle the maximum capacity of the operating pump or pumps. It also should be suitable under conditions of minimum return flow to storage. Note that minimum flow is the difference between the capacity of the operating pumps and maximum system requirements. Provide a three-valve bypass around the control valve to allow manual operation should the valve require maintenance. If a self-contained unit is used as a combination pressure regulator and pump relief valve, do not specify a bypass.

Temperature-Regulating System. The temperature-regulating system maintains the temperature of the oil leaving the PHS set at that corresponding to the optimum atomizing viscosity. As shown in Fig. 4.3, only one temperature control system is required for a duplex heating set, because one heater is always on standby while the other is in operation (or if personal preference requires, each heater can have its own control system). If two heaters are in operation with one on standby, each heater should have its own control system. For these cases, position thermal probes at the oil outlet of each heater, before any shutoff valve, and pack probe wells with a suitable heat-transfer compound.

A suitable condensate trap is recommended for the heaters' condensate-outlet header. Traps should be protected by Y-type strainers and should be installed with a three-valve bypass. As for the temperature control system, only one trap is required for a duplex set unless personal preference dictates otherwise.

Valves and Fittings. All valves and fittings on the oil side of pumping, heating, and straining sets must be of steel or ductile-iron construction, according to ANSI B 31.1 as amended. Since ductile-iron valves and fittings are not available, it is apparent that all these items must be of steel. This applies to all the standard shutoff, throttling, and check valves, as well as relief valves, pressure-regulating valves, etc.

Pumps, heat exchangers, oil strainers, and flow meters, when required, do not fall within the jurisdiction of B 31.1, so pumps, oil strainers, and flow meters with cast-iron casings are still acceptable—unless their use is prohibited by other factors. Steam and condensate piping, valves, and fittings can be made of bronze or steel or any other suitable material within the limitation of B 31.1 for pressure piping.

Piping Layout

Many operating problems that are generally blamed on the PHS set can be minimized or eliminated by paying close attention to piping layout. This is another seemingly insignificant area where experience and careful engineering pay off.

Basic Characteristics. Suction conditions are most important in pipe selection, yet they are seldom mentioned in specifications. A characteristic of a well-designed system is relatively straight suction piping, which slopes gradually upward from the storage tank to the pump suction. Sudden bends or restrictions are avoided, and, in general, piping is at least one size larger than pump suction. Another characteristic of a well-designed system is a low pressure drop, to preclude use of a pump with high-suction-lift capabilities. Still another characteristic: Suction lines and transfer or forwarding pumps (if required) are sized to satisfy the capacity of the main fuel pumps. This is often overlooked, and the suction lines and forwarding pumps are sized for the discharge capacity of the set, resulting in main-pump flow starvation.

On the discharge side of the pumps, pressure-drop restrictions are less stringent. To illustrate: The pressure drop in the equipment and piping from the pump discharge to the outlet of the discharge strainer may be as high as 25 psi (172 kPa) without excessively penalizing the pump/drive selection.

Design Details. Oil piping 2 in (50.8 mm) in diameter and smaller can be socket-welded or threaded. For 2½-in (63.5-mm) diameter and larger, butt-welded or flanged connections are recommended. Where threaded piping and fittings are used, the pipe should be at least schedule 80.

Use gate valves for shutoff or isolation service. For throttling service, such as bypasses around control valves, specify globe valves with a flow coefficient C_V as close as possible to that for the control valve. When two or more pumps discharge into a common header, install a check valve immediately downstream of each pump discharge, to protect any pump that is down for maintenance even if the stop valve is inadvertently left open. Since the system may require check valves in vertical lines with upward flow, as well as in horizontal lines, use swing, ball, or spring-loaded lift checks. Never install a check valve in a vertical line with downward flow. Finally, provide isolation valves between any pressurized line and any pressure instrument.

Testing and Operation

Although testing of each individual component included in each PHS set is done by the manufacturer of that equipment, shop testing of the assembled set usually is limited to a visual examination and a pressure check for leaks. The reason is that it is not possible to check properly, in the shop, under all the operating conditions for which the set was designed.

A Checklist. Here is a checklist for operation and initial testing of a typical PHS set, which can be used to verify certain design specifications. But before taking any action, be sure to first review carefully the manufacturer's instruction book.

First, prior to start-up:

- Check coupling alignment.
- Check lubrication of motors, if applicable. Pumps normally are lubricated by the fluid pumped.
- Check rotating elements for free movement by turning over by hand.
- Check drive for correct direction of rotation, as indicated by the arrow on the pump casing.
- See that the electric power supply to motors, electric heater, and controls agrees with specifications.
- Prime the discharge side of the pump, thoroughly venting the casing. It is a good idea also to prime the suction side, although it will prime itself if there is oil in the casing and the unit can be vented. If the pump casing does not include a connection for priming, this connection should be included in the pump-discharge piping ahead of any valves. Never start pumps dry.
- Make sure pump suction and discharge valves are open, as well as oil inlet and outlet valves on the heater.

Now start one pump and:

- Completely vent all air from the PHS set at high points in the system.
- Make a visual inspection for oil leaks that may have been caused by improper handling during transportation and installation.
- Set the thermostat on the electric heater (if provided) for the desired oil temperature, and close heater contact.
- Check the discharge-pressure gage. If the pressure is not at the desired level and if the set has a combination pump-relief and pressure-control valve, the pressure can be changed by adjusting spring tension. Where a separate pressure control valve is installed, it should be adjusted to the desired pressure. Verify that pump relief valves are properly set. The control-valve pressure can be raised temporarily to check these settings. Regarding heater or thermal relief

valves, remember that they are for thermal relief only and are not sized for full pump capacity, so exercise care in checking these valves.

- Put one of the steam heaters in service when steam is available. Isolate the spare heater.

Early Precautions. On initial start-up, it is advisable to close the gate valve ahead of the fuel-oil-heater temperature control valve and to slowly introduce steam through the bypass globe valve, to raise the oil temperature gradually. Avoid coking of oil in the heater by initiating oil flow before admitting steam. When normal operating conditions are established, steam is sent through the control valve, and the bypass is closed. The oil temperature is adjusted to the desired temperature by changing the dial or spring tension. For details, see the instruction book for the particular valve.

After an initial break-in period specified by the vendor, and at regular intervals, it is advisable to clean oil, steam, and condensate strainers, which may have collected scale or other foreign matter from the system piping.

If problems occur during start-up, or after the PHS set is in operation, refer to the troubleshooting section of the instruction book. Also, check actual operating conditions against conditions for which the unit was designed. For problems not readily solved in the field, one should advise the manufacturer of the following:

- Viscosity characteristics of oil
- Oil inlet temperature and/or pressure
- Oil outlet temperature and/or pressure
- Steam conditions for oil heaters
- Description of operation as related to the operating problem
- Size and description of piping arrangement from storage area to point of use, as well as in the size and arrangement of the oil-return line
- Electrical characteristics

ACKNOWLEDGMENT

P. F. Schmidt, fuel oil consultant, and E. R. Perez, Diamond Power Specialty Corp.

CHAPTER 3.5
WOOD FUEL AND WOOD HANDLING

Roger B. Bloomfield, P.E.
Bloomfield Associates, PC
Concord, NH

INTRODUCTION

For engineers whose association with wood firing has been restricted to the family-room fireplace, this chapter gives an introduction to available wood fuels and their handling and combustion to ensure clean, efficient operation. The many equipment options are identified along the way. For engineers who have industrial wood-firing experience, the chapter offers a review of time-proven technology, plus a look at new equipment and concepts and some ideas on how to operate existing facilities better.

RESOURCE AVAILABILITY

Forests cover more than 1.1×10^6 mi^2 (2.9×10^6 km^2) in the contiguous 48 states, and the vast amount of wood they contain is a viable fuel for industry in at least some sections of the country. Although it is virtually impossible to determine just how much wood is available nationwide for steam production (Fig. 5.1), the information would be of little value to the engineer responsible for establishing a plant's fuel supply network. Far more important is an intimate knowledge of wood resources in the vicinity of the plant, because the low-density, high-moisture nature of this fuel generally makes it uneconomical to transport beyond 40 to 50 mi (64 to 80 km).

Surveying Available Resources

Thus, the engineer's first step in deciding whether a plant should burn wood is to conduct a comprehensive survey of available resources, including the following:

Wastes from Wood-Process Plants. Mills producing lumber, plywood, veneer, and other wood products generate substantial quantities of waste that are suitable for fuel—for example, slabs, edgings, bark, sawdust, sander dust, planer shavings, debris from log unloading and handling operations, etc. Wastes from furniture plants are another potential fuel source.

What makes these mills attractive as fuel suppliers is that they produce more waste material than they need to satisfy their own energy requirements. To illustrate, only about one-half of the logs going into a sawmill come out as lumber. This is in sharp contrast to pulp-and-paper mills, which require more energy than they produce in the form of waste to sustain boiler operations.

The Forest. Wood and bark suitable for fuel include residue from logging operations—such as treetops, limbs, and cull logs—as well as other on-the-ground or standing fiber. The volume of noncommercial timber, which includes *rough* and *rotten* trees and salvageable dead trees, may amount to 40 percent or more of the forest inventory, particularly in the eastern United States.

By definition, *rough trees* are live trees of commercial species that do not contain at least one 12-ft (3.7-m) sawlog of suitable diameter, or two noncontiguous sawlogs each 8 ft (2.4 m) or longer, and they have defects attributed to roughness or to poor form. Rotten trees are those having more than 50 percent decay. Another possibility is the tree's stump and root system.

Estimating U.S. Wood Reserves

Region	Forested land		Average forest inventory		Annual growth rate	
	1000 acres	1000 ha	tons/acre*	Mg/ha	tons/acre†	Mg/ha
1	86,606	35 075	110	246	6.5	15
2	99,877	40 450	100	224	6.5	15
3	211,884	85 813	120	269	8.5	19
4	339,948	137 679	190	426	9	20
Total	738,315	299 018				

*About 40% of the total tonnage in most regions is noncommercial wood.

†Morbark Industries claims that these Forest Service estimates of annual growth rate could double—possibly triple—with environmental thinning of the nation's forests. Culled trees of no commercial value could be chipped for fuel.

Sources: U.S. Dept. of Agriculture Forest Service and Morbark Industries Inc.

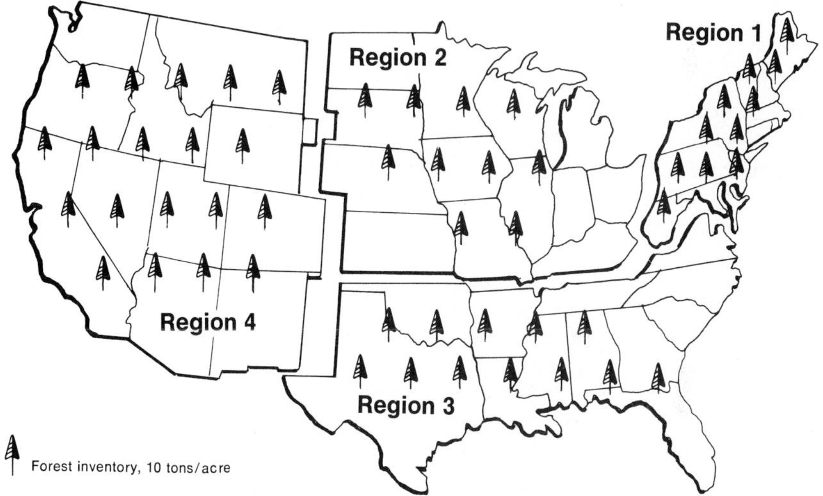

FIG. 5.1 Wood reserves by regions across United States. Note: 10 tons/acre = 22.4 Mg/ha.

Prepared Wood Fuels. Companies selling biomass fuels in pellet and briquette form operate plants in several parts of the country. Information on these firms, and others offering specialized equipment for handling and burning wood, can be obtained through organizations such as the Forest Products Research Society, Technical Association of the Pulp and Paper Industry, and Solar Energy Research Institute.

Estimating Future Resources

Conducting the survey is the easiest part of the overall evaluation. The next step is to determine what impact current and possible future trends in the pulp-and-paper and forest-products industries will have on the plant's fuel supply. This is an extremely difficult task, and it requires good intuitive judgment. Qualitative conclusions are about all one can expect from this analysis. Even if all the variables were identified that could adversely affect deliveries, it would be impossible to quantify them, because no reliable data exist on which to base the calculations.

Mill Wastes. In general, future availability of mill wastes must be considered uncertain. Improved technologies being introduced into lumber mills—such as thinner saw blades—will curtail production of sawdust and planer shavings. Greater use of these wastes in composite materials, and in pulping, will add to the shortfall. The movement by large pulp-and-paper mills to buy available mill wastes to reduce their consumption of fossil fuels is another mitigating factor. Complicating the picture even more, some sawmills are adding cogeneration facilities with the idea of converting their extra waste to electricity for sale to utilities.

Thus, in evaluating the vitality of local pulp-and-paper and forest-products industries, questions such as these must be answered favorably:

- Can a long-term contract be obtained for mill wastes?
- Do mills operate year-round?
- Will projected decreases in housing starts and other market factors force the closing of lumber and plywood mills?
- Will greater use of particleboard and other composite materials adversely influence plywood production?
- Can the old mills locally compete favorably with new ones in other areas?

Forest Residue. If the engineer does not have confidence in the long-term availability of wood waste from surrounding mills, she or he should consider forest residue and noncommercial timber. Chipping in the woods of treetops and limbs and of noncommercial species has been demonstrated and is being done in several areas. Commercially available whole-tree chippers make it possible to slice into uniform pieces entire trees up to about 2 ft (0.6 m) in diameter, to dispose of dirt and grit, and to blow finished chips into waiting vans (Fig. 5.2).

Even with this equipment, it is no small matter to establish a reliable and economical forest-oriented supply system. Trees are difficult to harvest in some places on a continuous basis. Rain, soft ground in the late spring, and heavy snows in winter may affect boiler operations if the storage capability is limited. Figure that it takes a pile 200 × 300 × 25 ft (61 × 91 × 7.6 m) high to supply a 100,000-lb/h (12 600-g/s) boiler for 1 mo.

Whole-Tree Chips. Several large wood fuel supply companies have been operating for some years in the northeast and midwest; in some cases, the firms were established expressly to satisfy a demand for whole-tree chips for boiler fuel. Some of the wood fuel supply firms were originally formed years ago to supply sawmill chips and whole-tree chips to paper mills for feedstock, and they merely expanded their operations to supply biomass fuels to other customers.

State or federal foresters can help provide information about yields of biomass fuels per acre (hectare, ha) of forest and about trends in price. Generally, foresters (both public and private) tend to be cautiously enthusiastic about whole-tree chipping for industrial fuel because the materials chipped and removed from the woodlot are the "weeds"—deadfalls, shrubs, unsuitable species, crooked and

FIG. 5.2 Skidder hauls logs and slash to chipper.

stunted trees, as well as the tops and limbs of marketable timber. One whole-tree chip supplier likened the process to weeding a garden and then burning the weeds to make steam and electricity.

Several intensive studies of short- and long-term availability of wood fuel taken by whole-tree chipping have been conducted by the U.S. Department of Agriculture. One such study of a northern New England area indicates that, over the long run, a typical woodlot acre will yield between 3.5 and 1.5 tons/yr of biomass fuel [10.4 and 4.4 Mg/(ha e yr)]. Since the woody biomass that is chipped for wood fuel is, by definition, unsuitable for pulp or timber, it is generally believed that demands by the paper or timber industries will not have a great effect on wood fuel availability.

Whole-Tree Chip Availability. As an example of wood fuel availability, assume that a proposed wood-fired steam plant has an area within a 25-mi (40-km) radius from which to draw. If one-half of that area consists of cities and towns, lakes, rivers, and highways; if one-half of the remaining area is unsuitable for chipping operations because of terrain or type of woodlot; if one-half of the remaining section consists of woodlots owned by persons or institutions not wishing to log their property—then there are available for wood fuel (whole-tree chip) recovery some 157,000 acres (63 585 ha).

If the amount of wood fuel that can be taken each year ranges from 1.5 to 3.5 tons/acre (3.36 Mg/ha to 7.84 Mg/ha), then there are available between 235,000 and 549,000 tons (213 000 and 498 000 Mg) of wood fuel annually. Since each ton of green whole-tree chips is capable of replacing approximately one barrel of no. 6 fuel oil, the woods within 25 mi (40 km) of the proposed plant can provide the equivalent of 10 to 23 × 10^6 gal (37.9 to 87 × 10^6 L) of no. 6 oil each year.

Choosing a Course of Action

If, after all known factors relating to supply are taken into consideration, the engineer is still not confident that there is enough wood to sustain boiler operations for many years to come, then the engineer should consider this alternative: Design fuel-handling, combustion, and pollution control systems that are flexible enough to accommodate two or more fuels. All the wood or bark that can be bought on an as-available basis is burned, and one switches to other energy sources when forest fuels cannot be obtained.

FUEL PROPERTIES

The chemical and physical characteristics of wood and bark must be known in detail before work can begin on the design of fuel-handling, combustion, and pollution control systems (Fig. 5.3). Although laboratory analysis shows that most species of wood and bark have approximately the same chemical composition on a dry basis (Table 5.1), the moisture content can extend over a broad spectrum. Heating value, size consist, and other properties influencing plant design also may vary so much that one consultant says that the only consistent property of wood is its inconsistency.

Principal Characteristics

The principal characteristics of wood are expressed in a proximate analysis, which shows the exact chemical composition of a fuel without reference to the physical form in which the compounds appear. It gives the power engineer a good picture of a fuel's behavior in the furnace. The analysis is relatively simple, involving the determination of the percentage of moisture, ash, and volatile matter and the calculation of the percentage of fixed carbon by difference. Since the percentages of these four variables total 100, the fixed carbon can be found easily

FIG. 5.3 Hog fuel.

TABLE 5.1 Fuel Properties: Comparing Different Types of Bark and Wood to a Range of Coals

Fuel characteristics	Chemical composition, % by wt (dry basis)										
	Bark				Wood			Coal			
	Pine	Oak	Spruce[a]	Redwood[a]	Redwood	Pine	Fir/Pine[b]	Lig[c]	Sub[d]	Bit[e]	Bit[f]
Proximate analysis											
Volatile matter	72.9	76.0	69.6	72.6	82.5	79.4	75.1	44.1	39.7	35.4	16.0
Fixed carbon	24.2	18.7	26.6	27.0	17.3	20.1	24.5	44.9	53.6	56.2	79.1
Ash	2.9	5.3	3.8	0.4	0.2	0.5	0.4	11.0	6.7	8.4	4.9
Ultimate analysis											
Hydrogen	5.6	5.4	5.7	5.1	5.9	6.3	6.3	4.6	5.2	4.8	4.8
Carbon	53.4	49.7	51.8	51.9	53.5	51.8	50.7	64.1	67.3	74.6	85.4
Sulfur	0.1	0.1	0.1	0.1	0	0	0	0.8	2.7	1.8	0.8
Nitrogen	0.1	0.2	0.2	0.1	0.1	0.1	2.4	1.2	1.9	1.5	1.5
Oxygen	37.9	39.3	38.4	42.4	40.3	41.3	40.2	18.3	16.2	8.9	2.6
Ash	2.9	5.3	3.8	0.4	0.2	0.5	0.4	11.0	6.7	8.4	4.9
Heating value											
Dry basis, Btu/lb (kJ/kg)	9030 (21 040)	8370 (19 500)	8740 (20 364)	8350 (19 455)	9220 (21 480)	9130 (21 273)	8795 (20 492)	11,084 (25 825)	12,096 (28 184)	13,388 (31 194)	15,000 (34 950)

[a]Logs stored in saltwater.
[b]Sanderdust.
[c]Texas lignite.
[d]Wyoming subbituminous B.
[e]Illinois bituminous (high-volatile A).
[f]West Virginia bituminous (low-volatile).
***Sources:* Babcock & Wilcox Co., Combustion Engineering Inc., Coen Co.**

once the other three are known. It is also customary to determine separately the total amount of sulfur contained in the wood, as well as its heating value.

Proximate and ultimate analyses are not conducted routinely on receipt of fuel at wood-fired plants (they *are* in coal-fired facilities) because wood waste, unlike coal, is not sold based on its heating value or other properties. Rather, transactions are based strictly on weight or volume. Units of measurement used most often are the ton (megagram) and the *unit*. The latter is the amount of uncompacted wood waste that will fit into a 200-ft^3 (5.7-m^3) container.

Moisture Content

The moisture content of bark and wood usually influences the design of both the firing equipment and the steam generator more than any other property. Sander dust and furniture plant scraps, which contain the least amount of moisture of the wood fuels—less than 10 percent on a wet basis—allow the highest boiler efficiencies. Bark from hydraulically debarked logs or from trees in areas with high rainfall, and sawdust from mills using water-cooled saws, may contain 65 percent moisture or more. At such high levels, combustion becomes unstable, and the fire goes out. *Hog fuel*—the term used to describe the mixture of wood and bark that is burned to produce steam in most plants—normally contains 45 to 55 percent moisture on a wet basis.

Note that there are two ways to express moisture content—on a wet or on a dry basis. Engineers usually calculate the moisture as a percentage of the as-received weight (wet basis), while people in the wood products industries and wood technologists express it as a percentage of the dry weight. Moisture known on one basis can be converted to the other by using Fig. 5.4. For example, 50 percent moisture on a wet basis and 100 percent moisture on a dry basis mean the same; that is, every pound of fuel contains a half-pound of water and a half-pound of bone-dry wood.

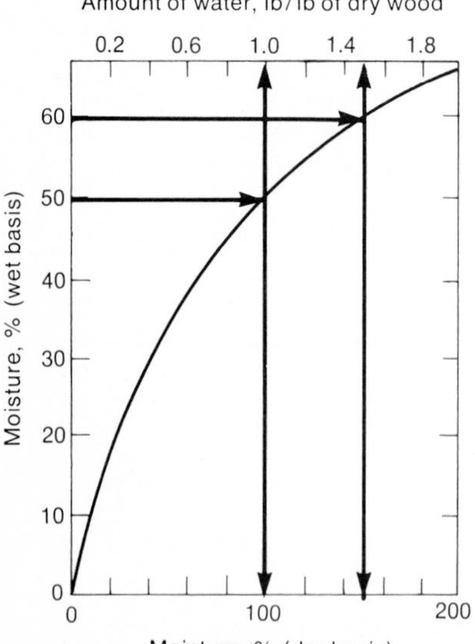

FIG. 5.4 Moisture content of wood.

Ash Content

Ash is the noncombustible mineral matter left behind when a fuel burns completely. It may differ from ashes, as the power engineer knows them, because ashes taken from a furnace sometimes contain unburned carbon. The ash content of wood is low, generally less than 1 percent. By contrast, the ash content of bark from most softwood (evergreen) species extends up to 3 percent, while that for hardwood (broad-leafed) species typically ranges from 2 to 5 percent (Table 5.2).

Besides the ash contained in the bark on the standing tree, harvesting and handling of logs frequently contribute more dirt, rock, and sand (Fig. 5.5). The

TABLE 5.2 Ash Properties: Comparing Bark to Coal

Fuel and ash characteristics	Bark				Coal			
	Pine	Oak	Spruce*	Redwood*	Lig†	Sub‡	Bit§	Bit¶
Ash content (dry basis), %	2.9	5.3	3.8	0.4	12.8	6.6	17.4	12.3
Sulfur content (dry basis), %	0.1	0.1	0.1	0.1	1.1	1.0	4.2	0.7
Ash constituents								
Silicon dioxide (SiO_2), %	39.0	11.1	32.0	14.3	41.8	24.0	47.5	60.0
Aluminum oxide (Al_2O_3), %	14.0	0.1	11.0	4.0	13.6	20.0	17.9	30.0
Iron oxide (Fe_2O_3), %	3.0	3.3	6.4	3.5	6.6	11.0	20.1	4.0
Calcium oxide (CaO), %	25.5	64.5	25.3	6.0	17.6	26.0	5.8	0.6
Magnesium oxide (MgO), %	6.5	1.2	4.1	6.6	2.5	4.0	1.0	0.6
Sodium oxide (Na_2O), %	1.3	8.9	8.0	18.0	0.6	0.2	0.4	0.5
Potassium oxide (K_2O), %	6.0	0.2	2.4	10.6	0.1	0.5	1.8	1.5
Titanium oxide (TiO_2), %	0.2	0.1	0.8	0.3	1.5	0.7	0.8	1.6
Manganese oxide (Mn_3O_4), %	Trace	Trace	1.5	0.1				
Sulfite (SO_3), %	0.3	2.0	2.1	7.4				
Chloride (Cl), %	Trace	Trace	Trace	18.4				
Other compounds, %	4.2	8.6	6.4	10.8	15.7	13.6	4.7	1.2
Ash-fusion temperatures, °F (°C)								
Initial-deformation								
Reducing environment	2180 (1193)	2690 (1477)			1975 (1079)	1990 (1088)	2000 (1093)	2900+ (1593+)
Oxidizing environment	2210 (1210)	2680 (1471)			2070 (1132)	2190 (1199)	2300 (1260)	2900+ (1593+)
Softening								
Reducing environment	2240 (1223)	2720 (1493)			2130 (1166)	2180 (1193)	2160 (1182)	
Oxidizing environment	2280 (1249)	2730 (1499)			2190 (1199)	2220 (1216)	2430 (1332)	
Fluid								
Reducing environment	2310 (1266)	2740 (1504)			2240 (1227)	2290 (1254)	2320 (1271)	
Oxidizing environment	2350 (1288)	2750 (1510)			2290 (1254)	2300 (1260)	2610 (1432)	

*Logs stored in saltwater.
†Texas lignite.
‡Wyoming subbituminous.
§Illinois bituminous (high-volatile).
¶West Virginia bituminous (low-volatile).
Source: ***Babcock & Wilcox Co.***

amount of additional noncombustible material that clings to the bark depends on the logging methods, type of soil in the forest, method of transportation (wet or dry), handling at the plant, and other factors. Compared to the ash burden in most coal-fired boilers, however, relatively little noncombustible material is discharged from bark-fired boilers.

FIG. 5.5 Woodwaste here has too many rocks.

Qualitatively, bark ash differs from coal ash. Although there is no such thing as a typical analysis, bark ash generally is high in calcium oxide, sodium oxide, and potassium oxide compared to coal ash. Conversely, coal ash usually contains more silicon dioxide and aluminum oxide than bark ash.

These differences may be significant in some cases where bark and coal are burned in combination. The reason is that if the bark has a high concentration of calcium oxide, this compound could act as a flux and reduce the fusion temperature of the coal ash to the point where slagging problems are possible. Also, high concentrations of sodium and potassium oxides may lead to superheater fouling.

At least one very large powerplant fired with purchased wood fuel and coal in the northeast has reportedly found that the sulfur oxide emissions have been reduced from the anticipated values because of the combination wood and coal firing. That is, the sulfur in the coal combines with wood-ash constituents and is removed from the furnace with bottom ash.

Heating Value

In judging fuel values, the heating value plays a basic part because in buying fuel actual energy units are being bought. When an oven-dry wood or bark sample is burned in a bomb-type calorimeter filled with oxygen under pressure, the fuel's *higher heating value* is measured. This assumes that the latent heat of water vapor contained in the combustion products is absorbed in the boiler. Since water vapor in the flue gas is not cooled below its dewpoint during normal boiler operation, this latent heat is not available for making steam.

Woods of different species have approximately the same heating value on a moisture- and resinfree basis, about 8300 Btu/lb (19 340 kJ/kg). But since resin has a heating value higher than that of the dominant cellulosic material (about 16,900 Btu/lb, or 39 380 kJ/kg), resinous woods—such as Douglas fir and pine—contain about 9000 Btu/lb (21 000 kJ/kg). Hardwoods, such as oak, have heating values near 8300 Btu/lb (19 340 kJ/kg). Heating values for nonresinous barks may extend up to about 9800 Btu/lb (22 830 kJ/kg); softwood barks range from 8800 to 10,800 Btu/lb (20 500 to 25 200 kJ/kg).

Particle Size. The particle size of wood waste can extend from 39.4 to 197 μin (100 to 500 μm) sander dust to sawmill slabs several feet in length. Dry, finely divided wood fuels are stored dry and burned in suspension, as is pulverized coal. Oversize dry and wet material generally is comminuted until the largest dimension does not exceed about 2 in (50 mm), and then it is burned on a grate. Green planer shavings and sawdust are burned either entirely on a grate or partially in suspension and partially on a grate.

Particle size and shape also influence the total energy content of any fuel shipment. A substantially greater heating value is obtained when bark is purchased on a unit basis than when sawdust or planer shavings of equivalent moisture content are purchased.

Comparing Wood to Fossil Fuels. A meaningful method for assessing the value of an alternative fuel is to compare its cost, based on dollars per 10^6 Btu, to that of the fuel being burned. The following example shows how Fig. 5.6 can be used for

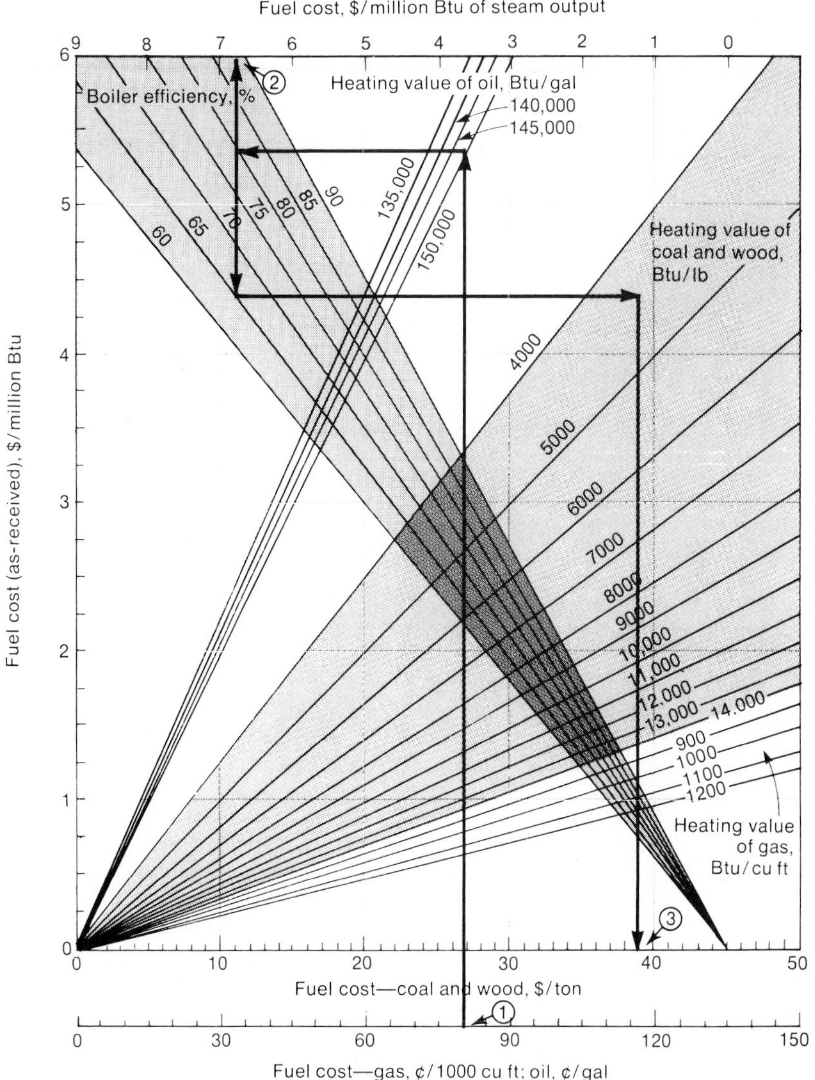

FIG. 5.6

that purpose. Note: Metric conversion factors are Btu/gal × 0.28=kJ/L; Btu × 1.055 = kJ;Btu/lb × 2.33 = kJ/kg;Btu/ft^3 × 373 = kJ/m^3;ton × 0.907 =Mg; and gal × 3.79 = L.

EXAMPLE A plant is burning 150,000 Btu/gal fuel oil, which costs $0.80 per gallon. The boiler efficiency is 80 percent. What would be the equivalent as-fired value of 4500 Btu/lb wood if the boiler efficiency dropped to 65 percent when hog fuel was burned?

SOLUTION First, determine the value of steam produced by burning oil. Enter Fig. 5.6 at point 1 ($0.80 per gallon) and extend a vertical line until it intersects the curve for 150,000-Btu/gal oil. Draw a horizontal line from this point to the curve representing 80 percent boiler efficiency, and then go up vertically until the abscissa is intersected, where one finds that it costs $6.67 to produce 1 × 10^6 Btu worth of steam from oil. Next, draw a vertical line from point 2 to the 65 percent efficiency curve and extend a horizontal line from that point until it crosses the curve representing a heating value of 4500 Btu/lb for the hog fuel. Now construct a vertical line to the lower abscissa and find the value of wood—$39 per ton in this case (point 3).

Note that this analysis does not consider transportation, preparation, handling, pollution control, and other costs associated with burning a particular fuel, some or all of which may be substantial. To illustrate: Factoring these penalties into the foregoing example would reduce the value of wood by 50 percent or more.

Finally, the heating values used in Fig. 5.6 for solid fuels are as-fired values. If fuel moisture is expressed on a dry basis, convert to the as-fired heating value by using Fig. 5.4.

FUEL PREPARATION

Wood waste generally is shipped via truck from the source of supply to the powerplant, strictly because that is the most economical way to transport bulk materials over the short distances normally traveled. Rail transport is often preferred when the wood must be hauled more than about 150 mi (242 km). Barges sometimes are used in areas such as the Pacific northwest, where plants and forests have access to navigable waterways. Barges carry the biggest payloads, 600 units or more; railcars hold 30 to 40 units; the largest trucks transport 15.

Transportation

Transportation is one of the largest components of fuel cycle cost for plants that buy hog fuel from outside sources. Estimates are that it can cost $1.50 per mile ($0.93 per kilometer) or more to haul wood and bark in a 25-ton (22.7-Mg) van. With diesel fuel prices on the rise, expect this rate to increase. Most plants rely on contract haulers or the seller to deliver hog fuel.

Key Equipment. Three types of trucks are used to haul wood waste: dump trucks, self-unloading vans, and conventional semitrailers. The first two have the advantage of dumping directly onto the storage pile, possibly eliminating the need for unloading facilities. A bulldozer or front-end loader is essential, however, to shape the pile and compact the fuel. The payload of these vehicles is considerably less than that of a semitrailer, and they are more expensive to operate. Self-unloading vehicles may be economical when the plant and the fuel source are within a few miles of each other or when the plant requires only a few shipments per day.

Semitrailers are unloaded by hydraulic dumpers; they may be either coupled to or uncoupled from the cab (Fig. 5.7). These dumpers have a maximum tilt angle of 62° and can discharge an entire truck in 3 to 5 min.

Railcars and barges are usually unloaded with the same type of equipment used to discharge coal, such as rotary car dumpers and clamshell buckets.

Unloading Systems. Systems to unload the fuel should be designed to minimize the amount of time a delivery vehicle spends at the plant, since demurrage charges can accrue quickly. A practical problem with truck transportation is that these vehicles arrive at the plant on a random schedule and often come in groups of two or more. A rule of thumb for sizing a truck unloading system that operates a nominal 12-h day is: Design it to handle one-half the daily volume in a 4-h period. Thus, if one expects twenty 20-ton (18-Mg) trucks per day, the system should be able to move 50 tons/h (45 Mg/h) of wood waste from the unloading station to storage.

Designing a wood waste unloading and handling system can be a challenge to even the most experienced engineer. At a minimum, equipment must be capable of operating continuously, under extremely dusty conditions, with a very abrasive material. Remember, wood waste can vary widely in density, moisture, ash content, and other characteristics. The physical idiosyncrasies of the fuel also in-

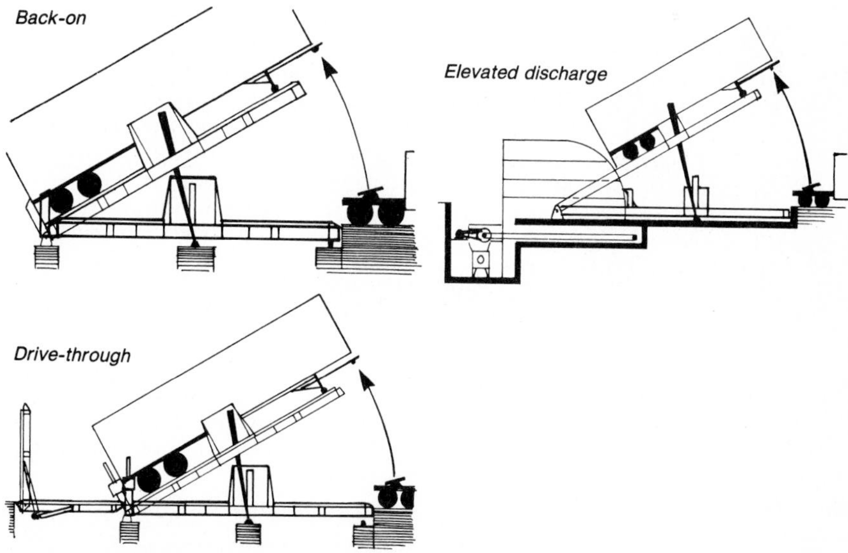

FIG. 5.7 Semitrailers are unloaded by hydraulic dumpers.

fluence design. For example, redwood bark breaks into long stringy segments, which are very difficult to handle and reduce in size.

In-Plant Handling

In many modern plants that burn a significant amount of hog fuel, wood waste is discharged from the transport vehicle, after weighing, into a live-bottom hopper and then metered to a screen, which splits the stream into two or more fractions. Oversize material is sent to a hog mill for comminution, while particles of acceptable size bypass the mill. The screen sometimes also separates dirt from the wood; it is shipped to a landfill.

Tramp iron usually is removed ahead of the hog by a magnet, although some plants opt for a metal detector. After the wood and bark have been sized, and large noncombustible objects eliminated, the resulting hog fuel is conveyed to storage, either outside on a pile or in a covered bin.

Scales

In some areas, particularly the east, wood waste is purchased on a weight rather than a volumetric (unit) basis. When this is the case, a scale should be installed at the plant to verify the quantity delivered. Trucks and combination truck/rail scales, as well as conveyor belt scales, are available for this purpose.

Conventional truck scales are preferred in most wood-fired plants. The weight of a given load can be transmitted from the scale to its weight indicator mechanically—through a system of levers or electrically—through load cells. Weight indicators may be of the dial, registering-beam, or digital type.

Screens

Most screens selected for wood handling systems are of the disk or vibrating type. Disk screens (Fig. 5.8), the most popular, are designed primarily for removing oversize material from the incoming stream. This minimizes the power requirements for hogging fuel as well as the production of fines, which add to the flue-gas particulate loading. Advantages of disk screens include

FIG. 5.8 Disk screen.

- Nonclogging, self-cleaning operation.
- High capacity compared to other types of screens. Equipment is available to process more than 200 units/h with a drive system rated less than 20 hp (15 kW).
- Variable product size. Disk spacing can be adjusted in the field to accommodate changes in fuel requirements over the plant lifetime.

Another way to reject slivers is to use a vibrating or shaker screen, which is essentially a perforated, vibrating flat plate. Vibrating screens are generally not selected to process the entire incoming wood waste stream because of their smaller capacity compared to disk screens and their tendency to blind (that is, to have their holes plug up) during processing of unsized wet hog fuel.

Alternatives to Screens

Other methods for separating grit and dirt from the fuel stream include

- *Flotation*: This is a simple procedure. Wood waste is dumped in a flume, and bark and wood are collected from the surface of the water as dirt and gravel settle to the bottom. A disadvantage to flotation is that the wood becomes saturated, reducing its net heating value.
- *Air separation*: Sophisticated air classification systems may be justified for removing noncombustibles from hog fuel if these materials would cause intolerable operating problems.

Hogs

Oversize wood waste is comminuted by a hog, which is a shredding machine serving the same purpose as a crusher in a coal preparation system. Hogging with either a knife hog or a hammer mill, the preferred types, can be done either before wood waste is conveyed to storage or after it is retrieved from storage. Most designers prefer the former approach.

Unsized material can be extremely difficult to handle, particularly if it is compacted in storage. Hogs generally are adjusted to discharge product with a top size of 2 to 3 in (50 to 75 mm), which is a good compromise among power cost for shredding, ease in fuel handling, and furnace performance in stoker-fired combustion systems.

Before buying a hog, the engineer should know the

- *Maximum size of pieces of wood or bark to be handled by the machine*: It influences the size of the hog-inlet opening as well as other parameters.
- *Nature of the wood waste*: Some types of wood and bark require special consideration by the designer.
- *Quantities of wood waste requiring comminution*: Both average and maximum flow rates are needed.
- *Required product size*: This factor has significant impact on hog capacity. The larger the allowable product size, the greater the capacity of a given machine and the lower the power consumption for a specified through-put.

Hammer Mills

These hogs can handle light tramp iron [nominally ¼ in (6 mm) and smaller], small bolts, steel strapping, etc., without incurring damage. (Knife hogs are vulnerable.) Further, these machines normally are protected against heavy tramp metal by a shear-pin arrangement. Figure 5.9 shows how a typical hammer mill works. Wood waste is gravity-fed through a large opening in the top of the hog,

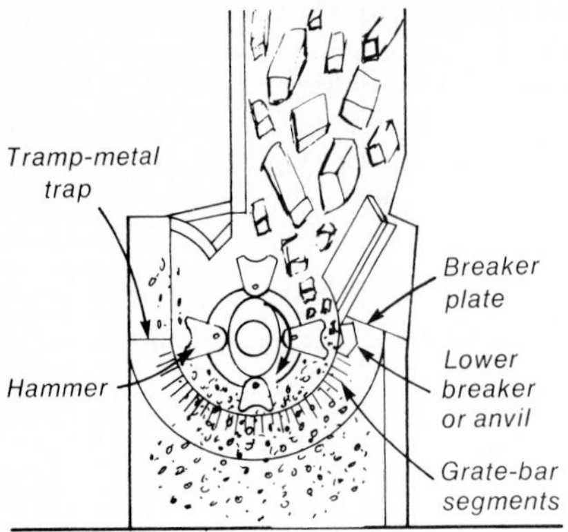

FIG. 5.9 Hammer mill hogging wood waste.

chopped between hammers and the breaker plate, and then ground between the hammers and the screen at the bottom of the unit.

Two types of mills are shown in Figs. 5.10 and 5.11. The first has heavy notched hammers designed primarily for normal wood residue grinding. The other (Fig. 5.11) is designed for processing bark. In operation, rotating teeth (hammers) pass through rectangular pockets formed by the stationary anvils on the side of the machine, breaking up large pieces of incoming wood waste. Final sizing takes place as hammers work the bark against the curved particle-sizing screen, which fits underneath the rotating element.

CONVEYING SYSTEMS

Two types of conveying systems are used in wood-fired powerplants: mechanical and pneumatic.

Mechanical Conveyors

Most energy systems engineers are familiar with the different kinds of mechanical conveyors—drag, flight, belt, screw, vibrating, and variations of these—having previously applied them in coal-handling operations. The success of the wood-handling system hinges in large part on selection of the proper type of conveyor for each job. Providing sufficient load-carrying capacity is important, too.

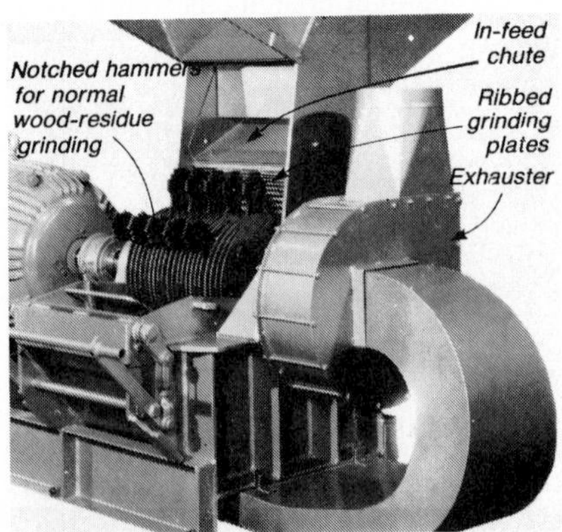

FIG. 5.10 Notched hammers grind wood in this mill.

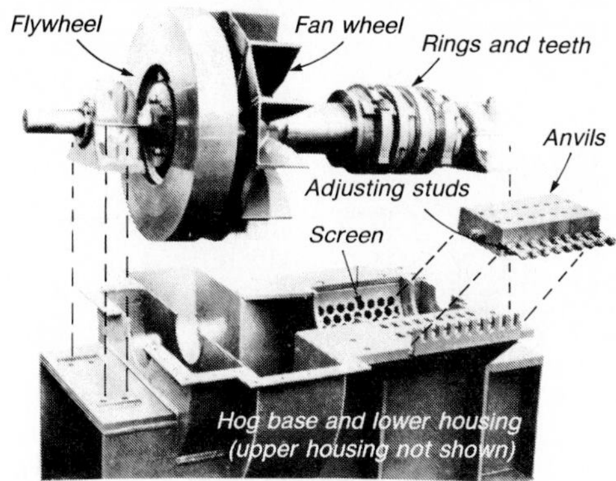

FIG. 5.11 Hammer mill is arranged to process bark.

Belts. The most popular method for transporting hog fuel involves belts. They are less costly to operate than other types of conveyors, can carry material of practically any size (from sawdust to large pieces of wood and bark), can traverse almost any distance, have high capacities, and are very reliable. Although they are not maintenance-free, repairs are generally easy to make and can be handled by most mechanics.

Here are some tips from consultants on how to design a successful transport system using belts:

- Keep conveyors as close to the ground as possible, to minimize the cost of support structures.
- Install cleaners after the head pulley, to wipe small particles of wood from the belt and prevent housekeeping problems.
- Hold the angle of incline to 12 to 15°, particularly in areas where icing occurs. Consider using ribbed belts to reduce slippage where the layout dictates steep inclines.
- Enclose the conveyor to avoid fugitive-dust problems while handling fines as well as to prevent ice formation.
- Use feeders to meter material onto the belt, thereby providing uniform load along the conveyor.

Drag Conveyors. As Fig. 5.12 shows, drag conveyors handle wood waste well and are used for such specialized jobs as reclaim from hoppers and storage piles, besides general conveying service. Most have box-link chains in single or multiple strands or chains with pusher bars. Drag conveyors work well on inclines up to about 18° and on properly designed concave and convex runs, also on horizontal runs. They have modest power requirements and can take more abuse (such as overloading) than other conveyors, while handling a wide range of piece sizes.

Past wear problems have been sharply reduced by lining the bottoms of troughs with ultrahigh-molecular-weight polyethylene or Teflon. The relatively high cost of chain conveyors often limits them to runs of about 50 ft (15 m), but some plants have units stretching 200 ft (60 m) or more.

FIG. 5.12 Drag conveyor.

Flight Conveyors. This type of conveyor (Fig. 5.13) serves well on inclines up to about 30° as well as on horizontal runs where there is little head room. Discharge can be at the end or through a gate (or series of gates) in the trough. One manufacturer says 50 ft (15 m) is a practical maximum length for these units, as with chain conveyors, because of their cost.

Screw and Vibrating Conveyors. Screw (Fig. 5.14) and vibrating (Fig. 5.15) conveyors generally are used as feeders, rather than for moving material between two distant points. Screw feeders find their greatest application in retrieving hog fuel from storage bins and in metering it to distributors on the boiler front. Problems have been reported when these are used to convey unsized or stringy material. Screw conveyors are compact, inexpensive over short distances, clean, and versatile; specifically, they can transport fuel horizontally, on an incline, or vertically. Augers often are hard-faced and stress-relieved to prolong life. Vibrating feeders require little power and usually are installed to regulate the flow of wood waste to belt conveyors or to hogs.

Pneumatic Conveyors

Much more flexible in arrangement than mechanical conveying systems, pneumatic conveying works best on finely ground, clean fuel such as sawdust and

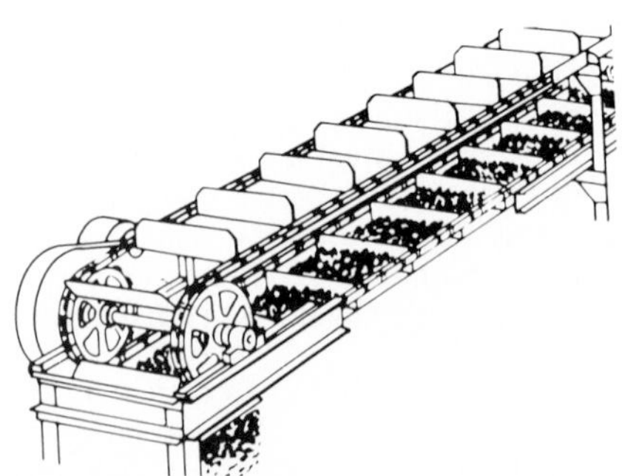

FIG. 5.13 Flight conveyor.

FIG. 5.14 Screw conveyor.

sander dust; also, it is usually cheaper to install. Transport piping can run vertically, horizontally, or on an incline, and, like the tube-type belt conveyor described earlier, it needs only simple support structures.

Major Components. Pneumatic systems typically have only four major components: a positive-displacement blower, transport piping, a rotary air lock to inject fuel into the pipe, and a cyclone to separate the wood from the conveying air at the point of use. The last item is eliminated when the system reclaims hog fuel from storage, because the combined fuel/air stream is injected directly into the furnace.

Consultants say that pneumatic transport usually becomes economical when straight runs exceed 500 ft (150 m). Even shorter systems may be cost-effective if turns are required. The reason is that mechanical conveyors usually cannot negotiate changes in direction unless two or more conveyors are installed, which boosts the cost substantially. The actual economic breakdown for any installation depends heavily on electric power costs and environmental requirements.

Pneumatic systems require about 10 times the horsepower of belt conveyors. Further, if the cyclone used to separate fuel and conveying air cannot capture

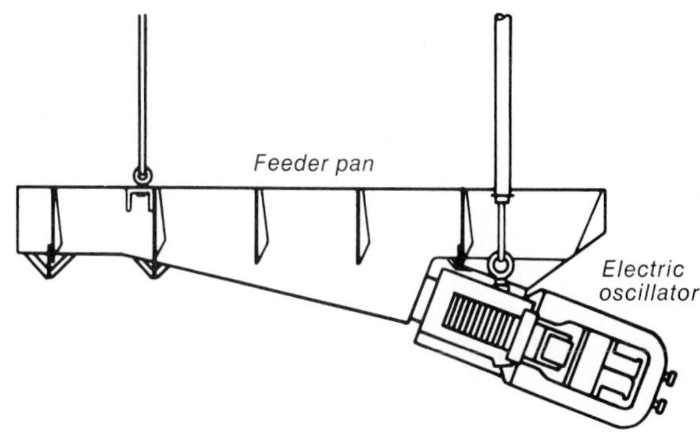

FIG. 5.15 Vibrating conveyor.

enough of the fine particulates to meet existing air pollution codes, a secondary dust collector must be added to filter the cyclone's exhaust stream.

Combined Systems. Recognizing the limitations and benefits of both mechanical and pneumatic conveying, as well as the increasing value of wood waste as a fuel for power boilers, designers are beginning to recommend combined systems, at least in large plants. There are substantial benefits to handling dry, fine fuels pneumatically and coarse or wet material by belt.

First, and most obvious, fugitive emissions from the wood yard are reduced dramatically. Second, boiler performance is enhanced by firing the powdery material in suspension with a solid-fuel burner, which is similar in concept to a pulverized-coal nozzle, and by burning the remaining hog fuel on a grate. Suspension firing of fines allows wood-burning boilers to respond to load swings quickly, without the need for auxiliary fuels, and minimizes the carry-over of combustible material.

DRYERS

Before any fuel can burn, the moisture it contains must be evaporated. Virtually all steam generators designed to fire wet wood and bark have furnaces that can drive off, without difficulty, all the water from hog fuel containing up to 50 to 55 percent moisture. But when the moisture level hits about 57 to 58 percent—as is often the case in the Pacific northwest—traveling-grate, spreader-stoker-fired boilers, the backbone of most large industrial powerplants, cannot produce enough heat on the grate to sustain combustion. The reason is that the temperature of the primary air forced up through the fuel bed to evaporate the moisture is limited to about 500°F (260°C) by grate metallurgy. Thus, although hotter undergrate air could prevent a furnace blackout, it would reduce the grate life dramatically.

Avoiding Moisture

There are several ways to avoid the problems of high moisture:

- Burn oil or gas over the grate to help evaporate the water. This is the customary approach, but one that is losing favor because of skyrocketing fuel costs. A similar method is to burn a dry, pulverulent form of wood, such as sander dust, in suspension with a solid-fuel burner. This option, of course, is limited to plants with a reliable supply of low-moisture (10 to 15 percent or less) wood waste.
- Burn a low-moisture fuel on the grate in conjunction with the hog fuel. Coal is the obvious choice for the supplementary fuel, and it is a good one. If it is required on a continuous basis, however, separate coal-handling and storage facilities must be installed, the boiler equipped with coal distributors, and pollution control equipment provided capable of handling the flyash from both fuels.

- Select a grate capable of handling higher-temperature combustion air. Stationary water-cooled pinhole grates, which accommodate air temperatures up to about 600°F (316°C), can support the combustion of wood waste containing more than 60 percent moisture. Several types of mass-burning stokers also are available for high-moisture fuels.
- Install a dryer to remove some moisture from the fuel before the fuel enters the furnace. This option offers several benefits, in addition to promoting stable combustion. It decreases excess-air requirements and improves efficiency both by reducing the amount of heat lost up the stack and by minimizing carry-over of carbon into the flue-gas stream. It also helps to curtail particulate emissions, facilitates control of the combustion process, and allows production of more steam than is possible by burning a wetter fuel. The last point is particularly important in the case of boilers converted from a relatively high-quality fuel (such as coal) to hog fuel.

Two Dryer Types

There are two basic types of dryers for removing moisture from hog fuel: one uses mechanical energy to wring the water out, and the other uses hot gas to evaporate the water.

Hydraulic Presses. Installed primarily in pulp-and-paper mills on the Pacific Coast of Canada, hydraulic presses squeeze water from bark that may contain up to 70 percent moisture. Applications for presses are limited, because they cannot reduce the moisture level below about 55 percent. They also consume large amounts of power and require continual maintenance. And water squeezed from the bark must be treated prior to its release to the environment.

Hot Gas. Hot-gas dryers can be either the fired or the unfired type. The latter are receiving most of the attention from powerplant designers today because, by using boiler flue gas as the drying medium, they offer an improvement in steam cycle efficiency—that is, the dryer can extract some heat that normally is discharged up the stack in the flue gas. Flue gas leaving the dryer is about 225°F (107°C), or about 100°F (55.6°C) below stack temperatures for large industrial boilers with air heaters and economizers.

Two kinds of hot-gas dryers for powerplant applications are the rotary-drum (Fig. 5.16) and cascade types. In both systems, the entire hog fuel stream is passed through the dryer and is then separated into fine and coarse fractions. Fines are conveyed pneumatically to the boiler and are burned in suspension. Coarse particles are transported by mechanical means to the equipment that distributes fuel onto the grate. This arrangement is not necessarily the optimum one for all systems; in some cases it may be more desirable to combine the coarse and fine fractions and feed them into the boiler together.

Two other hot-gas dryers are the hot hog and the hot conveyor. Essentially, the latter is a vibrating conveyor with holes in the tray that allow flue gas to contact the fuel. Hot hogs are said to have high power requirements and maintenance costs, and they are limited in the amount of moisture they can remove. Hot conveyors have low capacities and also are hampered by high maintenance cost.

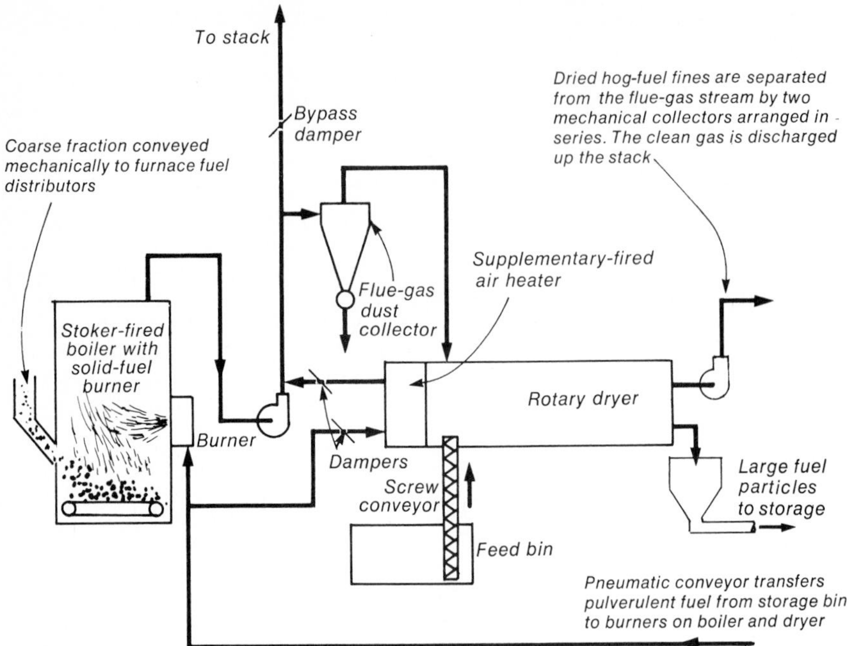

FIG. 5.16 Rotary-drum dryer.

Dryer Design

Progressive engineers generally agree that flue-gas dryers should be considered in the design of hog-fuel-fired steam systems whenever the fuel moisture exceeds about 55 percent. When the economic climate is bleak, they can be justified in many installations on the basis of efficiency gains alone.

When one is specifying a system, one of the first design criteria to establish is the moisture content of the fuel as it leaves the dryer. Some research in this area is worthwhile before concrete decisions should be made. Knowledgeable engineers say that less moisture is not always better, that there is a moisture level—possibly around 35 percent—that gives an optimum balance among dryer cost, furnace and dryer performance, system efficiency, and problems (such as dusting and explosions) associated with handling dry, pulverulent fuels.

STORAGE AND FEED

Hog fuel is stored outside in huge piles or inside in bins or silos designed to hold enough material to sustain boiler operations through a long weekend. High-moisture wood waste usually is kept outside, while dry fines are protected from wind and rain in silos. Some plants have both outdoor and indoor storage for wet hog fuel, relying on the pile for reserve capacity in the event that supplies are interrupted and on the bin to supply fuel to the boiler automatically.

Equipment Groups

Although it may appear that there are as many methods for storing and reclaiming hog fuel as there are plants burning it, the equipment available for these jobs can be split into four groups:

- Sophisticated stacker-reclaimers, to build and recover fuel automatically from storage piles containing tens of thousands of units
- Conventional mechanical and pneumatic conveyors and mobile equipment to build and recover fuel from storage piles of virtually unlimited size
- Storage bins, designed to hold more than 1500 units, which can be filled and reclaimed automatically
- Storage silos and small bins, designed to hold up to about 200 units, which can be filled and reclaimed automatically

The size of an outdoor storage pile should reflect the confidence operators have in their fuel suppliers and the geographic locations of their plants. Facilities in areas subjected to heavy snows, and in areas where forests are inaccessible for weeks at a time because of poor ground conditions, often stockpile enough wood waste to last 4 wk or more at full steaming capacity, that is, when they rely on hog fuel to supply all their power. Such piles are enormous, being nearly an order of magnitude larger than a coal pile containing an equivalent amount of energy. Even when they are stacked 20 ft (6 m) high, they sprawl across acres of land.

Stacker-Reclaimers

Storage piles are built easily by stacker-reclaimers like the one shown in Fig. 5.17. These machines operate on a first-in–first-out basis, eliminating the possi-

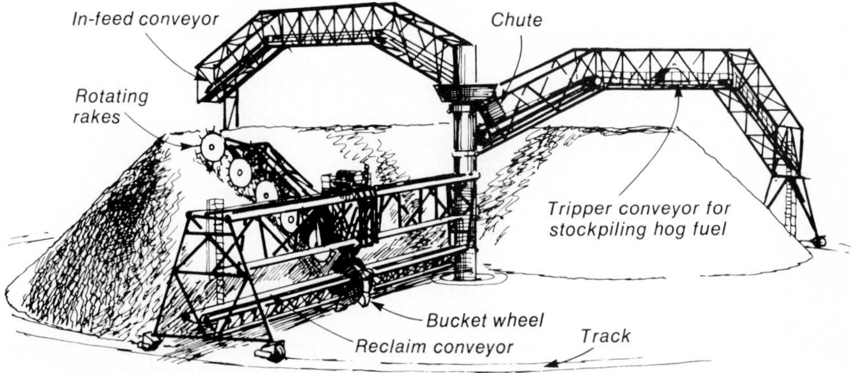

FIG. 5.17 Stacker-reclaimer.

bility of fuel deterioration, which some experts say can occur—in some species of wood and bark, at least—when wood waste is exposed to the elements for more than about 6 months. Most stacker-reclaimers require no operator in constant attendance and usually make less noise and create less dust than the bulldozers often used to work piles.

The circular-pile stacker-reclaimer shown in Fig. 5.17, a new entry in the marketplace, can store up to 60,000 units on a plot 600 ft (183 m) in diameter. Smaller versions of the arrangement are designed to maintain storage inventories as low as 200 units. Stockpiling is accomplished with a tripper conveyor, which pivots about a vertical support column. Fuel can be delivered to the stacker by pneumatic or mechanical conveyors.

Mechanical and Pneumatic Conveyors

Most storage piles are built with, and reclaimed by, mechanical and pneumatic conveyors working in concert with a bulldozer or front-end loader. The recommendations for building wood and bark piles are virtually the same as those for piling coal. The key criteria include the following:

- Select a well-drained area for the pile, where the ground is solid, not porous. Dress the area with paving materials or a compacted layer of wood waste that will never be reclaimed.
- Build piles in the shape of a cone or a topless pyramid, with the steepest sides possible, to prevent water from penetrating the mass of hog fuel.
- Install a temperature monitoring system to detect fires. One source says 200°F (93°C) is the point at which action should be taken, such as cutting out the affected area or injecting carbon dioxide.
- Build up the pile in successive layers, each not more than about 2 ft (0.6 m) thick. Thoroughly compact each layer before adding to the pile.
- Disperse fine particles [those smaller than ¼ in (6 mm)] throughout the pile. They present a greater fire hazard than coarse material and should not be concentrated in a particular area.

Reclaim-system design usually is based on one of four basic concepts:

- Drag-chain conveyors that extend about 40 ft (12 m) into one edge of the pile and have an adjustable shear gate and variable-speed drive system for regulating the reclaim rate (Fig. 5.18). One consultant claims that bridging can be avoided by providing 450 ft^2 (42 m^2) of active reclaim area under the storage pile.
- Traveling screws under the pile, which feed hog fuel onto a reclaim belt, blending the wood waste as they feed.
- Reciprocating pusher blocks, powered by a variable-speed hydraulic drive system.
- A screw rotating about a fixed point under the pile (Fig. 5.19). Material is transported by the screw to the center of the circle swept by the reclaim device and is then dumped onto an out-feed conveyor located in a trench below the screw.

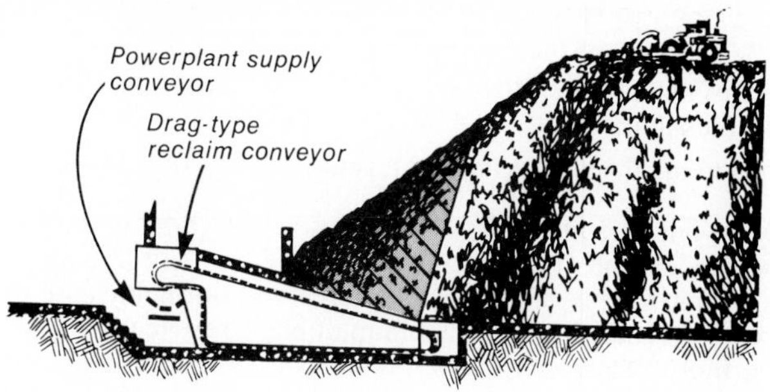

FIG. 5.18 Drag-type reclaim conveyor.

Bins and Silos

Most large bins and some of the smaller silos have negatively sloped walls to prevent bridging. When handling coarse hog fuel, they also are equipped with live bottoms. Both types of enclosed storage structures are capable of storing wood waste varying in consistency from dry sander dust to wet hog fuel, and they can be insulated to prevent freezing problems.

When the possibility of an explosion exists, as it does with sander dust, blowout panels are provided, and a steam smothering system often is installed. Material is delivered to bins and silos by a variety of pneumatic and mechanical conveying systems and is metered from them with equipment like that described in Fig. 5.18, as well as with other types of augers and digger-feeders.

Fuel Feed

The boiler fuel-feed system, which transfers wood waste from storage to furnace distributors, is a critical part of any hog-fuel-fired powerplant. It should be designed to accomplish these primary objectives:

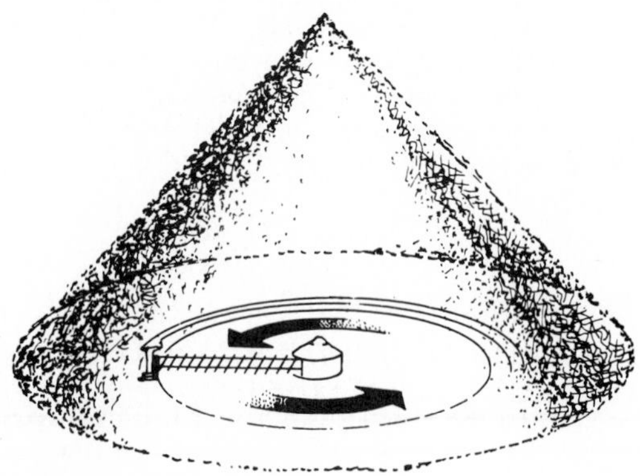

FIG. 5.19 Rotating auger.

- Reliably deliver a controlled amount of fuel to each feeder consistent with steam demand
- Be capable of biasing fuel flow to one or more furnace distributors, to correct for system imbalances
- Prevent tramp air from entering the furnace via the fuel chute

Many wood-burning plants have live-bottom surge hoppers on the boiler to maintain the fuel inventory needed for operating flexibility. These hoppers typically hold a 20-min supply of hog fuel. Where live-bottom bins are used, the most popular type of discharge system is the multiple-screw feeder.

Once hog fuel is delivered to the boiler front, it must be distributed evenly, and without segregation by size, among the various feed chutes supplying the furnace distributors. The fuel mixture in each chute must have approximately the same consistency, to ensure an even fire across the grate and to prevent blowholes, which contribute to inefficient combustion and particulate carry-over.

Poor flow control and segregation are two characteristics of vibrating distributors that limit their effectiveness in plants designed to burn hog fuel as the primary fuel. They are more suitable for boilers designed to burn wood waste as a secondary fuel.

PILE BURNING

Wood is a valuable, renewable source of energy and must be burned cleanly and efficiently, not merely incinerated. Design of a successful combustion system requires detailed knowledge of the fuel's physical and chemical characteristics, plant load profiles, clean-air laws, and other factors. The optimum system strikes a balance among such variables as degree of fuel preparation, capital and operating costs, energy conversion efficiency, and particulate carry-over.

Combustion Fundamentals

The burnable portion of wood is composed of the same three elements as the combustible material in fossil fuels—carbon, hydrogen, and sulfur. First, consider the burning of carbon to carbon dioxide. From elementary chemistry, the weights of the elements entering the reaction must equal the weight of the product. In the following equations, the first line represents the chemical reaction; the second, the respective molecular weights; and the third, the proportioning of these molecular weights, with 1 lb (0.454 kg) of carbon as the base:

$$C + O_2 \rightarrow CO_2$$
$$12 + 32 = 44$$
$$1 \text{ lb } (0.454 \text{ kg}) + 2.67 \text{ lb } (1.2 \text{ kg}) = 3.67 \text{ lb } (1.65 \text{ kg})$$

The last equation shows that it takes 2.67 lb (1.2 kg) of oxygen to burn 1 lb (0.454 kg) of carbon to produce 3.67 lb (1.65 kg) of carbon dioxide. Carbon originally is a solid, oxygen a gas, and the product is a gas. When hydrogen is burned,

$$2H_2 + O_2 \rightarrow 2H_2O$$
$$4 + 32 = 36$$

$$1 \text{ lb } (0.454 \text{ kg}) + 8 \text{ lb } (3.6 \text{ kg}) = 9 \text{ lb } (4.05 \text{ kg})$$

Here the combustion product is in the form of water vapor, which condenses when flue gas cools to near atmospheric temperature.

Finally, sulfur burns to sulfur dioxide:

$$S + O_2 \rightarrow SO_2$$
$$32 + 32 = 64$$
$$1 \text{ lb } (0.454 \text{ kg}) + 1 \text{ lb } (0.454 \text{ kg}) = 2 \text{ lb } (0.908 \text{ kg})$$

The oxygen required to complete all these reactions comes from both the fuel and the atmosphere. For combustion calculations, consider air as being a mixture of oxygen and nitrogen; the traces of water vapor and various other gases can be ignored.

The actual burning process can be divided into three phases: evaporation of moisture, distillation and burning of volatile matter, and combustion of fixed carbon. Here is what happens: After moisture has been evaporated, heat is absorbed by the fuel, raising its temperature and driving off the volatile elements, which burn, sustaining the combustion reaction. When most of the volatile elements are gone, the highly reactive surface of the fixed carbon that remains burns in the presence of oxygen.

Combustion Systems

Combustion systems generally are classified by the way in which they burn wood waste—that is, in a pile, in a thick bed (*semipile burning*), both in suspension and in a bed (*semisuspension burning*), and completely in suspension. Pile burners, including Dutch ovens, fuel cells, and similar equipment, often are selected where a high-moisture fuel must be burned. Pile burners are relatively simple, quiet-burning systems, which have a high thermal inertia to carry them through minor interruptions in fuel supply and fluctuations in moisture content.

Dutch Oven. Shown in Fig. 5.20 is a Dutch oven, the first combustion system

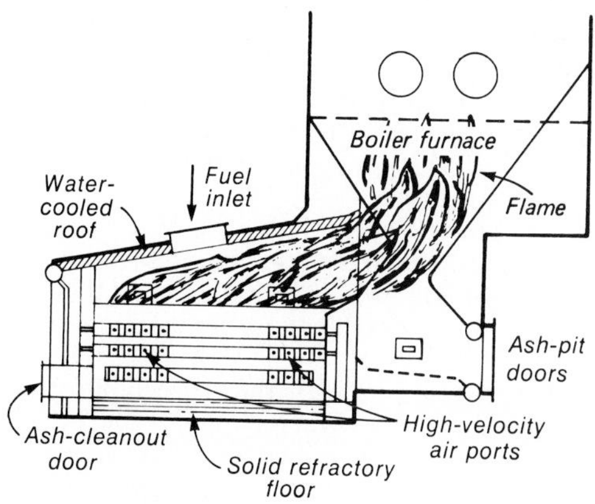

FIG. 5.20 Dutch oven.

used to incinerate bark and until recently the one most often specified for that service. It is essentially a refractory combustion chamber placed adjacent to or under the boiler furnace. Disadvantages of these ovens include high capital cost, high refractory maintenance requirements, poor load-following capability, and manual ash removal.

The floor of a modern Dutch oven often is constructed of solid brick, and all the hot [300 to 600°F (149 to 316°C)] combustion air is admitted through high-velocity ports in the side and rear walls. Fuel is fed through the furnace roof, preferably via an air lock feeder in the supply chute. The conical pile that is formed burns on the surface and around the edges. Combustion of volatile gases is completed in the boiler furnace.

Most Dutch ovens are zoned into two or three compartments, so one section can be shut down for manual cleanout of ash without shutting down the steam generator. In some systems, that portion of the load lost during ash removal is picked up by auxiliary burners in the boiler furnace.

Refractory-Lined Fuel Cells. Normally cylindrical in shape, these fuel cells differ from Dutch ovens in that hot flue gas exits through the top of the chamber rather than through the back or side. Depending on the steam requirements, from one to three cells are installed under the firebox of a shop-assembled boiler. Up to 60,000 lb/h (7560 g/s) of steam can be produced with three cells; cell turndown ratios of up to 5:1 are possible. Since a large furnace opening is required to accommodate the cells, an A-type boiler is a possible choice.

Hogged wood waste containing up to 50 percent moisture [particles smaller than 3 in (76 mm) are recommended to prevent materials handling problems] is piled at the bottom of the cell, on a water-cooled cast-iron grate or refractory floor. When the grates are installed, about 30 percent of the total combustion-air requirement enters the cell from under the grate as primary air; the remainder (secondary and tertiary air) is admitted through ports above the fuel pile. Up to 100 percent excess air may be necessary when a 50 percent moisture fuel is fired. Control of fuel and airflow can be automatic.

Cyclone Furnace. Wood waste enters the cyclone furnace from underneath the grate via a screw feeder. The pile spreads out from the centrally located retort onto grate bars; ash is collected around the periphery of the chamber.

Preheated primary air [up to 600°F (316°C)], at 6 in H_2O (1494 Pa), is fed through the grate, while cold, high-pressure secondary air [125 to 200 in H_2O (31 to 50 kPa)] enters tangentially at the top of the furnace. The whirling effect produced by the secondary air helps dry fuel particles on the surface of the pile and holds charcoal particles in suspension by centrifugal force until they are burned out. Low-pressure [6-in-H_2O (1494-Pa)] tertiary air ensures complete burnout of the combustible gases—primarily carbon monoxide and hydrogen—produced in the oxygen-starved lower furnace.

The main advantage of the cyclone furnace is that it can burn as much as 20 tons/h (18 Mg/h) of wood waste containing up to 68 percent moisture, with minimal carry-over and low carbon loss—all without an auxiliary burner. The disadvantages of the design are the same as those for a Dutch oven.

Wet-Cell Burner. As shown in Fig. 5.21, the wet-cell burner features fully automatic operation, including continuous ash removal. The combustion system can hold excess-air requirements under ideal conditions to 10 percent over a 5:1 turndown range. Units capable of producing 150×10^6 Btu/h (44 MW), and of going

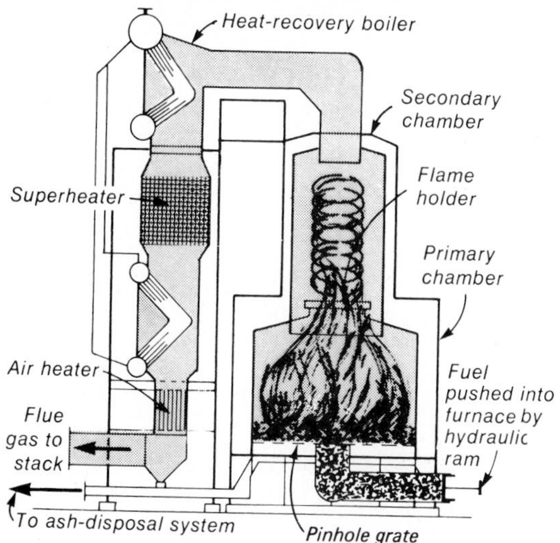

FIG. 5.21 Wet-cell burner is an alternative to Dutch oven.

from minimum to maximum load (and vice versa) in 15 s, are being planned. Exit-gas particulate loadings of 0.02 to 0.06 gr/dry standard ft^3 (0.05 to 0.14 g/m^3) are claimed when operation is at moderate loads.

The wet-cell burner can be integrated with waste-heat boilers for production of steam and/or electricity, as Fig. 5.21 shows. Modular systems of this type are available in capacities up to 100,000 lb/h (12 600 g/s).

Conventional Boilers. Pile burning also is possible in furnaces of conventional packaged boilers rated less than about 35,000 lb/h (4410 g/s), by using a combustion system like that described in Fig. 5.22. The variable-speed screw feeder delivers fuel to the furnace quietly, holding emissions to low levels.

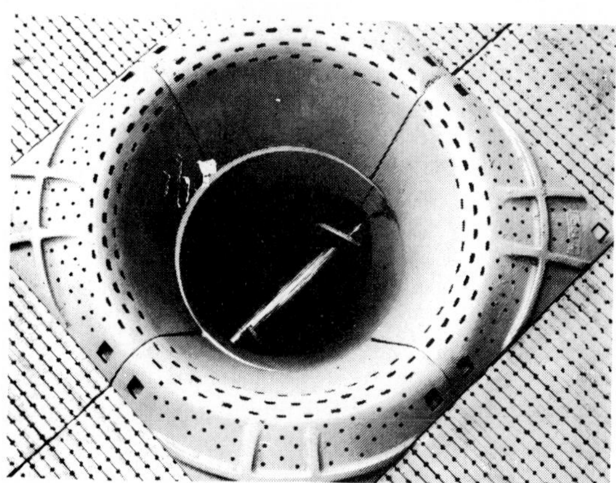

FIG. 5.22 Underfeed stoker.

The underfeed stoker in Fig. 5.22 was developed to handle a wide range of sized wood waste, including bark with 50 percent moisture or more. Although the underfeed principle is the same as that used for burning coal, retort design is different to permit the feeding of large volumes of hog fuel into the furnace, where a conical fuel bed is formed. Overall thermal efficiencies from 60 to 75 percent are possible, depending on the wood's moisture content and other variables, such as sizing.

Semipile Combustion

Semipile combustion systems are designed to burn hog fuel containing 60 percent moisture or more without auxiliary firing and with a minimum of particulate carry-over. But the relatively low grate heat-release rates required to achieve these objectives [250,000 to 300,000 Btu/h•ft^2 (788 to 945 kW/m^2)] make boilers that house semipile combustors expensive because of the additional furnace area needed to achieve a given steaming rate.

Pinhole Grate. The most popular semipile system in North America is the inclined water-cooled pinhole grate (Fig. 5.23), which is used in boilers rated up to about 300,000 lb/h (37 800 g/s). It consists of parallel grate tubes inclined at 55° from the horizontal and covered by stationary alloy grate castings. An advantage of this system, and others like it, is that the large quantity of wood on the grate permits brief outages of conveying equipment and allows rapid load pickup.

Here is how the system works: Unhogged wood waste—pieces as large as 4 × 18 in (102 mm × 457 mm) in cross section are permissible—is dropped from a feeder conveyor onto the upper portion of the grate, where drying takes place. Volatiles are distilled off and burned as the material slides down the grate. Final combustion occurs on the essentially flat bottom surface. Steam output is controlled by varying the airflow.

Multigrate System. Another system for semipile burning of wood waste containing up to 65 percent moisture consists of (1) a stationary grate with three different slopes, to accommodate the change in the natural angle of repose of the fuel as it dries and burns on its way down the grate, (2) a reciprocating floor grate, and (3) a dump grate for continuous, automatic ash removal. Units of this type are not yet installed in North America, but several are operating successfully in Europe. Their capital cost is high compared with costs for competitive systems.

The fixed-grate portion of the system is a finned welded wall, with air holes in the fins. The reciprocating grate has alternately fixed and moving grate bars made of cast steel containing 25 to 28 percent chromium to withstand temperatures up to 1920°F (1050°C). Fuel is fed to the furnace via chutes or feed screws located across the width of the grate under the ignition arch.

SUSPENSION BURNING

Semisuspension and suspension firing are far more familiar to powerplant engineers than pile burning, since both methods are applied in the combustion of fossil fuels, process wastes, and shredded municipal refuse. The grates (pinhole, traveling, and vibrating), the feeders (pneumatic and mechanical), and the burners used for all these fuels and for wood waste have many similarities.

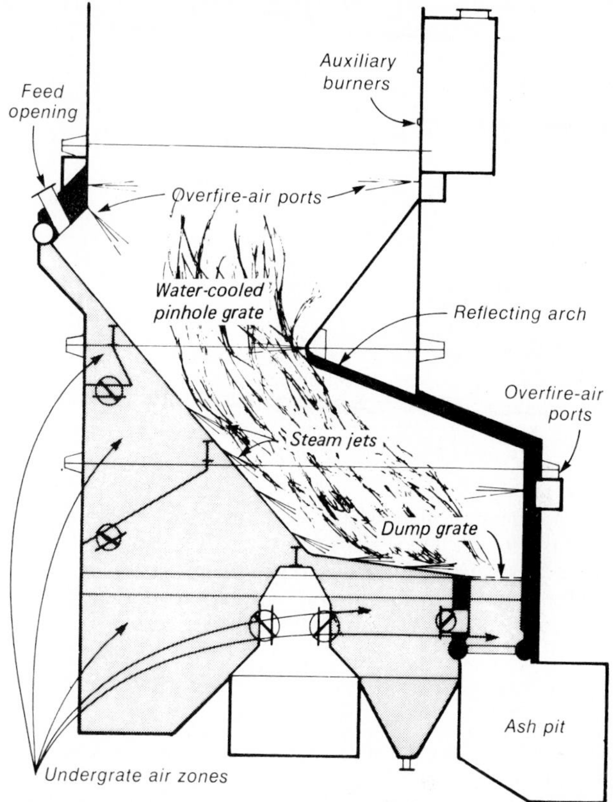

FIG. 5.23 Inclined water-cooled pinhole grate.

Semisuspension Firing

Pinhole grates (Fig. 5.23) of the flat air-cooled and slightly inclined (15°) water-cooled types are selected for the majority of hog-fuel-fired boilers producing up to about 250,000 lb/h (31 500 g/s).

Wood waste, normally hogged until the largest dimension does not exceed about 3 in (76 mm), is injected into the boiler by pneumatic (Fig. 5.24) or mechanical distributors. The acceptable top size can range anywhere from 2 to 4 in (51 to 102 mm). Mechanical feeders rarely see service today because their moving parts are subject to fouling by stringy bark, jamming by tramp iron, and wear problems. Pneumatic distributors, by contrast, have no moving parts and are highly reliable.

Pneumatic Distributors. The unit in Fig. 5.24 shows how the pneumatic distributor works. Air under pressure enters the inlet, passes the air control damper, and flows into the distributor nozzle. High-velocity air emerges from the nozzle at the distributor tray. The trajectory of the hog fuel, which flows by gravity down the angled chute from the feeder above, is controlled by raising or lowering

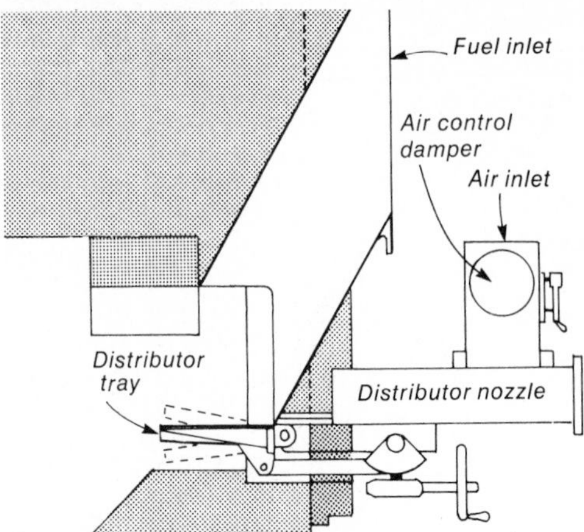

FIG. 5.24 Pneumatic distributor.

the tray with a regulating handwheel. Small particles entering the furnace are consumed in suspension; larger pieces burn on the grate.

Designers locate fuel distributors as low in the furnace as possible to minimize particulate carry-over. When distributors are properly designed and installed, the entire grate is covered with a fairly even blanket of fuel.

Mechanical Distributors

Flat Air-Cooled Grates. These grates are more economical than water-cooled grates for boilers producing up to about 25,000 lb/h (3150 g/s) of steam, simply because no water circulation system is required. Fixed air-cooled grates, however, are limited to a heat release rate of only 600,000 Btu/h • ft^2 (1.9 MW/m^2) and a primary-air temperature of about 400°F (204°C), even when ductile-iron grate bars are used.

Another point to remember is that there probably are as many air-cooled grate designs as there are manufacturers, so be sure to verify the service record preferred with users before signing a purchase order. Large gaps between grate bars and humping are two problems to look out for in operating plants.

Water-Cooled Grates. Grates cooled by water permit heat release rates up to about 1×10^6 Btu/h • ft^2 (3.15 MW/m^2) and primary-air temperatures to 600°F (316°C) or more. Thus, they allow a much smaller furnace for a given steam demand than do air-cooled grates, and they can accommodate hog fuel containing up to 55 percent moisture. Both types of grates have a simple design and have no moving parts—features contributing to high reliability and low maintenance.

Auxiliary burners are sometimes specified for boilers with pinhole grates to guarantee full steaming capability when a section of the grate is being cleaned or when very high moisture fuel is burned. Grates would be protected from furnace radiant heat under such operating conditions by the water-cooling system.

Traveling Grates. Traveling-grate spreader stokers (Fig. 5.25) feature continuous ash discharge, accurate load control, rapid response to changes in steam demand, and the ability to burn coal on the grate with wood. They are almost

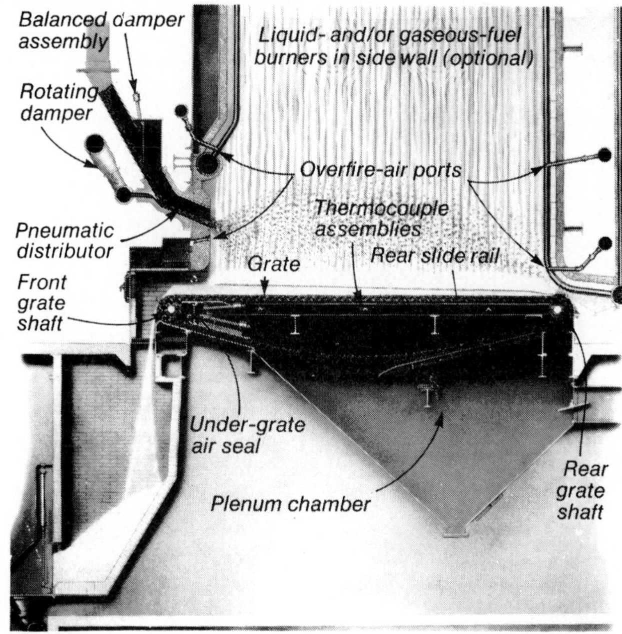

FIG. 5.25 Traveling-grate spreader stoker.

a unanimous choice for boilers rated 250,000 lb/h (31 500 g/s) and above. A disadvantage of the traveling grate is its high cost compared to the pinhole grate.

Feeders for these grates are the same as those for pinhole grates. The top size of fuel particles recommended by most manufacturers is between 2 and 4 in (51 and 102 mm); however, they stress the importance of a good distribution of particle sizes, from fines to the top size, for proper operation.

Air temperatures vary with fuel moisture content up to these limits: 450°F (232°C) for cast-alloy grates and 550°F (288°C) for ductile-iron grates. Grate heat-release rates extend from about 1×10^6 Btu/h • ft^2 (3.15 MW/m^2) for 55 percent moisture fuels to 1.2×10^6 Btu/h • ft^2 (3.8 MW/m^2) for 35 percent moisture.

The ability of the traveling grate to burn both coal and wood provides the fuel flexibility needed by companies with a tenuous supply of hog fuel. Expensive oil or gas is not required for steam production when wood is unavailable, as it is for boilers with pinhole grates. (Pulverized coal has been burned in steam generators with pinhole grates, but only rarely.)

Water-Cooled Vibrating Grates. This grate design provides another option for semisuspension thin-bed burning when continuous ash discharge is necessary. Higher primary-air temperatures are possible than with the spreader stoker; thus, the grate has the capability of burning higher-moisture fuels without supplementary fuel. Allowable grate heat-release rates extend up to 1×10^6 Btu/h • ft^2 (3.15 MW/m^2).

Overfire Air

Most boiler and stoker manufacturers claim that the total amount of excess air required for semisuspension thin-bed burning of wood—using either pinhole, traveling, or vibrating grates—can be held to 25 to 35 percent at the furnace exit. For contingency purposes, however, the combustion air system should be sized to provide at least 50 percent excess air when the poorest fuel expected is burned.

In a scientific sense, no one knows for sure what percentage of the total air requirement should be introduced under the grate and over it. Overfire air provides the oxygen and turbulence needed for complete combustion of volatiles distilled off the fuel bed, as well as of small carbon particles carried into the furnace by the under-grate air. Some engineers argue that 75 percent of the total air, or more, should be admitted as undergrate primary air; others say 40 percent or less.

In actual practice, the ratio of underfire to overfire air depends on the characteristics of a given installation, including fuel moisture content, grate design, firing technique, etc. When a thick bed is maintained on a pinhole grate, for example, one company's experience indicates that maintaining hot primary air at barely substoichiometric levels and injecting the remaining air over the grate—hot, and at a pressure high enough to provide the degree of turbulence and penetration necessary for good combustion—offer the best overall performance.

However, no off-the-shelf automatic control system is yet available that has the degree of sophistication required to maintain the substoichiometric condition on the grate and to burn, over the grate, the carbon monoxide produced by the two-stage combustion process. One benefit of this firing method is that load swings can be accommodated quickly by increasing or decreasing the amount of underfire air.

The leading manufacturer of traveling-grate spreader stokers recommends 25 to 40 percent of hot, high-pressure [25 to 30 in H_2O (6.2 to 7.5 kPa)] overfire air for its equipment. The actual percentage depends on conditions. But designers also insist that the primary-air system must still be sized to handle the full-load rating. The extra capacity gives flexibility for changing fuels, loads, and firing conditions. It is not provided to permit operation without overfire air, that is, all the high-pressure air that enters the furnace through the fuel distributors, the overfire-air ports, and the fly-carbon reinjection system (when used).

In some steam generators, additional overfire-air ports are installed high in the furnace, for example, near the bottom of the boiler bank in a top-supported unit. They are effective for particulate burnout and flame stabilization under adverse fuel conditions.

Reinjection

Another debate rages over the merits of fly-carbon reinjection for spreader-stoker-fired boilers. The trade-off is between increased efficiency and lower flue-gas dust loadings. Particulates leaving the furnace contain a substantial quantity of carbon—enough that boiler efficiency can drop by as much as 7 percent (Fig. 5.26). The largest pieces of char fall into hoppers under the boiler bank and air heater, and smaller flakes are captured by the mechanical dust collector.

Although many engineers agree that combustible material collected in the boiler hopper should be reinjected into the furnace, few are in favor of reinjecting

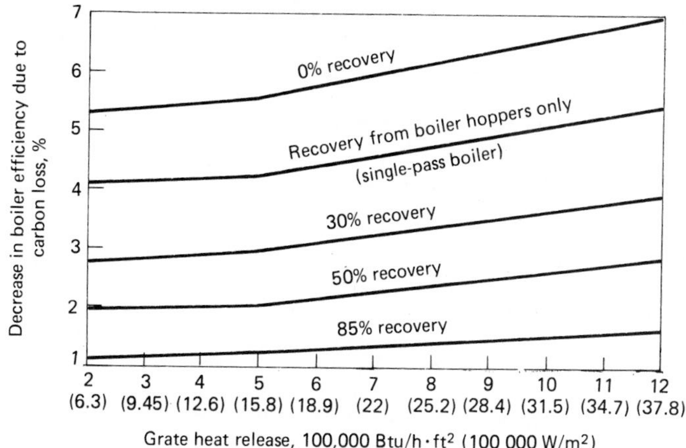

FIG. 5.26 Reinjection of fly carbon gives high efficiency.

carbon from the air-heater hoppers, and still fewer want to reinject fine char from the mechanical collector. What displeases engineers about reinjection is the nature of the fly carbon itself. The devolatized particles have a very hard structure, with combustion characteristics similar to coke—that is, high temperatures and a considerable amount of time are required to burn them out. In many furnaces, they are reentrained in the flue-gas stream before they burn up completely.

Engineers who think that disposal is a good alternative to reinjection should be aware that some particles are so reactive that they glow on contact with air and can start fires in disposal pits and dumps. If there is a consensus today, it is to reinject the large pieces of char from the boiler and air-heater hoppers and to dispose of the remainder.

For trouble-free operation, ensure that a positive flow of fly carbon to the feeder exists by making hopper sides adequately steep (minimum slope: 60°). Ideally, a uniform layer of fly carbon should be deposited on the fuel bed. The best way to do this is to use many injection ports—for example, 15 to 20 nozzles 4 in (102 mm) in diameter—equally spaced across the width of the furnace.

Two alternative approaches to fly-carbon reinjection are receiving attention:

- Sluice fly carbon from the sand classifier to a wet screening system, which collects the large particles and returns them to the fuel conveyor; fines are diverted to a settling basin.
- Collect the char dry, and burn it in an auxiliary combustion chamber specifically designed and sized to handle it. Then duct the flue gas from the auxiliary burner to the boiler furnace.

Suspension Firing

Efficient, clean combustion with suspension burners is possible only with clean, dry, finely divided wood waste, such as sander dust. When they are installed and operated properly, these firing systems burn so cleanly that they can be used on packaged boilers designed for oil and/or gas. For example, a furniture plant recently converted its D-type packaged boiler to pulverized wood by merely changing burners. Although the capacity has dropped from the 60,000 lb/h (7560 g/s) rating on oil/gas to 40,000 lb/h (5040 g/s) with wood, particulate emissions are comfortably below the state standard of 0.4 lb/10^6 Btu (0.17 kg/MJ) without a dust collector of any kind.

Two basic types of suspension burners are available for use on steam generators: the cyclonic and solid-fuel designs.

Cyclonic Burners. Essentially a cylindrical furnace, this type of burner is designed to mix fuel and air in the correct proportion and to complete combustion before the swirling particles of wood reach the end of the refractory chamber, as Fig. 5.27 shows. Although they are suitable for use in boilers, cyclonic burners are more apt to be used on direct-fired dryers and kilns, where their relatively high cost can be justified by the need for clean flue gas at the burner outlet. Units operating on −⅛-in (−3.2-m) particles are available from at least one manufacturer, in capacities up to 40 × 10^6 Btu/h (11.7 MW) with a 5:1 turndown ratio.

Experience with cyclonic burners reportedly is satisfactory in the many plants where they are operated in the nonslagging mode. But in some of the few installations designed to remove ash as molten slag, operating and maintenance problems have been severe. Variations in the amount and composition of dirt in the fuel, as well as variations in fuel density, upset the fuel/air ratio, caused flame-temperature fluctuations that overheated the refractory, and produced a molten slag that eroded the refractory in a relatively short time.

Solid-Fuel Burners. This burner design mixes the air and wood in the correct proportion and ignites the combustible mixture (Fig. 5.28). Burnout of fuel particles is completed in the boiler furnace. Solid-fuel burners also are being specified for many stoker-fired boilers today, instead of oil/gas burners, to improve the combustion of wet wood waste on the grate. The suspension burner in Fig. 5.28 combusts pieces smaller than ¾ in × ⅛ in × ⅛ in (19 mm × 3.2 mm × 3.2 mm), containing less than 12 percent moisture.

The dual-air-zone scroll-feed burner pneumatically injects dry-wood fines in an annular scroll discharge between two oppositely rotating combustion airstreams. The turbulent mixing provided by the register gives stable ignition of the fuel/air mixture. Although the wood fuel may provide a self-sustaining flame

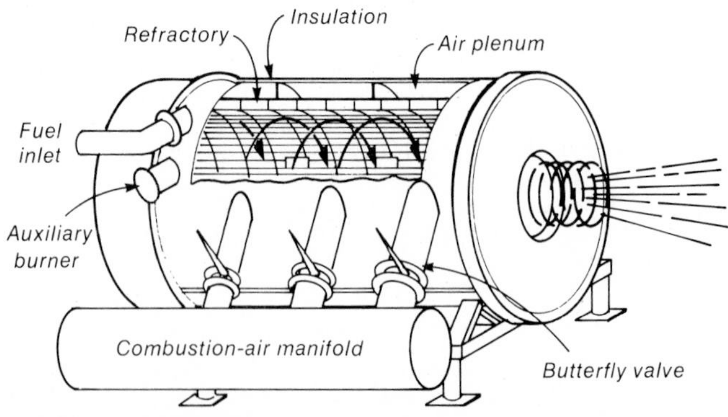

FIG. 5.27 Cyclonic burner.

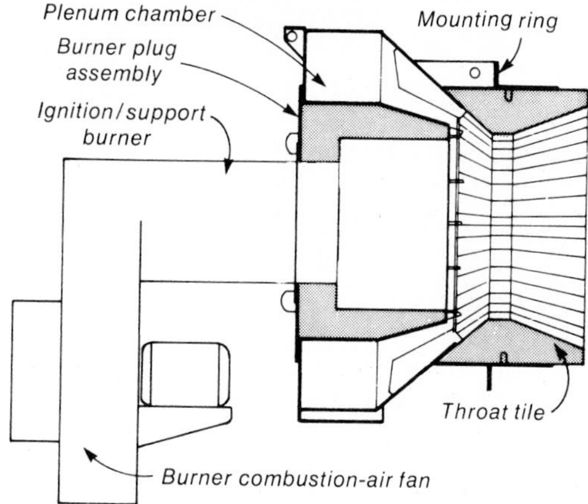

FIG. 5.28 Standing pilot ensures reignition if fuel flow is interrupted.

without auxiliary fuel, a standing pilot is recommended. An interruption in fuel flow and a malfunction of the flame safeguard system could create a hazardous situation. The standing pilot, which provides up to 10 percent of the heat input, provides safe reignition if interruption does occur.

FLUIDIZED-BED COMBUSTION

Although it is only recently that most powerplant engineers have become familiar with fluidized-bed combustors, these devices have been used by some petroleum companies for over three decades to incinerate refinery waste. Other industries have applied the technology to roasting, calcining, drying, sizing, and similar operations.

Principles of Operation

The simplest type of fluidized-bed combustor consists of a cylindrical, refractory-lined furnace with (1) a perforated air distribution plate at the bottom of that chamber, (2) a hot-gas outlet at or near the top, and (3) ports for introducing fuel. During combustion, air, which is admitted to the furnace through the distribution plate, fluidizes a bed of hot, granular, inert material, such as sand. Wood waste (or almost any other fuel) burns in suspension when it is injected into the bed, which resembles a boiling liquid because of its high turbulence.

Primary Functions. Primary functions of the inert material are to

- Disperse incoming fuel particles throughout the bed
- Heat the fuel particles quickly to the ignition temperature

- Act as a flywheel for the combustion process by storing a large amount of thermal energy
- Provide sufficient residence time for complete combustion

Wood is pyrolyzed much faster in a fluidized bed than on a grate, because of its intimate contact with the hot, inert material. Also the charred surface of the burning fuel is continuously abraded by inert particles, increasing the rate of char formation and oxidation. And volatile gases and air are continuously mixed, increasing the rate of gas-phase combustion reactions.

To engineers working for companies experienced in burning wood and bark, the outstanding benefit of fluidized beds is their ability to combust, with relative ease, problem fuels such as high-ash wood waste and slow-burning char. It is difficult to justify fluidized-bed steam generators economically, however, for fuels that can be burned efficiently by more conventional combustion systems.

Two Systems. A fluidized-bed combustion system that has burned wood waste is shown in Fig. 5.29. In another, a refractory-lined combustor installed in a West Coast pulp-and-paper mill works on log yard waste—a very poor fuel. For each pound (kilogram) of dry fiber, it can handle 0.67 lb (kg) of rock and other inert materials, 2 lb (kg) of water, 0.1 lb (kg) of fly carbon, and 0.1 lb (kg) of inert fines from the fly ash. The 183,000 actual ft^3/min (86 400 L/s) of 1800°F (982°C) flue gas produced by this unit is ducted to a modified spreader-stoker-fired boiler, where it generates 300 psig/500°F (2067 kPa/260°C) steam. Theoretically 100 × 10^6 Btu/h (29.3 MW) can be recovered by the system.

The packaged system in Fig. 5.29 is designed to handle only hogged wood waste—it must be smaller than a 2-in (51-mm) cube—which contains 55 percent moisture or less. Like the previous unit, its refractory-lined combustion chamber is separate from the boiler. And, as with other fluidized beds, the turndown ratio is limited to 2:1 to 3:1. The Scotch marine heat recovery boiler serving this system can produce up to 60,000 lb/h (7560 g/s) of steam. Wood combustion is self-sustaining, after a brief start-up period with oil or gas (given a separate preheat

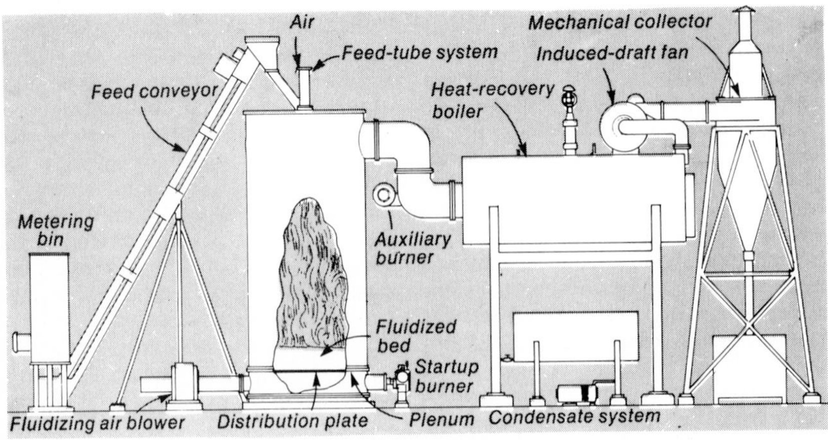

FIG. 5.29 Fluidized-bed combustor burns hogged wood waste.

burner). The boiler has an auxiliary burner for standby service or, if necessary, to supplement bed output.

Many such systems are operating across the country. Engineers at one company, with six in service at various plants, say they consider the design the only true commercial fluidized-bed combustor because of its predictable operating behavior. Experience has been good at these facilities, except for some start-up problems associated with the fuel handling equipment and the unfamiliarity of plant personnel with equipment that burns anything except oil or gas.

BOILER DESIGN

Wood-fired boilers are similar in many respects to the steam and hot-water generators used in industrial plants that burn coal and shredded municipal refuse. Capacities, for example, extend from a few thousand pounds per hour (grams per second) for the smallest fire-tube designs to more than 500,000 lb/h (63 000 g/s) for the largest watertube units—just as they do for coal.

Type vs. End Use

Packaged firetube and watertube steam generators, in sizes up to about 50,000 lb/h (6300 g/s), are the backbone of small plants (Figs. 5.30 and 5.31). Suspension

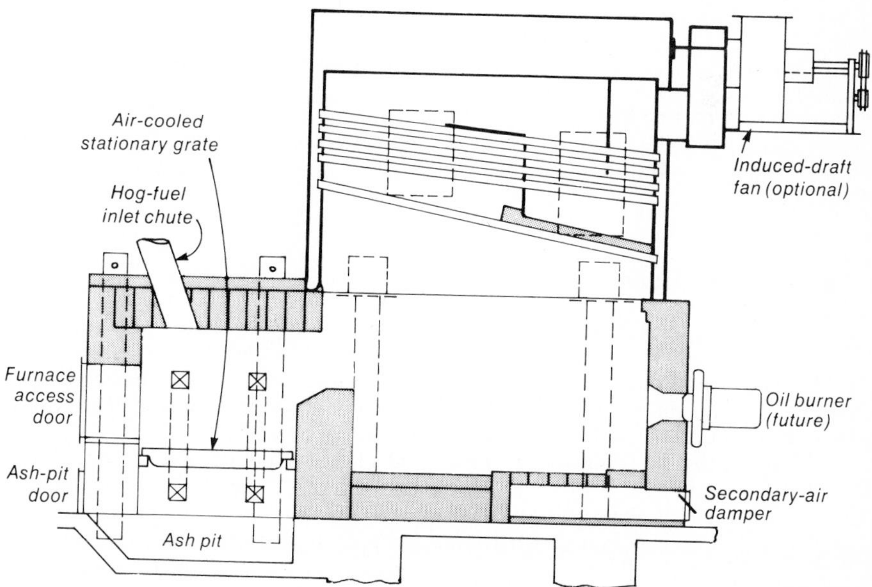

FIG. 5.30 Packaged firebox boiler with refractory furnace.

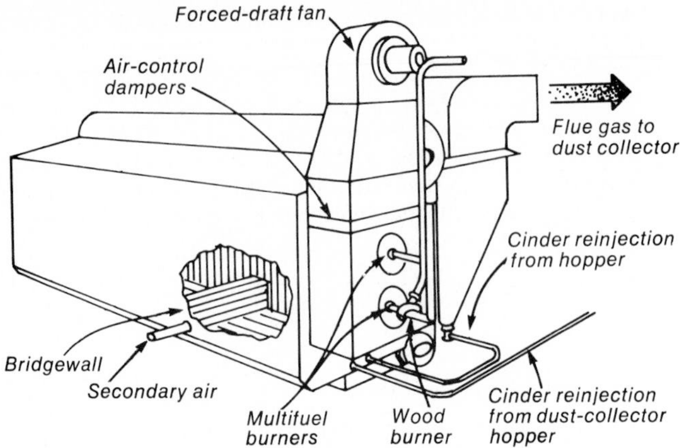

FIG. 5.31 Packaged watertube boiler with suspension firing.

burners, Dutch ovens, fuel cells, air-cooled stationary and dump grates, water-cooled stationary grates, fluidized beds, and gasifiers are the combustion systems usually specified. Selection of the firing method for a particular boiler depends, of course, on the type and quality of wood waste available.

Water-cooled pinhole, vibrating, and traveling grates normally are associated with field-erected steam generators (Fig. 5.32); pile and semipile combustion systems are specified occasionally. Suspension burners firing oil/gas or dry wood waste often are installed in a side wall or in the rear wall, to provide backup or supplementary capacity, to help burn out carbon-bearing particulates before they leave the furnace, to permit rapid response to load change, etc.

Many traveling-grate spreader-stoker-fired boilers purchased today are designed for burning coal together with wood waste on the grate, or for firing pulverized coal and hog fuel. Stoker coal and wood are burned in steam generators rated up to about 400,000 lb/h (50 400 g/s); pulverized coal and wood are not usually considered below about 300,000 lb/h (37 800 g/s), because of the high cost of such systems.

Furnace Construction

Construction of the furnace varies with the size and type of boiler and with the buyer's budget. Tube-and-tile walls (Fig. 5.33) are often specified for small plants, because of their low cost. This design is satisfactory for steam generators producing up to about 200,000 lb/h (25 200 g/s); above that, expansion of the unit becomes so great that refractory maintenance is prohibitive.

Engineers restrict furnace heat-release rates for hog-fuel-fired boilers to about 20,000 Btu/h • ft^3 (207 kW/m^3), allowing ample room to complete combustion. They generally hold flue-gas velocities under 50 ft/s (15 m/s) and to 30 to 40 ft/s (9 to 12 m/s) in multiple-pass (baffled) units, to prevent damage from sand erosion and cutting. Design parameters for boilers burning clean wood fuels such as sander dust are much less restrictive and, in some cases, may approach the heat-release rates and tube-bank velocities allowed for oil-fired units.

Tubular air heaters are installed on almost all steam generators firing wet

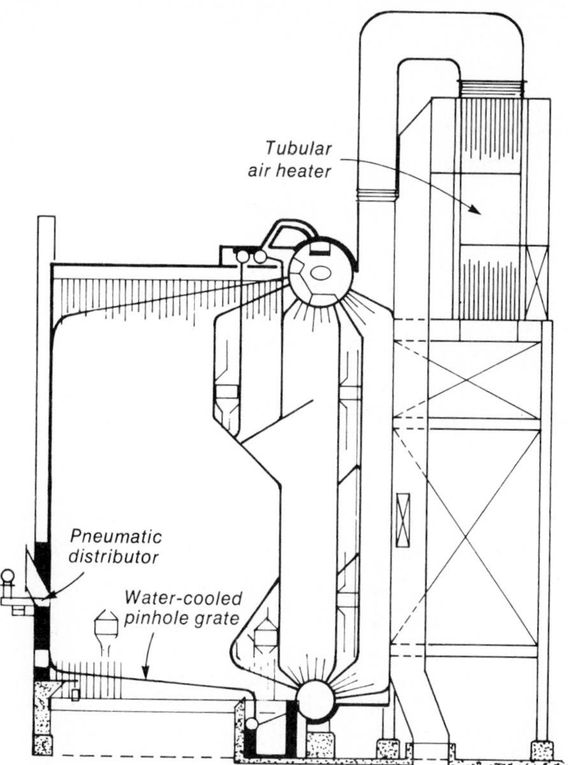

FIG. 5.32 Field-erected boiler with water-cooled pinhole grate.

wood to help evaporate moisture. The regenerative air heaters often specified for coal-fired plants are never used with wood, because char is easily trapped in them and fires may result.

Efficiency

Engineers involved with wood-fired boiler projects should be careful when comparing the predicted performance of boilers, whose efficiency depends very much on the fuel quality and whether the efficiency calculations use *high heating value* (HHV) or *low heating value* (LHV). Here is what may be expected for properly specified, designed, installed, and operated steam generators: Burning sawmill residue and whole tree chips with an HHV of 4250 Btu/lb (9900 kJ/kg) as fired—U.S. manufacturers use HHV, Europeans LHV—wet-basis moisture content of 50 percent, ash content not to exceed 1.0 percent per dry pound (kilogram); an energy-out/energy-in ratio of 65 to 68 percent may be expected with an air heater or feedwater economizer to lower stack temperatures to 300°F (149°C) or so.

By burning the same fuel without heat recovery, for instance in a 500-hp (4905-kW) HRT (horizontal return tubular) boiler or small underfeed stoker firetube unit, a boiler efficiency of 52 to 60 percent may be expected. Some boiler

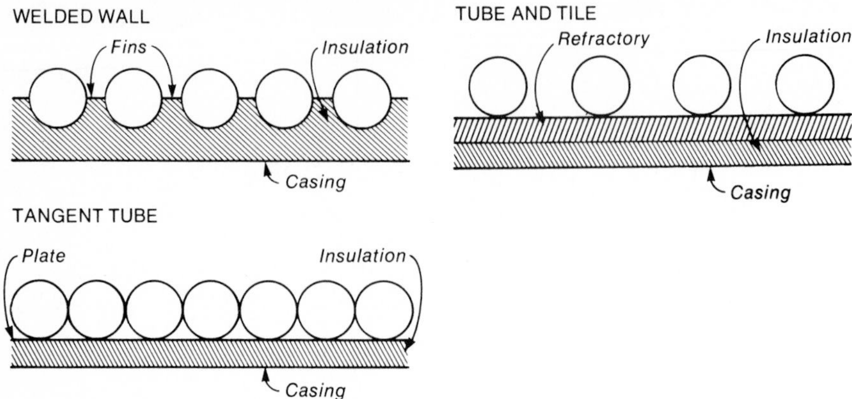

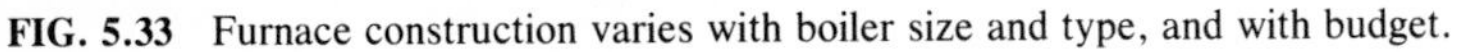

FIG. 5.33 Furnace construction varies with boiler size and type, and with budget.

manufacturers claim higher efficiencies, but careful examination of the claim usually reveals low-moisture-content fuel or calculations based on LHV.

By burning bone-dry (10 percent moisture content or less) wood fines in suspension with a burner that permits close control of excess air in a packaged boiler with heat recovery, fuel-to-steam efficiencies exceeding those for no. 6 fuel oil may be possible. This is because wood fuel contains a large amount of oxygen chemically bound (40 percent by weight for wood vs. approximately 12 percent by weight for coal) and thus requires less air for complete combustion than does coal or even oil. In fact, the theoretical flame temperature of bone-dry wood fuel exceeds 3600°F (1982°C). The high oxygen content and the high flame temperature of sander dust are why extreme caution is called for in handling, storing, and firing it.

SUGGESTIONS FOR FURTHER READING

1. Adams, P. J., et al., "Overview of Wood-Refuse Firing," American Power Conference, April 1979.
2. Andrew, J. G., "Fluidized-bed Combustors—An Overview," *Hardware for Energy Generation,* Forest Products Research Society, January 1979.
3. Archibald, W., "Woodwaste Feed Systems to Boilers," *Wood Residue as an Energy Source,* Forest Products Research Society, September 1975.
4. Arola, R. A., "Wood Fuels—How Do They Stack Up?" *Energy and the Wood-Products Industry,* Forest Products Research Society, November 1976.
5. Christensen, G. W., "Wood Residue Sources, Uses, and Trends," *Wood Residue as an Energy Source,* Forest Products Research Society, September 1975.
6. Corder, S. E., "Fuel Characteristics of Wood and Bark and Factors Affecting Heat Recovery," *Wood Residue as an Energy Source,* Forest Products Research Society, September 1975.
7. Dalton, C. Larry, *State-of-the-Art of Woodwaste Fuel-Handling to the Furnace,* Technical Association of the Pulp & Paper Industry, Engineering Conference, November 1979.

8. DeHoff, G. B., "Design Considerations Affecting Fixed Grates," *Hardware for Energy Generation,* Forest Products Research Society, January 1979.

9. DeRoy, G. L., *Mechanical Ash-Handling System Promises Lower Costs, Simplified Operation,* McDowell-Wellman Co., 1979.

10. Ferguson, C. E., *Storage Equipment, Bins, and Unloaders,* Miller Hofft Inc., 1979.

11. Flick, R. A., "Pulping Industry Experience with Control of Flue-Gas Emissions from Bark- and Wood-Fired Boilers," *Energy and the Wood-Products Industry,* Forest Products Research Society, November 1976.

12. Hall, E. H., et al., *Comparison of Fossil and Wood Fuels,* EPA Industrial Environment Research Laboratory (EPA-600/2-76-056), March 1976.

13. Hoff, E. B., "Handling of Forest Products Fuels," *Energy and the Wood-Products Industry,* Forest Products Research Society, November 1976.

14. Johnson, N. H., "Woodwaste Burning on a Traveling-Grate Spreader Stoker," *Hardware for Energy Generation,* Forest Products Research Society, January 1979.

15. Junge, D. C., *Design-Guideline Handbook for Industrial Spreader-Stoker Boilers Fired with Wood and Bark Residue Fuels,* Oregon State University, February 1979.

16. Levi, M. P., *Wood Fuel for Small Industrial Energy Users,* Solar Energy Research Institute (SERI/TR-8234-1), October 1979.

17. MacCallum, C., "The Sloping Grate as an Alternative to the Traveling Grate on Hog-Fuel-Fired Boilers," *Hardware for Energy Generation,* Forest Products Research Society, January 1979.

18. Paley, B. K., *The Modern Bark- and Wood-Refuse-Fired Boiler,* Bailey Control Applications Workshop, March 1976.

19. Porter, S. M., and R. W. Robinson, "Waste-Fuel Preparation System," *Energy and the Wood-Products Industry,* Forest Products Research Society, November 1976.

20. Suda, S., *Effect of Coal and Bark on Specifying and Designing Industrial Boilers,* Paper Industry Boiler Seminar, sponsored by Babcock & Wilcox Co.

21. Towne, R. S., *Materials-Handling Alternatives for Energy Conservation in the Forest-Products Industry,* American Institute of Chemical Engineers, annual meeting, 1976.

22. Walton, F. H., and D. R. Moody, "Energy Recovery from Log Yard Waste and Flyash Char," *Energy and the Wood-Products Industry,* Forest Products Research Society, November 1976.

CHAPTER 3.6

WASTE FUELS: THEIR PREPARATION, HANDLING, AND FIRING

J. D. Blue, D. R. Gibbs, T. W. Montgomery, L. A. Kreidler, R. W. Kronenberger, and J. R. Strempek

Babcock & Wilcox Co.

Barberton, OH

INTRODUCTION

Energy consumption in the United States in 1985 was almost 75×10^{15} (quadrillion) Btu (79×10^{15} kJ). Most of this energy was generated by the usual fossil fuels, hydroelectric power, or nuclear power. One of the fastest-growing sources of energy accounted for about 0.1×10^{15} Btu (0.1055×10^{15} kJ). This is the waste-to-energy source fueled by *municipal solid waste* (MSW).

The real growth in the U.S. waste-to-energy industry began in the 1970s. Prior to that, the disposal of municipal waste was limited to open dumps, sometimes burning it there or in some sort of refractory chamber in an attempt to obtain more complete burning in a more aesthetic manner. Local concerns with the smoke and ash effluents, coupled with increasingly strict federal and state standards, spelled the demise of this means of disposal. At the same time, the price of energy and land increased dramatically. Also, concerns for land and water pollution multiplied, while the federal government encouraged development of alternative energy sources. These disparate elements provided the motivation and favorable economics to begin generating power from municipal waste. The Arab oil crisis, PURPA (see Chapter 6, Section 2) and its attractive electric rates, and federal tax benefits made the economics of energy production even more favorable for burning refuse as a fuel.

WASTE TO ENERGY—A PERSPECTIVE

Before we go into the details of working with waste fuels, here is an overview of the kinds of usable wastes and the waste-to-energy techniques found in powerplant practice today, with emphasis on the level of boiler performance to expect.

Nature of Municipal Solid Waste

MSW is the combined residential and commercial waste materials generated in a given municipal area; their collection and disposal are usually provided by local government or private haulers. Industrial wastes are generally hauled in a separate procedure and are not discussed here. Formerly landfilled, MSW is now the energy source of most waste-to-energy plants, whether the waste is burned as it comes off the truck, called *mass burning,* or by processing the MSW via size reduction and material recovery to produce *refuse-derived fuel* (RDF).

Tables 6.1, 6.2, and 6.3 show the heterogeneous nature of MSW fuel. The variance in constituent composition, weights, moisture content, and density provides a challenge to the system designer. Ranges of weight and moisture percent-

TABLE 6.1 Ranges of Weight and Moisture Percentages*

Refuse item	Composition, %	Moisture, %
Corrugated boxboard	1.32–6.81	8.59–50.23
Newspaper	8.88–21.35	9.60–34.87
Magazines, books	2.05–3.74	7.23–26.27
All other paper	19.78–24.77	18.60–33.53
Plastics	2.00–6.82	3.62–19.65
Rubber, leather	1.22–2.60	3.57–18.42
Wood	1.18–6.58	8.09–24.98
Textiles	2.24–8.92	9.14–36.64
Yard trimmings	0.26–33.33	21.08–62.20
Food waste	7.23–16.45	52.35–73.45
Fines, −1 in (−25 mm)	2.83–11.75	10.10–43.00
Metallic	6.81–11.08	2.57–10.83
Glass, ceramics, etc.	7.13–23.06	0.59–6.00
Composite		16.77–42.10

*Elmer R. Kaiser, "Physical-Chemical Character of Municipal Refuse," *Combustion,* February 1977.

TABLE 6.2 Ranges of Inorganic Contents, in Weight Percent, Dry Basis*

Refuse item	%
Corrugated boxboard	2.01–3.57
Newspaper	1.31–2.96
Magazines, books	11.91–19.01
All other paper	4.98–28.94
Plastics	3.72–10.72
Rubber, leather	4.12–24.99
Textiles	1.84–3.17
Wood	1.56–5.62
Yard trimmings	5.59–30.08
Food waste	4.59–21.87
Fines, −1 in (−25 mm)	53.00–66.72
Metallic	90.49
Glass, ceramics, etc.	99.02
Composite	30.56–35.91

*Elmer R. Kaiser, "Physical-Chemical Character of Municipal Refuse," *Combustion,* February 1977.

TABLE 6.3 Approximate Densities of MSW Constituents

Refuse item	lb/ft^3	kg/m^3
Magazines	35	560
Paper	3–5	48–80
Cardboard	7	112
Corncobs	11	176
Green grass	3	48
Meat scraps	15	240
Rubber	44	704
Shoe leather	20	320
Vegetable food waste	14	224
Wood chips	15–25	240–400
Hardboard	33	528
Plastic bags	0.75	12
Plastics	2–7	32–112
Textiles	9–11	144–176
Cast iron, steel	450–490	7200–7840
Sand	90–117	1440–1872
Glass bottles, unbroken	22	352
Glass bottles, broken (1½ in or 38 mm max.)	67	1072
Aluminum, elemental	11	176
Aluminum cans (single-can basis)	2–3	32–48

ages, inorganic (ash) content, and approximate densities for some of the constituents are given in the tables. The composition of MSW varies from municipality to municipality and from season to season. Indicative of this variance is a recent study conducted by the National Bureau of Standards on the chlorine content of

MSW based on samples taken at Baltimore County, Maryland, and Brooklyn, New York. The total chloride content (by mass) was 0.45 and 0.89 percent, respectively. Not only was the variance wide, but also the component contributing the largest fraction to the chlorine content was different—paper in Baltimore and plastics in Brooklyn. Systems and equipment design must reflect the varying composition of MSW and should be based on representative sampling for the area involved, not on average or typical data.

Not all materials are acceptable for use as waste fuels unless the facility is designed specifically to accommodate them. Not acceptable as MSW feed for waste fuels are (1) materials that may cause a waste-to-energy facility to violate an air- or water-quality effluent standard or (2) items, when processed, that could cause harm or damage to personnel or hardware. Unacceptable materials are explosives, pathological and infectious wastes; radioactive wastes; poisons; concentrated acids and bases; human and animal remains; bulk quantities of paints, solvents, and other highly volatile materials; large amounts of coal; white goods (refrigerators, stoves, air conditioners, etc.); bathtubs and sinks; bulk quantities of ferrous and nonferrous metals; incinerator residues; concentrations of heavy metals; oil sludges; excavation wastes; and any other concentrations of materials that may place the facility in violation of Environmental Protection Agency (EPA) rules. This is largely an administrative rather than an engineering concern.

Many waste-to-energy plants, both the mass-burning and RDF types, install equipment to process oversized bulky waste (OBW) for use as waste fuels or for ferrous recovery. The purpose of OBW processing equipment is to reduce in size materials that have value as waste fuel or as recyclable material, so they may be handled by other processing or material-handling equipment. Typical OBW materials are household furnishings (tables, chairs, davenports, dressers), mattresses, bed springs, rolled carpets, tires, timber, empty drums, light-gage scrap metals, demolition debris, bundled paper and corrugated boxes, and loose roofing materials.

Boiler Performance Characteristics

The two major design areas of MSW energy recovery systems are combustion and materials handling. The well-designed combustion process should provide economical heat recovery, maximum reliability, and minimum environmental impact. It is covered first.

Refuse Fuel Analysis. Considerable engineering judgment is required to properly design a refuse burning and heat recovery system, because the chemical analysis and calorific value of the fuel may vary from day to day, week to week, and location to location. These values (that is, the fuel analysis and higher heating value) have a significant impact on the design of the boiler and associated equipment. The calorific value determines the heat input for a given quantity of fuel, while the chemical analysis determines the quantity of air required for combustion and the resulting flue-gas quantity and quality.

During the early stages of a project, it is in the engineer's best interest to specify the refuse analysis and calorific value so that she or he may have a common base for comparison of the various offerings. However, the proper relationship of the analysis and heating value to each other must be maintained, to achieve the proper design. If the calorific value is uncharacteristically high for the analysis, the air and flue-gas systems will be undersized for the probable operating range of the unit. If the heating value for a given analysis is too low, the fuel-burning capacity of the furnace will be limited and the contracted quantity of refuse will not be processed.

The best cross-correlation between analysis and calorific value is obtained from the air quantity required to combust 10,000 Btu (10 550 kJ) of a fuel. This value is remarkably consistent for most fuels, for example, for natural gas varying from 7.23 to 7.46 lb/10,000 Btu (3.11 to 3.21 kg/10 000 kJ). This same rationale can be applied to refuse.

When refuse is considered as a fuel, three of its components are critical: mineral matter, moisture, and organic matter. Although mineral matter is the root cause of most potential boiler problems, it has little participation in the combustion process, except for metals that may be partially oxidized. Likewise, moisture does not participate in the combustion process, but does contribute to the total moisture in the flue gas. The concern for the combustion process is, therefore, limited to organic matter.

Table 6.4 presents a typical analysis of the various components of refuse (newspaper, plastic, wood, yard trimmings, food waste, etc.). The derived theoretical air values are shown. The required pounds (kilograms) of air per 10,000 Btu (10 000 kJ) are reasonably consistent, varying from 7.2 to 7.3 (3.10 to 3.14) for wood and wood products, to 7.35 (3.16) for food and yard trimmings, to 7.64 (3.29) for plastics. Test data from a number of RDF and MSW fuels on operating units show similar results. Here the majority of the data were centered about 7.5 (3.23) with one-half of the data in the relatively small range of 7.3 to 7.7 (3.14 to 3.31).

Table 6.5 shows three example analyses provided for design purposes. Column 3 exemplifies the problem that can be anticipated if the heating value is not correlated to the chemical analysis. If the analysis in column 3 were used, the combustion fans, boiler, and flue-gas cleanup equipment would be undersized, and the desired tonnage of refuse could not be processed.

In summary, engineers of refuse units should design for a range of heating values and analyses. This range will account for the following:

- The inherent heterogeneous nature of the fuel
- Fluctuations due to seasonal trends
- Forecasted projections that show a steady trend of increasing heating values in the future

Throughout this range, the theoretical air must fall between 7.3 and 7.8 lb/10,000 Btu (3.14 and 3.36 kg/10 000 kJ). If the air required falls outside this range, a heating value should be inferred from the analysis by using a theoretical-air requirement of 7.5 lb/10,000 Btu (3.23 kg/10 000 kJ).

Boiler System Efficiency. The operating efficiency of a refuse-fired incinerator is dependent on the fuel analysis, final exit-gas temperature, excess air, radiation heat loss from the setting, and heating value remaining in the residue and flyash. Table 6.6 shows a typical fuel analysis of raw refuse and RDF compared to bituminous coal. Actual refuse analyses, of course, can vary widely from the values given. Unit efficiencies are also shown to depict, in part, the impact of fuel analysis on performance.

The exit-gas temperature is limited by the acid dew point of the flue gas, which is in the range of 300 to 350°F (149 to 177°C) when a regenerative or tubular air

TABLE 6.4 Analyses of Refuse Components, Percentage by Weight

Refuse item	Component wgt, %	C	H	O	N	Cl	S	Water	Ash
Corrugated boxboard	5	36.79	5.08	35.41	0.11	0.12	0.23	20.00	2.26
Newspaper	12	36.62	4.66	31.76	0.11	0.11	0.19	25.00	1.55
Magazines	3	32.93	4.64	32.85	0.11	0.13	0.21	16.00	13.13
All other paper	23	32.41	4.51	29.91	0.31	0.61	0.19	23.00	9.06
Plastics	3	56.43	7.79	8.05	0.85	3.00	0.29	15.00	8.59
Rubber, leather	2	43.09	5.37	11.57	1.34	4.97	1.17	10.00	22.49
Wood	3	41.22	5.03	34.55	0.24	0.09	0.07	16.00	2.82
Textiles	3	37.23	5.02	27.11	3.11	0.27	0.28	25.00	1.98
Yard trimmings	10	23.29	2.93	17.54	0.89	0.13	0.15	45.00	10.07
Food wastes	10	17.93	2.55	12.85	1.13	0.38	0.06	60.00	5.10
Fines, −1 in (−25 mm)	10	15.03	1.91	12.15	0.50	0.36	0.15	25.00	44.90
Metallic	7								
Glass, ceramics, etc.	9								
	100								

TABLE 6.5 Effect of Three Refuse Analyses on Combustion Air Requirements*

Constituent	1	2	3
Ash	25.63	9.43	21.30
S	0.33	0.27	0.20
H_2	3.38	4.72	3.85
C	23.45	33.47	21.70
H_2O	31.33	19.69	26.42
N_2	0.19	0.37	0.40
O_2	15.37	31.90	25.92
Cl_2	0.32	0.15	0.21
Total, %	100.00	100.00	100.00
HHV, Btu/lb	4174	5501	4713
(kJ/kg)	(9725)	(12 815)	(9723)
Air, lb/10,000 Btu	7.71	7.49	5.76
(kg/10 MJ)	(3.31)	(3.21)	(2.47)

*Values in percentage except where noted otherwise.

TABLE 6.6 Analyses of MSW and RDF Compared to Bituminous Coal

	Analyses, % (by weight)		
Constituent	MSW	RDF	Bituminous coal
C	27.9	36.1	72.8
H_2	3.7	5.1	4.8
O_2	20.7	31.6	6.2
N_2	0.2	0.8	1.5
S	0.1	0.1	2.2
Cl_2	0.1	0.1	0
H_2O	31.3	20.2	3.5
Ash	16.0	6.0	9.0
HHV (wet), Btu/lb	5100	6200	13,000
(kJ/kg)	(11 880)	(14 450)	(30 290)
	Losses per fuel, %		
Loss items	MSW	RDF	Bituminous coal
Dry gas loss	10.1	6.3	6.2
Moisture in fuel loss	14.5	11.0	4.2
Moisture in air loss	0.2	0.2	0.2
Unburned combustibles	3.3	2.2	2.5
Radiation	0.5	0.5	0.3
Unaccounted	1.5	1.5	1.5
Total losses	30.1	21.7	14.9
Efficiency	69.9	78.3	85.1

heater is used as the final heat trap. The exit-gas temperature may also be limited by economics as the flue-gas temperature approaches the incoming feedwater temperature if the last heat trap is an economizer, or to the saturation temperature at the operating pressure of the unit if no economizer or air heater is used.

Air heaters are not generally selected for mass-burning units since ambient air is needed to cool the stoker grate. Stokers firing RDF can use hot air, thus air heaters are often installed to ensure good operating efficiency. Excess air values for mass-burning units are generally about twice those of RDF units. This higher excess air is necessary to minimize the effects of maldistribution of combustion air caused by the uneven fuel loading on the stoker. Typical excess-air values range from 80 to 100 percent for mass-burning units and from 35 to 50 percent for RDF-fired units. The remaining heating value in the ash and other residue removed from the boiler system is highly variable and heavily dependent on how the unit is operated. Unburned combustible losses range between 1 and 3 percent of the total heat input because of the variable nature of the fuel supplied.

Handling the Raw Waste

The materials handling of MSW begins at the curbside of the home or business generating the refuse. In some cities, separation of the material to recover recyclables such as newspaper, ferrous metals, aluminum, and glass is provided for. The logistics of the municipality may favor spotting of the transfer station around the city, where smaller collecting trucks can dump their loads. Compactor trucks then transfer the MSW to the resource recovery plant (Fig. 6.1). The ma-

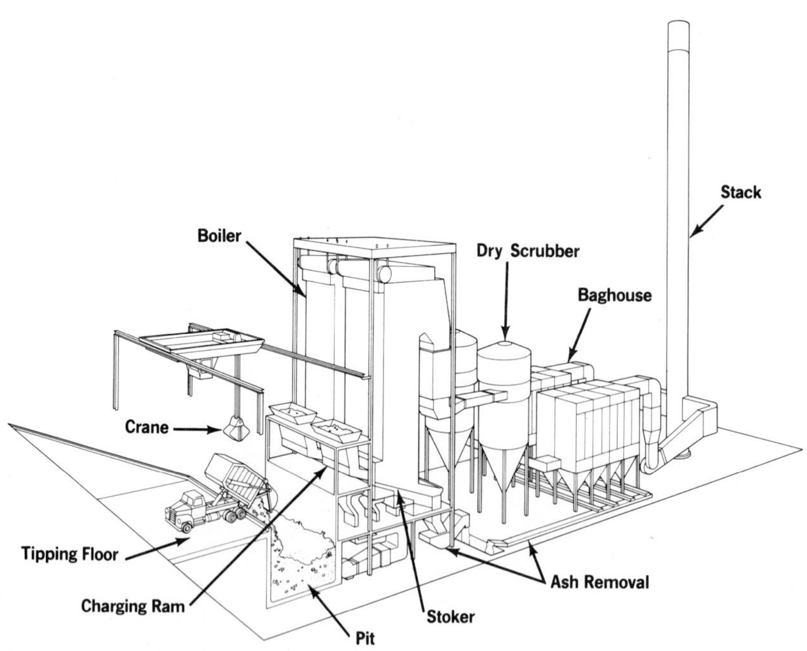

FIG. 6.1 Typical mass burning refuse-to-energy system.

terials handling system for mass burning is discussed here, for the RDF system later.

Receiving or Tipping Floor. On the incoming roadway to the resource recovery plant, the MSW loaded truck is weighed before it proceeds to the receiving or tipping floor area. The tipping floor is adjacent to the refuse storage pit that allows trucks to maneuver to unload into the pit. The area is normally enclosed and kept under slight negative pressure by putting a forced-draft fan intake duct over the pit; by drawing air from the pit area, dust and odor are kept to a minimum.

Tipping floor entrance and exit doors should allow at least 25 ft (7.6 m) horizontal and vertical clearances, should be at opposite ends of the building, and should be protected by barriers. Traffic flow through the area should be arranged to put the pit on the driver's left as he or she enters the tipping area, to allow better visibility.

Careful consideration must be given to the number of tipping bays and the width of the tipping floor, since both have a direct influence on traffic flow. The number of bays should reflect the anticipated daily volume of refuse delivered as well as the hours of delivery. The width of the tipping floor should be able to accommodate the largest vehicle that will enter the facility, which would normally be a transfer trailer. Large facilities of more than 1500 tons/day (1360 Mg/day) should have a tipping floor width of at least 125 ft (38 m) of clear space. A front-end loader moves rejected material to one side of the tipping floor for disposal and for general housekeeping purposes.

MSW Storage Pit. The refuse pit is a heavy, reinforced-concrete bunker designed to hold a designated volume of MSW. The size of the pit will depend on the quantity of refuse delivered each day, the number of days that deliveries will be made per week, the capacity of the combustion system, and the desired storage capacity. For example, if refuse is to be delivered for only 5 days and the plant will operate continuously over the weekend, as well as operate over a 3-day holiday weekend, it will probably be desirable to have 4 days' storage capacity.

Once the dimensions of the pit are determined, the effective volume of the pit must be calculated to determine whether it matches the desired days of storage. Determination of the effective pit volume must take into account the interior dimensions of the pit, the density of the refuse, and whether the refuse will be stacked (sloped).

The refuse pit capacity is calculated as

$$C = VD/2000$$

where C = refuse pit capacity, tons (Mg)
V = effective volume of refuse input, ft^3 (m^3)
D = average density of refuse, lb/ft^3 (kg/m^3)—use approximately 20 lb/ft^3 (320 kg/m^3)

Once the capacity is determined, it can be divided by the daily designed throughput of the plant in tons (megagrams) per day to arrive at the storage time in the pit in days. Conversely, the size of the required pit V can be similarly found, given the other factors.

Crane Design. In the modern mass-fired resource recovery plant, there will be one full-capacity spare crane to allow continuous operation of the facility during crane maintenance. Each refuse crane should be designed with the capability to feed all furnaces at rated capacity. These cranes should be designed for continuous severe service and have the ability to withstand high ambient temperatures. The pit itself will be fitted with water spray nozzles in the event of fire there.

Refuse crane design criteria are influenced by several factors not directly related to the mechanical capability of the crane to feed the furnace at a given rate. The crane is a basic critical part of the refuse flow system, which originates at the source of refuse generation and terminates at the disposal of residue and the energy delivery system. Furnace feeding is generally a continuous operation, 24 h/day, 7 days/wk. So the crane system must be designed with minimum unscheduled outage probability and the capability to feed all furnaces at rated capacity. Also, there must be slack time for the crane operators to mix the incoming refuse in the pit—especially during times of wet weather and when large amounts of difficult-to-burn materials such as grass and leaves are received.

The grapple system is important in determining the crane's duty cycle. Two basic types of grapple are the clamshell and the orange peel. The latter features longer cable life, the ability to pick up more refuse per pound of bucket, and smaller hoisting motions because of its lighter weight. The clamshell is more durable and requires a smaller stoker hopper opening.

The crane operator's ability is critical in determining the availability and effectiveness of the refuse crane; maintenance is also important. Thus, it is imperative that a crane operator training program as well as a comprehensive preventive maintenance program be instituted prior to commercial operation. These programs may be run by the crane vendor or the plant operator, but they are absolutely essential in bringing the plant onstream on a timely basis.

COMBUSTION SYSTEM DESIGN

The method of incinerating refuse fits into two major categories—prepared RDF and mass-fired (MSW). Within these two broad divisions, several technologies are available for the incineration of refuse, some of which are spreader stokers, reciprocating grates, roller grates, rotary combustors, kilns, gasifiers, and fluid beds. RDF is sometimes cofired in existing power boilers, but this design is not part of the technology considered here.

RDF and mass firing are distinguished by the mode of refuse preparation. In the mass-burning technique, the refuse is used in its as-received state, where it is fed directly into a large storage pit from a tipping floor. Large objects, noncombustibles, and hazardous materials are removed either manually from the tipping floor or remotely by crane from the refuse pit prior to burning. Mixing of the various constituents may occur in the refuse pit at the crane operator's discretion. The refuse is then fed onto a reciprocating grate stoker, the combustible portion is burned, and the residue is dropped into an ash pit for reclamation or disposal.

In the prepared-refuse technique, as-received material is processed in any number of schemes to yield a high-quality shredded fuel and other salable or recyclable by-products. Hazardous and large, bulky materials and noncombustibles are removed prior to the processing system. The RDF is then fed into a furnace to burn on a traveling grate stoker, fluidized bed, rotary combustor, or other suitable means.

Refuse is moderately volatile fuel and is readily combusted with any of the above-listed technologies. As a fuel it is heterogeneous, however, and difficult to handle, and it contains variable amounts of water and ash. In addition to the concerns for hang-ups and abrasion in the fuel handling, fuel feed, and ash removal systems, refuse incineration produces significant potential for corrosion, erosion, slagging, and fouling, which presents a challenge to the boiler designer.

Plant and Unit Sizing

Refuse plant sizing is generally expressed in tons per day (megagrams per day) of as-received refuse. The design of the plant, the determination of the number of incinerators, and the size of the units are dictated by the redundancy desired, the sizes commercially available, the capacity to handle peak refuse receiving periods, and the capacity to catch up from lost production periods caused by outages. Different operators may view the aforementioned design options differently as they relate to plant availability, keeping low *operating and maintenance* (O&M) costs, and the guarantee of refuse processing commitments.

Of significance is the heat input of refuse to the unit. Since the as-fired Btu (kilojoule) value of refuse can vary widely, proper evaluation of the Btu and tonnage ranges is important in arriving at a properly sized unit. The heat input is set by the design refuse rate and heating value. Since the furnace is a "Btu machine," the fuel-burning capacity of the furnace will be inversely proportional to changes in the refuse heating value. Proper consideration should be given to unit sizing so that the contracted quantity of refuse can be processed without overloading the furnace.

To further complicate the selection, the Btu value of refuse has been steadily increasing with time and is expected to continue to do so. In fact, projections for future RDF heating values are in the range of 7000 to 7500 Btu/lb (16 300 to 17 500 kJ/kg) with 5200 to 5500 Btu/lb (12 100 to 12 800 kJ/kg) forecasted for mass refuse. Present-day design levels are 6500 and 4500 Btu/lb (15 150 and 10 500 kJ/kg), respectively.

The above considerations may be further explained by demonstration of a simplified case. A project required that 365,000 tons/yr (331 000 Mg/yr) of refuse, averaging 4800 Btu/lb (11 200 kJ/kg), as-received, be guaranteed on an annual basis. The heating value is expected to range up to 5200 Btu/lb (12 100 kJ/kg). In the evaluation of the overall plant design and sizing, it is determined that the annual capacity factor for the refuse incinerator(s) will be 80 percent.

In this example, the capacity factor is defined as the percentage of time the incinerator(s) will operate at rated capacity on an annual basis. Therefore, to assume that 365,000 tons/yr (331 000 Mg/yr) can be incinerated, the incinerators must be sized at a rate of 456,000 tons/yr (413 590 Mg/yr). On the basis that the plant will contain two mass-fired incinerators, the base size of each would be 625 tons/day (567 Mg/day) or 26 tons/h (24 Mg/h) each. At 4800 Btu/lb, the hourly heat input to each incinerator would be 250×10^6 Btu/h. At 5200 Btu/lb, the average hourly heat input would be 270×10^6 Btu/h. The final unit sizing would then be 625 tons/day (567 Mg/day) at a heat input of up to 270×10^6 Btu/h (79 000 kW).

A similar evaluation can be made for an RDF facility. This is equally important for an RDF or a mass-burning plant because the sizing of all the balance of plant components will be driven by this initial analysis. Otherwise, there can be extreme overloading of the facility with the potential reduction of availability, higher operating and maintenance costs, and negative impact on landfill life and project financial health.

Mass Firing

In the mass firing concept, the refuse is incinerated in essentially the as-received state. Large objects and toxic and explosive materials, if present, are removed for alternative disposal. Figure 6.2 shows a typical mass-fired unit.

Charging Hoppers. Refuse is fed to a charging hopper in mass firing. The hopper is designed to permit the unobstructed plug flow of refuse down the hopper to the feed device. The hopper volume should be sized to provide a continuous supply of refuse to the feeder between charging intervals. The height of refuse in the hopper should be such as to effectively eliminate the infiltration of tramp air.

Charging hoppers should also be designed to resist abrasion and fire. Usually the hoppers are made of carbon steel with refractory liners in the lower areas, where the potential for fire exists and elevated temperature from furnace radiation is possible. Water-cooled charging hoppers, which eliminate the need for large quantities of refractory, are available. Sprinklers can be installed for fire protection.

Slide gate and guillotine isolation dampers may also be installed for maintenance, shutdown, and emergency provisions. Within the foregoing considerations, hopper design and sizing will vary from vendor to vendor.

Combustion Systems. Many mass firing combustion systems or technologies are in place today that effectively incinerate refuse. Among these are rotary combustors, starved-air combustors, rotary kilns, reciprocating grates, and roller grates. The technologies range from those developed in Europe and Japan to those developed either alone or jointly with North American interests. Because of their vast number, it is not possible to cover each one in detail; two are briefly described here.

The combustion system is defined to include the charging hopper, feeder, grate, undergrate/overfire air, combustion controls, and the furnace design and configuration. These components are interrelated and interdependent for an effective and controllable mass firing combustion system.

The most important requirement for incineration is a steady, controllable feed of refuse onto the grate, which is generally accomplished with a hydraulically driven ram located in the bottom of the charging hopper (Fig. 6.3). These rams must be rugged and designed for ease of maintenance. High-duty wear zones and parts subject to furnace radiation should be equipped with high-wear-resistant, easily replaceable liners. Ram speed and stroke should be adjustable to suit the needs of the boiler combustion control system.

Grate Systems. Once on the grate, the refuse dries, is heated to ignition temperature, and burns, and the ash is conveyed to the discharge end. Grate systems range from horizontal to inclined and from reciprocating to roller types. In general, however, all types are designed with grate motions to convey the refuse from inlet to discharge while mixing or tumbling the refuse to enhance combustion and carbon burnout. Some include steps or drop-offs to mix and expose the refuse to radiant heat and air for effective combustion (Fig. 6.4).

Generally, the grates are arranged in sections along the length, front to rear; they are independent for effective control of the combustion process, refuse, and

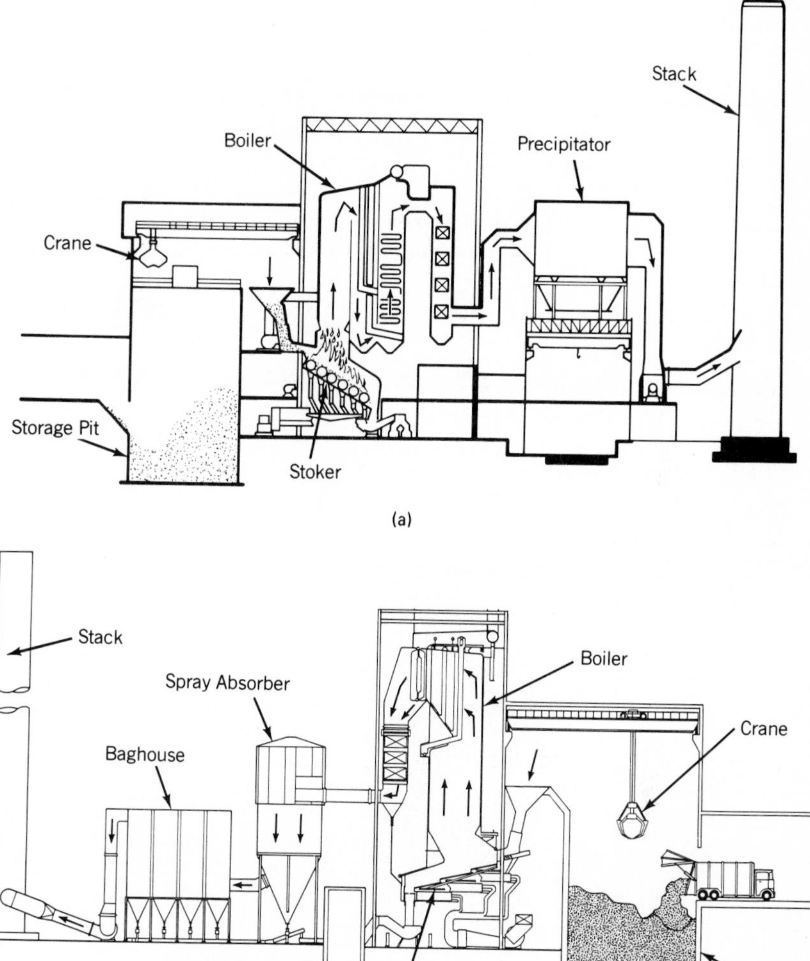

FIG. 6.2 Cross section of mass-fired units: *(a)* Multiple pass; and *(b)* single pass. *(Courtesy American Ref-Fuel Corp.)*

ash bed. The undergrate air is usually partitioned such that the airflow to the drying, combustion, and burnout zones is independently measured and controlled (Fig. 6.5). The number of air zones along the length will vary with different vendors but generally is from three to five.

Air admission through the grates varies depending on the type and design. In general, the grate air resistance should be high enough to provide good air distribution and be designed to cool the grate effectively, thus providing for longer life and lower maintenance. Since grates operate in a severe-duty environment, grate

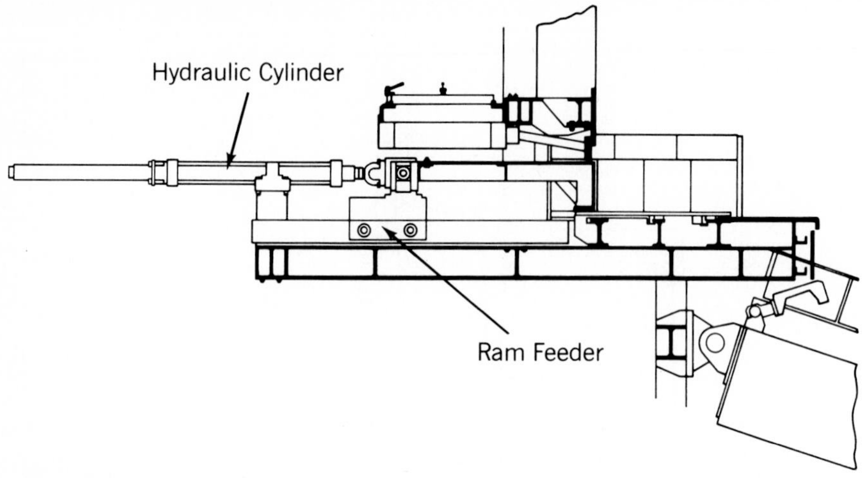

FIG. 6.3 Hydraulic ram feeds refuse onto grate. (*Courtesy Von Roll Inc.*)

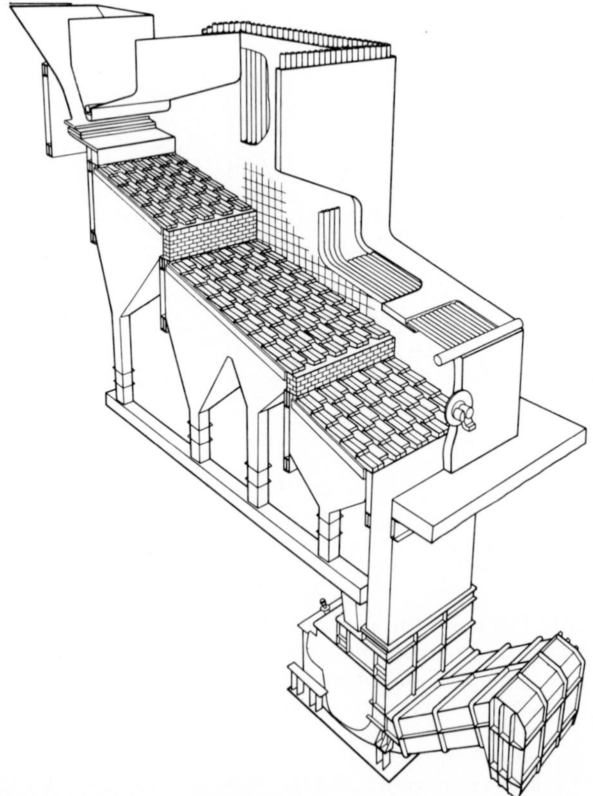

FIG. 6.4 Grate system has dropoffs to mix and expose refuse to radiant heat and air. *(Courtesy Riley Stoker Co.)*

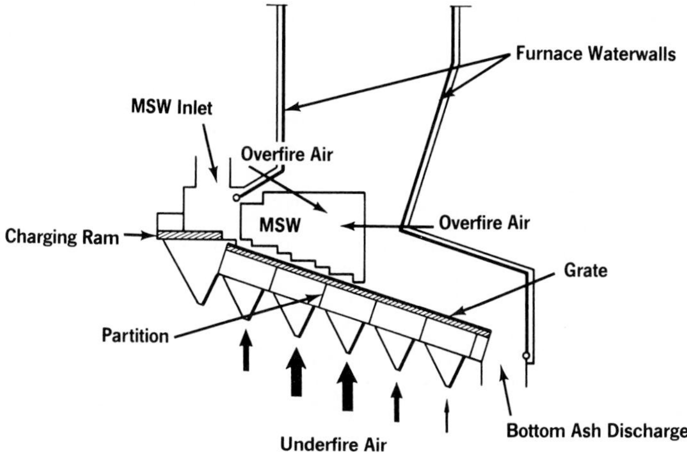

FIG. 6.5 Undergrate air is partitioned for better control. *(Courtesy Von Roll Inc.)*

materials should be high-alloy, high-heat, and high-wear-resistant materials; 23 to 25 percent chrome alloys are not uncommon. Grate vendors predict a 3- to 5-yr grate life when they are properly operated and maintained.

In addition to underfire air, which is introduced under the grate, an effective overfire air system is necessary to reduce the potential for corrosion and slagging and to provide for combustion efficiency. The location of overfire air nozzles is shown in Fig. 6.5. The number, size, and distribution of overfire air nozzles are set by the grate and/or the boiler designer, based on both theoretical factors and experience, to provide turbulence, mixing, and distribution of high-pressure air with combustible gases in the lower furnace.

Rotary Combustor. Technologies other than reciprocating- or roller-type grates are available for mass incineration of refuse. One is depicted in Fig. 6.6, which shows a water-cooled rotary combustor and afterburning grate system. A ram feeding system charges the waste to the elevated end of the combustor barrel. The slightly inclined, water-cooled combustor barrel rotates slowly, causing the waste to tumble and advance as combustion proceeds. Since the combustor is water-cooled, corrosion and clinker formation are reduced.

A forced-draft fan draws air from the waste storage area and supplies preheated combustion air through the ducts and the window hopper to the multiple-zone wind box located around the combustor barrel, where a controlled amount of preheated combustion air [at 450°F (232°C)] is forced into the combustor barrel. This air flows through the air holes in the steel webs connecting the combustor-barrel watertubes. Regulated distribution of air to the multiple zones controls the combustion process. Because the combustor walls are cooled by water instead of combustion air, the combustor requires only 50 percent excess air. The water used to cool the combustor is part of the boiler water circuits.

The tapered end of the rotary combustor extends through the front wall of the waterwall radiant section of the connected boiler. Small ash particles drop between the tubes of the tapered section onto a water-cooled afterburning grate, which permits additional burning of any remaining combustibles. The residue drops from the grate through the residue collection chute and into a water-filled

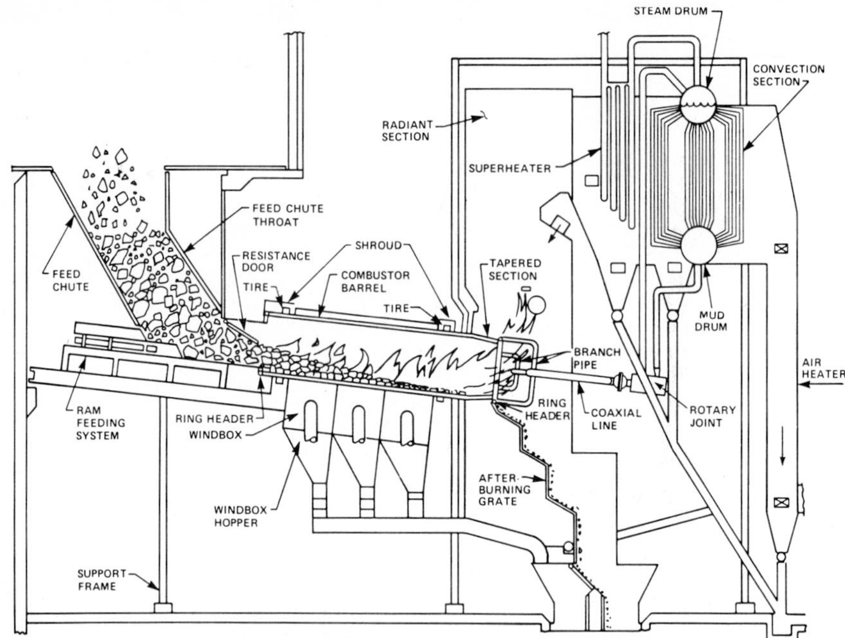

FIG. 6.6 Water-cooled rotary combustor and afterburning grate system. *(Courtesy Westinghouse Electric Corp.)*

conveyor trough. Large pieces advance to the end of the tapered section and fall through the residue chute into the ash conveyor. The small amount of siftings, which fall through the air holes in the combustor webs, are conveyed to the ash conveyor.

Combustion gases flow from the combustor barrel through the boiler radiant section, the screen tubes, the superheater, and the convection sections of the boiler. To maximize energy recovery and expedite combustion of high-moisture waste, the combustion gases leaving the convection bank section pass through a tubular heat exchanger that preheats the incoming combustion air. A forced-draft fan draws air for combustion from the waste storage area. This arrangement maintains these areas under a slight negative pressure to preclude odor or dust emissions from the structure. Steam coils at the forced-draft fan outlet preheat the combustion air to a set temperature to control flue-gas acid condensation in the air heater.

RDF Firing

Combustion technologies available today to fire RDF are (1) the overfeed and traveling grate system (sometimes referred to as *semisuspension firing*) cofiring in combustion with other fuels such as pulverized coal, and, to a lesser extent, (2) the fluidized-bed combustion system. RDF firing may be favored when rapid load response in the steam user's system is needed, higher boiler efficiency is desired, material recycling is practiced, or truck traffic to a refuse plant must be minimized.

Figure 6.7 shows a perspective of a typical RDF-fired system. It consists of a series of RDF feeders, air-swept spouts, traveling grate, and boiler. In the combustion process, the larger RDF falls to the grate, where it is burned with its ash discharged to the ash removal system. The lighter RDF is burned in suspension above the grate.

Feeders and Stoker. Many variations of feed arrangements and devices have been used on RDF units, several of which are derivations of those applied to feed wood- and bark-fired units (see Chapter 5, Section 3). The feed system has to be able to handle streamers, sand, bulky waste, and ferrous, nonferrous, and glass materials. Because of these extreme physical requirements, many of the early RDF systems were unreliable and troublesome and needed frequent maintenance and (ultimately) modifications.

Out of this early experience came the overfeed-type supply system. In this concept, the RDF is distributed to a bank of feeders located across the unit width in sufficient number to satisfy full-load requirements. The overfeed concept provides for RDF delivery to the feeders in an amount about 5 to 15 percent over that needed to satisfy boiler demand. Thus, the feeder surge hoppers are always full, the excess refuse is discharged to a conveyor, and it is returned to the RDF storage area. Individual feeders, each having a surge capacity, utilize a variable-speed ram to feed an inclined discharge conveyor whose speed is based on steam

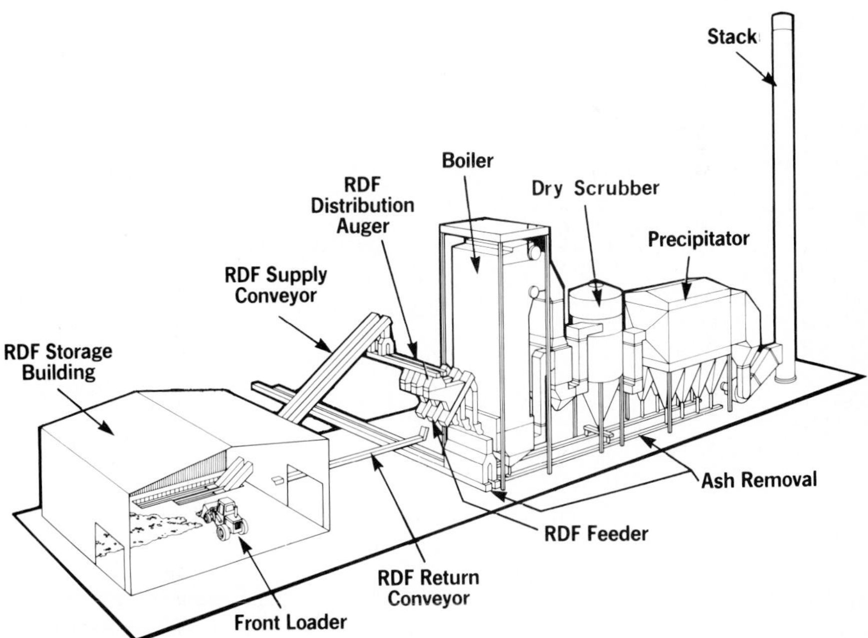

FIG. 6.7 Typical RDF refuse-to-energy system.

demand. Each feeder can be independently controlled to bias and control refuse distribution side to side and to modulate load to meet demand on 100 percent RDF firing.

Each feeder supplies RDF to an air-swept spout. Pulsed air is introduced through the spout in combination with underspout air nozzles to provide furnace front-to-rear depth of throw and fuel distribution. The air-swept spouts should be sized large enough to avoid pluggage; a minimum of 30 in (760 mm) wide and 18 in (460 mm) high is preferred.

Figure 6.8 shows a typical traveling grate for RDF firing. The overall grate size is set so as not to exceed 750,000 Btu/h• ft^2 (2.36 MW/m^2) of effective grate area. The stoker length is determined by the ability to distribute the RDF uniformly front to rear; the length has been up to 19½ ft (5.9 m). With the grate release rate and grate length determined, the stoker width may be found on an area basis if the resulting RDF feed and grate operating speeds reflect levels consistent with successful operating experience. Additional stoker width may be necessary to provide adequate feeder capacity and slower feeder and grate speeds consistent with acceptable wear levels.

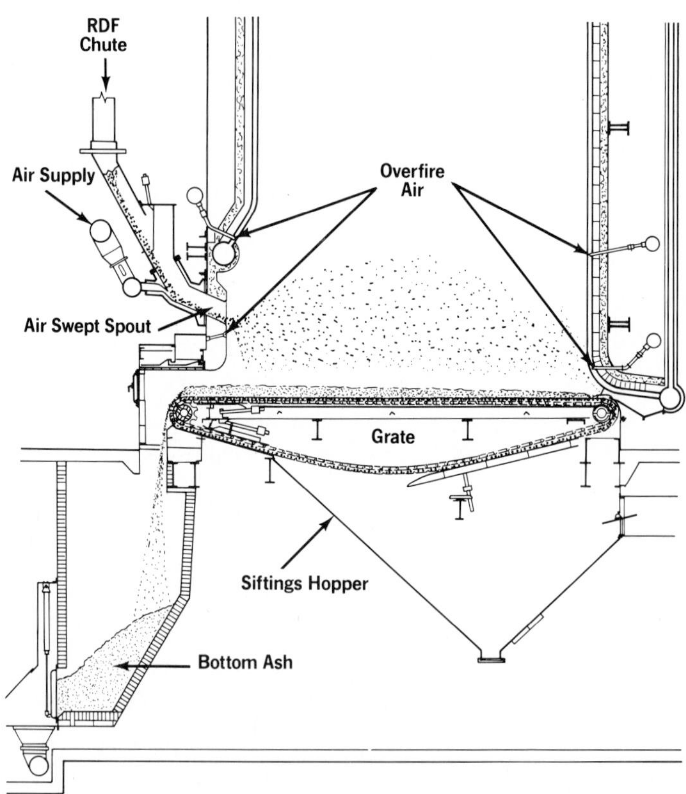

FIG. 6.8 Traveling grate for overfeed RDF firing. *(Courtesy Detroit Stoker Co.)*

In the combustion of refuse, aluminum will melt. In both mass and RDF firing, molten aluminum can solidify on the grates, plugging air holes and requiring periodic outages for cleaning. Stoker design should also provide enough flexibility to control ash-bed thickness of sufficient depth to allow molten aluminum to solidify as much as possible within the bed instead of on the grate.

Good stoker designs recognize the potential aluminum problem and have the ruggedness needed to shear the fused aluminum at parts moving relative to one another at seals and tuyeres. Similarly, the abrasive nature of RDF noncombustibles dictates the need for high-wear-resistant properties on moving grate parts, which should be consistent with requirements imposed due to thermal shock, strength, and impact resistance from potential slag falls. The design and arrangement of the RDF feed system may be affected by RDF received from the processing system. For this reason, communication between designers of the RDF processing system, feed system, boiler, and stoker is a must. The boiler-stoker designer should set specifications for RDF sizing.

RDF sizing requirements may vary between suppliers. Experience indicates that the size range should be biased toward the heavier side to control the quantity of suspension burning. A typical size distribution is −6 in + 4 in (−150 mm + 100 mm), 4 to 15 percent; +2 in − 4 in (+50 mm − 100 mm), 39 to 66 percent; −2 in + 1 in (−50 mm + 25 mm), 20 to 25 percent; and −1 in (−25 mm), 8 to 18 percent.

A more even distribution of fuel on the RDF grate and a more consistent fuel size permit operating at somewhat lower excess air than with mass firing. Overfire air nozzles in the front and rear walls are provided for an RDF unit as with a mass-burning unit for good combustion and emissions control. Heated air is often used under the grate and for overfire air. The nature of the fuel, coupled with suspension burning, generally permits selection of a narrower unit.

Furnace Sizing. The resource recovery furnace is usually a water-cooled, membrane-wall enclosure. It contains the combustion process and provides for sufficient surface to cool the products of combustion to temperature levels suitable for passage into the convective surfaces, with enough residence time for burnout of entrained materials. As noted, the design and internal arrangement of furnaces will vary with suppliers and technologies. In some designs, there are "furnace extensions" in the form of open passes that cool the gases before they enter the convection sections. Likewise, the criteria used to size the furnace and open passes will vary with the technology and suppliers.

The furnace exit-gas temperature is basically fixed by furnace size for a given heat input capacity. Significant changes can be effected by altering the fuel input, although small changes are possible by deviating the excess air from the design value. These deviations, however, are limited due to the increased potential for corrosion, slagging, or erosion. Therefore, it is important that conservative parameters set the furnace size. To avoid corrosion and fouling of convection surfaces, relatively low furnace exit-gas temperatures [1300 to 1600°F (705 to 820°C)] are recommended with relatively long residence times to complete combustion.

The interface between stoker and the lower furnace design is very important. The physical shape and arrangement of the lower furnace vary between mass- and RDF-fired units and are chiefly dictated by stoker dimensions. The lower furnace includes the overfire air system; thus the design and arrangement of both must complement each other to promote adequate mixing for rapid burnout of combustion products. This step ensures that the remaining furnace surface will be effective in reducing the gas temperature to design levels and will also provide an oxidizing environment to reduce the potential for slagging and corrosion in the upper furnace.

Combustion Air Requirements. Mass-fired units are typically designed to operate between 80 and 100 percent excess air. Although earlier RDF designs were based on 35 percent excess air to the stoker, experience indicates that higher levels may be required to control furnace slagging. To provide this flexibility, RDF units should be designed to permit operation up to at least 50 percent excess-air levels at the stoker. Air leakage and cooling air are additive to the stoker air requirements and will result in higher values at the unit outlet. Overfire air requirements vary depending on the particular stoker supplier. Experience shows that overfire air systems have operated at air quantities in the range of 40 percent of total stoker air requirements. Based on this, the overfire air system should not be sized for less than 40 percent excess air for good combustion burnout and slagging control.

Flow modeling of overfire air systems is often desirable to guide design. To boost effectiveness, arches may be added to reduce the overall depth of penetration required and to provide improved fuel/air mixing for complete burnout. Figure 6.9 shows an arrangement of an RDF-fired unit, with arches designed and arranged to maximize the effectiveness of the overfire air system.

Furnace Corrosion Protection. Refuse fuels contain significant quantities of chlorine, which can lead to corrosion of pressure parts. Chlorine contents of up to 1.8 percent and more (by weight) on a dry basis have been observed in refuse analyses. In the combustion process, the chlorine may exist as gaseous hydrochloric acid or low-melting-point chlorides, such as zinc chloride, which can cause significant corrosion when they are in contact with carbon-steel tube materials.

Operating experience on mass-fired waterwall units has indicated that the lower-furnace tube metal should be protected from the corrosive flue gas. Designs using pin studs and coatings of silicon carbide refractory have effectively eliminated corrosion of the lower carbon-steel walls over the range of today's operating units. The design and arrangement of the lower-furnace combustion system are important when the extent of refractory coverage is defined. Typically, the area of coverage encompasses all four walls up to about 30 ft (9 m) above the grate, at which elevation there is reasonable assurance that an oxidizing environment is predominant. The need for and the type and location of corrosion protection will vary with the technology and the supplier.

Silicon Carbide. When silicon carbide is selected, it is vital that the quality and physical characteristics of the silicon carbide refractory be maintained through proper application and curing. Lack of control in these processes will result in spalling, deterioration, and increased maintenance. The refractory material should have high thermal conductivity rates, to minimize the reduction in effectiveness of the water-cooled surface it is protecting. However, such characteristics may reduce its resistance to erosion, as experienced along the grate line, because of the scrubbing action of the refuse fuel and ash as it moves along the grate to the ash discharge. Silicon carbide materials with increased erosion resistance are suitable for these zones. Figure 6.10 shows a typical arrangement of stud pattern and size that has been used successfully to retain the silicon carbide refractory.

Maintenance and refractory life are affected by the refractory surface temperature. A higher temperature will increase the potential for slagging and may re-

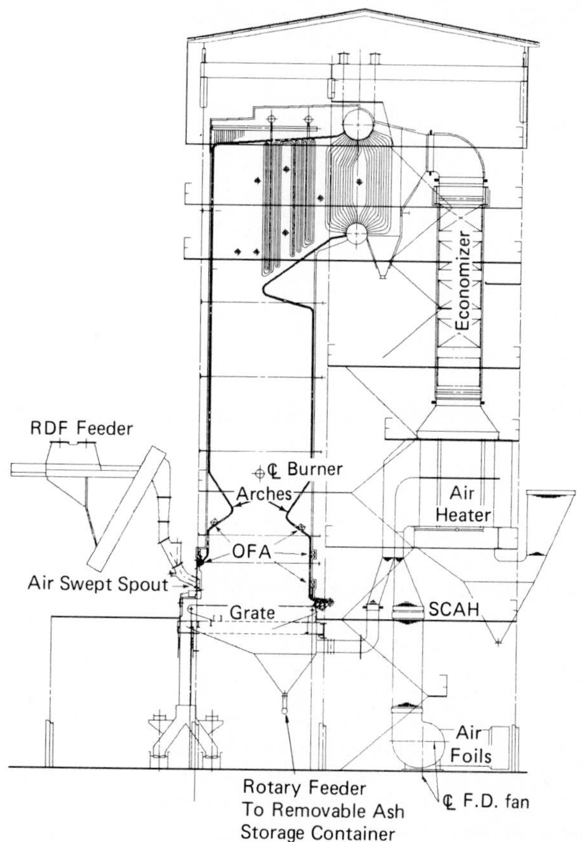

FIG. 6.9 RDF boiler with arches to boost effectiveness of overfire air system.

duce the refractory life because of spalling; higher temperatures may also result if the refractory chemically reacts with flue-gas constituents to reduce the material integrity. Stud design, density, and application can help by yielding lower surface temperatures.

RDF units now in service are operating at pressures that range from 250 to 1250 psig (1720 to 8610 kPa). With one exception, these units were supplied without lower-furnace wall protection, although one did have curtain air capability. The lower-pressure unit [250 psig (1720 kPa)] has operated for many years without significant furnace wall corrosion. Corrosion has been reported in various degrees on some of the units operating at 700 to 750 and 1200 psig (4820 to 5170 and 8270 kPa). A unit was retrofitted with studs and a silicon carbide refractory on the lower-furnace wall panels. It has been reported that the refractory has stopped the corrosion but the furnace is still subject to slagging.

Lower-Furnace Protection. Experience shows that, contrary to earlier beliefs, corrosion may occur on the unprotected lower wall panels of RDF-fired units if the lower furnace is not protected. Experience also indicates that the traditional

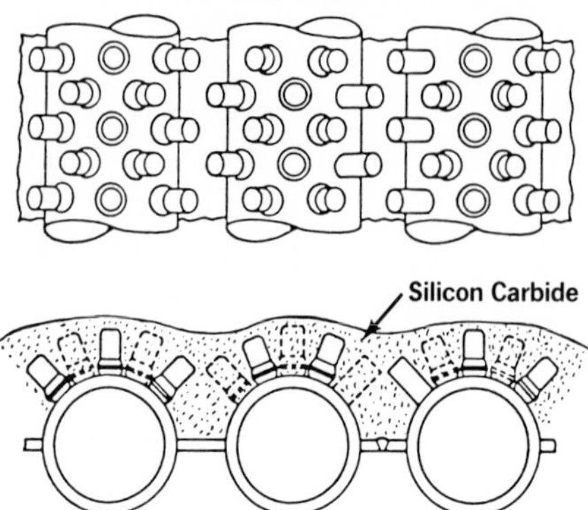

FIG. 6.10 Typical stud arrangement anchoring silicon carbide.

mass-fired studs and refractory protection may not be a practical solution to the corrosion potential for RDF units because of the potential for slagging. Field work has identified the lower-furnace corrosion mechanism to be caused by chlorine. Laboratory analysis indicated that this corrosion is partially due to HCl acid, with some evidence of a molten salt ($ZnCl_2$) attack. Corrosion rates may vary and be inconsistent around the periphery of the furnace. Thicker-wall carbon-steel tubes with adequate corrosion allowance will undoubtedly suffice to provide reasonable life in a major portion of the lower walls in medium- to low-pressure units. It may be desirable, however, to consider corrosion protection for these units in localized areas that have typically shown higher corrosion rates, and in higher-pressure units with their correspondingly higher tube-metal temperatures. The physical impact of raw fuel feed on removing the products of corrosion undoubtedly enhances the corrosion mechanism.

Developmental Work. Developmental work is underway to identify alloys that resist corrosion in the lower-furnace wall area. One unit has been overlaid with a chrome-nickel alloy. This selection was based on laboratory corrosion studies and the results of field test panel work, which showed the alloy to be effective in resisting chloride corrosion.

Field experience further shows the importance of water quality as it relates to tube corrosion. Poor water quality can result in internal deposition on tube surfaces, which insulates the internal surface of the tubes, resulting in higher-than-normal outside tube-metal temperatures. Recently, in an operating refuse unit, internal deposition ranging up to 20 gr/ft^2 (14 g/m^2) resulted in a carbon-steel tube corroding to failure in the equivalent of about 6-wk operation on refuse. This rapid corrosion highlights the fact that the corrosion rate is metal-temperature-dependent, as evidenced by laboratory studies of HCl corrosion vs. temperature (Fig. 6.11).

RDF TYPES AND PROCESSING EQUIPMENT

Refuse-derived fuel is the combustible or organic fraction of municipal solid waste that has been prepared for use as a fuel in boiler (or other energy recovery

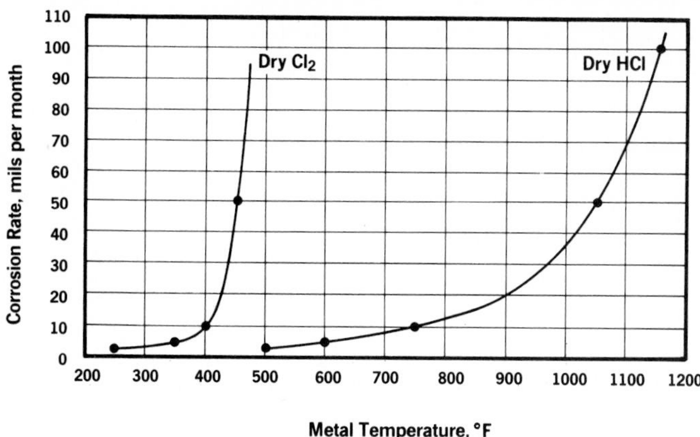

FIG. 6.11 Corrosion of carbon steel in chlorine and hydrogen chloride. Note: °C = (°F − 32)/1.8.

systems) by mechanical processing methods. Included with RDF is a small inorganic fraction, or ash, which is a variable percentage of that in the incoming MSW, depending on the design of the processing system. This section discusses the types of RDF and process equipment used in its preparation.

Types of RDF

The different types of RDF represent the fuel fraction that is derived from MSW to suit the needs of the combustion system, the material recovery system, or both. The forms of RDF that are usable as boiler fuels are described in Table 6.7, which may not be all-inclusive, because some plants use processing system designs that fall in between or beyond the classes described.

Table 6.8 compares the characteristics of raw MSW with three grades of RDF most commonly selected as stoker-fired industrial boiler fuel. The comparison ignores the reduction in moisture content caused during processing of MSW to RDF and the increase in −1-in (−25-mm) fines content caused by shredding, but does provide a theoretical comparison of the changes to the individual constituents, in percentage by weight. Important conclusions can be drawn about the three RDF grades.

Coarse RDF. Production of coarse RDF recognizes the advantages to be gained in combustion of a more consistent, smaller-size fuel over that of raw MSW. Advantages of lower excess air, smaller flue-gas cleanup equipment, the potential for lower unburned carbon loss, possible reduced emissions, and better steam flow control often cement the decision to burn RDF over raw MSW.

With coarse RDF, all the heating value of the incoming MSW is sent to the boiler-combustor. Since all the ash that is bound in the organic fraction, as well as that in the inorganic fraction, is sent to the boiler-combustor, a lower Btu/lb (kJ/kg) value in the fuel results as compared with prepared RDF. For the same energy production, coarse RDF leads to higher throughputs and more ash removal from the combustion system than for prepared RDF.

TABLE 6.7 Classification and Application of RDF*

Class	Form	Typical use	Description
1	Raw MSW	Mass-burning incinerator	Municipal solid waste as a fuel in an as-discarded form without oversized bulky waste.
2	Coarse RDF	100% RDF stoker-fired industrial boiler	MSW processed to coarse particle size with or without ferrous-metal boiler separation, such that 95% by weight passes through a 6-in (150-mm) square mesh screen. Also called *crunch and burn.*
3	Prepared RDF	100% RDF stoker-fired industrial boiler	MSW processed to produce a particle size such that 99% by weight passes through a 6-in (150-mm) square mesh screen. Stringers [material greater than 9 in (225 mm) in one dimension] are minimized. The normal particle size falls within the 2 in × 2 in to 4 in × 4 in (50 mm × 50 mm to 100 mm × 100 mm) range. Ferrous recovery of at least 90% of the incoming MSW is specified, as is removal of the glass, grit, sand, and dirt fractions.
4	Recovery prepared RDF	100% RDF stoker-fired industrial boiler	The same as class 3 with the following additions: • Processing to remove aluminum • Processing to remove other nonferrous metals • Processing to prepare recovered ferrous, nonferrous, and glass fractions for the resale market • Processing to return the fine-fraction combustibles to the RDF fraction
5	Fluff RDF	Up to 20% RDF, 80% coal, suspension-fired utility boiler; dedicated or cofired fluid-bed boiler	Shredded fuel derived from MSW processed for the removal of metal, glass, and other entrained inorganics; particle size of this material is such that 95% by weight passes through a 2-in (50-mm) square mesh screen.
6	Densified RDF	Stoker-fired industrial boilers designed for coal	Combustible waste fraction densified (compressed) into pellets, slugs, cubettes, briquettes, or similar forms. Fluff RDF free of glass and grit is used as feed to densifying equipment.

*Also see ASTM E 38.01, E 776 to E 791, which also includes such RDF forms as powders, liquids, and gases.

TABLE 6.8 Waste Fuels for Dedicated MSW and RDF Applications

Composition, % (by weight)[a]	Raw MSW	Coarse RDF	Prepared RDF	Recovery prepared RDF
Corrugated boxboard	5	5	6.4	6.2
Newspaper	12	12	15.4	14.9
Magazines, books	3	3	3.9	3.7
All other paper	23	23	29.5	28.7
Plastics	3	3	3.9	3.7
Rubber, leather	2	2	2.6	2.5
Wood	3	3	3.9	3.7
Textiles	3	3	3.9	3.7
Yard trimmings	10	10	12.8	12.5
Food waste	10	10	12.8	12.5
Fines, −1 in (−25 mm)	10	10	1.3[b]	6.0[c]
Metallic[d]	7	7[e]	2.6[f]	0.8
Glass, ceramics, etc.	9	9	1.2[b]	1.1
Composite	100	100	100	100
Composite Btu/lb (kJ/kg)	4451 (10 370)	4451 (10 370)	5360 (12 490)	5447 (12 690)
Composite ash/lb (kg), %	24.4	24.4	10.5	9.1
% of Btu (kJ) in incoming MSW to RDF	100	100	94	98
% Yield of RDF from MSW	—	100	78	80

[a] Artificially assumed that shredded RDF will not increase the amount of fines −1 in (−25 mm). In reality, after shredding 25% by weight may be in fines, −1 in (−25 mm) (including glass, ceramics, etc.).
[b] 90% removal.
[c] Of the 90% fraction removed, 90% Btu (kJ) content is recovered as paper.
[d] 80% metallics are assumed to be ferrous.
[e] No ferrous removal.
[f] 90% removal of ferrous.

Note: Neither MSW nor RDF is a homogeneous fuel. For design purposes, actual analysis of representative samples of the MSW in question should be performed. Removal and recovery efficiencies are based on the equipment used and its location within the processing system. Information presented above is based on MSW of a single composition to which various assumptions were applied to demonstrate points. Values shown should not be considered absolute, since they will vary with the specific processing equipment used as well as the actual MSW incoming composition.

Fluff RDF and densified RDF were excluded because of their very specialized use.

Prepared RDF. By producing prepared RDF, recognition is made of the deleterious impact of some of the fuel constituents on components of the entire plant system. By removing portions of the fines, glass, ceramics, sand, and grit, the abrasion of moving parts of conveyors, shredders, stokers, and ash removal equipment is decreased. Similarly, abrasive-erosive wear of stationary components at transfer points, chutes, liners, and housing is reduced. It is thought by some that furnace-boiler slagging is caused by melting glass in 100 percent RDF-fired boilers, although conclusive evidence is not available at this time. Removal of the ferrous fraction provides a reduction in the impact loading on various components in the system. The removal of the ferrous fraction decreases both equipment wear in the system and chances of pluggage caused by long, stringy wire items, which often pass through shredding equipment intact.

By removing the inorganic fractions, as indicated in Table 6.8, the percentage by weight of the organic constituents increases. These derivative changes also occur, as compared with raw MSW or coarse RDF:

- More Btu per pound (kilojoules per kilogram) of RDF.
- Less ash per pound (kilogram) of RDF.
- Fewer Btus (kilojoules) are produced from the incoming MSW.
- The yield of RDF produced from the incoming MSW is decreased. Note that most of this reduction comes through the removal of the inorganic fraction, which makes little contribution to the heating value.

Recovery Prepared RDF. By producing recovery prepared RDF, recognition is given to two important considerations:

- Materials recovered from MSW may be recyclable; thus an income stream may be realized and landfill expense decreased.
- Combustible materials may be recovered from the fines fraction removed with prepared RDF and returned to the fuel fraction.

Preparation of recovery prepared RDF provides more Btus per pound (kilojoules per kilogram), less ash per pound (kilogram), more Btus (kilojoules) from the incoming MSW, and increased yield, as compared with prepared RDF. Here the majority of the metallic fraction will have been removed, including aluminum, zinc, copper, and brass. Besides the recyclable advantage, removal of these low-melting-point metals has the advantage of reducing maintenance on stoker grate bars and stoker seals. In particular, aluminum has a tendency to plug grate-bar air holes by melting and resolidifying on the back side in a cooler temperature zone.

RDF Processing Equipment

Early experiences at RDF installations has shown MSW to be more difficult to process than plant designers originally thought. Much of the equipment selected was developed for other industries, in particular the mining and minerals processing industries. The characteristics of MSW are substantially different from minerals that are solid and essentially rectangular, cubic, or spherical. MSW has flat, long and stringy, irregular, and interlocking shapes, as well as other characteristics that make it difficult to handle, such as nonhomogeneity, abrasiveness, varying moisture content, a tendency to compact during storage, and putrescibility. Selection of RDF processing equipment has been on a trial-and-error basis, and many modifications have been made to accommodate the extremes of MSW characteristics.

Typical System Arrangement. MSW is deposited by refuse collection trucks or transfer trailers on a tipping floor in a storage building. From storage, material is moved to an apron conveyor, which transports the MSW to the input of the RDF processing equipment. Unacceptable wastes on the tipping floor are removed by grapple to a disposal point. MSW moves via an apron conveyor to the first processing step, which is to break open the bags to expose the contents. While the system in Fig. 6.12 shows a trommel to provide this function, other systems may rely on flail mills, knives, rotating discs, or hammer mills as the first processing equipment.

By placing a trommel or disc screen before the shredder, less glass and grit are embedded in the combustible fraction, abrasion on the shredders and down-

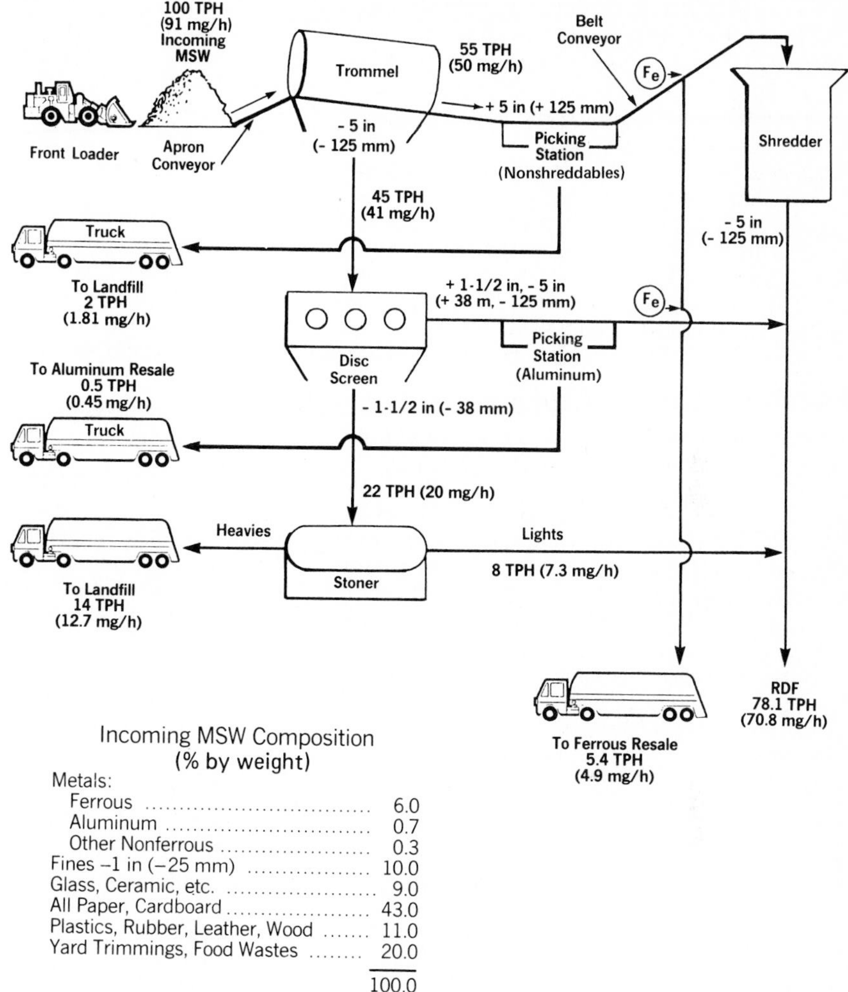

Incoming MSW Composition
(% by weight)

Metals:	
Ferrous	6.0
Aluminum	0.7
Other Nonferrous	0.3
Fines −1 in (−25 mm)	10.0
Glass, Ceramic, etc.	9.0
All Paper, Cardboard	43.0
Plastics, Rubber, Leather, Wood	11.0
Yard Trimmings, Food Wastes	20.0
	100.0

FIG. 6.12 RDF system flow and mass balance diagram.

stream equipment is reduced, and refuse that meets fuel size limits is not shredded, resulting in more uniform final RDF size, less fines, and lower shredder power.

A picking station is shown downstream of the trommel in Fig. 6.12. Its purpose is to provide for manual removal of undesirable items, such as explosives, propane and gasoline containers, and other deleterious items that may have been missed on the tipping floor. Magnetic separation on both the trommel overstreams and understreams removes the ferrous materials, which are of no value for fuel but may have a resale value. The oversized ferrous-free material is conveyed to a shredder to provide the size reduction needed to meet the RDF size requirements.

The undersized material that passes through the trommel openings is further classified by size with a disc screen to remove the fines fraction. The oversized material passes an aluminum hand-picking station under a magnetic separator and is combined with the shredder discharge to be conveyed to RDF storage. The undersized material is conveyed to a residue load-out point.

Equipment Descriptions. Key processing equipment consists of conveyors, screens, shredders, air classifiers, and magnetic separators.

Kinds of Conveyors. A pan or apron conveyor has a series of interlocking metal pans mounted on chains that are driven by terminal sprockets. It is designed for impact loading of raw refuse, which is pushed from the tipping floor onto the recessed pan conveyor; from there, the conveyor transports raw MSW to the first processing equipment. A belt conveyor consists of an endless rubber or treated-fabric belt, terminal pulley, and idlers or rollers, and it may be provided with the means to set the belt tension and to change direction. Belt conveyors transport processed RDF and ash.

A vibrating conveyor moves material in a horizontal direction by repeatedly giving it an impulse upward and in the direction of motion desired (Fig. 6.13). The conveyor can withstand the impact loading below shredders and tends to even out the material flow rate to downstream equipment.

Drag conveyors consist of a single chain or two parallel chains with flights, or plates, connected at intervals. As the chains move, the flights drag material over a stationary surface. Although they have been used as in-feed and distributing conveyors, their best application appears to be for ash conveying (see Section 4, Part 1, Chapter 6).

Screw conveyors contain a steel helix mounted on a shaft suspended in bearings, usually in a U-trough. As the shaft rotates, the material is moved by thrust at the lower part of the helix and is discharged through openings in the trough bottom or at the end. They have typically seen service in RDF plants to move the fuel out of storage bins and hoppers, to fluff material in hoppers, to feed combustors, to remove flyash, or to supply multiple feeders to a boiler.

Pneumatic conveyors for RDF applications consist of surge bins, rotary feed locks, blowers, and cyclone separators to convey small-particle-size material in suspension. Their key application has been to feed −1½-in (−38-mm) size RDF to utility-type boilers for burning in suspension with coal. The majority of the glass and grit fraction of the RDF should be removed to control abrasion of the piping and elbows housing the pneumatically conveyed material. Pneumatic conveyors have not been successful when they were applied to coarse RDF fuel.

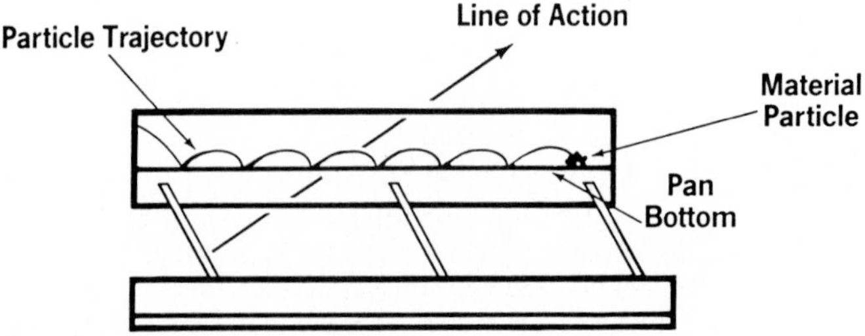

FIG. 6.13 Typical material flow on vibrating conveyor.

Size Separation with Screens. A trommel consists of a rotating cylindrical screen fabricated from ⅝- to ¾-in (16- to 19-mm) thick carbon-steel plate with holes of appropriate diameters to size-separate the material inserted (Fig. 6.14). When it is located ahead of a shredder, the holes are generally of a single size. Trommel screens are oriented with about a 5° downward slope and rotate at 8 to 12 r/min to lift the MSW and aid the forward travel. Spikes and lifters may be installed inside the trommel to break bags and to increase the tumbling action.

Disc screens consist of a series of rotating, parallel, horizontal shafts to which are attached interspaced, staggered disks (Fig. 6.15). Undersized material falls through the gap between discs, and oversized material is agitated and moved along across the top of the discs, which all rotate in the same direction.

A vibrating screen has a perforated flat plate, which is subjected to an amplitude action similar to a vibrating conveyor. Undersized material drops through the holes, and oversized material is conveyed off the end. The use of vibrating screens in RDF processing systems is limited to that fraction which is dry and has been shredded, sized, air-classified, and magnetically separated. When they are used elsewhere in the system, vibrating screens tend to blind and/or stratify the material.

Shredder Types. Shredders selected for refuse processing include flail mills, hammer mills, and shear types. Shredders reduce the RDF material to a size compatible with the other plant systems, such as distributor, fuel feed, grate, ash handling, etc.

Vertical hammer mills have the rotor shaft mounted vertically, and the refuse flows parallel to the rotor axis (Fig. 6.16). Hammers are affixed to the rotor, and

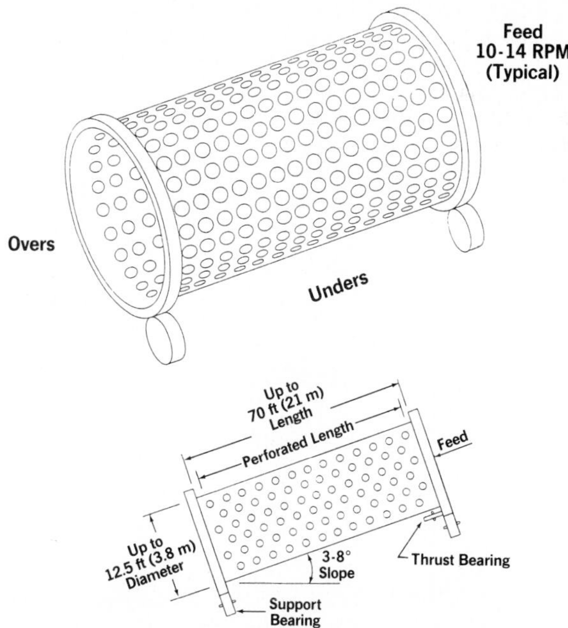

FIG. 6.14 Trommel is rotary screen for size separation.

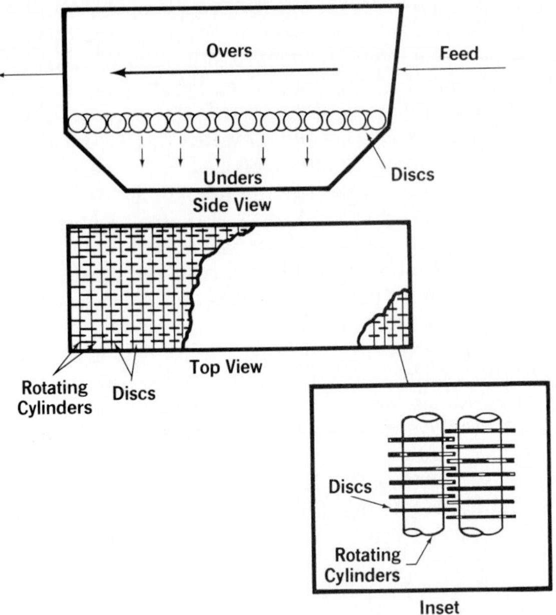

FIG. 6.15 Disc screens for size separation.

both rotor and hammers are enclosed in a heavy-duty housing, which may be lined with hard metal members. Refuse exits from the bottom between the lowest row of hammers and breaker bars. Vertical hammer mills are excellent for volume reduction prior to landfill. They have also been used for RDF production in many installations, but they have a tendency to discharge rags and similar materials with little size reduction, particularly when the hammers are worn.

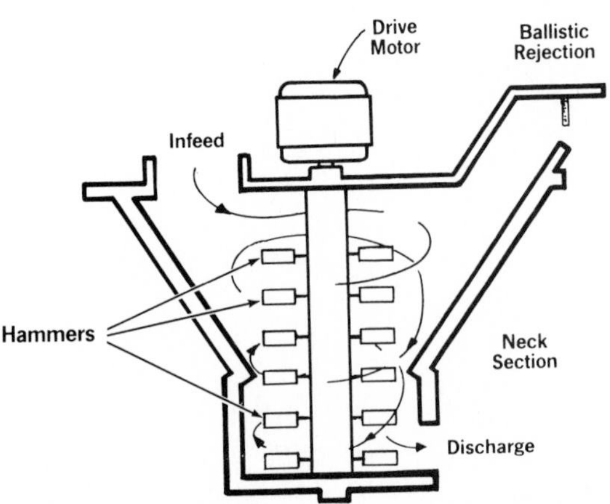

FIG. 6.16 Vertical hammer mill for shredding operations.

Horizontal hammer mills have the rotor shaft mounted horizontally, and the refuse flows perpendicular to the rotor axis. As with the vertical mills, the hammers are affixed to the rotor, and both rotor and hammers are enclosed in a heavy-duty housing that may be lined. Refuse exit is from the bottom through a set of cast-iron grate bars, which provide size classification. The mills may have high dust blowback at the inlet caused by rotor windage; baffling at the shredder inlet can correct this problem.

Shear shredders are low-speed, high-torque machines that impart a cutting action to the refuse. Although they operate with lower power requirements than hammer mills and have less explosion potential, their use for RDF production has been limited. Shear shredders have been criticized for producing large quantities (up to 15 percent of total RDF) of long, spearlike objects. Also, when difficult-to-shred material is encountered, throughput may suffer. Flail mills see service in some systems to break open bags and to provide primary size reduction, but not final size reduction in RDF processing systems.

Air Classifiers. Air classification is a technique that separates the light, mainly combustible fraction from the heavy, mainly noncombustible fraction of RDF. In air classifiers, there is a trade-off between fuel quantity and quality as set by the air velocity (Fig. 6.17). Other characteristics include the need for uniform feed and air cleanup equipment, increased horsepower, and high abrasion-induced maintenance. Air classifiers of this type are a part of most RDF facilities, producing fluff fuel for suspension burning in utility boilers.

An air knife is often used to clean up a stream of recovered ferrous material. Lighter tramp material is diverted outward from the free-falling material stream, whereas the heavier metallic fraction is less affected.

A "stoner" classifier has been developed to separate stones from grain. It re-

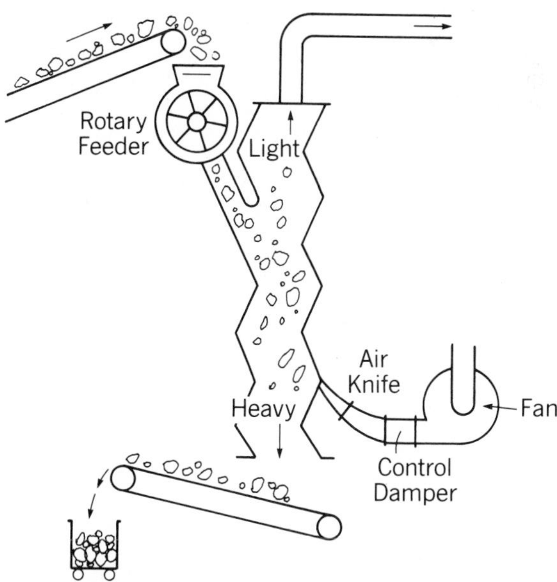

FIG. 6.17 Air classifier has zigzag configuration to help separate RDF fractions.

covers the combustible light fraction from the heavy fraction, after secondary size separation provided by a disc screen or trommel. This device imparts a fluidizing action on the light fraction and a vibrating conveyor-type action on the heavy fraction.

Magnetic Separators. To remove iron and steel from refuse, magnetic separators of the belt (Fig. 6.18), drum (Fig. 6.19), and conveyor pulley types are available, with either permanent magnets or electromagnets. In addition to providing a resalable product, ferrous material removal reduces wear on processing equipment and on boilers and grates.

Single-drum and single-belt (Fig. 6.18) magnets have experienced operational problems. The ferrous materials buried under a thick layer of refuse either are not picked up or, if the magnet is powerful enough, are picked up with an overburden of combustible, tramp materials. To correct this, both drum and belt separators are available in two stages (Fig. 6.19). The first stage is powerful enough to pick up the ferrous and tramp materials, which are conveyed and then dropped in an area where no magnetic attraction exists, away from the main refuse stream. The second, less powerful magnet then separates the ferrous material out of a densified ferrous stream, without tramp material pickup.

FINAL CONSIDERATIONS

Besides the front-end technologies of fuel preparation and handling, processing equipment, and combustion design, ash removal, storage of fuels, recovery markets, and health and safety should also be considered.

Ash Removal Systems

When refuse is burned in a waste-to-energy plant, the resultant ash takes two forms—light ash (called *flyash*) and coarse ash, which comes off the stoker. The flyash portion is entrained in the gas stream until it is removed by a particulate collection device or falls out in the hoppers below the boiler, economizer, and air heater. Stoker ash consists of ash from the fuel and slag deposits on the grate, from the furnace walls, and from superheater surfaces in some designs. Boiler ash deposits fall by gravity into hoppers, either naturally or after removal by soot

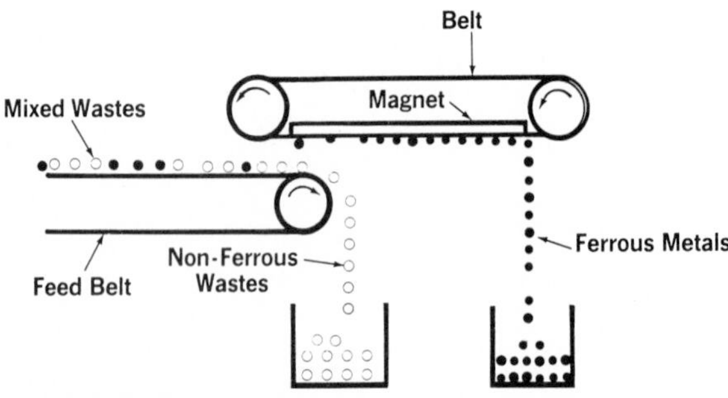

FIG. 6.18 Belt magnetic separator.

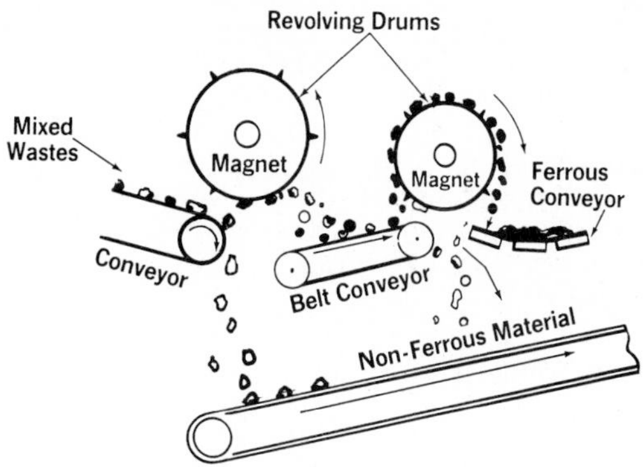

FIG. 6.19 Double-drum magnetic separator.

blowing or rapping. Stoker ash is removed from the stoker discharge chute and from the sifting hoppers.

The amount of ash produced in a boiler depends on the amount of ash in the raw fuel and the amount of front-end processing. For a mass-fired unit, which has little or no fuel preparation, the total ash in the as-burned fuel is near to that in the as-received fuel. In the RDF unit, the total ash in the as-burned fuel is significantly less. The split between flyash and stoker ash is 20 percent/80 percent for the MSW unit and 50 percent/50 percent for the RDF unit. This difference in the splits is due to the degree of preparation of the fuel before it is burned and to the method of feeding.

Because of the degree of preparation the as-received MSW undergoes, the consistency of the ash varies from boiler to boiler. Considering this, it is necessary to choose the proper ash conveying system for the specific conditions in each waste-to-energy plant. This is particularly true for the stoker discharge, which can include oversized bulky waste (OBW) that is not removed prior to firing in a mass-fired unit. The OBW may be shear-shredded in some mass-fired units to reduce its size before it is sent to the furnace. The designer of the ash conveying system also needs to consider whether the stoker will be kept dry, so that it can be processed for material reclamation, or whether it will be combined with flyash for final disposal.

Because of these and other considerations, ash removal systems must be well understood. See Section 4, Part 1, Chapter 6, "Ash Handling Systems," for the components and systems used to handle ash.

MSW and RDF Storage

Storage of both MSW and RDF is required—for MSW, because trailers bringing waste are not evenly spaced throughout the day; for RDF, because power production and RDF processing facilities have different schedules. Similarly, the amount of storage should reflect RDF processing-system redundancy to accommodate throughput variations.

Storage times for MSW are minimized, if possible, in the capacity range equivalent to 1 to 3 days. Storage time for RDF is usually acceptable for a longer period, typically up to 5 days, because RDF characteristically does not attract insects and rodents as readily as MSW. Types of storage and retrieval equipment found in operating plants include the tipping floor and front-end loader, live-bottom floor with conveyors, pit and overhead crane, pit and hydraulic rams, live-bottom bin and conveyors, conical bin with drag buckets, live-center bin, and bin with screw unloaders.

Selection of a storage and retrieval method should be made carefully because of the tendency of refuse to compact and agglomerate, to decompose, to be abrasive, and to contain long, stringy items.

Resource Recovery Markets

Both domestic and export markets are available for materials recovered from MSW. The more common resale items are ferrous metals, aluminum, glass, and ash. Other materials that have been recycled include nonferrous metals, paper, cardboard, and textiles. Although the volatility of scrap markets make analysis difficult, the potential income from them is worthy of consideration.

Health and Safety

Explosions can occur in and around shredders when flammable or explosive materials present in the as-received waste or generated during processing are ignited. Hammer mills in particular are susceptible to explosions, but all MSW shredders should be studied with respect to explosion potential. System design should (1) isolate shredders from personnel by explosion-resistant blockhouse protection, (2) provide explosion venting if the shredder is not located in an isolated outside area, and (3) furnish other explosion protection consisting of water sprays and/or inert-gas or steam injection.

The building walls and roof near a shredder should be designed to take explosion relief into account. Ignition sources must be prohibited in the vicinity of an operating shredder, and combustible-gas detectors are recommended. Fire-extinguishing equipment should be provided in both MSW and RDF storage areas. A key factor in explosion and fire prevention, however, is careful inspection of incoming waste materials.

Handling and processing refuse are extremely dusty, so the plant design should include dust collection systems for continuous removal at transfer points and at appropriate equipment locations.

For odor control, provide adequate ventilation in building designs. When possible, give consideration to taking the boiler's forced-draft-fan suction from the RDF floor. However, a filtering system may be necessary to keep dust from plugging the steam-coil air heater and similar surfaces.

ACKNOWLEDGMENT

Illustrations in this chapter are courtesy of Babcock & Wilcox Co. unless noted otherwise.

SUGGESTIONS FOR FURTHER READING

1. Alter, H., *Materials Recovery from Municipal Waste, Unit Operations in Practice,* Dekker, New York, 1983.
2. Blue, J. D., and J. R. Strempek, "Considerations for the Design of Refuse-Fired Waterwall Incinerators," *Energy from Municipal Waste,* Power Conference, Washington, D.C., October 1985. Proceedings published by *Power* magazine.
3. Churney, K. L., A. E. Ledford, S. S. Burce, and E. S. Domalski, *The Chlorine Content of Municipal Solid Waste from Baltimore County, MD, and Brooklyn, NY,* Dept. of Commerce, National Bureau of Standards, NBS IR 85-3213, October 1985.
4. Clunie, J. F., A. Leidner, and J. T. Hestle, "The Importance of Proper Loading on Refuse-Fired Boilers," *ASME National Waste Processing Conference,* Orlando, Fla., June 1984.
5. Daniel, P. L., J. D. Blue, and J. L. Barna, "Furnace-Wall Corrosion in Refuse-Fired Boilers," *ASME National Waste Processing Conference*, Denver, Colo., June 1986.
6. Gibbs, D. R., and L. A. Kreidler, "From Hamilton to Palm Beach—The Evolution of Dedicated RDF Plants," *Conference on Processed Fuels and Material Recovery from Municipal Solid Waste,* Gershman, Breckner & Bratton, Inc., December 1986.
7. Hainsworth, E., J. L. Mayberry, and R. P. Piscitella, *Energy from Municipal Solid Waste: Mechanical Equipment and Systems Status Report,* EG&G Idaho, Inc., for Dept. of Energy, DOE Contract DE-AC07-76 ID01570, March 1983.
8. Hasselriis, F., *Thermal Systems for Conversion of Municipal Solid Waste,* vol. 4; *Burning Refuse-Derived Fuels in Boilers—A Technology Status Report,* Argonne National Laboratory for Dept. of Energy, ANL/CNSV-TM-120, March 1983.
9. Hepp, M. P., R. M. Nethercutt, and J. F. Wood, "Considerations in the Design of High Temperature and Pressure Refuse Fired Power Boilers," *ASME National Solid Waste Processing Conference,* Denver, Colo., June 1986.
10. *Resource Recovery State-of-the-Art: A Data Pool for Local Decision-Makers,* Institute for Local Self-Reliance, Washington, D.C., June 1986.
11. Turner, W. D., *Thermal Systems for Conversion of Municipal Solid Waste,* vol. 2: *Mass Burning of Solid Waste in Large Scale Combustors,* Argonne National Laboratory for Dept. of Energy, ANL/CNSV-TM-120, December 1982.

SECTION 4

POLLUTION CONTROL

PART 1

AIR POLLUTION CONTROL

CHAPTER 4.1
CLEAN AIR ACT LEGISLATION AND REGULATED POLLUTANTS

Theodore C. Koss and Robert J. Jonardi
NUS Corporation
Gaithersburg, MD

ABBREVIATIONS

BACT Best available control technology
CO Carbon monoxide
CO_2 Carbon dioxide
EDF Environmental Defense Fund
EPA U.S. Environmental Protection Agency
GEP Good engineering practice
HC Hydrocarbons
LAER Lowest achievable emission rate
NA Nonattainment area

NAAQS	National ambient air quality standards
NESHAP	National emission standards for hazardous air pollutants
NO	Nitric oxide
NO_2	Nitrogen dioxide
NO_x	Nitrogen oxides
NRDC	National Resources Defense Council, Inc.
NSPS	New-source performance standards
O_2	Oxygen
O_3	Ozone
PAN	Peroxyacetal nitrate
Pb	Lead
PSD	Prevention of significant deterioration
SIP	State implementation plan
SO_2	Sulfur dioxide
TSP	Total suspended particulates
VOC	Volatile organic compound

INTRODUCTION

From 1970 to the present, Congress has considered and passed a number of laws governing air quality that have had significant impact on the electric utility industry. This legislation and the EPA regulations that implement the legislation have had pronounced effects on plant design, operation, and costs, and will continue to have a substantial effect in the future. Indeed, many observers feel that proposed acid rain legislation, primarily targeted toward coal-fired powerplants, will be the next offspring of the Clean Air Act legislation of the 1970s.

The first portion of this chapter reviews sections of the Clean Air Act and the ensuing EPA regulations that affect fossil-fueled powerplants. The major topics covered are (also see Abbreviations): NAAQS, NSPS for stationary sources, stack height regulations, interstate and international pollution abatement, the PSD of air quality, and requirements for NA. Several other sections of the act are treated in a more cursory manner. The second portion of this chapter deals with the kinds of air pollutants released from powerplants.

CLEAN AIR ACT

As the 1960s drew to a close, it became increasingly clear to Congress that no effective mechanism existed for controlling air pollution (except, perhaps, for smoke control in those localities that had undertaken the effort) and that a completely new attack on the problem was needed. The resulting 1970 Clean Air Act (passed as amendments to the 1963 Clean Air Act) was so far-reaching that it effectively commenced the modern era of air pollution control.[1]

The Clean Air Act amendments of 1970 introduced the philosophical premise that the most efficient means of air pollution control would result from a mixture of two ideas. NAAQS, designed to protect public health and welfare, was to be established at threshold levels below which no adverse effects would occur. Emissions standards, based on control technology, would be imposed to bring pollution concentrations below ambient standards and to keep them there.

The 1970 amendments had mandated attainment of NAAQS by 1975. While some modest progress had been made, these standards were not met in many parts of the country, including almost all urban areas. The major purposes of the 1977 Clean Air Act amendments were, therefore, to provide additional and stronger influence in relation to PSD for areas that had attained the standards and to improve air quality in areas where pollutant concentrations had not attained the standards.[1]

Air Quality Control Regions

Section 107 of the Clean Air Act requires the EPA to publish a list of all geographic areas not in compliance with NAAQS as well as those attaining NAAQS. Areas not in compliance with NAAQS are termed *nonattainment areas*. Areas meeting the NAAQS are referred to as *attainment areas*, and these are subject to PSD regulations. Areas which have insufficient data to make a determination are unclassified, but these are treated as being attainment areas until proved otherwise. The designation of an area is made on a pollutant-specific basis. The geographic regions established for designating the air quality status with respect to compliance with NAAQS are known as *air quality control regions*.

National Ambient Air Quality Standards

Section 109 of the Clean Air Act of 1970 mandates that the EPA establish ambient standards for certain criteria pollutants based on the latest scientific information regarding all identifiable effects a pollutant may have on the public health or welfare. The achievement of these standards represents the primary goal of the 1970 Clean Air Act. Subsequently, EPA promulgated regulations which set NAAQS for SO_2, TSP, NO_2, CO, HC, photochemical oxidants, and Pb. Two classes of ambient air quality standards were established: (1) primary standards that define levels of air quality that EPA has judged as necessary to protect public health and (2) secondary standards that define levels of air quality for protecting the public welfare (such as vegetation and aesthetic values). The Clean Air Act amendments of 1977 established timetables for periodically reviewing existing NAAQS and adopting new standards. Consequently, NAAQS for photochemical oxidants were reviewed and, in 1979, were restated as ozone (O_3),[2] while NAAQS for HC were revoked in 1983.[3] A new standard for inhalable particulates (PM_{10}) has recently been proposed but not promulgated.[4] Table 1.1 summarizes NAAQS in effect as of this writing. States were required by the 1970 Clean Air Act and subsequent 1977 amendments to develop plans to attain and maintain NAAQS.

New-Source Performance Standards

The NSPS program was authorized by Congress in the Clean Air Act of 1970 and codified in Section 111 of the act. By emphasizing new, rather than

TABLE 1.1 National Ambient Air Quality Standards*

Pollutant	Averaging time	Primary standard,† $\mu g/m^3$	Secondary standard,† $\mu g/m^3$
Sulfur dioxide (SO_2)	Annual	80	None
	24 h	365	None
	3 h	None	1,300
Particulate matter	Annual‡	75	60
	24 h	260	150
Carbon monoxide (CO)	8 h	10,000	10,000
	1 h	40,000	40,000
Ozone (O_3)	1 h	235§	235§
Nitrogen dioxide (NO_2)	Annual	100	100
Lead (Pb)	Calendar quarter	1.5	1.5

*From 40 CFR 50.4–50.12.

†Not to be exceeded more than once a year (except where noted), other than those based on annual averages or annual geometric mean.

‡Annual geometric mean.

§The ozone standard is attained when the expected number of days per calendar year with maximum hourly average concentrations above 235 $\mu g/m^3$ is equal to or less than 1.

existing, sources of pollution in this control program, Congress believed that NSPS would prevent new air pollution problems in the short term and result in long-term improvements in air quality. The older and dirtier existing plants would supposedly be replaced with new ones. Regulations promulgated to date to implement Section 111 have focused primarily on large, new sources of SO_2, particulate matter, NO_x, and VOCs, or hydrocarbons. NSPS have been viewed as "technology-forcing" provisions that provide a statutory incentive to develop new control technology to meet emission limitations.

In amending the act in 1977, Congress authorized the EPA administrator to propose regulations for any category of stationary sources that "causes, or contributes significantly to, air pollution which may reasonably be anticipated to endanger public health or welfare." Congress also reinforced the importance of Section 111 by requiring that the EPA administrator list every category of major stationary sources that is a significant contributor and had not already been listed. A *major source* was defined as one that has the potential to emit 100 tons (90.7 Mg) annually of any air pollutant, after reductions achieved by pollution control equipment are taken into account.[5]

In establishing NSPS, Congress, in the 1977 amendments, treated utility boilers with a capacity greater than 250×10^6 Btu/h (73.25 MW) differently from other sources in the regulatory scheme, due to the fact that electric generating plants constitute by far the largest source of SO_2 in the country. The act now requires a "percentage reduction in emissions" in addition to the maximum emission limit that still remained in effect. The determination of actual levels to be used was left for EPA to make in its rule-making procedures.

EPA is responsible for establishing, reviewing, and revising NSPS. In response to the 1977 amendments, EPA revised the NSPS for fossil-fuel-fired steam generators to provide for new emission standards for particulate matter, SO_2, and NO_x. Subpart D standards for fossil-fuel-fired steam generators promulgated prior to the 1977 amendments and subpart Da standards promulgated after 1977 are summarized in Table 1.2. As they are applied to SO_2 emissions from coal-burning units, the percentage reduction requirements are 90 percent for medium- and high-sulfur (eastern) coals and 70 percent for low-sulfur (western) coals. The reduction is calculated on the basis of the amount of uncontrolled SO_2 which would be emitted in the absence of any controls. This percentage reduction requirement virtually mandates the use of flue-gas desulfurization on all affected coal-fired boilers regardless of sulfur content.[1]

The NSPS apply not only to newly constructed sources, but also to existing sources that undergo substantial modification. This has been done to make it im-

TABLE 1.2 Summary of New-Source Performance Standards Applicable to Coal-, Oil-, and Gas-Fired Units

Subpart D: Fossil-fuel-fired steam generators for which construction commenced after Aug. 17, 1971

Source*	Pollutant	Allowable emission rate	
		$lb/10^6$ Btu	$kg/10^6$ kJ
Coal-fired boiler	SO_2	1.2	0.516
	NO_x	0.7	0.3011
	Particulate	0.1	0.043
Oil-fired boiler	SO_2	0.8	0.344
	NO_x	0.3	0.1299
	Particulate	0.1	0.043
Gas-fired boiler	NO_x	0.2	0.086
	Particulate	0.1	0.043

Subpart Da: Electric utility steam generation units for which construction commenced after Sept. 18, 1978

Source*	Pollutant	Allowable emission rate†		Reduction in potential emissions,%†
		$lb/10^6$ Btu	$kg/10^6$ kJ	
Coal-fired boiler	SO_2	1.2	0.516	90
		or	or	
		0.6	0.258	70
	NO_x	0.6	0.258	65
	Particulate	0.03	0.013	99
Oil-fired boiler	SO_2	0.8	0.344	90
		or	or	
		0.2	0.886	0
	NO_x	0.3	0.129	30
	Particulate	0.03	0.013	70
Gas-fired boiler	SO_2	0.8	0.344	90
		or	or	
		0.2	0.086	0
	NO_x	0.2	0.086	25
	Particulate	0.03	0.013	—

*Boilers greater than 73 MW, with capacity greater than 250×10^6 Btu/h (73.25 MW). See 40 CFR 60, subparts D and Da, for details on other fuels.

†Both column 3 and column 4 requirements must be met.

Adapted from Reference 5.

possible to evade the application of NSPS by piecewise modification of an existing plant to the point where it is a rebuilt plant.

National Emission Standards for Hazardous Air Pollutants

Section 112 of the 1970 Clean Air Act requires the EPA administrator to publish, and occasionally revise, a list of hazardous air pollutants for which she or he intends to establish emission standards and to establish emission standards for these pollutants. A *hazardous air pollutant* is one "to which no ambient air quality standard is applicable and which...causes, or contributes to, air pollution which may reasonably be anticipated to result in an increase in mortality or an increase in serious irreversible, or incapacitating reversible, illness." The *national emission standards for hazardous air pollutants* are commonly referred to as NESHAP. Hazardous emission standards must provide "an ample margin of safety to protect the public health" and may apply both to new and existing sources.

Although some pollutants have been found to be hazardous at only high levels of exposure with no effects below a certain level, scientists have not been able to identify thresholds for carcinogens below which effects do not occur. For carcinogens, EPA has taken the position that any level of exposure may pose some risk of adverse effects, with the risk increasing as the exposure increases. Hence, risk assessment (that utilizes mathematical dose-response models and population exposure estimates) is a major element in the implementation of section 112.[6]

As EPA promulgates additional NESHAP, and as state programs controlling toxic pollutants become better established, the evaluation of ambient impacts and the control of hazardous air pollutants will be of increasing importance to the electric utility industry.

International and Interstate Air Pollution

Since air pollution can move freely across state or international boundaries, the potential exists for sources in one state or country to affect the air quality in another. Interstate or international pollution exists when sources located near a geographical boundary emit air pollutants that are transported across the boundary. Transboundary pollution also occurs when pollutants are transformed and transported across political borders over longer distances, producing such problems as acid deposition, regional haze, and excessive ozone concentrations.

If EPA determines that emissions from a given state endanger public health or welfare in a foreign country, section 115 of the Clean Air Act as amended in 1977 authorizes the EPA administrator to formally notify the governor of the state that its implementation plan is inadequate and must be revised. This section applies only to those foreign countries that give essentially reciprocal rights to the United States.

In sections 110 and 126 of the Clean Air Act, Congress provided certain federal remedies for specific interstate pollution problems. Specifically, a state may petition the EPA administrator to find that a major pollution source in another state either prevents it from attaining or maintaining NAAQS or interferes with required measures to prevent deterioration of air quality or to protect visibility. The EPA administrator, after a public hearing, must grant or deny the petitions. If he or she finds one or more of these situations to exist, the EPA administrator may prohibit the construction or operation of the offending sources or may impose on them compliance schedules designed to reduce emissions sufficient to remove the interstate injury.

Regulatory Developments and Court Decisions. A section 126 petition was filed in 1979 by Jefferson County, Kentucky, with respect to SO_2 emissions from a nearby powerplant in Indiana. In its first-ever determination under section 126, EPA found that (1) a 3 percent contribution by an out-of-state source to violations to a state's ambient air standards "does not constitute prevention of attainment and maintenance of NAAQS" and (2) in an NA (where PSD requirements are not applicable) an out-of-state source cannot interfere with PSD measures in violation of section 126.[7]

In 1980 and 1981, Pennsylvania, New York, and Maine petitioned EPA under section 126, claiming that sources in the midwest were interfering with their ability to meet NAAQS for SO_2 and TSP and were causing acid rain. This petition alleged much broader impacts than the one-county, one-powerplant Kentucky complaint. In addition, Maine and New York alleged impacts on visibility in the form of regional haze.

In March 1984, six northeastern states and a number of major environmental groups sued EPA to take action to control acid rain under sections 115 and 126 of the Clean Air Act. The suit charged that EPA violated the Clean Air Act by failing to rule on the section 126 interstate air pollution petitions filed in 1981 to seek stricter air pollution controls in the midwest, and it also asked the court to order EPA to require reduced emissions in midwestern states under section 115.[8]

EPA denied the 1980 and 1981 petitions in December 1984, on the grounds that the petitioning states did not adequately demonstrate that out-of-state sources were interfering with measures required to prevent significant deterioration or to protect visibility, or that out-of-state sources prevented the petitioning states from attaining or maintaining NAAQS. Also, since sections 110 and 126 apply only to pollutants for which EPA established NAAQS or which are included in the PSD or visibility requirements, EPA ruled that they cannot apply directly to acid deposition or regional haze.[9]

In response to the 1984 suit, a federal court ordered EPA in July 1985 to begin the process of requiring states to cut emissions that have caused acid rain in Canada, under section 115. EPA has had until April 1986 to notify governors about what changes will have to be made to *state implementation plans* (SIPs).[10]

Stack Heights

Prior to and since the issuance of the 1970 amendments, many electric utilities have used tall stacks and other dispersion techniques to control ambient concentrations at ground level. However, section 123 limits the manner in which tall stacks and dispersion techniques may be considered in setting emission limits. This section of the act has been the source of a great deal of controversy and a court challenge involving EPA, the environmental community, several states, and the electric power industry.

Section 123 of the Clean Air Act, as amended in 1977, regulates the manner in which dispersion of pollutants from a source may be considered in setting emission limitations. Specifically, the act requires that the degree of emission limitation not be affected by that portion of a stack which exceeds *good engineering practice* (GEP) or by "any other dispersion technique." It defines GEP, with respect to stack heights, as "the height necessary to ensure that emissions from a stack do not result in excessive concentrations of any air pollutant in the immediate vicinity of the source as a result of atmospheric downwash, eddies, or wakes which may be created by the source itself, nearby structures or nearby terrain obstacles."

Section 123 further provides that GEP stack height not exceed 2.5 times the height of the source unless a demonstration is performed showing that a higher stack is needed to avoid "excessive concentrations." The statute delegates to EPA the responsibility for defining such key phrases as *excessive concentrations, nearby,* and *other dispersion techniques.*

Regulatory Developments and Court Decisions. In February 1982, EPA promulgated a final regulation limiting stack height credits and other dispersion techniques.[11] This regulation was challenged in the U.S. Court of Appeals for the D.C. Circuit by the Sierra Club Legal Defense Fund, Inc., the Natural Resources Defense Council, Inc., and the Commonwealth of Pennsylvania. In October 1983, the court issued its decision, ordering EPA to reconsider portions of the stack height regulation, reversing certain portions, and upholding other portions.[12] Generally, the court required EPA to rewrite the definitions of *excessive concentrations, nearby terrain,* and *dispersion techniques*; to reevaluate GEP formulas for stack heights; and to delete credits for plume impaction.

In response to the court mandate, EPA issued a final regulation in July 1985.[13] The July 1985 regulation has a number of technically complex provisions. Two of the most important aspects are as follows:

1. EPA disallowed dispersion credit for "plume impaction," which previously allowed sources to raise their stacks to avoid causing high pollution concentrations on elevated terrain downwind of the source. The July 1985 regulation requires that sources near elevated-terrain features reduce their emissions through constant controls rather than through the use of dispersion techniques.
2. EPA also expanded its definition of a prohibited dispersion technique. The July 1985 regulation focused more directly on practices that, as with tall stacks, increased the height at which pollutants are released into the atmosphere, and it limited the reliance on such practices as substitutes for emission controls.

The July 1985 regulation required that states revise their implementation plans and emission limitations for affected sources to reflect the provisions of the stack height regulation by May 1986 (nine months from regulation promulgation). It is expected that this rule will result in more stringent SO_2 emission limitations for some sources, primarily coal-fired powerplants. EPA has estimated that the July 1985 regulation could result in potential reductions in SO_2 emissions of up to 1.7×10^6 tons/yr (1.54×10^6 Mg/yr).[14]

Measures to Prevent Economic Disruption or Unemployment

Section 125 of the Clean Air Act stipulates that a state governor (with federal approval), or the EPA administrator (with state approval), or the President of the United States may prohibit major fuel-burning sources from using any fuels other than those locally or regionally available, if such use would result in local economic disruption or unemployment. Under these circumstances, the utility operator may be required to enter into long-term contracts for local coal and to contract for additional emissions control equipment as the state or EPA determines necessary to comply with the act. The intent of this section is to restrict the use of western coal in eastern states, where high-sulfur coal is available but not usable without scrubbers.

Prevention of Significant Deterioration

The 1970 Clean Air Act did not adequately address the degradation of clean air areas of the country where ambient air quality standards were being met. The act allowed air quality in these areas to degrade up to the level of NAAQS, which certainly did not conform with one of its primary purposes—"to protect and enhance the quality of the nation's air resources. . . ." As the result of a number of court cases, EPA promulgated a complex set of PSD regulations for source construction in clean air areas to satisfy the nondegradation philosophy of the act.[1]

The 1977 Clean Air Act amendments ratified the EPA program with respect to PSD as set forth in part C of the act (sections 160 to 169). In fact, as mentioned earlier, a major purpose of these amendments was to strengthen the nondegradation concept. The purposes of part C are (1) to protect health and welfare, "notwithstanding attainment and maintenance of all national ambient air quality standards"; (2) to preserve, protect, and enhance air quality in areas of national or regional scenic, natural recreational, or historic value; (3) to ensure consistency between economic growth and clean air; (4) to ensure that a source in one state will not interfere with a plan to prevent significant deterioration of air quality in any other state; and (5) to ensure that decisions regarding increased air pollution are based on evaluation of all consequences and that public participation is included in the decision-making process.

One of the most important provisions of part C was the establishment of three classes of areas for those air quality control regions that have ambient air quality levels better than any national primary or secondary air quality standard. Class I allows the least amount of degradation, while class III allows the greatest amount. Mandatory class I areas include international parks, national wilderness areas, and national memorial parks exceeding 5000 acres (2025 ha), as well as all areas that had been redesignated as class I under EPA regulations that existed prior to the enactment of the act. All other applicable areas were categorized as being class II areas unless they were redesignated. The maximum allowable increases in particulate matter and sulfur dioxide for class I, II, and III areas are listed in Table 1.3. These are the only two pollutants controlled through the use of increments under these provisions of the Clean Air Act.

The impact of class I areas on powerplant siting is a function of their size, location relative to existing sources, and regional meteorology. Class I areas, in fact, constitute a severe restriction to powerplant siting within an exclusion zone determined by plant- and area-specific factors. The use of conservative air quality modeling techniques recommended by EPA in assessing the SO_2 ambient impact of sources results in the exclusion of large new coal-fired powerplants from any areas within approximately 31 mi (50 km) of class I areas, unless the plants install extremely stringent SO_2 controls.

Regulatory Developments and Court Decisions. In the last few years, EPA's PSD program has been a significant hurdle in the construction of new facilities and in

TABLE 1.3 Federal PSD Increments, μg/m³

Pollutant	Averaging period*	Class I	Class II	Class III
SO_2	3 h	25	512	700
	24 h	5	71	182
	Annual	2	20	40
TSP	24 h	10	37	75
	Annual	5	19	37

*For any period other than the annual period, the applicable maximum allowable increase may be exceeded once per year at any one location.

the modification of existing facilities. The "PSD permit" is an approval which must be obtained prior to the construction of certain new or modified facilities and which requires extensive technology justification and detailed ambient air quality studies. Not surprisingly, EPA regulations implementing the PSD program not only have been very complex but also have changed frequently.

On December 14, 1979, the U.S. Court of Appeals for the District of Columbia forced an extensive modification of these regulations with its decision in *Alabama Power Co. v. Costle.*[15] On August 7, 1980, EPA issued revised PSD regulations in response to the court's directives.[16] The key aspects of these revisions were as follows:

1. PSD provisions (such as BACT, monitoring, and modeling) were extended to include "all pollutants regulated under the Act," rather than restricting the application of PSD to only the major pollutants emitted by the sources.
2. Consideration is given to fugitive emissions (including fugitive dust) to determine whether a source construction or modification is major and thus subject to PSD for 28 different source categories (including powerplants).
3. In the revised PSD regulations, the term *potential to emit* is defined as the capacity of a source to emit a pollutant at its maximum design capacity, after reductions achieved by pollution control equipment are taken into account.
4. All modifications to major sources in a clean air area that result in net increases in emissions above a *de minimis* level are subject to PSD review.

Concerning item 1, whereas the old regulations subjected a source to PSD review if it had the potential to emit a major amount of SO_2 or particulates, the August 1980 regulations subject sources to PSD review of *any* pollutant regulated under the act. Pollutants regulated under the act are the criteria pollutants (i.e., pollutants for which criteria documents have been prepared and for which NAAQS have been established), pollutants regulated by NSPS, and pollutants regulated by NESHAP.[1]

PSD Permit Requirements. In general, a PSD permit application must contain the following basic components:

1. An evaluation of alternative control devices and techniques demonstrating that BACT will be applied to the affected source
2. An analysis of existing measured ambient air quality in the area
3. A dispersion modeling analysis demonstrating that emissions from the project

in conjunction with other nearby sources will not cause a violation of NAAQS or PSD increments

4. An assessment of the source's impact on soils, vegetation, and visibility
5. An analysis of the air quality impact associated with indirect growth created by the project

The basic control technology requirement for new sources is the application and evaluation of BACT to determine an emissions limitation based on the maximum degree of pollutant reduction, by taking into account energy, environmental, and economic impacts. BACT must be applied to all new major sources for those pollutants for which there will be a net significant increase in emissions. This level has been defined by the "significant emissions rates" established by EPA. The evaluation of the proposed air pollution control system must include an analysis of alternative control systems capable of achieving a higher degree of emissions reduction.

Visibility Protection

Section 169A declares as a national goal the remedying of existing, and prevention of future, man-made visibility impairment in mandatory class I areas. In December 1980, EPA promulgated visibility protection regulations which established visibility requirements for SIPs for the 36 states where the mandatory class I areas are located.[17] The visibility regulations required these states to develop a visibility monitoring program and a visibility new-source review program.

In December 1982, the *Environmental Defense Fund* (EDF) filed a citizen suit alleging that EPA failed to promulgate visibility SIPs for the 34 states that failed to submit SIPs to EPA. In April 1984, EPA entered into a settlement agreement with EDF which required EPA to implement plans to protect visibility for the deficient states.[18] EPA proposed rules for these states in October 1984 which addressed visibility monitoring strategies and visibility new-source review.[19] The remaining SIP requirements for visibility to be addressed by EPA include a long-term strategy, a review of existing sources causing impairment in mandatory class I areas, and provisions for consideration of integral vistas in permit review and existing impairment control.[20] An *integral vista* is a view from within a mandatory class I area to a panorama or landmark that lies outside the boundaries of the area. The concept of protecting integral vistas has been a controversial issue in the Clean Air Act authorization debate. As consideration of visibility becomes incorporated into SIPs, states will play an increasing role in the protection of the mandatory class I areas.[20]

Nonattainment Areas

In the discussion on PSD, it was pointed out that the 1970 Clean Air Act incongruously allowed air quality in clean areas to degrade to the level of NAAQS. A similar situation existed for dirty areas in that the 1970 act never addressed the problem of industrial growth in areas where ambient standards were not being attained. To remedy this, EPA issued an interpretive ruling in 1976 which placed stringent conditions on permits for new or modified sources in NAs.

Section 129 of the 1977 amendments added a new part D (sections 171 to 178), "Plan Requirements for NAs," to the Clean Air Act. The principal provisions of part D can be summarized as follows: An offset policy was promulgated (approving EPA's approach in its interpretive ruling) requiring that all major new-source construction or modification in an NA be accompanied by emissions offsets (i.e., corresponding decreases in emissions elsewhere in the area) to mitigate any increase in emissions arising from the new source, such that the total emissions in the area do not increase and so that ambient pollutant concentrations do not increase. Also, the use of emissions control technology based on the *lowest achievable emissions rate* (LAER) and compliance by all other facilities owned by the applicant and located in the state are required.

The term *lowest achievable emissions rate* is defined as the rate of emissions which reflects (1) the most stringent emissions limitation which is contained in the SIP of any state for such class or category of source, unless the owner or operator of the proposed source demonstrates that such limitations are not achievable; or (2) the most stringent emissions limitation which is achieved in practice by such class or category of source, whichever is more stringent. In addition, the SIP submitted to treat NAs was to provide for attainment of NAAQS (or "reasonable progress" toward achieving attainment) by December 31, 1982. As with the PSD regulations, the NA regulations are complex and have undergone frequent changes.[1]

EPA's Bubble and Emissions Trading Policies. In late 1979 EPA issued its *bubble policy*.[21] The term arises from the concept whereby one places an imaginary "bubble" over the plant and views all the pollutants as exiting from a single hole in the bubble, rather than from individual stacks. Under the bubble policy, a source operator is permitted, under certain circumstances, to adjust individual point source discharges so that the total of all the discharges is less than or equal to the sum of all the individual point source limitations set forth in the various regulations.[1]

In April 1982, EPA published its proposed Emissions Trading Policy Statement.[22] This policy encourages states to use emissions trades to achieve more flexible and rapid attainment of NAAQS at lower cost than would otherwise be possible. One major change in this policy that differs from the original 1979 bubble policy was that trading was no longer limited to attainment areas only. However, the court decision in *NRDC v. Gorsuch* negated this proposed use of the nonattainment bubble.[23] As of early 1986, the issue of whether EPA should allow bubbles in areas not in compliance with Clean Air Act regulations continues to be an area of debate.[24]

Permit Requirements. A major new stationary source, or major modification to an existing source, located in an NA must obtain new-source permits in accordance with EPA's NA regulations. The new-source review of a major source located in an NA requires (1) the achievement of LAER, (2) certified compliance by all other sources owned or operated by the applicant in the state, (3) emissions offsets greater than one-for-one (or as otherwise provided in an approved SIP), and (4) emissions offsets providing a net air quality benefit to the NA. Major stationary sources are subject to review under the offset ruling only if they have the potential to emit 100 tons/yr (90.7 Mg/yr) or more for the pollutant(s) for which the area is designated nonattainment.[25]

Standardized Air Quality Modeling

The Clean Air Act has provided a significant incentive for the use of air quality dispersion models to predict ambient air concentrations. As regulations were developed in response to the 1970 and 1977 amendments, models became the major evaluation tool for enormously expensive and important decisions. Allowable emissions limits for existing sources and decisions about the acceptability of new sources in a given area are based on calculations from an EPA-sanctioned dispersion model. Models have played a prominent role in studies demonstrating conformance with NAAQS (section 109 of the act), in the stack height regulation (section 123), in interstate pollution abatement (section 126), in the PSD program (sections 160 to 169), and in the NA program (sections 129 and 171 to 178). In section 320, the act requires EPA to provide a forum for discussions on modeling by conducting a modeling conference every 3 years, with special attention to modeling necessary to carry out the PSD program.

KINDS OF AIR POLLUTANTS

Presently, coal combustion supplies more than 40 percent of U.S. electrical needs. Natural gas and petroleum had been replacing coal prior to the oil embargo of the early 1970s because of their cleaner, more stable combustion characteristics. Oil combustion was especially attractive in eastern U.S. locations because of cheaper water transportation costs and the much lower particulate emissions levels. However, the oil embargo and the subsequent upward surge in foreign crude oil prices—and the depletion of domestic reserves of oil and gas—stopped the replacement of coal by oil. Coal use has increased substantially since 1973. Coal combustion will continue as the largest source of electrical energy for many years into the future. Since coal is not a clean fuel, the growth in coal utilization, with attendant potential increases in pollutant emissions, has been an issue of concern from the environmental viewpoint.[26] In the remainder of this chapter, a qualitative discussion of the types of air pollutants released from powerplants is presented.

Table 1.4 provides an overall look at national emissions estimates of TSP, SO_x, NO_x, and CO for 1978.[27] Table 1.4 shows that the electric utility industry is the predominant source of SO_x emissions, but an insignificant contributor to CO emissions totals. Electric utility fuel combustion contributes about one-third of the annual United States NO_x emissions. Other air releases from powerplants include CO_2 and hydrocarbons.

TABLE 1.4 National Estimates of Emissions for 1978, 10^6 tons*

Source category	TSP	SO_x	NO_x	CO
Electric utilities				
Coal	2.6	17.5	5.6	
Oil	0.2	1.9	1.4	
Natural gas	<0.1	0	0.9	
Total	2.8	19.4	7.9	0.3†
Percentage of grand total	20	65	31	0.3†
Other sources	11.0	10.4	17.8	112.2†
Grand total	13.8	29.8	25.7	112.5

*To obtain 10^6 megagrams, multiply by 0.907.
†Estimates based on 1977 data, from Reference 28.

Sulfur Oxides

One of the most troublesome air pollutants in terms of quantity produced and difficulty and cost of control is SO_2. It is emitted in large amounts by coal- and oil-fired powerplants, industrial boilers, steel mills, and ore processing facilities associated with the mining and metallurgical industries.[1] Table 1.4 shows that coal-fired powerplants produced an estimated 59 percent of the nation's SO_2 emissions in 1978, while oil-fired plants produced 6 percent. Total United States SO_2 emissions from electric utilities decreased 16 percent from 1973 to 1982, despite a substantial increase in coal use during this period. This reduction in utility SO_2 emissions was due both to the use of lower-sulfur coals and to the operation of scrubbers.[29]

In the combustion of most coals, more than 90 percent of the coal sulfur is converted to gaseous SO_2. About 1 to 2 percent of the emitted sulfur oxides is in the form of primary sulfates (SO_4). Oil-fired boilers generally convert over 90 percent of available fuel sulfur to gaseous SO_2 emissions. However, high flame temperatures used in the combustion of oil promote the formation of primary sulfates. Tests have shown that about 7 percent by weight of sulfur oxide emissions from oil combustion is emitted as primary SO_4.[30]

In the atmosphere, SO_2 may undergo chemical or photochemical transformation to sulfuric acid or SO_4. The acid rain debate hinges in part on the hypothesis that acid rain is caused by these oxidation products of SO_2. The SO_2-SO_4 transformation is one that has received much attention in the past. During the 1960s, air pollution regulatory and research activities focused on SO_2. It is now felt that particulate sulfates pose a more severe health hazard, particularly in the presence of other pollutants.[1] Also, SO_4 is often the predominant constituent in secondary particulates associated with visibility impairment.

Particulate Matter

Particulate matter is the general term used for a broad class of chemically and physically diverse substances that exist as particles (either liquid droplets or solids) over a wide range of sizes. Particles originate from a variety of stationary and mobile sources. They may be emitted directly or be formed in the atmosphere by transformations of gaseous emissions such as SO_x, NO_x, and VOC. The major chemical and physical properties of particulate matter vary greatly with time, region, meteorology, and source category.[4]

Nationwide emissions of particulate matter (not including fugitive emissions) have decreased since 1950. Likewise, particulate emissions from stationary fuel combustion decreased during this period due to increased use of oil and natural gas prior to the oil embargo and due to installation of control equipment and conservation efforts following the embargo.[30] Table 1.4 shows that fossil-fuel-fired plants produced an estimated 20 percent of the nation's 1978 particulate emissions.

Several factors affect the quantity and characteristics (size and composition) of particulate matter emissions from stationary sources. Examples of such factors are source type, operating conditions, fuel characteristics, and type of emissions control equipment. Modern particulate control devices are very efficient in removing large particles, so that most particles emitted are less than 2.5 μm in diameter.[30] The overall mass of these small particles is, however, small.

Although coal combustion particles consist primarily of carbon, silica, alumina, and iron oxide, trace elements (such as chlorine, iron, and mercury) are also included. Although a large percentage of the trace elements in raw coal remains in the bottom ash after combustion, about 15 percent of total particulate emissions released to the atmosphere consists of trace metals. Although coal combustion contributes most of the trace elements associated with particulate emissions, oil combustion is the source of 50 to 80 percent of cadmium, cobalt, copper, nickel, and vanadium emissions.[30]

Impairment of visibility is perhaps the most noticeable and best documented effect of particles in ambient air. Pollution-derived effects on visibility can be classified into regional haze and visible plumes. Regional haze is relatively homogeneous, reduces visibility in every direction from the observer, and can occur on a geographic scale ranging from an urban area to multistate regions. Visible plumes of smoke, dust, or colored gas obscure the sky or horizon relatively near their source of emission.[31]

Nitrogen Oxides

The predominant molecular form of NO_x emitted from coal-fired powerplants is NO; the remainder, NO_2. Nitrous oxide is rapidly oxidized in the atmosphere to NO_2 by combination with O_2 and by reactions involving the hydroxyl (OH) radical. Nitrogen dioxide is considered more toxic than NO and is also involved in atmospheric reactions that produce secondary pollutants. On absorbing light in the ultraviolet range, NO_2 dissociates in the atmosphere to NO and atomic oxygen. This atomic oxygen may react with oxygen molecules to produce O_3 or may combine with hydrocarbon radicals (formed by absorption of light) to produce oxidized organic molecules. In addition, NO_2 itself may combine with hydrocarbon radicals to produce secondary pollutants, most notably PAN. Ozone, PAN, and oxidized hydrocarbons are the major components of urban smog. Also, the oxidation of NO_x to nitric acid or nitrates, followed by wet or dry deposition, results in the removal of NO_x from the atmosphere. As shown in Table 1.4, powerplants are not significant sources of NO_x compared with industrial and transportation sources.[32]

Carbon Monoxide and Carbon Dioxide

The presence of CO in boiler flue gas is a result of incomplete combustion of fuel. Consequently, the quantity of CO emitted is a function of boiler efficiency, which varies with boiler type, size, and age. Large coal-fired utility boilers, under conditions of relatively constant load, emit significantly less CO per ton (megagram) of coal burned than do smaller industrial boilers that are often operated intermittently.[32] As was mentioned earlier, powerplants are minor contributors to nationwide CO emissions.

Traditionally, CO_2 has not been considered a pollutant. However, since CO_2 plays a major role in determining the earth's climate by absorbing outgoing infrared radiation from the earth, there has been considerable concern in recent years that the warming of the atmosphere caused by increased CO_2 concentrations resulting from human activities will have undesirable consequences in the future. Atmospheric CO_2 levels have increased approximately 25 percent from 1800 to 1985, an increase attributable mainly to human influences, first from deforestation during the massive global expansion of agriculture and now primarily from fossil-fuel burning. A significant portion of the overall increase has occurred in the last few decades, during a period of unprecedented fossil-fuel use. A doubling of the atmospheric CO_2 level of 1800 may produce a global average temperature warmer than that of any period in the last 100,000 years or more, and the CO_2-induced warming (greenhouse effect) could be augmented significantly by increasing levels of other atmospheric trace constituents.[33]

Two of the most important possible impacts from CO_2 warming are coastal flooding and agricultural disruption. A significant increase in atmospheric CO_2 might lead to the disintegration of the west Antarctic ice sheet and the eventual disappearance of the entire Arctic Ocean ice pack in the summer, resulting in coastal flooding. With a global climate warming, some crop-producing regions would probably deteriorate while others would become more favorable.[34]

Hydrocarbons

The quantity of HC in boiler exhausts is also dependent on combustion efficiency in the boiler. HC in flue gas represents unburned fuel and may consist of compounds not initially present in the fuel but formed during the process of combustion. HC emissions may be either gaseous or solid. Powerplants are insignificant contributors to nationwide HC emissions.[32]

REFERENCES

1. J. P. Bromberg, *Clean Air Act Handbook,* Government Institutes, Rockville, Md., 1983.
2. "National Primary and Secondary Ambient Air Quality Standards—Revisions to the National Ambient Air Quality Standards for Photochemical Oxidants," EPA, *Federal Register* (44 FR 8202), Feb. 8, 1979.
3. "National Primary and Secondary Ambient Air Quality Standards," EPA, *Federal Register* (48 FR 628), Jan. 5, 1983.
4. "Proposed Revisions to the National Ambient Air Quality Standards for Particulate Matter," EPA, *Federal Register* (49 FR 10408), Mar. 20, 1984.
5. D. Pahl, "EPA's Program for Establishing Standards of Performance for New Stationary Sources of Air Pollution," *Journal of the Air Pollution Control Association,* 33(5): 468–482, 1983.
6. S. Tabler, "EPA's Program for Establishing National Emission Standards for Hazardous Air Pollutants," *Journal of the Air Pollution Control Association,* 34(5): 532–536, 1984.
7. "EPA's First Final Order on Interstate Pollution Sets Action Yardsticks," *Inside EPA,* Inside Washington Publications, Washington, Feb. 12, 1982, p. 6.
8. "Northeastern States, Environmentalists Sue EPA to Control Acid Rain," *Inside EPA,* Inside Washington Publications, Washington, Mar. 23, 1984, p. 12.
9. "Interstate Pollution Abatement, Final Determination," EPA, *Federal Register* (49 FR 48152), Dec. 10, 1984.
10. "EPA Must Order States to Cut Emissions to Reduce Acid Rain in Canada, Court Says," *Environment Reporter—Current Developments,* Bureau of National Affairs, Washington, Aug. 2, 1985, p. 541.
11. "Final Stack Height Regulations," EPA, *Federal Register* (47 FR 5864), Feb. 28, 1982.
12. *Sierra Club v. Gorsuch,* nos. 82-1384, 82-1412, 82-1845, 82-1889, D.C. Circuit, 1983.

13. "Stack Height Regulation; Final Rule," EPA (40 CFR Part 51), *Federal Register* (50 FR 27892), July 8, 1985.

14. "EPA Issues Final Smokestack Height Rules," EPA, Press Release, EPA Office of Public Affairs, Washington, June 27, 1985.

15. *Alabama Power Co. v. Costle,* 13 E.R.C. 1225, D.C. Circuit, 1979.

16. "Requirements for Preparation, Adoption, and Submittal of Implementation Plans; Approval and Promulgation of Implementation Plans," EPA, *Federal Register* (45 FR 52676), Aug. 7, 1980.

17. "Subpart P—Protection of Visibility," 40 CFR Pt 51, secs. 300–307.

18. *Environmental Defense Fund, Inc., et al. v. Ruckleshaus,* Order and Decree, signed by Judge Aguilar, Apr. 20, 1984, Case no. C82-6850 RPA. U.S. District Court for the District of Northern California. Notice of settlement appeared in *Federal Register* (49 FR 20647), May 16, 1984.

19. "State Implementation Plans for Visibility, New Source Review and Monitoring Strategy; Proposed Rulemaking," EPA, *Federal Register* (49 FR 42670), Oct. 23, 1984.

20. B. Polkowsky, "Implementation Plans for Visibility Protection, Pt II," Paper presented at 78th Annual Meeting of Air Pollution Control Association, Detroit, June 16–21, 1985.

21. "Air Pollution Control; Recommendation for Alternative Emission Reduction Options within State Implementation Plans," EPA, *Federal Register* (44 FR 71780), Dec. 11, 1979.

22. "Emissions Trading Policy Statement; General Principles for Creation, Banking, and Use of Emission Reduction Credits," EPA, *Federal Register* (47 FR 15076), Apr. 7, 1982.

23. *NRDC v. Gorsuch,* 685 F. 2d 718, D.C. Circuit, 1982.

24. "EPA Hears Conflicting Views on Use of CAA Bubbles in Nonattainment Areas," *Inside EPA,* Inside Washington Publications, Washington, Feb. 14, 1986, p. 2.

25. "Emission Offset Interpretive Ruling," 40 CFR Pt 51, Appendix S.

26. D. Randerson (ed.), *Atmospheric Science and Power Production,* DOE/TIC-27601, prepared by Dept. of Commerce for Dept. of Energy, Springfield, Va., 1984.

27. *National Air Pollutant Emission Estimates, 1970–1978,* EPA, EPA-450/4-80-002, PB82-232687, Office of Air Quality Planning and Standards, Research Triangle Park, N.C., 1979.

28. *Air Quality Criteria for Carbon Monoxide,* Informatics, Inc., EPA-600/8-79-022, PB81-244840, prepared for EPA, Research Triangle Park, N.C., 1979.

29. E. H. Pechan and J. H. Wilson, Jr., "Estimates of 1973–1982 Annual Sulfur Oxide Emissions from Electric Utilities," *Journal of the Air Pollution Control Association,* 34(10): 1075–1078, 1981.

30. *Air Quality Criteria for Particulate Matter and Sulfur Oxides,* EPA, EPA-600/8-82-029, Environmental Criteria and Assessment Office, Research Triangle Park, N.C., 1982.

31. *Review of the National Ambient Air Quality Standards for Particulate Matter; Assessment of Scientific and Technical Information,* EPA, PB82-177874, Research Triangle Park, N.C., 1982.

32. R. C. Christman, J. Haslbeck, B. Sedlik, W. Murray, and W. Wilson, *Activities, Effects, and Impacts of the Coal Fuel Cycle for a 1000-MWe Electric Power Generating Plant,* NUREG/CR-1060, prepared for the U.S. Regulatory Commission by Teknekron Research, Inc., 1982.

33. J. R. Trabalka (ed.), *Atmospheric Carbon Dioxide and the Global Carbon Cycle,* Dept. of Energy, DOE/ER-0239, Carbon Dioxide Research Div., Washington, 1985.

34. *Global Energy Futures and the Carbon Dioxide Problem,* Council on Environmental Quality, Washington, 1981.

CHAPTER 4.2
COAL CLEANING

James D. Kilgroe and Jerry Strauss
Environmental Protection Agency
Research Triangle Park, N.C.

INTRODUCTION

Coal is a readily combustible substance containing significant quantities of inorganic elements such as iron, aluminum, silica, and sulfur. These elements occur primarily in ash-forming minerals associated with the coal and are incorporated to a lesser degree into the organic coal structure itself. The relative amounts of contaminants, the manner in which they are included in the coal assemblage, and the degree to which they can be removed vary widely with different coals.

Coal preparation, or beneficiation, is a series of operations that remove mineral matter (ash) from coal. Preparation relies on different mechanical operations to perform the separation, such as size reduction, size classification, cleaning, dewatering and drying, waste disposal, and pollution control. Mechanical preparation of coal is also referred to as *physical coal cleaning* (PCC), or simply *coal washing*. Coal preparation processes are designed mainly to provide ash removal, energy enhancement, and product standardization. Sulfur reduction is achieved because the ash material removed contains sulfur (in pyritic form). Preparation plants are not currently designed to optimize sulfur removal.

The degree of desulfurization achieved by crushing and specific-gravity separation depends on the sulfur content of the raw coal, the chemical form of the sulfur, and the coal techniques used. Although each commercial coal preparation plant is different, all are generally capable of reducing potential sulfur emissions from 20 to 40 percent. Besides reducing sulfur levels, preparation reduces sulfur variability in the product coal and thus sulfur and ash variability in powerplant emissions.

Coal mines are increasing their use of washing plants because modern continuous-mining techniques increase the amount of rock and other noncombustible materials mixed in with the coal. Also, as the mines are exhausted of good-quality coals, washing is necessary to maintain product quality. Preparation plants are usually located at the mine and often blend several different-quality coals to meet customer fuel specifications. The more than 520 coal preparation plants operating in the United States today (almost all at or near the coal mine) can process 400×10^6 tons/yr (363×10^6 Mg/yr) of raw coal. In 1984, about 350×10^6 tons (317×10^6 Mg) of coal were delivered to utilities from the major eastern and midwestern coal-producing states. More than one-half of this coal was cleaned to some extent.

Coal preparation as a precombustion control technology for the boiler feed is quite different from combustion or postcombustion control options, which are strongly integrated into each particular boiler facility. One important result is that the cost impact of coal preparation on boiler operations is much less than the cost impact of boiler-specific control options. The lack of any additional equipment requirements at the boiler facility nearly eliminates any capital expenditures and considerably reduces operating and maintenance concerns. Whether cleaned coal is used does not preclude the boiler operator from using any other combustion or postcombustion control options. Preparation normally increases delivered fuel costs by 10 to 25 percent.

Use of coal preparation for moderate sulfur dioxide emissions control is receiving considerable support from the federal government and state agencies, because of its relatively low cost, widespread application potential, and the flexibility retained for instituting other controls. The *Environmental Protection Agency* (EPA), *Department of Energy* (DOE), *Electric Power Research Institute* (EPRI), and several states (including New York, Ohio, and Kentucky) have put considerable resources into evaluating the desulfurization potential of current and future preparation technologies. Much of the interest is spurred by the need to consider further controls on SO_2 emissions from coal-fired powerplants. Coal preparation is particularly well suited to such considerations, because it avoids the high costs associated with retrofitting pollution control equipment on existing boilers.

FUNDAMENTAL PRINCIPLES

Coal is a complex mixture of materials that originate from peat and detrital material deposited in swamps or bogs. Besides its organic constituents, coal contains significant quantities of minerals and inorganic elements. The relative abundance of organic and inorganic constituents varies widely between coals of different types and locations in a single coal seam. The variability of coal properties is a function of the environment and chemical composition of the coal vegetable matter and the subsequent geological history of the consolidated coal-burning material.

Coal Properties

The classification of coal by rank provides information useful in assessing its properties as a fuel. Rank is related to the degree of coalification—the process of deposition, burial, and metamorphosis. Age, volatile matter, fixed carbon, bed moisture, and oxygen are indicative of rank, but no one item completely defines it. The *American Society for Testing and Materials* (ASTM) classification system includes four types of coals: anthracite, bituminous, subbituminous, and lignite.[12] In turn, each class is divided into groups. For the higher-ranked coals, the ordering is based on the fixed-carbon and volatile-matter content. Fixed carbon increases and volatile matter decreases with increasing rank. Lower-ranked coals are ordered in terms of their calorific value as determined on a moist, mineral-matter-free basis; the higher heating value decreases with decreasing rank. The location of coal fields and the rank of coals in these fields are shown in Fig. 2.1.

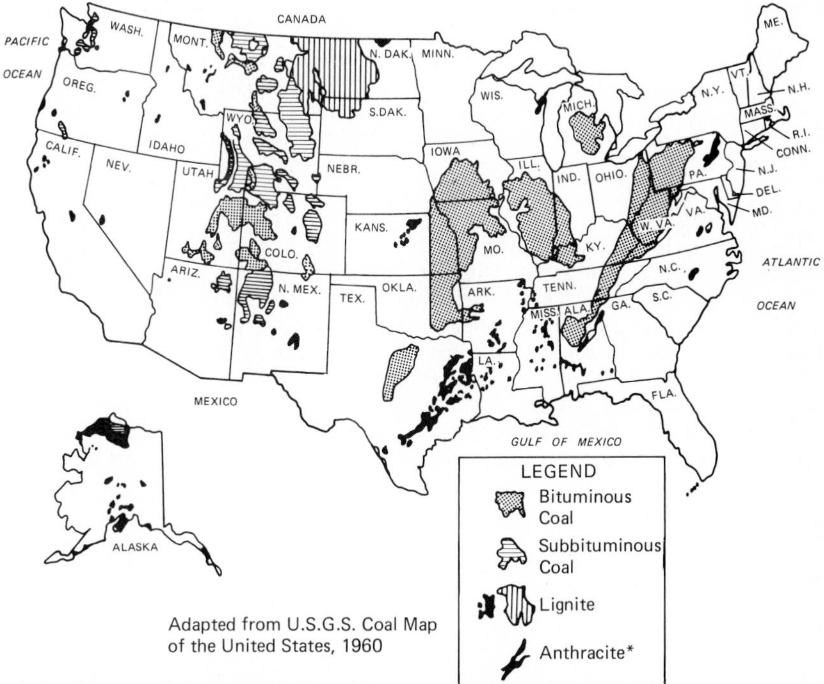

FIG. 2.1 Coal fields of the United States. *(Adapted from U.S. Geodetic Survey coal map of U.S., 1960.)*

A strong correlation generally does not exist between coal rank and the quantity and characteristics of the coal mineral matter. The mineral content of coal is most frequently a function of the depositional environment, the subsequent exposure of sediments, and the presence of mineralized groundwater or surface water. Coal rank and degree of metamorphosis are mainly functions of the temperature, pressure, and time of coalification.

Minerals and elements associated with the original organic matter and the early depositional phases of coal formation and coalification are fine-grained and intimately interspersed in the coal. This inorganic material is often called *inherent minerals*. By contrast, the minerals that form during the later stages of the coalification process are often large-grained and distinct from the coal's organic structure. These minerals are often called *extraneous minerals*. They can be removed by conventional physical coal-cleaning techniques; inherent minerals cannot.

The density or specific gravity of coal minerals is an important property which is used in separating the organic and mineral components of coal. The specific gravity of coal macerals (that is, organic matter) ranges from 1.2 to 1.7. The specific gravity of pyrite is 5.0 while the specific gravity of other minerals, such as clays and carbonates, generally ranges from 1.7 to 2.2.

Constituents of Concern

Coal constituents of concern for powerplant engineers and environmentalists include ash, sulfur, nitrogen, and various trace elements. The in-seam ash content of U.S. coals ranges from less than 2 to more than 50 percent. The effects of these components on boiler fouling and slagging are discussed in Section 1. The sulfur content of U.S. coals ranges from less than 1 to more than 7 percent. Sulfur is contained in three principal forms: mineral sulfur, organic sulfur, and sulfate sulfur. Mineral sulfur, generally in the form of pyrite (Fe_2S), can range from microscopic inclusion in the coal to large mineral fragments several inches in size. Organic sulfur, comprising from 30 to 70 percent of the total sulfur content of most U.S. coals, is an integral part of the coal matrix and can be removed only by chemical modification of the coal structure. Sulfate sulfur is present in very small amounts and usually can be removed by water washing of the coal.

Nitrogen is usually associated with the organic structure and can be removed only by chemical means. More than 40 trace elements have been identified in coal (see Table 1.3, Section 3). Most are distributed as accessory elements in coal minerals. The amounts of trace elements and their partitioning during coal use, while not discussed here, are the subject of several studies.[5]

Cleaning Techniques

The two most important unit operations used in physical coal preparation are size reduction and the subsequent separation of the organic and mineral components of the *run-of-mine* (ROM) coal. Size reduction through breaking or crushing generally causes the coal to fracture along the boundaries between the organic structure and extraneous mineral components. The degree to which the mineral components are liberated depends on coal properties, size reduction techniques, and the degree of size reduction. Crushing the coal to fine-particle sizes liberates more mineral particles, but physical separation of fine particles is less efficient and more costly.

Coal cleaning can be defined as the classification of coal particles by differences in the physical, chemical, or electrical properties. Physical coal-cleaning technology employs a wide variety of separating phenomena that depend on the properties of the particle (size, shape, density, surface characteristics), properties of the separating medium (density, viscosity, etc.), type of cleaning equipment, and operating conditions of the cleaning equipment.

Chemical coal-cleaning technology uses chemical reagents to selectively remove mineral or organic components from the coal structure. Electrical separation depends on differences in the electrostatic or electromagnetic properties of the coal particles. Although chemical and electrical separation processes are being developed, none is used commercially in the United States.

Coarse- and Medium-Coal Cleaning. Different physical cleaning techniques and equipment are often used for particles of different size ranges. Three ranges are commonly defined: *coarse* [4 to ⅜ in (102 to 9.5 mm)], *medium* [⅜ in to 28 mesh (9.5 mm to 0.589 mm)], and *fine* [less than 28 mesh (0.589 mm)]. Equipment selected to clean coarse- and medium-size particles generally employs one or more of these phenomena: gravity separation, centrifugal separation, stratification, hindered settling, or differential acceleration. In all cases, the size, shape, and density or specific gravity of the particles are important. In gravity separation, particles are placed in a fluid medium of controlled density. The difference between the gravitational and buoyant forces results in separation of particles in accordance with their densities relative to the separating medium.

Three types of gravity separation techniques are used in cleaning coarse- and medium-size particles. In *centrifugal separation,* the more dense particles are forced to the peripheral boundary layer of a cyclonic flow path, where they are segregated from the bulk flow. Particles that are nearly the same density as the separating medium remain entrained in the bulk flow. In a related separation technique, called *consolidated trickling,* particles are separated because of the mechanical interference between particles of different sizes and shapes. When large and small particles are placed in a container with a fluid medium and the container is agitated in a horizontal direction, the large particles stay together at the top of the particle bed while the small particles move downward through the voids between large particles.

In *differential acceleration,* the rate at which particles of different densities reach their terminal velocity is used to achieve separation. By changing the vertical direction of water flow in a controlled volume, differential displacement of particles will occur according to their density, size, and shape. For particles of a given size and shape, the more dense ones will be accelerated by the upward water flow at a lower rate. In hindered settling, particles are separated due to the combined phenomena of all three techniques.

Fine-Coal Cleaning. Fine-size coal is generally cleaned by techniques that rely on the differences in surface properties of mineral and organic coal particles. One technique operates on this mechanism: The organic particles are hydrophobic and will adhere to small air bubbles introduced into a slurry of fine coal, water, and selected chemical reagents. Air bubbles containing the fine-coal particles rise to the surface, where they form a froth, a process called *froth flotation.* Another technique that relies on selective adhesion is *oil agglomeration.* The fine organic particles in a coal-water slurry can be selectively wetted by small oil drops that are stirred into the slurry. The oil-coated coal particles agglomerate to form larger particles which can be separated from the fine mineral particles.

WASHABILITY

Laboratory washability tests are used to assess the sulfur and ash reduction potential of coal. Washability is a function of the inherent properties of a specific coal and washability-test technique. Two types of techniques are commonly used: those for specific-gravity separation and those for froth floatation.

Washability Tests

The potential for reducing the ash and sulfur content of a coal by specific-gravity separation can be determined by laboratory *float-sink tests*. In these tests, a coal sample is crushed and separated into various size ranges. Each size fraction is float-sink-tested in organic liquids of standardized specific gravity. The weight, percentage of ash, heat content, total sulfur, and sometimes pyritic sulfur content of each density and size fraction are then found (Table 2.1 shows an example). These data are used to determine the mineral and desulfurization potential and to specify the design and operating conditions for physical coal preparation equipment.

The effects of crushing and specific-gravity separation are similar for nearly all coals. The weight recovery (yield), energy recovery, ash content, and sulfur content decrease with decreasing specific-gravity separation. At low specific gravities, only those particles that are relatively free of embedded pyrite or mineral particles will float. Coal particles containing moderate amounts of mineral and pyrite particles sink along with pure mineral particles. This produces a cleaner

TABLE 2.1 Float-and-Sink Data for Coal*

Specific gravity		Dry basis, %			Cumulative results					
					Recovery (float), %			Reject (sink), %		
Sink (1)	Float (1)	Weight (2)	Ash (3)	Sulfur (4)	Weight (5)	Ash (6)	Sulfur (7)	Weight (8)	Ash (9)	Sulfur (10)
—	1.30	35.3	3.35	0.72	35.3	3.35	0.72	100.0	18.42	1.29
1.30	1.35	24.5	6.51	0.92	59.8	4.65	0.80	64.7	26.64	1.60
1.35	1.40	8.2	12.46	1.14	68.0	5.59	0.84	40.2	38.93	2.01
1.40	1.45	4.3	17.25	1.46	72.3	6.29	0.88	32.0	45.68	2.24
1.45	1.50	3.3	22.29	1.74	75.6	6.99	0.92	27.7	50.13	2.36
1.50	1.55	2.9	26.50	1.92	78.5	7.70	0.95	24.4	53.93	2.44
1.55	1.60	2.5	30.71	1.91	81.0	8.40	0.98	21.5	57.57	2.51
1.60	1.70	3.8	36.16	2.74	84.7	9.63	1.06	19.0	61.05	2.59
1.70	1.90	3.6	44.66	3.26	88.4	11.07	1.15	15.3	67.16	2.55
1.90	—	11.6	74.23	2.33	100.0	18.43	1.29	11.7	74.23	2.33

*Sample size: ¼ in (6.3 mm) by 100 mesh (0.149 mm), or ¼ × 100.

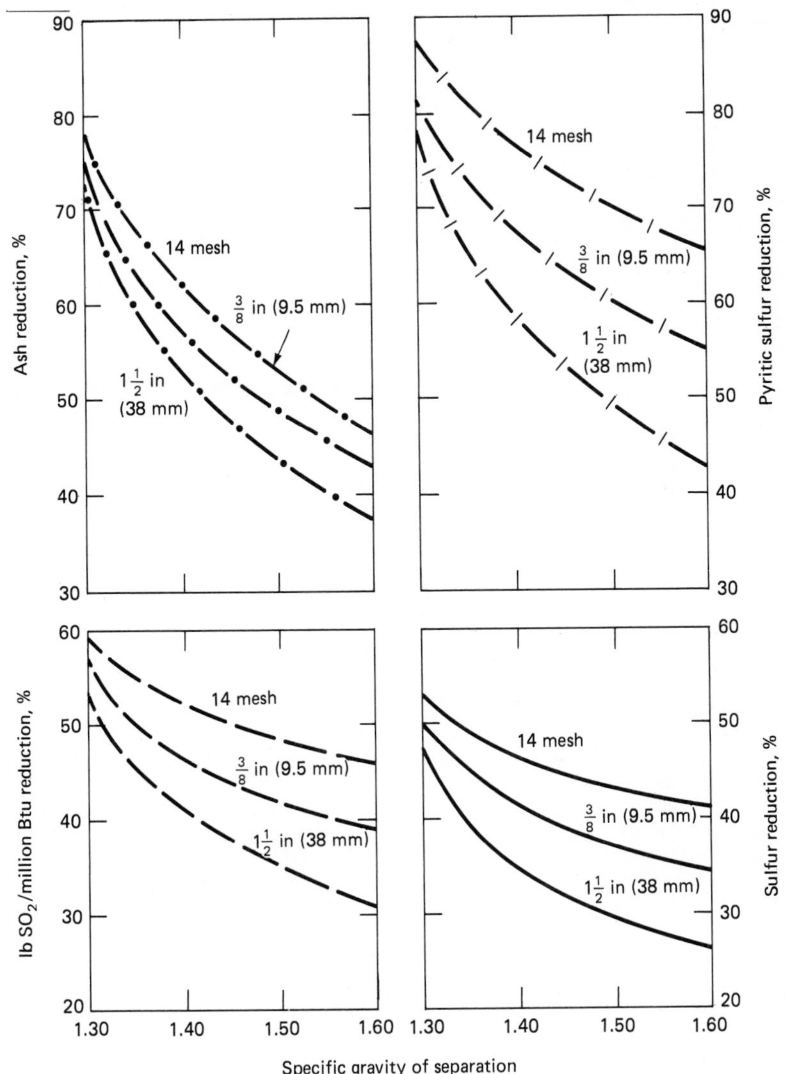

FIG. 2.2 Effect of crushing top size and specific gravity of separation on the reduction of ash, pyritic sulfur, total sulfur, and lb $SO_2/10^6$ Btu ($\times$ 0.43 = kg $SO_2/10^6$ kJ) with 227 coal samples from northern Appalachian region.[15]

product but at the expense of high energy losses to the sink (that is, refuse) fraction.

Crushing the coal to finer particle sizes liberates a higher fraction of the mineral and pyrite particles from the organic coal structure. This liberation leads to improved separation of the mineral and organic components during the subsequent float-sink tests. The general effects of coal-particle top size and specific-gravity separation are shown in Fig. 2.2.

Washability Curves

With data generated from float-sink tests, a series of five curves can be drawn, with an option for others depending on the requirements of the study. These curves are used to assess the cleaning potential of the coal by specific-gravity separation techniques. The principal curves include specific gravity (yield), cumulative-float ash, cumulative-sink ash, elementary ash, and ±0.10 specific-gravity distribution (Fig. 2.3). If samples are analyzed for sulfur content, a cumulative-float sulfur curve may also be included.

The specific-gravity curve is plotted directly from the cumulative-percentage (by weight) float data and specific-gravity fractions (column 5 versus column 1 of Table 2.1). This curve shows the theoretical yield of washed product from the raw coal at any specific-gravity separation. For example, at 1.56 gravity separation, a 79 percent yield could be obtained. As the specific gravity is reduced, the yield is reduced. This curve illustrates that washing at lower gravities generates more reject material that contains increasing amounts of heat content per unit.

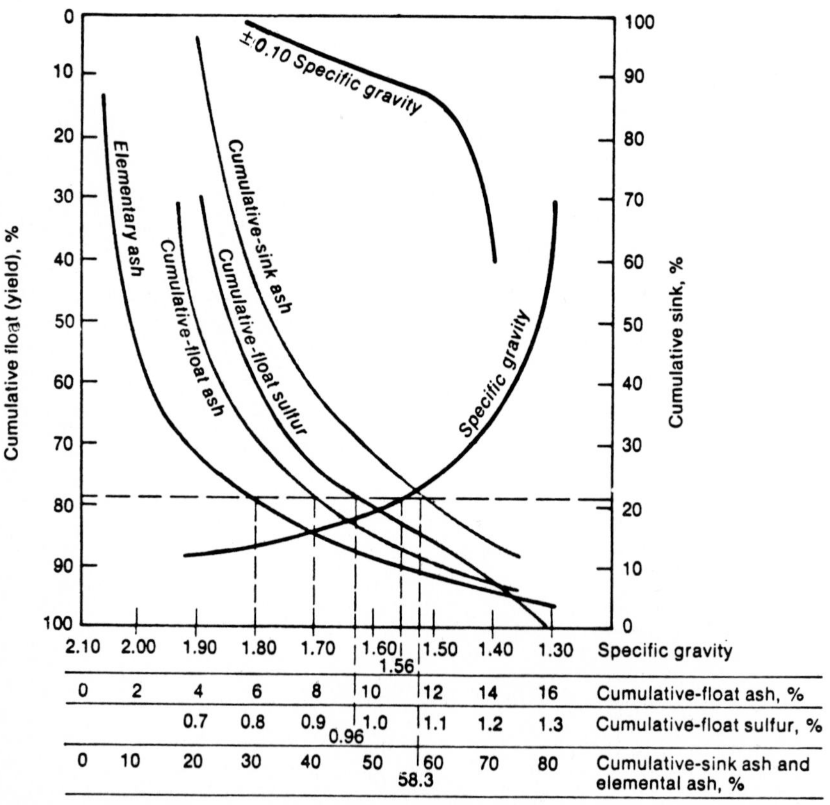

FIG. 2.3 Washability curves, derived from float-sink data.

The cumulative-float ash curve is plotted directly from the cumulative-percentage (by weight) float and cumulative-percentage ash float data (columns 5 and 6), and it shows the theoretical percentage of ash in the washed product at any given yield or washed product. The flatter the curve, the more ash that can be removed from the raw coal with the least penalty in yield and energy recovery. Also, since the specific-gravity and cumulative-float ash curves have common ordinate values, it is possible to determine what the separating gravity would be for a desired ash content.

The cumulative-float sulfur curve is plotted from the cumulative-percentage (by weight) float and cumulative-percentage sulfur float data (columns 5 and 7), and it shows the theoretical percentage of sulfur in the washed product at any given yield of washed product. For example, at 79 percent yield, the sulfur content of the clean-coal fraction would be 0.96 percent. The flatter the cumulative-float sulfur curve, the greater the sulfur reduction potential at a specific yield and heat-content recovery.

The cumulative-sink ash curve shows the theoretical ash content of the refuse at any yield of washed product. The elementary-ash curve is a derivation of the cumulative-percentage ash in the float material and is intended to show the rate of change of the ash content at different specific gravities or yields. The slope of the elementary-ash curve indicates the relative ease of separation of the coal from the refuse. Steep slopes indicate relatively small ash differences for large differences in yield; a shallow slope indicates an easy separation.

The ±0.10 specific-gravity distribution curve shows the percentage by weight of the coal that lies within ±0.10 specific gravity unit at any given specific gravity. Here is the commonly held definition of near-gravity material. It is used as a guide for determining the lowest practical specific gravity to wash a particular coal, and it indicates the degree of difficulty of separation. Less than 7 percent of near-gravity material constitutes a simple separation, while separation at a specific gravity with greater than 25 percent content of near-gravity material is difficult.

The meaning of this distribution curve has changed recently with the introduction of heavy-media cleaning equipment that yields high separation efficiencies. Although a separation at 25 percent of near-gravity material would be difficult for a jig or a concentrating table, heavy-media processes can make separations at well over 50 percent of near-gravity material.

Performance Criteria

A number of criteria are traditionally applied to evaluate the performance of coal-cleaning equipment. The various criteria are usually classified according to the degree to which they are dependent on the characteristics of the coal being cleaned. Definitions for these dependent criteria are provided in the box.

Some equipment performance criteria are substantially independent of coal properties. They are generally found from the separation distribution curve. This curve is a plot of the percentage of each gravity fraction of the coal feed recovered in the clean-coal product versus the median of the specific-gravity fraction. The *specific gravity of separation* is defined as the specific gravity of the material in the raw feed that is divided equally between clean coal and refuse. A typical distribution curve for fine coal is shown in Fig. 2.4. Independent criteria include:

DEFINITIONS

Ash Error The numerical difference between the actual and theoretical ash content of cleaned coal at the yield of cleaned coal obtained.

British Thermal Unit (Kilojoule) Recovery The ratio, expressed as a percentage, of the total calorific value of the clean coal divided by the total calorific value of the feed coal.

Emissions Parameter The total heat-specific SO_2 emissions which result when coal is completely combusted and the sulfur is converted entirely to SO_2. The emissions parameter is normally expressed in pounds of SO_2 per 10^6 Btu (kilograms of SO_2 per 10^6 kJ).

Emissions Reduction The ratio, expressed as a percentage, of the differential emissions parameter (raw-coal emissions parameter minus the clean-coal emissions parameter) divided by the raw-coal emissions parameter; the percentage reduction in the coal emissions parameters due to cleaning.

Misplaced Material The sum of the sink material in the cleaned coal and the float material in the refuse, expressed as a percentage of the raw coal.

Recovery Reduction The ratio, expressed as a percentage, of the yield of cleaned coal to the yield of the (washability test) float coal of the same ash content.

Weight Yield The ratio, expressed as a percentage, of the weight of clean coal divided by the weight of the feed coal.

Yield Error The difference between the yield of coal actually obtained and the theoretical yield at the ash content of the cleaned coal obtained.

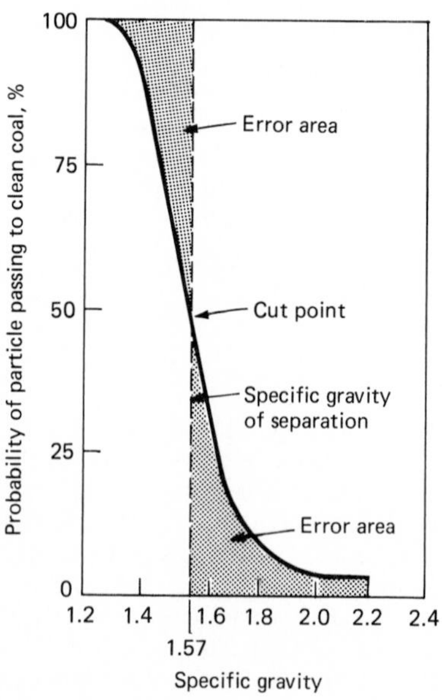

FIG. 2.4 Typical distribution curve for fine coal.

- *Probable error:* An indication of the slope of the distribution curve equal to one-half of the specific-gravity difference between the 25 and 75 percent ordinate values on the curve
- *Error area:* The area between the actual distribution curve and the theoretically perfect distribution curve, which is a step function of the specific gravity of separation (see Fig. 2.4)
- *Imperfection:* The probable error divided by the specific gravity of separation

PREPARATION EQUIPMENT

Coal preparation operations can be classified as comminution, sizing, cleaning, and dewatering. The major types of coal preparation equipment for each of these operations are discussed now.

Comminution

The major objective of comminution in U.S. preparation plants is to reduce the ROM coal to sizes suitable for cleaning and to liberate mineral impurities. In addition, crushing is usually required to meet the product coal-size specification imposed by the market. Production of fines in crushing is not desirable, mainly because of the difficulty in cleaning, dewatering, and handling fine-size coal.

Crushing in a preparation plant is done in one or two stages, depending on the feed and product coal sizes, coal characteristics, type of crusher selected, and process economics. For efficient operation with hard coals (that is, coals with low grindability indices), a reduction ratio of 4 is generally considered reasonable, and up to 10 may be used. (A *reduction ratio* is the ratio of feed to product size.) For friable coals, reduction ratios of as much as 40 may sometimes be economical. Major types of crushers found in coal preparation plants include rotary breakers, roll crushers, and hammer mills.

Sizing

Sizing of coal is accomplished by screens or hydraulic classifiers. Air classifiers are used by a few facilities. Screens in coal preparation separate refuse and fines prior to crushing, separate raw coal into size fractions for further processing, remove fines prior to cleaning, and remove water. Screens accomplish a classification of particles on the basis of size alone, the acceptance or rejection of the particles being made by a screening surface. Screens are often vibrated to give stratification by size, which enhances efficiency. For finer sizes of coal, screens are designated by mesh size, that is, the number of openings per inch.

Hydraulic classifiers have a wide size range, with hydrocyclones classifying coal as coarse as 24 μin (600 μm). The principle of operation of these devices is based on the difference in particle specific gravities as well as sizes. As their name implies, air classifiers operate with air as a fluid and classify particles with sizes ranging from 0.2 to 6 μin (5 to 150 μm). Air classifiers do not operate efficiently if the feed material is wet and are not widely used for wet sizing of coal in preparation plants.

Cleaning

Cleaning is the step in which coal is separated from its impurities. The separation takes place in water, in a dense medium, or in air. Mechanisms for cleaning operations are based either on the specific gravity or on the surface-property differences between coal and its impurities. Examples of equipment using water as the separation medium include jigs, concentrating tables, hydrocyclones, and froth flotation cells. Table 2.2 provides characteristics of major wet-coal cleaning devices.

Jigs are the most widely used coarse-coal cleaning equipment in the United States. However, they separate coal at relatively high specific gravities and with some misplaced material. Dense-medium cleaning devices generally use a mixture of finely ground magnetite (−325 mesh) dispersed in water; they perform sharper separations than devices using water or air as the separation medium. Devices currently found in preparation plants include dense-medium vessels and dense-medium cyclones. The specific gravity of the medium in these devices can be adjusted by the amount of magnetite added and is normally set between 1.3 and 1.7.

Hydrocyclones are separating devices that clean medium- to fine-size coal. They apply centrifugal force to send the product material through the center vortex finder to the overflow. Efficient separation is hindered by very fine size coal and ash that remain with the product material, resulting in a large percentage of fine coal as misplaced material in the overflow. The hydrocyclone is not as efficient a separator as other devices that clean the same-size fraction (that is, concentrating tables or heavy-media cyclones). Therefore, two-stage recirculation of overflow material is often recommended to raise the device's efficiency to acceptable levels.

Dry concentration methods use air as the medium. Selection of this separation method reached a peak in 1965 in the United States and has decreased in popularity since then. The predominant pneumatic cleaner in the industry at present is the air jig.

Froth flotation has become an economical necessity in many steam-coal beneficiation processes because of rising fuel prices, environmental regulations restricting refuse removal, and increasing fines content of ROM coal. As already mentioned, chemical reagents establish a water-repelling surface on the material to be floated. These flotation reagents come in three types and are most often used in combinations of two or more. They have these characteristics:

- *Frothers* change the surface tension of the water to prolong bubble life.
- *Collectors* or *promoters* induce contact between air bubbles and coal particles by forming a water-repellent covering over the material to be floated, while selectively ignoring other material.
- *Modifying agents* enhance the flotation process under special conditions where material selection needs further improvement.

TABLE 2.2 Characteristics of Major Wet-Coal Cleaning Devices

Device	Medium	Principle of separation	Sizes typically handled*	Unit capacities	Specific-gravity range	Cleaning efficiency	Comments
Coarse-coal jig	Water	Stratification	6 in × ¼ in (152 mm × 6.4 mm)	High	1.60–1.90	Good	Typically used to clean coarse coal at relatively high specific gravities. Efficiency decreases rapidly at sizes below ⅛ in (3.1 mm).
Fine-coal jig	Water	Stratification	½ in × 28M (12.7 mm × 0.589 mm)	High	1.60–1.90	Good	Reported to be as efficient as concentrating tables for handling compatible size. High unity capacity.
Hydrocyclone	Water	Centrifugal forces in ascending vortex	½ in × 100M (12.7 mm × 0.149 mm)	High	1.60–1.90	Poor	Poor cleaning efficiency, but can be used as "scalpers" ahead of more efficient devices.
Concentrating table	Water	Stratification and hindered settling	⅜ in × 200M (9.5 mm × 0.074 mm)	Low	1.50–1.80	Good	Provides good separation when near-gravity material is less than 10% of feed. Requires very low water-to-solids ratio.
Heavy-medium vessel	Fine magnetite in suspension	Float and sink	6 in × ¼ in (152 mm × 6.4 mm)	High	1.30–1.80	Excellent	Efficiency of separation near theoretical, even when feed contains large amounts of near-gravity material.
Heavy-medium cyclone	Fine magnetite in suspension	Float and sink plus centrifugal forces	⅜ in × 28M (9.5 mm × 0.589 mm) 1½ in × 0 (38.1 mm × 0)	High High	1.30–1.80 1.30–1.80	Excellent Very good	Excellent separating efficiency over wide range of specific gravities. Raw feed must be deslimed. Can replace multiple cleaning circuits. Requires no raw-coal desliming.
Flotation	Water	Particle attachment	28M × 0 (0.589 mm × 0) 100M × 0 (0.149 mm × 0)	High Low	N/A N/A	N/A N/A	Not effective for oxidized coal or for removal of pyritic sulfur. More expensive than other cleaning methods.

*M = mesh.

TABLE 2.3 Fine-Coal Dewatering and Handling Equipment[a]

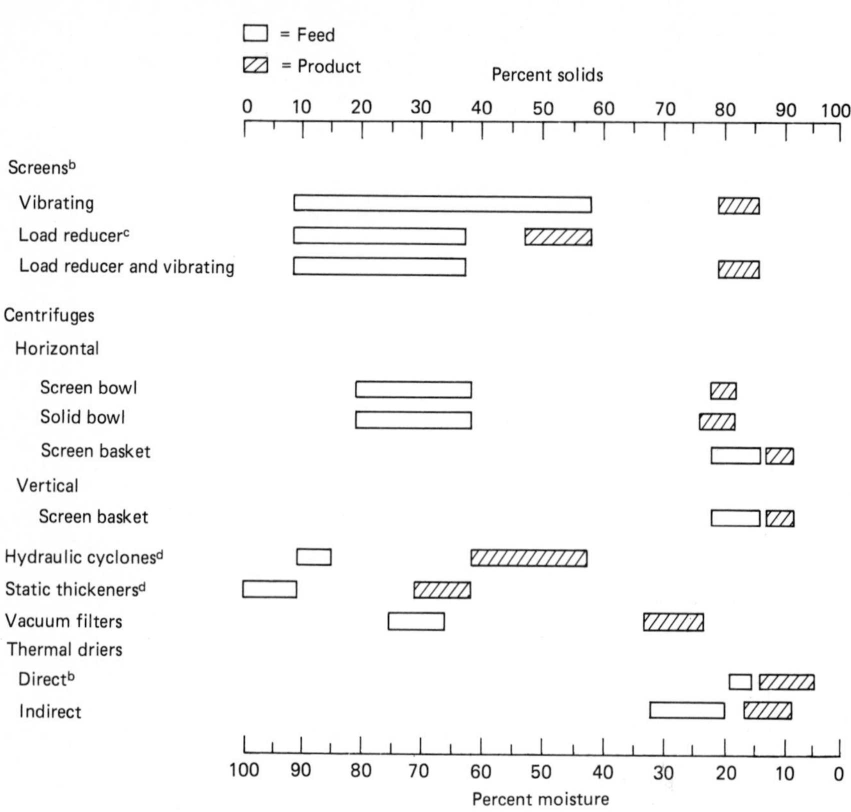

[a]Used on coal 28M (0.589 mm) × 0 unless otherwise indicated.
[b]⅜ in × 0 (9.5 mm × 0).
[c]Stationary screen, sieve bend, vor-siv, etc.
[d]Primary dewatering.

One major deficiency of flotation is its difficulty in selectively rejecting pyrites. To eliminate them from the clean-coal concentrate, it is generally necessary to reclean the froth concentrate one or more times by using a modifying agent to depress the clean coal. Then pyrite is floated with a normal reagent, skimmed, and discarded.

Dewatering

Cleaned coal and refuse streams from wet cleaning need dewatering to meet product specifications and refuse landfill-disposal requirements. Excessive moisture in the cleaned coal and refuse is undesirable because it creates handling problems, increases transportation costs, reduces the heating value of the cleaned coal, and boosts boiler fuel requirements.

Selection of dewatering equipment for a certain application depends on the initial moisture content from the cleaning equipment, the desired level of moisture after dewatering, and the size consist of the coal. Moisture can be removed by two types of equipment: mechanical dewatering devices and thermal driers. Mechanical dewatering devices can be classified as those that do not produce a final product (for example, cyclones and thickeners) and those that do produce a final product (for example, screens, centrifuges, and filters). The performance of dewatering equipment is highlighted in Table 2.3.

PROCESS DESIGN

Design Factors

The design of a coal preparation plant depends on a number of factors: in-seam coal characteristics and mining method, coal washability properties, desired clean-coal properties, and preferences of the design engineer. Coal from seams inherently low in ash content that also are carefully mined to minimize the amounts of dilution material in the ROM coal may not need cleaning. Examples are many surface- and underground-mined coals which do not contain partings and are mined by traditional blast techniques. If the coal seam has partings or is mined by techniques that result in significant amounts of in-seam or extraneous dilution material (for example, continuous miners or long-wall mining), then some degree of preparation may be required before the coal can be marketed.

When coal is cleaned, the intensity or level of cleaning will depend on its washability characteristics, the product specifications defined by the boiler operator, and the cost of cleaning. If the amount of additional ash or sulfur that can be removed does not increase significantly with the intensity or level of cleaning, then rudimentary cleaning which removes coarse dilution material and extraneous mineral matter is recommended. If the sulfur or ash reduction potential of a coal is large, if it contains substantial quantities of fine mineral matter, or if it must meet stringent product specifications, then the coal is often intensively cleaned.

Viewed from the designer's perspective, the design alternatives often represent trade-offs between capital and operating costs and yield. For example, it is possible to produce the same coal quality with a plant incorporating baum jigs and hydrocyclones as with a plant that uses dense-medium vessels, dense-medium cyclones, and froth flotation. Although the last-named plant will have higher capital costs and operating expenses, it will produce coal at a lower net cost per ton (megagram) of product because its yield will be significantly better than the jig/hydrocyclone plant. Plant designs also vary substantially depending on the preferences of the design engineer. Some firms that design and construct preparation plants also manufacture preparation equipment.

Cleaning-Plant Classification

Cleaning plants are classified by the type of circuit used (jig, concentrating table, heavy medium, etc.). Because of the large number of different circuit combinations, plants are also classified by the degree or intensity of coal cleaning. Although there are a number of different descriptors and associated definitions of the level of cleaning, the authors prefer a system based on (1) the number of different coal-size ranges cleaned in separate circuits and (2) the number of product streams. In the following definitions, alternative descriptors that generally apply to similar definitions are shown in parentheses:

Level 0 (Level A, Absence). Coal is shipped as mined.

Level 1 (Level B, Breaking). Mine debris and large noncombustible impurities are removed. Unit operations include crushing and sizing.

Level 2 (Level C, Coarse Beneficiation). Coarse- and medium-size coal is cleaned in a single circuit (usually a jig or heavy-media vessel). The fine coal is not cleaned but is either recombined with the coarse and medium coal for shipment or is discarded as refuse.

Level 3 (Level D, Deliberate Beneficiation). In this case, all the feed is wetted, and coarse- and medium-size coal and fine-size coal are washed in separate circuits. All the washed coal is then mechanically dried. At this cleaning level, fine-coal material (the most difficult and expensive to dewater) can be excluded from the product, with separate cleaning circuits for the coarse- and medium-size coal. The process generally results in a low yield and sometimes high cost. Typical level 3 plants have heavy-media washers to clean coarse- and medium-size coals together, plus flotation cells or tables to clean the fine coal.

Level 4 (Level E, Elaborate Beneficiation). The coarse-, medium-, and fine-size coals are cleaned in separate circuits. The coals from these circuits are combined to form a single product. A level 4 plant may typically have heavy-media vessels, heavy-media cyclones, and flotation cells to handle the three fractions of coal, respectively.

Level 5 (Level F, Deep or Multistream Cleaning). Coarse, medium, and fine coals are cleaned in a number of different circuits. Coal from the circuits is combined to form two products: a deep-cleaned coal and a middling coal. The former is a metallurgical- or compliance-grade coal, which is low in ash and sulfur. The middling coal has a higher ash and sulfur content. Diagrams depicting the levels of cleaning are shown in Figs. 2.5*a* through *c*.

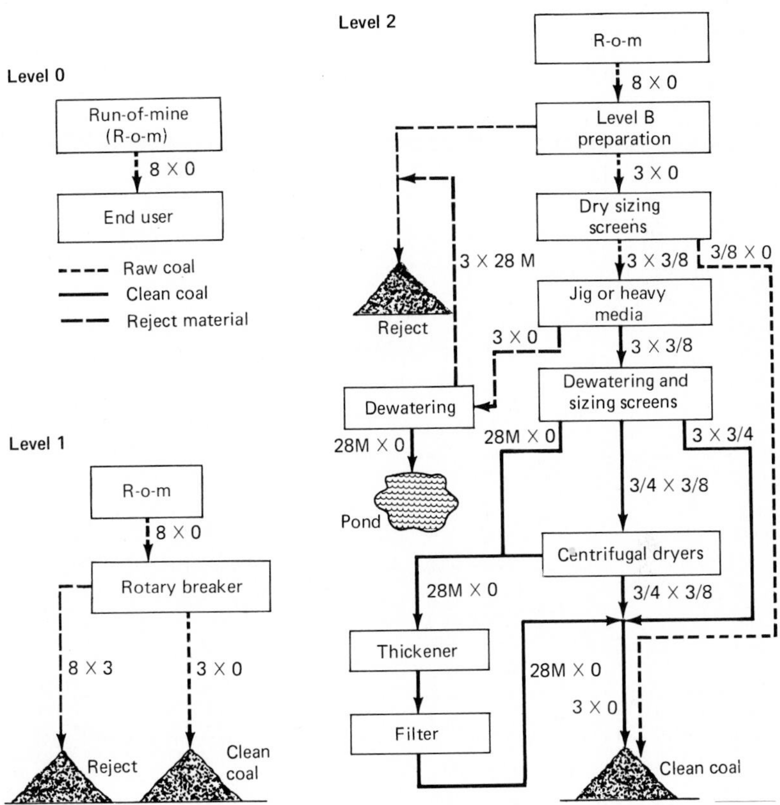

FIG. 2.5*a* Levels of coal-cleaning plants (levels 0, 1, 2). (Power *magazine)*

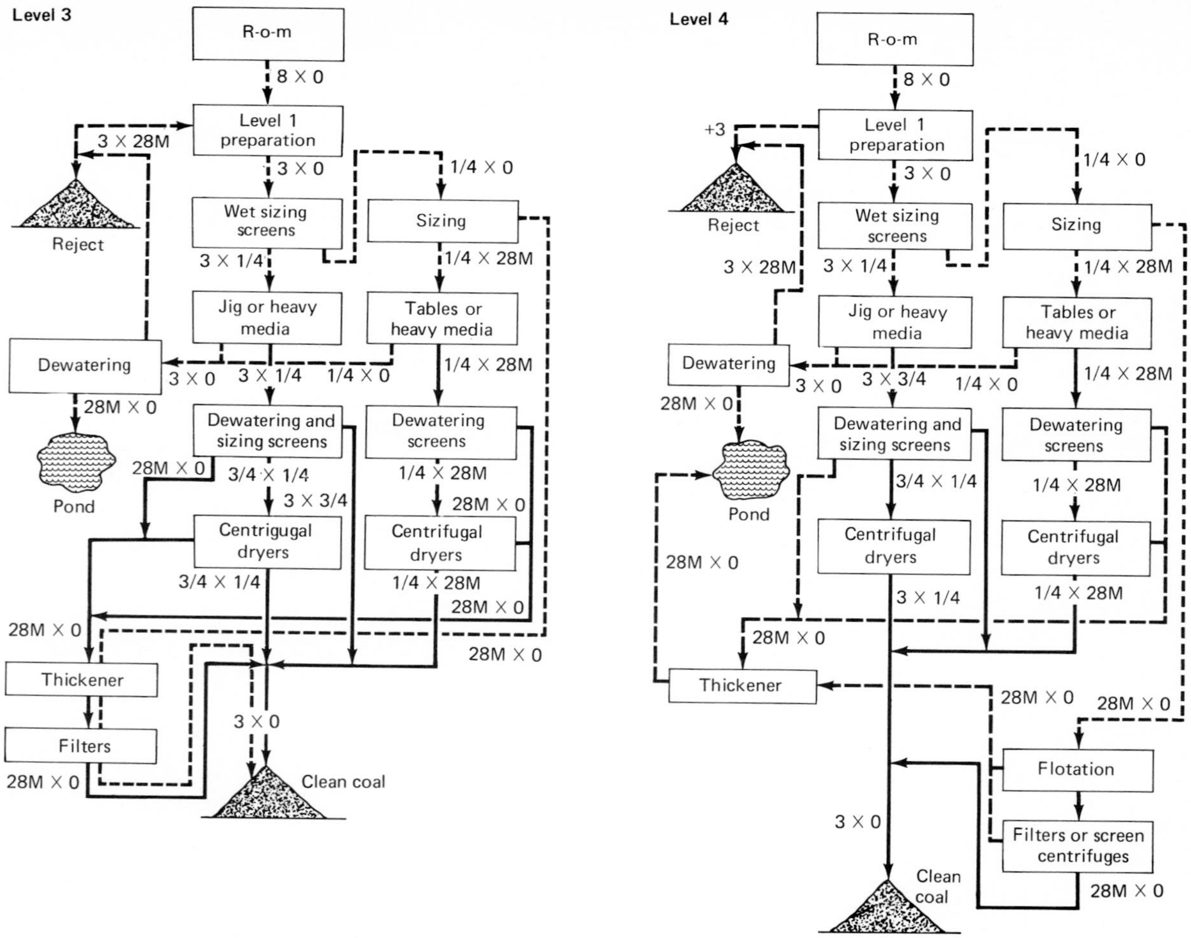

FIG. 2.5*b* Levels of coal-cleaning plants (levels 3, 4). (Power *magazine*)

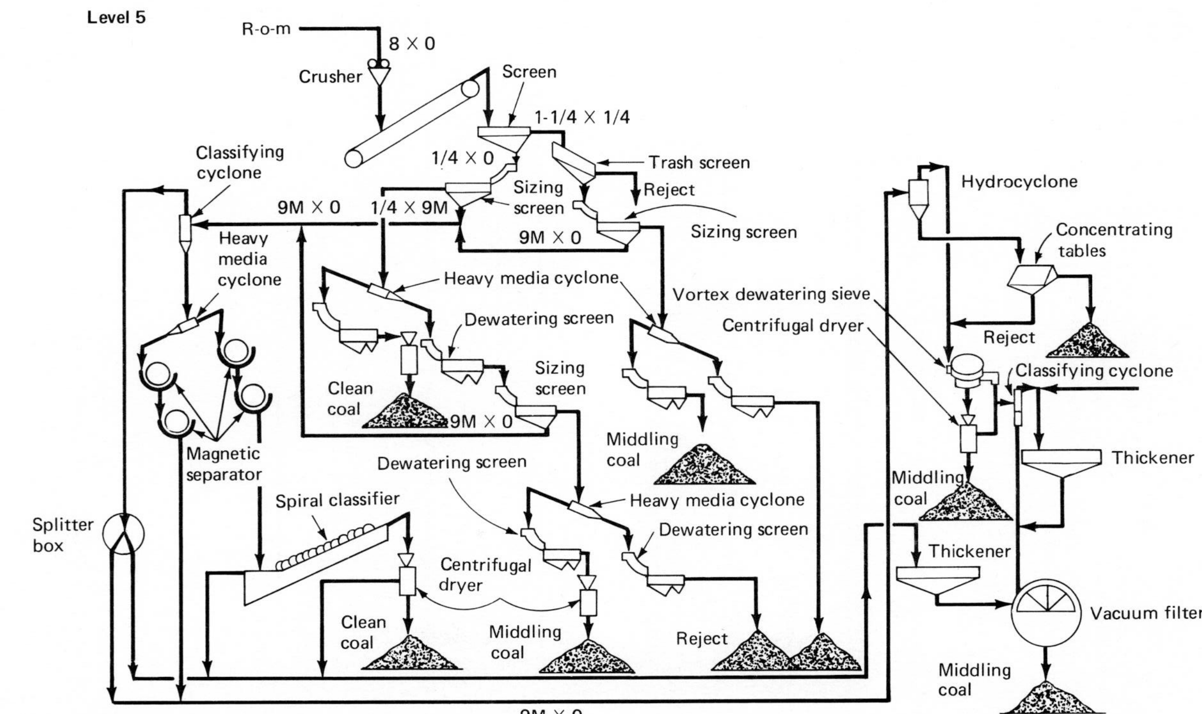

FIG. 2.5*c* Levels of coal-cleaning plants (level 5). (Power *magazine*)

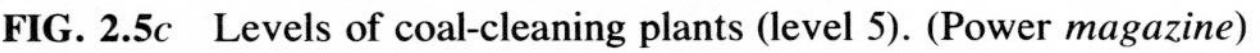

COST OF CLEANING

The cost of coal preparation is a function of the level of preparation, percentage of material discarded as refuse, coal washability, and plant size. Preparation costs are not strongly influenced by the coal's sulfur content. Costs for coal preparation can vary widely, and the costs for each plant must be considered individually with respect to location, terrain, coal properties, preparation specifications, and coal production rate.

Capital Cost

The total capital investment for a coal preparation plant is 2 to 2.5 times higher than the total direct costs. The total direct cost for a typical large preparation plant is estimated to range from $10 million to $30 million, depending on the level of preparation. Direct capital costs include raw-coal storage and handling equipment and preparation-plant equipment.

Indirect costs include those for engineering, construction, and field expenses. The range of capital costs for major line items for a 600-ton/h (544-Mg/h) plant is shown in Table 2.4. Thermal drying is not included in the table and would cost an additional $2 million per 100 tons/h (91 Mg/h) dried. Costs for larger or smaller plants may be estimated, since direct costs are proportional to the ratio of plant capacities raised to the 0.8 power.

TABLE 2.4 Capital Costs for a Preparation Plant*

Costs	Million dollars
Direct	
Processing equipment	3–8
Coal receiving and storage facilities	5–14
Other	2–5
Other costs	
Engineering, design, construction, interest	13–33
Total	23–60

*600-ton/h (544-Mg/h) plant, 1986 $.

Annual Cost

Annual costs include material costs, operating and maintenance expenses, and capital charges. Amortization costs and the costs of material mined and discarded as refuse have the greatest effect on annual costs. Amortization accounts for 10 to 40 percent of the annual cost, and coal waste materials may account for 15 to 70 percent. Total annual costs are often expressed in terms of dollars per ton (megagram) of cleaned coal, the value added to the coal product by preparation. For a 600-ton/h (544-Mg/h) plant, the costs of preparation, excluding discarded material, may range from $1.70 to $4.00 per ton ($1.54 to $3.63 per megagram) depending on the level of preparation (see Table 2.5).

For large preparation plants, the annual costs of preparation are reduced be-

TABLE 2.5 Annual Costs of Coal Preparation, 1986 $

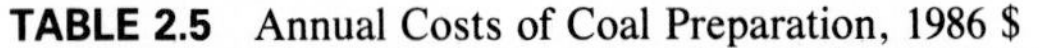

Cost elements	$/ton	$/Mg
Direct labor	0.49–0.70	0.44–0.64
Equipment and supplies	0.28–0.70	0.25–0.64
Utilities and chemicals	0.15–0.31	0.14–0.28
Refuse disposal	0.06–0.70	0.05–0.64
Overhead	0.35–0.61	0.32–0.55
Annualized capital costs, taxes, insurance, G&A	0.70–1.62	0.64–1.47
	2.03–4.64	1.84–4.22
Coal lost to refuse	1.00–10.5	0.91–9.52
Total	3.03–15.14	2.75–13.74

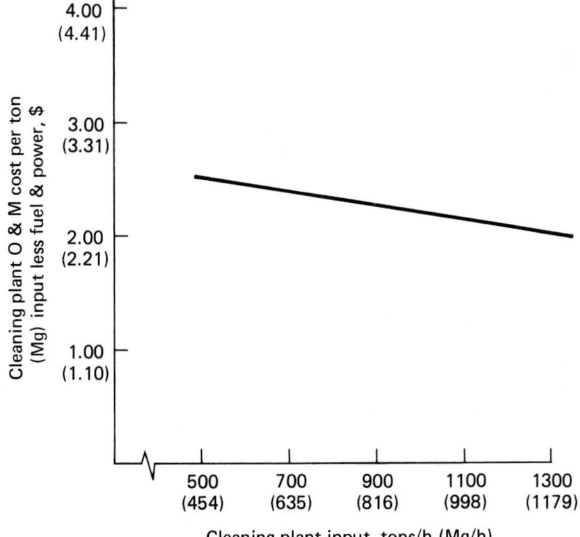

FIG. 2.6 Operation and maintenance (O&M) cost per unit of coal input (not including energy cost), 1986 $.

cause of the economies of scale (Fig. 2.6). Exclusive of coal reject costs, the annual value-added costs for a 1200-ton/h (1088-Mg/h) plant are estimated to range from $1.50 to $3.00 per ton ($1.36 to $2.72 per megagram). Including the cost of coal rejects [assuming mined material costs $25 per ton ($22.68 per megagram)], total annual costs can range from $3.00 ton to $15.00 per ton ($2.72 to $13.60 per megagram). In the worst case, the coal is of poor quality, and more than 40 percent of the mined material must be discarded as refuse. Although preparation adds substantially to the costs of these coals, they probably could not be sold unless they were cleaned.

Coal preparation is one of the most cost-effective SO_2 removal technologies that can be applied. One EPA study showed that preparation costs for a high-sulfur coal range from \$0.02 to \$0.10 per pound (\$0.009 to \$0.045 per kilogram) of SO_2 removed. For low-sulfur coals, which have considerably less pyritic sulfur content, the removal efficiency costs increase to about \$0.35 per pound (\$0.16 per kilogram) of SO_2 removed. To put the cost-effectiveness values into perspective, a 90 percent efficient FGD system costs from 2 to 5 times as much as preparation per pound (kilogram) of SO_2 removed for high-sulfur coals. For low-sulfur coals, *flue-gas desulfurization* (FGD) systems are more cost-effective for sulfur removal.

Preparation Benefits

The benefits derived from coal preparation fall into three general areas: coal handling, coal combustion, and pollution control. Transportation reductions occur if the coal is prepared at the mine site. Preparation reduces the coal's ash content and increases its energy value. This reduces the amount of coal needed to meet boiler energy requirements, thereby reducing coal handling at the powerplant.

Of the reasons advanced for the use of cleaned coal, perhaps the greatest single benefit (other than the control of emissions) is in the area of boiler fireside performance. Fireside problems result in forced and scheduled outages. They greatly affect the cost of boiler operation and maintenance, the capacity factor, and the availability of the generating facility. By modifying the coal characteristics that contribute to these problems (for example, slagging, fouling, and corrosion), coal preparation can favorably affect the economics of coal use. Typical cost benefits from improved boiler operations are estimated at \$2.60 per ton (\$2.36 per megagram) of cleaned coal (18 percent operation and maintenance, 79 percent availability, 3 percent boiler efficiency), but can range from \$0.50 per ton (\$0.45 per megagram) to as high as \$8 per ton (\$7.31 per megagram). Recently, several major utilities have demonstrated that increased boiler availability and reduced maintenance costs alone justified a switch to cleaned coal.

Coal preparation reduces ash disposal costs, since both bottom ash and flyash amounts decrease when cleaned coal is burned. Particulate control equipment may not function as efficiently with this cleaner fuel, however, especially electrostatic precipitators requiring gas conditioning to meet particulate standards. However, several studies have shown that in many cases the combination of coal preparation and FGD is less costly than FGD alone. Coal preparation, by removing 30 to 40 percent of potential SO_2 emissions, may allow partial scrubbing of flue gases to meet emissions regulations, rather than full scrubbing.

The box summarizes the costs and benefits of coal preparation and provides typical values for some benefits. In many cases, especially for high-sulfur, high-ash coals, preparation pays for itself through the benefits derived.

Increased Cleaning Constraints

Any legislative or regulatory strategy that called for increasing the amount of coal constraints would face a number of barriers. These constraints are institutional, technological, and regulatory in nature.

A major institutional constraint is the reluctance of state *public utility commissions* (PUCs) to allow utilities to pay the higher fuel costs associated with cleaned coal. The PUCs and some utilities continue to believe that the cheapest coal produces the cheapest electricity. Since benefits are site-specific and not

COSTS AND BENEFITS FROM COAL PREPARATION

- F.O.B. mine coal cost
  - Coal-cleaning plant operating and maintenance cost
  - Coal-cleaning plant capital amortization
  - Heat energy lost during cleaning process
  - Crushing and screening cost
  - Payment to United Mine Workers trust funds
- Transportation cost
- Ash disposal cost
- Pulverization cost
- Utility plant maintenance cost
- Emissions control costs
- Boiler availability
- Boiler efficiency

easily predicted (particularly future boiler availability and maintenance costs), the savings at the busbar from using a clean fuel are not easily quantified. Several major utilities, however, now believe a cleaned coal produces the cheapest electricity.

Another financial constraint is that coal preparation plants are not considered to be pollution control facilities like FGD retrofits. As a result, there are no tax incentives to either utilities or coal companies for lowering SO_2 emissions by cleaning coal.

The major technological constraint for cleaning coal is the limitation in removing organic sulfur. Thus, cleaning is less effective on low-ash, high-organic-sulfur coals. They are generally found in the southern Appalachian coal region. Although greater sulfur removal is achieved by grinding to fine sizes, cleaning and dewatering problems make this type of preparation more costly and less attractive for commercial plants. For certain coals, cleaning may actually worsen slagging characteristics or cause coal-handling problems.

INDUSTRY TRENDS

Developing circumstances are making coal cleaning more economical. These circumstances include

- Higher coal prices and transportation costs
- Diminishing coal quality because of less selective mining techniques
- Reduction of easy-to-mine coal reserves
- The need to increase availability and capacity factors at existing boilers

- Stringent air quality standards
- Lower costs for improving fuel quality versus investing in extra pollution control equipment, and a projected 20 percent increase of consumption by eastern and midwestern utilities in the next 10 years

Increase Likely

Discussions with coal and utility industry personnel support a relatively optimistic outlook for coal preparation. They believe an increase in cleaning is likely because

- The high ash content in ROM coal is pushing utilities toward more stringent boiler specifications.
- Current research on the benefits of cleaning is expected to indicate considerable savings to utilities.
- Tight markets are forcing coal companies to be competitive by offering better fuel products.

To compete in today's coal market, most large coal companies are cleaning high- and medium-sulfur coals and improving mining techniques. To offer better products to their customers, several coal companies are investigating modifications at all their preparation plants. Blending good- and poorer-quality coals at the preparation plant to give an average mix that meets product specifications is becoming more common.

Overall, utilities are not against using cleaned coal. They are attempting to obtain more information (as are state regulatory agencies) about its advantages. EPRI is critically testing the premise that the lowest-cost coal produces the lowest-cost electricity. Some arguments against increased cleaning cite the availability of low-sulfur, low-ash western coals; stable oil prices; decreased metallurgical coal demand; and slow orders for new coal-fired units.

Research Activities

Areas that have seen recent research activity are fine-coal cleaning, preparation circuits, automatic controls, and productivity improvements. The R&D area of greatest emphasis is improvement of washing and dewatering equipment for fine coal. This is a direct result of the high cost of coal that penalizes coal losses to refuse and the increased fines production from continuous-mining equipment. Studies are being performed by DOE, EPRI, and several states to further develop advanced coal-cleaning processes. The objective for most advanced cleaning technologies is to remove ash and sulfur from very fine size coals.

The long-term outlook for increased coal preparation may depend on the ultimate ability to produce a low-ash (less than 3 percent) and low-sulfur (less than 0.5 percent) coal suitable for use with coal-water mixtures in converted oil-fired boilers and synthetic-fuel facilities. The problem? Large-scale research is needed to produce this type of fuel, and physical cleaning alone cannot reduce the sulfur to 0.5 percent for most coals.

SUGGESTIONS FOR FURTHER READING

1. M. K. Buder, et al., *Impact of Coal Cleaning on the Cost of New Coal-Fired Power Generation,* EPRI CS-1622, Electric Power Research Institute, Palo Alto, Calif., Mar. 1981.
2. J. Buroff, et al., *Technology Assessment Report for Industrial Boiler Applications: Coal Cleaning and Low Sulfur Coal,* EPA-600/7-79-178c (NTIS PB 80-174055), Environmental Protection Agency, Research Triangle Park, N.C., May 1976.
3. J. A. Cavallaro, M. T. Johnston, and A. W. Deurbrouck, *Sulfur Reduction Potential of U.S. Coals: A Revised Report of Investigations,* EPA-600/2-76-091 (NTIS PB 252965), Environmental Protection Agency, Industrial Environmental Research Laboratory, Research Triangle Park, N.C., Apr. 1976.
4. *Evaluation of Conventional and Advanced Coal Cleaning Techniques: Final Technology Document,* EPA-600/4-84-239, Environmental Protection Agency, Industrial Environmental Research Laboratory, Research Triangle Park, N.C., Dec. 1983.
5. H. J. Gluskoter, et al., *Trace Elements in Coal: Occurrence and Distribution,* EPA-600/7-77-064 (NTIS PB 270922), Environmental Protection Agency, Industrial Environmental Research Laboratory, Research Triangle Park, N.C., June 1977.
6. H. R. Hazard, *Influence of Coal Mineral Matter on Slagging of Utility Boilers,* EPRI CS-1418, Electric Power Research Institute, Palo Alto, Calif., June 1980.
7. E. C. Holt, Jr., *An Engineering/Economic Analysis of Coal Preparation Plant Operation and Cost,* EPA-600/7-78-124 (NTIS PB 285251), Environmental Protection Agency, Office of Energy, Minerals and Industry, Washington, July 1978.
8. *Impact of Cleaned Coal on Power Plant Performance and Reliability,* EPRI CS-1400, Research Project 1030-6, Electric Power Research Institute, Palo Alto, Calif., Apr. 1980.
9. G. R. Isaacs, R. Ressl, and P. Spaite, *Cost Benefits Associated with the Use of Physically Cleaned Coal,* EPA-600/7-80-105 (NTIS PB 81-113953), Environmental Protection Agency, Industrial Environmental Research Laboratory, Research Triangle Park, N.C., May 1980.
10. J. D. Kilgroe and R. C. Lagemann, "The Technology and Costs of Physical Coal Cleaning for Controlling Sulfur Dioxide Emissions," paper prepared for Third Seminar on Desulfurization of Fuels and Combustion Gases, Salzburg, Austria, May 1981.
11. J. D. Kilgroe, "Combined Coal Cleaning and FGD," *Proceedings: Symposium on Flue Gas Desulfurization,* Las Vegas, March 1979. In hardcover: Vol. 1 (Research Triangle Institute), EPA-600/7-79-167a (NTIS PB 80-133168), Environmental Protection Agency, Industrial Environmental Research Laboratory, Research Triangle Park, N.C., July 1979.
12. J. W. Leonard (ed.), *Coal Preparation,* 4th ed., American Institute of Mining, Metallurgical, and Petroleum Engineers, New York, 1979.
13. D. C. Nunenkamp, *Coal Preparation Environmental Engineering Manual,* EPA-600/2-76-138 (NTIS PB 262-716), Environmental Protection Agency, Industrial Environmental Research Laboratory, Research Triangle Park, N.C., May 1976.
14. B. Onursal and L. C. McCandless, "Assessment of Coal Cleaning Technology: An Evaluation of Froth Flotation Circuits for Removal of Sulfur from Coal," draft report, EPA Contract 68-02-2199, Environmental Protection Agency, Industrial Environmental Research Laboratory, Research Triangle Park, N.C., July 1980.
15. B. Onursal, et al., "Assessment of Coal Cleaning Technology: An Evaluation of Dense-Medium Cyclone Circuits for Removal of Sulfur from Final Coal," draft report, EPA Contract 68-02-2199, Environmental Protection Agency, Industrial Environmental Research Laboratory, Research Triangle Park, N.C., Dec. 1980.

16. P. J. Phillips and R. M. Cole, "Economic Penalties Attributable to Ash Content of Steam Coals," presented at the Coal Utilization Symposium, AIME Annual Meeting, New Orleans, Feb. 1979.

17. P. J. Phillips, *Coal Preparation for Combustion and Conversion,* EPRI AF-791, Electric Power Research Institute, Palo Alto, Calif., May 1978.

18. E. Stach, et al., *Stach's Textbook of Coal Petrology,* 2d ed., Gebruder Borntraeger, Berlin-Stuttgart, West Germany, 1975.

19. T. W. Tarkington, F. M. Kennedy, and J. G. Patterson, *Evaluation of Physical/Chemical Coal Cleaning and Flue Gas Desulfurization (TVA),* EPA-600/7-79-250 (NTIS PB 80-147622), Environmental Protection Agency, Industrial Environmental Research Laboratory, Research Triangle Park, N.C., Nov. 1979.

20. T. D. Wheelock and R. Markuszewski, "Physical and Chemical Coal Cleaning," presented at the conference on the Chemistry and Physics of Coal Utilization, West Virginia University, Morgantown, June 1980.

CHAPTER 4.3
CONTROLLING POLLUTION DURING COMBUSTION

Jason Makansi and Robert G. Schwieger
Power *Magazine*
New York, NY

INTRODUCTION

Over the years, many techniques have been advanced to control a variety of pollutants during combustion of fuel in the boiler. Of those attempted, two basic methods have generally been accepted and are in commercial service today. Fluidized-bed boilers are becoming the choice for removing sulfur dioxide from high-sulfur coal, while low-NO$_x$ burners are being selected to reduce the emissions of oxides of nitrogen, which may be derived from nitrogen contained in the combustion air or in nitrogen chemically combined in the fuel.

Both techniques are attractive because of their adaptability to a range of fuel types, boiler capacities, and operating conditions without seriously compromising efficiency or performance. Also, the economics of both is competitive with that of other types of pollution control at other stages in the steam generating process.

FLUIDIZED-BED BOILERS

Fluidized-bed boilers produce steam from solid fossil and waste fuels by using a tachnique called *fluidized-bed combustion* (FBC). These abbreviations are used in this section: *atmospheric fluidized-bed* (AFB) combustion and *circulating fluidized-bed* (CFB) combustion.

FBC Fundamentals

In a typical fluidized-bed combustor, solid, liquid, or gaseous fuel or fuels together with inert material—for example, sand, silicia, alumina, or ash—and/or a sorbent such as limestone are kept suspended through the action of the primary air distributed below the combustor floor (Figs. 3.1 and 3.2).

To appreciate what makes this concept so attractive compared to other firing techniques, the function of two combustion parameters—turbulence and temperature—must be understood.

Turbulence. Turbulence is promoted by fluidization making the entire mass of solids behave much as a liquid. Improved mixing generates heat at a substantially lower and more uniformly distributed temperature—typically 1500 to 1600°F (816° to 871°C)—than a stoker-fired unit or a pulverized-coal burner. Thus, a properly sized combustion chamber will release heat at a rate equivalent to that

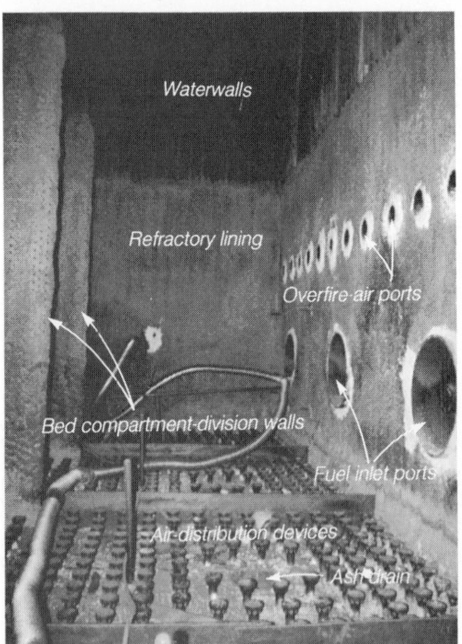

FIG. 3.1 Combustor internals include fluidizing-air nozzles, fuel-feed ports, secondary-air ports, bed divider walls, and waterwalls lined at the bottom with refractory.

FIG. 3.2 Wind box in FBC unit is located directly beneath the bed.

of a conventional boiler, but at a lower temperature and theoretically without any loss in efficiency.

The fluidizing mechanism, or added turbulence, offers several advantages: less volatilization of alkali compounds, reduced chance for hot spots on boiler and shell surfaces, less sensitivity to the quantity and nature of ash in the fuel, and a smaller furnace volume. Extra volume is not required to allow the ash to cool before flowing through the convective passes. These advantages are the essential ingredients of the improved fuel flexibility often attributed to AFB units.

Temperature. Advantages offered by the lower temperature are focused on emissions control. First, operating temperatures are well below the formation point for thermally induced nitrogen oxides. Staged combustion can be applied to minimize fuel-bound nitrogen oxide formation as well. Second, by including a suitable sorbent as part of the bed material, sulfur dioxide and other acid gases are absorbed, eliminating the need for downstream control. In fact, the operating temperature range for FBC units is such because this is where the sulfur-absorption reactions are thermally balanced. Thus, performance trade-offs tend to be involved in maintaining high sulfur dioxide removal efficiency together with high combustion efficiency.

Another major advantage of the lower combustion temperature, especially when coal is the fuel, is that it avoids the appreciable slagging, fouling, and other fireside problems associated with pulverized-coal- and stoker-fired units.

As in any boiler, a suitable heat sink must be provided in the combustion area to control temperature. This takes the form of the traditional waterwall enclosures, but also, in several designs, boiler tubes that are located within the bed

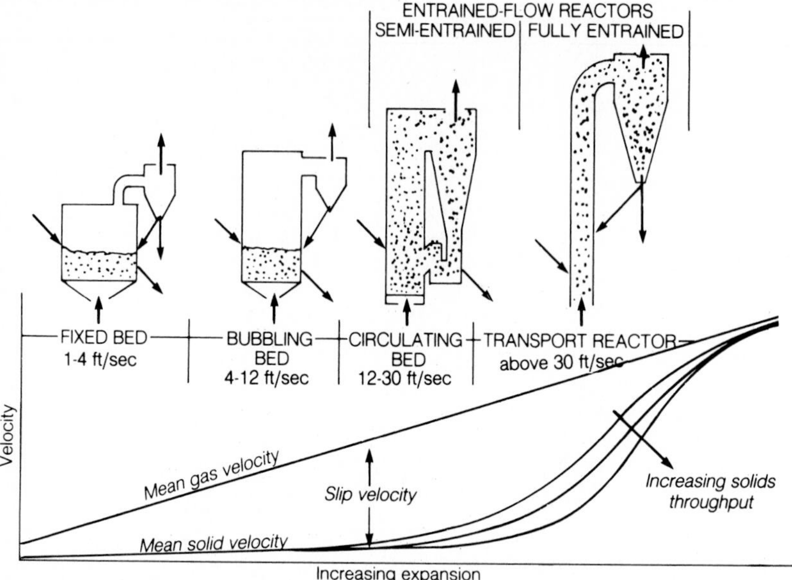

FIG. 3.3 Turndown is accomplished in a bubbling bed by slumping sections of the bed so that heat transfer is effectively cut off to tubes in that region. Note: ft/s × 0.3048 = m/s.

itself, oriented either vertically or horizontally. Both firetube and watertube designs are available; the latter are more popular for larger sizes.

Steaming Rate. Steaming rate is controlled by manipulating the primary bed parameters: height, temperature, inventory, and superficial gas velocity. In practice, velocity changes account for most of the control. For example, by reducing the fluidization velocity, less heat-transfer surface is exposed to the bed (Fig. 3.3). When air is shut off completely, the heat transfer is effectively zero, offering quick turndown capability. Turndown and steam rate control can also be accomplished by reducing the amount of bed material or by regulating fuel feed, as with other firing systems.

Once the hot gases leave the combustor area, they see convective heat-transfer surface, air heaters, and particulate cleanup devices that are similar or even identical to components used in boilers for years.

Basic Operation

Major categories of AFB boilers include bubbling-bed units, CFB units, and the two-stage bubbling bed, with designs that, technologically, fit between categories. This last point is emphasized because several suppliers have modified their bubbling-bed designs to compete with the popularity of CFBs, at least for the larger sizes. Essentially, this is done by increasing the recycle rate.

Air Velocity. The fundamental distinguishing feature of all FBC units is the velocity of air through the bed (Figs. 3.4 and 3.5). Bubbling-bed units have lower

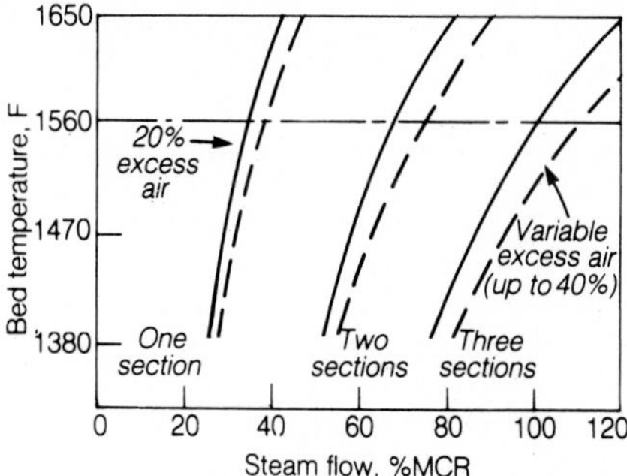

FIG. 3.4 The circulating bed can be thought of as an extension of a bubbling bed by dramatically increasing the recycle rate. Note: (°F − 32)/1.8 = °C. Also, MCR = maximum continuous rating.

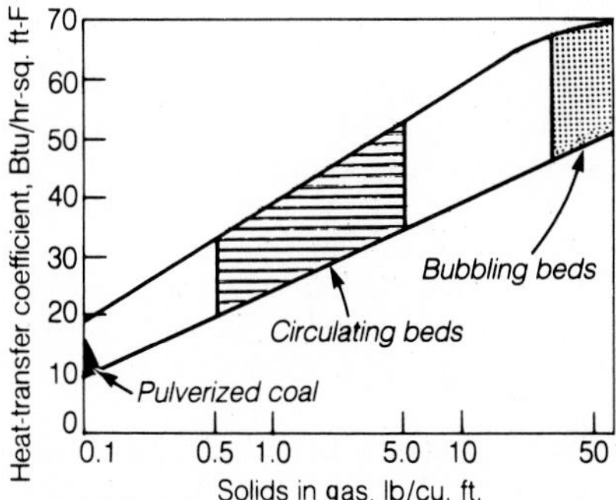

FIG. 3.5 The amount of solids in the gas directly affects the transfer of the combustion heat to the boiler tubes. Note: $lb/ft^3 \times 16 = kg/m^3$ and $Btu/(h \cdot ft^2 \cdot °F) \times 5.68 = W/(m^2 \cdot °C)$.

fluidization velocities, around 5 to 12 ft/s (1.52 to 3.66 m/s). The goal is to prevent solids from carrying over, or elutriating, from the bed into the convective passes. CFBs work with velocities as high as 30 ft/s (9.14 m/s), and they actually promote solids elutriation.

Most bubbling-bed units, however, require at least partial recirculation or reinjection of up to 90 percent of the solids escaping the bed to obtain satisfactory performance without substantially increasing the size of the combustor. Thus, a CFB merely maintains a continuous, high-volume recycle rate by removing solids in a mechanical separator and allowing the hot combustion gases to pass through the convective section.

Some of the advantages ascribed to CFBs over bubbling beds include the longer residence time to increase combustion efficiency and improve absorption of sulfur dioxide and other acid gases, capability to use less prepared fuel and absorbent, and elimination of many problems associated with underbed and/or overbed feeding of the fuel. One common disadvantage is the greater fan power needed to maintain the higher velocity through the bed.

Bed Circulation. Some bubbling-bed units have the fuel and air distribution configured so that a high degree of circulation takes place within the bed. One design accomplishes internal circulation with a tube in the center of a cylindrical bed that divides the lower combustor into two parts. Primary air and secondary air are divided between the two parts. The unequal force between the tube and the annular area causes the bed to circulate. Among the potential advantages are a more uniform bed temperature, simple feed into the central tube, and improved ability to accommodate a wide variety of low-grade fuels.

The concept behind two-stage bubbling-bed units is to separate the combustion and absorption processes so each can be optimized without the design compromises inherent in a single-bed approach. For example, a higher combustion temperature can be attained in a lower bed since absorption is no longer the limiting factor.

CFB Design

Recent activity in FBC design has focused on developing the CFB boiler. Compared to pulverized-coal- and stoker-fired units, the process is less advanced so that important variations still exist among vendor offerings. Areas where differences are significant include the flue-gas velocity through the unit, thermal-loading philosophy, hot-solids separation and reinjection, solids circulation rate, power requirements, fuel and limestone preparation requirements, amount of refractory in the combustor, presence or absence of an *external heat exchanger* (EHE), structural support method, and amount of solids inventory.

Major Components. CFBs include these major components: a refractory-lined combustor bottom section with fluidizing nozzles on the floor above the wind box, an upper combustor section usually with waterwalls, a transition piece including a hot-solids separator and reentry downcomer that may be all or partially lined with refractory, and a convective boiler section.

Primary combustion air is supplied as the fluidizing medium, while secondary air is supplied several feet (meters) above the combustion floor to effect staged combustion (Fig. 3.6). Usually, the split between primary and secondary air is 50-50.

Parameters such as the circulation rate and velocity affect the erosion potential of a particular design, the heat-transfer coefficient, and the required amount of boiler surface. To illustrate, a lower solids inventory and circulation rate could allow the unit to operate at a higher velocity with less threat of erosion.

The most fundamental structural parameter with CFBs is whether the boiler is bottom- or top-supported. The top-hung type is generally more expensive, but it avoids the expansion joint between the combustor and the hot collector.

External Heat Exchanger. Individual designs are distinguished primarily by the presence and duty of the EHE and the hot-solids separation and reinjection devices. The EHE in Fig. 3.7 is essentially a refractory-lined box containing an air distribution grid and an immersed tube bundle designed to cool material from the separator (typically a cyclone) before it is returned to the combustor. Fluidizing velocity is low, less than 3 ft/s (0.91 m/s), while solids density and heat-transfer

FIG. 3.6 Secondary-air ports achieve a staged combustion effect to minimize nitrogen oxides emissions.

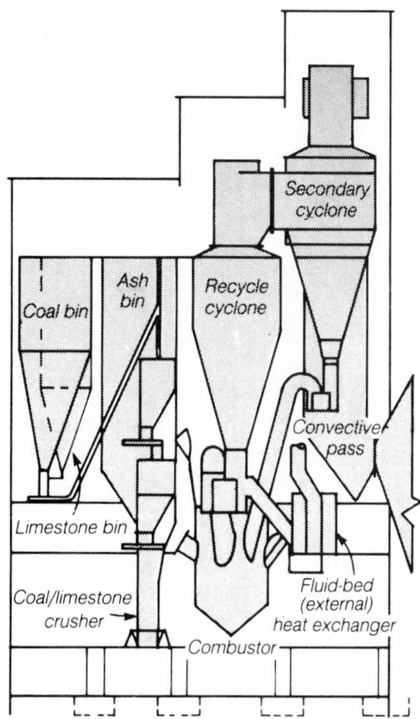

FIG. 3.7 An external heat exchanger provides flexibility in absorbing heat, an important feature when many different types of fuels are fired.

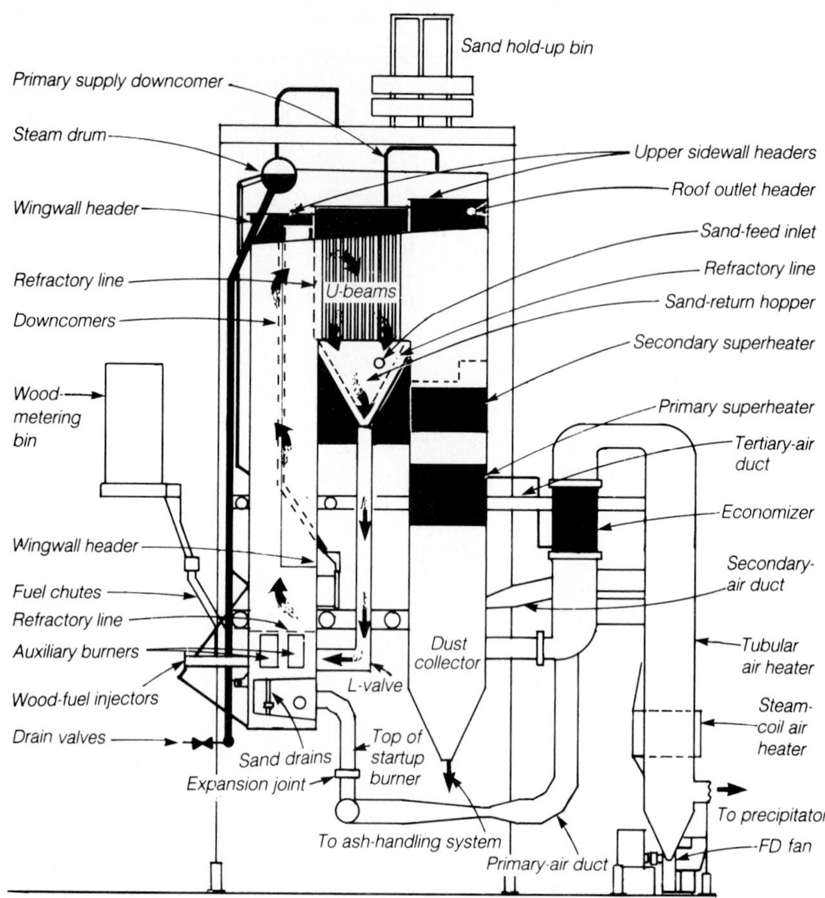

FIG. 3.8 The solids-separation device used with this FBC design is a departure from refractory-lined cyclones. It relies instead on particle impaction principles.

coefficients are high. Little or no combustion takes place in the box, according to theory, because the recycled solids contain little carbon. Erosion is thought to be negligible because of the very low velocities.

One advantage of the EHE is that it can easily compensate for variations in the heat absorption rate caused by changes in fuel properties and load conditions. Heat transfer in the EHE increases when absorption in the combustor decreases, and vice versa. Thus, the solids inventory in the combustor can be varied along with the heat-transfer ratio between the combustor and the EHE, without moving the primary-secondary air ratio away from the optimum point.

A disadvantage of the EHE is that it complicates the design and operation of the unit. For reheat steam cycles, however, the EHE is a possible location for reheat tube surface.

Solids Separators. For solids separation, most units feature one or more refractory-lined cyclones to keep the solids circulating. At least one supplier has integrated the cyclone with the combustor and waterwalls. That is, steam- or water-cooled walls are used as the sides and roof of the cyclone, adding structural and thermal stability, reducing high-temperature ductwork and expansion joints, and improving thermal efficiency.

A departure from the use of cyclones is the U-beam particle separator. It is comprised of U-shaped bars installed as a staggered array in a horizontal section, where the gas makes a 90° turn exiting the combustor (Fig. 3.8). As the gas stream decelerates by about 50 percent compared to the combustor shaft velocity, particles impinge on the bars and fall into a hopper. The reason behind this design is to avoid the use of thick refractory in a high-abrasion environment, such as a cyclone, where velocities are often quadrupled. This section also adds a second or two of residence time, which helps in converting carbon monoxide to carbon dioxide to reduce emissions and improve combustion efficiency.

In addition to the U-beams, this design calls for a multicyclone separator following the economizer. Solids captured here can be either recycled or purged. The rationale given for the backup collector is that it can eliminate potential material inventory problems that might result when the source of fuel is switched, such as from a high-ash to a low-ash coal.

Solids-Reinjection Device. Called an *L-valve,* a *J-valve,* a *loop seal,* a *Fluoseal,* or a *Sealpot,* a solids-reinjection device essentially provides a simple nonmechanical hydraulic barometric seal against the combustor shell. Most are large-diameter steel pipes with refractory lining. Inside, there are no internals, but there are nozzles for admitting compressed air that keeps the material fluidlike. Further design details are often proprietary.

In most CFB designs, especially those without an EHE, the convective pass is similar to that for other boilers. The superheater may be located in the dense-phase flow at the top of the furnace (Fig. 3.9) or in the convective section downstream of the solids separator. Low velocities and favorable flow patterns can minimize chances for erosion in the upper furnace. Downstream in the convective pass, the solids loading is much reduced so that erosion is of little concern at typical design velocities.

Sootblowers originally were not thought to be required with most CFBs, but are supplied as a precautionary measure; sometimes the ports that feed the sootblowers are inserted in case they are added later. In any event, sootblowing requirements are drastically reduced over stoker- and pulverized-coal-fired units.

Fuel Feeders. The combustor temperature is regulated by controlling the solids returning from the separator, either through the EHE if it exists or through the solids return leg. Control during fuel or load changes can also be accomplished by changing the ratio of primary to secondary airflow. As Fig. 3.10 shows, fuel is fed based on total boiler steam flow and airflow signals, while the limestone feed generally follows the fuel-feed signal if the two solids feed streams are independent. In some cases, they are fed together. During start-ups and shutdowns, the temperature differential between the inside and outside of the refractory must be carefully monitored to avoid cracking and damage.

A CFB boiler may be equipped with gas recirculation, especially when it is designed for fuels with substantially different heating values. This helps achieve a proper energy balance between the furnace and the convective passes.

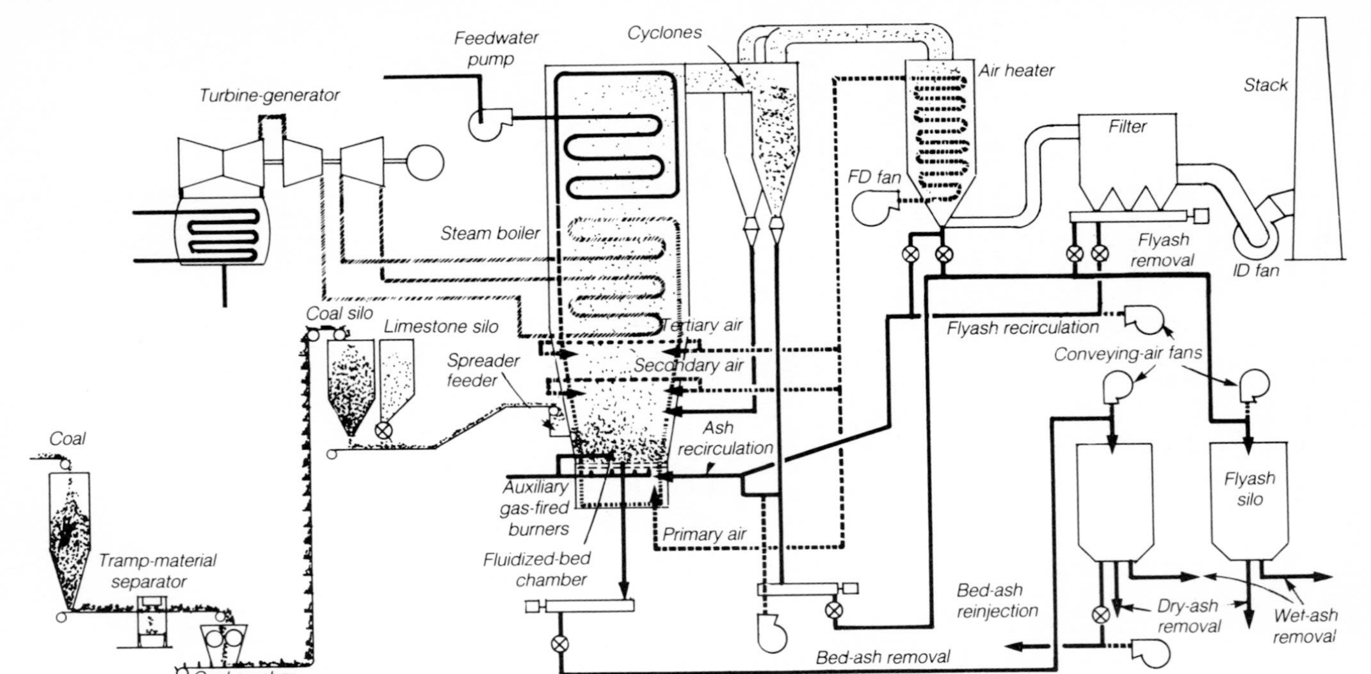

FIG. 3.9 CFB unit with superheater in the dense-phase flow and heat-transfer surface upstream of the cyclones.

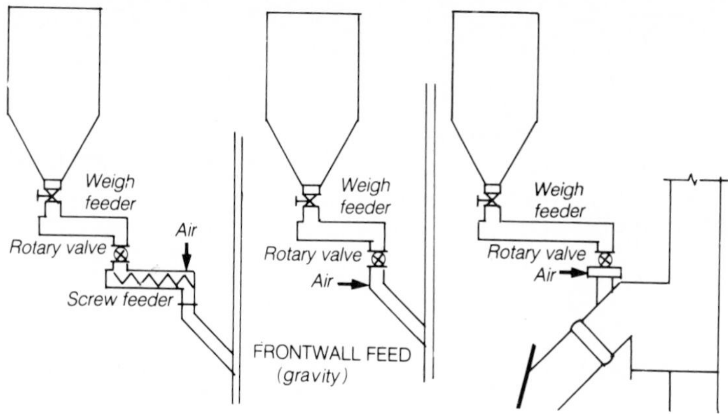

FIG. 3.10 Several types of fuel-feed systems are used to inject coal and other fuels into fluidized-bed combustors.

How ash is removed from the bed to control the solids inventory and to purge the system can be another distinguishing feature of CFBs. Since ash discharge is a key factor in controlling the combustor temperature, most approaches are proprietary. Points at which ash may be removed, separately or in combination, include the bottom of the combustor, the economizer hopper, the EHE, or the recycle leg to the combustor. Because the ash is initially at the combustor temperature, it can be used to heat feedwater or preheat combustion air, increasing the overall thermal efficiency of the boiler plant. Once the ash is cooled, it is removed by conventional means.

Bubbling-Bed Design

A design controversy associated with scaling up bubbling-bed boilers is whether to feed the fuel over the bed or under it. The choice has important implications for combustion efficiency (Fig. 3.11), boiler control, and emissions control, especially sulfur dioxide absorption.

Fuel Feeding. One utility uses an overbed feed system, but with this variation: Overfire air ports above the coal feeder inlets to the furnace disburse the incoming coal from the discharge area. A drawback with overbed feeders, one similar to spreader stokers, is that they are limited in their throwing distance and often require a long, narrow furnace.

Another utility employs an underbed feed system specifically designed to avoid the extensive plugging and erosion problems found with underbed feed in its pilot plant. First, coal is crushed to less than ¼ in × 0 (6 mm × 0) and dried with flue gas to less than 6 percent moisture. Then the fuel is passed through a fluidized bottle splitter with a central inlet and fuel lines arranged concentrically around the inlet. Blowers pressurize the bottles enough to transport the coal-sorbent mixture to the furnace. Each bottle is analogous to a burner and can be used to control load in the same way as cutting a burner in and out.

Ash Removal. Ash removal from bubbling beds can be another distinguishing feature of the various designs. Generally, it is accomplished through drain ports

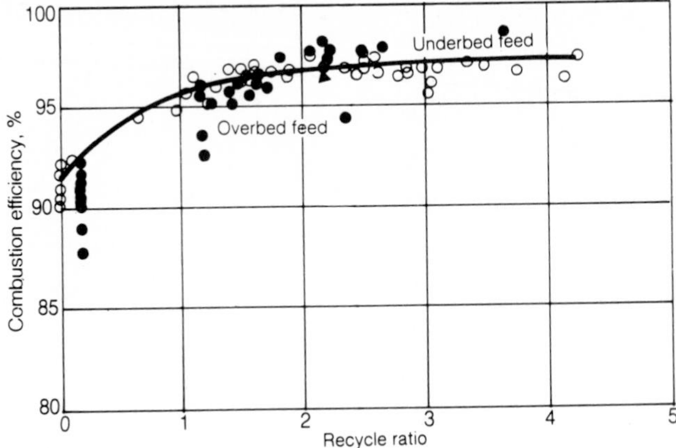

FIG. 3.11 Underbed feed is superior for large bubbling-bed units, at least in terms of combustion efficiency.

in each bed compartment. Location—along the sides, in the corners, or in the center—is a vendor preference and reflects each one's attempt to deal with the problem of tramp and agglomerated material that can block the fluidizing-air ports.

LOW-NO_x BURNERS

Pressed by the tight regulatory climate that exists today, vendors have refined burner designs to short-circuit NO_x emissions without adversely affecting performance or thermal efficiency or increasing furnace corrosion rates. Indeed, these improved burner designs may well replace furnace-oriented combustion modifications for both new and retrofit utility and industrial powerplants. With gas and oil fuels, low-NO_x burners appear capable of reducing oxides of nitrogen concentrations by 40 to 65 percent compared with conventional burner designs. Similar reductions are being demonstrated on full-size coal-fired units.

Basic Burner Operation

Low-NO_x burners control mixing of fuel and air in a pattern that keeps the flame temperature down and dissipates heat quickly. Whereas conventional burners mix secondary air with the primary fuel-air stream as soon as they are injected into the furnace, creating a high-intensity combustion process, low-NO_x burners establish distinctly separate primary and secondary combustion zones, thereby staging the flame at the burner.

Basically, the same techniques for NO_x reduction apply to the burner as were used in the furnace. Thus, almost all burner modifications rely on some form of low-excess-air staged combustion and/or internal flue-gas recirculation to reduce NO_x emissions. Overfire air in combination with low-NO_x burners offers even further reduction.

Although the type of fuel fired generally dictates hardware selection, several burner parameters affect NO_x emissions:

- Geometry of the burner and fuel-injection system
- Method of fuel injection
- Velocity and degree of swirl of the combustion air
- Division of total air among primary, secondary, and tertiary streams

Promising Applications

Of the many low-NO_x burner designs on the commercial market, two distinct types have prevailed—the dual-register and the distributed mixing burners. Variations of each have appeared, depending on vendor ingenuity.

Dual-Register Burners. With regard to coal fuels, dual-register pulverized-coal burners are available which produce a limited-turbulence, controlled-diffusion flame. One design operates as follows: A venturi mixing device, located in the coal nozzle, provides a homogeneous mixture of coal and primary air to the burner (Fig. 3.12). Secondary air is introduced through two concentric air zones surrounding the coal nozzle. Airflow to each zone is controlled independently by the inner and outer registers. Adjustable spinner vanes in the inner air zone provide the proper amount of swirl to control coal-air mixing during combustion. This delayed-mixing feature allows combustion to begin at the burner throat and end within a defined but variable zone in the furnace.

Benefits of the dual-register burner over staged combustion include these:

- An oxidizing environment is maintained in the furnace, minimizing slagging and reducing the potential for furnace-wall corrosion when high-sulfur coal is burned.

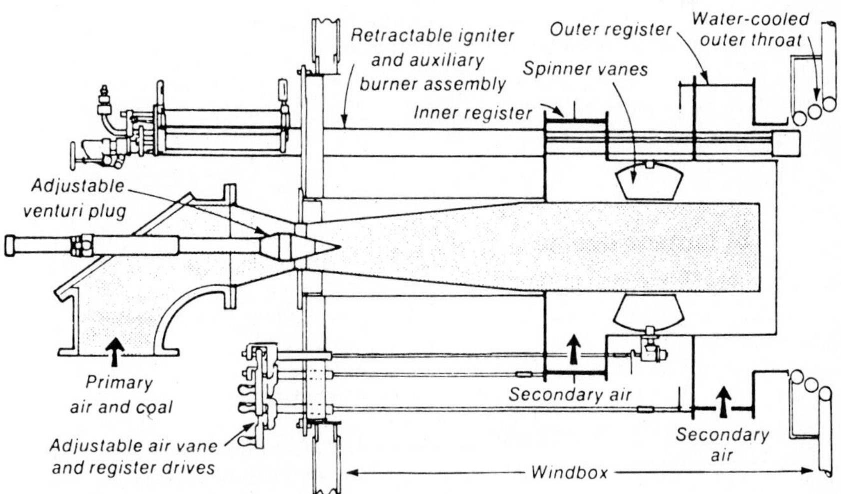

FIG. 3.12 Low-NO_x burners delay mixing and allow combustion to begin at the burner throat.

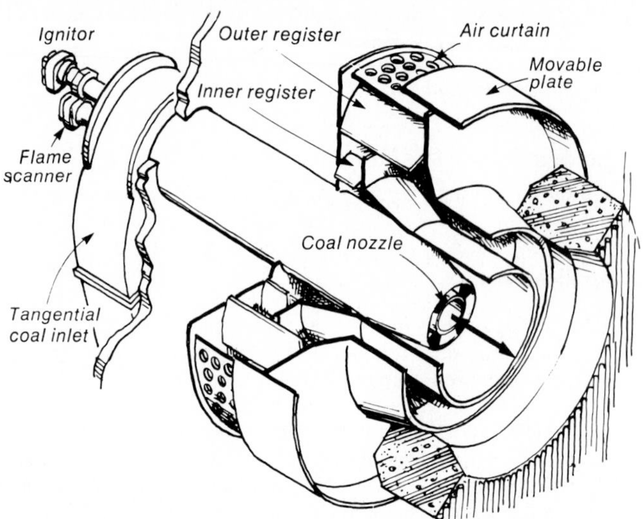

FIG. 3.13 Dual-register burner produces a limited-turbulence, controlled-diffusion flame.

- Carbon burnout is more complete.
- Less air is admitted to the furnace.

Another type of dual-register burner has been introduced (Fig. 3.13), whose distinctive aspects include:

- A flame-splitting distributor tip, which causes a flower-petal flame pattern to minimize mixing between the coal and primary air.
- A perforated air plate that measures airflow, improves secondary-air distribution, and balances air on a burner-to-burner basis.
- A tapered-annulus coal nozzle with a movable axial inner sleeve. For a constant flow of primary air, the nozzle velocity can be varied to minimize shear-induced turbulence and thus NO_x emissions.

A variation of the staged-combustion principle has been tested by another burner manufacturer. What it does is recirculate flue gas above and below the coal nozzle to maintain a reducing atmosphere at the nozzle during primary combustion; this action minimizes infusion of secondary air into the coal stream. In this design, the coal nozzle has a divergent tip, which acts as a flame holder for maintaining ignition close to the nozzle. Thus, devolatilization and combustion of the coal within the fuel-rich zone are improved.

In still another variation, an adjustable inner sleeve has been added to the coal nozzle of Fig. 3.13 to provide a controlled-flow, split-flame burner.

Distributed Mixing Burners. The distributed mixing burner has application for both new and existing wall-fired pulverized-coal watertube boilers. The burner introduces a portion of combustion air to the coal flame through tertiary-air ports located around the burner periphery. Injecting large amounts of combustion air

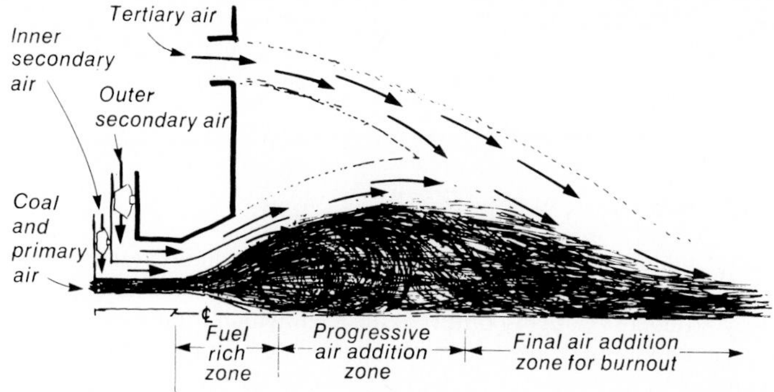

FIG. 3.14 Distributed mixing burner introduces combustion air through tertiary-air ports located around burner periphery.

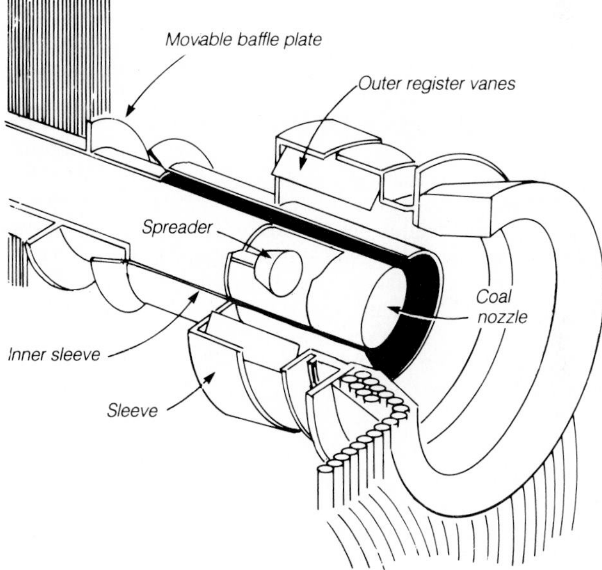

FIG. 3.15 Coal spreader was added to burner to improve carbon burnout.

through these ports is intended to induce staged combustion conditions nearer the burner than is possible with other units (Fig. 3.14).

Major problems for retrofitting the distributed mixing burner include fitting the tertiary-air ports around the burner with adequate clearances and possible revisions to the structural supports in the wind box.

A coal spreader (Fig. 3.15) has adjustable outer-air registers, closure plates, and spreader position. Reductions in NO_x emissions of 35 percent can be achieved with minimal changes in overall performance.

A secondary benefit with the modified burners is flexibility. The boiler can be fine-tuned to trade off between NO_x emissions and unburned carbon. If a pulverizer is grinding poorly, for example, the burner can be adjusted to accommodate the coarser coal. Or the NO_x emissions can be minimized at the expense of boiler efficiency, or vice versa.

Industrial Experience

Liquid fuels, which are the main energy supply for industrial boilers, present even greater problems than coal for NO_x control. Liquid fuels are normally introduced by atomizers of many different designs, which produce a wide variety of fuel-spray characteristics, such as spray angles, spray type, drop size, etc.

In general, by narrowing spray angles, longer flames are produced with delayed mixing. This can also be accomplished by reducing the number of atomizing holes or by repositioning them. Atomizers have been tested with fuel-injection holes of different diameters, which create fuel-rich and fuel-lean combustion zones. With staged combustion, finer atomization not only provides an earlier release of the fuel nitrogen into the fuel-rich zone, but also improves mixing and/or vaporization of the oil.

Poor mixing in a flame, however, leads to smoke or excessive particulate emissions. An important procedure is the careful matching of atomizer performance and fuel distribution with airflow.

The use of low-NO_x burners to retrofit industrial oil- and gas-fired boilers is limited, because their larger flame volumes generally require additional clearance to complete combustion. Extensive testing with single burners of the type normally selected for utility boilers indicates that changes in burner block, nozzle geometry, and swirl-vane angles may decrease NO_x emissions significantly.

One industrial-commercial design—the so-called self-recirculation burner (Fig. 3.16)—uses gas recirculation and two-stage combustion, all within the burner. Its key feature is the creation of a strong recirculation eddy in the burner throat, which draws combustion reaction products from the furnace to gasify the fuel stream. Primary airflow is usually around 30 percent of stoichiometric conditions. The result is that gas leaving the burner throat is very rich in hydrogen

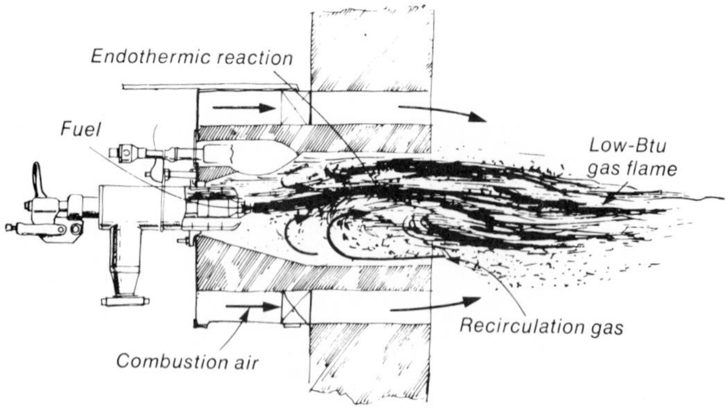

FIG. 3.16 Industrial burner for oil- and gas-fired units uses gas recirculation and two-stage combustion with the burner.

and carbon monoxide. This gas is mixed with secondary air in the furnace for complete combustion.

SUGGESTIONS FOR FURTHER READING

Bersani, A. A., and J. F. Laukaitis, "Operation and Maintenance Experience with the Shamokin and Wilkes-Barre Fluid-Bed Boilers," American Institute of Chemical Engineers, national meeting, New Orleans, Apr. 1986.

Butterworth, S. C., et al., "Operation and Performance of an Industrial Cogeneration Plant with an AFPC Boiler," ASME Industrial Power Conference, Chicago, Oct. 1985.

Herzog, G. P., et al., "Sohio's Experience with the Installation and Operation of a Bubbling-Bed Boiler," ASME Industrial Power Conference, Chicago, Oct. 1985.

Liu, E. H., et al., "Design Considerations on Fuel Flexibility of Circulating Fluidized-Bed Boilers," ASME Joint Power Generation Conference, Portland, Ore., Oct. 1986.

Proceedings of the First Annual Council of Industrial Boiler Owners Fluidized-Bed Boiler Seminar, Washington, D.C., Dec. 1985.

Proceedings of the Second Annual Council of Industrial Boiler Owners Fluidized-Bed Boiler Seminar, Novi, Mich., Dec. 1986.

Ruettu, S., et al., "Development of a CFB Combustor with a Horizontal Cyclone," Ninth International Conference on Fluidized-Bed Combustion, Boston, May 1987.

Shedd, R. W., "Emissions from Fluidized-Bed Boilers," annual meeting of the Air Pollution Control Association, Minneapolis, June 1986.

Stringfellow, T. E., and P. A. Ireland, "Fluidized-Bed Combustion and Its Role in Acid Rain Compliance," *Power* Magazine Conference, Washington, D.C., Oct. 1986.

CHAPTER 4.4
ELECTROSTATIC PRECIPITATORS AND FABRIC FILTERS

Jason Makansi
Power *Magazine*
New York, NY

INTRODUCTION

Controlling particulates from industrial and utility powerplants has grown from humble beginnings to a complicated engineering exercise. The reasons for this have as much to do with existing and potential regulations as with state-of-the-art technology in powerplant engineering and the design of the primary collection devices—*electrostatic precipitators* (ESPs) and fabric filters.

Powerplants are changing. Conventional pulverized-coal-, cyclone-, and stoker-fired boilers will still be installed in central stations, but they will be expected to handle a variety of different fuels and fuel combinations, undergo cycling duty, and be compatible with existing and even proposed environmental regulations. New plants entering service produce flue-gas streams that have different characteristics.

Existing plants will undoubtedly have to be retrofitted to comply with new and existing air emissions regulations. To illustrate the latter: Many plants are operating at less than full capacity. Under part load, they are in compliance. But depending on the condition of the particulate collection equipment, they may be out of compliance when a return to full load is warranted.

One piece of legislation is the Environmental Protection Agency's (EPA) new standard for fine particulates, called *PM 10* (particulate matter under 10 μm), promulgated to protect the public from respirable particles. Not as close to law but still a foregone conclusion in the eyes of many experts is the perennial acid rain legislation. For different reasons, both acid rain and PM 10 could require extensive retrofits to existing ESPs or, in some cases, replacement with fabric filters.

On the industrial side, cogenerators and others are turning to solid and solid-waste fuels and the systems that can burn them with high efficiency and environmental compatibility (see Section 2, Chapter 7). Fluidized-bed boilers, cocombustion of fossil and biomass fuels, and refuse and refuse-derived-fuel (RDF) combustion are all part of today's industrial powerplant. Not only do ash and flue gas have different characteristics from single-fuel plants, these characteristics could be changing from season to season, day to day, even hour to hour. Processes that remove other contaminants from the flue gas, such as SO_2 and acid-gas scrubbers, often compound particulate collection problems.

How are ESPs and fabric filters being designed and constructed today to accommodate these new regulations and powerplant technologies? The thrust of this chapter is to identify and describe the solutions this question raises.

ELECTROSTATIC PRECIPITATORS

These devices eliminate dust or other fine particles from flue gas by charging the particles inductively with an electric field and then attracting them to highly charged collector plates, from which the particulates are removed and disposed of. How this principle is applied to modern ESPs in utility and industrial applications is explained below.

Operating Fundamentals

An ESP consists of a hopper-bottomed box containing rows of plates forming passages through which the flue gas flows (Fig. 4.1). Centrally located in each passage are emitting electrodes energized with high-voltage, negative-polarity direct current provided by a *transformer-rectifier* (T-R) set. The voltage applied is high enough to ionize the gas molecules close to the electrodes, resulting in a visible corona. Flow of gas ions from the emitting electrodes across the gas passages

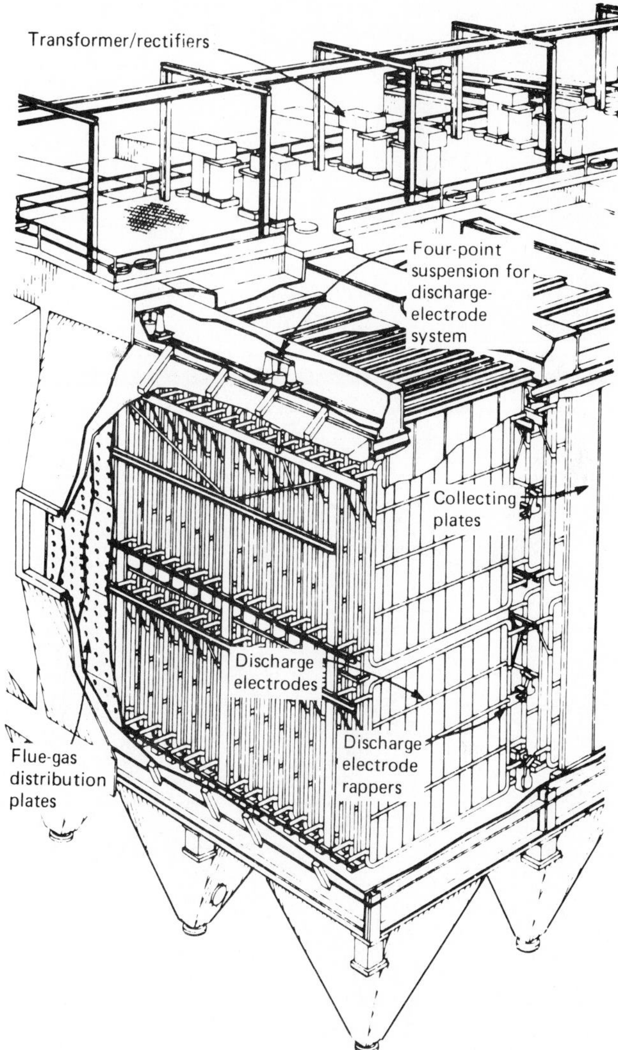

FIG. 4.1 Electrostatic precipitator (ESP) collects flyash from boiler flue gas.

to the grounded collecting plates constitutes what is called *corona current* (Fig. 4.2).

When passing through the flue gas, the charged ions collide with, and attach themselves to, flyash particles suspended in the gas. The electric field forces the charged particles out of the gas stream toward the grounded plates, and here they collect in a layer. The plates are periodically cleaned by a rapping system (Fig. 4.3), to release the layer into the ash hoppers as an agglomerated mass.

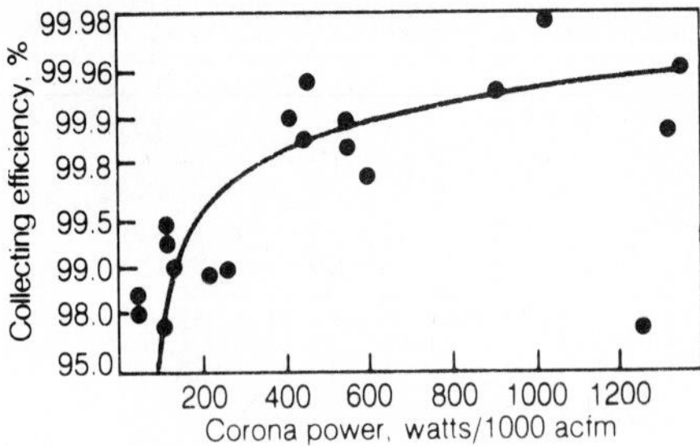

FIG. 4.2 Collection efficiency of ESP depends on applied corona power. Note: acfm × 0.472 = L/s.

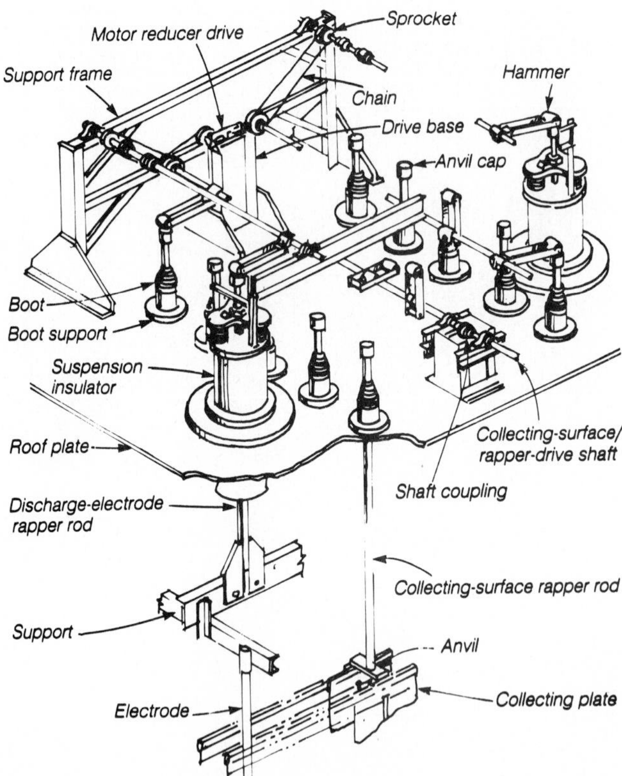

FIG. 4.3 Mechanical rapping system periodically dislodges accumulated ash into hoppers.

In most cases, the ESP is located after the air heater or economizer and is referred to as a *cold-side installation*. In special cases, it is located before the air heater to take advantage of the higher temperature and smaller gas volume. Here, it is called a *hot-side installation*.

Process Efficiency. The efficiency of the process can be approximated by the well-known Deutsch equation

$$\text{Eff} = 1 - \exp\left(-\frac{wA}{V}\right)$$

where exp = the base of the natural log = 2.718
w = precipitation constant, ft/s (m/s)
A = collecting plate area, ft^2 (m^2)
V = gas volume handled, ft^3/s (m^3/s)

Details of the equation are available in standard references on ESPs. Of importance is that A/V is often referred to as the *specific collection area* (SCA), the most fundamental ESP size descriptor. Efficiency increases as the SCA increases. Also important is that efficiency increases as w increases. The value of w increases rapidly as the voltage applied to the emitting voltage is increased. But the voltage cannot be increased above that level at which an electric short circuit, or arc, is formed between the electrode and ground.

Even in well-aligned ESPs, the maximum voltage applied to the emitting electrode without arcing is limited by the properties of the flyash. During operation, a layer of ash is deposited on the collecting plate surface. This layer has an electrical resistance and an associated voltage drop equal to the product of the current density and the resistivity of the flyash (Fig. 4.4).

Flyash Resistivity. Flyash resistivity plays a key role in dust-layer breakdown and the resulting limitations on ESP performance. It is highly dependent on the temperature and chemistry of the gas stream and the chemical composition of the particle itself. Flyash resistivities below about $2 \times 10^{10}\ \Omega \cdot \text{cm}$ do not generally limit corona-current density and ESP voltage to the point of affecting perfor-

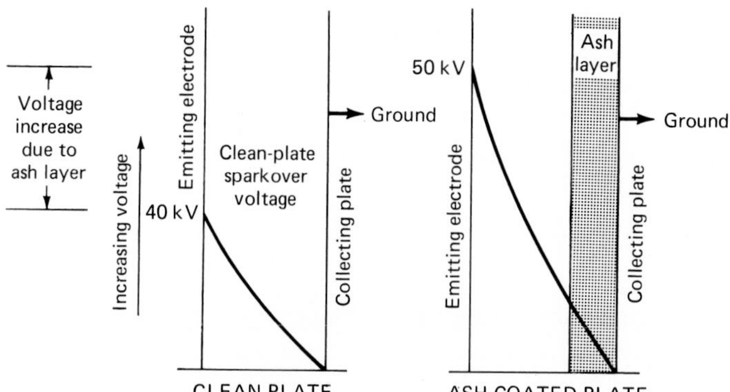

FIG. 4.4 Flyash layer builds up on plates, increasing voltage gradient across plates and also ESP power requirements.

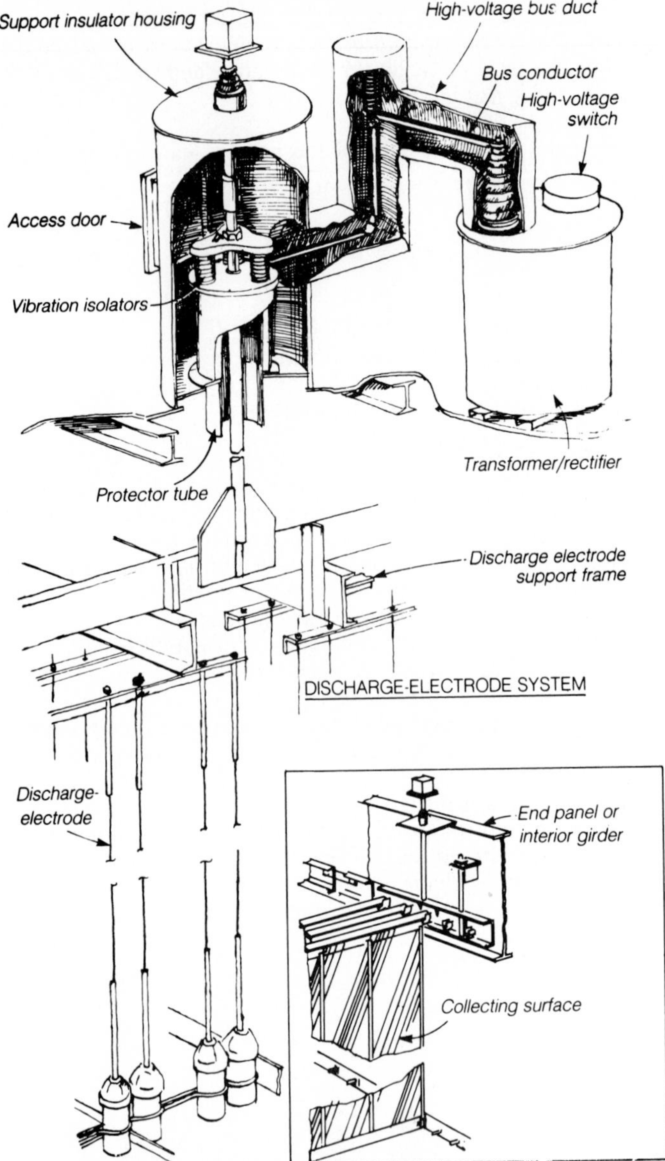

FIG. 4.5 Mechanical characteristics of so-called American design ESP.

mance. Above this value, sparking and arc-over can occur at lower voltage, and severe back corona will result when it is higher than about 1×10^{12} $\Omega \cdot$ cm. Back corona happens when positive ions are generated in the flyash layer and migrate toward the emitting electrodes, neutralizing negatively charged particles en route.

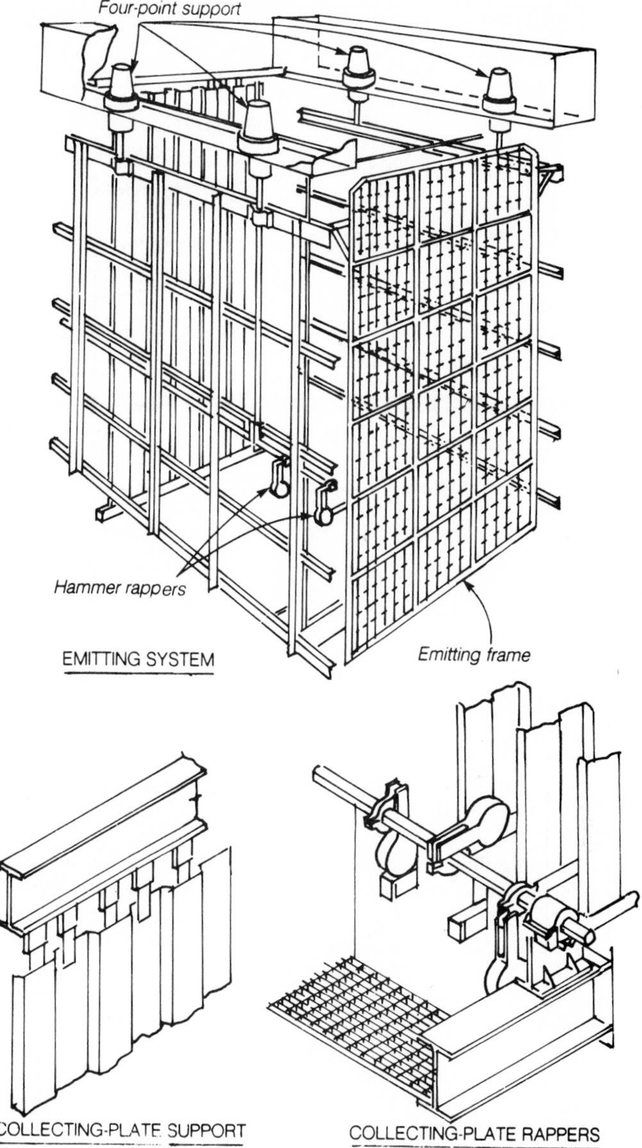

FIG. 4.6 Mechanical characteristics of so-called European design ESP.

High-resistivity ash suffers from power limitations at temperatures below 600°F (316°C). This relationship is the design basis for hot-side installations.

Adverse Influences. Mechanical aspects are as important to ESP performance as ash characteristics (Figs. 4.5 and 4.6). Three primary deficiencies in operating units are gas sneakage, flyash reentrainment, and flue-gas distribution. Flue gas

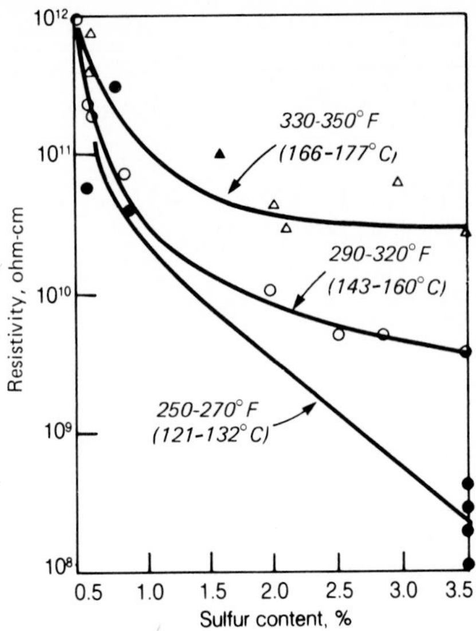

FIG. 4.7 Sulfur content of coal dramatically affects flyash resistivity.

that bypasses the effective regions of the ESP greatly increases the outlet dust loading. Reentrainment occurs when individual dust particles escape from the agglomerated mass falling into the hoppers. Many strategies have been imposed to avoid reentrainment, including rapping-cycle timing adjustments, off-line rapping, conditioning, and others. Finally, when the flue-gas velocity distribution is uniform throughout the entire cross section, the maximum collection ability of the unit is realized.

From the foregoing discussion, it is clear that the precipitation process is dramatically affected by what is happening upstream regarding the quality and type of fuel (Fig. 4.7), the combustion process, and the preceding components including the air heater, mechanical collectors, pulverizers, boiler, and *flue-gas desulfurization* (FGD) system. Any upstream factors that affect the moisture content of the coal or the sulfur compounds in the gas, the chemical constituents of the ash, and/or the temperature of the gas will also affect ESP performance. Likewise, mechanical changes to the fuel, such as pulverization and other preparation, can often be correlated to ash parameters such as resistivity and particle-size distribution.

Precipitator Enhancements

Difficulties in collecting high-resistivity flyash and fine particulates have had profound effects on suppliers and users of ESPs. In some cases, the prospect of difficult-to-collect flyash has resulted in specifications for very large units or unacceptable increases in ESP power consumption. In extreme cases, this has caused users to turn to fabric filters. For upgrading an existing ESP that is out of compliance, enlarging the box is an option, although it is rarely practical or even desirable.

Fortunately, better commercial options exist today, which are described in the following paragraphs. Like so many technologies, none of these are new concepts. They have just been more rigorously applied and commercialized in recent years, to overcome prior technical limitations and maintain competitiveness with fabric filters.

Pulse Energization. Pulse energization is a modification to conventional ESP power supplies. Essentially, a high-voltage pulse is superimposed on the base voltage to enhance ESP performance during operation under high-resistivity conditions. When it is applied to a simple wire or rod discharge electrode, the density of emitting points is greatly increased. The duration of the pulse is short enough to avoid the formation of sparks between the discharge electrode and the electrical ground.

Commercial high-voltage pulse power supplies fall into two generic categories: One type uses spark-gap switching in the high-voltage circuit to produce a narrow waveform with a fast rise time; the other type uses a string of thyristors, sometimes followed by a pulse transformer, to switch the pulse and produce a sinusoidal waveform. Suppliers have advanced their own theories about the optimum pulse-waveform and repetition rate.

In a retrofit application, installing a high-voltage pulse power supply requires minimal downtime. And during continuous operation, the system needs little maintenance or supervision. However, these advantages are somewhat obscured by the relatively high cost compared to other enhancements and by some past reliability and maintainability problems.

Intermittent Energization. To lower the cost of pulse energization, ESP engineers are turning to its derivative, *intermittent energization* (IE). Conventional ESP controls normally apply unfiltered full- or half-wave rectified *direct-current* (dc) voltage to the discharge electrodes. When IE is used, the voltage to the ESP is turned off during selected periods. This step allows a longer period between each energization cycle, which limits the potential for back corona.

One reason the technique is effective is that only a small fraction of the current actually charges particles, so it is theoretically possible to reduce current, and thereby power consumption, without reducing the charge on the particle. Another reason is that the average voltage in the ESP does not decrease dramatically when selected pulses to the primary side of the T-R set are blocked.

IE's most immediate advantages are that it significantly reduces ESP power consumption and can also reduce emissions at the same time, depending on the ash and the application (Fig. 4.8). Commercial tests have shown reductions in power consumption of over 50 percent. Some installed ESP controllers can be reset for IE at no cost and little service interruption. For many others, the new solid-state circuits can be retrofitted to the existing T-R sets, at a cost about 90 percent less than for full pulse energization. In fact, IE has been designed into the automatic controls of some new precipitator offerings. Although IE is effective in reducing emissions, side-by-side tests have confirmed the superiority of full pulse energization. In either case, both reach their potential only when the ESP is in good mechanical and electrical condition.

Wide Plate Spacing. Primarily thought of as a good way to reduce the capital cost of an ESP, wide plate spacing is the result of designers using components recently available to reevaluate the basic dimensions of the unit. Accompanying the work in this area is the potential to increase the thickness of discharge electrode wires. Like other design concepts, wide plate spacing has been more

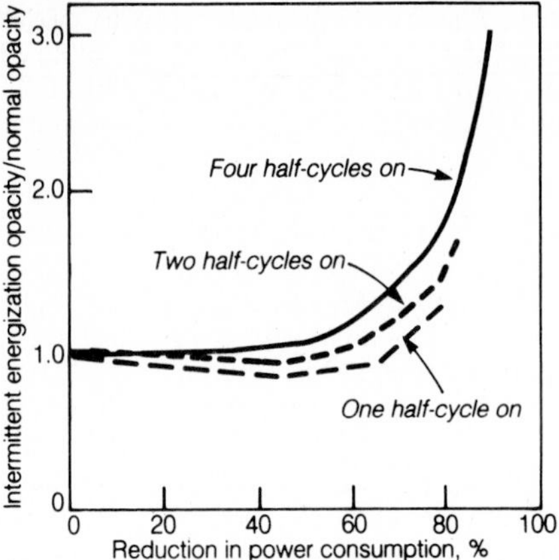

FIG. 4.8 Intermittent energization can reduce power consumption without affecting opacity.

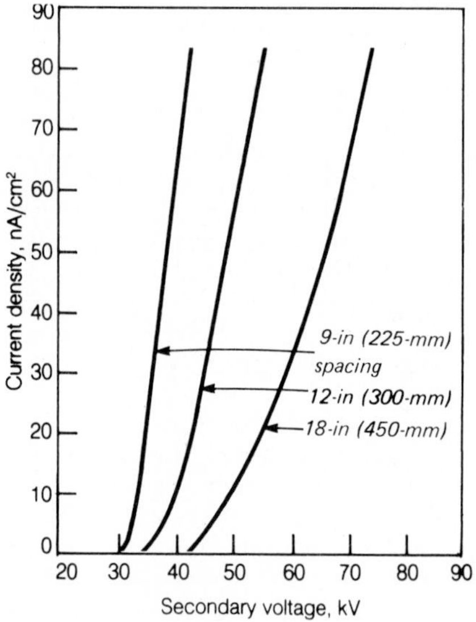

FIG. 4.9 Plate spacing affects critical ESP voltage/current relationships.

broadly applied in Europe and Japan. In the United States, industrial designers have specified wider plate spacings than their utility counterparts for years.

Electrode spacings are commonly 9 to 12 in (225 to 300 mm), and the goal is to widen them to 20 in (500 mm) or more (Figs. 4.9 and 4.10). Theoretically, this increase is possible without degrading performance because the calculated value of migration velocity in the Deutsch equation increases in direct proportion to

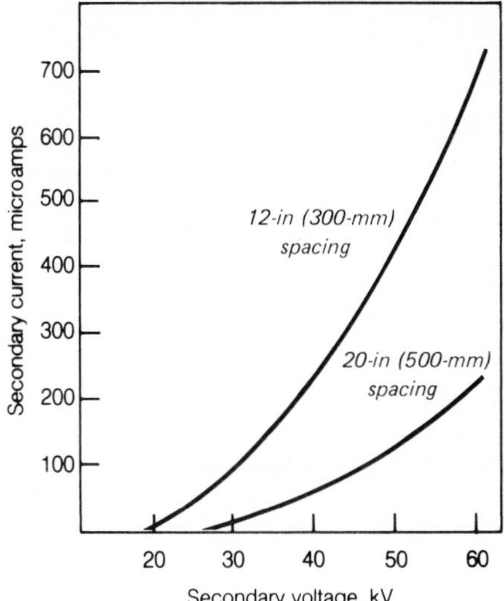

FIG. 4.10 Electrode spacing as wide as 20 in (500 mm) is possible as long as current density increases also.

plate spacing, as long as the current density is increased as well. Thus, equivalent efficiency is maintained in spite of the lower SCA.

In practical terms, wide plate spacing is possible today because of the availability of power supplies that can reliably provide the additional power and current density, especially to the thicker-diameter electrode discharge wires. Advantages associated with the new dimensions are several. Capital cost reductions have been estimated between 10 and 20 percent. Because there are fewer discharge electrodes, collection plates, and rapper system components, maintenance costs should go down. Access to internal parts is better because more space exists between plates, and alignment is easier because the tolerances are greater.

Wide plate spacing can be considered along with the latest electrode designs for rebuilds of existing units. For new units, specifiers need to reconsider the requirements for a minimum SCA, since offering wide plate spacing without a reduction in SCA often means an uncompetitive unit.

Microprocessor Controls

Application of microprocessor controls has probably had the greatest impact on ESP technology over the past few years because (1) it can truly integrate ESP operation with the balance of plant it serves (Fig. 4.11), (2) it enables designers to cut back on the design margins used to guarantee performance under worst-case conditions, and (3) it has enjoyed a more rapid mutual acceptance rate between users and suppliers, particularly utilities, than other enhancements.

As in so many other computer applications, controls for ESPs are limited only by the creativity of the programmer and the ability of the customer to understand the control concepts and pay for the hardware and software. Systems available today can sense changes in the shapes of voltage-current curves and avoid back

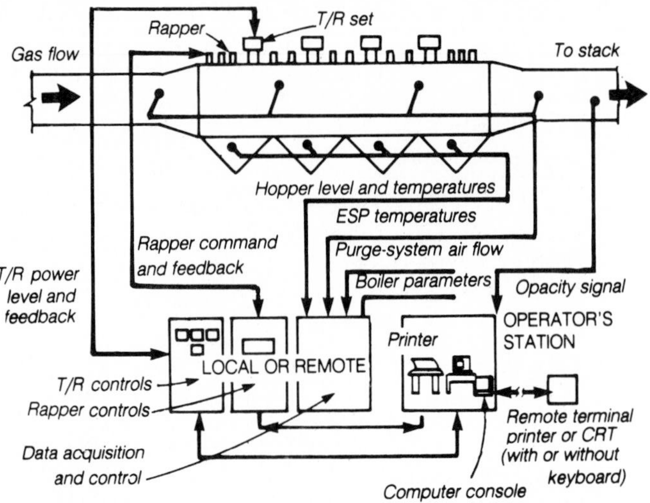

FIG. 4.11 Microprocessor-based controls link ESP operation with that of powerplant.

corona, use opacity or other variables to control energy consumption, implement state-of-the-art control algorithms for power savings or performance improvement, and control T-R sets, rapping mechanisms, and ash hoppers.

Compared to older controls, spark detection, for example, is on the order of a fraction of the half cycle in which it occurs. Parameters that previously were preset, such as quench time, recovery rate, setback hold, and gradual rise rate, can now be varied automatically. All these benefits accrue with a substantial reduction in the amount of control hardware.

As far as the balance of the plant is concerned, modern controls are especially useful for plants operating with a variable fuel source or undergoing much cycling duty. If wide variations in particle resistivity, particle size, or ash loading are expected over the remaining life of the plant, then the unit can be designed for worst-case conditions, but operated to provide only the amount of power needed based on real-time or close to real-time conditions.

Controls can be programmed to respond to changes in boiler load and related variations in such parameters as flue-gas temperature and volume and sootblowing cycles. Rapping schedules can be matched to changes in the characteristics of the fuel being burned in the boiler. The degree of rapping and its frequency can be set at maximum levels during low-temperature boiler start-up, then automatically reverted to a normal operating mode when flue gas is safely above the acid dew point. If an ESP has a high degree of electrical sectionalization, collecting plates can be deenergized before rapping ensues, greatly enhancing plate cleaning and minimizing rapping reentrainment losses. *Power-off rapping,* as this technique is called, was previously unavailable without modern controls.

Flue-Gas Conditioning

A different idea is to condition the flue gas or alter those flyash characteristics that affect ESP performance. The technique has been applied commercially to both hot-side and cold-side installations. Its commercial application rose with the switch to low-sulfur coal in the previous decade to meet SO_2 regulations at existing central stations. Over 200 utility boilers are equipped with some form of conditioning in the United States. There are virtually no industrial boilers equipped with gas conditioning, especially those units fired with fuels other than coal. Other types of additives have been used, however.

Basically, the objective of conditioning is to modify the electrical resistivity of the ash and/or its physical characteristics. The conditioning takes the form of changing the surface electrical conductivity of the dust layer deposited on collecting plates, increasing the space charge on the gas between the electrodes, and/or increasing dust cohesiveness to enlarge particles and reduce rapping reentrainment losses.

***SO_3* Injection.** The most common conditioning agents are sulfur trioxide (SO_3), ammonia (NH_3), and sodium compounds. By far the most widely applied for cold-side ESPs is SO_3; for hot-side units, it is sodium compounds. With the latter, adding a compound like sodium sulfate to the coal has been commercially demonstrated to have a consistent level of success in preventing ESP performance deterioration. Besides potential problems with increased slagging of boiler tubes, however, specific fuel constituents interfere with the process and affect the amount of sodium required and the subsequent economics.

The most popular apparatus for SO_3 injection is an elemental-sulfur burner that produces SO_2, which is then catalytically converted to SO_3 in amounts that depend on the boiler load and variations in the fuel (Fig. 4.12). Sulfur trioxide is injected after the air preheater in amounts that are generally less than 20 *parts per million* (ppm). Some developers have been pursuing a thermal decomposition system using ammonium sulfate.

Disadvantages of SO_3 injection systems include the possibility of plume color degradation under conditions of high gas temperatures, high ambient humidity,

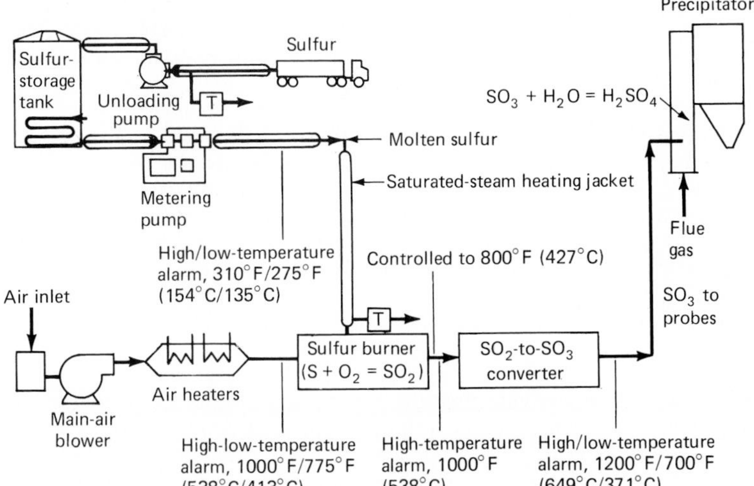

FIG. 4.12 Sulfur-burning system is popular SO_3-conditioning technique for controlling ash resistivity.

and low ambient temperatures. This problem is aggravated when combined with overinjection of SO_3.

SO_3-NH_3 Injection. Combined SO_3-NH_3 conditioning has been used in a few special cases. The SO_3 adjusts the resistivity downward, while the NH_3 modifies the space-charge effect, improves agglomeration, and reduces rapping reentrainment losses.

In a limited number of cases, proprietary compounds have been tried successfully to improve ESP performance, most notably when the unit's operating life is short and the capital costs of more conventional systems are prohibitive. It is difficult to comment generically about these agents because the development is often on a case-by-case basis and the process chemistry is not made available.

Some experts in ESP operation are convinced that conditioning combined with a modern ESP design of moderate SCA can provide the equivalent performance of a fabric filter in terms of total outlet emissions and fine-particulate emissions for conditions of high-resistivity ash.

Advanced Concepts

Other improvements to the basic ESP unit have been proposed and tested, especially to accommodate the retrofit market, but in most cases these are not as far along in their commercial application or acceptance for U.S. powerplants. Some are presented here.

Revised Operating Method. The concept of the *revised operating method* (ROM) is to cut rapping reentrainment losses by stopping the gas flow during rapping of the plates. It is especially suited for large units with four or more chambers arranged in parallel. It can be thought of as an off-line rapping technique. Gas flow is stopped successively in a single chamber by double-louver dampers with air-seal injection. Among other criteria, ROM requires absolutely reliable gastight dampers that can handle the frequent operating cycle and a flue system designed for automatic redistribution of the gas flow while any one chamber is closed off.

A recent full-scale demonstration of ROM showed it capable of reducing rapping reentrainment losses by 80 percent or more and halving the outlet dust loading compared to normal ESP operation.

Self-Cleaning Electrodes. One Japanese firm has developed a moving-electrode ESP for reducing rapping reentrainment losses. Collecting plates—each connected to a link chain drive mechanism—pass continuously through rotating brushes to remove the collected dust. In coal-fired boiler applications in Japan, the moving-electrode ESP is actually a second stage to a conventional ESP.

Commercial units are smaller and consume less power than units with conventional electrodes. Of particular importance to powerplant designers is the effect of the highly abrasive ash on the drive system.

Two-Stage Precipitators. Two-stage ESPs have been proposed for collecting high-resistivity ash. The underlying philosophy is that particle charging can be carried out effectively in a very short section of the ESP while collection can be handled in the larger portion, where high electric fields and low current density are more easily maintained. One design makes use of a temperature-controlled electrode that is based on the relationship between flyash resistivity and temperature.

SO_2 Removal. One area where ESP development has accelerated is in adapting the units to handle the scrubber and nonscrubber SO_2 removal technologies—fluidized-bed boilers, spray-dryer FGD, *limestone-injection multistage burners* (LIMBs), furnace sorbent injection, and in-duct sorbent injection (see Chapters 3 and 5 in this section, Part 1). In some cases, with retrofits in particular, the desire is to prevent ESP performance deterioration. In others, it is to allow the ESP to compete with fabric filters in these services.

Sorbents change the flue-gas stream in two primary ways: (1) As a result of the absorption reactions upstream, the dust loading is appreciably higher and must be reduced at comparable efficiency; and (2) there can be orders-of-magnitude increases or decreases in the particle resistivity. Some research with the spray-dryer ESP combination showed that particulates were easier to collect than the flyash alone. Particles were found to be larger, and the temperature, solids composition, and added moisture combined to lower the resistivity of the ash significantly. The research could ultimately support installation of spray dryers upstream of existing ESPs, many with low SCA values, as an acid-rain-control retrofit strategy.

Many of the enhancements discussed earlier can be brought to bear on the desulfurization applications. There are other ones being developed, including flue-gas-stream cooling and humidification. However, the overall performance and economics compared to fabric filters loom as a large question mark that is still years away from resolution.

Mechanical-Electrical Upgrades

A mechanical-electrical upgrade of the existing ESP components is often all that is necessary to make a unit perform adequately. Areas for study include repairing discharge and collecting electrodes; replacing electrodes (Fig. 4.13), rappers, and controls with state-of-the-art components; increasing electrical sectionalization; minimizing air in-leakage; improving gas flow distribution; and minimizing corrosion of internal parts.

Electrode Replacement. A basic upgrade is to replace old weighted-wire electrodes with rigid-electrode designs (Figs. 4.14 and 4.15). Benefits include an increase in performance with little or no increase in power consumption and higher reliability. Such a replacement may be more cost-effective than it appears because plate spacings can be increased and gas passages decreased, and each rigid electrode may replace more than one hanging wire.

Collector Plate Repair. ESP suppliers and several independent service-oriented companies have developed new techniques for repairing distorted collector plates. Warped and damaged plates interfere with electrode clearances, gas flow distribution, and rapping-force application. In extreme cases, untreated flue gas may sneak through the unit, greatly degrading performance, or, worse, may be diverted through the hoppers where it reentrains collected flyash. Warped plates are caused typically by rapid, excessive, or abnormal changes in temperature; high hopper-ash levels; or improper installation.

Spacers have been used in many installations to maintain proper clearances between plates and wires. They take many forms and installation procedures, but essentially they accomplish a low-cost mechanical upgrade that in the best case may avoid plate replacement or at least postpone it to a later time.

Rapping System and T-R Set Upgrades. Upgrading rapping systems is necessary when electrode fouling suppresses corona power. Upgrading can take the form of replacing rappers with ones that impart higher forces, increasing the number of

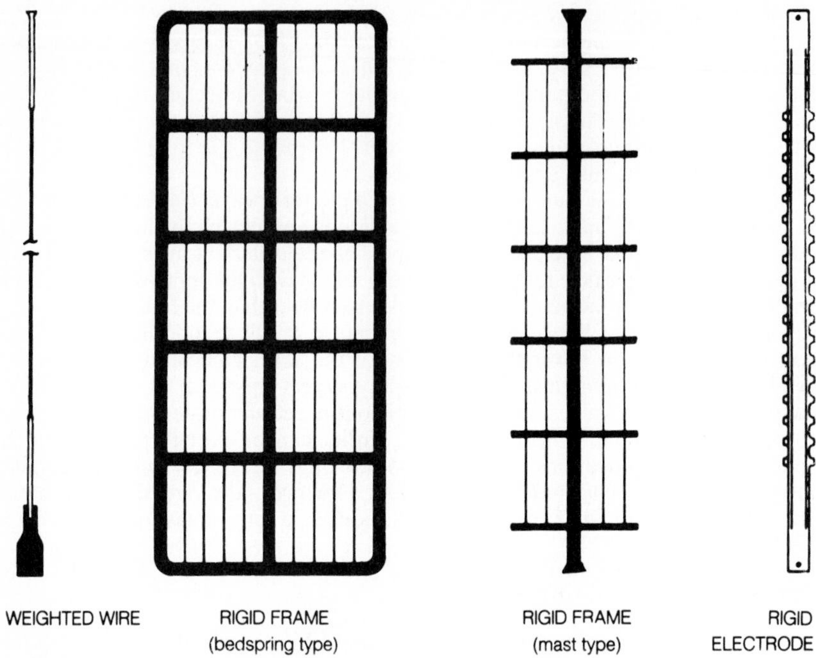

FIG. 4.13 Rigid discharge electrodes come in a variety of designs.

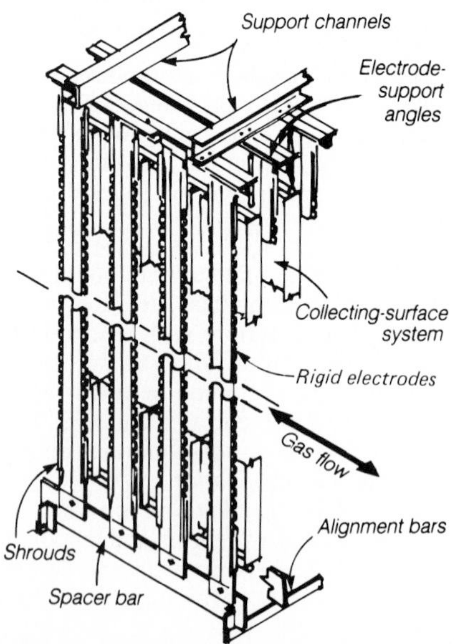

FIG. 4.14 Rigid-type discharge electrodes being retrofitted into ESP, replacing weighted-wire design.

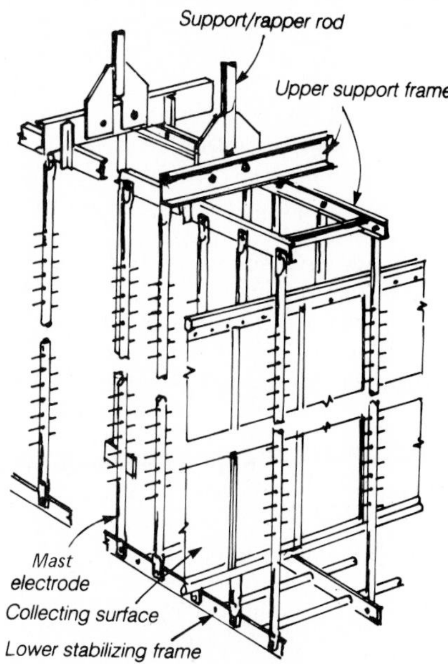

FIG. 4.15 Mast-type discharge electrodes being retrofitted into ESP.

rappers, and/or reducing the size of the electrode sections being rapped. The opposite problem—rapping that is too harsh—causes excessive reentrainment losses and is remedied by properly tuning the rapping cycles.

Upgrading the T-R sets can also improve performance, especially when the unit is collecting a flyash with higher or lower resistivity than it was designed for. T-R sets that were originally oversized as a design margin are less stable when operating at much less than their rated capacity. In this case, the strategy may be to decrease the ratio of the T-R sets. Units with small T-R sets that operate continuously at maximum rating may benefit from increasing the rating of the sets or their number. The latter alternative is preferred, because this increases the electrical sectionalization of the unit.

FABRIC FILTERS

The emergence of fabric filters for boiler flue-gas applications has doubled the pollution control opportunities for many powerplants. Clearly, the momentum today is with fabric filters—if not market share, at least for new plants. They can be as equally sensitive to flue-gas and flyash characteristics as ESPs, although in different ways, as the following delineates.

Three Basic Types

Fabric filters come in three basic types: reverse-air, shake-deflate, and pulse-jet units. They are distinguished by the cleaning mechanism and by their *air-to-cloth*

(A/C) value, a fundamental fabric-filter descriptor denoting the ratio of the flue-gas flow to the amount of fabric, or filtering surface area.

Reverse-Air Units. Reverse-air fabric filters are characterized by low A/C values and by the flow; dirty gas flows from inside the bags to outside (Fig. 4.16). They make use of what is called *low-energy cleaning,* which means that filter airflow is periodically reversed, causing significant bag motion that releases the collected dust.

Large, multicompartment reverse-air and shake-deflate units are designed with suitable valving to isolate one or more compartments while the others service the full rated load. In reverse-air units, additional ductwork, fans, and valves are included to return a portion of the cleaned outlet gas to serve as the reverse air for a compartment taken off-line for cleaning. Conventional poppet valves serve as the outlet manifold and reverse-air dampers.

A large majority of U.S. utilities use reverse-air units, mostly because they are deemed the economic choice for the huge sizes required for central stations, generally larger than 300 MW.

Shake-Deflate Units. A shake-deflate unit, another low-A/C type, collects dust on the inside of the bags as in the reverse-air design. To clean the bags, the top end is shaken by a drive linkage adjusted for the optimum frequency and amplitude of the shake (Figs. 4.17 and 4.18). It is typically an off-line cleaning technique. Recent evaluations indicate that shake-deflate units offer economic advantages over reverse-air units; operating experience with several of them in utility service has been good.

Pulse-Jet Units. In pulse-jet units, gas flow is from outside the bag inward (Fig. 4.19). Typically A/C ratios are at least double those of reverse-air units (Fig. 4.20), offering a more compact installation in most cases. Cleaning is accomplished with a high-pressure burst of air into the open end of the bag. A high-

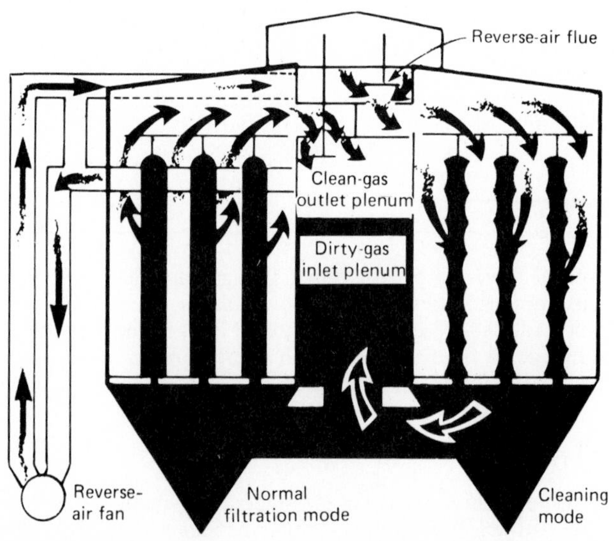

FIG. 4.16 Reverse-air fabric filter is often the conservative choice.

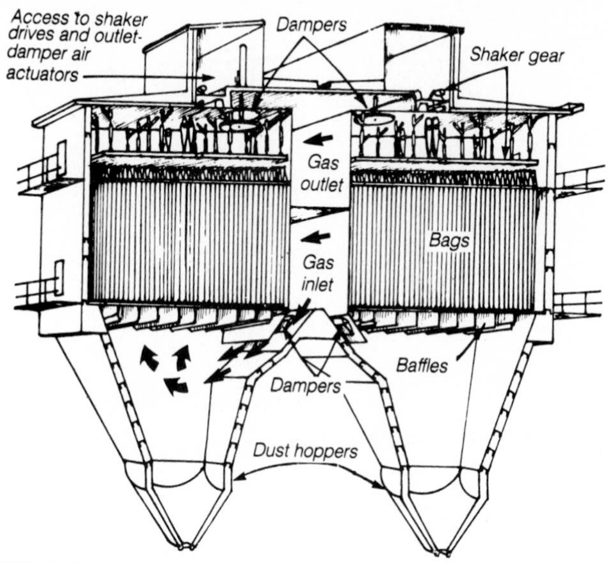

FIG. 4.17 Shake-deflate fabric filter.

pressure line, maintained by a compressor, runs above the bag openings. At appropriate times, a valve in the line opens to direct a pulse to the open bag. The units can be cleaned on- or off-line.

Pulse-jet units are preferred for industrial boilers, although reverse-air units are also widely applied. There are indications, however, that utility preferences may change, and pulse-jet units may be more of a competitor for the smaller boilers and new combustion technologies targeted for future central stations.

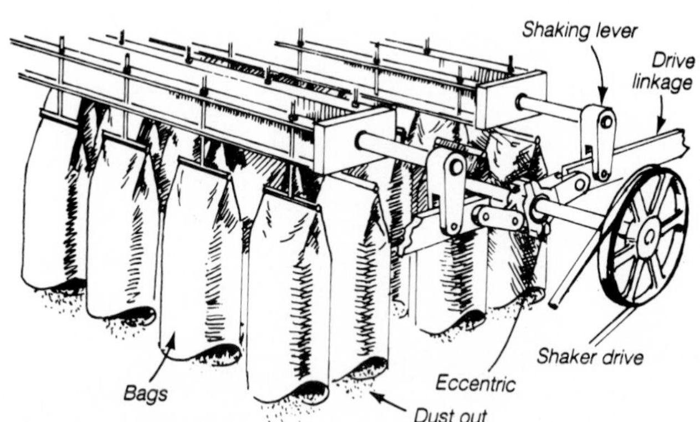

FIG. 4.18 Bags are shaked to remove collected dust in shake-deflate design.

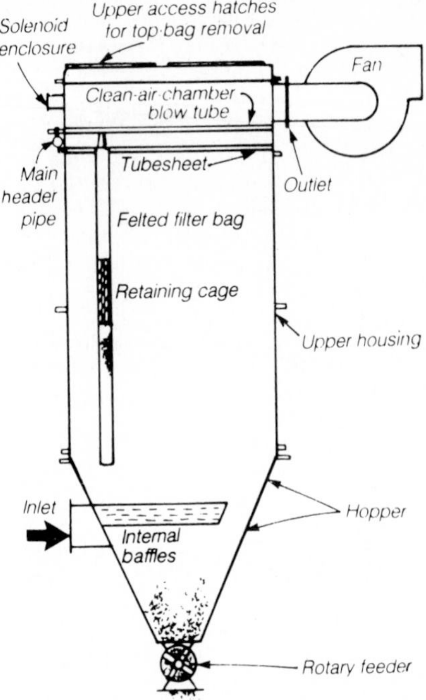

FIG. 4.19 Pulse-jet fabric filter has traditionally seen service in industrial boilers.

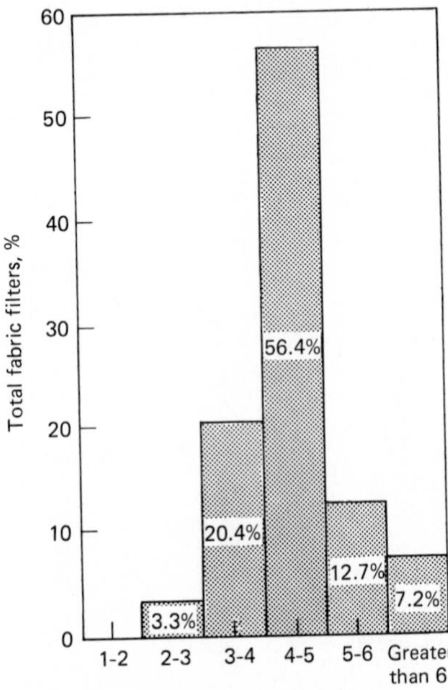

FIG. 4.20 Air-to-cloth (A/C) values for pulse-jet units. Note: acfm/ft^2 $\times$ 0.005 = m^3/s • m.

Flue-Gas Flow Patterns

Getting flue gas into and out of the fabric filter is an important design consideration. The unit must minimize pressure drop, maintain appropriate temperature and velocity profiles, and distribute the ash-laden flue gas evenly to the individual compartments and bags. Obviously, the importance of flow distribution increases as the size of the unit increases.

General recommendations are these: (1) Velocity through the unit should be minimized without allowing ash to fall out; (2) inlet and outlet manifolds should be tapered for constant velocity; (3) turning vanes should be used in the outlet manifolds and hopper to ensure ash flow balance among the compartments and prevent entrainment of ash from the hopper; and (4) internal structural supports in the ductwork should be avoided.

Tensioning of bags is another design parameter that has received more attention, especially for reverse-air units. Bag suspension mechanisms must be able to apply a predetermined force, or tension, to the bag so that it will clean efficiently with a minimum amount of wear. Also, the mechanisms must compensate for operating conditions, such as increased bag weight from collected ash and variable flows and temperatures brought about by changes upstream to the boiler load.

The need for adjustable tensioning devices and other bag supports varies with the type of unit, application, and supplier. Hopper design and the flyash evacuation system should ensure that collected dust is not reentrained into the bags and that the ash level never gets too high. Ash reentrainment is more of a concern for bottom-inlet unit designs, but it has been mentioned often as an operating problem among users.

Systemwide Effects

Of critical importance to fabric-filter design and operation are the upstream factors: how coal-ash chemistry and combustion in the boiler affect the resultant fly-ash-laden flue-gas stream. Particle-size distribution and ash loading are the most critical factors. Different fuels and combustion systems produce ash with different characteristics. Stoker-fired systems produce ash with high carbon content, moderate load, and relatively large size distribution; pulverized-coal-fired units produce high ash loads and small size distributions; cyclone-type boilers produce low ash loads with a very fine size distribution; and bubbling fluidized-bed boilers exhibit ash with high carbon content and high loads. Mechanical dust collectors upstream of the fabric filter can drastically alter the size distribution and loading of the ash.

Sulfur content of coal has been correlated to fabric-filter operation. It has been demonstrated conclusively that the cohesiveness of ash from high-sulfur-coal-fired plants is greater than from western low-sulfur-coal-fired plants. Fabric filters are extremely sensitive to the condensation of acid. Maintaining temperature in the unit above the acid dew point is critical in high-sulfur coal applications, especially when the plant is operating at reduced loads for significant periods. Some form of temperature control is essential. A corollary to this precaution is to minimize any temperature drop in the fabric filter caused by air infiltration or poor insulation.

Gas and ash characteristics affect other parameters whose importance has only recently been recognized. One of these is dust cake weight. Over time, bags accumulate a residual dust cake—ash that is never dislodged during the cleaning cycle. This is one reason why the pressure drop tends to increase gradually over time, so users should interpret pressure-drop guarantees carefully. Research shows that heavier cakes tend to accumulate in high-sulfur coal units and those that undergo cycling and frequent shutdowns. Residual dust cake weights affect the bag suspension and tensioning mechanisms, among other things.

Filter Enhancements

One of the most recognized and accepted enhancements to fabric-filter technology is the application of sonic energy to improve bag cleaning and to reduce pressure drop. Conversely, gas conditioning, another enhancement, is still at the prototype stage.

Sonic Energy Technique. The sonic energy technique got its start after early users looked for solutions to the gradual rise in pressure drop over several months of operation. For a variety of reasons, bags became increasingly difficult to clean back to original design pressure drops. Sonic horns were then applied and found to work well, reducing pressure drops anywhere from 20 to 50 percent and above (Fig. 4.21). Over the past several years, virtually all reverse-air units have included sonic horns. They can also be effective, but perhaps less so, in shake-deflate and pulse-jet units.

The effect of the sound wave action on the dust cake is to create a shear force at some boundary layer within the cake. The depth of the boundary layer and the corresponding level of cleaning depend on the dust, flue-gas, and bag-material characteristics. Frequency and power are the most significant horn parameters affecting bag cleaning. For a given value of sonic pressure, low frequencies are the most effective in terms of dust-cake removal and pressure-drop improvement (Fig. 4.22). Sonic frequencies above 250 to 300 Hz are not effective.

Horns are usually mounted above the bags in the clean-air plenum, if space permits, for reverse-air units. They are spaced to distribute the sound energy uniformly among the compartments. At least one utility has installed and tested horns mounted in the reverse-air manifolds. This placement helps reduce the number of horns required, but at the same time they may have to be operated more frequently, increasing compressed-air costs.

Gas Conditioning. Gas conditioning is another fabric-filter performance enhancement, although it is not yet a commercial reality. Results from one extensive pilot program and a commercial installation burning lignite show that partic-

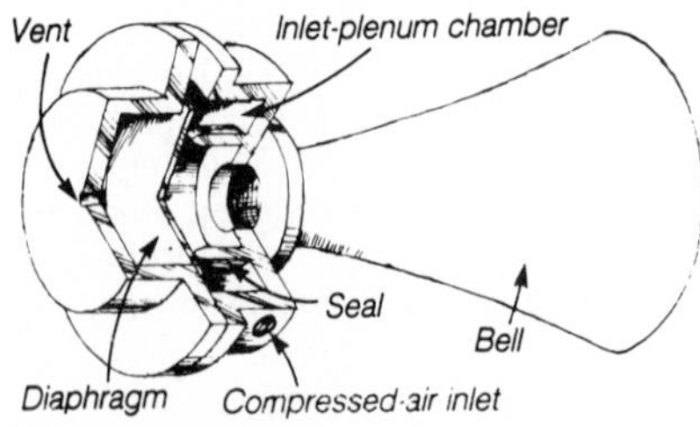

FIG. 4.21 Sonic horn is diaphragm-driven pneumatic design.

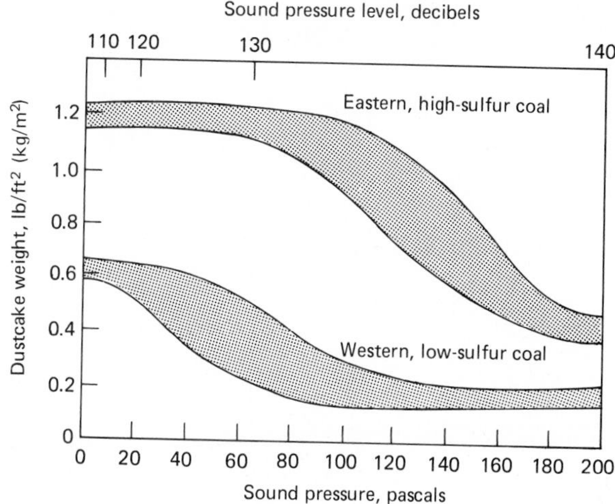

FIG. 4.22 Sonic horn effectiveness depends on sound pressure and properties of the fuel, like fuel content. Note: $lb/ft^2 \times 48 = Pa$.

ulate removal in a fabric filter is enhanced by conditioning the flue gas with low concentrations (10 to 15 ppm) of ammonia and/or sulfur trioxide.

Thus, conditioning may be an effective means of controlling fine-particulate emissions and reducing pressure drop, especially when low-rank fuels such as lignite are fired. The physical and chemical mechanisms explaining the phenomenon are complex, and the procedure is far from optimized in terms of temperature, cleaning cycle, fabric, and ash-flue-gas characteristics. It has been postulated that conditioning increases the cohesivity of the ash.

Filter Material Selection

Operating costs of fabric-filter installations have two major components: flange-to-flange pressure drop and bag life. In one sense, they are related because proper bag-material selection can often relieve pressure-drop problems. Thus, the bag itself can be considered the heart of the fabric filter.

Woven Fiberglass. The cleaning mechanism has a lot to do with the bag material, at least historically. Reverse-air units have traditionally been accompanied by woven fiberglass bags with Teflon B finish because they are an inexpensive choice for the low-energy cleaning application. In Australia, woven homopolymer acrylic materials are used extensively, but their temperature limitations are much lower than those of fiberglass. Pulse-jet units require stronger materials, such as heavier types of woven fiberglass or nonwoven fabrics like felts (Fig. 4.23).

Over 90 percent of the bags in U.S. utility service are the woven fiberglass variety. Woven and felted fiberglass and Teflon fluorocarbon fiber make up the majority of industrial pulse-jet and reverse-air units. However, these figures belie the complexity of fabric selection today. Fabrics are much more sensitive to the ash and flue-gas characteristics than the filtration community originally thought. About the only fact known for sure is that the fabric selected will not last the life

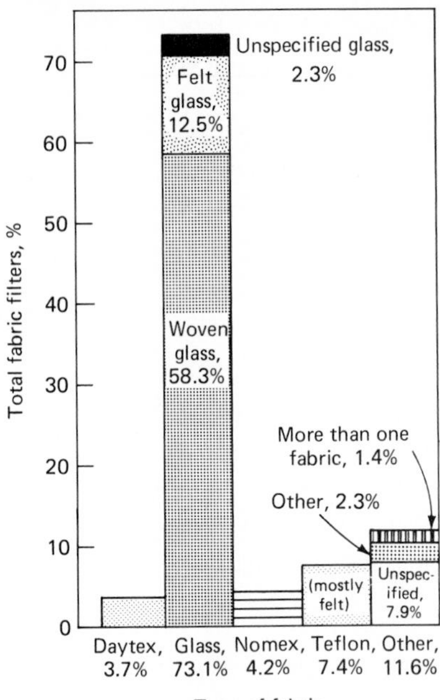

FIG. 4.23 Different fabric materials are selected for pulse-jet fabric filters.

of the unit. Beyond technical superiority, the ultimate choice must compromise the relative first cost and the anticipated replacement cycle.

Fabrics cannot be judged alone. How the fabric is constructed and what finish or coating is applied are equally important. A finish is required for any one or combination of these reasons: (1) to protect the fibers against acid and alkali attack, (2) to lubricate the fibers and improve abrasion resistance, and/or (3) to enhance the particulate release properties.

Texturization of the yarn is one area where fabric construction helps fabric-filter performance. It is a percentage that refers to the amount of texturized yarn on the surface of the fabric. Data from pilot units show that it has a significant impact on residual dust-cake weight and emissions.

In essence, the bag acts as a surface on which a dust cake is formed; it is not the primary filtering medium. Therefore, materials of bag construction are chosen for temperature and chemical resistance, mechanical stability, and ability to collect the particulates in question as a cake and then release the cake during the cleaning cycle. Bag performance typically is rated in terms of its permeability, cleanability, and durability.

Acrylic Materials. Acrylic materials can be considered as an alternative to fiberglass if they can be adequately protected from temperatures higher than 300°F (149°C), such as after a spray-dryer FGD system. At approximately 50 percent the cost of fiberglass, they provide interesting economic benefits even if they have to be replaced more often.

Regarding temperature limitations, Nomex is a material that fits between the acrylics and fiberglass and Teflon. It has not seen wide application in the United

States, but has been used on many utility-sized pulse-jet units in Europe and Canada. More cost-effective materials are being sought for pulse-jet units. One is a 75/25 percent blend of Teflon fiber and fiberglass needle-punched onto a substrate of 100 percent Teflon. Another is Ryton, similar to Teflon and fiberglass but with improved flex life and lower temperature limitations. A third is a felted Daytex, Ryton fiber on an improved Teflon fluorocarbon substrate; its strength makes it a strong candidate for on-line cleaning applications.

Gore-Tex is a special material that has solved some difficult collection problems, especially when particle penetration is the cause of the pressure drop increasing to abnormally high levels. The Gore-Tex material is laminated onto the regular glass weave, providing a nonstick, microporous surface for holding the dust cake only on the surface, so the cake can be broken and released more readily. In some cases, Gore-Tex bags have lengthened the period between cleaning cycles since it cleans to a lower pressure drop than other materials. This attribute could extend bag life, reduce the unit's downtime, and offset the significantly higher capital cost of the materials for bag construction.

Range of Applications

Fabric filters are being applied with increasing frequency at some of the latest utility and industrial powerplants—those fired by biomass, solid-waste fuels; those equipped with spray-dryer FGD systems or sorbent injection processes; and those using fluidized-bed boilers. Except for the U.S. utility market, the majority of these units are using, or plan to use, pulse-jet collectors.

In some cases, notably refuse-fired powerplants, the choice of a fabric filter may be based more on institutional or political influences than on technical or economic criteria. Because it is perceived as the alternative providing the most insurance against present and future regulations, plants with fabric filters for emissions control are sometimes more readily accepted by the public and public officials who must approve and fund the plant.

In most cases, there are important technical considerations. To illustrate, units following sorbent injection and spray-dryer processes become more of an integrated environmental control system because the fabric filter provides an appreciable degree of SO_2 removal. Following are some of the technical issues for a range of applications.

Fluidized-Bed Boilers. Products from SO_2 absorption reactions, especially calcium sulfate, become cementitious when combined with moisture, which makes blinding of the bags more of a concern. Also, the finer particles associated with *circulating fluidized-bed* (CFB) combustion tend to work their way into the interstices of the fabric, causing pressure-drop problems. The carbon content is greater in ash from bubbling boilers because the combustion efficiency is lower than pulverized-coal-fired units or CFBs. If ash were allowed to build up in hoppers through improper evacuation, it would create a fire hazard.

Fabric-filter manufacturers have been designing very conservative units when they are coupled with fluidized-bed boilers, mainly because not enough information is available from commercial experience. Low-A/C, pulse-jet collectors are often specified with a heavy-bulk, woven fiberglass bag material with a Teflon finish.

SO_2 Removal Systems. It is generally agreed that fabric filters can provide a key advantage in desulfurization. They are able to collect unused reagent from the process and absorb more SO_2 while it is still on the bags, or collect it for later recycling back to the absorber. In the spray-dryer case, SO_2 removal capability may be 25 percent or more above that accomplished in the spray dryer. Even with a higher particulate loading, the fabric filter is little affected by having the spray dryer upstream, unless a process upset sends wet, sticky particles to the collector. It is imperative to maintain temperatures above the water and acid dew points, especially during start-ups and shutdowns. Maintaining the dust layer on the bags by carefully controlling the cleaning cycle is another objective during operation.

One potential problem area in desulfurization service is corrosion of the clean side of the compartments. Some installations have had to implement new design concepts to protect against corrosion. New concepts include flue-gas reheat between the spray dryer and the fabric filter, wall temperature monitoring, increased insulation, reduction of air in-leakage, thermal gaskets on hopper flanges, air cavity elimination in casings by adding wall stiffeners, and isolation of individual casings with guillotine dampers.

In several sorbent injection processes, the fabric filter is intended to serve as the primary desulfurizing medium (Fig. 4.24). Extensive tests with sorbent injection (sulfur compounds) have revealed little impact of this technique on fabric-filter operation. A change in the cleaning cycle may be required to compensate for variations in pressure drop or the increased particulate loading. One area that deserves further research for desulfurization applications is choosing the fabric finish that assists in providing the most efficient SO_2 removal in terms of dust-cake buildup and cake release properties.

Refuse Plants. As with coal-fired boilers, the recent trend in particulate collection for refuse-fired plants has been to install fabric filters instead of ESPs, although there is every indication that ESPs can provide equivalent service. Until recently, fabric filters were deemed unsuitable for this service because frequent temperature excursions would create fire hazards. More frequent application of

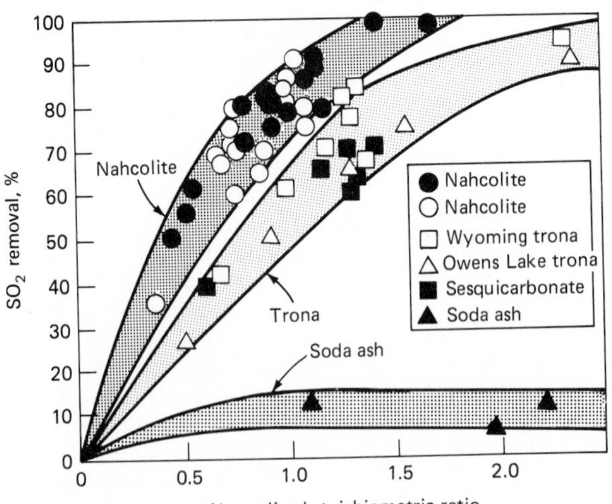

FIG. 4.24 SO_2 can be removed with a variety of sodium compounds.

scrubbers for acid-gas control—providing a continuous quench of the flue-gas stream—helps favor fabric filters.

There is mounting evidence that the bulk of gaseous, trace heavy metals from RDF plants is emitted with the fine-particulate fraction of the flyash. Thus, the fabric filter circumvents problems, meeting fine particulates and public concern over trace-metal emissions. Another aspect to consider is the effect of acid gas on the filter fabric. To illustrate, high hydrofluoric acid levels may preclude the selection of fiberglass bags. Some fabric substrates are more dimensionally stable in the face of certain constituents in the flue-gas stream.

Wood-Fired Plants. Fabric filters are rarely applied to wood-fired powerplants, but this could change as a few industries look to the benefits of being the good neighbor with the clear stack. One major concern is the fire hazard as glowing char particles inevitably find their way back to the units, even if a mechanical collector is located upstream. Another concern is the variable nature of the fuel. Moisture levels and contaminants like lignin in the wood contribute to this concern.

At least one U.S. industrial wood-fired boiler contains a fabric filter, which has these characteristics: reverse-air, low A/C, positive pressure, extensive fire suppression, and sophisticated ash handling. Still, several fires have been experienced as well as abnormal pressure drops. In fact, the pressure-drop problems were severe enough to warrant installing a whole new module after correction measures failed.

SUGGESTIONS FOR FURTHER READING

"Conference on Electrostatic Precipitator Technology for Coal-Fired Powerplants," *Proceedings,* Electric Power Research Institute (EPRI) Report no. CS-2908, Apr. 1983.

Donovan, R. P., *Fabric Filtration for Combustion Sources,* Marcel Dekker, New York, 1985.

Dry SO_2 Particulate Removal for Coal-Fired Boilers, EPRI Report no. CS-2894, June 1984.

Economics of Fabric Filters and Electrostatic Precipitators—1984, EPRI Report no. CS-4083, Apr. 1983.

EPA/EPRI Fifth Symposium on the Transfer and Utilization of Particulate Control Technology, EPRI Report no. CS-4404 (four volumes) and EPA Report no. EPA-600/9-86-0086, Feb. 1986.

Fluid Dynamic Design Guidelines for Utility Fabric Filter Systems, EPRI Report no. CS-3811, Oct. 1984.

Gallaer, C. A., *Electrostatic Precipitator Reference Manual,* EPRI Report no. CS-2809, Jan. 1983.

Investigation of Ammonium Sulfate Conditioning for Cold-Side Electrostatic Precipitators, EPRI Report no. CS-3354, Feb. 1984.

Sodium Conditioning for Improved Hot-Side Precipitator Performance, EPRI Report no. CS-3711, Oct. 1984.

CHAPTER 4.5
FLUE-GAS DESULFURIZATION

William Ellison
Director
Ellison Consultants
Monrovia, MD

INTRODUCTION

A wide variety of flue-gas cleaning processes exist for controlling SO_2 emissions. Site-specific considerations, economics, and other criteria govern the selection and manner of application of any one of these processes. The processes proposed for commercial use in the 1980s and 1990s may be ranked (high, medium, low) with respect to the major individual considerations reviewed in Tables 5.1 and 5.2.

Site-Specific Considerations[1]

Wet-limestone systems, the most common, have been installed across the country at plants of all sizes firing low-sulfur subbituminous coals, high-sulfur coals,

TABLE 5.1 Assessment Relative to Cost and Performance

	Criterion*				
FGD process	Operating cost	Capital cost	SO_2 removal	Reliability record	Commercial use
Limestone					
Natural oxidation	M	M	M	M	H
Forced oxidation	M	M	M	M	H
MgO-lime	M	M	H	H	H
High-calcium lime	M	M	M	M	M
Dual-alkali					
Lime	M	M	H	H	M
Limestone	L	M	H	—	—
Dry scrubbing	H	M	L	H	H
LIMB	M	L	L	—	—
Dry injection	M	L	L	—	—
Wellman-Lord	H	H	H	M	M
Regenerable MgO	M	H	H	M	L
Dowa	M	M	M	H	L
Aqueous carbonate	H	H	M	—	—
Sulf-X	M	H	M	—	—

*H = high, M = medium, L = low.

TABLE 5.2 Assessment with Respect to Flexibility of Application

	Criterion*				
FGD process	High sulfur	Low sulfur	Retrofit ease	Waste management	SO_2/NO_x removal
Limestone					
Natural oxidation	H	H	M	L	L
Forced oxidation	H	M	M	L	L
MgO-lime	H	L	M	L	M
High-calcium lime	H	M	M	L	L
Dual-alkali					
Lime	H	L	M	L	M
Limestone	H	L	M	L	L
Dry scrubbing	M	H	M	M	M
LIMB	L	H	H	M	M
Dry injection	L	H	H	L	L
Wellman-Lord	H	L	L	M	M
Regenerable MgO	H	M	M	H	M
Dowa	H	M	M	M	M
Aqueous carbonate	H	M	M	H	L
Sulf-X	H	M	M	H	H

*H = high, M = medium, L = low.

and Texas lignites. Limestone is an abundant raw material with deposits in many areas of the United States, and it is economical for use as a *flue-gas desulfurization* (FGD) reagent at powerplants located in the vicinity of such sources.

Wet-lime systems have been installed at powerplants of all sizes firing both low- and high-sulfur coals. The plants are predominantly in the Ohio River Valley, where a major source of magnesia-buffered lime affords unsaturated-$CaSO_4$-mode operation in high-sulfur coal service. Others are at specific locations in the west, where the relative cost of lime delivered to plant sites is preferable to that of limestone. Some high-sulfur coal-fired plants in the midwest have selected wet sodium-base dual-alkali systems with lime postprecipitation, an unsaturated-$CaSO_4$-mode operation, to achieve high SO_2 collection efficiencies and reduced FGD system maintenance costs.

Dry-scrubbing FGD systems have typically been selected at powerplants firing low-sulfur coals (bituminous and subbituminous)[2] or North Dakota lignite. Dry systems are thought to be most economical at units firing lower-sulfur coals, while wet systems have typically been most economical at units firing high-sulfur coals—the dividing line being at approximately the 1 percent coal sulfur level.[3] To accommodate waste dewatering operations using thickeners and vacuum filters, wet FGD systems generally require greater plot space than dry FGD systems. However, alternative use of hydroclones and/or centrifuges for wet-FGD waste dewatering substantially offsets this difference in space requirements.

The operation of wet-FGD processes may be affected by a number of plant and site-specific factors, including coal-ash analysis, makeup water quality, and the possible reuse of available plant wastewater streams in the FGD system. For FGD systems in northerly climates, weather enclosures should be provided to facilitate repair and protect against freezing.

Economics

Minimizing the overall cost of ownership and operation provides a principal basis for the selection and manner of application of FGD processes. In any specific application, particular conditions may exist that strongly affect the economic attractiveness of individual processes, such as the following:

- *Coal sulfur concentration:* With high-sulfur coal, processes that introduce low-cost reagents such as limestone will generally be found attractive in cost evaluations.
- *Solid-waste volume:* At utility plants firing bituminous coal, processes such as limestone forced oxidation, which improve the characteristics of solid waste and offer simple and versatile means of managing and disposing of this output, are increasingly favored because they reduce system cost.
- *Boiler size:* Sodium carbonate reagent with low-cost, clear-liquor scrubbers is suitable for relatively small industrial boilers, since waste is generally disposable as a totally water-soluble liquid effluent.

In a comprehensive economic evaluation of available FGD processes, comparative December 1982 levelized busbar costs of SO_2 removal (the added cost to produce power as a result of the FGD system) for two new 500-MW coal-fired generating units were developed.[4] For throwaway waste systems firing high-sulfur (4 percent) coal, levelized busbar costs in mills per kilowatthour were as follows:

natural-oxidation limestone, 18; forced-oxidation limestone, 16; natural-oxidation lime, 20; dual-alkali lime, 17; and dual-alkali limestone, 16.

For regenerative systems on high-sulfur coal, the busbar costs in mills per kilowatthour were Wellman-Lord, 26; MgO, 19; Dowa, 16; aqueous carbonate, 30; Sulf-X, 20. For throwaway waste systems firing low-sulfur (1 percent) coal, the busbar costs in mills per kilowatthour were natural-oxidation limestone, 8; dry scrubbing with lime, 7.5; dry injection with nahcolite, 8; dry injection with trona, 7.

WET THROWAWAY FGD

The most common application of flue-gas desulfurization in the United States has been, and continues to be, through the use of systems incorporating wet collectors that scrub and water-saturate the flue gas, yielding a sulfurous solid waste that is discarded.

Lime and Limestone Scrubbing

Lime or limestone alkali reagents are commonly used in wet-slurry scrubbing systems to absorb SO_2 and to yield a calcium-base solid-waste mass (see box).

REAGENT PREPARATION FOR FGD SCRUBBING

Limestone reagent is reduced in size in wet grinding systems so that it will chemically react in the FGD operation. Typical closed-circuit arrangements incorporate a weigh feeder, ball mill, slurry sump and pumps, and liquid-cyclone classifiers with instrumentation and controls—all oversized particles in the mill product being recycled until they are ground sufficiently to meet FGD reagent feed criteria.[5] The ball-mill size is determined by reagent feed capacity considerations, size analysis of the limestone supplied to the mill, required size of product to be supplied to the FGD operation, and resistance of the limestone supply to grinding, that is, its bond work index.[6]

For any individual mill, throughput capacity is reduced 10 percent by an increase in feed size from 0.25 to 0.75 in (6.3 to 19 mm). Feed larger than 0.75 in (19 mm) is not recommended for ball mills. The capacity is reduced to 65 percent if product requirements are changed from 200 to 325 mesh. The degree of limestone grind has a pronounced effect on FGD system characteristics, including internal scale control and mist-eliminator fouling as well as reagent utilization and reactivity. Most limestone preparation systems are designed to yield a limestone supply that falls in the range of 60 to 80 percent passing 325 mesh.

Lime scrubbing systems are supplied with quicklime (CaO) that is hydrated in a slaker which mixes a set proportion of lime and water under appropriate conditions of agitation and temperature for dispersing soft hydrated particles formed and yielding a milk of lime.[7] Detention, paste, and ball-mill slakers are used. Grit consisting of sand and other solid impurities in the quicklime feed is generally separated from the slaked lime product, typically by gravity settling or by screening. The viscosity of the lime milk is reduced by water dilution, if necessary, to achieve adequate degritting.

Process Using Limestone Slurry. In this process, a slurry of precipitated SO_2 reaction product and limestone is circulated, generally in a spray tower, to absorb SO_2 and convert it to a calcium sulfite/sulfate solid waste.

System Description. The forced- and natural-oxidation FGD processes are available. In the latter, the FGD process most commonly selected in the United States in the 1970s, the catch is converted primarily to calcium sulfite solids accompanied by calcium sulfate formed through oxidation of sulfite by oxygen contained in the inlet flue gas. In Fig. 5.1, dedusted flue gas is contacted countercurrently by slurry of calcium sulfite, calcium sulfate, and limestone (calcium carbonate) solids to absorb SO_2, forming dissolved sulfite and sulfate, some of which precipitates in the absorber to form suspended solids. Entrained scrubbing slurry is removed from the scrubbed gas prior to discharge from the plant stack.

The balance of dissolved sulfite and sulfate created in the absorption step is reacted with the limestone reagent feed in the effluent hold tank, reducing dissolved calcium and sulfate to a low level of supersaturation and precipitating additional reaction-product waste. Suspended solids in a bleed stream from the recirculating slurry loop are discarded—generally after primary dewatering in a thickener followed by final dewatering in a vacuum filter. Dewatering prepares the solids for stabilization by intermixing with dry flyash for optimum control of water pollution in disposal by landfill.

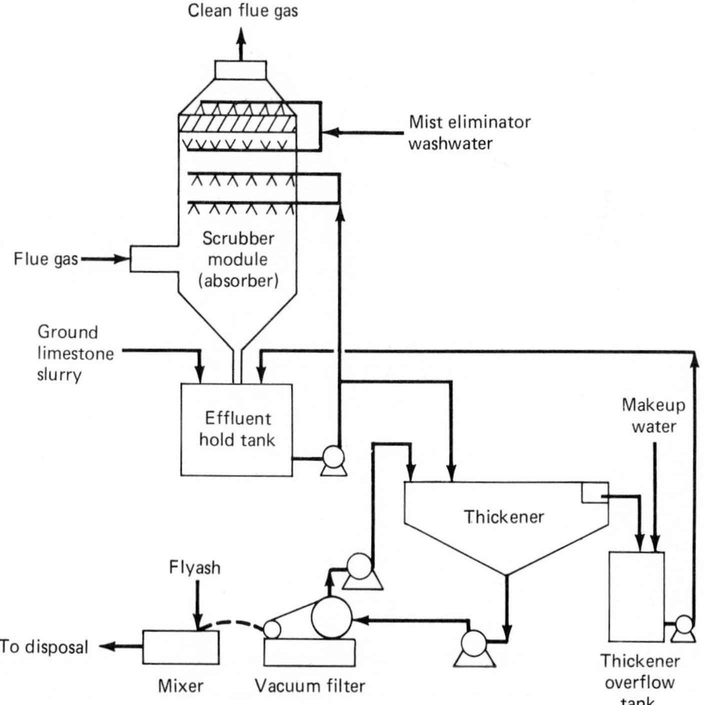

FIG. 5.1 Limestone natural-oxidation process.

In the *forced-oxidation* (FO) FGD process, commonly selected in the United States during the 1980s over the *natural-oxidation* (NO) process, compressed air is introduced to the system to render the solid waste as gypsum (calcium sulfate) or a gypsumlike mass that has better settling, filtering, and structural or physical properties than solids of predominantly calcium sulfite. As shown in Fig. 5.2, an oxidation tank is interposed between the absorber slurry outlet and the recirculating tank, the compressed air contacting the slurry at the low-pH point of the system, where calcium sulfite solubility and reaction with oxygen are the greatest. Gypsum solids may be dewatered to a filter cake for landfill disposal, or the scrubber bleed stream may be pumped directly to the disposal site, by using a method of wet stacking similar to that used in the phosphate fertilizer industry.

Advantages and Disadvantages. The systems offer comparative simplicity at relatively low cost. In the case of FO, it is readily adapted to either throwaway or by-product utilization. The growing use of organic acid additives to solubilize limestone reagent substantially improves SO_2 removal and offers potential for economies in system design.

A recent evaluation[8] of limestone slurry scrubbing alternatives for high-sulfur utility plant service indicates these levelized costs, in mills per kilowatthour: NO operation with ponding of waste, 14.42; FO with discarding waste as filter cake, 15.08; and FO mode with adipic acid additive, 13.31. Generally operating in a supersaturated $CaSO_4$ mode with limited soluble sulfite alkalinity, limestone slurry scrubbing systems have had major reliability and maintenance shortcomings, particularly in high-sulfur coal applications where typical average availability is less than 80 percent.

Equipment Design. A myriad of details, some of them described in Table 5.3,[9] go into the development of an engineered FGD system design. These system components are critical to system performance and reliability:

Absorber Spray towers, open vessels in which scrubbing slurry from atomizing nozzles contacts the gas stream, are most often selected for slurry scrubbing systems. In a conservative design,[9] flue-gas velocity is limited to roughly 8 ft/s (2.5 m/s) to prevent excessive carry-over of slurry to the mist eliminator (Fig. 5.3). In typical spray towers,[6] the pressure imparted to the scrubbing slurry discharged from the spray nozzles, together with the velocity of the incoming gas stream, produces fine liquid droplets that serve as sites for SO_2 absorption.

Nozzle pressures of 15 to 20 psig (103 to 138 kPa) are normally established in spray towers to produce droplets 0.1 to 0.16 in (2.5 to 4 mm) in diameter. Droplets in this size range provide sufficient surface area for SO_2 absorption, and the entrainment problems normally associated with smaller droplets are minimized. Relatively low gas-phase pressure drops, approximately 1 to 4 in H_2O (250 to 1000 Pa), are normally encountered because the spray tower includes no internal parts other than spray headers.

The number of spray headers (spray banks) through which slurry is fed to the nozzles varies with the amount of SO_2 loading and the required SO_2 removal efficiency. One to six spray headers are installed in limestone spray towers in commercial FGD systems operating on coal-fired flue gas. The towers must include enough spray nozzles to provide a spray zone of uniform density. Placement of nozzles so as to provide a considerable overlap of slurry spray reduces the problems associated with nozzle failure, which could create a path of least resistance to gas flow. Nozzles should be fabricated of abrasion-resistant material such as silicon carbide or ceramic.

High *liquid/gas* (L/G) ratios improve SO_2 removal by increasing the surface

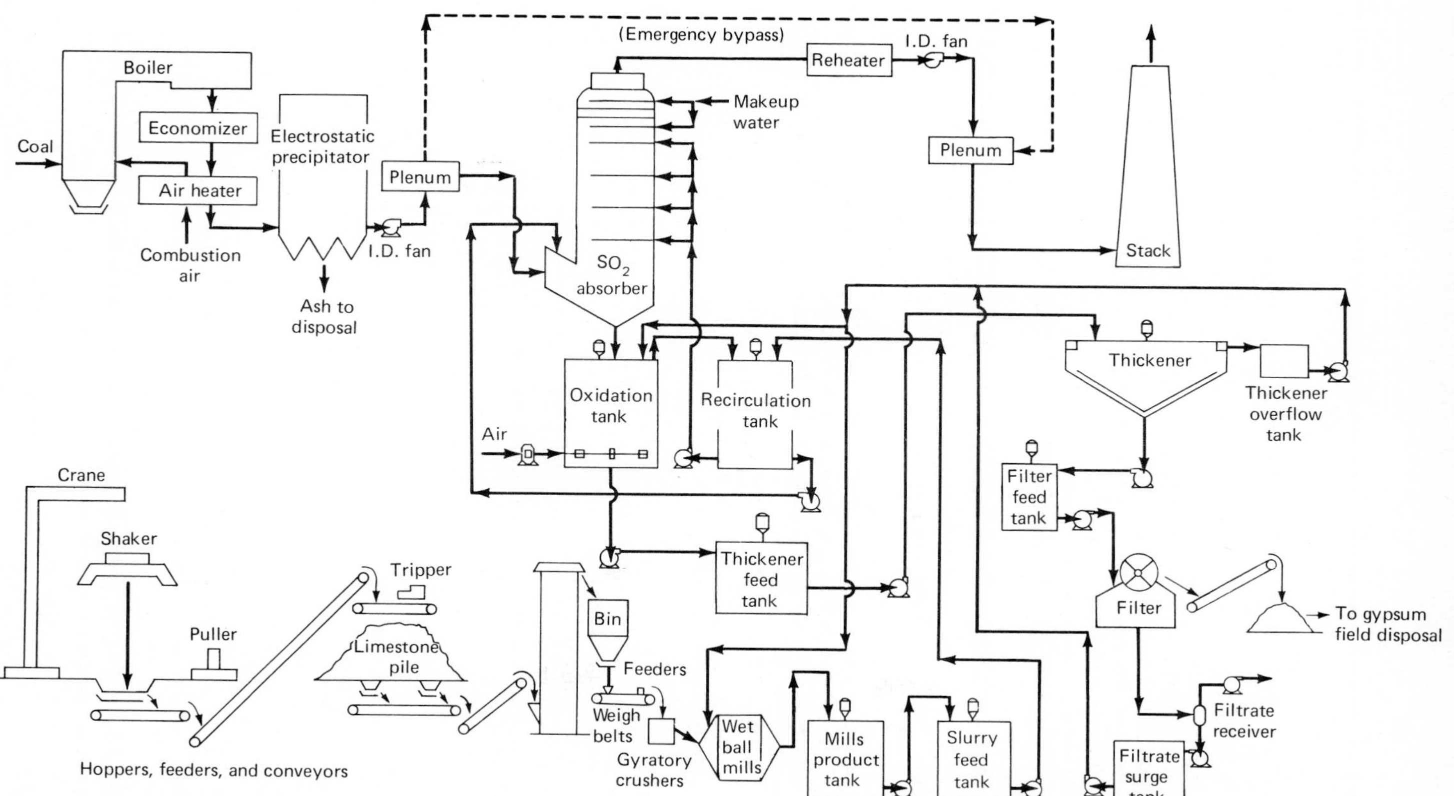

FIG. 5.2 Limestone forced-oxidation process.

TABLE 5.3 Reliable Design Features of Mechanical Equipment

Dampers—Double louver and guillotine
• Dual chain and sprocket drives, one on each side of blades on guillotine (avoid screw jacks) • Alloy blade construction on absorber outlet, carbon-steel blade construction on absorber inlet, alloy 625 seals on all dampers • Seal air system (alloy 625) including dampers, fans, and ductwork designed for 5 in H_2O (1245 Pa) over duct pressure • 200% maximum torque on operators to allow full closure under maximum fan pressure and/or dirty conditions • Accessibility of moving parts (outside of gas stream), such as linkages, bearings, motor drivers, drive mechanisms, and other maintenance items • Aerodynamic blade shape for double louver to minimize flow resistance and pressure drop • Bearings to withstand high temperature, self-lubricating, self-cleaning, self-aligning, corrosion-resistant, and gastight for long life and to minimize maintenance • Manual handwheel to override drive failures
Pumps—Slurry service
• Split pump casing to enable maintenance and replacement of parts • Grease-lubricated, antifriction, heavy-duty roller bearings, with minimum L-10 life of 100,000 h • Rubber liners with durometer of 35 to 50, natural rubber used only at temperatures less than 130°F (54°C), sufficient lining thickness to allow for erosion of liner [minimum ⅛ in (3.2 mm)], easily replaceable liners • Tip speeds for impellers less than 5000 ft/min (25.4 m/s) and pump speeds less than 1200 rpm • Wearing parts removable without disturbing pedestal, bearing, and drive assembly • Through-bolts for casing • No asbestos packing or gaskets • No rubber exposed to oils, solvents, or hydrocarbon
Valves—Slurry service
• Full port design (pinch, knife gate, butterfly, or eccentric ball); no areas where solids collect on packing; minimize pressure drop and wear points • Valve materials—rubber-lined bodies and internal parts (stainless-steel knife gates) • Flanged connections • Enclosed cast-iron body design (pinch valve) to prevent spills • Internal parts accessible and replaceable while valve body in-line • Field-repairable • Position indicators to prevent overtorquing • Spare parts in stock (e.g., knife gates, sleeve liners) • No slurry control valves
Induced-draft fans
• Sized for 110% of gas flow and 121% of gas pressure at maximum continuous rating of boiler • Located after particulate removal equipment and ahead of absorbers, dry service • Self-contained lubrication system and thermocouples on bearings • Vibration monitoring system • Inlet screens and silencers • Tip speeds less than 22,000 ft/min (112 m/s)

TABLE 5.3 Reliable Design Features of Mechanical Equipment *(Continued)*

Agitators
• Chemical- and erosion-resistant linings on wetted parts • Shaft removable without disturbing speed reducers or motor • Shaft operating speed with impeller—below 80% of first resonant frequency • Motor and couplings protected from slurry • Coupling designed to transmit 200% of full torque and 200% of axial load • Minimum torque per unit volume—1 in • lb/gal (0.03 J/L) (for recycle tank) • Minimum L-10 bearing life of 100,000 h • Grease-lubricated bearings of antifriction type, no pressed-on bearings • Solid shafts, coupling for every 20-ft (6-m) length • Top entering mixers; side entering not recommended • Abrasion-resistant tank bottom or wear plate for slurry service • Ability to run agitator in reverse by switching rotor leads; ability to start when buried in slurry • Helical and/or spiral bevel gearing; no worm gears
Thickener and thickener rakes
• Sized for 25% to 45% (35% average) by weight-hydrated solids • Sized at 110% of full-load slurry flows • Torque measurement for rake arms • Automatic rake-lifting mechanism, activated when overtorquing sensed (set at 50% of design torque) • Solids loading—30 ft^2/ton • day (3 m^2/Mg • day), size not based on use of polyelectrolyte additive • Redundant underflow pumps • Drain from thickener to emergency pond • Rake arms equipped with variable-speed gear mechanism • Rake arms—steel truss construction only (rubber-lined) with no support tie rods—fixed-type (not hinged) • Walkway to tank center, access to drive machinery for maintenance and repairs • V-notch design weirs • Limit switches for raised and lowered positions of rake mechanism • No pumps in thickener column
Belt conveyors
• Maximum belt conveyor speed of 300 ft/min (1.5 m/s) • Synthetic belt material (moisture- and mildew-resistant) • Gear-driven drive motors over 300 hp (224 kW) • Drives with safety guards and adjustment between motor and reducer • Walkways along conveyors • Outdoor portions of belt conveyors equipped with semicircular hood covers • Hood cover materials—minimum 14-gage corrugated aluminum, stiffened with ¼-in (6-mm) angles or plates • Troughing angle of all load-carrying idlers equal to 35° • Return idlers to be self-cleaning • Troughing training idlers to automatically maintain belt alignment • Labyrinth-type seals to protect bearings from dirt and moisture • Access to grease pipes of idler bearings from conveyor walkway

TABLE 5.3 Reliable Design Features of Mechanical Equipment *(Continued)*

Radial stackers
• Mast type with electric power drive capable of vertical movement of boom arm • Radial movement up to 120° arc • Designed for sludge at 40% (wet) to 80% (dusty) hydrated solids by weight • Maximum elevation above grade based on minimum angle of repose of 40°
Ball mills
• Designed for 110% capacity at full load • Two manholes per compartment • Slurry product—minimum fineness of 80 percent through 200-mesh screen • Ball-mill slurry-circuits piping cross-connected between redundant mills • Sumps sized for flooding condition during improper mill shutdown • Ball loading and handling system • Slurry density monitoring and control system • Bearing lubrication system • High-lift pumps to be an integral part of ball mill • Trunnion bearings of self-aligning type with replaceable bearing inserts
Vacuum filters
• Designed according to solids composition, solids content, particle-size distribution, and filter-cake dryness • Rubber-lined drum, vat, and equipment exposed to slurry and filtrate • Bypass piping around filters to be used for low-sludge and high-ash flows • Drains for cleaning and flushing filter vat to remove accumulated grit • Filter cloths with demonstrated nonblinding characteristics (such as multifilament nylon) • Air blow-back from service air or separate blowers to release filter cake from cloth
Pug mill mixers
• Designed for complete range of ash-to-sludge ratios which can occur • Maximum 60-s residence time • Motor and gears sized for "dirty" conditions • Wet-scrubber-type dust collectors discharging back into process • Abrasion-resistant liner and Ni-hard paddles • Regular flushing and washing • Direct flyash feed to mixer through rotary valves, no screw conveyors • Provision to accept thickener underflow sludge directly for mixing with flyash • Access panels on sides and top for cleanout

Source: Stone and Webster Engineering Corp.

area for mass transfer and by reducing the high liquid-film resistance to SO_2 absorption. Moreover, high L/G ratios reduce the potential for scale formation in the scrubber by preventing any instantaneous $CaSO_4$ supersaturation beyond 30 percent of normal solubilities.

Reaction Delay Tank Also known as an *effluent hold tank* (EHT),[6] the reaction delay tank must provide sufficient retention time to relieve supersaturation of the liquor and thereby minimize scaling. The EHT may be a separate tank or form the bottom part of the scrubber vessel that serves as a reservoir. This is a critical design factor: Size the EHT so that it provides enough holdup time to ensure optimum limestone utilization and precipitation of gypsum. Required holdup time is a function of (1) the degree of supersaturation allowed to take place during SO_2 absorption and (2) the tank design.

High-Calcium. High-calcium quicklime, containing less than 5 percent MgO, has found application primarily when it is available locally as a by-product carbide lime. High-calcium lime systems, which operate in the supersaturated-$CaSO_4$ mode, have not reached the above 90 percent availabilities sustained in magnesia-buffered operations, and are no longer being applied in new, high-sulfur FGD installations.

Other Throwaway Wet Processes

Dual-Alkali. A number of high-sulfur coal installations have used dual-alkali FGD systems to avoid the presence of suspended solids in the recirculating scrubbing medium and to gain the benefits of an unsaturated-$CaSO_4$ mode of operation. The flue gas is contacted with a clear liquor, typically of dissolved sodium sulfite/sulfate, the absorbed SO_2 converting sulfite to bisulfite. A bleed stream from the absorber is reacted with calcium-base reagent to regenerate sodium sulfite, precipitating a calcium sulfite/sulfate sludge that is thickened, vacuum-filtered, and water-washed for recovery of sodium values prior to disposal of the solid waste in a landfill.

Lime-Based. All early commercial installations of dual-alkali FGD have used slaked high-calcium quicklime as the sacrificial alkali for sulfite regeneration.

Process Operation Flyash is removed in an electrostatic precipitator, after which flue gas is scrubbed with a recirculating sodium-sulfite-rich liquor, absorbing SO_2 to form sodium bisulfite. A liquor purge stream taken at the absorber outlet is reacted with slaked lime in a postprecipitation tank to form a calcium sulfite/sulfate sludge. It is dewatered to a cake that is freshwater-washed to limit the loss of sodium in the discarded solids. The reaction with lime regenerates dissolved sodium sulfite, which is recycled to the absorber.

Advantages and Disadvantages Because of the high concentration of dissolved sodium sulfite in the recirculating scrubbing medium, dual-alkali systems operate unsaturated with respect to calcium sulfate as do magnesia-buffered lime scrubbing systems. Installations of each of these unsaturated-mode processes, invariably in high-sulfur service, have achieved very high system availability (over 90 percent) as contrasted with the more common supersaturated-mode lime/limestone scrubbing processes that have been applied in the United States.

The dual-alkali system provides high SO_2 removal efficiency at low liquid/gas flow ratios. The process has the disadvantage of requiring makeup of soda ash to maintain dissolved sodium concentration in the scrubbing liquor. Because of the fine-grained, predominantly sulfite solids in the solid waste, the dual-alkali method, like other natural-oxidation processes, generally requires fixation of the sludge solids by intermixing of flyash and lime additive.

Economics A recent evaluation of available FGD processes by the *Electric Power Research Institute* (EPRI)[4] indicates that the levelized busbar cost for dual alkali in high-sulfur coal service is closely comparable to that of limestone slurry scrubbing systems. The demonstrated higher availability of existing dual-alkali installations may thus provide a significant economic edge favoring selection of dual alkali in some high-sulfur applications.

Limestone-Based. Dual-alkali systems relying on limestone reagent instead of lime have been undergoing testing for purposes of commercial application in the United States.

Sodium-Scrubbing Type[19] Beginning in the late 1970s, extensive efforts have been made to develop sodium-scrubbing dual-alkali systems using limestone regeneration. The process has been found to be technically feasible. Subsequent field demonstration activity employs lower pH than lime-regenerated systems and limits the regeneration fraction—the percentage of bisulfite in the scrubbing bleed solution that is regenerated in the regeneration section—to only 35 to 60 percent. Economic evaluation by the *Tennessee Valley Authority* (TVA) indicates that limestone-regenerated dual alkali has direct costs about 13 percent lower than those of forced-oxidation limestone scrubbing, primarily because of lower maintenance and power costs. Other advantages include high availability, high flow-turndown and load-following capability, and reduced disposal space requirements.

Dowa Process Used commonly in Japan in smelter and contact-sulfuric-acid plants, this dual-alkali process employing a 3 to 3.5 pH liquor of $Al_2(SO_4)_3 \cdot Al_2O_3$ (basic aluminum sulfate) in the gas-scrubbing loop is offered for commercial use in coal-fired boiler service. The SO_2 absorption forms dissolved $Al_2(SO_4)_3 \cdot Al_2(SO_3)_3$ that is converted in an external oxidation tower to $Al_2(SO_4)_3 \cdot Al_2(SO_4)_3$, which is bled off to a limestone postprecipitation step, yielding potentially usable gypsum and regenerating the $Al_2(SO_4)_3 \cdot Al_2O_3$ reagent.[19] Before chloride purge is discarded from the system, it is treated with an excess of limestone to precipitate aluminum ions as aluminum hydroxide, which is recovered for reuse.

The Dowa process offers unsaturated-$CaSO_4$-mode dual-alkali operation and the use of limestone instead of lime. Because of low operating pH, it also achieves an extremely high level of limestone utilization with a minimum solid-waste quantity—typically 10 percent less than for forced-oxidation limestone. Thus, in utility plant applications with high-sulfur coal, the Dowa process has been estimated to have levelized busbar costs more than 10 percent less than either natural-oxidation or forced-oxidation wet-limestone scrubbing systems.[4] The principal disadvantages are a potentially high liquid-purge rate and process complexity compared to the simplest limestone slurry scrubbing systems.

Flyash Scrubbing. Coal from some sources, including Minnesota and North Dakota mines, may have a sufficiently high alkali content (CaO, MgO, Na_2O, and K_2O) to make flyash an attractive reagent in slurry scrubbing operations—substituting for some or all of the limestone that would otherwise be required to achieve target SO_2 removal.[20] Metals such as iron and manganese dissolve in the recirculating scrubbing slurry and catalyze oxidation of absorbed SO_2, typically converting 90 percent of the catch to sulfate, reducing SO_2 vapor pressure, and yielding an unusually low absorber pH (3.5 to 5), with good control of internal chemical scaling.

When it is economical to do so for purposes of meeting particulate emissions and stack opacity requirements, a venturi scrubber can be inserted upstream of the absorber to collect the flyash and assimilate it in the wet-scrubbing system. Thus flyash scrubbing, when applicable, has advantages similar to those of forced oxidation together with quite low reagent cost. Its overall cost may not be attractive, however, unless the venturi scrubber meets stack opacity requirements at a reasonably low flue-gas pressure drop.

Clear-Liquor Scrubbing. Typically, clear-liquor scrubbing utilizes sodium-base alkali (sodium carbonate or sodium hydroxide) to absorb SO_2 in a comparatively clear solution of sulfite, bisulfite, and sulfate. (The SO_2 catch is discarded as a clear-liquid effluent purge stream, usually after aeration to convert sulfite forms to sulfate.) Except for one or two installations in the arid west near an economical supply of soda ash (sodium carbonate) reagent, utility plants have not capitalized on this simple, trouble-free FGD technique because of (1) an inability to dispose of the comparatively large output of liquid waste and (2) the high cost of sodium alkalis such as soda ash.

Clear-liquor scrubbing is the typical means of SO_2 removal in smaller industrial boiler applications, the wet collection of flyash and SO_2 often being carried out simultaneously in a venturi scrubber that is pH-controlled to prevent corrosion and to limit SO_2 emissions to a target level. In some industrial applications, soluble alkali wastes are used to meet all or part of the FGD reagent requirement. In some industries, including some types of pulp and paper mills, sulfite liquor formed by FGD may be recovered for use in plant processes.

DRY THROWAWAY FGD

Dry FGD systems are of great value in some applications because they remove SO_2 without saturating the flue gas with water, and they yield a dry waste product.

Dry Scrubbing

Dry-scrubbing systems, generally installed without prior dedusting of flue gas, typically contact the flue gas with slaked lime slurry in a spray dryer, thereafter collecting the dry SO_2 reaction product along with flyash in a fabric filter or an electrostatic precipitator. In lieu of a spray dryer, a circulating-fluid-bed contacting device has also been used commercially,[21] primarily in high-sulfur applications.

Process Operation. The process passes a flue gas containing flyash and SO_2 through a spray-dryer type of absorber (Fig. 5.4), where it is contacted with a finely atomized slurry or solution of reagent feed. The SO_2 is absorbed into the alkaline droplets as water is simultaneously evaporated. The reactions taking place are nearly the same as in wet-lime or clear-liquor scrubbing.

Solid sulfites and sulfates formed in the dryer along with flyash and excess reagent exit the dryer with the scrubbed gas and are collected in either a high-efficiency electrostatic precipitator or a fabric filter. Additional absorption may take place in the dust collector. Recycling a portion of the collected solids to the reagent feed system can improve reagent utilization, solids drying, and feed-slurry handling. Dry scrubbing produces no liquid wastes. Solid waste is a dry mixture comprising the boiler flyash catch and the sulfite/sulfate SO_2 reaction product.

Besides having potential for SO_2 removal efficiency approaching that of wet scrubbing, dry scrubbing offers a simplified system without the need for solid-waste dewatering equipment, large rubber-lined recirculating slurry pumps and piping, or flue-gas reheat facilities. Unlike with common wet scrubbing, the spray-drying action achieves in excess of 90 percent SO_3 removal, thus greatly lessening potential downstream corrosion tendencies. The principal disadvantages are lack of the means of achieving commercial utilization of SO_2 catch and high operating cost because of the expensive reagent, typically quicklime.

Theory and Design.[22] Design considerations for the spray dryer include the size, flue-gas conditions (temperature, moisture, SO_2 content), type of atomization

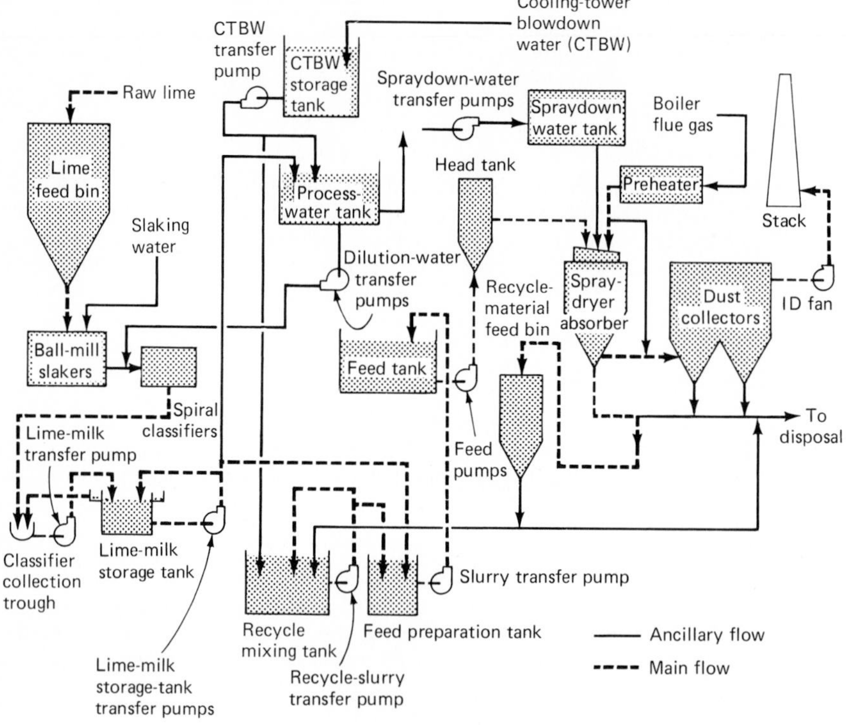

FIG. 5.4 Dry-scrubbing process. (Power *magazine*.)

(see box), types of gas dispersers, and approach to saturation temperature. The size of the spray dryer is a function of the flue-gas flow rate and spray-down temperature (difference between inlet and outlet temperatures). The spray-dryer volume must allow for adequate flue-gas and reagent mixing and must achieve sufficient flue-gas residence time to allow for high SO_2 removal and to produce a dry product (nominally 10 s for a lime-based process).

The amount by which the outlet temperature exceeds the flue-gas wet-bulb (saturation) temperature inversely influences SO_2 removal efficiency and solid-

ATOMIZER FUNCTIONS IN DRY SCRUBBING

The function of the atomizer is the production of large numbers of fine droplets from the reagent feed slurry. Both rotary atomizers and dual-fluid nozzles are used in dry scrubbing, although most large installations rely on rotary atomizers. The degree of droplet atomization, a function of the velocity and diameter of the atomizer wheel, influences SO_2 removal and solids moisture content. Internal devices such as a compound gas disperser achieve intimate mixing of the entering flue gas and the atomized droplets, preferably by introducing the inlet gas both above and below the atomized slurry cloud formed.

waste moisture content. This temperature approach is generally no less than 18°F (10°C), to permit control of wetness downstream of the spray dryer.

Fabric filters are selected for 90 percent of dry-scrubbing installations and contribute more to SO_2 removal efficiency than electrostatic precipitators in this service because of the effectiveness of available alkalinity in the solids collected on the bags. Fabric filters are also preferred in dry scrubbing because their collection performance is less sensitive to variations in spray-dryer operation.

Dust collection devices downstream of a spray dryer operate close to the water dew point temperature and thus may require additional insulation to minimize heat loss and cold spots. Typically, insulation is specified to achieve a maximum of 10°F (5.6°C) gas temperature drop across the dust collector. Also, special care must be taken around all access doors.

A majority of lime-based systems employ ball-mill slakers for lime slaking, although paste, detention, and tower-mill slakers are also used. Ball-mill slakers offer the advantage of smaller absorbent particle size and eliminate the need for a grit removal system.

Limestone Injection into a Multistage Burner

Early FGD field evaluation beginning in the 1960s of dry-limestone injection into utility boilers gave unsatisfactory results because of poor SO_2 removal (15 to 50 percent) and substantial loss of feed caused by dead-burning and nonuniform distribution. As applied, it also caused other major operational problems, such as tube fouling and impairment of precipitator performance. More recent use of low-NO_x burners with reduced peak-flame temperature and enlarged fuel-rich furnace zones provides an improved means for overcoming these limitations and successfully applying dry-limestone FGD by injecting the sorbent with either the staged air or the pulverized coal.[23]

Early LIMB Performance. The *limestone injection into a multistage burner* (LIMB) installation is comparatively simple to retrofit to coal-fired boilers of varied size and is low in cost compared to available FGD alternatives. Results from bench- and pilot-scale testing by the EPA indicate that limestone calcination conditions that enhance the effective surface area of the calcine will at the same time improve capture of SO_2. In so doing, it will be possible to create momentarily 6100 ft^2 (20 m^2) or more of surface area per ounce (gram) of limestone. Maximizing the presence of high-surface-area lime, in part by selecting an optimized limestone-feed grind (particle size), is vital in optimizing LIMB performance.

Also, favorable temperature conditions for up to approximately 1 s after calcination must be provided for sulfation reaction—that is, 1560 to 2370°F (850 to 1300°C)—through appropriate choice of limestone feed location. This selection may be made from several of the combustion air injection locations, e.g., auxiliary air and overfire air, or the coal injection location.

Process Operation. A general schematic of LIMB is shown in Fig. 5.5 as it may be applied in the late 1980s after completion of test and demonstration activities.[24] It will serve as a combined NO_x/SO_x control system for a pulverized-coal-fired powerplant. Coal is fired in low-NO_x burners to minimize NO_x emissions, and SO_x emissions are reduced by a combination of in-furnace capture and downstream cleanup. The sorbent could be (1) mixed with the coal prior to the pulverizer, or mixed with the coal after grinding and fed to the furnace through

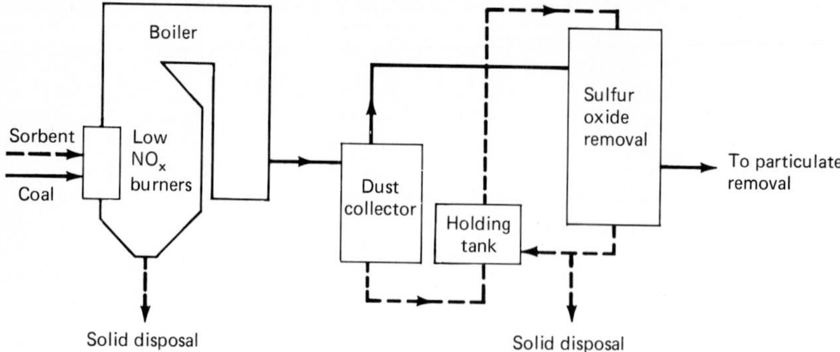

FIG. 5.5 Generalized LIMB process.

the burner, or (2) mixed with one of the combustion airstreams (secondary or tertiary staged air) and then injected into the furnace.

The addition of sorbent will increase the total solids loading in the furnace and may also cause problems from slagging and fouling. A dust collector downstream of the air heater removes the sorbent, which could then be sent to a wet or dry contactor to reduce flue-gas SO_2 content further.

Advantages and Disadvantages. A major advantage of LIMB is its low capital cost, which may (1) make it feasible and attractive for retrofit to existing sources and (2) provide an optimum means for substantial reduction of SO_2 emissions from utility boilers having limited remaining lifetimes. The disadvantages of LIMB include uncertainties as to its feasibility for continuous high-efficiency SO_2 removal and concerns about its possible impact on boiler reliability.

Waste Product. For a specific low-sulfur bituminous coal,[25] the calculated composition of solid waste collected in the flue-gas dust collector will vary as a function of the ratio of the limestone feed to the coal sulfur molar value (see Table 5.5). Due to its high lime content, the solid waste may be marketable for specific industrial uses.

TABLE 5.5 Waste Products from LIMB Operation[24]

Constituent	Composition, %	
	Ca/S = 2	Ca/S = 4
Flyash	61.1	45.4
CaO	21.0	33.6
$CaSO_4$	17.3	20.5
$CaCl_2$	0.5	0.4
CaF_2	0.1	0.1

Dry Injection

Dry injection is an FGD method particularly tied to low-sulfur coal service requiring only 70 percent SO_2 removal. Finely divided dry sorbent such as nahcolite ($NaHCO_3$) or Trona ($NaHCO_3$, Na_2CO_3) from Owens Lake, California, is injected in the air heater or directly into the flue-gas duct upstream of particulate collection equipment. The dry sorbent reacts with flue-gas SO_2 in the duct and in the particulate collection equipment.

In a full-scale demonstration on a 22-MW pulverized-coal-fired boiler equipped with a fabric filter,[26] technical feasibility for low-sulfur service using nahcolite has been proved. The SO_2 removal of 70 percent was consistently achieved with over 90 percent alkali utilization, and no adverse impacts on the performance of the fabric filter were experienced because of nahcolite injection. However, at low-load conditions with the flue-gas temperature less than 275°F (135°C) at the fabric filter, SO_2 removal is impaired unless provision is made to predecompose the bicarbonate by subjecting the nahcolite to an intermediate temperature of 450 to 500°F (230 to 260°C).

In an assessment of alternative FGD technologies for low-sulfur coal,[4] dry injection of sodium compounds has been found (1) to require less than 25 percent of the capital investment for wet- or dry-scrubbing methods and (2) to have a levelized busbar cost approximately equal to these conventional methods. Potential disadvantages of dry injection in some site-specific cases are the possible long shipping distances from western U.S. sources of these minerals and possible problems in managing the disposal of the water-soluble throwaway waste material, which is intermixed with the flyash catch.

REGENERATIVE FGD

These FGD processes regenerate the alkaline reagent and convert the SO_2 catch to a usable chemical by-product. To minimize the impact of tramp components of the dedusted flue gas on product quality and FGD operations, hydrogen chloride and other constituents collectible at very low pH are generally removed from the inlet gas in a separate prescrubber system upstream of the SO_2 absorption process.

Wellman-Lord Process

The Wellman-Lord process offers the most highly demonstrated regenerative technology in the world.

Process Operation. In the absorber (Fig. 5.6), flue-gas SO_2 reacts with sulfite ions in a concentrated aqueous sodium sulfite/bisulfite liquor.[27] Subsequently, bisulfite in the spent absorber solution is regenerated to active sulfite by thermal decomposition using steam in a double-effect evaporator-crystallizer, driving off a concentrated SO_2 gas product. The regenerated sodium sulfite solution is returned to the absorber, while the SO_2 product is cooled to remove water vapor and is compressed for further processing. The yield is either concentrated sulfuric acid or elemental sulfur for sale. A small proportion of collected SO_2 oxidizes to the sulfate form and is converted in a crystallizer to sodium sulfate solids that are marketed as a salt cake.

Advantages and Disadvantages. The process offers a proven means of achieving these key advantages: minimal solid-waste production and alkaline-reagent consumption. Also, because of soluble sulfite scrubbing, the process is capable of high SO_2 removal rates without internal chemical scaling. It is a complex process,

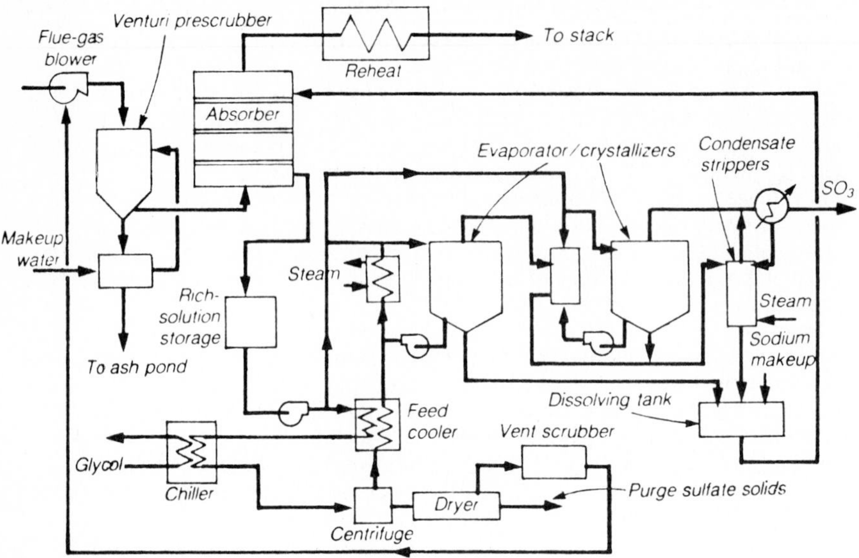

FIG. 5.6 Wellman-Lord process. (Power *magazine*.)

however, with high plot area, energy, and maintenance requirements. The process has no provision for regeneration of sulfate formed by incidental oxidation, which adds to the complexity of waste management. Its economics are not attractive in common coal-fired applications; in high-sulfur utility plant service, the process has been estimated to have a levelized busbar cost more than 40 percent greater than that of limestone throwaway waste scrubbing.[4]

Regenerative Magnesia Scrubbing

The MgO process is in commercial service in the U.S. utilities industry.

Process Operation. After low-pH prescrubbing (Fig. 5.7), the flue gas is scrubbed with a magnesium sulfite/sulfate slurry, the liquid phase of which is a magnesium sulfite/bisulfite solution containing a high level of dissolved sulfate.[28] A bleed stream of scrubber slurry is centrifuged to form a wet cake containing 75 to 90 percent solids. The cake is then dried at 400°F (204°C) in a fuel-fired rotary dryer to form a dry, free-flowing mixture of magnesium sulfite and sulfate.

This mixture is decomposed to SO_2 and MgO in a direct-fired fluidized bed or rotary-kiln calciner maintained at about 1800°F (982°C) to achieve decomposition of a substantial proportion of the magnesium sulfate. The regenerated MgO is reused in the absorber after slaking. The SO_2 product gas is generally washed and quenched and fed to a contact-sulfuric-acid plant to produce concentrated sulfuric acid by-product.

Advantages and Disadvantages. The capability of the process to regenerate the tramp sulfate produced by oxidation effects greatly simplifies waste management. Other advantages include high SO_2 removal efficiencies (up to 99 percent), min-

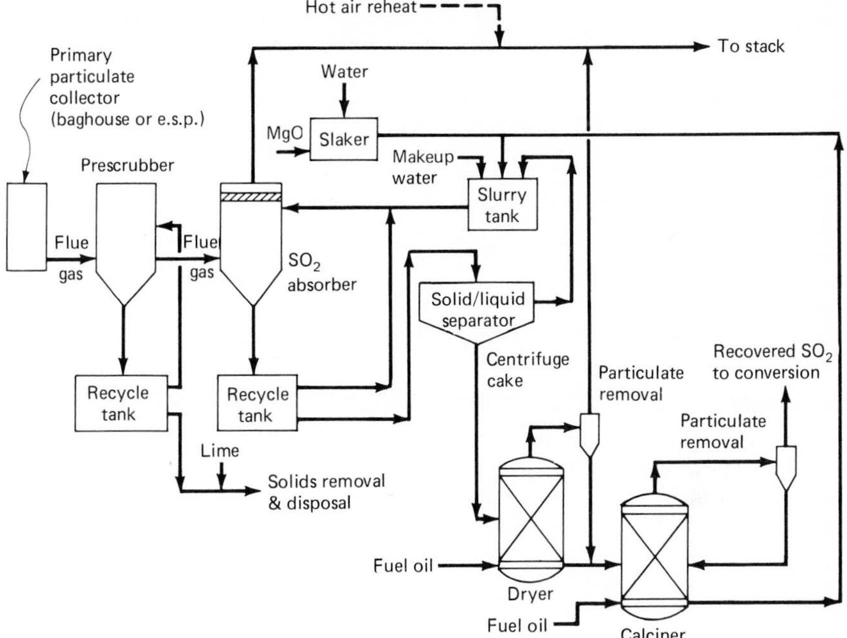

FIG. 5.7 Regenerative magnesia-scrubbing process. (Power *magazine*.)

imum impact of fluctuations in inlet SO_2 levels on removal efficiency, and very low chemical-scale potential. Moreover, its reported economics[4] are more favorable than for other available regenerative processes, its levelized busbar cost in high-sulfur applications being only about 5 to 15 percent greater than that for limestone throwaway-waste scrubbing. The principal disadvantage of the process is its complexity, including the need for direct-contact thermal regeneration of product solids and introduction of a contact sulfuric acid plant to produce a potentially salable by-product.

Other Regenerative Processes

Three additional regenerative processes have undergone demonstration and seen service at commercial-scale installations.

Ammonia Processes. Numerous ammonia-based scrubbing processes have been developed and marketed in the United States.[28] Most of these involved once-through ammonia scrubbing in which an ammonia-laden solution is contacted with flue gas, producing a mixed ammonium sulfite/sulfate liquor. The ammonium sulfite is then oxidized to sulfate, and the ammonium sulfate is recovered from the liquor by evaporation of water—normally with flue-gas heat prior to the scrubbing step. Ammonia scrubbing offers the advantages of high SO_2 removal efficiency; a simple, well-defined chemistry developed through years of use; and a relatively uncomplicated process configuration. The principal disadvantage is high ammonia consumption; thus a well-established market for the ammonium sulfate by-product is necessary as well as close control of the absorbers to avoid ammonium-particulate plumes and to minimize ammonia blowdown.

Aqueous Carbonate Process. This process treats dedusted flue gas in a sodium-carbonate-based dry-scrubbing system to produce sodium-sulfite-rich solids. This product is converted to elemental sulfur via a series of steps,[28] including chemical reduction to sodium sulfide melt by coke or coal in a molten-bed reactor at 1800°F (982°C); quenching of the melt in a tank to form a 220°F (104°C) liquor of sodium sulfide and carbonate, subsequently cooled and filtered; carbonation of the liquor by CO_2 with water releasing H_2S and forming a slurry of sodium carbonate/bicarbonate, subsequently decomposed thermally to produce a sodium carbonate solution for reuse in flue-gas scrubbing; and conversion of the H_2S to elemental sulfur by-product in a Claus plant.

Advantages include use of efficient sodium-base dry scrubbing instead of wet scrubbing and production of a desirable elemental sulfur by-product without the need for costly chemical reductants. Complications such as the impact of ash contamination on molten-bath operation have been found in a 100-MW demonstration to be of limited concern with acceptable wear rate. Economics appear to be disadvantageous—the process has been estimated[4] to have a levelized busbar cost more than 10 percent greater than that of the Wellman-Lord process.

Citrate Process. After prescrubbing, dedusted flue gas is scrubbed with sodium citrate solution in a packed tower. Collected SO_2 in a bleed stream of spent liquor is converted to elemental sulfur by steps that include reaction with H_2S (formed from a principal portion of the final sulfur product) in a turbine-agitated regeneration reactor system to yield crystalline elemental sulfur; separation of the sulfur froth by air flotation; and heating and decanting to yield molten sulfur. Sulfate is removed and discarded by crystallizing it from a citrate solution sidestream, which is later reused.

Advantages include high SO_2 removal efficiency, scale-free absorber operation, and the ability to accommodate substantial fluctuations in inlet SO_2 concentrations. The many complexities of the process, however, and the need to consume a reducing gas and produce and use toxic H_2S, appear to detract significantly from the economics and practicality of the process.

FUTURE FGD TECHNOLOGY

Trends in FGD technology embrace system selection, design, and operation. New process developments embrace short- and long-term activities.

FGD Process Selection

As of 1983, some 16 percent of coal-fired generating capacity was controlled by FGD. By 1993 this figure is projected to rise to 34 percent through construction of new powerplant units. Based on announced contract awards for such plants in the design and construction stage, an industry preference for throwaway-type limestone slurry scrubbing systems continues to grow, based most often on forced-oxidation designs.

At the same time, an increasing proportion of lime-based dry-scrubbing systems are being selected for low-sulfur coal applications.[17] A growing preference for selection of forced-oxidation operation of limestone scrubbing systems is based on (1) the greater ease and lower cost in management of its more completely dewatered gypsumlike solid waste and (2) the benefits of increased limestone utilization, less water use, improved process chemistry, minimized internal gypsum scaling, and the ability to tolerate a high chloride concentration in the scrubbing slurry without a significant decrease in SO_2 removal.[29]

Design and Operation

Electric utility companies and FGD system suppliers continue to draw on extensive field experience accumulated in the 1970s and 1980s. This experience is helping to establish comprehensive specifications that address process design considerations and to specify types and manufacturers of auxiliary equipment based on good performance in similar FGD environments.[29] Materials applications having a history of numerous and catastrophic failures are receiving major attention in specifications. Many laboratory and field tests of alloys and linings have provided the designer with extensive information for specifying the proper materials of construction throughout the FGD system.

Moreover, guarantees on materials of construction are increasingly being offered by FGD system suppliers. Utility companies have also been strengthening their technical expertise, and chemical engineers, chemists, and metallurgists are now being regularly included in the staffing of FGD operations. Extensive laboratories have been set up to monitor and control process chemistry, serving to optimize FGD operations and lower operating and maintenance costs.

New Process Developments

In the near term, commercial advancements taking place as a result of technical developments are not reflected in the use of substantially new FGD processes, but rather take the form of refinements in limestone-based FGD design.[29] In this connection, EPRI has performed vital work to enhance the knowledge of FGD chemistry, improve equipment reliability, and lower operating and maintenance costs. Design guidelines for use by utilities, architect-engineers, and system suppliers have resulted in significant benefits to utilities.

In the long term, through R&D activities within EPA and EPRI, increasing emphasis is being placed on integrated environmental control wherein FGD and other pollution control functions are engineered into new powerplant designs in such a manner that overall economy and effectiveness are maximized. At the same time, major long-range new-technology development within the government sector, notably the *Department of Energy* (DOE),[30] is focused on FGD processes that can potentially lead to integration of future control requirements for NO_x, particulates, and SO_2.

The focus of this new process development is to bring to commercial fruition those combined SO_2/NO_x removal concepts that would be expected to gain a potential savings of at least 20 to 25 percent compared to in-tandem selective catalytic reduction and limestone scrubbing. Processes of growing interest include fluidized-bed copper oxide, moving-bed copper oxide, the Lurie sodium aluminate scrubbing process, electron beam irradiation, the Noxso process, glow-discharge irradiation, the Sulf-X process, and zinc oxide spray-drying.

REFERENCES

1. Paul R. Predick, Written communication, Sargent and Lundy, Chicago, Ill., June 27, 1984.
2. N. N. Dharmarajan, R. D. Forbus, and P. R. Predick, "FGD System Selection for Coleto Creek Power Station, Unit 2," American Power Conference, Vol. 46, Illinois Institute of Technology, Chicago, Apr. 1984.
3. Paul R. Predick, "Dry and Wet Flue Gas Desulfurization Systems," American Power Conference, Vol. 44, Illinois Institute of Technology, Chicago, Apr. 1982.
4. J. B. Reisdorf, R. J. Keeth, J. E. Miranda, R. W. Scheck, and T. M. Morasky, "Economic Evaluation of FGD Systems," Electric Power Research Institute, EPA/EPRI Flue Gas Desulfurization Symposium, New Orleans, Nov. 1983, p. 5.
5. *Rock Talk Manual,* Kennedy Van Saun Corp., no. K1082, Denver, 1982, pp. 71–73.
6. D. S. Henzel, B. A. Laseke, E. O. Smith, and D. O. Swenson, *Limestone FGD Scrubbers: Users Handbook,* Environmental Protection Agency, EPA-600/8-81-017, Apr. 1981.
7. *Lime FGD Systems Data Book,* Electric Power Research Institute, EPRI FP 1030, May 1979.
8. T. A. Burnett, R. L. Torstrick, and F. A. Sudhoff, "Economic Evaluation and Comparison of Alternative Limestone Scrubbing Options," Electric Power Research Institute, EPA/EPRI Flue Gas Desulfurization Symposium, Hollywood, Fla., May 1982.
9. L. N. Davidson, R. J. Mongillo, S. Ou, and C. P. Wedig, "Designing, Operating, and Maintaining a Highly Reliable Flue Gas Desulfurization System," 76th Annual Meeting and Exhibition of the Air Pollution Control Association, Atlanta, June 1983.
10. Richard T. Egan and William Ellision, "Developments and Experience in FGD Mist Eliminator Application," Electric Power Research Institute, EPA/EPRI Flue Gas Desulfurization Symposium, Hollywood, Fla., May 1982.
11. Envirotech Corp., *Sludge Dewatering for FGD Products,* Electric Power Research Institute, EPRI FP-937, May 1979, pp. 2–42 to 2–51.
12. Richard A. Arterburn, Written communication, Krebs Engineers, Menlo Park, Calif., June 13, 1984.
13. "Guidelines for Slurry Pump Specifications for Utility and Process Power Applications," Warman International Inc., Madison, Wis., revised Nov. 9, 1984.
14. D. S. Henzel and D. H. Stowe, "Application of the Thiosorbic Process: The Most Widely Used Reagent for High Sulfur Coal," annual meeting, Air Pollution Control Association, New Orleans, June 1982, pp. 2, 3.
15. J. Z. Abrams, R. M. Sherwin, and D. T. Berube, "Dolomitic Lime in Flue Gas Desulfurization," 82-JPGC-Pwr-38, Joint Power Generation Conference, American Society of Mechanical Engineers, New York, 1982.
16. Donald H. Stowe, Jr., D. S. Henzel, and C. J. McCormick, "Lime vs. Limestone: A Technical and Economic Assessment," National Lime Association FGD Symposium, Denver, Sept. 1983.
17. B. A. Laseke, Jr. and M. T. Melia, "Trends in Commercial Applications of FGD," Electric Power Research Institute, EPA/EPRI Flue Gas Desulfurization Symposium, New Orleans, Nov. 1983, pp. 16, 19.
18. Gerald A. Hollinden, C. D. Stephenson, and J. G. Stensland, "An Economic Evaluation of Limestone Double Alkali Flue Gas Desulfurization Systems," Electric Power Research Institute, EPA/EPRI Flue Gas Desulfurization Symposium, New Orleans, Nov. 1983.
19. Paul S. Nolan and D. O. Seaward, "The Dowa Process: Dual-Alkali Flue Gas

Desulfurization with a Gypsum Product," *Proceedings of Seminar on Flue Gas Desulfurization,* Canadian Electrical Association, Ottawa, Sept. 1983.

20. David Nixon and Carlton Johnson, "Particulate Removal and Opacity Using a Wet Venturi Scrubber—the Minnesota Power and Light Experience," Peabody Process Systems, Stamford, Conn., ca. 1980.
21. Rolf Graf, "First Operating Experience with a Dry Flue Gas Desulfurization (FGD) Process Using a Circulating Fluid Bed (FGD-CFB)," Lurgi Corp., River Edge, N. J., May 1985.
22. J. R. Donnelly, Written communication, Niro Atomizer Inc, Columbia, Md., June 13, 1984.
23. D. C. Drehmel, G. Blair Martin, and James H. Abbot, "Results from EPA's Development of Limestone Injection into a Low NO_x Furnace," Electric Power Research Institute, EPA/EPRI Flue Gas Desulfurization Symposium, New Orleans, Nov. 1983.
24. P. Case, M. Heap, J. Lee, C. McKinnon, R. Payne, and D. Pershing, "Use of Sorbents to Reduce SO_2 Emissions from Pulverized-Coal Flames under Low-NO_x Conditions," progress report, Environmental Protection Agency, EPA-600/S7-82-060, Jan. 1983.
25. M. Y. Chughtai and S. Michelfelder, "Direct Desulfurization through Additive Injection in the Vicinity of the Flame," Electric Power Research Institute, EPA/EPRI Flue Gas Desulfurization Symposium, New Orleans, Nov. 1983.
26. L. J. Muzio, T. W. Sonnichsen, R. G. Hooper, G. P. Green, H. G. Brines, and N. D. Shah, "Demonstration of SO_2 Removal on a Coal-Fired Boiler by Injection of Dry Sodium Compounds," Electric Power Research Institute, EPA/EPRI Flue Gas Desulfurization Symposium, Hollywood, Fla., May 1982.
27. Jason Makansi, "SO_2 Control: Optimizing Today's Processes for Utility and Industrial Powerplants," Special Report, *Power,* Oct. 1982, pp. S-18–S-20.
28. Richard R. Lunt and J. S. MacKenzie, "Longer-Term Options for Reducing SO_2 Emissions: Regenerable Scrubbers," *Acid Rain Sourcebook,* McGraw-Hill, New York, 1984.
29. Stephen D. Jenkins, Written communication, Tampa Electric Co., Tampa, July 31, 1984.
30. John E. Williams, "Status of the DOE Flue Gas Cleanup Program," Electric Power Research Institute, EPA/EPRI Flue Gas Desulfurization Symposium, New Orleans, Nov. 1983, pp. 2, 5.

SUGGESTIONS FOR FURTHER READING

Ando, Jumpei: *SO_2 Abatement for Coal-Fired Boilers in Japan,* EPA-600/7-83-028, EPA, Research Triangle Park, N.C., May 1983.

Arterburn, R. A.: *The Sizing and Selection of Hydroclones,* AIME, Littleton, Colo., 1982.

Bitsko, R., et al.: *Operating Experience with the United/PECo Magnesium Oxide Flue Gas Desulfurization Process,* United Engineers & Constructors, Philadelphia, July 1983.

Black and Veatch Consulting Engineers: *Lime FGD Systems Data Book,* 2d ed., EPRI CS-2781, EPRI, Palo Alto, Calif., Mar. 1983.

Black and Veatch Consulting Engineers and Radian Corp.: *The Limestone FGD Systems Data Book,* EPRI CS-2949, EPRI, Palo Alto, Calif., Mar. 1983.

"Control of Pollution from Combustion Processes," Chapter 22, *Chemistry of Coal Utilization,* Wiley, New York, 1981.

Ellison, W., K. B. Hinerman, and J. D. Maddox, *Salable FGD Chemical Production through National SO_2 Emission Inventory Reduction,* Air Pollution Control Association Annual Meeting, San Francisco, June 1984.

Ellison, W., *Experience with Wet Lime/Limestone FGD Systems,* McGraw-Hill's Second International Conference on Acid Rain, Washington, D. C., Mar. 1985.

Ellison, W., *Control of Environmental Impact of Air Pollution Control Residuals,* paper no. 85-JPGC-APC-9, ASME, Joint Power Generation Conference, Milwaukee, Oct. 1985.

Ellison, W., J. Leimkuhler, and J. Makansi, "West Germany Meets Strict Emission Codes by Advancing FGD," *Power,* Feb. 1986.

Ellison W., J. Leimkuhler, and J. Makansi, "Strict European NO_x Code Brings SCR into the Limelight," *Power*, Aug. 1986.

Ellison, W., and J. A. Ratafia-Brown, *Flue-Gas NO_x Control: Recent Operating Experience with Commercial Systems and the Potential of Advanced Technologies,* McGraw-Hill's Third Annual Conference on Clean-Coal Technologies and Acid Rain, Washington, D. C., Oct. 1986.

Ellison, W., *Assessment of SO_2 and NO_x Emission Control Technology in Europe,* EPA 600/2-88-013, Environmental Protection Agency, Research Triangle Park, N. C., Feb. 1988.

Ellison, W., and E. Hammer, "FGD-Gypsum Use Penetrates U. S. Wallboard Industry," *Power,* Feb. 1988.

Makansi, Jason: "SO_2 Control: Optimizing Today's Processes for Utility and Industrial Powerplants," Special Report, *Power,* Oct. 1982.

Michael Baker, Jr., Inc.: *FGD By-product Disposal Manual,* 3d ed., EPRI CS-2801, Jan. 1983, pp. 4-12–4-13.

Midulla, R. J., and J. D. Peterson: *Design and First Year Operational History of Pollution Control Systems at Seminole Units 1 and 2,* Seminole Electric Cooperative Inc., Tampa, Fla., Apr. 1984.

Oldshue, James Y.: *Fluid Mixing Technology,* McGraw-Hill, New York, 1979.

Singer, Joseph G.: *Combustion Fossil Power Systems,* Combustion Engineering, Windsor, Conn., 1981, pp. 17-25–17-66.

Slack, A. V., and G. A. Hollinden: *Sulfur Dioxide Removal from Waste Gases,* 2d ed., Noyes Data Corp., Park Ridge, N.J., 1975.

CHAPTER 4.6
ASH-HANDLING SYSTEMS

Douglas M. Rode
C-E Power Systems
Combustion Engineering, Inc.
Windsor, CT

INTRODUCTION

An important adjunct to any pollution control scheme is the ash-conveying system. The need for various ash-conveying processes is created by the numerous boiler-imposed variables such as characteristics of fuel fired, discharge points, and site limitations. The subject of ash conveying is simplified by segregating it into two broad categories: (1) *Bottom ash* is that portion of ash which falls to the bottom of the furnace or from the stoker discharge, and (2) *flyash* represents that fraction which is airborne from the furnace proper and precipitates out in the heat-recovery and pollution control equipment.

The distribution of bottom ash and flyash is predicated on the fuel-firing system and the quantity of fuel being burned. As illustrated in Fig. 6.1, the total design quantity of ash exceeds 100 percent of the ash in the fuel burned in the furnace. This excess represents the ash system's conveying tolerance because the actual distribution of ash is not an absolute.

For each category, the available types of systems are briefly defined. The ad-

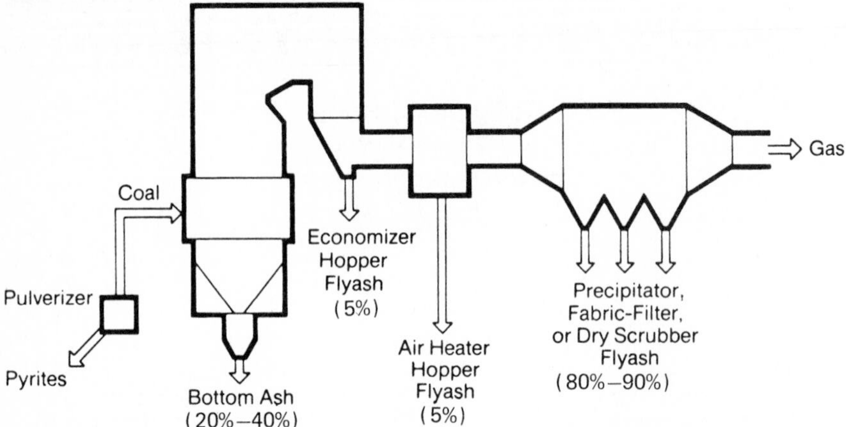

FIG. 6.1 Ash must be collected and transported from at least the five points shown. In each location, the physical and chemical characteristics of the ash vary. *(Combustion Engineering, Inc.)*

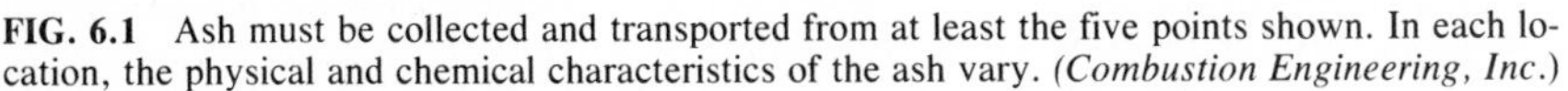

vantages and/or disadvantages of each process as well as comparisons regarding power consumption are given.

BOTTOM-ASH CONVEYING

Figure 6.2 highlights the available process systems for bottom-ash conveying. Note that when economizer ash is handled wet, it is considered a part of the bottom-ash conveying system. Pyrites, which are the rejects from the pulverizers, are disposed with the bottom-ash system.

Important considerations regarding the type of system to be employed are the ash characteristics, ash quantity, and final disposal determination.

Vacuum Pneumatic System

Industrial boiler applications which produce less than 350,000 lb/h (44 100 g/s) of steam typically use a vacuum pneumatic system. The ash quantities are generally small to allow for manageable gravity-assisted emptying of a bottom-ash hopper. This hopper, constructed of carbon steel and internally lined with refractory and insulation for wear and heat protection, is located beneath the furnace opening. The furnace seal is maintained through the use of boiler seal plates that form a water seal with the hopper seal trough, which extends around the upper perimeter of the hopper.

Alternatively, the hopper is generally suspended from the stoker support steel, eliminating the need for a seal trough, because there is no furnace expansion in that area. The ash is gravity-fed through vertical lifting gates arranged along the length of the hopper. Steep hopper sidewall angles in excess of 60° and the use of poke holes helping to promote flow through the ash discharge gates are incorporated in the hopper design.

Hopper Configurations. Since the bottom-ash conveying line serves as a branch line to the entire vacuum pneumatic system, a single-roll clinker grinder or a siz-

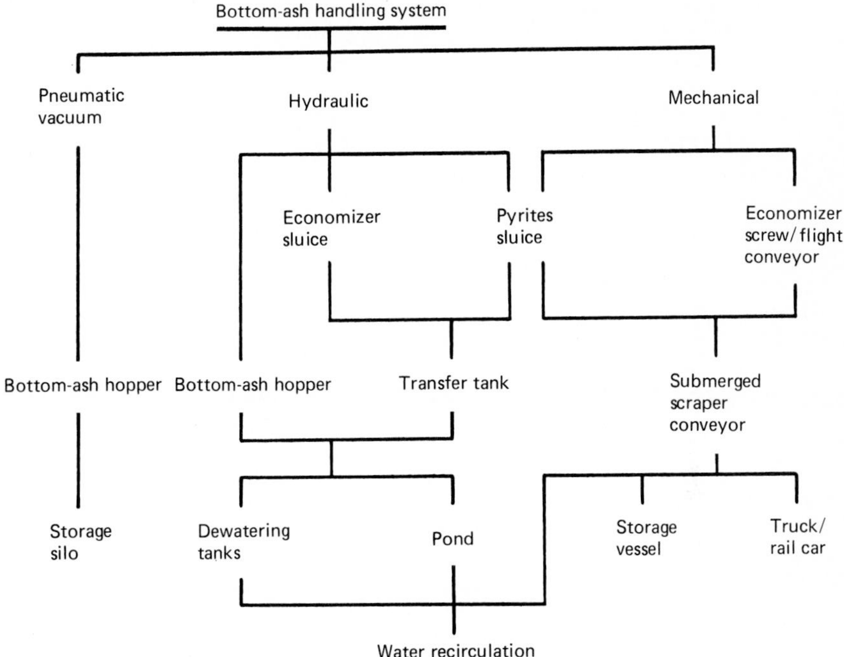

FIG. 6.2 Available processes for bottom-ash conveying.

ing grid is placed at the discharge of the ash gate to prevent oversized particles from entering the pipeline and causing pluggage problems. When it is located beneath a stoker discharge, the potential exists for burning material to fall from the end of the grate into the hopper. To address this situation, some operators have opted for a water-impounded bottom-ash hopper. Until recently, the use of a water-impounded ash storage hopper has been the most common method employed.

This style of hopper is constructed to provide a deep pool of water to quench and absorb the impact of the falling ash from the furnace. The use of water nozzles on its sloped walls provides a more positive means of extracting the ash from the hoppers. Depending on the furnace dimensions, the hopper configuration may be a flat, stepped horizontal bottom, a single-vee shape, a double-vee or "W" or, for large utility units with a center wall, a triple-vee hopper. The hopper is intermittently emptied by a hydraulic sluicing operation.

Discharge Arrangement. For all these configurations, the discharge arrangement consists of a sluice gate discharging into a single- or double-roll clinker grinder. The hydraulic conveying force is developed by either a jet pump, which is a venturi device with no moving parts, or a centrifugal slurry pump. Centrifugally cast pipe with a minimum Brinnell hardness of 280 is used to transport the slurry to either a pond or dewatering tanks.

Although bottom ash has been defined as a nonhazardous waste by the *Resource Conservation and Recovery Act* (RCRA), construction of any pond must address both the potential of leachate and the possibility of overflow of water.[1]

If the bottom ash is to be transported off the site, a minimum of two dewatering tanks is necessary. As one dewatering tank receives the ash and water slurry from the bottom-ash hopper, the ash in the other tank is dewatered through decanting screens, which conduct the precipitated water from the ash as they run vertically from the top of the tank to the cone-bottom discharge. The dewatered ash is discharged through an outlet gate, similar to the bottom-ash hopper sluice gate, to a vehicle.

Mechanical System: "SSC"

Because of the increasing importance of water use and power consumption, a device for the continuous removal of boiler-furnace bottom ash has been successfully reintroduced to the United States. Normally called a *submerged scraper conveyor* (SSC), it differs from the conventional system which stores ash in a bottom-ash hopper before it is intermittently sluiced through a pipeline. Although used in the United States before the 1930s, the SSC passed from the scene with the advent of large pulverized-coal-fired units and the increased use of oil-fired boilers. The SSC has been used extensively in Europe, however, because of water use and power consumption concerns.

Two Separate Compartments. The reinstitution of the SSC in the United States in the 1980s warrants a detailed description, as shown in Fig. 6.3. The SSC configuration consists of two separate compartments. The ash, which falls from the stoker grate or furnace, is quenched by 140°F (60°C) water contained in the upper trough or compartment of the SSC and settles to the bottom of the trough.

The steel scraper flights are located between the two parallel lengths of chain traveling along the upper-trough bottom. The flights transport the ash to and up the dewatering slope located at the head end of the SSC. At the top of the dewatering slope, the flights reverse direction, dumping the conveyed ash. The flights and chain return through the dry lower compartment, which has open sides to facilitate inspection while the equipment is in operation. A reversing roller, located at the tail end of the SSC, is equipped with a take-up adjuster to maintain tension in the conveyor chains.

Conveyor Chain and Dewatering Slope. The chain commonly used in the SSC is chrome-nickel-alloy abrasion-resistant ship's chain that is annealed and carburized to obtain a 550 to 600 Brinnell hardness in the interlink. Experience has shown that the typical service life of such a chain is about 12,000 h on high-silica, high-ash coals and 25,000 h on medium-ash bituminous coals. Service life is a function of chain lengthening, caused by wear in the interlink, which results in a mismatch of chains and driving sprocket and not breakage attributable to thinning. The flights are simple structural angles attached to the chains by fittings provided by the chain manufacturer.

The dewatering slope can be any angle between 25° and 45° with the horizontal, with 35° a commonly used angle. The physical configuration depends on the unobstructed space available, structural-steel design considerations, and method of final disposal from the SSC discharge. The dewatering slope floor is arranged in a herringbone fashion to promote drainage of water back into the upper trough of the SSC. Although the final water content of the dewatered ash is a function of the ash particle size and the chemical analysis of the ash, it normally ranges from 15 to 30 percent by weight. This is equal to the ash discharged from a dewatering tank.

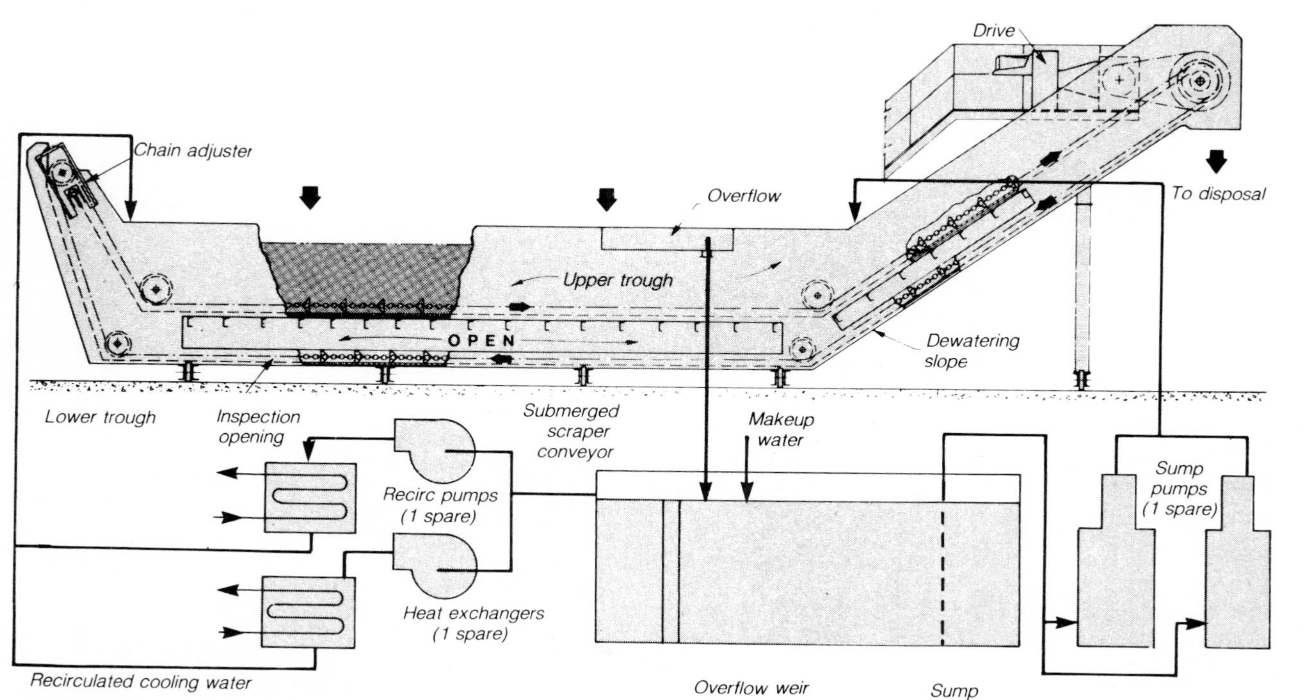

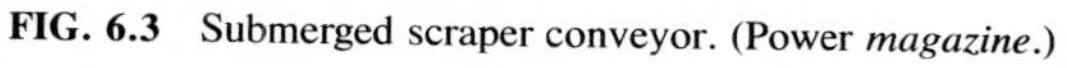

FIG. 6.3 Submerged scraper conveyor. (Power *magazine*.)

Speed and Mobility. The speed of the SSC ranges from 2 ft/min (0.01 m/s) to a maximum of 18 ft/min (0.09 m/s) with the normal speed less than 10 ft/min (0.05 m/s), mainly to inhibit wear. This adjustable speed can be achieved by using a hydraulic drive consisting of a pump, flexible hoses, and a drive unit, which is coupled to the SSC head shaft, or a mechanical arrangement utilizing a variable-speed drive unit and a worm gear coupled to the head shaft.[2]

One feature of an SSC is that it can be mobile when it is fitted with travel wheels, supporting the entire conveyor on rails, thereby enabling easy accessibility into the furnace during unit shutdown. The ash discharge from the SSC can be dumped directly into a truck or temporary storage area consisting of a three-sided bin. If physical limitations restrict this scheme, belt conveyors can transport the SSC discharge to a storage silo, which is similar to a dewatering tank but with fewer decanting elements.

Auxiliary Subsystems

Bottom-ash auxiliary subsystems include economizer and pyrite removal. Removal of ash from the high-temperature zone of the economizer hopper has been found to reduce the incidence of sintering of ash in this zone. If the calcium oxide content of the ash is greater than 15 percent, the ash will characteristically exhibit pozzolanic and cementitious properties in the presence of water and create concrete.[3] For these applications, the economizer ash must be conveyed dry either pneumatically or mechanically. If this is not the case, the economizer ash drops into temporary water-filled storage tanks, which are either hung from the hoppers or bottom-supported with a seal trough to account for expansion. Intermittently, this ash is sluiced from these tanks to a transfer tank for eventual transportation to the pond or dewatering tanks.

Pyrite Removal. This same transfer tank may be utilized to receive pyrites, which are sluiced from the individual storage hoppers at each pulverizer. The individual hoppers must be designed to withstand 50 psig (345 kPa) according to the requirements of *National Fire Protection Association* (NFPA) no. 85F, paragraph 2-6.1, when pressurized mills are employed, which is typical for utility applications.

Removal of the material from each hopper is accomplished through the use of jet pumps (smaller versions of the ones used for bottom-ash hoppers), through either individual centrifugally cast pipelines or a header system using the same pipe material. When the quantity of mill rejects is small, typically from suction mills used on industrial-size boilers, the pyrites may be removed manually with a wheelbarrow.

Water Recirculation. One possible subsystem involved with either hydraulic or mechanical bottom-ash removal is that for water recirculation. To minimize water makeup, a closed-loop, water recirculation system is used. For the hydraulic system, a combination of settling and surge tanks precipitates the ash carry-over from the dewatering tanks and acts as a water-inventory tank. Water recirculation pumps, usually designated to operate at principal pressure levels, convey the water to the various points of application. The surface area of these tanks is sufficient to provide cooling of the water. The mechanical SSC system, which does not require sluicing water, incorporates the same principle.

Because of the lower water quantities, the settling and surge functions are usually combined in a two-chamber tank or sump. Since only low-pressure water is

TABLE 6.1 Bottom-Ash Process Advantages and Disadvantages

Pneumatic	Hydraulic	Mechanical
	Advantages	
Not a separate system but a branch of vacuum removal system	Storage time up to 12 h available with bottom-ash hopper	Reduced power consumption
		Reduced water consumption
No need for dewatering tanks	Only process which sluices directly to a pond	Simpler operational control system
		Less clearance required from grade to furnace; less structural steel, fuel piping, ductwork, and building costs
		Easy external visibility of ash removal process
		No need for dewatering tanks
		Mobility to clear furnace outlet for access when SSC equipped with wheels
	Disadvantages	
Manual assistance to remove the ash from the hopper typically required	Consumes most horsepower due to conveying and supplying water requirements	Less storage time available in upper trough than a bottom-ash hopper
	Longest erection span	

required by the SSC, a single-pressure recirculation pump is used in conjunction with plate-type heat exchangers, which are used to cool the recirculated water. The size, complexity, and power consumption required by each type of water recirculation system affect any comparison between hydraulic and mechanical conveying systems.

Tables 6.1 and 6.2 summarize the discussion of bottom-ash conveying systems.

FLYASH CONVEYING

Figure 6.4 graphically shows the alternative processes available for the conveying of flyash. A major division exists between pneumatic and mechanical conveying systems.

Pneumatic System, Vacuum

The most common pneumatic system employed is the vacuum system (Fig. 6.5). Its popularity is attributed to its generally lower cost and understandable operation, but it is limited by the conveying distance and site altitude. The sequential

TABLE 6.2 Power Consumption Comparisons

	600-MW coal-fired steam generator						Industrial-size steam generator			
	Conventional system with jet pumps		Conventional system with slurry pumps		Continuous removal system		Conventional system with jet pumps		Continuous removal system	
	High-pressure sluice pumps	Low-pressure pumps	Slurry pumps	Low-pressure pumps	SSC	Low-pressure pumps	High-pressure sluice pumps	Low-pressure pumps	SSC	Low-pressure pumps
Bhp (kW)	615 (481)	20 (15)	65 (48.5)	20 (15)	10 (7.5)	20 (15)	175 (131)	8.5 (6.3)	1 (0.75)	2 (1.5)
Time per shift (min)	90	480	90	480	480	480	90	480	480	480
kWh/day	2,064	358	218	358	179	358	588	152	18	36
Installed hp (kW)	700 (522)	50 (37.3)	150 (112)	50 (37.3)	15 (11.2)	50 (37.3)	200 (149)	10 (7.5)	5 (3.8)	5 (3.8)
Energy charge										
kWh/day	2,422		576		537		740		54	
× $50/kWh • day, $	121,100		28,800		26,850		37,000		2,700	
Vs. continuous removal, $	94,250		1,950		Base		34,300		Base	
Capability charge										
Installed hp (kW)	750 (560)		200 (149)		65 (49)		210		10	
× $600/hp, $	450,000		120,000		39,000		126,000		6,000	
Vs. continuous removal, $	411,000		81,000		Base		120,000		Base	
Total auxiliary-power cost advantage, $	505,250		82,950		Base		154,300		Base	

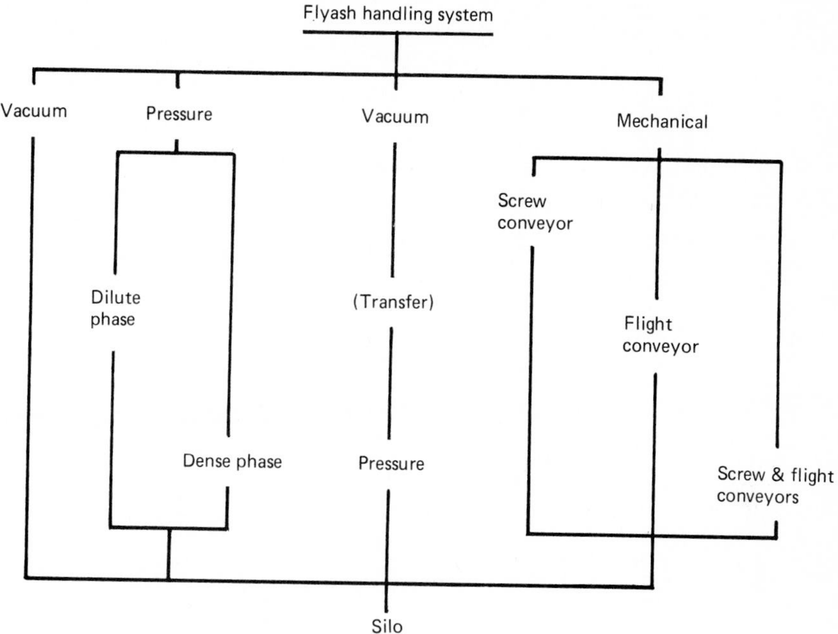

FIG. 6.4 Alternative processes available for conveying flyash.

regulation of flow rate and quantity of flyash leaving each hopper is achieved by the use of a flyash intake valve of the swing-disk or knife-gate design, both of which are automatically actuated by the system's control logic. The system's vacuum is created by mechanical blowers, water exhausters, or steam exhausters.

To protect the mechanical blowers, the ash is separated from the conveying medium by a cyclone separator and fabric filter, operating in series and discharging into a storage silo. Because the fabric filter is normally pulse-cleaned with compressed air, it is recommended that an inert medium be used for handling high-carbon-content flyash which results from the firing of wood products, a common practice for industrial units.

When a pond is used for storage, it is normal to use a water or steam exhauster to create the system vacuum. High-pressure water or steam, injected into a series of nozzles, causes a vacuum to be effected downstream and a slurry of ash and water to result, which allows for the ash to be transported by gravity to the pond. Although it is a simple device with no moving parts, the exhauster's inherent inefficiencies require high-pressure water to be supplied.

Pneumatic System, Positive-Pressure

The second type of pneumatic conveying system is a positive-pressure system (Fig. 6.6). Instead of the flyash being sucked from the hoppers, a pressure system blows the ash down the pipeline. The feeding apparatus consists of an airlock feeder which combines the hopper-sealing, storage capacity, and system-loading

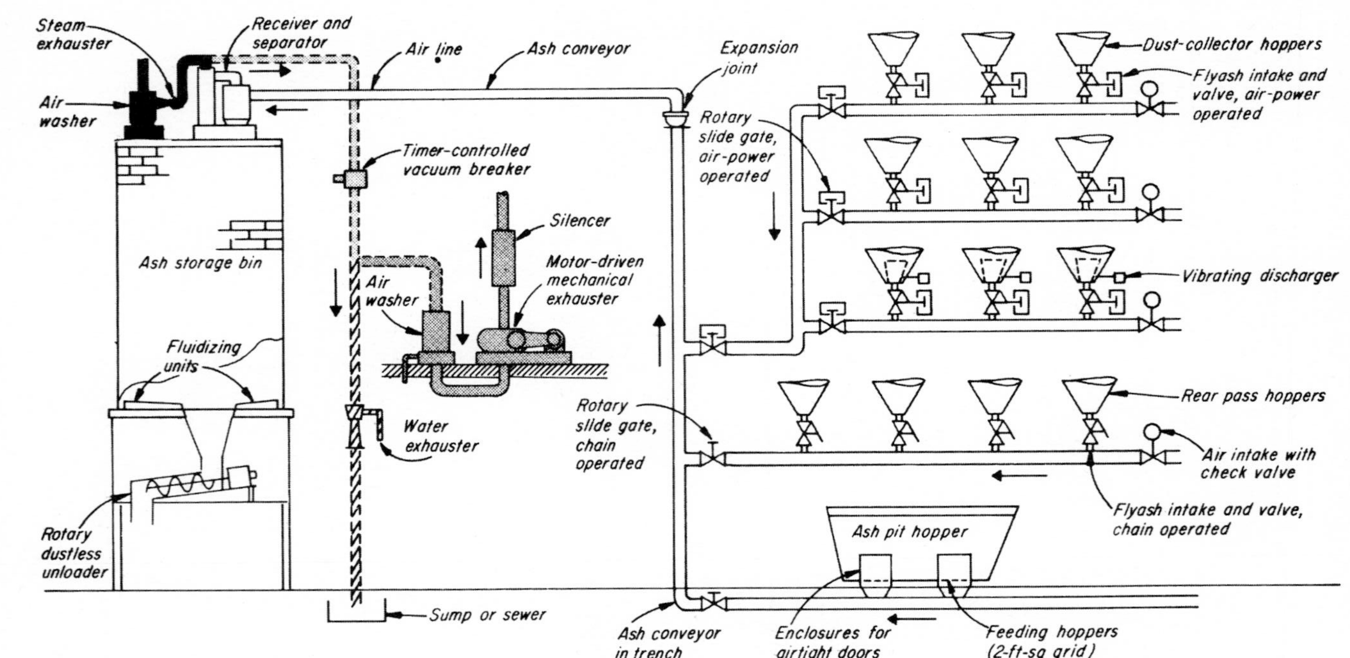

FIG. 6.5 Vacuum pneumatic system. (Power *magazine*.)

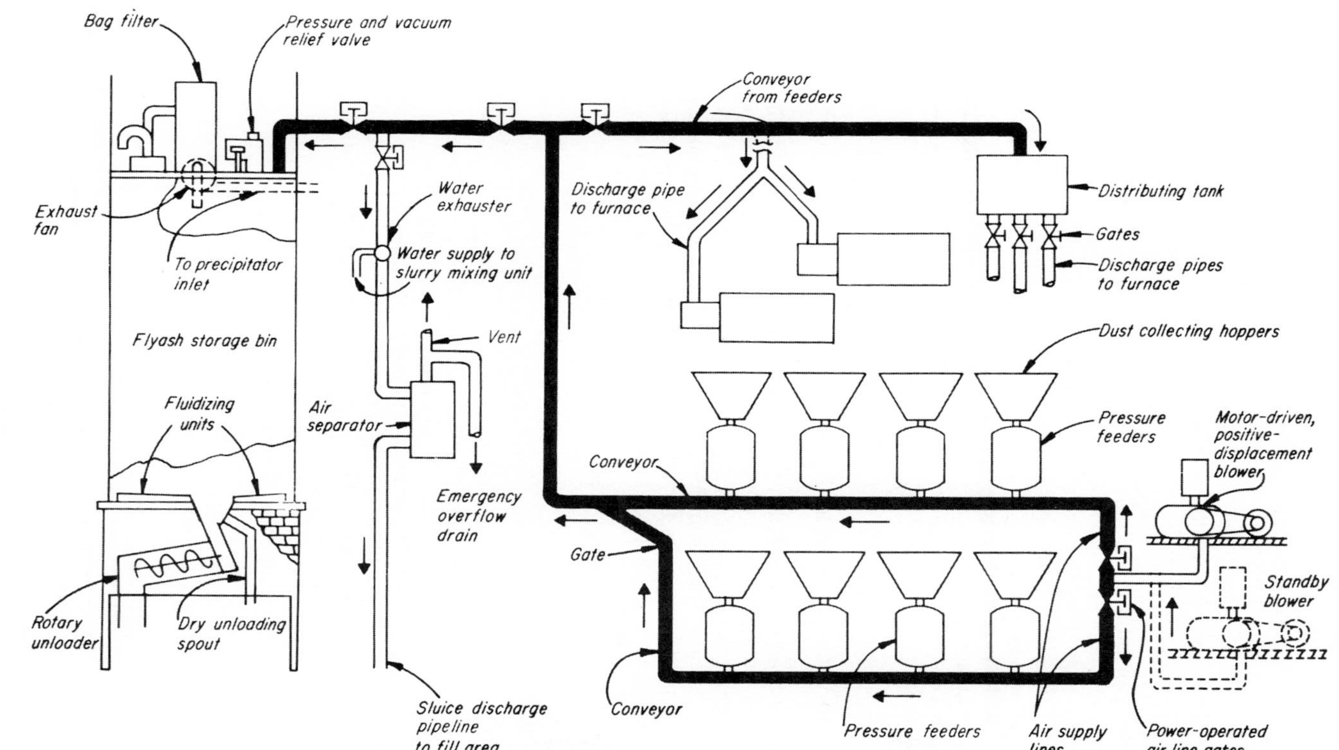

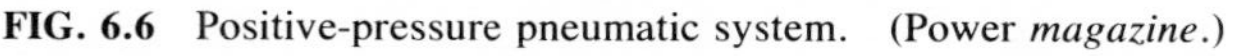

FIG. 6.6 Positive-pressure pneumatic system. (Power *magazine*.)

functions. The airlock feeders are normally grouped to permit a staggered cycle of filling, storing, and emptying of ash into the pipeline. If the conveying pressure exceeds 15 psig (103.4/kPa), the airlock feeders must be designed per the *American Society of Mechanical Engineers* (ASME) Code for Unfired Pressure Vessels. Because there is no limitation on pressure, theoretically there is no conveying length limitation on pressure systems.

Because the ash is blown into the silo, there is no need for the ash separation equipment required by the vacuum system. However, the conveying air and ash-displaced air in the silo must be vented through a fabric filter to prevent environmental discharge.

The ash-to-air ratio in the conveying mixture classifies the type of pressure system. A conventional limiting value for the pounds (kilograms) of ash being conveyed per second per square foot (square meter) of pipe area for dilute-phase conveying is 100 (488). When this value is exceeded, saltation takes place in the pipeline, which is referred to as *dense-phase conveying*. There is a practical rule of thumb which suggests that dense-phase conveying implies that a decision has been made to incorporate compressors [rated for 100 psig (689 kPa)] instead of blowers; low conveying velocities, less than 2000 ft/min (10 m/s); and small pipe sizes—generally less than 6 in (152 mm) in diameter.

Combination Vacuum, Pressure System

When there are numerous ash discharge points, in excess of 30, and the storage silo is not in close proximity to the pollution control equipment, a frequent solution is to combine the merits of a vacuum system to a transfer point and utilize a pressure pneumatic system from there to the silo. This combination reduces the capital and maintenance costs for a straight pressure system, since less equipment is required on a per-hopper basis, but at the expense of higher power demands.

Mechanical System

There is one process which does minimize power consumption while affording other operational advantages—mechanical conveying. A mechanical conveying system continuously removes all the flyash simultaneously from all hoppers, which effectively eliminates the problems of hopper pluggage due to compaction and deaeration. The design of the mechanical conveyor must account for the density of material being conveyed and for a physical arrangement that does not have the flexibility of a pipeline.

To be analogous to a pneumatic branch line, each mechanical conveyor should be located beneath the hoppers in a single field of a precipitator and perpendicular to the gas flow through the precipitator. The effect would be uniform loading and field responsibility, since each conveyor would be responsible for serving a single field, thus ensuring greater precipitator availability.

Two Types. Two types of mechanical conveyors are used to convey flyash. Screw conveyors, with internal hanger bearing supports, typically operate with their troughs less than 15 percent full, to avoid wear and damage to the internal hanger bearings. The screws are usually equipped with a hard surface to restrict wear. Flight conveyors transport ash in a dusttight enclosure through the use of single- or double-strand chain and flights. Flight conveyors are sized either on a volumetric basis or on an en masse basis, in which the material is conveyed in bulk without agitation. For some applications, a combination of screws and flights produces the most desirable results.

Storage Bin. Associated with mechanical conveying can be the use of a three-sided concrete bin which receives conditioned ash from the mechanical conveyors after they pass through a pug mill (a double screw or paddle mixer). Equipped with a roof, such a bin can store ash at grade for eventual loading into trucks or can accept immediate truck loading. This arrangement, favored in some industrial applications, eliminates the problems of storage of ash in elevated silos from freezing and ash pluggage.

Tables 6.3 and 6.4 compare the types of flyash conveying processes as to their advantages and disadvantages, and by power consumption characteristics.

Storage and Discharge Equipment

One common component of any flyash system is the storage device. For all pneumatic systems, this storage device is in the form of a cone-bottom silo, for small flyash quantities, or a fluidized, flat-bottom silo. The cone-bottom silo with angles of 60° or greater is generally reserved for storing ash which cannot be fluidized due to its characteristics, such as high carbon content associated with wood products, which can cause fires and explosions. For coal-fired applications, the flat-bottom silo is equipped with several outlets from which spoke-configured fluidized troughs extend from the silo outlet. Fluidization assists in the removal of ash because the fluidizing air causes the ash to assume the flow properties of a fluid.

Silo discharge equipment is designed to remove the flyash in a dry state through the use of a telescoping discharge chute or, in a moistened condition of

TABLE 6.3 Flyash Process—Advantages and Disadvantages

Vacuum pneumatic	Pressure pneumatic	Mechanical
	Advantages	
External indication of hopper conditions via strip charts Lower overall cost Least height required below hoppers Ash system leakage is inward	Long-distance conveying achieved for least cost Smaller pipe sizes generally used Less power consumed than for vacuum system Multiple hopper evacuation on a staggered cycle	Continuous conveying; therefore, no storage in any hopper Reduced power consumption Simpler operational control system Easiest maintenance due to readily available access Low conveying velocities minimize wear
	Disadvantages	
Storage in hoppers	Pressure leaks out, causing fugitive-dust problems Dense phase requires dry and oil-free air System venting required to prevent blow-back Storage in hoppers	Restricted physical arrangement

TABLE 6.4 Power Consumption Comparisons

	600-MW coal-fired steam generator				Industrial-size steam generator	
	Vacuum system	Dilute-pressure system	Dense-phase system	Mechanical system	Vacuum system	Mechanical system
Bhp (kW)*	330 (246)	175 (131)	110 (82)	125 (93)	37 (28)	9.3 (6.9)
Time per shift (min)	240	240	240	480	240	480
kWh/day	2,953	1,566	984	2,237	331	166
Installed hp (kW)	600 (448)	400 (298)	250 (187)	280 (209)	60 (45)	13.5 (19)
Energy charge						
kWh/day × $50/ kWh • day, $	147,650	78,300	49,200	111,850	16,555	8,300
Vs. continuous removal, $	35,800	−33,550	−62,650	Base	8,255	Base
Capability charge						
Installed hp (kW) × $600/ hp, $	360,000	240,000	150,000	168,000	36,000	8,100
Vs. continuous removal, $	192,000	72,000	−18,000	Base	27,900	Base
Total auxiliary-power cost advantage, $	+227,800	+38,450	−80,650	Base	+36,155	Base

*Conveying power to storage facility only.

15 percent moisture by weight, through an ash conditioner. The device employed is predicated on the type of vehicle used to transport the flyash.

Prudent Selection

The design parameters for ash receiving and conveying systems must be determined and evaluated in conjunction with the design of the boiler and particulate control equipment. Prudent selection of the ash-conveying process will enhance the availability of the unit, instead of constipating the system because the ash-conveying considerations were originally ignored.

REFERENCES

1. *Federal Register,* 47 (224): 52290–52309, Nov. 19, 1982.
2. J. C. Fleming and D. M. Rode, "Ash Removal from Industrial Boilers—The Changing Scene," *Power,* Sept. 1982, p. 92.
3. J. G. Singer, *Combustion,* 3d ed., Combustion Engineering, Inc., Windsor, Conn., 1981, chapter 8, p. 19.

SUGGESTIONS FOR FURTHER READING

Bosso, A., and J. E. Horne: "Southwestern Public Service Company Pioneering in Continuous Bottom Ash Removal," presented at the Frontiers of Power Conference, Stillwater, Okla., Oct. 11–12, 1982.

Fleming, J. C., and D. M. Rode: "Ash Removal from Industrial Boilers—The Changing Scene," *Power,* Sept. 1982.

Foley, G. F., and F. W. Deininger: "Bottom Ash Handling from the '50's to the '90's," Electric Power Research Institute, Coal and Ash Handling Conference, St. Louis, Oct. 27, 1980.

Gretz, C. R., "Bottom Ash Handling Methods: A Look at Drag Conveyors for U.S. Applications," presented at the Coal Technology Conference, Houston, Nov. 6–8, 1979.

Joint Technical Committee of the American Boiler Manufacturers Association and the Industrial Gas Cleaning Institute, Inc.: "Design and Operation of Reliable Central Station Flyash Hopper Evacuation Systems," American Power Conference, Chicago, Apr. 21, 1980.

Shulof, G. F., and S. T. Hinckley: "Innovations for Conservation in Generation," Missouri Valley Electric Association Annual Engineering Conference, Kansas City, Apr. 16, 1980.

Singer, J. G.: *Combustion,* 3d ed., Combustion Engineering, Inc., Windsor, Conn., 1981, chapter 8.

Singer, J. G.: "Design for Better ESP/Fabric-Filter Hopper Operation and Maintenance," presented at the Air Pollution Control Association 76th Annual Meeting and Exhibition, Atlanta, June 19–24, 1983.

Singer, J. G., and A. J. Cozza: "Design for Continuous Ash Removal: Alternative Concepts in Ash Handling," American Power Conference, Chicago, Apr. 23, 1979.

PART 2

WATER POLLUTION CONTROL

CHAPTER 4.1
LEGISLATION AND POLLUTANT SOURCES

James K. Rice
Consultant
Olney, MD

ABBREVIATIONS

ACC	Acid copper chromate	
As	Arsenic	
BAT	Best available technology	currently available
BCT	Best conventional technology	currently available
BPT	Best practical technology	currently available
CCA	Chromated copper arsenate	
Cd	Cadmium	
Cn	Cyanide	
Cr	Chromium	
Cu	Copper	

CWA	Clean Water Act
EEI	Edison Electric Institute
EPA	U.S. Environmental Protection Agency
EPRI	Electric Power Research Institute
FAC	Free available chlorine
Fe	Iron
FGD	Flue-gas desulfurization
GW	Gigawatts
Ha	Hectare
Hg	Mercury
K	Potassium
LOD	Limit of detection
Mg	Magnesium
Mn	Manganese
Ni	Nickel
NPDES	National pollutant discharge elimination system
NRC	Nuclear Regulatory Commission
NRDC	National Resources Defense Council
NSPS	New-source performance standards
O&G	Oil and grease
P	Phosphorus
Pb	Lead
PCB	Polychlorinated biphenyl
PSES	Pretreatment standards, existing sources
PSNS	Pretreatment standards, new sources
SA	Settlement agreement
Se	Selenium
TDS	Total dissolved solids
TOX	Total organic halogen
TRC	Total residual chlorine
TSS	Total suspended solids
TVA	Tennessee Valley Authority
UWAG	Utility Water Act Group
Zn	Zinc

INTRODUCTION

Where does this country stand in its quest to control the release of pollutants into the nation's water courses? Taking as a starting point the effluent guidelines for steam-electric plants promulgated by the U.S. *Environmental Protection Agency*

(EPA), this chapter reviews the evolving legislation and status of today's requirements for controlling point sources and examines the spectrum of pollutants contained in waste streams. Although steam-electric plants are the reference point here, common problems of industrial applications and utilities relative to boilers and cooling towers could eventually lead to standardized limitations across the entire industrial horizon.

Also, state regulatory agencies act apart from federal agencies, but may well refer to EPA's steam-electric guidelines in setting their own emissions standards. Thus, this chapter will be valuable for those engineers engaged in steam generation throughout industry. It can be used as a practical reference for waste-stream treatment problems and to help the engineer place in perspective new systems yet to come.

EVOLVING LEGISLATION

On November 19, 1982, the EPA, under the authority of the *Clean Water Act* (CWA)[1] and in compliance with the terms of a *settlement agreement* (SA),[2] promulgated effluent limitations guidelines, pretreatment standards, and *new-source performance standards* (NSPS) for the steam-electric power generating point-source category.[3] A description of the agency's study methodology, data-gathering efforts, and analytical procedures supporting the regulation is contained in a companion development document.[4]

EPA's previous rule-making efforts, which resulted in the promulgation of effluent-limitation guidelines in 1974,[5] emphasized the achievement of *best practicable* control *technology* (BPT) currently available by July 1, 1977. A subsequent court challenge by the *Natural Resources Defense Council* (NRDC) resulted in the SA, which required EPA to develop a program and to meet a schedule for promulgating (1) effluent-limitation guidelines and standards of performance and (2) water quality criteria for 65 priority pollutants and classes of pollutants for 21 major industrial categories, including the steam-electric point-source category.

Clean Water Act of 1977

Many of the elements of the SA were subsequently incorporated into the CWA of 1977. Also, the 1977 act established *best conventional technology* (BCT) for pollutant control to replace the *best available technology* (BAT) for conventional pollutants, e.g., *total suspended solids* (TSS) and *oil and grease* (O&G). EPA published revised BCT methodology for determining the reasonableness of BCT limitations, but as yet has not issued BCT for the steam-electric power industry. Until the BCT limitations are issued, the 1974 BPT limitations for TSS and O&G continue. The previous limitations for BAT, NSPS, *pretreatment standards for existing sources* (PSES), and *pretreatment standards for new sources* (PSNS) were all amended by the 1982 regulations. Table 1.1 summarizes both the 1974 and the 1982 limitations on the discharge of pollutants from steam-electric powerplants.

Note from Table 1.1 that no limitations were issued for four types of wastewaters: nonchemical metal-cleaning wastes, *flue-gas-desulfurization* (FGD) wastewater, runoff from materials storage and construction areas (other than

TABLE 1.1 Effluent Limitations, 1974 vs. 1982

Pollutant	October 1974 BAT, mg/L Max.[a]	October 1974 BAT, mg/L Avg.[b]	November 1982 BAT, mg/L Max.[a]	November 1982 BAT, mg/L Avg.[b]
All discharges				
pH (except once-through cooling)	6.0–9.0		6.0–9.0	
PCBs	No discharge		No discharge	
Low-volume wastes[c]				
TSS	100	30	100*	30*
O&G	20	15	20*	15*
Bottom-ash transport water[c,d]				
TSS	8	2.4	100*	30*
O&G	1.6	1.2	20*	15*
Flyash transport water[c]				
TSS	100	30	100*	30*
O&G	20	15	20*	15*
Chemical metal-cleaning wastes[e]				
TSS	100	30	100*	30*
O&G	20	15	20*	15*
Copper	1.0	—	1.0	—
Iron	1.0	—	1.0	—
Boiler blowdown				
TSS	100	30	—	—
O&G	20	15	—	—
Copper	1.0	—	—	—
Iron	1.0	—	—	—
Once-through cooling water				
Free available chlorine	0.5	0.2	—	—
Total residual chlorine	—	—	0.2	—
Cooling-tower blowdown				
Free available chlorine	0.5	0.2	0.5	0.2
Zinc	1.0	1.0	1.0	1.0
Chromium	0.2	0.2	0.2	0.2
Phosphorus	5.0	5.0	—	—
Other corrosion inhibitors	Case by case		—	—
Other 124 priority pollutants (in added maintenance chemicals)	—	—	No detect. amt.	
Coal-pile runoff				
TSS (promulgated in 1980)	—	—	50	—

[a]Maximum for any one day.
[b]Average of daily values for 30 consecutive days.
[c]Probable BCT limits shown with an asterisk; BAT withdrawn 1982.
[d]Concentration/12.5. Use for mass limit set in 1974 BAT.
[e]Divided into nonchemical and chemical categories, 1982.

coal-pile runoff), and thermal discharges. EPA has reserved putting limits on these wastes for a future rule making.

The 1982 Amendment

Surveys of the industry's discharges were made by EPA in the course of its 1982 rule making. They determined for once-through cooling water, for example, that

112 of the priority pollutants—mostly organic compounds but including 15 inorganic elements or compounds—were not detected by section 304(h) analytical methods. Where the pollutants were detected, they were present from sample contamination or present in amounts too small to be effectively reduced by the technologies available. Table 1.2 lists all the priority pollutants and notes which were excluded from specific national regulation for each of the discharges.

Guideline Specifications. The 1982 effluent-limitation guidelines specify both a concentration and a mass limitation. The latter is obtained by multiplying the former by the discharge flow—not by multiplying a mass per unit of production by the number of units produced per unit of time, as is commonly done for many other industrial discharges. EPA has recognized that discharges from steam-electric powerplants do not lend themselves to expression in terms of production units; that is, they are not expressible as pounds (kilograms) per day per megawatt of power produced. This condition arises from the facts mentioned above regarding the influence of fuel type, cooling system, etc., which results in plants having an identical generating capacity but fired with different fuels, for example, coal versus gas, discharging different pollutants at different concentrations in streams with different flow rates.

While EPA has recognized the absence of a relationship common to all powerplants between the mass of pollutants discharged and the units of power produced, the issue remains as to what flow should be selected at each plant as the basis of the permitted mass limitations. This is an especially important issue for plants in either load-following or load-peaking service. The industry has urged the use of historically expected flows at maximum-design electrical production for each unit. Such a basis takes into account the requirements contained in the charter of most public utility companies to meet electrical demand at any given moment insofar as it is practicable.

Industry Recategorization. In the 1982 regulations, EPA also recategorized the industry, treating it instead as a single subcategory with separate limitations for each type of waste, some of which were further differentiated on the basis of plant size, for example, greater than 25 MW. Table 1.3 lists the process discharges contained in each waste stream type. For waste streams other than noncontact once-through cooling water, reclassification by waste stream types resulted in little change from the 1974 regulations. These wastes are combined in each plant in ways that are dictated by site considerations, type of fuel, and receiving-water quality, and operational requirements. Thus, while ash sluice water may be treated by itself in one plant, it is combined with other internal streams in another plant.

Power Industry Study

To better understand the nature of its discharges, the power industry—through its research and development arm, the *Electric Power Research Institute* (EPRI)—assembled and analyzed the available information on the characteristics of discharges from fossil-fuel-fired steam-electric powerplants.[6] The study included an analysis of the concentration and flow data contained on the *national pollution discharge emissions standard* (NPDES) permit application form 2C from a representative sample of 100 steam-electric powerplants.

TABLE 1.2 Priority Pollutants Regulated by Each Waste Category*

Pollutant	OTCW	CTBD	LOVOL	CHCLN	COLPL
001 Acenaphthene	N	Y	N	N	N
002 Acrolein	N	Y	N	N	N
003 Acrylonitrile	N	Y	N	N	N
004 Benzene	N	Y	N	N	N
005 Benzidine	N	Y	N	N	N
006 Carbon tetrachloride (tetrachloromethane)	N	Y	N	N	N
007 Chlorobenzene	N	Y	N	N	N
008 1,2,4-trichlorobenzene	N	Y	N	N	N
009 Hexachlorobenzene	N	Y	N	N	N
010 1,2-dichloroethane	N	Y	N	N	N
011 1,1,1-trichloroethane	N	Y	N	N	N
012 Hexachloroethane	N	Y	N	N	N
013 1,1-dichloroethane	N	Y	N	N	N
014 1,1,2-trichloroethane	N	Y	N	N	N
015 1,1,2,2-tetrachloroethane	N	Y	N	N	N
016 Chloroethane	N	Y	N	N	N
018 Bis(2-chloroethyl) ether	N	Y	N	N	N
019 2-chloroethyl vinyl ether (mixed)	N	Y	N	N	N
020 2-chloronaphthalene	N	Y	N	N	N
021 2,4,6-trichlorophenol	N	Y	N	N	N
022 Parachlorometa cresol	N	Y	N	N	N
023 Chloroform (trichloromethane)	N	Y	N	N	N
024 2-chlorophenol	N	Y	N	N	N
025 1,2-dichlorobenzene	N	Y	N	N	N
026 1,3-dichlorobenzene	N	Y	N	N	N
027 1,4-dichlorobenzene	N	Y	N	N	N
028 3,3-dichlorobenzidine	N	Y	N	N	N
029 1,1-dichloroethylene	N	Y	N	N	N
030 1,2-trans-dichloroethylene	N	Y	N	N	N
031 2,4-dichlorophenol	N	Y	N	N	N
032 1,2-dichloropropane	N	Y	N	N	N
033 1,2-dichloropropylene (1,3-dichloropropene)	N	Y	N	N	N
034 2,4-dimethylphenol	N	Y	N	N	N
035 2,4-dinitrotoluene	N	Y	N	N	N
036 2,6-dinitrotoluene	N	Y	N	N	N
037 1,2-diphenylhydrazine	N	Y	N	N	N
038 Ethylbenzene	N	Y	N	N	N
039 Fluoranthene	N	Y	N	N	N
040 4-chlorophenyl phenyl ether	N	Y	N	N	N
041 4-bromophenyl phenyl ether	N	Y	N	N	N
042 Bis(2-chloroisopropyl) ether	N	Y	N	N	N
043 Bis(2-chloroethoxy) methane	N	Y	N	N	N
044 Methylene chloride (dichloromethane)	N	Y	N	N	N
045 Methyl chloride (dichloromethane)	N	Y	N	N	N
046 Methyl bromide (bromomethane)	N	Y	N	N	N
047 Bromoform (tribromomethane)	N	Y	N	N	N
048 Dichlorobromomethane	N	Y	N	N	N
051 Chlorodibromomethane	N	Y	N	N	N

TABLE 1.2 Priority Pollutants Regulated by Each Waste Category* *(Continued)*

Pollutant	OTCW	CTBD	LOVOL	CHCLN	COLPL
052 Hexachlorobutadiene	N	Y	N	N	N
053 Hexachloromyclopentadiene	N	Y	N	N	N
054 Isophorone	N	Y	N	N	N
055 Naphthalene	N	Y	N	N	N
056 Nitrobenzene	N	Y	N	N	N
057 2-nitrophenol	N	Y	N	N	N
058 4-nitrophenol	N	Y	N	N	N
059 2,4-dinitrophenol	N	Y	N	N	N
060 4,6-dinitro-o-cresol	N	Y	N	N	N
061 N-nitrosodimethylamine	N	Y	N	N	N
062 N-nitrosodiphenylamine	N	Y	N	N	N
063 N-nitrosodi-*n*-propylamin	N	Y	N	N	N
064 Pentachlorophenol	N	Y	N	N	N
065 Phenol	N	Y	N	N	N
066 Bis(2-ethylhexyl) phthalate	N	Y	N	N	N
067 Butyl benzyl phthalate	N	Y	N	N	N
068 Di-*N*-butyl phthalate	N	Y	N	N	N
069 Di-*n*-octyl phthalate	N	Y	N	N	N
070 Diethyl phthalate	N	Y	N	N	N
071 Dimethyl phthalate	N	Y	N	N	N
072 1,2-benzanthracene (benzo(a) anthracene)	N	Y	N	N	N
073 Benzo(a)pyrene (3,4-benzo-pyrene)	N	Y	N	N	N
074 3,4-Benzofluoranthene (benzo(b) fluoranthene)	N	Y	N	N	N
075 11,12-benzofluoranthene (benzo(b) fluoranthene)	N	Y	N	N	N
076 Chrysene	N	Y	N	N	N
077 Acenaphthylene	N	Y	N	N	N
078 Anthracene	N	Y	N	N	N
079 1,12-benzoperylene (benzo(ghi) perylene)	N	Y	N	N	N
080 Fluorene	N	Y	N	N	N
081 Phenanthrene	N	Y	N	N	N
082 1,2,5,6-dibenzanthracene (dibenzo(,h) anthracene)	N	Y	N	N	N
083 Indeno (,1,2,3-cd) pyrene (2,3-o-pheynylene pyrene)	N	Y	N	N	N
084 Pyrene	N	Y	N	N	N
085 Tetrachloroethylene	N	Y	N	N	N
086 Toluene	N	Y	N	N	N
087 Trichloroethylene	N	Y	N	N	N
088 Vinyl chloride (chloroethylene)	N	Y	N	N	N
089 Aldrin	N	Y	N	N	N
090 Dieldrin	N	Y	N	N	N
091 Chlordane (technical mixture and metabolites)	N	Y	N	N	N
092 4,4-DDT	N	Y	N	N	N
093 4,4-DDE (p,p-DDX)	N	Y	N	N	N
094 4,4-DDD (p,p-TDE)	N	Y	N	N	N
095 Alpha-endosulfan	N	Y	N	N	N

TABLE 1.2 Priority Pollutants Regulated by Each Waste Category* *(Continued)*

Pollutant	OTCW	CTBD	LOVOL	CHCLN	COLPL
096 Beta-endosulfan	N	Y	N	N	N.
097 Endosulfan sulfate	N	Y	N	N	N
098 Endrin	N	Y	N	N	N
099 Endrin aldehyde	N	Y	N	N	N
100 Heptachlor	N	Y	N	N	N
101 Heptachlor epoxide (BHC-hexachlorocyclohexane)	N	Y	N	N	N
102 Alpha-BHC	N	Y	N	N	N
103 Beta-BHC	N	Y	N	N	N
104 Gamma-BHC (lindane)	N	Y	N	N	N
105 Delta-BHC (PCB-polychlorinated biphenyls)	N	Y	N	N	N
106 PCB-1242 (Arochlor 1242)	Y	Y	Y	Y	Y
107 PCB-1254 (Arochlor 1254)	Y	Y	Y	Y	Y
108 PCB-1221 (Arochlor 1221)	Y	Y	Y	Y	Y
109 PCB-1232 (Arochlor 1232)	Y	Y	Y	Y	Y
110 PCB-1248 (Arochlor 1248)	Y	Y	Y	Y	Y
111 PCB-1260 (Arochlor 1260)	Y	Y	Y	Y	Y
112 PCB-1016 (Arochlor 1016)	Y	Y	Y	Y	Y
113 Toxaphene	N	Y	N	N	N
114 Antimony	N	Y	N	N	N
115 Arsenic	N	Y	N	N	N
116 Asbestos	N	Y	N	N	N
117 Beryllium	N	Y	N	N	N
118 Cadmium	N	Y	N	N	N
119 Chromium	N	Y	N	N	N
120 Copper	N	Y	N	N	N
121 Cyanide, total	N	Y	N	N	N
122 Lead	N	Y	N	N	N
123 Mercury	N	Y	N	N	N
124 Nickel	N	Y	N	N	N
125 Selenium	N	Y	N	N	N
126 Silver	N	Y	N	N	N
127 Thallium	N	Y	N	N	N
128 Zinc	N	Y	N	N	N
129 2,3,7,8-tetrachloro-dibenzo-*p*-dioxin	N	Y	N	N	N

**Notes:* N = No.
Y = Yes.
OTCW = once-through cooling water.
CTBD = cooling-tower blowdown.
LOVOL = low-volume wastes.
CHCLN = chemical metal-cleaning wastes.
COLPL = coal-pile runoff.

Limitations reserved for future rule making on nonchemical metal-cleaning, FGD, and materials storage and construction runoff waste types.

UWAG Data Base. A data base from the foregoing sample had been assembled by the *Utility Water Act Group* (UWAG)[7] to supplement the limited data available in the development document as well as those subsequently made available in the EPA *Treatability Manual,*[8] which included brief descriptions of each major industry's discharges, the treatability of each pollutant, the available technologies, and the means of estimating the costs of treatment.

TABLE 1.3 Process Discharges Contained in Each EPA Waste Stream Type

Waste stream type	Waste sources
Once-through cooling	Once-through cooling water passed through the main condenser in one to two passes, then to waste
Cooling-tower blowdown	Cooling-tower recirculating water that is discharged to waste
Flyash transport water	Wastewater from transporting flyash to the ash disposal area
Bottom-ash transport water	Wastewater from transporting bottom ash to the ash disposal area
Low-volume wastes	Boiler blowdown, wet-air-scrubber pollution control systems, ion-exchange water treatment, water evaporator blowdown, water clarification wastes, cooling-tower basin cleaning, house service water, laboratory and sampling drains, floor drains, roof drains
Chemical metal cleaning	Wastewater from cleaning of boiler watersides and steam sides, of feedwater heaters, condensers, etc., with chemical solvents
Nonchemical metal cleaning	Wastewater from cleaning gas sides of boiler, from cleaning of air preheaters, coolers, condensers, etc., with water to which no chemicals have been added
Coal-pile runoff	Drainage from coal storage area
Materials storage runoff	Drainage from materials storage areas other than coal and from construction areas
Flue-gas desulfurization	Blowdown from flue-gas desulfurization systems
Thermal discharges	Added heat contained in once-through cooling, cooling-tower blowdown

The concentration of each pollutant, the average flow rate of each discharge, and thus the mass of each pollutant discharged are a function of the type of fuel fired at each plant and the type of cooling system. Also, the flyash transport system used, whether dry or wet, and the materials of construction of the main condenser tubes influence the types and mass of each pollutant discharged. Other influences include coal-pile runoff and area runoff, when these are combined with other process streams prior to discharge.

Complications in Analysis. Since substantial quantities of pollutants may enter the plant in the intake water, representative sampling and careful analysis are important, although both operations complicate the understanding of a plant's contribution to its discharges. Sampling and analysis are especially important for iron, copper, zinc, and organic pollutants, particularly pesticides in the makeup to cooling towers.

TABLE 1.4 Flow-Weighted Concentrations in Powerplant Discharges from RP 1851

Pollutant	Average concentration, mg/L	Relative standard deviation, %	Number of data points	Range, mg/L	
				Min.	Max.
TSS	32	91	68	0.34	119
O&G	3.3	68	68	0	17
Ammonia	0.355	103	69	0.005	1.9
Antimony	0.037	150	71	0	0.45
Arsenic	0.041	130	68	0	0.23
Beryllium	0.005	65	69	0	0.01
Cadmium	0.005	148	71	0	0.05
Chromium	0.019	96	72	0	0.13
Copper	0.045	155	71	0	0.77
Iron	0.744	94	70	0.001	5.2
Lead	0.035	119	70	0	0.40
Manganese	0.861	196	70	0.006	5.0
Mercury	0.001	241	68	0	0.01
Nickel	0.039	100	70	0	0.22
Phosphorus	0.302	215	67	0.005	5.8
Selenium	0.016	143	68	0	0.08
Zinc	0.076	87	67	0.005	0.32

The relative standard deviation of analytical methods that are available and approved by EPA is typically in the range of 20 to 40 percent for elements such as iron and copper and is substantially greater for the organic pollutants. Thus, any determination of the contribution of the plant to the discharge over that in the intake must take the analytical variability into account. In the referenced EPRI study, discharge of a pollutant was deemed significant only when its concentration was analytically different at the 95 percent confidence level. Table 1.4 is drawn from the data in the EPRI RP 1851 report[6] and illustrates the difficulty in generalizing the discharge characteristics, even by fuel type.

Powerplant Discharges

In practice, the complexity of powerplant discharges has led to considerable effort on the part of both EPA and industry to agree on equitable means of calculating limits on the discharges from combined treatment facilities. For example, it is not clear how to calculate the limits on the discharge of an ash pond that treats both flyash sluice water and coal-pile runoff, plus a portion of the cooling-tower blowdown used to sluice the flyash to the pond. Each of these streams has different effluent limitations.

Rainfall on the pond adds significantly to the combined flow through the pond (thus the hydraulic load), yet is an unpolluted and unregulated stream that for all practical purposes cannot be diverted and must be combined with whatever streams are treated. Resolution of the effect of rain that falls on ash-pond surfaces is one of the important considerations in the general issue of how to interpret both concentration and mass limitations on ash-pond discharges. Table 1.5 illustrates the effect of various rainfall rates on the TSS in the discharge from model-plant ash ponds of varying levels of efficiency. It becomes apparent that

TABLE 1.5 Effect of Rainfall on TSS in Ash-Pond Discharge, mg/L

Rainfall		Discharge		Pond efficiency factor		
in/day	mm/day	gal/min	L/s	$n = 1/5$	$n = 1/3$	$n = 1/2$
0	0	4,100	259	3.5	17.2	52.5
1	25.4	4,883	308	4.2	18.4	52.4
2	50.8	5,665	358	5.0	19.7	52.6
4	101.6	7,230	456	6.6	22.1	53.4
8	203.2	10,361	654	10.1	26.5	55.2

Note: Based on model 1000-MW coal-fired plant with 4100-gal/min (259-L/s) average ash sluice water flow, 41.5-acre (16.8-ha) combined bottom-ash/flyash pond, zero material storage area runoff, zero other non-rainfall-related flows.

Pond efficiency factor n is inversely related to relative pond performance; for example, 1/5 corresponds to a highly efficient pond, 1/2 to a poorly efficient one.

ash ponds serving only ash sluice water, which are able to meet average limitations for TSS during dry-weather flows, will meet concentration limitations during wet-weather conditions. EPA's final guidance on combined waste streams has not been published to date.

Interlaboratory Precision and Bias

EPRI has assembled and analyzed all the available data on the precision and bias of analytical methods approved by EPA and appearing in 40 CFR Part 136, also known as the section 304(h) methods.[9] Knowledge of interlaboratory precision and bias is necessary to determine with stated confidence whether effluent pollutant concentrations are in compliance with the limitations contained in the NPDES discharge permits.

It is also necessary to characterize the discharges realistically so that treatment systems able to meet the limitations can be designed. In this respect, a key industry position has been that the determination of compliance with effluent limitations must employ the same analytical methodology, and thus the same precision and bias, as that used in the development of the limitation itself.

Interlaboratory precision and bias are also required to establish the *limit of detection* (LOD) of the methods. LOD is essential, for example, in the case of maintenance chemicals added to cooling-tower systems, where EPA specifies that the addition of any such chemical shall not result in detectable amounts in the cooling-tower blowdown of any of the 126 priority pollutants, except Cr and Zn. As an alternative to monitoring the blowdown for the 124 pollutants, EPA allows demonstration through engineering calculation that any amounts contained in the maintenance chemicals would be discharged in the tower blowdown at concentrations below the LOD of the part 136 methods used for their analysis. Table 1.6 is based on the EPRI report and shows the best current estimate of interlaboratory precision and bias for the most appropriate section 304(h) method for a number of pollutants of interest.

Water Quality Criteria and Standards

Pursuant to the provisions of section 304(a)(1) of the CWA, EPA developed water quality criteria for the priority pollutants. Criteria documents were published for

TABLE 1.6 Interlaboratory Precision and Bias Data from RP 1851

Parameter	Method	Concentration, μg/L	RSD, %
Antimony	FAAS	37	12.9
Arsenic	GHAAS	41	21.7
Chromium	GFAAS	19	29.6
Copper	GFAAS	45	15.7
Mercury	CVAAS	1.3	41.4
Nickel	GFAAS	39	22.4
Selenium	GHAAS	39	27.8
Zinc	GFAAS	76	33.4

RSD = relative standard deviation
AAS = atomic absorption spectrometric
FAAS = flame AAS
GFAAS = graphite furnace AAS
GHAAS = gaseous hydride AAS
CVAAS = cold vapor AAS

64 priority pollutants or classes of pollutants on November 28, 1980.[10] EPA proposed revised water quality criteria for As, Cd, Cr, Cu, Cn, Pb, Hg, and ammonia and proposed new criteria for chlorine in February 1984.[11] Table 1.7 shows the present or proposed criteria for priority-pollutant elements.

In November 1983, EPA consolidated its rules governing the development, review, and approval of water quality standards by the states in a new water quality standards regulation. Guidance for applying the new regulation was contained in a companion *Water Quality Standards Handbook*[12] and in *Technical Support Manual: Waterbody Surveys and Assessments for Conducting Use Attainability Analyses.*[13]

TABLE 1.7 EPA Water Quality Criteria for Priority-Pollutant Elements, μg/L

	Freshwater		Seawater	
Element	Maximum	30-day avg.	Maximum	30-day avg.
Arsenic*	140	72	120	63
Cadmium*	4.5	4.5	38	12
Chromium (+ 6)*	11.0	7.2	1200	54
Copper*	15.7	10.8	3.2	2.0
Lead*	64	2.5	220	8.6
Mercury*	1.1	0.2	1.9	0.1
Nickel	1800	96	140	7.1
Selenium	260	35	410	54
Zinc	320	47	170	58

Notes:

Wherever the criterion is expressed as a function of hardness in the water, a total hardness of 100 μg/L has been used to arrive at the above numerical values.

The criteria for elements with an asterisk were proposed revisions in February 1984; the remainder are as promulgated in November 1980.

Purpose and Impact. Water quality standards are to protect health and welfare; to enhance the quality of water for the protection and propagation of fish, shellfish, and wildlife; and to provide recreation in and on the water. Other purposes of the standards are to promote agriculture, industry, and navigation. The 1983 regulation also describes the dual role of the standards in establishing water quality goals for a specific body and in serving as the regulatory basis for establishing water-quality-based treatment controls and strategies beyond that level of treatment required by sections 301(b) and 306 of the CWA, that is, beyond technology-driven effluent limitations.

The importance to powerplants of the water quality criteria and water quality standards regulation lies in the potential impact of water quality standards in the receiving water on the technology-driven effluent limitations contained in each NPDES permit. The regulation contained in 40 CFR part 122.62 states that whenever a state water quality standard for a pollutant cannot be met by the limitations imposed by the applicable effluent-limitation guidelines, the permitting authority (generally the state) is allowed to establish a more stringent limit for that pollutant in the NPDES permit.

Thus, as the states put in place revised water quality standards to reflect EPA's water quality criteria, especially those for the toxic (priority) pollutants, effluent limitations for these pollutants may become significantly lower than those contained in existing permits or as specified by the 1982 effluent guidelines. Equally important, limitations derived from water quality standards for toxic pollutants not now regulated by the present guidelines or contained in past permits may be placed in both existing and newly issued permits.

Making Allowances. As noted in Table 1.1, only pH, TSS, O&G, Fe, Cu, Cr, Zn, and *total residual chlorine* (TRC) are regulated under BAT in powerplant discharges considered as a whole. Yet, as Table 1.4 shows, As, Ni, Cd, Pb, and Se appear in significant numbers of discharges at concentrations greater than water quality criteria. However, allowance must be made for mixing-zone dilution (low flow is generally used at present for such calculations). Such allowable dilution may eliminate for many plants the necessity of treatment for the Table 1.4 elements, even though they may be present in some of their discharges.

Nonetheless, the impact may be much greater than these general figures would indicate, since effluent limitations on each individual waste stream type may regulate none, or only one or two, of the toxic elements. For example, ash sluice water contains limitations for none of the toxic pollutants, while cooling-tower blowdown contains limitations for only Cr and Zn, plus no toxic pollutants in any added maintenance chemicals.

EPA Guidance Document

EPA has published a guidance document[14] to assist state and regional EPA permit writers in preparing NPDES permit limitations. These limitations are designed to protect the aquatic ecosystem in receiving waters primarily beyond the edge of the mixing zone, but including acute toxicity limitations within the mixing zone outside the initial dilution area. Derivation of these toxicity-based limitations relies heavily on the use of both acute and chronic effluent toxicity tests.

Although EPA has proposed a two-tiered document for assessment of the toxic impact of a discharge, only the first tier, *screening,* seems likely to be required of most coal-fired plants and plants with cooling towers. Such screening may entail whole-effluent toxicity testing by using two test species, a fish and a daphnid, on as many as six grab samples of each discharge once per quarter for a year. If any of these tests result in greater than 50 percent mortality, more testing is required and the second tier, *definitive data generation,* is started.

Although few powerplant effluents meeting BAT are likely to enter the second tier, EPA notes that in the first-tier testing false positives are more desirable than false negatives. With few data available on the application of acute and chronic toxicity tests to the wide range of powerplant effluents, it is not possible to predict the number that might require second-tier testing.

Toxic Treatment

If an effluent is found to be toxic and to require further limitation because of its impact on receiving waters, the development of treatment procedures is likely to require a great deal of new effort. Little experience with existing technology is available to help reduce the measured toxicity explicitly. Since chronic toxicity is of equal concern—even the shortest chronic life-cycle tests require 7 days to conduct—toxicity performance testing will be expensive and time-consuming. Finding surrogate parameters that will faithfully mimic toxicity testing may also be an expensive undertaking specific to each case.

From the direction that regulations governing the discharge of toxic pollutants have taken, it appears that steam-electric powerplants will be faced with new, possibly extensive, and generally unfamiliar monitoring requirements. For some plants, such as those with copper-alloy main-condenser tubes in recirculated cooling systems, the results of effluent toxicity testing may require main-condenser tube replacement to secure compliance. Compliance for plants with flyash ponds may require retrofitting dry flyash transport systems.

Steam-electric powerplants are now faced with another major round of pollution control requirements—not from the requirements of recently issued effluent-limitation guidelines this time, but rather from a new and generally untried in-stream aquatic toxicity requirement. Not unexpectedly, however, and as was true in the past, coal-burning powerplants with cooling towers in nonarid areas of the country will be faced with the most problems.

POLLUTANT SOURCES

Effluent limitations for conventional, nonconventional, and priority pollutants are given in the effluent guidelines or are found in Table 1.4, which itemizes discharges from powerplants. Remember, the concentrations in this table were derived as if all the discharges from a plant (except once-through condenser cooling water) were combined and the average concentration reported. Viewing the plant as a "bubble" in this fashion appears to be realistic, because of the diversity of water management designs stemming from a variety of requirements. EPA, however, has categorized the industry's discharges by waste stream type, as previously shown in Table 1.3. In the following coverage, each of these pollutant sources is examined in turn.

Once-Through Cooling Water

The principal pollutant, and the only one specifically regulated in this source, is total residual chlorine, which results from the addition of chlorine at some point ahead of the main condenser or at the inlet to an auxiliary cooling system. The chlorine controls the biological growth in the system, most often the growth of biofilm on the tube surfaces in the main condenser. The discharge of TRC is limited by both a concentration maximum and a maximum duration of each chlorination event.

In 1982, it was estimated that plants representing about 277 GW of U.S. electrical capacity use once-through cooling. Of such plants, 73 percent employ chlorine for biofouling control, or 42 percent of the total 1982 steam-electric generating capacity. A study of the duration of chlorine additions, based on practices reported in 1980,[16] showed that approximately 70 percent of those plants which chlorinate do so for 1 h/day or less, and 95 percent do so for 2 h/day or less.

EPA's concern with the discharge of TRC relates to both the toxicity of TRC to biota in receiving waters and the generation of organic halogen compounds (TOX) from the reaction of *free available chlorine* (FAC) with natural organic compounds present in most surface waters. Several studies have been performed on the TOXs generated by powerplants during chlorination of once-through cooling water.[17,18,19] These have shown that typically 4 μg/L of chloroform is generated by the process and is 12 to 40 percent of the TOX so generated. It has been estimated that the chloroform discharged from chlorination of cooling water by steam-electric powerplants in 1982 was 32 tons (29 Mg), while that discharged by publicly owned treatment works was 1800 tons (1633 Mg).[20]

In its development document,[4] EPA reports on a study undertaken by a utility to determine the concentration of several trace elements in the influent and effluent cooling water from eight coastal generating stations. These data (summarized by EPA) are shown in Table 1.8 Note that only Cu and Zn increased in excess of 0.1 μg/L.

Cooling-Tower Blowdown

When a steam-electric plant rejects condenser heat to an evaporative cooling tower (see Section 1, Chapter 9), the dissolved and suspended matter in themakeup water must be removed from the system. The large volume of air drawn through the tower may contain gases and particulate matter; these will be scrubbed from the air by the water in the tower and must also be removed.

TABLE 1.8 Metal Pickup from Condenser by Once-Through Cooling Seawater, μg/L

	Median influent concentration		Net concentration change (effluent − influent)	
Metal	Dissolved	Particulate	Dissolved	Particulate
Cadium	0.06	0.006	0.034	0.005
Chromium	0.16	0.200	− 0.010	0.097
Copper	0.80	0.320	0.21	0.10
Nickel	0.44	0.160	0.10	0.004
Lead	0.14	0.24	0.04	0.07
Zinc	0.20	0.48	0.09	0.17

Part of the suspended matter from the makeup water and air may precipitate and settle out in the cooling-tower basin. It can be removed as sludge at infrequent intervals—possibly during the annual plant outage—by mechanical methods. The remaining suspended matter and all the dissolved matter are removed continuously in two ways: (1) partly by drift, in the tower exhaust, (2) but mostly by cooling-tower blowdown. Drift losses in modern mechanical-draft towers average 0.005 percent of the cooling-water circulating rate (0.002 percent in natural-draft towers), while in older and less efficient towers it can be as much as 0.2 percent of the circulating rate.

The combined removal through drift and blowdown is determined by the maximum concentration allowed for each of the dissolved and suspended substances in the circulating water. The controlling factor is the possible fouling and corrosion of condenser tube surfaces. Since drift loss is fixed for a given tower, the blowdown flow is varied to achieve the desired concentrations. Thus, for a tower with 0.005 percent drift, blowdown may vary from 65 percent of the makeup for a saltwater system to 20 percent of makeup for the typical freshwater tower. Where a portion of the recirculating cooling water is continuously treated to remove the deposit-forming substances—called *sidestream treatment*—the blowdown may be as low as 3 to 5 percent of the makeup. In this case, total dissolved matter in the blowdown may be as high as 10 000 to 30 000 mg/L.

Chemicals as Pollutant Sources. In addition to substances brought into the system in the makeup and in the air, other chemicals may be added continuously, or at intervals, to control corrosion and deposits. In excess, the chemicals may themselves be pollutants. Most common among these is chlorine (usually as chlorine gas in solution), added for the control of biofouling in the condenser. Biological growths also occur in the cooling tower in ways that hinder heat transfer. Since control of biological growths in both the condenser and the tower require that 0.3 to 0.5 mg/L of FAC be maintained in the system for 1 h or more, usually once per day, the blowdown will likewise contain the same pollutant, for which there is an effluent limitation (see Table 1.1).

Organic chemicals may be added for biological control, to supplement or possibly replace chlorine. Most of these chemicals are purchased as proprietary blends of several agents, under a trade name or number. The label on such compounds must bear the generic chemical names and their percentages of the total composition. Some of the generic chemicals used in these microbiocides appear on the priority-pollutants list in Table 1.2 and are subject to effluent limitations.

Corrosion Inhibitors. A variety of methods are used to control corrosion of metal cooling-system components. In the steam-electric power industry, highly corrosion-resistant tube metals are commonly selected, tube sheets may have cladding, and inlet structures and water boxes may feature cathodic protection. Sometimes, however, the most economical choice is to select the lowest-cost alloy that meets steam-side conditions and to add corrosion inhibitors to the cooling water. Combinations of chromate and zinc or of zinc and polyphosphates are the most common inhibitors, together with phosphonates and polymers to assist inhibitor action.

Unlike many other industrial categories, where tower blowdown is generally considered to be combined with other plant waste streams, steam-electric powerplants with cooling towers and incorporating corrosion inhibitors must meet effluent limitations that are quite restrictive. Thus, the use of chromate- and zinc-containing corrosion inhibitors is generally limited to those plants in arid areas where zero discharge is practical.

Maintenance Chemicals. Chemicals for microbiological, corrosion, and scale control are classed by EPA as chemicals added for cooling-tower maintenance. With the exception of Cr and Zn (see Table 1.1), which are separately regulated, none of the 126 priority pollutants may be present in cooling-tower blowdowns in detectable amounts. Instead of monitoring for such priority pollutants in the discharge, however, EPA allows the use of engineering calculations to demonstrate that these maintenance chemicals are not present in the discharge in detectable concentrations.

Release of corrosion products to the circulating water from system construction materials constitutes another pollution source in tower blowdowns. The concentration of corrosion products that can theoretically build up in the circulating water and appear in the tower blowdown depends on the number of cycles of concentration and corrosion rates. For example, a corrosion rate of 0.2 mil/yr (0.005 mm/yr) for an Admiralty-tubed condenser in a cooling tower operating at six cycles of concentration would result theoretically in a copper concentration of 0.6 mg/L in the blowdown, 1.4 mg/L at 14 cycles. How much actually does appear in the blowdown depends on the effectiveness of dispersing agents, either those naturally present or those added as deposit control chemicals, in preventing the hydrous oxide corrosion products from settling in the tower basin or redepositing elsewhere in the system.

Tower Materials. Many cooling towers are constructed of wood that has been treated with preservatives to reduce its susceptibility to fungus attack. *Acid copper chromate* (ACC), *chromated copper arsenate* (CCA), creosote, and pentachlorophenol are preservatives commonly selected; all contain or consist principally of substances on the priority-pollutants list. The extent to which they leach into the cooling water is low, although it is likely to be a significant amount in new or newly rebuilt towers using lumber so treated.

The fill in large natural-draft cooling towers is frequently asbestos cement. The amount of asbestos per cubic foot (meter) of fill can vary from 2 to 3 lb (32 to 48 kg), depending on the type of fill (film or splash) selected. Asbestos fibers may be released to the water by physical disintegration from freeze-thaw cycles, by chemical dissolution of the cement because of low-pH cooling water, and by biological action.

The U.S. *Nuclear Regulatory Commission* (NRC)[23] surveyed 18 towers using asbestos fill. The analyses of samples of both intake and blowdown showed no detectable asbestos in the makeup and a wide variation of asbestos fibers in the tower blowdowns. The concentration of asbestos fibers was above detection limits (generally 6×10^4 fibers per liter) in only 3 of the 18 blowdowns. In these three, the concentrations varied from below detection to as high as 160×10^6 fibers per liter. Fibers were detected in basin sediments in instances where they were not detected in the blowdown.

Where raw surface water is the source of makeup to the cooling tower, still other substances may enter, such as insecticides and herbicides from upstream agricultural runoff. After concentration in the tower, these chemicals could reach levels that exceed effluent limitations imposed by water quality standards.

Fuel-Related Wastes

The ash generated by fossil-fuel-fired powerplants varies in quantity and quality depending on the type of fuel, its source, the method of combustion, and the point in the boiler from which it is obtained. Ash formed from the combustion of fuel oil consists primarily of oxides and salts of vanadium, nickel, iron, plus organometallic compounds and carbon (soot). Sulfuric acid and traces of other metals may also be present. Table 1.9 shows the range of ash composition that might be expected in a plant where magnesium oxide is added to the furnace to control corrosion by the ash.

TABLE 1.9 Oil and Flyash Composition

Constituent	Weight, %
Silica, SiO_2	1–4
Iron oxide, Fe_2O_3	2–8
Vanadium pentoxide, V_2O_5	7–14
Calcium oxide, CaO	0–1
Magnesium oxide, MgO	20–45
Sulfate, SO_4	12–40
Other oxides (Ni, Cr, Na, etc)	1–4
Loss on ignition (presumably carbon)	2–20
Specific gravity, typically 2.64	

Coal-fired plants produce substantially larger quantities of ash than oil-fired plants—300,000 tons/yr (272 000 Mg/yr) versus 2000 (1800) for a 1000-MW steam-electric plant is typical. The quantity of ash varies widely depending on the source of coal: 8 percent for a western coal to as much as 18 percent for some eastern coals. Ash quality likewise depends on the source and whether it is bottom ash or flyash. Tables 1.10 and 1.11 from an EPRI report[24] illustrate the bulk-coal and bulk-ash compositions for 11 powerplants burning coal from midwest and Appalachian regions. Also shown in Table 1.11 are data from another study[25] for 23 ashes collected from plants in 21 states.

The amount of sluice water needed to convey ash varies considerably with ash quality and system design. The typical concentration of ash in sluice water routed to the settling pond is 5 percent by weight. Flows range from 1200 to 10,000 gal/day • MW (4550 to 37 900 L/day • MW); typical rates are 5000 gal/day • MW (18 950 L/day • MW) for combined coal-ash pond effluent and 800 gal/day • MW (3030 L/day • MW) for oil. The amount of matter that dissolves in sluice water—or remains in it as suspended solids in the ash-pond discharge—is again a function of many factors. Oil flyash generally contains more soluble matter, up to 40 percent, so the sluice water will contain substantial quantities of dissolved iron, nickel, chromium, and vanadium sulfates at low pH, along with suspended siliceous and carbonaceous particles. The dry ash is of low density (10 to 20 lb/ft^3, or 160 to 320 kg/m^3), so it settles with difficulty.

Bottom Ash. Bottom ash becomes a source of pollution when it is transmitted from the furnace bottom to bins and a settling pond, or to a high-rate settler. From a coal-fired furnace it is a vitreous combination of metal oxides and silica, thus generally has low solubility, is relatively coarse (only 5 percent may be finer than 0.08 mm), and settles rapidly.

Flyash. Flyash becomes a source of pollution when it is transmitted by water from the electrostatic precipitator or fabric filter to the settling pond. Coal flyash can have solubilities of several percent, particularly when a significant amount of

TABLE 1.10 Coal Analyses Average for 11 TVA Plants, EPRI EA2588, μg/g*

Constituent	Mean	*N*	Range	Coefficient of variation
C, %	66.4	13	55.8–82.8	9.6
S, %	2.42	13	0.56–4.14	50
Ash, %	17.4	13	12–21	15
Heat value, Btu/lb	11533	13	9468–12,737	7
(kJ/kg)	(26 872)		(22 064–29 677)	
Na	465	13	250–1225	52
Mg	2223	13	1120–3400	28
Al, %	2.11	13	1.12–3.39	31
Cl	795	13	34–2290	99
K, %	0.39	13	0.08–0.62	36
Ca, %	0.40	12	0.11–1.04	76
Sc	4.19	13	1.50–7.20	37
Ti	1118	13	613–1660	29
V	49.3	13	12–84	41
Cr	21.7	13	7.0–29	27
Mn	51.5	13	16.6–97	46
Fe, %	1.54	13	0.36–2.62	47
Co	5.45	13	1.70–11	44
Ga	5.46	13	2.60–11	40
As	12.1	13	3.5–24	49
Se	2.73	13	0.8–7.10	56
Br	7.01	13	0.54–17	82
Rb	31.6	13	15–42	29
Sr	117	12	42–302	64
Mo	1.90	4	1.0–2.6	35
In	0.030	5	0.028–0.033	5.9
Sb	1.02	12	0.30–1.90	54
I	1.71	9	0.83–4.6	71
Cs	1.43	13	0.30–2.10	33
Ba	246	13	92–700	75
La	11.1	13	5–21	40
Ce	19.0	13	8.5–30	36
Sm	2.08	12	0.8–3.8	40
Eu	0.39	13	0.11–0.65	37
Tb	0.31	13	0.19–0.40	27
Dy	1.46	12	0.7–2.1	32
Yb	1.27	12	0.3–7	146
Lu	0.25	6	0.2–0.3	18
Hf	0.79	13	0.44–1.20	26
Ta	0.28	9	0.19–0.40	27
W	0.73	12	0.42–1.40	38
Th	4.44	13	1.5–21	114
U	1.93	13	0.95–2.98	31

*Except where indicated otherwise.

TABLE 1.11 Coal Ash Analyses for 11 TVA Plants, EPRI EA2588, μg/g*

	TVA study				National study			
Element	Mean	*N*	Range	Coefficient of variation	Mean	*N*	Range	Coefficient of variation
Na	2620	22	1340–6000	42	5860	23	1180–20 300	98
Mg, %	1.33	22	0.84–2.39	28	1.95	23	1.15–6.08	53
Al, %	12.4	22	8.13–19.6	24	10.8	23	5.91–14.3	23
Cl	106	2	63–150	57	83.8	15	13–460	144
K, %	2.31	22	1.51–3.26	20	1.55	23	0.15–3.47	54
Ca, %	2.16	20	0.33–6.82	73	5.50	23	0.60–17.7	96
Sc	26.6	22	18–38	24	16.2	23	6.0–38	45
Ti	6610	22	5050–8650	18	5630	23	2760–8310	28
V	313	21	161–644	41	213	23	68–442	42
Cr	131	22	34–170	24	132	23	43–259	41
Mn	259	22	79–560	46	199	23	58–543	64
Fe, %	9.86	22	2.52–21.9	57	7.82	23	2.71–29	72
Co	41.7	22	22–82	44	27.4	23	4.9–57	55
Ni	90	10	35–132	28	33	23	1.8–115	86
Cu	138	15	57–262	60	240	23	45–616	75
Zn	348	16	60–832	67	77.9	23	14–406	126
Ga	47.8	22	15–99	56	76.1	15	13–230	99
As	150	22	30–513	82	85.6	23	2.3–312	98
Se	16.1	14	4.6–53	83	8.43	23	1.2–17	59
Br	5.5	4	1.5–9.5	63	5.3	23	0.3–21	92
Rb	194	21	139–240	14	165	23	36–300	47
Sr	724	19	218–2240	70	1330	23	59–3860	81
Mo	34.2	15	7–110	85	21.4	23	6.5–41	53
Cd	3.0	15	0.28–10.2	99	0.86	23	0.10–3.9	124
Pb	92	15	41–136	34	40.3	23	3.1–241	154
In	0.23	14	0.12–0.38	34	0.31	11	0.10–1.10	96
Sb	9.4	22	1.9–20	65	3.96	23	0.8–13	74
Cs	8.9	22	7–12	15	8.2	23	1.5–18	51
Ba	1220	22	345–3380	65	1740	23	570–6920	93
La	70.5	22	37–106	30	60.9	23	33–104	29
Ce	121	22	69–172	28	180	23	74–300	32
Sm	12.4	21	7–19	31	12.5	22	5.4–24	34
Eu	2.53	22	1.5–3.6	26	—	—	—	—
Tb	1.66	22	0.7–4.2	55	—	—	—	—
Dy	9.03	22	5.9–13	27	—	—	—	—
Yb	5.63	22	2.5–11	43	—	—	—	—
Lu	1.68	13	1.2–2.6	29	0.87	23	0.5–1.5	29
Hf	4.84	22	3.1–7.0	21	—	—	—	—
Ta	1.40	21	0.9–2.0	23	1.31	23	0.5–2.6	37
W	5.59	22	2.0–11	50	6.74	23	2.19–21	56
Th	19.5	22	11–28	28	43.5	23	13–68	33
U	14.1	22	7.9–21	32	7.33	23	0.8–19	47

*Except where indicated otherwise.

TABLE 1.12 Leachability of Flyash, EPRI EA2588 (Table 3.6), Percent Extractable in River Water*

Constituent	Ash Code C2	C4	E2	F2	G2†	H2	I2	I3	J2	J3	K3†
Ca	12	10	10	15	16	13	12	21	2.7	7.8	—
Mg	0	−0.8	−1.9	−0.19	1.4	−1.3	0.90	1.8	−1.8	−1.2	2.7
Na	3.3	4.4	8.0	6.9	5.7	7.3	3.4	15	1.9	7.0	12
K	1.3	2.0	2.3	2.3	1.0	2.2	1.5	2.9	1.0	2.3	1.5
SO_4	42	64	55	85	80	69	71	84	52	61	(118)‡
Al	0.03	0.06	0.09	0.06	0.02	0.11	0.02	0.03	0.27	1.6	0.002
Fe	1.5×10^{-4}	3.5×10^{-4}	1.8×10^{-4}	-7.3×10^{-5}	3.0×10^{-4}	2.2×10^{-5}	-1.7×10^{-4}	0.01	2.3×10^{-5}	4.5×10^{-4}	1.0×10^{-3}
Mn	0.05	0.07	−0.04	0.01	0.56	−0.01	0.58	3.1	−0.11	−0.09	0.77
Ni†	0.25	0.35	1.1	1.1	–	0.84	0.71	—	0.73	0.77	—
Cr	0.87	2.0	9.3	3.5	0.20	13	0.80	0.01	3.3	8.5	0.13
Cu	−0.13	0.16	−0.53	−0.35	—	−0.14	−0.12	2.2	0.04	0.09	—
Zn	0.47	0.11	0.03	0.009	—	0.01	0.01	4.8	−0.15	−0.08	—
Cd	0.25	0.02	0.004	0.10	—	−0.06	1.3	22	−0.15	−0.04	—
Pb	0.02	0.01	0.005	0.03	—	0.005	1.2	0.16	0.08	0.04	—
Se	27	—	—	36	42	9.0	10	7.2	—	20	50
As	18	3.4	− 0.17	3.9	25	3.0	3.4	0.65	1.8	0.43	2.2
Mo	58	(150)‡	93	35	—	54	24	46	—	24	65
pH	9.2	9.8	11.2	8.8	8.4	10.1	8.3	5.5	10.5	10.1	7.7
E.C.§	324	409	1016	706	212	637	484	646	441	720	312

*Except where indicated otherwise.

†Values uncorrected for composition of intake water; corrected values would be lower.

‡Suspect value.

§Electrical conductivity, μS/cm.

limestone is in the coal. In addition, the very fine size of the flyash particles (typically 90 percent are finer than 0.08 mm, and 5 percent are finer than 0.003 mm) and the presence of hollow particles, which will float, make the sedimentation process for flyash much more difficult than for bottom ash.

The bulk of the matter dissolved from flyash particles consists of calcium, magnesium, potassium, sodium, and iron sulfates and hydrous oxides. The trace elements that are leached come largely from the surface of the particles. The fraction of each element that appears on the particle surface, and is available for leaching, is characteristic of that element and of the section of the gas pass in the boiler from which the ash particle was obtained. Table 1.12 from the EPRI report[24] shows the results of the extraction of nine flyash samples expressed as the percentage of each element leached after a contact time of 4320 min in the river water used for sluicing.

Trace Elements. Trace elements that appear in the ash-pond overflow are a complex function of the sedimentation characteristics of the pond, the pH, and the oxygen-reduction potential at the sediment/water interface at the bottom of the pond. Table 1.13 from the EPA development document[4] presents data on average trace-element concentrations in overflows from 20 combined ash ponds. The elements were gathered from plant monitoring reports after the implementation of BPT, and are thus presumably representative of current operations. Table 1.14 presents a range of values for ash-pond discharges from 11 coal-fired powerplants obtained from 1973 to 1975, which was generally prior to implementation of BPT and thus may not be representative of current discharges.[26]

Flue-Gas Desulfurization

Most, if not all, FGD systems installed at eastern powerplants are designed to have zero aqueous discharge. Dissolved matter extracted from the flue gas is designed to depart from the system in the liquor that constitutes the surface moisture in the dewatered waste sludge. Problems that arise and necessitate some

TABLE 1.13 Combined Ash-Pond Overflow Data, EPA Development Document (Table V-33), μg/L

Trace metal	Flyash ponds*			Bottom-ash ponds†			Combined ponds‡		
	Min.	Max.	Avg.	Min.	Max.	Avg.	Min.	Max.	Avg.
As	10	66	29.2	7	70	21.1	3.5	416	67
Cd	3.5	26.9	11.8	2	16.3	9.7	0	82	18.7
Cr	5	15.2	10.2	4	41.7	15.6	2.5	84.2	30.4
Cu	20	209	84.8	5	70	36.9	0	130	59
Fe	1055	8138	4011	657	10 950	3410	80	2600	664.6
Pb	10	200	59.4	10	60	25.5	0	100	40.1
Hg	0.1	1.8	0.6	0.4	1.7	0.8	0	65	3.9
Ni	33	100	61.1	13.3	1345	191.4	0	100	49
Se	2	7.8	4.4	2	10	6.7	1.7	68.3	23.6
Zn	50	1139	358.4	10	302	131.9	10	293	94.9

*Data for four facilities.
†Data for nine facilities.
‡Data for 20 facilities.

TABLE 1.14 Characteristics of Ash-Pond Discharges, mg/L*

Parameter	Flyash pond Min.	Flyash pond Avg.	Flyash pond Max.	Bottom-ash pond Min.	Bottom-ash pond Avg.	Bottom-ash pond Max.
Flow, gal/min (L/s)	3100 (196)	6212.5 (392)	8800 (555)	4500 (284)	16,152 (1020)	23,000 (1451)
Total alkalinity (as $CaCO_3$)	—	—	—	30	85	160
Phen. alkalinity (as $CaCO_3$)	0	0	0	0	0	0
Conductivity (μmho/cm)	615	810	1125	210	322	910
Total hardness (as $CaCO_3$)	185	260.5	520	76	141.5	394
pH	3.6	4.4	6.3	4.1	7.2	7.9
Dissolved solids	141	508	820	69	167	404
Suspended solids	2	62.5	256	5	60	657
Aluminum	3.6	7.19	8.8	0.5	3.49	8.0
Ammonia (as N)	0.02	0.43	1.4	0.04	0.12	0.34
Arsenic	<0.005	0.010	0.023	0.002	0.006	0.015
Barium	0.2	0.25	0.4	<0.10	0.15	0.30
Beryllium	<0.01	0.011	0.02	<0.01	<0.01	<0.01
Cadmium	0.023	0.037	0.052	<0.001	0.0011	0.002
Calcium	94	136	180	23	40.12	67
Chloride	5	7.12	14	5	8.38	15
Chromium	0.012	0.067	0.17	<0.005	0.009	0.023
Copper	0.16	0.31	0.45	<0.01	0.065	0.14
Cyanide	<0.01	<0.01	<0.01	<0.01	<0.01	<0.01
Iron	0.33	1.44	6.6	1.7	5.29	11
Lead	<0.01	0.058	0.2	<0.01	0.016	0.031
Magnesium	9.4	13.99	20	0.3	5.85	9.3
Manganese	0.29	0.48	0.63	0.07	0.16	0.26
Mercury	<0.0002	0.0003	0.0006	<0.0002	0.0007	0.0026
Nickel	0.06	0.11	0.13	0.05	<0.059	0.12
Total phosphate (as P)	<0.01	0.021	0.06	<0.01	0.081	0.23
Selenium	<0.001	0.0019	0.004	<0.001	0.002	0.004
Silica	10	12.57	15	6.1	7.4	8.6
Silver	<0.01	<0.01	<0.01	<0.01	<0.01	<0.01
Sulfate	240	357.5	440	41	48.75	80
Zinc	1.1	1.51	2.7	0.02	0.09	0.16

*Except where indicated otherwise.

aqueous discharge are related to the buildup of chlorides in the recycled absorbing-liquor slurry. Chlorides upset the close balance between sulfate and sulfite that is needed to prevent supersaturation and scaling by calcium sulfate in the absorber.

Excessive carry-over of flyash into the absorber circuit apparently adds to the problem by catalyzing the oxidation of sulfite to sulfate, further disturbing the balance. These tramp contaminants must be removed from the system, either as fluid retained in the dewatered sludge or as aqueous blowdown. Such blowdown streams are principally saturated solutions of calcium sulfate, calcium sulfite, and sodium chloride, with traces of metal from the ash. In its 1982 effluent-limitations document,[3] EPA reserved putting restrictions on FGD discharges pending a future rule making.

Coal-Pile Runoff

Some of the rain that falls on coal storage piles runs off the surface, but some infiltrates the pile. The portion that infiltrates may percolate through all or a fraction of the pile, adding to the moisture content of the coal. Rainfall may erupt from the sides farther down from the top (interflow), emerge at the base of the pile (base flow), or percolate into the ground beneath the pile. The fraction of total rainfall that becomes each of these resultant flows, and their relation to time as measured from the start of the rainfall event, is a complex function of the manner in which the pile is constructed, the previous rainfall on the pile, and the rate and duration of the immediate event.

An EPA and *Tennessee Valley Authority* (TVA) study[27] reports that 800 to 2400 yd^3 (612 to 1836 m^3) of coal storage is required for every megawatt of rated capacity. A survey jointly sponsored by *Edison Electric Institute* (EEI) and EPA[28] showed for the 81 plants responding an average of 500 tons (454 Mg) stored per megawatt of capacity, an average height of 36 ft (11 m), an average density of 0.63 ton/yd^3 (0.75 Mg/m^3), an average slope of 0.63, and an average storage period of 2.8 months.

The reaction of coal minerals with water and oxygen in the air is similar to that which takes place in coal mines, causing contaminated drainage. Water quality of the interflow and the base flow from the pile during and after rainfall is largely determined by the amount of these oxidation products stored in the pile since the previous rainfall. A certain amount is generated during the event itself.

The metallic sulfide-bearing minerals that predominate in coal are pyrites and marcasites, both iron sulfide ores. Their oxidation, which is catalyzed, results in the production of ferrous sulfate and sulfuric acid. Below pH 5, certain iron bacteria accelerate the formation of additional oxidation to ferric iron. The low pH from the sulfuric acid formed dissolves many other complex sulfides and sulfosalts, releasing some of or all 10 metals (aluminum, zinc, copper, etc.). Table 1.15

TABLE 1.15 Coal-Pile Drainage Analyses, EPA Development Document (Table V-74, 75)

Plant		pH	Acidity, mg/L (as $CaCO_3$)	Sulfate, mg/L	Dissolved solids, mg/L	Suspended solids, mg/L	Fe, mg/L	Mn, mg/L
J	Range	2.3–3.1	300–7100	1800–9600	2500–16K	8.0–2300	240–1800	8.9–45
	Mean	2.79	3400	5160	7900	470	940	28.7
	N	19	18	18	18	18	19	19
E	Range	2.5–3.1	860–2100	1900–4000	2900–5000	38–2270	280–480	2.4–10.0
	Mean	2.67	1360	2780	3600	190	380	4.13
	N	6	6	6	6	6	6	6
E*	Range	2.5–2.7	300–1400	870–5500	1200–7500	69–2500	62–380	0.88–5.4
	Mean	2.63	710	2300	2700	650	150	2.3
	N	14	14	14	14	14	14	14

*Discrete storm.

shows the results of analysis of coal-pile runoff over a 6-month period, as reported by EPA/TVA for two plants.[4]

Chemical Metal-Cleaning Wastes

The waterside of steam generators, feedwater heaters, and associated piping is cleaned with chemical solvents periodically. Plant operators determine the intervals between cleanings on the basis of boiler type, its characteristics, and the condition of operation. In general, most large steam generators are chemically cleaned at 3- to 5-year intervals. The volume of wastes resulting from each cleaning depends on the fill volume of the particular boiler as well as on the type of cleaning process used. Boiler volumes range from 20,000 to 100,000 gal (75 800 to 379 000 L), with most single-stage solvent/neutralizer systems producing three to four volumes of waste per application.

The metal content of the waste-cleaning solution will be largely a function of the materials of construction of the feedwater system, since it is the corrosion products released from these surfaces and transported to the steam generator that must be removed. Iron is the principal metal in spent boiler-cleaning solvents, with varying amounts of copper, nickel, zinc, chromium, calcium, and magnesium, depending on the specific feedwater/steam-generator system and its condition.

The particular chemical-solvent composition is determined by several factors: materials of construction of the unit being cleaned, composition of the deposit to be removed, and the available means of spent-solvent disposal. Where hydrochloric acid with copper complexers is used, sometimes preceded with ammonium bromate to oxidize and remove metallic copper, the spent solvent is generally treated on the site, either separately or in combination with other metal-cleaning wastes. The neutralized, clarified waste is discharged subject to the effluent limitations in Table 1.1 or the more stringent limitations derived from water quality standards. Table 1.16[29] shows the results of the treatment of several different types of waste by the addition of lime, with or without sulfide.

Organic solvents, such as ammoniated EDTA, hydroxy acetic-formic acid, or ammoniated citric acid, are incinerated by many operators in the furnace of an operating coal- or oil-fired boiler. In coal-fired units, the metals added to the ashes by incineration are but a small fraction of the total of such metals already in the ashes; in oil-fired units, the contribution may be significant. Where wet-ash

TABLE 1.16 Raw and Treated Spent-Solvent Analyses, APC 1981 (p. 1158)

		Effluent treatment				
Waste	pH	TSS, mg/L	Fe, mg/L	Cu, mg/L	$Ca(OH)_2$, g/L	Na_2S, g/L
Hydrochloric acid	10.0	—	0.39	—	2.4	—
Hydrochloric acid plus ammonium bromate	8.5	—	0.88	0.09	1.3	0.07
Hydroxyacetic/formic acid	11.2	—	0.82	—	1.2	—
	11.0	—	0.94	0.05	1.3	—
	9.3	4.0	0.25	—	1.8	—
Ammoniated EDTA	11.4	—	0.55	0.5	1.4	0.64

TABLE 1.17 Gas-Side Wash Wastewater Analyses, UWAG Report (Sec. IV)

	Gas-side wash wastewater analyses			
Pollutant*	Average air-preheater wash wastewater composition, four coal-fired units, mg/L	Gas-side wash wastewater composition (furnace), one coal-fired unit, mg/L	Average air-preheater wash wastewater composition, five oil-fired units, mg/L	Gas-side wash wastewater composition (furnace), one oil-fired unit, mg/L
Antimony	0.80	0.1	0.55	0.5
Arsenic	0.70	0.01	0.14	0.01
Beryllium	—	—	—	—
Cadmium	0.03	0.04	0.03	0.05
Carbon, organic	—	—	—	—
Chromium, total	1.31	0.03	6.6	1.4
Copper	0.89	0.06	9.66	90.0
Iron	2485.00	0.02	893.3	210.0
Lead	0.11	0.05	0.13	0.50
Manganese	—	—	—	—
Mercury	—	—	—	—
Nickel	1.75	0.21	115.33	120.0
Selenium	0.13	—	—	—
Silver	0.02	—	—	—
Thallium	0.35	—	—	—
Vanadium	2.23	0.2	429.33	130.0
Zinc	2.25	0.13	2.93	7.1
Total	2495.57	0.95	1458.00	559.56
Priority-pollutant metals (excluding iron)	10.57	0.93	564.7	349.56
Priority-pollutant metals only (excluding iron and vanadium)	8.34	0.73	135.37	219.56

*All metals reported as soluble metals.

sluicing is used, the metals may or may not be leachable by the sluice water. In either case, proper control of the incineration process can achieve up to 98 percent retention of the metals within the waste ash.

Nonchemical Metal-Cleaning Wastes

In some boilers, fireside deposits are removed by high-pressure hosing with water, occasionally with a small amount of alkali added where the deposits are highly acidic. These deposits can be a source of the same metals and other pollutants that are contained in the boiler ash. Those from the furnace resemble bottom ash while those from the superheater/reheater resemble flyash. Deposits from the air preheater, however, tend to contain a higher percentage of the more volatile elements in the ash as well as more soluble matter in general. The quality of the waste wash water thus depends not only on the characteristics of the fuel and the manner of combustion, but also on where the wash water originates in the gas passes of the steam generator.

Obtaining samples representative of the total wash-water waste is difficult. Table 1.17 is from a report by UWAG[30] and shows the results of analyses of air preheater and furnace waste wash water from both coal- and oil-fired boilers. Table 1.18 is based on the same report and reveals the average volumes and frequencies of wash-water waste per megawatt for the furnace, superheater, economizer, and air preheater for both coal- and oil-fired units.

The UWAG report also shows that 84 percent of coal-fired capacity treated the gas-side wash-water wastes as low-volume and not metal-cleaning wastes, whereas only 48 percent of oil-fired capacity treated gas-side wastes as low-volume, 52 percent as metal-cleaning. The distinction is important because the effluent limitations on low-volume wastes contain limits only on pH, TSS, and O&G, while those on chemical metal-cleaning wastes contain, in addition, limits of 1.0 mg/L on both Fe and Cu.

Low-Volume Wastes

EPA defines *low-volume wastes* to mean those from all sources taken collectively as if from one source, except wastes for which specific limitations are otherwise established. Wastewaters include those from ion-exchange water treatment systems, evaporator blowdown, boiler blowdown, laboratory and floor drains, cooling-tower-basin cleaning wastes, and drains and losses from recirculating house-service-water systems. At this time, and until EPA promulgates a specific regulation, low-volume wastes also include FGD blowdown streams.

From Demineralization. Ion-exchange regeneration processes may be used in a steam-generating plant, both for makeup water treatment and for polishing condensate, which is returned to the boiler as feedwater. When a fixed-bed ion-exchange resin system is selected, it is regenerated periodically; if a powdered

TABLE 1.18 Frequency and Volume of Gas-Side Waste Wash Water

Source*	Unit size, MW	Flow rate		Volume per wash		Washes per year
		gal/min • MW	L/s • MW	gal/MW	L/MW	
Furnace						
Coal, 213	113	4.0	0.25	3439	13 000	1.9
Oil, 110	121	2.9	0.18	2921	11 800	3.9
Superheater, reheater						
Coal, 205	113	3.5	0.22	3291	12 500	1.0
Oil, 152	114	3.2	0.20	3181	12 100	2.7
Economizer						
Coal, 211	148	3.5	0.22	2759	10 100	1.0
Oil, 144	123	3.2	0.20	1178	4500	2.9
Air preheater						
Coal, 213	114	3.9	0.24	3817	14 500	1.4
Oil, 152	114	2.9	0.18	1962	7500	3.3

*Fuel (coal or oil), number of units.

ion-exchange system is used, the exhausted resin is discarded and replaced with fresh resin. With regeneration, the waste volumes, composition, and frequency of regeneration are determined by the particular resins used and the cycle in which they are operated. The regeneration frequency for condensate polishing systems will vary from one mixed bed per day to one per week. For makeup demineralizers, the frequency will generally vary from one train per shift to one train per 1 to 2 days. The amount of regenerant waste produced depends on the size of the units in the system and the manner in which the rinse water is handled. Some multiple-bed ion-exchange demineralizers use rinse recycling techniques that may halve the total volume of rinse discharged to waste.

When the ion-exchange process is one of sodium-cycle softening, the principal waste is a 0.5 to 1.0 percent solution of a mixture of Na, Ca, and Mg chlorides. Only where water quality standards establish effluent limitations for *total dissolved solids* (TDS) will sodium-cycle softening regenerants be a source of pollutants.

In a water treatment system featuring ion-exchange demineralization, the regenerants will contain such metals as Fe, Al, and Mn from the clarified water source, besides Na, K, Ca, and Mg. The regenerant will also contain anions such as sulfates, chlorides, nitrates, and most of the soluble organic matter in the clarified supply. The combined regenerant wastes from a single cation/anion train thus are principally a source of TDS, plus either free NaOH or free H_2SO_4 and some trace elements and organic matter, depending on the quality of water being demineralized and the resin combinations used.

From Fixed-Bed Polishing. Fixed-bed polishing of condensate or feedwater to high-pressure boilers produces wastes of considerably different character than those from demineralization of makeup. Except for start-up and periods of high condenser leakage, regeneration of the entire capacity is done only once a week to once a month. Most important, however, the composition of the regenerant is significantly different. The resin bed may remove 20 μg/L of Fe from the condensate, while removing from 10 to 100 μg/L of Na, Ca, and Mg (total).

Since most plants using condensate polishing add ammonia for condensate pH control, the polishers also remove 200 to 400 μg/L of ammonia—at least up to the ammonia breakpoint in each cycle. Since the effluent quality requirements of the polisher are much higher than those for makeup demineralization, several times as much regenerant is used per unit of dissolved matter removed.

The net effect is this: While Ca and Mg salts in the regenerant are low, Fe, Cu, and Zn will be relatively high, and the amount of ammonia higher still. Typically, for a system operating on the H/OH cycle, the concentrations in the combined cation/anion spent regenerant plus two volumes of rinse might be as follows: ammonia, 500 to 5000 mg/L; Fe, 10 to 100 mg/L; Cu, 5 to 50 mg/L; Zn, 2 to 10 mg/L; Ca and Mg together, 10 to 100 mg/L; plus 2000 to 5000 mg/L of Na and 5000 to 10 000 mg/L of sulfate. Thus, the ammonia, Cu, and Zn may be a significant source of pollutants, although the volume of discharge—10,000 to 20,000 gal/day (37 900 to 75 800 L/day) for a 500-MW plant, for example—is small compared to ash sluice water of 2 to 4 $\times$ 10^6 gal/day (7.58 to 15.16 $\times$ 10^6 L/day), or cooling-tower blowdown of 1 to 2 $\times$ 10^6 gal/day (3.79 to 7.58 $\times$ 10^6 L/day). For those systems operating with an $H,NH_3/OH$ cycle, the concentrations of Fe, Cu, Zn, Ca, and Mg will be several times greater, while the average volume per day of waste will be several times smaller.

From Powdered Resin. Polishing of condensate with powdered ion-exchange resin generates a different type of waste. The resin is not regenerated when it is exhausted, but is simply flushed to waste and replaced with new resin. The waste resin slurry will contain exchanged Ca, Mg, Na, plus ammonia, Fe, Cu, and Zn, filtered or exchanged during the service cycle of the resin. Most such powdered exchange systems operate on the H,NH_3/OH cycle, so that the normal service-cycle length is 20 to 40 days.

The volume of waste slurry is about one-third the volume encountered in regenerating fixed-bed polishers. Instead of containing largely dissolved matter, the slurry will be 5000 to 10 000 mg/L of suspended solids containing 50 to 500 mg/L of Fe, 20 to 200 mg/L of Cu, and 10 to 50 mg/L of Zn, depending on the materials used in the condenser and low-pressure feedwater heaters. The suspended resin solids as well as the contained metal oxides thus are sources of both conventional and toxic pollutants.

From Boiler Blowdown. Boiler blowdown was specifically regulated by EPA in the 1974 effluent-limitation guidelines. On the basis of more information, EPA concluded that boiler blowdown was sufficiently similar in characteristics to the other specific sources of low-volume wastes that there was no need to regulate it as a separate waste stream.

The volume and composition of blowdown vary depending on the age, size, and pressure of the boiler; on the composition of the makeup water; on the tightness of the condenser; and on the chemistry used. Large, modern, high-pressure drum-type steam generators will have blowdown rates of 0.1 to 1.0 percent of the steam flow. Since better than 90 percent of the entering metal oxides are deposited within the boiler, such rates are largely for the control of the dissolved matter present. Blowdown thus contributes only small amounts of Fe, Cu, Ni, and Cr to the plant wastes.

Even where discharges of phosphate might be limited by water quality considerations, modern high-pressure boilers operating on phosphate chemistry discharge only 0.2 to 1.0 lb/day (0.091 to 0.454 kg/day) of phosphate at concentrations of 1 to 2 mg/L as P, which must be compared to the 5-mg/L P limit contained in the 1974 effluent limitations (abolished in 1982) on cooling-tower blowdown at flows generally 100 to 200 times larger. Only in older, low-pressure boilers will blowdown possibly become a significant source of TDS and phosphate in the plant discharge.

REFERENCES

1. *The Federal Water Pollution Control Act Amendments of 1972,* Pub. L. 92-500, as amended by the *Clean Water Act of 1977,* Pub. L. 95-217.
2. *Natural Resources Defense Council v. Train,* 8 ERC 2120, D.D.C., 1976, modified at 12 ERC 1833, D.D.C., 1979.
3. *Federal Register,* 47 FR 52290 et seq., Nov. 19, 1982.
4. *Development Document for Final Effluent Limitations Guidelines, New Source Performance Standards, and Pretreatment Standards for the Steam Electric Point Source Category,* U.S. Environmental Protection Agency, EPA 440/1-82-029, Nov. 1982.

5. *Federal Register,* 39 FR 36186 et seq., Oct. 4, 1974, and 40 FR 7095, Feb. 19, 1975.
6. "Aqueous Discharges from Steam Electric Power Plants, Data Evaluation," RP 1851, interim report, Electric Power Research Institute, Palo Alto, Calif., Oct. 1982.
7. The Utility Water Act Group (UWAG) is a consortium of 73 electric power generating companies representing more than 50 percent of the electric power generating capacity of the United States. Edison Electric Institute, the National Rural Electric Cooperative Association, and the American Public Power Association also are members.
8. *Treatability Manual,* vols. 1 to 5, EPA 600/8-80-442a–e, Environmental Protection Agency, July 1980.
9. "Aqueous Discharges from Steam Electric Power Plants, Analytical Methods, Precision and Bias Data," draft report, RP 1851-2, Electric Power Research Institute, Palo Alto, Calif., Nov. 1982.
10. *Federal Register,* 45 FR 79318 et seq., Nov. 28, 1980.
11. *Federal Register,* 49 FR 4551 et seq., Feb. 7, 1984.
12. *Water Quality Standards Handbook,* Office of Water Regulations and Standards, Environmental Protection Agency, Dec. 1983.
13. *Technical Support Manual: Waterbody Surveys and Assessments for Conducting Use Attainability Analyses,* Office of Water Regulations and Standards, Environmental Protection Agency, Nov. 1983.
14. "Technical Support Document for Water-Quality-Based Toxics Control," Draft, Office of Water Regulations and Standards, Environmental Protection Agency, Nov. 1983.
15. J. K. Rice and S. D. Strauss, "Water-Pollution Control in Steam Plants," *Power,* 121 (4), Apr. 1977.
16. *Biofouling Control Assessment—A Preliminary Data Base Assessment,* EPRI CS-2469, Electric Power Research Institute, July 1982.
17. *Organohalogen Products from Chlorination of Cooling Water at Nuclear Power Stations,* NUREG/CR 3408, PNL-4708, Nuclear Regulatory Commission, Oct. 1983.
18. *Chlorine Minimization/Optimization for Condenser Biofouling Control: Final Report,* EPA-600/170-80-143, Environmental Protection Agency, Aug. 1980.
19. "The Proposed Limitations and Standards for Once-Through Cooling Water and Cooling Tower Blowdown," comments of UWAG on EPA's proposed effluent limitations for steam-electric generating point-source category, Sec. III, Hunton & Williams, Washington, Jan. 1981.
20. P. M. Cumbie, T. A. Miskimen, and J. K. Rice, "Environmental Impacts of Chlorine Discharges: A Utility Industry Perspective," 5th Conference on Chlorination, Williamsburg, Va., June 1984.
21. *Federal Register,* 45 FR 68328 et seq., Oct. 14, 1980.
22. "Dechlorination Technology Manual, Final Report," RP 2300-3, Electric Power Research Institute, May 1984.
23. *Asbestos in Cooling Tower Waters, Final Report,* NUREG/CR-0770, ANL/ES-71, Nuclear Regulatory Commission, Mar. 1979.
24. *Leachability and Aqueous Speciation of Selected Trace Constituents of Coal Flyash,* EPRI EA 2588, Electric Power Research Institute, Sept. 1982.
25. A. K. Furr et al., "National Survey of Elements and Radioactivity in Flyash—Absorption of Elements by Cabbage Grown in Flyash-Soil Mixtures," *Environmental Science & Technology,* vol. 11, Nov. 1977.
26. *Design of a Monitoring Program for Ash Pond Effluents,* EPA-600/7-79-236, TVA PRS-41, Environmental Protection Agency, Nov. 1979.

27. *Characterization of Coal Pile Drainage,* EPA-600/7-79-051, TVA PRS-42, Environmental Protection Agency, Feb. 1979.

28. "Report on Phase I—Planning Study to Model and Monitor Coal Pile Runoff," TRC-Environmental Consultants Inc, Edison Electric Institute, Washington, July 1980.

29. W. Willersdorf and R. O'Connell, "The Selection, Design and Operation of a Mobile Waste Facility for Treatment of Boiler Cleaning Solutions," *American Power Conference Proceedings,* Vol. 43, Illinois Institute of Technology, Chicago, 1981.

30. "The Proposed Limitations and Standards for Metal Cleaning Wastes," comments by the Utility Water Act Group on EPA's Oct. 13, 1980, *Proposed Effluent Limitations and New Standards of Performance for the Steam Electric Generating Point Source Category,* Sec. IV, Hunton & Williams, Washington, Jan. 1981.

CHAPTER 4.2
KEY TREATMENT SYSTEMS

Sheldon D. Strauss
Power *Magazine*
New York, NY

INTRODUCTION

Aside from drinking and sanitary purposes, water has a number of uses in industry. The first is cooling: Every day, huge amounts of water go into removing unused heat from powerplant condensers, compressors, and diesel engines; steel mills, oil refineries, and chemical plants consume large quantities. Powerplants are also big consumers in another area—generation of steam to produce electricity. Other tasks for water are heating; transporting and processing materials, such as pulp and paper; as a raw material that goes into a finished product, such as canned goods and beverages; and to pass along domestic and industrial wastes.

Depending on the source, water can sometimes be used without treatment. But more often, it contains impurities that must be removed, reduced, or stabilized. If all water carried the same impurities, treatment could zero in on a nearly standardized prescription for each use. But pure water never occurs in nature, and impurities vary widely. The closest approach to purity is rain, but even rain contains enough dissolved oxygen and carbon dioxide to make it corrosive. Thus, water treatment is essential.

This chapter covers key systems, equipment, and techniques in water treatment—clarification, filtration, deaeration, partial demineralization, and combinations of these. Ion exchange (full demineralization) is covered in the next chapter.

CLARIFICATION

Water clarification has two goals: to settle out those larger suspended particles that are readily settled and to condition smaller colloidal (suspended) particles to make them settle out, so that the subsequent filtering operation is not impeded. A sedimentation basin takes care of the first. Once the coarser matter is removed, the finer solids can be handled by equipment of moderate size.

The basin may take the form of a pond, reservoir, or tank large enough to permit suspended sediment to settle while passing through. More than likely, a fabricated unit would be equipped with a means for removing settled sludge periodically. The facility is usually sized for settling of the heavier solids in a matter of hours, compared with days or weeks for removal of finer particles in a facility of this type.

From the basin, the overflow proceeds to the clarifying station. The key to clarification is the addition of chemicals to agglomerate suspended solids, shortening the detention time for settling to a few hours at most. The basic goal of agglomeration is to overcome the factors that tend to prevent finely divided matter from joining together, and to encourage further aggregation and growth to a size large enough to settle out of suspension.

The primary factor is electric charge. All particles of turbidity and color have an electric surface charge, usually negative. The charge attracts a compact layer of oppositely charged ions (counterions) in very close proximity and a more diffuse layer farther away (Fig. 2.1). The combination of the two layers forms an

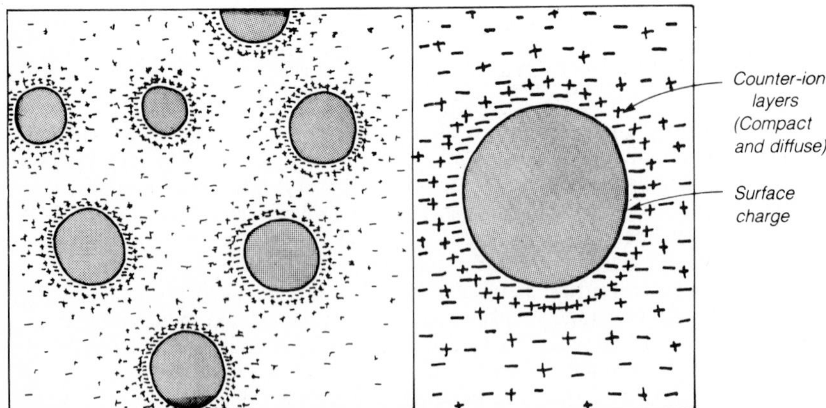

FIG. 2.1 Electrical surface charges present on all solids suspended in water attract ions of opposite charge to form dual counter-ionic layer—one compact, one diffuse.

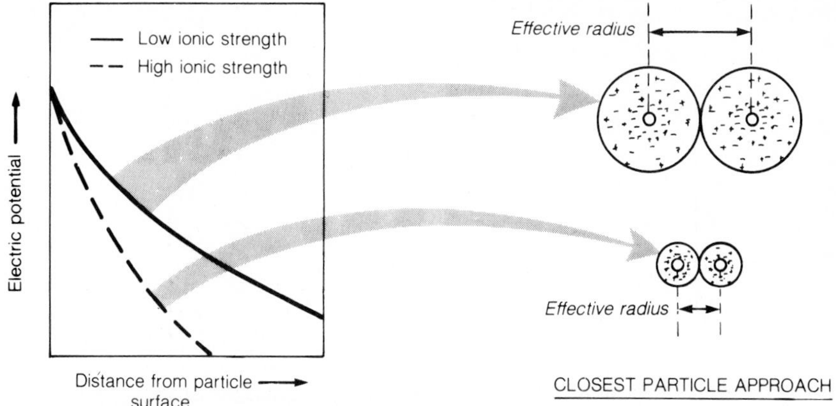

FIG. 2.2 Surrounding electrostatic zeta potential thus formed (see Fig. 2.1) creates repulsion between particles with same charge sign. Repulsive effect varies with ionic content of water.

electrostatic potential—called the *zeta potential*—around the particle, which repels other particles similarly charged (Fig. 2.2).

Coagulants: Physical Mechanisms

Chemicals called *coagulants* are added to the water to break down the zeta-potential barrier. They introduce ions of opposite charge—positive in this instance—which act to compress the double electrical layer, thus reducing the barrier to agglomeration. Actually, a second force exists, the van der Waals force, which is attractive in nature; it is the algebraic sum of these two forces—the net repulsive force—that must be overcome to coalesce individual particles. This can be considered as neutralization of particles, to allow attachment to one another.

Metallic salts are often selected as coagulants. They introduce positive metal ions, carrying two or more electric charges. Thus, they increase the concentration of counterions in the diffuse layers, enabling a thinner layer to maintain overall electrical neutrality in the vicinity of the particle. In this way, the range of the repulsive force between proximate particles decreases, and the barrier is reduced. With proper choice and amount of coagulant, the barrier is removed completely, allowing contact between particles.

Compounds of aluminum, especially alum and sodium aluminate, are the most widely used coagulants (Table 2.1). Besides neutralizing the electrostatic barrier, they react with the alkalinity of the water to form a tough aluminum hydroxide precipitate, or *floc,* which is stable over the pH range 5.7 to 7.5. The finer suspended particles are trapped in this floc. The hydroxide dissolves at pH above 7.5 and below 5.7. Iron salts precipitate a hydroxide floc over a much broader pH range—5.0 to 11.0—and, therefore, are often preferred for coagulating the precipitates found in softening reactions.

Evidently, successful coagulation depends on both pH and the water analysis after addition of chemicals. When the water contains natural alkalinity, alum hydrolizes into aluminum hydroxide, liberating sulfuric acid. This reacts with alkalinity to form calcium sulfate and CO_2. Each *part per million* (ppm) of alum con-

TABLE 2.1 Compounds Commonly Selected for Coagulation

Name	Chemical formula	Commercial strength	Grades available	Suitable handling materials	Type of feeder	Alkalinity needed, ppm*
Aluminum sulfate (filter alum)	$Al_2(SO_4)_3$ d $18H_2O$	17% Al_2O_3	Lump, powder, granules	Lead, rubber, silicon, iron (wet); concrete, iron (dry)	Dry or solution	0.45
Sodium aluminate (soda alum)	$Na_2Al_2O_4$	55% Al_2O_3	Crystals	Iron, steel, rubber, plastic, concrete, ceramic	Dry or solution	—†
Ferrous sulfate (copperas)	$FeSO_4$ d $7H_2O$	55% $FeSO_4$	Crystals, granules	Lead, rubber, tin, wood (wet); iron, steel (dry)	Dry or solution	0.36
Ferric sulfate	$Fe_2(SO_4)_3$	90% $Fe(SO_4)_3$	Powder, granules	Lead, rubber, 316 SS, plastic, ceramic	Dry or solution	0.75
Ferric chloride	$FeCl_3$ d $6H_2O$	60% $FeCl_3$	Crystals	Rubber, glass, plastic	Solution	0.92
Bentonite	—	—	Powder	Iron, steel	Slurry	—

*Per ppm compound.
†Sodium aluminate is alkaline.

sumes 0.45 ppm of natural alkalinity, suppressing the pH level. To counteract this effect and ensure coagulation, lime or soda ash is added at the rate of 0.4 ppm lime (90 percent calcium hydroxide), or 0.5 ppm soda ash, per ppm of alum. The quantity actually added is closely controlled to give the best pH.

Polymers: Chemical Mechanisms

Another set of chemicals also assists coagulants, but through a chemical rather than a physical mechanism. These are *polyelectrolytes*—synthetic organic polymers of high molecular weight, containing chemical groups that can interact with surface sites on colloidal particles. When a polymer molecule attaches to a particle in this way, the rest of it extending out into the bulk of water, other particles—brought to the vicinity by gentle stirring—can become attached (Fig. 2.3).

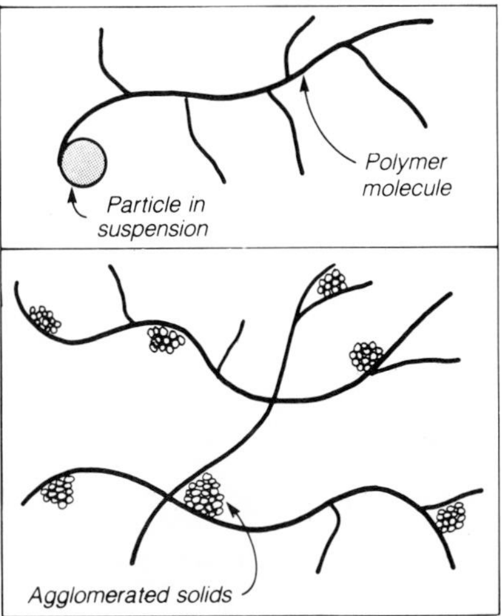

FIG. 2.3 Long-chain polymers attach to charge sites on floc and enhance floc growth by chemical bridging.

This is the "bridging" effect—and the basic mechanism by which coagulant aids, also called *flocculants,* help pinpoint floc to grow into larger floc. Such aids act with surprising speed and are particularly helpful with water that is cold or high in color content.

Polymers are cationic (positively charged), anionic, or nonionic. Slightly anionic polyacrylamides with molecular weights up to 20 million generally appear to be the most effective, but some cationic and nonionic types have been found useful in certain cases. Because the optimum charge and molecular weight vary with the situation, tests are needed to determine the best choice.

Two developments have attracted interest in recent years. One is the selection of polyelectrolytes as primary coagulants. These are cationic polyamines with relatively low molecular weights (less than 500,000) and very high charge densities. They become adsorbed on particle surfaces, neutralizing the repelling negative charge. To some extent, they also bridge particles, reducing the need for inorganic coagulants.

The other development is direct filtration. This technique dispenses with a clarifier per se, introducing water directly into a filtration system. Chemical coagulant is injected slightly upstream of the system and in smaller quantities than those used with clarifiers, to avoid interference with filter operation. Polymers, usually cationic, are preferred to inorganics for this application since they do not introduce additional solids loadings. A short mixing period is needed to provide for the optimum degree of chemical reaction, and dilution water may be needed to ensure adequate dispersion of the polymer.

Equipment Designs

Circular basins have a rotating vertical shaft with radial arms and rake blades (Fig. 2.4). These scrape settled solids in toward a central discharge hopper for continuous removal. Clarified water flows out the collecting flume at the far end and into a clear well. Storage in the clear well allows the system to equalize flows and override variations in raw-water turbidity.

Detention times in circular basins can be relatively long. To shorten the time and improve the precipitated material—that is, make it larger and more settleable—use is made of the *solids-contact principle*. This is essentially a catalytic effect, obtained by contacting newly precipitated material with precipitates

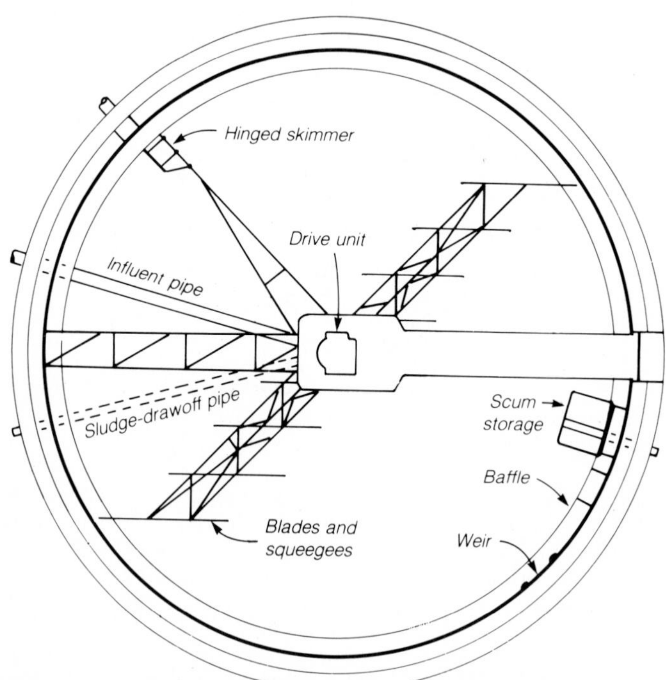

FIG. 2.4 Circular clarifier has central feed and sludge scrapes.

formed earlier. The effect increases the efficiency of clarification and enables reduction of the size of the facility. Retention time is reduced from 6 h or more to 1 or 2 h.

Solids-Contact Clarifiers. Most modern clarifiers embody the solids-contact principle. They take one of two basic forms: the "sludge-blanket" unit or the "sludge recirculation" unit. Both types combine coagulation, flocculation, and hydraulic separation (settling) in a single vessel. In such equipment, flash mixing takes place at the top center, the flocculation region is below this, and the sludge blanket forms in an area concentric to both, in the lower portion of the vessel. The clarified-water zone is above the blanket. In a typical clarifier (Fig. 2.5), water and chemicals are introduced together at the top and then move downward to the mixing zone. Here, they are slowly agitated to make continued contact with sludge particles formed previously. Agitation continues until water and precipitates pass through a constricted opening at the bottom and into the sludge-blanket zone.

The entering mixture moves upward through the sludge blanket, which serves two functions: (1) It entrains or adsorbs particulates from the rising water, and (2) it furnishes nuclei for continued particle agglomeration. As the particles in the blanket become coated with the finer floc particles, or as the latter agglomerate with those forming the blanket, the larger particles soon become too heavy to be supported by the up-flowing water and fall back to the mixing zone. Thus, fine, suspended particles flow continuously from the mixing zone into the blanket, and there is a counterflow of larger particles in the opposite direction. An equilibrium condition exists, with approximately the same concentration in both regions.

In Fig. 2.5, the cross-sectional flow area increases vertically upward, which serves to reduce the flow velocity as the water rises. The effect is to delineate more sharply the point where the larger and heavier particles are no longer sup-

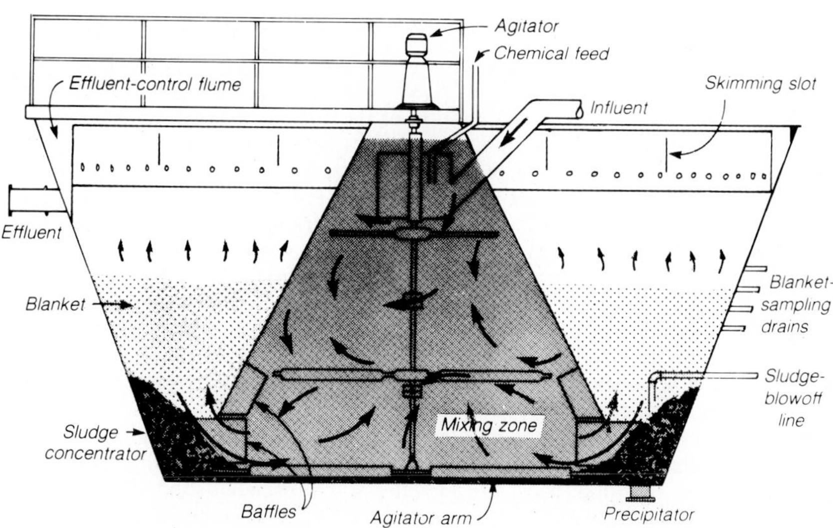

FIG. 2.5 Solids-contact clarifier.

ported, giving the blanket a well-defined upper surface. Proper choice of influent flow rate, which is a function of its characteristics and the chemicals added, ensures that the sludge remains below the treated-water collectors and that the effluent water maintains its clarity.

Conditions of operation can change, of course, and care is important to ensure proper performance (Table 2.2). The rise rate of the clarified water is about 1 to 1.5 gal/min d ft^2 (0.68 to 1.02 L/s d m^2). With a 1- to 2-h retention time, and depending on influent conditions and quality of operation, the unit generally produces water with 5 to 10 ppm suspended solids from most available water sources.

TABLE 2.2 Troubleshooting a Sludge-Blanket Clarifier

Symptom	Possible causes
Sludge blanket too high	Chemical dosage incorrect
	Flush-back pressure too high
	Sludge blowdown low
	Flow rate high or fluctuating too rapidly
	Agitator speed too high, or incorrect rotation
Sludge blanket too low	Chemical dosage incorrect
	Sludge blowdown high, flow rate low
High turbidity in effluent	Agitator speed too high, or incorrect rotation
	Chemical dosage incorrect
	Sludge-blanket level incorrect
	Rapid temperature change
Agitator overload	Sludge density too high
	Oil level in reducer low
	Motor amperage incorrect
	Excessive sand, silt, or foreign matter in water or chemical supply (drain unit, if necessary)

Source: "Operating Instructions for Water Conditioning Equipment," Permutit Co., Paramus, N.J.

Solids-Recirculation Clarifiers. Operation of the second broad category of clarifiers is not tied to a stabilized sludge blanket. It achieves intimate solids contact by recirculating preformed sludge with incoming water in a central cylinder or cone. In Fig. 2.6, influent water enters the lower portion of a rapid-mix zone, or draft tube. Under the action of an impeller, it is drawn upward and mixed with treatment chemicals (both coagulant and flocculant), plus some recirculated precipitates. Impeller speed must be kept low, so shear forces do not break up large floc particles.

The mixture of raw water and sludge rises through the tube, leaving at the top to enter the concentric flocculation compartment. There, the mixture is stirred gently, while it is flowing downward, to provide sufficient time for completion of the flocculation process. By the time the water passes into the outer settling zone, the larger and heavier precipitates are formed and separate rapidly from the water. While the lighter, clarified water then rises toward the top, it constitutes just a portion of the total. Most of the water and suspended solids that move to the bottom of the flocculation compartment enter the draft tube at the lower end. Recirculation there provides positive solids contact regardless of the sludge inventory level.

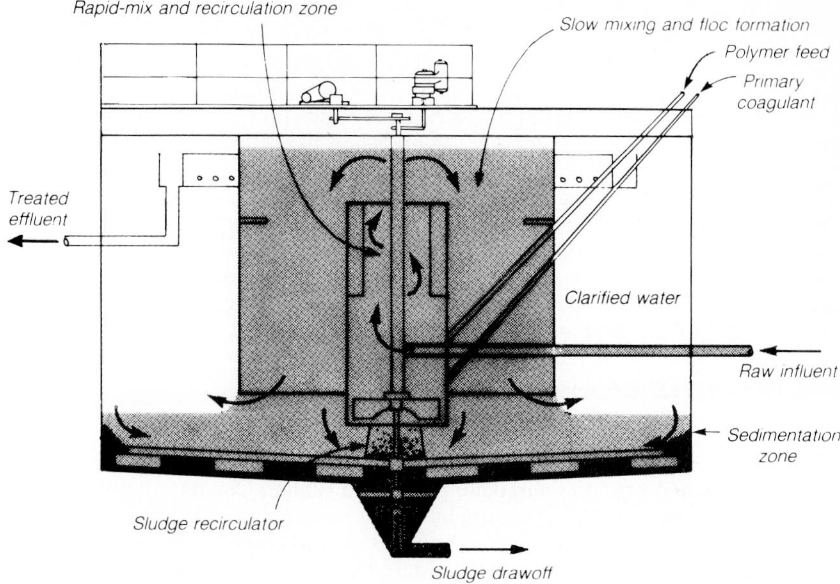

FIG. 2.6 Solids-recirculation clarifier.

Settled precipitates are moved continuously along the floor toward the center of the unit by a rotating scraper. Under this action, accumulated sludge is moved to the sludge pit, where it is concentrated by thickening pickets. Sludge inventory is maintained at the predetermined level by automatic blowdown of excess material from this point.

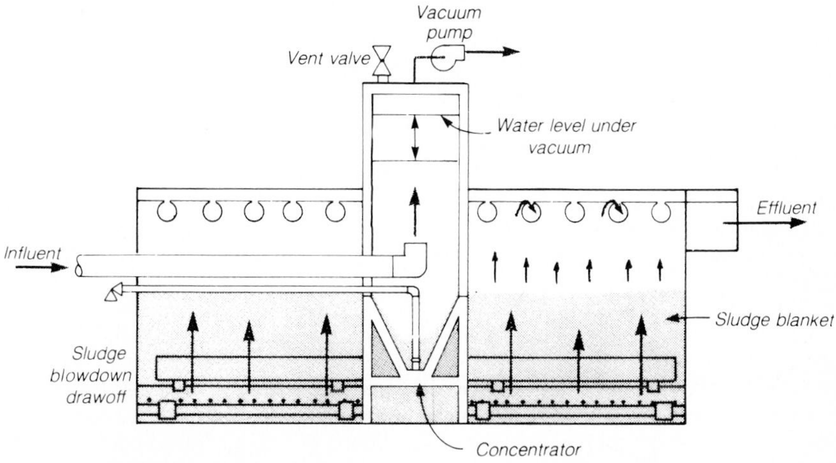

FIG. 2.7 Pulsed-type clarifier.

Pulsed-Type Clarifiers. Pulsed-type clarifiers are a variation on the sludge-blanket type (Fig. 2.7). They move newly flocculated water up through the blanket in a cyclical, pulsating flow, to avoid formation of irregularities followed by channeling of water through the bed. When raw water and chemicals enter the vacuum, the pressure is lowered, forcing the water up to a predetermined level. A level sensor opens the main vent valve when the proper level is reached, restoring atmospheric pressure.

The resulting hydraulic head causes a surge into the distribution pipes at the bottom of the vessel. The water-chemicals mixture is discharged evenly from the pipe system into the flocculation zone, where the hydraulic energy is converted to gentle turbulence. This stirring action mixes the new floc with existing flocculated sludge. The surge slows when a lower-level limit is reached; a sensor closes the air-vent valve, the pump evacuates the chamber, and the cycle repeats. Excess sludge flows over the top edge of a V-shaped concentrator, which thus establishes the depth of the blanket; sludge is drawn out only from the concentrator.

External-Recirculation Clarifiers. Variations from the basic designs sometimes lie in the method used to recirculate preformed sludge. One design differs significantly, recirculating sludge externally to the vessel (Fig. 2.8). As the sludge is formed at the bottom, it is removed by pipes spaced along the scraper arm under the pressure head of the overlying water. Depending on existing water conditions, a fraction is reintroduced at the central reaction well by a recirculating pump located in the sludge sump. The method is intended to provide better sludge control without disrupting the floc. A second variation dispenses with both the mix-zone impeller and the sludge pump. Instead, it uses the hydraulic head of incoming water to draw previously formed precipitates up into the draft tube to contact the influent directly.

FILTRATION

Even with clarifiers operating at optimum levels under anticipated conditions, additional solids-removal equipment is required downstream. It serves to reduce the remaining suspended solids to the parts-per-billion level required for boiler operation, and it provides backup to the clarifiers in case of severe upsets.

Filter development has progressed so far that, in some cases where water has moderate turbidity, addition of a small amount of coagulant allows the clarifier to be dispensed with entirely. But in preparing boiler feedwater, filtration is generally needed to reduce clarified water in the range of 2 to 10 Jackson turbidity units (JTU) to a fraction of 1 JTU. This applies as well to condensate return, where the slightest corrosion or in-leakage can contaminate boiler water intolerably.

Filtration is the process of placing a pervious (porous) barrier across flowing water to remove matter held in suspension. It can be accomplished simply through mechanical straining at the surface of the barrier or by removal throughout the depth of the medium. Surface filtration is generally suitable only for water extremely low in suspended solids, since operating cycles would otherwise be too short. For makeup preparation, however, more of the medium depth must be brought into play for practical operation.

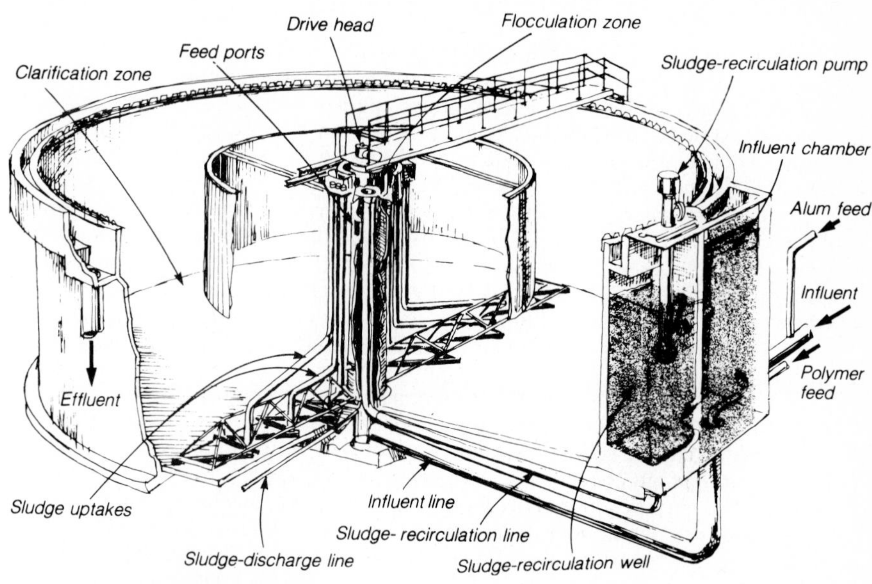

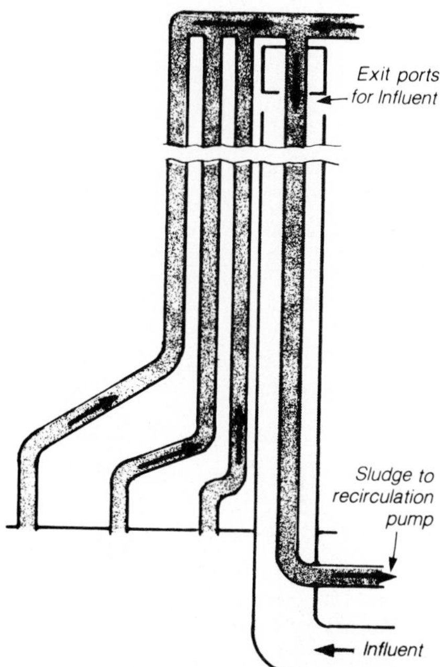

FIG. 2.8 External-recirculation clarifier.

Sand Filters

Granular media are the accepted standard for medium-depth filtration. Silica sand, the most common, has been the standby for over a century. The sand filter had its inception in England, and still finds application abroad in its original form. It features a bed of fine sand—3 to 5 ft (0.9 to 1.5 m) thick—supported by a gravel subfill. Overlying water percolates down through the sand and is collected and removed by the underdrain system.

Proper performance does not begin until a layer of silt and bacterial matter has developed in the top inch or so of sand. The unit then produces exceedingly clear water, but at a rate no faster than 0.12 gal/min d ft^2 (0.08 L/s d m^2). This continues until the limiting head loss, 3 to 5 ft (0.9 to 1.5 m), is reached. The upper layer, where all the filtering action takes place, is then scraped off, washed externally, and restored to the bed. Adaptations of early sand filters feature circular tanks with built-in scrapers (Fig. 2.9) or in-situ cleaning by reverse-flow backwashing.

The effectiveness of slow filtration depends on the nature of the water being processed. Highly successful in Europe, where the absence of chemical costs offsets the high capital cost, slow filters are not well suited to domestic application because of the high concentrations of very fine clay often present in U.S. waters. Because the clay penetrates deep into the sand body, cleaning is far more difficult. The development of granular-media filters in the United States was based on pretreating the water and modifying the media used. With preconditioning provided by coagulation, whether it is done in a clarifier or in-line, filters take a va-

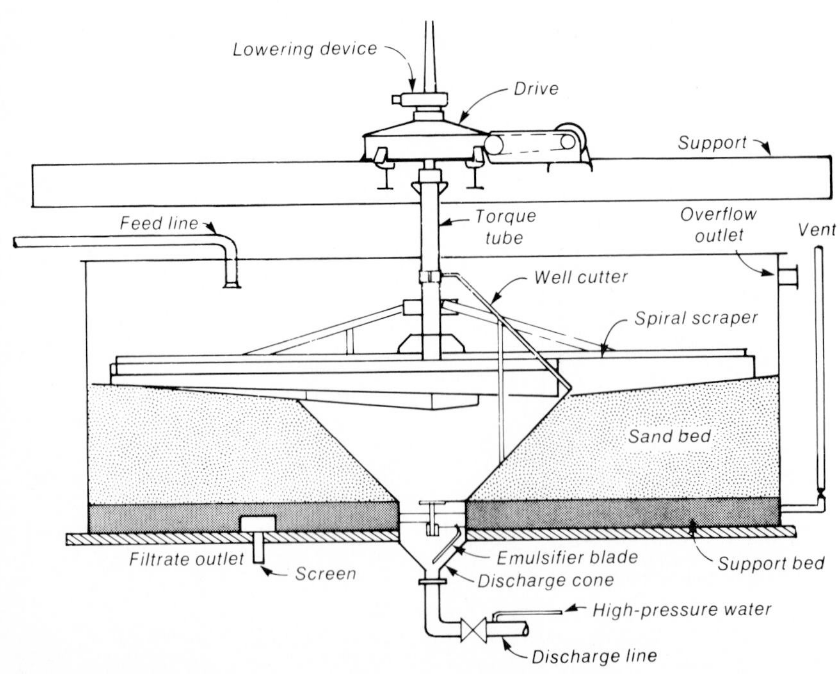

FIG. 2.9 Sand filter.

riety of forms and provide filtration rates in the range of 1 to 4 gal/min d ft^2 (0.68 to 2.72 L/s d m^2).

Rather than being limited to a surface effect, solids removal in these units takes place within the bed volume (Fig. 2.10). Several mechanisms contribute to filtering efficiency. The larger particles are removed within the interstices of the medium by mechanical straining; finer particles must be first transported close to the sand-grain surfaces and then attached to them (or to other particles).

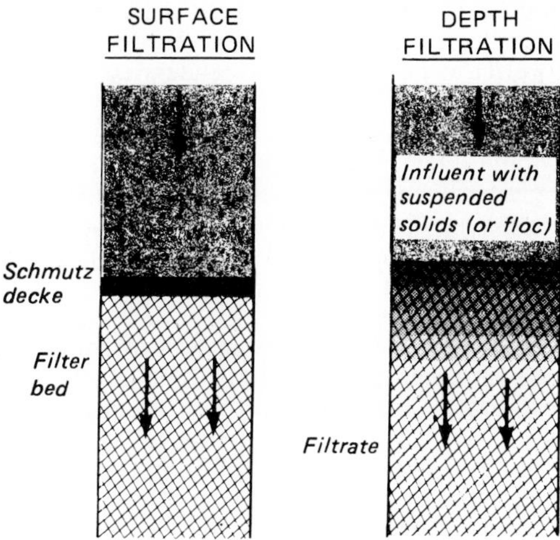

FIG. 2.10 Solids-removal zone progresses through media in depth filtration versus surface type.

The attachment mechanisms involve simple adsorption, electrostatic effects, chemical bridging, or a combination of these. Both surface and depth reactions operate during a run, and different transport and attachment mechanisms may become dominant under varying conditions. The physical characteristics of the medium, flow rate, particle size and density, and water temperature all have a bearing on performance. Changing conditions can account for changing quality and head-loss patterns.

Dual-Media Filters

Even with pretreatment of the water, no more than the first several inches of media participate in the filtering action when the bed comprises average-grade silica sand, normally from 0.3 to 0.8 mm. Clogging of these layers impedes penetration and terminates the cycle rather quickly. Studies in the 1960s aimed at circumventing this problem showed that grain coarseness was the primary parameter governing the effectiveness of solids removal from a given treated water. The result of this work established the practice currently favored, namely, selection of two or more different granular media for filtration.

Coarse and Fine Media. When a coarser granular layer is superimposed over the bed, it removes the larger solids selectively, allowing much deeper and more uniform penetration by the finer particles. This action reduces head loss, allowing higher flow rates and longer operating cycles before cleaning is necessary. One design featuring this approach has two grades of sand, fine and coarse. A variation aims at achieving the theoretically ideal granular filter, one that is graded continuously from coarse to fine in the direction of flow. To simulate this condition, average-grade sand of low uniformity coefficient is specified to obtain a greater variation in size between the smallest and largest grains. In normal practice, these grain sizes are commonly used: anthracite, 0.8 to 1.2 mm; sand, 0.5 to 0.8 mm; garnet, 0.4 to 0.6 mm; and magnetite, 0.3 to 0.4 mm.

Crushed anthracite coal is generally preferred as the second medium. More porous and more angular than sand, it removes the larger matter effectively and allows passage of smaller suspended solids to the deeper sand layers and for longer periods before the collected material renders continued operation inefficient.

A Third Medium. Some filter designs include a third medium for particular applications—ground garnet. Finer and denser than sand, it is placed at the bottom of the bed to provide extra polishing action. Another variation goes a step further by incorporating megnetite. In a more recent development, the bed is topped with a low-density anthracite in larger granules for use in treating industrial wastewater or municipal secondary effluent.

For most applications, however, a combination of anthracite and sand provides sufficient depth filtration to reap the full value of dual-media design. The combination enables filtration rates to be increased from 2–4 to 8–10 gal/min d ft^2 (1.36–2.72 to 5.44–6.79 L/s d m^2) or higher and provides considerably higher solids-holding capacity at relatively low cost. For an existing facility, similar improvements can be obtained by replacing 2 to 6 in (50 to 150 mm) of the bed with up to 8 in (200 mm) of crushed anthracite. Capacity can be increased further by replacing more of the sand, however, this should not be done if it degrades the effluent quality too much.

Gravity Filters

Gravity filtration, featuring downward flow, characterizes most industrial filters. Usually of concrete and either square or rectangular, gravity filters may also be cylindrical steel structures (Fig. 2.11). They may be open or closed at the top. Filtration rates vary with the model, ranging from 4 to 8 gal/min d ft^2 (2.72 to 5.44 L/s d m^2). The maximum height is about 12 ft (3.6 m), but enough freeboard must be provided to prevent loss of media during backwash. Flow-rate controllers are advisable to prevent ejection of trapped solids during sudden changes in influent rate.

Vertical and Valveless Types. Vertical gravity filters may contain a single medium or two. To avoid surface plugging and provide suitable holding capacity, coarser media are selected and in greater depths—1 to 2 mm in diameter, 4 to 8 ft (1.2 to 2.4 m) deep. Such media provide longer service runs and do not require excessive backwash volumes. An added advantage derives from the use of underdrains with larger strainer openings, since there is less likelihood of clogging. Disadvantages are that deeper beds may incur greater structural costs and that filtrate quality may not be as good as with multiple media.

The valveless filter offers something different in gravity units: It stores water in a compartment above the filter for use as backwash; the water level rises

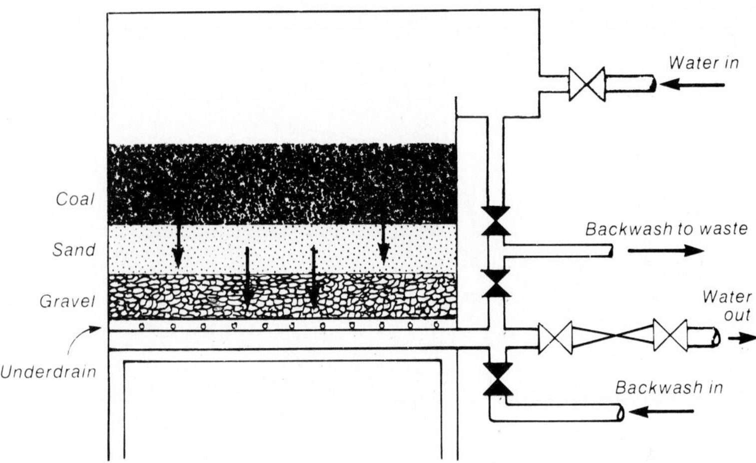

FIG. 2.11 Gravity filter.

slowly in the backwash pipe in the course of operation (Fig. 2.12). When a preselected level (rather than pressure drop) is reached, it overflows to waste. Stored water then flows up through the bed and out the pipe; exposure of the end of the siphon breaker stops the action. Single- and double-valve designs are alternative options available with integral backwash storage.

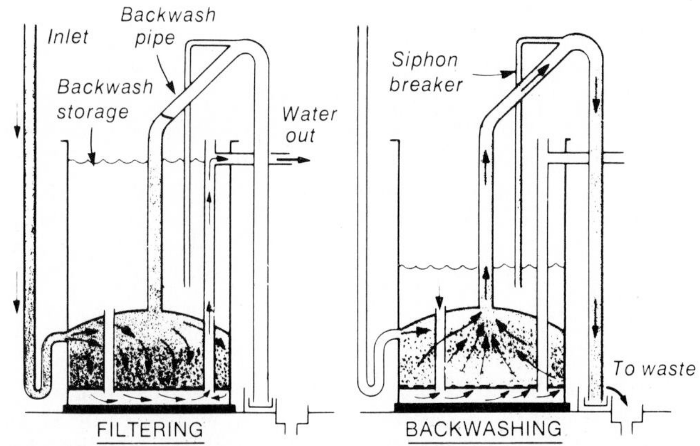

FIG. 2.12 Valveless gravity filter.

Concrete Construction. Concrete construction offers first-cost advantages when the filter is set in the ground, because this location avoids pumping requirements to raise feedwater to an elevated flow-divider box. This type of box is used in designs featuring multiple compartments arranged around a central control module. Compartmental designs reduce headroom, inlet head, and waste-handling requirements. While each compartment acts as an individual filter, all are served by

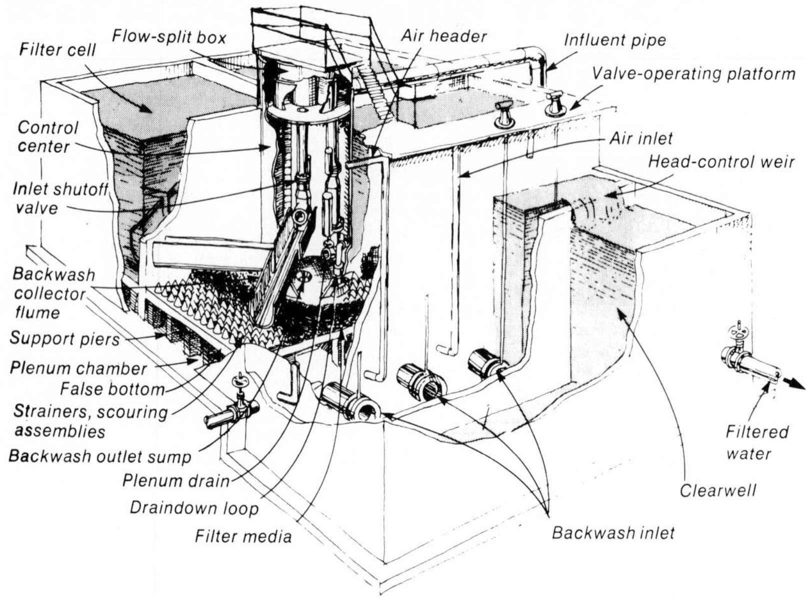

FIG. 2.13 Concrete construction is feature of this multiple-cell filter.

a common backwash chamber. Only one cell in the battery is washed at a time, so less total storage water is required.

Compartmentalization brings additional savings in headroom and reductions in waste-pipe size. Figure 2.13 depicts a modern commercial unit featuring four-cell design, dual media, air scouring, and automatic backwash cycling triggered when the pressure drop across the bed exceeds a preselected value.

The low headroom required for gravity units can be a disadvantage in some industry applications. Operating cycles terminate when the pressure drop in the bed equals the gravity head. For this reason, many users opt for pressure-operated filters; besides longer service runs, these filters operate at rates as high as 8 gal/min d ft^2 (5.44 L/s d m^2), may not require repumping of water, and are readily automated.

Pressure Filters

These types of filters have operating and internal features similar to those of gravity units, except that process water flows under pressures of 50 to 100 psig (345 to 689 kPa). The tanks are made in short vertical or horizontal designs and are shaped like cylinders with dished heads. Vertical units are 4 to 5 ft (1.2 to 1.5 m) high, with diameters ranging from 3 to 11 or 12 ft (0.9 to 3.3 or 3.6 m); maximum dimensions of horizontal units are 50 ft (15 m) in length and 10 ft (3 m) in height. Media rest on gravel subfill or steel plate (fitted with strainers), the latter to preclude possible sand-gravel upsets under surges of water or air (Fig. 2.14).

The horizontal configuration is suitable for indoor areas and for large sizes to effect cost savings. Although the large filtering surfaces require high backwash

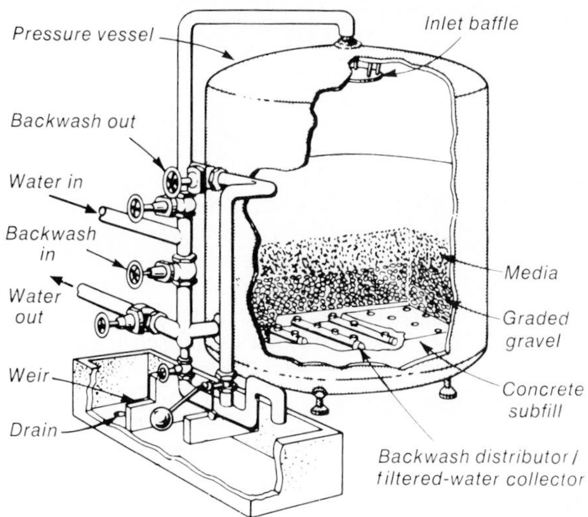

FIG. 2.14 Pressure filter.

flows, compartmental design can compensate somewhat, allowing one section to be washed at a time.

Cleaning Techniques

Maintaining a clean filter bed is important. Washing restores filter efficiency when effluent quality deteriorates or when excessive head loss reduces filtration rates. Thus, either pressure drop or effluent turbidity can be the criterion for terminating a run to wash the media. In downflow units, fittings and construction for backwashing force clear water through the bed in the upflow direction and then to waste via wash channels in the upper part of the vessel.

Backwashing. Backwashing must be forceful enough to expand the bed by at least 15 to 20 percent, so that trapped solids may be loosened and removed. Flow rates, therefore, are about 12 to 15 gal/min d ft^2 (8.15 to 10.19 L/s d m^2) for sand media and 8 to 12 gal/min d ft^2 (5.44 to 8.15 L/s d m^2) for anthracite. Excessive backwash rates should be avoided. Strong streams may channel through several paths, causing unequal cleaning and swirling motions within the bed. Surface sludge can then be carried into the deeper layers and compacted under this motion, producing mud balls.

To avoid mud balls, surface washing and air scouring are recommended. In surface washing, high-pressure water jets from fixed or rotating nozzles mounted above the bed are used to break up encrusted matter at the surface. Another design features an additional distributor within the bed, about a foot below the top; perforations direct water upward to the overlying layer. After the initial rinse water is discarded, the filter is ready for service with 5 to 10 min of backwashing.

In air scouring, pressurized air is blown up into the bed from a header-lateral distribution system mounted above the gravel subfill. The scrubbing action of the air loosens solids attached to the media granules. These are then flushed out under backwash flow at about 8 gal/min d ft^2 (5.44 L/s d m^2), after as little as 4 min of scouring in some designs.

Upflow Washing. Upflow washing is well suited to multimedia filters. It tends to reclassify the media, leaving the larger-grained but less dense medium at the top of the bed and the smaller but denser grains at the bottom, ready to start a new service run. Upflow backwashing imposes a problem, however, in single-medium systems of uniform density. At the end of the backwash cycle, the bed is graded with the finest grains at the top and the coarsest at the bottom—an arrangement bringing all the disadvantages of surface filtration.

One filter design, embodying a single medium in a deep bed and featuring large holding capacity, avoids the graded bed by flowing service water vertically upward. A media-retaining grid permits operation at 6 to 10 gal/min d ft^2 (4.08 to 6.79 L/s d m^2) without loss of media. Raw water can be used for washing, also done in the upward direction, but at a higher rate. Gradation at the end of washing is from fine at top to coarse at bottom, the proper sequence for upflow service. The unit in Fig. 2.15 features two support structures: a series of parallel bars atop the media to resist hydraulic forces and a steel retention screen below the bed to prevent movement of support rock into the filter medium. This design accommodates injection of polyelectrolyte slightly upstream to provide in-line clarification as well as filtration.

Air-Water Washing. A relative newcomer to filtration also merits mention—a uniform-media, downflow filter (Fig. 2.16). Backwash includes simultaneous air and water scouring, leaving the media thoroughly cleansed and unclassified. Using coarse sand provides high holding capacity. Including smaller sand grains within the interstices of coarser grains increases the surface area for attachment

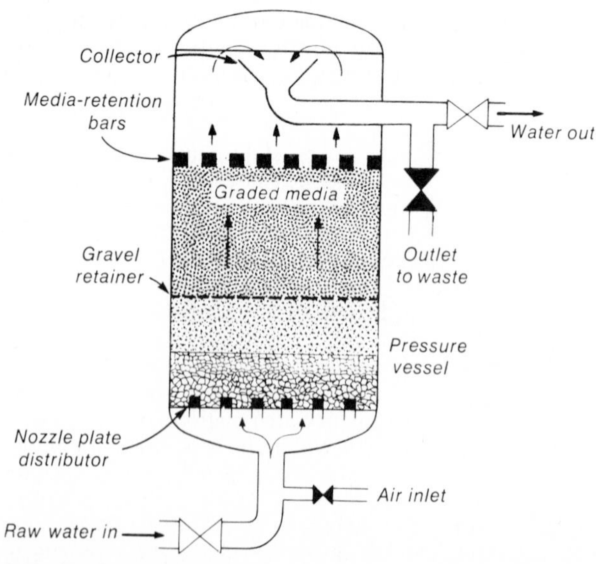

FIG. 2.15 Upflow filter.

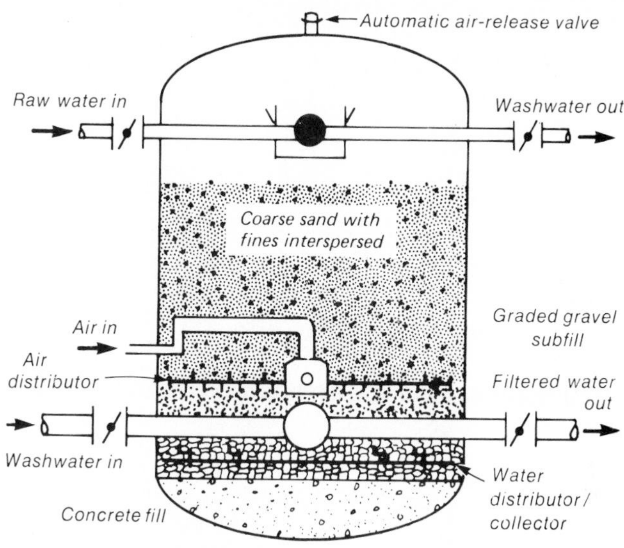

FIG. 2.16 Uniform-media filter.

of solids while retaining capacity for long runs. Designed to operate at 12 gal/min d ft^2 (8.15 L/s d m^2) or higher, the unit is suitable for direct filtration; appropriate chemical treatment ensures high-quality effluent.

Other Filtering Techniques

Although media filtration accounts for the bulk of suspended-solids removal in industrial steam systems, several other techniques have carved out niches for specialty applications. These include polishing service, equipment protection, and nuclear service.

Precoat Filters. Precoat filtration involves deposition of a thin layer of filtering material on a supporting substrate (Fig. 2.17). Although the initial filter-aid precoat may be no more than ⅛ in (3.2 mm) thick, exceptionally small passages retain minute particles, below 1 μm. Substrates are wire-wound, fiber-wound wire screen, or porous-cloth septa; the shape is either tubular or flat leaf.

Filter aid is applied in place just prior to the service run. Retention of suspended solids during service creates a pressure differential; when the differential becomes too large to continue operation, the precoat must be discarded. In tubular configurations, the precoat is backwashed from the substrate in-line, and the filter is returned to service after another precoat application. Leaf-type elements are removed for external cleaning, or cake is sluiced off in place. Precoat units find wide use as polishing filters following a roughing stage. Nuclear condensate is a special application. Powdered ion-exchange resin is used for combined filtration and demineralization.

Magnetic Filters. Magnetic filters and etched-disk units are also finding nuclear application. The former comprises vessels containing ¼-in (6.4-mm) metal

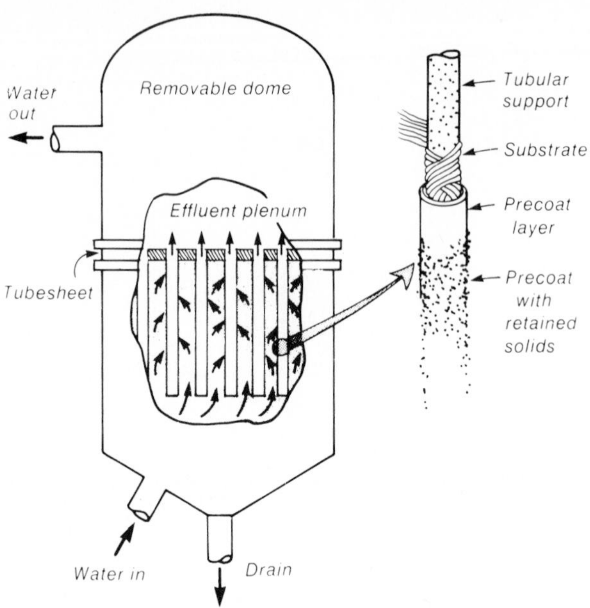

FIG. 2.17 Precoat filter.

spheres, which are magnetized by an applied magnetic field. Installed in condenser lines, they remove metallic oxide impurities down to 0.1 μm with 90 percent efficiency or better, affording protection to upstream ion-exchange demineralizers and nuclear steam generators.

Etched-disk filters have specialized application to streams containing radioactive suspended solids. Comprising stacks of disks with etched surfaces, they provide filtration for particles of submicrometer size and up. Primary applications are spent-fuel storage pools and floor drains, offering the advantage of minimizing waste volumes.

Ultrafilters. Ultrafilters and filter cartridges are two more protective devices. Most popular cartridge types embody either nylon yarn wound over a cylindrical core or metallic screening with pleated paper filter attached. Filters in the range of 3 to 10 μm are selected for reactor-water cleanup in nuclear plants, or to protect reverse-osmosis units from particulates in any service application.

Ultrafilters comprise extremely fine hollow-fiber membranes, with diameters of about 90 μm and pore openings of 0.001 to 0.02 μm. Assembled in bundles, they are suited to removal of colloidal silica from feedwater to protect boiler tubes and upstream turbines.

DEAERATION

Dissolved gases in a powerplant's water supply can produce corrosion and pitting, and they must be removed to protect piping, pumps, boilers, and condensate lines. Predominant gases are oxygen and carbon dioxide, together with hydrogen sulfide and ammonia. Their removal is based on raising the water temperature to the saturation point for the existing pressure. Eliminated are CO_2, H_2S, and NH_3 through the introduction of oxygen, which is precisely the opposite of deaeration.

Oxygen removal is accomplished in deaerators of two types, pressure and vacuum. When heated water is needed, as for boiler feed, steam deaerators are selected. Otherwise, the choice is vacuum units, for example, in cooling applications.

Steam Deaerators

Steam deaerators break up water into a spray or film, then sweep the steam across and through them to force out dissolved gases. The oxygen content can be reduced below 0.005 cm^3/L, which is near the limit of chemical detectability.

The typical deaerator has a heating and deaerating section, plus storage for hot deaerated water. Usually, a separate tank is provided above, alongside, or underneath to hold 10-min storage at rated capacity. Designs fall into two broad categories, spray and tray, or combinations of these (Figs. 2.18 and 2.19). Entering steam meets the hottest water first, to thoroughly scrub out the last remaining fraction of dissolved gas. The latter is carried along with steam as it flows through the deaerator. The direction of steam flow may be across, down, or counter to the current. The steam picks up more noncondensable gas as it goes through the unit, condensing as it heats the water.

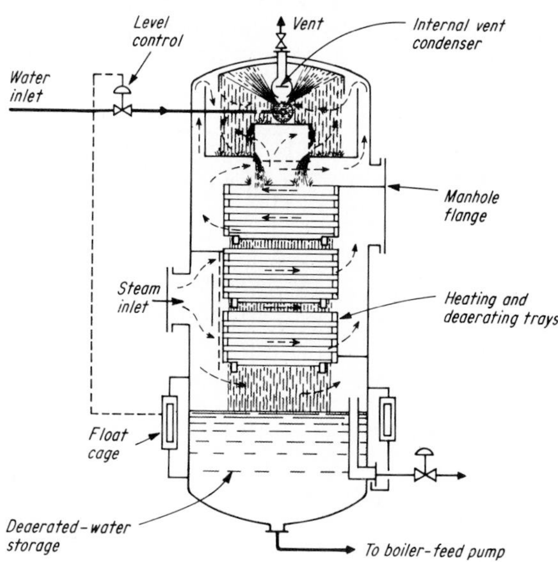

FIG. 2.18 Spray-type steam deaerator.

The bulk of the steam condenses in the first section of the deaerator (the top, if the unit is vertical) when it contacts entering cold water. The remaining

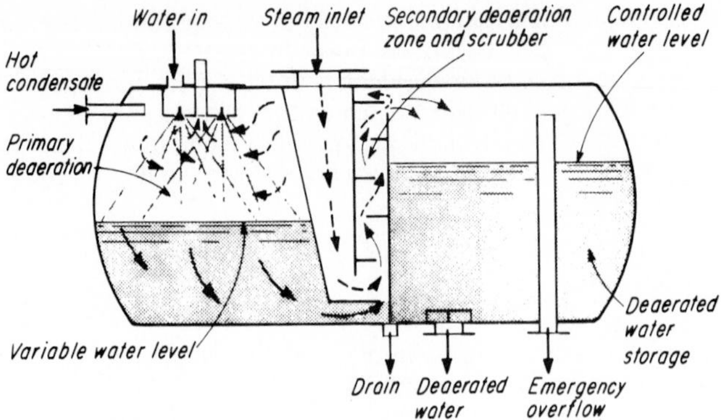

FIG. 2.19 Tray-spray steam deaerator.

mixture of noncondensable gases is discharged to atmosphere through a vent condenser.

Vacuum Deaerators

Vacuum deaerators use steam-jet ejectors or mechanical vacuum pumps to pull the required vacuum (Fig. 2.20). The degree of vacuum depends on the water temperature, but is usually in the range of 29 in Hg (98 kPa). The efficiency of O_2 and CO_2 removal does not compare with that of steam deaerators. Oxygen residuals run about 0.2 ppm; CO_2 will vary from 2 to 10 ppm, depending on the initial content and temperature of the water. Vacuum

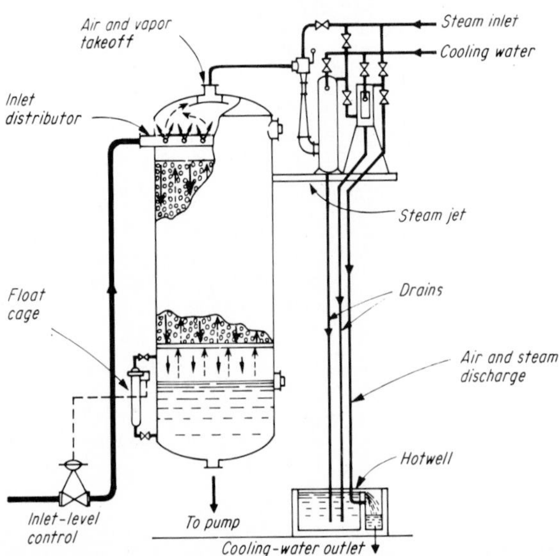

FIG. 2.20 Vacuum deaerator.

deaerator units are often used in demineralizing systems both to reduce chemical operating costs and equipment size and to protect anion-exchange resins from possible oxidation damage.

PARTIAL DEMINERALIZATION

The next phase of water treatment is removal of dissolved salt and mineral solids, which tend to ionize in solution. Ion-exchange demineralizers are the choice for the more common water sources, rivers and streams, which are relatively low in solids content. Increasingly, however, industry has been, and will be, turning to less pure water—brackish water from a deep well, for example, with several thousand ppm total dissolved solids, or estuary water with 10,000 ppm solids or more.

To apply ion exchange to water of such quality for significant throughput would require large equipment and exorbitant chemical costs. A more practical approach is partial demineralization to remove the large majority of dissolved solids, followed by polishing in an ion-exchange demineralizer. Special techniques for brackish waters involve separation by membrane and distillation processes, as described below. Ion exchange is covered in the next chapter.

Membrane Processes

Membrane processes (Fig. 2.21), while not so well established as distillation, have matured to a point where some are particularly suited to special conditions. The two leading processes are covered here.

Electrodialysis. *Electrodialysis* (ED) has seen the longest development. It differs from other membrane techniques and thermal evaporative processes in that it moves the dissolved minerals away from the water, rather than vice versa. Since the quantity of minerals is far less than that of the water containing them, there are practical advantages.

For ED, the basis of operation is the passage of water between paired sets of parallel membranes, with an electric field applied perpendicular to the direction of flow (left to right in Fig. 2.22). The field moves positive impurity ions toward the negative electrode and negative ions in the opposite direction. The membranes are selectively permeable, the one in each pair closer to the negative electrode (*C*) allowing passage of the positive ions (cations), the other (*A*) favoring negative ions.

Actually, hundreds of pairs of membranes of alternating sensitivity are assembled into membrane stacks. Plastic spacers separate the membranes, directing flow of water through a tortuous path across the membrane faces. Present removal rates justify ED costs for water with solids concentrations up to those found in seawater, or 1000 to 35,000 ppm.

Reverse Osmosis. *Reverse osmosis* (RO) serves as the basis for a more varied group of membrane devices. The underlying principle is this: When two solutions of different concentrations are separated by a semipermeable membrane, solvent (water) will be transported from the dilute to the more concentrated side. Application of pressure to the concentrated side will produce a flow of solvent in the reverse direction. If the solution is that of salt in water, the choice of membrane

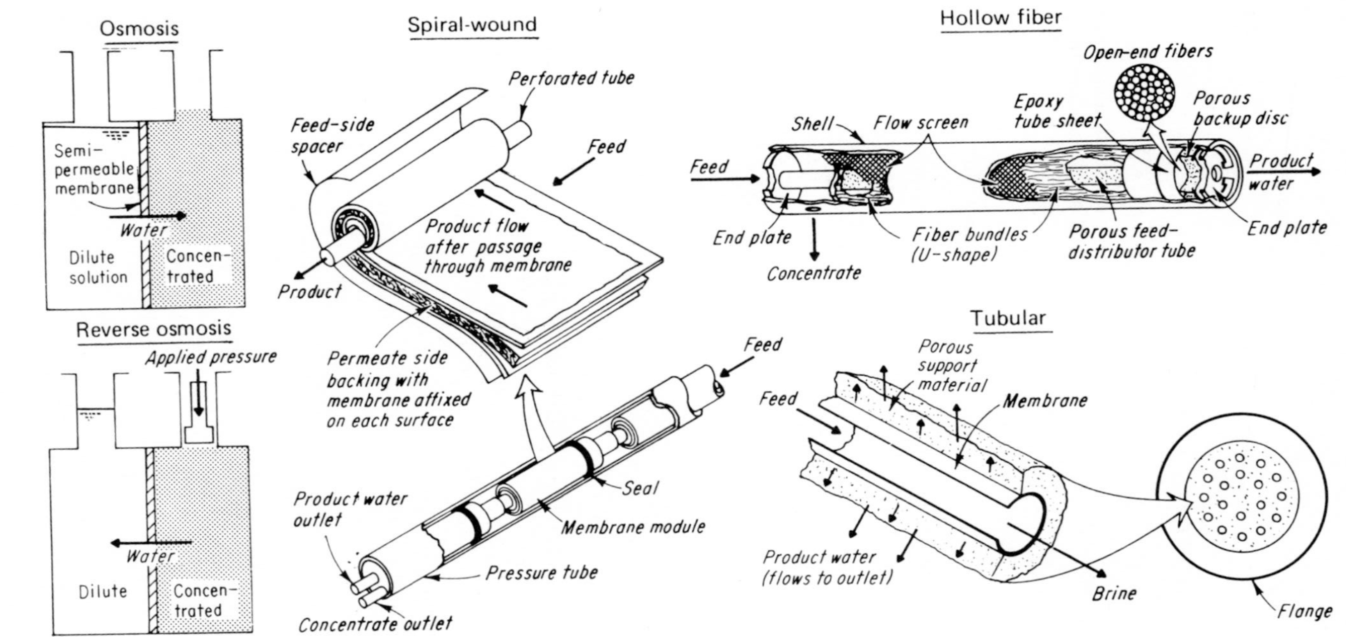

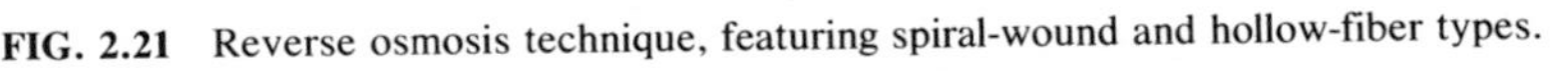

FIG. 2.21 Reverse osmosis technique, featuring spiral-wound and hollow-fiber types.

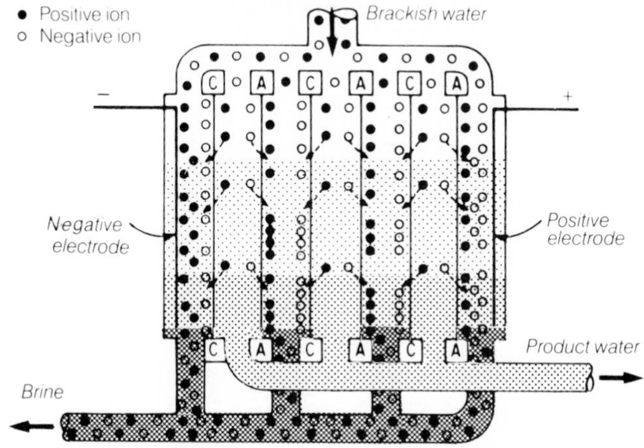

FIG. 2.22 Electrodialysis cell.

that is permeable to water but not to salt will result in a flow of water to the unpressurized side. The process is thus one of concentration, resulting in one stream more concentrated in salt than the original feed and one that is purer (more dilute). Figure 2.21 shows some designs available commercially.

RO costs for planned installations are comparable with those of ED for moderate concentrations of dissolved solids—1500 to 10,000 ppm. Moreover, RO systems remove ion-forming solids as well as nonionic matter, both colloidal and high-molecular-weight organic; ED applies only to ion removal, perhaps making charcoal adsorption a necessary adjunct. ED units require electrical and mechanical auxiliary equipment and somewhat more highly skilled operators. However, their initial cost is probably lower than for RO units in the low-solids range. Pressure requirements for ED are lower—perhaps 100 psi (689 kPa) compared with 400 to 600 psi (2756 to 4134 kPa) for RO, although the trend is downward for RO.

Evaporative Techniques

Another way to remove dissolved salts and minerals is via evaporation. Three distinct methods find application in powerplant practice.

Distillation. The favored evaporative technique is distillation, the purely mechanical process of removing dissolved and suspended solids by vaporizing water with heat. The process is carried out in a pressure tank traversed by steam coils, with a moisture separator at the outlet end. The vapor thus removed is then condensed to form purified water. Another type, the flash evaporator, operates with its chamber under partial vacuum (Fig. 2.23). Incoming water is preheated sufficiently to cause water to flash into vapor on entering the chamber. Long-tube evaporators are a variation, in which evaporation takes place in long vertical tubes rather than vacuum chambers.

Multiple-effect systems feature two or more units in series, the vapor from one stage being used to heat the next stage at lower pressure and then being con-

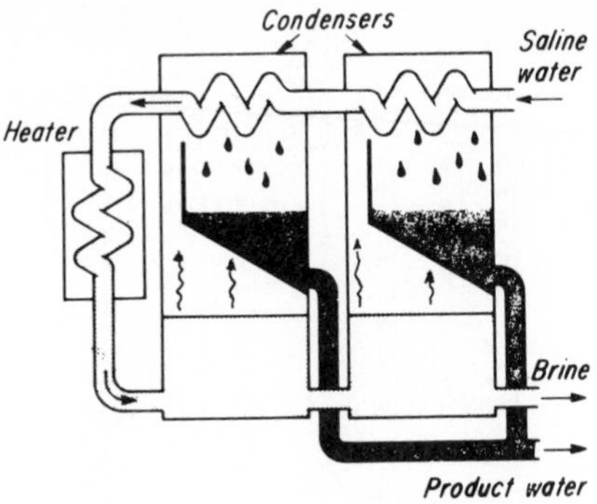

FIG. 2.23 Flash distillation unit.

densed. Such systems are generally required to produce potable water. Boiler feedwater applications favor single-effect units for greater output per unit area of tube surface.

Vapor Compression. In vapor compression, the equipment needed comprises a starting heater, heat exchanger, boiling chamber, and vapor compressor (Fig. 2.24). Hot water entering the evaporation chamber from the heat exchanger is boiled at atmospheric pressure. The resulting steam is fed to the compressor, which raises pressure about 3 psi (21 kPa). Compressed steam flows back to the boiling section and condenses, giving up heat to boil more water.

Vacuum Freezing. A quite different process is vacuum freezing (Fig. 2.25). Here, freezing of saline water (more than 10,000 ppm dissolved solids) produces ice crystals consisting of pure water coated with brine. The brine has a much lower freezing point and thus can be washed off the ice crystals. The major ap-

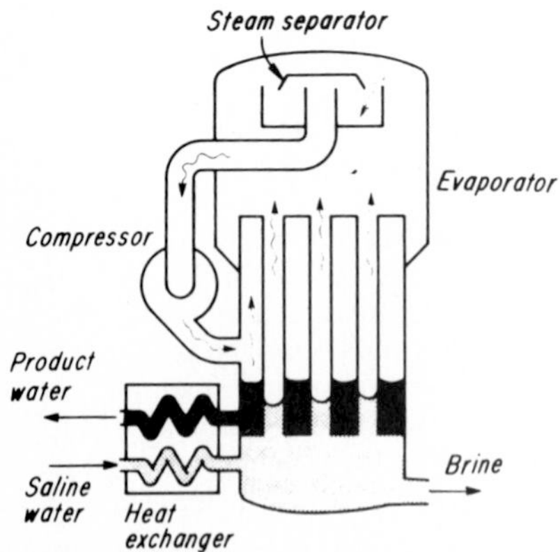

FIG. 2.24 Vapor compression unit.

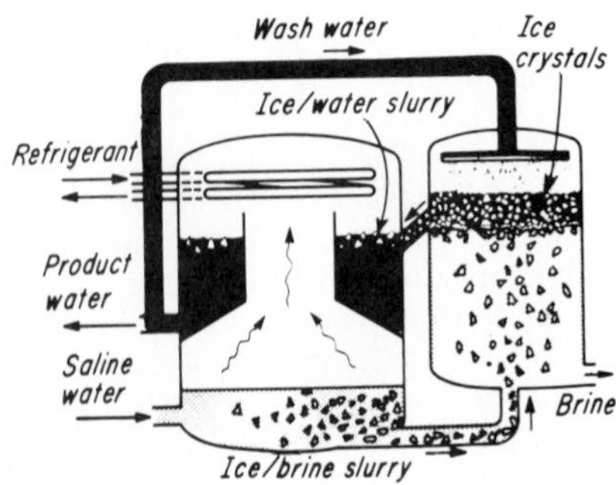

FIG. 2.25 Vacuum freezing unit.

plication is to ocean water or to estuary water suffering incursion from the sea (up to 30,000 ppm).

Comparison of Methods

Of the foregoing methods, only distillation is well established, plants having been built in the million-gallon-per-day (million-liter-per-day) range. However, they all have disadvantages. One is the production and disposal of brine. While this is no problem in sea-bordering regions, it can add considerable expense to treatment at an inland site. Another cost penalty stems from the large quantity of heat required to change the physical state of the water. This drawback is magnified by an escalating cost of energy. The most favorable economies exist in plants or areas where waste heat is available as a by-product of some other activity. A prime example is steam generation, in which dual-purpose plants are designed from the outset to produce both desalinated water and electricity. Plants of this type are coming on-line, particularly overseas in places like the Middle East.

Both RO and ED provide less expensive separation than evaporation units. For contained ions that are predominantly monovalent (sodium, chloride), membrane systems reject 85 to 95 percent of solids. Where divalent ions predominate (calcium, magnesium, sulfate), larger particle size and charge enable 98 percent rejection with RO and 90 percent with ED. On the whole, water with up to 10,000 ppm dissolved solids can be purified up to 90 percent, in some cases to 99 percent. Both ED and RO systems are offered for throughputs as high as several million gallons (liters) per day. Evaporative (and freezing) processes can produce purer water, provided carry-over of solids in spray droplets is inhibited. But the requirement of heat makes these processes costly.

As in all water treatment systems, equipment selection depends on water analysis and system needs. Where high product recovery is necessary to reduce waste brine volume, it may be feasible to apply a membrane process to the waste stream followed by evaporation. Most units must be tailored to meet particular situations. For example, pretreatment may be advisable to protect RO membranes; some require that pH be adjusted for slight acidity (5.5 to 6.5). When chromates are removed from recycled water, a polyamide membrane, which requires high pH for best results, is recommended. Prior treatment may be necessary for iron removal for all membrane types.

COMBINATION SYSTEMS

Settlers, clarifiers, media and precoat filters, and ultrafilters are used alone and in combination to treat a variety of waters, reducing the levels of those impurities that cause scaling, silting, and corrosion of boiler surfaces, piping, and turbine blades. Coagulation and filtration also remove impurities that may foul expensive ion-exchange resins and RO membranes. RO and cooling-water systems require the reduction of the concentrations of slightly soluble salts (such as calcium sulfate) and silica. These impurities are concentrated on the brine side of RO membranes and in cooling waters. The result is supersaturation and scaling.

Combinations of liquid-solid separator operations are designed to maximize the efficiency of suspended-solids removal and to expand the capability of the plant to include removal of scaling species. The three most common separation equipment systems are

- *Direct filtration:* This group includes all filtration of source waters without the use of presettling or clarifiers for preliminary solids removal. Both precoat and media filters are commonly selected.
- *Clarification and filtration:* This pairing represents the most commonly found treatment scheme for raw water. High-rate solids-contact clarifiers are followed by media filters; precoat filters can also be selected.
- *Polishing filtration:* This type of filtration is used to trap colloids in boiler makeup and, more often, in condensate.

Direct Filtration

Direct filtration with media or precoat filters treats waters containing relatively low levels of turbidity. Precoat filters using perlite, diatomaceous earth, or other filter aids produce high-quality effluent without prior coagulation of influent colloids and suspended solids. Other advantages are low waste volume (less than 1 percent of the treated water) and low cost. Precoat filters are generally limited to treating waters containing little more than 5 to 10 ppm suspended solids. Appreciably higher turbidity will rapidly lead to formation of an impervious and compressible filter cake, quickly producing a large pressure drop and ending the filter run.

Direct filtration of more turbid waters is possible with granular-media filters. Deep-bed granular-media filters have been demonstrated to filter natural, coagulated turbidity at a rate of 8 to 12 gal/min d ft^2 (5.44 to 8.15 L/s d m^2) for downflow designs and 6 gal/min d ft^2 (4.08 L/s d m^2) for upflow filters, producing very high effluent quality. Direct filtration is capable of filtering water containing 200 ppm of suspended solids, removing 90 to 98 percent of all particles greater than 2 μm when sufficient coagulant is added upstream. This approach to pretreatment is finding increasing acceptance for treating waters that do not require substantial reduction of reactive silica, hardness, or organic constituents.

Highly colored surface waters have traditionally been treated with alum in clarifiers and filtered. In-line coagulation and filtration have proved effective and may be a viable alternative. Careful pH control is necessary here for proper coagulation.

An alternative to in-line coagulant addition is the use of a preflocculation, or floc-forming, tank. Configured to encourage intimate mixing of treatment chemicals and raw water to accelerate floc growth, these tanks afford a flexibility not possible in an in-line system. They facilitate color removal by adsorption on aluminum hydroxide floc, as well as iron and manganese removal by oxidation, precipitation, and filtration.

Clarification and Filtration

Combined clarification and filtration is recommended where filters alone would not be adequate to remove high turbidity loadings and where the removal of some dissolved impurities would be advantageous (Fig. 2.26). For example, power-

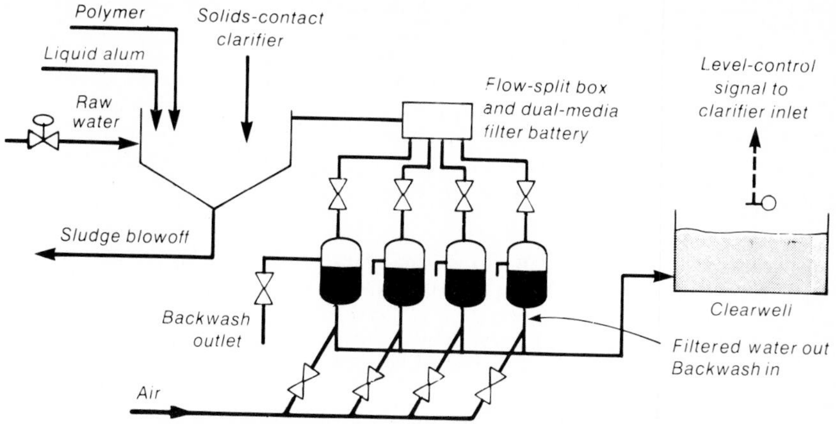

FIG. 2.26 Clarifier-filter combination.

plants situated on rivers exhibiting spates of high turbidity during spring rains and/or melts would be well advised to include a high-rate clarifier, to ensure adequate solids removal performance. Also, savings of regenerant chemicals are made possible by preceding a demineralization system with a cold-process, lime-soda softening clarifier. Where less than 2- to 5-JTU turbidity is required, clarifiers alone are not enough. The great majority of filters added after clarifiers are the granular-media type.

Some surface waters show periods of extreme levels of suspended solids, often far in excess of 1000 ppm. Although high-rate units can often accommodate these loadings, a settling basin may be used to remove the bulk of the larger and faster-settling suspended particles before they reach the clarifier. Such a basin also equalizes the influent water quality during highly variable periods, minimizing the need for timely operator reactions to changes of influent quality dictating treatment modifications.

Polishing Filtration

Polishing systems include both precoat filters and ultrafilters. Precoat formulations containing powdered ion-exchange resin may be used, to simultaneously remove dissolved and suspended solids from steam condensate. Fibrous material may be added to increase the holding capacity and thus run lengths. Although ultrafiltration systems still do not have a well-defined niche in steam plants, these membrane filters have been shown to be effective traps for colloidal particles. High levels of turbidity may tend to form a compressible gel layer on the ultrafilter membrane, reducing permeate flux. Many commercial units now on the market have provisions for backwashing.

ACKNOWLEDGMENT

The author would like to thank *Power* magazine for the use of some of the material that appears in this chapter.

SUGGESTIONS FOR FURTHER READING

Bozeka, C.G., and F.J. Pocock, "Waterside Corrosion Control in Industrial Boilers," Corrosion '85, National Association of Corrosion Engineers, Boston, Mass., March 1985.

Handbook: Impurities in Water and Means for Their Removal, Drew Chemical Corp., Boonton, N.J., 1972.

Handbook of Industrial Water Conditioning, 8th ed., Betz Laboratories, Inc., Trevose, Pa., 1980.

Kemmer, F.J., *The Nalco Handbook,* McGraw-Hill Book Co., New York, 1979.

Nolan, L.L., "Precoat Overlay Improves Demineralizers," *Power,* January 1980.

Nordell, E., *Water Treatment for Industrial and Other Uses,* 2d ed., Reinhold Publishing Co., New York, 1961.

Puri, V.K., "Pretreatment of Water for Cooling and Steam-Generating Systems," 21st Annual Liberty Bell Corrosion Course, Philadelphia, Pa., 1983.

Strauss, S.D., "Water Treatment—A Special Report," *Power,* June 1974.

Tunnell, H.R., "Shallow Bed Rapid Sand Filtration for Water and Wastewater," International Water Conference, Pittsburgh, Pa., 1975.

Wagner, J.N., "Solids Contact Devices—Past, Present, and State of the Art," 15th Annual Liberty Bell Corrosion Course, Philadelphia, Pa., 1977.

Weber, Jr., W.J., *Physicochemical Processes for Water-Quality Control,* Wiley Interscience, New York, 1972.

Zirps, G.T., "Magnetic Filters Enhance Plant Performance," *Power,* May 1977.

CHAPTER 4.3
ION EXCHANGERS

Leonard J. Lefevre
Larkin Laboratory
Dow Chemical Co.
Midland, MI

INTRODUCTION

Ion-exchange resin processes have become a vital part of the operation of fossil-fuel and nuclear powerplants. Modern steam-generator designs using high-pressure drum-type boilers or once-through operation mandate the use of high-purity water makeup and the "polishing" of the recycled condensate to keep cation and anion concentrations at very low levels, in some instances at levels below 1 part per billion (ppb). These low levels of ions, such as sodium, chloride, and sulfate, are necessary to control corrosion. Specific requirements are dependent on the chemistry and metallurgy of the system involved and the equipment vendor's recommendations for the system. Groups such as the *American Society of Mechanical Engineers* (ASME) are involved in studies aimed at further defining specific needs.

These requirements are effectively met only by ion-exchange demineralization systems. However, a wide variety of ion-exchange systems are available, combining several available resin types and a variety of equipment design approaches for optimum use of these resins.

Ion exchange is also used to demineralize water used in stator-cooling recirculation loops, nuclear-plant deborating circuits, fuel pool demineralization, cooling-tower water dealkalization, and softening. Each of these areas is treated in this chapter.

Makeup Water Demineralization

As shown in Fig. 3.1, water demineralization by ion exchange is essentially a two-step process: (1) the removal of dissolved cations (Na^+, K^+, Ca^{++}, Mg^{++}, etc.) by exchange with hydrogen ions (H^+) on a cation-exchange resin (see Common Ion-Exchange Terms in box) and (2) the removal of the remaining anions (Cl^-, SO_4^-, HCO_3^-, $HSiO_2^-$, etc.) by exchange with the hydroxide ion (OH^-) on an anion-exchange resin or by sorption of the mineral acids formed in step 1 on a weak-base resin operating in the free-base form. These steps are in this order to prevent precipitation problems, which may occur if magnesium ions in raw water were contacted with the hydroxide form of a strong-base resin, and to provide a lead bed having the more rugged and less expensive resin to take the greater physical and chemical strain inherent in this position. Filtration of particulates, with the subsequent need for cleaning, and oxidation due to residual chlorine in feedwater contribute to this strain.

Although it is basically a two-step operation, each "step" can use combinations of resins and processes. For example, weak-acid resins can be used to economically supplement a strong-acid resin in the cation removal step, if the influent water analysis is appropriate. While strong-acid resins remove cations from both neutral and alkaline salts, weak-acid resins effectively remove only divalent cations associated with alkalinity in the raw water.

Weak-Acid vs. Strong-Acid Resins. The value of the weak-acid resin use lies in the very efficient utilization of acid regenerant, especially compared to the effi-

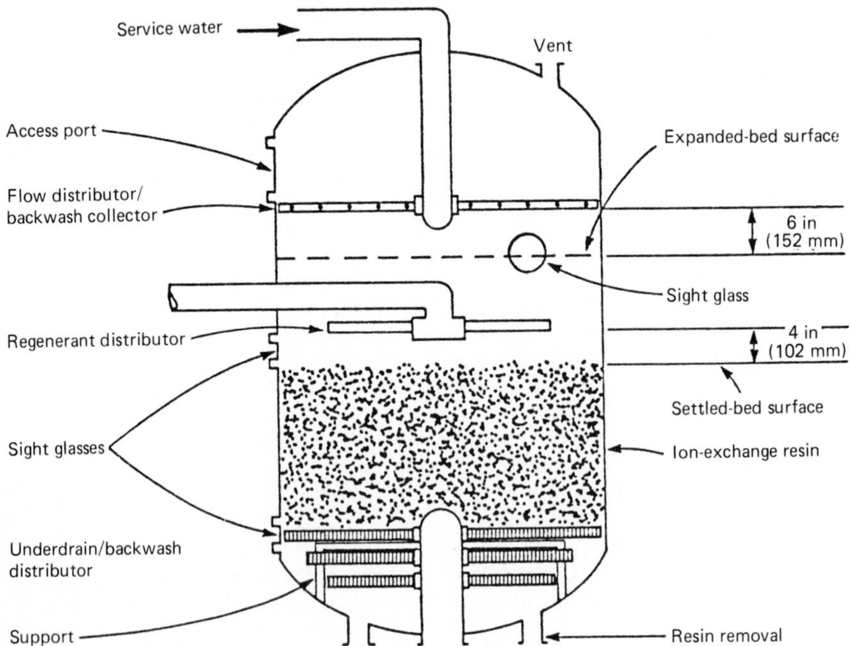

FIG. 3.1 Makeup demineralizer vessel. Indicated is proper location of elements to ensure adequate freeboard above bed and fluidization of contaminating particulates. (Power *magazine*)

COMMON ION-EXCHANGE TERMS

Alkalinity May exist in three forms—caustic (or hydrate), carbonate, and bicarbonate. Bicarbonate and carbonate may exist together as may carbonate and caustic. Alkalinity is commonly expressed as P and M *parts per million* (ppm) as calcium carbonate ($CaCO_3$). "P" represents the titration with a standard acid solution to a phenolphthalein endpoint; "M" represents the titration to a methyl orange endpoint. These values are used to calculate the caustic, carbonate, and bicarbonate alkalinities.

Attrition Breakage and wear of resin particles.

Bed The ion-exchange resin contained in a column. (*See* ion-exchange bed.)

Bed depth The depth of the ion-exchange resin in a column.

Bed expansion The separation and rise of ion-exchange resin particles in a column during backwashing.

Breakthrough That volume of effluent where the concentration of the exchanging ion in the effluent reaches a predetermined limit. This point is usually the limit of the exhaustion cycle and the beginning of the regeneration cycle.

Channeling The creation of isolated paths of less resistance in the ion-exchange resin bed caused by the production of air pockets, dirt, and other factors which cause uneven pressure gradients in the bed. Channeling prevents the liquid being processed from uniformly contacting the entire resin bed.

Column operation The most common method for employing ion exchange in which the liquid to be treated passes through a fixed bed of ion-exchange resin.

Countercurrent regeneration Application of regenerant to an ion-exchange bed in the flow direction opposite the direction of the water being treated. Reduced leakage results.

Cross-linkage Binding of the linear polymer chains in the matrix of an ion-exchange resin with a cross-linking agent that produces a three-dimensional, insoluble polymer.

Degasifier Also called a *decarbonator.* Used to reduce the carbon dioxide content of water from hydrogen cation exchangers. Also reduces CO_2 to about 8 ppm but saturates water with air. Common problem in connection with demineralizing is pollution of water with air contaminants, especially around industrial areas and powerplants. Adequate air filters are needed, with regular maintenance required.

Degasifier, vacuum Actually a deaerator. Reduces oxygen as well as CO_2 and all other gases to very low level. Preferred as a means of CO_2 reduction when boiler-water makeup is demineralized. Eliminates water pollution and reduces corrosion problems when water is transferred through steel equipment. Use usually results in longer anion-exchange resin life.

Demineralization (deionization) The removal of ionizable constituents and silica from a solution by ion-exchange processing. Normally performed by passing the solution through the hydrogen form of a cation-exchange resin and through the hydroxide form of a strong-base anion-exchange resin, either as a two-step operation or as an operation in which a single bed containing a mixture of the two resins is employed.

Distributor A piping configuration, in an ion-exchange vessel, designed to uniformly spread or collect solution flow through the resin bed.

Freeboard The space provided above the resin bed in the column to accommodate the expansion of the resin particles during backwashing.

Hydrogen cycle Cation-exchange resin operation in which the regenerated ionic form of the resin is the hydrogen form.

Inert resin Nonfunctional spherical beads with hydraulic properties selected to position the inert resin between the cation and the anion resins of mixed beds by backwash separation prior to regeneration.

Interstitial volume The space between the particles in an ion-exchange resin bed. Also termed the *void volume.*

Leakage (hardness, sodium, silica, etc.) Caused by incomplete regeneration of the exchanger bed. Since complete regeneration is usually inefficient, most ion-exchange processes operate at one-half to two-thirds of the ultimate capacity of the ion exchanger. Operating cost is a factor, and effluent water quality (tolerable leakage) greatly affects this.

Monosphere The Dow Chemical Co.'s designation for its line of resins having very uniform particle-size distribution.

Pressure drop (head loss) The loss in liquid pressure resulting from the passage of the solution through the bed of an ion-exchange resin.

Regenerant The chemical used to convert an ion-exchange resin to the desired ionic form for reuse.

Regenerant crossover Contact of cation-exchange resin with anion-exchange resin regenerant (and the reverse), usually at the interfacial distributor of a mixed bed.

Regeneration The displacement from the ion-exchange resin of the ions removed from the process solution. Performed by passing through the bed a solution containing the ion desired on the resin.

Regeneration level The amount of regenerant used per cycle. Commonly expressed in pounds per cubic foot (kilograms per cubic meter) of resin.

Rinse The passage of water through the ion-exchange resin bed to flush out excess regenerant.

Sphericity The unbroken state of ion-exchange resin beads.

Strong-acid cation-exchange resin When regenerated with acid, cation-exchange resin exchanges hydrogen ions for metal cations, forming acids of the salts present in the water. Usually a sulfonated copolymer of styrene and divinyl benzene.

Thoroughfare regeneration Use of regenerant effluent from a strong-acid or strong-base resin as regenerant for a weak-acid or weak-base resin, respectively, used in series during the operating cycle.

ciency possible with strong-acid resins. Used ahead of a strong-acid resin, weak-acid resins remove calcium and magnesium ions, which are associated with alkalinity in the influent water, reducing the load to the strong-acid resin. The waste regenerant acid from the strong-acid resin can be used to regenerate the weak-acid resins, resulting in significant reduction in the amount of regenerant which would be required if only a strong-acid resin were used. In some cases it is practical to layer the weak-acid cation resin on top of the strong-acid resin in a single bed.

The acid generated from cation exchange is removable, in step 2, by several methods. In addition, different types of anion-exchange resins may be used in some methods. Mineral acids may be removed by sorption of the acid on a weak-base resin and efficiently regenerated from the resin by neutralization with any of several base chemicals. These resins have little affinity for the ions of neutral or basic salts, thus are regenerable with regenerant effluent from strong-base resins. (See "Thoroughfare regeneration" in box.) The regenerant requirement is nearly the theoretical value required for neutralization of the sorbed acid. Several types of weak-base resins are commercially available.

Dissolved CO_2 Removal. Dissolved CO_2 generated from the raw-water alkalinity when the water is passed through the cation-exchange resin may be removed by mechanical means, such as decarbonation by air blowing or degasification under vacuum, or by exchange on strong-base anion-exchange resins. Removal of CO_2 by mechanical methods is the most economical process when water quantities being treated and the alkalinity level in the raw water are relatively high.

When very low levels of CO_2 are required in the product water, or when dissolved silica must be removed, strong-base resins must be used, either alone or following the mechanical methods. A wide variety of strong-base resins are available commercially (see Table 3.1). These may be grouped, generally, as styrene-based type I and type II resins and acrylic strong-base resins. Each type is available in gel (gelular) and macroporous (macroreticular) forms.

Several swollen porosity ranges are available in the gel resin group. Various combinations of degasification, weak-base resin, and strong-base resin, including layered beds of weak-base on top of strong-base resin, can be considered for economic optimization (see Table 3.2).

TABLE 3.1 Representative U.S. Ion-Exchange Resins for Softening and Demineralization[1]

	Manufacturer (trade name)			
	Dow Chemical (Dowex)	Ionac Chemical (Ionac)	Rohm & Haas	
Resin type			Amberlite	Duolite
Gel cation S.A.[2]	HCR-S	C-249	IR120	C20
	HCR-W2	C-298	IR130	C20J
	HGR	C-250	IR122	C20X10
	HGR-W2	C-299	IR132	C225X10
Gel cation	650C	650C	650C	650C
W.A.[8]	CCR-2	CNN	IRC-84	C-433
Gel anion	SBR	ASB-1	IRA400	A109
S.B.I[3]	SBR-P	ASB-1P	IRA402	A101D
	11	A642	—	—
	550A	550A	550A	550A
Gel anion S.B.II[4]	SAR	ASB-2	IRA410	A104
Gel anion	WGR	A305	—	A30
W.B.[5]	WGR-2	A305	IRA47	A340
Gel anion W.B.A.[6]	—	—	IRA68	—
Gel anion S.B.A.[7]	—	—	IRA458	—
M.P. cation S.A.	MSC-1	CFP110	200	C26
M.P. cation W.A.[8]	MWC-1	CC	IRC51	C464
M.P. anion S.B.I	MSA-1	A641	IRA900	A161
M.P. anion S.B.II	MSA-2	A657	IRA910	A162
M.P. anion W.B.	MWA-1	AFP-329	IRA94	A368

[1]Special grades are available for nuclear and condensate polishing applications.
[2]Strong acid.
[3]Strong-base type I.
[4]Strong-base type II.
[5]Weak base.
[6]Weak-base acrylic.
[7]Strong-base acrylic.
[8]Weak acid.
Note: M.P. = macroporous.

TABLE 3.2 Representative Ion-Exchange Systems

Systems	Relative cost (1 = lowest) First	Operating	Expected quality, μs
SA(⇊)-SB(⇊)	2	7	2
SA(⇅)-SB(⇅)	3	4	0.5
SA(⇊)-WB(⇊)	1	1	4
SA(⇊)-DG-SB(⇊)	4	6	2
WA(⇊)-SA(⇅)(⇆)-WB(⇊)-SB(⇅)(⇆)	8	3	0.5
SA(⇊)-SB(⇊)-MB	6	8	0.1
SA(⇊)-SB(⇊)-SA(⇊)-SB(⇊)	7	5	0.5
LBC(⇆)-LBA(⇆)	5	2	2

DG = degas
SA = strong-acid resin bed
WA = weak-acid resin bed
SB = strong-base resin bed
WB = weak-base resin bed
MB = mixed bed of SA-SB resins
(⇊) = cocurrent regeneration
(⇅) = countercurrent regeneration
(⇆) = thoroughfare regeneration, strong to weak resin
LBA = layered bed, anion WB/SB
LBC = layered bed, cation WA/SA

Countercurrent Regeneration

Another demineralization system variable to be considered in system design is the regeneration mode: (1) cocurrent application of regenerant to the bed (relative flow direction) and (2) countercurrent application. Cocurrent application is the historic and currently most widely used mode. However, the significant improvement in product water quality at a given regenerant dosage that can be realized by using countercurrent application is arousing considerable interest for new equipment. Retrofitting of existing equipment is also viable with some of the countercurrent techniques of application.

Ion Leakage. Key to obtaining the improved quality of product water when the countercurrent ion-exchange mode is used is prevention of ion leakage by reexchange of contaminating ions from the effluent end of the resin bed (by the acid formed from cation splitting of neutral salts in the case of cation exchange) during the service cycle. Reexchange is prevented by elimination of the contaminating ion from the effluent end of the column. This requires

- Regenerant relatively free of the contaminating ion.
- Regenerant displacement rinse water free of the contaminating ion.
- Maintenance of a packed bed to avoid migration of exhausted resin to the product end of the resin bed.
- Reduction in the frequency of backwash. This requires removal of filterable particulates coming to the resin bed and reduction in dissolved gases to below their solubility level at the temperature and pressure involved.
- A minimum bed-depth requirement of 5 ft (1.5 m) to optimize resin utilization.

Commercial Techniques. Commercial units designed for countercurrent regeneration use various techniques to maintain the resin bed in the packed condition during the upflow portion of the operation. This includes application of (1) a blocking flow of water, (2) an air block, (3) a combination cocurrent-countercurrent flow through discrete zones in the resin bed, (4) an inflated bladder in the freeboard space of the bed, (5) a bed operated in the full condition, and (6) resin lifted hydraulically against an upper screened distributor. The quality of the effluent obtained depends on how well the design and operation meet the criteria noted above.

Mixed-Bed Resin Demineralizers

For the ultimate in water quality from an ion-exchange demineralization train, a mixed bed of cation- and anion-exchange resins operating in the H^+ and OH^- forms, respectively, is often used as a polishing unit. With anion and cation resins acting together, the leakage of ions associated with individual bed operation is reduced to very low levels. The leakage due to reexchange in a single resin column is eliminated since the immediate formation of water by the H^+ ion and OH^- released from the resins removes the driving force responsible for the reexchange, even if some reexchangeable ions are present. Mixed-bed resin design and regeneration practices are critical to obtaining the optimum capacity and leakage characteristics. The resins chosen for mixed-bed operation must be matched so hydraulic separation and subsequent remixing can be done effectively.

Hydraulic Separation. Hydraulic separation is obtained by the uniform application of water upflow through the resin bed (backwash) at a rate sufficient to expand the cation portion of the bed approximately 20 percent for approximately 10 to 15 min. When flow is stopped, the cation and anion resins settle into layers. In internally regenerated equipment, the regenerant outlet distributor is positioned at the interface formed between the settled layers. In vessels having separate regeneration tanks for each resin, the anion transfer distributor is positioned at or near the cation-anion interface.

Regenerant crossover, the contact of anion-exchange resin with cation-resin regenerant (or the reverse), must be minimized to optimize capacity and reduce leakage in subsequent cycles. Particular attention should be given to reducing the contact of anion-exchange resin with sulfuric acid, since this results in long rinse times to remove sulfuric acid formed by hydrolysis of the bisulfate anion taken up by the anion resin. If it is not rinsed, poor water quality from the mixed bed may result.

Minimizing Crossover. The crossover problem at the distributor may be minimized by the addition of an inert resin, to the mixture of resins, which has hydraulic properties such that it is positioned between the cation- and anion-exchange resins following the backwash operation. Thus, it fills the regenerant crossover zone at the regenerant outlet distributor with material which does not react with either regenerant chemical.

The inert concept requires the use of matched sets of cation, inert, and anion resins. It is important that the finest cation resin and the coarsest anion resin do not mix, even without the inert resin present. To optimize this separation, at least one resin manufacturer has introduced a line of resins having very uniform particle-size distribution.

Stator coolant demineralization is a special case where the mixed-bed demineralizer is removing metals, predominantly copper, which enters the cooling water during recirculation. Hot water in contact with copper-laden resin tends to catalyze resin degradation reactions with resin degradation components tending to foul the resin bed. Resin pairs which resist the degradation and/or the subsequent resin fouling are specified for optimum service life of these resin beds.

Condensate Polishing

Recycle condensate contains both particulate products of corrosion, various oxides of iron commonly called *crud,* and contaminating ions from in-leakage of cooling water and air into the condensate. In certain systems, corrosion-inhibiting compounds, such as ammonia and hydrazine, may also be present. To protect the system from the impact of these contaminants on the water chemistry in the steam generator and turbine, condensate-polishing demineralizing systems can be used. Both deep-bed and powdered-resin systems are used, depending somewhat on the major contaminating specie present or likely to be present.

Powdered-Resin Systems. In the absence of significant ionic load to the polisher system, crud may be removed by using powdered-resin systems. These consist of candle- or septum-type filters coated with a thin layer [approximately ¼ in (6 mm) thick] of a mixture of fine-particle-size cation-exchange resin (hydrogen form) and anion resin (hydroxide form). The pressure drop across the resin layer is controlled by the amount of each resin in the mixture and resin pretreatments designed to control resin agglomeration characteristics and therefore the void fraction of the layer. On start-up of the filter, resin is precoated onto the septum with recycle until the desired layer has formed. Units are taken off line, and resin is flushed from the septum and processed for appropriate disposal when a pressure drop or an ion-leakage endpoint has been reached. The ion-exchange capacity is limited.

Deep-Bed Polishers. Deep-bed condensate polishers (see Fig. 3.2) provide equivalent filtration capability to powdered-resin filters and significantly greater ion capacity. These systems usually consist of several mixed-bed polishing units operating in parallel at flow rates in the range of 50 to 60 gal/min • ft^2 (34 to 41 L/s • m^2). The high flow rates evolved to keep equipment sizes manageable at the large condensate return volumes involved, 10 to 20 gal/min • MW (0.6 to 1.2 L/s • MW), and to cause crud filtration to be in-depth filtration as compared to bed-surface filtration. Resin-exchange reaction kinetics become important at these flow rates, particularly if ion levels increase in the condensate because of cooling water in-leakage.

Polisher Regeneration. Regeneration of the polishers is done in equipment separate from the polisher vessel for several reasons. The external regeneration system allows isolation of regenerant chemicals from the direct flow path of the condensate, reducing chances for contamination. Also, since the run time on a polisher is significantly longer than the time required for regeneration, one regeneration station can accommodate several polishers and thus save capital.

The high flow rate and in-depth crud filtration result in a buildup of pressure drop across the system with time. As a result, pressure drop across the system is one of the determining run-length factors (ion "breakthrough" and convenience

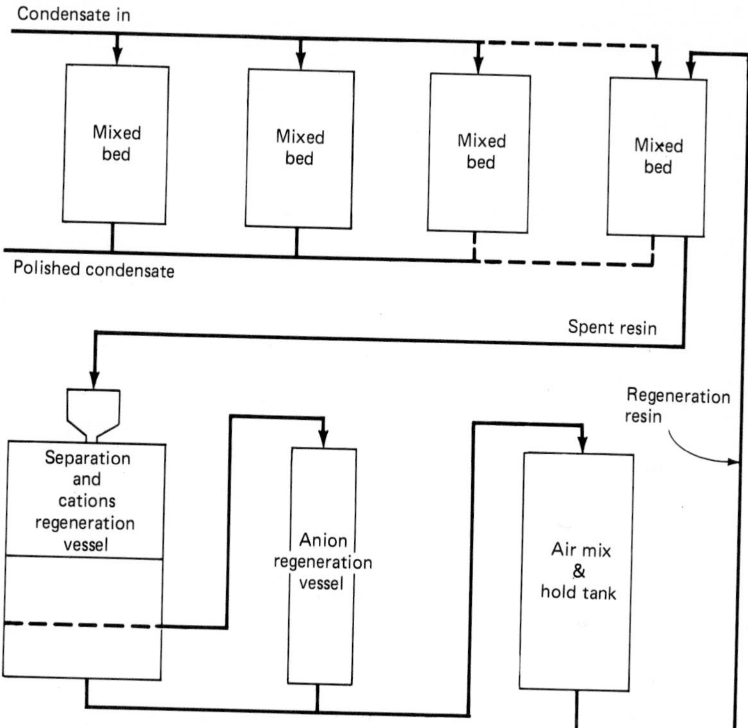

FIG. 3.2 Utility condensate-polishing flowsheet.

of the operators are others). The crud-loaded or ionically exhausted resin is then sluiced to the external regeneration unit, where several stages of cleanup are performed. Air scouring, followed by backwash or downflow purging, and ultrasonic cleaning are among the cleaning techniques practiced. These resin transfer steps and cleanup procedures require the use of physically stable, premium-grade resins for acceptable life.

Regeneration generally (but not always) follows the physical cleanup before the resin is ready to be replaced into the polishing vessel.

Resin Separation. Separation of the resin is necessary, and for optimum operation of the polisher, as a demineralizer, separation must be optimized. This is particularly critical in systems operating with *all-volatile treatment* (AVT) through the ammonium break or on the ammonium cycle, to control sodium leakage levels and sulfate leakage levels caused by regenerant crossover.

Several approaches have been commercialized to aid in this optimization. Resin manufacturers have prepared special grades of resins for condensate polishing operations, which are physically stable and are sized to minimize crossover following backwash (see Fig. 3.3). Ambersep resins and Dowex Monosphere resins are examples.

Equipment vendors have devised techniques to maintain the separation inherent in the resin pairs involved and/or to minimize the impact of regenerant cross-

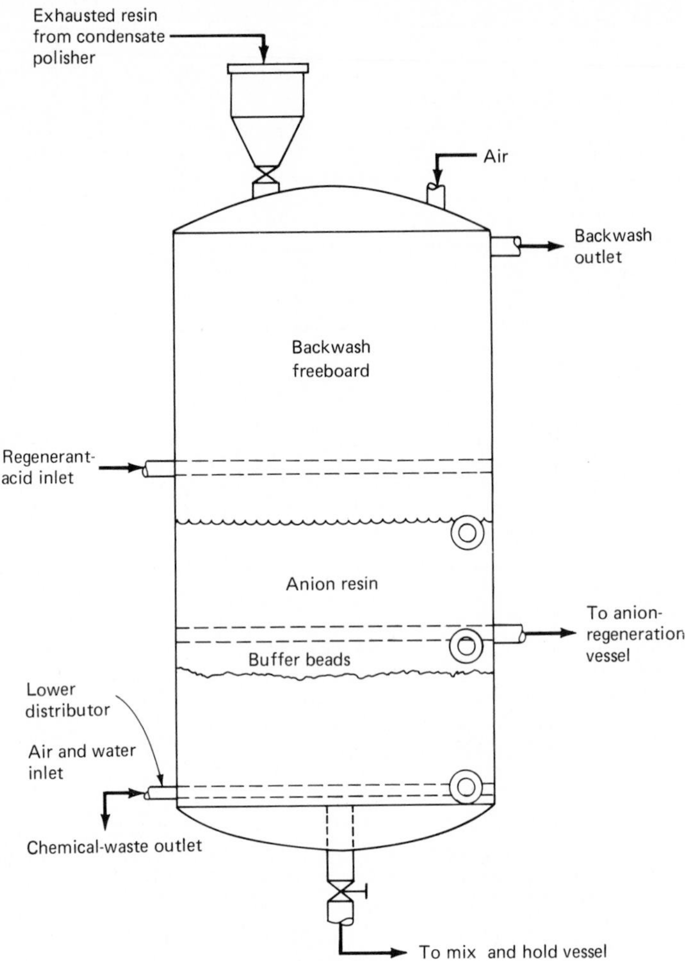

FIG. 3.3 Separation and cation regeneration vessel.

over. These include techniques which (1) isolate the crossover zone, using the resin in that zone only during the separation operation (Belco Co.); (2) transport cation resin from the interfacial zone into the anion regeneration tank, where the cation resin is removed by flotation of the anion resin from it by using 16 to 20 percent regenerant sodium hydroxide (Seprex-Graver); and (3) similar cation resin transport followed by conversion of the cation resin to a form which does not throw sodium during the polishing cycle, such as

- Calcium form by liming (Infilco Degremont and Permutit)
- Ammonium form (Ammonex-Cochrane)

At least one equipment vendor is promoting the use of separate resins, as opposed to mixed beds, for condensate polishing, thereby eliminating the separation problem (Tripol-Permutit Boby).

Short H^+-OH^- cycles with systems using AVT at relatively high pH (9.3 to 9.6) and the preference to maintain the ionic protection provided by the H^+-OH^- cycle relative to operating in the ammonium cycle have led to condensate-polishing mixed beds being preceded by H^+-form cation-resin beds, which remove the pH-adjusting additive (i.e., $NH_4{}^+$) and which do most of the particle filtration.

Nuclear Powerplant Applications

In addition to the use of ion-exchange softening and demineralization of makeup water and condensate polishing which they have in common with fossil-fuel plants, nuclear powerplants use ion exchange for several critical functions. Demineralization mixed beds are used to purify the water in the fuel pool and the reactor coolant. Radioactive waste is both concentrated and immobilized for burial by using single-bed and mixed-bed resin combinations for cesium, cobalt, and iodine removal, among others. Combinations of regenerated resin systems and regenerant concentration by evaporation are also used.

In pressurized water reactors, boric acid solution concentrations used to moderate neutron density are modified by pickup of boric acid on a strong-base anion-exchange resin. This unit may also be used as a boric acid sink since the resin capacity for boric acid varies with the temperature of the water flowing through it. The capacity of a strong-base resin for boric acid is also a function of the boric acid concentration in the solution being treated, due to the formation of complex borate anions at higher solution concentrations. These variations in capacity with concentration and temperature have resulted in operating schemes optimizing resin use in controlling boric acid concentrations for use as a neutron absorber (shim).

Cooling-Tower Water Treatment

Finally, ion-exchange softening and/or weak-acid resin dealkalization of makeup water to cooling towers can be used to reduce the chemical treatment requirements for scale control and to increase the cycles of concentration possible in the tower. The total dissolved-solids reduction and softening of the dealkalization process and the effluent use of regenerant acid are of particular value in this operation.

SUGGESTIONS FOR FURTHER READING

Applebaum, S.B., *Demineralization by Ion Exchange,* Academic Press Inc., New York, 1968.

Auerswald, D. C., and F. M. Cutler, "Two Year Study on Condensate Polisher Performance at Southern California Edison," *Proceedings, 43d International Water Conference,* Pittsburgh, Pa., Oct. 1982.

Brown, C.J., "Water Deionization by Recoflo Short-Bed Ion Exchange," International Water Conference, Pittsburgh, Pa., 1986.

Emmett, J.R., and P.M. Grainger, "Ion-Exchange Method in Condensate Polishing," International Water Conference, Pittsburgh, Pa., 1979.

Fisher, S., "'Extractables' in New Resins: A Further Look at What They Are and What Becomes of Them," International Water Conference, Pittsburgh, Pa., 1986.

Fisher, S. A., "Ion Exchange Part III—Resin Analysis," *Power,* Oct. 1982.

Gottlieb, M., and F.X. McGarvey, "Fundamental and Basic Theories Regarding the Proper Use of Modern Day Ion-Exchange Resins," International Water Conference, Pittsburgh, Pa., 1979.

Kunin, R., and J. Barrett, "Twenty Years of Macroreticular Ion-Exchange Experience," International Water Conference, Pittsburgh, Pa., 1979.

Lefevre, L. J., "Ion Exchange Part I—System Design and Operation," *Power,* Aug. 1982.

Nitti, N.J., "A New Condensate-Polishing Technique for Medium-Pressure Boilers," International Water Conference, Pittsburgh, Pa., 1979.

Rios, J., and M. Maddagiri, "Use of Two Bed Polishers in Side Stream to Reduce Introduction of Insoluble and Soluble Solids into PWR Steam Generators," *Proceedings of American Power Conference,* 47:1066–1075, 1985.

Selby, K. A., "Ion Exchange Part II—Operations and Supervision," *Power,* Oct. 1982.

Strauss, S. D., and R. Kunin, "Ion Exchange—Key to UltraPure Water for High Pressure Fossil and Nuclear Steam Generation," *Power,* Sept. 1980.

Thompson, J., and A.C. Reents, "Counterflow Regeneration," International Water Conference, Pittsburgh, Pa., 1966.

CHAPTER 4.4
KEY TREATMENT CHEMICALS

James A. Baumbach
Nalco Chemical Co.
Oak Brook, IL

INTRODUCTION

Utility and industrial powerplants rely on water as a primary process fluid for generating power, for cooling condensers and heat exchangers, and for drinking purposes. All waters, regardless of their source, require some conditioning before, during, and sometimes after use. The type of conditioning depends on the ultimate application. When water enters the plant, it is typically clarified or softened by a precipitation process. Chemicals are added in a stepwise process as the water proceeds from the initial source to the area where it will be used for other applications.

Water for boiler feedwater is further treated to remove as many dissolved solids and potential scale-forming minerals as possible. After the water is demineralized and deaerated, the internal chemicals are added to prevent scale and corrosion inside the boiler. The steam and condensate systems are treated with different chemicals to prevent corrosion by carbonic acid and pitting by oxygen.

Chemicals used in cooling systems are quite different from those used in water preparation or in the boiler system. Open cooling circuits often receive whatever water is available or that needs to be concentrated or reused prior to disposal. Many cooling systems are made up of combinations of water sources that can vary with seasonal changes—lakes, rivers, wells, and plant process waters and wastewaters. Treatment programs usually contain a number of different chemicals, all added to the circuit at the same time.

WATER PREPARATION

Chemicals can be classified by their application in various parts of the plant. As the following paragraphs explain, each application requires different chemicals and treatment technologies. Also see Fig. 4.1.

Clarification and Softening Chemicals

Clarification and softening remove suspended solids and color from the water (Table 4.1). The softening process also reduces hardness, alkalinity, and some

TABLE 4.1 Common Clarification and Softening Chemicals

Chemical	Formula	Equivalent weight	Application
Aluminum sulfate (alum)	$Al_2(SO_4)_3 \cdot 18H_2O$	111.0	Coagulant, removes suspended solids and color
Ammonium alum	$Al_2(SO_4)_3 \cdot (NH_4)2SO_4 \cdot 24H_2O$	151	Coagulant, removes suspended solids and color
Bentonite	Varies	Varies	Flocculation aid
Calcium chloride	$CaCl_2$	55.5	Alkalinity reduction
Calcium hydroxide (hydrated lime)	$Ca(OH)_2$	37.1	Softening, reduces Ca and Mg
Calcium hydroxide magnesium oxide (bolomitic lime)	$Ca(OH)_2 \div MgO$	Varies	Softening and silica removal
Calcium oxide (burnt lime)	CaO	28.0	Softening, reduces Ca and Mg
Calcium sulfate (gypsum)	$CaSO_4 \cdot 2H_2O$	86.1	Alkalinity reduction
Ferric chloride	$FeCl_3 \cdot 6H_2O$	91.0	Coagulant, removes suspended solids and color
Ferric sulfate	$Fe_2(SO_4)_3$	66.7	Coagulant, removes suspended solids and color
Ferrous sulfate (copperas)	$FeSO_4 \cdot 7H_2O$	139	Coagulant, removes suspended solids and color
Magnesium oxide	MgO	20.2	Silica reduction
Sodium aluminate	$Na_2Al_2SO_4$	82	Hot or cold process coagulation
Sodium carbonate (soda ash)	Na_2CO_3	53	Softening, reduces Ca and Mg
Sodium hydroxide (caustic soda)	NaOH	40	Softening, reduces Ca and Mg
Sodium silicate (activated silica)	Varies		Flocculant, promotes floc formation

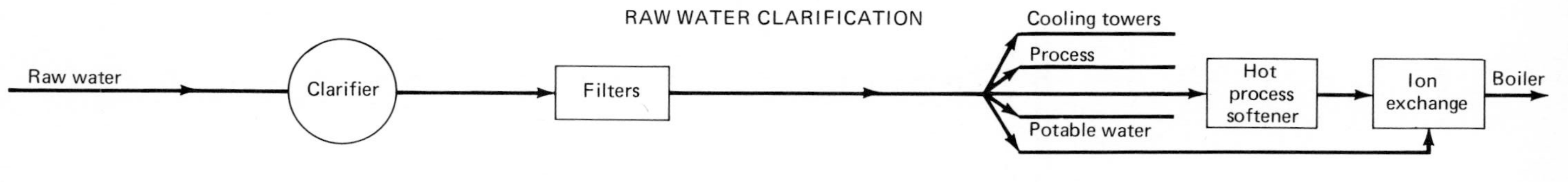

FUNCTION	Suspended solids removal			Direct filtration. Suspended solids removal without prior clarification		Filter aid Improves filter efficiency		Filter cleaners Clean organic flocculants from filters	Hot process softening Dissolved hardness and suspended solids removal		Ion exchange The exchange of unwanted ions for wanted ions	
EFFICIENCY (Typical)	Influent 5-200 NTU Effluent 2-5 NTU			Upflow 5-10 NTU 0-5 NTU	Gravity-downflow 5-10 NTU 0-5 NTU	3 NTU 0.5 NTU		60-90%	Influent 150 ppm hardness Effluent 20 ppm hardness		99.9%	
GENERIC NAME	Coagulants	Clay	Flocculants	Coagulants	Coagulants	Coagulants	Flocculants	Cleaners	Flocculant	Sodium aluminate	Resin cleaner	Regenerant

COLD LIME SOFTENING

Raw water	Cold process softener			Filters			
						Process Cooling towers Potable water Ion exchange	
FUNCTION	Dissolved hardness and suspended solids removal			Filter aid Improves filter efficiency; extends life; minimizes plugging	Filter cleaner Clean scale from filters	Ion exchange The exchange of unwanted ions for wanted ions	
EFFICIENCY (Typical)	Influent 150 ppm hardness Effluent 35-50 ppm hardness			In 3 NTU Out 0.5 NTU	80-95%	99.9%	
GENERIC NAME	Inorganics Sodium aluminate	Coagulants	Flocculants	Coagulants	Cleaners	Cleaner	Regenerant

FIG. 4.1 Treatment chemicals for water preparation.

silica. After clarification, the water may be used as is for makeup to the cooling system. If the water has been suitably disinfected such as by chlorination, it may be used for potable purposes. Cooling systems will have a separate chemical treatment program to prevent corrosion, deposition, and biological problems.

Chemicals may be added to the water after the initial conditioning step to minimize corrosion in the piping before the water reaches its final destination. (This is especially important for desalinated water because of its highly corrosive nature.) If the water is also used for potable service, any chemicals added must be safe for drinking and must not contribute scale-forming materials to steam generators.

For boiler feed, the clarified or softened water will require additional treatment to remove scale-forming salts, carbonate alkalinity, and dissolved oxygen. However, after all the pretreatment, the boiler water and steam may require additional chemicals to prevent the last remaining impurities from depositing on hot surfaces and corrosive gases from corroding steam lines.

Defining Clarification. *Clarification* is the process of removing suspended solids, turbidity, color, and colloidal materials from large volumes of water. The process uses coagulation, flocculation, and sedimentation to rapidly settle out the solids. Each step in the process is distinct and requires special attention to produce the desired results.

During the initial stages of clarification, solid particles coming in with the raw water bear electrostatic charges which keep them suspended in the water. *Coagulants* are chemicals that are added to the water to break down or neutralize the charged layers surrounding the particles. Coagulation, therefore, is the process of overcoming the repulsive charge on suspended particles so that they can come closer to each other.

Salts of highly cationically charged metal ions, such as alum (aluminum sulfate) and iron, or organic cationic polymers are frequently used as coagulants to neutralize the charged particles. Under proper water conditions they may also hydrolyze to form metal hydroxides. The hydroxide produces a floc that traps particles. During flocculation, the destabilized particles agglomerate and form large particles.

Each stage in the clarification process has an optimum time to produce the best results. During coagulation, the chemicals are rapidly mixed into the water to get complete distribution and charge neutralization. Slower mixing is used during flocculation to allow floc to grow uniformly and agglomerate. Finally the water reaches a slow-moving stage, and the floc is allowed to settle in a sedimentation basin.

The clarification process achieved with metal salts can usually be improved by the addition of polyelectrolytes (Table 4.2). In some systems these highly charged cationic polymers may be used as the primary coagulants. By partial or total re-

TABLE 4.2 Polyelectrolytes Used in Water Conditioning

Polyelectrolytes	Molecular weight	Application
Polyacrylamide	1 to 10 million	Nonionic flocculant
Polyacrylamide (anionic)	1 to 10 million	Anionic flocculant
Poly(DADMAC)	50,000–150,000	Cationic primary coagulant, coagulant aid, and flocculant

placement of metal salts, the polymers neutralize particle charges without adding more solids to the water. When fewer solids are used for coagulation, there is less sludge to dispose of and less dissolved solids remain in the water that must be removed to make boiler feedwater. Other organic polymers which are nonionic or slightly anionic can act as coagulant aids or flocculants by bridging between floc particles. Since the floc settles faster, sedimentation rates increase significantly. Proper clarification requires on-site testing of the raw water to produce the quantity and quality of water needed.

Defining Softening. *Softening* is the process of precipitating hardness, alkalinity, silica, and other materials by adding lime, soda ash, and/or sodium aluminate.

Lime Softening. Calcium hardness can be removed from water by the addition of lime, lime and soda ash, or lime and sodium aluminate. Calcium is precipitated as the calcium carbonate, $CaCO_3$, salt:

$$\underset{\text{[Lime]}}{Ca(OH)_2} + \underset{\text{[temporary hardness]}}{Ca(HCO_3)_2} \rightarrow 2CaCO_3 + 2H_2O$$

At high pH (about 10), $CaCO_3$ may carry a negative charge which resists flocculation; therefore, a coagulant such as sodium aluminate can be added to enhance precipitation.

Partial Lime Softening. The amount of calcium removal by lime depends on the carbonate alkalinity in the water. If the water contains more calcium than alkalinity, all the bicarbonate alkalinity will be removed or partially converted to carbonate alkalinity. If the calcium is less than the alkalinity, a large reduction in the calcium concentration can be achieved. Some of the magnesium can also be removed by lime:

$$Mg^{++} + \underset{\text{[Lime]}}{Ca(OH)_2} \rightarrow Mg(OH)_2 + Ca^{++}$$

Complete Lime Softening. Additional calcium can be removed by using soda ash. The reaction is

$$Ca^{++} + \underset{\text{[Soda ash]}}{Na_2CO_3} \rightarrow CaCO_3 + Na^{+}$$

The practical limit of calcium removal by this method is 20 to 35 mg/L.

Warm- and Hot-Process Softening. The softening process can be improved by heating the water. The reaction rates increase, and the solubilities of $CaCO_3$ and $Mg(OH)_2$ are much less. In hot-process softening, it is possible to reduce calcium to 8 mg/L and magnesium to 2 to 5 mg/L.

Other Water Conditioning Chemicals

Magnesium compounds such as magnesium oxide, magnesium sulfate, magnesium carbonate, or dolomitic lime are used to reduce silica. For highly alkaline waters, where alkalinity exceeds the hardness, gypsum (calcium sulfate) or calcium chloride may be used to reduce the alkalinity. Chlorine and potassium permanganate may be introduced to remove iron and other heavy metals.

BOILER TREATMENT

Boilers are extremely sensitive to deposits in high-heat-transfer zones; even the slightest insulating scale can cause rapid overheating and tube failures. The high water temperatures also make the boiler systems susceptible to corrosion by dissolved oxygen, organic-chemical-breakdown products, acids, and excess caustic. Proper treatment of boilers begins with good water preparation and sound operational practices. Chemicals are used to control trace impurities that enter the boiler with the feedwater. Unchecked, these impurities cause deposition or corrosion. As feedwater quality deteriorates or mechanical equipment fails, chemicals become less and less effective, and treatment programs must be changed to meet the new conditions.

Chemicals are applied in three general locations: the preboiler system (chemical deaeration), inside the boiler (internal treatment), and the condensate system. See Fig. 4.2. A complete program must be carefully selected to match the boiler system pressure and feedwater and operating conditions.

Pressure and Water Quality Considerations

Internal boiler treatment is usually determined on the basis of the operating pressure. Low-pressure systems, < 350 psig (2400 kPa gage), can use a variety of

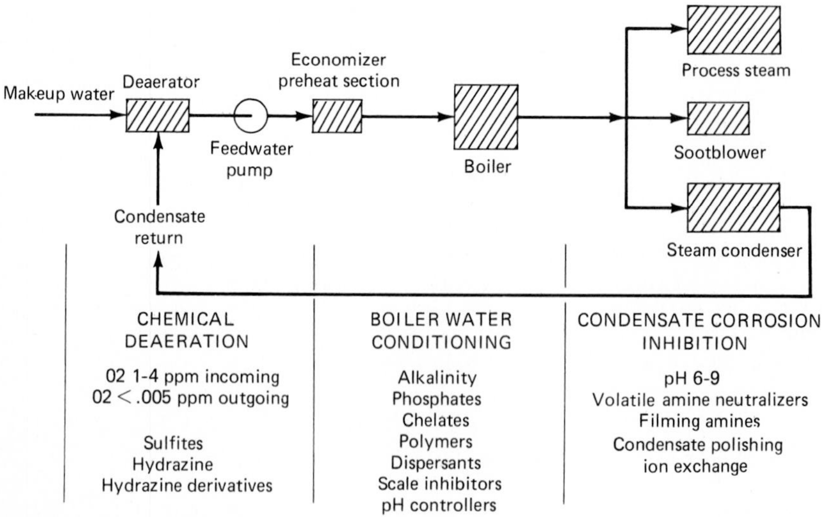

FIG. 4.2 Key boiler-treatment chemicals.

treatment chemistries ranging from coagulation (hardness precipitation and sludge conditioning) to hardness chelation and coordinated phosphate programs. Intermediate-pressure boilers of 350 to 1000 psig (2400 to 6900 kPa gage) can be treated with chelates, coordinated phosphate, congruent phosphate, or all-organic programs. As pressures increase, boilers become less tolerant of dissolved solids in the water. High-pressure systems of 1000 to 2000 psig (6900 to 13 800 kPa gage), are often treated with coordinated phosphates or *all volatile treatment* (AVT). In very-high-pressure plants of 2400 to 3200 + psig (16 560 to 22 080 kPa gage), supercritical programs have to be AVT.

Water quality plays an important role in determining which program will provide the best results. When fresh makeup water is added to replenish losses from steam and condensate leaks and boiler blowdown, impurities can come into the boiler. If hardness coming in with the makeup is greater than 1 mg/L (as $CaCO_3$) and the boiler is low-pressure, then a sludge-dispersant or chelate program should be used. When the hardness is a trace to 1 mg/L (as $CaCO_3$) and the boiler pressure is low to intermediate, then an all-organic, chelate, or coordinated phosphate program can be used. For systems with very low makeup and high-quality condensate return, a coordinated phosphate or all-organic program is preferred. An important consideration in all boiler treatment applications is that the chemicals applied should be harmless to feedlines and boiler internal parts. A mismatched or poorly applied program can cause considerable damage.

Chemical Deaeration

There is always some residual oxygen in untreated feedwater. It comes from makeup water added to the deaerator or from leaks in the suction side of feedwater pumps. If dissolved oxygen gets into the feedwater, it reacts with the primary passive-surface magnetite (Fe_3O_4) to form a nonprotective hematite (Fe_2O_3), which allows pitting to occur in the economizer and/or feedwater heaters.

Boilers operating below 1000 psig (6895 kPa gage) may be treated with catalyzed sodium sulfite to scavenge or remove traces of oxygen. A sulfite residual is maintained in the boiler to remove any remaining unreacted oxygen from the hot feedwater. At high boiler pressures, the sulfites break down to corrosive gases and are no longer used (Table 4.3).

Hydrazine, one of its derivatives, or certain volatile organics can be used to control the effects of oxygen in high-pressure systems. The primary reaction of these agents in the boiler is to form and repair the magnetite surface. In this application, hydrazine contributes no dissolved solids to the system. Since hydrazine has potential health problems associated with handling the chemical, substituted hydrazines or volatile organic reducing agents may be used instead.

Internal Boiler Programs

Scale and corrosion are potential problems for all boiler systems. Chemicals added to the feedwater or fed directly to the boiler are called *internal treatments* (Table 4.3).

Corrosion Control. Boiler corrosion can be minimized by maintaining high pH in the boiler water. The method for controlling pH is to adjust incoming carbonate

TABLE 4.3 Internal Treatment

Typical chemicals	Formula or abbreviation	Recommended application	Limitations
Antifoam	Synthetic organics	High solids waters or when foaming occurs	Overfeed may cause foaming
Crboxymethyl-cellulose	CMC, natural organic	Dispersant, deposit conditioner	Difficult to monitor, may form carbonaceous deposits
Caustic	NaOH	Alkalinity addition for pH control. Coagulation program in low-pressure boilers	Adds free OH alkalinity and may cause caustic gouging, embrittlement, or corrosion
Chelates	NTA, EDTA	Sequester hardness in low-solids, low-hardness boiler water. For low- to medium-high pressures	Becomes corrosive in presence of dissolved oxygen and at high concentrations
Disodium phosphate	Na_2HPO_4, DSP	pH control in coordinated phosphate programs	Precipitates hardness and may form scale on heater tubes
Ethylenediamine tetraacetic acid	EDTA	Sequester hardness in low-solids, low-hardness boiler water. For low- to medium-high pressures	Becomes corrosive in presence of dissolved oxygen and at high concentrations
Hexamethylenediamine-tetraphosphonate	HMDA	Control hardness scale in low-hardness boiler water	Difficult to monitor
Hydrazine	N_2H_4	Metal-surface passivator and oxygen control. Medium- to high-pressure boilers and AVT programs	Slow-acting at low temperatures. Carcinogenic vapor
Hydroxyethliden-diphosphonate	HEDP	Control hardness scale in low-hardness boiler water	Difficult to monitor
Lignins	Natural organics	Control sludge in coagulation programs for low-pressure boilers	May cause carbonaceous deposits
Monosodium phosphate	NaH_2PO_4, MSP	pH control in coordinated phosphate programs	Precipitates hardness and may form scale on heater tubes
Nitrilotriacetic acid	NTA	Sequester hardness in low-solids, low-hardness boiler water. For low to medium-high pressures	Become corrosive in presence of dissolved oxygen and at high concentrations
Polymers	Synthetic organics	Control hardness scale, iron oxides. Sludge dispersant	Difficult to monitor
Starch	Natural organic	Control oil in coagulation programs at low pressures	Forms carbonaceous deposits
Soda ash	Na_2CO_3	Precipitate calcium in low-pressure coagulation program. Alkalinity control	Forms heavy scale if not conditioned and dispersed. Adds CO_2 to condensate
Sodium hydroxide	NaOH	Alkalinity addition for pH control. Coagulation program in low-pressure boilers	Adds free OH alkalinity and may cause caustic gouging, embrittlement, or corrosion
Sulfite	Na_2SO_3	Oxygen scavenger	Adds solids to boiler water. May cause acidic condensate
Tripolyphosphate	$Na_5P_3O_{10}$, STPP	Phosphate and dispersant source for coagulation programs	Precipitates hardness and may form scale on heater tubes
Trisodium phosphate	Na_3PO_4, TSP	pH control in coordinated phosphate programs	Precipitates hardness and may form scale on heater tubes

or hydroxide alkalinity. The alkalinity may be present in the makeup water or added in the form of caustic or basic sodium phosphates. The type and quantity of alkalinity allowed in the boiler depend on the operating pressure and quality of feedwater. Note that alkalinity must be controlled with care to prevent possible scaling and corrosion damage due to excessive caustic in the boiler.

Scale Control. Trace quantities of mineral salts and other substances can accumulate inside the boiler and eventually form deposits on high-temperature surfaces. Typical contaminants are calcium, magnesium, iron, silica or silicates, phosphates, sulfates, oils, and organic elements. These contaminants may be removed before the water enters the boiler, or they must be treated inside. Chemicals that work inside the boiler are called internal treatments.

Precipitation and Sludge Conditioning Treatments. Hardness and alkalinity in the feedwater concentrate rapidly inside the boiler. Without treatment they will form a hard insulating scale. Phosphates and sludge conditioning polymers can be used to precipitate the hardness as a soft fluid sludge of basic calcium phosphate and magnesium hydroxide.

Precipitation programs require relatively heavy blowdown to hold down the sludge accumulation. However, not all the sludge is removed by blowdown, and the boiler must be cleaned periodically to remove tube deposits. Precipitation programs are suitable only for low-pressure boilers.

Chelate Programs. Certain organic chemicals, called *chelates,* have the ability to stabilize hardness against precipitation at the elevated temperatures inside the boiler. Chelates such as NTA (nitrilotriacetic acid) and EDTA (ethylenediaminetetraacetic acid) form soluble complex ions with calcium and iron. Chelates may require supplemental polymer dispersants to prevent nonchelatable magnesium and silicate scale.

A well-controlled chelate program can produce very clean boiler surfaces. Since the chemicals are applied with a slight excess of chelate, careful monitoring is necessary to prevent overfeeding chemical and producing chelate attack in feedwater lines, economizers, steam drums, and separators. In typical boiler system conditions, an overfeed can cause corrosion by dissolving the protective magnetite surface layers. Underfeeding the chemicals may result in scale formation. Chelate programs are also sensitive to trace quantities of oxygen in the feedwater. The chelates are rapidly oxidized to aggressive organic acids which corrode metal surfaces.

Coordinated Phosphate Programs. High-pressure boilers with high heat-transfer rates require feedwater with little or no feedwater solids. As pressures increase, the boiler becomes susceptible to caustic attack by free caustic alkalinity. To maintain high boiler-water pH without adding free caustic, phosphate chemicals can be used in a special coordinated phosphate program. The application introduces disodium phosphate in combination with tri- or monosodium phosphate to adjust pH without forming free hydroxide ions.

Coordinated phosphate programs must be monitored consistently and with great care. Successful operation depends on a reliable source of extremely pure feedwater. Traces of hardness can precipitate the phosphate and upset the pH/free-caustic balance. Another source of concern is the proper ratio of sodium to phosphate. An imbalance can affect the solubility of sodium phosphate and cause "phosphate hide-out," or even deposition of phosphate salts.

All-Organic Treatment. Recent developments in polymer technology have produced boiler treatment programs as effective as chelates without the application problems typically associated with chelates. The polymers solubilize hardness and disperse solids to keep generating tubes clean. They also act as mild reducing agents to maintain the magnetite surface intact.

All-Volatile Treatment. The *all-volatile treatment* (AVT) program is especially designed to control boiler corrosion and carry-over deposition in superheaters and turbines and in once-through boilers. The program uses only hydrazine, ammonia, or other neutralizing amines to regulate boiler water and steam pH; no dissolved solids are added to the system. However, the program requires ultra-high-purity feedwater and precise boiler control to minimize pH excursions that could be caused by feedwater contamination.

Condensate Treatment

Steam carries corrosive gases such as carbon dioxide and oxygen through the steam lines to traps, condensers, receivers, etc. Wherever steam condenses, carbonic acid and dissolved oxygen form highly aggressive solutions. Treatment for preventing condensate problems includes decarbonization and deaeration of the feedwater, pH control with ammonia and volatile organic amines, and filming metal surfaces with hydrophobic oils (Table 4.4).

The treatment program is selected on the basis of system operation. Volatile amines can be used to regulate condensate pH in medium- to high-pressure turbine plants. Volatile amines or ammonia and hydrazine may be paired with AVT programs, but care must be taken to minimize possible damage to copper and admiralty condensers. Filming chemicals are useful for protecting iron surfaces

TABLE 4.4 Condensate Inhibitors

Typical chemicals	Formula or abbreviation	Typical application	Limitations
Ammonia	NH_3	Volatile gas, pH control. Use with AVT programs	Aggressive to copper and copper-alloyed condensers
Cyclohexylamine	$C_6H_{11}NH_2$	Volatile organic amine for pH control. Low- to high-pressure systems with extensive distribution	Strips out during deaeration. May meet FDA specifications
Diethylaminoethanol (DEAE)	$(C_2H_5)_2 \cdot NCH_2 \cdot CH_2OH$	Volatile organic amine for pH control. Low- to high-pressure systems with extensive distribution	May meet FDA specifications
Hydrazine	N_2H_4	Volatile gas, pH control, oxygen scavenging. Use with AVT programs. Condensate oxygen scavenger	Breakdown products are aggressive to copper and copper-alloyed condensers
Morpholine	C_4H_8ONH	Volatile organic amine for pH control. Medium- to high-pressure systems with limited distribution	High solubility limits range in steam lines. FDA-approved
Octadecylamine		Surface filmer, controls oxygen corrosion. Medium- to high-pressure systems. Chemical added to steam lines	May cause organic deposits in traps and dead lines

against oxygen attack, but their detergent action can remove iron oxides and create organic/oxide deposits.

COOLING-WATER TREATMENT

Cooling systems provide large quantities of heat-transfer fluid to condensers and process equipment. The fluid, usually water, is corrosive to most metals used in transmission lines and many heat exchangers. The water also contains dissolved minerals and gases, suspended solids, process contaminants, and biological matter. Without treatment of the water, there may be a loss of cooling, plugged tubes, destruction of expensive equipment, and serious process contamination.

Treatment Programs

Treatment programs for cooling systems are generally added at the tower basin or the system reservoir. See Fig. 4.3. The chemicals used are divided into three basic groups: corrosion inhibitors, scale inhibitors, and biocides. Complete treatment programs include chemicals from each group. Products for cooling systems obtained from water treatment suppliers often contain blends of chemicals from one or more groups. However, whatever the final combination of products used in the system, it should cover all groups.

Treatment programs vary according to the type of cooling system and water source. Cooling systems are typically open recirculating (evaporative cooling), once-through, or closed circuits. Since all systems are subject to corrosion by water and its dissolved solids and oxygen, they must be protected by corrosion inhibitors. Each system has special treatment requirements tailored to its operation.

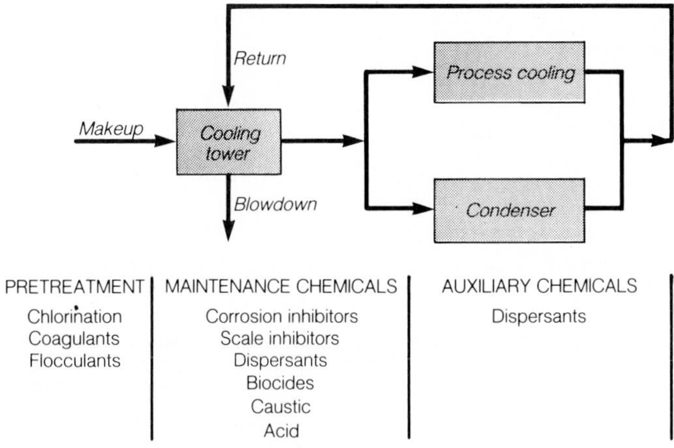

FIG. 4.3 Key cooling-water chemicals.

Systems Treated. *Open systems* are the most complicated because they cool by evaporation, which causes the dissolved and suspended solids to concentrate until they exceed the solubility limits of certain minerals. The systems must be treated with scale inhibitors and dispersants to prevent insulating scale from depositing on heat-transfer tubes. Biocides are required to control organic fouling and corrosion from bacterial activity.

Although *once-through systems* have less tendency to corrode or scale up than evaporative systems, some treatment may be required to keep them operating properly. For treatment to be effective, a minimum amount of chemical is required. However, the proper treatment may be too costly because once-through cooling uses very large quantities of water. There are no evaporation or significant water losses in *closed systems,* so they use much less water. Often they are filled with high-quality softened or demineralized water and require only corrosion and biological control.

Cooling systems consume large volumes of water. The makeup may consist of river, lake, or well water, seawater, desalinated or demineralized water, waste or process water, etc. Every water source has its own type of contamination and corrosivity which requires special consideration. Moreover, when the cooling system blowdown is discharged, there may be environmental regulations that restrict the chemicals used in treating the water. The treatment program must be safe to use and technically sound, to produce the desired results.

Treatment Stages. There are several stages of treatment during the normal life of a cooling system. During start-up, the system is prepared by flushing to remove debris left from construction. Then it is degreased and cleaned to remove oil and mill scale and passivated to prevent rapid initial or "flash" corrosion. Cleanup and passivation should be completed before the system comes on-line.

When the system is operational, it is treated with a maintenance program. Start-up of the maintenance program involves pretreating or "high-leveling" the regular product to satisfy the "system demand" for the corrosion inhibitors. After a period of 3 or 4 days, as determined by the uptake of chemicals, the treatment is added at normal use dosages.

Under normal operating conditions, treatment chemicals should be added on a continuous basis. However, during upsets, such as water-source changes, process leaks, airborne contamination, acid spills, etc., the program may require supplemental dispersants, corrosion inhibitors, or even a complete revision in the treatment chemicals.

Corrosion Control Chemicals

Corrosion is a chemical process where a metal reacts with water and reverts to its natural oxide state. Note: All metals used to fabricate cooling systems can corrode by pitting. In some cases if the metal is covered by a deposit, corrosion by differential aeration or bacteria may be so rapid that tubes or walls are perforated in a matter of days.

Minimizing Corrosion. Some common methods for minimizing corrosion of metals in contact with water are coatings (epoxy or rubber), plating (galvanizing with zinc), sacrificial anodes (magnesium or zinc), and chemical additives. Each method is used for specific applications. For example, coatings are often used to protect water transmission lines or pipes. Galvanizing prevents corrosion in small sheet-metal systems such as air conditioners. Sacrificial anodes protect individual condensers and exchangers with a reactive metal electrically attached inside the shell of the unit. Chemicals are added to the water to reduce its corrosivity and/or develop a protective barrier on the metal surface.

Two Reactions. The corrosion of metals requires at least two reactions (electrochemical half-reactions) to describe the process. One reaction is the loss of metal (anodic) and is typically shown as

$$Fe \rightarrow Fe^{++} + 2e^{-} \qquad \text{(anode)}$$

The Fe^{++} is soluble in water and leaves the metal surface. Once it is lost, the metal cannot be recovered, and it often deposits as iron oxide on heat-transfer surfaces. The other reaction is a reduction reaction where something accepts electrons from the metal (cathodic). This is typically shown as the conversion of dissolved oxygen to hydroxide:

$$\tfrac{1}{2}O_2 + 2e^{-} + H_2O \rightarrow \tfrac{1}{4}OH^{-} \qquad \text{(cathode)}$$

Corrosion inhibitors are designed to minimize metal losses by affecting both anodic and cathodic reactions. The function of a chemical as an anodic or a cathodic inhibitor is determined electrochemically; however, its application is governed by combinations with other chemicals or alteration of the water chemistry. See Table 4.5.

Alkaline Water Treatment

The corrosivity of water can be reduced significantly by increasing pH while maintaining calcium carbonate in solution. Special indices (Langelier saturation and Ryznar stability) have been developed to relate water corrosivity to the solubility of calcium carbonate. According to the indices, corrosion decreases as the pH increases until calcium carbonate starts to precipitate.

Calcium carbonate has a very low solubility in neutral-pH water. When acid is added to water, the pH drops and the carbonate is converted to bicarbonate. Calcium bicarbonate is very soluble. When the pH increases, as in a cooling tower where air scrubs carbon dioxide out of the water, the bicarbonate forms carbonate ions. When sufficient carbonate is present, calcium carbonate precipitates:

$$\underset{\substack{\text{Soluble} \\ \text{(low pH)}}}{Ca^{++} + 2HCO_3^{-}} \rightarrow \underset{\substack{\text{insoluble} \\ \text{(high pH)}}}{CaCO_3} + CO_2 + H_2O$$

If calcium carbonate precipitates, scale forms, reducing heat-transfer rates and increasing corrosion rates. However, by stabilizing the water with special phosphates or organic compounds, the normal solubility limit can be exceeded without scale formation. Corrosion rates are then reduced to very low levels.

Some programs use stabilized zinc and chromate at high pH to help protect

TABLE 4.5 Corrosion and Scale Inhibitors

Typical chemicals	Formula or abbreviation	Recommended application	Limitations
Aminotri (methylene-phosphonic) acid	$N(CH_2PO_3H_2)_3$, AMP	Control calcium scales at high pH. Partially inhibits corrosion of steel	Can decompose during chlorination. High calcium may form deposits with chemical
Benzotriazole	BZT	Copper corrosion inhibitor. Used with copper and alloyed condensers. Formulated in products	Expensive but effective at low concentrations
Boric acid	H_3BO_3	Combined with other inhibitors for closed-system pH control. Helps remove organic deposits	Absorbs radiation, avoid use in nuclear power plants
Caustic	NaOH	Increase pH and alkalinity in soft-water systems	Very difficult to control. May cause scale formation at high pH
Chromate	CrO_4	Opened- or closed-system corrosion inhibitor. Protects most metals at appropriate dosages	Highly toxic to aquatic life. Severely limited or prohibited for environmental reasons. May be corrosive at wrong dosage
Hexametaphosphate	$Na_6P_6O_{18}$	Corrosion inhibitor for once-through and potable systems. Pretreatment and passivation chemical	Breaks down to ortho phosphate which can form deposits. May be environmentally restricted
1-hydroxyethane-diphosphonic acid	$CH_3C(OH) \bullet (PO_3H_2)_2$, HEDP	Control calcium scales at high pH. Partially inhibits corrosion of steel	Can decompose during chlorination. High calcium may form deposits with chemical
Lignosulfonate	Natural organic	Dispersant for iron oxides and silt	Reacts with cationic biocides. May cause corrosion if overfed. Biological nutrient
Lime	$Ca(OH)_2$	Adds calcium and alkalinity to control corrosion in soft water	Very difficult to feed and control. May cause scale if not used properly
Mercaptobenzotriazole	MBT	Closed-system corrosion inhibitor for copper and copper-alloyed exchangers	Easily decomposed in oxygenated waters
Molybdate	$Na_2MoO_4 \bullet 2H_2O$	Nontoxic corrosion inhibitor mostly used in closed systems	Expensive chemical, requires high concentrations to protect steel
Nitrate	$NaNO_3$	Mild corrosion inhibitor found in closed systems with boron-nitrite treatments	Not usually added to systems to inhibit corrosion

Nitrite	$NaNO_2$	Closed-system corrosion inhibitor for steel and mixed metal systems. Normally used with borates	Biological nutrient, forms corrosive solutions when attacked by bacteria. May form carcinogenic compounds
Phosphate	Na_3PO_4	Open, closed, once-through, and potable-water system corrosion inhibitor. Pretreatment or passivation chemical	Ortho phosphate readily forms deposits without dispersants. Environmental restrictions may apply
2-phosphonobutane-1,2,4-tricarboxylic acid	PBTC	Controls calcium scales at high pH. Partially inhibits corrosion of steel	Difficult to monitor
Polyacrylic acid	PAA	Typical polymeric material for dispersing suspended solids in open systems	Numerous polymer variations may have specialized applications
Polymaleic anhydride	PMA	Polymeric material for dispersing suspended solids in open systems	Numerous polymer variations may have specialized applications
Polyolester	POE	Organic phosphate for controlling calcium scales in open systems	Requires good biological control
Polyphosphate	$K_4P_2O_7$	Corrosion inhibitor for once-through and potable systems. Pretreatment and passivation chemical	Breaks down to ortho phosphate which can form deposits. May be environmentally restricted
Silicate	$Na_2O \cdot 2SiO_3$	Corrosion inhibitor for once-through and potable systems	Slow rate inhibitor can form deposits if overfed
Soda ash	Na_2CO_3	Alkalinity and pH control for soft waters. Provides corrosion protection when calcium is present	May cause calcium carbonate precipitation if not controlled properly
Sulfuric acid	H_2SO_4	Controls pH and alkalinity	Very corrosive and dangerous to handle. Very good control is necessary to prevent damage
Tannin	Natural organic	Dispersant for iron oxides and silt	
Tolyltriazole	TT	Copper corrosion inhibitor. Used with copper and alloyed condensers. Formulated in products	Expensive but effective at low concentrations
Zinc	Zn	Corrosion inhibitor for open, closed, once-through, and potable-water systems	Heavy metal, toxic to fish but not humans. Forms deposits if not carefully controlled

against pitting. All alkaline treatments require good control of chemicals and pH and alkalinity.

Phosphate Treatment

Phosphate chemicals have been used to control corrosion and deposition in water systems for many decades. They prevent corrosion of mild steel surfaces and the precipitation of soluble iron and manganese in once-through and potable-water lines.

Phosphates prevent corrosion by forming a corrosion-resistant barrier on the metal surface. The barrier may be a thin film containing various combinations of phosphate, iron oxides, hardness, and alkalinity. If iron is exposed to water without treatment, it will rapidly corrode. The corrosion products are nonprotective and with time will build up an undesirable oxide scale. Phosphates help control the corrosion products and allow the protective barrier to develop.

When they are used in recirculating cooling systems, phosphates are effective as long as they do not precipitate as calcium phosphate scale. The potential for scale formation is present when the phosphate concentration exceeds 3 to 5 ppm in the water. High skin temperatures and low flow rates in heat exchangers also promote scaling. Any program containing phosphates should have a strong dispersant to control calcium phosphate deposition.

Phosphates may be used alone or in combination with heavy metals to prevent corrosion. Polyphosphates and phosphate-zinc combinations can be used in potable-water systems. Phosphates combined with chromates reduce the amount of heavy-metal chemicals used in the water.

Heavy-Metal Treatment

Chemicals such as chromates and zinc are classified as heavy metals. In the water treatment industry, the classification implies that they may be toxic or restricted in some way. Technically, molybdates are also heavy metals but are considered nontoxic to fish or the environment. Note: Since large quantities of cooling water may be blown down from the system, any application of chromates or zinc should be checked for compliance with local and federal environmental restrictions.

Chromates. Chromates are the most widely accepted corrosion inhibitors for cooling systems. They provide the best corrosion protection for most metals and are very easy to apply. Chromates are oxidizing agents that cause metals to rapidly form a strong protective oxide layer. If chromates are used alone, they must be applied at very high concentrations to overcome pitting tendencies. Such applications are usually limited to closed systems. Any blowdown must be treated before disposal.

To reduce the amount carried in the water, chromate is usually blended with zinc and/or phosphate. Severe discharge limitations or laws may prohibit the use of chromates at any concentration.

Zinc. Zinc-based treatments are frequently applied at high pH by using chemicals with stabilizers to prevent precipitation of $Zn(OH)_2$. The treatments require good control to achieve the desired results. The chemicals prevent corrosion at high pH by combining the lower corrosivity of the water with zinc's ability to precipitate a protective hydroxide at the metal cathode where the localized pH is significantly higher.

Zinc chemicals such as $ZnCl_2$ should never be overlaid or simply added on top of an "alkaline-side" program (i.e., pH > 7.0). Zinc must be prestabilized in a concentrated form to prevent precipitation. Specially formulated products for potable-water systems may contain zinc and/or phosphate. When it is applied properly, zinc is considered safe for human consumption. However, there may be environmental restrictions regarding the discharge of zinc-containing effluent to commercial fishing waters.

Zinc and magnesium metals are also used as sacrificial anodes to prevent corrosion on surfaces which are electrically connected to the anode. The effective range of sacrificial anodes is limited to the equipment where they are installed. They are generally used for extra protection of special equipment.

All-Organic Treatment

In recent years, environmental pressures have caused industrial water users to seek safer chemicals for their cooling systems. Discharge regulations may restrict heavy metals, phosphates, *biological oxygen demand* (BOD), *chemical oxygen demand* (COD), biocides, pH, and/or dissolved solids.

All-organic chemical programs have been developed to help plants cope with effluent restrictions. The chemicals may contain various combinations of organic phosphates, anionic polymers, surfactants, copper-alloy inhibitors, and filming agents. In most cases the organic chemicals alone are not sufficient corrosion inhibitors to provide complete protection. Usually they are used in systems that can operate at high pH and that have enough hardness and alkalinity in the water to lower the corrosivity.

Deposit Control Chemicals

The control of deposits is critical to producing good results in any cooling system. Deposits are formed from one or more of the following: precipitating mineral scale, tubercles and migrating iron oxides, settling solids from suspended dirt and organic matter, leaking process materials, equipment oil, and biological slime. Deposits reduce the heat-transfer efficiency of condensers and exchangers, and if the deposits are not removed, they can cause serious damage to metals through localized corrosion or under-deposit pitting. Under-deposit corrosion by pitting is rapid and uncontrollable. Deposits can destroy all metals including expensive corrosion-resistant alloys.

Scale Inhibitors. During the cooling process, the recirculating water concentrates by evaporation. The normally soluble salts reach their saturation point and begin to precipitate. If the system is not treated, insoluble calcium and silicate minerals will form hard, insulating scale on heat-transfer surfaces. Typical scales are calcium carbonate, calcium phosphate, calcium sulfate, silica, and magnesium silicate.

Inhibitors prevent scale by interfering with the regular crystalline growth of the minerals precipitating in concentrated cooling water. Most scale control chemicals are organic phosphates or synthetic polymers. See Table 4.5.

Dispersants. Suspended solids, dust, and fine particles settle out in low-flow areas. The solids also adsorb treatment chemicals and interfere with corrosion inhibition and scale control. Most dispersants are negatively charged synthetic polymers or natural organics. These chemicals impart a charge to the surface of particles, which causes them to repel one another and reduces or eliminates settling. Dispersants are often combined with scale inhibitors to improve deposit control. The combination is useful when the system is operating at or above the mineral solubility limits and there are suspended solids in the water.

Antifoulants. Cooling-water deposits may contain complex mixtures of scale, sediment, biological matter, and/or oil. Sometimes even the corrosion inhibitor contributes to the deposit.

Antifoulants are chemicals that penetrate and loosen the organic component and emulsify the oily layers. They are usually nonionic or cationic polymers or alkyl amines.

Biological Control Chemicals

The cooling-water environment supports and promotes the growth of bacteria, algae, and fungal masses. Other creatures such as clams can also take up residence in the cooling circuit and in condenser head boxes and tubes. Without effective controls, biomasses can cause severe pitting-type corrosion, loss of heat transfer, plugging, and fluid friction (increased pumping energies).

Biocides. Biocides are chemicals that kill living organisms in cooling systems. They can be broadly classified as *oxidizing* or *nonoxidizing*. The former type destroys the organism by oxidizing the organic-cell walls. This biocidal mechanism cannot be defeated by the organism becoming immune. However, the chemicals also react with and are consumed by most organic materials, such as oil, and by reducing agents such as sulfides that contaminate the water. Chlorine, an oxidizer, is the most widely used and accepted biocidal treatment for both industrial cooling and potable-water applications. See Table 4.6.

Nonoxidizing biocides are frequently used to supplement chlorination programs. The mechanism of biological toxicity varies from chemical to chemical. However, bacteria may develop immunities for most of these biocides if they are used continuously, without periodic changes of chemical.

Many nonoxidizing biocides are specially formulated for microbiological control under conditions where chlorine may lose some of it effectiveness. During oil or hydrocarbon leaks, the nonoxidizing biocide will continue to control bacteria, while chlorine may be consumed by oxidizing the organics. Sludge, oil, and slime deposits shelter bacteria from oxidizers, but penetrating nonoxidizers can get into the deposit and kill the bugs.

Biodispersants. Biodispersants are chemicals commonly used with biocides to increase penetration and removal of organic matter fouling the heat-exchanger surfaces. They are usually nonionic polymers with little or no toxicity. The biomass dispersant ability and nontoxicity of some biodispersants make a chlorination program highly effective and relatively safe for most environmental regulations.

Wood Preservatives. Some cooling towers may contain wood construction materials. Although the wood may have been treated with a preservative before it was

TABLE 4.6 Biological Control

Typical chemicals	Formula or abbreviation	Typical application	Limitations
1,3-Dichloro-5,5-dimethylhydantoin		Slow release chlorination type. Sanitizer for swimming pools and small cooling systems	Less active than chlorine, very expensive for large systems
2,2-Dibromo-3-nitrilopropionamide	DBNPA	General biocide for low-pH cooling systems. Can be deactivated by increasing effluent pH	Not applicable in high-pH systems. Expensive chemical
Acrolein	$CH_2{=}CH{-}CHO$	General biocide effective in organic contaminated waters. Deactivated by sodium sulfite	Flammable, volatile, lachrymator. Dangerous to handle
Bis (tri-*n*-butyltin) oxide	$(H_9C_4)_3{-}Sn{-}O{-}Sn(C_4H_9)_3$	Films out on surfaces, provides long-term surface protection. Effective on algae and fungi	Very toxic to aquatic life, has severe discharge limitations. Foams
Bis (trichloromethyl) sulfone		Wide-pH-range biocide for slime and algae	
Bromine chloride	BrCl	Oxidizing biocide effective at high pH. Unaffected by ammonia	More expensive than chlorine
Bromine salts	$CaBr_2$, NaBr	Oxidizing biocide effective at high pH. Unaffected by ammonia	More expensive than chlorine. Requires chlorine for production
Calcium hypochlorite	$Ca(OCl)_2$	Chlorine type, dry chemical. Small cooling systems	Adds calcium to the water. Expensive for large systems
Chlorinated phenols		Highly effective, surface-active wood preservative	Extremely toxic to fish, and generally objectionable. No longer available in United States
Chlorine	Cl_2	Most common oxidizing biocide, effective for all biological control	Difficult and dangerous to handle. Chlorine consumed by organics and ammonia
Chlorine dioxide	ClO_2	Oxidizing biocide effective in neutral to high-pH cooling waters. Retains activity in ammoniated waters	Requires hypochlorite and chlorine to make the chemical. Explodes at concentrations above 15%

TABLE 4.6 *(Continued)*

Typical chemicals	Formula or abbreviation	Typical application	Limitations
Copper sulfate	$CuSO_4$	Control algae in ponds	Not recommended for recirculating systems. Copper causes pitting of steel surfaces
Dodecylguanidine hydrochloride	$n\text{-}C_{12}H_{25}NH\text{—}C(NH)\text{—}NH_3Cl$	Control of bacteria and algae. Protects wood. Surfactant	Can form insoluble salts with phosphates and sulfates
Methylene bis (thiocyanate)	$NCS\text{—}CH_2\text{—}SCN$	Broad-spectrum biocide. Controls anaerobic bacteria	Low water solubility. Breaks down at pH above 8
Ozone	O_3	Sanitizing biocide for small systems. Electrically generated	Impractical for large systems
Quaternary ammonium salts		Broad-spectrum biocide. Effective in high-pH systems	May precipitate anionic polymers. Tendency to foam
Sodium dichloroisocyanurate		Slow-release chlorination type. Sanitizer for swimming pools and small cooling systems	Less active than chlorine, very expensive for large systems
Sodium dimethyldithiocarbamate	$Na[(CH_3)_2N\text{—}CSS]_2$	Broad-spectrum biocide. Effective for pH 7 and up	
Sodium hypochlorite	$NaOCl$	Chlorine type, liquid. For small cooling systems	Expensive and increases pH. May cause $CaCO_3$ precipitation

shipped to the tower manufacturer, years of operation leach out the preservation chemicals and the wood becomes vulnerable to fungal attack. Special biocides containing organotin compounds can be used to coat the wood and help protect it.

Cleaners and Passivation Chemicals

Cooling-system cleanliness and adequate surface protection are critical to reliability and long-term satisfactory operation. Significant damage can occur during hydrostatic testing if the water used is very corrosive. Prior to start-up, the system should be cleaned and passivated. Mill scale and iron oxide should be removed by a noncorrosive cleaner followed by a phosphate or nitrite passivation. A clean system will tend to stay clean, but a dirty system will foul and corrode very rapidly.

SUGGESTIONS FOR FURTHER READING

General

Butler, G., and H. C. K. Ison, *Corrosion and Its Prevention in Waters,* Robert E. Krieger Publishing, Huntington, N.Y., 1978.

Drew Principles of Industrial Water Treatment, Drew Chemical Corp., Boonton, N.J., 1977.

Faulkner, L. L., and S. B. Menkes (eds.), *Corrosion and Corrosion Protection Handbook,* Marcel Dekker, New York, 1983.

Handbook of Industrial Water Conditioning, Betz Laboratories Inc., Trevose, Pa., 1980.

Makansi, J., "Protecting Today's Systems for Long-term Reliability," *Power,* Apr. 1983.

The Nalco Water Handbook, McGraw-Hill, New York, 1979.

Water Preparation

American Water Works Association, Inc., *Water Quality and Treatment, A Handbook of Public Water Supplies,* McGraw-Hill, New York, 1971.

Cohen, J. M., and S. A. Hannah, "Coagulation and Flocculation," *Water Quality and Treatment,* 3d ed., McGraw-Hill, New York, 1971.

Faust, Samuel D., and Osman M. Aly, *Chemistry of Water Treatment,* Butterworth Publishers, Woburn, Mass., 1983.

Knoppert, P. L., G. Oskam, and E. G. H. Vreedenburgh, "An Overview of European Water Treatment Practice," *Journal of the American Water Works Association,* Nov. 1980.

Operation of Wastewater Treatment Plants, Manual of Practice, no. 11, chaps. 14 to 20, Lancaster Press, Lancaster, Pa., 1976.

Powell, Sheppard T., *Water Conditioning for Industry,* McGraw-Hill, New York, 1954.

Strauss, S. D., "Removing Suspended Solids from Water," *Power,* June 1981.

Water Treatment Handbook, Degremont, Rueil-Malmaison, France, 1973.

Boiler Treatment

Barker, P. A., "Water Treatment for Steam Generating Systems," *Industrial Water Engineering,* Mar./Apr. 1975.

McCoy, J. W., *The Chemical Treatment of Boiler Water,* Chemical Publishing Co., New York, 1981.

May, L. M., L. R. Gess, and J. A. Kelly, "Performance of Oxygen Scavengers in Steam Generating Systems," *Corrosion/83,* Anaheim, Calif., Apr. 1983.

Singer, Joseph J. (ed.), *Combustion: Fossil Power Systems,* Combustion Engineering, Windsor, Conn., 1981.

Steam: Its Generation and Use, Babcock and Wilcox Co., Barberton, Ohio, 1978.

Cooling-Water Treatment

Franco, R. J., and J. A. Baumbach, "Ensuring Success for Your Cooling-Water Treatment Program," *Power,* Mar. 1982.

McCoy, J. W., *The Chemical Treatment of Cooling Water,* Chemical Publishing Co., New York, 1974.

Strauss, S. D., and P. R. Puckorius, "Cooling Water Treatment for Control of Scaling, Fouling, Corrosion," *Power,* June 1984.

White, G. C., *Handbook of Chlorination,* Van Nostrand & Reinhold, New York, 1972.

CHAPTER 4.5
ZERO-DISCHARGE WATER SYSTEMS

Roger M. Jorden
President
Water Management Associates
Steamboat Springs, CO

INTRODUCTION

As the name implies, zero-discharge water management practice involves no return of plant wastewaters to surface streams, including contained salts and minerals. Roughly 33 coal-fired stations and one nuclear plant have zero-discharge water-wastewater systems in the United States. In practical terms, this means that all waters brought onto the plant site for use are evaporated from the cooling tower, up the stack, or from ponds or waste piles; and the residual salts from makeup water, plus any added chemicals, as well as residue from ash, etc., are contained and disposed of on-site.

Zero discharge is practiced largely in the western United States (Fig. 5.1), but extends to a water-scarce portion of Florida. Solar evaporation is a major factor for making zero discharge within economic reach today. In areas where water is plentiful and downstream salinity is not a problem, zero discharge simply does not make sense.

From a global view of water management, zero discharge accomplishes both water conservation and salinity control. In the western United States, no laws

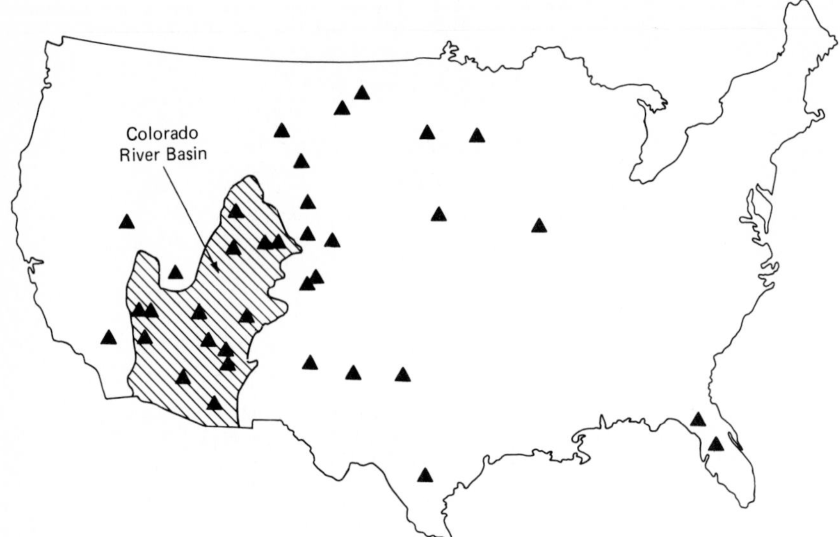

FIG. 5.1 Location of operating zero-discharge stations.

specifically mandate zero discharge; however, salinity control regulation or environmental permitting activities indirectly lead to the adoption of zero discharge (see box).

Experience at existing zero-discharge plants has shown that this is truly an "easier-said-than-done" thing. This is not because water management practice is inherently difficult to accomplish, but because of a combination of characteristics of a zero-discharge water system and the way this system's priorities typically shake out in the overall process of designing, constructing, and operating a very large, complex, integrated facility like a modern coal-fired powerplant.[1]

Reference to Fig. 5.2 dramatically illustrates that achieving zero discharge really is a tough job. For Craig Station the total plant makeup flow 10,922 gal/min (689 L/s) must ultimately be reduced to 44 gal/min (2.78 L/s) going to the evaporation pond. This represents 0.4 percent of the makeup flow, or a volume reduction by a factor of $\frac{1}{248}$.

Outlined below is an example of zero-discharge consideration in which the impact of wastewater routing is calculated (see also Fig. 5.2).

IMPACT OF WASTEWATER ROUTING What is the difference in flow rate of water entering the evaporation pond if it is alternately routed through the high quality sump versus directly through the low quality sump?

Assume: (1) Wastewater flow equal to 285 gal/min (18 L/s) and (2) concentration factors are cooling tower equals 12 and desalting equals 33.

CALCULATED RESULTS High quality flow is concentrated in cooling tower and in desalting

QE = final flow to evaporation pond

$$QE = \frac{285 \text{ gal/min}}{12} \times \frac{1}{33} = \frac{23.7}{33} = 0.7 \text{ gal/min (0.04 L/s)}$$

If direct to evaporation pond from low quality sump

$$QE = 285$$

The impact is then as follows:

$$0.7 \text{ \$ Impact} = 285 \text{ gal/min (18 L/s)}$$
$$\text{Impact} = 407\text{-fold cost difference}$$

Thus, a cruel feature of zero-discharge systems is that they are unforgiving of design or operational mistakes which result in recyclable wastewater blending with the concentrated wastes in the final disposal facility. The evaporation pond level is the ''bottom line'' in these systems. A full pond signals the need for change, for which there are two major avenues: control the excess flow from within the plant or build additional evaporation ponds. The first avenue is usually the best choice, but methods as well as costs are very situation-specific. The general cost for added evaporation ponds dramatically illustrates the penalties for a full pond. The

SALINITY CONTROL WITHIN COLORADO RIVER BASIN

Regulations limit the discharge of dissolved salts to 2000 lb/day (908 kg/day) to a surface stream at an industrial plant site. This does not mandate zero discharge; however, in a practical sense it does. Let us see how.

Assume:

- *Total dissolved salts* (TDS) of the makeup water of a Colorado River Basin plant are 500 mg/L.
- Total plant water needs are 8 gal/min • MW (0.5 L/s • MW).

Question:

How many megawatts are required to result in 2000 lb/day (908 kg/day) of dissolved salts from the makeup water? Also, how big a plant can be built and still discharge all the salts coming in via the makeup water and yet stay under 2000 lb/day (908 kg/day)?

Calculation:

$$\frac{2000 \text{ lb/day}}{8 \text{ gal/min} \cdot \text{MW} \times 500 \ \dfrac{0.012 \text{ lb/day}}{\text{gpm} \times \text{mg/L}}} = 42 \text{ MW}$$

Meaning:

1. The largest powerplant for which desalting of plant effluent does *not* have to be practiced is 40 to 50 MW.
2. For plants less than 40 to 50 MW, three possible options exist:

- Discharge up to 2000 lb/day (908 kg/day) if state regulations allow and if TDS are below the limit, and use desalting for treating the remaining wastewater flow for discharge. Retain all residual salt over 2000 lb/day (908 kg/day) on-site.
- Use desalting to treat all water for discharge (many local regulations and permitting activities lead to this as a minimum) and retain all salt residue on-site. Note: This choice normally makes less sense economically than the next one.
- Select zero discharge if expensive desalting on-site is the alternative, that is, \$2/1000 gal to \$10/1000 gal (\$0.53/1000 L to \$2.65/1000 L). Reuse that desalted water because it is superior in quality to the raw water supply. Retain all salts on-site.

In general terms, salinity control requires desalting. From this point, local regulations and economic choices generally lead to the selection of zero discharge.

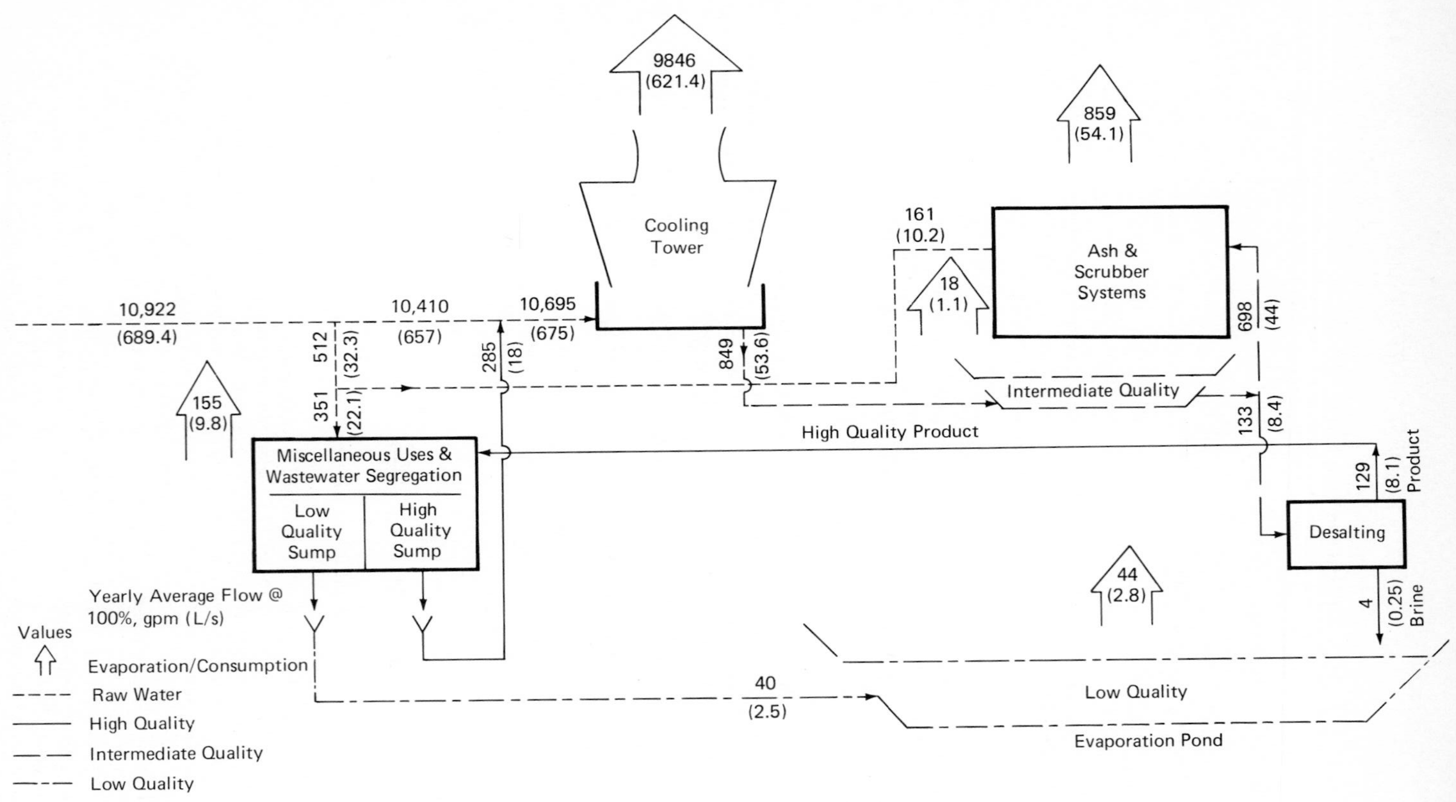

FIG. 5.2 Craig Station design water balance.

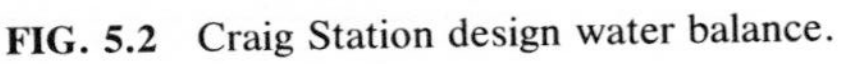

box below shows that an increase of 10 gal/min (0.63 L/s) for this case would cost about $1 million—severe, indeed.

Thus, in the real world it appears that there are many strikes against having a successful zero-discharge water system. In fact, this is correct.

So what is really necessary to do this seemingly difficult undertaking? This chapter addresses the engineering approach for design of a zero-discharge facility. The type of facility is assumed to be a modern coal-fired station with evaporative cooling in an arid climate. It focuses on overall system conceptual design and design criteria for the more crucial detailed features.

Before that, let us first put this unforgiving water management practice in perspective. A missing link in the engineering design process, plant operation personnel (and their attitudes), and some general features of water systems are items that a workable design and operating practice must take into consideration to achieve success.

FINAL WASTEWATER DISPOSAL COST MAGNITUDES

An ordinary household garden hose has a flow of roughly 10 gal/min (0.63 L/s). For an evaporation pond to dispose of this volume of flow continuously requires about 10 acres (4.05 ha) of pond area (less in Arizona, more in Wyoming). In the late 1980s that would cost about $1 million to construct. Thus, the knowledgeable and prudent designer and operator must go to extreme lengths in preventing fugitive wastewater flows with reusable quality from mixing with nonrecyclable waters. Picture 10 gal/min (0.63 L/s)—not much of a flow—as a $1 million problem if that flow goes to the wrong place.

NATURE OF THE PROBLEM

Water, much like people, is said to "have a mind of its own." A keen appreciation of how that "mind" works represents a design challenge. The problem can be divided into two parts—people and nature. Briefly we list features, things, realities, or ideas that designers should try to accommodate. This should begin by reflecting on the past.

Lack of Feedback

The technically intensive design-construction-operation process for achieving acceptable performance from complex entities such as a powerplant or a commercial aircraft is highly involved, but divisible into discrete tasks. An essential component of this process at any level is feedback of performance data to improve new designs or correct nonfunctioning ones. "Learn from our mistakes, lest history repeat itself." This is especially crucial for success in the detailed engineering phase.

Feedback is essential because of the engineer's inability to anticipate the extremes of operating conditions, long-term effects, and details which imperfect sciences cannot predict in real-world applications. It rids the engineer of half-baked ideas. If procurement practices do not get in the way, feedback can even help to overcome the significant problems of differing vendor qualities. The lack of feedback is an open invitation to disaster.

In the aircraft industry, feedback is alive and well, in part because of the dramatic form that failures can take. For powerplants, feedback has been a crucial part of improved developments over the years for subsystems directly affecting the power production train—boilers, turbines, generators, and boiler feedwater treatment, for example.

Incredulous as it may seem, feedback is largely inactive, if not totally absent, when it comes to wastewater systems and the engineering design, specification, and procurement of detailed features like pumps, valves, sumps, controls, etc. This is an industrywide problem for the cases involving large zero-discharge stations and major architect/engineers.[2] The lack of this very fundamental element of improving on past experience manifests itself largely in failures in seemingly minor features. Failures in numerous cases of well-engineered conceptual designs result from inadequate detail component performance. The overall wastewater system in many cases would work satisfactorily if only the little items did not malfunction, e.g., sump pumps, controls, etc.

The causes for the breakdown of the feedback loop in the engineering-construction-operation process are, in fact, unknown. The important and positive approach is to make feedback an integral part of the design process in wastewater systems (see box). To understand its causes is the means to correcting its nonoccurrence.

Fortunately, it is not an insurmountable task to make feedback active. A critical analysis of the causes suggests that the needed ingredients are awareness, commitment, resources, and effective management. For success, some fundamental stumbling blocks must be corrected, such as (1) poor operation and maintenance documentation procedures that do not provide a clean basis for relating cause-effect of failures and (2) the characteristic "knee-jerk" reaction evoked by words like *design deficiency* that prohibits any prospect of feedback. It is likely, but difficult to prove, that the total costs are significantly cheaper when feedback is healthy and active.

People

Without specific training that emphasizes the concise control necessary for operating a zero-discharge system, some commonly held feelings and attitudes about

FACTORS CONTRIBUTING TO POOR QUALITY IN WASTEWATER SYSTEMS

- *Slow system response time* In the absence of a strong monitoring program during start-up, problems may appear months or even years later, when the architect-engineer is no longer around.
- *Causes may be diffuse* It is often difficult to relate cause and effect. Finger pointing and blaming the other side are paths too easy to follow, especially after a malfunction.
- *Cost overruns* These usually lead to a shift in priorities, limiting of funds, and diminished attention to details, especially in peripheral systems like wastewater systems, which do not produce megawatts.
- *Lack of an effective quality assurance program* An inadequate program basically results from a lack of accountability in massive projects spanning many years, poor personnel continuity, and diffused decision-making documentation.

water compound the problem. For example, consider these comments: "Why should I get excited about a thing which the city delivers to my house, exactly when I need it, and it costs me less than a penny per ton?" "It's cheaper than dirt." "My neighbor washes his driveway with it every weekend." "Nature provides it, and the leftovers flow back to the river." With false economics for water pricing, municipal sewer systems, gutters, and storm drains, such attitudes do not have a direct, perceptible effect—just increased taxes. In a zero-discharge system, a combination of improperly trained personnel who lack the awareness of the delicate balances of a zero-discharge system plus an "unaware" design can only lead to disaster. Both training and design—see below—are key to dealing with this.

In the large and impersonal environment of a modern powerplant, accountability for water misuses or abuses is virtually undetectable. Recognize that sources and sinks are omnipresent, that is, hose bibs and drains are everywhere. It can become seductively wonderful that all leaks, forgotten hoses left running, spills, wash-down liquid of any kind, so conveniently disappear down the nearest drain.

Water and wastewater systems are largely hidden from sight—"out-of-sight, out-of-mind." Unless floor drains are color coded, only infrequently does anyone know the fate or impact of water disappearing so conveniently down a particular drain. Additionally, a nonfunctioning sump pump, or even ones experiencing a temporary inflow that exceeds their capacity, may lead to overflow which is going undetected. This is usually the case since sump systems are underground. An important practical question for the designer (and then the operator) is: Can this system be monitored, or is anyone even checking?

Defective pump seals can introduce large quantities of fresh water to a scrubber or ash system, for example. This will displace makeup from a recycled water-wastewater source such as cooling-tower blowdown. Without provisions for proper detection equipment, it is impossible to know the amount of water introduced unwisely and unknowingly into the zero-discharge system.

Proper training is a means of overcoming these factors, but creative design features enhance the likelihood of success.

Water and Zero-Discharge Systems

Water obeys a strict set of physical and chemical laws which are well known. However, when it is introduced into the environment of a large powerplant, especially a zero-discharge one, it is helpful to key on some particular features for improving design.[3]

Water Is Essential. Water is an invaluable, versatile, and inexpensive medium (working fluid, coolant, transporting agent, and cleaning agent). Unfortunately, it can also scale, corrode, erode, and explode.

Evaporation and Dissolved Chemicals. Water evaporates (or is consumed) in the process of use—100 percent ultimately for a zero-discharge station, as Table 5.1 shows. However, all the ions stay behind with the remaining liquid, and their concentration increases in direct proportion to volume reduction (in the absence of scaling or chemical precipitation). Also, additive chemicals or other ions may dissolve as the water comes in contact with flyash, flue gas, or during many of the cleanup operations. The consumed water must be replaced, bringing with it more dissolved ions.

TABLE 5.1 Craig Station Design Water Balance for Major Components*

Component	Makeup to plant		Recycle use, gal/min (L/s)	Total use, gal/min (L/s)	Blowdown, gal/min (L/s)	Consumption	
	gal/min (L/s)	Percent				gal/min (L/s)	Percent
Cooling towers	10,410 (657)	95.312	285 (18.0)	10,695 (675)	849 (53.6)	9846 (621.4)	90.15
Scrubbers	136 (8.6)	1.245	588 (37)	724 (45.6)	0 (0)	724 (45.6)	6.63
Ash system	25 (1.6)	0.229	110 (6.9)	135 (8.5)	0 (0)	135 (8.5)	1.24
Boiler/demineralizer	142 (9.0)	1.300	129 (8.1)	271 (17.1)	200 (12.6)	71 (4.5)	0.65
Service water	203 (12.8)	1.859	0 (0)	203 (12.8)	119 (7.5)	84 (5.3)	0.77
Domestic water	6 (0.4)	0.055	0 (0)	6 (0.4)	6 (0.4)	0 (0)	0.00
Pond evaporation	0 (0)	0.000	62 (3.9)	62 (3.9)	0 (0)	62 (3.9)	0.57
Total	10,922(689.4)	100.000	1174 (73.9)	12,096(763.3)	1174 (73.9)	10,922(689.4)	100.00

*Yearly average flow rate (gal/min or L/s) for 1233 MW at full load.

Such continued increases in concentration ultimately limit the extent that water can be recycled in a particular use, because of the chemical scaling and corrosion that inevitably result. In fact, the engineering of zero-discharge water systems is *chemical engineering*. An interesting sidelight is that the chemical engineering discipline has been largely foreign to utilities and powerplant architect/engineers until recent years; at the decision level, this lack of expertise still exists today.

System Approach. The extent, complexity, multiplicity of uses and of waste sources, supply and demand swings, chemical requirements for uses, and chemical changes with use mean that the entire water and wastewater system must be designed (and operated) as an integrated entity. In the design of other plant subsystems, such as the demineralizer, boiler, ash, and *flue-gas desulfurization* (FGD) systems, the water system's in's and out's must be carefully documented as any design changes occur.

The ideal situation is that water system impacts are considered in the decision-making process for design and selection of these "independent" subsystems. The design must encompass all sources and sinks of water, including all uncontaminated runoff and contaminated runoff from such areas as the coal pile and sludge disposal pond.

Site-Specific Designs. Designs must be tailor-made for each site, largely because of variations in raw water chemistry, individual plant subsystems' requirements, site and major component topography, site evaporation rates, and land availability and cost.

Water System Demands. The water system must accept *all* subsystem upsets and whatever happens to come down the thousand or more floor drains from the entire plant, including oil and chemical spills. For best results, the wastes must be segregated and routed according to chemical quality. The designer is indeed challenged to anticipate what may happen. The system must be able to accept both flow and chemical swings to achieve dampening of both. Flexibility for detecting chemical quality, temporary isolation, and optimal wastewater rerouting is desirable.

More Critical Components. If a hose is left running into a floor drain, the effect can vary from minimal to near disaster. If, for example, the drain leads to the higher-quality wastewater system which is recycled to cooling-tower makeup, the effect is very minimal; that is, this displaces direct makeup to the cooling tower from the raw water system and has almost no cost impact.

However, if that same flow is a tributary to the evaporation pond, the impact can be estimated at about $100,000 per gallon per minute (0.063 liter per second)—severe indeed. Figure 5.2 dramatically illustrates that the difference is a 400-fold impact. The exact impact will depend on the actual chemistry, cooling tower, desalting operation, and wastewater routing. Design details must be sensitive to such criticality.

Slow Response Time. This characteristic results from the large surge volume necessary to handle swings in the supply-demand equation (for both volume and chemistry). In addition to having an effect on the design process, as discussed previously, it likewise interferes with plant operation. Anything which blurs the perception of cause and effect needs to be changed or overcome through design features. For example, use color coding of floor and equipment drains, visual and/or instrument access of critical system points, and a full-functioning water management training program available at start-up.

Start-up Is Critical. Several zero-discharge systems have been overloaded because of excess wastewater flows accumulated during start-up. A host of factors may contribute to this, and some may be eliminated, dealt with, or minimized by a variety of approaches—full-functioning monitoring equipment, start-up water management procedures covering *all* site subcontractors, facilities flexible enough to maximize wastewater routing and recycle by quality, etc. Design detail features and their actual timing for completion (full function) are critical. Rapid devastation of equipment, contingency pond volumes, and morale can have long-lasting ill effects.

Integral Water and Waste Disposal. Integrating water and waste disposal is expensive but essential. The trick is to engineer the natural evaporation of water (and still control scaling and corrosion within each individual water-using subsystem) and to concentrate the salts, typically until the final residual can be economically evaporated in a solar evaporation pond, as Fig. 5.2 illustrates. Additionally, a variety of desalting and crystallization equipment are available. In concept, there are an infinite variety of ways of concentrating residual chemicals to a solid form for final disposal. In practice, there is a finite list of commercial approaches which the designer is challenged to select from and optimize in a unique configuration for a particular site.

System Is Unforgiving. Problems resulting from improper design and/or operation may be slow in appearing. However, small abuses from multiple sources, accumulating in the wrong places for extended periods, can inevitably lead to a full evaporation pond. A number of factors can mask pending doom: seasonal swings in supply, demand, and surplus on hand; "fires" in more critical plant subsystems; or lack of properly functioning monitoring equipment. However, abuses not corrected will eventually manifest themselves.

SYSTEM DESIGN AND INGREDIENTS FOR SUCCESS

This section embraces conceptual design, zero-discharge fundamentals, critical water system activities in the powerplant, and such features as quality assurance and water management.

Conceptual Design

Conceptual design, as used here, refers to the process of selecting major components that comprise the overall water and wastewater system. This must include the consideration and detail necessary to maintain the performance and reliability within each of the water-using subsystems to control scaling, corrosion, fouling, etc. Conceptual design, however, is not intended to extend to the all-important activity of detailed design specification procurement of "minor" components such as pipes, valves, pumps, sump structures, controls, etc. Also, it does not encompass questions of equipment redundancy. Conceptual design covers the engineering development of the major-component configuration layout. These two design stages are, in fact, not totally separable because of their interdependence and iterative interaction. Effectiveness in *both* conceptual design and detailed design is prerequisite for success.

In general, the objectives of the conceptual design process are to (1) meet the water use requirement of each and every plant subsystem both for supply and for waste disposal; (2) maintain acceptable performance and reliability in the water system itself and in each of the plant's subsystems that it serves, including control of scale, corrosion, fouling, etc.; and (3) minimize the total evaluated cost.

The design process itself is iterative. In general, it deals with all station waters (raw water, wastewater, site runoff, and evaporation) and all chemicals and chemical changes in the water system (chemicals dissolved in the makeup, additives or chemicals dissolved during the use of water). The designer must identify each and every plant water use, its flow and chemical requirements for makeup, and all discharge flows and their associated chemistry.

Also, the designer must integrate the many water uses—their differing quality needs, anticipated wastewater flow and varying composition, and differing treatment requirements—into a functioning system that can be operated to produce a small final wastewater flow. Additionally, the designer has to provide the operating flexibility to allow the balancing of supply and demand in order to handle swings and upset conditions.

Cornerstones of Success

Individual cases may involve adoption of special features. In general, however, the major features of a functioning zero-discharge system evolve from application of these fundamental concepts:

- Conservation—minimize water use.
- Maximize recycle—or minimize blowdown from individual water-using systems. This can be approached in several ways: (1) Continue use of a water until satisfactorily approaching the corrosion or scale limits; (2) treat water, such as sidestream softening of cooling-tower circulating water to control blowdown flow; (3) apply additive chemicals to control scaling, corrosion, or fouling.
- Maximize water reuse. Because of the water quality requirements for the various plant uses (ranging from ultrapure to rather poor quality), it is possible to find uses for some blowdown flows. This is accomplished in two ways: (1) Cascading reuse, such as boiler blowdown to cooling-tower makeup, cooling-tower blowdown to ash system makeup, cooling-tower blowdown to a portion of the FGD makeup; and (2) treatment and reuse, such as desalting of cooling-tower blowdown to upgrade water to a higher use like demineralizer makeup, lime softening of cooling-tower blowdown for makeup to FGD.
- Maintain wastewater segregation according to chemical quality. Because the designer cannot anticipate all wastewater quality variations and the inevitable upset conditions, it is important to provide flexibility for isolation and for optimal routing of wastewater flow, that is, multiple ponds, chemical monitoring, and versatile conveyance systems. Also, overflow direction for equipment,

drains, sumps, and ponds is crucial. It is important to protect the final disposal facilities from accepting overflows by default of absolutely any potentially recyclable water.

- Treat the entire system as an integrated entity. Every water-using plant subsystem which is changed during the design process (for example, selection of a different bottom-ash system) may require a commensurate change in the design of the overall water system.
- Carefully document all critical design assumptions so that they can be incorporated in the quality assurance program and in system operation.

Critical Activities

Some individual plant water systems are critical and bear special mention in guiding design activities.

Raw Water Supply. The chemistry of the supply water is a major factor in determining the actual system design details. Obviously it is important to properly characterize the water. Unfortunately, some plants have apparently been designed on inadequately characterized raw water, and significant problems have resulted.

Each type of raw water supply presents a special problem for proper characterization, whether it be rivers, lakes, wells, or recycled waters such as sewage plant effluent. Figure 5.3 suggests a logical method to use in systematically characterizing a water supply. This activity should attempt to identify the design chemistry and the ranges in chemistry associated with that particular water supply. It should encompass anticipated long-term changes from such things as other industrial development, upstream diversions, etc.

In general, rivers present greater chemical variations because of seasonal changes. Groundwater has some special problems if multiple wells are tapped because long-term pumping may lead to significant individual well production and chemical quality changes. Likewise, municipal wastewater poses a number of unique problems, which are best dealt with through pilot studies.

Also, water-sampling programs or pilot testing must be carefully planned. The following examples illustrate the types of precautions necessary to overcome potentially large cost increases:

- The plant's raw water pond was placed such that it received almost the entire plant site runoff including oil spills and ash system residue, and it even intercepted some hard groundwater seams. The result—the raw water chemistry experienced a 20 percent higher Ca^{++}. Cooling-tower blowdown, likewise, was 20 percent higher than design. This effect rippled through the system.
- Pilot testing at one site completely overlooked the need for silica control. This oversight occurred because large volumes of the source water were transported to a laboratory site and then stored in an open tank. Apparently silica was removed during storage as a result of diatom (algae) growth, which utilizes the dissolved silica to form internal skeletons. The result—the system was built; it failed because of silica scale and had to be heavily modified.
- A water system design for a station drawing on a well field was based on water samples from exploratory wells. After plant start-up, it was discovered that the chemistry from the long-term well field was significantly better and did not re-

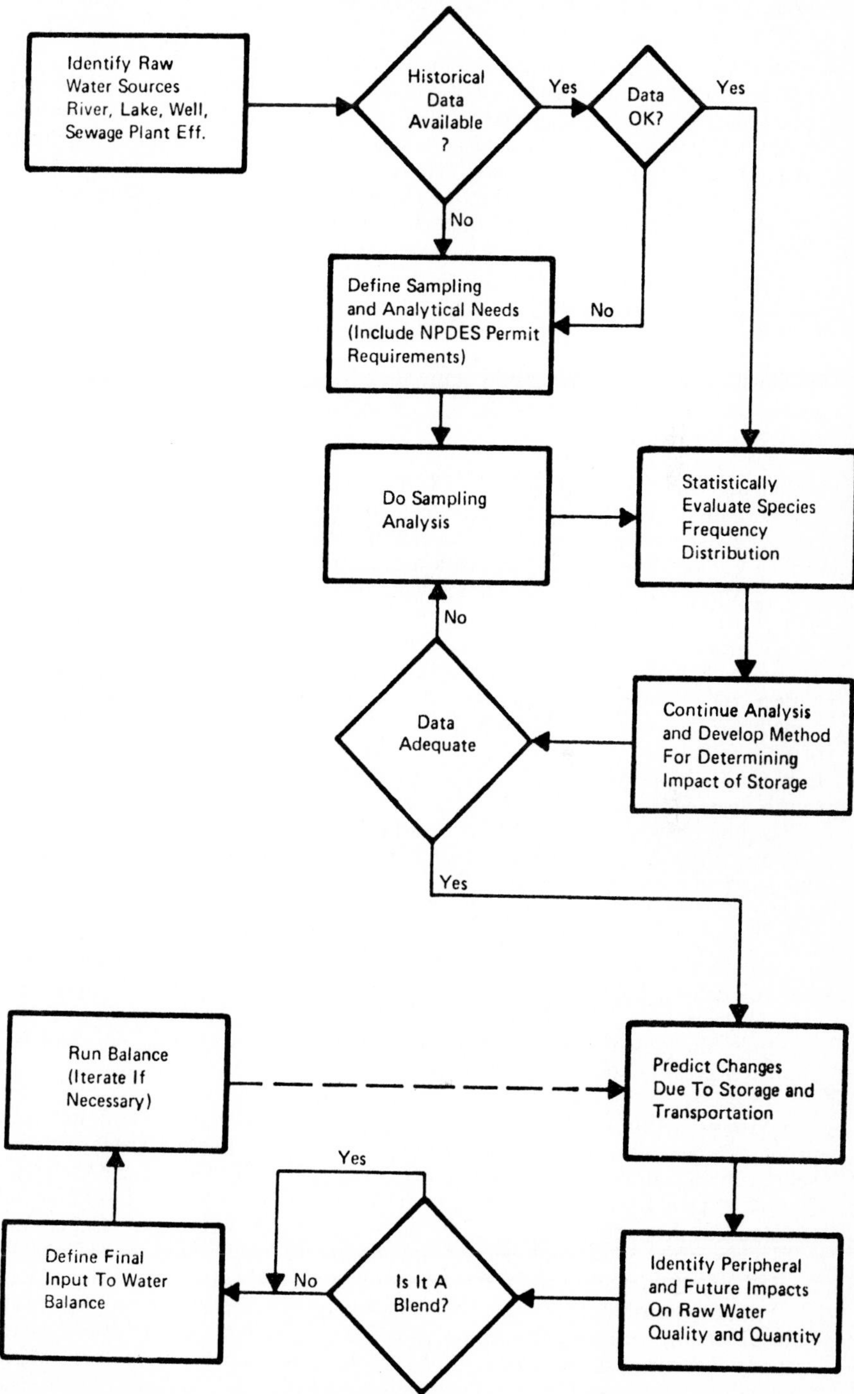

FIG. 5.3 Logic diagram showing analysis and sample collection for water data.

quire treatment. The result—an expensive but unnecessary treatment system was installed.

- The design of a major station treatment facility was based on a pilot study for treatment of municipal wastewater and was conducted at the sewage treatment plant site. The powerplant water treatment system was selected based on pilot study performance and assumed no additional chemical changes to the water. After construction was completed, two major differences became apparent: (1) The plant site was 30 mi (48 km) away from where the pilot testing was conducted, and (2) the treated effluent was stored in an open reservoir. The result—significant biological growth occurred both during transport in the 30-mi (48-km) open channel and during storage of the treated water. This growth altered the water chemistry on both ends, and off-design conditions required change.

Obviously, water quality is not a static look-it-up-in-the-handbook item like the properties of a particular stainless alloy. Mother nature has numerous tricks that must be guarded against.

Circulating-Water System. As Fig. 5.1 and Table 5.1 clearly indicate, the circulating-water system is the major consumer for stations with wet evaporative cooling. Likewise, this subsystem can be one of the major generators of wastewater for the station. (Where cooling lakes are used, this becomes a different case.) Thus, proper design of the circulating-water system offers major leverage for design of the entire water and wastewater system.

For a given heat load and makeup water chemistry, the minimum flow rate of cooling-tower blowdown is set by controls on chemical scaling, principally within the condenser tubes. Acid addition provides some latitude for blowdown flow control because of carbonate reduction caused by off-gassing of CO_2. Scale inhibitors allow further reduction of the required blowdown flow rate. (However, some question the advisability of planning on inhibitors at the design stage.)

Fortunately, lime softening is a versatile process that sees service in either the makeup or the sidestream mode to selectively control almost all the major scaling species (calcium, magnesium, silica, carbonate, phosphate, and suspended solids).[4,5]

A particularly promising extension of lime softening involves its integration with some advanced combustion technology, that is, fluidized-bed combustion and solvent injection. Both can use $CaCO_3$ for SO_2 capture. The sludge by-product from lime softening is largely $CaCO_3$. Dovetailing the circulating-water system's blowdown flow control with softening solids disposal, SO_2, and NO_x control may offer some real cost savings.

Ash System. Almost universally, ash systems have proved to be a significant problem in zero-discharge water systems. This system, like the water system itself, probably suffers from a low priority accorded to ash systems and the lack of feedback. Generally, ash systems are designed as a wastewater-consuming subsystem, but they often wind up to be net producers of wastewater. Experiences with ash system failures are many.[2] Beyond barely functioning, the systems frequently result in nearly unbearable messes and severely affect the water system.

A major contributor to water imbalances appears to be inadvertent freshwater makeup to the components tributary to the ash system. Any indirect system makeup directly displaces the use of recycle water makeup. This comes from unmonitored pump seals, sump flush water, "emergency" makeup, boiler seals, clinker grinders, etc. The displacement defeats both raw water conservation and recycle maximization.

Installation of flow monitors in each pump-seal supply line may not be necessary. Reliance on numerous weld-on meter saddles that will accept a flow monitor for periodic checking is another less expensive approach.

FGD System. Early designs of FGD systems projected these to be consumers of wastewaters. However, like the ash system, the earlier systems turned out to be net producers of wastewater. Later designs—the so-called dry scrubbers and liquid scrubbers—appear to be better consumers of wastewater. Actual system performance data of "in" and "out" monitoring should, in theory at least, be available today as a more appropriate source of design information than that provided by vendors.

Final Disposal. For handling the nonrecyclable waste, two major avenues are open:

- Evaporation ponds.
- Treatment and recycle, employing desalination processes. Many types exist, and the majority produce a concentrated brine and a high-quality product water suitable for recycle.

The optimal choice between evaporation ponds and desalting is largely dependent on site land costs and evaporation rates. In addition to total evaluated costs, considerations of performance, reliability, and operational preferences may affect the selection. Care should be taken to obtain actual site solar evaporation rates and to account for salinity effects.

Pond surface area is provided in relation to the desired evaporation rate, while volume or depth is provided for surge capacity or to store the residual salts. An evaporation pond should receive salts at the rate of input from raw water, plus or minus any additional chemicals from dissolving or water treatment processes. The major variable is the raw water chemistry. To protect against reduction of the pond volume from inflow of suspended solids, the capability of upstream solids retention may be advisable.

By and large the residual salts are composed of the major, innocuous constituents of water (calcium, magnesium, sodium, carbonate, sulfate, and chloride), but will include trace elements. The classification and required handling procedures of the ultimate residual solid waste will relate to the specific chemical constituents. For example, if hexavalent chromium is used for corrosion control in a cooling-tower system, it could lead to classification as an "RCRA hazardous waste" and all the difficulty which that entails.

The evaporation-pond water level is the "bottom line," literally and figuratively speaking, for a zero-discharge system. Obviously, facilities for protecting the evaporation pond from receiving recyclable water are an important design consideration. Other considerations should include monitoring equipment and multiple ponds as well as a flexible conveyance system for rerouting recyclable waters.

Other Features

Quality Assurance. At the Craig Station (Table 5.1), operating experience showed that the evaporation-pond design flow rate of 22 gal/min (1.4 L/s) for two units was being exceeded by over 600 gal/min (38 L/s). A detailed inspection of the piping and sump system revealed several problems. For example, 140 gal/min (8.8 L/s) of freshwater to cool a blower motor was being directed to the evaporation pond via a hub drain that was, fortunately, visible. (Reference to Fig. 5.2 emphasizes the 400-fold impact difference that such a mistake creates.) Repiping promptly corrected this significant foul-up. A more systematic approach, however, is to have a good quality assurance program. This advice is especially valid because much of the piping in wastewater systems is buried and out of sight. Consequently, some improper piping (unlike in the above case) may be virtually beyond detection. Doing it once correctly is much cheaper.

The quality assurance program should center on the documentation evolving from design assumptions. The quality assurance effort should be sharply focused on the drain and wastewater system, and it should span from site planning through the start-up phase. Personnel continuity and clear, direct accountability combined with effective management are the overriding keys for success.

Water Systems Management. Management, of course, is under the direct responsibility of station operation. Efforts at developing a management program in the design phase, however, help to stimulate the inclusion of features and details that enhance the ability to effectively manage a system. The monitoring system should, at a minimum, focus on all the critical locations in water and wastewater systems. It should allow complete in and out monitoring of important subsystems such as ash and the FGD system, and it must allow flow monitoring of each and every individual flow to the evaporation ponds, including overflows.

REFERENCES

1. R. M. Jorden, *Design and Operation Checklists for Zero Discharge Power Plant Water Systems,* EPRI CS-4045, June 1985.
2. *EPRI Zero Discharge Symposium Papers,* Denver, Sept. 1981. This collection of 27 papers, available from the Electric Power Research Institute, is an invaluable source of detailed information on zero discharge.
3. R. M. Jorden and J. Maulbetsch, "Zero Discharge Concept Grapples with Reality," *Power,* June 1982, pp. 107–111.
4. J. T. Aronson et al., *Design and Operation Guidelines Manual for Cooling Water Treatment,* EPRI CS-2276, Mar. 1982. This manual includes design calculation procedures, rules of thumb, and procedures for establishing raw water quality.
5. J. K. Laughlin et al., "Study of Saline Water Use at the Jim Bridger Power Plant," U.S. Bureau of Reclamation, Denver, May 8, 1985.

CHAPTER 4.6
SOLID-WASTE MANAGEMENT

Earl H. Rothfuss
Baker Engineers
Beaver, PA

INTRODUCTION

This chapter addresses the management of large-volume solid wastes—bottom ash, flyash, and scrubber by-product—at coal-fired powerplants. It approaches solid-waste management from the perspective of a utility manager charged with overseeing and coordinating this management program at a particular coal-fired powerplant. Solid-waste management programs are normally team efforts with experienced specialists assigned specific technical issues. The chapter is intended as a guide to the person(s) responsible for managing the program and directing the work of these contributing specialists.

For a review of waste characteristics of bottom ash, flyash, and scrubber by-product, see Chapter 5, Flue-Gas Desulfurization, and Chapter 6, Ash Handling Systems, Section 4, Part 1. Also, many elements of solid-waste management rely on complex engineering technologies, such as filter design, pump selection, slope stability and settlement analysis, open-channel hydraulics, seepage net calculation, etc. Because of space limitations, these topics are not detailed here but are covered in other sources such as textbooks and publications of government agencies. See the References at the end of the chapter for suggested reading regarding specialized technical information.

DISPOSAL METHODS

In simple terms, all disposal methods are either wet or dry depending on the physical condition of the material delivered to the disposal area. As the names imply, wet disposal involves the handling and placement of the material in slurry or liquid form, while dry disposal requires the receiving and placement of the waste as a bulk solid material. Dry disposal systems are generally referred to as *landfills*; wet disposal systems, *impoundments* or *reservoirs*.

In many cases, a wet disposal impoundment becomes, in effect, a dry disposal landfill with the settlement of the waste solids and the recycling or release of the excess water or supernatant. Complete designs of any wet disposal system must include the facilities for removal, treatment, and either recycle or release of the inevitable supernatant. Both wet and dry systems must provide for the management of precipitation falling on the disposal area.

Wet Disposal

Wet disposal requires the safe containment of the waste slurry until it has settled and stabilized or fixated. Impermeable barriers must be provided to contain the full volume of waste slurry anticipated during the life of the disposal site. The configuration and general arrangement are controlled by the terrain, geology, hydrology, and other features of the site. The simplest approach is to construct a dike around the perimeter of the site with materials from off-site locations (Fig. 6.1). Where the geology and hydrology of the site permit, it is often less costly to excavate the dike material within the future disposal area, creating a site that is partially diked and partially excavated or incised in the site (Fig. 6.2).

Another option is to excavate for the entire disposal volume and to use the removed soil for other purposes (Fig. 6.3). Often, a decision on this point is based on the overall fill and borrow needs of the entire powerplant project. Sloping sites offer the possibility of taking advantage of the terrain to provide portions of the containment, as with a sidehill disposal area diked on three sides (Fig. 6.4) or a valley impoundment diked only on one side (Fig. 6.5).[4]

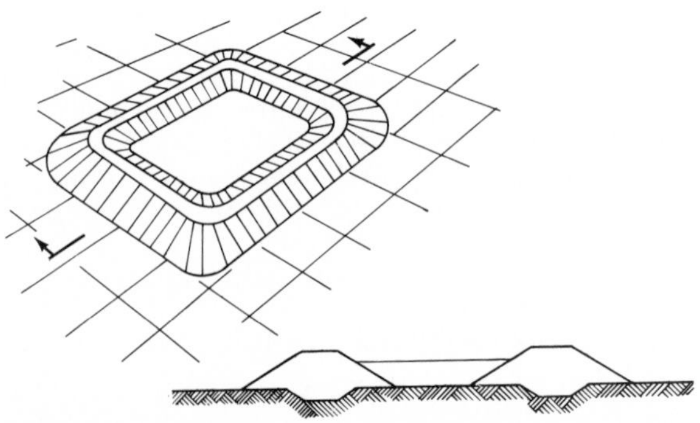

FIG. 6.1 Diked pond constructed above grade.

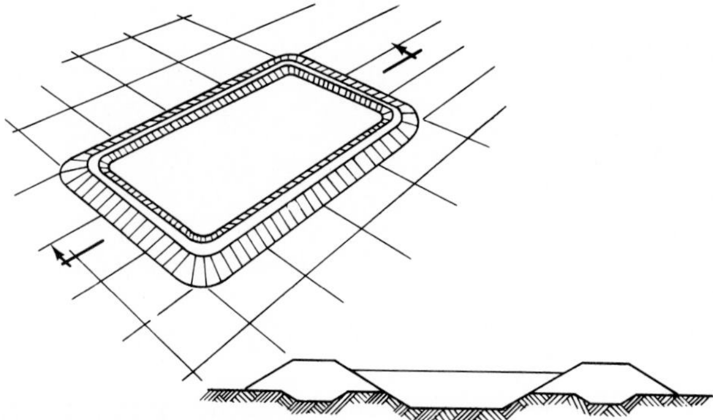

FIG. 6.2 Diked pond partially excavated below grade.

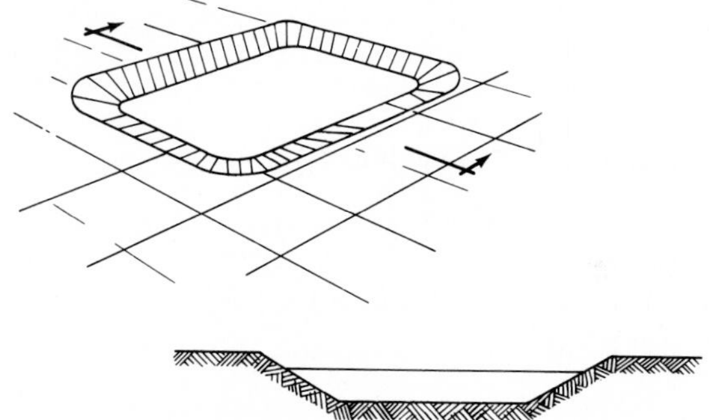

FIG. 6.3 An incised disposal pond.

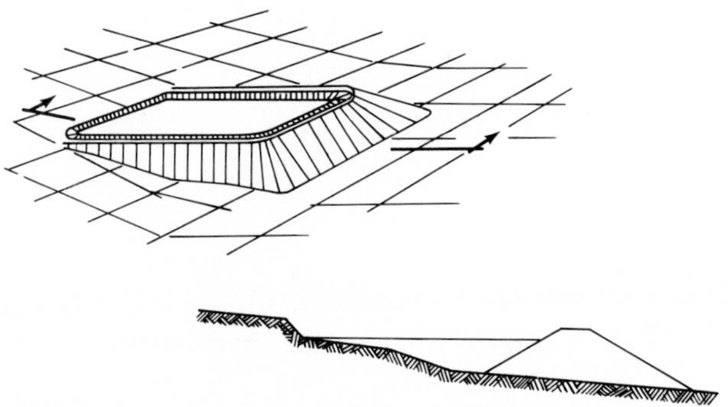

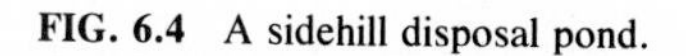

FIG. 6.4 A sidehill disposal pond.

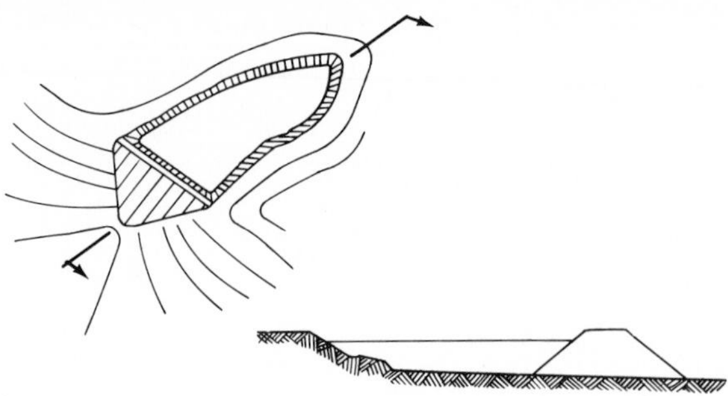

FIG. 6.5 A cross-valley pond configuration.

Dry Disposal

A similar, somewhat shorter list of configuration choices is available for landfills of dry-waste materials. The simplest method of landfill construction is a heaped fill in which the waste material is placed and compacted in lifts to form a mound or artificial hill (Fig. 6.6). This approach is generally used in flat terrains.

For greater stability and reduced visual impact, the designer should take advantage of sloping terrain whenever possible. A sloping site allows the construction of a sidehill landfill (Fig. 6.7), and a valley site leads naturally to a valley fill configuration (Fig. 6.8). In practice, a landfill site may employ more than one of these configurations as it is developed over the life of the powerplant. Being alert to such options can greatly extend the life of a disposal site.

In new waste disposal facilities, to serve new or existing plants, there is a trend toward dry disposal. This technique leads to smaller volumes for the same useful life, more options for site or material reclamation, and higher reliability. The increasing popularity of dry methods of collection (dry scrubbers, fabric filters, etc.) further reinforces this trend.[4]

Combined Disposal

The distinction between wet and dry disposal, as well as between disposal and utilization, is often blurred in practice. Some plants are successfully using dry

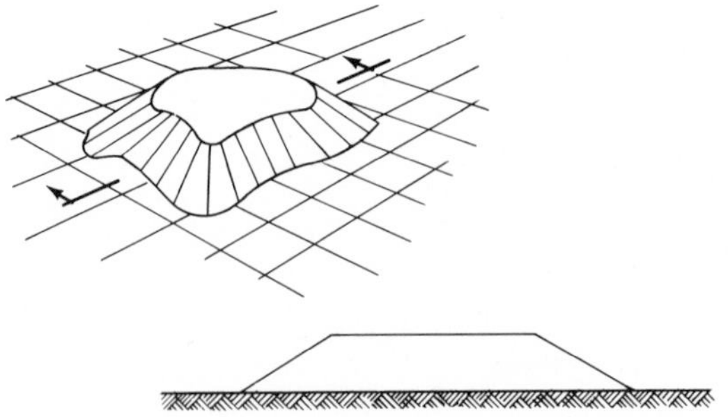

FIG. 6.6 A heaped landfill configuration.

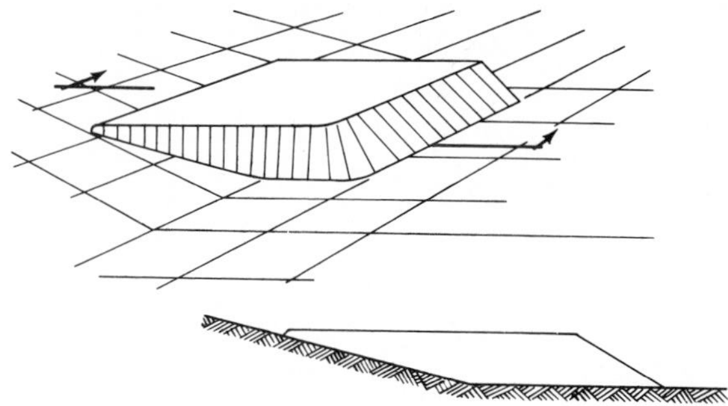

FIG. 6.7 A sidehill landfill.

wastes (mainly ash) to construct dikes for containing wet wastes [mainly *flue-gas desulfurization* (FGD) products], or in blending dry materials (ash) with wet (FGD products) to produce stable materials suitable for disposal in a landfill.

Disposal Method Selection

Two main considerations guide disposal method selection. Clearly, the disposal methods must be responsive to the waste collection methods used in the powerplant itself. The chosen system must either handle the wastes as received or include the facilities to alter them to the chosen disposal method. To provide valid comparisons, the cost tabulations should include all the costs associated with each method, including treatment, transportation, and disposal. Estimates should include both operating and capital costs, and consistent rules should be applied to both estimating and financial calculations. Some other factors influencing the selection include terrain, soil types, hydrology, and regulatory preferences.

TREATMENT METHODS

In powerplants with wet collection systems, disposal design often includes processing to produce a solid. Three considerations influence this decision: waste volume reduction, need for stability and better handling, and liquid recovery for reuse. Cost is the primary deciding factor, and treatment is introduced when its cost is less than the potential savings for disposal and/or recovery.

The treatment method can be considered in three broad categories:

- *Dewatering:* Physically separating water and solids to recover the water and/or increase the solids content of the product
- *Stabilizing:* Adding dry solids to the slurry to increase the solids content of the product
- *Fixating:* Adding an agent to cause a chemical change in the product to bind the water into the solids to produce a physically dry solid

Many designs include specific methods from more than one category, either as a multistep process or as one step combining more than one method. For example,

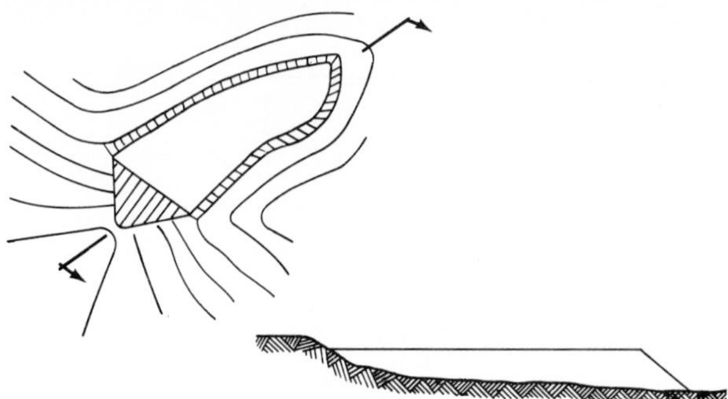

FIG. 6.8 A valley-fill disposal configuration.

many plants treat scrubber slurries with one or more dewatering steps followed by stabilization or fixation. The designer must select the most appropriate treatment or sequence of treatments.[4,12]

Dewatering

Dewatering encompasses a wide variety of mechanical devices to separate water and particulate solids (Table 6.1). The terminology makes a distinction between clarifying done to recover clean water for reuse, or release, and thickening done to increase the solids content in the slurry. Some of these methods are described here. This survey is incomplete, however, and the equipment suppliers are continually bringing new and/or improved models to market. The designer is encouraged to contact as many vendors as possible for up-to-date product data and design assistance.[4]

Settling Ponds. Often used alone with ash or limestone scrubber slurries, settling ponds are the simplest dewatering method. The pond is sized to provide a large volume so that flow velocities are very low and gravity acts to settle the solids. The liquid fraction is drawn off at a clear well and the solids are collected during periodic cleaning of the pond bottom. Care must be taken in pond design to prevent short-circuiting of the flow between the inlet and the clear well.

Dewatering Bins. Dewatering bins include a wide variety of devices ranging from the simple to the complex. The simplest are the hoppers with a single sloping, perforated plate to separate granular materials such as bottom ash from the slurry water fraction. In more complex units, multiple plates collect the heavy solids from scrubber sludges on a moving belt at the bottom of the bin (Fig. 6.9). The plates provide a large surface area in a small volume to settle the solids and to separate particles from the moving fluid.

Thickeners. Thickeners operate much as ponds or bins, relying on gravity to separate high-specific-gravity solids from water. Typical thickeners are large cylindrical tanks with a center column (Fig. 6.10), which supports the drive mechanism for two or more long radial raking arms extending from the bottom of the shaft. These arms carry a series of plows to stir the material at the bottom of the tank, which slopes toward the center. The plows push the settled solids toward the underflow discharge point.

TABLE 6.1 Leading Dewatering Methods

Dewatering method	Range of solids concentrations	Advantages	Disadvantages
1. Settling ponds	10–50% FGD slurry 20–70% ash	• Simple operation • Not sensitive to inlet solids content • Low maintenance costs • High reliability	• Substantial land area • Unpopular with regulatory agencies • Solid removal difficult since pond must be shut down for cleaning
2. Dewatering bins	15–25% FGD slurry 25–75% ash	• Reduced land area • Relatively simple maintenance • Clear water produced • Attractive first-stage treatment	• Low slurry product solids • Sensitive to inflow characteristics • New technology
3. Thickeners	20–45% FGD slurry	• Reduced land area • High throughput rates • Established technology	• Complicated operation controls • Higher capital cost • Higher maintenance cost
4. Cyclones	35–65% FGD slurry	• Low space requirements • Relatively low cost • Recover high portion of large particles • Low solids content in liquid fraction	• More complicated operation • Do not recover fine particles • Inefficient with feeds over 15% solids • Susceptible to abrasion and corrosion
5. Centrifuges	40–65% FGD slurry	• Low space requirements • Accept variation in inflow • High product solids content • Established technology	• High liquid content in solids fraction • Do not produce clear liquid • High cost • High maintenance • Subject to abrasion and corrosion
6. Vacuum filters	35–65% FGD slurry 60–75% ash	• Low space requirements • High product solids content • Consistent product quality • Established technology	• High cost • High maintenance • Complicated operation • Do not produce clear liquid

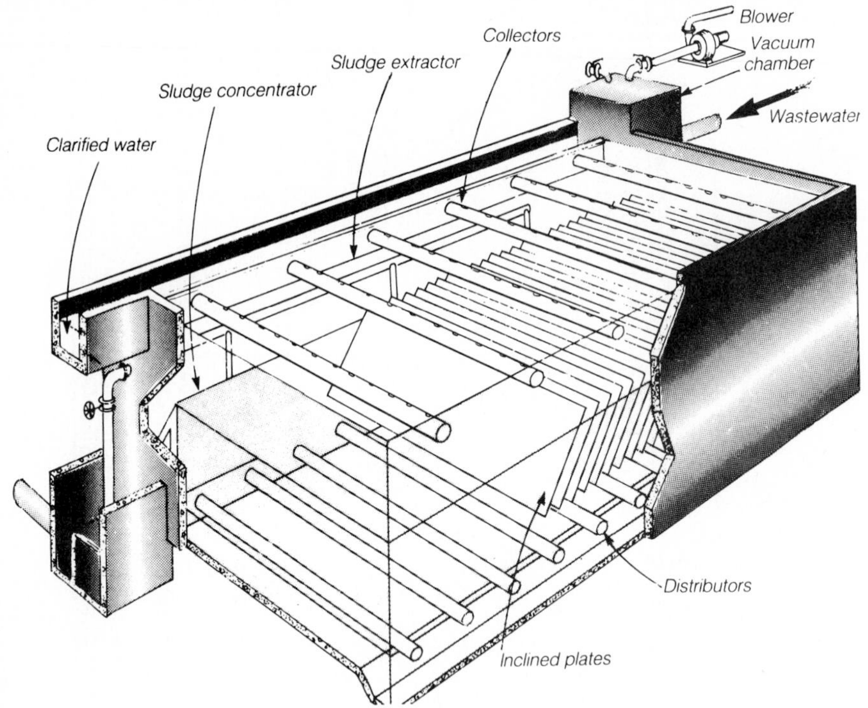

FIG. 6.9 Plate settler performs clarification among a bank of parallel inclined plates. (Power *magazine.*)

The center column also contains a feedwell that delivers the inflow slurry to the midpoint of tank depth. The initial separation is made by gravity, and the clear flow exits over a weir around the top perimeter of the tank. The thickened solids are compacted by the rakes and discharged by a pipe at the bottom center of the tank. Thickeners are sized according to inflow volume, solids, loading, settling characteristics, and target outflow solids concentrations.

Recommended surface loading for granular solids is 300 to 4000 gal/ft^2 • day (12 240 to 163 200 L/m^2 • day); for slow-settling solids, 800 to 2000 gal/ft^2 • day

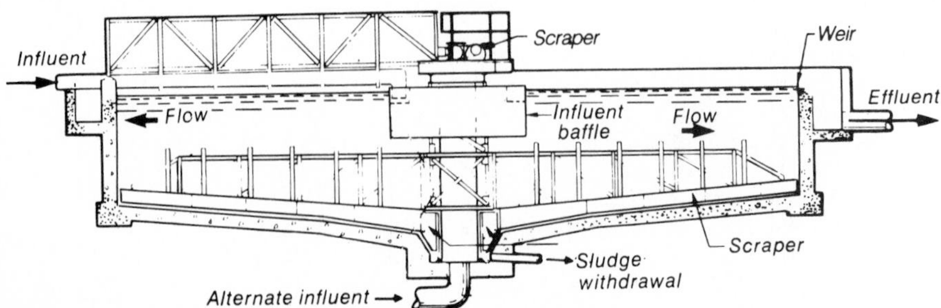

FIG. 6.10 Thickener gives clarifier sludge additional compaction under effect of gravity. (Power *magazine.*)

(32 640 to 81 600 L/m^2 • day); for flocculated solids, 1000 to 2000 gal/ft^2 • day (40 800 to 81 600 L/m^2 • day).

Tank depths are between 8 and 14 ft (2.4 and 4.3 m). The tanks are usually constructed with steel sides and a concrete bottom. The feedwell and mechanical components are generally mild steel with heat-treated or alloy steels being used in the gear mechanisms. To ease lubrication and prevent premature wear caused by particle intrusion, the drive components are normally placed in dusttight, cast-iron housings. Because of varying pH levels and chloride content, the submerged elements are coated with epoxy or rubber to prevent corrosion.

Cyclones. In some sedimentation systems, the liquid cyclone is used in place of the tank-type thickener. Cyclones promote free vortex separation that removes solids from slurries by the combined effects of centrifugal force and liquid shear (Fig. 6.11). The pressure energy of the slurry liquid pumped into the cyclone changes to velocity energy at the inlets and into rotational energy as the liquid moves down and through the cyclone. This rotation causes a natural vortex that moves the heavier solids particles to the outside of the vortex and the lighter liquid particles to the center. The heavier solids are discharged from an orifice in the cyclone's tip.

A cyclone is fundamentally a clarifier, a device that removes particles from water. Cyclones will separate and collect particles down to a given size; the finer particles remain with the liquid overflow. The separation limit is controlled by forces such as particle mass and shape, liquid viscosity, and cyclone feed concentration. Generally, collection of finer particles requires larger equipment and produces a higher final moisture content in the dewatered sludge. In sum, a very low portion of solids falls into the liquid fraction, but a high portion of liquid falls into the solids fraction. Cyclones are ineffective in recovering solids particles finer than 5 μm and are quite inefficient when the feed contains more than 15 percent solids.

Centrifuges. Found in a wide variety of dewatering applications, centrifuges are normally a secondary dewatering step that follows an upstream thickener. They are basically compact, high-intensity settling basins; centrifugal force is the principal cause of solid-liquid separation. Centrifuges can produce 1500 to 4000 times the acceleration force of gravity—the limiting force in settling basins. Centrifuge reliability and efficiency are improving, and they accept variations in feet (meter) rates and solids contents. The solids fraction is consistent and can reach 65 percent solids concentration. They are not effective in producing a clear liquid fraction and are subject to abrasion and corrosion.

The most common type of centrifuge is the horizontal, solid bowl design in which slurry is pumped into a spinning cylindrical tank (Fig. 6.12). Centrifugal force pushes the solids to the outside wall of this tank, where they are extracted by a concentrically mounted screw conveyor. The liquid fraction is discharged at a port

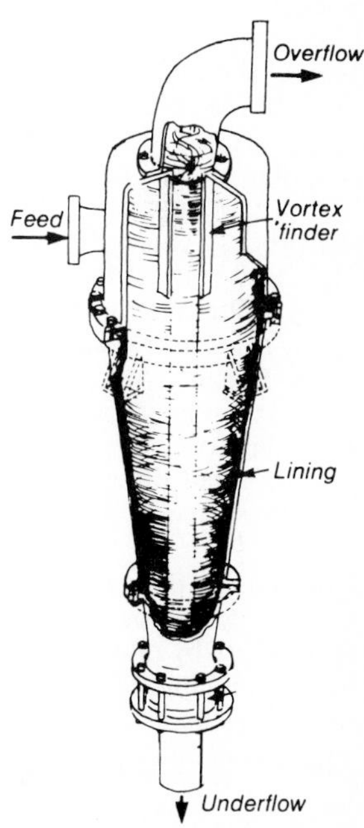

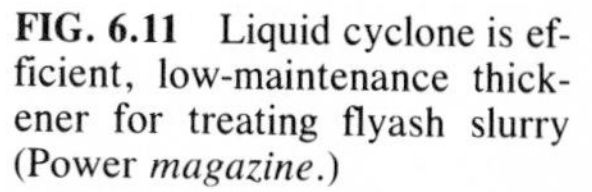

FIG. 6.11 Liquid cyclone is efficient, low-maintenance thickener for treating flyash slurry (Power *magazine*.)

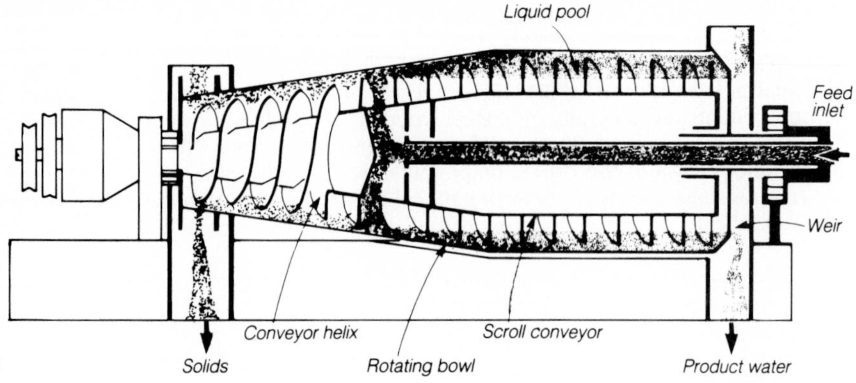

FIG. 6.12 Centrifuge is commonly used for dewatering scrubber or ash slurries. (Power *magazine*.)

near the machine's centerline. Centrifuge selection is a search for a balance between solids-fraction moisture content and solids recovery. The major problem in centrifuge operation is the disposal of the liquid fraction, which is high in solids content.

Vacuum Filters. Two types of vacuum filters—drum and belt—are commonly selected for second-stage dewatering of slurries in powerplants. The rotary-drum filter is the most popular type since it is usually the least expensive (Fig. 6.13). The drum carrying the filter media is divided into sections. Feed is held in the tank at the bottom of the unit. As the drum rotates, each section passes in turn through the feed holding tanks. From within the drum, the vacuum draws slurry to the drum's face. As the drum rotates, the vacuum draws the filtrate through the filter media until a high-solids filter cake remains on the outside of the filter. The drum continues to turn past a scraper, which removes the filter cake so that the process can repeat. Drum filters are available in sizes offering 3 to 800 ft^2 (0.28 to 74 m^2) of filter area.

Belt filters are improved versions of the rotary-drum filter. The filter medium lifts from the drum after the dewatering portion of the cycle and passes over a small-diameter roller to remove the cake. This small roller completely discharges the cake without a scraper, thus increasing filter cloth life. Since the belt is lifted from the drum, the drying time is only a fraction of the total cycle. The installed cost of a belt filter is typically 30 percent higher than that for a drum filter with the same capacity.

Dewatering Aids. Very often chemicals are added to aid floc formation in scrubber sludges. Flocs of sludge particles are removed more efficiently and more completely by most thickening and filtering processes. Chemicals selected to promote sedimentation are generally known as *flocculants,* which are long-chain molecule chemicals from a broad group known as polymers. Depending on their charge condition, polymers will be anionic (positive), cationic (negative), or nonionic (neutral).

Anionic polymers are generally more effective with FGD slurries. Effective dosages range from 0.001 to 0.02 lb (0.000 5 to 0.01 kg) of polymer per ton (megagram) of slurry solids. As an example, polymers have been found to reduce

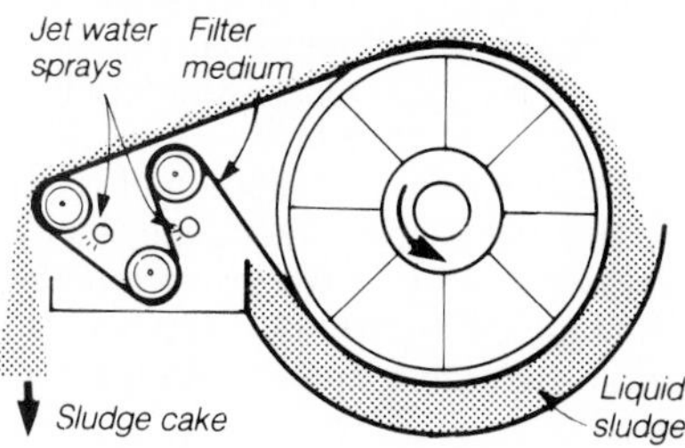

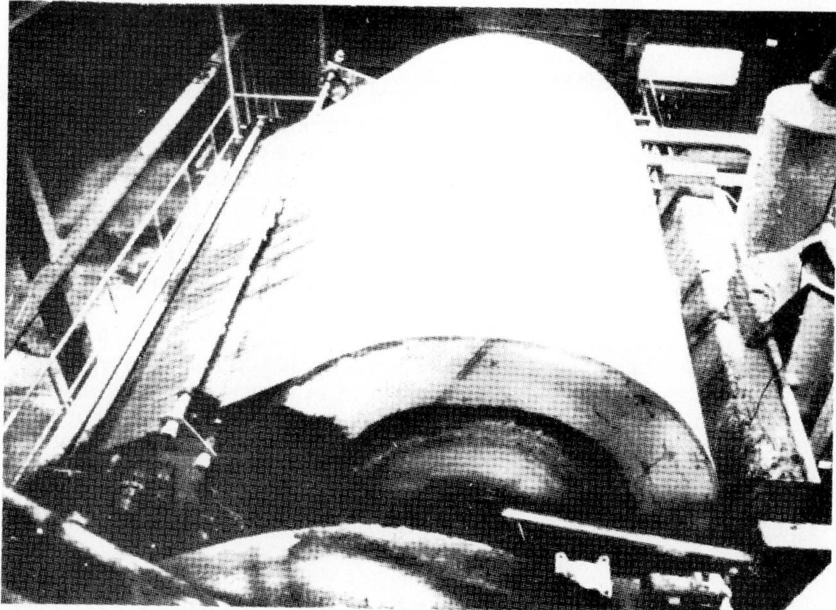

FIG. 6.13 Vacuum dewatering technique concentrates sludge by sucking it against moving cloth on rotating drum. Cake is separated from belt by scraper, as photograph shows. (Power *magazine*.)

required thickener surface by 75 percent. The addition of polymers (also called *electrolytes*) to slurries has proved effective in many situations. A wide variety are currently available, and polymer effectiveness is highly sensitive to the point and method of application. Thus, laboratory and/or pilot tests and vendor assistance are essential in polymer selection and system design.

Stabilization and Fixation

Frequently, FGD slurries require additional treatment (even after two mechanical dewatering steps) to be suitable for safe and efficient disposal. Although there may be no specific legal or regulatory requirement for further treatment, stabilization or fixation may be necessary for operating and/or safety considerations at the disposal area. Both impart greater physical stability to the processed waste material, making it easier to haul and place in a landfill and making the landfill itself less susceptible to failure or erosion.[4]

As used here, *stabilization* means the addition of relatively dry material to the slurry to increase the combined solids content and to change its physical characteristics without a chemical reaction. *Fixation* means the addition of material to the slurry to produce a chemical reaction to alter the physical characteristics of the product. In practice, a treatment process often amounts to a combination of stabilization and fixation.

Since flyash and bottom ash are generally granular materials that dewater easily, stabilization and/or fixation is most often required for treatment of scrubber products. Stabilization and/or fixation is commonly the final treatment step after one or more mechanical dewatering steps.[12]

Stabilization—A Dry Process. Stabilization is generally the simple mixing of a relatively dry solid such as soil or flyash with the slurry material to be treated. Physical stabilization is most applicable where dry disposal is advantageous. The proportions are chosen to optimize the moisture content of the resulting mix so that it is a stable solid and can be compacted to a maximum density in the landfill. The addition of the dry material spreads the water entrained in the slurry through a larger weight of solids and improves the particle size distribution so that closer packing can be obtained in the disposal area. These produce increased shear strength, lower permeability, and lower combined volume. However, the material is subject to erosion, saturation, and leaching.

Stabilization can be reversed since it is not a chemical process. If a mixture is rewetted or saturated, it may fluidize, causing a rapid decrease in shear strength and possible structural failure. The mixing may be done in large mechanical mixers with conveyors or trucks to haul the resulting solid to the disposal or storage area. In a few cases, the mixing is done by interbedding alternate layers of slurry and dry material in the storage area and mixing the layers with harrows or rollers. In either case, care should be taken to protect the product from large volumes of water, saturation, and erosion by proper placement, drainage, sloping, and soil-vegetation covering.

Four Fixation Processes. Many powerplants burn coal that produces alkaline flyash. Mixing such ash with the wet scrubber product yields a chemical reaction combining stabilization and fixation. So far four fixation processes have been identified:

- *Mixing with alkaline flyash:* Depending on the chemistry of the coal, the flyash may contain sufficient calcium oxide (CaO) to produce a chemical reaction when dry ash is mixed with scrubber slurry. Western utilities use this method with brown coals and lignite. When adequate CaO is present and ash and sludge are mixed in a suitable ratio, compressive strengths close to that of concrete have been obtained. This method reduces product permeability, which enhances placement efficiency and minimizes water-related problems.
- *Mixing with lime and flyash:* Where the coal does not produce an alkaline flyash, lime may be added with the flyash to provide the CaO needed for the fixation reaction. The lime is added at about 4 percent of the weight of the combined ash and sludge solids. Generally, the ash and sludge are mixed in equal proportions by weight of solids. The resulting cementitious reaction provides

the necessary physical properties for disposal or utilization. Permeabilities as low as 10^{-7} cm/s are obtained after 90 days of in-place curing. The cured material has good compressive strength and angles of repose and is thus suitable for structural fills and sites that will accommodate buildings after the landfill is complete.

- *Mixing with blast furnace slag:* Ground and basic blast furnace slag may produce the cementitious reaction needed to fix the scrubber sludge. This process works under water, so the slurry can be pumped to an impoundment or an interim curing pond where the reaction will continue. At the pond, the material is allowed to cure, the supernatant is removed, and the material is excavated as a dry solid and taken to a landfill. In other cases, the slurry is mixed with flyash and ground slag to produce a stable, fixed material suitable for immediate placement on a structural fill. The ground slag is added at a rate equal to 5 to 10 percent of the weight of solids in the slurry. In some cases, lime is added to raise the pH of the mixture to 10.5. The treated sludge can attain an unconfined compressive strength of 4.5 tons/ft^2 (44 Mg/m^2) after 30 days of curing. Permeability can be as low as 10^{-6} cm/s; thus, the material is stable and suitable for structural fills.
- *Mixing with portland cement:* A group of utilities have experimented with the use of portland cement to fix both scrubber sludges and flyash slurries. When portland cement is added at about 5 percent of the sludge solids weight, the process may be cost-competitive in some areas. The resulting materials cure under water and produce materials suitable for structural fills.

DISPOSAL SITE SELECTION

Selecting a site for the storage or disposal of powerplant wastes generally requires balancing a large number of competing goals related to costs and impacts. This balancing act often requires establishing a series of compromises, so that (1) no aspect of the environment or the public is subjected to unacceptable insult and (2) costs remain in the bearable range. Success in this effort depends on both the procedure and the criteria used for site selection.

Those responsible for selecting disposal sites must recognize that their work is the subject of intense public concern and will, therefore, inevitably be the object of close scrutiny by regulatory agencies and public interest groups. The procedure and criteria outlined here address the most common concerns (public safety, environmental impact, and project cost) in an organized, rational manner. The responsible engineers still must modify both the criteria and the methods to the specific conditions at their search location.

Site Selection Criteria

Site selection criteria are conveniently divided into two categories: exclusionary criteria and ranking criteria. Exclusionary criteria are features which, if present, remove an area or site from further consideration. Many exclusionary criteria would make it impossible or prohibitively expensive to construct the disposal site. Other criteria imply severe public or environmental impacts that would make the site permitting process unacceptably long or difficult. Properly developed and applied, exclusionary criteria eliminate sites in the early stages that would rank very low in the evaluation process if they remained on the list of candidates.

Exclusionary criteria also guide the search for candidate sites by eliminating areas (such as bodies of water, urban areas, national parks, etc.—see Table 6.2) that are unlikely to contain attractive candidates. A list of typical exclusionary criteria should be reapplied at each step of the selection process as better information is obtained about the candidate sites so that unsuitable sites are removed from the process as early as possible. Thus, the search effort can focus on the most attractive candidates.

Ranking criteria are features of a candidate site that provide comparisons of the suitability or desirability of one site relative to the other candidates. A suggested list of ranking criteria is presented in Table 6.3 (engineering criteria) and Table 6.4 (environmental criteria). Although there is significant overlap, the engineering criteria address features primarily affecting project cost, while the environmental criteria address those features primarily impacting public safety, public convenience, and ecological resources.

For each ranking criterion, Tables 6.3 and 6.4 provide a potential set of scoring standards. Five scores (0 to 4) are possible, each with specific standards, as a reasonable number of rating levels for each criterion, based on the limited, preliminary information that will be available during most of the site screening process. Fractional scores may be used whenever appropriate, especially late in the selection process when more detailed data are available for each candidate site. The scoring standards should be adjusted to suit each search area inasmuch as

TABLE 6.2 Recommended Criteria to Apply to Exclude Areas or Sites from Further Investigation

- Beyond a predetermined distance from the powerplant
- Within federal parks, forests, wildlife areas, and listed historic sites or districts
- Within military reservations
- Within state parks, forests, wildlife areas, recreation areas, natural areas, historic sites and districts, and archaeologic sites
- Within county parks or other areas with protected or dedicated land use
- Within urban areas as defined by a well-developed street pattern and/or a high density of structures
- Within or immediately adjacent to saltwater wetlands, brackish water wetlands, or freshwater wetlands
- Within an identified 100-yr flood plain
- Within or immediately adjacent to a critical habitat area as defined by federal or state agencies as habitat threatened or endangered species
- Valleys adversely affecting major streams, rivers, lakes, or other large water bodies
- Draining to or adjacent to surface water intakes, water supply reservoirs, or major water wells
- Within major federal, state, or county highway rights-of-way
- Cemeteries
- Within private or semiprivate recreation areas such as camps, golf courses, ski areas, etc.
- Requirements imposed by state or local regulations
- Minimum siting requirements, such as disposal volume, arising from plant design

TABLE 6.3 Typical Ranking Criteria Used in Preliminary Engineering Evaluation of Candidate Sites

Engineering criteria	Weighting factor	Ranking criteria: Score = 0 (lowest)	Score = 1 (low)	Score = 2 (medium)	Score = 3 (high)	Score = 4 (highest)
Site development[a]						
New haul road construction required, mi (km)	0.1	Over 2 (3.2)	1.2–2 (1.9–3.2)	0.8–1.2 (1.3–1.9)	0.4–0.8 (0.6–1.3)	Less than 0.4 (0.6)
Sedimentation pond construction difficulty	0.1	All excavated ponds	—	Equal combination of excavation, embankment ponds	—	All embankment (valley) ponds
Run-on diversion channel construction required, linear ft (linear m)	0.1	Over 24,000 (7315)	20,000–24,000 (6096–7315)	16,000–20,000 (4877–6096)	12,000–16,000 (3658–4877)	Less than 12,000 (3658)
Haul route[b]						
One-way distance, mi (km)	0.1	Over 16 (25.6)	12–16 (19.2–25.6)	8–12 (12.8–19.2)	4–8 (6.4–12.8)	Less than 0.4 (0.6)
One-way travel time, min	0.2	Over 40	30–40	20–30	10–20	Less than 10
Potential soil problems[c]						
Average depth to water table, ft (m)	0.1	Less than 5 (1.5)	5–10 (1.5–3)	10–20 (3–6)	20–30 (6–9)	Over 30 (9)
Predominant soil type	0.1	Sand and gravel	Silt and sand	Silt and loam	Clay and silt	Clay
Potential for future development (after landfill closure)[d]	0.1	Very poor	Poor	Fair	Good	Very good
Stability of pile based on gen eral slope[e]	0.05	Heaped pit	Sidehill and heaped	Valley and heaped	Sidehill fill	Valley fill
Upstream drainage area, acres[f] (hectares)	0.05	Over 500 (202)	300–500 (122–202)	150–300 (61–122)	80–150 (32–61)	Less than 80 (32)

[a]Comparison of features directly related to project capital costs.

[b]A comparison of features related to the operating costs of the project; also, a relative measure of the number of haul route users and neighbors impacted by selection of a site.

[c]Addresses the relative cost of groundwater protection.

[d]A comparison of the relative possibilities for development of the candidate site after waste disposal operations are complete.

[e]For a landfill, pile and stability compare the potential for stability problems because of the configuration of the pile at each candidate site. If wet disposal were proposed, this criterion would have to be modified to address the length (cost) and stability of the impoundment dikes.

[f]A relative measure of the project capital investments for run-on diversion facilities and of the public concern over local flooding and erosion.

TABLE 6.4 Typical Ranking Criteria Used in Preliminary Environmental Evaluation of Candidate Sites

Environmental criteria	Weighting factor	Ranking criteria: Score = 0 (lowest)	Score = 1 (low)	Score = 2 (medium)	Score = 3 (high)	Score = 4 (highest)
Public safety[a]	0.24	High	Moderate	Low–moderate	Low	Very low
Aquatic resources[b]	0.22	Pristine	Very good quality, habitat, and use	Good quality, habitat, and use	Fair quality, habitat, and use	Poor quality, habitat, and use
Terrestrial ecology[c]	0.15	Very good existing on-site habitat	Good existing on-site habitat	Fair existing on-site habitat	Marginal existing on-site habitat	Poor existing on-site habitat
Noise impact[d]	0.12	Very high	High	Moderate	Low	Very low
Aesthetics[e]	0.10	Highly visible	Visible	Visible from some points	Not visible from frequented areas off-site	Not visible
Zoning compatibility[f]	0.07	Requires major zoning change	Requires moderate zoning change	Requires variance from zoning	Compatible with existing zoning	Compatible with existing zoning
Air quality[g]	0.06	High	Medium-high	Medium	Low-medium	Low
Impact on historic or archaeologic resources on-site[h]	0.04	Undisturbed with evidence of on-site resources	Undisturbed with evidence of nearby resources	Undisturbed with no evidence of nearby resources	Disturbed with evidence of nearby resources	Disturbed with no evidence of nearby resources

[a] A relative comparison of the potential for accidents caused by on-site operations and hauling. Such impacts could arise from trespassers [number of residences within 1000 ft (305 m) of the site], pedestrian accidents [number of residences within 200 ft (61 m) of the haul route], vehicular accidents (current traffic load on haul route and haul route quality).

[b] A relative assessment of the potential for water quality degradation in the receiving streams. It is considered to be greater in areas with pristine or "special" use streams and less in areas where streams are subject to industrial and municipal discharges.

[c] A relative estimate of the impact on plant and animal communities that would result from site development. Woodland and mature fields are considered a greater loss than young fields, and young fields a greater loss than quarries or strip mines.

[d] Noise impact considers both hauling and site operations in terms of relative impacts on neighbors (especially schools and hospitals) within 200 ft (61 m) of the haul route and 1000 ft (305 m) of the site perimeter.

[e] A relative comparison of the extent to which the completed disposal area will be visible from nearby highways and communities. It is generally a function of disposal area configuration, with a heaped pile being much more visible than a valley fill.

[f] A measure of the degree to which the proposed use would be consistent with current actual or allowable activities on the site and on neighboring properties [within 1000 ft (305 m)].

[g] Air quality considers fugitive dust from the site and exhaust from the haul vehicles in terms of their relative impacts on site neighbors [within 1000 ft (305 m)] and haul route neighbors [within 200 ft (61 m)], and it includes consideration of site configuration and current traffic counts.

[h] A relative measure of the changes of site development, which may damage such resources on the site or diminish the attractiveness of neighboring resources because of changes in drainage or topography.

several of those suggested may be inappropriate, or even unattainable, in a particular search area.

For each ranking criterion, Tables 6.3 and 6.4 also provide a weighting factor expressed in decimal fractions. The score for each site for each criterion is multiplied by the weighting factor for that criterion, to obtain a rating. The final score for each site is the arithmetic total of all ratings for that site.

The weighting factors are suggested values only, and the engineer is invited to modify the criteria in response to local concerns as well as the type of baseline environmental data available for all sites in the search area. The data in all three tables are not represented as exhaustive or universally applicable. Criteria may be added, substituted, or deleted.

Site Selection Process

The process outlined here proceeds from the general to the specific, from a large search area through a series of increasingly shorter candidate site lists to a final recommended site. It has two objectives: (1) to identify and select the most preferred site in terms of the best possible balance of the many competing concerns impacting the program and (2) to make efficient use of resources for the search program by focusing investigative efforts on the best prospects.

The site selection process presented here has nine defined, sequential steps. The reality is not so simple; the steps will often overlap in time, be repeated, and/or be reordered. Regardless, the process consists of these elements:

1. *Criteria definition:* A list of exclusionary and ranking criteria (with scoring and weighting factors) is developed for the specific area of the search and the design parameters of the powerplant.
2. *Search area definition:* Generally, the search area is a circle centered on the plant. Often, segments of the circle can be eliminated as obviously violating one of the exclusionary criteria.
3. *Map study:* By working from large-scale maps of the search area [U.S. Geological Survey (USGS) maps or county planning maps], the area is outlined. Large excluded areas are delineated to reduce search time. The balance of the search area is examined for map-identifiable candidate sites, making sure no exclusionary criteria are violated.
4. *Candidate list:* By working from the maps, a first list of candidate sites is made. These sites are those for which the maps show no indication of violations.
5. *Field reconnaissance:* Since access to private properties is not usually available for this element, reconnaissance must be conducted from roads and other public areas. The object here is to obtain the maximum information about each candidate site in terms of both exclusionary and ranking criteria.
6. *Site screening analysis:* First, any more indications of exclusionary criteria violations are applied to eliminate unsuitable candidates. The remaining sites are scored according to the ranking criteria, and the weighting and summary calculations are completed for each candidate site.
7. *Sensitivity analysis:* As a check on step 6, this analysis is recommended. It

should vary the weighting factors systematically and delete each criterion, one at a time, to test the resulting changes in the rank order of the candidate sites.

8. *Short list of sites:* Based on the results of steps 6 and 7, a short list of finalist candidates is prepared. This short list should include two to four of the very best sites in terms of the ranking criteria for the specific project. This short list should also pass the "commonsense" test. That is, do the sites on the short list belong there based on the best overall professional judgment of the project team?

9. *Detailed study:* Finalist sites are then subjected to detailed study, by starting with checks of ownership records and negotiation of access privileges with current landowners. On-site studies include control surveys for aerial mapping as well as environmental and geological surveys—walk-over surveys of flora and fauna, checks for ecologically sensitive historic-archaeologic areas, and a preliminary drilling program for geologic structure and groundwater data.

   The sites should be evaluated in terms of both exclusionary and ranking criteria, and the ranks should be reconfirmed. A preliminary design and a life-cycle cost estimate (acquisition, development, operation, and closure) should be proposed for each finalist candidate site.

10. *Repeat process:* All the information assembled in step 9 should be used to repeat the site ranking process. The criteria, weighting factors, and scoring procedures should be revised to reflect the site-specific environmental concerns and cost data. A final site selection is then made by using the highest-ranking available site.

Data Sources

In most locations, published data sources are readily available to assist in the early stages of the searching and screening process.[4] These published resources are especially helpful, since most of the early work can be done efficiently only in the office; reconnaissance work usually must be carried out without trespass on private property. The commonly available sources include:

- *U.S. Geological Survey maps:* Detailed topographic and land use maps are available covering much of the continental United States at workable scales.
- *State geological survey maps:* Many state surveys publish their own copies of the USGS maps and provide overlays indicating geologic structures, groundwater levels, soil types, and other data.
- *State environmental resources inventories:* Many states publish, in a variety of forms and levels of detail, inventories of resources showing parks and natural areas, streams and water uses, vegetation covers, and similar items.
- *Soil conservation service maps:* The state and county offices of the Department of Agriculture's *Soil Conservation Service* (SCS) often have maps, following the USGS formats, showing soil types, land uses, streams, and like data.
- *County planning agency data bases:* Many county governments include planning agencies that publish maps and other information indicating current land use, transportation routes, current zoning, and projected development. Munic-

ipal planning and zoning agencies often provide the same information, in greater detail, within their jurisdictions.

- *Highway department traffic data:* Both state and county highway departments maintain current data on highway design (lanes and intersections), signals and signs, highway capacities, and average daily traffic volumes.

DISPOSAL SITE DESIGN

Disposal site design primarily relies on geotechnical and hydrologic-hydraulic engineering. Regardless of the process or disposal method selected, the design process starts with a detailed study of the disposal site selected for the project. After the site study procedure is described, the major design steps for a landfill-type disposal site can be outlined. Those major features of impoundment design that are different from landfill are discussed afterward.[4]

Site Investigation

The disposal site design starts with site investigations, which begin from the work done to select the disposal site and use the information collected during the site selection studies. Large-scale mapping, regional soil and water data, and site reconnaissance notes are useful input to the design investigations. They are intended to provide information needed for final design and permitting of the disposal site. These investigations cover six tasks:

Topographic Mapping. Large-scale regional maps (such as USGS maps) for site selection are not adequate for final analysis and design. More detailed and current mapping at much smaller scales is required for disposal area design and permitting. Such mapping is generally produced by using recent aerial photography, since disposal site size makes manual mapping in the field impractical. The sequence of work to produce detailed mapping includes the following steps:

- Definition of the area to be mapped, covering the proposed disposal area, all drainage control proposed for the project, access roads, and such adjacent areas as may be helpful in defining drainage, property ownership, etc.
- Control survey to establish photograph-identifiable points and to tie those points to the appropriate state plane grid to (1) establish horizontal and vertical dimensional control over the map and (2) determine the legal definition of property lines for deeds and permits.
- Aerial photography covering the entire area to be mapped at an appropriate scale and with appropriate forward and side overlaps between individual photographs to allow stereoscopic analysis of photographic pairs.
- Map plotting from the aerial photographs by using pairs of photographs and special plotting equipment to project the stereoscopic effect onto the plotting surface, allowing the operator to plot both horizontal and vertical data.

Aerial photography and mapping for these purposes require extensive precision equipment and trained, experienced operators. These services are obtained from contractors specializing in this work. Their product will include both complete sets of aerial photographs and maps of the project site.

Site Reconnaissance. Using the best available mapping, the design team should conduct a detailed reconnaissance of the proposed disposal site, including accessible adjacent areas and routes for access roads and drainage courses. If detailed topographic mapping is available, it should be used for this purpose. The reconnaissance will provide the designers with an orientation to the whole site, including major terrain features, drainage patterns, and access. It will also allow initial planning of the stream sampling and subsurface drilling locations.

Environmental Inventory. Following as a logical extension of the site reconnaissance, environmental specialists on the design team should initiate a site environmental inventory, which will characterize the proposed disposal site in terms of existing land use (farm, forest, meadow, etc.), surface condition (slope, wetland, pond, etc.), and vegetation. Specific vegetation types, wildlife habitat, and wildlife observations (animal sitings, tracks, etc.) are noted. Special emphasis is placed on evidence of endangered or threatened species. The information is recorded in organized field notes and on copies of site maps for incorporation into the permit applications.

Surface Water Studies. Any significant surface water bodies should be sampled and tested for the parameters specified in the appropriate regulations. Surface water samples should be analyzed for the parameters shown in Table 6.5. State or local regulations may require additional testing. Often testing is required for constituents known to appear in powerplant wastes to establish background data for monitoring programs during disposal site operations. Where appropriate, stream flow readings may be needed. Testing may be advisable at different times of the day or months of the year to establish reliable data.

Subsurface Investigations. A typical subsurface investigation program has three objectives: assessment of foundation conditions at the site, estimation of available borrow material for construction and final cover, and initiation of groundwater studies. A typical program might include the following:

- Test pits at possible on-site borrow locations to estimate quantity and quality of material available for construction of dams and other structural fills for the project and for cover of the waste materials as disposal progresses

TABLE 6.5 Suggested Analytical Parameters for Water Samples

pH	Arsenic
Specific conductance	Barium
Alkalinity	Boron
Acidity	Vanadium
Iron (total)	Chromium (total)
Calcium	Lead
Magnesium	Mercury
Sulfate	Selenium
Sulfite	Silver
Chemical oxygen demand (COD)	Total dissolved solids (TDS)

- Shallow borings—up to 50 ft (15 m)—to assess the load-carrying capacity of those areas of the site where significant foundation loads may be imposed during construction
- Deep borings—up to several hundred feet (meters)—to allow measurement and testing of the groundwater under the site, both to provide data on existing conditions and to initiate monitoring during construction, disposal, and closure

Appropriate caution and conservation should be applied to the data obtained from any subsurface exploration. Experienced geotechnical engineers should be employed to reduce the inherent uncertainties and to develop design parameters with adequate safety margins.[4,17]

Groundwater Studies. These studies usually proceed from the deep boring phase of the subsurface investigation, when at least some of the borings are developed as monitoring wells (Fig. 6.14). One or more monitoring wells are usually located hydraulically upgradient of the disposal area to protect the owner from blame for groundwater contamination by others. Similarly, several wells are located downgradient to sample across the plume of groundwater that may be impacted by leachate flows from the disposal site (Fig. 6.15). The screen zone at the bottom of each well unit is positioned and sized to draw water from the full depth of the aquifer being monitored.

Unless specified site or regulatory conditions dictate more frequent monitor-

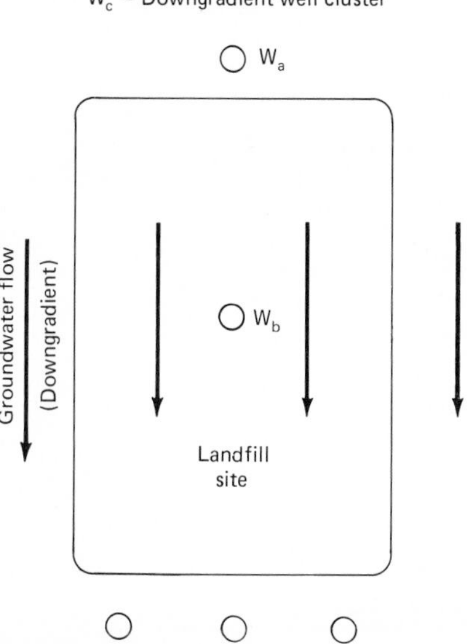

FIG. 6.14 Typical pattern for groundwater monitoring well locations.

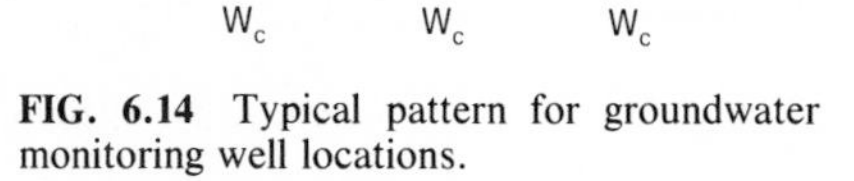

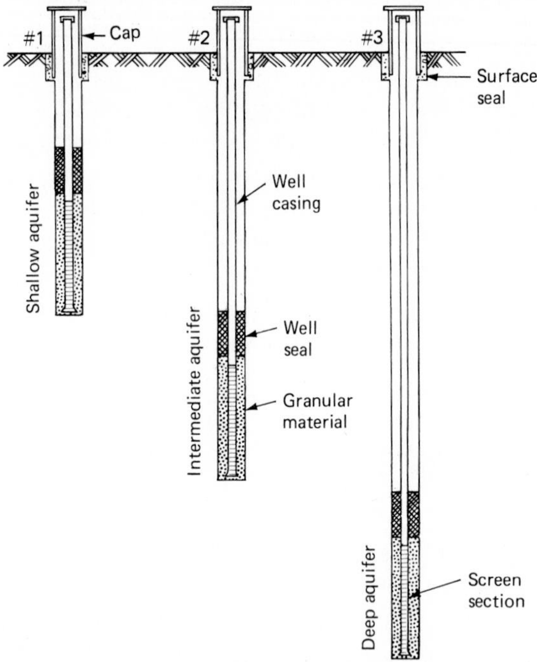

FIG. 6.15 Typical monitoring well cluster to sample multiple aquifers. Well seals prevent water from a high aquifer flowing down well between casing boring wall to sampling zone of deeper well.

ing, wells are sampled quarterly starting with the date of initial use. The early samples provide a basis of comparison to monitor the impact (if any) from disposal site construction and use. Generally, the leachate quality is compared to drinking water standards (Table 6.5). Testing should be conducted according to EPA-endorsed procedures. More testing may be required to accommodate local or state regulations and/or permit conditions.[4,26]

Landfill Design

Landfills are the common solution to disposal and, under current regulations, are more likely to be selected for future disposal sites. Landfill designs must satisfy three basic objectives (Fig. 6.16):

- The landfill must be stable at all times during construction, operation, and closure—and in all weather conditions.
- Surface water management must protect water quality and prevent flooding caused by storms.
- Groundwater must be protected by preventing infiltration and collecting leachate for treatment and disposal.[4,12]

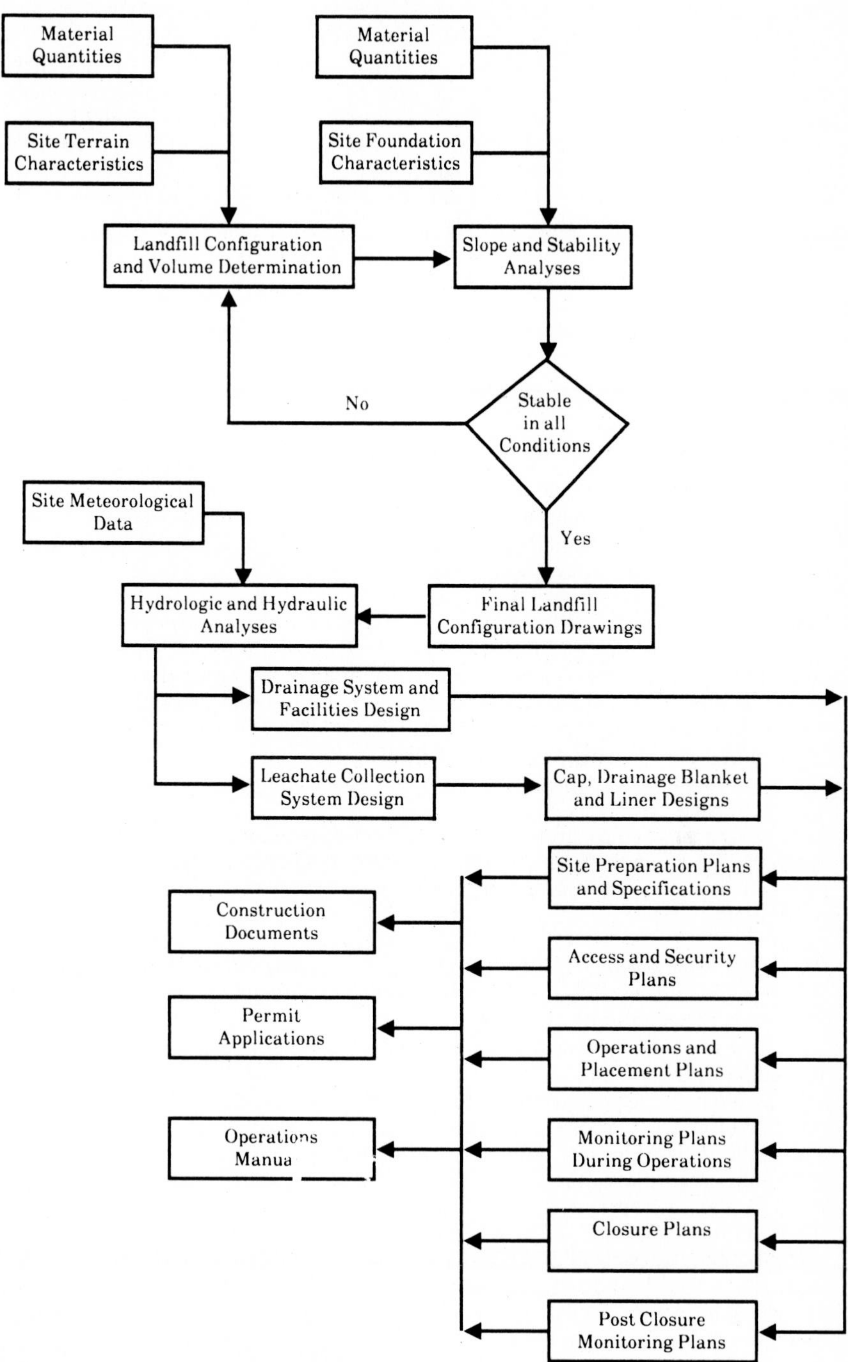

FIG. 6.16 Work flow for landfill design tanks.

Configuration. After the landfill site has been selected, the material quantity projections and site terrain are applied to develop the landfill configuration. The material quantity projection, estimated with assistance from the plant designers, considers coal-burning rate, plant utilization factors, coal quality, flue-gas cleaning system performance, and by-product chemistry and moisture content.[4]

Once the material quantities are established, the site topography is used to develop a landfill configuration, with adequate volume for the anticipated material plus all liner, intermediate cover, and final cover material needed to prepare, operate, and close the landfill. Usually several configurations and landfill boundaries are tried before a final layout is developed that accommodates all the material and makes appropriate utilization of the site.

Stability. The landfill designer's next responsibility is to demonstrate the stability of the landfill in all weather conditions and at all phases of development. Stability analysis must include the landfill materials and the underlying natural soils. In most landfills, potential failure surfaces will exist both within the landfill and extending to the landfill and foundation soils. For this reason, stability analysis input data must include structural data for both the by-products and existing costs. Required data will include compressibility, shear strength, internal friction angles, and other information related to stability and strength. Stability analysis of landfills (or structural fills) is a complicated, demanding technical exercise calling for experienced geotechnical engineers.[4,8,12,16,17,24–26]

General Arrangements. When a final configuration has been devised that provides adequate volume and is stable in all design conditions, the development of landfill drawings begins. Usually, the final landfill configuration drawings are presented as plans and sections showing site preparation, phases of by-product and intermediate cover placement, and completed landfill and final cover-closure arrangements. Development of all three drawing sets can begin when the final configuration is made. Completion of these drawings requires completion of the remaining design steps.

After the final configuration is established, the focus shifts to drainage management, in which both precipitation and process water must be considered. Drainage management affects landfill stability and may require changes in landfill configuration.

Drainage Management. The technical title for drainage management is *hydrology and hydraulics*. Like stability analysis, it is a sophisticated technology, a distilled discussion of which follows.

For most projects, hydrology and hydraulics begins with a thorough review of applicable federal, state, and local meteorological data. The regulatory review will establish the design storm parameters—duration and intensity—to be applied to sizing drainage channels and retention basins. By using data from the National Atmospheric and Oceanic Administration's weather bureau, local data on average annual precipitation, runoff, and specific storm events can be obtained for weather observation stations near the landfill site. In many cases, data from a single station very close to the project will be available; in some cases, however, it will be necessary to interpolate data from two or three stations in the general area.[1,26]

The regulations establish a design storm for the project. It is expressed in total rainfall in a specified number of hours for the most intense storm recorded in a specified number of years. For example, a 3-h, 25-yr storm is the total rainfall in a 3-h period for the greatest storm expected to occur once in 25 yrs.[1]

Separate runoff calculations are made for the land areas above (upgradient) the landfill and for the landfill itself. To the extent possible, the upgradient flows are diverted around the landfill and discharged via natural stream channels without contacting the by-product material. Any run-on entering the landfill plus direct runoff from the landfill surface is collected for prerelease treatment.

For both runoff volume calculations, the same methods are used. The general procedure is as follows:

- Using a planimeter, estimate the area contributing to the runoff being calculated.
- Estimate the runoff coefficient for the contributing area. This coefficient is the proportion of rainfall that runs off the area as surface flow and is determined by considering soil types and vegetative cover (Table 6.6).
- Size the retention basins for holding the runoff from the design storm by multiplying the contributing area by rainfall and the runoff coefficient.
- Size the drainage channels and culverts for peak flow from the design storm. Peak-flow calculations involve contributing areas of specific channel points and concentration multiplied by the peak flows coupled with rainfall and runoff coefficient.
- Design the actual channels and culverts for open-channel flow, using Chezy's and Manning's equations.[1,6] Worked together, these equations estimate the water-carrying capacity of a channel based on cross-section geometry, slope, and interior surface roughness.

After the flows are calculated for each segment of the surface drainage system, detailed design of the drainage facilities begins and includes these critical elements:[1,6]

- Channels and basins must be located, sized, and arranged with due consideration for existing terrain to minimize earth work during site development.
- Planning must include runoff control and sedimentation collection during construction. In most cases, a soil erosion and sedimentation control plan for construction must be submitted to and approved by the local office of the SCS (Department of Agriculture) before construction can begin.

TABLE 6.6 Typical Runoff Coefficients

Watertight roof surfaces	0.75–0.95
Asphalt parameters	0.80–0.95
Concrete pavements	0.70–0.90
Gravel or macadam pavements	0.35–0.70
Impervious soils	0.40–0.65
Impervious soils with turf	0.30–0.55
Slightly pervious soils	0.15–0.40
Slightly pervious soils with turf	0.10–0.30
Moderately pervious soils	0.05–0.20
Moderately pervious soils with turf	0.00–0.10

- Basins must be sized for peak storm storage plus settled sediments, and operations planning must include provisions for periodic sediment removal to maintain the basin volume.
- Design details such as basin inlets and outlets, emergency spillways, channel culverts, and energy dissipaters are addressed at this time.
- Embankments for the retention basins are designed in accordance with applicable regulations.
- Site preparation designs and specifications must meet requirements for the interior surfaces of the runoff diversion and collection channels (Table 6.7).

Leachate Collection.[4] To protect groundwater and to preserve landfill stability, a leachate collection system is installed under the whole landfill. It should drain to the toe line of the fill, and channels should be provided to carry it from there to one or more of the retention basins.

Many leachate collection systems consist of a network of perforated pipes arranged to discharge at the landfill toe. Experience shows that these pipe systems are subject to crushing and sediment plugging and become impossible to clean or repair after material placement starts. Current design practice provides a blanket of granular material laid over the entire landfill "footprint," as shown in Fig. 6.17. The granular material may be sand, gravel, or bottom ash.

Leachate, along with landfill surface runoff, is routed to one or more retention basins. The basin provides for settlement of suspended solids and a weir or clear well allows discharge of water that meets the regulations. In a limited number of cases, additional treatment, such as filtration to remove solids, acid or alkali treatment to modify pH, and removal of specific chemicals, is required to satisfy the regulations.

Cap and Liner Design.[4,21] To isolate landfills, caps and liners are typically included in the program design. The cap, or final cover, is designed to minimize percolation into the landfill, while the liner contains any leachate and routes it to the landfill through the drainage blanket. A well-designed cap will reduce the volume of water percolating into the landfill. Minimizing the water contacting the by-product materials minimizes the potential problems of leachate quantity and quality.

Proper cap design accounts for both permeability and sloping. The cap should include a layer of low-permeability material (10^{-6} cm/s or less) to minimize percolation into the landfill body. The cap should also provide finishing grading plans to promote runoff and eliminate standing water. Runoff should be routed

TABLE 6.7 Comparison of Methods for Interior Surface Preparation of Runoff and Diversion Channels

	Relative costs		Relative performance	
Method	Initial construction	Operating maintenance	Hydraulic conductivity	Channel stability
Planted turf	Low	Moderate	Moderate	Moderate
Geofabric with turf	Moderate	Moderate	Moderate	High
Rip-rap	High	Low	Moderate	High
Paving	High	Low	High	High

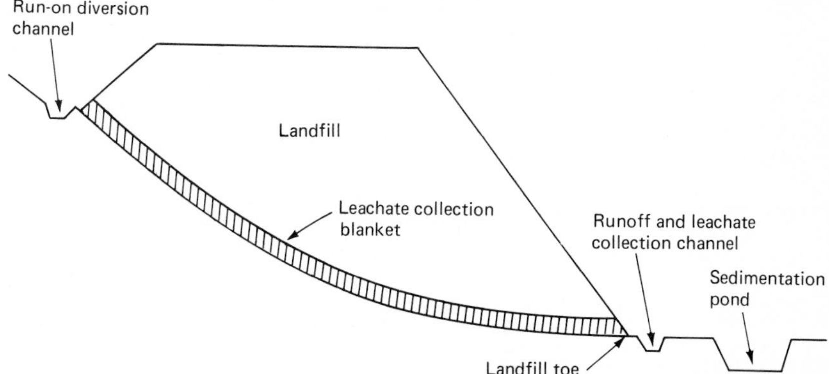

FIG. 6.17 Sidehill landfill with leachate collection blanket and drainage management facilities.

through collection channels to one or more retention basins. Since the runoff has no contact with by-product materials, it can normally be discharged to natural surface water streams after sediment settling in the retention basins.

Design Documents. Design document preparation usually begins with the topographic mapping. By using a photographic process, topographic map sheets are screened to reduce line intensity and provide background sheets for the design drawings. With this approach, a complete design package includes four sets of site maps: topographic maps showing initial conditions, plan drawings showing site preparation, plan drawings showing by-product placement sequence, and plan drawings showing final contours and site closure. For clarity, separate drawings should be prepared for each design phase rather than combining phases on a single set.

By using the analyses and designs compiled earlier, the landfill project plans are prepared. Then they are assembled into packages for site preparation construction, permit applications, and operations manuals. In a typical landfill program, site preparation is done by an outside construction contractor and operations services are performed by utility personnel or a service contractor, so separate document packages are needed for site preparation and operations.

Impoundment Design[4]

Because of typically higher capital costs, greater process management problems, and more difficult permitting, waste impoundments are not selected as often as waste landfills. Increasing regulatory strictures and rising public awareness should bolster the trend toward landfills for disposal of coal-derived by-products. However, some by-product collection processes and powerplant locations make wet disposal preferable, and will dictate the design, permitting, construction, and operation of an impoundment facility. Furthermore, at some plants, dry ash has been used to construct the impoundment for scrubber slurry, thus making dam erection (at least in part) an operating program rather than a capital cost.

Figure 6.18 shows the major tasks required to design a waste impoundment

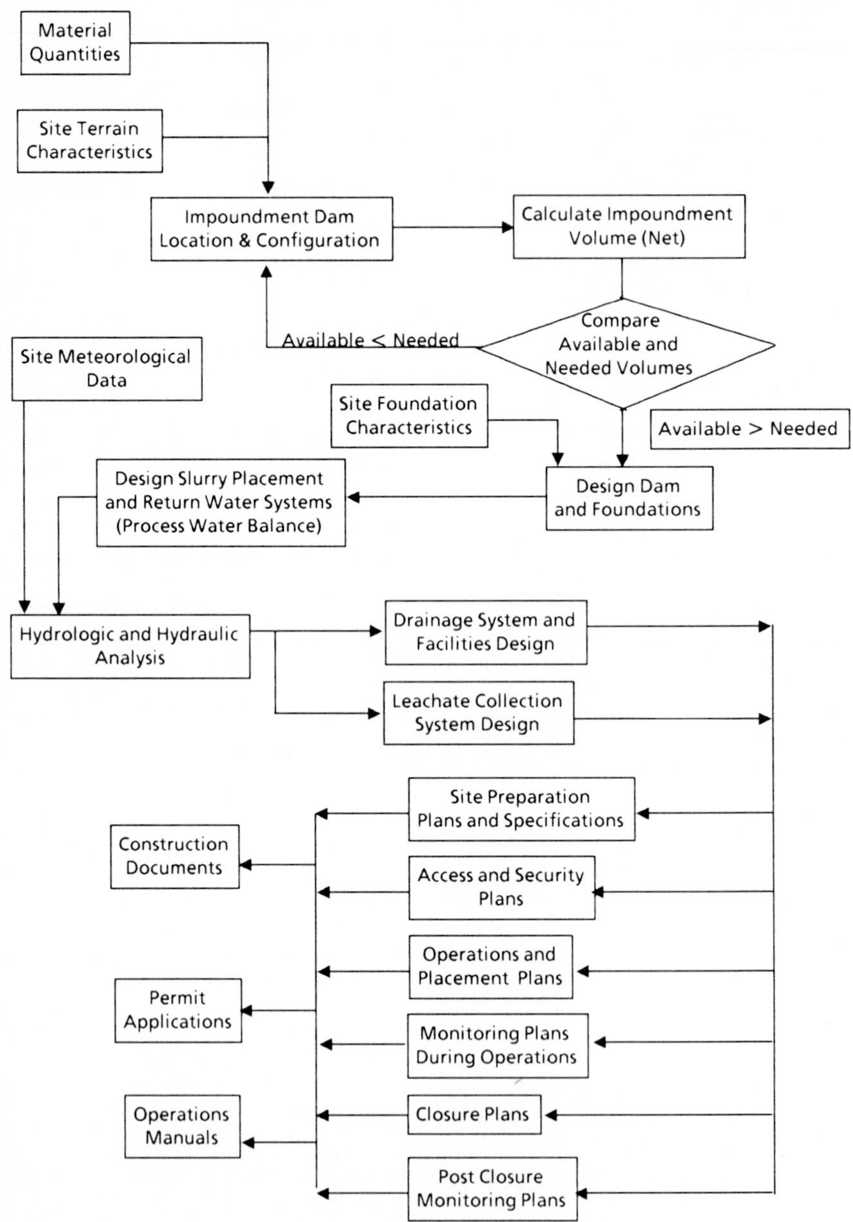

FIG. 6.18 Work flow for impoundment design tanks.

facility. Many of the tasks and work products are similar to those required for landfill design. This discussion focuses on those work elements that are different.

Impoundment Configuration. Selection of the impoundment configuration takes maximum advantage of existing terrain features to define the impoundment containment. By using the best available topographic maps of the site, trial locations for the dam or containment dikes are made, as are trial crest elevations. The slope and toe lines of the dams/dikes are plotted on the map, and the appropriate contour lines are sketched across the faces of the dams/dikes.

By using a planimeter, the net disposal volume of the impoundment is estimated. If plans include obtaining borrow material for dam/dike construction from within the impoundment, this volume should be added to the net disposal volume. Although precise water balance calculations are not possible at this time, an allowance in net volume should be made for storm and process waters. The resulting net-volume estimate is compared to expected by-product volume, settled in place. This becomes an iterative process where dam/dike locations and crest elevations are adjusted until the required net volume (including freeboard) is obtained.

General Arrangements. When the locations and configurations at the dams/dikes have been finalized, general arrangement drawings can be developed. By using the topographic maps, general plans for dams/dikes, pipelines, drainage, access roads, and other support facilities can be made.

Foundation Drilling Program. Final locations and configurations for the dams/dikes allow development of the structural drilling program for impoundment foundations—an extension of the subsurface exploration program outlined under landfills. Borehole locations in the footprint of the dam/dike are chosen to obtain information about two critical issues: bearing capacity of soils under the dam/dike and permeability (or depth to impermeable stratum) of soils under the dam/dike.[26]

Generally, these foundation borings are concentrated along the proposed crest line and toe lines of the planned dam/dike, since these are the areas of greatest stress on the foundation soils. Boring programs should include careful logging of field work and collection of samples for laboratory analysis. Laboratory programs should include compression, shear, and permeability testing. All results should be plotted on boring logs and soil profiles through the site parallel and perpendicular to dam/dike centerlines.

Dam/Dike Design. Impoundment dams/dikes are constructed from earth materials and rely on gravity for stability and strength. Dam/dike profiles are of three types (Fig. 6.19). As a variation on type 3, in a few projects the mass of material downstream of the impervious upstream face is constructed of carefully placed and compacted flyash.[12,26]

Regardless of the materials of construction, the dam/dike must be designed to be stable during all phases of construction and impoundment operation. As a general guide to dam/dike design, these proportions are suggested: upstream slope 3:1 (horizontal:vertical), downstream slopes 2.5:1, and crest width ⅕ times the maximum vertical height.

In addition to standard stability calculations, estimates of seepage and pore pressure must be made. All dams leak; the variable is the seepage rate. Seepage calculations should also include seepage flows through the soils under the dam.

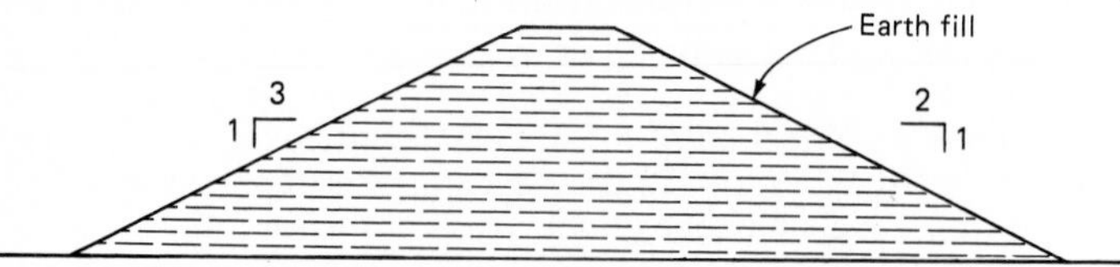

Type 1: Homogeneous earth fill construction

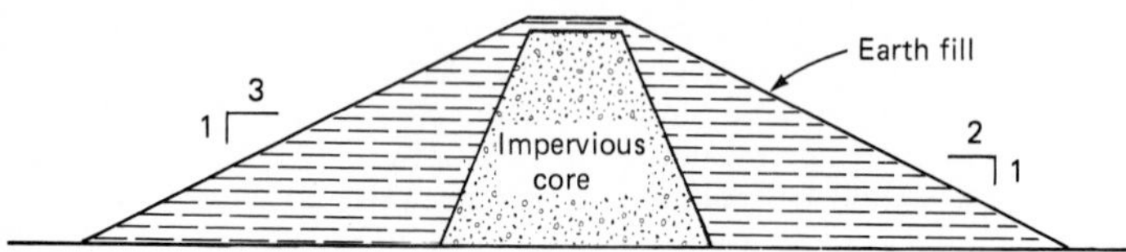

Type 2: Impervious core with earth fill

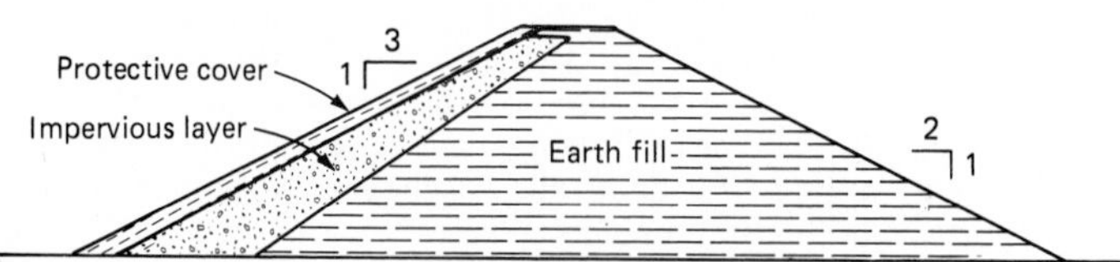

Type 3: Impervious upstream face with earth fill

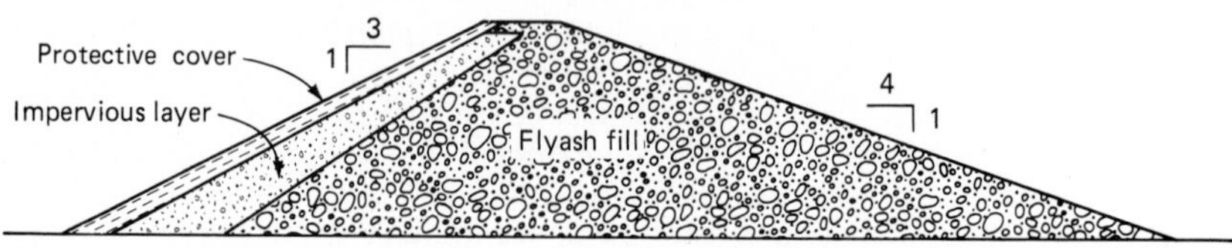

Type 3: Alternate—Impervious upstream face with flyash fill

FIG. 6.19 Typical gravity earth-fill dams.

Seepage-related pore pressures can be controlled, to some extent, by using a rock toe or filter blanket (Fig. 6.20). These devices reduce the pore pressures above the upper limit of seepage and thus increase the physical stability of the downstream face of the dam. Bottom ash may be substituted for some of or all the rock and sand materials in these drainage aids.[12,26]

Dam Foundations. The soils under the dam must be analyzed for compressive strength (settlement), shear strength (dam stability), and permeability (seepage). These analyses start with the data collected during the programs mentioned earlier under Subsurface Investigations and Foundation Drilling Program.

If compressive and shear strengths are not adequate, the engineer should consider remedial actions to increase soil strength. These actions include removal and replacement or recompaction of the inadequate soils, grouting, and dynamic in-place compaction. If the soils under the dam are pervious, several options should be considered (Fig. 6.21). If an impervious layer is known to exist at a shallow depth below the bottom of the dam, a cutoff trench can be excavated through the pervious soils under the dam and backfilled with an impervious soil material (Fig.

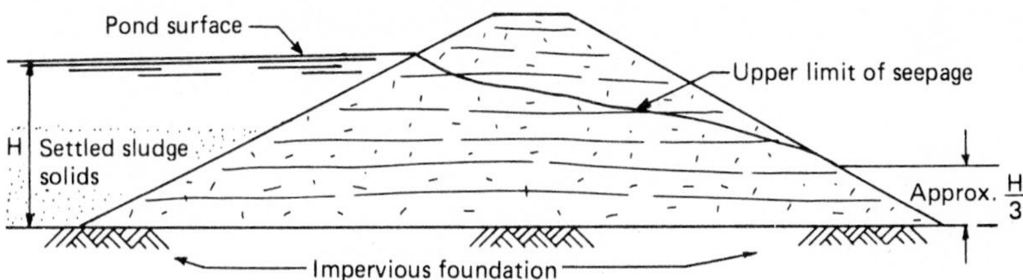

(a) Completely homogeneous

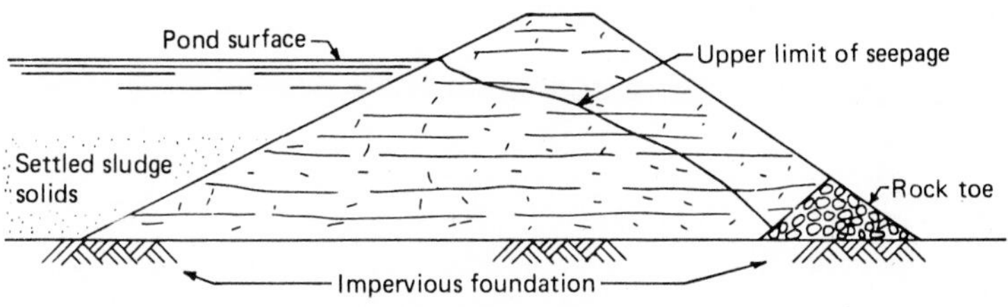

(b) With rock toe

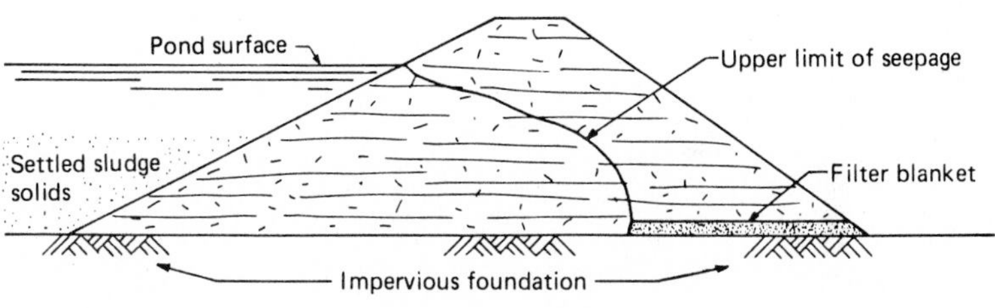

(c) With filter underdrain

FIG. 6.20 Seepage patterns and modifiers for earth-fill dams.

6.21*a*). If the depth to the impervious strata is greater, grouting or sheet piling can be added to control seepage under the dam (Fig. 6.21*b* and *c*). In many cases, adequate seepage control is achieved in a cost-effective manner by using a combination of these methods.

Slurry Placement and Return-Water Systems. The project plan must include facilities for slurry placement and clear water return. Typically, the slurry pipelines extend out onto the impoundment surface and are supported by floats to rise with the water surface and to move about the impoundment as required. The pipelines normally discharge below the water surface to minimize suspended and dissolved solids at the surface of the reservoir.

To minimize impacts on local water quality, most slurry systems have a return-water loop. In a typical arrangement, surplus water from each waste processing step (thickeners, filters, settling ponds, and impoundments) is returned to

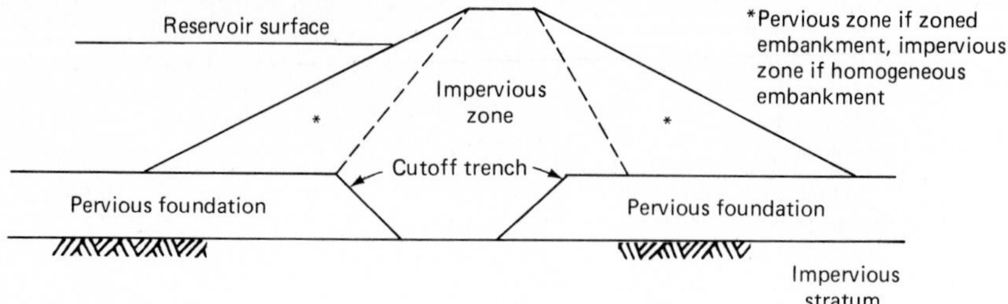

(a) Cutoff trench for shallow pervious foundations

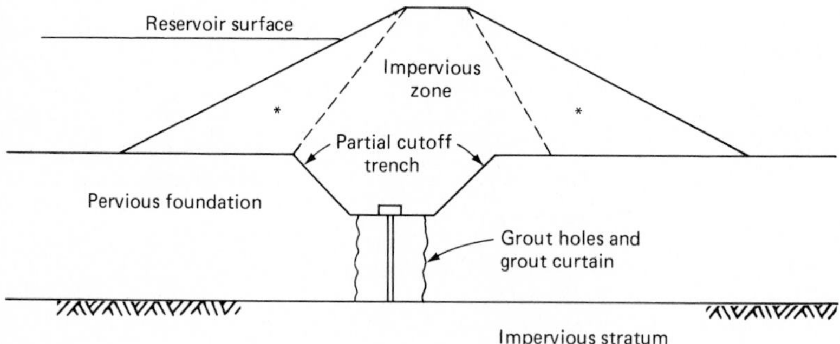

(b) Partial cutoff trench with grout curtain

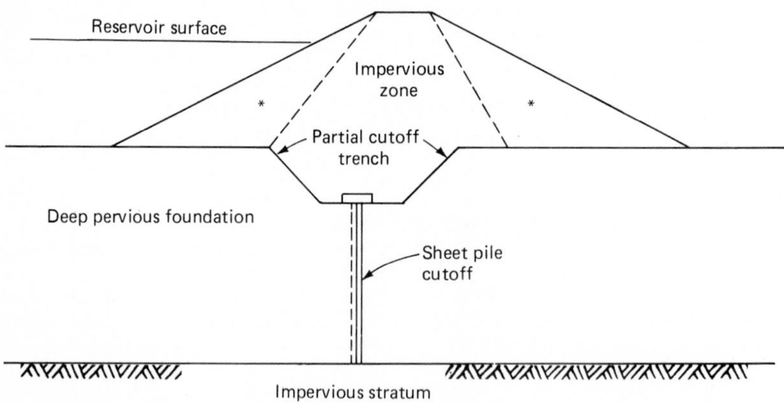

(c) Partial cutoff trench with sheet pile barrier for deep pervious foundation

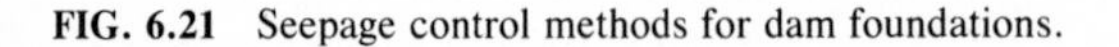

FIG. 6.21 Seepage control methods for dam foundations.

the scrubber makeup water supply. At most slurry impoundments, return-water intakes are located at the reservoir surface away from the slurry outlet. The intake and return-water pumps are usually on a raft designed to move up and down with the water surface. This arrangement minimizes "short-circuiting" of slurry or high-solids-content water directly from the slurry outlets to the return-water intakes.

Drainage Management. By using the methods outlined under drainage management for landfills, rainfall directly on the reservoir and runoff from the upgradient areas are estimated for peak storm events and for an average year. If possible, runoff from upgradient areas should be routed around the impoundment, just as with landfills. Runoff that must enter the impoundment and rainfall directly on the reservoir surface are managed through the reservoir.

An overall water balance can be calculated for the whole impoundment by considering these plus or minus factors:

- Surplus process water, that is, the water released as the slurry solids consolidate at the bottom of the reservoir (plus)
- Return-water demands at the powerplant (minus)
- Rainfall on the reservoir and upgradient runoff routed to the reservoir (plus)
- Evaporation from the reservoir surface (minus)

Both rainfall and evaporation data are obtained from the records of the nearest weather station. Normally, the sum of rainfall and evaporation will leave a surplus of water to be discharged to the preexisting water channels. A separate intake should be provided for this surplus water discharge over the dam to the stream channel. It should be located near the return-water intake and use the same raft-mounted pump and intake system.[1,29]

To satisfy regulatory requirements, the water discharged to existing stream channels must be monitored for water quality. If necessary, appropriate predischarge treatment systems must be provided. The most common problem is pH adjustment. Most ash and sludge slurries are alkaline, and the supernatant must be treated with acid to meet the pH limits for free discharge.

Leachate Management. For impoundments, leachate collection comes through the toe drains or drainage blankets of the dam or dikes. These flows are routed to the stilling basin or treatment area for the supernatant discharge system. Quality monitoring and treatment will also be applied to these flows.[4,13,26]

Design Documents. The lists of design documents and packages are the same for an impoundment project as for a landfill, although the contents of each document are different.

TRANSPORTATION

As used here, the word *transportation* includes all the material handling procedures required to move the waste products from their areas of generation through treatment (if any) to their final placement in the disposal or long-term storage area. Typically (but not universally), the waste material travels in a series of transportation segments separated by accumulation, storage, or treatment facilities. Each segment must be designed separately to respond to the conditions prevailing at that point. Thus, the planner will develop a series of transportation concepts based on economics and other factors. The available options, reviewed here in terms of their applicability, advantages, and disadvantages, will help in developing the concepts.

Options Available[4]

For both wet (slurry) and dry materials, seven basic transportation modes are available:

- *Pneumatic systems:* Applicable to dry, fine-grained materials only, these systems commonly are selected for short moves inside the powerplant. They are flexible in operating rates, are easily routed around obstacles in the plant, and have low environmental impact. They are also capital-intensive and limited in effective travel distance.
- *Pipelines:* Pipelines handle slurries within the plant just as pneumatic systems handle dry materials. In addition, since their effective range is virtually unlimited, pipelines often see service for long-distance transportation. They are easily routed around obstacles and over steep terrain; moreover, they are reliable and generally produce relatively low environmental impacts. Pipelines are relatively inflexible in operating rates, because fluid velocities must be in the turbulent-flow range to prevent settlement of solids within the pipe. Pipelines are also costly to build and require careful, if not extensive, maintenance.
- *Conveyors:* Effective for both dry materials and fixed or stabilized sludges, conveyors are a widely applied, proven means of material transport. They mainly see service for short transport segments; in some instances, however, their range has been extended to 20 mi (32 km). Because of their high reliability and capacity, conveyors are most appropriate where large volumes of waste must be moved short distances. Used with bridges, conveyors have also been successful where the haul route crosses steep terrain. They are costly to construct and have significant environmental impact because of their ungainly structure.
- *Aerial trams:* A variation on conveyors, aerial trams are occasionally selected when relatively small quantities of material must be moved over adverse terrain. Their capital cost is less than that of conveyors, and they are more easily routed around obstacles; because of their lighter structure, their environmental impact is also less. Aerial trams have high operating costs and generally low capacities.
- *Trucks:* Among vehicles legally operable on public highways, this mode of transportation is common when the powerplant and disposal area are not contiguous; in fact, trucks are probably the most common single method of waste transportation. Alone among all the transport modes, trucks do not require ownership or lease rights for rights-of-way. Their advantages are great flexibility in the face of changes in waste production rates (changes in work hours or fleet size are easily arranged) and low capital costs (some leasing arrangements eliminate all capital costs to the utility). Disadvantages are high operating costs and significant environmental impact, especially on the public highway portions of the haul route. Trucks are normally selected to haul dry or fixed stabilized wastes; however, with specially designed bodies or with sealed tailgates, they have been used to carry sludges with some success.
- *Off-road vehicles:* Both large-capacity off-road trucks and construction equipment, such as modified scrapers and front-end loaders, are available to transport waste products where the disposal area can be accessed from the powerplant on private rights-of-way in which weight and vehicle size restrictions do not govern. Sometimes the vehicles have been modified to increase travel speed or carrying capacity or to make them suitable for carrying slurries. Where conditions allow their use, off-road vehicles offer flexibility, reliability,

and moderate to low capital costs. Their operating costs may be high because of vehicle and roadway maintenance. They may also be responsible for substantial environmental impact.

- *Railroad:* Transport of waste products by rail is feasible in situations in which the wastes can be returned to the mine that provided the fuel. These special circumstances have arisen at mine-mouth plants and/or at very large plants where huge volumes of fuel and waste are involved. Rail systems are dependable and flexible in terms of operating rates. High capital and fixed operating costs limit the desirability of rail systems to plants whose large volumes result in acceptable unit costs. Railroads also require right-of-way ownership and often entail major impact to the environment.

Advantages and Disadvantages

Tables 6.8 and 6.9 summarize the principal conditions of applicability and the more prominent advantages and disadvantages of each transportation mode just described.

UTILIZATION[2]

As a general concept, utilization is attractive to engineers confronted with the problems and costs of waste management. The concept grows more attractive as the cost of disposal creates opportunities to subsidize the utilization and improve the competitive position of the waste materials vis-à-vis other materials with the same uses. The reality of utilization so far has not lived up to its apparent attractiveness. Only minor amounts of FGD products are utilized for any purpose. Although significant portions of bottom ash (25 percent) and flyash (14 percent) are utilized, the majorities of each are still sent to landfills. The value of utilization actually accomplished is limited by the extremely low value and low technology of the utilization practices actually applied.

Table 6.10 presents a general summary of the utilization markets assessed in terms of their potential value and outlook. The attractive markets involve high volumes of bottom ash and flyash in low-value, low-technology applications. The applications for FGD products are still in the study stage and are generally restricted by the high costs of the required technology.

Ash Utilization

Currently, only bottom ash (including boiler slag) and flyash are utilized in any significant quantities. The actual uses are mainly bulk consumption of the wastes with minimal processing or refining. As a result, the material values derived from utilization are very low, and the principal advantage is avoidance of disposal costs.

Utilization is also limited by the economics of transportation. The product values are so low that transportation distances are limited to areas where the producing powerplant is the closest source, unless the utility chooses to subsidize the transportation costs by direct payments to haulers or users.

Table 6.11 lists areas where significant quantities of bottom ash and flyash are currently utilized. The major areas are ones in which the ash material can be

TABLE 6.8 Applicability of Transportation Modes

Transportation mode	Material type			Haul distance			Waste volume			Terrain	
	Dry granular	Fixed or stabilized sludge	Slurry	Short within plant	Long within plant	Long out of plant	Small	Medium	Large	Steep	Normal
Pneumatic	1	4	4	1	2	3	1	1	3	1	1
Pipeline	4	4	1	1	1	1	1	1	1	1	1
Conveyor	1	1	4	1	1	1	2	2	1	1	1
Aerial tram	1	1	3	3	2	2	1	1	2	1	1
Truck	1	1	3	4	3	1	1	1	1	3	1
Off-road vehicle	1	1	3	4	1	3	1	1	1	3	1
Railroad	1	1	3	4	3	3	4	4	1	3	1

Note: 1 = broadly applicable, 2 = sometimes applicable, 3 = applicable in special circumstances, 4 = not applicable.

TABLE 6.9 Advantages and Disadvantages of Transportation Modes

Transportation mode	Costs			Reliability	Flexibility	Environmental impact	Public impact
	Capital	Fixed operating	Variable operating				
Pneumatic	H	L	L	M	L	L	L
Pipeline	H	L	L	M	L	L	L
Conveyor	H	M	L	H	M	M	M
Aerial tram	H	L	L	M	H	M	M
Truck	L	L	H	H	H	H	H
Off-road vehicle	L	M	H	H	H	H	H
Railroad	H	H	M	H	H	H	H

Note: H = high, M = moderate, L = low.

TABLE 6.10 Utilization Market Assessment Summary

Utilization markets	Conventional materials	Type by-product*	Potential by-product volume	Technology requirements	Market value	Major advantage	Major disadvantage	Utilization outlook
Cement	Cement	BA, FA	Moderate	Moderate	High	Cost savings	Quality control	Good
Concrete and construction materials	Sand, gravel, and stone	BA, FA	Moderate	Moderate	Low	Cost savings	Quality control	Good
Bituminous pavements	Sand and gravel, stone	BA, BS, FA	High	Low	Low	Processing economics	Product acceptance	Moderate
Structural fill and fill materials	Soil, stone, sand and gravel	BA, BS, FA	High	Low	Low	Urban and industrial proximity	Product acceptance	Good
Soil stabilization	Lime, cement	FA	Low	Moderate	High	Cost savings	—	Moderate
Deicer/anti-skid	Salt, sand, and gravel	BA, BS	Moderate–high	Low	Low–moderate	Noncorrosive	—	Good
Blasting grit	Stone, sand, and gravel	BS	Moderate–high	Moderate	Moderate	Cost savings	—	Good
Roofing granules	Stone, sand, and gravel	BA, BS	Moderate	Moderate	Low	—	—	Good
Grouting	Cement	FA	Low–moderate	Moderate	High	Cost savings	Ash quality	Good
Mineral wool	Furnace slag, wool rock	BS, FA	Low	Moderate	Moderate	Market proximity	Atypical furnace	Moderate
Agriculture	Ag-lime fertilizers	FA, FGD	High	Low	Low	—	Replacement ratio	Poor–moderate
Metals recovery†	Natural ores	FA, FGD	High	High	High	—	Costs, residue volume	Low
Sulfur recovery	Natural sulfur	FGD	High	High	Moderate	—	Costs	Low
Gypsum	Natural gypsum	FGD	High	Moderate	Moderate	—	Product acceptance	Low

*BA = bottom ash; BS = boiler slag; FA = flyash; FGD = flue-gas desulfurization products.
†Includes aluminum, titanium, iron, and silica.
Adapted from *Coal Combustion By-products Utilization Manual,* vol. 1: *Evaluating the Utilization Option,* Table 4-1, Electric Power Research Institute, EPRI CS-3122, Feb. 1984.

TABLE 6.11 Reported Ash Utilization in the United States in 1983, 10^3 short tons (10^3 megagrams)*

	Utilization area	Bottom ash or boiler slag utilized	Flyash utilized	Total ash utilized
Aggregate	63 (57)	63 (57)	126 (114)	
Blasting grit	1,239 (1124)	NA	NA	1,239 (1124)
Cement additive	490 (444)	1,618 (1468)	2,108 (1912)	
Concrete admixture	126 (114)	3,097 (2809)	3,223 (2923)	
Concrete block	45 (41)	314 (285)	359 (326)	
Dam construction	0 (0)	155 (141)	155 (141)	
Fill material	700 (635)	391 (355)	1,091 (990)	
Grouting	NA	NA	541 (491)	541 (491)
Hazardous waste fixation	NA	NA	63 (57)	63 (57)
Ice control	577 (523)	43 (39)	620 (562)	
Roadway construction	595 (540)	227 (206)	822 (746)	
Roofing granules	686 (622)	NA	NA	686 (622)
Other	19 (17)	203 (184)	222 (201)	
Total utilized	4,540 (4118)	6,715 (6091)	11,255 (10 208)	
Total produced	18,089 (16 407)	48,310 (43 818)	66,399 (60 224)	

*Adapted from *Coal Combustion By-products Utilization Manual*, vol. 1: *Evaluating the Utilization Option*, Table 2-1, EPRI CS-3122, Electric Power Research Institute, Palo Alto, Calif., Feb. 1984.

substituted for sand or gravel. This central fact shapes and controls the economics and practices of utilization. The strong, granular nature of bottom ash makes it an attractive material for use as blasting grit, controlled fills, and roofing granules. Both the chemical and physical nature of flyash make it suitable for some applications in the manufacture of portland cement and in the preparation of concrete mixes.

FGD By-product Utilization[2]

Utilization of FGD by-products is more studied than practiced. Several options have been explored in depth, but implementation so far has been deferred for economic reasons. Some of the more significant methods studied to date are as follows:

- *Agriculture*: Generally FGD by-products contain large amounts of lime and thus may be suitable for substitution for natural agricultural lime. However, such use has been blocked in practice by two facts: (1) available lime is lower in the FGD by-product than in the natural material, so more is required for similar effectiveness; and (2) trace elements in FGD by-products may have an unacceptable impact.
- *Metals recovery*: Both flyash and FGD by-products contain aluminum, titanium, iron, silica, and other metals in very low concentrations. The technology is not fully developed yet for recovery of trace elements, so this approach does not appear cost-effective at current metals prices. Furthermore, the final waste volume is not significantly reduced by the extraction of trace amounts of these metals.

- *Sulfur recovery*: FGD by-products are also rich in sulfur that may be recoverable in its elemental, thus marketable, form. The obstacles are the incomplete development of the recovery technology, the low market price of natural sulfur, and the large volume of by-product remaining after the sulfur is extracted.
- *Gypsum*: Many FGD systems that are operated with forced oxidation can be made to produce gypsum suitable for drywall and other construction uses. This is a potentially large-volume method of utilization, but actual implementation continues to be limited by the price of natural gypsum and by trace materials in the by-product gypsum that may bleed through to the finished wall surfaces.

Site Utilization[4]

Some utilities have determined that the best way to utilize waste products is in the construction of engineered fills. Disposal sites are chosen, in part, as developable locations that are hindered by terrain problems. The by-products are placed in systematic, compacted lifts designed to recontour the site to correct the terrain problems.

After completion, the site is ready for commercial or light industrial development. The advantage comes from the increased value of the site at completion, not from reduced disposal costs during site operations. However, a delayed advantage is better than no advantage. The ingenuity of utility engineers is challenged in site selection and disposal area design, but the rewards can be well worth the extra effort.

ACKNOWLEDGEMENT

The author would like to thank Baker Engineers and the Electric Power Research Institute for the use of Figs. 6.1 to 6.8, 6.14, 6.17, and 6.19 to 6.21.

REFERENCES

1. *Handbook of Steel Drainage and Highway Construction Products,* American Iron and Steel Institute (AISI), Washington, D.C., 1971.
2. Michael Baker, Jr., Inc. (Consulting Engineers), *Coal Combustion By-products Utilization Manual,* vol. 1: *Evaluating the Utilization Option,* Electric Power Research Institute, Palo Alto, Calif., 1984.
3. Michael Baker, Jr., Inc. (Consulting Engineers), *Engineer-Evaluation of Projected Solid-Waste Disposal Practices,* vol. 1, Electric Power Research Institute, Palo Alto, Calif., 1982.
4. Michael Baker, Jr., Inc. (Consulting Engineers), *FGD By-product Disposal Manual,* 3d ed., Electric Power Research Institute, Palo Alto, Calif., 1983.
5. H. R. Cedergren, *Seepage Drainage and Flow Nets,* Wiley, New York, 1967.
6. V. T. Chow, *Open-Channel Hydraulics,* McGraw-Hill, New York, 1959.
7. *Hydraulic Design of Spillways,* Engineer Manual, EM 1110-2-1603, Corps of Engineers, Department of the Army, Washington, D.C., 1965.

8. *Stability of Earth and Rockfill Dams,* Engineer Manual, EM 1110-3-1902, Corps of Engineers, Department of the Army, Washington, D.C., 1960.

9. E. D'Appolonia Consulting Engineers, Inc., *Engineering and Design Manual, Coal Refuse Disposal Facilities,* Mining Enforcement and Safety Administration, Department of the Interior, Washington, D.C., 1975.

10. *Design Charts for Open-Channel Flow,* Federal Highway Administration, Department of Transportation, Washington, D.C., 1973.

11. *Hydraulic Design of Improved Inlets for Culverts,* Federal Highway Administration, Department of Transportation, Washington, D.C., 1974.

12. GAI Consultants, Inc., *Ash Disposal Reference Manual,* Electric Power Research Institute, Palo Alto, Calif., 1979.

13. M. E. Harr, *Groundwater and Seepage,* McGraw-Hill, New York, 1962.

14. J. A. Havers and F. W. Stubbs, Jr. (eds.), *Handbook of Heavy Construction,* McGraw-Hill, New York, 1971.

15. F. M. Henderson, *Open Channel Flow,* Macmillan, New York, 1969.

16. R. C. Hirschfield and S. J. Poulos, *Embankment-Dam Engineering,* Wiley, New York, 1973.

17. Roy E. Hunt, *Geotechnical Engineering Techniques and Practices,* Chap. 8: "Slope and Embankment Stability," McGraw-Hill, New York, 1986.

18. H. W. King and E. F. Brater, *Handbook of Hydraulics,* McGraw-Hill, New York, 1963.

19. G. A. Leonards (ed.), *Foundation Engineering,* McGraw-Hill, New York, 1963.

20. R. K. Linsley and J. B. Fraxini, *Water-Resources Engineering,* McGraw-Hill, New York, 1972.

21. R. J. Lulton et al., *Design and Construction of Covers for Solid Waste Landfills,* EPA-600/2-79-165, Environmental Protection Agency, Research Triangle Park, N.C., Aug. 1979.

22. *Soil Erosion and Sedimentation Control Manual,* Pennsylvania Department of Environmental Resources, Harrisburg, Pa., 1973.

23. *Engineering Field Manual for Conservation Practices,* Soil Conservation Service, Department of Agriculture, Washington, D.C., 1969.

24. Donald W. Taylor, *Fundamentals of Soil Mechanics,* Chap. 16: "Stability of Slopes," Wiley, New York, 1965.

25. K. Terzaghi and R. B. Peck, *Soil Mechanics in Engineering Practice,* 2d ed., Wiley, New York, 1967.

26. *Design of Small Dams,* A Water Resources Technical Publication, Bureau of Reclamation, Department of the Interior, Washington, D.C., 1965.

27. *Earth Manual,* A Water Resources Technical Publication, Bureau of Reclamation, Department of the Interior, Washington, D.C., 1974.

28. *Soil Mechanics Design Manual,* Chap. 7: "Slope Stability and Protection," NAVFAC DM-7.1, Naval Facilities Engineering Command, U.S. Navy, Alexandria, Va., 1982.

29. *Rainfall Frequency Atlas of the United States,* Technical Paper no. 40, U.S. Weather Bureau, Department of Commerce, Washington, D.C., 1963.

PART 3

NOISE CONTROL

CHAPTER 4.1
POWERPLANT NOISE AND ITS CONTROL

Robert M. Hoover and Robert D. Bruce
Hoover Keith & Bruce Inc.
Houston, Tex.

INTRODUCTION

Fossil-fuel and nuclear powerplants use equipment that has the potential for causing excessive noise exposure for plant personnel or annoyance and noise complaints from nearby residential areas. For neighbors, the plant operation should not produce any noise in the community that might be considered distracting or annoying. For in-plant personnel, noise should be controlled to avoid the risk of hearing damage and interference with speech, communication, and alarm systems.

Neighborhood noise complaints have been caused by *forced-draft* (FD) and *induced-draft* (ID) fans, *primary-air* (PA) fans, transformers, safety and control valves, deaerator vents, motors, boiler feed pumps, precipitator rappers and vibrators, coal car shakers, start-up blowdown operations, and sound paging systems.[1,2] Within the plant, dominant noise sources are FD and PA fans, boiler feed pumps, and control valves. Other significant noise sources can be air compressors, seal air blowers, motors, deaerators, coal mills, coal car shakers, and the various components of the turbine-generator system.[2,3]

To control the noise of these sources, engineers should apply appropriate noise control measures during the design and specification stages of powerplant equipment. They may first make use of various noise control prediction procedures for specific pieces of equipment. The predicted noise levels may be compared with appropriate noise criteria, and if necessary, various noise control remedies such as mufflers, enclosures, or lagging can be incorporated to meet the criteria.

In estimating in-plant noise, another factor is the acoustical environment of the powerplant. In an enclosed plant, the spaces are typically live and reverberant; that is, in any given area of the plant, the noise level is controlled by a number of sources, not just the nearest noise source. In addition, the various operating levels of the plant are usually acoustically interconnected via open gratings, hatches, or stairways. Thus, a loud boiler feed pump at a basement level may be quite audible on the turbine floor. For an open powerplant, most of the areas become less reverberant, but neighborhood complaints due to equipment noise increase.

TERMS USED

Some acoustical and noise control terminology and noise criteria are reviewed in the following paragraphs. These terms are important for a full understanding of noise and its control.

Basic Parameters of Sound

The basic parameters of sound are its magnitude, its frequency content, and its duration. The magnitude of a sound is most often measured in decibels (dB). The sound pressure level L_p in decibels is 20 times the logarithm of the ratio of the pressure of the sound to the reference pressure, which is 20×10^{-6} Pa. Sound pressure level L_p is a function of the sound power of a source, the distance from the source, and the environment in which the source is operating. The sound power level L_w of a source is relatively independent of source location or characteristics of the space in which the source is operated; L_w is 10 times the logarithm of the ratio of the sound power of the source to 10^{-12} W.

The frequency content of a sound can be determined by weighting networks that filter the acoustical signal and by narrowband filters. The most common weighting network used in the analysis of powerplant noise is the A-weighted network. The A-weighted sound level is the L_p that has been filtered through the A-weighted network (Table 1.1). The most common narrowband filters are (1) the constant-bandwidth filter in which the bandwidth is constant for all center frequencies and (2) the constant-percentage bandwidth type in which the bandwidth is a constant percentage of the center frequency.

The most common of the constant-percentage bandwidth types is the octave-band filter. The octave-band center frequencies normally used are 31.5, 63, 125, 250, 500, 1000, 2000, 4000, and 8000 Hz. In many instances, the sound consists of tonal components superimposed on a broadband noise where a tone is a sound

TABLE 1.1 Frequency Response of A-Weighting Network

Frequency, Hz	A-weighting relative response, dB
10	− 70.4
12.5	− 63.4
16	− 56.7
20	− 50.5
25	− 44.7
31.5	− 39.7
40	− 34.6
50	− 30.2
63	− 26.2
80	− 22.5
100	− 19.1
125	− 16.1
160	− 13.4
200	− 10.9
250	− 8.6
315	− 6.6
400	− 4.8
500	− 3.2
630	− 1.9
800	− 0.8
1,000	0
1,250	+ 0.6
1,600	+ 1.0
2,000	+ 1.2
2,500	+ 1.3
3,150	+ 1.2
4,000	+ 1.0
5,000	+ 0.5
6,300	− 0.1
8,000	− 1.1
10,000	− 2.5
12,500	− 4.3
16,000	− 6.6
20,000	− 9.3

that has a single pitch.[4] A narrowband constant-percentage filter or a constant-bandwidth filter is used for analysis of tonal components.

Another important parameter of sound is its duration. This can be quantified by noting the hours the source operates, by statistically evaluating the percentage of time that L_p is above a particular level or within a particular range, and by reporting the level that is exceeded for a certain percentage of the time. Two other methods use the equivalent sound level L_{eq} and the day-night sound level L_{dn}. Here L_{eq} is defined as the sound level of a steady sound having the same energy as the time-varying sound. The day-night sound level is defined as

$$L_{dn} = 10 \log \frac{1}{24}\left[15 \times 10^{\left(\frac{L_d}{10}\right)} + 9 \times 10^{\left(\frac{L_n + 10}{10}\right)}\right]$$

where $L_d = L_{eq}$ for the daytime period (7 a.m. to 10 p.m.). and $L_n = L_{eq}$ for the nighttime period (10 p.m. to 7 a.m.).

Sound Propagation

At a receiver location outdoors L_p can be calculated from

$$L_{p\theta} = L_w + DI_\theta - 20 \log r - A - 1 \qquad [-1 = -11 \text{ in SI}]$$

where $L_{p\theta}$ = L_p in direction θ
L_w = L_w of source, dB re 10^{-12} W
DI_θ = directivity index of the source in direction θ; DI is 3 dB for hemispherical radiation, such as a source resting on a flat plate
r = distance from source to receiver, ft (m)
A = sum of attenuations caused by atmospheric absorption, weather conditions, barriers, mufflers, enclosures, etc.

Typically, these calculations would be performed in each of the octave bands. This equation illustrates that for every doubling of the distance from the source, the L_p will decrease by 6 dB. Attenuation values for atmospheric absorption, weather conditions, and barriers can be determined from the literature.[3,4] Indoors L_p can be estimated from

$$L_p = L_w + 10 \log \left(\frac{Q}{4\pi r^2} + \frac{4}{S\overline{\alpha}} \right) + 10 \qquad [+10 = 0 \text{ in SI}]$$

where Q = directivity of source
S = total area of surfaces of room, ft² (m²)
$\overline{\alpha}$ = average Sabine absorption coefficient for the room, including objects

For large values of r, the distance term becomes small and L_p in the space is controlled by the reverberant energy. In analyzing an indoor noise problem, calculations are typically performed in octave bands of frequency.[4]

Noise Criteria

Noise criteria (NC) are available to the power industry for both in-plant noise and noise in the community. In this discussion, both of these types are covered.

In-Plant NC. The purposes of in-plant NC are to

- Reduce the likelihood of employees incurring *noise-induced permanent threshold shift* (NIPTS), or hearing loss
- Create an acoustic environment where speech in person or over the telephone is more readily understood
- Reduce annoyance caused by particular sounds

TABLE 1.2 OSHA Noise Exposure Regulations

Exposure sound level, dB(A)	Allowable time at sound level, h
90	8
95	4
100	2
105	1
110	0.5
115	0.25

Note: No exposures to continuous sounds greater than 115 dB(A) are allowed under the regulations. If the exposure is to a variety of sound levels (typical in industry), then the *daily noise dose* (DND) determines the acceptability of a noise exposure:

$$\text{DND} = \frac{c_1}{t_1} + \frac{c_2}{t_2} + \frac{c_3}{t_3} + \ldots \frac{c_n}{t_n}$$

where c_n = actual exposure time at a sound level and t_n = allowable exposure time at that sound level. The DND should be less than or equal to 1.0.

These purposes can be achieved through careful selection and application of the criteria. At present, the *Occupational Safety & Health Administration* (OSHA) has the responsibility of regulating the noise exposure of employees in the United States. The basic purpose of this regulation is to reduce the amount of NIPTS that employees will have when they retire after 30 or 40 yr of service (Table 1.2). If their daily noise exposure is greater than 85 dB(A), OSHA requires all employees to be in a hearing protection program.[5]

A number of corporations have developed more restrictive goals than required by OSHA. Key ones are

- To allow 85 dB(A) for a time-weighted exposure rather than the 90 db(A) allowed by OSHA
- To limit the noise levels throughout the plant to less than 85 db(A) in all locations where employees are exposed
- To provide hearing protection to maintenance people working in noisier areas

Companies have set these more restrictive goals because of their concern for employees and to protect themselves from future lawsuits seeking compensation for NIPTS. Also, it is imperative that all employees be able to hear and understand warning signals and announcements.

Community NC. The selection of community NC requires careful deliberation. First, ambient sound levels in the community must be determined: typically, L_{eq} or L_{dn} values are found; in addition, measurements of L_p in octave bands of frequency are made. Then local and state noise regulations are identified and analyzed. In some locations, powerplant noise must be controlled such that the sound levels after the plant is in operation are not greater than the ambient levels. In other locations, no regulations govern the noise radiated from a powerplant. In

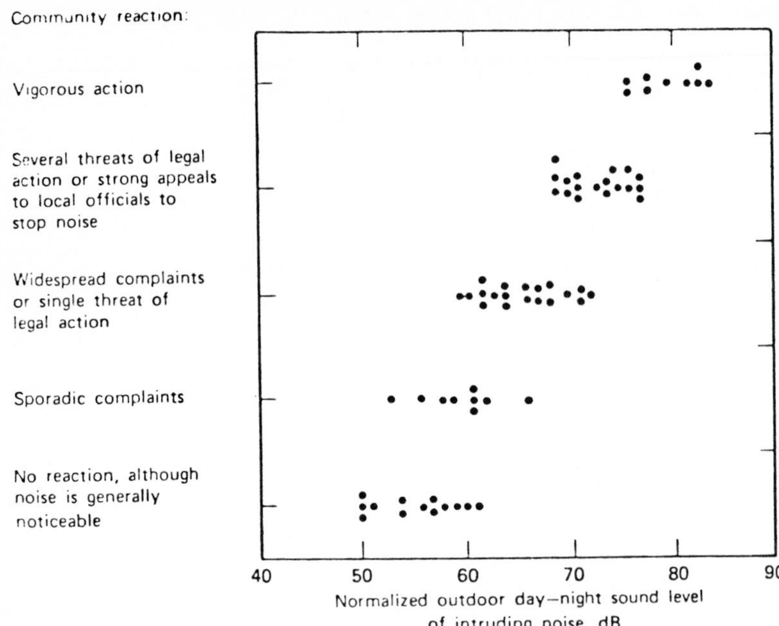

FIG. 1.1 Community reaction versus normalized day-night sound level with powerplant data added.

either situation, the plant engineer will do well to consider the community NC, taking into consideration the potential for noise complaints.

During the past 30 yr, communities around powerplants and other industrial facilities have complained about the noise from certain operations. As a result, it is possible to evaluate the potential for complaints from noise by using a factor called the *normalized outdoor day-night sound level* (Fig. 1.1).[1] Since the powerplant operates 24 hr/day and the noise level from the plant is fairly continuous, the day-night sound level is about 6 dB greater than the measured or calculated L_{eq}. The normalized day-night sound level is determined by modifying the day-night sound level, drawing on the information in Table 1.3.

An example of the use of the criteria to assist in the evaluation of complaints may be useful. Suppose that a fan with a tone is to be added to an existing powerplant (+ 5 to L_{dn} for tone). No complaints have been received from the neighborhood about existing plant noise. No experience with the tonal sounds of the fan (+ 0) exists. The neighborhood can be considered a noisy urban residential community (− 5). The intruding day-night sound level will be 65 db(A). The noise will occur year-round (0). Thus, the normalized day-night sound level will be $65 + 5 + 0 - 5 + 0 = 65$. From Fig. 1.1, a normalized outdoor day-night sound level of 65 dB will result in widespread complaints or a single threat of legal action.

If the plant desires no reaction from the community, the intruding sounds should be reduced to a normalized L_{dn} of 50 or to a sound level within 5 to 10 dB of the existing ambient sound level, whichever is lower. Since this approach does

TABLE 1.3 Corrections to Add to Measured Day-Night Sound Levels of Intruding Noise to Determine the Normalized L_{dn}

Type of correction	Description	Correction*
Seasonal	Summer (or year-round operation)	0
	Winter only (or windows always closed)	−5
For outdoor noise level measured in absence of intruding noise	Quiet suburban or rural community (remote from large cities, industrial activity, and trucking)	+10
	Normal suburban community (not located near industrial activity)	+5
	Urban residential community (not adjacent to heavily traveled roads and industrial areas)	0
	Noisy urban residential community (near relatively busy roads or industrial areas)	−5
	Very noisy urban residential community	−10
For previous exposure and community attitudes	No prior experience with intruding noise	+5
	Community has had some previous exposure to intruding noise, but little effort is being made to control it	0
	Community has had considerable previous exposure to intruding noise, and the noisemaker's relations with the community are good	−5
	Community is aware that operation causing noise is very necessary and it will not continue indefinitely; this correction can be applied for an operation of limited duration and under emergency circumstances	−10
Pure tone or impulse	No pure tone or impulsive character	0
	Pure tone or impulsive character present	+5

*Amount of correction to be added to measured L_{dn}, in decibels.

not measure ambient sound levels and does not consider in detail the frequency characteristics of the sounds, it should be used only to obtain an initial estimate of the community response. A more detailed approach is normally undertaken.[3]

NOISE SOURCES

Presented here are the principal acoustical characteristics of some of the most important noise sources in current state-of-the-art powerplants. For more de-

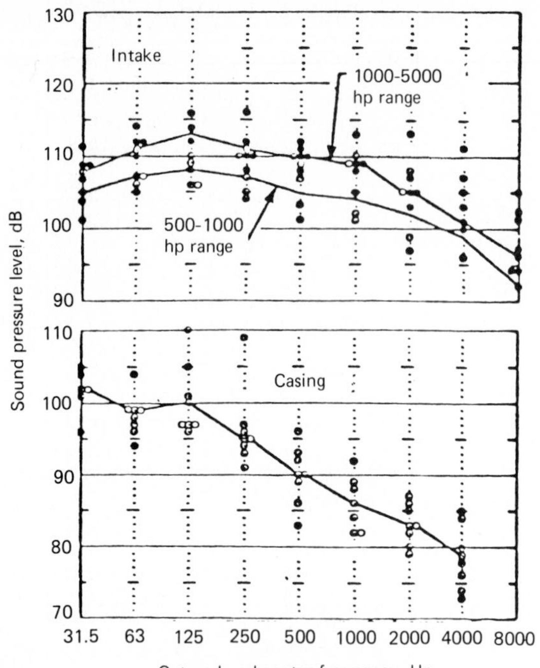

FIG. 1.2 Sound pressure levels for fan intake and casing. (*Institute of Electrical and Electronics Engineers, Inc.*)

tailed information on the noise characteristics of these and other plant noise sources, the reader is referred to the literature.[3,6,7]

Draft Fan Noise

Draft fans may be either centrifugal or axial and produce both broadband and tonal sounds. The broadband sound is generated by interaction of turbulent flow with the fan impeller, casing, and blades. The tonal sound is created by the fan blades as they rotate by obstacles in the airflow and produce pressure pulsations at the blade passage frequency and its harmonics. For a centrifugal fan, the pulsations are primarily generated at the fan cutoff, which occurs at the junction of the fan casing and the discharge evasé. For an axial fan, the obstacles may be control or stator vanes on either side of the rotor or support struts located in the airstream.

In a typical powerplant application, the blade passage tone of a centrifugal fan will be in the range of 120 to 200 Hz; for axial fans the tone is typically above 200 Hz. Generally, the tonal components of the fan sound will be submerged in the broadband noise for fans operating at low blade-tip speeds. As the speed and pressure are increased, however, the level of the tonal components will increase relative to the broadband noise and become dominant.

FD and PA Fan Noise. Both FD and PA fans draw air from the atmosphere and supply combustion air to the boiler. As such, these fans are always provided with discharge breeching between the fan outlet and the boiler. For centrifugal fans, the air inlet on either one or two sides of the fan may open directly to the surrounding space or be provided with an inlet duct to draw the air from above the fan or from some remote location. Normally, the fan casing, discharge, and inlet ducts are not insulated.

With an open, unducted inlet, the noise levels in the vicinity of the fan will be controlled by the fan intake noise. This noise is typically in the range of 105 to 120 dB(A) within a distance of two or three diameters of the intake opening. If the intake is ducted to some remote location, the nearby noise levels will be controlled by casing and breeching radiated noise and possibly by motor and drive system noise.

For FD fans, in a balanced-draft boiler application, the fan noise will be broadband for either the ducted or nonducted installation. However, as the pressure rise across the fan is increased to the range of 25 inH_2O (6000 Pa) or higher, the noise will be dominated by the sound at the blade passage frequency of the fan. This also holds true for PA or FD fans serving pressurized boilers.

Typical octave-band L_p values measured near FD centrifugal fan installations serving balanced-draft boilers are shown in Fig. 1.2 for both open air intake and casing noise.[7,8] These data show that the noise near the open intakes is relatively broadband with octave-band levels at or about 110 dB. For the casing noise, the spectra show some evidence of tonal components in the 125- and 250-Hz octaves, but in most cases the noise is only slightly tonal. In Fig. 1.3, however, spectra at the air intake and near the fan casing are given for one pressurized FD fan. These

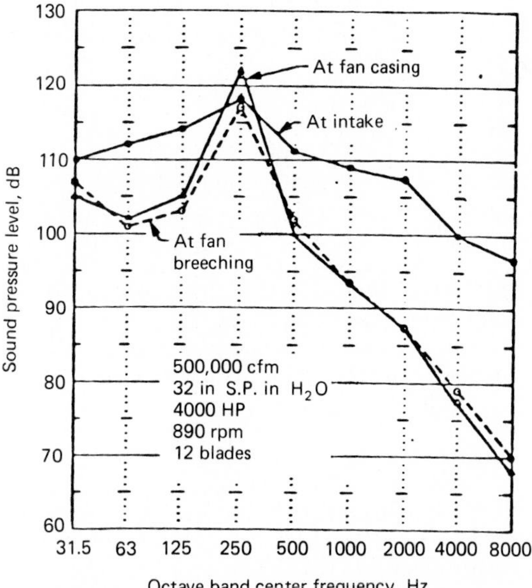

FIG. 1.3 Sound pressure levels for forced-draft fan. Note: cfm × 0.472 = L/s; in × 25.4 = mm. (*Institute of Electrical and Electronics Engineers, Inc.*)

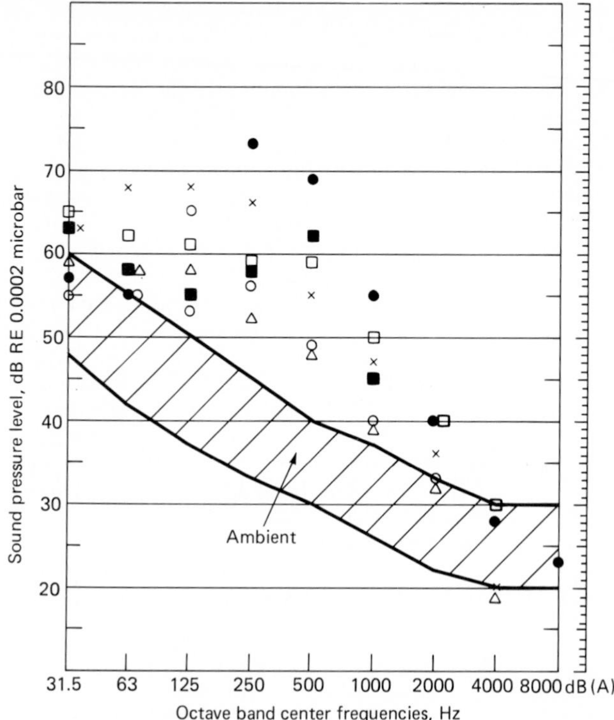

FIG. 1.4 Sound pressure levels for induced-draft fan measured in community locations 2000 to 5000 ft (600 to 1500 m) from boiler stacks. Comparison is made with typical range of ambient L_p's in rural and suburban communities.

data, with the highest level in the 250-Hz octave band, show the presence of a dominant blade passage tone at the fan casing and open air intake. For this fan, the level of the tone (180 Hz) near the fan casing is comparable to the level at the open air intake.

Axial fans in FD applications may generate levels similar to those of centrifugal fans, shifted upward somewhat in frequency. An inlet muffler is usually provided because of the higher pitch of the noise; its single inlet can easily be fitted with a muffler. For similar reasons it is also the practice to insulate both fan casing and discharge breeching acoustically.

ID Fan Noise. ID fans differ acoustically from FD fans in that both the intake and discharge are ducted to and from the fan, and the fan casing and breeching are provided with thermal insulation. Thus, the sound levels near an ID fan for either the centrifugal or axial type will be considerably lower than those of the typical untreated centrifugal FD fan. However, the sound of the fan can propagate via the discharge breeching to the boiler stack and radiate from the top of the stack to surrounding community areas. This may lead to noise complaints.

Examples of ID fan L_p values measured where community noise complaints were registered are shown in Fig. 1.4, along with a typical range of L_p values in

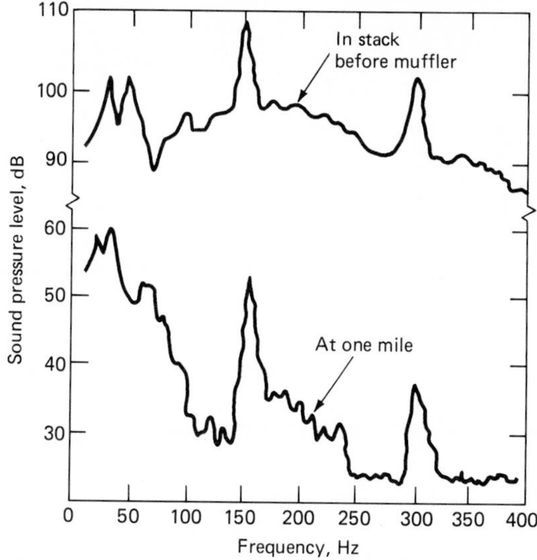

FIG. 1.5 Sound spectrum for induced-draft fan. Note: mile × 1.61 = km. (*Institute of Electrical and Electronics Engineers, Inc.*)

the community. In this figure, the range of the ID fan noise exceeds the background L_p by the greatest amount in the 250- and 500-Hz octave bands. In the 125-, 250-, and 500-Hz octave bands, the noise levels are controlled by the sound at the blade passage frequency or its harmonics. In the other octave bands, the noise consists of broadband noise produced by the turbulent airflow through the fan assembly.

Figure 1.5 presents a comparison of the sound spectrum measured inside a 30-ft (9-m) diameter stack served by four 5000-hp (3730-kW) centrifugal ID fans with a simultaneous spectrum measured in a residential area 5000 ft (1500 m) from the stack.[9] Similarly, Fig. 1.6 shows a spectrum measured inside the discharge breeching of one axial ID fan and the corresponding spectrum measured at 4500 ft (1350 m) from the stack with eight fans in operation. These data clearly show that sound at locations remote from the boiler stack has the same tonal characteristics as sound in the stack or in the discharge breeching. Thus, clearly if ID fan sound in community locations is to be controlled, some measure must be taken to control the sound radiated from the boiler stack.

Prediction of Draft Fan Noise

Information is available to enable the estimation of draft fan noise levels and the determination of noise control requirements for installations during the design stage. One may draw on past experience with similar installations, obtain data from fan manufacturers, or use available noise-level prediction methods. Fan makers can normally supply octave-band L_w information for a selected operating point, usually close to the point of maximum static efficiency. If other operating points (lower static efficiency) are anticipated, L_w data should be requested for those points. Information on the variation of L_w with operating points has re-

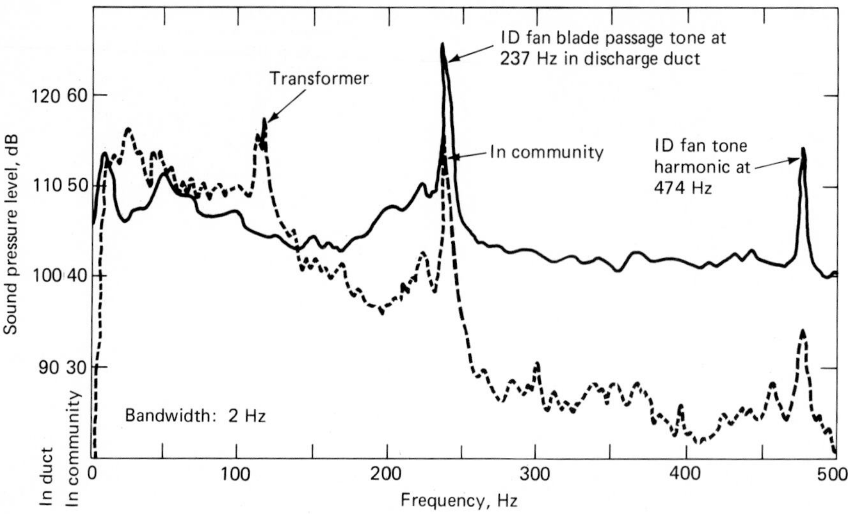

FIG. 1.6 Narrow-band spectrum of induced-draft fan noise in discharge duct and in community at 4500 ft (1350 m).

cently been provided in a study of ID fans.[10] The study shows that L_w may increase for low load operation by as much as 6 to 8 dB for a centrifugal fan.

Also note that L_w data developed by fan manufacturers are based largely on extrapolations from data obtained under laboratory conditions on relatively small fans. Thus, when fan makers submit data, they may increase the test L_w data by a safety factor to allow for field conditions. This factor, which should be known and taken into account, emphasizes a remaining difficulty in predicting actual field performance. There is currently no method for quantitatively predicting the effect of various field conditions on fan L_w values, especially for centrifugal ID fans. Axial fan octave-band L_w values at low frequency may increase by up to 4 to 6 dB if the inlet flow is distorted.[10]

Basic Approach. The basic approach to estimating L_w at the intake or discharge of a draft fan in the early design stages of a system, possibly before a fan manufacturer is involved, is to calculate the octave-band L_w values by using the equation in Table 1.4 and the data in Tables 1.4 and 1.5, given for airfoil, backward-curved, and inclined-blade types.[3,11] The octave-band L_w values are shown to be dependent on the specific L_w for the fan blade type from Table 1.5, the gas flow rate, the pressure across the fan, and the static efficiency. It is suggested that this methodology be applied in the absence of manufacturers' data when the operating point is known.

If the specific power levels in Table 1.5 are used, then corrections for the blade passage tone of 3 dB and corrections for static efficiency (from Table 1.4) are to be applied. However, when one is using the specific power levels from Edison Electric Institute, no general correction is made for the blade passage tone or static efficiency, since the values were derived from actual in-plant fan installations.[3] For an FD fan serving a pressurized boiler, however, an additional 5 dB should be awarded to the octave containing the blade passage frequency. For centrifugal ID fans, 10 dB should be added to the octave band containing the blade passage frequency if operation at low load is anticipated.

TABLE 1.4 Equation for Estimating Fan L_w (dB re 1 pW)

$$L_w = K_w + 10 \log \frac{Q}{Q_1} + 20 \log \frac{P}{P_1} + c$$

where L_w = estimated sound power level of fan (dB re 1 pW)
K_w = specific sound power level
Q = flow rate, ft^3/min (L/s)
Q_1 = 1 ft^3/min (0.472 L/s)
P = pressure, in H_2O (Pa)
P_1 = 1 in H_2O (249 Pa)
c = correction factor for point of fan operation

For off-peak operation:

Static efficiency, % of peak	c, dB
90–100	0
85–89	3
75–84	6
65–74	9
55–64	12
50–54	15

Institute of Electrical and Electronics Engineers, Inc.

TABLE 1.5 Specific Sound Power Levels (dB re 1 pW) for Airfoil, Backward-Curved, and Backward-Inclined Fan Blades*

Octave-band center frequency, Hz	Organization	
	EEI†	ASHRAE‡
31.5	31	—
63	33	32
125	35	32
250	34	31
500	33	29
1000	33	28
2000	29	23
4000	25	15

*Values are for either inlet or outlet of fan.

†From Edison Electric Institute data, which already include some effect of blade passage tone; but add 5 dB to octave containing blade passage frequency to forced-draft fans serving pressurized boilers.

‡From *American Society of Heating, Refrigerating & Air-Conditioning Engineers Handbook,* 1984 *Systems Volume,* Chap. 32, "Sound and Vibration Control." Add 3 dB to octave band containing blade passage frequency.

TABLE 1.6 Specific Sound Power Levels (dB re 1 pW) for Casing-Radiated Sound, Centrifugal-Type, Induced- or Forced-Draft Fans

Octave-band center frequency, Hz	Casing, dB
31.5	30
63	27
125	26
250	22
500	17
1000	15
2000	11
4000	7
8000	0

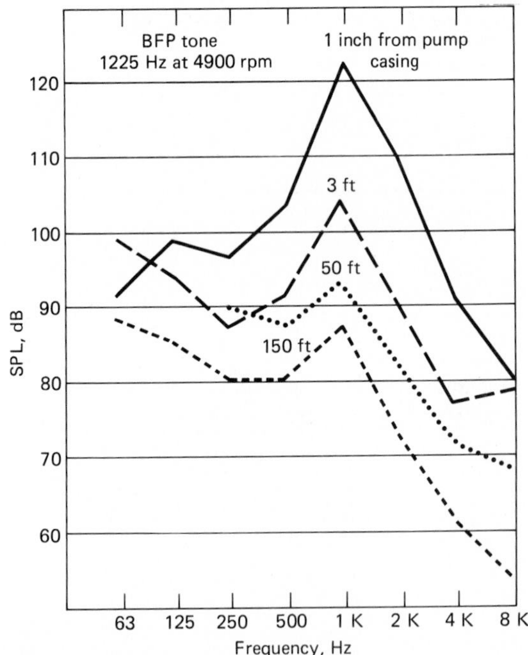

FIG. 1.7 Octave-band sound pressure levels at various distances from boiler feed pump. Note: in × 25.4 = mm; ft × 0.3048 = m. *(Institute of Electrical and Electronics Engineers, Inc.)*

Making Allowances. The foregoing procedures provide a method for estimating L_w at the intake of an FD or a PA fan or the discharge of an ID fan. With these predicted L_w values, the L_p values in the vicinity of the open inlets of an FD or a PA fan can be estimated by allowing for hemispherical spreading modified by the appropriate acoustical characteristics of the nearby space, if enclosed.

For ID fans, L_p at some distance can be calculated by allowing for hemispherical propagation from the top of the stack to the point of observation, the directivity of the stack opening for sound propagation along the ground plane, molecular absorption, and wind and thermal gradients in the atmosphere.[3,4] Also, for multiple fans discharging into one stack, or for multiple stacks, 10 log N, where N is the number of identical fans operating at or near the same load point, should be added to L_p.

Finally, even with the foregoing tools and information, uncertainty still exists in the industry about these power-level prediction procedures, especially for ID fans, since many installations without noise control measures have not been the source of noise complaints. While the absence of complaints may in part be caused by a lack of nearby neighbors, it appears that some installations are quieter than would be predicted. The equation in Table 1.4 may also be used for the prediction of L_w radiated directly through uninsulated fan housings and breechings. The specific power levels for casing L_w values are given in Table 1.6.[3]

Boiler Feed Pump Noise

Even though they may not be driven by the main turbine, boiler feed or reactor feed pumps are often installed at one or two levels below the turbine floor. Independent of its location, however, a loud feed pump with a typical tonal sound can usually be heard throughout several levels of the plant.

Two Examples. One example is shown in Fig. 1.7,[7,12] in which a peak in the sound spectrum occured at a frequency of 1225 Hz when the pump was operating at a speed of about 4900 rpm. The level of the tone was as high as 122 dB next to the pump casing and about 104 dB at 3 ft (1 m). The level of the tone was 93 dB on the opposite side of the turbine hall directly behind the main generator. This tone is characteristically heard throughout other areas of the plant.

Another example is shown in Figs. 1.8 and 1.9. In Fig. 1.8, the spectrum shows a peak in the 500-Hz octave band of about 95 dB about 3 ft (1 m) from the pump. The narrowband sound spectrum in Fig. 1.9 shows that it is made up of a tone at about 610 Hz and four harmonics. This spectrum is characteristic of many

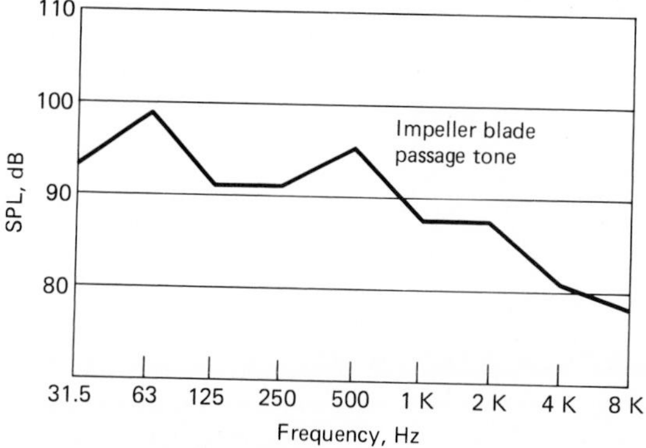

FIG. 1.8 Octave-band spectrum near boiler feed pump. (*Institute of Electrical and Electronics Engineers, Inc.*)

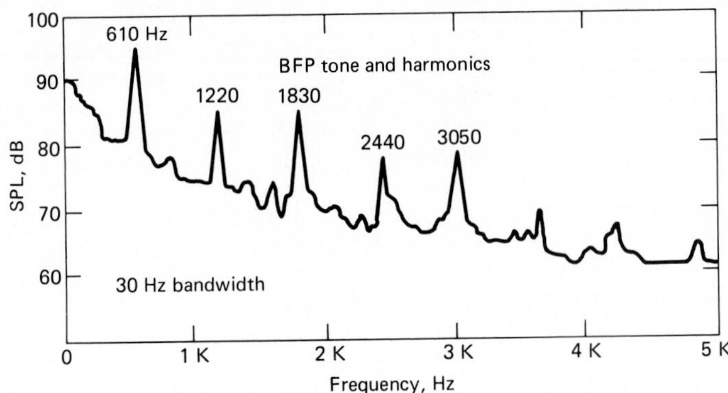

FIG. 1.9 Narrow-band spectrum near boiler feed pump. (*Institute of Electrical and Electronics Engineers, Inc.*)

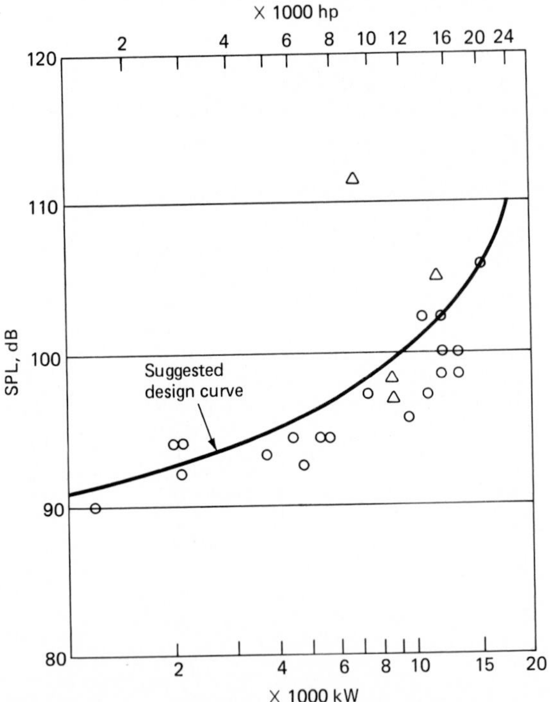

FIG. 1.10 A-weighted sound pressure levels at 3 ft (1 m) versus power rating for boiler feed pumps. (*Institute of Electrical and Electronics Engineers, Inc.*)

TABLE 1.7 Estimated Sound Power Levels of Boiler and Reactor Feed Pumps

Pump power rating		Sound power level, dB	
hp	kW	Overall	A-weighted
1,300	1000	108	104
2,700	2000	110	106
5,300	4000	112	108
8,000	6000	113	109
12,000	9000	115	111
12,600	9500	113	112
16,000	12 000	115	114
20,000	15 000	119	118
24,000	18 000	123	122

Octave-band correction, dB (subtract from overall)

Octave-band center frequency, Hz	Pump rating	
	A*	B†
31	11	19
63	5	13
125	7	15
250	8	11
500	9	5
1000	10	5
2000	11	7
4000	12	19
8000	16	23

*A = 1300 to 12,000 hp (1000 to 9000 kW).
†B = 12,600 to 24,000 hp (9500 to 18 000 kW).
Institute of Electrical and Electronics Engineers, Inc.

boiler feed pumps and produces a sirenlike sound. Although these examples of pump tonal sounds show potential for causing excessive noise exposure for personnel, it is fairly common for these tonal sounds to be so annoying that some retrofit noise control measures are often applied primarily to reduce annoyance.

Prediction of Feed Pump Noise. For the prediction of the sound level of these pumps, the expected A-weighted sound level is given in Fig. 1.10 as a function of pump power rating.[3] Note that this prediction curve is based primarily on data obtained at relatively loud installations and is thus fairly reliable for a loud pump installation.

Also, for more detail, Table 1.7 gives the overall and A-weighted L_w and a procedure for calculating L_w octave-band spectrum for a boiler feed pump.[3] With these empirical data, a design engineer can estimate the spectrum and level of a pump fairly reliably if the pump indeed produces pronounced tones. This is emphasized by Fig. 1.11, which shows a range of data taken on 14 boiler and steam-generator feed pumps.[3] The upper portion of this figure shows how the pumps with tones produce octave-band levels of about 100 to 100 dB, as compared with 90 to 95 dB for pumps that do not generate pronounced tonal components.

Because the tonal sound produced by a boiler feed pump can raise the sound level and the associated annoyance and risk of hearing damage significantly, it is important to note what causes the generation of the tone. A tone is associated with the rotation speed of the impeller and the number of impeller

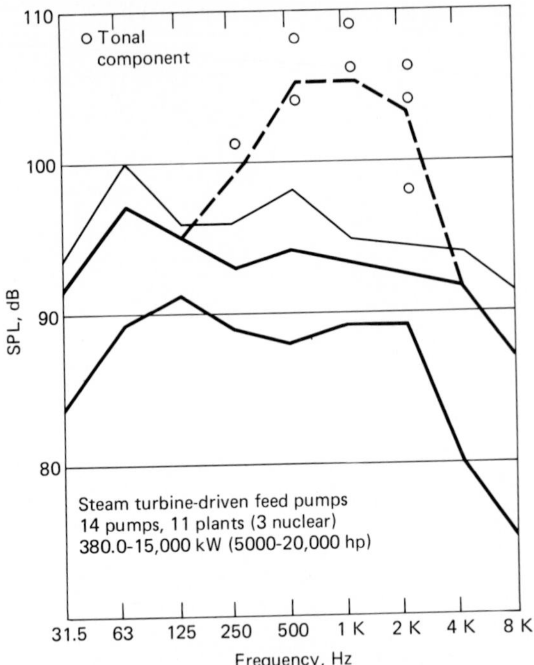

FIG. 1.11 Octave-band sound pressure levels at 3 ft (1 m) for 14 boiler and reactor feed pumps. (*Institute of Electrical and Electronics Engineers, Inc.*)

blades; this is the blade passage frequency. In the spectrum for the pump of Fig. 1.9, a fundamental tone occurs at about 610 Hz. This pump was driven at a speed of 6100 rpm by a 10,000-hp (7460-kW) turbine and has six impeller blades. Thus, the fundamental tone occurs at the blade passage frequency of 610 Hz (6100/60 × 6).

Steam Admission and Safety Valve Noise

Valve noise is prevalent in powerplants. Within the plant, excessive noise for personnel exposure often is associated with steam desuperheaters, deaerator systems, gas regulators, and steam-turbine admission valves. Excessive community noise is often caused by boiler vents, main power control valves, and safety valves. With these valve systems, the dominant levels of noise, which radiate from downstream piping or openings to atmosphere, occur in the frequency range of 500 to 8000 Hz. Steam admission and safety valves are covered here.

Steam Admission Valves. Measurements have been made at several fossil-fuel powerplants of the sound levels produced by the main turbine-generator steam admission valves.[3,7] These data show that the valve noise spectrum is dominated by noise primarily in the 2000- and 4000-Hz octave bands. Also, the measurements show that the sound level is a strong function of load, or more precisely, the throttling of the valve. Thus, in the range of 25 to 50 percent open, the A-

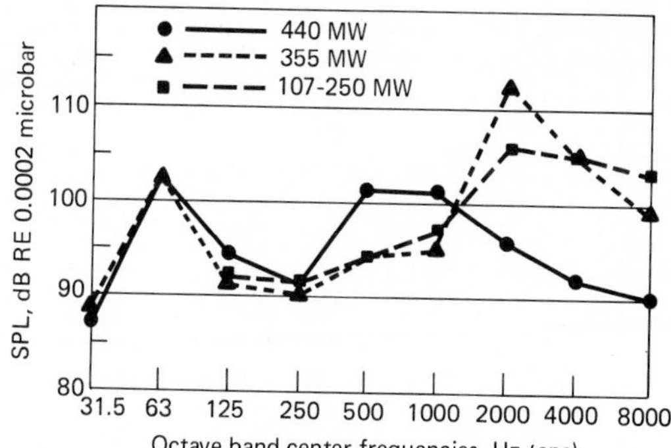

FIG. 1.12 Steam admission valve sound pressure levels at 3 ft (1 m) versus generator load. (*Institute of Electrical and Electronics Engineers, Inc.*)

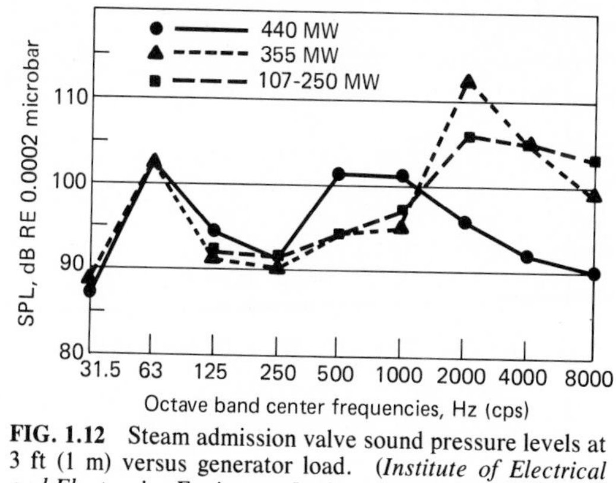

weighted sound level can be on the order of 10 dB(A) higher than at the full-load or wide-open position of the valves.

An example of the measured data is shown in Fig. 1.12, where the A-weighted level at 1 m (3 ft) from the valve casings ranged from 104 dB(A) at near full load (440 MW) to 113 dB(A) at an intermediate load of 355 MW. In the chart, the spectrum in the octave bands of 250 Hz and above is controlled by valve noise. The maximum level in the spectrum in the 63-Hz octave band is controlled by other sources, such as the turbine-generator system. This particular set of four valves was about 100 percent open at the 440-MW load point and 75 percent open at the intermediate load of 355 MW. Table 1.8 gives estimated octave-band L_w values for fossil-fuel plant steam admission valves for three different ranges of valve settings.[3]

Safety Valves. Boiler safety valves have been the source of many community noise complaints. Only a few power companies, however, have provided mufflers for these valves because of their anticipated infrequent operation, the costs associated with supplying mufflers for a large number of valves, and the cost of

TABLE 1.8 Estimated Sound Power Levels of Steam Admission Valves

Octave-band center frequency, Hz	Sound power level, dB		
	Under 60%*	60–80%*	90–100%*
31	—	—	—
63	—	—	—
125	—	—	—
250	98	95	88
500	100	97	90
1000	105	102	95
2000	115	112	105
4000	113	110	103
8000	105	102	95
Overall	118	115	108
A-weighted	119	114	109

*Approximate valve setting, relative to fully open.
Institute of Electrical and Electronics Engineers, Inc.

structural support. Each boiler also has a main power control valve, which is frequently assigned to steam pressure relief in reducing the generating load. Here, because of anticipated frequent use, a muffler or a quiet trim valve is generally provided.

Figure 1.13 shows octave band L_p values measured at two distances away from a 3-in (76-mm) boiler drum safety valve with an upstream pressure of 2950 psi (20 300 kPa) and a temperature of 1000°F (538°C).[7] At 300 ft (90 m), the corresponding A-weighted sound level was 112 dB(A), and at 3400 ft (1035 m) in a residential community the sound level was 82 dB(A), where the daytime and evening ambient could range from 40 to 45 dB(A). Because noise complaints are often received when a steady state noise 10 to 15 dB(A) above the ambient level intrudes into a community, a valve noise 40 dB above ambient will certainly lead to complaints. Although the sound may have a duration of only a few minutes or less, the sudden increase in noise level has a startle effect, somewhat offsetting its infrequent operation.

As noted earlier, only a few power companies have provided mufflers for these valves, mainly because of cost. However, the size and cost of the mufflers could be reduced by modification of the noise control goal. Because the typical community intrusion can be 40 to 50 dB(A) higher than the ambient, when mufflers have been installed, noise reductions on the order of 40 to 50 dB(A) have been specified. However, reductions on the order of 20 to 25 dB(A) would be adequate for infrequent valve operation; this would reduce the cost of mufflers, possibly leading to increased usage.

Finally, with regard to safety valves, it is common practice to provide a bias cut at the end of the valve discharge pipe. Tests indicate that no significant di-

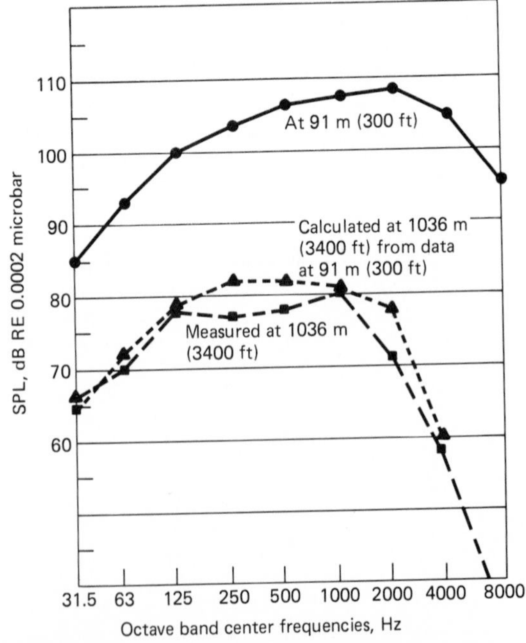

FIG. 1.13 Typical safety valve sound pressure levels. (*Institute of Electrical and Electronics Engineers, Inc.*)

rectivity benefit can be realized by orientation of the bias cut, but the cut did serve to reduce the sound by 5 dB at high frequencies for the vent tested.

Transformer Noise

Transformers at fossil-fuel or nuclear powerplants are potential sources of community noise complaints, although the problem is greater at power distribution facilities. Transformer sound consists primarily of tonal components at multiples of twice the power line frequency (60 Hz), such as 120, 240, 360 Hz, etc. The fundamental tone at 120 Hz is the most troublesome of the components, because sound at this frequency can propagate with favorable weather conditions over long distances with little attenuation beyond that due to spherical divergence. Also, a typical residential structure provides only a moderate degree of protection from sound intrusion at this frequency. Thus, complaints from residences have been received about transformer noise at distances of 3000 to 10,000 ft (1 to 3 km) from the plant.

NEMA Standard. For many years, the transformer industry has used the National Electrical Manufacturers Association (NEMA) sound-level rating to designate the sound of a transformer.[13] This rating system requires only the determination of the average A-weighted sound level at a distance of 1 ft (0.3 m) from the wall surfaces of the transformer. The standard has been useful and has facilitated the specification and marketing of transformers. It is common practice to specify a transformer to have either a standard NEMA rating or to have a lower, or quieter, rating. The quieter rating is obtained by various methods, such as using a larger than normal core or providing double-wall construction. The cost for such lower ratings has been reported to be about 1 percent of the transformer cost per decibel of reduction in the A-weighted sound level up to reductions of 10 dB, and 2 percent per decibel for reductions greater than 20 dB.[14]

Although the NEMA standard is useful, it has one major weakness that is important, especially when residential areas are located at some intermediate distance, such as 1000 to 10,000 ft (300 to 3000 m) from the plant. At these distances, the component at 120 Hz is the most important aspect of transformer sound, and the close-in A-weighted sound level is not an adequate indication of the level of the tone at 120 Hz at either the close-in or residential location. Thus, when the potential is estimated for noise complaints at some significant distance from the transformer site, it is advisable to concentrate on the sound level at 120 Hz and not on the A-weighted level.

Empirical Procedure. Here is an empirical procedure for predicting L_p at 120 Hz for sound caused by a transformer. Designed according to normal NEMA sound standards, the procedure has two steps. First, the A-weighted sound level at 100 ft (30 m) is estimated as follows:

$$L(100\text{ ft}) = 40 + 8.5 \log R$$

where $L(100\text{ ft})$ = A-weighted sound level at 100 ft (30 m) and R = total maximum (forced cooling) rating of the transformer, in MVA. Then, to obtain the average L_p of the tone at 120 Hz and several of its harmonics, add these numbers in decibels to $L(100\text{ ft})$:

Tone (Hz) :	120	240	360	480
dB :	15	3	− 6	− 10

To estimate transformer tone levels at downwind distances other than 100 ft (30 m), spherical divergence (−6 dB per doubling of distance) may be assumed. To determine whether a particular level at either of the tones is acceptable, a comparison is made between the tonal level and the octave-band ambient residual level during a quiet period, such as between 2:00 a.m. and 4:00 a.m., in the vicinity of the nearest homes. The tonal level should be in the range of 0 to 5 dB below the appropriate octave-band level to ensure that no complaints will be received.

Electric Motor Noise

Electric motors are generally not major noise sources in the powerplant; however, the sound of large induction motors that drive boiler feed pumps, circulating and condensate water pumps, and draft fans can range from 85 to 95 dB(A) in personnel areas. Also, the motor sound usually includes middle- and high-frequency tones that can cause neighborhood annoyance. Thus, there have been community noise complaints caused by circulating water pumps, which are often located outdoors and sometimes near plant boundaries.

Generally, the noise of these motors radiates from the ventilation or cooling-air openings. Narrowband analysis of motor noise usually reveals either a single dominant tonal component, believed to be of either magnetic or aerodynamic origin, or a group of several tones separated in frequency by 120 Hz caused by rotor-slot magnetic interactions.

Design Options. Although motor manufacturers do not publish as much information on the sound levels of large motors as they do for small motors [200 hp (149 kW) or lower], they do have a number of design options to control sound, including speed selection, internal mufflers, and acoustical treatment in the airflow paths of motors provided with WPI or WPII weatherproof enclosures. Thus, if a user determines a need for noise control, L_p at a given distance should be specified.

Also, especially if community complaints are a concern, the specification should call for some sound level limit for any tonal components. Within a plant it may be adequate to specify an A-weighted sound level at some distance, but for circulating-water-pump motors that may be a disturbance in residential areas, the allowable level of any tones should be specified for some particular distance and direction of interest.

Compliance Measure. In the electric motor industry, IEEE Standard no. 85, Test Procedure for Airborne Noise Measurements, is generally used in conducting compliance tests to meet specifications. In this procedure, however, the motor is

TABLE 1.9 Estimated Sound Power Levels of Electric Motors*

Octave-band center frequency, Hz	Sound power level, dB				
	Col. 1	Col. 2	Col. 3	Col. 4	Col. 5
31	94	88	88	88	86
63	96	90	90	90	87
125	98	92	92	92	88
250	98	93	93	93	88
500	98	93	93	93	88
1000	98	93	96	98	98
2000	98	98	96	92	88
4000	95	88	88	83	78
8000	88	81	81	75	68
Overall	106	102	102	102	100
A-weighted	104	101	101	100	99

Col. 1: 1800 and 3600 rpm and for all TEFC motors.
Col. 2: 1200 rpm.
Col. 3: 900 rpm.
Col. 4: 720 rpm and lower.
Col. 5: 250- and 400-rpm vertical motors.

*Applies to induction motors rated between 1000 and 5000 hp (746 and 3730 kW). Includes drip-proof and WP-I, WP-II enclosures (with no acoustical specification by customer).

Notes: For motors rated above 5000 hp (3730 kW), add 3 dB; for motors rated between 200 and 1000 hp (149 and 746 kW), subtract 3 dB in all octave bands. For motors rated under 200 hp (149 kW), use the speed ranges above, and reduce 200 hp (149 kW) L_w values further by 10 log 200/hp dB. For high-efficiency design motors rated at 1200, 1800, or 3600 rpm and 250 hp (187 kW) or smaller, use the above relationships and reduce the calculated sound power levels by an additional 7 dB.

Institute of Electrical and Electronics Engineers, Inc.

not under load, which may reduce the levels of the tones generated relative to those produced under load. Consequently, especially for outdoor circulating-water pumps and possibly draft fan installations, the specifications should call for compliance in the field under load with regard to control of the tonal components.

From field measurements made on a number of motors under load conditions in powerplants, some estimates have been made of the typical L_w values for motors without acoustical treatment. These data, given in Table 1.9,[3] may be referred to in estimating the L_w values of new motor installations as a guide for the preparation of purchase specifications.

Turbine-Generator System Noise

Turbine-generator systems occupy a major part of the powerplant, in a highly visible area, and thus the noise produced by these units is of interest. Noise sources in a turbine-generator system include the steam admission valves, turbine units, generator, exciters, brush gear, and shaft-driven boiler feed pump and its coupling to the turbine shaft. Also, in the turbine hall there may be other control valves or cross-over steam lines which generate sounds that may appear to be caused by the turbine-generator system.

Figure 1.14[3] shows representative data for turbine generators without steam admission valves on the turbine floor or shaft-driven boiler feed pumps. In the

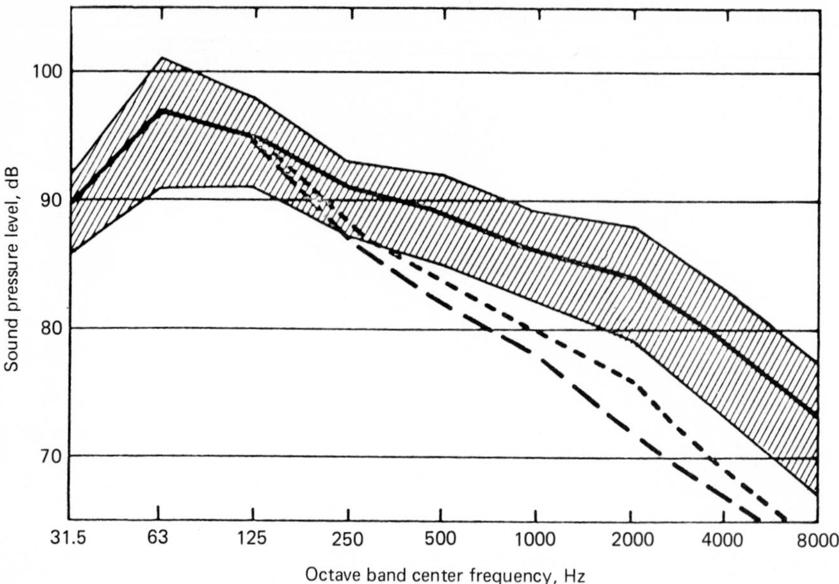

FIG. 1.14 Octave-band sound pressure levels at 3 ft (1 m) from steam turbine-generator units. (*Institute of Electrical and Electronics Engineers, Inc.*)

figure, the shaded band represents the range of octave-band sound levels measured around 10 turbine-generator systems in the 400- to 1100-MW range. The dashed curve represents data, 84 dB(A), for one 400-MW unit with acoustical treatment in the turbine hall and noise control measures applied to the unit. The solid curve is the expected value for indoor installations, 92 dB(A), with no acoustical treatment in the turbine hall and no noise control measures applied to the turbine-generator system. The dotted curve can be used for outdoor units.

Noise control measures for turbine-generator systems include sound-absorptive treatment for the turbine hall, which reduces the cumulative effect of several units located in one long turbine hall. Other options include large, removable, acoustically treated enclosures sometimes provided for the admission valve area and the high-pressure turbine unit, enclosures for couplings, and mufflers for cooling air openings for exciters and brush gear. Typically, enclosures for this equipment are options provided by the turbine manufacturer.

NOISE CONTROL MEASURES

The first step in controlling the noise of powerplant equipment is to include noise as a factor in selecting the type of equipment and the operating conditions or procedures to be used. For new equipment, the next step is to specify the maximum noise level that the equipment will be allowed to produce at some defined distance and operating conditions. This procedure will work toward obtaining equipment that will meet the specified level even though some add-on noise control measures may be provided by the supplier. The specified level may be in terms of the A-weighted level, octave-band levels, and/or the maximum sound level at a blade passage tone or any other tonal component.

For both new and existing equipment, a variety of noise control measures may be applied. One form of noise control is to modify the equipment operation, such as reducing speed or adjusting load control procedures. However, the primary noise control measure, used mostly in powerplants with existing equipment, is to modify the propagation path between the source and the receiver. This may mean installing mufflers, enclosures, barriers, lagging, and sound-absorptive materials.[3,16]

Mufflers

In powerplant applications, mufflers in ducts or pipes reduce the sound that is transmitted along these carriers without significantly impeding the flow of air or gas being conducted. Mufflers that depend on sound-absorptive material to reduce the level of sound as it propagates along muffler air passages are *dissipative* devices. Mufflers that depend on tuned cavities and/or expansion chambers are called *reactive mufflers*. Some mufflers incorporate both dissipative and reactive elements.

The most common type of muffler used in powerplant applications is the dissipative, parallel-baffle design. It consists of baffles that split an existing duct or pipe into an array of parallel paths for the gas flow. Each baffle contains some dissipative material, which is protected with cloth, screen, and perforated metal to prevent erosion or loss of material caused by high air velocity in the passages. A traditional baffle is shown in Fig. 1.15 (top), along with another type (bottom) in which the sound-absorptive material does not face the gas flow. It consists of a group of cavities with some sound-absorptive material on the cavity walls, and it is designed for flows containing particles that might clog the perforated facing of the conventional baffle. Typical application is in the breeching or stack of ID fans, where flyash may deteriorate the traditional baffle.

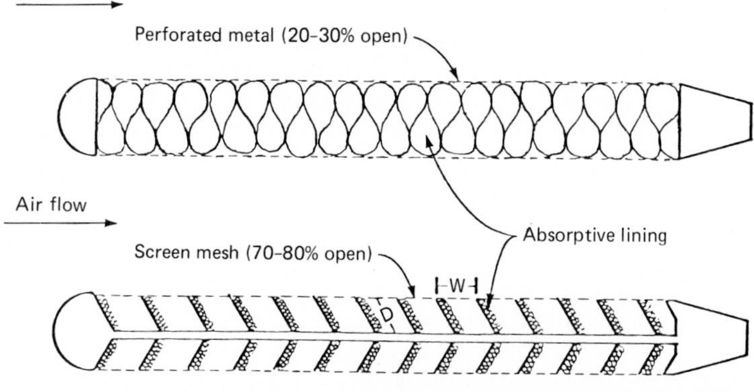

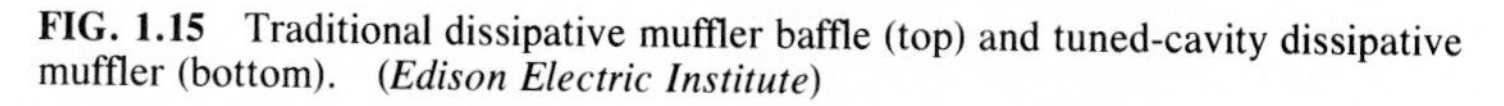

FIG. 1.15 Traditional dissipative muffler baffle (top) and tuned-cavity dissipative muffler (bottom). (*Edison Electric Institute*)

Acoustical Performance. The acoustical performance of mufflers as a function of frequency depends on several factors. For the standard dissipative muffler, it depends on the thickness of the baffle, the properties of the dissipative material, and the width of the airspace between the baffles. For the cavity type of muffler, it depends primarily on the depth of the cavity; in particular, the baffles provide maximum insertion loss at that frequency where cavity depth is about one-fifth of a wavelength. As a result, the baffles are referred to as *tuned dissipative units*.

Generally, mufflers using the dissipative type of baffle will provide significant noise control over a wide range of frequencies, with peak performance between 200 and 2000 Hz. They are applicable to broadband noise sources such as fans and steam vents of various types, including steam blowdown operations. They are also used to control exhaust (frequency above 100 Hz) and inlet noise from gas turbines.

Acoustical performance is most commonly stated in terms of *insertion loss* (IL) in decibels, where IL is the reduction in sound level that should be observed at some selected location when the muffler is installed in the sound propagation path. The IL is usually specified as a function of frequency for each of the normal octave or one-third octave bands. This method of specifying muffler performance is most appropriate with noise control of an existing fan, since the sound level can be measured before and after the muffler is installed to obtain the IL.

For a new installation, where a "before muffler" condition is not observable, two other methods are used. One is to specify the allowable noise level at some distance from the muffler inlet or discharge. Here, however, the muffler vendor will require some knowledge or information pertaining to the sound power of the fan. The other method is to specify the *noise reduction* (NR), defined as the difference in the noise level (in decibels) as measured at the entrance and exit of the muffler. This method is usually applicable to relatively large mufflers, where it is feasible to carry out measurements at both entrance and exit.

Aerodynamic Performance. The aerodynamic performance of a muffler is normally specified by naming the allowable pressure drop in in H_2O (Pa) for a specified air or gas volume flow in cubic feet (meters) per second. When these factors are specified, it is also common practice to specify the muffler in terms of its *dynamic insertion loss* (DIL). This is the IL of the muffler with the given flow conditions.

Besides acoustic and aerodynamic performance, several other factors should be considered when one is purchasing a muffler. These include the noise produced by flow through the air passages (muffler self-noise) and the sound transmission paths that may result in limiting the effective performance of the muffler by flanking. Other factors are the materials selected for fabrication of the muffler, structural design, and cost.

Lagging

Lagging is the application of a resilient material with an outer impervious cover around pipes, ducts, or equipment surfaces for the control of sound radiated by these elements or surfaces. The lagging may consist of an inner layer of glass fiber, mineral wool, or a closed- or open-cell polyurethane foam covered by a metal skin, a troweled mastic, or a lead-metal laminate.

Essentially, lagging forms the second wall of a double-wall construction scheme. For lagging to be effective, the outer impervious cover must be as heavy as possible and have no rigid ties to the sound-radiating surface that comprises the first wall. To specify the acoustical performance of a lagging treatment for an existing installation, the lagging IL should be specified for the various octave bands of concern. For a new installation, it may be necessary to specify the allowable sound level at some specific distance from the duct, pipe, or equipment after the lagging is applied and the equipment is operating.

Enclosures

Enclosures may range from a complete, fully accessible room to a tight-fitting boxlike structure that may even follow the contour of the noise source. They also may be partial enclosures, such as some telephone booths with only three sides.

An enclosure consists primarily of some impervious side panels, combined with a top and bottom construction that may or may not be of the same material as the sides. And almost all enclosures have a variety of openings. These may be deliberate for airflow, ventilation, or access. They may also be necessary for clearances around moving parts, such as drive shafts, or as unwanted gaps in the construction, such as under doors or around pipe penetrations. Finally, all effective enclosures should have some form of interior sound absorption.

Construction Materials. A variety of construction materials may be chosen for the sides and top or bottom of the enclosure, ranging from concrete to steel or plywood. Generally, the higher the mass of the enclosure material, the greater the potential noise reduction of the enclosure. Also, in practice, relatively high-mass-enclosure construction in the range of 10 to 50 lb/ft^2 (50 to 250 kg/m^2) is needed for low-frequency sound sources, whereas construction weighing 2 to 4 lb/ft^2 (10 to 20 kg/m^2) may be suitable for high-frequency sound sources. Other factors entering into the choice of panel construction are the desired stiffness of the structure, ease of fabrication, availability of materials, allowable total weight for the supporting structure, maintenance, and cost.

Although the ultimate noise reduction of an enclosure is determined by panel material characteristics, including its mass, the actual performance is almost always determined by the enclosure openings. In some instances, these openings are effectively controlled by installing mufflers in air passages, by providing resilient gaskets for doors and window perimeters, or by caulking construction joints. However, construction joints, door gaskets, and openings around rotating shafts usually reduce the acoustical performance of the enclosure. The degree of degradation is often quite large; for example, with a wall construction capable of providing an NR of 30 dB, an opening equal in size to 0.1 percent of the enclosure surface area can reduce the NR to 27 dB. Likewise, openings equal to 1 percent of the enclosure area can reduce the NR to about 20 dB, and a 10 percent opening would lower the NR to about 10 dB.

Acoustical Performance. The acoustical performance of an enclosure for an existing noise source can be described by specifying the noise level in octave bands that must be achieved at specific locations nearby. If the source operates at various speeds or loads, the operating conditions under which the enclosure is to be evaluated must be defined. This type of specification is the most easily understood; however, performance evaluation can be hampered if the background noise caused by other sources in the plant is not more than 6 dB (and preferably 10 dB) lower than the specified levels for the enclosed source. If it is known that the background noise will be too high for a proper evaluation, then the specified levels outside the new enclosure can be raised by about 3 dB, to allow for the somewhat equal contributions of the background and the enclosed source.

To design an enclosure for a noise source not yet installed or operating, it is first necessary to estimate the noise levels to be produced by the source. These levels should then be compared with the selected noise criteria to determine the NR required. Then, to settle on the material for the enclosure envelope, it is necessary to define the *transmission loss* (TL), in decibels, of the material. For an enclosure that is to have sound-absorptive materials on the full ceiling and wall surfaces, the TL should be chosen to be about 10 dB greater than the required NR. If no sound-absorptive materials are to be provided, the difference should be 20 dB; and if only partial treatment is scheduled, then the difference should be 15 dB. As a guideline, note that the TL of a construction generally increases with mass and frequency.[17,18]

Barriers

A noise control barrier in powerplant applications is usually in the form of a wall that intercepts the direct sound path between a source and a potential receiver. A barrier provides NR in the shadow zone created on the receiver side of the barrier. Barrier NR depends on (1) the height of the barrier relative to the line of sight between source and receiver and (2) distances between barrier, source, and receiver. If the distance from barrier to receiver is large compared to the barrier-source distance, the IL provided by the barrier is proportional to the square of the height above the line of sight and inversely proportional to the barrier-source distance and the wavelength of sound.

Under ideal conditions, the barrier can provide ILs on the order of 15 to 20 dB. In practice, however, barrier performance seldom exceeds an IL of more than 10 dB and is often in the range of 5 to 10 dB. One reason for low performance in outdoor applications is that often many of the receivers are located a considerable distance from the barrier. At relatively long distances of 1000 ft (300 m) or more, atmospheric bending and scattering of the sound into the shadow zone diminishes barrier performance. Also, in downwind directions, the sound wave is essentially bent over the barrier when the wind velocity exceeds 10 knots (18 km/h).

In spite of its limitations, a barrier can provide sufficient NR in many cases to alleviate a noise problem if certain conditions are met. First, the barrier wall should be impervious and have a density of about 4 lb/ft^2 (20 kg/m^2). Second, the barrier height has to be sufficient to extend significantly above the line of sight and must extend horizontally to block the line of sight to all expected receivers. Alternatively, the barrier may be two- or even three-sided. Finally, reflecting objects or nearby structures must be provided with suitable sound-absorptive treatment.

Sound-Absorptive Treatment

When a noise source is placed in a room with acoustically live surfaces, the noise level at some moderate distance from the source will be larger than if the source were outdoors in a free field. This increase in noise level is caused by reflections from the interior room surfaces; these reflections make up what is termed *reverberation*. Sound-absorptive materials are added to room surfaces to reduce the reverberation. This addition also reduces the noise levels at those locations where the reverberation of the untreated room would cause an increase in noise over that which would be observed outdoors.

Conventional sound-absorptive materials include glass or mineral fibers and polyurethane open-cell foam. These materials are available in blanket and board form and in acoustical ceiling or wall panels. The blanket and board materials are normally applied to room surfaces in thicknesses of 1 to 4 in (25 to 100 mm). Here, the greater thicknesses are required to reduce reverberation at low frequencies (below 200 Hz). Also, it is sometimes desirable to cover the fibrous blankets or boards with a neoprene spray-on coating, a fibrous-mat layer, a glass fiber cloth, or screen and perforated metal.

The NR that can be attained with the addition of sound-absorptive materials to the interior surfaces of a room is a function of the noise spectrum; the shape and size of the room; the type, thickness, and extent of the treatment; and the location of the auditor relative to the noise source or sources. Under the most favorable conditions, the NR can be 6 to 8 dB in the middle- and high-frequency ranges; in most cases, the NR achieved will be on the order of 5 dB.

The application of sound-absorptive materials to the boundaries of a room provides other benefits. The treatment reduces the distraction caused by reverberant sound and enables occupants to localize sound sources, thus aurally identifying the equipment that is operating. Also, the treatment reduces the cumulative buildup of sound from multiple sources. For example, when ceiling and wall treatments are applied to a large turbine hall with a low, flat ceiling and containing several generating units, the sound level in any one area is dominated by the nearest unit and is not influenced significantly by the other units.

ACKNOWLEDGMENTS

The section *Noise Control Measurers* is adapted with permission from *Electric Power Plant Environmental Noise Guide*, Chap. 7, vol. 1, 2d ed., Edison Electric Institute, Washington, D.C., rev. 1984.

REFERENCES

1. R. M. Hoover, "Study of Community Noise Complaints Caused by Electric Power Plant Operation," *Noise Control Engineering,* 6(2):74, Mar./Apr. 1976.
2. *Noise Control at Fossil Fuel Power Plants: An Industrywide Assessment of Costs and Benefits,* EPRI Report CS-3262, Electric Power Research Institute, Palo Alto, Calif., Dec. 1983.
3. L. N. Miller, E. W. Wood, R. M. Hoover, et al., *Electric Power Plant Environmental Noise Guide,* 2d ed., Edison Electric Institute, Washington, D.C., 1984.
4. L. L. Beranek (ed.), *Noise and Vibration Control,* McGraw-Hill, New York, 1971.
5. *Title 29, Code of Federal Regulations,* Pt 1910.95, Government Printing Office, Washington.
6. A. M. Teplitzky, "Power Plant Coal Conversion Noise Analysis," *Proceedings of Inter-Noise 83,* Institute of Acoustics, Edenburgh, U.K.; p. 103, July 13, 1983.
7. R. M. Hoover, "Characteristics of Power Plant Mechanical and Electrical Equip-

ment,'' IEEE Tutorial Course, *Engineering Control of Power Plant Noise,* Institute of Electrical and Electronics Engineers, Inc., New York, 1984.

8. E. W. Wood, ''The Prediction, Measurement, and Control of Power Plant Draft Fan Noise,'' *Symposium Proceedings: Power Plant Fans—The State of the Art,* EPRI Report CS-2206, Electric Power Research Institute, Jan. 1982, pp. 4-30 to 4-49.
9. W. J. Frederick, J. R. Shadley, and R. M. Hoover, ''Induced-Draft Fan Noise Measurements Solve Community-Relation Problem,'' *Power,* Aug. 1980, p. 62.
10. I. Ver and E. W. Wood, *Induced Draft Fan Noise Control,* Eseerco Technical Report EP 82-15, Empire State Electric Energy Corp., New York, 1984.
11. ''Sound and Vibration Control,'' Chap. 32, *ASHRAE Handbook, Systems Volume,* American Society of Heating, Refrigerating & Air-Conditioning Engineers, Atlanta, 1984, p. 32.3.
12. R. M. Hoover and E. Makay, ''Boiler Feed Pump Sound and Its Control,'' *Symposium Proceedings: Power Plant Feed Pumps—The State of the Art,* EPRI Report CS 3158, Electric Power Research Institute, Palo Alto, Calif., July 1983, pp. 5-44 to 5-72.
13. *Transformers, Regulators, and Reactors,* NEMA Standard no. FR1-1974, National Electrical Manufacturer's Association, Washington, D.C., 1974.
14. I. Ver and D. W. Anderson, *Characteristics of Transformer Noise Emission,* Eseerco Project EN-2, Empire State Electric Energy Research Corp., New York, 1976.
15. E. W. Wood, ''Transformer Sound Predictions,'' *Proceedings of 47th International Conference of Doble Clients,* Doble Engineering Co., 1980, Watertown, Mass., p. 6-201.
16. A. M. Teplitsky, ''Community Noise Emissions from Enclosed Electric Power Plants,'' *Noise Control Engineering,* 6(1):4, Jan./Feb. 1976.
17. *Compendium of Materials for Noise Control,* Pub. no. 75-165, NIOSH, Dept. of Health, Education, & Welfare, Washington, D.C., 1975.
18. M. I. Schiff, ''Use Barriers and Enclosures to Reduce Power-Plant Noise,'' *Power,* 122(8):85, Aug. 1978.

SECTION 5

PLANT ELECTRIC SYSTEMS

CHAPTER 5.1
ELECTRICAL INTERCONNECTIONS

John Reason
Power *Magazine*
New York, NY

INTRODUCTION

Plant electric systems apply to in-plant industrial generating plants that are interconnected with electric utility networks or systems. Key elements of plant electric systems are the distribution equipment (Section 5, Chapter 2), cable and bus duct (Section 5, Chapter 3), and *alternating-current* (ac) motors and generators (Section 5, Chapters 4 and 5). Tying these elements together so they will interface with one another and with the electric utility is accomplished with electrical interconnections, the subject of this chapter.

Even though the utility enjoys great economies of scale, there are several possible reasons for generating electricity in-house:

- *Cogeneration* means the simultaneous production of heat and electricity from the same powerplant (Chapter 7, Section 2; also see box). It is the most efficient way to generate electricity and may use as little as one-half the fuel that the utility does. But the economic operation of a cogeneration plant depends even more heavily on the ability to interconnect, because it is subject to the upswings and downswings in the plant's heating needs and its electrical demand.
- *Small power production* is a term used by the *Federal Energy Regulatory*

Basic Metering Arrangements With Cogeneration

Two basic metering arrangements are used when excess in-plant power is sold to the utility (see illustration). These arrangements allow either the sale of excess power only or the sale of all generated power to the utility with simultaneous purchase of all the plant's power needs. The latter is generally known as a *buy-all, sell-all arrangement.*

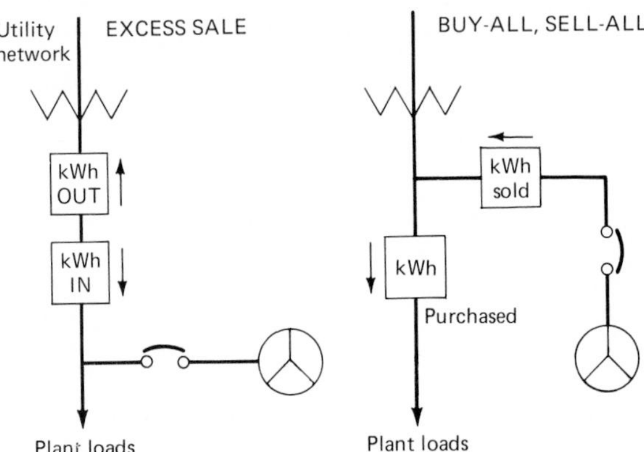

An excess power sale requires only one tie line connecting the customer to the utility. Two watthour meters are installed in the line by the utility, each ratcheted to operate in opposite directions of power flow. During normal operation, power generated in-house flows directly to plant processes and is supplemented by utility power, which is recorded on the in-meter. When the plant is generating more power than it needs, usually in the evening and on weekends, power flows into the utility network and is recorded on the out-meter. Both meters are usually time-of-use devices; with both, the amount paid for purchased power and the amount billed for power consumed vary with the time of day the power is delivered.

A buy-all, sell-all metering arrangement is designed to take advantage of state or federal regulations that require the utility to pay at its avoided cost for purchased power. For many utilities, the avoided cost is higher than the standard industrial rates at which they sell power. Thus, it becomes advantageous to connect the in-plant generator directly to the utility network through one meter and to sell all the generated power at the avoided cost. Then plant power is purchased at standard industrial rates through another meter.

Remember, the customer may also be looking for capacity payments as well as kilowatthour payments. This is particularly true of small power producers who have a constant supply of fuel to be disposed of or a constant hydroelectric power capacity. With the buy-all, sell-all arrangement, a capacity contract can be written for the full capacity of the generator, whereas with the excess-power sale, capacity payments can be made only for excess capacity, which is usually zero.

Penalty payments are levied by the utility if the plant load draws a power factor below a certain level. But a synchronous generator is an excellent way of improving the power factor and can even be used to inject VARS into the utility network. With a buy-all, sell-all arrangement, the plant gets charged for the VARS it uses, but the plant does not get paid for the VARS it generates. So if the plant has a poor power factor, the situation favors the sell-excess arrangement. Although no mainland utility presently includes flow of VARS in its contract, the practice has been common in Hawaii for years. See page 5.14 for more information on VARS.

Commission (FERC) to describe power generation from renewable power sources, including low-head hydroelectric, refuse and biomass combustion, wind, and solar. Electricity generated by these methods may or may not be cheaper than by conventional means, but because the fuels are renewable, the small power producer is given special authority to interconnect with the utility.

- *Peak shaving* is another way to cut the cost of purchased electric power. A peak-shaving generator does not itself produce cheap power; in fact, the kilowatthours it generates may be quite expensive. Because of the high-demand component of electric rates, however, a peak-shaving generator can be operated for short periods each day to cut down on the peak power level drawn from the utility. A peak-shaving generator does not have to be interconnected, but it works much more effectively if it is.
- *Dispersed storage and generation* is an expression coined by the *Institute of Electrical and Electronic Engineers* (IEEE) in an attempt to clear up the semantic confusion between interconnection and cogeneration. It has introduced semantic confusion of its own, mainly because dispersed storage is an advanced concept that is still waiting for the development of suitable storage batteries.

In 1979, FERC moved to encourage cogeneration and the use of renewable fuels by ruling that (1) regulated utilities must agree to interconnect with qualified cogenerators and small power producers and (2) the former must purchase the latter's excess power. The price that utilities must pay for the power they purchase, originally set at the full avoided cost, has since been challenged in court.

In any event, the need for cogeneration and small power production is now a fact of the times, and the interconnection problems can and will be solved, given adequate cooperation between the utility and its new customer-generator.

SWITCHING

In industrial powerplants, switching is achieved with circuit breakers. They are the key to power system reliability.

Circuit Breakers in Operation

Some newly interconnected customers are surprised because their main circuit breaker repeatedly drops out and shuts down the whole plant. They always throught that the main circuit breaker was the last resort to safety and that it operated only when all else had failed. Now, with all the protection that the utility insists on, the main circuit breaker often opens for no apparent reason.

For the utility, these trips represent normal and correct functioning of the network protective system. The customer often sees them as harassment by utility engineers who would rather not have the customer's generator connected in the first place. Either contention may be partly or wholly true. It is certainly true that when a customer also becomes a generator of electricity, the customer must take some responsibility for the continuous operation of the entire network to which it is connected. Certainly, he or she must understand something about how the utility network is designed and operated to provide the highest possible availability of power.

One common misconception is that the customer's generator is so small compared with the vast utility network that it could not possibly affect the network's operation. If the generator is at the end of a feeder, however, or on the secondary side of a distribution transformer serving a small group of customers, then its capacity may be a significant part of the locally available capacity. Its potential ability to disrupt other customers may be serious.

At the same time, it is often true that utility personnel are insensitive to the needs of the interconnected customer. For the utility to isolate a faulty or questionable generator, especially a small one, is no problem. There is always excess spinning reserve to take up the lead. The utility's position is that if the customer wants to become part of the generating network, the customer must accept both the benefits and the burdens. Clearly, no utility will be able to maintain that attitude. Public pressure for widespread interconnection is so great, and the capacity needs of some utilities so desperate, that ways must be found to preserve the network's integrity while providing interconnected customers with reasonable security of service.

How Circuit Breakers Work

A circuit breaker is an automatic switch that is specifically designed to open when a heavy current is flowing (Fig. 1.1). Its contacts are closed mechanically against the action of a heavy spring and are held closed by a latch. The circuit

FIG. 1.1 A circuit breaker is designed to open and isolate an electric circuit under load. *(Courtesy Gould-Brown Boveri Inc.)*

breaker opens when an external signal, usually from a protective relay circuit, releases the latch. This tripping signal may be an indication that the current flowing in the circuit is too heavy, that the voltage is too high or too low, or that one of many other fault conditions exists.

If a current is flowing when the circuit breaker opens, an arc is drawn out between the contacts. The circuit breaker action is to extinguish that arc as quickly as possible. This usually takes between three and eight cycles, depending on the design of the circuit breaker, the operating voltage, and the size of the current. During these several cycles, severe damage may be done to the electric components where the fault occurred.

Unlike a current-limiting fuse, the circuit breaker is not designed to limit damage at the fault, but rather to protect the continuity of power to the remainder of the network. In the case of a low-grade fault, however, circuit breaker operation may very well be fast enough to prevent the fault arc from growing and doing major damage.

A large utility network is capable of feeding thousands of amperes into a fault in the customer's system, so the main circuit breaker must be capable of opening the line to the plant against that short-circuit current. Hopefully, it will never have to operate, because another plant circuit breaker closer to the fault will open first and shut off the current flowing into the fault. This is called *coordination*.

When customer generation is introduced, the problem is magnified several times. A fault may be fed from the utility network, from the customer's generator, or from both—sometimes from opposite directions. In the latter case, at least two circuit breakers must open before the arc is extinguished.

If one of these circuit breakers is controlled by the utility and the other by the customer, it becomes obvious why special relay coordination is essential.

Special Problems

Closing a circuit breaker may present special problems. In a normal plant circuit, fed only from the utility, all the plant has to do before closing a circuit breaker is to make sure that everything that is not supposed to start is already switched off. Before a circuit breaker is closed to connect two live generating systems, both voltages and frequencies must be the same, and in particular the two ac waveforms must be exactly in phase. If they are not, with the inertia of several large generators behind it, the utility system will jerk the plant generator into phase in a fraction of a second, shearing the generator shaft or much worse.

The business of correctly closing a circuit breaker that connects to ac systems is known as *synchronizing,* and it may be done manually, under relay supervision, or fully automatically.

A special problem for interconnected customers is that after a utility circuit breaker has opened automatically to clear a fault, it may automatically reclose. This happens because most faults in a utility network are of a transient nature. They might be caused by lightning or by a tree branch falling across an overhead line and setting up an arc to ground. Once the fault current has been interrupted and the arc extinguished, it is usually safe to reclose the circuit breaker and to continue operations without physically inspecting for the cause of the fault. Automatic reclosers may operate as many as 5 times, repeatedly attempting to close the circuit breaker, until the fault clears. After the preset number of operations, the recloser locks out, isolating the line for manual intervention.

Most often, manual intervention is not necessary, and all the customer sees is a flicker on the line. But the effect on an interconnected generator can be traumatic, and it must be protected against this event. Conversely, if the customer's generator remains connected during the time that the reclosing circuit breaker is attempting to clear the fault, it will feed the fault, keeping the arc alive. This situation is especially possible if the customer's generator is large in comparison to the capacity of the feeder to which it is connected.

Coordination and Testing

Circuit breakers can be electrically interlocked if necessary to ensure that one is not closed when the other is also closed, or vice versa, but it is unlikely that the utility will agree to physically interlock its circuit breakers with the customer's. The best way to ensure safe operation of circuit breakers is through correct selection and setting of the relays that control them. Correct coordination ensures that when problems occur, the right circuit breaker opens to correct them. It also ensures that circuit breakers do not close when doing so would cause damage or an accident.

Testing of circuit breakers is an important part of the interconnection agreement. The utility will want a chance to approve the plant's interconnection circuit before the plant goes on line, and the utility will also want reasonable assurance that the circuit is maintained regularly, so that it keeps working properly. When the interconnection circuit is ready, the utility usually sends inspectors with the correct instrumentation to see that the equipment works as planned. They will probably do a trip test, which involves injecting signals into the relay circuit that simulate various kinds of faults and checking circuit breaker operation. After that, periodic maintenance is up to the customer, who will be required to conduct tests at specific intervals and to keep a log of these tests. This procedure may include relay settings before and after the test. Usually such tests must be done at a time when the plant can be shut down.

GROUNDING

Three-phase power systems may be operated grounded or ungrounded. In either system, the power is carried on three conductors or phases. The grounded system may also include a fourth neutral conductor, which carries only the unbalanced current, if any. The windings in transformers, generators, and other three-phase equipment may be connected in wye or delta form. Generally, the delta connection is used in ungrounded systems, and the wye in grounded systems, with the common or neutral point connected to ground and/or the neutral conductor. The neutral conductor operates essentially at ground potential if the three phases are balanced.

Three Types of Systems

The grounded system provides two voltage levels: line-to-line voltage (480 V in the common 480-V system) and line-to-ground or line-to-neutral voltage (Fig. 1.2). The line-to-ground voltage is $1/\sqrt{3} = 0.57$ times the line-to-line voltage. Typical four-wire systems provide 480/277 V for power and lighting or 208/120 V.

If a ground fault develops on one of the phase conductors, a ground protection

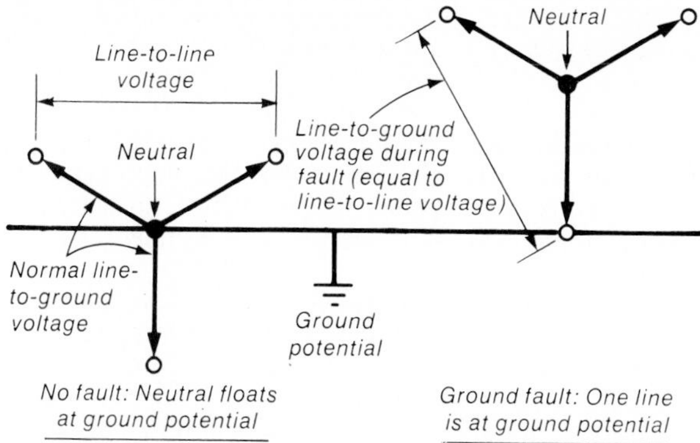

FIG. 1.2 No fault versus ground fault for a three-phase power system.

relay immediately detects the large unbalanced current in the neutral conductor and trips a circuit breaker, isolating the fault and cutting off power to the process.

The ungrounded system has a fixed line-to-line voltage only. In the event a ground fault develops on one conductor, that conductor falls to ground potential, while the potential of the other two conductors rises to the line-to-line voltage above ground. If the fault current is small, which is often the case, the system can continue to operate until the ground fault can be located and repaired. This is why the ungrounded system is preferred by many plant operators.

Problems arise only if the ground fault is left unrepaired—eventually, a second ground fault occurs on another phase in another part of the system. The result is a much more destructive line-to-line fault (at the higher line voltage), which damages equipment in two parts of the system.

An additional problem that can occur with ungrounded systems is overstressing of insulation. Even with a low-level fault, the two unfaulted conductors are raised above their rated potential. If a high-level fault occurs, transient voltages are generated as the fault arc strikes and restrikes, and in the circuit breaker as it operates to clear the fault. These transient voltages impose much higher stresses on the conductor insulation of the ungrounded system.

The resistance-grounded system is preferred by many industries. In this case, the neutral conductor is grounded, not solidly, but through a resistance. The size of this resistance is selected so that if a ground fault occurs, the current flowing through the neutral conductor will be large enough to trip the ground fault relay, but not so large that the fault arc can do serious damage, such as destroying motor laminations.

Utilities prefer large ground fault currents that ensure unambiguous operation of ground fault relays. Another reason that utilities prefer solidly grounded systems is that it allows them to use grounded-neutral lightning arrestors, which are less expensive and more effective than ungrounded arrestors.

Transformer Connections

Choosing a grounded or an ungrounded distribution system can be a source of conflict. Over the years, as noted, utilities have grown to prefer solidly grounded systems. Industrial systems may be operated grounded or ungrounded, depending largely on the preference of the electrical engineer in charge and the organization of the maintenance. Even where an industrial system is grounded, however, the neutral point is often connected to ground through a resistance or reactance.

These different approaches have been adopted to suit the different operating conditions of utilities and industrial companies. The approaches work well until the utility and the industrial company try to generate in parallel. That is when problems surface, and most focus on the transformer that connects the customer to the utility power supply.

Transformer primary and secondary windings can be connected to the three power conductors in either wye or delta configurations. Transformers may be supplied in wye-wye, wye-delta, or delta-delta connections (Fig. 1.3). Normally, these connections cannot be changed once the transformer is designed because this would alter the transformer ratio. The critical difference between them is that wye-connected windings can be grounded whereas delta-connected windings cannot.

Wye-delta connected transformers (Figs. 1.3*a* and 1.4) with the grounded wye on the utility side are the easiest to apply for interconnection, assuming that the customer is satisfied with an ungrounded system. It is particularly important, however, if the utility circuit breaker opens to clear a fault and isolates the customer's system, that the customer's circuit breaker opens, too. If it does not, and the customer's generator is large compared to the capacity of the utility feeder, then the customer's generator will continue to feed the fault. Since the customer's system is ungrounded, its conductors may be subject to overvoltages as the utility circuit breaker, with its reclosing mechanism, repeatedly tries to clear the fault and the arc keeps restriking.

Delta-wye connected transformers (Fig. 1.3*b*) with the customer side wye-connected and the utility side ungrounded can cause the same problem for the utility if a fault occurs on the customer's side. In this case, the utility will almost certainly insist on installing an additional grounding transformer or open-delta connected potential transformers to detect a ground fault, and the customer will be expected to pay for these additions.

Wye-wye connected transformers (Figs. 1.3*c* and 1.5) might appear at first glance to be the most suitable arrangement since it provides a ground on both sides, but this is not so. One reason why a utility prefers a solidly grounded sys-

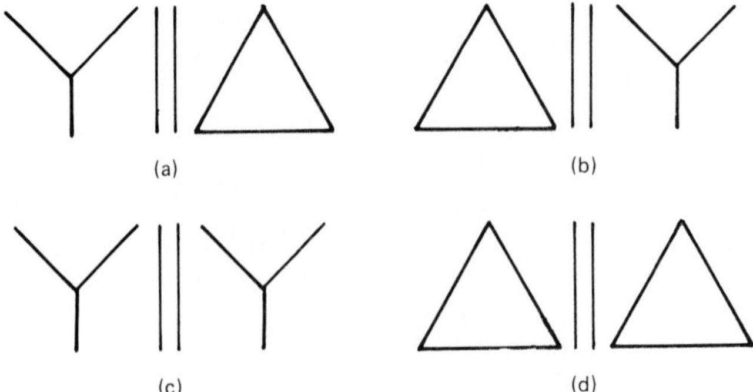

FIG. 1.3 Transformers may be connected: (*a*) wye-delta; (*b*) delta-wye; (*c*) wye-wye; and (*d*) delta-delta.

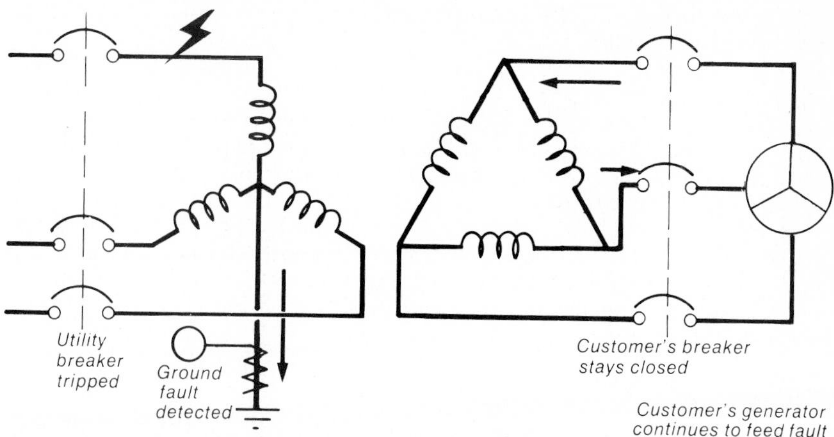

FIG. 1.4 Wye-delta connection is best for most installations, but if the customer's is larger, it may continue to feed a fault on the utility side of the transformer.

tem is because a ground fault on such a system produces a large ground fault current through the neutral leg, which ensures unambiguous operation of the ground fault relay. When both utility and customer sides are grounded, the fault current is divided in some indeterminate ratio between the ground legs on either side of the transformer. This imprecision makes it difficult to determine the setting on the ground fault relay and to coordinate it with other relays in the network—a situation that is anathema to the utility protection engineer. A delta-connected tertiary winding installed in the transformer to absorb generator harmonics makes the problem of coordination even worse.

Delta-delta connected transformers (Fig. 1.3*d*), in which neither the utility nor the customer side is grounded, are common when there is no generation on the customer side, in which case they are quite suitable. Once a generator is inter-

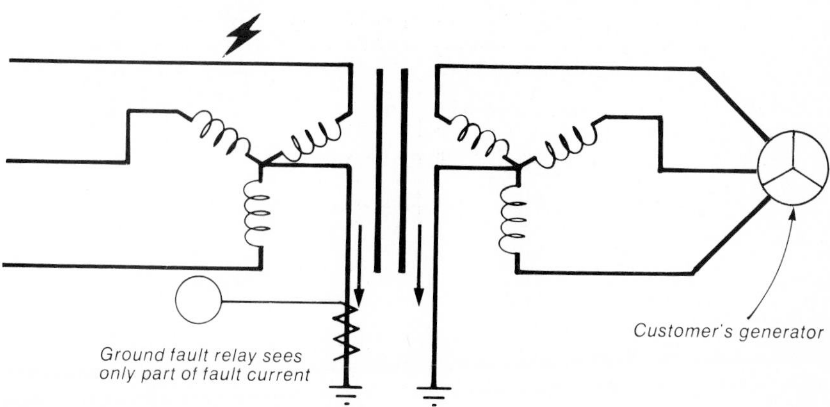

FIG. 1.5 Wye-wye connection is least desirable because ground fault current is divided between two grounding legs.

connected, however, some way must be found of introducing a ground connection at the transformer.

A grounding transformer is commonly selected to introduce a ground leg into a previously ungrounded delta system (Fig. 1.6). Since the grounding transformer is required to carry current only during a ground fault, and then only for the short time needed to operate the circuit breaker, its rating may be considerably less than that of the main power transformer, and thus its cost is much less than that of a new power transformer.

A zigzag grounding transformer has four terminals (Fig. 1.7), three of which are connected to the three power phases and the fourth to ground. It carries practically no current unless there is a fault, in which case the fault current flows down the fourth grounded leg to operate the ground delay.

Harmonics Inhibit Grounding

One of the advantages of a delta-connected system when a generator is installed is that the problem of harmonics is eliminated. Any ac generator puts out a waveform that is slightly distorted, and these distortions can be represented largely as third-order harmonics (180 Hz). A small- to medium-size generator can be expected to generate worse third-order harmonics than a utility-side unit, because special core design features can be incorporated into large generators to eliminate harmonics.

Third-order harmonic waveforms on the three phases of an electric system are in phase with one another. So when the three phases are joined in a wye connection, the harmonic currents reinforce one another in the grounded neutral leg. This may be detected as a ground fault current by an overcurrent relay in the neutral leg.

A delta-connected system does not have the same problems with third-order harmonics because the harmonic currents are effectively short-circuited by the delta, in which they generate circulating currents. (Note that ac generators are almost universally supplied with wye-connected windings to avoid the internal

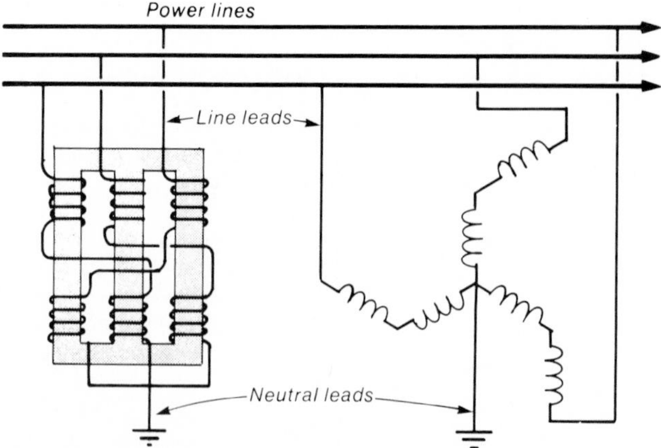

FIG. 1.6 Any transformer will double as a grounding device if its primary is wye connected and its secondary is a closed delta.

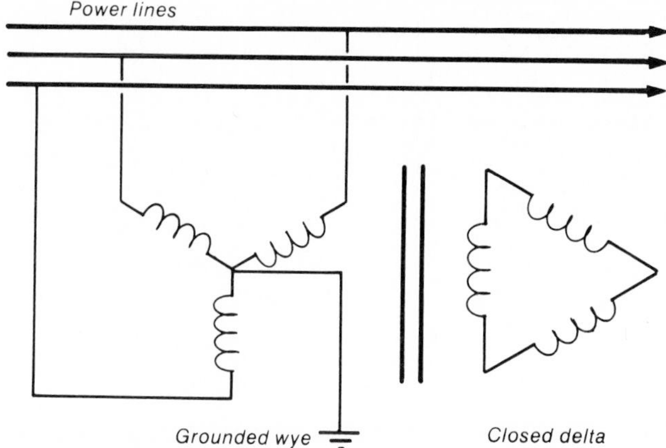

FIG. 1.7 Zigzag grounding transformer has primary and secondary windings connected as shown, with the secondary neutral point grounded.

heating caused by such circulating currents.) If a customer prefers not to abandon his or her grounded system, another solution to the problem of harmonics may be a grounded-wye transformer connection with a tertiary delta-connected winding in the transformer in which the third-order harmonic currents can flow. Clearly, this solution is economical only if a new transformer must be purchased anyway. Note that tertiary delta windings further complicate the problem of ground fault relay coordination and may be strongly resisted by the utility.

A ground resistor or reactor in the transformer's neutral leg may keep the third-order harmonic currents below the setting of the ground fault relay. In a wye-wye connected transformer, a resistor on the customer side and solid grounding on the utility side may also solve the problem of relay coordination.

VOLTAGE REGULATION

To control the flow of reactive power, voltage regulators and load tap changers are selected. As shown in Fig. 1.8, when the current lags on the voltage waveform, it transmits less power than when it is in phase. The apparent power is found by multiplying voltage by current, but the real power is less than this by the power-factor percentage. It is convenient to divide the apparent power into two vector components—*real power* and *reactive power*. Both components are important in power generation and must be metered and controlled separately.

The power factor becomes harder to understand once an in-plant generator is interconnected and operated in parallel with the utility network. This is because the power factor (or reactive power flow) is intimately tied in with the voltage control of the plant generator and the voltage of the utility supply. Understanding the problems involved is now much more than a matter of avoiding the utility's power-factor penalty.

The power factor is a way of representing the extent to which alternating current drawn by the plant is out of phase with the voltage. It is expressed as the

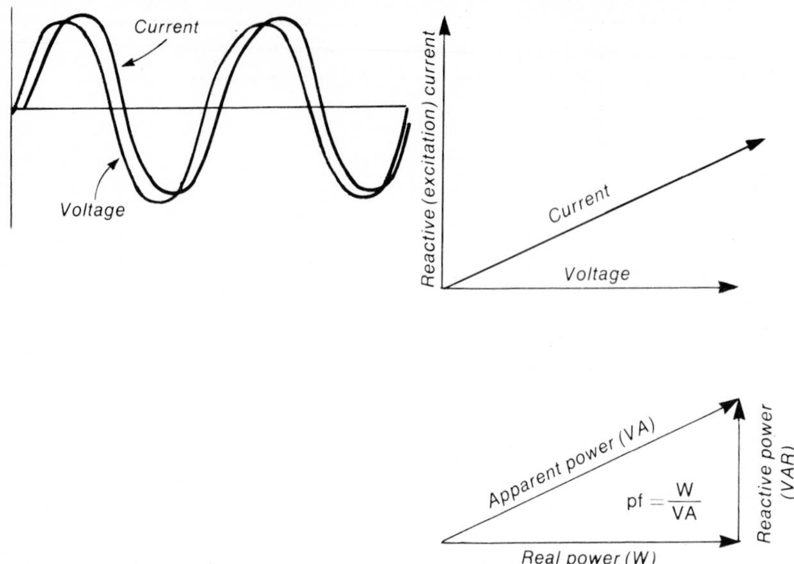

FIG. 1.8 Apparent power consists of real and reactive power.

ratio (or percentage) of real power (watts) to apparent power (volts × amperes). Most industrial plants draw a lagging current and have a power factor somewhere between 70 and 95 percent.

When in-plant generation is involved, the concept of power factor becomes less useful because it is difficult to tell whether the power factor is leading or lagging. (A power factor that is lagging from the point of view of one generator may be leading to another generator.) That is why utilities talk about *reactive current*—that component of total current that is 90° out of phase with the voltage. This leads to the concept of reactive power, normally expressed in *volts-amperes reactive* (VARS).

Flow of Watts and VARS

Generator engineers are concerned with the flow of watts and the flow of VARS. The power output of a generator is controlled by varying the torque applied to its shaft by the prime mover. The VARS output is controlled by varying the field excitation. This function can be valuable in an industrial plant. If an in-plant generator is overexcited, it produces VARS as well as watts, and these VARS flow into the plant's motors to provide their excitation current. This flow reduces the amount of VARS that the motors draw from the utility system. It is exactly the same as installing power-factor correction capacitors to correct a poor power factor in a plant. (The generator functions as a synchronous capacitor.)

The voltage at which the utility supplies power to a plant is seldom constant, although it is supposed to be held within ±5 percent of the nominal value. In practice, it varies with the load on the utility's transmission line and the distance of the plant from the nearest substation.

An in-plant distribution system in which there is no generation is designed to accommodate this variation in voltage. Static taps on the plant's main transformer usually provide adequate adjustment and are set so that the average voltage at the plant's load center is never too far from the rated voltage of the equipment.

Voltage Regulator

An in-plant generator coupled to a system with a varying voltage must have its excitation constantly adjusted according to the plant voltage. This is done automatically by the voltage regulator. A potential transformer produces a signal proportional to the voltage, and the regulator adjusts the excitation to a level at which the generator would produce the same voltage if it were operating in isolation. This excitation level is then trimmed to control the flow of VARS from the generator. To do this, an instrument known as a *cross-current control transformer* senses the flow of reactive current from the generator and sends a trimming signal to the voltage regulator.

If adjacent plant equipment draws a badly lagging power factor, additional control of the generator excitation may be needed in the form of a power-factor controller. This is a unit that measures the plant power factor and trims the generator VARS output to ensure that the plant does not draw excess VARS from the utility. The dynamic range of the generator's voltage regulator must be able to cover the full range of supply voltage and control VARS flow at the same time.

Load Tap Changer

VARS control gets to be more of a problem when the size of the in-plant generator is large compared with the capacity of the utility intertie. In such a case, it is preferable to let the in-plant generator set the plant voltage, while the utility intertie acts only to cover peaks. Problems occur if the tie voltage drops while the plant generator tries to maintain the voltage at the rated level. The result is that VARS flow from the in-plant generator into the utility network, adding to line loss and voltage drop, even though they are flowing in the opposite direction to the power.

The solution is a load tap changer on the main plant transformer or voltage regulators in the utility tie line (Fig. 1.9). A load tap changer is a motor-driven switching device with multiple contacts that can adjust the transformer ratio in response to variations in the incoming voltage. A voltage regulator is an autotransformer connected in series with the main transformer. The choice is a matter of economics.

With the load tap changer in place, the utility supply voltage is automatically adjusted to the nominal plant value so that voltage regulators on the in-plant generators can control the flow of VARS while holding a constant plant voltage.

PROTECTIVE RELAYING

Most electrical engineers have no more than a casual understanding of protective relaying, and most powerplant engineers are only dimly aware of its existence. It

FIG. 1.9 Three voltage regulators can take the place of a load-tap-changing transformer. *(Courtesy Siemens-Allis Inc.)*

is a foreign subject, talked and written about with foreign symbols, that the average power user does not need to understand.

For the utility engineer, however, protective relaying is a way of life. The utility's relay engineers spend all day figuring possible combinations of faults that might occur in the network and designing relay circuits to protect against them. In the process, relay engineers have developed a highly systematic approach to relaying and an efficient shorthand for designing circuits and specifying relays.

The purpose of protective relays is to detect unsafe or out-of-limit conditions in a power system and to trip appropriate circuit breakers. The prime intent of the protective relay circuit is to isolate the faulty section of the circuit so that the remainder of the network can continue to deliver power without interruption. If the circuit breakers can be opened fast enough to prevent damage to electrical components, so much the better.

The most commonly needed relays in an interconnected system are overcurrent, overvoltage and undervoltage, overfrequency and underfrequency, differential, and relays that detect the direction of power flow. The relay engineer shows each of these relay types (along with many others) on a relay diagram, not by its name or function, but by a number.

Instrument Transformers

Instrument transformers transmit current, voltage, or frequency signals to the protective relays (Fig. 1.10). Two types of instrument transformers are the cur-

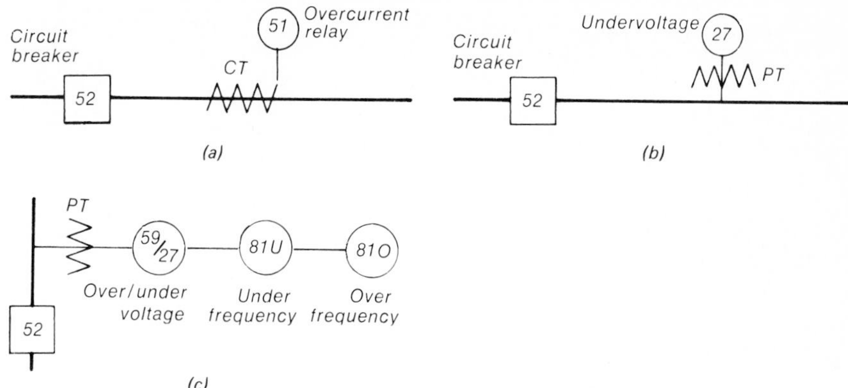

FIG. 1.10 Instrument transformers may be: (*a*) current (CT); or (*b, c*) potential (PT).

rent and potential. They are needed because the high voltages and currents in a power circuit must be reduced to values that can be conveniently handled by instruments and protective relays.

A *current transformer* (CT) uses the power conductor as its primary winding, and it produces a small current in its secondary winding that is proportional to the power line current. Normally, 5 A at the secondary is equivalent to full-load current flowing in the primary.

A *potential transformer* (PT) has its primary connected between two of the three power lines or between one phase and ground, and it produces a voltage at its secondary proportional to the line-to-line or line-to-ground voltage, respectively. Normally, 120 V at the secondary is equivalent to the rated voltage at the primary.

Most power circuits are three-phase and may use three or four conductors, but they are specified by using one-line diagrams. At each measurement point, there may be one, two, or three instrument transformers and up to three relays. But only one transformer and one relay are indicated on the relay diagram.

Ground Faults

A ground fault on one phase of a grounded three-phase circuit produces an unbalanced current in the grounded leg of a wye-connected transformer or generator (Fig. 1.11*a*). The ground current may be quite small initially and will go undetected by the regular overcurrent relay, which is set to read hundreds of thousands of amperes. But a sensitive ground overcurrent relay can be set to detect current flowing in the neutral leg and to trip the appropriate breaker.

A ground fault in an ungrounded circuit can be detected only as an imbalance in the line-to-line voltages (Fig. 1.11*b*). This is done with three voltage transformers connected between the three phases. The secondaries of these transformers are connected in what is known as an *open delta*. If the three phases are balanced, no voltage appears around the delta. As soon as the circuit becomes unbalanced due to a ground fault, a voltage appears across the open delta and can be detected by a voltage relay.

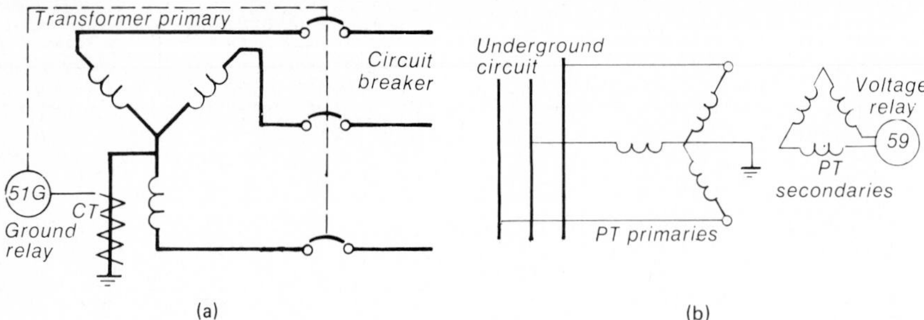

FIG. 1.11 Ground fault: (*a*) on one phase of a grounded three-phase circuit; and (*b*) in an ungrounded circuit.

Types of Relays

Directional relays are vital in interconnected systems to control the flow of power (Fig. 1.12*a*). They are usually set to detect a value of current, but since current flows both ways in an ac circuit, they detect direction in terms of power flow. Two instrument transformers are needed: a CT to measure magnitude and a PT to polarize the relay to read in one direction only. A directional relay is always needed to ensure that power does not flow into the interconnected generator, causing it to act as a motor.

A synchronizing relay supervises the manual synchronization of an interconnected generator (Fig. 1.12*b*). An operator manually adjusts the speed and fre-

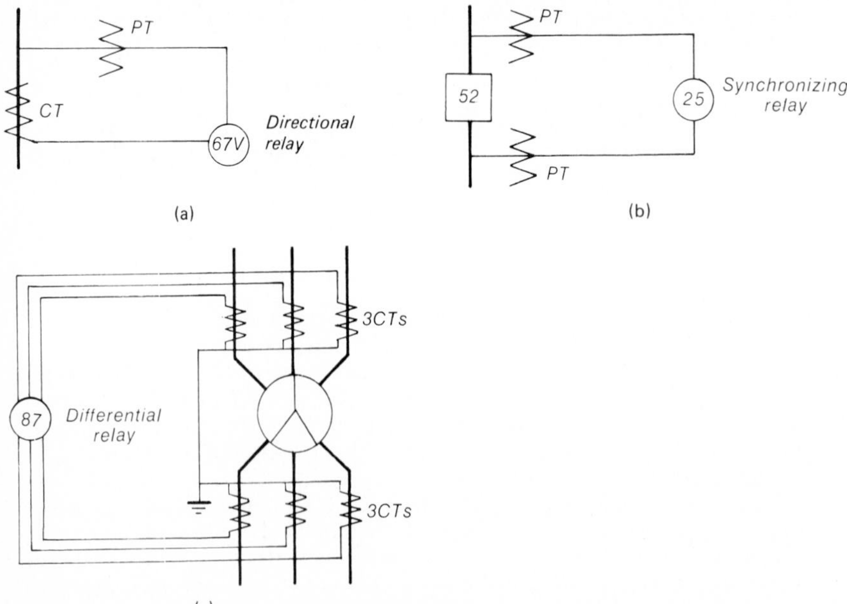

FIG. 1.12 Protective relays may be: (*a*) directional; (*b*) synchronizing; or (*c*) differential.

quency of the in-plant generator, using a synchroscope to judge when the generator is in phase with the utility system, at which point the operator closes the circuit breaker. But the damage that could be inflicted by closing the circuit breaker out of synchronism is so great that a supervisory synchronizing relay is usually mandatory. This relay needs two PTs, one on each side of the circuit breaker.

An automatic synchronizer makes the frequency adjustment by itself. All the operator has to do is to push a button; no separate protective relay is needed.

Differential relays are needed to detect internal faults in the windings of transformers and generators (Fig. 1.12*c*). They do this by looking at the current flowing into and out of each phase winding. Any imbalance in these readings indicates there is a fault between windings, and the circuit breaker must be opened before the damage gets worse. A differential relay needs six CTs, two for each phase.

SPECIFIC PROTECTION MEASURES

Once a generator is interconnected and operated in parallel with the utility network, it becomes part of a vast, sophisticated machine. So it needs sophisticated protection against what the network can do to the generator and what the generator can do to the network. How much protection is needed for a specific situation is controversial.

When a large company that employs its own electrical and relay engineers wants to interconnect with the utility system, knowledgeable engineers from both the company and the utility must sit at a table and decide on the need for each interconnection component and each protective device. They also must agree on who will pay for the capital and installation cost and for maintenance. Nothing is standard, and such contracts typically take 6 months to complete. As the number of interconnections increases and the technical sophistication of the average customer-generator decreases, there is a need to bring some standardization into the protection package and to reduce its cost.

Forward-thinking utilities have produced printed guidelines that attempt to delineate exactly what equipment is needed for different types of interconnection. Some of these publications are quite specific, others quite general.

The following situations pose major interconnection problems that must be handled:

Main Circuit Breaker Trips

Isolation of the customer's generator occurs if the utility circuit breaker opens for any reason (such as for clearing a fault). This leaves the generator coupled to the full plant load and to any adjacent customers connected downstream of the tripped circuit breaker. In such an event, it is almost inevitable that the generator will be grossly overloaded and will immediately start to drop in both voltage and frequency, with possible damaging results. Thus, in any interconnected system, the customer's main circuit breaker must trip whenever utility power is interrupted.

Usually underfrequency or overfrequency relays can cause the main circuit breaker to trip. Typically, these are set at 59 and 61 Hz so that the circuit breaker trips as soon as the generator output reaches these frequency limits. But it is just possible that the customer's governor works better than expected, so that the generator maintains exactly 60 Hz for a few seconds after being isolated. For this reason, most utilities are also insisting on overvoltage and undervoltage relays in addition to the frequency relays.

Once the main circuit breaker has opened, the customer is responsible for what he or she does with the generator output. Hopefully, the customer will be able to shed nonessential loads and continue to operate essential loads on her or his own generator. Usually this means opening the generator circuit breaker first, then shedding nonessential loads, next reconnecting essential loads to the generator. With suitable high-speed relaying, it may be possible to shed nonessential loads quickly enough to avoid interrupting generator output.

Main Circuit Breaker Reclosures

Automatic reclosing of utility circuit breakers can be catastrophic for the customer-generator. If the in-plant generator has fallen well out of phase by the time the utility circuit breaker recloses, the resulting shock can be enough to shear the generator shaft, not to mention the electrical damage that will be done. There is no way that plant circuit breakers can operate fast enough in this situation to prevent damage.

The only sure protection is to keep the in-plant generator off the line until the utility system has stabilized. The utility may or may not agree to inhibit automatic reclosing if the customer's circuit breaker is still closed. Normally, inhibiting is done with a voltage relay on the customer's side of the circuit breaker, which prevents reclosing if the line is live.

Reclosing of the customer's circuit breaker presents further problems. Clearly, this circuit breaker must not close onto a dead utility line or onto a live line to which it is not synchronized. If the utility will not provide an interlock between circuit breakers or agree to inhibit automatic reclosing, then a time delay is needed to ensure that the utility system has stabilized.

Synchronizing equipment is also needed across both the customer's main circuit breaker and individual generator circuit breakers. It can be either an automatic synchronizer or a synchronism check relay. In the latter case, the generators are synchronized and the circuit breaker is closed manually, but the check relay prevents reclosing out of synchronism. If the plant has become disconnected from the utility and is running on its own generators, the main circuit breaker's synchronizer allows the plant to be reinterconnected without stopping the generators.

Safety Assurances

Hazards to line workers are a legitimate concern of utilities. Every utility has established a strict operating procedure to ensure that power lines being worked on are locked out and dead. This procedure must now be extended to ensure that a customer will not inadvertently or out of ignorance reenliven a line that is presumed dead. Understandably, utilities are insisting on the right to inspect a customer's facilities to make sure that procedures are being followed correctly. These procedures usually include keeping a log of regular maintenance, which must be available for utility checks.

Proper Relay Settings

Relay settings are an endless source of misunderstanding between the utility and the customer-generator. This is largely because the utility's standard approach is to drop a generator offline at the moment trouble is detected, while the customer's main interest is uninterrupted power for the plant. Thus, even if the utility has published protection requirements and has inspected the relaying on an interconnected system, there may be no clear-cut agreement on where relay trip limits should be set.

A typical problem is voltage surges on the utility's line, caused by the switching of banks of power-factor correction capacitors. Often these surges are sufficient to trip the undervoltage relays and drop out the customer's plant along with the generator. The utility is not about to forewarn the customer of these surges. In at least one case, the customer discovered the cause of repeated plant trips only after realizing that they were occurring at the same time every day.

Conversely, relay coordination is becoming a big headache for utilities. A small interconnected generator does not present too many problems here, but as the generator gets larger, phase overcurrent relays that are a normal part of any generator must be capable of coordination with utility relays. The utility may insist on voltage-controlled overcurrent relays.

What constitutes a large or a small generator? Southern California Edison Co. has published two sets of relay requirements—for generators under and over 200 kVA (Fig. 1.13). Georgia Power Co. classifies all generators over 5 MVA as large and all others as small. Hawaiian Electric Co. has set 500 kVA as the limit for

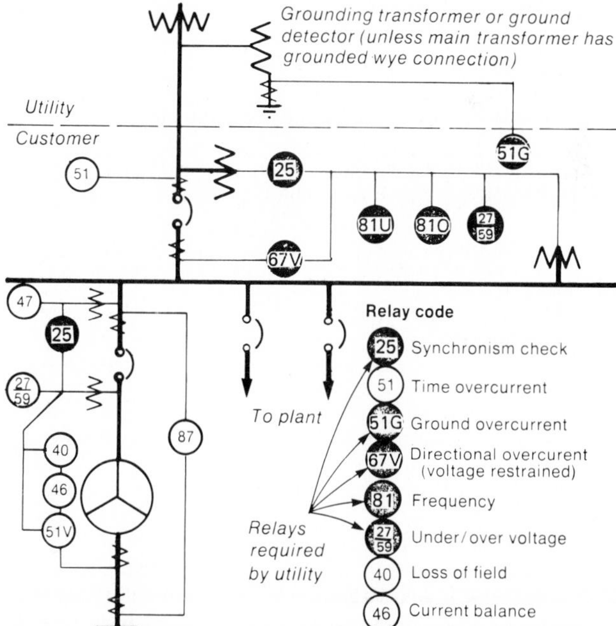

FIG. 1.13 Basic protection scheme suggested by Southern California Edison Co. for total generation over 200 kVA. *(Courtesy Southern California Edison Co.)*

small generators. From the point of view of the relay engineer, however, the importance of the generator's size depends more on its output relative to the capacity of the feeder to which it is interconnected than on its absolute kilovoltampere rating. Right now, customers that are interconnected to relatively small utility feeders, rather than to a utility substation, are having trouble merely staying online.

Relay Specifications

Utility-quality relays are a must for most interconnection jobs. No real standard exists to differentiate a utility- from an industrial-quality relay, but there is no doubt in the utility engineer's mind about which is which. Basically, the utility-quality relay is a much heavier, more costly piece of equipment in which trip settings can be adjusted accurately to known values. The relay has operation indicators showing that it has tripped and why it has tripped, and its guts can be removed for servicing without breaking the connection. By comparison, the industrial relay is a packaged, solid-state device with simple dials to adjust settings. Accurate coordination of industrial relays can be accomplished only by trial and error.

Cost is the big problem. A utility-quality relay costs between $400 and $1200, and a complete protection package, containing everything the utility would like to have, can cost $45,000 or more. For the customer-generator between 200 and 1000 kW, this cost can be a real burden.

Design and specification of the relay protection package are almost always the customer's responsibility. Even if the utility has published guidelines on protection needs, the customer will be expected to come up with the detailed design and submit it to the utility for approval.

Standards and conventions for interconnection protection have had little chance to become established. But one convention that is already universal is that the customer pays for all necessary protection. Even if the utility prefers to install and own the protection package, the customer will pay for the initial cost as well as a monthly maintenance fee. The customer may even be billed separately for the electric power used by the relaying system. Also the customer pays for any changes and additions needed to the existing switchgear and relaying.

EMERGENCY SYSTEMS

When the lights go out, emergency generators sometimes fail to start. They have been sitting in the basement for months, even years, representing thousands of dollars in capital investment, but when the time comes, they may fail to perform their vital function. The answer, of course, is frequent maintenance and regular exercising under load.

There are four ways in which interconnection and operation of generators in parallel with the utility supply may streamline the exercising routine: (1) Generators can take up in-house loads without the usual momentary interruption associated with a transfer switch. (2) Any magnitude of exercising load can be applied to the generators without the need to select actual loads for transfer. (3) Generators can double as peak-shaving equipment, either for the utility's peak or for the customer's. (4) In the case of large generators, revenue from power fed back into the network may be used to offset the cost of exercising.

Transfer Switches

Most emergency generating systems are operated at 480 V and are coupled to the emergency loads, when needed, by means of transfer switches. These systems are carefully designed to eliminate the possibility of the emergency generators ever being paralleled with the utility supply. This arrangement has been dictated partly by traditional utility resistance to interconnection and partly by the need to keep the emergency system totally separate. Some codes even require that no transformer be interposed between the emergency generator and its loads.

One problem with the transfer method, whether done with a transfer switch or with interlocked circuit breakers, is that one source must be dropped before the other is picked up. This causes a brief power interruption to vital systems. The result is that hospital administrators, and other building managers concerned with the continuous operation of their facilities, may actually resist regular exercising of emergency generators.

Peak Shaving

As power costs continue to soar and utility resistance to interconnection is tempered by experience, more and more emergency generating systems are being built for parallel operation with the utility. But the cost is high. All the protective relays used with a continuous interconnected generator are needed, and usually additional circuit breakers must be purchased. So the emergency generating system must be either very large or very important before interconnection is justified.

Peak shaving may very well be the cost cutter that justifies interconnection. Peak shaving with transfer switches can be achieved by switching selected loads over to the emergency generator as the total plant load nears its peak, although this is a cumbersome method that depends on having switchable loads of the right size. Ideally, the load shaved should be carefully selected to minimize generator run time for the maximum reduction in peak kilowatt load. An interconnected generator can be automatically loaded to keep the total plant load below a predetermined kilowatt level, whatever the total plant load.

Network Systems

One might think that, because an emergency generator operates only when the utility power supply fails, there is not much opportunity for operating the systems in parallel when an emergency actually occurs. But this is not the case. Most facilities that need emergency generation, such as hospitals, telephone exchanges, or computer facilities, also have multiple-feeder network systems to ensure maximum possible reliability of utility-supplied power, even before the emergency generator must be called into action.

It is unlikely that all the utility feeders will fail at once. Individual feeders may fail because of problems at local substations, and their loads usually can be assumed by the other feeders. In the event of a total utility blackout, the feeders usually go dead one after the other as the utility system goes down.

ACKNOWLEDGMENT

The author would like to thank *Power* magazine for the use of some of the material that appears in this chapter.

CHAPTER 5.2

IN-PLANT ELECTRIC DISTRIBUTION

Staff Editors
Power *Magazine*
New York, NY

INTRODUCTION

In-plant electric distribution is essentially matching electrical needs with the electrical realities of generation. This means a careful analysis of the plant load, always a time-variable, multifaceted entity in modern industrial facilities, against the practicalities of generation, both by the utility and by the plant itself. In this chapter, distribution basics are covered, beginning with load planning and closing with secondary substations, the final element in the distribution hierarchy. In between, coverage of such pragmatic topics as system voltage selection, power supply options, and circuit arrangements fleshes out the fundamentals of electric distribution within the industrial plant and other buildings.

LOAD PLANNING

Since the end purpose of the electric distribution system is to handle present and future loads, planning starts with the load—its magnitude, nature, and characteristics. The load size should be examined first because both supply and distribution arrangements depend on it. The nature of the load—lighting, motors, welding, heating, computers, etc.—is important from the standpoint of distribution. Motor and lighting loads are often supplied by the same feeder. Some questions must be answered: Are loads intermittent or continuous? If continuous, how costly is an interruption? How far would it pay to go in providing extra reliability?

Estimating load accurately is not easy, especially for a new plant where the utilization equipment also is in the planning stage. The designer can only work from data at hand, then check and supplement them with figures from similar plants. The designer starts by considering light and power separately and then combines them to determine demand in various areas, since both often are fed from the same substation. Load figures are most often expressed as load density in voltamperes per area (VA/ft^2 or VA/m^2).

Lighting and Power Loads

Lighting loads are not difficult to estimate. Three factors are considered: type of light source, its intensity, and luminaire height. Lighting load density will vary from 1 VA/ft^2 (11 VA/m^2) in storage areas to 15 VA/ft^2 (161 VA/m^2) or more for office work and precision manufacturing. Outdoor lighting for shipping areas, protection, and aesthetic effects should not be overlooked. Much estimating help can be found in the literature issued by major lamp and luminaire manufacturers.

Power load estimating is tougher. Usually, the best way to start is to look first at major loads such as big synchronous motors, furnaces, welders, etc. These influence demand strongly. They should be dealt with early in system design since their size and operating schedules can usually be predicted closely. For large induction and 0.8 *power factor* (PF) synchronous motors, it can be assumed that the kilovoltampere demand equals the horsepower (kilowatt) rating. For 1.0-PF motors, kVA = 0.8 × hp (kW) can be assumed. Often the larger machines will operate alternately, so diversity factors should be investigated and taken advantage of (see box).

Experience plus a good overview of the plant's activities will be necessary for estimating demands of smaller machines on a load-density basis. In any one plant, density may vary from zero in storage areas to 40 VA or more in concentrated machine areas. Typical load densities for representative industries (and other types of buildings) may be found in handbooks and manufacturers' literature. A study of loads in an existing plant with similar processes will aid considerably in estimating loads in a new plant.

Total Load Estimates

Total load estimates must be made after the loads of individual machines and plant areas have been determined. The individual loads are combined, and factors are applied to take advantage of the characteristics of the loads, as the box explains. The total maximum demand determines the system capacity for which the designer must provide. Selection of demand and diversity factors (such as load density) must be based on knowledge of plant conditions, on past experience, and on data from similar plants or buildings.

Factors To Apply In Estimating Total Load

The *demand factor* is the ratio of the maximum demand on a system to the total connected load of the system (see illustration). The *diversity factor* is the ratio of the sum of individual maximum demands of various parts of the system to the maximum demand of the system as a whole. The *load factor* is the ratio of the average load over a period to the peak load occurring in that period.

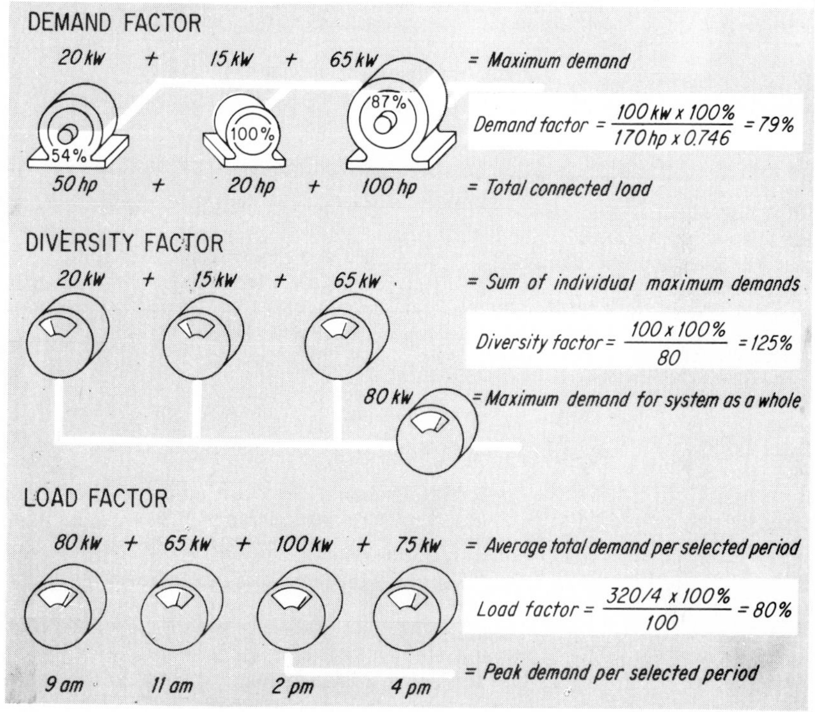

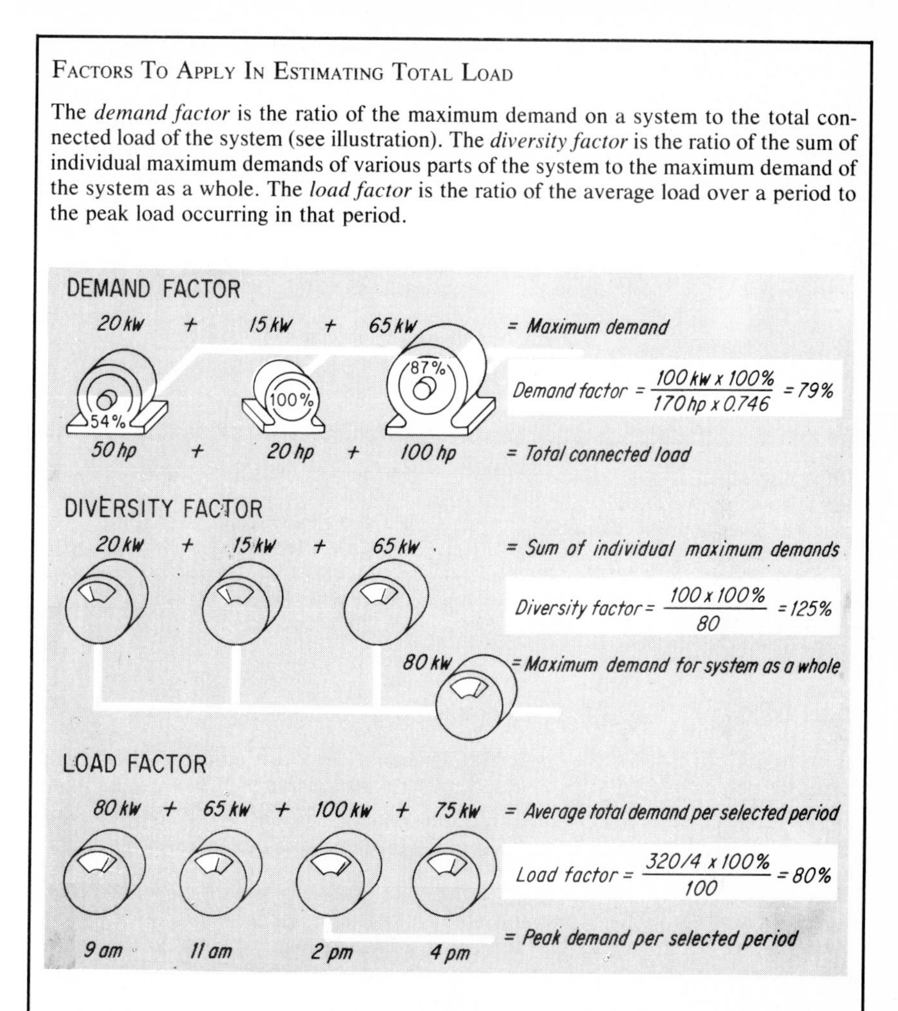

Diversity factor may be applied to determine the rating of the main substation transformer and service entrance equipment. In determining the capacity of major feeders and risers, demand factors may be applied. A load factor should be applied only when a detailed knowledge of the operating characteristics of the load can be matched to the thermal capacity of the distribution equipment. Total load is multiplied by demand and load factors and divided by the diversity factor, which normally has a value greater than 1.

For example, the demand of an arc furnace is 100 percent while the resistance welder has a demand of only 20 percent. A diversity factor of 1.0 is used by some engineers to provide ample system capacity to take care of inevitable growth. Capacity for more than 100 percent in the main system may be beneficial if unexpected expansion of facilities is likely in the future.

The demand factor will vary considerably with different types of loads (Table 2.1). For example, the demand of an arc furnace is 100 percent while the resistance welder has a demand of only 20 percent. A diversity factor of 1.0 is used by some engineers to provide ample system capacity to take care of inevitable growth. Capacity for more than 100 percent in the main system may be beneficial if unexpected expansion of facilities is likely in the future.

TABLE 2.1 Demand Factors for Various Types of Loads

Load type	Demand factor, %
Arc furnaces	100
Arc welders	30
Induction furnaces	80
Lighting	100
Motors	
For general-purpose equipment*	30
For semicontinuous processes†	60
For continuous operations‡	90
Resistance ovens, heaters, furnaces	80
Resistance welders	20

*Machine tools, cranes, elevators, ventilation, compressors, pumps, rolling mills, etc.
†Paper mills, refineries, rubber processes, etc.
‡Textile mills, chemical plants, petrochemical plants, etc.

Future Loads

Future loads often lead to large errors, because they are either ignored completely or not estimated wisely. Load growth should not be guessed at; plans should be committed to paper for what to do if it doubles, triples, etc. Fortunately, modern distribution equipment lends itself to easy additive growth if provision is made in the basic plan.

Space is often a limiting factor. At the outset, space for additional substations, primary switchgear, generators, and other major equipment should be reserved, but planned to occupy areas that are not in prime demand for production equipment. Enough spare ducts, conduits, etc., should be provided that added feeders can be run through the plant without difficulty.

The electric system should be flexible enough to accommodate changes in manufacturing processes (or in occupation of nonindustrial buildings). Unit substations, easily added to or moved, are a big help in maintaining flexibility. So is a plug-in busway, which simplifies shifting production equipment. From an overall point of view, future electrification of apparatus now using steam or fuel should be anticipated, if possible.

Special Loads

Loads such as from computers may require extra attention. Power requirements of typical large computer systems run from about 200 to 500 kVA, with air conditioning requirements of from 25 to 75 kVA. High-level lighting may also be needed in computer rooms. Even more important is the quality of the electric supply needed to keep the computers functioning properly. For typical units, the voltage may not vary more than $\pm$ 3 percent, and the three line-to-line voltages must remain balanced to within 2 percent with the computer operating. The frequency may vary ± 1 Hz. Also, the waveform must not be distorted by harmonics and must be free of high-voltage transients. Before computers are installed, their manufacturer should be contacted well in advance to find out electric requirements and limitations.

Since computers in general are likely to be damaged by system transients that may otherwise go undetected, it would be wise to feed the computer from a separate transformer and to provide the machine with its own grounding circuit. In some cases, further isolation will be needed: the computer may be supplied from a battery and inverter or from a motor-generator set that isolates it from line transients.

Computers are also sensitive to power interruptions of only a few cycles; programs in some cases will be interrupted and have to be started over. The power supply that provides isolation (buffering) should also give uninterruptible power—transferring the load from the normal source to the alternative source (battery or engine) with no disturbance beyond a slight distortion of the waveform for a fraction of a cycle. It is less costly to install special power supplies at the start than to install them when trouble develops.

SYSTEM VOLTAGE SELECTION

The voltage selected for the greatest effectiveness of the distribution system must be chosen from one of the standard system voltages accepted by electric utility companies and manufacturers of electric equipment. These voltages are established by *American National Standards Institute* (ANSI) Standard C84.1, Voltage Ratings for Electric Power Systems and Equipment (60 Hz). Table 2.2, derived from the Standard for systems supplying both light and power, covers the voltages that may be selected for in-plant electric distribution.

Nominal Voltages

Nominal system voltage identifies the system; actual voltages may lie within voltage range *A* or *B* in the table. Except under unusual conditions, utilities are expected to maintain service voltages within range *A;* the plant should design and operate its system so that most utilization voltages lie within range *A*. Properly designed and rated utilization equipment will give fully satisfactory performance throughout this range.

Range *B* includes voltages both above and below range *A*, which result from practical design and operating conditions—on the utility and user systems—that are limited in extent, frequency, and duration. Utilization equipment will ordinarily give acceptable performance at range *B* voltages, although not as good as in range *A*.

Utilization Voltages

Utilization voltages offering the most advantages for a typical industrial plant are 480 V or 480Y/277 V (Fig. 2.1). These systems have been adopted almost universally for plants where most of the load consists of integral-horsepower motors

TABLE 2.2 Standard System Voltages for Three-Phase, In-Plant Distribution*

Nominal system voltage		Voltage range *A*			Voltage range *B*		
		Minimum		Maximum	Minimum		Maximum
Three-wire	Four-wire	Utilization voltage	Service voltage	Utilization and service voltage	Utilization voltage	Service voltage	Utilization and service voltage
	208Y/120	**191Y/110**	**197Y/114**	**218Y/126**	**184Y/106**	**191Y/110**	**220Y/127**
	240/120	**220/110**	**228/114**	**252/126**	**212/106**	**220/110**	**254/127**
240		220	228	252	212	220	254
	480Y/277	**440Y/254**	**456Y/263**	**504Y/291**	**424Y/245**	**440Y/254**	**508Y/293**
480		**440**	**456**	**504**	**424**	**440**	**508**
600			(Limits not established)			(Limits not established)	
2,400		2,160	2,340	2,520	2,080	2,280	2,540
	4,160Y/2,400	3,740Y/2,160	4,050Y/2,340	4,370Y/2,520	3,600Y/2,080	3,950Y/2,280	4,400Y/2,540
4,160		**3,740**	**4,050**	**4,370**	**3,600**	**3,950**	**4,400**
4,800		4,320	4,680	5,040	4,160	4,560	5,080
6,900		6,210	6,730	7,240	5,940	6,560	7,260
	8,320Y/4,800		8,110Y/4,680	8,730Y/5,040		7,900Y/4,560	8,800Y/5080
	12,000Y/6,930		11,700Y/6,760	12,600Y/7,270		11,400Y/6,580	12,700Y/7,330
	12,470Y/7,200		**12,160Y/7,020**	**13,090Y/7,560**		**11,850Y/6,840**	**13,200Y/7,620**
	13,200Y/7,620		**12,870Y/7,430**	**13,860Y/8,000**		**12,540Y/7,240**	**13,970Y/8,070**
	13,800Y/7,970		13,460Y/7,770	14,490Y/8,370		13,110Y/7,570	14,520Y/8,380
13,800		**12,420**	**13,460**	**14,490**	**11,880**	**13,110**	**14,520**
	20,780Y/12,000		20,260Y/11,700	21,820Y/12,600		19,740Y/11,400	22,000Y/12,700
	22,860Y/13,200		22,290Y/12,870	24,000Y/13,860		21,720Y/12,540	24,200Y/13,970
23,000			22,430	24,150		21,850	24,340
	24,940Y/14,400		**24,320Y/14,040**	**26,190Y/15,120**		**23,690Y/13,680**	**26,400Y/15,240**
	34,500Y/19,920		**33,640Y/19,420**	**36,230Y/20,920**		**32,780Y/18,930**	**36,510Y/21,080**
34,500			33,640	36,230		32,780	36,510

*The more popular systems are shown in boldface type.

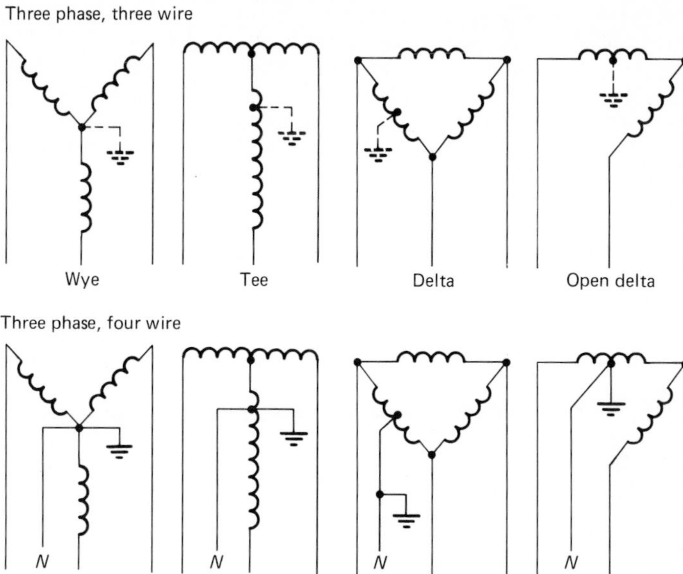

FIG. 2.1 Transformer secondary connections available to supply the standard system voltages.

between 10 and 500 hp (7.5 and 375 kW). A few larger motors can be operated at higher voltages such as 2300 or 4000 V, obtained directly from the primary distribution system or through unit substations.

Convenience outlets and 120-V equipment can be operated from circuits supplied by local dry transformers. Large commercial buildings with 277-V fluorescent lighting and air conditioning generally have adopted 480Y/277 V systems. Although there are some economies in a 600-V distribution system, suitable standard utilization equipment is not readily available.

In plants where the load consists of many small motors [less than 10 hp (7.5 kW)], hand tools, or laboratory-type apparatus, 208Y/120 V systems are advantageous. The clothing and electronic manufacturing industries are typical examples. Although rare today, in some cases utilities may have only 208Y/120 V to offer from dense, downtown networks. Where the load is predominantly electric furnaces, 240 V may be advantageous because of furnace voltage ratings. Furnace loads are usually spot loads in a large plant, however, and can be supplied from a separate load center substation. Even though a sizable portion of the total plant load may consist of 240-V furnaces, 240 V is seldom economical for general distribution.

Load Center Distribution

Load center distribution occurs at a voltage above 600 V to substations strategically located with respect to the load. It should be considered for plants with total loads above 1000 kVA, especially where long feeder runs are involved. The actual decision depends on what service voltages the utility has to offer for the plant's size load. Where the utility has available a voltage between 2 and 15 kV (without extra charge), selection of a primary distribution voltage is not difficult.

Since distribution at the supply voltage eliminates a primary substation transformer, this is usually the most economical course to follow.

There is generally no reason for transforming voltages of the order of 13.8 kV down to 2400 or 4160 V for distribution throughout the building. Where it is desirable to operate motors at a lower medium voltage, transformation can be done at a load center substation. (In most cases, a 2300-V motor with a substation costs less than a 13,200-V motor with appropriate switchgear.) If the utility offers an option between 13.8 kV and an existing lower primary voltage for a plant expansion or addition, serious consideration to the higher voltage for the new system is recommended. Although operation with two distribution voltages may entail some inconvenience, as the plant grows and older equipment becomes obsolete, the lower voltage can be eliminated in favor of the economy and flexibility of the higher voltage.

A special case occurs where the utility offers 34.5 kV as a service voltage. Increasingly, plants are selecting 34.5 kV for the distribution voltage. Although distribution equipment is available at this voltage, it is usually more expensive than 13.8-kV apparatus. Sometimes 34.5-kV circuit breakers cost 2 to 4 times as much as their 13.8-kV counterparts, and a 34.5-kV secondary substation may cost 35 to 60 percent more than a 13.8-kV unit. With long cable runs, however, 34.5 kV may have an economic advantage.

Primary Substation

To transform down to a medium distribution voltage, a primary substation is desirable both technically and economically where the utility supply voltage is greater than 34.5 kV and frequently *at* 34.5 kV. Exceptions are large steel mills, chemical plants, and other large industrial facilities having large, widely scattered loads.

With a primary substation, the choice of a primary distribution voltage is influenced only by considerations in the plant. While the trend is generally toward higher voltages, studies have shown that either of two voltages—say, 4.16 versus 13.8 kV, or 2.4 versus 4.16 kV—will perform equally well in a majority of cases. For larger plants (above 20,000 kVA), the 13.8-kV system is likely to have more advantages.

In some cases where there is a concentrated motor load at 2300 V, it may be preferable to distribute power at 2.4 kV. The economies gained by using 2300-V motors and controls can outweigh the economies lost.

POWER SUPPLY OPTIONS

The type of electric power service available depends mainly on two factors: the location of the plant on the utility system and the size of its load. If the plant's energy requirements represent a sizable load to the electric utility, chances are that it will have several options from which to select the service best suited to system needs. Smaller plants may have no choice.

Low-Voltage Networks

Low-voltage networks are prevalent in dense urban areas (Fig. 2.2). A choice between 208Y/120 V or 480Y/277 V may be available, or the plant may have to accept one or the other of these voltages. The great advantage of the utility network is the continuity of service. The main disadvantage is that the network provides a very large available short-circuit current, which complicates protection. Protection can be enhanced by purchasing power through more than one service, although multiple services are more expensive. In most network areas today, large power users can obtain service at the utility feeder voltage, or they can have the network extended inside their buildings.

Outside cities, small plants are usually served from pole lines. Depending on load size, service may be obtained from a pole-top transformer shared with other customers. Or the load may be large enough to justify serving the plant alone from a single transformer, or even at the utility's primary distribution voltage. If primary distribution is installed at a plant and the utility distribution voltage is suitable, a transformer can be eliminated.

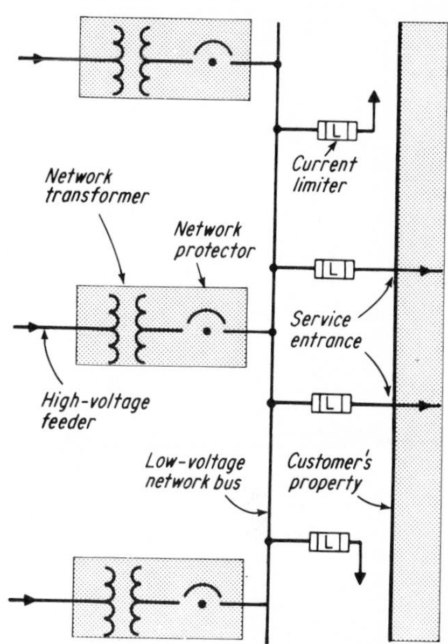

FIG. 2.2 Utility network supplies power from a low-voltage bus shared by many customers. Failure of one or more feeders or transformers doesn't interrupt service.

Primary Substations

Primary substations are in order when the plant is large enough to be served at the utility's transmission or subtransmission voltage (69 kV and above). Figure 2.3 shows a typical primary substation with a tap (*A*) off the high-voltage line. Other customers (*B*) may be served from this tap, but preferably not. The entire substation (below *C*) may be owned by the utility or by the customer, or the ownership may be divided. A common divided-ownership arrangement is for the customer to accept service at the transformer (*D*). Among plants surveyed with loads of 5 MW or more, between 50 and 60 percent own their substations, between 20 and 30 percent of the substations are utility-owned, and between 10 and 20 percent have divided ownership.

Where a choice of ownership is offered, the many pros and cons can be settled only by an economic study. Rates are often more favorable when the user owns the substation, but this advantage is offset by tying up capital that might be invested more profitably elsewhere. Control over design and operation of the substation is valued by some companies, but they must have personnel versed in high-voltage technology.

Weatherproof equipment makes outdoor installation the general rule for primary substations. The unit is usually located adjacent to the main building(s) being served. Some engineers prefer to locate the high-voltage switchgear and transformer outside the plant and the medium-voltage switchgear inside it. Where corrosive fumes or conducting residues may settle on high-voltage insulators, indoor design is recommended.

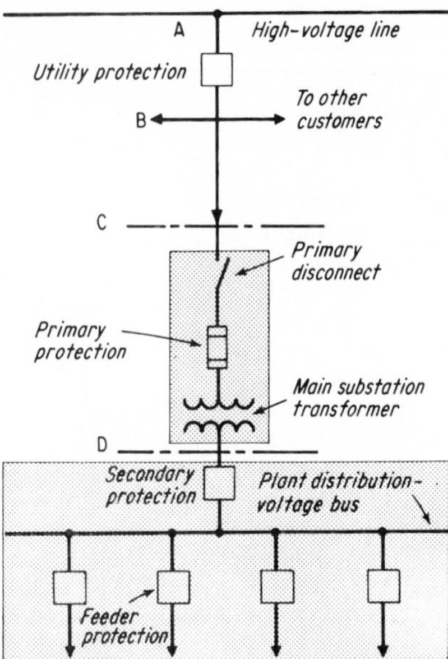

FIG. 2.3 Primary substation is most common where plant distribution is at medium voltage (2.4 to 13.8 kV), and utility transmission is at voltage above 15 kV.

System Reliability

System reliability cannot exceed that of the utility supply. Details of outage expectancy based on past experience are available for the asking from the utility. Two utility supplies from different high-voltage lines increase reliability. The cost of the second supply varies greatly, depending on the location of the alternate high-voltage line. Lightning protection should be investigated; the plant may want and need more than is supplied.

Voltage stability required by the plant may be greater than that maintained by the utility; thus, the utility's regulation data should be checked, if possible. To improve voltage regulation, a load-tap-changing transformer in the primary substation may be the answer. Conversely, if voltage swings are critical to only a portion of the plant load, it may be more economical to select feeder regulators.

Short-circuit current available from the utility should be checked carefully, not only for the present, but for the foreseeable future. Utility engineers can be questioned as to what plans the utility has for expansion of its system that will lead to an increased available short-circuit current. Otherwise, the plant runs the risk that protective gear adequate at the time of installation may become dangerously undersized when greater interrupting capability is required.

CIRCUIT ARRANGEMENTS

The many possible circuit arrangements can be classified broadly into five types: radial, primary selective, primary loop, secondary selective, and network.

Radial System

Of the five system types, the radial scheme is the most often selected (Fig. 2.4). The basic radial system has feeders branching out from a single source without any interconnections. The system may distribute at one voltage throughout the plant, but the load center arrangement is usually more efficient and adaptable. Primary feeders at medium voltage carry power to secondary unit substations, which transform the voltage to the utilization level. Like most load center systems, the radial system can be expanded readily by extending the medium-voltage feed or adding other feeders. Reliability can be increased (within economic limits) by splitting the load and adding feeders.

The radial system has several advantages deriving from its relative simplicity. First costs are lower because there is no duplication of equipment. If sufficient substation capacity is installed, the radial system has enough flexibility to take care of practically any shifting of loads. Voltage regulation is inherently good because of the short secondary feeders. Safety is another benefit of radial system simplicity, since the possibility of switching mistakes—especially during an emergency—is cut to a minimum.

The major drawback of the radial system, and the main reason for considering alternatives, is the shutdown of service when a component fails or is deenergized for maintenance. Should a primary cable or transformer be removed from service, a large part of the load will be without power until the key component is restored to service.

Primary Selective System

This system duplicates primary cables and thus increases continuity in the primary distribution circuit. The flexibility of this arrangement to handle shifting or growing loads is the same as for the radial system, given the same reserve transformer capacity for both.

Two primary feeders, each big enough to carry the entire load, go to each substation (Fig. 2.5); with proper switching equipment on the incoming side, either can be used to supply power. This switching, plus primary cable, usually increases feeder costs compared with radial or secondary selective systems. With

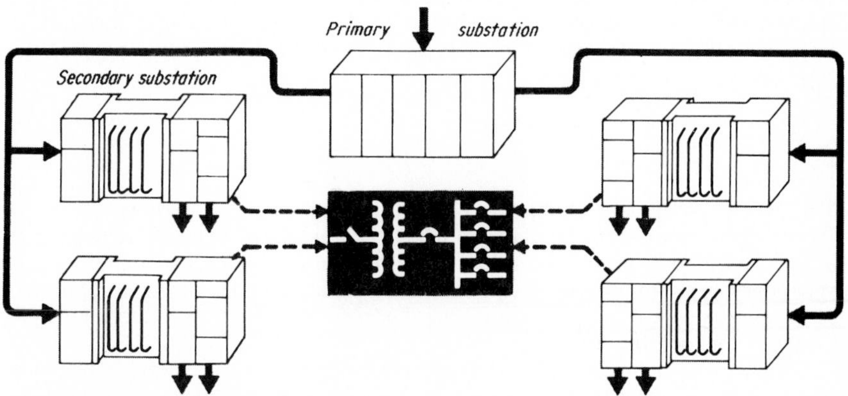

FIG. 2.4 Radial system.

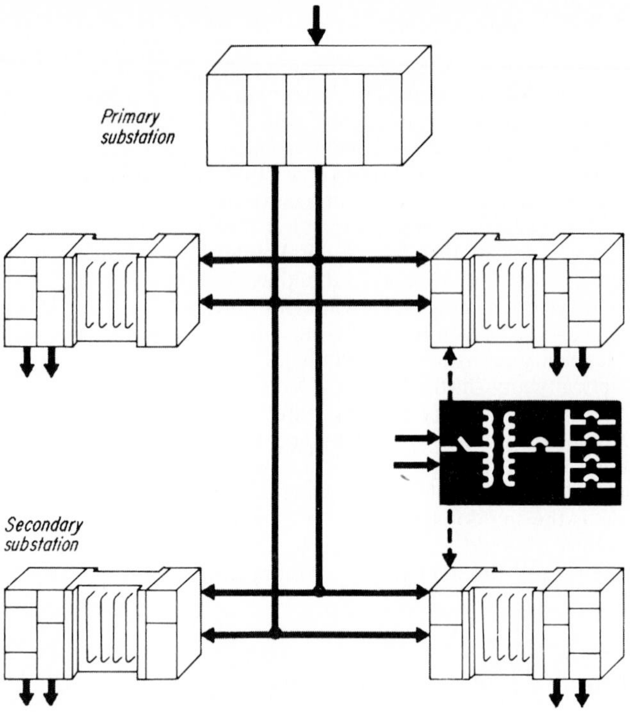

FIG. 2.5 Primary selective system.

adequate circuit breakers for transfer, however, this system is as safe as any; where cable failures are more frequent than average, service reliability can be increased.

If a fault occurs in a transformer, its associated primary circuit breaker opens and interrupts service to one-half the load. Where load-interrupter switches instead of circuit breakers are selected, the transformer is disconnected by opening the switch. Reclosing the primary feeder circuit breaker then restores service to the rest of the load. But without secondary ties, service to the area normally fed from the faulty transformer cannot be restored until the transformer is repaired or replaced.

Primary Loop System

The primary loop system, having primary feeders arranged peripherally rather than radially, may offer advantages when the centers of load are relatively far apart. The loop is ordinarily sectionalized by load-interrupter switches at each secondary substation. Two circuit breakers (one for each leg of the loop) may be at the source end, or a single circuit breaker may be feeding two load-interrupter switches for the loop ends.

When a transformer or primary feeder fault occurs in either loop arrangement, the primary feeder circuit breaker(s) will open to interrupt service to all loads served from that loop. The fault can be located by opening all load-interrupter switches and closing them one at a time in sequence. To avoid the possibility of closing the switch on a fault, the primary circuit breakers should be open during switching. When the fault has been located, it can be isolated by leaving the associated switches open. The loop is now open, but other unit substations are still supplied power through one or the other of the loop legs. A typical loop arrangement costs more than a radial system, but less than a primary selective scheme.

Secondary Selective System

This system offers somewhat greater flexibility and continuity than the radial system at only slightly higher cost. By using a double-ended substation, or two substations side by side, all switching and interconnections are at one spot for any one pair (Fig. 2.6). The secondary buses are tied together so that either unit substation can supply the entire load. The tie can be made between the two single-transformer substations or through the double-ended substations. The tie breaker is normally open, and the system operates as two parallel radial systems entirely independent of each other beyond the power supply point.

In another arrangement, the two substations are separated in order to be closer to their load areas. The proximity reduces the secondary feeder cable needed, but savings may be partially offset by the need for extra tie cable. Nev-

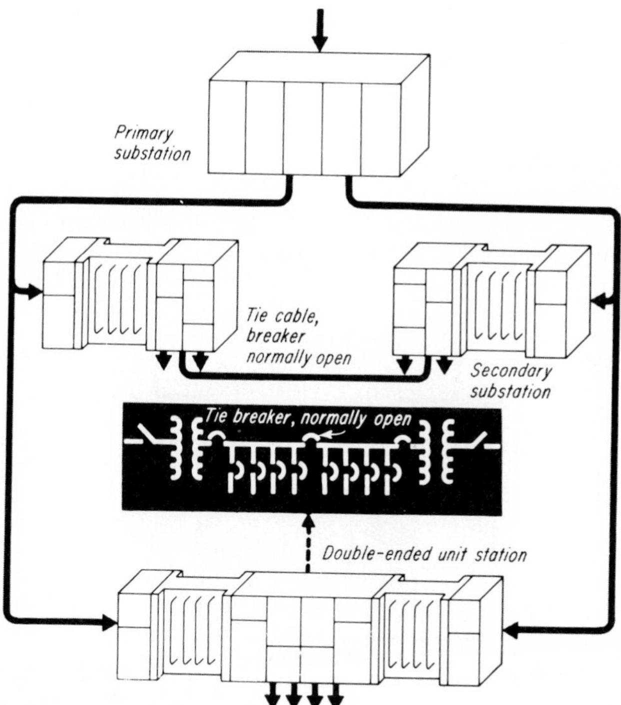

FIG. 2.6 Secondary selective system.

ertheless, many engineers prefer this arrangement for a new plant: As the load grows, another transformer can be added to each substation, allowing 100 percent growth in any load area.

Network Systems

Network systems are designed to give uninterrupted secondary power, even if a primary distribution feeder or secondary transformer fails. All secondary substations have their low-voltage sides tied together. In Fig. 2.7, the medium-voltage feeds are set up like those in straight radial schemes. As load is added anywhere on the secondary circuit, it is shared by the system collectively. Load is equalized on all transformers feeding the network, and any transformer or primary feeder can be taken out of service for maintenance without interrupting service to any load. The network can be adopted for the whole plant, or it may be a spot network.

Although networks are more costly than other systems, the extra cost is justified if maximum continuity is essential. The primary feeders may have a radial arrangement, but a primary selective feed gives added reliability and is the form most frequently used in industrial plants. Besides continuity, a network usually leads to improved voltage conditions, with less effect felt from current surges such as those generated in starting large motors.

When any network arrangement is considered, the available short-circuit currents on the secondary buses should be looked into. Since all transformers in secondary unit substations are paralleled, additional available short-circuit current may call for circuit breakers having greater interrupting capacity than those required for one of the simpler systems.

SECONDARY SUBSTATIONS

Location of secondary substations must be carefully thought out, since they are key equipment in the load center scheme (Figs. 2.8 and 2.9). Three building levels

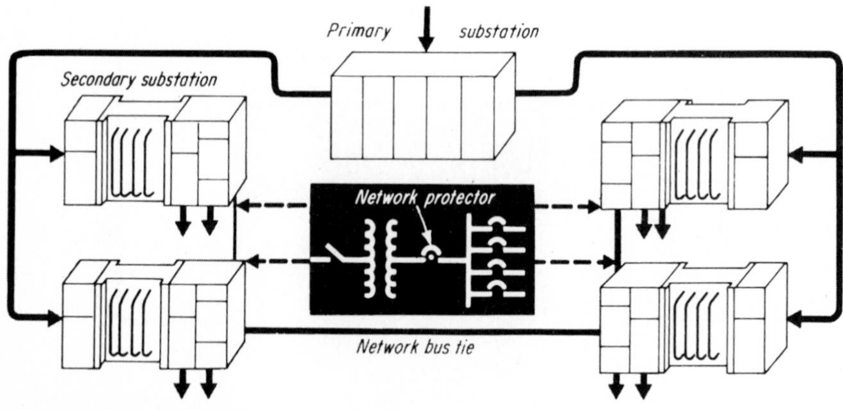

FIG. 2.7 Network system.

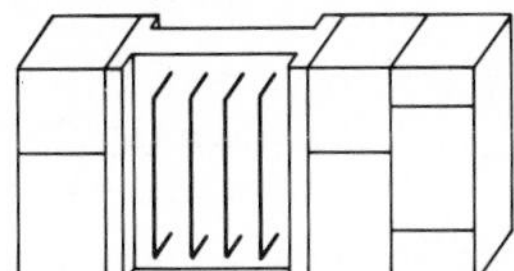

FIG. 2.8 Single-ended substations have primary section, transformer, and secondary switchgear, in that order. They are always used with radial and primary selective systems.

are considered: floor, balcony, and roof. If plant layout indicates low-cost areas indoors at load centers, these are desirable locations. Few consider placing the substations immediately outdoors. Balcony (or mezzanine) areas free floor space, but call for steel structures with expanded metal flooring placed high enough to keep the substations out of normal traffic. Substations are often suspended in roof trusses, generally slightly above the lowest structural member.

When primary distribution is run overhead and busway is installed throughout the plant, a roof location is often popular. Another advantage to roof location is avoiding high ambient temperatures or dirt and fumes in production areas.

Economical size is another consideration when the secondary substation is planned. As a rule, the smaller the substation, the higher the cost per kilovoltampere. As the number of substations increases in a given plant area, the length of primary feeder cable needed to supply them increases. At the same time, however, the length of secondary feeder cable required decreases. The basic cost per kilovoltampere of the substation and the cost of both primary and secondary cable must all be considered in determining the most economical substation rating. Other factors influencing substation rating are the following.

Substation Rating Factors

Higher primary voltage may require substations with larger kilovoltampere ratings, so that a greater kilovoltampere value per primary feeder can be handled without unduly complicating the substation overcurrent protection. In general, higher voltages justify substations with higher kilovoltampere ratings, especially in systems with primary voltages of 15 kV and above.

Large spot loads can sometimes justify larger unit substations. For example, a single furnace, large oven, or welder may justify substations on the order of 2000

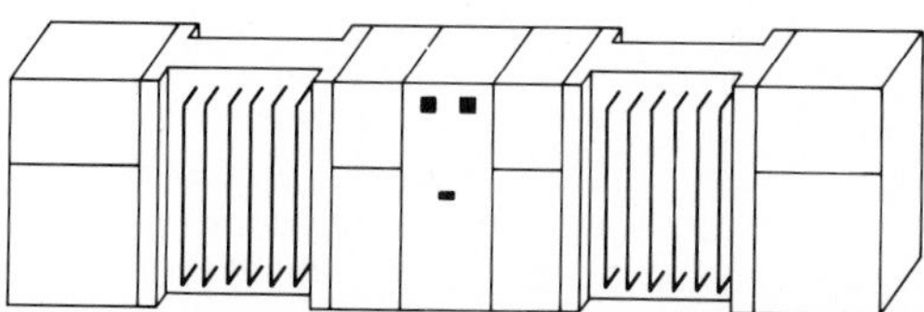

FIG. 2.9 Double-ended substations are often used with secondary selective and network systems. Two transformers feed two sections of low-voltage switchgear, with a normally open breaker between.

to 2500 kVA at a 480-V secondary, since no secondary distribution offsets the economies of size. For general factory areas, two 1000-kVA unit substations will nearly always provide a lower overall system cost than one 2000-kVA substation.

Space available for unit substations sometimes dictates larger substation kilovoltampere ratings. Sufficient space may be available at one location for the larger unit, while space at two locations may not be available for the smaller units. Conversely, other areas may allow space for the two smaller units but not for the larger one.

Air Interrupter Switches

Switching the primary input of the transformer makes wide use of air interrupter switches. The simplest is the three-pole, two-position (open-closed) switch with all three poles operated simultaneously by a handle in front of the switch compartment and providing a visible air break. With two separate incoming lines, the three-position (line 1/open/line 2) air interrupter selector switch provides an added factor of reliability by allowing the operator to switch from one feeder to the other, in case of line failure, or to the open position for maintenance. The selector switch is a dead-break device that is mechanically interlocked so that it cannot be operated unless the interrupter switch is open.

Liquid-filled interrupters are available for either oil- or askarel-filled transformers. The switch is a three-pole device, either two- or three-position—or, as an alternative, open/closed/cable-grounded. Oil cutouts may be paired with liquid-filled or dry-type transformers. The three-pole, two-position cutouts are gang-operated through a handle.

An air-filled terminal chamber may be furnished where a substation will be protected by an individual metal-clad circuit breaker. If plans call for a second substation to be added to the primary feeder at a later date, thought should be given to equipping the first substation with an interrupter switch at the time of installation.

Load-interrupter switchgear is available from several manufacturers, at both primary and secondary voltages. Switches are manually operated and should be of a quick-make, quick-break design, so that it is impossible to tease the switch into any intermediate position. The typical primary switch has arc chambers and spring-loaded auxiliary switch blades to ensure fast load current interruption. Mechanical interlocks prevent opening the door when the switch is closed or closing the switch when the door is open. Fault protection is provided by current-limiting fuses.

Transformers

The type of transformer is influenced by location and environment; liquid-filled transformers are installed where they can effectively be located outdoors, dry (both sealed and vented) units are used indoors. Insulation of transformers is frequently class F. This provides for an average temperature rise not to exceed 80°C and is based on a 40°C ambient temperature with 100 percent of the rated nameplate load connected. Class H insulation is available in dry transformers, permitting smaller units for a given rating. These transformers are designed for maximum temperature rises of 115 and 150°C.

Transformers are designed for various nominal voltage classes. The primary voltage applied to a 15-kV transformer must not exceed 15 kV; it is normally used with a system voltage of 13.8 kV. Each nominal operating voltage class has a corresponding withstand voltage, or *basic impulse level* (BIL). For example, a 15-kV-class transformer may have a BIL of 50 kV; a 25-kV unit may have 95 kV; a 35-kV unit, 125 kV.

The use of fans to increase transformer capacity is common practice, although disapproved of by some engineers. A typical new dry-transformer design is provided with forced-air cooling, which increases the rating to 133 percent of the self-cooled continuous rating without exceeding a 150°C rise and a 180°C hot-spot rise. The transformer is equipped with a winding temperature relay, which indicates winding hot-spot temperature. Alarm contacts are furnished to signal overload conditions. A separate set of contacts turns on the fans when the self-cooled rating is exceeded.

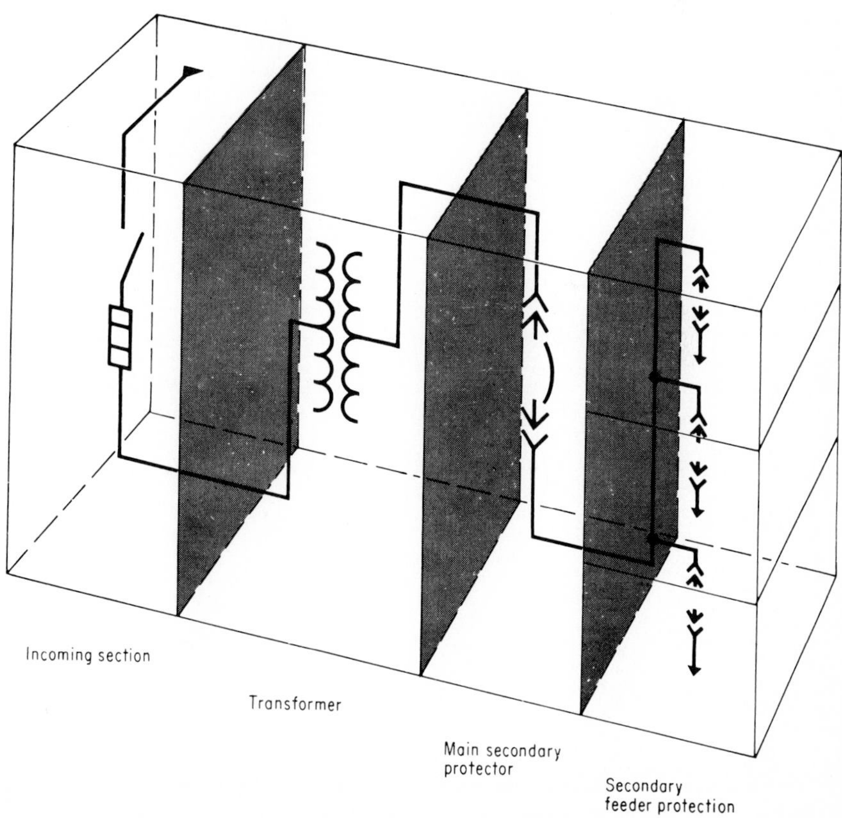

FIG. 2.10 Incoming section between primary feeder and transformer may have circuit breaker, load-interrupter switch, fused cutout, selecter switch, or a solid connection.

Substation Compartmentation

The low-voltage section of the substation consists of several compartments (Fig. 2.10). Ordinarily, there is one for the main circuit breaker, and as many others as are needed for feeder circuit breakers, motor control, instrumentation, etc. Compartmentation is a major factor in low-voltage-switchgear design, and it has made a significant contribution to system reliability.

The major elements in the low-voltage section are isolated—incoming line, main bus, circuit breakers, instruments, cable connections—by metal barriers. Operating mechanisms are in the forward portion of the enclosures, with the bus arrangements in the rear. Typical circuit breakers (and some fuse units) are draw-out types; the circuit breaker connects to the bus through pressure contacts and is automatically disconnected when it is withdrawn.

ACKNOWLEDGMENT

The authors would like to thank *Power* magazine for the use of some of the material that appears in this chapter.

CHAPTER 5.3
CABLE AND BUS DUCT

Ralph H. Lee
Consultant
Wilmington, DE

INTRODUCTION

Cable and bus duct are the key conductor systems that carry electric energy from its source(s) to the components in which this form of energy is used. Since electricity is also instrumental in providing many of the other services supplied by powerplants, such as steam and compressed air, loss of electric energy could bring about a complete shutdown of the plant and the systems it supplies. Thus the reliability of electric systems, including conductors, is paramount. Conductor reliability is achieved by a combination of these factors:

- Quality of components

- Suitability for exposure in service
- Skill in installation
- Proper loading (limited overloading)
- Adequate overload protection
- Protection from physical damage
- Diversity of supply (routing)
- Maintenance, including testing

Cable, the more widely used of the conductor types, is discussed first.

CABLE

General

Cable, basically, is one or more strands of metal, encased in insulating material. The term *cable* is also applied to multiple individual insulated conductors assembled together. The insulating material will be, for voltages higher than about 2 kV, completely encased in a grounded electrostatic shield, to keep the voltage stresses uniform all around the insulation and to provide safety, should anyone touch the insulation. For protection of the relatively fragile shield, and to provide enclosure of the conductors of multiple-conductor cable, an outer jacket is a general component.

Cable Construction

In North America, most cables of no. 8 *American wire gauge* (AWG) and larger are stranded; smaller sizes are solid unless flexibility is of great importance. In stranded cable, individual strands are not larger than 0.128-in (3.25-mm) diameter, for flexibility and ease of installation in conduit. With this, the number of strands is one of 1, 7, 19, 37, or 61, up through 1000-MCM (1000 circular mils, or 507 mm^2) size.

In the rest of the world, where conduit is not used as predominantly, the conductors may be solid, or single-strand, up through quite large sizes. Cables are quite generally self-supporting between cable straps at suitable spacings.

Conductor Materials. The conductor materials are copper or aluminum, the former being predominant for sizes below about no. 2 AWG, with aluminum becoming widely used for feeders, etc., of larger size. Copper cable is almost universally annealed, or dead soft, for easy flexibility. Aluminum, however, has given problems in the *electrically conductive* (EC) grade, owing to cold flow away from terminal clamping fittings.

Consequently, the EC aluminum used for wiring has been generally the ¾ hard-drawn temper, which is springy and hard to bend. It also tends to break where bent when it is placed under a screw terminal. A newer alloy type, containing about 0.5 percent iron, has displaced the EC type for circuit sizes (no. 8 AWG and smaller). This alloy type is also available in larger sizes, if requested, although EC will be supplied unless alloy is specified.

Copper Advantages. Copper has greater conductivity than aluminum, size for size, but copper's 3.3 times greater specific gravity, and about double the cost per pound (kilogram) compared to aluminum, makes the latter economically attractive. On a cost-per-ampere basis, copper wire is about 4 times as costly as aluminum.

Aluminum has received much criticism because it has been widely used with connectors suitable for copper only. In the past 10 years, however, connecting means suitable for aluminum, or copper interchangeably, have been developed and are performing well. More details are included in Connection Methods, later in the chapter.

Insulation

600 V and Under (Low Voltage). The principal insulation for cables in this voltage class is *polyvinyl chloride* (PVC), having good electrical and physical properties. *National Electrical Code* (NEC) classifications include types T, TW, and THW, which have basically different temperature ratings. PVC is receiving criticism because in fires the insulation gives off HCl in gaseous form, which is purported to have an anesthetic effect on people.

Another widely used type has a thin PVC coating, covered with an even thinner nylon coating. Types TFN, THWN, and THHN are in this class. The coating material, being hard and slippery, reduces friction in pulling cable into conduits, and the smaller diameter makes greater cable fills in conduit possible.

Cross-linked polyethylene (CLP or XLP) also receives much use in this class. The type designation is XHHW. It is slightly larger than THWN and THHN. Coatings of this type have to be "painted" on; coding can be done by printing letters or stripes on the insulation exterior.

Ethylene-propylene rubber (EPR) insulation is used on much of the low-voltage cable intended for powerplants, particularly nuclear generating stations. In nuclear plants, unusually large quantities of cables are used in cable trays, and the resistance of this type to burning is advantageous. Fires of even moderately flammable cables in large quantities in trays have been responsible for a number of long-time shutdowns of nuclear stations. IEEE has a number of standards covering cables for nuclear powerplants.

All types of low-voltage cables can be assembled in multiconductor cables. A number are tray cable, metal-clad cable, armored cable, and power-limited tray cable.

Over 600 V. Cables in the 600-V-plus class are almost universally solid-dielectric-insulated, that is, insulated with an extruded plastic or rubberlike solid insulation. The laminated-type cables (paper-oil-lead and varnished-cambric types), long nonpareil, have largely dropped out of use except among utilities, which can keep an adequate corps of cable splicers for the special techniques required for this material.

Solid dielectric insulation has one inherent weakness: There must be no air (voids, etc.) in the dielectric path between the high-voltage conductor and the grounded shield. To accomplish this, two semiconducting layers are applied, one over the conductor and the other over the outside of the insulation. The inner layer, called the *conductor shield* or *screen* in modern practice, is a thin extrusion [about 0.015 in (0.381 mm)] over the "peaks" of the conductor strands and filling the "valleys" between the strand peaks.

Conductor Shield. The conductor shield provides a smooth cylindrical surface on which the insulation is extruded with little or no air inclusion. Without the conductor shield, the air voids in the valleys between strands would break down from dielectric stress and arc internally, with the arc burning into the insulation, resulting in insulation failure, sometimes called *corona cutting*. Older cable types utilized semiconducting tapes, helically or longitudinally applied over the conductor, under the insulation. This leads to inferior performance caused by folding or "skips" in the tape and strands of semicon fabric tape extending into the insulation.

On the outside of the insulation, the semiconducting layer is extruded on, for best performance, or taped helically, which gives adequate performance. The tape will always have a spiral triangular cross-sectional void along the overlap between adjacent turns. For lower-voltage ranges, this is tolerable, but on voltage surges and for the higher voltages, it can be responsible for corona cutting and failure. The metal portion of the shield, either spiral metal tape or spaced spiral wires, is applied over the outer semiconducting layer. The semiconducting layer is the actual electrical shield, and the metal tape or wires are simply the electric current drain path and distribution conductor since the longitudinal current capability of the semiconducting layer is very limited. Where the metal path becomes broken or corroded apart, the current through the semiconducting layer quickly overheats and destroys adjacent insulation, causing cable failure.

Major Insulation Types. The modern medium-voltage insulated cables utilize, almost universally, one of two major insulation types: EPR and CLP. Another type is silicone rubber. EPR, a rubberlike material, is a cross-linked combination of ethylene and propylene with a number of additives. It looks like, and performs very much like, the rubber-type insulations in use for many decades. Its performance has been excellent, especially in freedom from "treeing," a weakness of CLP cables. (*Treeing* is the slow progressing of microscopic channeling through insulation, usually from some void or point on or within the insulation.)

XLP or CLP, a milky-appearing hard material, is polyethylene cross-linked by using dicumyl peroxide and other additives. The cross linking overcomes the melting temperature [221°F (105°C)] of pure polyethylene, although above this temperature it is softer than EPR. Its cost is slightly lower than that of EPR, making it more widely used for distribution lines but not favored where the highest reliability is desired.

Silicone rubber is a rather soft, rubbery material occasionally used where abnormally high or low ambient temperatures are encountered. Its electrical performance is excellent, but its low physical strength makes protection against any physical stresses obligatory. It is frequently used near boilers, steam and hot-water lines, stacks, etc.

Operating Temperature Ratings. Both EPR and XLP have continuous operating temperature ratings of 194°F (90°C); silicone rubber, 251°F (125°C). The temperatures permissible during the very short times involved with system faults are 482°F (250°C) for all types. The *Insulated Cable Engineers Association* (ICEA) has Standard P-32-382, which interrelates available fault current, cable operating temperature, cable size, conductor type, and fault-device clearing time. Where a medium-voltage cable is used to supply a relatively small load, such as a 125-hp (93-kW) motor, the cable size may have to be larger than that required for motor overload current, just to keep the cable insulation from being destroyed in the event of a fault at the load.

All these cable types have an emergency overload temperature rating. For

EPR and XLP, it is 266°F (130°C); for silicone rubber, 302°F (150°C). In an emergency, the cables may be operated at these temperatures for 36 h, five separate times in any year, but an average of only once each year. Under this loading, there will be no detectable shortening of life of the cables. For an EPR or XLP cable, operating at 194°F (90°C) in a 104°F (40°C) ambient temperature, the loading to reach 266°F (130°C) is 126 percent of rated ampacity. A more extensive coverage of this and short-time ampacities is found in IEEE Standard 242 (Buff Book), Chapter 8, Protection and Coordination of Industrial and Commercial Power Systems.

Much has been made of high operating temperatures available in modern cables. With the higher operating temperature of the conductors, the conductor resistance is increased, so I^2R losses of the cable are increased. With the increasing cost of energy, the loading of cables requires scrutiny to determine the optimum economic choice.

Effects of Contaminants. All these insulations exhibit excellent resistance to moisture, ozone, and corona. XLP is not affected by petroleum products. Water can, however, affect other parts of cables, such as shielding, semiconducting tapes, and even the conductors themselves. As mentioned above, the presence of water has been traced to the slow progress of treeing in this cable type. This form of insulation degradation was discovered only after translucent insulation of polyethylene and XLP types became available, making this phenomenon visible.

Progressive treeing in XLP insulation normally takes years, usually 5 to 15 yr, to completely penetrate the insulation, thus causing failure. Tests have shown that this does not occur in EPR insulation, or that if it does, it progresses so slowly as not to be a factor in normal cable life.

Jackets and Sheaths

The insulation or shield of medium-voltage cable generally requires the protection of some outer covering. Such a covering is called a *jacket* or *sheath,* the latter normally being applied to the lead covering of paper-oil-insulated cable. The sheath type can develop internal pressures of up to about 50 psi (345 kPa) under periods of heavy loading. Other than lead, in order of popularity, the jacket materials are

- Polyvinyl chloride
- Polychloroprene (Neoprene)
- Chlorosulfonated polyethylene (Hypalon)

All these serve well as mechanical and waterproofing coverings. Neoprene has one possible problem: Contact with the creosoted portion of a creosote-treated pole is likely to cause softening of the jacket.

At splices and terminations, an effective overlap or seal to prevent water entrance is necessary. These are shown in the figures to be found later in this chapter.

Wiring Methods

Cable cannot, except for emergencies, be simply laid on the ground or floor, or even suspended from building steel or ceilings. The latter approach is permitted for construction and reconstruction periods, but none are permitted or desirable for permanent installations. There are three reasons for this:

- Cable, while excellently insulated, needs physical protection from damage.
- Cable is inherently flexible and so needs support at relatively close intervals.
- Should persons damage cable insulation, they could receive an electric shock.

So cable is normally enclosed in what is called a *wiring system*—really an enclosure and support system. This includes conduit, cable tray, and underground or aboveground runs.

Conduit. Conduit is the best known and probably the oldest wiring system. It consists of standard size pipe, with fittings and pull boxes suited for the pulling in of cables. The conduit is strong enough to require support at only relatively wide spacings and to suitably resist mechanical impingement from personnel and light equipment-handling machines. It is, however, rather a costly method and not adaptable to modification or addition.

Surprisingly, too, conduit is rather poor from a reliability standpoint, since low-voltage cable insulation can be damaged by the pulling-in process and by being scraped across sharp corners of conduit fittings and pull-box edges. In case of fire, the conduit protects the cable for a short time, but ultimately becomes an oven and permits melting of the cable insulation, preventing the pulling out of the wires. So, after a moderate-duration fire, the conduit must be sawed apart and replaced.

Steel and aluminum are the metals from which conduit is made. Steel is stronger, but causes more voltage drop in the contained conductors, due to greater reactance. Aluminum is susceptible to galvanic corrosion where moisture and other metals or earth can contact it. Lighter walls are available, too—a half-thickness type called *intermediate metal conduit* (IMC) and an even thinner-walled *electrical metallic tubing* (EMT). IMC is suitable for powerplant use, but not EMT because of its fragility and minimal grounding characteristics.

Cable Tray. A relative newcomer except in utility generating stations, cable tray has come into its own in industrial and even large commercial use. It simply consists of a surface with side rails, angle and cross fittings, and, if desired, cover plates. It is adaptible to changes or addition, since new cables can be laid in the trays. Trays are simply supported at intervals of 10 to 20 ft (3.05 to 6.1 m), depending on side-rail strength and physical loading. Cables are laid or pulled into the tray, which can be filled up to a depth of approximately 3 in (76 mm) *regardless of the height of the side rails*. That depth limitation is necessary because the cable "bed" can dissipate its self-generated heat only through the top and bottom—the sides are effective only for about 3 in (76 mm) into the sides. The ampacity of cables in the tray is the same as for cables in conduit in the 1984 NEC, but will probably be derated to 90 percent of the in-conduit ampacity of the Neher-McGrath ampacities.

Cable tray is rather like an air duct, in that it cannot be bent as readily as conduit. And it is too large to connect directly into a motor terminal box. Its great use is as a throughway or main wiring path out through a load area, with conduits projecting from its sides to the individual loads. There are standard fittings (T & B Co.) to attach conduits to the side rails of trays; these also supply a bonding-earthing function. Tray cable (NEC 340) can be run from the tray into and down a conduit for support to a motor conduit box, pushbutton, instrument, etc. Trays, of course, can be fitted to power centers, or vertical wireways can interconnect between tray and center. Also trays can be run vertically as well as horizontally; cables need to be tied into the former for security.

With approved splices and side rails, cable trays are suitable for "equipment grounding conductors" or the "green wire" in the electrical grounding system. As such, they serve exceedingly well, actually better than a single grounding conductor carried in the tray. The latter tends to throw cables out of the tray by magnetic repulsion, in the event of a heavy ground fault. The fault return current, carried by the tray side rails, tends to balance between the two sides and neutralizes the magnetic thrust.

Voltage classes should not be mixed in a single-cable tray, but separators can be used to segregate instrument wiring from power wiring. Additional separation may be required to eliminate magnetic induction from power circuits into instrument cables.

Underground and Aboveground Runs. Cables between buildings can be run in underground ducts or even directly buried. In the latter case, the cable runs need to be marked to minimize the likelihood of dig-ins. Also, cable which is directly buried must never be operated at over 149°F (65°C), even though it is a higher-temperature rated cable. Higher surface temperature than 140°F (60°C) [9°F (5°C) drop in insulation] will cause the natural moisture in the ground to migrate away from the cable, reducing the thermal conductivity of the soil. This, in turn, causes the cable temperature to rise. The process results in burnup of the cable insulation and is called a *thermal runaway*.

Aerial cable, supported on poles, is not recommended in the factory-assembled form, because of catastrophic difficulties in installation. If, however, it is field-assembled by being raised as three single conductors into a field-spinning unit, then its reliability is excellent with little difficulty in installation.

Ampacity

Ampacities of AWG and MCM wire sizes, of both copper and aluminum for 600 V and lower, and 2000 up to 35,000 V, are most readily found in NEC, the 1984 edition. Similar data are included in IEEE Standards S 135-1 and S 135-2. Note that the 600-V and lower ampacities in the 1984 NEC are based on the "Sam Rosch" ratings and are to be replaced by the "Neher-McGrath" ratings. All the IEEE ampacities are based on the Neher-McGrath ratings, the more exact method for determining ampacity.

When the United States converts to the metric system, all wire sizes will be based on square-millimeter sizes. Tables are presently prepared for this conversion, but no ampacity tables are yet on hand. When metrication occurs, the corresponding ampacities will be published promptly. For comparison, 1 mm^2 = 0.507 MCM, or 1000 MCM = 507 mm^2. Most of the rest of the world has been using the metric system for many years, so ampacity values are readily available in the standards of those countries.

Voltage Drop

Cable conductors have inherent resistance; additionally, they have inductance. Their inductance is relatively low, especially in comparison to the widely spaced open wiring used in overhead pole lines. Probably the most convenient source for resistance and reactance of cables in conduit is Tables 8 and 9 of the 1984 NEC. Table 8, for resistance, is applicable to all voltage cables, on direct current. Table 9, for reactance (and ac resistance), is directly applicable to 600-V cables only. For higher-voltage cables (because of the greater insulation thickness, causing greater spacing between the conductors), the inductance will be increased by the factors of Table 3.1.

The net impedance of a cable run is the square root of the sum of the squares of resistance and reactance. Tabular values are in ohms per 1000 ft from line to neutral, so the voltage drop (amperes times impedance) will be per wire to neutral and will need to be multiplied by 1.73, to obtain line-to-line voltage drop.

Rule-of-Thumb Method. A quicker rule-of-thumb method for voltage-drop estimation is that there will be a 1-V drop per 35 ft (10.7 m) of each conductor if the current loading is that permitted by the 1984 NEC. Where Neher-McGrath ampacities are used, as for *medium-voltage* (MV) cable, the voltage drop in each conductor will be increased to 1 V per 30 ft (9.1 m) of length. Another rule is that cable will incur a 4 percent voltage drop at cable ampacity rating in a length in feet equal to the circuit voltage in volts. This will convert to a 4 percent voltage drop at cable ampacity rating in a length in meters equal to 0.328 times the circuit voltage.

Equipment normally is designed to operate at 96 percent of source (transformer) voltage. It will operate satisfactorily at as low as 90 percent of rated voltage and 86 percent of source transformer voltage, but performance drops and heating increases with reduced voltage. Overvoltage causes a decrease in the power factor and an increase in heating.

Splices

Where cable runs are longer than the shipping lengths, or equipment is relocated such that the cable supplying it must be lengthened, splices are required. Until recently the field-made splices were not reliable for several reasons:

- Incompatible material was employed in splicing, such as Duxeal over the connector to smooth its surface (the oil from Duxeal is incompatible with rubber insulating tape).
- The connector and adjacent short, bare conductor lengths were not covered

TABLE 3.1 Inductance Multipliers*

	Cable voltage rating, kV			
Cable type	5	15	25	35
3-1/c cables†	1.1	1.2	1.3	1.4
1-3/c cable‡	1.0	1.2	1.3	1.4
1-3/c cable§	1.0	1.3	1.4	1.5

*From National Electric Code, Table 9.
†Applies also to steel-armored cable.
‡Magnetic conduit; applies also to steel-armored cable.
§Nonmagnetic conduit; applies also to aluminum-armored cable and cable in trays.

with semiconducting material. This left possible air voids in contact with the rubber insulating tape, allowing corona cutting of the insulation.

- Splicers experienced with paper-oil-lead cable would apply tape without tension and then try to "snug it up" at the end of each turn. Rubber and plastic cable tape must be applied with uniform tension all around the splice, or else there will be thick spots and thin spots. Also, tape applied without enough tension will not bond turn to turn, leaving internal voids.
- Splicers did not use a heated roller after each layer of tape, incurring voids in the tape buildup.
- Splicers did not taper the ends of the factory insulation sharply enough, so the tape buildup would not bond to it, allowing longitudinal failure.
- Occasionally, a mechanical connector was used, with the splice normally lasting from 1 min to 1 day. The outside of the connector needs to be as smooth a cylinder as possible, or made so by wrapping with semiconducting tape. (Both are necessary.)
- The ground lead was not suitably sealed, allowing water to enter into the shield structure. Flat grounding lead, embedded in the HV tape and with rubber cement added sparingly, makes a better seal than braided or round conductor.
- Bonding jumpers were not used, so the shield braid or knit tape would be required to carry any fault current, usually failing through overheat.
- Semiconducting tape was not used over either the connector and adjacent cable or the insulating tape, or both, incurring corona cutting.

In the past 10 yr, factory-prefabricated parts for field assembly of splices have been available. If they are properly assembled, these are successful. However, great care must be taken to be sure the longitudinal matching of splice part and cable factory shield is correct. Without this, open surface of factory insulation can occur, with rapid corona cutting and failure. Factory-prefabricated splices eliminate the need for precise taping skill. Still, great care is necessary in preparation of the cable for the splice.

As in terminations, factory-prefabricated splices must be ordered, with cable insulation and jacket dimensions given quite accurately, and require time for delivery. If this time is not available, tape splices can be made by using the design shown in Fig. 3.1. Splices made using this design have been found to operate without problems for the life of the cable.

Terminations

Every cable has two ends or, where branches are involved, three or more. For low voltage, 600 V and below, simple removal of the insulation (and application of a terminal if it is aluminum) will suffice.

Higher Voltages. For higher voltages, it is necessary to

- Leave a length of factory insulation between the conductor end and shield end
- Provide stress relief at the end of the shield
- Provide necessary protection for exposed factory insulation

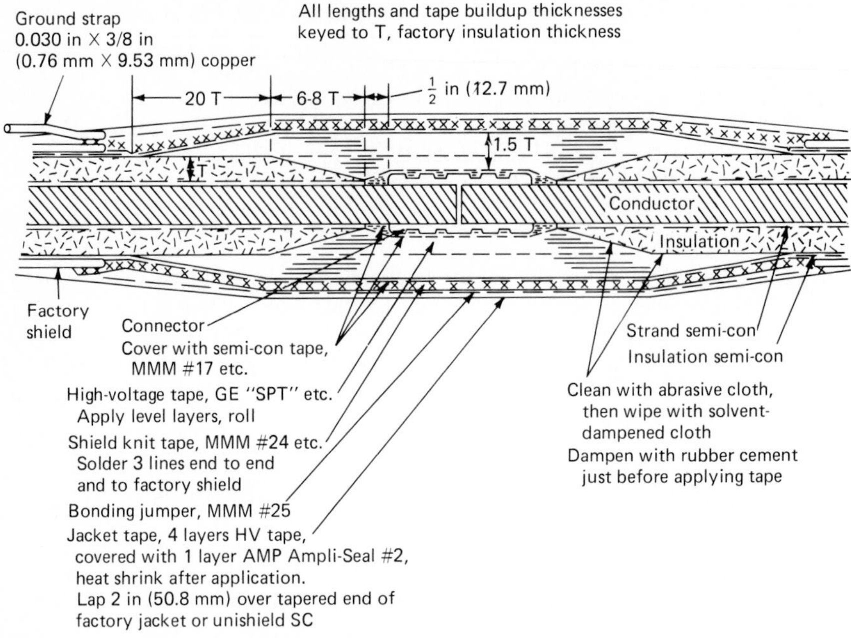

FIG. 3.1 Straight splice, solid dielectric cable.

- Provide a seal between terminal and factory insulation to keep precipitation out of the conductor
- Bring out the grounding connection from the end of the shield

High-voltage cable terminations were, at one time, a major failure item. They were either the "pothead" type or a hand-taped "stress-cone" termination. Both require a high degree of skill and utmost care in application and completion. Recently a number of factory-manufactured termination types have become available. These reduce the degree of skill required in application, but care must still be taken to avoid damaging the cable in preparation for applying the termination, as well as the application itself.

Three Classes. Replacing the obsolete term *pothead,* there are now three classes of termination:

- *Class 1:* Seals the factory cable from pressure, moisture, and contamination and provides stress control. (This class is the functional equivalent of the old "pothead.") It is suitable for all locations.
- *Class 2:* Seals against moisture and contamination, but not against pressure; provides stress control. These are slip-on types with overlap over factory jacket. They are suitable for moist, contaminated locations, but not for paper-oil-lead cable.
- *Class 3:* Provides stress control only (old stress cone) and is suitable only for clean, dry locations.

Stress control is a means of minimizing the stress concentration at the end of the factory shield. Conventionally, this has been done by simply increasing the insulation thickness progressively by wrapping it with tape, as shown in Fig. 3.2. For voltages up through 35 kV, this has been a linear taper, not exceeding a slope of 1:16 (for higher voltages, the slope has been logarithmic—a smaller angle next to the shield end, more at the apex of the slope). This is known as a *stress cone.* Recently, a material called *voltage control tape* has become available, which is simply wrapped spirally along 1 in (2.54 mm) of the factory shield and then along a length of the factory insulation beyond the factory shield end. The length to be covered is about ½ in (1.27 mm) per kilovolt to ground of the cable. This performs very well, needing only a weather covering for protection.

Figures 3.3 and 3.4 illustrate class 2 and class 3 terminations. These perform well in proper application environments. However, just as with any other insulators in unusual contamination conditions, it is prudent to use a higher-voltage rating of the class 2 terminator where these conditions prevail.

Applications. For suitable application, all these factory-prefabricated units call for precise diameters of factory insulation and jacket during ordering, with the customary delays between ordering and delivery. If an emergency arises and such time is not available, the fabrication of a stress cone termination as shown in Fig. 3.2 may be resorted to (even outdoors) with satisfactory performance and life. This can be either a class 3 for clean, dry areas or class 2 for weather or contamination exposure, when the factory insulation end is sealed and covered with silicone tape. This construction has proved very satisfactory in service, even in severe contamination.

In any of these terminations, there must be no bends in the stress cone area. Moderate bends are permissible between the stress cone and terminal in flexible

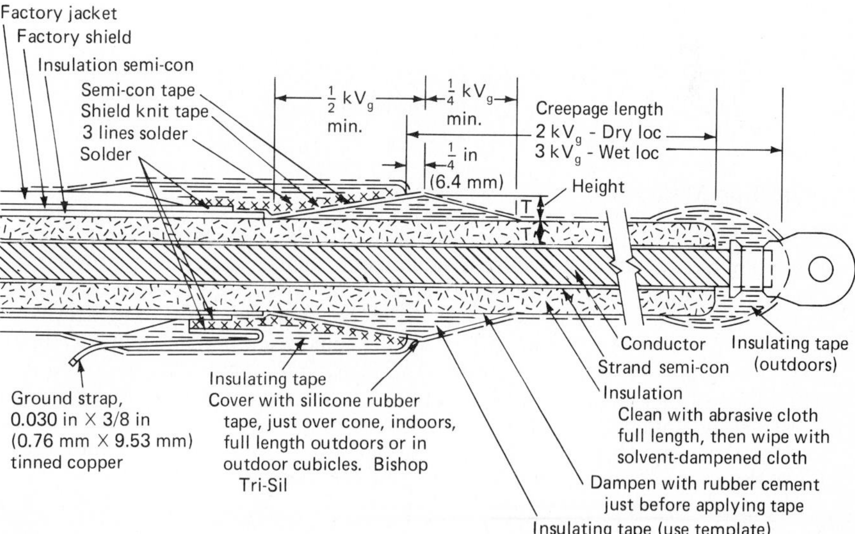

FIG. 3.2 Taped termination, solid dielectric cable. Note: ¼ kV_g, ½ kV_g, 2 kV_g, and 3 kV_g = ¼ in (6.4 mm), ½ in (12.7 mm), 2 in (50.8 mm), and 3 in (76.2 mm) per kilovolt to cable ground, respectively.

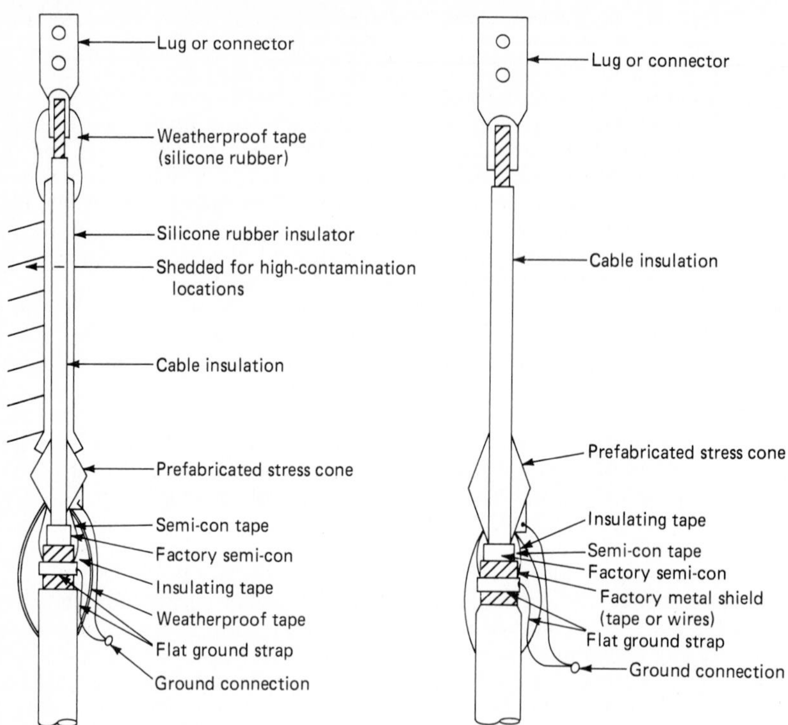

FIG. 3.3 Class 2 terminator.

FIG. 3.4 Class 3 terminator.

material and below the stress cone in undisturbed cable, but care must be taken not to introduce any bending that might affect the stress cone zone.

Connection Methods

Cable itself is an excellent material. The major problems are encountered largely at the point of connection of cables to equipment, such as switches, circuit breakers, etc. The conductors of cables are made up of "rods" of the metal, each rod carrying its share of the cable current. For connections, there is generally a clamping of part of or all the surface of the conductor bundle, and the inner strands must transfer their current into and through the outer strands to the terminal inner surface. With copper, this poses little problem; but with aluminum, the insulating aluminum oxide surface layer on each strand militates against this interstrand action. This, actually, is the principal criticism against aluminum cable. With proper attention, however, the problem is readily overcome.

Compression Connectors. There are two general types of connectors, compression and mechanical. In the former, the conductor end is placed within a connecting tube or compatible design, and the whole is compressed until some deformation of all the conductor strands occurs. For aluminum the tube is normally filled with the "joint grease" before the conductor is inserted; the excess squeezes out as the joint is compressed. Grease fills all the space in the connection not filled by metal, keeping air (oxygen) away from the deformed fresh surfaces and preventing formation of aluminum oxide.

Mechanical Connectors. The mechanical type of connector generally involves a setscrew, or stirrup device, which simply squeezes the conductor in one dimension. Thus a top setscrew deforms the top 25 percent of the cable surface, with little deformation on the bottom. Thus, with aluminum particularly, the only deformed surface is that in contact with the setscrew, and almost all the current comes out via the screw. Grease is desirable in this type as well as in the compression type.

In comparison, the voltage drop in a compression connection is about one-quarter that of a similarly rated mechanical type. So the surface area of the latter has to be larger to dissipate the additional heat generated. Also, the failure of properly sized and applied compression connectors is almost unheard of, whereas the failure of mechanical connectors is considerable. One mechanical type has a historic failure rate of about 1 percent per year. Yet, the necessity of needing a compression tool to apply the compression terminals when only a wrench is required for mechanical connections is the principal reason for the widespread use of the latter.

Solid-Aluminum Connectors. Connectors for medium- and large-size cables are predominantly solid aluminum, cast or extruded, and plated with tin. This combination is suitable for both copper and aluminum. One may not, however, drill out a (cable) hole to use such a connector for a larger size; this removes the tin and will produce galvanic failure of the connector.

Aluminum has another characteristic, which must be overcome to ensure good connections. The metal oxidizes within seconds of exposure to oxygen. Few people have ever seen pure aluminum; all have seen aluminum oxide. The aluminum oxide has a high electrical resistance and so is not desirable in an electric connection. A good connection breaks up or distorts the surface of the metals being connected. This should be done submerged within the body of a grease, which bars oxygen from the newly exposed surfaces. With no oxygen to turn the surfaces into resistive aluminum oxide, a good connection results. All aluminum connections, compression or mechanical, should have a complete internal coating of oxygen-barring grease.

Terminal Mounting. Adequacy of mounting a terminal on the bus is of prime importance. One terminal type, mounted with a small flathead screw collinear with the setscrew, but directly below the cable end, has a particularly high failure rate. This type is customarily supplied as the terminal of molded-case circuit breakers or switches in the 125- to 1600-A range. A basic problem here is that the terminal mounting screw tightness can never be checked. It is rare that the wire will ever be removed, as is required to check tightness; usually the only check is to try to wiggle the terminal by the cable or to inspect the cable insulation for burn-off or overheat.

Cable Damage

Although cable is flexible and is made to be pulled into place, it can be damaged by

- Bending shielded cables more sharply than to a radius of 12 times the individual cable's outside diameter, even momentarily, and for even an inch of its length. This can cause separation of helically wrapped shielding, with distortion on restraightening. Actually, the restraightening after overbending on installation is more damaging than the overbending itself. Unshielded cables can be bent to an inside radius of no less than those listed in Table 3.2.
- Pulling at a tension greater than 0.008 times the circular-mil (square-millimeter) area of copper conductors and 0.006 times the same area of aluminum conductors.
- Too sharp a bend for the pulling tension, causing side-wall compression damage to cable. This is the compression between conductor and the side of the insulation against the inside surface of the duct (or sheave). The amount of pulling tension in pounds (kilograms) divided by the duct bending radius in feet (meters) is the criterion called the *side-wall pressure*. Weak rubberlike (silicone rubber) insulation will be damaged by over 100 lb/ft (149 kg/m), while the harder rubbers or plastics (EPR and XLP) will withstand 300 lb/ft (447 kg/m). Paper-oil-lead cable will stand 1000 lb/ft (1490 kg/m).

Installation

Techniques in installing cable embrace gripping and pulling. When cable is installed underground, manholes are important.

Gripping Cables. It is safest, and obligatory with cable over 2-in (50-mm) *outer diameter* (OD), to pull on both the conductors themselves and the jacket. The pulling end of the conductors should be stripped of insulation, formed into loops and bent and twisted over itself, and finally taped together to prevent pull-out. The jacket can be secured by using a cage (or Kellems) grip, with the downstream end of the cage secured with strapping to prevent slippage. Installed pulling eyes, of course, are available when pulling lengths are specified in the purchase of cable.

Pulling Cables. Always pull so that the longest straight section is at the pulling end. This arrangement, of course, places most of the bends at the end where the wire is fed into the duct. Consequently, the cable can be trained into the duct end and can even be "pushed" into that end. Since the bends contribute much of the tension required for pulling, this "pushing in" for the first bend substantially reduces the tension required. In fact, it is good design to locate manholes or pull boxes adjacent to a bend for this reason.

TABLE 3.2 Minimum Bending Radius Times the Outside Diameter of Unshielded Cable

	Outside diameter, in		
Insulation thickness, in*	1.00−	1.001–2.00	2.001+
0–0.155	4	5	6
0.170–0.310	5	6	7†
0.325 or more	—	7†	8†

*In × 25.4 = mm.
†Armored.

The required tension for pulling may be estimated from the following formulas:

For straight lengths:

$$T = W \times 0.5$$

where T = tension required for straight length, lb (kg)
W = weight of cable, total, lb (kg)
0.5 = estimate of friction coefficient, with lubricant

For curves alone:

$$T_C = T_1 \times e^{0.5a}$$

where T_C = tension required to overcome friction within curve or bend, lb (kg)
T_1 = tension at entering end of bend or curve, lb (kg)
e = 2.72
a = angle of bend, rad (angle of bend in degrees divided by 57.3)
0.5 = estimate of coefficient of friction

The tension in the curve will be the sum of the entering tension and T_C, by the above. This is the value to be used in determining the side-wall pressure.

All the components of a run should be added to determine the total pulling tension for the run. Check side-wall pressure limitations for practicality of installation, and the maximum tension for the cables. If these values are not met, reduce the length between pull points, or increase the radii of the bends.

If cable must be pulled past a point where it is to be terminated (switchgear, etc.), care must be taken not to overbend the cable, even temporarily, (1) in feeding the end into the terminating enclosures and (2) in hairpinning the cable loop during the feeding process. This is especially commonplace where oversharp bending occurs. A thorough cable installation handbook is issued, at no charge, by Anaconda-Ericsson Co. and goes into far greater detail than is possible here.

Manholes. A most common fault of manholes is their too small size. There needs to be a central space not less than 5 ft (1.52 m) in diameter, with a 1-ft (0.305-m) space along all walls for cable splices and bending entry into belled ducts. Probably a minimum size and shape would be a 7-ft (2.13-m) interior-dimension octagonal shape; four opposite faces would be for duct entrances, the four between these would be for splices, with inserts to support cables along these walls.

All underground installations are subject to water entrance, and many manholes and ducts remain essentially full of water all the time. Properly made and handled cables and splices are not damaged by submerged operation. However, any cable sheath damaged from improper installation and any splice not suitably made are likely to admit water and incur failure. For this reason, many users connect manholes to drainage systems or install pumps with float switches. Under these circumstances, flaws in cable jackets or splices may be tolerated and may not cause failures.

Hazardous Locations

It is good practice to locate the absolute minimum of electrical equipment in hazardous areas, because of the possibility of its becoming an ignition source and because of the much higher cost of equipment approved for such locations. In many, if not most, instances, the electrical equipment can be located outside of hazardous areas.

Two Classes. A number of locations in powerplants involve materials which fall into what are called (electrically) *hazardous areas*. Two key classes of hazards in these locations are covered here:

Class 1: Vapor-Liquid.

- Natural gas and methane (methane can be given off by coal in closed containers, such as bunkers, tunnels, and silos)
- Hydrogen
- Gasoline
- Liquified petroleum gas and propane
- Other liquids and gases unique to the location

Class 2: Dusts.

- Coal and coke (may also involve methane)

Requirements in Class 1 Areas. Class 1 exposure involves flammable liquids and gases, such that any sparking or temperature above the ignition temperature of the hazard could generate ignition, in either normal or overload operation of the equipment. This class is divided into two divisions, 1 and 2.

In division 1, the hazard may exist at any time, since the flammable material is or may be open to the atmosphere. In division 2, the hazard is less likely to exist at any given time, since it is normally confined to a closed vessel or pipe. Spaces beyond the prescribed extent of a division 1 location are normally classified division 2. The extent of these zones is detailed in NEC, NFPA 70, ANSI C-1 and C-2, and the National Electrical Safety Code.

Electrical equipment for such areas is designed and tested to contain, without rupturing, an explosion of the hazardous material involved. Most powerhouse hazardous materials, except hydrogen, are protected against by class 1, groups C and D components. However, there is very little equipment made which is suitable for a hydrogen environment. Ample ventilation to prevent hydrogen leakage concentration from reaching an ignitable level is probably the most suitable means.

Any electrical equipment would be classified class 1, group B. One other material, acetylene, may be kept for gas welding; it has such a high explosion pressure that there is no electrical equipment available to withstand it. The safe practice here is to store the acetylene outdoors, not less than 20 ft (6.1 m) from any electrical equipment.

Wiring methods for class 1, division 1, involve rigid and intermediate threaded-metal conduit, type MI (*mineral-insulated*) cable, and boxes and fittings having at least five threads of engagement with the conduit. Division 2 can also be served with enclosed gasketed busways and wireways and cable of MC, MV, TC, SNM, and PLTC categories, with approved termination fittings. These cables may be installed in trays.

Seals are required in conduits and cables to prevent them from acting as channels between components in division 1 areas or between division 1 and division 2, and division 2 and nonclassified areas. Some cables (TC and PLTC particularly) are available with internal filling material between individual conductors which, in sufficient lengths, serves the function of a seal fitting, such that the latter can be omitted.

Requirements in Class 2 Areas. Class 2 exposure involves dusts which may be flammable. Those found in powerplants are largely those of coal or its derivatives. The carbon dusts are conductive, not as highly so as metals, but conductive enough that a buildup on an insulating surface between circuit parts of different voltages is likely to initiate an electric arc, which in turn will ignite any dust buildup in the vicinity. This causes a double explosion, a bang-bang, in which the first one is at and near the source. The concussion from the first explosion throws out much of the dust which has settled, particularly on horizontal surfaces, into the air, where it can be ignited by glowing residue of the first explosion. This could readily involve the whole area in which dust has settled, with the pressure waves causing much building damage.

Equipment for this exposure is basically dusttight, and wiring methods involve tight closure of entry into equipment enclosures. In division 1, threaded rigid or intermediate-metal conduit and MI cable are approved types. Fittings are to be of types approved for this use. In division 2, additionally, cable types MC and SNM and approved fittings are specified.

An important addition in the 1984 NEC indicates that where the dusts have resistivity of under 100,000 Ω • cm, there shall be only division 1 locations. The reason is that dust can still build up, even though more slowly, in locations farther from its source, so the ignition hazard may be just as great, but possibly less frequent than in prime dust generation areas.

Testing

Like human beings and automobiles, equipment and cables require checkups at regular intervals to minimize spontaneous breakdowns. As in people going to a doctor's office or taking the car to a garage, cable has to be deenergized and out of service for its checkup.

When to Test. The required intervals for cable testing are 4 to 6 yr, so the cables will have to be out of service this frequently. If the powerplant is scheduled for absolute continuous operation, the cable system will have to be one of the redundant types, such as primary selective, secondary selective, or network. These systems are normally employed for a high-reliability facility such as a powerplant. Where the plant is shut down at intervals for maintenance of its nonelectrical components, the cables can be tested at these same periods. Likewise, where a cable is associated solely with a single component, it can be tested when that component is being checked out or reconditioned. This is especially true of low-voltage equipment, where dual supply is seldom provided because of the added expense.

Key Method. The most practical and by far the most widely used method of testing (away from the factory) is *dc high potential*. In this method, negative high-voltage direct current is applied in steps about equal to the ac *root-mean-square* (rms) voltage rating of the cable until the prescribed potential is reached. At each step, voltage is maintained for 1 min, and the current is recorded. From this and by using Ohm's law, $R = E/I$ is calculated and usually plotted graphically with test voltage as the abscissa. If the resistance remains reasonably constant (±20 percent) up to the maximum prescribed voltage, the cable is judged satisfactory. If, at any test voltage below the prescribed level, the resistance decreases 50 percent for a 20 percent increase in voltage, this indicates that the cable is faulty and likely to fail in the near future if it is left in service.

Holding the highest test voltage for 15 min, and watching the microammeter closely, is an important part of the test. Possibly after a few minutes, the meter indication will start to increase slowly and then rapidly, and failure on test will ensue rapidly. Actual resistance is not vitally important, except, of course, that all three conductors of any one circuit should measure within about ±50 percent of each other. Cable has resistance several magnitudes higher than that of the equipment to which it is normally connected, so the equipment should be disconnected to test the cable.

Precautions. Potential transformers and surge (lightning) arresters *must* be disconnected from the cable. Thus, only the cable is being tested. Other equipment may not be able to withstand test voltages entirely safe for cable, which may well be necessary to really test the cable. Actually, sound 15-kV cable will withstand a 250-kV dc test, but prescribed test potentials are only 25 to 40 percent of that.

To prevent the high voltage from ionizing the air at bare conductor ends of cable on test, and thus adding to the apparent cable leakage current, it is necessary to cover these ends with some insulating material hood. A glass jar or plastic bottle, partially sealed around the insulated portion of the cable, is a solution. Duct-seal tape provides a good seal. A plastic bottle must be kept about 1 in (25 mm) minimum from contact with the terminal, because less spacing will cause burn-through of the plastic.

Procedure. A detailed procedure for testing is found in IEEE Standard 400, Recommended Procedure for Making High Voltage Direct Potential Tests on Cable in the Field. Probably the most widely used standard test voltages are those of the ICEA. These were developed from in-factory test data to identify faulty cable in manufacture, and they have been modified for dc field use. This test information is shown in Table 3.3.

A group of utilities and users desiring higher reliability of cable systems service were able to have the following added in IEEE Standard 400: "Voltages up to 70 percent of system BIL for installation and maintenance testing may be considered in consultation with the suppliers of the cable and the accessories." Most long-time manufacturers will permit these values.

A relationship between percentage of rated working ac rms voltage to ground at which the dc resistance curve shows a 50 percent decrease for a 20 percent voltage increase is shown in Table 3.4. From this it may be seen that testing to

TABLE 3.3 ICEA Field-Test Voltages

System voltage, kV rms	System BIL, kV peak	Test voltages, kV dc to ground	
		Installation	Maintenance
5	75	35	25
15	110	55	40
25	150	80	60
35	200	100	75
46	250	120	90

TABLE 3.4 Relationship of Ac to Dc Parameters

Dc test potential for start of current runaway, % of normal ac rms working voltage		Estimated time before failure, yr
EPR, XLP	Butyl	
250	200	0
375	300	0.5
625	500	2
1000	800	6+

higher than the ICEA values may be necessary to ensure adequate future time before failure, or until the next test date.

After a 6-yr period of satisfactory tests, some users elect to leave the cable connected to equipment (switchgear, etc.) for testing and to simply test the cable at switchgear test levels, about 35 percent of system BIL. The leakage-current readings are not meaningful for the cable, since cable leakage current will probably be less than 1 percent of the total leakage current. The saving is in the time for removing (tape) insulation from the end connections and replacing it.

Actually, the many turns of thin (VC) insulating tape can be replaced by two layers of Bishop Electric Co. 5000 bus-seal tape. Of course, kneaded aluminum foil should be used to fill between and over the bolts and nut ends for easy disassembly. This changes a 2-h job into a 5-min job.

BUS DUCT

General

Bus duct is the name applied to a system of sizable metal bars supported and enclosed within a metal housing, suitable for

- *Type 1:* Carrying larger electric currents than are practical with electrical cables
- *Type 2:* Making load connections at numerous and changeable locations along the bus duct length

Type 1 is useful where it is necessary to transfer substantial current (e.g., over about 1000 A) from one point to another, but it is not practical or desirable to use higher-voltage cable and stepdown transformers. Type 2 is most suitable for a space where the utilization equipment is subject to rearrangement at relatively frequent intervals, such as an automobile assembly plant or machine shop.

Type 1 is widely known as *feeder duct,* type 2 as *plug-in duct*. In the feeder-duct type, the bars are, of course, covered with insulation and placed very close together for minimum inductance, giving low voltage drop. The plug-in type, because of receptacles at close centers, must use wider spacing, so the inductance and voltage drop are higher.

Conductors used are commonly copper or aluminum, the copper being smaller but of higher cost, the aluminum larger and less costly. Splicing problems may occur more frequently with aluminum than copper. Bus duct is commonly made in 10-ft (3.05-m) lengths, with other lengths as specials. There are various fittings—elbows, tees, crosses, etc.

In particular, feeder bus duct is associated with large power supplies, so it has high current-withstand capability. Commonly, this ranges from 50,000 to 200,000 A. If the power source has higher output fault capability than the capability of the bus duct, a current-limiting device to reduce source availability current to that of the bus duct is a necessity.

Construction

Bus duct construction differs by type. Feeder bus duct employs conductors, generally completely insulated by wrapping and placed as closely together as possible, to minimize impedance. Commonly, these are thin bars, wrapped with insulating material full length, placed flat against one another, and separated only at the ends for splicing to adjacent sections. Frequently, this splicing is accomplished by means of a single bolt, compressing all members together, along with the necessary insulating members.

For plug-in bus duct, the individual buses are arranged more conventionally, normally spaced at the separation of the plug phase connections and supported by insulators at spacings meeting the fault current capability currents. Tap receptacles are normally spaced at 3-ft (0.9-m) separations, with removable covers at each outlet. The bus manufacturer will supply the switches or circuit breakers and means to attach them to the buses. The switches are normally operable by means of a hook stick by a person standing on the floor below. Enclosing both types is a sheet metal, or perforated metal, enclosure to exclude contact by personnel or material.

Ratings and Voltage Drop

Standard ratings and voltage-drop values may be selected from Tables 3.5 and 3.6.

Resistance and Reactance

For determination of let-through fault current, the resistance and reactance of the various types of bus duct are tabulated in Table 3.7.

Short-Circuit Ratings

Buses in bus ducts, as in motor control centers and switchgear, can be damaged by currents beyond those for which the buses and supports were designed. The standard currents for various ratings of 600-V bus ducts are shown in Table 3.8.

Higher-Voltage Bus Ducts

Higher-voltage bus ducts follow the ratings and capacities outlined in Table 3.9.

Installation

Most bus ducts are suitable only for indoor, dry installation. However, even those designed for outdoor service have open ends for splicing as shipped. Not all

TABLE 3.5 Three-Phase Copper Bus Duct*

Current rating, A	Voltage drop per 100 ft (30 m), line to line at rated current — Load power factor, % lagging: 50	60	70	80	90	95	100
				Feeder bus duct, totally enclosed			
600	2.93	3.09	3.23	3.31	3.31	3.23	2.83
800	2.23	2.35	2.44	2.49	2.48	2.42	2.10
1000	2.39	2.66	2.92	3.15	3.33	3.39	3.29
1350	2.37	2.60	2.80	2.98	3.11	3.13	2.96
1600	2.43	2.56	2.67	2.73	2.72	2.66	2.31
2000	2.30	2.43	2.52	2.57	2.55	2.49	2.15
2500	2.18	2.29	2.36	2.40	2.37	2.30	1.96
3000	2.43	2.55	2.63	2.67	2.63	2.55	2.17
4000	2.29	2.40	2.49	2.53	2.49	2.42	2.07
5000	2.11	2.22	2.30	2.35	2.33	2.27	1.96
				Plug-in bus duct, totally enclosed			
225	2.36	2.46	2.54	2.56	2.52	2.42	2.04
400	2.60	2.66	2.70	2.66	2.54	2.40	1.90
600	5.11	5.04	4.89	4.62	4.11	3.67	2.38
800	6.02	5.96	5.80	5.50	4.92	4.42	2.92
1000	5.02	4.98	4.84	4.60	4.12	3.70	2.46
1350	3.94	3.94	3.84	3.68	3.32	3.01	2.06
1600	3.82	3.80	3.72	3.54	3.22	2.92	2.00
2000	4.36	4.82	4.68	4.42	3.94	3.52	2.30
2500	4.30	4.26	4.14	3.94	3.54	3.18	2.12
3000	3.94	3.90	3.80	3.60	3.24	2.90	1.92
4000	4.79	4.73	4.57	4.30	3.81	3.38	2.15
5000	3.78	3.78	3.62	3.40	3.02	2.70	1.76

*Values based on full loading at end, 167°F (75°C) busbar temperature.

these are suitable for moisture exposure, so all bus duct pieces need effective protection from water in shipping and storage. Dry indoor storage at the site is highly recommended.

Bus duct is supported rather like air duct, usually from drop rods. Duct pieces must be connected, "bolt end" to "slot end," since they cannot be connected the other way, except by special fittings. "Top" or "up" labels need to be observed. Even outdoor types may sustain leakage; in a few such installations, an umbrella-inverted channel over a bus duct had to be used to exclude moisture. Some outdoor types have weep holes that are fitted with screws for shipment and storage. These screws must be removed to provide drainage of condensation.

Installation must include provision for expansion. Bus will expand ½ to 1 in per 100 ft (4.17 to 8.33 mm per 100 m) between no load and full load. Hangers need to allow for such movement, and changes in direction of the bus duct run are useful here. Where direction changes are not possible, expansion lengths can be included in the main run.

TABLE 3.6 Three-Phase Aluminum Bus Duct*

Current rating, A	Voltage drop per 100 ft (30 m), line to line at rated current						
	Load power factor, % lagging						
	50	60	70	80	90	95	100
	Feeder bus duct, totally enclosed						
600	2.48	2.73	2.96	3.16	3.30	3.34	3.17
800	2.44	2.66	2.86	3.03	3.14	3.15	2.94
1000	2.39	2.66	2.92	3.15	3.33	3.39	3.29
1350	2.37	2.60	2.80	2.98	3.11	3.13	2.96
1600	2.45	2.67	2.87	3.04	3.14	3.15	2.94
2000	2.23	2.42	2.59	2.73	2.81	2.81	2.60
2500	2.18	2.35	2.51	2.63	2.70	2.69	2.48
3000	2.37	2.58	2.78	2.94	3.04	3.05	2.85
4000	2.24	2.42	2.59	2.71	2.79	2.78	2.56
	Plug-in bus duct, totally enclosed						
100	3.76	4.30	4.83	5.33	5.79	5.98	6.01
225	2.73	2.96	3.15	3.31	3.41	3.40	3.13
400	3.92	3.99	3.99	3.92	3.69	3.45	2.64
600	5.30	5.41	5.45	5.37	5.10	4.80	3.76
800	4.70	4.81	4.84	4.78	4.54	4.28	3.36
1000	3.75	3.85	3.39	3.86	3.70	3.50	2.79
1350	3.36	3.44	3.48	3.44	3.28	3.08	2.44
1600	4.31	4.59	4.61	3.52	4.27	3.99	3.07
2000	3.99	4.07	4.09	4.04	3.83	3.60	2.81
2500	3.24	3.30	3.32	3.27	3.10	2.92	2.27
3000	4.09	4.14	4.12	4.01	3.74	3.47	2.60
4000	3.50	3.55	3.55	3.47	3.26	3.04	2.31

*100 ft = 30.5 m.

TABLE 3.7 Bus Duct Resistance R and Reactance X, Line to Neutral*

Current rating, A	Feeder bus duct				Plug-in bus duct			
	Aluminum		Copper		Aluminum		Copper	
	R	X	R	X	R	X	R	X
100	—	—	—	—	20.1	5.0	—	—
225	—	—	—	—	6.74	3.45	4.44	3.94
400	—	—	—	—	3.20	4.33	2.31	2.76
600	2.56	0.99	2.28	1.68	3.03	3.80	1.92	4.35
800	1.78	0.81	1.27	0.98	2.03	2.52	1.78	3.80
1000	1.59	0.50	1.05	0.82	1.35	1.57	1.20	2.52
1350	1.06	0.44	0.76	0.65	0.88	1.06	0.75	1.44
1600	0.89	0.41	0.70	0.53	0.93	1.24	0.61	1.17
2000	0.63	0.31	0.52	0.41	0.68	0.86	0.56	1.24
2500	0.48	0.25	0.38	0.32	0.44	0.56	0.42	0.86
3000	0.46	0.21	0.35	0.31	0.43	0.62	0.32	0.66
4000	0.31	0.16	0.25	0.21	0.28	0.39	0.26	0.62
5000	—	—	0.19	0.15	—	—	0.17	0.69

*mΩ/100 ft (30 m) at 77°F (25°C). For 167°F (75°C), multiply the resistance (only) by 1.19.

TABLE 3.8 Short-Circuit Ratings*

Current rating, kA		Maximum short-circuit current ratings, kA	
Plug-in	Feeder	Symmetric	Asymmetric
0.1		10	10
0.225		14	15
0.4–0.8		22	25
1, 1.35	0.6, 0.8	42	50
	1, 1.35	75	85
1.6, 2, 2.5		65	75
	1.6, 2.0	100	110
3		85	100
	2.5, 3	150	165
	4, 5	200	225

*Higher short-circuit capacity can be obtained on request.

TABLE 3.9 Higher-Voltage Bus Ducts

Voltage, V_{rms}		Continuous current, A	Insulation withstand, kV			Max. current, kV asymmetric
Nom.	Max.		60 Hz, 1 min	Dc, 1 min	Impulse	
4.16	5	1200	19	27	60	19–78
13.8	15	2000	36	50	95	19–78
23	25.8	3000	60	—	125	58
34.5	38	—	80	—	150	58

Field Testing

Bus duct is related to switchgear from the high-potential testing standpoint. For 600-V and lower bus, megohmmeter testing is suitable, with 1000-V maximum test potential. Now 1 MΩ per 10 ft (3.05 m) of length should be regarded as the minimum resistance, but actually 100 MΩ divided by the number of 10-ft (3.05-m) lengths or fittings is the normally expected value.

For higher-voltage high-potential testing, the use of 75 percent of the factory test values for switchgear may be used. For resistance values, the manufacturer should be consulted.

CHAPTER 5.4
AC MOTORS

R. L. Nailen
Project Engineer
Wisconsin Electric Power Co.
Milwaukee, WI

INDUCTION MOTOR

To review briefly the basis for three-phase "squirrel-cage" induction-motor operation: Slots around the inner diameter or "bore" of the stator core contain individual coils, as Fig. 4.1 shows, interconnected so that the net effect of the three applied phase voltages is to produce a rotating magnetic field traveling around the bore at a "synchronous" speed determined by the applied frequency and the number of magnetic poles for which the winding is connected. Synchronous rpm N_s is equal to $120f/P$, in which f is the frequency in hertz and P is the number of poles.

Rotating Field

The field does not really rotate. Consider an ocean wave approaching the shore. The water does not travel in and out very much relative to the shoreline. Rather, at any given point, the water simply rises and falls as the wave crest approaches, then passes on. Similarly, in the motor, the magnetic field strength rises to a peak, then drops again, at successive points around the stator circumference.

Through induction, this field produces voltages in the squirrel-cage rotor con-

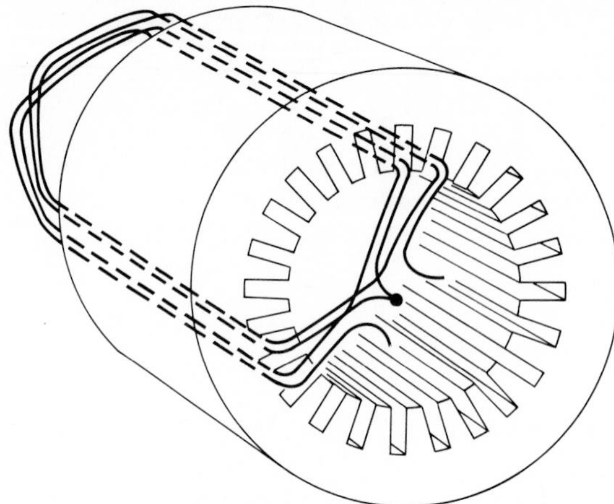

FIG. 4.1 Stator core showing two 2-turn coils in slots. Complete winding would fill all the slots.

ductors or bars which occupy the rotor periphery. Because those bars are short-circuited at their ends, these voltages produce bar currents which in turn surround each bar with a rotor magnetic field. Interaction of that field with the "rotating" stator field produces the torque to turn the rotor on its axis (Fig. 4.2).

No induced rotor voltages can exist without relative motion between the bars and the stator field. Hence the rotor always travels a little slower than the stator field, rather than in synchronism with it. This speed difference is called *slip*. Its value varies with the motor size, poles, rotor-bar resistance, and stator-winding arrangement. The frequency of the rotor bar current is the slip in percentage

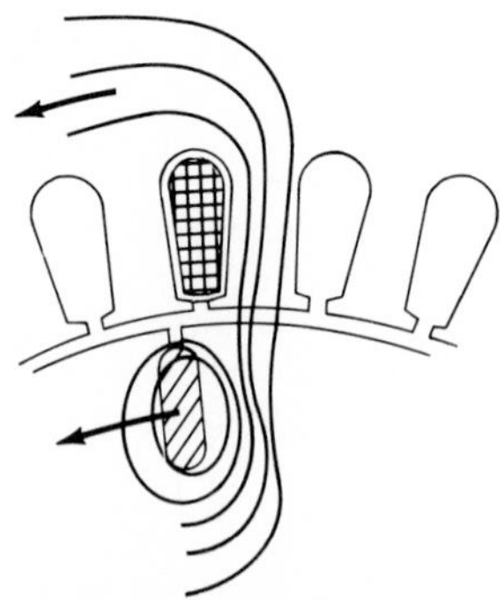

FIG. 4.2 Portion of induction motor stator (upper, with coil cross section shown in one slot) and rotor (lower, with one cage bar and slot). Lines show magnetic flux path. Note "crowding" of flux lines where stator field (moving in direction of upper arrow) and rotor field come together. This produces force on each rotor bar as shown by lower arrow.

times the line frequency. Percentage of slip is defined as

$$\frac{\text{Synchronous rpm} - \text{actual rotor rpm}}{\text{Synchronous rpm}} \times 100$$

and it ranges from ½ to 4 percent for most standard motors and up to 20 percent for special designs (at full load).

Torque Characteristics

Variations in rotor resistance and reactance (the "deep bar effect" in the cage) during acceleration result in a varying torque as the motor comes up to speed. Those variations occur because the rotor circuit frequency varies from a maximum of line frequency at zero speed to the much lower slip-frequency value at full speed. When different rotor-bar and slot shapes are used (Fig. 4.3), the torque versus speed relationship can be changed to suit different types of driven loads, as Fig. 4.4 shows.

Easy to extrude (or to cast in place for small rotors), aluminum is especially suited to special bar shapes. Wide experience shows that both aluminum and copper cages give long service life. Figure 4.5 illustrates common methods of fabricating and attaching the "short-circuiting" or end rings completing the circuit between cage bars at each end of a large (noncast) rotor.

FIG. 4.3 Typical rotor bar/slot shapes for squirrel-cage rotors. Double cage, at right, uses two separate conductors in each slot.

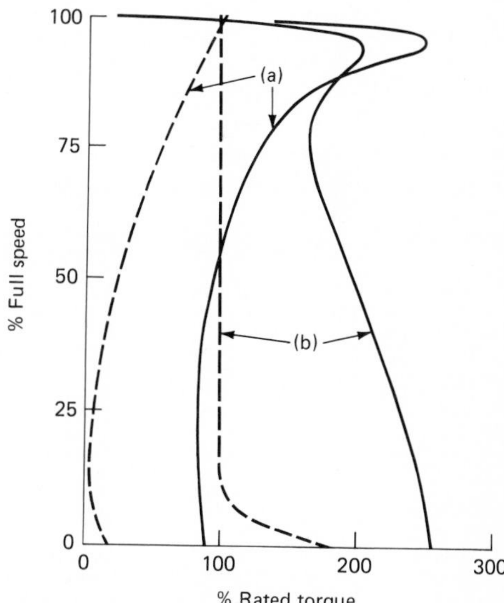

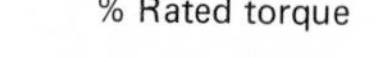

FIG. 4.4 Cage rotor design matches motor torque (solid curves) to what the load needs (dashed curves). At (*a*), pump drive uses slot like the one on the far left of Fig. 4.3. At (*b*), conveyor drive needs higher torque of double-cage rotor. Bar materials of different resistances, as well as shape, yield different curve shapes.

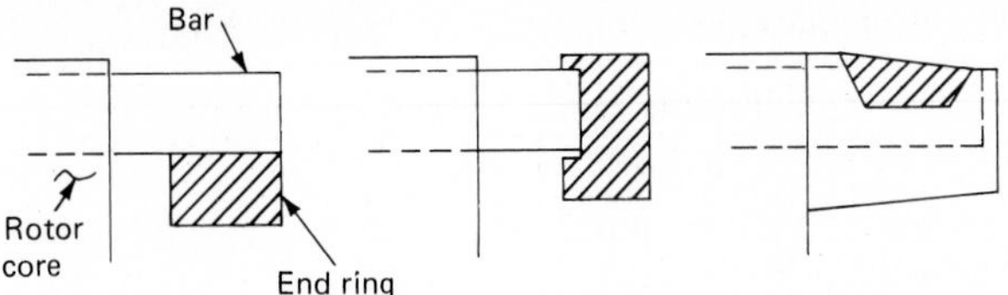

FIG. 4.5 Three common ways of fitting short-circuiting end rings to cage rotor bars. Version at right is aluminum using circumferential weld (shaded area) to connect bars.

The general shape of the speed-torque curve distinguishes several common motor types (Fig. 4.6). Locked-rotor, breakdown, and minimum or "pull-up" torques are standardized depending on the motor size and type. As the illustration shows, however, those values can be the same even with considerable variation in curve shape.

Applicable Standards

What is a "standard" motor? In this country, three agencies issue voluntary standards governing motor construction and operation. Most extensive documents are those of the *National Electrical Manufacturers Association* (NEMA). Although some apply to any motor, most standards deal with the so-called NEMA

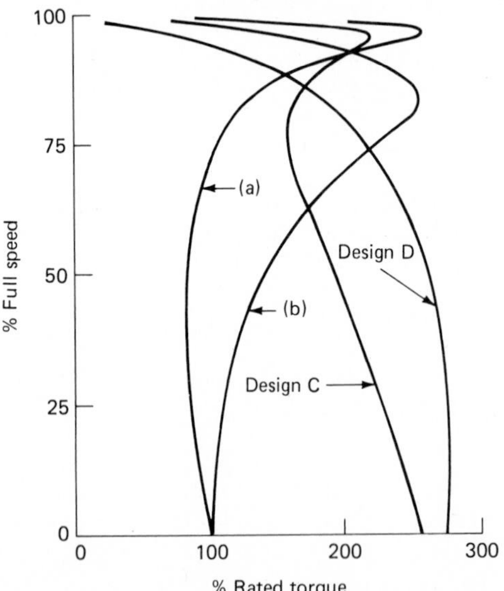

FIG. 4.6 Typical motor speed-torque curve shapes for standard NEMA Design B: large motor at (*a*), small motor at (*b*); for Design C, double-cage; and Design D, high-slip punch press drive. Lowest torque value at any point on the curve is called the "pull-up" torque value.

frame or integral-horsepower machines. Usually below 250 hp (187 kW), these may be as large as 500 hp (373 kW) depending on the speed (the fewer magnetic poles in the stator winding, the smaller the motor size for the same horsepower).

Published as MG1-1987 (latest year of complete revision), NEMA standards include, in part,

- Overall dimensions
- Limits on locked-rotor current and torques
- Standard horsepower (kilowatt) and voltage ratings
- Enclosure definitions
- Noise, balance, and vibration levels
- Temperature-rise ratings

The standards are less well defined for larger motors. Frame sizes are not fixed; neither are locked-rotor currents or specific noise levels.

A second standards-making body concerned with rotating machines is the *Institute of Electrical and Electronic Engineers* (IEEE). It issues no standards for complete motors. Rather, IEEE publications deal with insulation behavior, testing procedures, and motor circuit protection.

The *American National Standards Institute* (ANSI) adopts some standards of both IEEE and NEMA. It has also published some of its own, notably ANSI C50.41-1982 (defining certain application and performance conditions of powerplant auxiliary drive motors).

It is always most economical to design any drive so that a standard motor can be used. To do so, however, requires any motor user to be familiar with these agencies and their publications. Motors built by different manufacturers will not be the same even though they are built to the same standards. Each supplier decides in what way the standard objectives will be met.

Service Factor

In matching standard motor horsepower ratings to actual loads, questions often arise about the "service factor." According to NEMA Standard MG1-1.43, "The service factor of an alternating-current motor is a multiplier which, when applied to the rated horsepower, indicates a permissible horsepower loading which may be carried under the conditions specified for the service factor." Those conditions include nameplate voltage and frequency.

Some users expect too much of a service factor. It is intended only to allow for occasional overload caused by process variations, ambient-temperature swings, and the like. It can be a hedge, too, against intermittent voltage unbalance. Thus, motor-winding temperature rise at the service factor overload exceeds the normal insulation system limit, usually by 18°F (10°C). Such overheating, if infrequent, has little effect. But if the overload is continuous, the overheating will persist (Fig. 4.7) and the insulation life can be cut 50 percent or more (Fig. 4.8).

If such loading is necessary, the better choice is a motor of higher horsepower (kilowatt) rating, rather than a smaller one with a service factor. Limited advantages of a service factor have led to its elimination from NEMA standards above 200 hp (149 kW). International standards allow no service factor for any size motor.

A service factor will not increase the accelerating or breakdown torque, either. A motor with a service factor will therefore not necessarily start a heavier

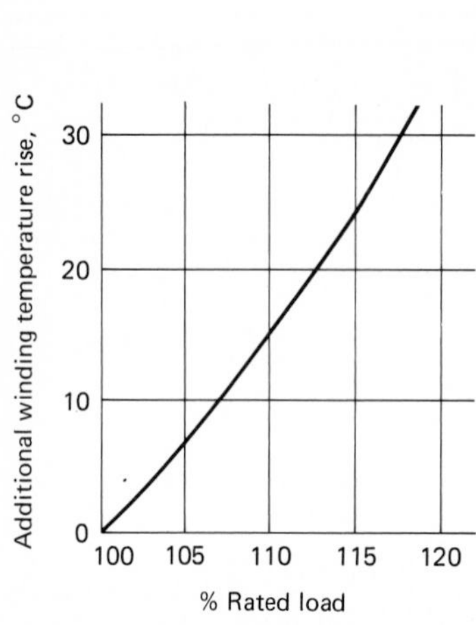

FIG. 4.7 Motor winding temperature rise versus continuous load.

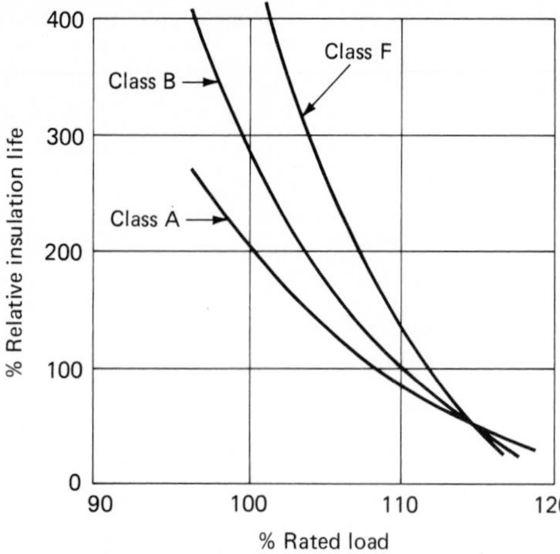

FIG. 4.8 Insulation life versus load for three standard insulation systems, showing the effect of continuous motor operation at 1.15 service factor load.

load or "ride through" high-torque peaks while running (as in crusher service). Finally, Fig. 4.9 shows that a service factor design generally offers no efficiency benefit.

"Premium Efficiency"

No trend in motor design or application since the 1960s can compare with the push for higher efficiency. All major manufacturers now offer "premium efficiency" motors, at least up to 250 hp (187 kW).

Improved efficiency commands a higher price. Payback analysis may justify extra investment now that electric energy cost has risen from $0.02 or $0.03 per kilowatthour to the $0.07 to $0.20 range.

In dealing with motor efficiency, designers recognize that "identical" motors are subject to manufacturing tolerances, resulting in some performance variation. Therefore, in arriving at nameplate efficiency ratings for standard motors, NEMA in 1980 adopted *nominal efficiency* values (Table 4.1), each indicating a range of efficiency. An earlier system used index letters rather than numbers. As NEMA Standard MG1-12.55 explains, "Variations in materials, manufacturing processes, and tests result in motor-to-motor efficiency variations for a given motor design; the full-load efficiency for a large population of motors of a single design is not a unique efficiency but rather a band. . . ." Nominal values, adds the Standard, ". . . shall not be greater than the average . . . of a large population of motors of the same design."

Each category in Table 4.1 also includes a minimum efficiency. Every motor of the same design should be "at least that good." Obviously, if a major operating-cost saving depends on a motor's efficiency, a guaranteed minimum figure is preferable to a nominal one.

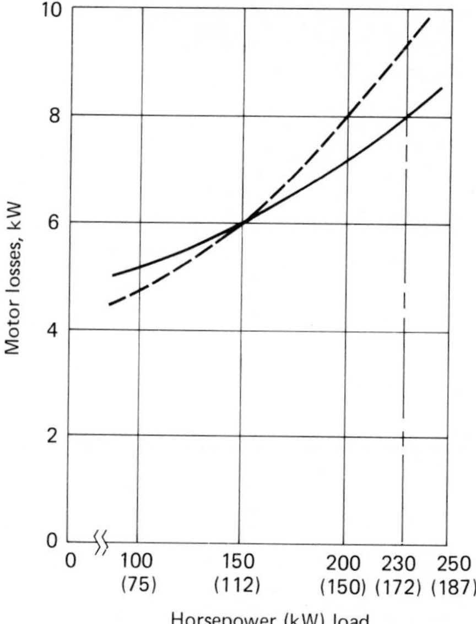

FIG. 4.9 Here a 230-hp load could be carried by either a 200-hp motor with service factor (dashed curve) or an underloaded 250-hp machine (solid curve). Typically, the latter choice is more efficient. Note: 1 hp = 0.746 kW.

TABLE 4.1 NEMA Nominal-Efficiency Ranges*

Nominal efficiency, %	Minimum efficiency, %
95.0	94.1
94.5	93.6
94.1	93.0
93.6	92.4
93.0	91.7
92.4	91.0

*Established for use on motor nameplates and the corresponding "minimum" figures. Complete table extends down to 74.0% value.

Reducing Losses. Many design and manufacturing techniques can reduce the five losses inherent in any induction motor:

- *To reduce stator and rotor I^2R loss:* Larger slots permit larger conductors. So that the consequent reduction in magnetic path areas between slots does not push magnetic field strength too high (plus magnetizing current and iron loss), this usually means a longer core stack.
- *To reduce core loss:* Better-quality lamination steel, thinner laminations, and—again—longer core stacks are used. Basic grades and thicknesses of lam-

ination steel have gone unchanged for many years. But options exist in manufacturing technique, such as heat treatment.

- *To reduce friction and windage:* Changes are made in bearings, fans, and air circulation systems.
- *To reduce stray load loss:* A great variety of slotting, winding, and manufacturing methods have been used with varying degrees of success.

Whatever the designer tries, he or she will act first on those losses which predominate. As Fig. 4.10 shows, these can vary considerably with motor size, speed, and enclosure. Because such practices are variable, different high-efficiency motor makes will not exhibit the same performance, as Table 4.2 shows.

TABLE 4.2 Efficiencies Available from Different Motor Manufacturers*

Manufacturer	Efficiency: Standard	Efficiency: Premium	Standard loss, kW	Premium loss, kW
A	0.910	0.940	9.2	6.0
B	0.918	0.954	8.4	4.5
C	0.920	0.935	8.1	6.5
D	0.925	0.945	7.6	5.5
E	0.921	0.950	6.8	4.9

*For a 125-hp (93-kW), four-pole open motor.

Efficiency Drawbacks. High efficiency can have drawbacks in other respects. For example, the starting current tends to be higher because the I^2R winding losses are most easily reduced by making a design magnetically "stronger"—that

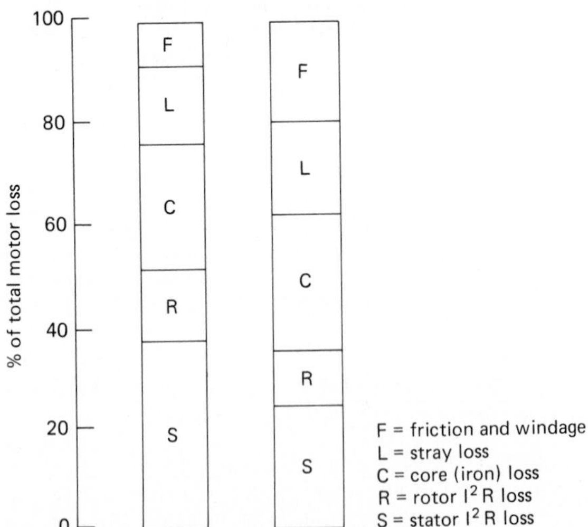

FIG. 4.10 Loss distribution for 10 to 50 hp (7.46 to 37.3 kW) motors—drip-proof at left, fan-cooled at right. To raise efficiency, designer would look first at stator I^2R loss for the drip-proof unit; iron or windage loss for the other.

is, by reducing the effective turns in the winding, resulting in higher torques and locked-rotor current. Lower rotor-cage resistance means lower slip, hence higher full-load speed. In a pump drive, the shaft power varies as the cube of the speed, so a higher-efficiency motor tends to drive the pump faster at a higher horsepower. The extra power, which must be paid for, is usually wasted in system losses.

A careful study of the entire operating system, then, should precede purchase of any premium-efficiency motor. The motor itself is often only a minor contributor to total drive losses (Fig. 4.11). Such a study should establish the actual operating horsepower needed. In large motors, efficiency versus output power may vary widely with such design options as accelerating torque (Fig. 4.12).

Equally important is the actual duty cycle. A premium-efficiency motor price justified for 80 operating hours a week may be meaningless if the motor runs only one-half that time.

Because newer premium-efficiency machines are larger and often cooler, users assume the machines are more tolerant of misuse. That is not necessarily true. The improved performance can quickly be lost if

- *Terminal voltage changes:* For example, if voltage drops only 5 percent, the loss in a typical 150-hp (112-kW), six-pole open motor will go up more than 1 kW, while efficiency declines nearly a full percentage point.
- *Voltages are not balanced:* A 2 percent unbalance can cut efficiency of a 500-hp (373-kW) machine by one-half a percentage point.

Enclosures

Whatever the motor type, correct application always depends on proper enclosure choice to suit the environment. The most common enclosure types (per NEMA MG1) are

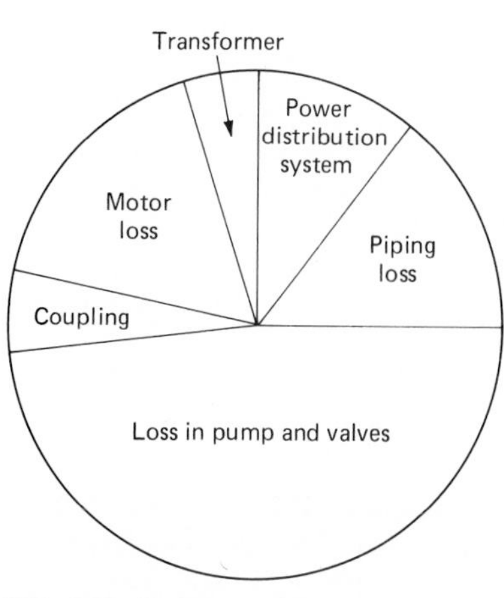

FIG. 4.11 In a pump drive, power loss in the motor alone may be minor.

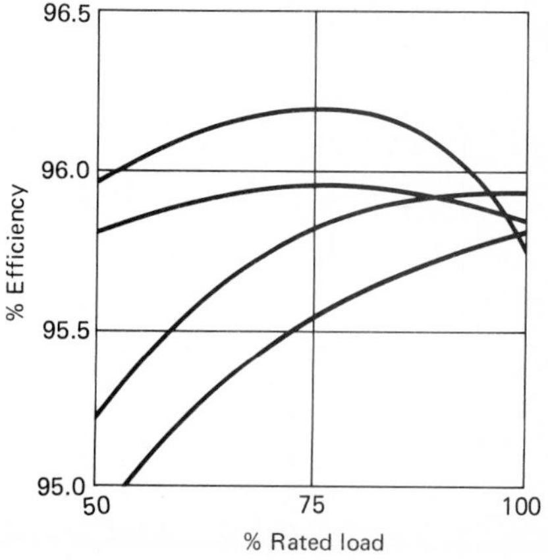

FIG. 4.12 Performance of four different designs, 4-pole motor. Which one offers the "best" efficiency depends on actual load. Motor worst at full load is best at 80% load.

- *Drip proof:* Having ventilation openings to let cooling air circulate freely, this enclosure is so constructed "that successful operation is not interfered with when drops of liquid or solid particles strike or enter the enclosure at any angle from 0 to 15 degrees downward from the vertical." This suits normal indoor service in reasonably clean, dry surroundings.
- *Weather-protected type II:* Intended for outdoor use, this enclosure excludes solid particles ¾ in (19 mm) or larger. It minimizes entrance of rain, snow, or airborne dirt. Air intakes allow high-velocity storm winds to blow harmlessly through or past the openings without driving into the motor interior.
- *Totally enclosed, fan-cooled:* Although neither airtight nor watertight, these machines allow no free circulation of outside air through the motor interior. Heat is dissipated either through a finned-frame surface across which a shaft-driven fan blows outside air or through an air-to-air heat exchanger. With some added features, most of these machines can be made explosion proof for hazardous atmospheres.

Not all motor ratings are available in all enclosures. Many other special constructions, such as water-cooled, exist for special applications.

Overload Protection

Protecting motors against overload damage involves selecting a protective device (relay, circuit breaker, or fuse) having a time-current rating curve that "fits" the motor's thermal damage curve. The latter is a plot of overload current versus the time that current may safely flow. Figure 4.13 is an example.

During acceleration, motor windings heat 30 to 50 times as rapidly as during running. Should the motor fail to rotate, the ratio is even higher because no fan action assists internal cooling. Therefore, motor protective equipment must act

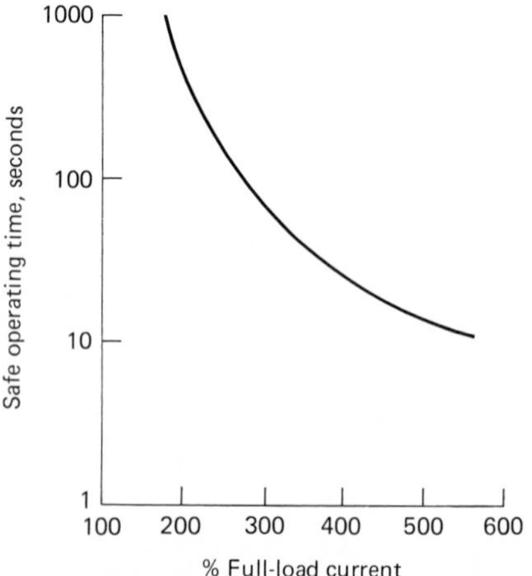

FIG. 4.13 Thermal damage curve for an induction motor.

quickly to get the machine off the line in time to prevent serious damage, yet not operate too soon on a normal start.

Adding an actual plot of accelerating current versus time to Fig. 4.13 results in Fig. 4.14. Note the "window" between the two curves. Without that margin, a "nuisance trip" would occur on every start.

Reduced-Voltage Starting. Reduced-voltage starting is often considered "easier" for the motor, because it reduces the starting current. Figure 4.14, however, shows that reducing the voltage will rapidly lower the accelerating torque to increase the starting time. Thus, a safe start at full voltage can become unsafe at a reduced voltage. Motor torque is usually assumed to vary as the square of that voltage. But that is not correct. On the average:

$$\text{Motor accelerating torque} \propto (\text{voltage})^{2.2} \qquad (4.1)$$

because of magnetic saturation in the stator and rotor.

Another misconception is that a 50 percent autotransformer tap for starting will produce 50 percent rated voltage at the motor. Because of line drop throughout the circuit, such a tap can result in only 40 percent rated voltage at the motor.

This relation between torque and voltage is valid only for sinusoidal voltage. Such a normal waveform will be present with conventional reduced-voltage starters like the autotransformer type. However, it will not be present with the new solid-state or "soft" starters, available for motors rated below 600 V and about 1000 hp (746 kW). During acceleration, these use *silicon controlled rectifiers* (SCRs) with variable firing angles to apply an effective voltage well below the rated value, having a "chopped" nonsinusoidal waveform. Cutting off part of each cycle is what lowers the effective value.

As a result, the motor torque will be even less than Eq. (4.1) would indicate.

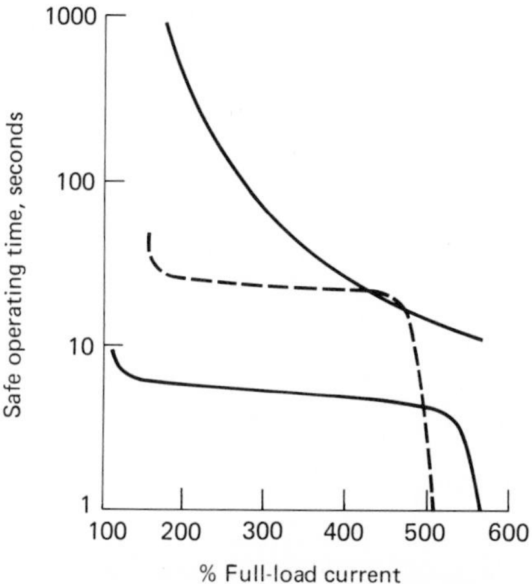

FIG. 4.14 Damage curve of Fig. 4.13 with actual time-current acceleration curves added. Solid curve is full-voltage acceleration; dashed curve is reduced voltage.

For any driven load requiring high accelerating torque, then, a solid-state starter may make acceleration impossible even though the ''measured'' motor voltage remains the same as it was with an autotransformer.

Required Shaft Torque. To be sure of adequate protection, investigate the load characteristics and power system behavior carefully beforehand. The most important single characteristic of any motor-driven machine is its required shaft torque. In specifying the drive motor, therefore, choosing the right motor torque is essential. Three separate values must be considered:

- *Locked-rotor torque (at zero speed):* This torque must overcome whatever ''breakaway torque'' the load needs to get underway.
- *Accelerating torque:* This is the torque that the motor develops during the range from zero to full rpm. If it does not exceed what the driven machine demands at all speeds, the motor either will fail to reach full speed or will overheat in doing so.
- *Full speed (rated) torque:* This torque must be sufficient to sustain the running load.

Not all motors are alike. Physics dictates that a high-speed, two-pole motor cannot exert as much accelerating torque as a six- or an eight-pole machine. NEMA standards reflect that; see Fig. 4.15. The shapes of the respective speed-torque curves will inevitably differ. Nor will a large machine exert as much torque as a small one, in percentage—again, see Fig. 4.15.

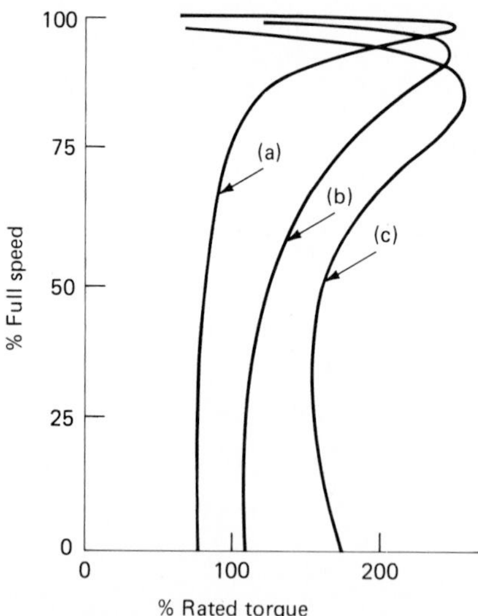

FIG. 4.15 How motor accelerating torque varies with speed and size. Because a 10-hp (7.46 kW) motor application worked well does not mean the same margins can be assumed for 1000 hp (764 kW). Curve (*a*) = large 2-pole motor; (*b*) = large 8-pole motor; and (*c*) = small 8-pole motor.

Rotor Heating. Heating of a squirrel-cage rotor during acceleration, which varies directly with drive inertia, is nonlinear with speed. Heating occurs much more rapidly during the low-speed portion of the acceleration than during the high-speed portion. Figure 4.16 shows this.

But rotor heat developed during each speed increment in Fig. 4.16 also varies with the amount of torque the load demands. Mathematically,

$$\text{Total rotor heat} \propto (\text{inertia}) \frac{T_M}{T_M - T_L}$$

Here T_M is the average motor torque during each interval, and T_L is load torque during that same interval. As the motor and load speed-torque curves approach each other, the ratio $T_M/(T_M - T_L)$ rises rapidly, and so does rotor heating. Although anyone could calculate the heat energy involved from this formula, only the motor designer (knowing the heat storage and heat dissipation capability of the parts) can determine the resulting temperature rises and stresses. Those determine whether the start is safe.

Two Device Types. Motor protection devices have usually been one of two types:

- *External current sensors:* Starter overload heaters, overcurrent relays, or magnetic circuit breaker trips are used.
- *Internal temperature sensors:* These have the advantage of detecting heating problems (such as air blockage) not dependent on current flow.

Solid-state electronics has brought radical changes to both methods. Compact black boxes now combine six or eight protective functions (such as overcurrent, undervoltage, or phase unbalance) into a single unit, controlled by an internal microprocessor. A separate input channel may monitor the motor winding temperature sensors. These controls can be programmed to "remember" how often the

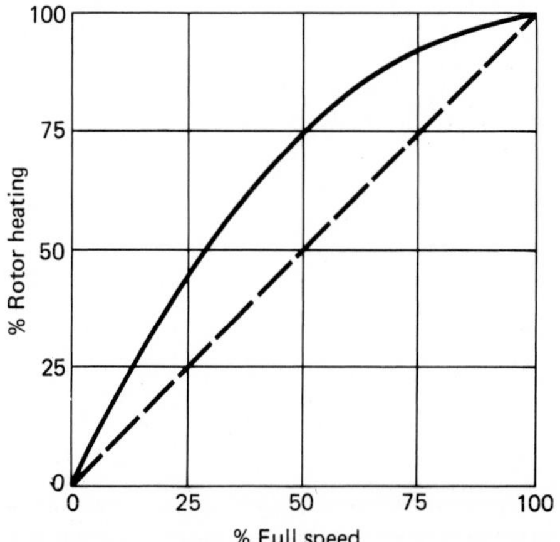

FIG. 4.16 Nonlinearity of rotor cage heating (solid curve) versus speed, during an induction motor start. Linear heating would produce the dashed line.

motor starts, forcing the interval between starts to increase further with each start.

Thermal Damage Curve. Whatever the protective device selected, properly coordinating it with the motor's thermal capability is the secret of adequate motor protection. That capability, too, is expressed differently today. The old practice was to pick a protective device current-time operating curve close to the motor's thermal damage curve (Fig. 4.13). The recently issued IEEE Standard 620, however, recognizes that such a thermal damage curve really includes two separate regions. Under running overload, current can increase only up to the breakdown point on the speed-torque curve. Further overload will stall the motor. Current then jumps at once to the locked-rotor value.

So a realistic thermal damage curve resembles Fig. 4.17. Two separate protective devices (or two separate protective functions within a single electronic control) are needed—one to cover running overloads, the other to monitor starts (Fig. 4.17). Now, when voltage is low or acceleration is long, the "bulge" in the acceleration curve can fall into the gap between the two protection curves, allowing effective protection to be applied. Nuisance trips are unlikely during a normal start.

SYNCHRONOUS MOTOR

Rotor Construction

Whereas the induction motor relies on slip for the rotor magnetization, without which no torque exists, in the synchronous motor a separate *direct-current* (dc)

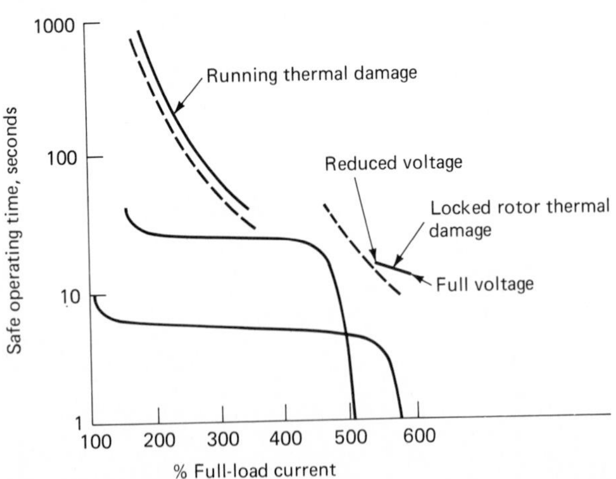

FIG. 4.17 Modification of Fig. 4.14 to reflect the actual two-part nature of the motor thermal damage curve. Protection for reduced voltage starting is made easier. Some electronic controls may combine the separate "running" and "starting" protection curves (dashed lines).

power source energizes the rotor directly. Both types of machine use the same stator construction and winding. But besides a cage of short-circuited bars, the synchronous machine's rotor circuit includes a set of wire-wound individual magnet poles (Fig. 4.18). In small motors, these may be permanent magnets.

The magnetized rotor poles are pulled around in synchronism, or "in step," with the stator field. To reach that operating condition, the motor initially accelerates as an induction machine, using the cage winding or damper bars fitted in slots on each pole (Fig. 4.18). During acceleration the dc field winding is short-circuited through a resistance called a *field discharge resistor*.

Near full speed, the resistor is disconnected, and the field winding is energized to bring the rotor into synchronism. This requires a "pull-in" torque that bears no relation to the pull-up torque of the induction motor.

Protection and Control

Synchronous-motor protection and control are more complex than for the induction-motor type. First, the field discharge resistor is needed to prevent the open-circuited field from developing destructively high voltage through induction from the energized stator.

Second, the field must be energized at just the right speed for the rotor to pull in to synchronism. Third, the machine must be deenergized if loss of field (a break in the field circuit) should occur during running, which would cause it to pull out of step and overheat.

Most industrial synchronous motors are of the salient-pole type, so called because each field pole projects outward from a central drum or "spider." When the combination of diameter and speed imposes too great a centrifugal stress on such poles, the solid-rotor design is used (Fig. 4.19).

A salient-pole machine having only a few poles, such as four or six, must have wide interpole gaps in the rotor circumference, limiting the room for damper bars. So in large sizes these are often built with solid pole heads or tips which

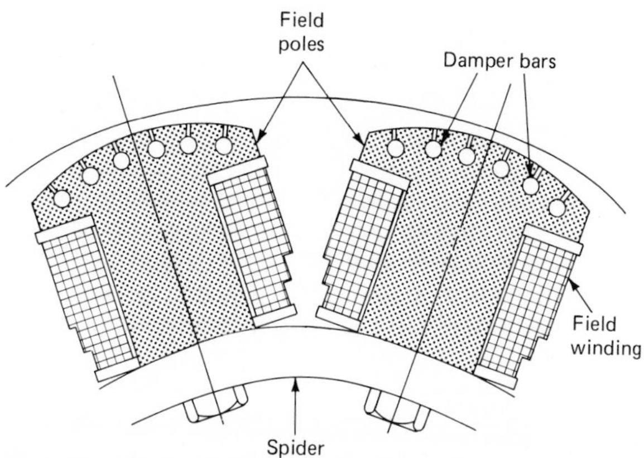

FIG. 4.18 Synchronous-motor field poles (lower) inside stator which is slotted and wound as in Figs. 4.1 and 4.2.

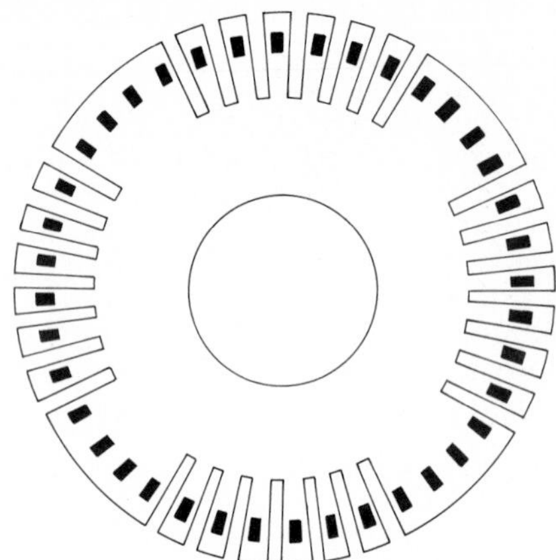

FIG. 4.19 Cylindrical rotor for large 4-pole synchronous motor. Deep slots contain field winding; damper bars (dark) lie between.

themselves act as short-circuited damper-winding "turns" to provide the needed accelerating torque. Otherwise, poles are laminated as would be the rotor of an induction machine.

Advantages and Disadvantages

Lack of slip gives the synchronous machine one of its greatest advantages: high efficiency. Whereas an induction motor might reach 0.96 full-load efficiency, for example, the corresponding synchronous unit may reach 0.97.

Another advantage is that field-current variation can control the synchronous-motor power factor. That power factor can be held constant with varying motor load, typical values being 0.80, 0.90, and 1.0—all leading rather than lagging as in induction motors, where it necessarily varies with load.

Offsetting these benefits are some drawbacks, such as

- Maintenance on the wound poles.
- The dc excitation system. Years ago, a separate dc generator, or one driven from the synchronous-motor shaft, was the field power supply. Now, most synchronous motors use "brushless exciters." Figure 4.20 illustrates a typical system. No brushes or slip rings are needed to connect an external power source to the rotating field, which reduces maintenance considerably.
- Greater risk of instability when terminal voltage dips or load inertia is high. In an unstable mode, the motor pulls out of step and has to be restarted.

Applications

Large synchronous motors are most often applied to compressors, particularly at low speed, where either the high efficiency or the power factor produces large operating-cost savings. Figure 4.21 shows typical motor-torque curves.

Although many exceptions can be found, the usual application of induction

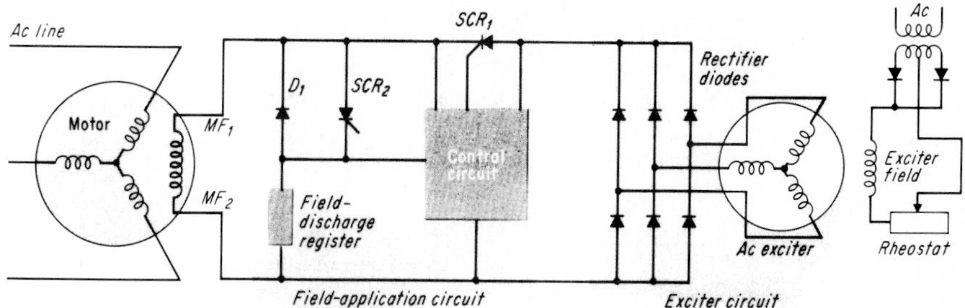

FIG. 4.20 Motor-mounted exciter gets its excitation from a small rectifier. Exciter ac output is fed to the recifier diode assembly; output is controlled by the exciter field. The solid-state control circuit applies dc to motor field through SCR1. Field discharge resistor is also shaft mounted and is controlled by diode D1 and SCR2. Motor fields (MF1, MF2) are connected at the correct rotor speed. *(*Power *magazine.)*

and synchronous motors in modern practice is as shown in Fig. 4.22. Induction motors at 3600 and 1800 r/min are quite common up to 20,000 hp (14 920 kW); synchronous motors are not.

VARIABLE-FREQUENCY DRIVES

Both induction and synchronous motors are thought of as constant-speed machines (the speed of the former varies only slightly with changing load). Yet in fan or pump drives, higher overall efficiency usually results from operation at different speeds, depending on flow. Flow rate changes may be frequent, and if the

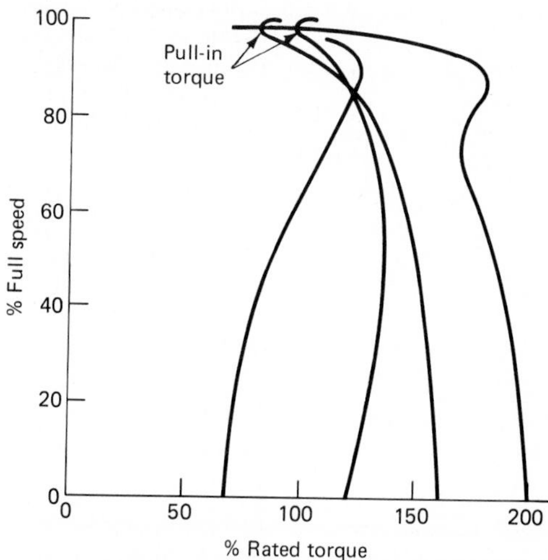

FIG. 4.21 Typical synchronous-motor speed-torque curves. As with induction machines, rotor bar/slot design can provide wide variation in accelerating torque.

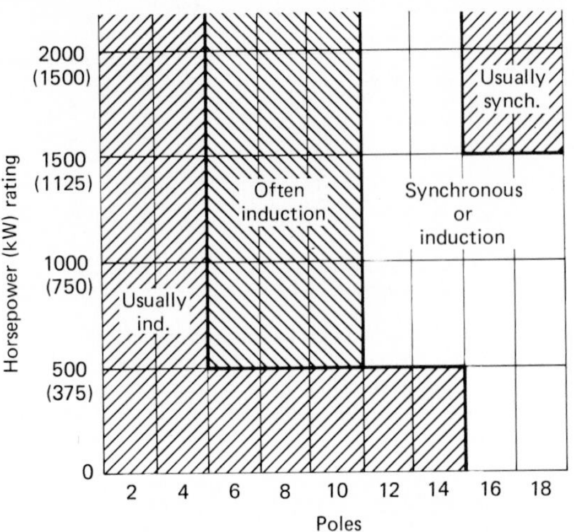

FIG. 4.22 Usual choice between induction and synchronous motors, depending on both horsepower and speed.

driven speed does not change, the flow can be varied only by throttling in the system, which wastes power (Fig. 4.23).

Modern electronics has made possible widespread use of variable-frequency power supplies. Both induction- and synchronous-machine rpm depends directly on the applied frequency. If that is varied, the speed will vary also. In Fig. 4.24, note how the different frequencies permit matching the motor output to the different operating points of the load.

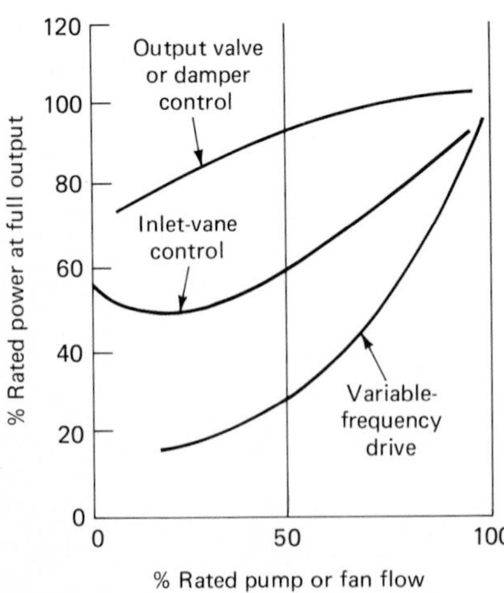

FIG. 4.23 How overall pump efficiency varies with the type of drive. Two upper curves apply with constant frequency, single-speed drive motor.

Because magnetic field strength (plus associated iron losses and magnetizing current) varies inversely as the frequency, lowering that frequency too much would push the field strength unacceptably high. So the applied voltage must be lowered along with the frequency. Logic circuits and variable-SCR firing angles in the power supplies can readily do that.

Three Types

Variable-frequency (sometimes called *variable-speed*) electronic drives come in three types:

- *Voltage-source inverter* (VSI)
- *Current-source inverter* (CSI)
- *Pulse-width-modulated* (PWM) design

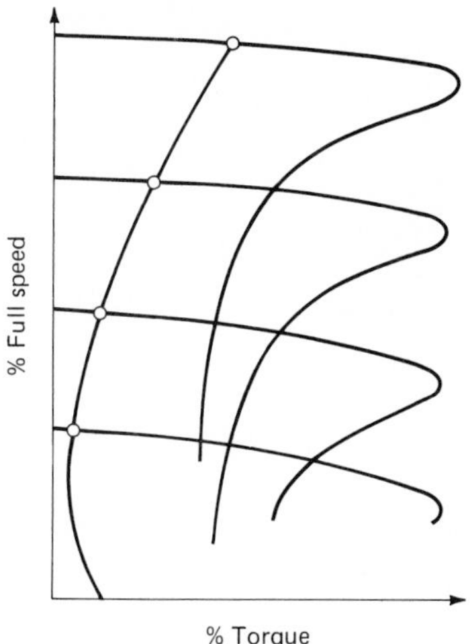

FIG. 4.24 For economical operation at any of the circled points on a pump or fan speed-torque curve, motor speed is varied by changing applied power frequency and voltage. Each frequency change shifts entire motor speed-torque curve up or down as shown.

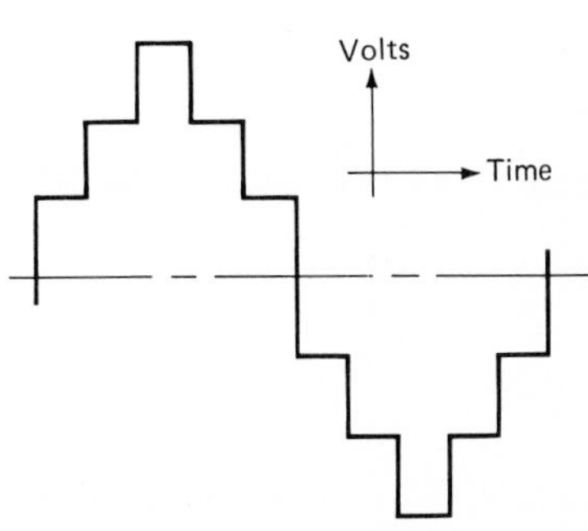

FIG. 4.25 "Stepped" voltage output of a VSI variable-frequency power supply.

In all three, a solid-state rectifier section converts the *alternating-current* (ac) line power to direct current. An inverter section then transforms that direct current to an ac output at adjustable voltage and frequency.

VSI voltage output varies as in Fig. 4.25; CSI current has a similar wave shape. The PWM design instead produces the voltage waveform of Fig. 4.26. Each system has its advantages and disadvantages, as Table 4.3 shows.

These power supplies are seldom economical for induction motors above 1500 hp (1119 kW); the 1- to 300-hp (0.746- to 224-kW) range is most popular. Thyristors require a reactive voltage in the motor windings to "commutate" the supply voltage, that is, to turn off and on at the proper phase positions to maintain the desired three-phase output sequence. In the usual inverter or induction-motor drive, this commutating voltage comes from a capacitor network across the thyristor output. Transistors, which are not subject to this limitation, are just becoming available for large, high-voltage motor circuits. And induction motors above about 1000 hp (746 kW) are difficult to wind for 460 V.

New Applications

The synchronous motor inherently produces a reactive commutating voltage by induction from the dc field. The inverter is automatically "load-commutated" by the motor winding itself.

Recently, therefore, variable-frequency synchronous-motor drives have appeared in two new applications:

- Replacement of dc rolling mill motors for steel mill service.

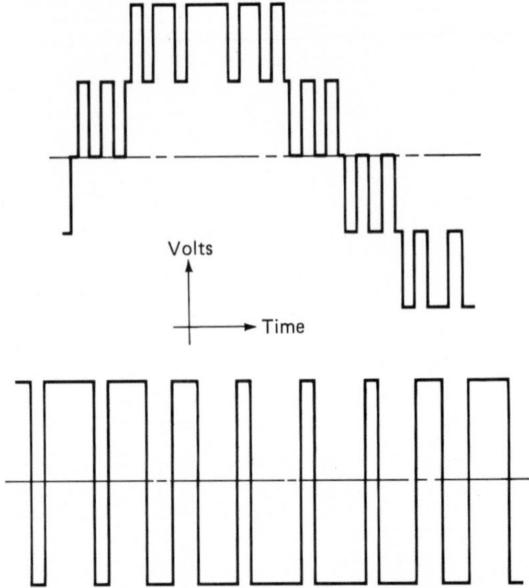

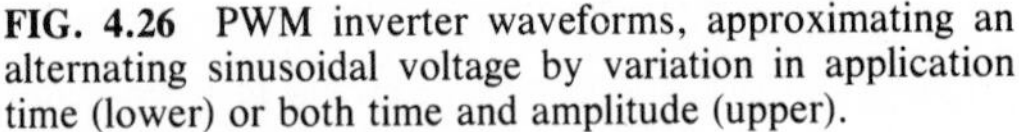

FIG. 4.26 PWM inverter waveforms, approximating an alternating sinusoidal voltage by variation in application time (lower) or both time and amplitude (upper).

- Powerplant draft fan or large compressor drives, from 1500 hp (1119 kW) to at least 60,000 hp (44 760 kW), replacing induction motors using either fluid couplings or multispeed connections.

BEARINGS AND LUBRICATION

Simple installation, ready availability, low cost, freedom from wear, and ease of lubrication have made antifriction bearings the universal choice in small industrial motors. Many larger machines, especially those driving vertical-shaft, high-

TABLE 4.3 Strong Points (+) and Weaknesses (−) of Three Variable-Frequency Inverters

Item	VSI*	CSI*	PWM*
Open-circuit protection	+	−	+
Short-circuit protection	−	+	−
Handle undersized motor	+	−	+
Handle oversized motor	−	+	−
Simplicity	+	+	−
Low cost	+	+	−
Regeneration capability	−	+	−
Stability	−	−	+
Efficiency at low speed	+	+	−

*VSI = voltage-source inverter, CSI = current-source inverter, and PWM = pulse-width-modulated design.

thrust pumps, also use antifriction (ball or roller) bearings because of their ability to sustain high axial (endwise) thrust loads.

Above about 2000 hp (1492 kW), however, two- and four-pole motors at least will most often have sleeve or journal bearings as standard. Depending on the journal diameter, these will be either pressure-lubricated (from a self-contained or remote oil supply system) or self-lubricated via oil rings. When the journal surface speed exceeds about 3000 ft/min (15.2 m/s), a cool flow of oil is needed—not to maintain lubrication as such, but to hold down the bearing temperature rise.

Trouble-free bearing service depends on proper alignment, journal surface finish and clearance, oil quality, and temperature. Thus, although sleeve bearings may wear quite slowly and are fairly easy to repair, maintenance cannot be neglected.

The normal lubricant for antifriction bearings is grease: a chemical "soap" acting as a carrier to keep its entrained oil within the bearing. For temperatures below about 300°F (148°C), normal petroleum greases are satisfactory, such as a lithium-base (to minimize moisture contamination) compound having an *extreme-pressure* (EP) additive. Synthetic greases accommodate more extreme temperature variations and water buildup in wet areas.

Ideal ball bearing lubrication, although too costly for most installations, is an oil-mist system, pressure-feeding atomized oil into the bearing. Vertical-motor thrust bearings of the ball or roller type need oil lubrication for two reasons:

- To maintain an oil film on bearing surfaces. In large bearings with high rolling speeds, grease tends to be pushed aside by the fast-moving parts and to overheat.
- To carry away heat generated by the thrust loading.

When the thrust exceeds the capacity even of available antifriction bearings, the plate-type or pivoted-pad bearing takes over. A wedge-shaped oil film of high load-carrying capacity forms between its individual shoes and a "thrust collar" or runner (Fig. 4.27). These bearings are quite costly and generate high internal losses (normally requiring water cooling of the lubricating oil).

In horizontal motors, sleeve bearings are ill equipped to sustain endwise thrust

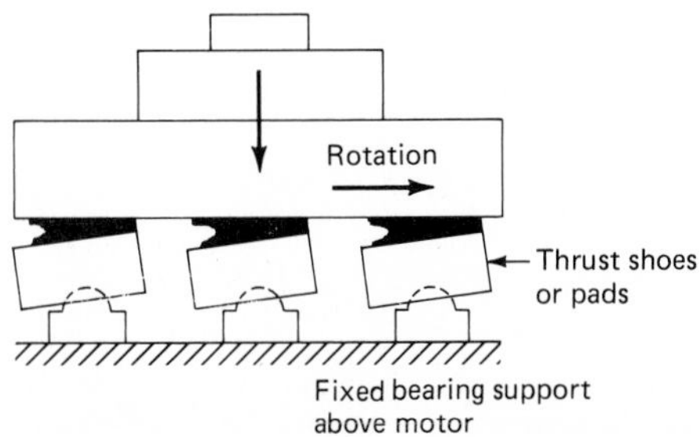

FIG. 4.27 Basic principle of a pivoted-pad thrust bearing in top end of vertical-pump motor. Dark areas are oil-film wedges that sustain thrust. Heavy arrow shows direction of downward thrust load.

imposed by some loads, or the heavy side pull of belt or chain drives. Such loading dictates roller bearings. Above a certain size and speed, however, grease lubrication is inadequate, and splash oil lubrication is then needed.

INSULATION SYSTEMS

Progress in stator winding insulation in recent years has occurred along several lines:

- Development of high-temperature, long-lasting magnet-wire coatings, such as the amide-imide synthetics, and fused glass or synthetic fibers.
- New forms, such as reconstituted paper, for the best high-voltage insulating material usable in windings—mica.
- Solventless impregnants, most often used as *vacuum-pressure-impregnation* (VPI) resins instead of conventional varnishes for the "dip and bake" treatment of windings.

Any coil, whether random-wound (below 600 V) or form-wound (see Fig. 4.28), inevitably includes voids or air spaces which will attract dirt and moisture or ionize under high voltage to break down the adjacent insulation. No insulation system develops its normal voltage or temperature rating until the entire winding is filled with an insulating liquid and cured solid by baking, to displace all internal air and moisture.

When it is applied to coils under vacuum, that liquid penetrates more effectively into the coil structure. The vacuum has already boiled off the trapped moisture. Finally, the pressure cycle of the VPI process drives the impregnant still more fully into the winding. Absence of any solvent in the liquid means that nothing will be driven off in the bake oven to form new voids.

Similar treatments exist for synchronous-machine field poles. An alternative

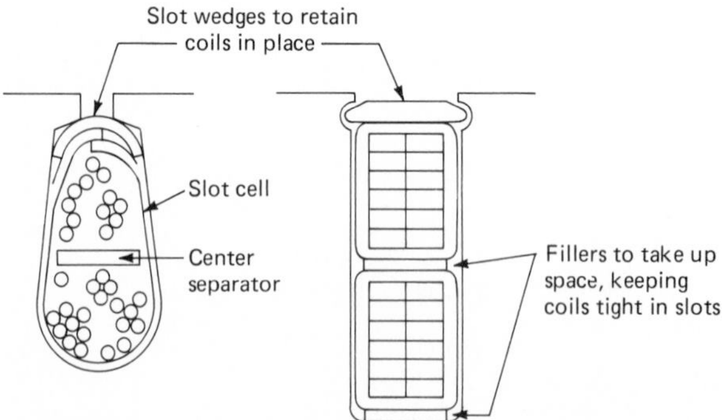

FIG. 4.28 Cross section of random-wound stator slot, left, shows how round wires are randomly packed into areas occupied by top and bottom coil sides, with insulating separator plus slot "cell" or liner. At right, similar sectional view shows rectangular wires of "formed" and taped coils used for higher voltages.

for the stator is the B-stage tape system used for many machines above 7 kV. A partially cured impregnant saturates the insulating tape with which coils are wrapped. Preheating the finished winding drives off moisture; then oven curing softens the resin in the tape enough to enable flow between layers for void-free bonding.

Either this treatment or the VPI system will withstand the IEEE Standard 429 test defining a sealed winding (one capable of withstanding high voltage under water).

Transient Surge

For any insulation system, one of today's most troublesome and elusive operating hazards is the transient surge, or "voltage spike." High-speed, automatic switching (especially involving capacitor banks) is the usual culprit. To use the hydraulic analogy, such a transient passing along the circuit toward a motor may be compared to a pressure wave along a water pipe. The wave moves rapidly until it encounters a sudden restriction. A destructive water hammer then exerts high pressure at that point.

Similarly, a voltage surge traveling along a circuit moves quickly until iron-embedded motor coils cause it to slow down abruptly. This "piles up" most of the surge magnitude across the first few turns in the winding. That can overstress the fairly thin insulation between those turns. In a 4000-V motor, an impulse peaking at 7000 V can damage turn insulation, even though twice that voltage between the winding and ground would do no harm.

An IEEE working group concluded in 1980 that motor windings have the surge-withstand capability expressed by Fig. 4.29. But even when it is able to

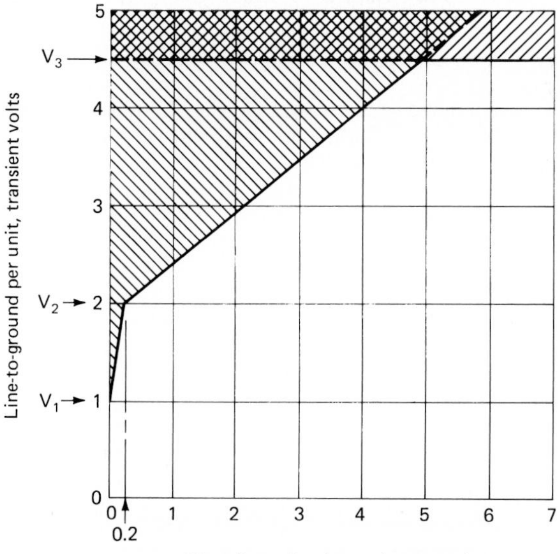

FIG. 4.29 Impulse or surge withstand capability of motor windings.

withstand an occasional transient peak, turn insulation becomes fatigued by long-term transient repetition (common in some applications). Unfortunately, surge failures are seldom identifiable as such. The initial turn-to-turn breakdown may escalate rapidly, with the resulting damage obscuring its true origin.

The curve in Fig. 4.29 can be interpreted with the following information.

$$V_L = \text{rated motor volts, kV}$$
$$V_1 = \sqrt{2/3}\ V_L = 1 \text{ per unit, kV}$$
$$V_2 = 2V_1, \text{ kV}$$
$$V_3 = 1.25\ (\sqrt{2}\ )\ (2V_L + 1), \text{ kV}$$

in which V_1, V_2, and V_3 are all *peak* values.

1. If the machine will see surges of either magnitude or rise time extending into the area shaded, wave-sloping capacitors should be supplied at the motor (usually 0.5 μF/day per phase).
2. If transient voltages in the area shaded will be present, surge arresters (metal oxide station class) should be supplied.

Some examples of the limits plotted (all values in kV) are:

Motor rating V_L	Peak V_1	Peak V_2	Peak V_3
2.3	1.88	3.76	9.9
4.0	3.27	6.54	15.9
6.6	5.40	10.80	25.0

Wave Front

The rise time of the traveling surge wave (see Fig. 4.30) is important. The shorter it is, the steeper the wave front—thus the more severe the "piling" effect on the

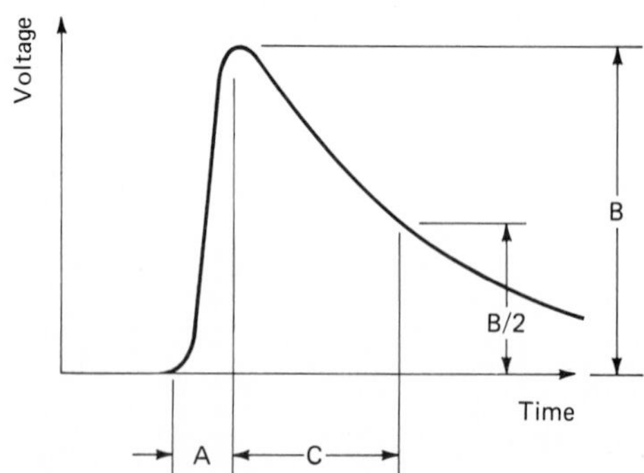

FIG. 4.30 Profile of transient voltage-surge wave: *A* is wavefront rise time; *B* is peak or crest magnitude; *C* is decay or fall time until voltage drops to half its original magnitude. Both *A* and *C* are usually measured in microseconds.

winding. If the wave front slopes gradually enough, the full surge magnitude will be spread out over many turns in several coils, reducing stress at each point to a safe value.

Surge protective devices at the motor itself increase that slope (the surge capacitor) or clip off the peak of the incoming transient (the surge arrester). Either or both may be needed, as Fig. 4.29 shows. If lead length between those devices and the motor winding exceeds 3 to 4 ft (1 m), protection is diminished because of the interaction between cable or motor inductance and the surge capacitor.

Although such protection is most often used for machines rated 2300 V or above, it is also available for 460-V machines. But the need is less frequent because of the greater surge-damping capability of low-voltage network systems.

SUGGESTIONS FOR FURTHER READING

Albright, H. E., "Application of Large High-Speed Synchronous Motors," *Trans. IEEE,* IA-16(1): 134–143, Jan./Feb. 1980.

Andreas, John C., *Energy-Efficient Electric Motors,* Dekker, New York, 1982.

Arnold, R. E., "NEMA Suggested Standards for Future Design of AC Integral Horsepower Motors," *Trans. IEEE,* IGA-6(2): 110–114, Mar./Apr. 1970.

Bartheld, R. G., "Motor Surface Temperatures in Hazardous Areas," *Trans. IEEE,* IA-14(3): 220–227, May/June 1978.

Bonnett, A. H., "Understanding Efficiency in Squirrel-Cage Induction Motors," *Trans. IEEE,* IA-16(4): 476–483, July/Aug. 1980.

Bowers, W. D., P. G. Cummings, and W. J. Martiny, "Induction Motor Efficiency Test Methods," *Trans. IEEE,* IA-17(3): 253–272, May/June 1981.

Douville, E. L., "Selection and Application of Variable Speed Motor Drive Systems," *Trans. IEEE,* IA-18(6): 698–702, Nov./Dec. 1982.

Gill, J. D., "Transfer of Motor Loads between Out-of-Phase Sources," *Trans. IEEE,* IA-15(4): 376–381, July/Aug. 1979.

Heidbreder, J. F., "Induction Motor Temperature Characteristics," *Trans. IEEE,* 77, pt 3, 800–804, 1958.

Jordan, H. E., *Energy Efficient Electric Motors and Their Application,* Van Nostrand Reinhold, New York, 1983.

Linders, J. R., "Effects of Power Supply Variations on AC Motor Characteristics," *Trans. IEEE,* IA-8(4): 383–400, July/Aug. 1972.

Merrill, E. F., "Should I Select a Service Factor Motor?" *Trans. IEEE,* IA-17(5): 458–462, Sept./Oct. 1981.

Owen, E. L., "Torsional Coordination of High Speed Synchronous Motors—Part I," *Trans. IEEE,* IA-17(6): 567–571, Nov./Dec. 1981.

Owen, E. L., H. D. Snively, and T. A. Lipo, "Torsional Coordination of High Speed Synchronous Motors—Part II," *Trans. IEEE,* IA-17(6): 572–580, Nov./Dec. 1981.

Schatz, M. W., "Overload Protection of Motors—Four Common Questions," *Trans. IEEE,* IGA-7(2): 196–207, Mar./Apr. 1971.

Weihsmann, P. R., and W. L. Subler, "Modern Lubrication Practices," *Trans. IEEE,* IA-16(4): 484–489, July/Aug. 1980.

Williams, A. J., and M. S. Griffith, "Evaluating the Effects of Motor Starting on Industrial and Commercial Power Systems," *Trans. IEEE,* IA-14(4): 292–305, July/Aug. 1976.

Woll, R. F., "Comparison of Application Capabilities of *U* and *T* Rated Motors," *Trans. IEEE,* IA-11(1): 34–37, Jan./Feb. 1975.

Working Group Report, "Impulse Voltage Strength of AC Rotating Machines," *Trans. IEEE,* PAS-100(8): 4041–4053, Aug. 1981.

CHAPTER 5.5
AC GENERATORS

Martin C. Hofheinz
H & G Engineering, Inc.
Stockton, CA

INTRODUCTION

Electric generators convert mechanical energy to electrical energy, which is more easily transmitted to remotely located points of application. The first large electric generating systems used *direct-current* (dc) generators, mainly because direct current was better understood than *alternating current* (ac). However, dc generators are limited to generating power at relatively low voltages, largely due to problems at their commutators.

As power networks developed, higher and higher voltages were required to transmit large blocks of power over longer and longer distances. Electric transformers can easily change the normally low voltage generated to the high voltages needed for efficient power transmission, and, of course, transformers only work on alternating current. In addition, ac generators, or *alternators* as they are commonly called, are so much simpler mechanically, so much more efficient, and require so much less maintenance than dc machines that all large generating plants output alternating current today. Although dc transmission lines can transport extremely large blocks of power very efficiently over long distances, the power is always generated as alternating current, transformed to the voltage required, rectified and transmitted as direct current, and then inverted back to alternating current at the point of application.

Mechanical Energy

The mechanical energy for driving the generator must be derived from a source with enough reliability and capacity to make it economically feasible to develop and transmit the energy electrically to the point of use. A small water supply running only during exceptionally wet years or located at a great distance from electrical consumers would probably not be suitable. Mechanical energy sources which cannot be moved, such as hydraulic turbines or even wind machines, must have the cost of transporting the energy produced (among other factors) taken into account when overall costs are calculated. Steam-turbine powerplants, however, can be located near a coal seam, lumber mill, or a reliable source of cooling water to save on transportation costs.

Some mechanical power may be obtained from sources more easily located near the point of utilization. Gas turbines and reciprocating gas or diesel engines fall into this category. Except for standby emergency power generators, even here it might be more economical to install large units and transmit the power to the point of use. Large powerplants will generally have better operating efficiencies than small ones, and it may be desirable to locate a large plant near the center of use and then distribute the power generated outward, assuming the fuel supply is transportable.

Each type of mechanical driver has its own peculiarities, and some have a sizable impact on the generator configuration. There are marked differences as to the engine output speeds available, the speed pulsations possible, the chances of overspeeds, etc. Normally, the generator shaft is horizontal and direct-connected to the driver (Fig. 5.1). Sometimes speed-changing gear boxes are installed between a high-speed turbine and a lower-speed generator. These allow the turbine to run at its most efficient speed, a speed that may be too high for the generator. Small hydraulic turbines usually have their shafts mounted horizontally; large hydraulic machines have their shafts direct-connected and vertically mounted. The

FIG. 5.1 Generator shaft direct-connected to driver.

generator may include special bearings to carry the thrust imposed by the water flowing through the turbine. Criteria like these for providing mechanical energy impose special designs on the generating machines.

Basic Machines

Two basic types of machines are used by electric utilities and industry for generating alternating current: the synchronous alternator and the induction generator. The synchronous alternator is much more commonly found in powerplant applications. It consists of a stationary stator with the output windings arranged around its periphery and a rotating, driven rotor with dc windings acting as an electromagnet (Fig. 5.2). The windings are fed from a dc source, either an external source or an exciter, or a source connected to the same rotor shaft.

The induction generator is much simpler and less expensive than the synchronous alternator, but it is really a special-use machine, finding application primarily at locations where electric power is to be delivered to an existing system. This speciality arises because the excitation for an induction generator is supplied from the energized power line to which it is connected. Without an energized line to connect to, the induction generator cannot function. The generator's stator is essentially the same as that of the synchronous alternator, but the rotor is the same as that of an induction motor (Fig. 5.3). It has solid conducting bars, usually copper or aluminum, embedded in the periphery of the rotor. The bars are connected at each end of the rotor.

The terms *ac generator, synchronous alternator,* and *synchronous machine* are all used to designate the synchronous ac generator. These machines are available in either single- or three-phase designs with almost any output voltage and frequency. Since single-phase machines are made only in very small sizes, how-

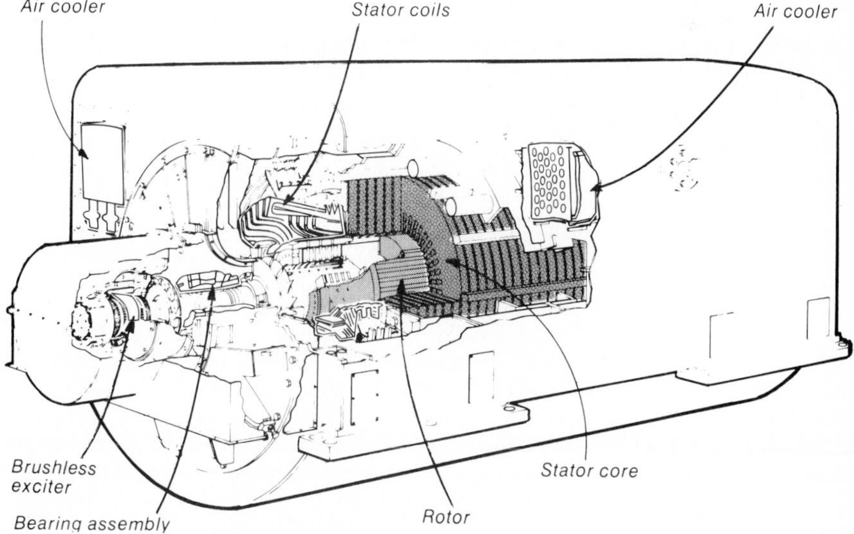

FIG. 5.2 Key elements of ac generator.

FIG. 5.3 Induction generator is simply an induction motor driven above synchronous speed.

ever, the following discussion is limited to three-phase machines, in the more common 480-V to 15-kV ratings.

SYNCHRONOUS GENERATORS

Synchronous generators are so named because an exact relationship exists among the number of electrical poles, the rotative speed, and the output frequency of the machine. The generated frequency is equal to the number of poles times the rotative speed (expressed in revolutions per minute) divided by 120. A two-pole machine rotating at 3600 r/min will have an output frequency of 60 Hz. Changing the speed of the driving engine or turbine will change the output frequency exactly as the speed is changed.

Depending on the rating, the operating speed, and the configuration of the mechanical driver, several basic constructions are available. However, all types of construction have a fixed mounted stator containing the output windings and a rotor supplying the excitation.

Stators

The rapidly varying magnetic flux in the stator iron causes hysteresis losses as the iron resists changes in the flux density. The varying magnetic flux also causes electric currents, called *eddy currents,* to flow in the iron laminations; losses also result from this current flow. The stator is built from thin laminations to minimize the electrical losses and of specially rolled silicon steel to minimize the hysteresis losses. For small machines, the laminations are circular, in the shape of the finished stator. For large machines, the laminations are punched as semicircles and then assembled into the finished circular stator (Fig. 5.4). Slots are punched for future installation of the windings. The winding slots are suitably insulated to provide both electrical insulation between the windings and the grounded stator and protection from abrasion damage to the windings by the stator iron. Windings are specified with the proper span, wire size, and amount of insulation required by the machine rating.

FIG. 5.4 Stator field winding.

For smaller machines, the windings are wound with loose coils of round wire, which are inserted into the slots provided in the stator, turn by turn, and fastened with slot wedges to prevent movement of the windings. To get as much conductor and stator iron as possible into the machine, large units are wound with square or rectangular wire, which is formed into rigid coils with insulation both between the individual wires and around the coils themselves. The coils are inserted into the stator slots, which have parallel walls to provide a snug fit between the coils and the stator iron; slot wedges hold the coils in place. Coil ends are connected into the proper groupings to provide the configuration of poles, voltage, and other parameters for which the machine is rated.

Rotors

Two basic types of rotors see service in synchronous alternators. High-speed machines (two- and four-pole) are built with round rotors; slots are cut into the rotor for the field windings. These alternators are referred to as *uniform-air-gap machines*.

Slower-speed machines have field poles that stick out from the rotor shaft, with the field winding wound around the projecting poles. The air gap obviously is not uniform. These alternators are called *salient-pole machines*.

Each pole on the alternator rotor has a winding through which direct current, usually at 63, 125, 250, or 375 V, is circulated to "excite" the field and create a magnetic field. The power required for field excitation is normally only a small percentage of the output, about 1 to 2 percent of the alternator rating. The dc excitation is obtained either from direct-connected machines driven by a prime mover or from separately mounted exciters that derive their power from other sources. The exciter output voltage level must be adjustable and have enough capacity to enable the alternator to produce rated voltage at rated output.

Exciters

Over the years, field excitation has been provided by three main exciter designs—rotating brush, rotating brushless, and static types.

Rotating Brush Type. Rotating, compound-wound dc designs were the only exciters used for many years. Exciters driven via a speed-increasing belt-and-pulley arrangement were sometimes specified so that less expensive, higher-speed exciters could be paired with slower-speed alternators. Direct current is delivered to the alternator rotor slip rings, which consist of two circular brass-alloy rings mounted on and insulated from the alternator shaft. Connections are made from the slip rings to the alternator field. Brushes riding on the slip rings are connected to the exciter.

The rotating brush exciter still sees service, but continual maintenance problems are associated with delivering large currents through slip rings, commutators, and brushes. These problems, together with the development of reliable, inexpensive semiconductors, make the brushless exciter the dominant choice today.

Rotating Brushless Type. The brushless exciter is simply a special type of alternator mounted on the same shaft as the main exciter (Fig. 5.5). It is special because its field, which must be excited with direct current, is stationary, and its ac output comes from the rotating parts. The output is rectified and connected to the main alternator's field by means of cables run along and fastened to the alternator shaft. Brushes, commutators, slip rings, and their maintenance are eliminated.

Static Type. As prices go down and the reliability and ratings of semiconductors go up, special cubicle-mounted controlled rectifiers, called *static exciters,* are becoming an increasingly popular choice. Their lower cost, reduced losses, reduced maintenance, and more flexible outputs also make them good choices for replacements of damaged rotating exciters.

A static exciter consists of an input transformer, *silicon controlled rectifiers* (SCRs), rectifier controls, and voltage regulator controls (Fig. 5.6). The complete assembly functions to rectify the incoming ac voltage into a properly controlled dc exciter voltage required by the alternator. Static exciter input may be connected to any convenient ac power source, such as station power (assuming it is available when the alternator is not running), but it is normally connected to the alternator output leads. Fuses and disconnect switches are installed between the alternator and exciter to protect against faults in the system.

Once an alternator field winding has had direct current passed through it, a small amount of residual magnetism remains. When the alternator is run again at rated speed, without excitation, an ac voltage of 2 to 10 percent of rated can be measured at the alternator's output terminals. This voltage is generated by the residual magnetic flux in the rotor acting on the stator windings. When it is connected to the alternator's output, the static exciter rectifies this residual ac voltage into direct current, which is applied to the alternator field windings. This ac-

FIG. 5.5 Brushless exciter (at top) is mounted on same shaft as main exciter.

tion further increases the excitation, which builds up until, in a very short time, the rated output voltage is obtained. Obviously, the correct connections must be made; if the output of the static exciter is in opposition to that of the residual voltage, no buildup will occur.

The exciter output is connected to the alternator field via the slip rings, which will require some brush and ring maintenance, but not as much as is required by the brush and commutator arrangement in a rotating exciter. Sometimes the residual magnetism is lost or it is desirable to reverse the direction of the residual magnetism. The field can be "flashed" by momentarily connecting a battery to the alternator field to establish some residual magnetism in the correct direction. On some static exciters this field flashing is done automatically every time the unit is started.

Static exciters also find application where the alternator must have special response characteristics, such as for starting abnormally large motors. The starting current of an induction motor is on the order of 6 times its normal full-load current. Starting a large motor (larger than one-half the generator load) causes the generator output voltage to drop, possibly enough to cause the motor starters to drop out. Reduced-voltage starters of several types are available to reduce the motor-starting current, but they are expensive and introduce time delays that may not be desirable. A static exciter can be provided with special "field forcing" equipment to give a quick increase in excitation in response to the demands of starting a large induction motor. Field forcing allows the generator to be smaller and less expensive than if standard equipment were used.

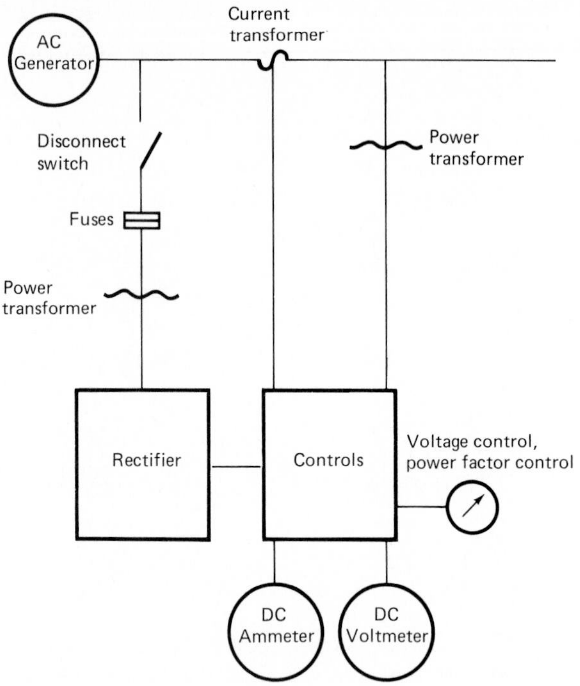

FIG. 5.6 Key elements of static exciter.

Motor-Generator Sets. Separately mounted dc generators driven by engines or ac motors are sometimes used as exciters. They are called *motor-generator* sets. The sets are occasionally specified as a replacement for a damaged direct-connected exciter. At one time, special types of motor-generator sets with voltage-regulating exciters were also used.

Voltage Control

The output voltage of an alternator running at rated speed, and not connected to other alternators, is a function of its excitation. Where the alternator is connected to other alternators, changing the excitation changes both the voltage level and the division of reactive power between the paralleled units. On a single unit, the level of field excitation can be changed manually with a rheostat, but this is practical only if the load and the voltage vary slowly. For alternators connected in parallel, it is virtually impossible to control the field excitation level manually to compensate for changes in reactive power flow.

Voltage regulators are used to regulate the excitation of an alternator—and thus the output voltage or reactive power output. The two basic voltage regulator types—the electromechanical and the static—work on completely different principles. Both types are connected, usually through potential transformers, to sense the alternator output voltage. The electromechanical regulator, by means of rocking arms, sliding contacts, and other linkages, changes the value of the resistance inserted into the exciter field in response to changes in the alternator output voltage. This varies the exciter field current, the exciter output voltage, the alternator field excitation, and ultimately the alternator output voltage. Because of maintenance problems with the mechanical parts, electromechanical voltage regulators are rarely used anymore.

Static voltage regulators predominate today; they are inexpensive, reliable, and almost maintenance-free. The devices work the same way that static exciters do, except on a smaller scale. They take power, via a step-down transformer, from the alternator output. The voltage regulator puts out rectified direct current, which is inversely proportional to the alternator output voltage. The direct current is fed to the exciter field. The exciter, which may be a rotating or static type, is the actual supplier of excitation to the alternator.

Voltage regulators usually sense only one phase, but for more accuracy, three-phase voltage sensing may be desirable. A voltage-adjusting rheostat, usually mounted on the front of the switchboard, changes the amount of the sensing voltage applied to the voltage regulator. On a static regulator, the rectified, adjusted, sensing voltage is compared to a constant-voltage diode. The regulator output voltage, which feeds the exciter, changes in response to differences between the two voltages. As is the case with static exciters, field flashing is often required with static voltage regulators.

Reactive Power Control

Any load fed by a generating plant has a characteristic power factor determined primarily by the electrical nature of the individual loads and the magnitude of power-factor correction introduced by devices added just for this purpose. Alternators must be operated to supply the load with the necessary kilowatts at whatever power factor the load has. When alternators are paralleled to supply the load, the real power and reactive power required may be divided between the paralleled alternators as desired (within their ratings, of course). Adjustment of real-power division is made with the governor controls on the prime mover.

Division of the reactive power is made by adjusting the excitation levels of the various alternators. The excitation level is changed by adjusting the voltage regulator output. Special voltage regulator circuits are needed, however, to ensure that the alternators continue to divide reactive power as desired. These circuits are called *parallel-compensation circuits*. They involve installing a current transformer in the phase opposite to the phases supplying the sensing voltage. The current transformer is connected across a resistor, which is connected in the voltage regulator sensing circuit. Since the voltage developed across the resistor is 90° out of phase with the sensing voltage, both alternators will tend to shed reactive power and thus will continue to divide as desired.

Parallel Operation

Except for an occasional noncritical standby machine, paralleling of generators is necessary to supply the loads. Since any machine is subject to unscheduled downtimes, it is normal to supply at least one-third spare capacity to allow for generators being out of service; that is, at least three machines would be provided to supply the *peak* load that could normally be carried by two machines.

To parallel two generators, the output voltages of each must have the same magnitude, frequency, phase rotation, and phase angle. The magnitude and frequency are obvious. The same phase rotation means that the voltages of the corresponding phases of each generator come to their peaks at the same time. The same phase angle means that the three phases come to their peaks at the same time.

Synchronizing means checking that the above four items are correct. Phase rotation is checked at the time of installation and, if correct, is not checked again at each synchronization, unless wiring changes have been made. Phase rotation meters and indicators may be used to check each generator, or a three-phase induction motor may be selected.

Manual Synchronization. A synchronizing panel will ordinarily contain two voltmeters, two frequency meters, and a synchroscope and/or synchronizing lights. One voltmeter and one frequency meter monitor the incoming machine, and the others monitor the running machine. The synchroscope pointer indicates the phase angle between the two generator voltages. The straight-up or 12 o'clock position indicates that the running and incoming machine voltages are in phase. The speed of the synchroscope pointer is proportional to the difference in frequency between the two alternators.

Synchronizing lights usually serve as backup to the synchroscope. They are connected across the circuit breaker contacts, and they go dark at synchronism. The connection between the two alternators is made when the synchroscope is rotating slowly (but not too slowly) in the clockwise direction and the pointer is at about 11:30 o'clock. When the pointer is rotating, it shows that the two frequencies are not exactly the same. Synchronization with the pointer rotating slowly clockwise will ensure that when the connection between the two units is made, a small amount of power will flow out from the newly connected unit and its reverse power relay will not trip erroneously.

Automatic Synchronization. Many types of automatic synchronizers have been designed to replace some of or all the manual synchronizing functions. The simplest are called *synch-check relays,* which simply check the two voltages and close a contact when the voltages are within certain limits for a certain length of time. On small machines, synch-check relays may serve to close the circuit breakers. On large machines, they are primarily backups in manual synchronizing schemes to block closing of the circuit breaker too far out of synchronism.

Highly accurate and reliable automatic synchronizing relays are available with adjustable ranges to monitor both synchronism and the voltage levels of the machines being synchronized. Even more elaborate synchronizing relays are available that automatically send out signals (1) to the prime mover for changes in the frequency and (2) to the voltage regulator for changes in voltage of the incoming unit, and then give a closing signal to the circuit breaker when all parameters are correct. Dead bus relays, sometimes included in the synchronizing relay itself and sometimes provided as a separate relay, allow connecting a machine to a dead bus, where the synchronizing relay itself would not give a close signal.

Generator Protection

Protective relays are included in all electric circuitry to protect equipment and maximize continuity of service. A vast array of protective devices is available to do this job. Fortunately, the *National Electrical Manufacturer's Association* (NEMA) has set up and published a system of function numbers, standards, and specifications for protective relays. NEMA standards are adhered to by virtually all manufacturers.

Many relays meeting NEMA standards are available in both standard speed and sensitivity and in the more expensive high-speed and/or special-sensitivity models. The latter are specified for the larger, more important, or more critical machines in the system. Most protective relays are now available in both electromechanical and static designs. The prices are about the same for the two types, but the static models are finding increasing acceptance, mostly because of their flexibility and the reduced maintenance associated with them.

Protective Relays. Protective relay schemes must provide coordination and backup between the various relays, as well as protection. Coordination means that the relays nearest a problem will operate first to remove the problem section without disturbing the rest of the system. Backup means that if the relays nearest the problem cannot or do not remove the problem section, then protective relays nearer to the source will operate to back up those relays that did not operate. When backup relays operate, it usually means that more of the system will be removed from service.

Protective relays may be connected to trip one or more circuit breakers, sound an alarm, or trip only if another event occurs. On 600-V and lower-rated circuit breakers, overcurrent devices normally are built into the circuit breaker, and they trip it directly. Other relays may be mounted separately and actuate the shunt trip on the circuit breaker. A battery, a capacitor trip, or a reliable source of alternating current is needed for this scheme. On circuit breakers rated at 2400 V and above, the relays are always separately mounted, are fed from current and/or potential transformers, and perform their trip or alarm function with the power obtained externally.

Here are some typical protective devices for the alternator itself. This treatment is by no means complete, nor are the schemes discussed the only ones used.

Overcurrent Protective Devices. NEMA device 50 represents instantaneous overcurrent protection, which can be provided either as a self-contained overcurrent trip in a 600-V circuit breaker or as a separately mounted relay fed from a current transformer mounted somewhere in the generator output circuit. Device 50 is set to trip instantaneously at a point above which all the other instantaneous relays on the feeders are set. Sometimes device 50 in a 600-V circuit breaker will be arranged with a very small time delay, to enable the circuit breakers further down the line to trip first. This is called a *selective trip*.

NEMA device 51 is a time-delay overcurrent relay. Again, it may be part of a 600-V circuit breaker overcurrent trip, a separate relay, or combined with NEMA device 50 in one relay. Device 51 comes in a multitude of different time-overcurrent curves, all designed for easier coordination or to protect a particular type of load or method of system operation. As a separate relay, NEMA device 51 may be provided with either voltage control or voltage restraint. The former means that the overcurrent contact can act only when the voltage on the circuit falls below a certain point. The latter means that the overcurrent-sensing element of the relay will trip at a point proportional to the voltage; with full voltage the relay will trip normally, and with no voltage it will trip at about 20 percent of normal current. These devices are both NEMA 51V. Usually, an instantaneous element is included in the same case, but it is not influenced by the voltage element.

As mentioned, the idea behind the 51V relay is coordination and backup. If a short circuit occurs at some distance from the alternator, the machine's current will rise, but the generator voltage, because of the voltage drops in the long feeder lines, will remain high. The voltage element keeps the 51V relay from tripping until a distant relay can act to clear the fault. When the fault is close to the alternator, possibly on the station bus, the voltage at the relay will drop, and the 51V relay acts to trip the alternator circuit breaker.

Undervoltage and Overvoltage Relays. Some alternators contain NEMA device 27 (undervoltage) and device 59 (overvoltage) relays. Undervoltage may be caused by a generator or voltage regulator failure or by a severe overload. Sometimes device 27 is included simply to ensure that the circuit breaker is tripped when the generator is shut down. If the voltage falls to zero in the voltage-sensing circuit of the regulator (perhaps caused by a blown fuse), the regulator acts as if the generator voltage were too low and tries to raise it, probably to the generator ceiling voltage. This can obviously damage loads and possibly the generator and its excitation equipment. Voltage regulators are available that will protect against this danger. An overvoltage relay NEMA device 59 can provide protection in many circumstances.

Differential Current Relays. Differential relays, NEMA device 87, act if there is an internal fault in the alternator. They operate by comparing the current going into a particular phase with that going out. If the currents are not identical, some immediate action must be taken. Internal leakage in the alternator probably means that the generator has suffered some winding damage already, and the problem now is to limit the damage. If the 87 relays are specified, it will usually limit the damage to that of the alternator windings and will protect against more serious and expensive damage to the stator laminations. The 87 relay not only should trip the output circuit breaker, but also should remove the field excitation. To do this, the 87 relays are connected to trip NEMA device 86, the lock-out relay.

Device 86 relays are multicontact, electrically tripped, hand-reset switches that trip the alternator circuit breaker and remove alternator field excitation. They must be reset by hand because the alternator damage indicated by device 87 relays must be investigated before the machine is restarted.

Directional Power Relays. When alternators are connected in parallel, there is always the possibility that a prime mover may, for some reason, lose power. The alternator connected to that prime mover will then run as a motor and drive the prime mover. Not only is this situation a waste of electrical energy, but also it can damage the prime mover. NEMA 32 directional power relays are connected in the generator output circuit to detect power flowing into the motoring machine and to trip the alternator circuit breaker.

NEMA type 81 over- and underfrequency relays may be used to detect, indirectly, over- and underspeed of the prime mover and take appropriate action. NEMA type 40 relays can protect the alternator field against open circuits, over- or underexcitation, etc. Many other types of protective relays are available, but mostly they are special-purpose devices, selected for special applications.

INDUCTION GENERATORS

As mentioned earlier, induction generators are special-use machines, and they are applied only where they may be connected to an existing power supply

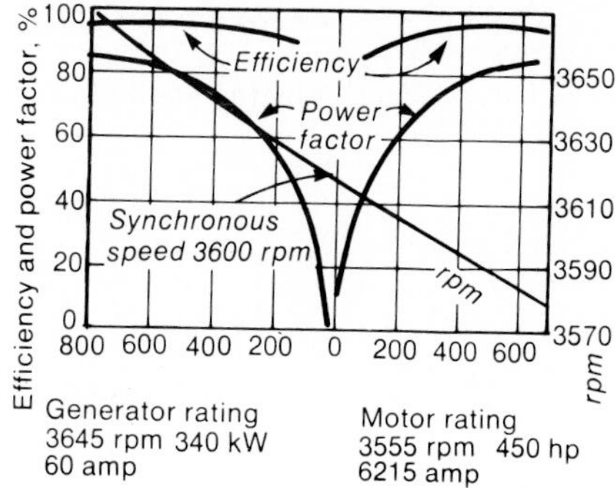

FIG. 5.7 Characteristics of induction generator.

which, by definition, has to be generated by synchronous alternators. The big advantages of induction generators are that compared to synchronous alternators they are inexpensive; do not require dc exciters, voltage regulators, and synchronizing equipment; and do not need to be synchronized (also see Fig. 5.7).

Line Connection

The normal way to connect an induction generator to the line is to bring it up to slightly above synchronous speed. A circuit breaker or contactor is closed, connecting the machine to the energized power line. If the power line is dead, nothing happens, of course, since the machine has not received excitation. No synchronizing procedure is required.

If the power line is energized, it provides the machine with the excitation required. The excitation is in the form of lagging-power-factor magnetizing current. The reactive power that this represents must be supplied either by the line or, more commonly, by capacitors connected across the induction generator terminals.

Since the excitation power is not controllable, neither the machine output voltage nor the output power factor is controllable. Only by raising or lowering the speed of the prime mover can the output power of the machine be controlled.

By their nature, induction generators are usually small. They generally see service in schemes where there are no station service requirements and where the necessary energized power line for exciting the generator is also present to carry away the generated power.

SUGGESTIONS FOR FURTHER READING

Beeman, D., ed., *Industrial Power Systems Handbook,* McGraw-Hill Book Co., New York, 1955.

Electric Transmission and Distribution Handbook, Westinghouse Electric Corp., Pittsburgh, 1964.

Hindmarsh, J., *Electrical Machines and Their Applications,* Pergamon Press Inc., New York, 1977.

Richardson, D.V., ed., *Handbook of Rotating Electric Machinery,* Reston Publishing Co., Reston, Va., 1978.

Say, M.G., *Alternating Current Machines,* Hulsted Press Ltd., Edinburgh, Scotland, 1983.

SECTION 6

INSTRUMENTATION AND CONTROL

CHAPTER 6.1
KEY MEASUREMENTS IN POWERPLANTS

J. A. Moore
Leeds & Northrup, A Unit of General Signal Corp.
North Wales, PA

INTRODUCTION

Operators of plants producing steam and electric power must know the current (real-time) value of process variables. The operators depend on measurements and indications of the variables to provide (1) information on which they can base decisions for making adjustments manually, (2) the real-time signals needed for closed-loop control, (3) information that is printed and presented in formatted form for management, and (4) process values that can be archived and displayed to indicate trends.

All instrumentation and controls measure and regulate either the flow of energy or the flow of material. Sometimes it is difficult to separate these two flows in describing the function of an instrument. Flow measurement of fuel provides figures that represent material flow; but because each unit of fuel has a heating value, the output of the transmitter whose signal is created by the flow measurement can also be calibrated in units of energy flow.

Two broad divisions of measurement and control exist in a power-producing installation. The first is associated with the production of steam; the second is concerned with the prevention of atmospheric and water pollution. As a broad generalization, it can be said that energy is measured and controlled in the steam power application, while material is measured and controlled in the pollution control application.

The variables that are of major importance for boiler operation include temperature, pressure, level, and flow. The ones that are of major importance for pollution monitoring and control include pH, conductivity, and gas analysis. These variables must be monitored constantly. Variables of secondary importance—demanding constant monitoring but not required for control—include turbidity, opacity, vibration, and flame detection.

All these variables are the subject of the first two chapters of Section 6.

TEMPERATURE

Temperature is probably the most important process variable affecting powerplant operation. It provides a measure of the operating conditions in every phase of power production. Chapter 1 in Section 1 defines temperature and the physics of heat and energy on which it is based.

Temperature Measurement: Reasons and Applications

The fact that temperature is a measure of the amount of energy contained in an amount of material gives rise to three principal reasons for temperature measurement in powerplant operations. Temperature-measuring applications in the plant stem from these reasons.

Three Reasons

- In conjunction with weight-rate measurements, temperature measurement allows calculation of heat balances. The same amount of energy that enters a system will leave the system, by the first law of thermodynamics, which states that energy can be neither created nor destroyed. Thus, by taking measurements of flow rates and temperatures, multiplying these rates by changes in temperature of input and output, and subtracting measurable values from known theoretical values, amounts of energy that cannot be measured can be calculated. Calculations of this sort help to determine performance and efficiency.
- A measurement of temperature infers the state of a process. Often the real variable of interest—percentage of reaction completion, for example, or the composition of a product of reaction—cannot be measured. The values of the desired variable can be inferred from temperature measurements, however, and so these are used as reference values for operation.
- Temperature measurement provides a means for sounding an alarm when op-

erating conditions are out of safe limits or when operating equipment fails. If process energy increases to a dangerous level, the temperature will also increase because of the increased energy. Comparison of a measurement of temperature to an acceptable limit provides a means of activating an alarm to notify the operator of a dangerous situation.

Applications. These temperature measurement applications cover the basic powerplant systems.

Fuel System. Temperature measurement in powerplants starts with the fuel, regardless of the type burned:

- Whether natural gas, oil, coal, lignite, or bark, the fuel temperature is an entry in overall heat-balance calculations. In the case of coal or lignite, the temperature of the coal-air mixture is the parameter.
- Temperature of high-viscosity fuel (Bunker C, no. 6 heating oil, or tar) is measured as a controlled variable. The fuel has to be heated to a controlled temperature so that it can be pumped.
- The temperature of piles of stored coal must be measured to detect spontaneous combustion.

For coal or lignite, temperatures are measured in mills and pulverizers:

1. Bearing temperatures of the grinding machinery are checked to detect bearing failure.
2. The primary-air temperature is measured, since it is the transport medium for pulverized coal. The coal-air mixture is controlled by adding tempering air to maintain a temperature slightly below the ignition temperature of coal.

Feedwater System. In the feedwater system, the temperatures of feedwater, condensate, and steam are measured for inputs to heat-balance calculations. At the boiler feed pumps, bearing temperatures are checked as well as feedwater inlet and outlet temperatures. Similar measurements are made at the condensate pumps.

Inlet and outlet temperatures are measured at the feedwater and condensate heaters, again for performance and efficiency calculations and also as a guide for preventive maintenance; that is, the cleaning of heater tubes is scheduled as a function of the temperature difference across the tubes.

Boiler Internals. Flame temperatures have been measured as a guide to combustion efficiency; the measured values determine and control the fuel-air ratio. In the boiler itself, boiler-tube temperatures determine the need for sootblowing and aid in leak detection. Waterwall temperatures are measured as a check for leakage.

Steam temperatures measured include those in the superheater, drum, and headers. Turbine-extraction and blowdown temperatures are also measured.

Air System. Air-temperature measurements are needed for heater control and to prevent condensation of acid-laden moisture in ducts, on stack surfaces, and in heaters.

The ambient-air temperature and the air temperatures at the inlet and outlet of the preheater and economizer are measured. The stack temperature is monitored.

Temperature-Measuring Devices

The important temperature-measuring devices covered here are thermometers, thermocouples, and resistance temperature detectors.

Thermometers. The simplest form of temperature measurement occurs when temperature produces a physical change in a material. If an element made of the material is inserted in a medium, the degree of change can be calibrated in such a way that the temperature of the medium is indicated. Such a device is called a *thermometer.*

The material physically affected by temperature most commonly selected for thermometers has been mercury, enclosed in a glass or metal tube so that its volume could be contained in a constant shape. When the tube is inserted into the medium to be measured, the temperature will expand or contract the volume of mercury and the top of the liquid will move to a position in a glass tube attached to the enclosure. A scale etched on the tube provides the temperature indication. Thermometers are always limited to local measurements.

Glass tubes have a regrettable fragility, so this type of thermometer has been generally replaced by the bimetallic variety. The expandable element for this device is a strip of composite metallic material made of two dissimilar metals fused to form a laminated strip (Fig. 1.1). Temperature causes expansion or contraction of the two metals, but to a different degree in each. The result of a temperature change is that the strip bends. An amplification of the bending effect is achieved by forming the strip into a helix, so that the tip of the strip rotates as the temperature changes. The helix is coupled to a shaft, usually with a pointer on the end, and as the shaft rotates, the pointer moves in an arc over a scale calibrated in degrees of temperature.

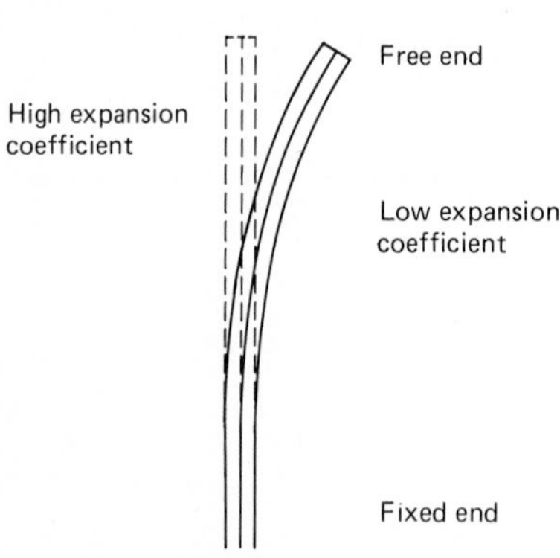

FIG. 1.1 Bimetallic element bends on temperature increase. *(Courtesy J. A. Moore.)*

Physical Characteristics

- *Range:* −200 to +1000°F (−130 to +540°C); 800°F (430°C) is the maximum temperature recommended for continuous operation.
- *Sensitivity:* Maximum sensitivity of 3 deg/°F (5.4 deg/°C) may be expected. Dial sizes are usually 3 or 5 in (76 or 127 mm). The greater the diameter, the longer the scale and the better the readability.
- *Accuracy:* Properly installed, an industrial-grade bimetallic thermometer will indicate to ±1 percent of the instrument span.
- *Materials of construction:* The helix enclosure, which is inserted into the process fluid, is generally 18-8 stainless steel, and it will withstand pressures up to 1000 psi (689 kPa). Special materials or coatings may be needed for the measurement of corrosive fluids; these increase the expense as well as the response time of the thermometer.

Limitations and Use. The main limitations stem from abnormal conditions. Shock or vibration can mechanically deform the helix. Prolonged exposure to temperatures at the upper end of the operating range can change the coefficients of expansion of the bonded metals, resulting in permanent shifts in calibration.

Thermometers are used for local measurements. They are inserted into pipelines through a threaded coupling, with the dial extending beyond insulation surrounding the pipe. If the dial is fixed, installation must be oriented so that the figures on the dial face can easily be read. On some models the dial head can swivel, so it can be turned to a plane at right angles to the viewer's line of sight.

Thermocouples. Using electrical transmission instead of mechanical transmission allows the readout device to be located remote from the sensor. The sensor most commonly found in powerplants for producing an electric signal as a function of temperature change is the thermocouple.

Theory. A thermocouple consists of two wires, made of dissimilar metals, with the wire ends joined to make a loop. When the junctions of the metals are at different temperatures, current flows in the wire. If one junction is kept at a constant temperature (the readout or reference end), the amount of current flow will vary as the temperature at the other junction (the measuring end) is varied. The *electromotive force* (emf) developed between the hot and cold ends of the wires can be read and calibrated as a function of temperature (Fig. 1.2).

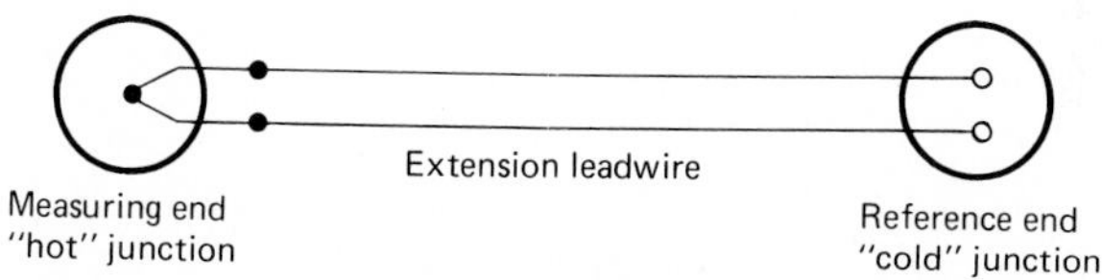

FIG. 1.2 Thermocouple circuit, with extended reference junction. *(Courtesy J. A. Moore.)*

Since a thermocouple acts as a small battery, two or more devices can be connected in series to develop an emf that is the sum of the individual emfs. When they are connected positive to negative, the emfs add; when connected positive to positive, they subtract (Fig. 1.3). This is one way of measuring temperature difference. The preponderant use of thermocouples, however, is for simple temperature measurement.

Ideally, the cold end of the loop is a constant-temperature ice-salt bath. This is fine for laboratory work, but impractical for an instrument in a powerplant. The practice for plant instrumentation is to include cold-junction compensation at the terminal board where the thermocouple wiring terminates. Compensation in-

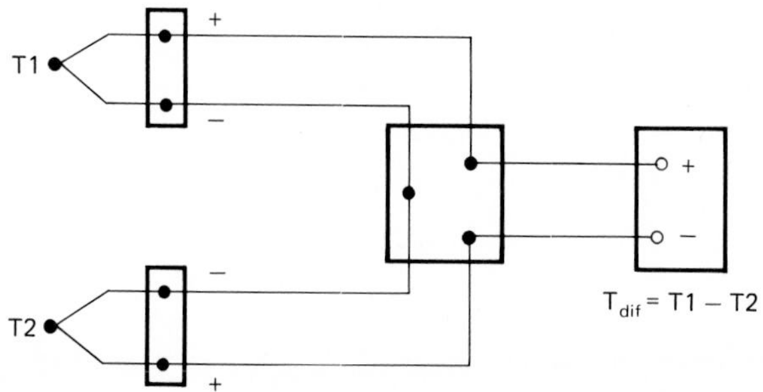

FIG. 1.3 Thermocouple connected in series opposition to produce temperature-difference measurement. *(Courtesy J. A. Moore.)*

cludes a circuit that introduces a small emf as temperature changes, so that the result is a constant temperature at the junction.

Materials. Thermocouple materials must be resistant to high temperatures, must have linear temperature-emf relationships, and must develop an amount of emf change per degree of temperature change to allow an accurate and reproducible measurement. The pairs of wires satisfying these requirements are listed in Table 1.1. They may be described further as follows:

- *Copper-constantan:* Constantan is an alloy of roughly 55 percent copper and 45 percent nickel. The combination oxidizes above 750°F (460°C) and is attacked by acid fumes.
- *Iron-constantan:* This material is subject to attack by sulfur, oxygen, and moisture.
- *Chromel-alumel:* Chromel is an alloy of 90 percent nickel and 10 percent chromium. Alumel is approximately 94 percent nickel, 3 percent manganese, 2 percent aluminum, and 1 percent silicon. The combination is resistant to oxidation, but must be protected from reducing atmospheres. It is affected by sulfur.
- *Platinum-platinum, rhodium:* There is no physical difference between types S and R. The principal limitation to the use of either is the high cost.

Many other combinations exist, but these are the thermocouple materials most frequently selected for powerplant applications.

Construction. Powerplant thermocouples are commonly of two types: the bare wire and the pencil. The bare-wire thermocouple is made of heavy-gage wire (no. 8 is a standard size), with the junction formed by twisting the two ends together and welding them. The wires will continue, twisted about each other, and terminate in a protective enclosure called a *head,* which contains terminals to which the wires are connected.

The pencil-type thermocouple uses small-gage wire inside a thin-walled stainless-steel sheath, ¼ in (4.2 mm) in diameter, which is packed with magnesium oxide. The junction is formed at the end of the wires and may (or may not) be welded to the sheath. The upper end of the sheath is fastened to a protecting head with terminals, as Fig. 1.2 shows.

The bare-wire thermocouple may have a range of lengths from a few inches (millimeters) to as much as 100 ft (30.5 m). The sheath diameter of the pencil-type thermocouple will support lengths up to about 2 ft (0.61 m), if the device is installed vertically.

TABLE 1.1 Thermocouple Materials

Material	Type	Temperature range, °F (°C)	Approx. millivolt change over 1000°F (540°C)
Copper-constantan	T	−300 to 700 (−185 to 370)	27.3
Iron-constantan	J	0 to 1400 (−18 to 760)	29.5
Chromel-alumel	K	0 to 2300 (−18 to 1260)	22.3
Platinum-platinum, 10% rhodium	S	0 to 2700 (−18 to 1480)	4.6
Platinum-platinum, 13% rhodium	R	0 to 2700 (−18 to 1480)	4.9

Extension Lead Wire. Ideally, thermocouple material should be connected to the instrument used for measuring the developed emf. To reduce the cost of long wiring runs, the actual thermocouple is made long enough to extend from the point of measurement to a junction at ambient temperature. Smaller-gage pairs of the same heavy-gage measuring wire complete the circuit.

As long as the junction points of the two dissimilar metals are at the same temperature, the temperature-emf relationship is not changed. Extension lead wires of platinum-platinum, rhodium are very expensive if the distance between the instrument and the point of measurement is considerable. Thus, alloys of copper and nickel for the negative element, and a copper wire for the positive element, are selected. These match the platinum-platinum, rhodium emfs closely in the measuring range.

Use of extension lead wires introduces some nonlinearity into the temperature-emf relationship, and this must be compensated for if linear scales are desired. Where temperatures are recorded, scales and chart paper can be made with nonlinear spacing. Digital instruments can provide linearization in the software of the input signal-conditioning circuitry.

Accessories. Two accessories used with thermocouples are protecting wells and transducers. Since the thermocouple junction must be located as close to the point of measurement as possible, it must be inserted into the process stream. To measure boiler tube temperature, the junction may be physically welded onto a pad that is, in turn, welded onto the tube. To measure the gas temperature in a furnace (or flame temperature), a bare-wire thermocouple can be positioned so that the junction is in the gas stream.

If the fluid to be measured by a bare-wire thermocouple is in a pipeline, however, a protecting well of stainless steel or ceramic must penetrate a threaded connection in the pipe wall, and the thermocouple must be inserted into the well (Fig. 1.4). The well maintains the integrity of the pipe wall, allows the thermocouple to be removed and replaced without having to shut off the flow of material passing through the pipeline, and protects the thermocouple from corrosive fluids or from oxidizing or reducing action of hot gases. Pencil-type thermocouples have the protection of the sheath, but even for these a separate well is often advisable.

Transducers are the second accessory used. Thermocouples develop millivoltages. To record temperatures, it is common to run the wiring all the way to the recorder terminal board on which the cold-junction compensation is mounted. Recorders are commonly designed to accept millivolt inputs. Present-day analog and digital instruments for control and special-purpose computation, however, are usually designed to convert voltages in a range of 1 to 5 V to their binary equivalents. Some digital devices will accept low-level (millivolt) inputs directly, but many do not. To make the conversion, a separate device must be introduced that will change the millivoltage of the thermocouple to a current signal, usually 4 to 20 mA (Fig. 1.5). This, in turn, is converted to 1 to 5 V by passing the signal through a 250-Ω resistor mounted on the input terminals of the digital device.

Some thermocouple protecting heads have converters built into them. More commonly, converters are housed in individual boxes or on printed-circuit cards plugged into relay rack mounts. The 4- to 20-mA output can be transmitted over shielded, twisted copper wire, thus reducing the cost by eliminating the need for long runs of expensive extension lead wire. The additional device required for the conversion is also an expense, however, as is the requirement to mount it.

Use. Thermocouples are selected for all the powerplant applications noted earlier, with the exception of bearing-temperature measurement that commonly

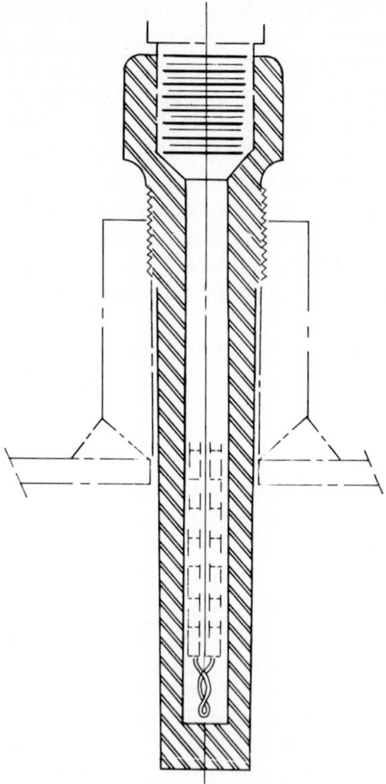

FIG. 1.4 Thermocouple well. *(Courtesy J. A. Moore.)*

uses resistance temperature detection. Thermocouples are always used for hot-gas temperatures and for high-pressure steam temperatures, and they may see service for heater and condenser inputs and outputs.

Resistance Temperature Detectors. Measurement of resistance as a function of temperature is possible because most metals, and some semiconductors, change in resistivity with temperature in a known, reproducible manner. As a metal becomes hotter, the molecules of which it is composed vibrate more rapidly, and

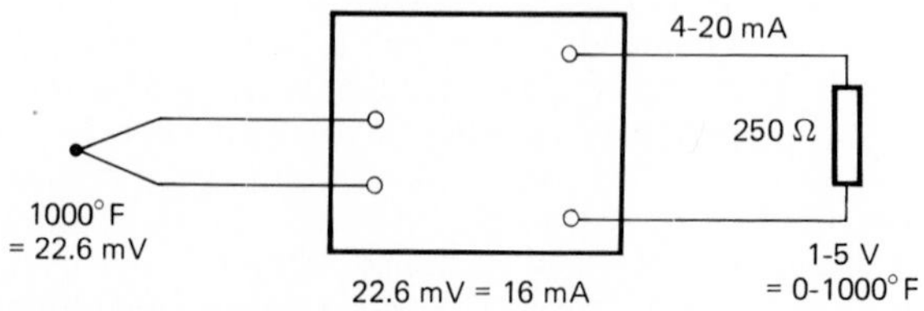

FIG. 1.5 Analog signal conversion. Note: (°F − 32)/1.8 =°C. *(Courtesy J. A. Moore.)*

the resistance to the passage of current increases. The relationship between temperature and resistance for a particular material is called the *temperature coefficient.*

The sensing element of a *resistance temperature detector* (RTD) is made from a length of wire that has a specific resistance at 0°C. The wire is coiled in a tube, usually in a double helix, or wound on a mandrel, stress-relieved (because strains in the wire cause resistance variations), and enclosed in a capsule that is filled with an insulating medium. The capsule is attached to a head that contains terminals to which transmission wiring can be attached (Fig. 1.6). The wires carry a current from the receiving instrument, and the passage of the current through the sensor resistance develops a millivoltage. The calibrating relationship, therefore, is temperature versus millivolts.

Three leads are attached, two to one end of the winding and one to the other end. This provides compensation for ambient-temperature changes and removes the effect of interconnecting-wire resistance from the measurement (Fig. 1.7).

Wire Sensing Element. The wire used for the sensing element of an RTD must have the following properties:

- High temperature coefficient
- High resistivity
- High purity when made; must remain pure during use

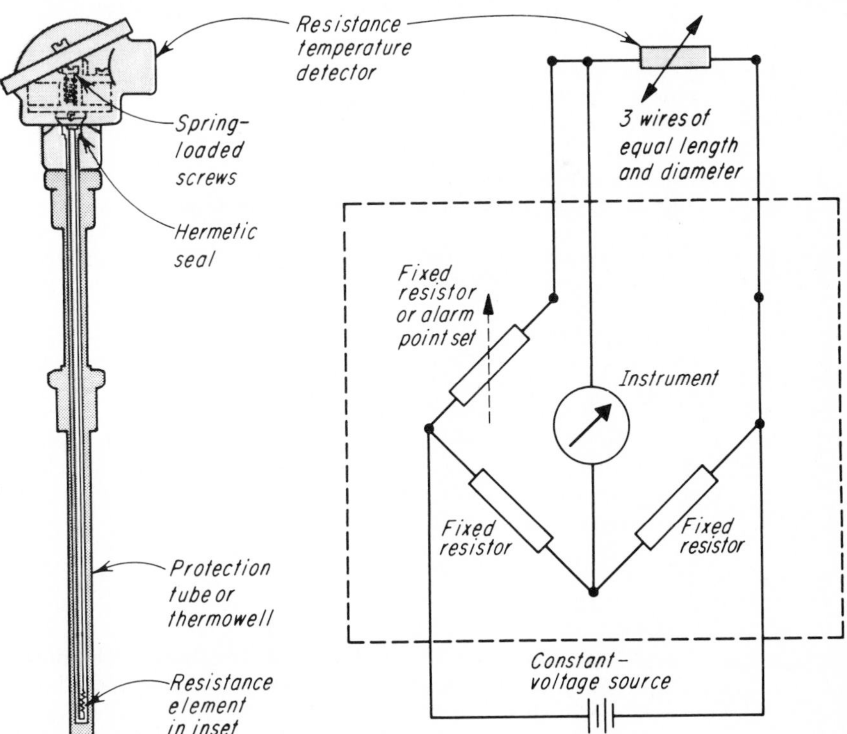

FIG. 1.6 Resistance temperature detector.

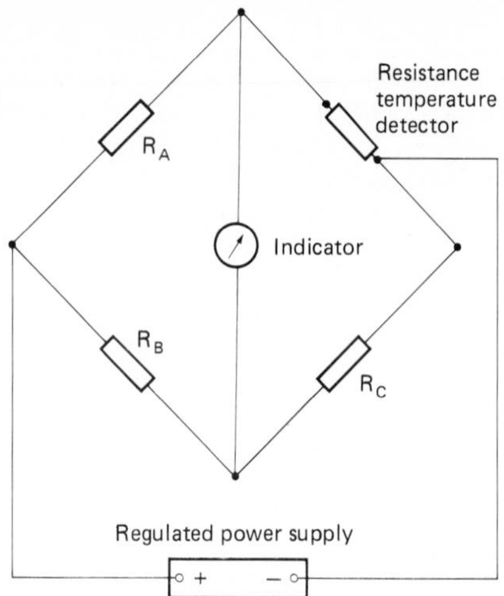

FIG. 1.7 Simplified bridge measuring circuit for temperature measurement using resistance. *(Courtesy J. A. Moore.)*

- Ductile enough to be drawn into a fine wire, but strong enough to permit forming and winding
- Very high melting point

Specifications. Copper, nickel, and platinum are the three metals most commonly selected for RTDs, with resistances of 10, 25, and 100 Ω.

Copper RTDs can be used over the range of −320 to 250°F (−196 to 121°C). The 10-Ω element has a sensitivity of 0.022 Ω/°F (0.0396 Ω/°C); the 100-Ω element, a sensitivity of 0.215 Ω/°F (0.387 Ω/°C). Accuracy is ±0.14°F (±0.25°C).

Nickel RTDs have a range of −100 to 300°F (−73 to 149°C). A resistance of 100 Ω of nickel produces a sensitivity of approximately 0.2 Ω/°F (0.36 Ω/°C). Accuracy is ±0.14°F (±0.25°C).

Platinum RTDs have two ranges: one covers temperatures from −297.3 to 32°F (−182.9 to 0°C), the other from 32 to 1112°F (0 to 600°C). A resistance of 100 Ω of platinum has a sensitivity of about 0.4 Ω/°F (0.72 Ω/°C). Precision platinum RTDs can be certified to ±0.006°F (±0.01°C).

Use. RTDs measure bearing temperatures and can measure liquid temperatures into and out of heaters, as well as fuel temperatures. They are always selected for relative-humidity measurements, one RTD sensing the wet-bulb, another the dry-bulb temperatures of an air sample.

The wet-bulb temperature is obtained by surrounding the element of an RTD with a wick that is kept wet by a water bath. As the air blows by the wick, the water evaporates to a degree dependent on the amount of water vapor in the air. Heat is required to evaporate water, and the only source is the air itself, so the air cools to the wet-bulb temperature. The combination of wet- and dry-bulb temperatures, read with a psychrometric chart, permits determination of the percentage of water vapor in the air relative to the amount that the air could hold if it were saturated with water vapor. This defines the relative humidity of the air.

FLOW

While temperature measurement is an obvious requirement where energy is being controlled, flow measurement is similarly associated with material control. Solids, liquids, and gases moving from one point to another are easily identified as material. The measurements of their flow rates can be used for material-balance calculations, for inventory control and accounting, and for determining production values. When the materials undergo changes in state or when the flows result in a release of energy, both material and energy control become a function of flow measurement. This is the case in powerplant flow measurement (also see box).

Mass Flow Mainly for Measuring Air-Gas Flow

Flow of material can be expressed as weight per unit of time, or as volume per unit of time. Usually weight is the important measurement, because the coefficients of energy are defined as Btu per pound (kilojoules per kilogram). Most flow measurements, however, are developed as a measure of volume, and so a conversion must be made if weight is really the variable required. (In the metric system, only mass is the considered variable, not weight; see schematic.)

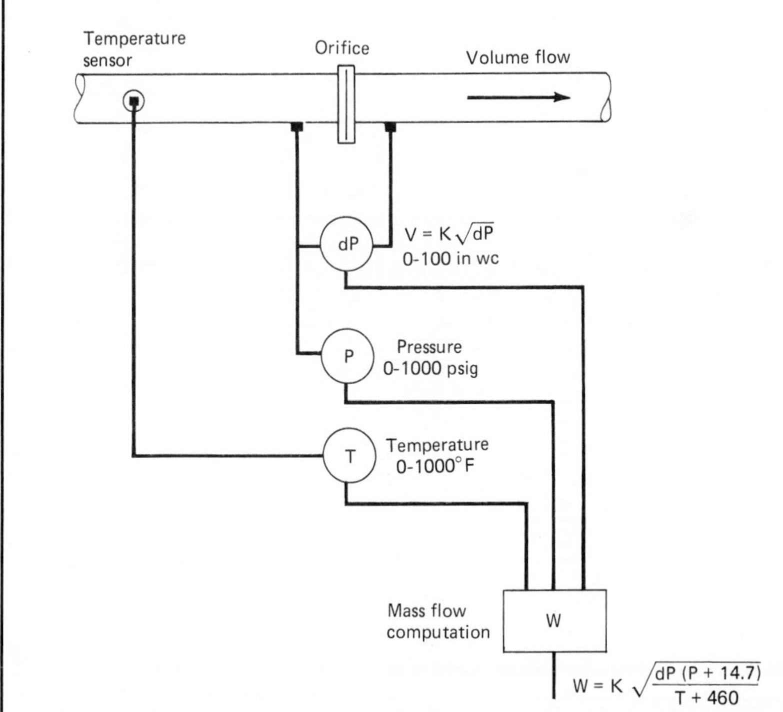

(Continued)

The conversion calculation for a gas, assuming that the specific gravity remains constant, is a function of temperature and pressure. Gas volume changes with temperature and pressure changes. It is a direct function of absolute temperature change and a reverse function of absolute pressure change. Imagine a rubber balloon filled with gas of such a volume that when the balloon rested on a beach at sea level, at a temperature of 60°F (15.6°C), the volume of gas was exactly 1 ft^3 (1 m^3). If the gas were dry air, the cubic foot (cubic meter) would weigh 0.0764 lb (0.0346 kg). If the temperature of the air became hotter, the volume of air in the balloon would increase to greater than 1 ft^3 (1 m^3), but the weight would stay the same; if the temperature of the air became colder, the volume of air would decrease, but the weight would still remain the same.

Pressure will also affect the volume of air in the balloon, but the change will be opposite to that caused by temperature. If the balloon were put into a chamber and the pressure increased, the volume of the balloon would become smaller; but if the chamber were evacuated, the air volume would swell and increase the size of the balloon. In every case, however, the weight of the air would still be 0.0764 lb (0.0346 kg). The volume of air for each of these combinations of temperature and pressure is called the *actual volume.*

The volume at sea level (1 atm of pressure) and 60°F (15.6°C) is a special combination, because all the other volumes are referenced to it. One cubic foot of gas at these conditions is called a *standard cubic foot* (scf). A volume of any gas has a weight that is specific to that gas. The volume-of-gas change as pressure changes is inversely proportional to the absolute pressure, while the volume change as temperature changes is directly proportional to absolute temperature.

Therefore, it is possible to convert any actual volume, if the temperature and pressure are known, to the equivalent volume in standard cubic feet. Since the weight of a standard cubic foot is known, the weight can be determined. This is an important conversion in powerplant measurements of preheated air, because the ambient temperature of the air will change from day to night and the temperature at the outlet of the preheater may change. The amount of moisture in the air is usually not considered.

Mass flow is used less frequently when liquid flow is measured, but it is still a factor to be considered. Liquids do not change significantly in volume as pressure changes, but they do have different densities, and density may change as temperature changes. If it is important to know weight flow, and the transmitter making the flow measurement is calibrated for volume flow, then the density and temperature of the flowing fluid must also be known. Relationships are not directly proportional for liquids as they are for gases, so the conversion must be made from information in a table of values.

Areas of Measurement

The three principal areas of flow measurement in steam power generation are the feedwater, fuel and air, and steam systems; consider all three to be measurements of energy as well as of material.

Feedwater System. The flow measurements associated with feedwater flow include

- Feedwater flow from each boiler-feed pump
- Condensate flow from each condensate pump
- Feedwater heater flows
- Total feedwater flow

- Total condensate flow
- Blowdown flow

All these flows use gallons per minute (liters per second) as the unit of measure and therefore represent material flows. Feedwater heater flow, in combination with the temperature differential of the entering and leaving water, is a measure of the Btus (kilojoules) added to the feedwater and therefore becomes a measure of energy flow. Condensate, since it represents recovered energy, is an important contributor to the energy balance around a boiler; the condensate flow rate in conjunction with condensing temperature and flowing temperature becomes a measure of energy flow. Similarly, blowdown flow, which represents lost energy, is as important to a heat-balance computation as it is to a material balance.

Fuel and Air Systems. Flow of oil for combustion fuel is measured, as is mass flow of air, natural gas, and other gaseous fuels. There is no dependable flow measurement for solid fuels such as coal, bark, or lignite. The measurement is usually made inferentially, by tachometer or feeder speed, or by a combination of primary airflow and damper position. Airflow is measured at individual fans and at selected locations in air ducts.

Fuel measurement is important for the information it gives about material flow as well as energy flow. Natural-gas billings are based on gas flow measurement, oil consumption is checked by integration of measured flow rates, and the ratio of steam produced to fuel consumed is an operator's measure of boiler efficiency. Every type of fuel, however, has a heating value, and so the material measurement is easily converted to one of energy (Fig. 1.8). The relation of air to fuel for combustion is expressed as a ratio of materials, and material flow units are important in order to display this ratio. The reason for using air, however, is to provide oxygen for combustion of the fuel, and the product of combustion is the energy needed to boil feedwater to convert it to steam. Air temperature is also measured so that the heat content of the air can be computed by combining it with a mass flow measurement of airflow.

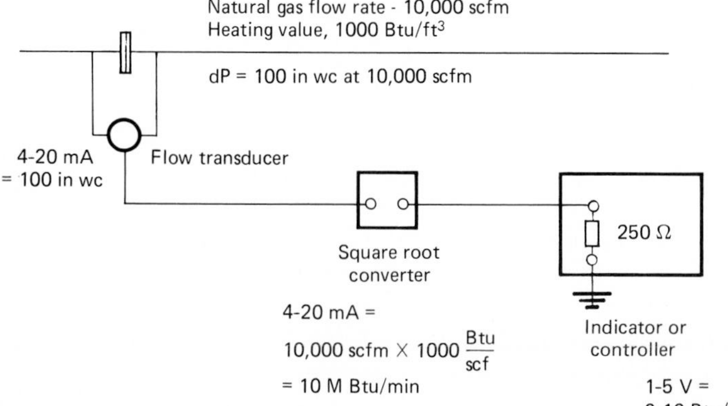

FIG. 1.8 Flow-of-energy computation from measurement of flow of material. Note: in × 25.4 = mm; Btu × 1.055 = kJ; scfm × 0.472 = L/s; Btu/ft^3 × 37.3 = kJ/m^3. *(Courtesy J. A. Moore.)*

Steam System. Steam flow measurement is also used to compute energy, in conjunction with steam pressure measurement, for use in heat-balance computations. Process-steam flows may be measured to produce a material flow figure. The flow of steam as a material also serves as an input to boiler combustion control systems that use boiler load information as feed-forward signals to the combustion control system in anticipation of changing fuel requirements.

How Flow Is Measured

Almost every conceivable branch of physics has been a basis for the design of flow-measuring devices. Each successful design has specific applications for which it is satisfactory, but there is no universal flowmeter. Most commonly found in powerplants for measuring fluid flow are devices that use differential-pressure loss in pipelines; variable-area meters also see service, as do positive-displacement devices. These are discussed below. Many other types of flowmeters are on the market—magnetic meters, acoustic meters, turbine meters, swirl meters, vortex shedding meters, and thermal meters, to name a few. Since they do not generally see service in powerplants, however, they are not reviewed here.

Differential-Pressure Measurement. If a restriction is placed in a pipe or duct through which fluid flows, the velocity of the fluid must increase as it passes through the restricted location. It will again flow at its original velocity, after the fluid flow pattern has increased to the original cross-sectional area of the pipe or duct. After the velocity has returned to normal, the pressure in the pipe or duct downstream of the restriction will be a measurable amount less than the pressure upstream of the restriction. Just downstream of the restriction, where the velocity is the greatest, there is an even larger loss of pressure. This difference in pressure occurs because the flow pattern through the orifice is not perfectly smooth. There is some eddying, and the turbulence is wasted energy; consequently, a net loss of pressure exists (Fig. 1.9).

The relationship between the flow rate (actual, not standard, flow rate) and *differential pressure* (DP) can be expressed as

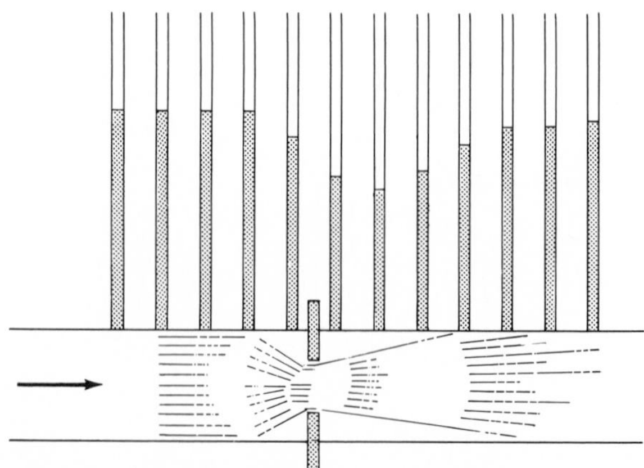

FIG. 1.9 Flow pattern through orifice. *(Courtesy J. A. Moore.)*

$$Q = kA \left(\frac{2ghp}{GT}\right)^{1/2}$$

where Q = flow
k = orifice coefficient
A = cross-sectional area of restriction
g = gravitational constant
h = head (differential pressure)
p = static pressure
G = specific gravity
T = absolute temperature

The coefficient k is a constant that is a function of the size and shape of the restriction. DP measurement is not linear with respect to the flow rate, which reduces its usefulness, but nevertheless it commonly finds application in powerplants and in chemical plants as well. The measurement is usually made in circular pipes, although it can be made in rectangular ducts. The restrictions in circular pipes are mostly orifice plates, flow nozzles, venturi tubes, and, less frequently, pitot tubes. Dall tubes and Gentile tubes also measure the flow rate by creating differential pressure, but they are not found to any extent in powerplants.

Orifice Plates. If a thin plate with a circular hole having a diameter smaller than the pipe's *inside diameter* (ID) is placed in the pipeline perpendicular to the flow path, a restriction is created (Fig. 1.10). The hole is called an *orifice,* and it normally will be concentric with the pipe. The ratio of orifice diameter to pipeline ID has a bearing on the value of the constant k in the flow equation. A popular ratio of orifice diameter to pipe ID is 0.6. The ratio may be smaller, and a differential pressure corresponding to the relationship expressed above will be produced; but the smaller the orifice diameter, the greater the permanent pressure loss. This is undesirable because a cost is associated with creating the pressure in the first place, and throwing it away across orifice plates is a small but constant operating expense. The diameter can be larger, saving money in pressure loss, but above 0.75 the flow equation becomes unreliable. Good practice is to keep the orifice ratio between 0.55 and 0.65.

The orifice plate will usually be ¼ in (6.4 mm) thick, although in large pipes with high-pressure gas passing through them, ½ in (12.7 mm) may be advisable to prevent distortion of the orifice from bending of the plate. This type of restriction is called a *sharp-edged orifice,* because the upstream face makes a sharp 90° angle with the bore. Halfway through the plate's thickness, however, the bore opens out at a 45° angle so that the downstream opening is beveled; it is also bigger than the upstream opening. The bevel decreases downstream turbulence and

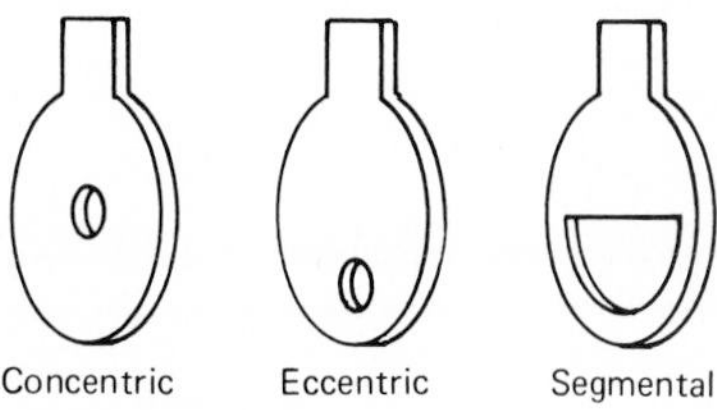

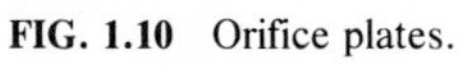

FIG. 1.10 Orifice plates.

shapes the flow so that it expands smoothly to fit inside the pipe wall again. The point where the pressure drops to its lowest level is called the *vena contracta*. It is a short distance beyond the downstream face of the orifice plate, where the diameter of the flowing fluid is actually smaller than the orifice bore. Exactly where the vena contracta falls will depend on the ratio of diameters of the orifice and the pipe ID.

The orifice plate is usually supplied with a tab, which extends outside the flanges that hold the plate in the pipe. On the tab will be inscribed the tag number of the orifice, its diameter ratio, and the pipeline size. A popular material for orifice plates in powerplants is type 304 stainless steel.

Openings are bored through the pipe wall, so that the upstream and downstream pressures can be measured by an instrument that will develop a flow signal. The openings, called *pressure taps,* are connected to the instrument by ½-in (12.7-mm) pipe or by copper tubing. Usually the instrument is an electronic transmitter (see page 6.28) located near the restriction.

Orifice plates are held in the line between special fittings called *orifice flanges*. These have openings for the pressure connections drilled through the flange (Fig. 1.11), threaded at the outside of the flange. They may turn 90° at the inside, perpendicular to the orifice-plate surface. This design is called corner taps. The outside diameter of the orifice plate is just small enough that the plate nests inside the bolts that join a pair of flanges, thus ensuring the concentricity of the orifice in the line. Orifice flanges are rated 300-lb ASA (or higher if pressure in the line requires it), but are supplied in 150-lb ratings only for very large pipe sizes. Orifice locations may vary to suit a particular location.

Flow Nozzles. A flow nozzle is a machined restriction containing an orifice hole, but continuing the hole through a short section of pipe machined and honed to a smooth, convergent surface that shapes the flow pattern and reduces turbulence (Fig. 1.12). The downstream end of the tube has the diameter of the vena contracta of an orifice of equal capacity. The flow is discharged more nearly parallel to the axis of the downstream pipe.

The diameter ratio of a flow nozzle is smaller than that of an orifice, and its

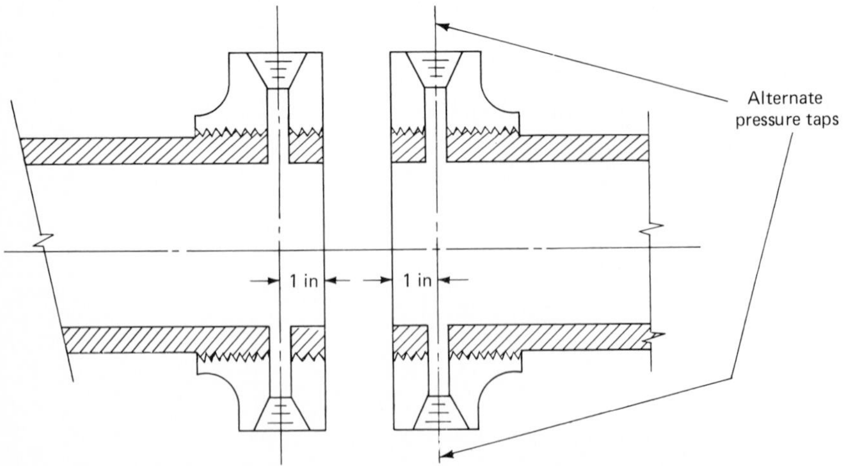

FIG. 1.11 Orifice flanges. Note: in × 25.4 = mm. *(Courtesy J. A. Moore.)*

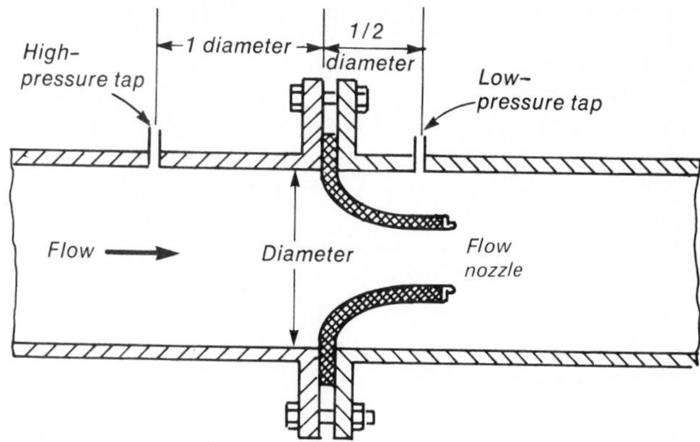

FIG. 1.12 Flow nozzle.

flow coefficient k is larger. The result is less permanent pressure loss, and the flow nozzle is more efficient than the orifice. It has application in powerplant instrumentation for feedwater flow and high-pressure steam flow measurements. A flow nozzle is about 10 times more expensive than an orifice plate.

Venturi Tubes. The venturi tube carries the advantages of the flow nozzle over the orifice to its ultimate limit by extending the cone of the nozzle in a smooth, gradually divergent shape until the original pipe size has been attained (Fig. 1.13). The intent is to match the normal shape of the fluid stream and to reduce pressure losses to a minimum. The body of the venturi tube is an entire pipe section, and the pressure connections are machined into the section.

The venturi tube is about 20 times as expensive as an orifice plate, but its design reduces pressure loss to such a degree that the savings in energy can offset the cost of an element and its installation. The venturi tube can handle 60 percent more flow than an orifice plate, with the same pressure drop, but the permanent loss will be much less. It is often used for high-pressure steam flow measurement.

Pitot Tubes. The pitot tube can be described as an element made of two pressure taps inserted through the wall of a pipe into the center of the flowing stream. Instead of depending on pressure loss across a restriction, the two taps measure a DP, which is a function of their orientation to the fluid flow. One tap, facing upstream, measures total static pressure and velocity head, while the second, facing perpendicular to the flow path, measures static pressure only. Because there is no reduction in pressure due to velocity change, the DP is much

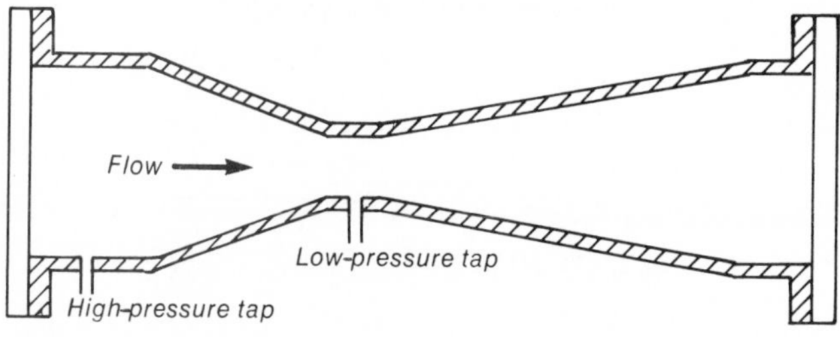

FIG. 1.13 Venturi tube.

lower than would occur across an orifice, but the same relationship of flow to DP exists.

Since the pitot tube measures at a point in the fluid path, it is necessary to position the tube so that its measurement is representative of flow. If there are elbows or branches upstream of the point of measurement, turbulence will be introduced and the flow pattern may be influenced so that the best point of measurement is not in the center of the pipe. Specially designed pitot tubes extend across the pipe section, with several sets of openings manifolded together so that the impact pressures are averaged (Fig. 1.14).

In another variation, the pitot-tube pressure openings are mounted on the end of a mast, which positions them inside a small venturi section that increases the fluid flow at the point of measurement. Pitot tubes more often see service for collecting flow measurements for test information than for process information.

Limitations. DP measurement is used for the majority of flow measurements in powerplants. Orifice installations are relatively inexpensive compared to those for which elements must be inserted into the pipeline. Also, for closed-loop control or for measurement over a small range of flow, they perform satisfactorily. DP systems are limited to flow ranges from 100 percent down to about 39 percent of full flow, because of the square root relationship between flow and DP (Fig. 1.15). If flow is at one-half of the calibrated range, the DP being measured is at one-quarter of its range. If flow is at one-third of its calibrated range, the DP is at one-ninth of its range, and so on.

Every transmitter for any measurement has some degree of error, although it may be very small. The square root relationship means that the error of measurement by the DP transmitter is occupying an increasing portion of the measurement as flow decreases. A standard commercial DP transmitter, with a range of 0 to 100 in H_2O (0 to 24.9 kPa), will have an error of ¼ in H_2O (62 Pa). When the flow has dropped to 50 percent, the error of ¼ in H_2O (62 Pa) still persists, but now it is for 25 in H_2O (6225 Pa), or one-quarter of the range.

As flow drops to one-third, the persistent ¼ in H_2O (62 Pa) now is part of the 11.1-in H_2O (2764-Pa) transmitter measurement, and its importance has increased from ¼ to 2¼ percent. It is for this reason that the recommendation is made not to use DP measurement for flow ranges that will consistently perform at less than 10 percent of maximum. This precaution can create a problem during start-up operations, because they are often conducted at low-flow conditions. Normal oper-

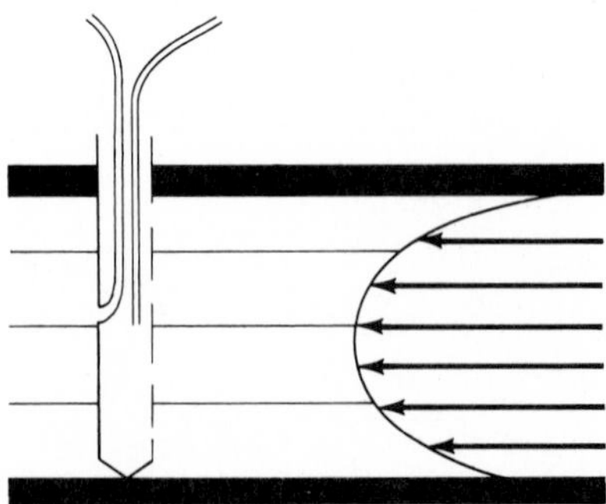

FIG. 1.14 Averaging pitot tube.

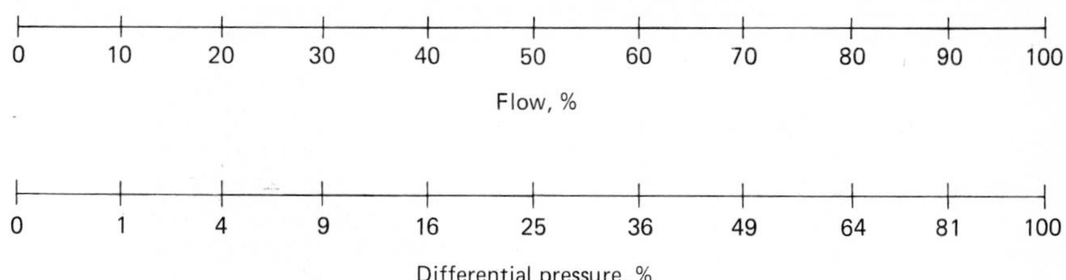

FIG. 1.15 Relationship between flow and differential pressure. Volume flow = $K\sqrt{DP}$. *(Courtesy J. A. Moore.)*

ating flows should be in the upper one-third of the calibrated flow range; for this application, DP measurement is quite satisfactory.

Another limitation of DP sensors is the straight length of pipe required before and after the sensor to ensure smooth flow patterns. The length will vary with the piping pattern. There should be the equivalent of two to four diameters of length after the elements, and anywhere from 10 to 30 diameters before it, which can result in some expensive piping (Fig. 1.16). If the accuracy of the measurement is important enough, the extra cost may be justified.

Note that even if the straight run is not the recommended length, the measurement will still be reproducible and can be used for reference purposes. In a closed-loop control system, this is no problem because the controller will find the correct output to make the measured process variable match setpoint, and the absolute value of the DP measurement is not important. If the measurement is intended for inventory purposes or for heat-balance calculations, however, the error is significant; here, the straight-run requirements and turndown ratio cannot be ignored.

Other Flow Measurement Techniques. Some flow measurement techniques operate on principles other than DP measurement and are not subject to the limitations discussed above. Of these techniques, only the variable-area meter finds significant application in powerplants.

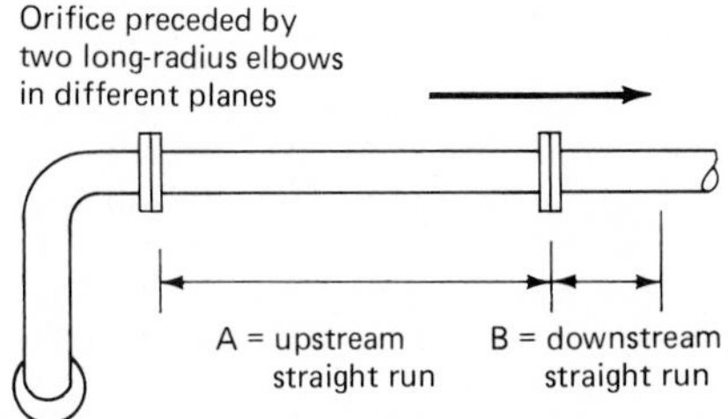

FIG. 1.16 Straight-run requirements for orifice flow measurement. With an orifice beta ratio of 0.6, *A* should be at least 16 (inside) pipe diameters and *B* should be at least 3.5 (inside) pipe diameters. *(Courtesy J. A. Moore.)*

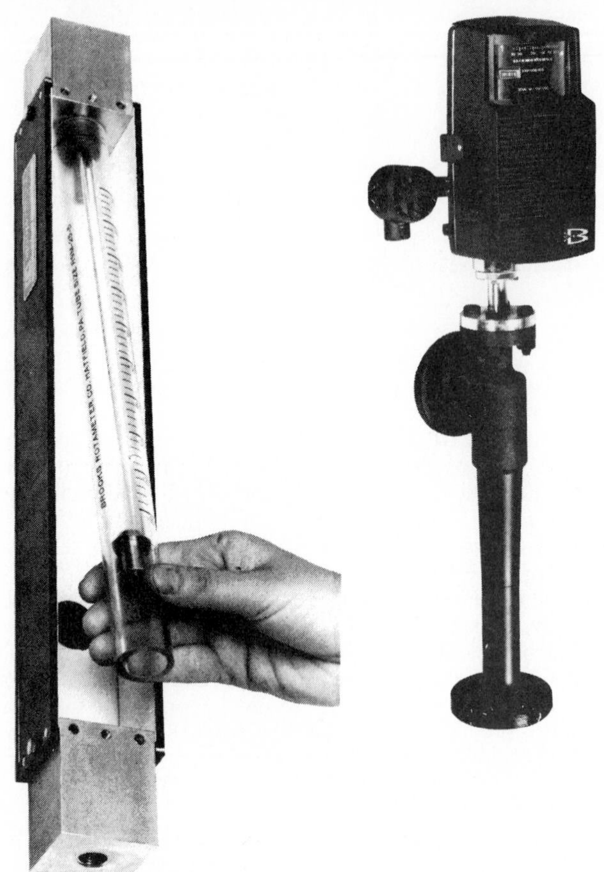

FIG. 1.17 Variable-area meters: glass-tube rotameter for local indication (left); magnetic rotameter for local and remote readout (right). *(Courtesy J. A. Moore.)*

Variable-Area Flowmeters. If a buoyant plummet is contained within a vertical tapered tube through which the fluid to be measured passes, from top to bottom, and if it is free to move up and down inside the tapered section as the flow rate changes, then upward and downward forces on the plummet will be in equilibrium, allowing the plummet to assume a specific position for every flow rate. The only force acting downward is the plummet's weight. The upward-acting forces are the buoyancy of the plummet, the pressure drop across the plummet (times its cross-sectional area), and the viscous drag of the fluid past the plummet.

Since the tube is tapered, the annular cross-sectional area for flow is variable, increasing as the flow increases; the pressure drop across the plummet is constant. The same relationship among flow, area, and DP given above governs the flow calculation, but area—the measured variable—varies as the first power, and so the flow is linear with position in the tube.

An extension of the plummet, extending outside the pipeline in which the mea-

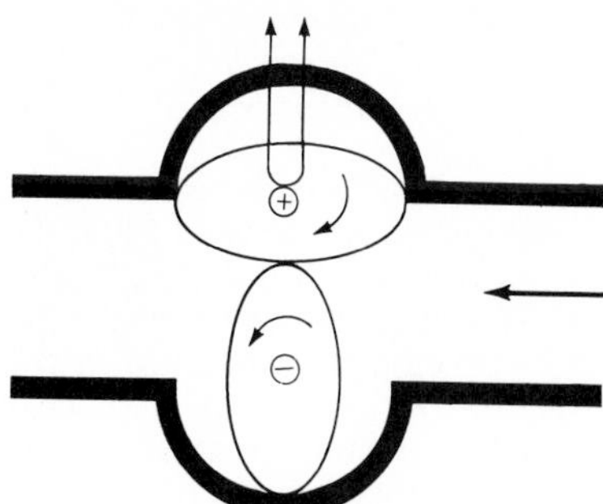

FIG. 1.18 Positive-displacement meter.

suring element is mounted, contains a magnetic device that can actuate a transmitter to develop an analog signal for transmission. This type of device is called a *rotameter* (Fig. 1.17). Rotameters are made in sizes for installation in pipe of ¼ to 6 in (6.3 to 152 mm), but ½- to 3-in (12.7- to 76-mm) sizes are more common. Advantages over orifices are that the flow relationship is linear, they can operate with reasonable accuracy down to 10 percent of flow range, and pressure drops are low. They can be designed so that the measurement is not affected by the viscosity of the measured fluid. An application of rotameters in powerplants is the flow measurement of heavy fuel oils, for example, no. 6 or Bunker C. Other applications are for local indication of gas and liquid flows in small pipes.

Positive-Displacement Meters. Another type of meter with powerplant application measures the *positive displacement* (PD) of the fluid flow. The measuring device acts as a bucket conveyor and consists of chambers that fill with fluid, move to the discharge, and empty. Rotary motion drives the chambers, and so the number of rotations in a given time becomes a measure of the amount of material moved. Chambers are formed by rotating vanes, nutating disks, intermeshing gears, water seals, and many other clever designs.

A powerplant application for PD meters, again for heavy fuel oil, relies on two oval gears rotating inside a housing with very close tolerances (Fig. 1.18). The pressure of the oil drives the gears, rotating them on shafts so that a quantity of oil is trapped between the gear teeth. The rotation transfers the trapped oil, discharging it after one full revolution. Magnetic pickups on the shaft develop pulses for each revolution, and these are counted to determine the flow rate.

PD meters for measuring gas flow can be wet (sealed with water) or dry, depending on the application. They are sometimes selected for measuring sample gases for analysis. Because PD meters are very accurate, natural-gas utilities use them for billing purposes. For this application, the dry-type PD meter is the choice. It consists of a gastight housing divided into two sealed compartments. Each compartment is again divided into two sealed chambers by flexible bellows made from gastight elastomers. Slide valves direct the flow through the chambers from inlet to discharge.

PRESSURE

The units in which pressure is expressed are a representation of the forces created by the gravitational pull on the molecules making up a column of air. In inch-pound units, if it were possible to weigh a column of air 1 in^2, extending from sea level up into the stratosphere, the weight would be about 14.7 lb. This is expressed as a pressure of 14.7 psi. Another way of expressing the same value is to call it 1 atm of pressure.

If the column of air could be balanced by a column of water 1 in^2, the water column would be about 34 ft tall. This volume of water, in this shape, also weighs 14.7 lb and exerts that amount of weight on a 1-in^2 surface. The same pressure can also be expressed in terms of the height of a water column, 34 ft H_2O, or 408 in H_2O. This is sometimes called a *head measurement* (Fig. 1.19). Pounds per square inch (psi), atmospheres, inches or feet of water (in H_2O or ft H_2O), and inches of mercury (in Hg) are all interchangeable units of pressure measurement. These are the units commonly found in powerplant measurements.

In SI units, 14.7 psi = 101.325 kPa. Conversion factors are: psi × 6.89 = kPa; ft H_2O × 2.99 = kPa; in H_2O × 249 = Pa; in Hg × 3.38 = kPa.

Defining Pressure Measurement

If gas at atmospheric pressure is enclosed in a housing and removed through a pump that can draw the molecules of gas out of the housing and discharge them to atmosphere, the pressure in the housing will be reduced. Pressures below atmospheric level are called *vacuums*. If it were possible to remove every last molecule (there are pumps that will very nearly do this), assuming that the atmospheric pressure when the gas was put into the housing was 14.7 psi (101.325 kPa), it could then be said that the pressure in the housing was a vacuum of 14.7 psi (101.325 kPa). One could also say that the pressure was −14.7 psi (−101.325 kPa). More commonly, however, this complete absence of molecules is defined as an absolute pressure of 0 psi (0 kPa). On this basis, atmospheric pressure is really 14.7 psi absolute (psia), or 101.325 kPa.

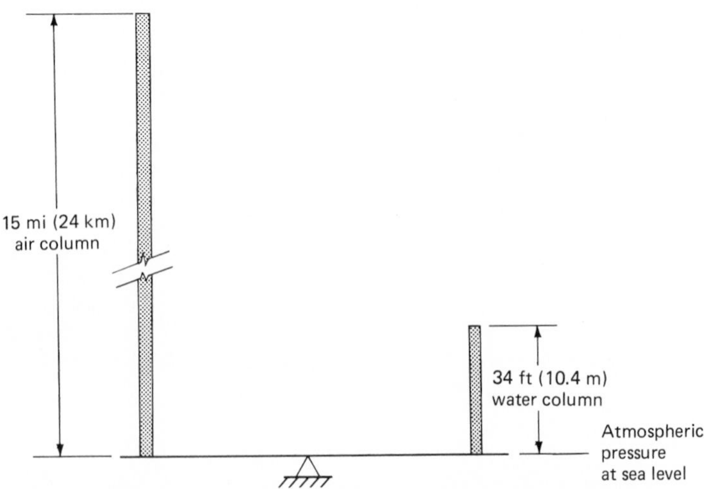

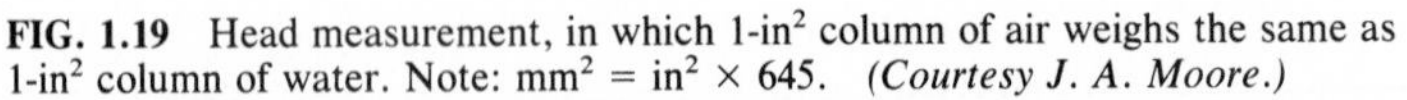

FIG. 1.19 Head measurement, in which 1-in^2 column of air weighs the same as 1-in^2 column of water. Note: mm^2 = in^2 × 645. *(Courtesy J. A. Moore.)*

Thus, there are two points of reference for measuring pressure: One is absolute zero, and the other is atmospheric pressure. Pressures based on the latter are called *gage pressures* (Fig. 1.20). Atmospheric pressure is an absolute pressure. If the vacuum pump in the example above reduced the pressure in the housing by 4.7 psi, it would be correct to talk about a vacuum of 4.7 psig (gage pressure) and equally correct to express the pressure as 10 psia (absolute pressure). Comparable SI units are 32.4 kPa gage and 68.9 kPa. Figure 1.21 shows a compound pressure-vacuum gage.

Use in Plants. Measurement of pressure is important in powerplants for two reasons: (1) Pressure is a direct indication of the energy level of steam produced from boiler operation and also of the electric power produced from steam-driven turbine-generators, and (2) pressure provides an indication of safe operating levels.

Pressure is a measure of energy in steam because there is a specific pressure associated with saturated steam at every temperature. Water that has been heated to 212°F (100°C) at atmospheric pressure will start turning into steam when one more Btu (kilojoule) is added to it; if the absolute pressure were raised a mere 0.3 to 15 psia (2.07 to 103.35 kPa), however, it would be necessary to heat the water to 213°F (100.6°C) before steam would begin to form. A higher pressure, 200 psia (1378 kPa) for example, would correspond to a boiling point of 381.8°F (194.3°C). Either pressure or temperature defines the heat content of saturated steam, but pressure is somewhat easier to measure.

The steam generated in a boiler is saturated. The measurement of steam pressure is used as the reference point for combustion control to maintain the header pressure at a desired level. It also serves as a reference for electric energy production via control of the steam pressure after it has passed through the various stages of a turbine. The pressure at which the steam is discharged is a measure of the energy that has been converted to mechanical rotation, as well as the amount of energy left in the steam for further expansion in other turbine stages or for condensate that will be returned to the boiler as feedwater.

From a safety point of view, pressure measurement helps determine whether

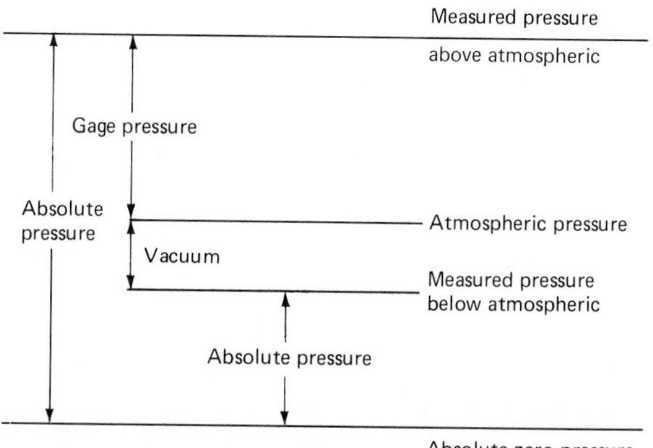

FIG. 1.20 Relationships among measured (gage) pressure, absolute-zero pressure, and atmospheric pressure. *(Courtesy J. A. Moore.)*

FIG. 1.21 Compound pressure-vacuum gage. *(Courtesy Perma-Cal Corp.)*

equipment is being operated in a manner that could damage either the equipment or the areas in which it is located. Powerplant equipment operates at pressures much higher than atmospheric, so it must be designed for safe limits of operation. Pressure measurement is a positive way of determining whether the limits are being exceeded.

Areas of Measurement. Pressure measurement starts in the fuel system, if gas or liquid fuel is burned. Pressure must be developed by pumping to move the fuel, and safety is also a consideration. Boiler drum and steam header pressures are important parameters to measure. Pressures are also measured to establish operating conditions for tanks, heaters, and pump suction and discharge points.

An entirely different kind of pressure measurement is made to determine conditions in the exhaust stack. The combustion of fuel takes place at a positive (above atmospheric) pressure in a combustion chamber, where air under pressure is combined with fuel under pressure. A negative (below atmospheric) pressure, induced by a fan as well as by natural draft, pulls the exhaust gases up the stack for discharge to the atmosphere. Somewhere between the combustion chamber and the stack, the pressure changes from positive to negative, and the location of this point for best combustion can be found by measuring the draft (another name for pressure in the stack) and controlling it. Draft is always measured as a gage pressure, because it is very close to atmospheric. A common range of measurement is −0.05 to +0.15 in H_2O (−12.45 to +37.35 Pa).

Measuring Instruments

Most pressure measurements are really measurements of differential pressure, with the pressure referenced either to atmosphere or to an absolute vacuum, although pressure can be measured as a direct force by piezoelectric crystals. Powerplants may select this type of device for acceleration measurement, but not usually for pressure measurement.

With DP measurement, if gage pressure is being measured, the differential is between atmospheric pressure and the pressure measured. There is only one piping connection; the other connection is made by exposing one side of the measuring device to the atmosphere. If the measurement is of absolute pressure, the reference side of the measuring device is exposed to a high vacuum, which approximates absolute-zero pressure.

The pressure transmitters most often selected for powerplants are the following:

- For positive gage pressures indicated locally, the bourdon tube or bellows
- For negative gage pressures indicated locally, the bellows
- For draft measurements that are transmitted, the slack-diaphragm electronic device
- For pressure measurements other than draft that are transmitted, the two-wire electronic transmitter. (Pneumatic transmitters are still in use in many plants, but they are being replaced as spare parts become unavailable, almost always by electronic devices.)

Bourdon Tubes. For measuring gage pressures, in positive ranges from 0 to 15 psi (103.35 kPa) up to 0 to 10,000 psi (68 900 kPa), and for negative ranges (vacuum) over 0 to 15 psi (103.35 kPa), bourdon tubes are used. A bourdon tube is made of flexible metal shaped into a flattened cross section and bent into a C shape, a spiral, or a helix (Fig. 1.22). One end is fixed and is connected to the pressure-inlet fitting. When the pressure increases, the tube tries to straighten and the tip of the other end moves. If the indication is local, the tip is connected through a linkage and sector gear to a pointer, which moves across a scale to indicate the pressure causing the movement. The accuracy of measurement with bourdon tubes is ±1 percent of span. They are made of brass or phosphor bronze for small ranges [up to 0 to 100 psi (0 to 689 kPa)] and stainless steel for high ranges. Monel tubes are selected for the measurement of pressure of saltwater or caustic solutions.

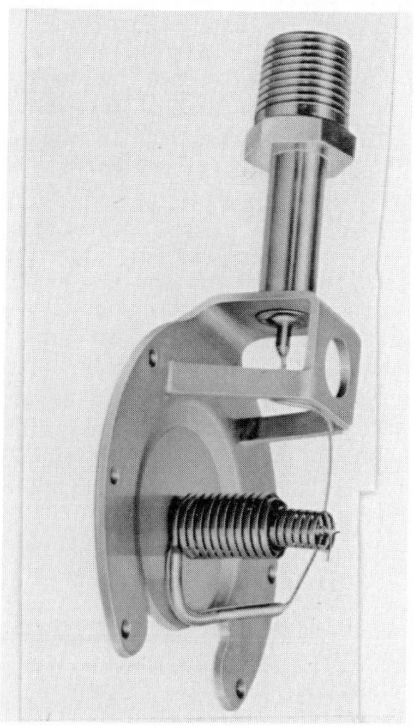

FIG. 1.22 Bourdon tube. *(Courtesy Perma-Cal Corp.)*

Bellows. A bellows is a flat capsule, or a stack of flat capsules, made of thin metal, usually phosphor bronze or brass (Fig. 1.23). Pressure is applied through a connection to the inside of the capsules and is balanced by a spring connected to the outside. The spring opposes the motion of the bellows, which expands from a zero position when the pressure increases, or contracts as vacuum increases. As with the bourdon tube, a linkage translates this movement to a pointer rotation, or a

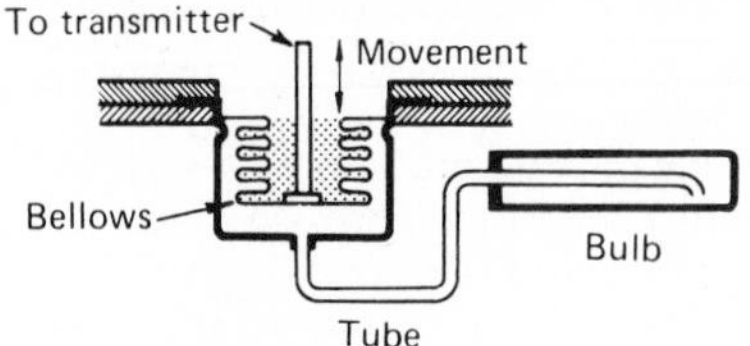

FIG. 1.23 Bellows. *(Courtesy Robertshaw Controls Co.)*

transduced signal. A pair of matched bellows, one of which is evacuated, makes possible absolute-pressure measurements. Bellows instruments measure low ranges, from 0 to 5 in H_2O (0 to 1245 Pa) up to 0 to 15 psig (0 to 103.35 kPa). Accuracy is ±1 percent.

Electronic Transmitters. Most pressure (or DP) measurements in powerplants develop an electric signal for transmission to a central control area (Fig. 1.24). The sensor consists of a thin flexible member, sandwiched between outer plates that are also flexible. The spaces on either side of the flexible sensing plate are usually filled with a noncorrosive liquid such as silicone oil. This capsule is mounted in a housing that has connections to which the process can be connected, so that the process pressure is in direct contact with the outer faces of the flexible plates that form the walls of the capsule.

If the sensor is measuring differential pressure, the higher-pressure side will be connected to one chamber of the housing, while the lower-pressure side will be connected to the other chamber. If the sensor is measuring gage pressure, the

FIG. 1.24 Electronic transmitter at remote location.

process connection will be made to the higher-pressure chamber, and the lower-pressure chamber will be open to the atmosphere. An absolute-pressure sensor will have the lower-pressure chamber sealed and evacuated.

When the differential pressure increases across the cell, the diaphragm on the high-pressure side is pushed toward the center, and the central flexible plate is distorted. It can be protected from overpressure by a supporting plate that stops motion before the elastic limit of the metal is exceeded. The signal, which varies with pressure variation, is developed by an electrical device associated with the flexing of the thin, central member of the capsule. Types of devices are capacitors and strain gages; inductive and reluctance elements also see service. The greatest degree of acceptance has been of the capacitive circuit devices.

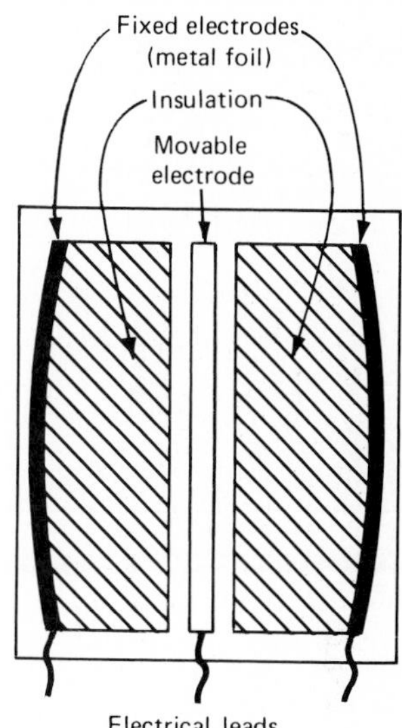

FIG. 1.25 Capacitive sensor.

Capacitive Sensors. A capacitive sensor contains cavities on either side of a central flexible member that are filled with insulating materials, such as glass or ceramic. In a typical design (Fig. 1.25), one surface of each insulating section is dish-shaped and has metal foil placed on it to serve as a capacitor plate. The central flexible member functions as a movable electrode, while the metal-foil coverings function as fixed electrodes. The electrostatic capacity between electrodes changes as pressure on the high side of the cell changes the spacing between the electrodes. The transducer is excited with high-frequency (10-kHz, for example) *alternating-current* (ac) carrier current, and the physical unbalance creates an electrical alternating voltage with an amplitude proportional to the difference in pressures.

Electronic pressure transmitters are calibrated for ranges from zero to a few psi (kilopascals), up to ranges of zero to several thousand psi (kilopascals). They can be turned down; that is, full electrical output can be obtained for a portion of the potential capsule span. Some accuracy error, expressed as a percentage of full capsule span, always exists. The greater the turndown, the greater this absolute error will be in terms of the measured span. For capacitive sensors, the figure is 0.2 percent. Turndown is limited to 10:1.

Strain-Gage Sensors. If a thin strip of metal is stretched within its elastic limits, the cross-sectional area decreases and the electrical resistance of the element increases (Fig. 1.26). A strain gage is made of four small-diameter wire grids arranged in an electrical bridge and bonded to a surface. If the surface can be distorted by pressure, two elements of the bridge will be in tension, while the other two will be in compression, and a millivoltage change occurs. The electrical value can be calibrated in terms of the measured pressure. A bridge-controlled oscillator system can transmit the electric signals from resistance strain gages to an amplifying circuit. Metals used for strain gages include nickel-chromium, copper-nickel, and platinum-tungsten alloys. Accuracies of ±0.25 percent of range are possible; turndown is limited to 5:1.

A serious limitation of the metal strain-gage bridge is the need for temperature compensation, since resistance change in metals can also be a function of temper-

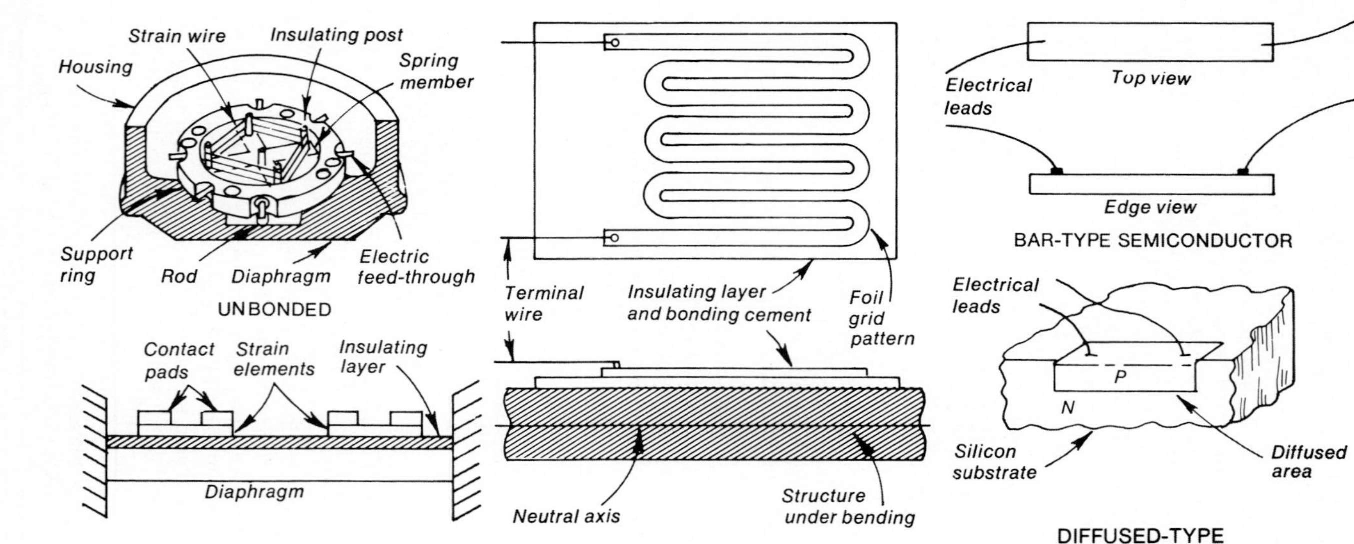

FIG. 1.26 Strain gages.

ature change. It is possible to provide some compensation or stabilization for temperature change by alloying, by heat treatment, and by the use of compensating resistors in the electric network. Nevertheless, drift caused by temperature change has limited the attraction of strain gages in powerplant measurements.

Variable-Reluctance Bridge. A variation of the same mechanical design—that is, a movable element between two isolating diaphragms separating high- and low-pressure chambers—capitalizes on magnetic coupling instead of the resistance of a strain gage or the capacitance between plates. There is a version of the linear variable transformer consisting of a single primary winding and two symmetrically spaced secondary windings. A ferrite disk forms a core, which moves between the two secondary windings to change the flux linkage between the coils, and coil inductances are varied.

The magnetically triggered movement is the result of differential-pressure change against the outer diaphragms. Changes in inductance are magnified and transduced to an analog signal for transmittal. Sensors designed to vary inductance have fast dynamic response and have better shock and vibration resistance than the strain-gage and capacitance types; they are, however, sensitive to temperature and stray magnetic fields.

Slack-Diaphragm Sensor. A suitable sensor for measuring stack draft, a variable with a range that is usually less than 1 in H_2O (249 Pa), is the slack diaphragm. The diaphragm is made from a thin sheet of plastic material (Buna-N-treated nylon or Mylar, for example) supported by clamping, but not stretched tightly, in a ring of metal. The central portion is also clamped between two circular plates of a diameter about one-third that of the external ring, so that the active diaphragm is an annular sheet that can move backward and forward as the differential pressure across it changes.

The high and low pressures are contained in the chambers of a housing, which is constructed so that the slack diaphragm forms one side of each of the two chambers. The central clamping plates are fastened to a lever arm which positions a capacitor to change current in a feedback loop, and the current becomes an analog signal of the draft measurement. Ranges as low as 0 to 0.2 in H_2O (0 to 50 Pa) can be measured with this device.

LEVEL

While temperature and pressure are associated with energy measurement, and flow measurement can represent quantities of either or both material and energy, level is usually an indication of material only. True, it is correct to refer to a level of energy, but the units of measure of that level are Btus (kilojoules). Level measurements in the context of this section are made to determine where the upper surface is of solid or liquid material contained in tanks and other vessels. From the measurement, it is possible to determine either the quantity of material contained in the vessel or whether too much or too little material is in the vessel.

Level can be measured as a dimension, that is, a specific distance above a reference point. Liquid level can also be measured as a pressure, by applying the concept of hydraulic head. In the first case, the measurement is absolute; in the second case, the reading will be expressed in inches of H_2O (pascals) or feet of H_2O (kilopascals) and must be converted to provide the required information.

The boiler drum level is probably the most useful of the level measurements in a powerplant; the feedwater control system is dedicated to maintaining a satisfactory drum level. The *American Society of Mechanical Engineers* (ASME) Boiler Code requires that there be direct reading indications of drum level, one at each end of the boiler; or that direct reading be replaced or supplemented by a hydraulic head measurement that mechanically actuates an indicator. The indicator can be remotely located from the boiler; an electrical transmission of level can be made to the control room. Levels of liquid fuel can be checked by measuring devices that will indicate whether they are above or below a limit, but not the amount above or below. Condensate and deaerator tank levels are measured continuously.

In this discussion, measurements of liquids and solids are reviewed separately; the division between mechanical and electrical sensors (another way to study level measurement) will be obvious.

Liquid-Level Measurement

The earliest method of measuring level, inserting a dipstick into a tank, is still used sometimes for determining the level of liquids in underground storage vessels. The next earliest method, direct visual observation of the liquid level in a gage glass mounted on the side of a tank, is still much in evidence. Sensors for measuring liquid at a single point rely on floats and probes. Sensors for continuous control over a range of levels make use of floats, displacers, hydraulic head, capacitance, ultrasonics, and radiation.

Gage Glasses. Flat glass and tubular gages provide visual indication of liquid levels in piping systems, vessels, and tanks (Fig. 1.27). This type of gage is required at both ends of every boiler operating in accordance with the ASME Boiler Code. Both flat glass and tubular gages can be supplied in welding-pad configurations

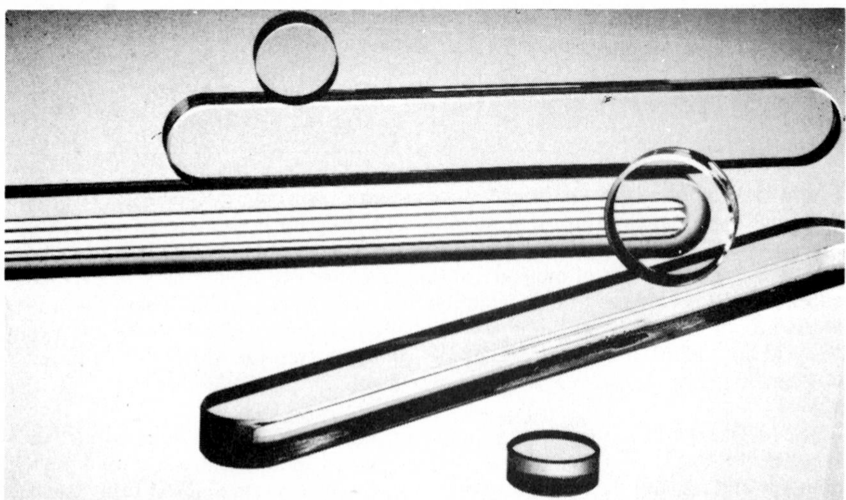

FIG. 1.27 Gage glasses.

for mounting directly to the metal of a vessel; in tubular designs they can also be mounted to pipe fittings.

Flat glass gages are available in reflex and transparent styles. Reflex gages have a prism glass window that refracts the light from the space above the liquid, reflecting it to the observer, so that the area appears silvery. Light striking the area of glass covered by the liquid passes through to the interior of the gage—it appears to be dark. The interface therefore is clearly marked.

Transparent gages have clear glass so that all the light which strikes the gage glass passes through into the vessel. This type of gage is best used when the liquid to be measured is colored or opaque. Reflex-style gages are available for pressures to 4000 psi at 100°F (27 560 kPa at 37.8°C); transparent-style gages are available for pressures to 6000 psi at 250°F (41 340 kPa at 121.1°C).

Floats. Metal or plastic floats can be used for either point or continuous-range measurements. Of the two classes of floats, one functions by staying out of the liquid, the other by being immersed in it. A ball float rides on top of the liquid, moving up or down as the level changes (Fig. 1.28). A displacer is cylindrical and becomes submerged as the level rises around it (Fig. 1.29).

The ball float can rotate a pivoted arm to which it is connected. If the arm extends through a seal in the wall of the vessel in which the float is housed, the motion can actuate a switch at a specific level as the liquid level changes. This is a single-point measurement. Alternatively, it can ride up a guide rod and transmit its position through the motion of a counterweighted cable riding over a pulley. This action provides a measurement over the entire length of travel.

The displacer, or immersed cylindrical float, operates on the principle that a body, immersed wholly or partly in a liquid, loses weight equal to the weight of the liquid displaced by the body. The float is almost stationary in this type of installation, being held by a short, flexible shaft connected through a flange that allows it to be mounted to a nozzle, providing access to the vessel. The shaft is packless, and a small amount of motion is imparted to a stiff tongue of metal that extends through a flattened cross section of the tubular connection, oriented so that only vertical motion is permitted. The tongue of metal actuates an electric or

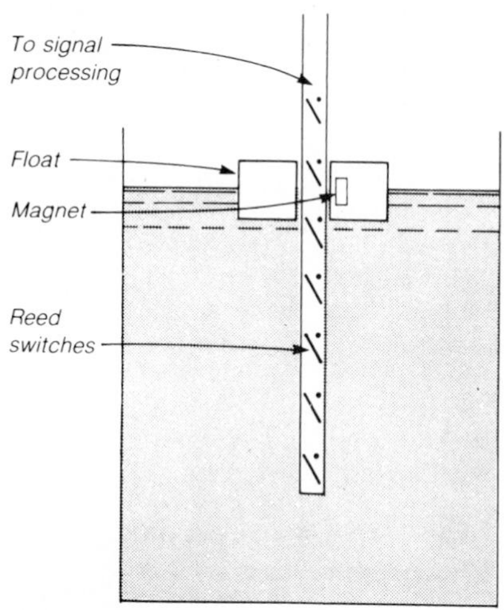

FIG. 1.28 Free float.

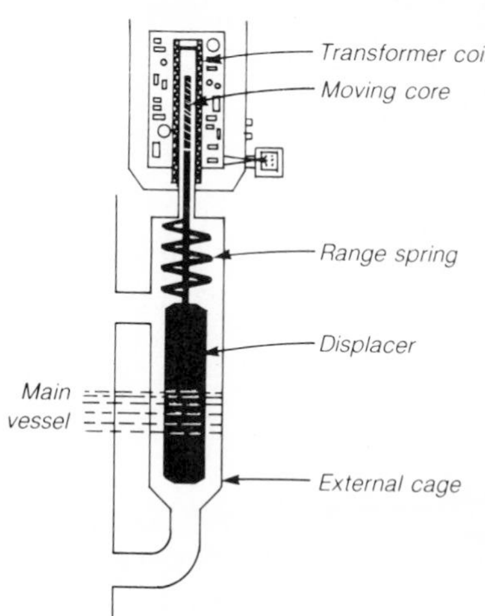

FIG. 1.29 Displacer float.

pneumatic transmitter, so that a continuous signal is generated corresponding to the length of the cylindrical float (Fig. 1.30).

Probes. Probes measure level by detecting when an electrical characteristic is changed as liquid comes in contact with them. They are mounted through the vessel wall or suspended from the top of the vessel (Fig. 1.31). They can be paired with a wide variety of processing equipment to control operating levels automatically or to detect critically high or low levels. Pump-up/pump-down circuits, in which a pump is turned on at low level to fill a tank and turned off at high level when the tank is full, are often activated by probes.

If the liquid in a vessel conducts electricity, it may be used to short-circuit a minute current to ground, actuating a solid-state detector. The liquid may act as a dielectric, changing capacitance so that a detector is actuated. Probes have no moving parts, so they require less maintenance than other types of level detectors. Probes work best when the level changes in the vessel are not rapid, but some designs include time-delay circuits, so that swirls or waves will not be recognized until the level has been maintained for a time long enough to represent a permanent condition.

Hydraulic Head. At any level below the surface of liquid in a vessel, a force exists equal to the weight of liquid above that level plus the weight (or pressure) of the va-

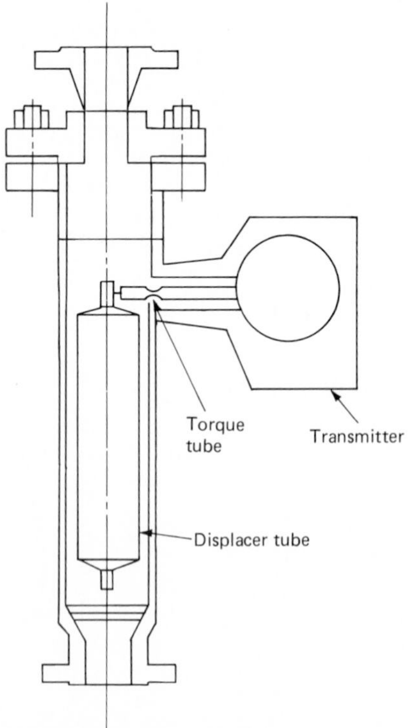

FIG. 1.30 Displacer, used for level measurement. *(Courtesy J. A. Moore.)*

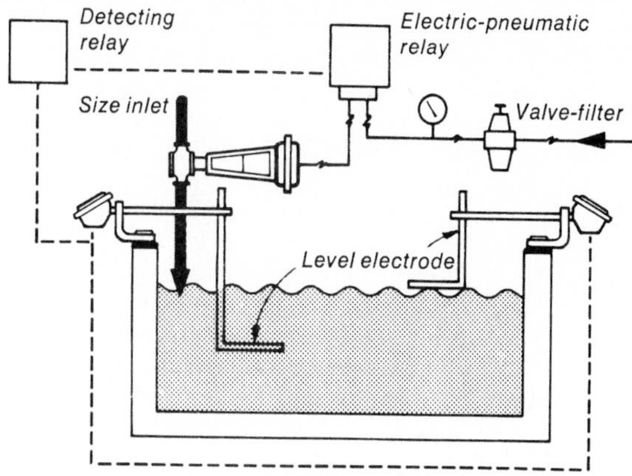

FIG. 1.31 Electrode-type probes.

por above the liquid. Measurement of this force provides an indication of the liquid level in the vessel. Called *hydraulic head,* it can be measured in several ways.

One approach is (1) to extend a small tube down from the top of the tank to a point representing the low end of the level to be measured and (2) to force air down through the tube at a slow rate, so that the air bubbles up slowly from the bottom of the tube. An equilibrium pressure at the tube bottom will be reached when the air pressure is exactly equal to the head of liquid it is displacing. If this equilibrium pressure is connected to the high-pressure side of a DP transmitter, with the low side connected to the vapor space above the liquid, it will represent the level of liquid in the tank. If the tank is open to the atmosphere, the low-pressure side of the transmitter may also be open to the atmosphere.

Instead of bubbling air into the tank, the point of level measurement can be accessed by a fitting through the tank wall to which the DP transmitter is connected (Fig. 1.32). If the tank is open to the atmosphere, the fitting is connected to the high-pressure chamber, and the low-pressure chamber is open to the atmosphere. As the level increases, the measured DP increases.

Most measurements in powerplants are for levels of closed tanks under pressure or vacuum (condensate or deaerator tanks, the boiler drum itself), so the low-pressure chamber of the DP sensor is connected to the fitting at the *bottom* of the tank, and the high-pressure side is connected to the *top* of the tank, above

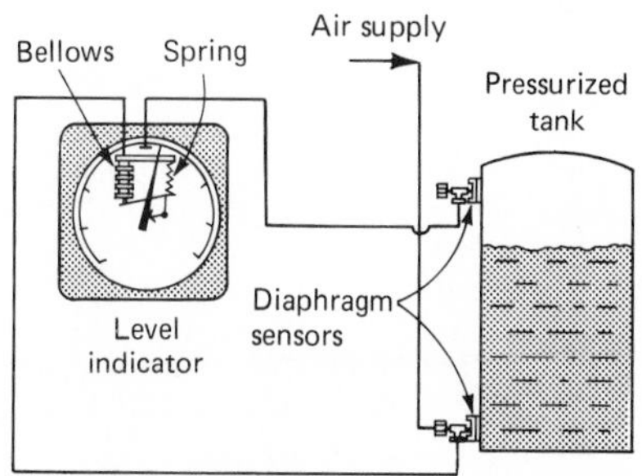

FIG. 1.32 Diaphragm sensors detect hydraulic head to measure level.

the upper limit of level. The reason for reversing the connection is this: The air above the liquid is full of water vapor, which will condense in the connecting line to the sensor, filling the line with water to the place where it is connected into the top of the tank. Thus, a permanent hydraulic head of water is in the connecting line, and it will always be higher than the hydraulic head because of the level itself. Thus when the tank is empty, DP will be maximum; when the tank is full, DP will be zero.

Noncontacting Sensors. If a sensor must come into contact with the liquid whose level it is monitoring, the sensor must be mounted in or on the vessel containing the liquid. This makes it difficult to inspect the condition of the sensor—and even more difficult to service it. For abrasive, corrosive, and viscous liquids that may harm the action of internally mounted sensors, an alternative is sensors that do not come into contact with the liquid. These are more easily maintained, but are substantially more expensive than contact-type sensors. Instruments most frequently encountered in noncontact measurement use nuclear radiation or sound as the means of detecting level change.

Radiation Gages. Gamma radiation from a source on one side of a tank is directed through vessel walls toward a radiation detector on the other side (Fig. 1.33). Radiation passing through liquids in the vessel will be attenuated and scattered to a greater degree than when it passes through empty space above the contents. Thus, a change in level changes the radiation received at the detector. This type of measurement usually acts to trigger a switch at high level.

Cesium 133 and 137 or cobalt 60 serve as radiation sources. Cesium 137 has a long half-life (30 yr) and is preferred to cobalt 60, which has a half-life of only 3 to 5 yr. Source holders and shipping containers must be protected with lead shield in accordance with Nuclear Safety Committee regulations to ensure that the radioactive field in occupied areas is less than the recommended maximum.

Ultrasonic Meters. Level of liquids can be measured by an ultrasonic echo-ranging technique. A sensor is mounted on top of the tank or vessel, with a clear view to the bottom. An ultrasonic wave generated by the sensor is bounced off the liquid and returns to the sensor (Fig. 1.34). The elapsed time for the echo's return is accurately measured by electronic circuits calibrated to a scale proportional to the tank level. Units should have temperature compensation.

The accuracy of the sound level detector is listed at 1 percent of range. A problem

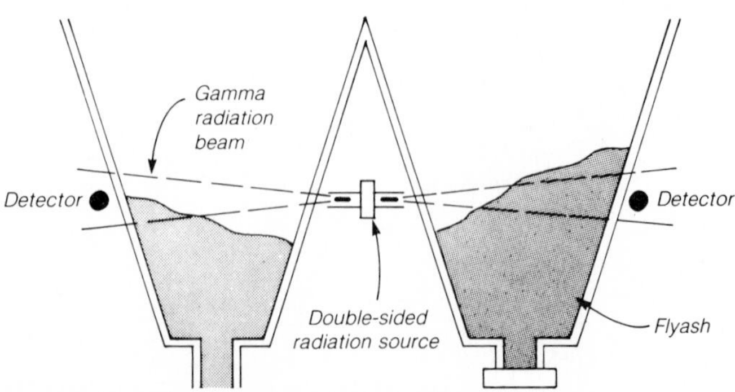

FIG. 1.33 Radiation gage.

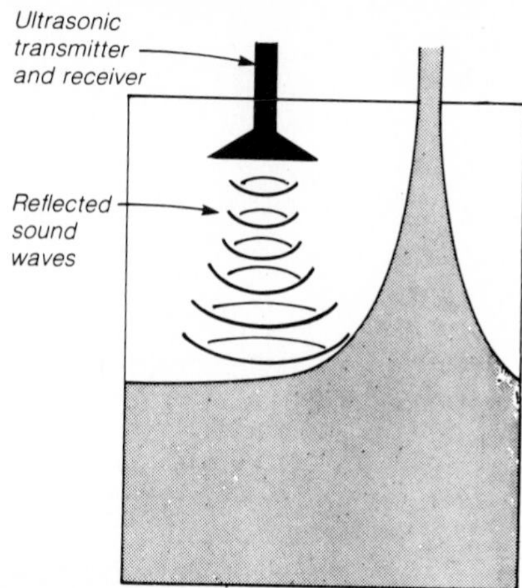

FIG. 1.34 Ultrasonic meter.

with this type of measurement has been interference from external acoustic and electrical noise; some circuits include statistical analysis to filter out interference by accepting inputs only in the range of the signal-sending component.

A variation of this technique detects high-frequency sound waves sent from one face to the other of a notch in the end of a probe. Liquid filling the notch as the level rises around the probe interrupts the sound waves and provides a point indication of level in the vessel in which the probe is mounted.

Solids-Level Measurement

Desirable solids-level measurements in powerplants include those of flyash, coal in pulverizers and grinders, and coal in chutes and bins. Solids-level measurement is not nearly as exact as liquid-level measurement, because solids build up a slope on the surface of the pile, hang up and "rat hole" when they flow out of bins and chutes, and in general present an irregular surface to a measuring device.

Sensors for measuring levels of solids depend on interruption of a light beam or sound "beam" by the solid; interference with a rotating paddle; pressure of the solid on a lever, diaphragm, or plate that will actuate a switch; proximity switches; or tilting switches actuated as the angle of repose changes.

One design measures the top surface of solid material by lowering a plummet into the vessel and detecting the level by the change in tension on the supporting cable when the plummet reaches the solid material (Fig. 1.35). Probes whose characteristics of vibration, capacitance, magnetic inductance, or resistance change when contacted by solid material are also available. These are all point measurements, useful for batching or alarming, but not continuous control. A degree of continuous control may be obtained with ultrasonic transmitters and sensors.

Electromechanical Devices. Point-type level controls can indicate inventory levels in bins. They can activate alarms and emergency shutdown when potentially dangerous clogs, overflows, or shortfalls are imminent. Electromechanical devices for point measurements include

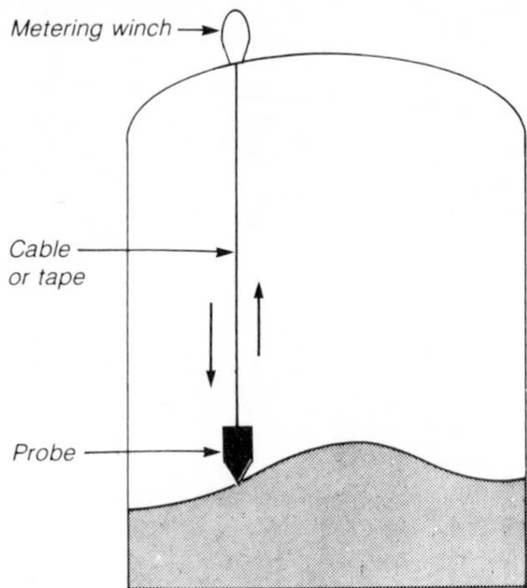

FIG. 1.35 Level is measured by detecting change in tension on cable when probe reaches material.

- Rotating paddles on the end of a short shaft. A torque measurement actuates an indicator when solids rise to a level that interrupts paddle rotation.
- A neoprene, Teflon, fiberglass, or stainless-steel diaphragm mounted in the straight or sloping side of a bin. When solid material reaches the diaphragm level, its weight depresses the flexible diaphragm and activates a mechanical contact (reed or mercury) switch, or a proximity switch, to complete an indicating or alarm circuit.
- Hopper switches mounted through the sides of bins contain probes that are excited by high-frequency oscillations. Material rising to a level that covers the probe will dampen the oscillations and create an indication of level.
- Tilting switches on spring-loaded levers that move when they are pushed by rising levels of material. They are mounted in bins and along the sides of conveyors.

Noncontacting Devices. Radiation gages and ultrasonic meters of the types described for liquid-level measurement are also applicable to the measurement of solids levels. They can be installed in train and barge unloading areas and in crusher-house chute systems and storage areas before the pulverizers, where stoppage of coal flow quickly is essential. Response times can be as fast as 0.5 s—important for eliminating coal backup in chutes and for turning off high-capacity vibrating feeders. Both types have been installed on the flyash collection hoppers of electrostatic precipitators and fabric filters.

MISCELLANEOUS

Here we cover the instrumentation for vibration monitoring and flame detection—both of increasing importance in powerplant measurements.

Vibration Monitoring

Continuous production of power depends on continuous operation of rotating machinery. Bearings wear and must eventually be replaced; the need for replacement is made evident by vibration. Slight misalignment in mounting also produces vibration, the detection of which is very important. Excessive vibration results in damaged equipment, decreased operating efficiency, increased operating cost, and inaccurate measurements. In the case of excessive vibration of a large turbine-generator, the results can be catastrophic.

Vibration monitoring keeps track of the condition of major pieces of rotating machinery: turbine-generators, electric motors, auxiliary turbines, boiler feed pumps, heater feed pumps, circulating-water pumps, condensate pumps, heater drain pumps, and forced-draft, induced-draft, and primary-air fans. Rotor unbalance is probably the most common of the vibration problems.

Three types of sensors are used to detect and measure vibration—accelerometers, velocity pickups, and displacement probes. The last two are most useful for detecting unbalance because their frequency response is within the range of most equipment running speeds, 500 to 5000 rpm.

Accelerometers. Accelerometers are most effective for monitoring high-speed phenomena, such as high-speed gear defects, and ball- or roller-bearing deterioration. Quartz (piezoelectric) sensors are used in one form of accelerometer. Quartz crystals act as a precision spring to oppose the acceleration forces of vibration and generate an electric charge proportional to the deflection. The crystals are very rigid, resisting deflection, so that they excel in dynamic pressure and force measurements. In motion-sensing accelerometers, they measure the force required to give a seismic mass the same motion as the structure to which the sensor is attached.

Velocity Pickups. Velocity pickups are most suitable for machines with low ratios of casing to rotor mass. The bearing-cap-mounted velocity pickup sensor is the most widely used because it is sensitive over a broad range. It most often sees service in the monitoring of smaller rotating machinery.

Velocity pickups and accelerometers are seismic devices: They do not measure shaft displacement directly, but rather measure the energy transferred from the shaft to bearing housings, where the pickup is generally located. Only displacement probes measure shaft displacement directly.

Displacement Probes. As a general rule, these probes have great accuracy, but are relatively expensive. They are the contacting (shaft riding) or noncontacting (proximity) types, mounted near or through the bearing (Fig. 1.36). A shaft rider consists of a spring-loaded rod assembly. The top of the rod is connected to a velocity pickup. It measures absolute vibration absolutely and exclusively.

Proximity probes do not contact the surface; they sit about 40 mils (1 mm) above and sense changes in a self-generated magnetic field, which is proportional to the gap distance. To measure the shaft absolute vibration, a proximity probe and velocity pickup are installed in tandem on the bearing housing (Fig. 1.37). This pairing provides both shaft-absolute and shaft-relative-to-bearing measurement.

Networks of vibration monitors can provide security for an entire installation. Not all measurements have equal importance, however. Turbines are protected by tripping them out on detection of high vibration; other devices are protected by alarms only.

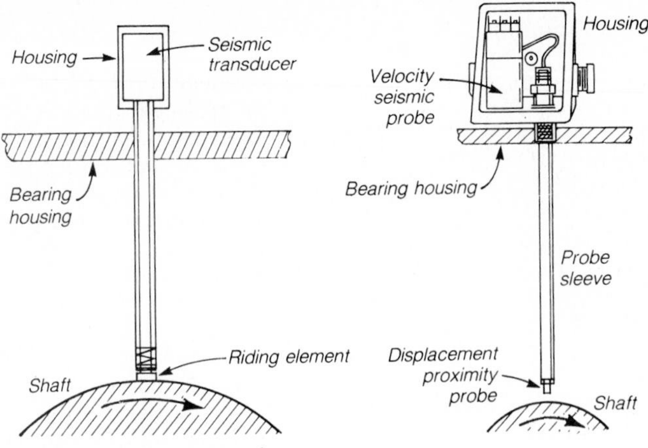

FIG. 1.36 Displacement probes.

Flame Detection

Detection of flame after firing is initiated is the keystone of burner safety systems. In addition to flame safeguards, a complete firing safety system should include the elements listed in Table 1.2.

After the firing of fuel in a combustion chamber has been established, and the refractory is heated to incandescence, loss of flame and reestablishing of firing are not serious; during start-up, however, fuel explosion is a potential hazard. Thus, detection of the presence of flame to verify ignition is a requirement for continued operation in burner safety systems.

Flame is an electromagnetic phenomenon. It contains ionized particles, components of infrared radiation caused by heat, visible radiation (mostly from incandescent carbon particles), and ultraviolet radiation. By applying these characteristics, four types of devices have been developed to detect the presence of flame: flame rods, photoelectric sensors, *infrared radiation* (IR) sensors, and *ultraviolet* (UV) sensors.

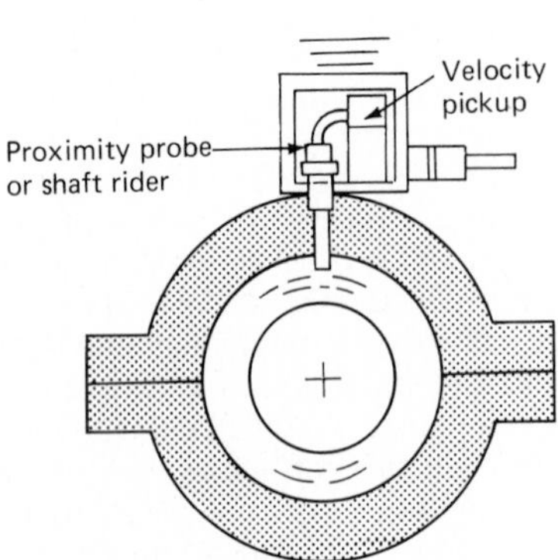

FIG. 1.37 Proximity probe and velocity pickup.

TABLE 1.2 Firing Safety System

Limit controls
High steam pressure
High water temperature
High and low gas pressure (gas-fired system)
Oil temperature and pressure (oil-fired system)
Minimum airflow
Water level
Atomizing steam (oil-fired system)
Operating controls
Fuel rate
Steam pressure
Water temperature and pressure
Air temperature
Interlocks
Low-fire start
Purge airflow
Fuel valves closed

Flame Rods. A flame rod is a stiff, heavy-gage metal rod mounted to project into the flame produced by gas or oil burners. Since flame is caused by a chemical reaction between fuel and oxygen, it liberates electrons (ionizes); consequently, it can conduct an electric current, alternating or direct. In the presence of a flame, a current passing through a rod in the flame will be short-circuited to ground. This type of probe does not require temperature buildup to function, and it proves flame at the ignition point. Flame rods are made of stainless steel and special metals such as Globar and Kanthal. They have a long life and operate in temperatures in excess of 2000°F (1093°C).

Limitations of flame rods are as follows: Carbon deposits can cause high resistance, blocking conduction. More seriously, any leakage can cause current flow and fool the flame safeguard circuitry. Unless precautions are taken, a short circuit caused by faulty insulation can simulate the presence of flame. The rods require frequent cleaning and become brittle. They cannot detect reliably when fuels high in sulfur are burned, because the flames have a low resistance. The low flame-rod current can result in nuisance shutdowns.

Photoelectric Sensors. Visible radiation in a flame is about 8 percent of the total band of wavelengths emitted. Photoelectric sensors are limited to oil flames because only these have enough visible radiation to make a cell effective. Photoelectric effects are produced by cells containing cesium oxide, lead sulfide, or cadmium sulfide. In a high vacuum, cesium oxide on the cathode emits electrons when light strikes it and acts as a rectifier. This phenomenon makes the cesium oxide cell useful for flame detection, because the rectification capability permits distinguishing between grounding caused by flame and spurious signals caused by high resistance or a short circuit to ground. The cell is also activated by light from hot refractory, however, and must have a view of the flame only. The temperature of the cell must remain under 165°F (73.9°C).

Infrared-Radiation Sensors. Almost 90 percent of the radiation emitted by a flame is in the infrared range of frequencies. IR sensors can sense this radiation from flame produced by burning oil, gas, or coal. The element most commonly selected is made of lead sulfide. The resistance of a thin film of this substance lowers in proportion to increasing radiation. Silicon photodiodes are also used. In either case, an amplifier, in combination with a bandpass filter tuned to eliminate the effect of flame flicker, generates the control signal.

Since the IR sensor does not come into contact with the flame, it avoids the deposition of carbon that troubles the flame-rod detector. Since its excitation source comes from a much greater amount of the available radiation, it is more responsive than the photocell. Nevertheless, there are limitations to the sensor's usefulness, many of which can be overcome by paying attention to three key application requirements:

1. The cell must have a good view of the flame. Hot refractory may emit more radiation than the flame itself, so the cell should be aimed to that part of the refractory that is as cool, is as far away, and has as small an area as possible (Fig. 1.38).
2. Sensitivity of the cell decreases as its temperature increases, so the cell must be protected from temperatures in excess of 150°F (65.6°C). The projection into the firing chamber should not exceed one-half the length of the cell.
3. Wiring should have twisted, short leads (50 ft, or 15 m); continuous ground returns; rigid conduit, firmly supported; and no power leads in the conduit.

Ultraviolet Sensors. Flame detectors that respond at wavelengths above 300 nm (IR detectors, for example) will respond to radiation from solid bodies, while UV detectors, which respond to wavelengths below 300 nm, will not. Thus, for some applications, UV detectors have advantages over IR detectors. In addition, all flames flicker and pulsate, and certain ranges of this phenomenon are more reliably monitored by UV sensors.

The lens of the UV detector is made of a special quartz-type glass, selected to pass the short UV frequencies but to block other short waves such as x-rays. The cathode of the UV detector emits electrons to the anode when high voltage is applied and the tube is exposed to UV light. The gas in the tube ionizes, and the tube becomes conductive. As it conducts, its potential drops sharply, electron flow stops, and ionization is quenched; then the cycle is repeated. The signal emitted from the tube is a random pulse, at a rate proportional to the UV radiation present.

Hot refractory is not a problem for this method of sensing, because most of the radiation is in the visible and IR bands until temperatures of the refractory exceed 2500°F (1371°C). This is not a problem either, because by the time this temperature is reached, any flame is self-sustaining.

Installation precautions should be observed: The sensor should have an unobstructed view of the flame and should preferably be sighted on the outer one-third of the flame area, where the pattern is irregular. The sensor must not sight the ignition spark used for gas pilots. If it cannot be sighted far enough out from the pilot to be beyond the spark, a barrier should be constructed to block the view. One possibility is to restrict the viewing angle by using a longer sighting tube. The electronics must be protected from temperature. A good rule of thumb is that it should be possible to lay a hand on the housing without discomfort. Wiring techniques noted above for IR detectors also apply to the UV type.

Because no detector is absolutely satisfactory, viewing heads that combine both UV and IR detectors have been designed and are being used. One approach

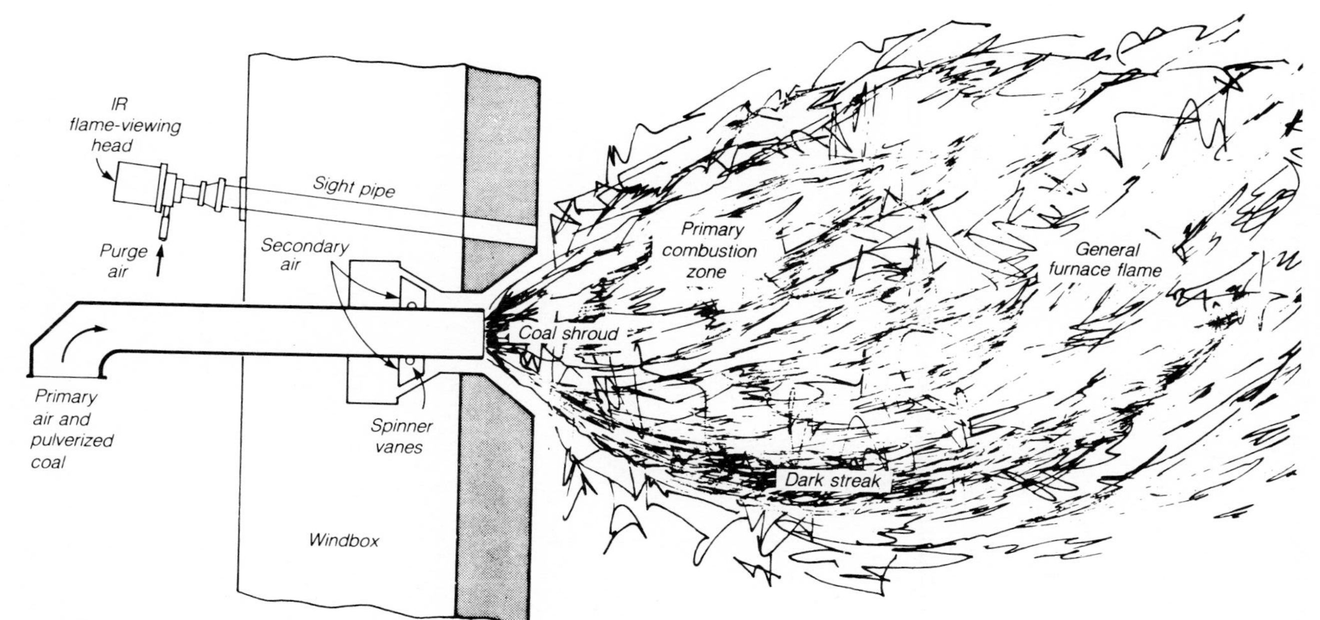

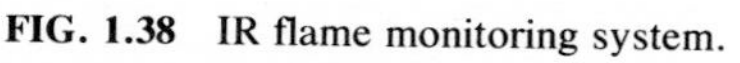

FIG. 1.38 IR flame monitoring system.

is to combine a UV tube and an IR solid-state photodiode and either combine the flame signals at the viewing head or transmit them separately.

ACKNOWLEDGMENT

Illustrations in this chapter are courtesy of *Power* magazine except where noted otherwise.

SUGGESTIONS FOR FURTHER READING

Benedict, R.P., *Fundamentals of Temperature, Pressure, and Flow Measurements,* John Wiley & Sons, Inc., New York, 1984.

Cho, C.H., *Measurement and Control of Liquid Level,* Instrument Society of America, Research Triangle Park, N.C., 1982.

Hougen, J.O., *Measurements and Control Applications,* 2nd ed., Instrument Society of America, Research Triangle Park, N.C., 1979.

Miller, R.W., *Flow Measurement Engineering Handbook,* McGraw-Hill Book Co., New York, 1983.

Sydenham, P.H., *Handbook of Measurement Science,* John Wiley & Sons, Inc., New York, 1983.

CHAPTER 6.2
POLLUTION INSTRUMENTATION

J. A. Moore
Leeds & Northrup, A Unit of
General Signal Corp.
North Wales, PA

INTRODUCTION

Pollution supports the descriptive adjectives *contaminated* or *impure*. In powerplant practice the incoming water, earmarked for cooling and as the raw material for making steam, must have its impurities removed to be usable. As water is used, it becomes contaminated and must again have its impurities removed, or at least limited, so that it can be discharged back into natural waterways.

At the same time that impurities are increasing in the water, impurities are introduced into the air above the plant—products of the combustion of the fuel that supplies the energy for making steam. It is important to keep these airborne impurities at a minimum, so that contamination surrounding the plant is limited. The instrumentation used for monitoring and controlling the pollution of water and air around powerplants is the subject of this chapter.

Reliability is an important characteristic of instrumentation for pollution monitoring, because instruments often must perform in harsh environments. State-of-the-art sensitivity is equally important, because measurements must often be made of trace amounts.

WATER POLLUTION

Treatment of water, plus its measurement to determine the need for and extent of treatment, is required for four different categories of powerplant water use:

- Source water from rivers, lakes, and other waterways
- Boiler feedwater and condensate
- Cooling water
- Wastewater to external waterways

See Section 1, Chapter 10, and Section 4, Part 2, Chapters 1 to 4 for in-depth coverage of the categories of powerplant water use and treatment recommended in each case. For review, also see Table 2.1.

Specific Measurements

Continuous measurements most useful for maintaining the purity of water are conductivity, pH, sodium ion, dissolved oxygen, hydrazine, and turbidity (also see Fig. 2.1). Other tests, from laboratory analyses of samples, may be conducted for these contaminants:

Aluminum	Ammonia
Calcium	Carbon dioxide
Chelant	Chloride
Chromate	Chromium
Copper	Hydrazine
Iron	Nitrate
Nitrite	Phosphate
Silica	Sulfate
Sulfite	Zinc

TABLE 2.1 Impurities in Natural Water Sources

Impurity	Harmful action	Treatment
Dissolved gases		
Oxygen	Metals corrosion	Deaeration, scavenging
Carbon dioxide	Metals corrosion	Deaeration, neutralization with alkali
Suspended solids		
Deposits in pipes and vessels	Carry-over to turbines	In incoming water: settling, coagulation, filtering Formed during process: blowdown
Dissolved solids		
Calcium and magnesium (hardness)	Scaling	Softening, demineralization
Sulfates, chlorides, nitrates, sodium	Corrosion, scaling, embrittlement	Demineralization
Silica	Scaling	Ion exchange

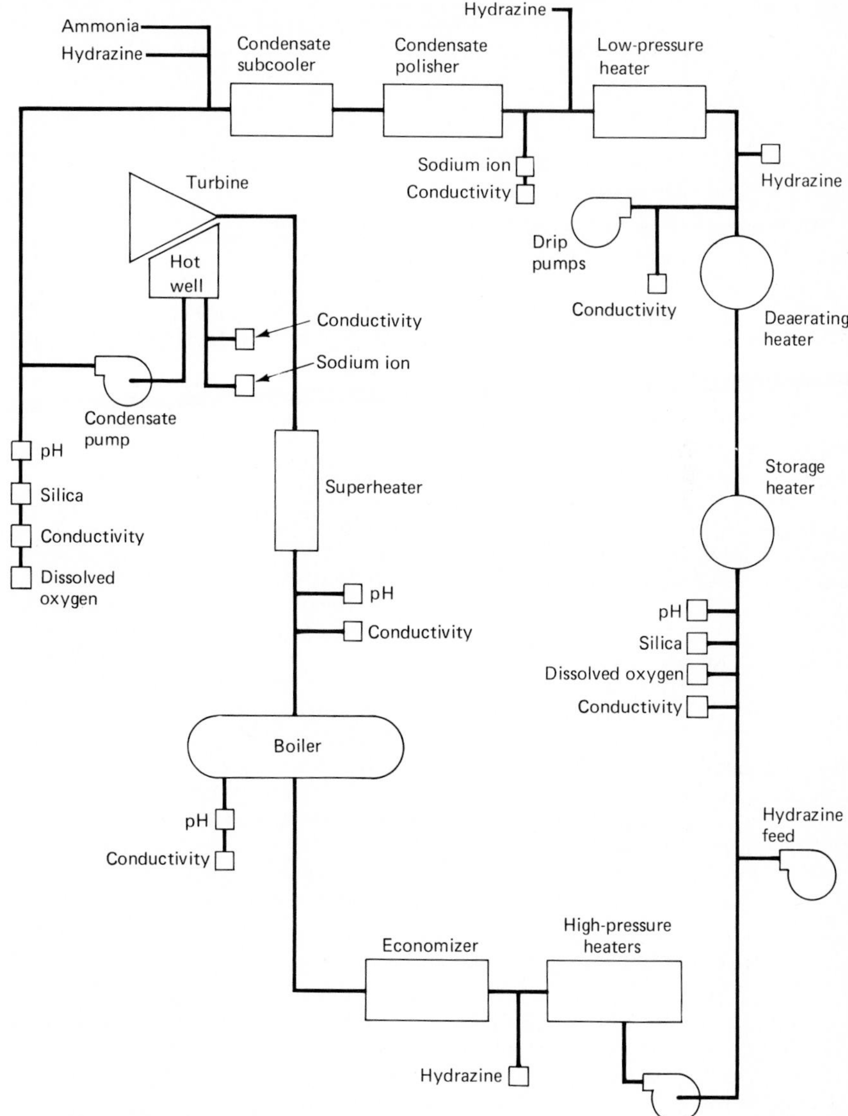

FIG. 2.1 Typical sampling points in high-pressure utility boiler. *(Courtesy J. A. Moore.)*

Conductivity. Conductivity relates to these activities:

- Monitoring continuous blowdown
- Monitoring influent and effluent of mixed-bed demineralizers
- Monitoring product-water quality
- Signaling exhaustion of the ion-exchange resin bed

Regeneration of a demineralizer is regulated from the conductivity measurement of rinse water. The regenerated unit may be placed back in operation when the final rinse-water conductivity shows that all the acid and caustic used in regeneration has been purged from the bed. Conductivity measurements can also

- Signal condenser-tube leaks at the hot well
- Show contaminated drips
- Monitor the solids content of polisher and deaerator
- Assure the operator that boiler water and steam are free of excess dissolved solids

Theory of Measurement. Electrolytic conductivity is a measure of the ability of a solution to carry electric current. Conductance is the reciprocal of resistance. The unit of conductance is the mho, or (in Europe) the siemens. The terms are equivalent.

Conductivity measurements are functions both of the shape of the measuring cell and of temperature. They relate to a cubic cell, 1 cm on a side, or a similar electrode length-area ratio. The reciprocal of the resistance of a solution in a cubic centimeter is the specific conductance, or conductivity, of the solution in mhos per centimeter. It is specific for the solution's composition and concentration measured. In practice, the conductivity of the most frequently measured solutions is so small that, to avoid the use of unwieldy decimals, a unit that is one-millionth of 1 mho/cm, the micromho per centimeter, has been adopted. The European equivalent is the microsiemens per centimeter.

The conductance of electricity through all forms of matter is associated with the flow of electrons. Applying a potential to a liquid conductor will cause current to be carried through the solution by dissolved particles having electric charges, called *ions*. To calibrate a commercial cell (which will not be 1 cm^3 in size or shape), the resistance of a standard solution R_c is measured in the cell to be calibrated, and it is compared to R_s, the measurement made in a standard cell at the same temperature. Calibration is usually done with a known concentration of KCl (potassium chloride) as a standard.

The ratio R_c/R_s is known as the *cell constant K*, and it can be used to convert readings made with the commercial cell to specific conductance:

$$\text{Specific conductance} = \text{Cell constant} \times \text{Measured conductance}$$

A large number of cell constants have evolved to measure extremes of conductivity without going to extremes of conductance.

Traditionally, conductance is measured with an *alternating-current* (ac) bridge (Fig. 2.2). The conductivity cell forms one arm of the bridge. The resistance of the solution is measured, and the reciprocal value is computed. Present-day models make the measurement with microprocessor-based electronics.

Electrode materials affect the conductive properties of conductivity cells. Commonly used materials are stainless steel and titanium; for high-accuracy measurements, platinum may be selected. Platinum is resistant to oxidation and eliminates dissolution of the electrode that may occur during measurement.

Conductivity in solutions occurs because of electrons carried by ions. Ion mobility increases with increasing temperature. Because temperature coefficients vary from solution to solution, special temperature compensators must be de-

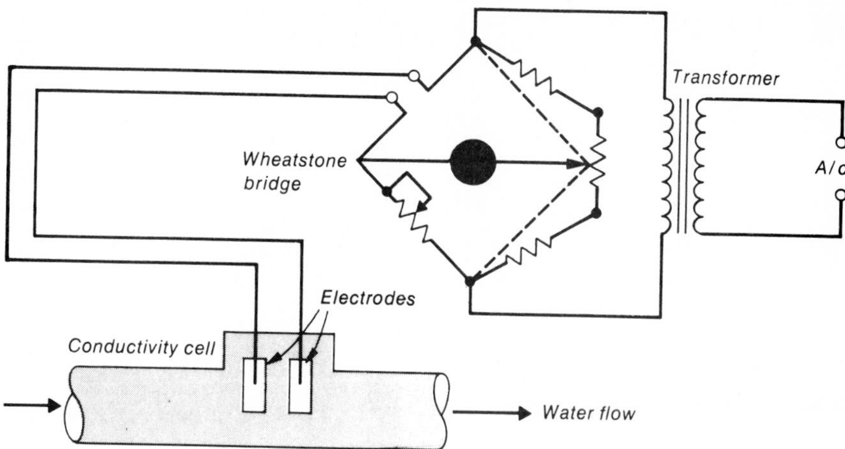

FIG. 2.2 Conductivity measured with cell electrodes and Wheatstone bridge.

signed for each solution. A typical automatic temperature compensator is made from a negative-coefficient thermistor having a characteristic curve with a slope similar to the curve of the solution to be compensated.

Conductivity cells can be inserted into the solution to be measured or can be built into an assembly through which the solution flows. Measurement in a well-mixed tank requires a fully immersed dip cell. Continuous measurement in a pipe requires an industrial flow cell assembly.

Measurements in High-Purity Water. In a conductivity measurement, current is conducted by whatever ions are in solution. The substance providing the ions is not identified—only the total ion activity is apparent. Each contribution, in turn, is affected by temperature, and each is affected in a different way:

- As temperature rises, the decreasing viscosity of water allows ions to slip through more easily. The conductivity-temperature coefficient of most solutions, at all conductivity levels, is the same, about 2 percent per degree Celsius.
- Absolutely pure water conducts current because of the presence of small amounts of H^+ and OH^- ions. The amount of water that ionizes into these ions changes dramatically with temperature. The greater the purity, the more significant the temperature effect caused by changes in ionization. Figure 2.3 illustrates how the effect of temperature is increased as water purity increases. At 77°F (25°C), the temperature coefficient at 0.055 micromhos/cm is about 5 percent.

Small amounts of any other ionized impurity—sodium, for example—also contribute to the conductivity measurement and have their own temperature relationships. This means that the conductivity measurement, to be reliable for measuring small amounts of material in high-purity water, must be performed at a specific temperature (usually 77°F, or 25°C) or else must be temperature-compensated—for both water ions and those of any other material.

To make this measurement at temperatures other than standard, follow these steps:

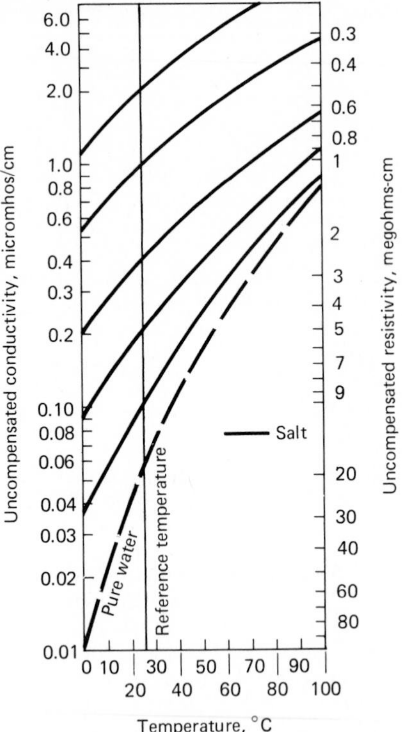

FIG. 2.3 Conductivity-temperature relationship in high-purity water.

1. Measure the temperature, and scale it to a voltage value that represents the conductivity of pure water at that temperature.
2. Subtract the pure-water value from the main reading, leaving only the conductivity contributed by the salt ions.
3. Make the correction for the salts, to the same temperature as used for the water correction.
4. Add the two corrected values.

This is difficult to do with analog equipment, but it can easily be programmed into a microprocessor-based, digital readout device.

Cation Conductivity. Conductivity of steam serves to detect the carry-over of solids. To measure conductivity, steam must be condensed, cooled, and measured at a constant temperature. Cation conductivity gives good results here. Cation-conductivity analyzers pass condensed steam through a small cation-exchange column operating on the hydrogen cycle, replacing other cations with more conductive hydrogen ions and removing the typical treatment chemicals (ammonia and amines). Subsequent reboiling of the condensate sample effectively removes CO_2, after which the specific conductivity is measured.

This technique greatly increases the sensitivity for impurity detection. However, the resulting acidic sample from the exchange column has a different temperature coefficient from that of salt-contaminated water, and the analyzer must be properly adjusted to take this fact into account.

pH for Acidity or Alkalinity. The pH is defined as the measurement of the hydrogen-ion, H^+, concentration in a particular solution. It is widely used as a measurement of the degree of acidity or alkalinity. In powerplants, pH is measured at the condensate pump discharge, at the deaerator outlet, and in boiler water and steam samples to maintain limits of acidity that prevent corrosion of copper and steel.

Theory of Measurement. When molecules of hydrogen and oxygen are bound together to form H_2O, the resulting molecule is said to be *associated.* A small fraction of the associated water breaks up into small fragments (dissociation). One of these fragments is the hydrogen ion, H^+; the other is called the *hydroxyl ion,* OH^-. The extent of the fragmentation is one H^+ or one OH^- to 5.5×10^8 water molecules. The concentration of either ion is 10^{-7}, and the product of the two concentrations is 10^{-14}.

The pH is a number indicating the quantity of hydrogen ions dissociated in a solution; it is also a relative index of acidity and alkalinity. The pH is the reciprocal of the logarithm of the hydrogen-ion concentration, and thus the pH of pure water (at 77°F, or 25°C) is 7. The measurement of pH is applied to control water treatment and serves as an index of water pollution. The relationships are illustrated in Fig. 2.4.

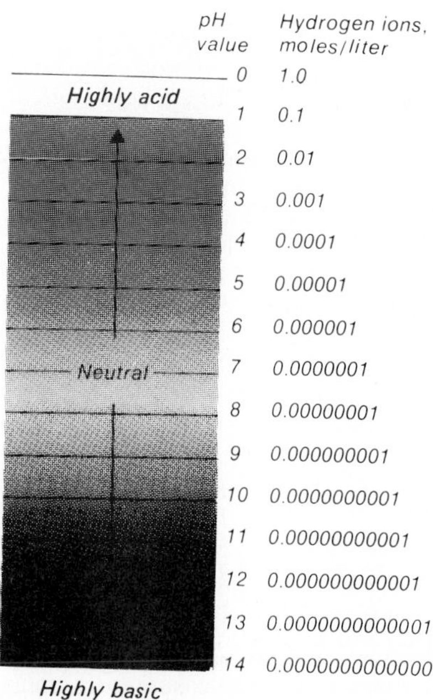

FIG. 2.4 The pH scale is a function of hydrogen-ion concentration.

Acids like hydrogen chloride (hydrochloric acid) contribute hydrogen-ion concentration to a solution. The product of H^+ and OH^- concentrations must remain at 10^{-14}, and so if the hydrogen-ion concentration increases to 10^{-3}, for example, the hydroxyl-ion concentration must decrease to 10^{-11}. The hydrogen-ion concentration in this particular solution is 10^{-3} *M*. And, of course, the reverse is true for an increasing hydroxyl-ion concentration. In the example, the pH is 3.

The pH is controlled by adding enough hydrogen ion or hydroxyl ion to the solution to neutralize the ion that is in excess. Remember that pH represents a logarithmic relationship between acidity and concentration. The amount of hydroxyl ion needed to change the pH from 5 to 6, for example, is 10 times that required for a change from 6 to 7.

Electrode Systems. The sensor commonly used to measure pH is an electrode assembly that develops an *electromotive force* (emf) proportional to the hydrogen-ion concentration. The assembly consists of a measuring electrode, reference electrode, and temperature compensator (recommended if the temperature at the measuring point is not constant).

The pH of an aqueous solution is defined in terms of the voltage which exists between a measuring element having a pH-sensitive glass membrane and a reference electrode immersed in the solution. The pH electrodes can be immersed directly into the solution, or they can be mounted in an assembly through which the solution flows. The three cell elements are usually mounted separately, but there are cell assemblies that combine all three into one mounting.

To measure the potential developed by ions in a solution, a closed circuit must be established with two conductors in contact with the solution. One will be sensitive to the ions present, and the other, the reference electrode, must be as insensitive as possible to all ions. It must produce a stable potential in any solution measured and be insensitive to its environment.

Acceptable reference electrodes include the silver–silver chloride electrode—a silver wire coated with insoluble silver chloride and placed in a solution of KCl saturated with silver chloride. There must be a liquid junction between the solution being measured and the solution inside the reference electrode. To keep process fluid out of the cell and prevent contamination of cell chemicals, most reference electrodes do not have an internal solution that "flows" out of the electrode enclosure. Silver–silver chloride electrodes contain a gel saturated with KCl and silver chloride. The liquid junction to the process is made through a porous medium such as wood, ceramic, or a polymer.

Figure 2.5 shows how potentials are developed in a hypothetical pH measuring cell that includes glass and reference electrodes. Glass electrodes are constructed from specially formulated glass membranes that respond to specific ions by means of an exchange of mobile ions at the membrane surface. For pH, the glass contains alkaline metals that make it conductive and respond to the hydrogen-ion concentration.

In Fig. 2.5, an internal electrolyte *A* is buffered and thus has a constant hydrogen-ion concentration. The electrolyte develops a concentration at the inner surface *C* of the glass membrane *M*, the active element in this measuring system. Hydrogen ions from the test solution develop a potential on the outer glass surface *D* proportional to the hydrogen-ion concentration of the solution. The result is a difference in potential between the external unknown solution and the known internal solution *A*. If the two solutions have the same pH, the differential will be 0 V. For every pH unit change, a specific output (in millivolts) will occur at a specific temperature.

A slow bleed of electrolyte solution *G* from the reference electrode through

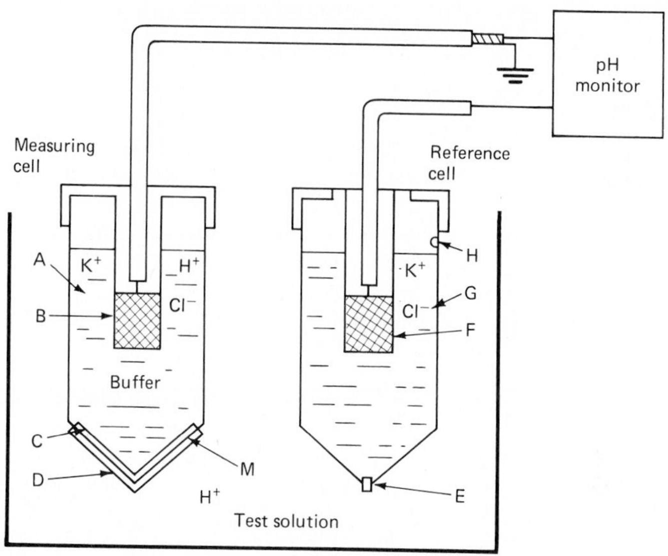

FIG. 2.5 The pH measuring system. *(Courtesy J. A. Moore.)*

junction *E* provides a "liquid wire" to the test solution, completing the electric circuit between the reference and measuring electrodes. Internal element *B* in electrode *A* and internal element *F* in electrode *G* theoretically generate identical potentials. The potential developed at the interface of the glass membrane and the test solution is transmitted to the pH monitor.

Temperature Compensation. Even though pH remains constant, electrode output will vary with temperature. This is called the *Nernst effect*. The variation can be compensated automatically by temperature-sensitive resistors mounted in assemblies in contact with the solution.

Also, the hydrogen-ion content, and so the pH of a solution, will vary as the temperature of the solution changes. The variation will not be significant if the ions are supplied by dissolved salts, but high-purity water is an exception to this rule because it contains no buffering ingredients. A more advanced type of temperature compensation, available in some microprocessor-based digital pH instrumentation, is recommended for obtaining reliable measurements.

Measurement in High-Purity Water. "Pure" water is a poor conductor, which puts constraints on how measuring equipment is applied, pH equipment in particular. In addition to the temperature effect mentioned above, these interferences should be considered:

- The pH measurement is sensitive to electrical noise and static electrical pickup.
- If the high-purity water flows through insulated tubing or plastic pipe, it can generate its own static charge. This is sometimes called *streaming potential.*
- The resistance must be low at the junction where the reference electrode maintains contact with the measured solution by means of a liquid junction or restricted interface. In high-purity water, the electrolyte salt fill can be washed away from the porous junction, leaving too high a resistance.

The following recommendations provide solutions to these problems:

1. Use a grounded, conductive-metal flow chamber.
2. Maintain a low sample flow—about 100 cm^3/min—to minimize streaming-potential effects.
3. Arrange the system so that the two electrodes are placed in parallel rather than series flow paths. This minimizes voltage differences between the electrode sensing ends.
4. Select combination electrodes with concentric measuring and reference sections; they can minimize streaming-potential effects.
5. Use a flowing junction reference electrode. Here, a continuous supply of electrolyte flows into the cell through a fill port and through the junction, at a slow rate. A slurry-filled reference electrode is another solution.
6. Provide temperature compensation that corrects for solution temperature changes as well as for electrode temperature response.

Sodium Ion. Measurement of sodium ion in high-purity water can detect condenser leaks, boiler carry-over, and breakthrough in ion-exchange equipment. It is often measured in the hot well to make sure the condensate has not picked up contaminants in the condenser. Measurement at the effluent end of the condensate polishing demineralizer detects breakthrough caused by resin exhaustion or channeling. Sodium-ion measurement is more sensitive than cation conductivity. It can give good results where ammoniated resins are used in demineralizers, a practice that decreases the sensitivity of conductivity measurement.

Sodium-ion measurement is made in an analyzer containing a glass electrode, a slurry-filled reference electrode, and a temperature compensator. The glass tip of the measuring electrode is a special formulation responsive to changes in sodium-ion concentration. Where the sample pH is less than 8.8, the sample at the analyzer must be treated to overcome the hydrogen-ion interference. Ammonia and amines are reagents used to increase the pH to permit sodium-ion measurement at low solids concentration levels.

Dissolved Oxygen. Dissolved oxygen is measured at the hot well to make sure the condensate has not picked up an excessive amount of dissolved oxygen in the condenser or in the first stages of the low-pressure heaters. It also indicates the efficiency of the deaerating process.

Methods available are based on galvanic principles, similar to fuel-cell design. One such design allows oxygen to diffuse through a semipermeable membrane into a polarigraphic cell. Another design also relies on this membrane, but balances regeneration of oxygen in the cell with the level of oxygen in the sample.

The probe for this second design consists of three electrodes. Two active electrodes (cathode and anode) are mounted on a supporting substrate and covered with an electrolyte. The third (reference) electrode also contacts the electrolyte to create the electromechanical potential. When the probe is immersed in a sample stream, oxygen diffuses through the membrane and is reduced at the cathode. Simultaneously, an equal amount of oxygen is generated at the anode. The diffusion continues until the oxygen tension on both sides of the membrane is at equilibrium. The current necessary to maintain them equal and in balance is converted by the electric circuitry of the analyzer to read out the concentration of dissolved oxygen in the solution.

Turbidity. Turbidity is the appearance of a liquid containing suspended particles so small that the solution appears "smoky," "hazy," or "milky" instead of clear. The relative concentration of suspended solids is measured by electro-optical means, that is, by shining a light through the liquid and comparing the amount absorbed to the amount scattered. In powerplants, turbidity measures suspended solids in raw water, filtered water, condensate, feedwater, cooling-tower effluent, and wastewater. Turbidity measurement can detect breakthrough of deionizers before reactive silica shows up.

The Jackson candle turbidimeter, based on light scattering, is the original laboratory source for the standard. The *Jackson turbidity unit* (JTU) is the standard unit of turbidity measurement, although solids concentration can be expressed in *parts per million* (ppm) also; 1 JTU = 1 ppm suspension of fuller's earth.

There is no direct method for measuring turbidity. The indirect method used is the determination of the ratio of light scattered to light transmitted (Fig. 2.6). Different analyzer designs determine light intensity for transmitted light, for forward-scattered light, and for light scattered at 90°. At low turbidity, the forward scatter is insignificant and transmitted light attenuation is minimal, so 90° scatter is the most effective. At high turbidity, both transmitted light and forward-scattered light have an effect on the 90° scatter measurement.

A turbidimeter that measures the right-angle scattering of light in a sample fluid is called a *nephelometer* and is sensitive to low turbidities. An *absorptometer* is a turbidimeter that measures light passing through a sample, often as a ratio of direct to scattered light, and is suitable for higher turbidity measurements.

Specific Techniques

Chemical analysis can be performed by collecting grab or composite samples and analyzing them in the laboratory, but this does not permit real-time control to be applied. On-line analysis of important but less commonly used variables is possible with automated colorimeters and suitable sampling systems.

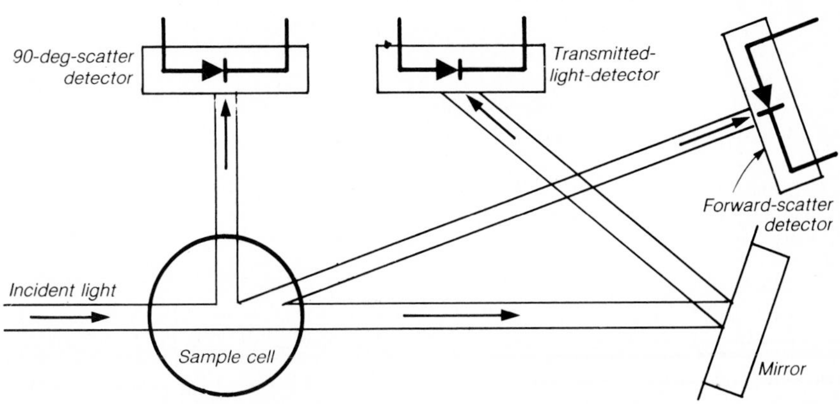

FIG. 2.6 Turbidity measurement.

Chemical Analysis. Colorimetric analyzers must measure and mix sample and reagents in precisely controlled ratios to ensure a continuous sample for the colorimetric cell. Capillary, gravity-flow analyzers may be used for trace analysis of silica. Most measurements are taken with a positive-displacement pump to sample, transport, and dispense process liquid and reagents. In one design, the pump cylinder serves as the colorimetric cell to pull in a sample, hold it until complete reaction with the color-producing reagent occurs, then flush it out and wipe the cell wall clean at the same time. This method allows semicontinuous analysis for hardness, chlorine, alkalinity, chelant, chromate, copper, hydrazine, and phosphate.

Silica. High-pressure boilers are vulnerable to accumulations of deposits. Reactive silica can be removed by ion exchange. Continuous monitoring of silica at the condensate pump and at the deaerator effluent tells the operator whether to reduce pressure as a result of either excessively high silica entering the system with makeup water or a condenser-tube leak. Monitoring of silica at the anion-exchanger outlet provides the most sensitive detection of resin exhaustion, since silica is the first contaminant to break through.

Hydrazine. Hydrazine is monitored at the effluent of the low-pressure heater and at the economizer inlet. When a residual amount of hydrazine shows at the economizer, it may be taken as conclusive proof that no dissolved oxygen is in the boiler water. Best control of hydrazine treatment depends on maintaining this residual amount within a zone. Too much reagent will decompose, yielding excessive ammonia and increasing system alkalinity.

Total Alkalinity. Excessive carbonate content promotes scale formation in cooling towers and recirculating-water systems. Too much hydroxide in boiler water can cause embrittlement of metals. Monitoring can detect these excesses.

Chelants. Chelants are reagents that chemically remove calcium and magnesium scale from the waterside of boiler surfaces. Insufficient addition of chelants permits accumulation of scale and sludge. Too much chelant is also unsafe, because the protective oxide films on iron- or copper-based metals will be attacked, opening the way for corrosion. Monitoring will alert maintenance people to incorrect chelant additions.

Chromates. One effective method of cooling-tower treatment for inhibiting corrosion hinges on maintaining an optimum concentration of hexavalent chromium in the system. Monitoring ensures that this concentration is maintained. Effluent treatment is mandatory, since chromium is highly toxic.

Phosphates. Phosphate is used as a scale conditioner in boiler and cooling-tower water. Effluent discharges, accurately monitored, limit phosphate additions to prevent excessive amounts from discharging to waterways, where the chemical can promote the growth of algae.

Sampling. Since water analysis relates to impurities, analyzers will require regular maintenance to remain effective. Problem areas and solutions include the following:

- Fouling occurs due to scale and solids deposits. Purge lines are recommended to divert flows containing large particles and filters to stop small particles. Two 0.9-μm filters in parallel should be added upstream, with valves to switch from a loaded filter to a clean one.
- Oversized sampling lines result in low flow velocities, which increase sample travel time, allowing the composition to change before the sample reaches the analyzer, and let solid materials drop out of suspension and block the flow

path. A bore of 0.166 in (4.2 mm) is recommended as an optimum diameter for a sample line.

- Variations in flow rate affect analyzer performance. Pressure reduction is recommended for flow regulation if the source pressure is above the analyzer's pressure limits.

AIR POLLUTION

Pollution of the air created by powerplant operation comes from undesirable products introduced with the fuel or combustion air and from products of incomplete combustion. If not removed, the pollutants become airborne, leaving the plant through the exhaust stack as gases and solid matter.

The amount of air pollution can be minimized by (1) efficient combustion—not creating pollutants in the first place; (2) removal during combustion via sorbent injection or fluidized-bed techniques; and (3) removing flyash and sulfur compounds from the exhaust gases with filters, precipitators, and scrubbers. See Section 4, Part 1, Chapters 3 to 6, for control details.

Basic Measurement Areas

To control air pollution, quantifying such pollutants as flyash, sulfur, and other stack exhaust products is the first step. Understanding the fuel-air ratio is also important.

Monitoring Flyash and SO_2. Wet process scrubbing is mostly a U.S. preference. Instrumentation is provided for the measurement and control of bypass and gas flows, and slurry preparation and density. Besides temperature and flow, pH and SO_2 measurements help to control scrubber operation. In lime systems, pH controls the lime feed rate; in limestone systems, control is a function of either pH or SO_2 concentration and boiler load.

Monitoring Stack Exhaust Products. To monitor the condition of exhaust gases, measurements are made of the amounts of SO_2 and NO_x. To control process conditions and to enhance boiler efficiency, the concentrations of O_2, CO, CO_2, and water vapor are also measured. The amounts of carbon and carbon compounds that are combustible are measured as part of determining the efficiency of combustion. Opacity measurements give a valuation of the amount of solid material in the exhaust gas. Carbon in flyash is measured also, as a measure of efficiency.

Two temperature measurements are helpful in the control of pollution:

1. In the furnace, the flame temperature has an influence on the formation of nitrous oxides.
2. At the stack, the temperature of stack gases must not fall to such a low value that water vapor in the gases condenses. If it does, it can dissolve SO_2 and SO_3 that may be present and create acids that will corrode the metal portions of the breeching and stack.

Opacity measurements contribute to efficient operation by indicating the energy that is lost through the stack in the form of unburned carbon. Also CO is a measure of inefficiency, because efficient combustion will convert it almost completely to CO_2.

Determining the Fuel-Air Ratio. Measurements of O_2 and CO are important in determining the amount of excess air (Fig. 2.7). These measurements can also help to trim controlled fuel-air ratios. With excess-O_2 measurement, the percentage of O_2 measured in the stack is matched to the setpoint of a proportional controller set at the average desirable value. Deviations from setpoint will create a controller output that, in turn, will adjust the fuel-air ratio.

Measurement of CO also serves as an indicator of combustion efficiency. There is a very sharp change in the relationship between CO concentration and efficiency between 200 and 300 ppm of CO. This is a function of burner operation and the burner's ability to convert all the energy in the fuel to usable energy. In Fig. 2.7*c*, the point of deflection (zero slope) is the most desirable position on the curve for operation.

Measurements of O_2 and CO both have their advocates as indicators of efficiency. The CO content of the exhaust gas is a direct measure of the completeness of the combustion process. And O_2 measurement is affected by air drawn into the stack by leaks in the ductwork and for this reason may be less represen-

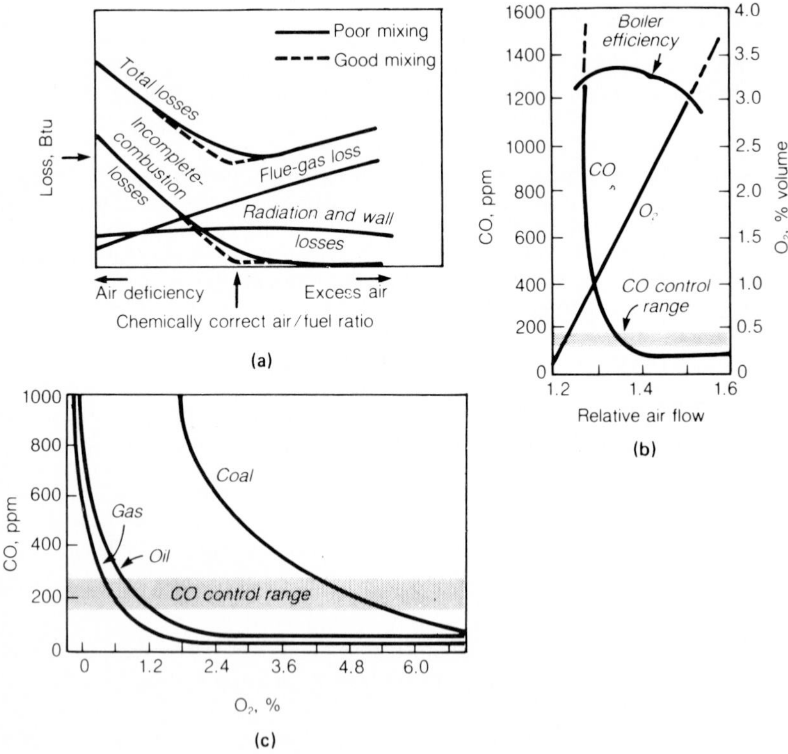

FIG. 2.7 Flue gas and combustion efficiency: (*a*) major system losses, (*b*) CO versus excess air, (*c*) CO versus O_2 control.

tative of the true burning condition than CO measurement. A spectroscopic analyzer that samples the entire gas stream by shining a beam of light across the stack is the usual technique for CO analysis.

Relative cost and payback are prime considerations in choosing the type of measurement for excess-air control. A system controlling on the basis of CO is inherently limiting combustibles from exiting through the stack; however, it is more expensive than O_2 measurement. Installed cost for CO analyzers is about 5 times that for O_2 analyzers, and maintenance costs are about twice as much. Nevertheless, energy savings should be at least 1 percent higher with CO measurement.

Combinations of the two measurements can be used to improve fuel-air ratio control. With solid-fueled boilers, O_2 measurement can determine the position on the CO curve where best operation takes place. Stack O_2 content varies in a relationship with stack CO content, as Fig. 2.7*c* shows.

Stack-Gas Sampling and Analysis

Stack-gas analyzers are either extractive or in situ. The terms refer to the way the gas sample is delivered into the analyzer. In extractive sampling, a gas sample is drawn out of the duct or stack, then passed through a sample conditioner, where it is prepared for analysis by a remotely located analyzer. In situ analysis features an analyzer mounted on the stack, with its sampling apparatus directly in contact with the stack gases.

Maintaining the sampling equipment has long been as much a problem as performing the analysis. Consequently, an analyzer that does not require sample transport and conditioning because it has its entire active element directly in the gas stream is welcomed by the industrial user. The in situ analyzer, which became available in the 1970s, provides this benefit and has been accepted for industrial and utility applications. Both types of sampling are discussed.

Extractive Sampling. In extractive sampling, a gas is drawn out of a duct or stack by an aspirator or a pump, and then it is passed through a sample conditioner on its way to the measuring instrument. Analyzers generally operate on a dry basis. Sample conditioning may include (1) filtration to remove particulates, (2) refrigeration to remove water vapor, (3) heating or insulation of lines to maintain proper processing temperatures, and (4) introduction of standard composition gases for calibration, so that zero and span adjustments for scaling and calibration can be made.

Extractive sampling starts with a probe inserted into the stack or duct (Fig. 2.8). The probe must include filters, a potential source of plugging. An air- or water-operated vacuum eductor draws the sample into the probe. Primary filters may pass up to 50-μm particles, while secondary filters may take out all but less than 1-μm particles.

The filtered sample enters a conditioning chamber mounted near the sampling point. Acid-mist condensables and entrained liquids are removed by a chiller. The sample is then heated, filtered to remove the last trace of particulates, and dried below its dew point (Fig. 2.9).

Extractive sampling allows location of the analyzer at a remote site, but an interval must necessarily lapse between the time at which the sample is pulled from the gas stream and the time at which analysis takes place. Thus, the reading is always somewhat behind the process. More seriously, sampling equipment is

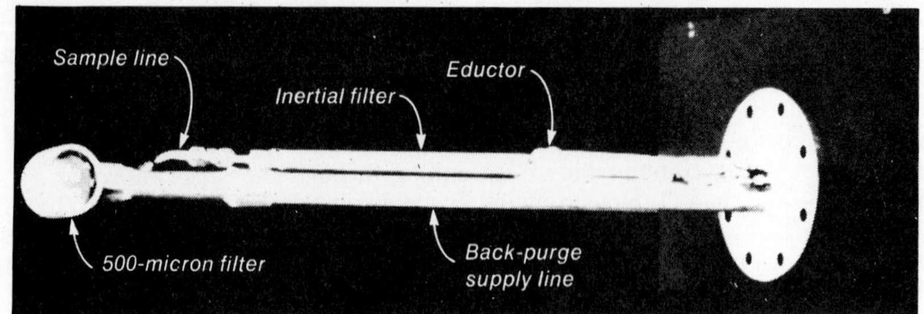

FIG. 2.8 Probe for extractive sampling.

vulnerable to plugging and loss of sample. Good consistent maintenance is required to make it work.

In Situ Sampling. When sampling for oxygen is done in place, the measurement element is mounted on a probe directly in the hot gas stream. Also, a sample can be drawn out of the gas stream and passed through an analyzer mounted on the side of the duct or stack.

In situ measurement of CO, CO_2, SO_2, NO_x, and unburned hydrocarbons combines a light source shining across the stack with a receiver-analyzer (Fig. 2.10). It is based on absorption spectroscopy—measuring in the ultraviolet, visible, and near-infrared portions of the optical spectrum. The molecules of each different material vibrate at specific frequencies, which cancel equivalent light frequencies in the light beam.

Detection of the absorbed frequencies in the spectrum from a narrowband source identifies the components and their concentrations. Maintenance is required to prevent fouling of windows at the transmitting and receiving ends of the light path, but this fouling can normally be minimized by a continuous air purge at both ends of the beam. The preferred location of sampling for O_2 is in the breeching; for other gases, on the stack.

Gas Analysis. Regardless of the sampling method, gas analyzer design follows a pattern that makes use of a measuring cell and a reference cell. The material in

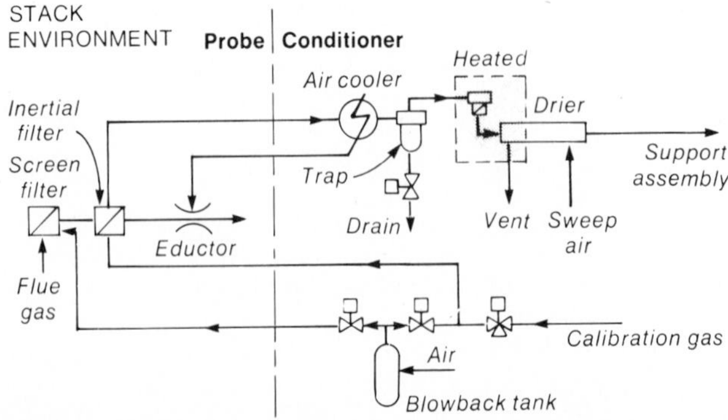

FIG. 2.9 Conditioning assembly is mounted on outside of stack.

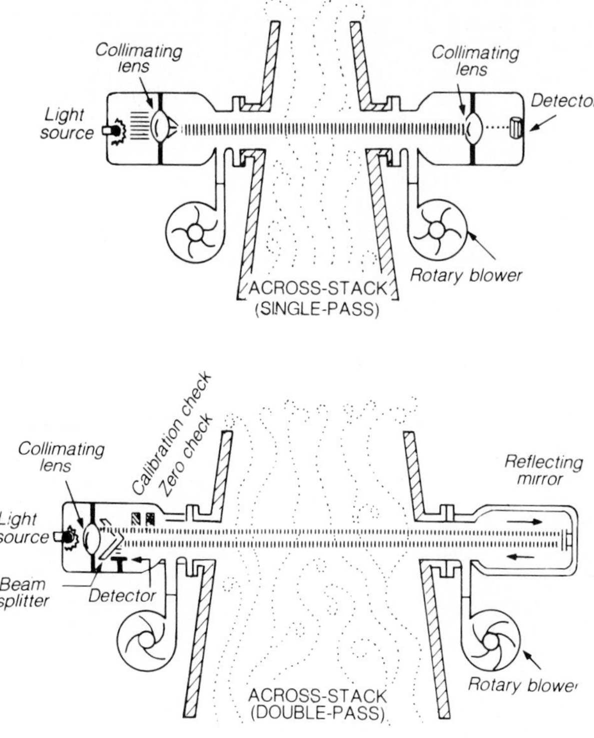

FIG. 2.10 In situ measurement, with single-pass (top) and double-pass (bottom) arrangement.

the reference cell is of standard composition, often air, and the results of the measurement in this cell are predictable. The same measurement is made in both cells, and the results are compared. The concept is no different from that of the Wheatstone bridge, which follows the same pattern in an electrical context. Differences in analyzers occur because of the physical principles invoked to make the measurements. Some are described below.

Paramagnetic Properties. A paramagnetic material is attracted by a magnetic field, while a diamagnetic one is repelled. Oxygen is one of the few gases that is paramagnetic (nitric oxide is another). The magnetization produced by a magnetic field of unity strength in a paramagnetic gas varies inversely as the absolute temperature. By properly combining a magnetic-field gradient and a thermal gradient, it is possible to induce and sustain a convective gas flow that is dependent on the percentage of oxygen in a gas sample containing no other paramagnetic gas. Changes in gas flow rate are measured by the effect produced on the resistance of a temperature-sensitive element.

Thermal Conductivity. The rate at which heat from a heated electrical element will be conducted through a gas mixture is a function of the composition of the gas. The thermal conductivity of the gas mixture is proportional to the product of the mole fraction of each gas in the mixture and its respective thermal conductivity. With a thermistor as the detector in a sample gas and another thermistor in a reference gas, when a constant temperature is maintained at the heat source, the difference in temperature of the two detectors is an indication of relative concentration when the thermal conductivities of the sample and the reference are known. Table 2.2 lists key gases and their thermal conductivities. These can be measured by this method or can be present as components in the background gas of a mixture being measured. Powerplant applications include measurement of CO_2 in stack gas and hydrogen purity in hydrogen-cooled generators.

Heat of Combustion. The concentration of combustible gases in a sample stream is converted to an electric signal by oxidation of these gases and measurement of the heat produced by the combustion, which takes place on the surface of a measuring filament. The filament is coated with a catalyst to permit controlled combustion at lower than ignition temperatures. It is one active arm of a Wheatstone bridge circuit; the other arm is an uncoated reference filament. Both filaments are exposed to the same pressure, composition, flow rate, and temperature of sample. Differences in resistance are a function of changes caused by combustion at one filament. The signal is proportional to the concentration of combustible material in the gas stream.

Leading Analyzers. Leading analyzer designs today for stack-gas analysis follow the same pattern of sample- and reference-cell measurement, but use photoconductive cells combined with optical filters for spectrographic analysis. Cell design depends on the absorption characteristics of the gases measured, since the various gases absorb different spectral wavelengths.

Spectroscopic Types. Spectroscopic analysis functions on the basis that gas molecules absorb energy at known wavelengths from beams of light transmitted through the gas. The spectrum is complex, and it is necessary to restrict light to a narrow bandwidth to avoid overlapping and interference. Analyzer cells may shine one or two beams of light. In either case, the pattern of measuring cell/reference cell is maintained. Cells with two beams will have one for reference and the other for the variable sample. Cells with one beam measure at two wavelengths—one for reference, the other for the unknown.

A *nondispersive infrared* (NDIR) analyzer may be used for CO and CO_2. One design focuses energy from an infrared source into two beams. One beam passes through a cell in which the sample gas flows; the other passes through a zero-reference cell. Each beam is passed and reflected intermittently by a rotating chopper, then focused by means of optics at the detector. The detector output is a differential pulse that is proportional to the absorption of the sample cell with

TABLE 2.2 Thermal Conductivities of Gases*

Component	Thermal conductivity	Component	Thermal conductivity
Air	1.000	Cl_2	0.342
O_2	1.028	SO_2	0.350
NH_3	1.040	H_2S	0.540
CH_4	1.450	Ar	0.665
He	5.530	CO_2	0.704
H_2	6.803	H_2O	0.771
CO	0.958	N_2	0.989

*Relative to that of air.

respect to the reference cell. The effect of the beam chopper in front of the infrared source generates an ac signal that helps minimize drift.

And SO_2 and NO_x can be analyzed by instruments that rely on ultraviolet light. The two gases are analyzed in sequence in the sample cell. A split-beam arrangement, with optical filters, phototubes, and amplifiers, measures the difference in light-beam absorption at two wavelengths—280 nm for SO_2 and 436 nm for NO_x. The light source is generally a mercury-vapor lamp.

Oxygen Analyzers. Analyses of O_2 and CO are the two most important exhaust-gas measurements for combustion efficiency control. In situ analyzers are almost always selected for the measurements. Usually CO measurement depends on infrared absorption. The most common O_2 analyzer design is based on a difference in O_2 partial pressures on the two sides of a zirconia wafer, or zirconium oxide cell. Palladium/palladium oxide is another possibility.

In one variation (Fig. 2.11), a difference in O_2 partial pressures on the two sides of a cell, which is heated and maintained at a constant temperature, creates ion migration that causes a proportional variation in an electric current conducted by the wafer, while the front side is exposed to the flue gas. Since the zirconia or palladium is affected only by oxygen, the back side sees a predetermined amount of O_2 in the instrument air as a reference gas. The difference in partial pressures generates a millivolt output that is representative of the oxygen level in the sample gas.

FIG. 2.11 Oxygen analyzer, positioner, and controller.

ACKNOWLEDGMENT

Illustrations in this chapter are courtesy of *Power* magazine except where noted otherwise.

REFERENCES

Gray, David M.: "Upgrade Your pH Measurements in High-Purity Water," *Power,* Mar. 1985, pp. 95–96.

Handbook of Industrial Water Conditioning, 8th ed., Betz Laboratories, 1980, Philadelphia, Pa.

Hunt, Robert C.: "Measurement of Conductivity in High-Purity Water," *Ultrapure Water,* July/Aug. 1985, pp. 26–29.

Nestel, William A., K. Anthony Selby, Lloyd E. Eater, and Robert M. Ricker: "Water Chemistry Control for Power Plants," *Power Engineering,* Apr. 1980, pp. 92–95.

Reason, John: "When It Pays to Monitor Flue-Gas CO," *Power,* Aug. 1981, pp. 37–43.

away from small packaged units to larger, more flexible, field-erected, multiple-burner units capable of firing combinations of fuels that, in many cases, include process waste, biomass, coal, etc. A variety of control strategies are available to implement waste-fuel firing; the one that is right for a given boiler will depend on its geometry, on fuel availability, and on the firing arrangement.

- *Furnace heat-balance control:* The heating value of fuels, particularly coal and waste fuels, varies considerably as they are introduced to the boiler. To make the necessary corrections to the fuel-air ratio to ensure safe and efficient operation, furnace heat-release calculations can be used to infer changes in fuel heating value.
- *Feedwater controls:* Industrial boilers and fossil-fuel-fired utility units relegated to cycling duty are subject to wide load fluctuations and require quick-responding control to maintain constant drum level. Multiple-element feedwater controls can help these swing units respond faster and more accurately to fluctuating demand.

In the area of burner management (also called *flame safety* and *burner control*), engineers are now able to design fail-safe systems to virtually any level of automation desired. New concepts in hardware design provide the additional advantages of plant diagnostics, system self-diagnostics, better operator control system communication, simultaneous installation and design, and simpler system modification.

Flame-scanning techniques have been developed for all types of fuels and fuel combinations. These systems are capable of greater flame discrimination, which has a direct influence on safe boiler operation. Flame scanner electronics has advanced to the point where signals now sense not only flame presence, but also flame quality. This capability has already led to their passive use in some areas of combustion control.

The type of combustion control and burner management system to select will depend on the size and type of boiler, the fuel or fuels fired, and the operating requirements of the plant. The proper selection will increase the efficiency, speed of response, and safety of the plant.

COMBUSTION CONTROL

In the operation of packaged or field-erected industrial or utility boilers, the system selected must provide the tightest, most efficient, and most economical control available. The methods and equipment used to control combustion will depend on the size and type of boiler, the fuel or fuels being fired, the number of burners, and boiler operating requirements. Clearly, the larger and more complex boiler will require a more complex control system; so will a boiler working under demanding load conditions. Thus, different types of control systems have been developed to accommodate different boiler types, sizes, and loading.

Control Types

There are three general types of combustion control schemes selected today: series, parallel, and series-parallel control (Fig. 3.1). In series control, variations in

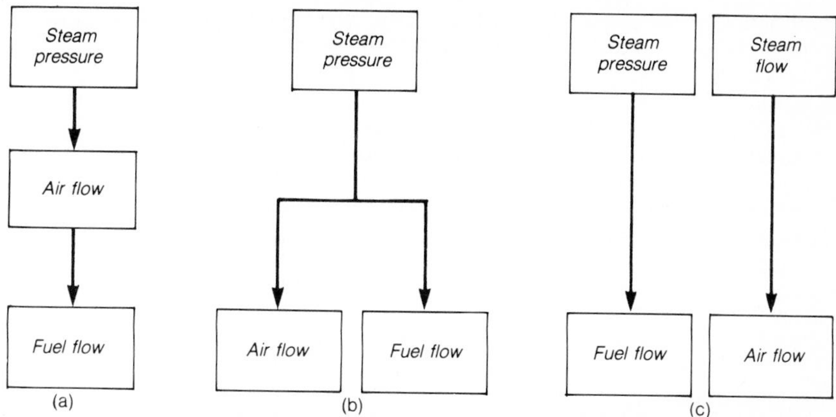

FIG. 3.1 Combustion controls input fuel and air in (*a*) series, (*b*) parallel, or (*c*) series and parallel.

steam header pressure (the master control signal) from setpoint cause a change in combustion airflow which, in turn, results in a sequential change in fuel flow. Systems using air as the controlled variable and fuel as the secondary variable are most suited to boilers that are required to pick up load rapidly and shed it slowly. The opposite is true when fuel is the controlled variable. This type of control is limited to small boilers up to about 100,000 lb/h (12.6 kg/s) having relatively constant steam load and burning fuel of constant Btu (kilojoule) value.

In parallel control, a variation in steam pressure simultaneously adjusts both the fuel flow and the airflow. This method is common to any size boiler firing liquid and/or gaseous fuels with a fairly constant Btu (kilojoule) value.

The series-parallel method finds application where the Btu (kilojoule) value of the fuel varies substantially or the Btu (kilojoule) input rate of the fuel is not easily monitored. Pulverized-coal-firing schemes, for example, are of series-parallel design. Here is how it works: Variations in steam pressure setpoints adjust fuel flow to the boiler. Since steam flow is directly traceable to heat release of the fuel, and since a relationship can be established between heat release and airflow requirements, steam flow can be used as an index of required combustion air. This relationship, however, holds true only at steady loads.

Control Hardware

To carry out the aforementioned schemes, the control hardware includes on/off, positioning, and metering systems.

On/Off Systems. On/off, or two-position, controls are still found in many industries, but are generally confined to firetube and the smallest watertube boilers, where the cost of more efficient positioning controls usually cannot be justified. Operation is analogous to thermostatic control; when pressure drops to a preset value, fuel and air are automatically fed into the boiler at a predetermined rate until the pressure has risen to its upper limit. The primary limitations of on/off controls are widely varying header pressure and inefficient operation through most of the boiler's load range.

The fuel-air ratio on a two-position control system is determined at the time of calibration. Since the combustion process is either on or off, however, fuel and combustion airflow are limited to a single value.

Positioning Systems. Positioning systems respond to changes in header pressure by simultaneously positioning the forced-draft damper and fuel valve to a predetermined alignment either through a common jackshaft and cam valve arrangement—called *single-point positioning*—or through adjustable cam positioners—called *parallel positioning*. These systems are suitable for single-burner boilers firing only one fuel at a time.

Single-Point Positioning. The single-point positioning, fuel-air ratio adjustments are set up through a manipulation of the cam valve and linkage angles, which is performed before the boiler goes into operation. The fuel-air ratio cannot be changed on line unless a trim mechanism is triggered. Since parallel positioning controls fuel and air separately through individual actuators and drives, the fuel-air ratio can be manipulated through manual or automatic adjustment.

The inherent safety and simplicity of single-point positioning systems have been their hallmark for many years, and they are still widely used on single-burner packaged boilers under 150,000 lb/h (18.9 kg/s). No method has proved more reliable than mechanically linking, matching, and locking fuel-air ratios over the load range.

Parallel Positioning. Parallel positioning with fuel-air ratio control is widely used on single- and multiple-burner boilers. It permits the operator to adjust the fuel-air ratio over the entire load range, through either a manual or an automatic bias- or ratio-adjustment station (Fig. 3.2). In bias adjustment, a constant is added to or subtracted from the signal to the fuel valve (or to the damper, if air leads fuel).

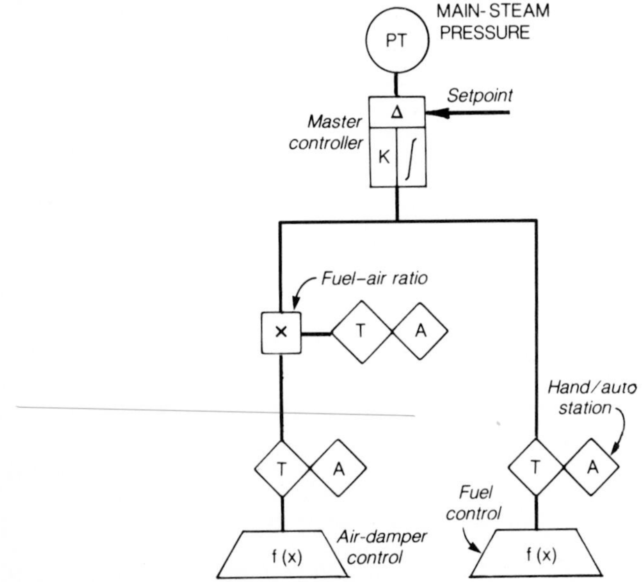

FIG. 3.2 Parallel positioning with fuel-air ratio control.

Bias adjustments are correct only at a particular firing rate, and therefore they must be changed with each new load demand. With ratio control, the operator can vary fuel flow in proportion to airflow over the entire load range.

The primary advantage of parallel positioning is the capability for adjusting fuel and air independently through the use of manual/automatic stations. This method is not recommended for gas or oil firing, however, because malfunction of either the fuel or air control element could allow unsafe firing conditions to develop. Such a system is adaptable to stoker firing of coal, wood, bagasse, etc., where a large reserve of unburned fuel resides on the grate and where changes in bed thickness do not have an appreciable effect on the fuel-air ratio.

Parallel positioning is also applicable to the firing of hard-to-meter fuels such as pulverized coal. Since steam flow is roughly a function of the fuel's Btu (kilojoule) input, combustion air can be consistently aligned with inferred fuel flow. This system has control functions that apply metering as well as position control concepts and that relate directly to the series-parallel control system in Fig. 3.1—with the only difference being a feed-forward signal from the steam-pressure controller to combustion-air control.

Metering Systems. In metering control, combustion is regulated in accordance with measured (not inferred) fuel flows and airflows. Metered flows serve as feedback signals to the controllers to ensure that flow corresponds to demand (Fig. 3.3). This method helps compensate for the effects of boiler slagging, barometric conditions, and changes in fuel quality on boiler performance. It also maintains combustion efficiency over wide load ranges and over long periods.

Both metering and positioning control systems use steam header pressure as their primary measured variable and as a basis for firing-rate demand. A master

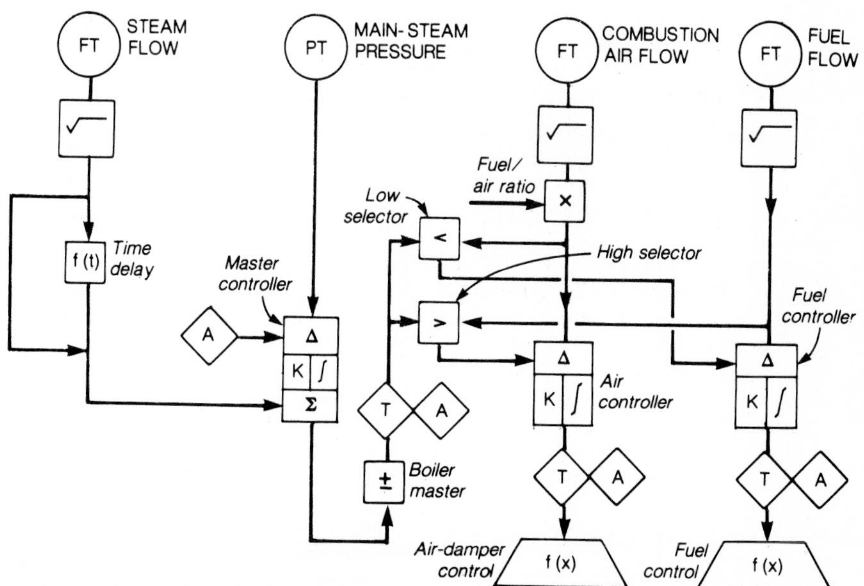

FIG. 3.3 Metering control systems regulate combustion in accordance with fuel flow and airflow to maintain efficiency over a wide load range.

pressure controller responds to changes in header pressure and, by means of power units or actuators, positions the forced-draft and/or induced-draft dampers to control airflow and the fuel valve to regulate fuel supply. The firing-rate demand signal may control more than one boiler where two or more are tied to a common steam header.

Two-Element Metering. If load changes are rapid, it is advisable to apply a steam flow feed-forward circuit to an existing header-pressure control scheme to provide swifter corrective action to fuel and air controls during extreme load changes. In this two-element, steam-pressure control arrangement, steam flow sets the initial demand for fuel and combustion air, and the steam setpoint pressure error generates the fuel correction or trim to the fuel and combustion-air control elements. This system improves the response of the boiler by anticipating a load change and thus minimizes upsets in the combustion rate and variations in outlet pressure.

In the control system shown in Fig. 3.3, the steam flow corrective signal is linearized and simultaneously sent to a summer and a signal-delay unit (optional), which inverts the signal and holds it for a fixed time before passing it to the summer. Meanwhile, the summer output (feed-forward) signal is passed on to the boiler master station, which passes the same signal on to the fuel and air controls. As the signal in the delay unit builds up, it cancels the effects of the steam flow signal to the summer, thereby permitting the control signal from the main-steam-header pressure controller to resume full control of combustion through the boiler master station.

Cross-Limit Metering. Although metering controls can be either series or parallel, parallel systems offer faster response to load changes because fuel flow and airflow corrections are made simultaneously. The master demand signal developed as a result of steam-pressure deviation from setpoint or steam flow changes establishes the setpoint flows required for fuel and air at their respective controllers. The fuel and combustion-air controllers develop corrective signals to the final control elements to establish proper flows that, in turn, feed back to their controllers.

The basic parallel metering system for single-fuel firing provides stable, accurate control of energy input and fuel-air ratio during steady-state operation, but does not guarantee absolute furnace safety during load changes. Therefore, most metering systems today use cross-limit control as an extra measure of furnace safety without sacrificing the efficiency inherent in parallel control systems.

In cross-limit metering control, maximum and minimum signal selectors provide a positive interlock to prevent fuel-rich conditions on a load change. Air always leads fuel on a load increase and lags fuel on a demand decrease. Under steady-state conditions, fuel flow and airflow controllers continuously hold measurement equal to setpoint. On steam demand increase, the low-selector module blocks the increase to the fuel controller and makes the controller setpoint equal to the actual airflow (Fig. 3.3). In this fashion, fuel flow cannot increase until after airflow has begun to increase. On steam demand decrease, the selecters prevent a decrease in airflow until fuel flow has begun to decrease. Although cross-limit metering controls provide a degree of safety, they also force excess-air levels higher than is optimum during load changes—a disadvantage of the system.

In combination systems where both a meterable fuel, such as gas or oil, and an unmeterable fuel, such as coal, are fired separately in the same boiler, metering of the coal is done inferentially by subtracting the metered gas input from the steam flow output on a Btu (kilojoule) basis.

Digital Systems

Acceptance of microprocessor-based digital control by more and more industries and utilities has had a noticeable effect on the way process control systems are normally defined. The line of separation between hardware that is used strictly for control, monitoring, or logic functions will become less obvious or even nonexistent in the future because all three can be performed within the same microprocessor-based network.

Plants built in the 1970s and before, and even some today, are generally outfitted with separate and distinct modulating control, logic, and monitoring systems—more than likely supplied by separate vendors. In these plants, modulating control is performed by discrete hard-wired analog control stations located in the control room. Plant logic, composed of on/off control and interlocks necessary for safe start-up and shutdown of a boiler, is made up of discrete solid-state, relay logic, or microprocessor-based devices. Monitoring and data acquisition are performed by a central computer hard-wired to many field inputs. A certain amount of information transfer between the three systems is necessary for safe and efficient plant operation, but communication has been limited by the amount of hardwiring necessary.

Digital equipment affords much easier intersystem communication because digital data transfer is simpler, and the natural distribution of control, logic, and monitoring functions within the plant lends itself easily to remote data processing and control. I&C engineers are currently developing ways of getting information from the modulating control system and plant logic system into the computer (and vice versa) via a single digital data link.

Most manufacturers will soon be offering plant control systems encompassing all three functions as an integrated package. Once vendors do begin to offer an integrated version of all three subsystems, users may find it difficult, if not impossible, to combine the hardware of two or more vendors because of conflicts in system philosophy and data transmission protocol.

Optimizing Combustion

The integration of new control philosophy into existing control systems to increase efficiency of the combustion process is probably the fastest-growing area of powerplant control. Increased operating and maintenance costs have forced utility and industrial managements to make the most of expensive fuel stocks by optimizing the efficiency of steam generation equipment. Probably the most cost-effective means of improving boiler efficiency is by using a control system that is capable of minimizing furnace excess-air levels.

Excess air is a necessary evil in boiler combustion processes. Although it is required to ensure complete mixing and optimum heat-release characteristics, it also adds significantly to heat loss. Excess air is also essential from a safety standpoint. Without it, the amount of O_2 at the burner might drop below the theoretical stoichiometric level during load changes, possibly leading to a boiler explosion.

By keeping excess air at the minimum level required for stable firing, effluent heat losses can be minimized. For boilers that are not currently operating at low excess-air levels, the potential efficiency gain by doing so is significant. For example, for every percentage reduction in excess O_2 at high excess-air levels, an improvement of about 1 percent in combustion efficiency is achievable. At lower

O_2 levels (less than 3 percent), a 0.5 percent improvement in combustion efficiency is possible for every 1 percent reduction in excess O_2.

The idea behind low-excess-air combustion control is to maximize boiler efficiency by operating at the theoretical point where both combustible energy losses and effluent energy losses are minimized (Fig. 3.4). This is commonly known as the *smoke point.*

Since the rate of energy loss because of excess combustibles is 6 times greater than that for excess air, it is wise from both a safety and an efficiency standpoint to operate the boiler as close to the smoke point as possible without falling below it (Fig. 3.5). The operating point will actually fall to the air-rich side of the excess-air curve—the extent is dependent on the boiler, its control system, firing equipment, and fuel burned.

Monitoring Flue Gas. The parameters most widely applied as the basis for low-excess-air trim are O_2 and CO. Other signals used primarily as secondary elements in trim control include CO_2, hydrocarbons, and opacity. Although most trim systems are based on O_2 as the primary element, this is not always the case.

Both O_2 and CO give important information about the combustion process and, when used in combination, provide a very accurate picture of firing conditions within the boiler. The CO measurement gives information that is clearly not available with just O_2 measurement. Here are some of the advantages of CO trim that are most often cited:

- *Accuracy:* Since the measurement is related directly to the amount of unburned fuel in the boiler, it can be applied to control completeness of combustion better than O_2.
- *Simplicity:* The control setpoint is independent of fuel type. Control is always performed at the breakpoint or knee in the CO curve, between 100 and 350 ppm CO (Fig. 3.6).
- *Security:* The CO signal is virtually unaffected by furnace air infiltration. However, O_2 readings are rendered almost useless if tramp air enters through the furnace casing or convection passes of balanced-draft boilers.

The benefits of monitoring CO are clear, but the fact remains that the CO signal is not always an accurate indicator of excess air. For example, a dirty burner or poor fuel mixing can cause a rise in CO level and associated increases in excess air without the operator being aware of it. Another problem is that if the

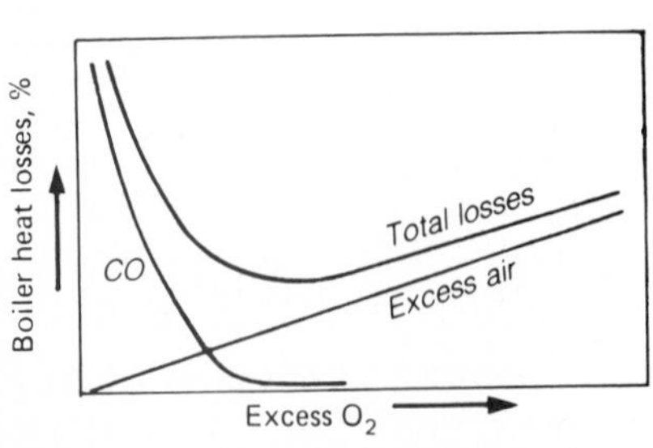

FIG. 3.4 Efficiency is maximized by locating the heat-loss minimum. Losses increase to each side of the correct firing point.

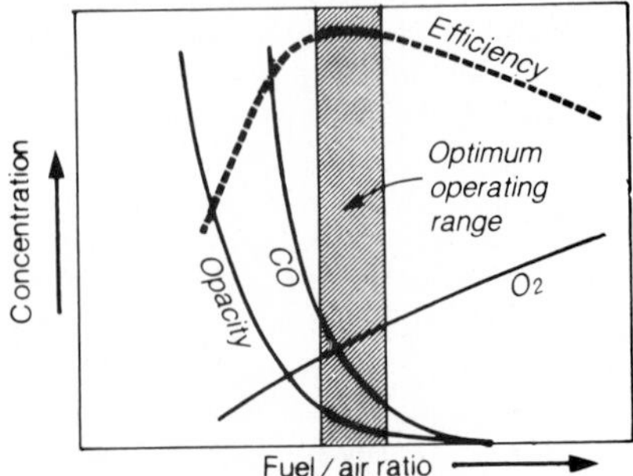

FIG. 3.5 The optimum operating point falls to the air-rich side of the efficiency curve, depending on boiler, control system, and fuel.

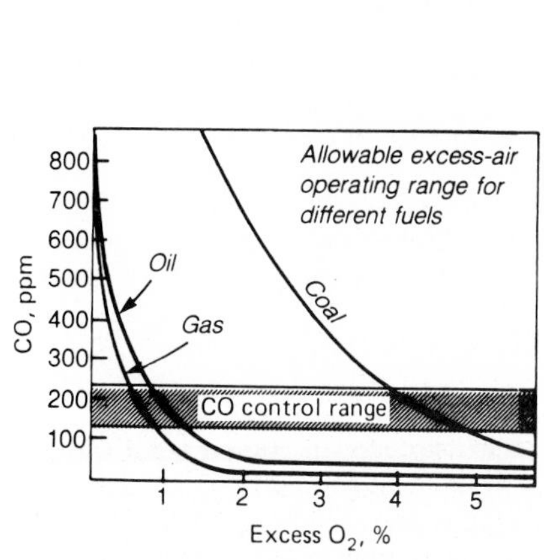

FIG. 3.6 Excess-air operating point varies for different fuels, whereas CO setpoint is always at knee or breakpoint in curve.

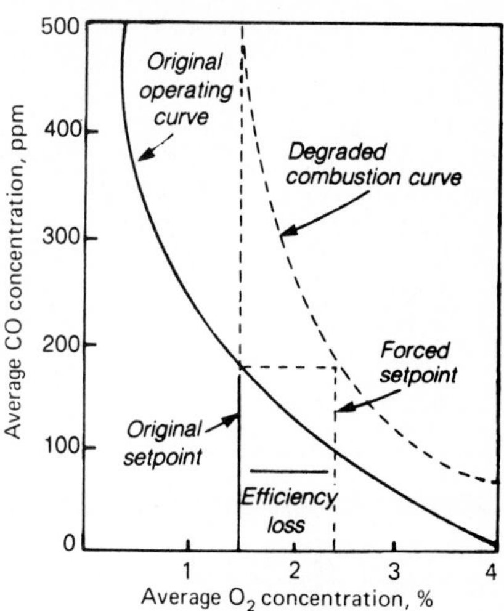

FIG. 3.7 Drift of operating point from original CO-O_2 curve indicates wrong fuel-air ratio, bad atomization, or change in fuel.

operator selects a high-excess-air firing condition, the CO concentration is liable to be undetectable.

Most control system vendors and consultants prefer to base excess-air trim on O_2 measurement and endorse CO measurement where it is economically justifiable. According to these engineers, a good O_2 measurement, whether it is tied into closed-loop control or is used simply as a basis for manual trim, can achieve 80 to 90 percent of the expected efficiency gain from low-excess-air firing. An additional 0.1 to 1 percent efficiency gain can be obtained by monitoring or controlling with CO, depending on the controller selected.

CO Monitoring. An application of CO monitoring that offers intrinsic advantages over just O_2 monitoring is the diagnosing of the firing characteristics of multiburner boilers. Here, the CO-O_2 relationship can be applied to evaluating combustion efficiency on a burner-by-burner basis. For a given firing rate, an increase in boiler CO level should accompany a predictable and repeatable decrease in O_2 level. The CO-O_2 curves identify the minimum O_2 and CO levels for reliable operation of the unit without unstable combustion, slagging, smoking, or other undesirable conditions.

Once baseline curves have been established for a given fuel and firing condition, deviations occurring during operation can be traced to a variety of combustion-related causes. Operating conditions that could lead to deviation from the CO-O_2 curve include improper fuel-air ratio to a burner, poor atomization, or a change in fuel properties (Fig. 3.7).

In multiple-burner boilers, each burner generally behaves as a separate combustion source with its own individual air requirements and products of combustion. Despite some gas mixing, combustion products generally travel from each burner to the boiler exhaust duct in paths displaying laminarlike flow patterns. Because of the discrete flow paths of burner effluents, the distribution of combustion products in the exhaust duct can generally be related directly to the firing characteristics of individual burners or burner groupings (Fig. 3.8).

By sampling flue gas in the duct with a multipoint extractive analysis system similar to that used in boiler performance testing, the firing characteristics of each burner can be identified. In addition to providing valuable diagnostic information about burner performance, this system solves some of the sampling-accuracy problems sometimes encountered if one relies on single-point sampling of O_2 and CO.

This concept is applicable to tangentially fired boilers as well as wall-fired designs. A boiler performance monitoring system of this type is currently operating on a tangentially fired 200-MW utility boiler; more systems are in the offing.

Excess-Air Trim Systems

Before an investment is made in an excess-air trim system, an overall inspection and tuneup of the boiler is recommended. Proper adjustment of existing equipment will ensure that efficiency of the boiler and operation of the control system are at their optimum level, which is important in making sure that the investment in added equipment is justified. Also, an inspection will provide good baseline data for a subsequent improvement in efficiency.

Prerequisites for Firing. Four prerequisites for efficient low-excess-air firing are

- Burners capable of low-excess-air firing
- An accurate and repeatable relationship between air and fuel over the load range
- Equal percentage bias-trim control over the entire load range

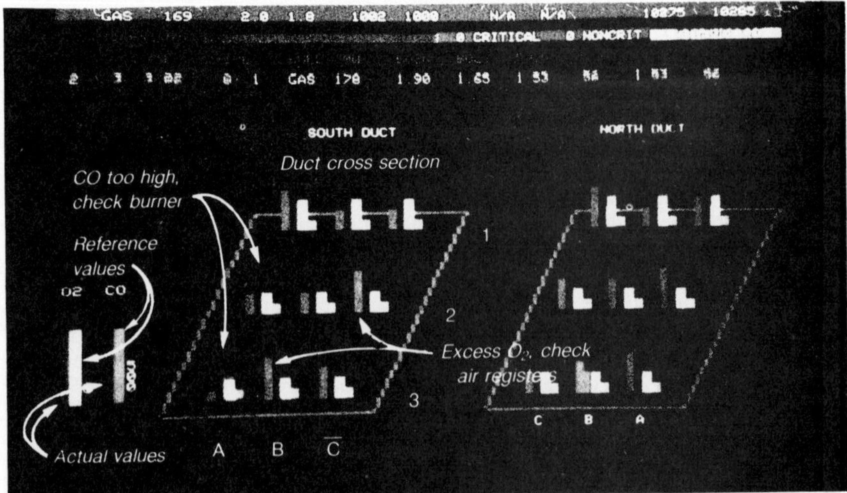

FIG. 3.8 Distribution of combustion products in exhaust duct can generally be related directly to firing characteristics of individual burners or burner groupings, as image on CRT screen shows.

- A control system capable of handling lead-lag functions, response delays, and other problems associated with modulating loads

Almost any excess-air trim system can operate the boiler at optimum efficiency during steady-state load conditions. The true colors of a control system, however, are shown during rapid load changes when fuel and air controls must act in harmony to keep the atmosphere within the boiler below a maximum allowable excess-air level. Because of the limitations of most control systems, few can attain boiler excess-air levels lower than 2 to 5 percent.

The problem is particularly acute on industrial boilers—which are typically subject to very large and frequent load swings—where operators often are forced to maintain a higher O_2 setpoint than the automatic trim control is capable of, simply because of limitations in the control system, which lead to smoking during changes. Thus, it is often the capability of the control system (or lack thereof), not the boiler, that is the limiting factor in how close excess-air levels can be run to the smoke point during load swings.

In choosing a metering system, make sure that the controller used to set the fuel-air ratio is capable of ratio-trim as well as bias-trim control. Ratio trim is the preferred control mode, since it allows the setting of a constant fuel-air ratio throughout the load range. This operation is commonly known as *equal-percentage trim* (Fig. 3.9). Bias trim is also necessary because the system should be calibrated with more excess air at lower firing rates; that is, the intersection point of the load curve on the airflow axis in Fig. 3.9 should be greater than zero.

The alternative control method is to let the analyzer bias airflow (Fig. 3.10). This is not considered good control practice, however, because every load change will introduce an error in the fuel-air ratio that will have to be recorrected by the O_2 trim control.

Applying O_2 Trim. Introduction of the in situ zirconium oxide sensor in the early 1970s has made O_2 trim commonplace on boilers of nearly every size. The testing ground for these instruments, however, was primarily confined to larger industrial and utility boilers, in which the economics of O_2 trim was initially more attractive.

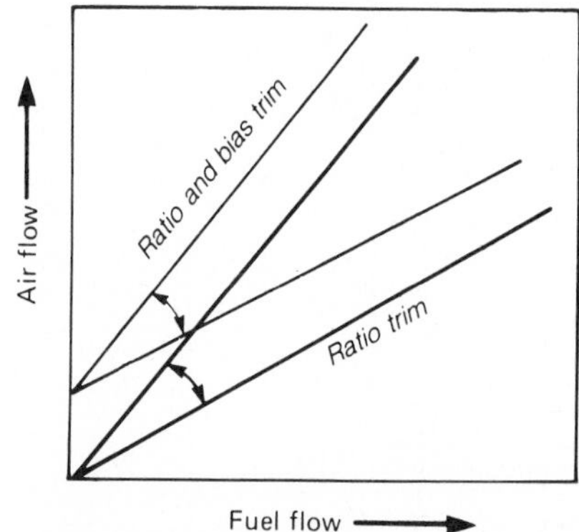

FIG. 3.9 Ratio trim allows setting of constant fuel-air ratio throughout the load range.

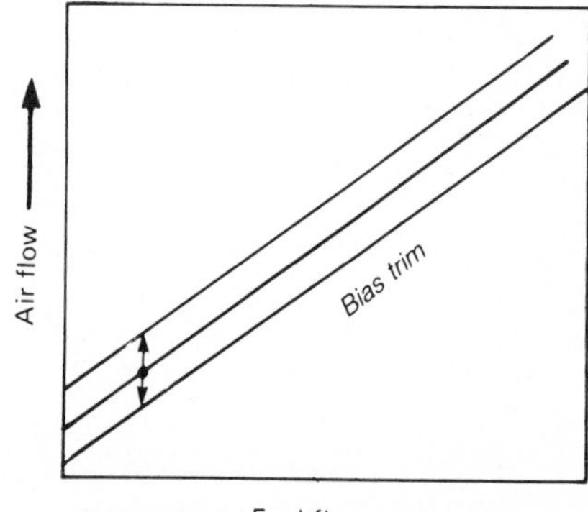

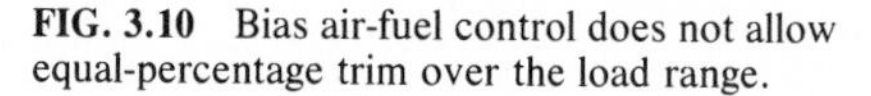

FIG. 3.10 Bias air-fuel control does not allow equal-percentage trim over the load range.

Modes of Operation. There are currently three operating modes for linking the flue-gas parameter on boilers with metering control systems: manually, semiautomatically, and automatically.

The most prevalent use of the signal from an O_2 analyzer is simply as a monitoring parameter that is recorded in the control room on a continuous or intermittent basis. Using steam generator load as a guide, the operator then manually adjusts the fuel-air ratio to the proper setpoint.

Manual control is generally the first logical step in the process of upgrading a control system to automatic trim control. It allows the operator to become familiar with and develop confidence in flue-gas monitoring instrumentation. Based on the fuel savings accomplished with manual trim, the economics of upgrading the control system to automatic trim can also be considered.

In the semiautomatic trim mode, the operator is able to program an O_2 setpoint and manually ratio or bias the setpoint to match the required fuel-air ratio at each firing rate.

In fully automatic trim control, the steam flow signal is run through a function generator that characterizes the optimum excess-air signal as a function of boiler load (Fig. 3.11). This signal then becomes the fuel-air ratio setpoint and is continuously compared with the actual oxygen measurement in the flue gas. The difference between the two signals is used for the variable setting of the fuel-air ratio relay. Since setpoint changes are automatic, boiler efficiency is optimized over the load range.

CO-O_2 Combinations. There are many ways in which a CO measurement can be used in combination with the O_2 signal to improve trim control. Generally CO supplements O_2 trim because it has the ability to fine-tune combustion in the vicinity of stoichiometric firing (Fig. 3.12). The CO signal generally is not used as a primary control element because it is subject to unusual transients during load

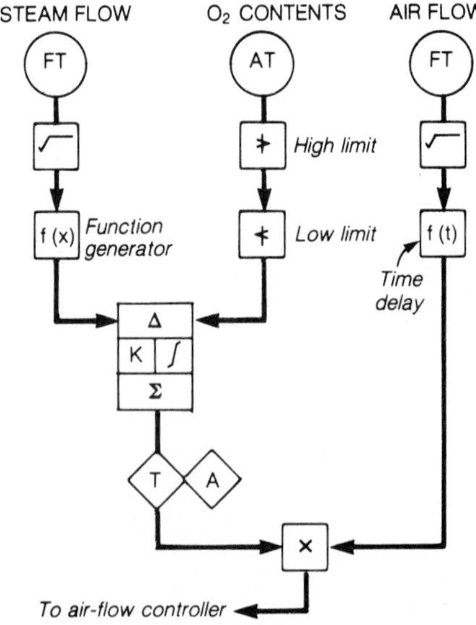

FIG. 3.11 In O_2 trim control, the steam flow signal determines the excess-air operating point.

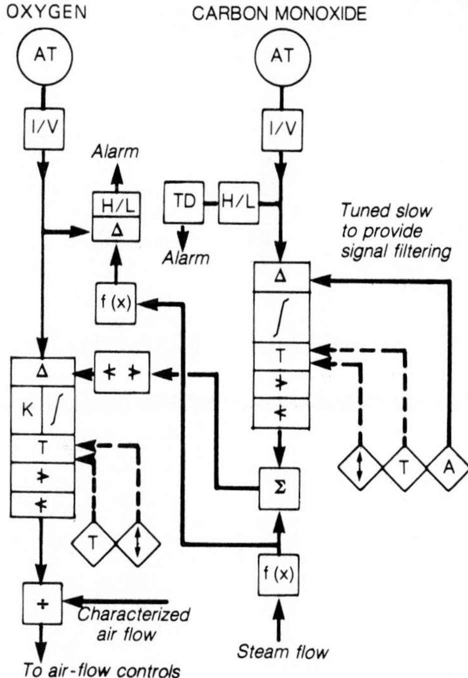

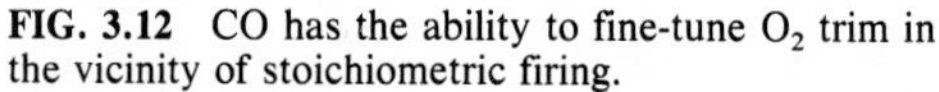

FIG. 3.12 CO has the ability to fine-tune O_2 trim in the vicinity of stoichiometric firing.

changes and limits the operator's ability to operate at high excess-air levels, if required. This type of O_2-CO trim control is best suited for high-carbon fuels like oil and pulverized coal.

Simpler O_2-based trim systems may use a less expensive and less sensitive combustibles monitor to provide an alarm setpoint to the fuel-air ratio controller. A high combustibles measurement would signal the effects of poor atomization, burner wear, improperly adjusted burner air registers, or O_2 probe failure.

A modified approach to combination O_2-CO-based trim is to control off of O_2 during load changes and switch over to CO control during steady-state conditions. This strategy makes good technical sense because O_2 is much more responsive over the air-rich range of the boiler than CO, and O_2 can provide more positive corrections during swings when excess-air levels are normally at their highest. When O_2 is the governing signal in boiler trim, the CO signal is used as a high limit for combustibles to ensure that levels do not exceed a nominal 1000 ppm.

When the load is constant and the O_2 signal has stabilized at a low excess-air level, CO takes over to refine the operating point to about 150 ppm. In this operating mode, high and low limits of O_2 ensure that the control system is not running blindly at high excess-air levels.

Several manufacturers offer complete trim systems that do not require any O_2 signal and instead control from an array of inputs—primarily CO, opacity, and combustibles. And CO has been used successfully as the primary element in trim control of gas-fired boilers.

Bed Firing. Bed firing of solid fuels such as coal and bark presents unique control requirements that can be met only by systems designed on a fuel- and boiler-specific basis. Because bed burning systems have no firing components to combine air and fuel precisely, normal low-excess-air firing techniques are not always adequate. Bed firing control schemes, however, do seek the same objectives as conventional optimization systems: the reduction of excess air and combustibles. These control systems generally rely on a combination of parameters, including CO, O_2, and opacity, to define the optimum firing point.

In both underfeed and overfeed stoker-fired boilers, combustion is confined primarily to the outer layer of the fuel bed. Air is introduced to the fuel at two points: below the bed as primary air and above the bed as overfire air. Boiler excess-air levels can be controlled by properly adjusting the two airflows as a function of fuel feed rate and flue-gas parameters such as CO and CO_2.

The CO signal is measured and controlled to a preset target by manipulating the overfire air damper. This is valid since studies indicate that CO is primarily a function of overfire air and is virtually independent of undergrate airflow. The CO target for stoker operation is about 200 ppm.

Undergrate airflow is generally controlled to maximize CO_2 formation and minimize unburned fuel in the ash. This is a complex control procedure because optimal levels of CO_2 vary with fuel Btu (kilojoule) value, moisture content, bed distribution, and flow rate.

A simple bed burning optimization procedure generally adjusts the undergrate air until CO_2 in the flue gas is maximized. The control-system search program automatically makes a change in undergrate airflow to determine if there is an increase in CO_2. If so, additional control moves are made in the same direction until a maximum CO_2 level is reached. This iterative process continually searches for a moving CO_2 peak.

The control strategy for maximizing base fuel flow (and minimizing expensive swing fuel flow) is to increase the fuel flow rate slowly and adjust overfire and underfire air until a boiler constraint is reached. Typical constraints include opacity, steam flow, drum level, fan speeds, and fuel flow.

Positioning Trim Systems. Most watertube packaged boilers with single-point or parallel-positioning control systems can save significant amounts of fuel by using automatic O_2 trim control. For units rated more than 200,000 lb/h (25.2 kg/s), installation of a CO controller and analyzer might also be economical. However, a CO monitor added to an existing O_2 trim system can usually improve boiler efficiency by only 0.1 to 0.5 percent. When choosing a trim control system, compare the net difference in expected operating point between a trim system using O_2 only, one using CO only, and a combination system.

For units that cannot justify the cost of a CO monitor, consider the installation of a much less expensive catalytic combustibles monitor. This instrument is not as sensitive as a CO unit, but it can provide valuable information about the combustion process if it is used as a combustibles high-limit override.

The development of automatic trim systems for single-point positioning controls has been approached from many angles by control manufacturers. The O_2 trim is difficult to implement because air and fuel controls are rigidly linked and refinement of the fuel-air relationship requires flexibility. The most widely accepted trim method is to break the jackshaft between the main control shaft and damper to insert a ratio control arm. In another, more accurate, method, the fuel valve pressure is manipulated (Fig. 3.13). On a well-tuned boiler, these systems can achieve performance that even metering systems cannot (Fig. 3.14).

A special problem exists when attempts are made to provide equal-percentage

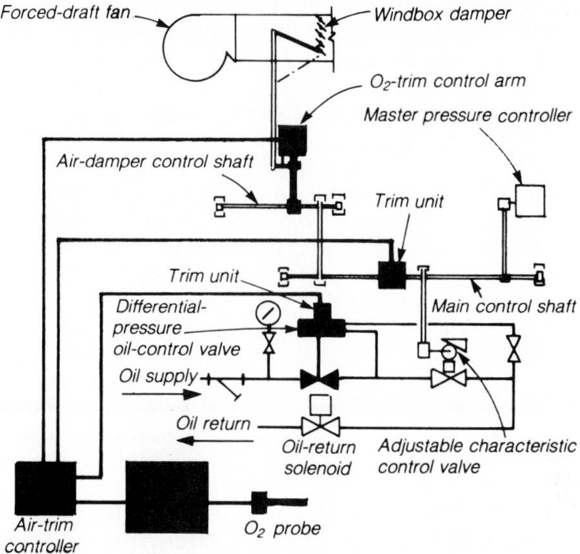

FIG. 3.13 The O_2 trim on jackshaft boilers can be performed on the air side, on the fuel side, or on the main control shaft.

trim capability for single-point positioning systems. The difficulty lies in the lack of any motion in the linkage system that is linearly representative of airflow or fuel flow and on which an equal-percentage device can be applied. Possibly the only way to provide the equal-percentage trim capability is to trim on the fuel valve.

Waste-Fuel Firing

As more and more industrial plants turn to burning by-products alone and in combination with premium fossil fuels, the control philosophies for waste-fuel firing

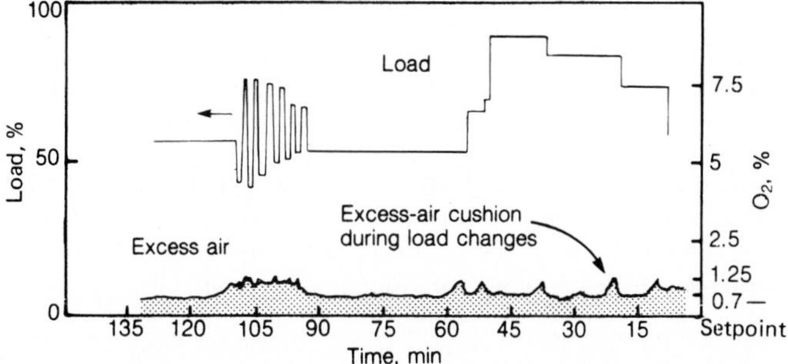

FIG. 3.14 Some trim systems offer accuracies not found even on parallel metering controls. Trace is from a 36,000-lb/h (4.5-kg/s) boiler trimming on fuel.

are making a transition from the simple approaches of the past to more complex multiple-fuel firing techniques. Instead of controlling combustion to eliminate waste products in the simplest and most convenient way, as was often done in the past, engineers are now taking more serious consideration of waste fuels for their Btu (kilojoule) value and their ability to displace oil and gas.

The practice in the past has been to base-load waste fuels manually, at some flow determined by the operator, then to respond to load demand changes by manipulating the prime fuel. Today, boiler optimization techniques are able, simultaneously, to displace as much prime fuel with waste fuel as possible, maximize overall boiler efficiency and capacity, and keep emissions within regulatory limits.

Since the aim is to minimize or eliminate the consumption of prime fuel, the waste fuel must often be manipulated to control steam header pressure. This makes controllability somewhat more difficult, because waste fuels typically are not uniform in Btu (kilojoule) value, nor are they easily metered. Simple, conventional feedback control loops cannot provide acceptable performance, so additional feed-forward techniques must be incorporated that use analytical or other measurements to anticipate and compensate for variations in the effects of different fuels.

Some of these functions can be performed best by analog control, while others can be done equally well by analog or digital control. The specific hardware configuration must be chosen on the basis of the application. The best control configuration usually contains both analog and digital elements, although analog-only and digital-only control systems have seen service.

Feed-Forward Control. The relatively poor dynamic response with waste fuel as the primary manipulated variable demands a fundamental change in the approach to steam pressure control. To begin, steam users can no longer ignore the steam producer's ability to meet changing demand, so better coordination between the two is necessary today. User demands under normal operating conditions must be stabilized. This means better scheduling and coordination of batch processes and limiting the rate of change of all types of process demands.

Under abnormal or emergency conditions, the system may need much more than feedback control of firing rate. Feed-forward control from different parts of the steam distribution system, and temporary steam venting and/or steam load shedding, will also be needed.

Figure 3.15 shows a feedback–feed-forward system used for steam pressure control of a boiler firing a solid-waste fuel. To establish the firing rate demand, a linearized measurement of boiler steam flow is fed through a summer, a static characterizer $f(x)$, a dynamic compensator $f(t)$, and a multiplier. The summer provides a feed-forward corrective trim from two additional measurements. Input C is a feed-forward derivative response from low-pressure steam flow to the process. Any sudden change in low-pressure steam is fed forward to the summer, where it is combined with the firing rate demand to reflect the process-steam load change. Thus, the effect of the load change is sensed almost immediately, without the usual delay while it ripples through the steam distribution system.

Input A in Fig. 3.15 provides feed-forward compensation for stored energy lost or gained as a result of changes in fuel heating value or any other boiler-side disturbance. At steady state, any change in fuel heating value or manual fuel supply causes a change in steam flow and steam pressure. A signal proportional to the pressure deviation from setpoint is fed forward and combined with the firing rate demand to minimize the effect of changing the Btu (kilojoule) value of the

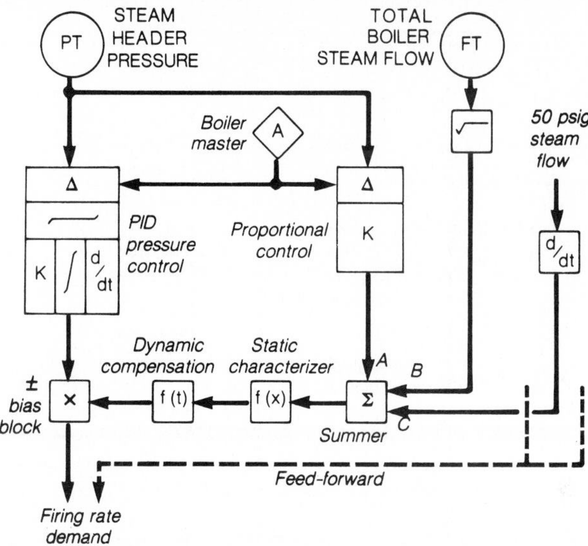

FIG. 3.15 Feedback–feed-forward system for steam-pressure control of a boiler firing solid-waste fuel. Note: psig × 6.89 = kPa.

fuel. This control also makes up for any surplus or deficiency of stored energy during load changes.

Combustion Control Guidelines. The best combustion control configuration for a given operation depends on how the waste and prime fuels can be fired in the boiler. Four possible combinations are (1) waste fuel fired alone; (2) controlled ratio of waste fuel to prime fuel; (3) waste fuel base-loaded, prime fuel modulated; and (4) prime fuel base-loaded, waste fuel modulated.

Waste Fuel Fired Alone. This method has the virtue of simplicity, but is not best in terms of continuous plant operation. It can be done only if the following conditions are met:

- Prime fuel is not needed to stabilize combustion.
- The dynamic response of the boiler with the fuel is compatible with the steam load pattern of the plant.

A parallel metering system is generally sufficient for this purpose. If direct measurement of fuel flow is not possible, an inferential measurement should be made to correct for fuel inconsistencies. Some factors in inferential measurement that have been used are belt scale output, fuel properties, and boiler heat balance.

Controlled Ratio of Waste to Prime Fuel. In this system, adjustments can be made to permit the two fuels to be fired simultaneously at any ratio. The system allows one fuel to be adjusted manually to provide base load, while the second fuel responds to load swings. This system generally is selected when a fixed ratio of waste to prime fuel is necessary to stabilize combustion. The ratio may be set manually or by measurable, or known, fuel characteristics.

In a system for fuel ratio control (Fig. 3.16), the input to a multiplier determines the proportion of the total firing rate demand that is transmitted to the

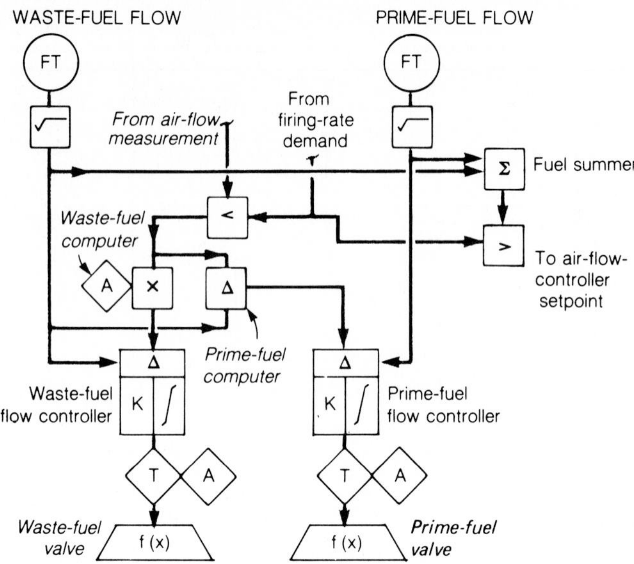

FIG. 3.16 In ratio firing, input to a waste-fuel multiplier determines what proportion of firing rate is from waste fuel.

waste-fuel feed. At steady state, the subtractor sees the difference between total and waste-fuel demands and calculates net prime-fuel demand.

This basic system can be adapted to improve response to load demands by rate-limiting waste-fuel demand. Load changes can thus be handled by the prime fuel, after which the waste-fuel firing rate ramps to the new steady-state demand.

The rate-limiting function may be one of these:

- Not required. *Application:* gaseous or liquid wastes.
- Required for both increasing and decreasing demand. *Application:* all types of solid-waste–burning units.
- Required on increasing demand only. *Application:* wet solid-waste firing where little fuel is stored in the unit.
- Required on decreasing demand only. *Application:* where a relatively large quantity of fuel is stored in the unit.

Waste Fuel Base-Loaded, Prime Fuel Modulated. The level at which waste fuel can be base-loaded depends on its availability and the capacity needed from the boiler. The capacity of a boiler firing waste fuel may be less than when it is firing prime fuel, and this may limit the amount of waste that can be base-loaded.

A simple manual loading station may be manipulated by the operator to set the base load, but a more desirable automatic control is shown in Fig. 3.17. With this configuration, the base-load demand is the lower of either the total firing rate demand or the base load set by the operator. In the event of a sudden loss of steam load, this ensures that the demand for waste fuel can be reduced accordingly. The subtractor calculates the difference between total firing rate and base load to find the net demand for prime fuel.

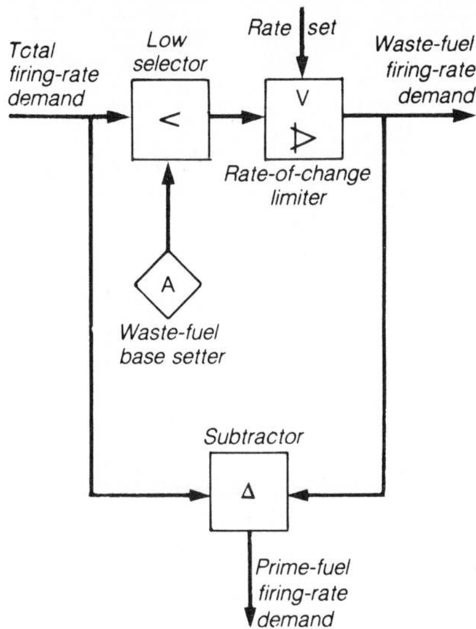

FIG. 3.17 This control system subcircuit sets the firing rate demand for prime fuel.

Prime Fuel Base-Loaded, Waste Fuel Modulated. This mode may be selected if a constant ignition-stabilizing source is needed or if the waste-fuel firing rate must be limited and load changes absorbed by the prime fuel (Fig. 3.18).

The prime-fuel base load is determined by the rate-of-change limit and the steam load pattern. If the rate limit applies to both increasing and decreasing demand, then the value of the base load must be such that the prime-fuel firing rate can be decreased sufficiently during transients. If this setting is left up to the operator, he or she will probably burn too much prime fuel.

The only way to optimize the amount of prime fuel burned is to calculate both the continuous running average load and the transient bandwidth. This requires a computer or microprocessor and cannot be done satisfactorily with analog control. The prime-fuel base-loaded mode is best applied where the waste fuel is rate-limited for increasing demand only. Here, the base load of the prime fuel is the minimum stable firing rate. Increasing transients are absorbed by the prime fuel, after which the waste fuel ramps up to return prime fuel to the minimum. Decreasing loads are absorbed by the waste fuel.

Preferential Firing. Another firing scheme that is similar to the ratio method cited earlier is preferential firing. Here, the prime fuel is allowed to vary in response to load demand, but is not allowed to exceed a predetermined limit. This operator-set limit is selected when a fuel purchase contract stipulates that the supplier may charge a premium if the user exceeds a certain rate of consumption. In other circumstances, fuel availability may vary arbitrarily, and a cutback controller may be introduced to generate the limit of primary-fuel use.

During normal operation, the firing rate demand signal is transmitted as a setpoint to the primary-fuel controller until the availability limit is reached (Fig. 3.19). As demand increases further, a computer calculates the difference between the firing rate demand and primary-fuel flow and transmits this difference as a setpoint to the waste-fuel controller.

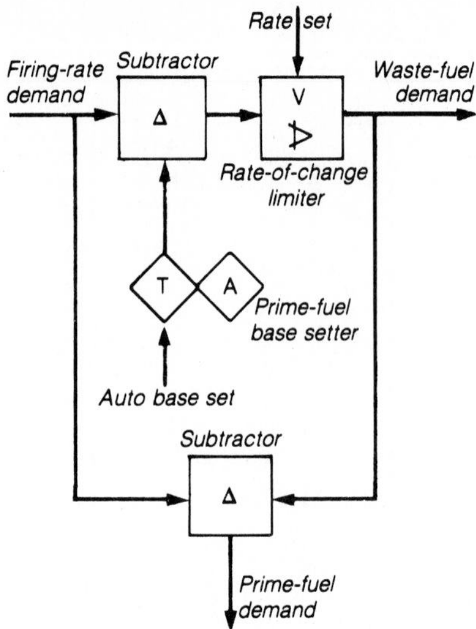

FIG. 3.18 Prime-fuel base load is subtracted from demand to obtain waste-fuel demand.

Multiple Firing. When more than two fuels are fired, it is often desirable to proportion fuels according to their availability. Various "participation setters" can be used to apply the desired percentage for each fuel (Fig. 3.20). This system can still be operated in base and swing modes by putting one or several fuels in manual.

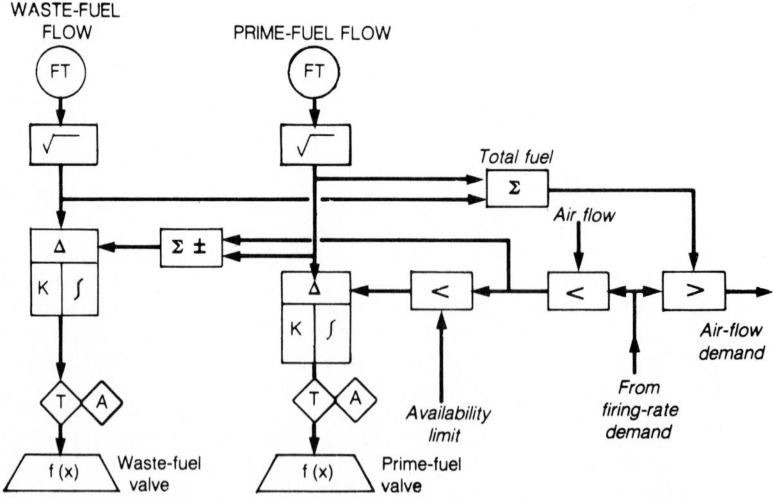

FIG. 3.19 In preferential firing, prime fuel may vary in response to load demand but may not exceed a predetermined limit.

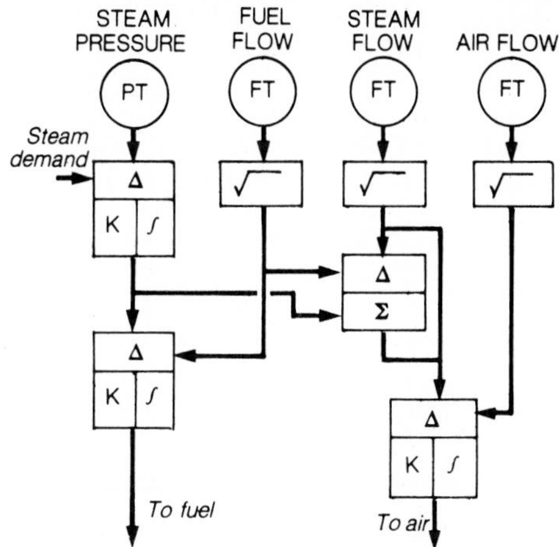

FIG. 3.20 When more than two fuels are fired, each can be loaded according to availability by ratio control or by base-loading certain fuels.

Each of the fuels must be accounted for on a Btu (kilojoule) basis. Inputs to the fuel controller include the master demand signal as a setpoint and the total fuel flow as a measured variable. The output of the controller is then split up by the various participation multipliers into the demand for each fuel. Each fuel has its own controller that compares demand to actual fuel feedback. The last fuel picks up whatever demand is remaining.

Control System Design

Swing-loaded industrial and utility boilers demand fast response of a control system, and feed-forward signals are one means of achieving this. Response time is quicker because variables upstream of a process inform primary controllers before each controller's input actually senses a change. Thus, the effects of outside disturbances are canceled before they have a chance to produce an upset.

Most pressure-based control systems use steam flow as the all-saving feed-forward signal, particularly in the control of firing rate and drum level. Although this type of system is adequate for most boilers, changing process conditions are creating plenty of new reasons for adding more feed-forward signals:

- To account for upsets that come from within the control system. For example, in coal or waste-fuel firing, a momentary change in Btu (kilojoule) value or the periodic operation of sootblowers will cause a reaction in steam flow and firing rate that is sometimes counter to what is really needed.
- To increase response accuracy, as in pressure-compensated level and temperature-compensated flow measurements.
- To improve control system safety by interlocking controllers with feed-forward and feedback signals, as in furnace draft controls.

Thus, the I&C engineer must design controls that are intelligent enough to distinguish between system-initiated events and external events and to take action against only the latter. The feed-forward techniques described in the following show how this control philosophy can be applied to the boiler control system.

Heat-Balance Correction Factor. Coal and waste fuels create a sizable dilemma for the controls engineer whose primary concern is maintaining combustion efficiency and boiler stability. These fuels are generally hard to measure and more often than not vary significantly in Btu (kilojoule) value. If the energy input to the boiler is not directly or indirectly measurable, then a metering control system will be hard put to proportion the correct airflow to fuel flow, especially if low excess air is required.

An unmetered fuel requires that a boiler maintain higher than normal excess-air levels to guard against the possibility of explosion. Several methods have been developed to measure the Btu (kilojoule) value of hard-to-meter fuels. Some are direct methods, but most infer heating value by calculating furnace heat release or mass flow into the boiler.

The Btu (kilojoule) content of waste gases such as refinery, blast furnace, and coke oven gases is easily metered; however, these fuels vary sufficiently in Btu (kilojoule) content to cause boiler upsets if heating value is not accounted for. Fortunately, heating value can be determined by sample combustion analyzers that provide an on-line correction to fuel-air ratio controls. In operation, a sample of the fuel is mixed in the analyzer's combustion chamber with a constant flow of air and then is burned. The products are analyzed for unused oxygen, and the relative airflow required to burn the gas is determined. The signal from the analyzer may be used either as feed-forward to a fuel-air ratio controller or as feedback to a gas-mixing station.

Unmeterable solid and liquid fuels (as well as pulverized coal) generally require some type of inferential measurement, based on belt scales, feeder speed, fuel properties, or boiler heat balance. One of the most widely used methods of inferring fuel flow is measuring boiler heat release, that is, treating the boiler as a calorimeter and balancing fuel input with energy output. This is the preferred method for firing pulverized coal but can be selected for other fuels as well.

In older pulverized-coal plants, the most common type of combustion control is the series-parallel system that bases airflow on steam flow. This type of heat-balance system provides accurate firing ratios during steady-state operation, but generally is not the best system for boilers involved in variable-load operation. The reason is that inaccuracies in the inferred fuel measurement may force the operator to increase excess air to ensure safe operation during load changes.

However, parallel metering systems are better suited for variable-load operation and can be adapted to pulverized-coal firing by inferring fuel flow. In some cases, the inferred fuel input is a crude approximation derived from available factors such as feeder speed, exhauster position, mill differential, and primary airflow. The accuracy of this measurement is affected by the moisture and Btu (kilojoule) value of the fuel and the fuel rate repeatability. The combustion control scheme shown in Fig. 3.21 can be applied to correct the inaccuracy of the inferred fuel measurement.

This heat-balance correction system can be further improved by increasing the accuracy of the calculation for boiler heat absorption (Fig. 3.22). The correction factor is obtained by first taking the difference between the temperatures of the economizer inlet and the superheater outlet and then multiplying this value by mass steam flow. When the correction factor is adapted to boilers with reheat, the energy correction of the reheat unit must be calculated and added to the equation.

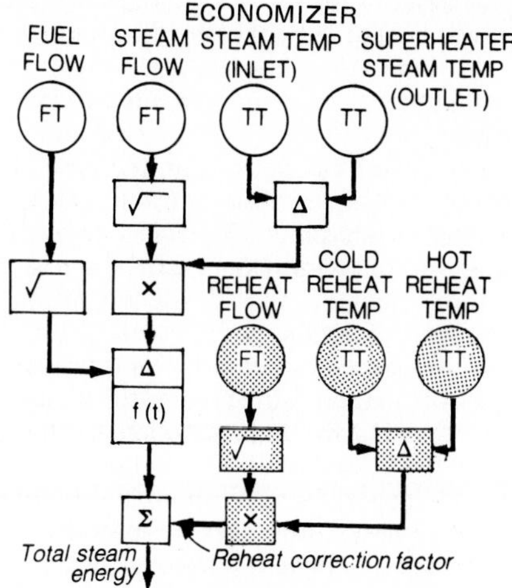

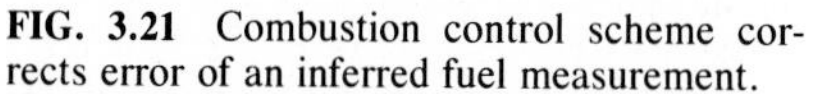

FIG. 3.21 Combustion control scheme corrects error of an inferred fuel measurement.

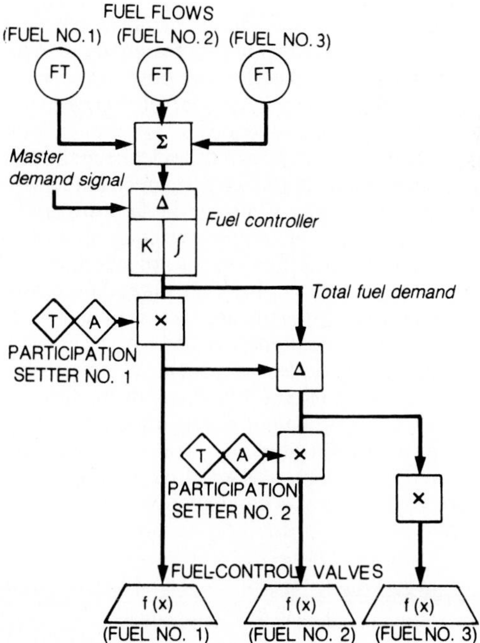

FIG. 3.22 Factor corrects for heat absorbed between economizer and superheater.

Feedwater Controls. Feedwater controls provide a mass accounting system for steam leaving and feedwater entering a boiler. A system for feedwater control must be designed to maintain the mass balance over expected boiler load changes, so that the level in the steam drum remains within the required limits for safe and efficient operation.

Just how this goal is achieved will depend on the type of boiler and its operating requirements. Boilers subject to frequent load changes will require more adaptive systems than steady-state units. Control system complexity is based on the number of measured variables initiating control action and includes single-element, two-element, three-element, and advanced control schemes to improve the accuracy of the final control action.

Single-element control bases feedwater flow on drum level; thus, the greater the drum level deviation from setpoint, the greater the control response to restore the level. Single-element systems are quite adequate for boilers that are not subject to rapidly changing loads and that fire a fuel with fairly constant Btu (kilojoule) value. These are generally small industrial boilers up to 100,000 lb/h (12.6 kg/s) intended for continuous processes or space heating.

Two good reasons why single-element control is not adequate for fluctuating demand conditions are that

- Response is very slow because a change in steam or feedwater flow takes a long time to show up as a level change.
- Pressure fluctuations in the steam drum cause water to increase and decrease in volume, resulting in a false level measurement. This phenomenon produces a control action counter to what is needed.

The two-element system overcomes these inadequacies by using steam flow change as a feed-forward signal. Response is faster because an increase or a decrease in flow rate is sensed immediately. The drum level signal serves as a feedback signal to provide trim control. This control system is capable of handling load changes that occur at a moderate rate.

Boilers that experience wide and rapid load changes require three-element control action—the third element is based on feedwater flow. This system is widely used on industrial and utility boilers. Steam flow acts as a feed-forward signal to balance the effects of expanding and contracting drum volume during load changes. The level and steam flow signals are summed and serve as an index or a setpoint to the feedwater flow. The feedwater flow measurement provides corrective action for variations in feedwater pressure and for the operating characteristics of the feedwater control valve.

Additional elements can be added to a feedwater control system to improve response accuracy. For example, the five-element feedwater control system in Fig. 3.23 is essentially a three-element configuration in which steam flow measurement is temperature-compensated and drum level measurement is pressure-compensated. Transmitters for blowdown flow and sootblower flow could be added to this system to make up seven-element feedwater control.

BURNER MANAGEMENT

Burner management systems supervise the admission of fuel and air into a boiler and provide all the on/off control and monitoring necessary for safe start-up, op-

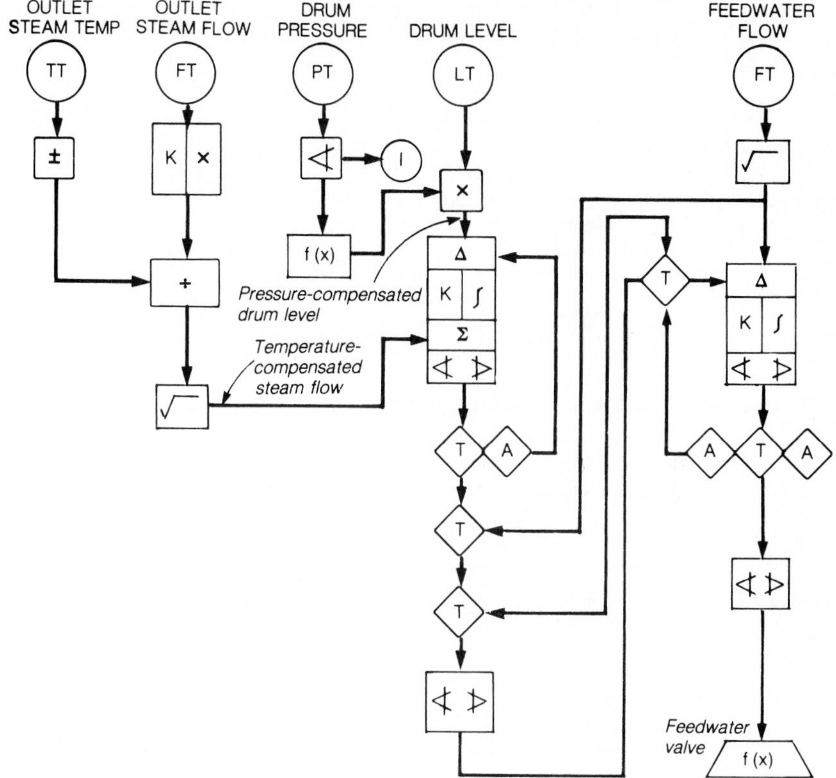

FIG. 3.23 Two elements have been added to a three-element feedwater control system to improve response accuracy.

eration, and shutdown of a fuel supply system. Although these digital-type controls generally are physically and functionally separate from the modulating control system, intersystem communication is a requirement. Also, because a furnace safety system must always have jurisdiction over firing system components, the designs of the two systems must be carefully coordinated.

The *National Fire Protection Association* (NFPA) has developed most of the groundwork for the various interlocks required in the start-up and shutdown of oil-, gas-, and coal-fired boilers, both single- and multiple-burner units. Other required interlock sequences are typically developed in conjunction with the user's consulting engineer, the boiler manufacturer, and the controls manufacturer. A typical interlock block diagram for unit start-up based on NFPA guidelines is shown in Fig. 3.24.

Basic Requirements

The basic requirements of a burner management system are safety, reliability, operational flexibility, and simplicity. The larger the unit and the greater the num-

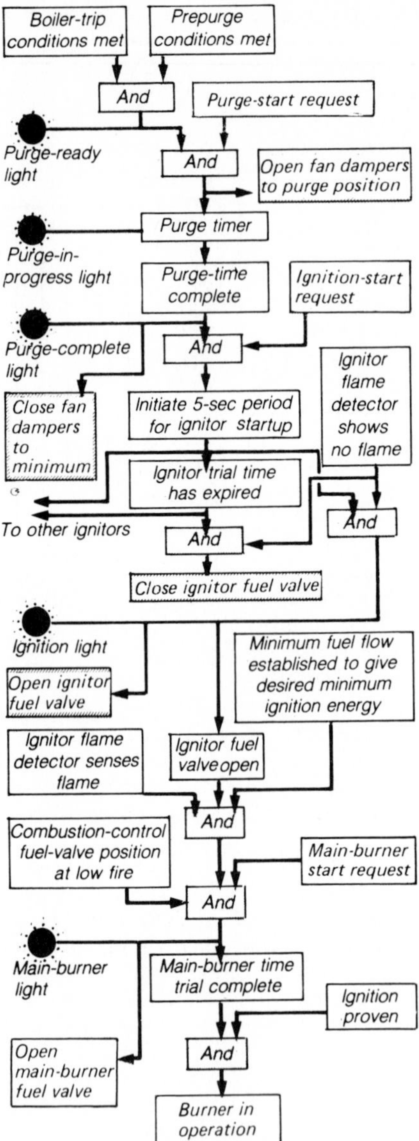

FIG. 3.24 Interlock diagram for a unit start-up shows complexity of the prepurge, purge, prelight, and light-off phases.

ber of burners, the more complicated the start-up and shutdown logic will be. There is abundant choice in the type of hardware and the degree of automation available from control system vendors, but the right type of system for a given unit will be defined by its size, operating requirements, fuels fired, and firing arrangement.

For boilers with less than four burners, a manual furnace supervisory system generally is adequate, although a trend is developing toward greater automation of the start-up of these units. Fully automated systems provide much greater safety at a comparable cost and allow a unit to be brought back on line faster and more reliably after a boiler trim.

With this system, the operator manually positions all fan dampers and valves for purge and burner start-up. The operator also must manually open the fuel safety shutoff valves. Advancement from one stage of start-up to the next is based on permissions obtained from an automatic monitoring system.

Boilers with four burners and up usually require at least some type of remote operating system, if not a fully automatic system. In the remote manual system, start-up and shutdown of each individual piece of equipment (air registers, fuel guns, fuel valves) are initiated by the operator from a control panel that may be located remote from the burner front. System logic ensures that operator commands are performed in the correct sequence, and logic intervenes only when it is required to prevent a hazardous condition.

In remote automatic systems, the prepurge, purge, prelight, and light-off phases of operation for each burner or burner level are performed at the push of a button. These systems are needed for the largest industrial and utility units typically having 16 or more burners. Systems of this type also may be operated in the remote manual mode if necessary.

The highest level of automation is similar to the automatic mode, but with load-initiated operation for peaking utility units and swinging industrial units. In load-initiated start-up, the burner management system automatically places in service or removes from service a related group of firing equipment based on load requirements. The system is designed to maintain stable flame conditions and make the most efficient use of firing system components.

Although the fundamental requirement of a burner management system is to protect plant personnel from injury, following not far behind are the needs for protecting equipment from damage and for optimizing plant availability. For this reason, safety interlocks must be capable of providing safety and avoiding unnecessary nuisance trips at the same time. How well these two functions are performed will depend largely on the type of logic chosen. The two systems available are the energize-to-trip and the deenergize-to-trip safety interlocks.

Safety Interlocks

Industrial boilers have traditionally operated on the deenergize-to-trip or fail-safe philosophy. The term *fail-safe* means this: When power to the control system is lost, fuel valves and other control devices return to their normal positions, leaving the boiler in a safe but inoperative condition. The prime advantages of a deenergize-to-trip system include lower initial cost and lower maintenance cost because the equivalent logic can be performed with less hardware than in the other type of system.

Utility boilers generally operate on energize-to-trip logic because these systems produce fewer nuisance trips, consume less power, remain operable on loss of power, and are more reliable. Because of basic utility industry requirements for availability and reliability, companies are usually willing to invest in this type of system to avoid costly downtime later.

Burner management logic can be accomplished with electromagnetic relays, hard-wired solid-state electronics, or microprocessor-based controls. Several factors determine which system is best for operation of a particular unit: the number of burners, the flexibility required, the plant's location, and the type of system installed on existing boilers.

If a fairly simple logic system is needed, expect to pay between 50 and 70 percent more for the latest microprocessor-based system than for a relay system. More complex logic with a large amount of input and output points (for example, pulverized-coal boilers) would be less expensive to implement with microprocessor-based controls, some of whose advantages are parallel installation and software design, expandability without major hardware or wiring changes, single or double redundancy, boiler operational diagnostics, and system self-diagnostics.

While microprocessor-based systems do have some inherent advantages over relays and solid-state systems, they are clearly not suitable for every installation, particularly for small plants without access to replacement parts or suitably trained maintenance people. These plants would probably be better off investing in a relay or solid-state system.

Flame-Monitoring Techniques. The flame detector is the key to the structure of all burner management systems. Some of the basic requirements of this instrument as it applies to furnace safety include

- Detection of the high-energy zone within the burner flame
- Ability to distinguish between igniter flame and main flame
- Discrimination between the source flame and adjacent or opposed flames and other background radiation

All flames emit some *ultraviolet* (UV) radiation and a considerable amount of *infrared* (IR) radiation (Fig. 3.25). The detectors available to monitor these flames include the UV-sensitive tube, the broadband IR lead sulfide cell, and the narrowband IR silicon diode detector (Fig. 3.26). Figure 3.27 shows spectral sen-

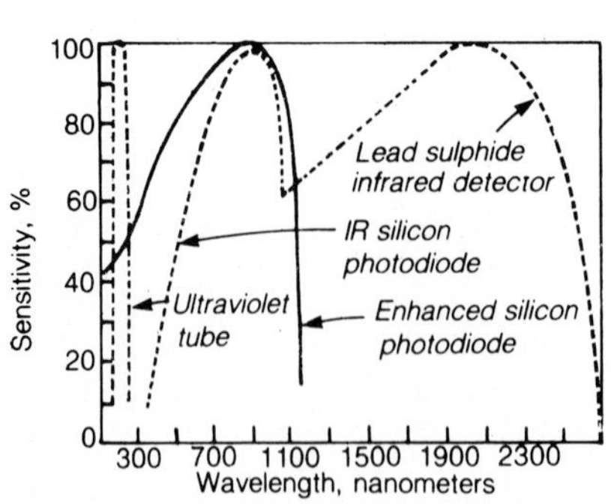

FIG. 3.25 All flames emit some UV radiation and a considerable amount of IR radiation.

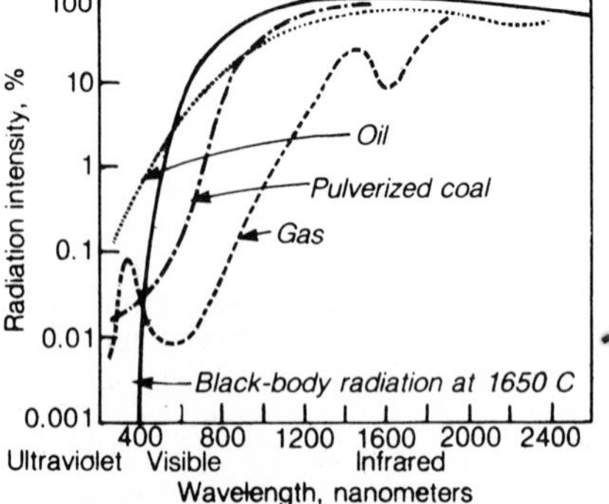

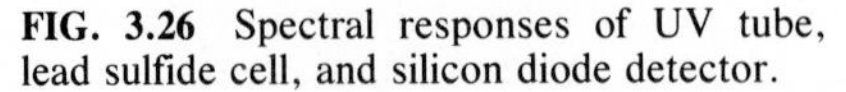

FIG. 3.26 Spectral responses of UV tube, lead sulfide cell, and silicon diode detector.

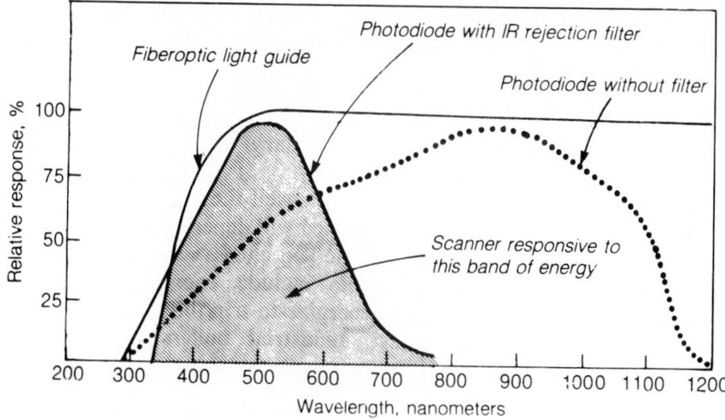

FIG. 3.27 Spectral response of a visible-light flame detector is obtained by introducing a special IR rejection filter. Monitor is suitable for oil and coal flames.

sitivity of a visible-light flame monitor developed specifically for oil and coal flames. A special IR rejection filter aids in flame discrimination.

Most of the technical problems have been solved that involve scanning the flames of boilers with up to four burners. These applications are rather simple since they do not have the problems with discrimination that the larger multiple-burner boilers have. An accepted method of scanning these oil- and/or gas-fired boilers is to use a UV detector aimed directly at the root of the flame.

For larger multiple-burner boilers, the flame detection technique depends on the fuel fired. UV detectors are very effective in monitoring hydrogen-rich gas and light-oil flames, and they provide excellent discrimination between adjacent and opposing burners and refractory (blackbody radiation) because the higher-UV-intensity zones of each burner generally do not overlap.

Before the UV detector was developed, there was virtually no way to design a scanning system that could differentiate between an igniter flame and a main flame. Now it is customary to use one UV cell to detect the igniter flame, which is typically oil- or gas-fired, and to watch the main flame with redundant detectors—the type based on the fuel fired.

With all its virtues, the UV scanner cannot provide reliable detection of heavy-oil and pulverized-coal flames, as well as flames of other high-carbon-content fuels. Unburned and partially burned hydrocarbons skirt the flame and absorb whatever UV radiation is emitted from the root of the flame. Since IR radiation is transmitted easily through this shroud of particles, some type of IR-sensitive or visible-light-sensitive detector is normally selected.

Scanning Coal Flames. Because UV detectors are unsuitable for pulverized-coal or heavy-oil flames, developers have found a way to use an IR detector to discriminate between adjacent flames. What has come out of research in this area is the *flame-flicker method,* where the high-frequency (ac) component of the radiation intensity is capitalized on to discriminate between flames. These detectors operate in the visible or near-infrared spectrum with a frequency range of 50 to 600 Hz. Various methods of signal processing using the ac signal component can now help scanners differentiate between (1) flames of different fuels, (2) flames of the same fuel, and (3) flame and hot refractory.

The flicker method works equally well with all types of waste fuels—black liquor, hog fuel, tail gases, trash, etc. UV detectors cannot scan these fuels for the same reasons as held for coal and heavy oil.

A special scanning problem arises for industrial users when a waste fuel is fired in combination with a prime fuel such as oil, gas, or pulverized coal. The boiler is brought up to load on prime fuel to establish a stable waste-fuel flame, after which the prime fuel is often shut off. To obtain permission from burner logic to restart the prime fuel while waste fuel is still firing, a scanner that is insensitive to the waste fuel is required. This requirement can be met by blocking out the flicker frequency characteristics of the particular waste fuel being fired.

Different boiler and burner arrangements require different philosophies in flame monitoring. Wall-fired units are monitored on an individual-burner basis—usually two main flame scanners and one igniter flame scanner are adequate. Should one of the main flame scanners indicate no flame, the igniter is removed from service along with the main burner.

Shutdown logic for tangentially fired boilers is similar to that for wall-fired units until a single fireball is present in the furnace, typically at 25 to 30 percent of rating. After this point, flame safety logic is transferred to fireball monitoring, and flame-failure protection is arranged on a furnace basis. If three out of four scanners indicate no flame on any given burner level, and if loss of ignition energy from associated igniters is confirmed, then fuel flow to that elevation of burners is shut off. If all elevations indicate loss of flame, the whole unit is tripped.

ACKNOWLEDGMENT

Illustrations in this chapter are courtesy of *Power* magazine.

SUGGESTIONS FOR FURTHER READING

Baur, P., "Combustion Control and Burner Management—A Special Report," *Power,* September 1982.

Dukelow, S.G., *Improving Boiler Efficiency,* 2nd ed., Instrument Society of America, Research Triangle Park, N.C., 1985.

Koska, T.J., and B.M. Redmond, "Nonlinear Measurement of Water and Steam in a Pressurized Vessel," ISA Power Symposium, Instrument Society of America, May 1981.

Lane, J.D., "Adaptive Control Techniques for Utility Boiler," Control Systems for Fossil-Fuel Powerplants Symposium, Electric Power Research Institute, Palo Alto, Calif., February 1987.

Martz, L., "Y-factor Correction of Combustion in Boilers," ISA Power Symposium, Instrument Society of America, May 1979.

Ryan, F.P., "Modernization—A Key to Improved Performance," ISA Power Symposium, Instrument Society of America, May 1982.

Schuss, J.A., et al., "Burner Management Specifications in an Environment of Technological Change," ISA Power Symposium, Instrument Society of America, October 1987.

Taft, C., "Performance Evaluation for Boiler Control System Strategies," Control Systems for Fossil-Fuel Powerplants Symposium, Electric Power Research Institute, Palo Alto, Calif., February 1987.

CHAPTER 6.4
KEY SYSTEMS AND COMPONENTS

J. A. Moore
Leeds & Northrup, A Unit of General Signal Corp. North Wales, PA

INTRODUCTION

Instrumentation to monitor and control power generation performs three basic functions: information gathering, control, and safety. Each instrument component must operate in accordance with its specifications, but only a combination of components—interconnected to accomplish monitoring, control, or safety tasks—is useful in the context of production. Such combinations are called *systems*.

Instrumentation systems work with the information that flows through them. Information flow through all industrial process systems, including those in powerplants, follows the broad divisions shown in Fig. 4.1.

During the last decade, great technological advances in instrumentation have

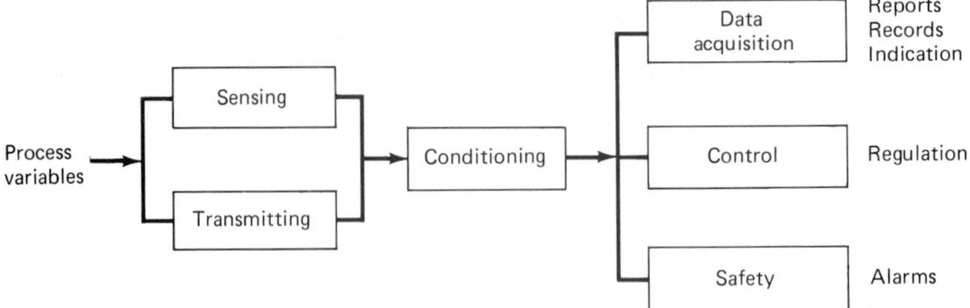

FIG. 4.1 Information flow through process system. *(Courtesy J. A. Moore.)*

been made. They are continuing today, although much of the equipment purchased previously is still in service. The older order can be called *analog* in character. It includes instrumentation based on pneumatic power and electronic designs that use *direct-current* (dc) signals. The new order can be called *digital* in character. Here, microprocessors and special-purpose computers combine to communicate over data highways.

The emphasis in this chapter is on digital equipment and local area networks, because that is the type of instrumentation that eventually is going to prevail. Older instruments will inevitably be replaced with newer designs (even though the equipment may have been operating satisfactorily for many years), simply because spare parts will become unavailable. Analog instrumentation is still a widely used form of measurement, however, and is also covered here.

SIGNAL GENERATION AND TRANSMISSION

Sensors create the measurements of process variables, while transmitters develop signals that interface the measurements to the instruments that will make use of them. In most cases, the signals developed are low-level millivoltages or capacitances. Electronic and digital systems require voltages, usually 1 to 5 V dc, although 0 to 10 V is sometimes specified. This means that transmission of signals really consists of three stages. In the first, low-level signals from sensors are changed to signals having a level that can be transmitted. This is sometimes called *transducing*. In the second stage, the transduced signals are moved from the process area to the location of the indicating and control instruments. This stage can be called *transmitting*. Finally, signals may have to be changed so that they interface properly with the equipment that will accept them. This stage is called *signal conditioning*.

Pneumatic Signals

Pneumatic transmission contains the three stages. The sensor produces a mechanical motion that is analogous to the process variable value; the motion is transduced by a nozzle-flapper and bellows assembly to a pneumatic signal that is transmitted. The signal ranges may be 3 to 15 or 3 to 27 psi (20.7 to 103.4 or 20.7

to 186 kPa) of air pressure. To become the input to an electrical system for measuring or control, the pneumatic signal must be conditioned by converting it to a voltage. This is done in a manner equivalent to that followed for electric signals, described below.

Electric Signals

Sensors producing electric signals for the significant process variables have been discussed in detail in Chapters 1 and 2 of this section. The transducers for these signals are generally bridge circuits or oscillators with chopper-stabilized amplifiers, producing a higher-level analog signal. Span and zero adjustments provide a means of matching the high and low ends of the outputs to the calibrated range of the sensor.

Typically, a two-wire transmitter produces a 4- to 20-mA current that is proportional to the value of the measured variable. A common circuit draws on 24-V dc power from the instrument panel. Carried over one of the wires to the transmitter, it becomes the source for a regulated supply voltage. Current from this source voltage passes through resistors and the emitter-collector junction of a transistor, producing a voltage that is fed back and matched against the value of the process variable analog.

An amplifier, comparing the two opposed signals, provides voltage to the base of the transistor. It does this by varying the current through the emitter-collector junction so that an equilibrium is established between the current flow and the signal value of the process variable. The 4- to 20-mA current returns to the signal common of the 24-V supply through the second wire of the two-wire transmitter, completing its circuit through a 250-Ω resistor to ground. This develops a 1- to 5-V signal that is introduced into the electronic-control or data-logging equipment through a very high impedance (1 MΩ or greater).

Electric Signals for Status

Discrete electric input signals are generated by interrupting ac or dc power or by generating pulses. The discrete open- and closed-circuit signals are created by contacts of switches, pushbuttons, or relays. For anything other than digital circuitry, the voltages do not have to be conditioned; 120-V, 60-Hz or 125-V dc inputs are most common. High-frequency pulses, generated by magnets in rotors or mounted on rotating shafts, are developed by tachometers and by some flowmeters. These are analog signals, since the frequency varies over a range of rotation or flow, and they have to be conditioned to fit the 1- to 5-V requirement of analog inputs.

FINAL CONTROL ELEMENTS

At the other end of the system, output signals to regulate material or energy flows that change process conditions have ranges of 3 to 15 or 3 to 27 psi (20.7 to 103.4 or 20.7 to 186 kPa) from pneumatic controllers or 4 to 20 mA from electronic or digital control systems. The majority of final-control-element operators are actu-

TABLE 4.1 Components for Powerplant Regulation

Output value	Regulator	Control value
3 to 15 or	Control valve	3 to 15 or
3 to 27 psi*	Damper operator	3 to 27 psi*
Electric pulse (forward-reverse)	Reversing motor	120 V, 60 Hz; 240 V, 60 Hz
4 to 20 mA	Control valve	3 to 15 or
	Damper operator	3 to 27 psi*
4 to 20 mA	SCR	4 to 20 mA
Discrete signal	Relay, solenoid valve, light	120 V, 60 Hz; 240 V, 60 Hz; 125 V dc

*In SI units, 20.7 to 103.4 or 20.7 to 186 kPa.

ated by pneumatic motors, however, and 4- to 20-mA signals that will be used must be converted to the corresponding range of air pressure. For varying the speed of motors, 4- to 20-mA signals can be transmitted directly through *silicon controlled rectifiers* (SCRs). Discrete outputs, in the form of ac or dc voltages, are supplied to operate solenoid valves, energize relays, and light indicating lamps. These conditions are summarized in Table 4.1.

Control Valves

Flows of steam, oil, water, gas, and sometimes air are controlled by throttling valves. Changes in controlled air pressure position the valve operator, a spring-loaded shaft (Fig. 4.2) that positions a plug in a port (Fig. 4.3) to throttle flow. To satisfy safety requirements, plugs may be contoured so that flow rates can be characterized at different shaft positions.

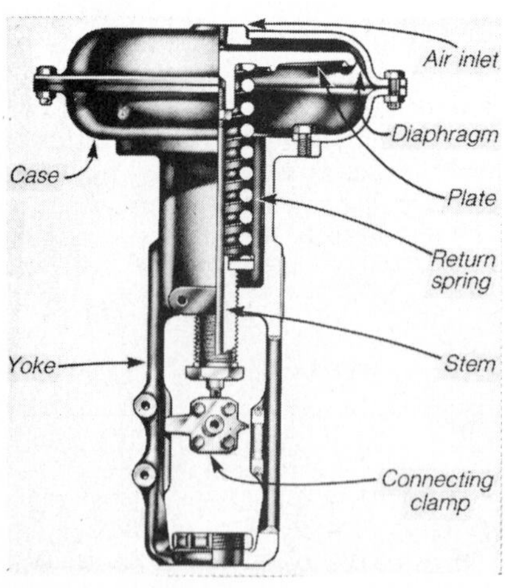

FIG. 4.2 Spring-loaded shaft in valve operator.

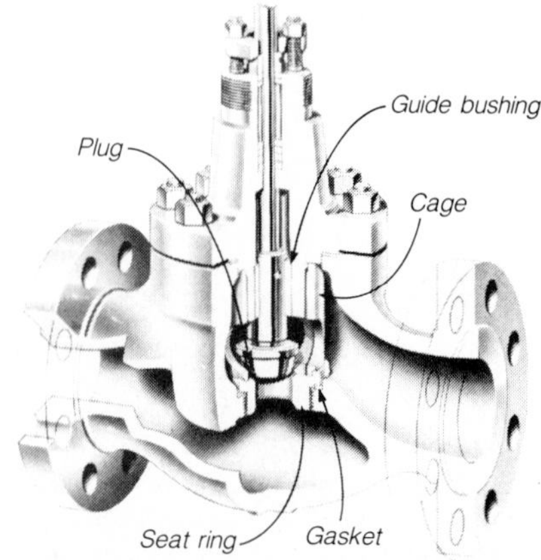

FIG. 4.3 Plug-in valve port throttles flow.

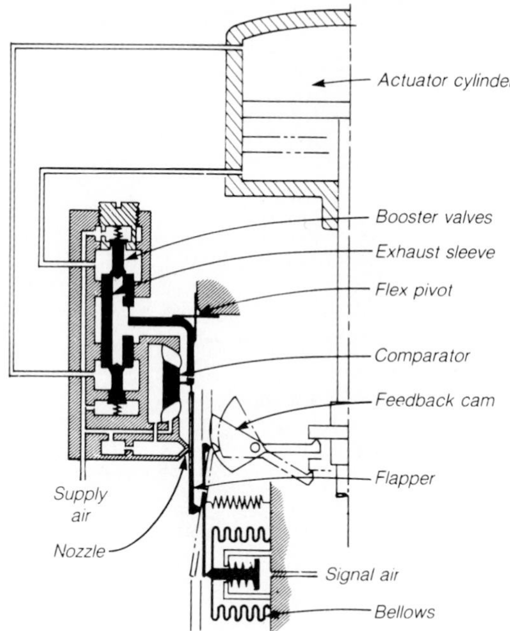

FIG. 4.4 Valve positioner.

A valve positioner is recommended for all valves having body sizes larger than 3 in (76 mm). It compensates for shaft friction, unbalanced forces across the plug, loading-spring nonlinearities, and valves sticking in one position.

The valve positioner is a device mounted on the drive motor that compares position transmitted from the shaft by a linkage to the value of the variable control signal (Fig. 4.4). A specific correspondence exists between the control signal and shaft position. If it is not attained, the valve positioner will actuate a pilot that applies high-pressure air to the operator to move it until the correct correspondence is reached.

Piston Operators

Flow of gases (through forced- and induced-draft fans, for example) is more often controlled by positioning the vanes of dampers than by actuating throttle valves. Positioners for large dampers must be able to develop high torque. Piston operators (Fig. 4.5) are specified for torque requirements up to 3000 ft • lb (4.08 kJ).

Pneumatic power positioners use pneumatic control signals to pilot a high-pressure air supply [50 to 100 psi (345 to 689 kPa)] to one side or the other of a double-acting cylinder to drive a piston that is linked to the damper operating lever. A feedback linkage from the positioner output linkage feeds the damper position back to the pilot for comparison with the control signal, producing the same action as the valve positioner just described.

FIG. 4.5 Piston operator.

Electric-Drive Motors

Control valves and damper operators can also be positioned by reversing electric motors, operating from forward-reverse control signals (Fig. 4.6). This type of operator is particularly useful for large torque requirements. As boiler capacities increase, coals with lower heating values are burned, and the addition of scrubber

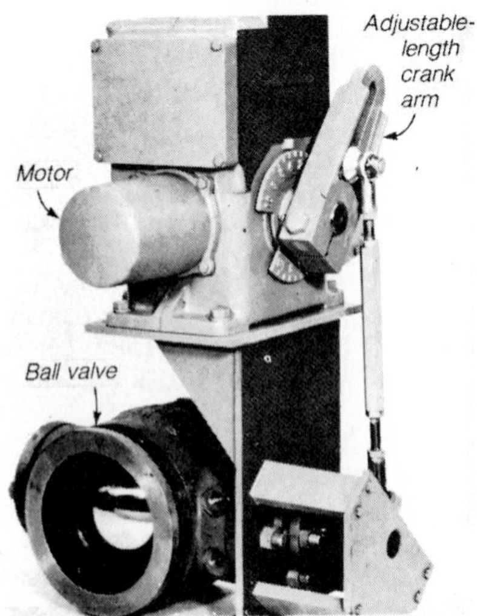

FIG. 4.6 Reversing motor on control valve.

systems increases draft loss. Thus, bigger fans with higher heads must be selected, increasing maximum torque requirements from 3000 to 7000 ft • lb (4.08 to 9.5 kJ) or more.

Motor drives employ gear reducers to transmit forces that will meet these torque requirements and at the same time execute the instructions of the control system. The holding force is provided by continuously energizing the drive motor, by providing a braking device (for high-efficiency gearing), or by using low-efficiency gearing that will not drive backward.

Position feedback is required to ensure that the unit has provided the control action called for by the control signal. Proponents of electric drives list as desirable these characteristics: high breakaway torque; fast, but accurate, response to signal changes; positioning accuracy to 0.25 percent; slewing speeds that give 15-s full stroking time; and retention of the last position in case of failure of the control signal or drive unit power.

Variable-Speed Control

Solids flow may be controlled by varying the speed of motors driving conveyor belts or rotating machinery such as pulverizers. A 4- to 20-mA control signal operates through an SCR to vary the field flux or armature voltage of dc motors. To regulate liquid flow, pump speeds can also be adjusted in this way. The current output signal can also control fan speed through the adjustment of hydraulic coupling devices (Fig. 4.7).

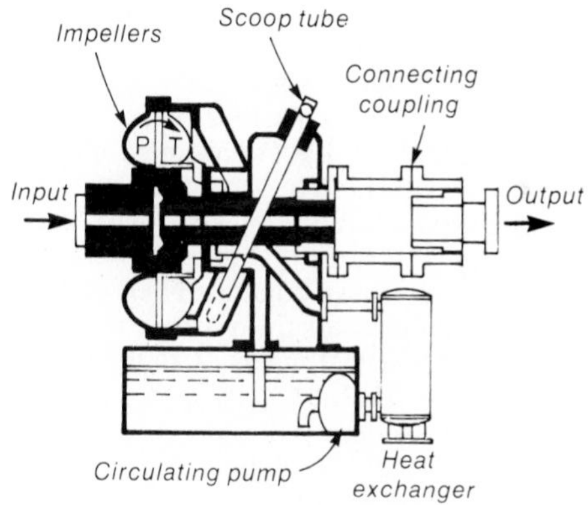

FIG. 4.7 Hydraulic coupling on feed pump drive.

On/Off Control

Discrete output signals in the form of voltages energize solenoids, relay coils, and indicating lights. This type of control, initiated by switches and pushbuttons on consoles, is found in burner control circuits and triggers annunciators in alarm systems (Fig. 4.8).

BETWEEN INPUT AND OUTPUT

This section of the instrumentation system, between input signal conditioning and output regulation, is undergoing revolutionary changes. Analog equipment is

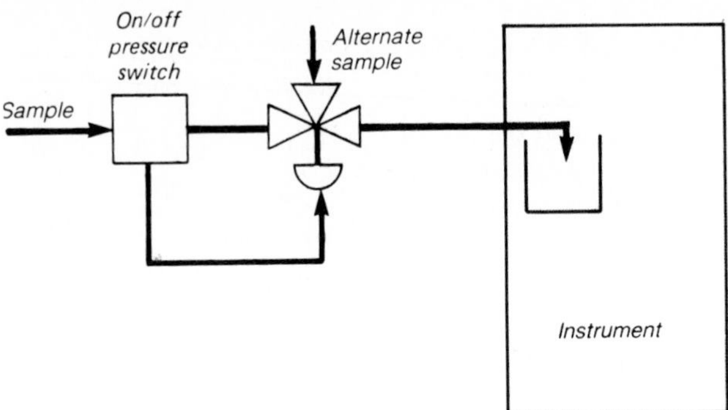

FIG. 4.8 On/off pressure switch in sample line.

reviewed first, as background for discussing these technological advances; also, it still sees frequent service after years of satisfactory performance. Existing analog installations for information retrieval, control, and safety have a mixture of pneumatic and electronic devices.

Analog Data Acquisition

Transmitted data from sensors are displayed for operator reference on panel-mounted indicators and recorders. Round-chart recorders display and store process data for one to four variables in 24-h segments, and strip-chart recorders can hold records of up to 16 variables for periods of several days to more than a week, depending on chart speed (see Recorders, next chapter).

Indicators will normally be designed to accept analog signals, either low-level millivolts or 4 to 20 mA, although draft indicators may be connected directly to the process (see Indicators, next chapter). Draft gages may have scales 12 in (305 mm) long, shaped in an arc that follows pointer travel. They are mounted high on the panel and are visible to the operator from across the control room. Other indicators will have horizontal or vertical scales, with the pointer responding to the rotary motion of a taut metal band. Indicators with 6-in (152-mm) scales can be expected to be about 2 percent accurate.

Information for reports is collected by hand, by a clerk with a clipboard, once an hour or once a shift. It is reviewed by management and retained with chart records for future reference.

Analog Control

The control portion of instrumentation has undergone a constant evolution, beginning with pneumatic equipment and continuing today with microprocessor-based hardware. Powerplants that have not yet made the conversion to digital control will have panel-mounted pneumatic or electronic control, providing *proportional, integral, and derivative* (PID) control functions for feedback control loops.

Electronic control equipment in powerplants usually follows a split-architecture design. That is, the operator will see, and have access to, panel-mounted control housings, each with a front plate that displays indications of pro-

FIG. 4.9 Manual/automatic (M/A) station.

cess variables and output control signals; has a switch that provides selection of automatic or manual modes of control (Fig. 4.9); and has knobs or pushbuttons for adjusting setpoint and output values.

The electronics portion of the controller includes the circuits producing error signals that compare process variable to setpoint—and the proportional band, reset, and rate circuitry and adjustments—with output signal generation. The electronics was developed by a transistorized operational amplifier and passive components mounted on printed-circuit cards (Fig. 4.10).

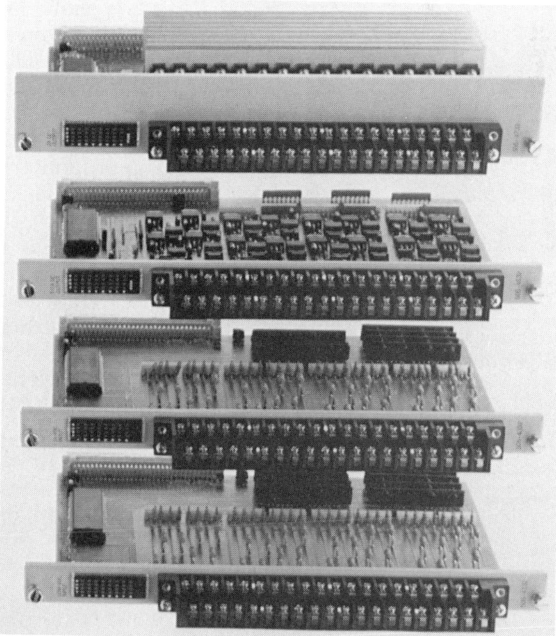

FIG. 4.10 Printed-circuit cards.

The printed-circuit cards will often be packaged with other cards that carry the comparators, computational functions, and relay logic that make up the balance of the process control system. The entire electronics package is transparent to the operator. Instrument technicians and engineers are responsible for tuning constant adjustments, calibration, and maintenance of this portion of the equipment, which is split off from the operator's panel and is quite often mounted in another part of the operating facility altogether.

Other control instruments include computing devices (such as square root extractors), integrators, lead/lag elements, functions generators, summers, multipliers, etc. These will be part of the control system for any but the simplest boiler and power generation systems. They may be designed into the electronics package that holds the PID controllers, or they may be supplied as separate, back-of-panel instruments. In the electronics package, they will be built onto printed-circuit cards and plugged into back panels that bus and interconnect all the signals for the entire system.

Analog Alarms and Safety Circuits

The principal alarm indication on powerplant panels is the annunciator, and the principal safety circuitry occurs in burner management systems for gas-fired boilers. Both annunciators and burner management systems depend on relay operation for routing control-action voltages, and burner management relies on timers and counters for making digital computations. This type of equipment is made up of discrete components, mounted in housings, with interconnections hard-wired to terminals.

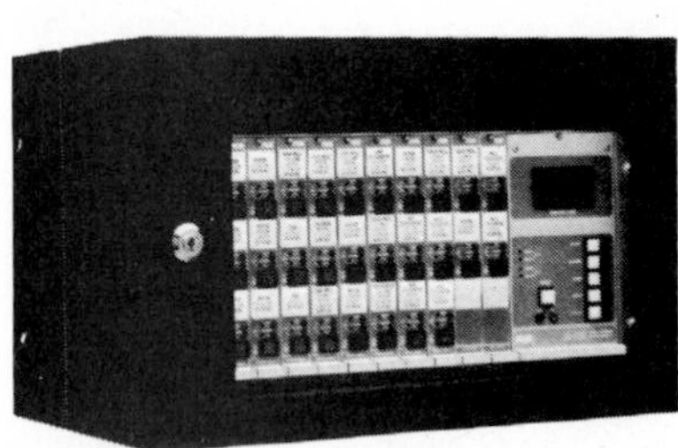

FIG. 4.11 Annunciator display.

Annunciator displays are a matrix of backlighted legend plates, panel-mounted for operator viewing (Fig. 4.11), which are normally activated by a contact closure that results from a process limit violation. The operator is alerted to an alarm condition by a horn and by the appropriate legend lighting up and flashing. A pushbutton on the panel is depressed to silence the horn and change the flashing indication to a steady lighted display.

The burner management system applies logical computations to perform batch or discontinuous operations on the basis of discrete signals or elapsed time. In gas-fired plants, this sort of logical design ensures that gas pilots and combustion burners are lighted after a sequence of purges of the combustion chamber, factoring in enough elapsed time to see that firing conditions have been safely established. Other sequential logic operations using relays can include pump-up/pump-down circuits (Fig. 4.12), actuation of solenoid valves for periodic blowdown, and sequential starting of pumps by operating motor control center relays.

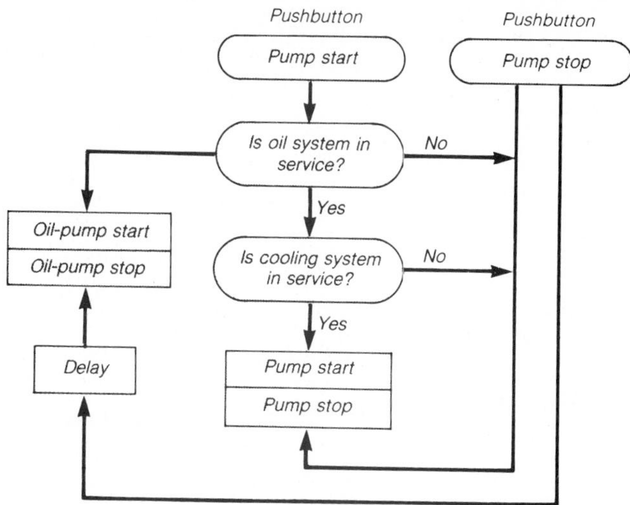

FIG. 4.12 Logic definition for pump start-up and shutdown.

Digital Equivalents

The following digital control devices are becoming state-of-the-art forms, suitable for replacing the devices just described to perform powerplant instrumentation and control:

- Data acquisition devices
- Data loggers
- Single-loop controllers
- Programmable logic controllers
- Distributed control systems
- Minicomputers
- Personal computers

These devices are really special-purpose computers that use microprocessor-based central processing units. Each can stand alone, performing specific parts of a plant instrument operation. In general terms, the data acquisition and logging devices accomplish the information-gathering function, while single-loop controllers, programmable logic controllers, and distributed control systems perform the control function. Minicomputers and personal computers have augmented the activities of the central control room. The power of microprocessor-based, integrated-circuit electronics is so great, however, that activities that were quite distinct in the description of analog equipment can be accomplished by more than one of the above-named digital control devices.

The significant change is that all the devices can work together in a single integrated-control scheme. This is possible because of a second technological advance (the microprocessor is the first)—the ability of digital information to be communicated among and shared by many devices. The communications networks that accomplish this are local-area networks and data highways. Since communications networks coordinate the system operation, it seems appropriate to review them first and then to look at the tasks of information gathering, control, and safety in the context of digital equipment.

DIGITAL COMMUNICATIONS

The key contribution of the digital communications concept is that operator functions for monitoring and control of process areas can be focused at one central point, and the information collecting and processing function can be distributed to the individual locations. The hardware at each location is modular, with each module designed to make use of integrated circuits. The information between modules, coded in 1s and 0s, can be transmitted serially over a pair of wires. All the local networks in turn communicate with the central operator area serially over a single path, again in coded 1s and 0s. These communications paths have been called *data highways,* and the interconnection of highways makes up a communications network.

Figure 4.13 shows a distributed system for a typical powerplant. It has a coal storage area; coal-fired boilers; facilities for ash handling, feedwater, and waste treatment; a scrubber; and turbogenerators for producing power. Operation is centralized, and there is a link to a plant computer system. The schematic illustrates the plant subsystems and applications that can make use of the types of digital devices available and that can be combined into one system by the communications networks that interface them to each other.

Networks: Two Kinds

The interconnections shown in Fig. 4.13 include two kinds of networks. Individual instruments in a local area are connected by a linear highway or by star configurations. The local networks connect to a global highway that provides communications between the central area and all the local networks. Global highways may have a linear formation or a ring formation (Fig. 4.14). An installation will normally include a redundant global highway to improve reliability of the system, since any fault in the highway can interrupt the entire process of communications. In smaller networks, the remote station connections may radiate from the central location in a star formation (Fig. 4.15).

Data Highways: Hardware

Digital signals between devices move as a string of pulses of energy, representing 1s and 0s. Each 1 or 0 defines the status of a binary digit, referred to as a *bit.* Single conductors (copper or aluminum) are a convenient means of carrying a string of information as bits of electric energy. Fiber-optic cables, carrying bits of

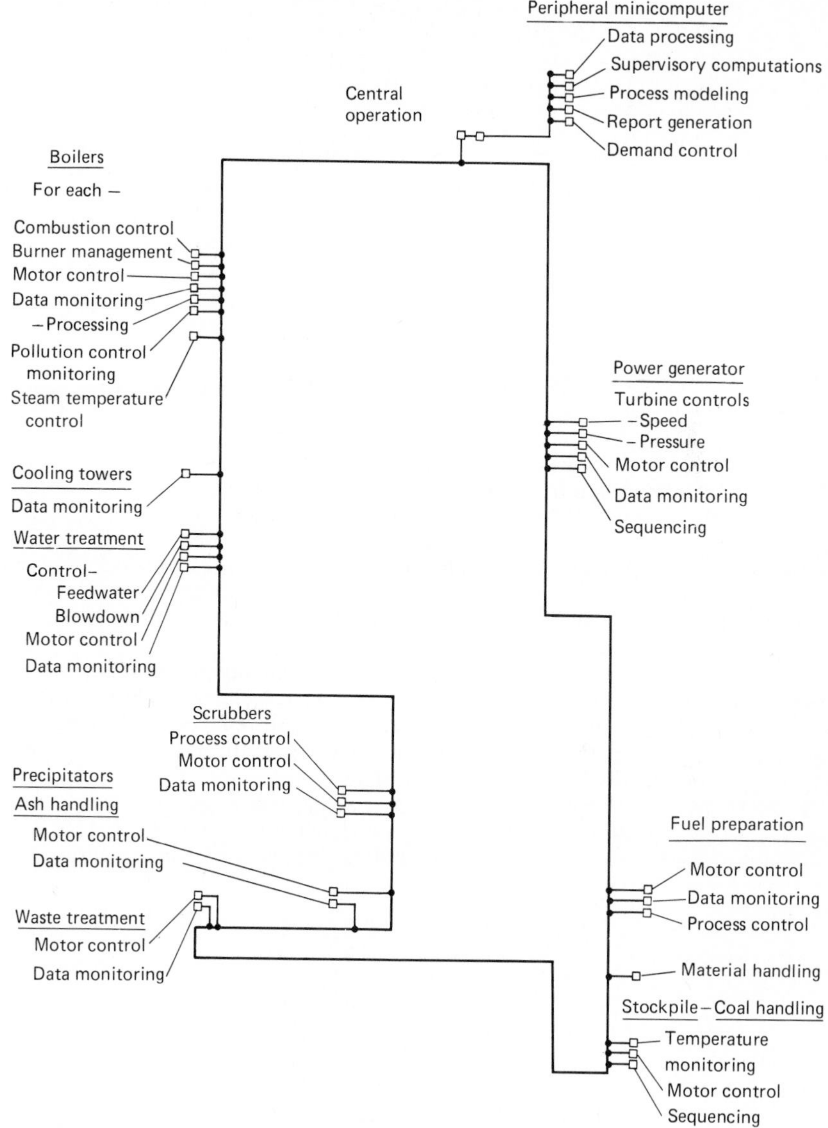

FIG. 4.13 Potential activities for data-highway-coordinated powerplant instrumentation. *(Courtesy J. A. Moore.)*

light energy, are also possible. Media for data highways include twisted pairs of wire, coaxial cable, or fiber-optic cable.

Multiple pairs of small wire (no. 20-24 AWG), each pair twisted and shielded, may be combined in a single cable for connections in a local network or star formation. Cable lengths of 985 ft (300 m) are practical. Number 16 gage wire,

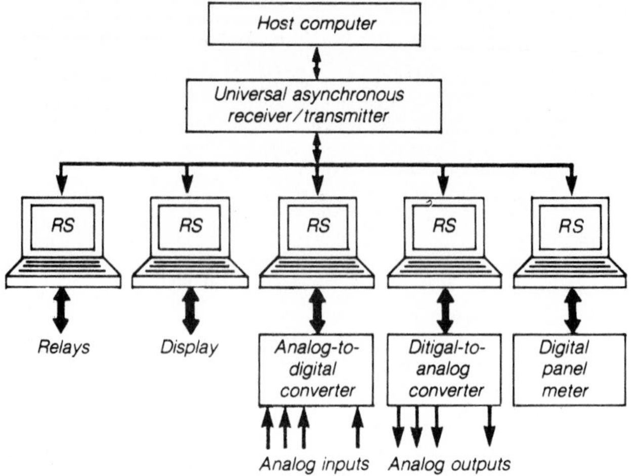

FIG. 4.14 Linear formation type of distributed control system.

twisted and shielded, can be used for linear and ring-shaped highways. Capacitive coupling between wires attentuates the signal, limiting the effective length of a twisted-pair highway to about 0.62 mi (1 km). The information traffic that a no. 16 gage twisted pair can handle is about 500,000 bits/s.

Coaxial cable is the most common medium for global highways. It is made up of one heavy central conductor separated by an insulating layer of plastic or foam from an outer sheath of braided conductor. The shield reduces electromagnetic interference, so longer transmission distances can carry 10^6 bits of data per second without the necessity of repeating the signal.

Fiber-optic cable is the third medium for transmission selected by some suppliers. The energy transmitted is light, at wavelengths of either 850 or 1350 nm. Fiber-optic cable is immune to electrical interference and carries very large

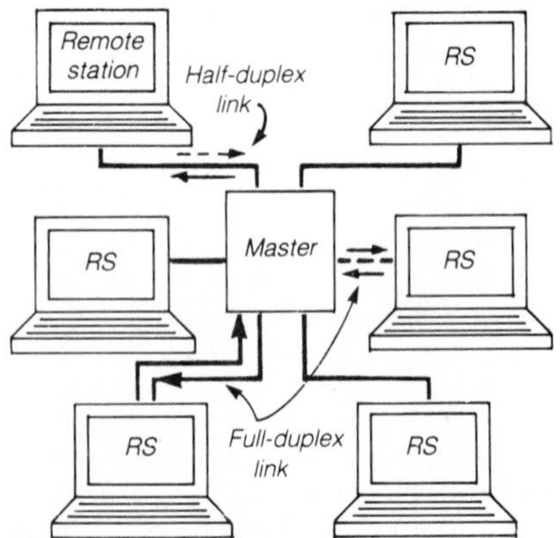

FIG. 4.15 Star configuration.

amounts of information—both desirable characteristics as the geography of systems and the amount of information transmitted become larger. Signals eventually attenuate, however, because of light dispersion, and must be repeated.

Installed costs are the same order of magnitude as for coaxial cable, but connections with other highways and branches are more difficult to make, because fiber-optic to electrical to fiber-optic transitions must be made. A typical cable for distributed-control-system highways has a diameter of 100 μm and carries 10^6 bits of information per second for a distance of 1.3 mi (2.1 km) before requiring signal repeating.

Data Highways: Software

Communications involving the moving of digital information are very much like systems for moving mail, or systems that handle automotive traffic. The mail analogy is useful for considering how data are extracted from a device, sent over a highway, and received by another device. Essentially, information stored in a sender's memory or records must be written in a form and language the recipient will understand, packaged in an envelope that meets postal regulations, and addressed properly.

Forming the Message. Information is stored in a computer's memory in strings of binary bits. When they are called to be put into a message, the bits are moved in parallel (side by side, all at the same time) to a staging area, where they are changed to a serial configuration (one at a time, one following the other) and combined with (preceded and followed by) other bit combinations that indicate they make up a message, and where the message is to go. The staging area is a chip and is sometimes called a *universal asynchronous receiver-transmitter* (UART).

The encoded message is then sent through a port and a *mo*dulator-*dem*odulator (modem) onto the data highway. A modem is a device that develops a carrier frequency which is varied by the message content. It may, for example, have a frequency of 2 Hz when representing a 1 bit, changing to 1 Hz to represent a 0 bit. Other types of modems change amplitude or phase to represent bit status.

The message leaves through a port, which is really a software address but physically becomes a plug and cable. The uses for the connections in the cable, the rate of change from one bit to the next (called the *baud rate*), the voltage levels of the status signals, and other operating parameters are defined by a published standard. It is RS-232C from the Electronics Industries Association.

Highway Standards. The parameters of RS-232C limit transmission between sender and receiver to about 50 ft (15.2 m). This is fine for slow devices such as printers that are sitting next to computers, but not for highway communications, which must handle a far greater volume of traffic over longer distances; consequently, other standards have been developed. For instrumentation highways the standard is RS-499, which defines pin designations and connector requirements. Two standards covering electric signal characteristics relate to unbalanced (RS-422) or balanced (RS-423) transmission lines.

The way that the serial bits are assembled in the message is called *protocol.* Communications using the RS-232C standard—to a printer or disk drive, for example—may relate to a code originally developed for transmission between teletype machines, called *ASCII,* for American Standard Code for Information

Interchange. This has 128 seven-bit combinations to represent upper- and lower-case letters, punctuation marks, and message instructions (message starting, message ending, etc.); for this standard, 9600 is a high baud rate.

By the time destination addresses, request to transmit and permission to transmit codes (called *handshaking*), and start and stop codes have been added in the UART, only about 40 percent of every transmission is useful message information; so the ASCII code is not a very efficient format. Other formats are available that can send a packet of information at one burst, including a paragraph or block of bits with a header, destination, and elaborate accuracy checks to make sure the message received is identical to the message sent. This increase in efficiency is necessary for data highways that may carry information at rates higher than 1 MBd. A standard format called *high-level data link control* (HDLC), developed by the *International Standards Organization* (ISO), is preferred by many suppliers (see box on following page).

Traffic Control. One more variable must be considered: how to manage traffic control. Here the analogy is to highway traffic, with rights-of-way, ramps to throughways, queuing, and multilane highways as the equivalent factors. Here are four ways that communications traffic may be controlled over a digital system highway:

- Each mode may include a highway director, a device that polls all devices, giving each client station a fraction of a second of highway use. Or it may run through a schedule of devices, allowing each one in sequence to send, and sharing unused time with the other stations on the network. Local-network directors (1) store information to be sent from a station and put it onto the highway when an opening occurs and (2) watch for messages coded to its station, pulling them off the highway and sending them to the station. They may also acknowledge that messages have been received.
- Another method allows any station on the highway to communicate when it has a message to send. Part of the name of this method is *multiple-access*. Here, courtesy must be observed, as at a polite board meeting. If a station is transmitting, all others wait until it has finished; if two stations start simultaneously, the lower-priority station will stop and let the other complete its transmission. This accounts for the rest of the name, *collision detection*. The complete name is *carrier sense multiple-access/collision detection* (CSMA/CD).
- A popular form of transmission, common to ring networks, uses a token, which is a coded signal that is passed from one station to the next in accordance with a schedule programmed when the system is configured. The station having the token may transmit for a timed interval, say 50 ms, then must pass the token to the next station.
- Another system functions as a bulletin board, posting its entire data base into a common block of shared memory accessible to every station. Information in the shared memory is broadcast periodically, like a stock market report, and each station listens, picking out the information it needs. To reduce the time for information exchange, this type of transmission is proposed for highways that encompass an entire plant.

Communications Standards. There are many standards for digital communications. A number of engineering groups are working to organize standards for different levels of communications (formatting, protocol, traffic control) into an ac-

HDLC Protocol Format

FLAG	ADDRESSES SEND/ RECEIVE	CONTROL	INFORMATION PACKET FIELD	CYCLIC REDUNDANCY CHECK	FLAG

Flags define the beginning and the end of transmission, so that the transmission can be synchronized with the reception.

Control defines the message-handling details.

The information packet can contain as many bytes of data as the time allocated to the transmission permits.

The *cyclic redundancy check* (CRC) is a system of error checking performed at both the transmitting and the receiving locations. In one form, each byte sent is divided into a polynomial expression, such as $g(x) = x^{15} + x^{12} + x^9 + 1$. The remainders are accumulated, creating a unique value. The same computation is performed on the bytes received, and the CRC bytes are compared to the generated CRC value. If the read CRC does not agree with the generated CRC, an error is assumed to have occurred, and the message is not accepted.

ceptable composite. This is badly needed, so that different types of digital control devices can communicate over one highway without interpretive programs having to be written for each separate one. A standard directed toward this objective that seems to be gaining acceptance is the *Manufacturing Automation Protocol* (MAP), which is strongly endorsed by General Motors Corp.

Data Acquisition

Earlier in this chapter, data acquisition by recorders, indicators, and counters was discussed. Data logging was a manual operation, in which the observed data were copied onto forms. The acquired data were processed by manual computation to obtain figures for operating reports and departmental records.

The first digital devices provided an intermediate step in data collection, allowing a number of records to be collected and displayed as seven-segment *light-emitting diode* (LED) indications or printed on narrow tape rolls (Fig. 4.16). These were data loggers. Microprocessor-based circuitry has expanded the number of inputs that can be handled and has added the capability of processing the data. Communications technology now permits real-time data, and performance values computed from it, to be displayed at every work station in the system and to be available in specifically formatted reports. These activities go beyond logging, and the equipment performing them is more properly referred to as *data acquisition and processing devices*.

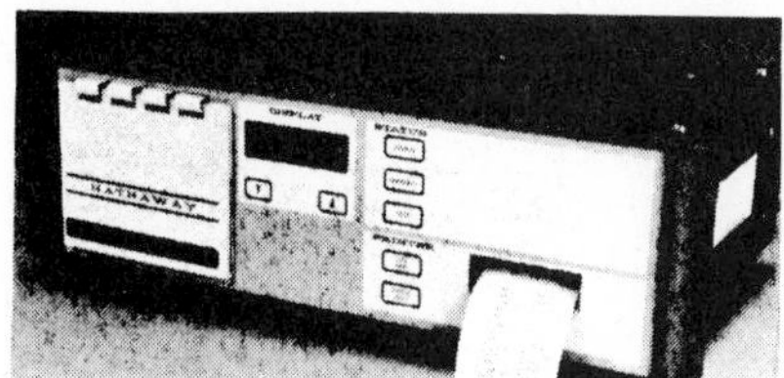

FIG. 4.16 Event recorder.

Signal Generation. To implement data acquisition to publication in report form, information flow from measurement through signal selection and conversion must be analyzed (Fig. 4.17). Most digital devices can accept low-level signals directly from sensors. Powerplant inputs typically accepted are millivolt signals from thermocouples and resistance thermometers; 1- to 5-V signals from differential-pressure sensors; pulse-counting inputs from tachometers; discrete inputs—120 V, 60 Hz and 125 V dc, for example—from limit switches and pushbuttons; and *binary-coded decimal* (BCD) inputs from other digital devices. By combining input modules, more than 1000 points can be handled by a data logger.

This information may require some conditioning—taking the square root of signals from *differential-pressure* (DP) transmitters or linearizing thermocouple signals, for example. Conditioning may be done externally with analog devices or

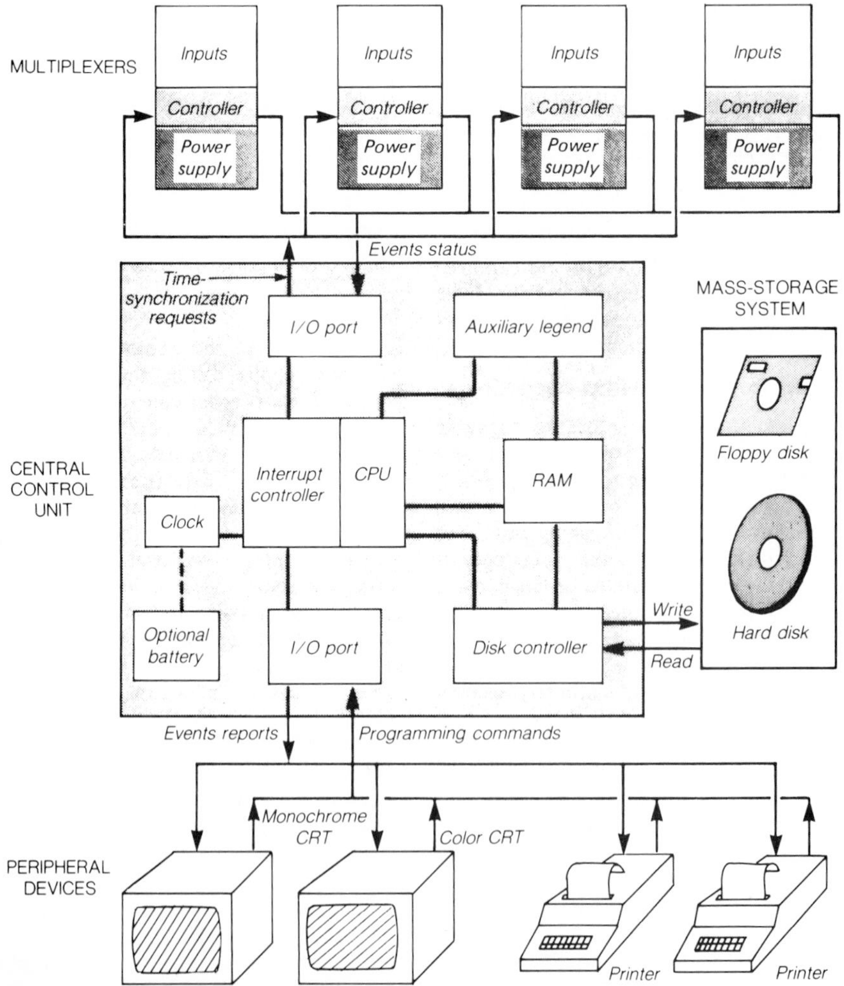

FIG. 4.17 Flow of information from measurement inputs to final deposition at CRT, printer, or storage device.

internally in the data acquisition circuitry itself. There must also be provision internally, however, for converting the combination of signal types listed in the preceding paragraph to a common form that can be handled by the circuitry which changes analog values to digital format. To do this, proportional gain amplifiers, with span and zero adjustments for creating range limits, are needed. All the parameters for these adjustments are created in software when individual points are defined by programming. To avoid interference from external noise, additional conditioning is required in the form of filtering and isolation circuitry.

An input value having been selected, it must next be converted to its digital format. An analog-to-digital converter will accomplish this. The resolution of conversion is a function of the number of bits in the digital equivalent produced from the analog value. A 12-bit word has a resolution of 1 part in 4000, or 0.025 percent; 13-bit resolution is common, and even 14-bit resolution (1 part in 16,000) is listed.

Information Processors. Today's data acquisition units may include processing capabilities, greatly enhancing the value of the information collected. The information collecting, multiplexing, conditioning, and conversion segments of a data acquisition system can send their data to a data processor that is part of a distributed control system; or the data can go to a processor that is dedicated to data logging and processing and does nothing else. In either case, if the necessary ports are included in the equipment design, acquired data can be put onto a data highway and sent to central locations for display.

The data can be printed at specific time intervals in specially formatted reports. They can be sent over a highway to a minicomputer at a higher level of the communications network, for use by management in process modeling or for any other purpose requiring real-time data. Because of its enhanced capabilities, data acquisition has a much greater place in the control system than when it was used for indication alone.

Recorders

Digital technology has brought several significant enhancements to strip-chart recording, still used for immediate display of trends as well as for record keeping. First, recorders have been designed so that the chart can serve as a combination printer and graph device. Instead of the continuous line produced by a pen or print wheel, integral numbers may be printed in columns and rows so that the values recorded can be read as numbers (Fig. 4.18).

Second, recorders can include a microprocessor and memory; thus each point of a multiple-point recorder can be programmed to have individual range, print-time interval, linearization, and alarm points, and the chart speed can be changed by programming as well. A small keyboard in the chassis allows information to be checked and changed readily. In many cases, the familiar scale has been augmented by an LED display of the variable value. Diagnostics continuously check instrument operation, and the same display can also indicate the status of test values.

Control Functions

Single-loop controllers and distributed control systems provide the digital-control-device equivalents to the control functions reviewed in the analog context

< < < ROCHESTER GAS & ELECTRIC > > >
– R. E. GINNA STATION –

WE FOR SHPT
9/26/83 10:57:10

SAMPLE COUNTING DATE	6/17/83	3:34:51
SAMPLE COLLECTION DATE	6/17/83	2:39:57
SHIPMENT DATE	9/27/83	

SAMPLE ANALYZED BY		N. KIEDROWSKI
DECAY TIME		102 DAYS 8.419 HRS.
ACQ. LIVE TIME		600 SEC.
SAMPLE VOL.		10 CC
GALLONS DRUMMED		30
DETECTOR NUMBER	2	
LINESHAPE FILE NO.	2	#2 ENERGY CALIB 12/8/82
EFF. FILE NO.	3	50 ML (1) – 12/22/82

SOLID WASTE ISOTOPE

NUCLIDE	UCI/CC		DECAY CORR.			
ENERGY		ABUNDANCE (%)		RATE (CPM)	% ERR	DELTA–E
CR-51	3.14E-04 +/– 6.73E-05		0.08			
320.0	3.14E-04	9.90		44.2	21.4	–0.1
MN-54	7.74E-04 +/– 8.14E-05		0.80			
834.8	7.74E-04	100.00		43.2	10.5	0.1

TOTAL ACTIVITY 2.60E-01 UCI/CC

ISOTOPE	MCURIES	% TOTAL ACTIVITY
CR-51	3.57E-02	0.12
MN-54	8.79E-02	0.29
CO-58	5.72E-01	1.93
FE-59	8.69E-02	
CO-60	2.79E+00	9.44
ZR-95	2.48E-02	0.08

FIG. 4.18 Printout in digital format.

discussed above. Control functions can also be performed by programmable logic controllers, data acquisition devices, and minicomputers; in the powerplant, however, the principal applications of these categories of instrumentation are other than for control.

The input and output ends of the control system have changed very little from those provided by analog equipment. Transmitters are beginning to appear with digital outputs that can be transmitted over highway networks, but they are not yet a standard device in powerplants. Most digital devices can accept low-level signals, and some can accept inputs in the form of high-frequency pulses, so less external input conditioning is required. The same regulating devices are still in service, and the only additional equipment required will be for converting 4- to

20-mA control signals to pneumatic ones, when a plant has replaced pneumatic controls with solid-state electronic hardware.

DISTRIBUTED CONTROL

Distributed control systems today are controlling boilers, scrubbers, and other pollution abatement equipment. The systems are frequently installed in retrofits as replacements for pneumatic and electronic analog instruments.

Two Characteristics

The adjective *distributed* describes two different characteristics of digital control instrumentation. First, the functions of control directly associated with the process being controlled are distributed about the operating facility, but in close proximity to the process areas. Signals from sensors and transmitters come to special-purpose computers, where the information is processed and stored. Control outputs and outputs to regulators are generated; alarm limits are monitored and alarms are generated, as necessary. All information about the process, as well as status information about the electronics at each remote location, can be sent over data highways to a central operating area for operator observation and action.

Second, the computing functions of the electronics system are distributed. This type of digital control can be, and has been, done by a mainframe computer. The mainframe can be programmed to access data, condition them, and manipulate them in almost any conceivable fashion. Putting all this responsibility into the software of a single machine, however, has always been a source of concern to management, mainly because of the threat to reliability. Also, preparing the programming to match the varied requirements of individual plants and installations is costly in time and money. Thus, mainframe computers have not become an accepted part of digital control, either in processing plants or in powerplants.

In distributed control, this responsibility is allocated among the special-purpose computers at remote locations. The burden of programming and the size of memory to support it are also distributed. Overall operating procedures, record keeping, and information distribution functions are assigned to other special-purpose computers in a central operating area. A minicomputer or a plant mainframe computer can also be included through data highway connections. Thus, the programmed functions are distributed among many devices, making the best use of the capabilities of each.

Information Flow

Information flow through a distributed control system is illustrated in Fig. 4.19. It begins at the input plugs or terminals of the remote units. Here, inputs are scanned and conditioned, and the digital equivalents stored in memory. After the inputs are processed, control signals are sent through output ports and out to the processing area to operate regulating devices.

Control action is on a timeshared basis, with the program cycling through a

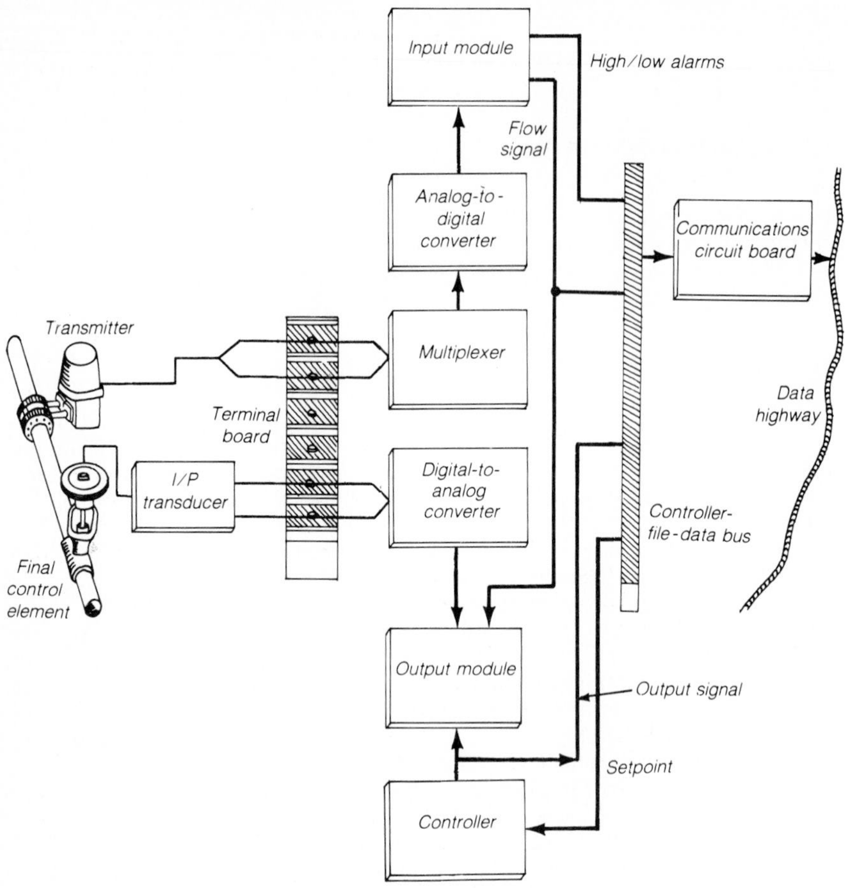

FIG. 4.19 Digital control system.

series of control loops. As a control loop is serviced, functions stored as preprogrammed subroutines (or algorithms) and common to all loops are called by the main program for matching with specific input and output addresses. Performance of one function may consume 40 ms.

The heart of the distributed control system includes processor circuitry that manages the transfer of information over data highways from the various remote areas and the programming of the time slots at each remote location. It also supports: (1) data bases where information for many process variables may be trended and archived and (2) information flow to peripheral equipment, including *cathode-ray tube* (CRT) monitors, printers, and departmental minicomputers at the next level of supervisory control.

Man/Machine Interfaces

The next chapter describes in detail the displays and operator interfaces for this type of equipment. It should be read in conjunction with the following. There will be keyboards associated with each operator station. Here, the operator can call up displays for any group of loops she or he needs to observe or must interact with. The operator can address any specific loop and observe process variable values, setpoint, and output control values, using the CRT as a guide. He or she can change the operating mode from automatic to manual, change the setpoint, and change output values (in manual mode). The operator can also acknowledge alarms. In some systems, she or he can call up operating instructions that define action to be taken to counter the specific type of upset creating the alarm condition.

Conversely, the keyboard can be locked to prevent the operator from taking certain actions. Loop tuning, parameter changing, and programming can usually be done only by the instrument technician or someone with access to a key that permits selection of these activities. In some systems, a separate keyboard may be installed for performing these activities.

Programming

Programming the loop functions, sometimes called *configuring,* is typically done with a high-level, process-oriented language that simulates the filling in of the forms developed for engineering design. It is popularly described as "fill-in-the-blanks" programming. By viewing the CRT display on which the form appears and moving the cursor (a blinking symbol on the screen acting as a position prompt) from blank to blank, the addresses of inputs and outputs, constant values, and limit values are filled in.

At the same time, the functions to be used in the loop are called up from the algorithm library by code number and connected in a manner very similar to drawing a flowchart with SAMA symbols. That is, by identifying the source of the signals, the output of one algorithm can be directed to the input of the next one. The example in Fig. 4.20 illustrates the types of algorithms and how they might be connected (not physically but via software) to implement a fuel-firing system that combines waste fuel with a prime fuel for best energy management.

This type of programming is normally done off-line; that is, by switching from operating mode to configuring mode, the operator station may be removed from its highway connection, to act as a piece of equipment that is no longer part of the operating system. The remote units will continue to control and to process data, as they have been; only the operator interface will be missing. This illustrates a significant strength of distributed control: Even if the highway were to be cut, the control action of the remote units would continue.

Other types of programming may also be done during off-line operation. These include the creation of group and graphic displays on the CRT, the definition of highway communications and trending of process variables, and the selection of variables to be measured and displayed. Some systems have minicomputers built into their processing sections for sophisticated computations, logic operations, and report formatting. These are also programmed off-line with FORTRAN, BASIC, or specially developed languages specific to their application.

Single-Loop Controllers

Some suppliers of distributed control equipment share circuitry of control loops (one microprocessor may accommodate 8 to 30 loops), while others, expressing

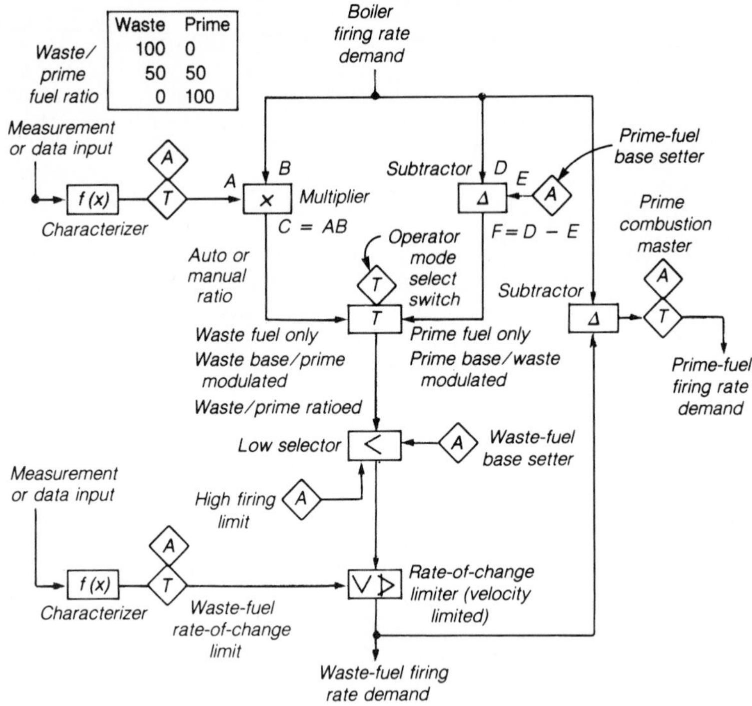

FIG. 4.20 Algorithms devised in programming of waste-fuel firing.

concern about single-loop integrity, specify a separate microprocessor for only one or two loops. A number of methods are available to increase the reliability of timeshared electronics:

- A duplicate set of electronics on ready standby, its data base constantly updated, to take the place of a controller file if it has a failure.
- Manual stations, displaying values of process variables and output signals, can be connected to allow the operator to shut down a process manually if the electronics becomes inoperable.
- Instead of going directly to a regulating circuit, the output signal can cascade into, or parallel, a dedicated single-loop controller, which can operate autonomously to control its loop if the operator so desires or if the shared control fails.

Single-loop controllers are also available with microprocessor-based circuits. In these controllers, the PID function is an algorithm stored on a chip. Some suppliers of digital single-loop controllers offer a large library of algorithms that can be combined by programming, so that a single housing can generate one or two outputs that are functions of many variables or computations. In one case, steam table functions can be created; an entire three-element drum level control can be programmed (Fig. 4.21); or a combustion control loop, including fuel-air ratio, cross-limiting, and oxygen trim, can be configured. Ports exist so that these devices can be combined on a data highway, with a CRT monitor and keyboard, or

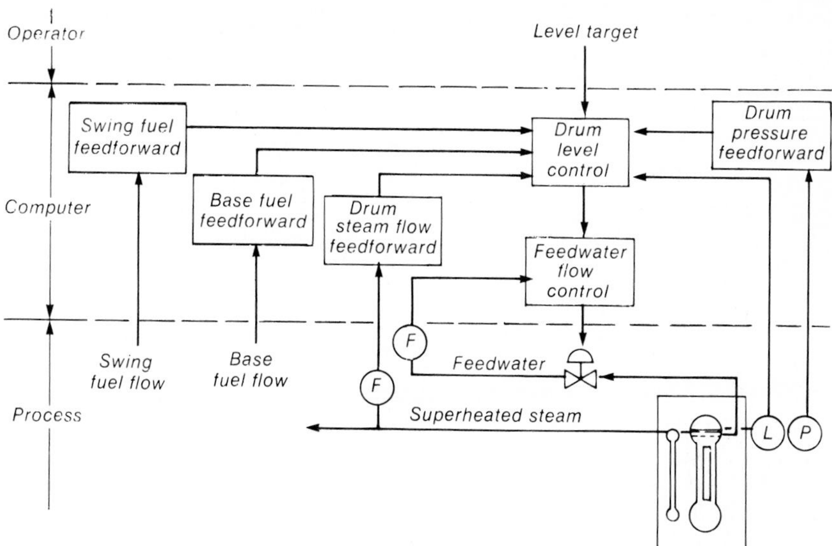

FIG. 4.21 Flow diagram for drum level control program.

even with peripheral computers. In this way, a distributed control system can be built by using single-loop controllers.

LOGIC OPERATIONS

Two types of logic operations easily handled by digital equipment must be implemented in powerplant instrumentation—batch, or sequential, operations and on/off control. Batch, or sequential, operations are a part of material flow control and occur in ash handling, coal handling, pollution control, water treatment, burner management, and interfacing between demand controls and electric switching systems. On/off control is required for every piece of electric operating equipment in the plant, from the largest induced-draft fan to the smallest solenoid shutoff valve. It is the type of control in effect when console-mounted switches and pushbuttons send discrete signals to switchgear circuit breakers and relays in motor control centers.

Programmable Logic Controller

As was mentioned earlier, logic functions have been handled by hard-wired relays, timers, and counters. The digital control device equivalent to these components is the *programmable logic controller* (PLC), which is becoming increasingly attractive for discrete control operations. The PLC has capabilities for PID control and computing functions, as well, and can implement many of the process control functions for which the distributed control system is more commonly used.

TABLE 4.2 PLC Specifications

Parameter	Value
Temperature, °F (°C)	
Operating	32 to 140 (0 to 60)
Storage	−40 to 140 (−40 to 60)
Humidity, relative, percent, noncondensing	5 to 95
Vibration, Hz	10 to 27 at 1.3 g 33 to 100 at 2.0 g
Memory, Kbytes	
Small	2
Medium	4 to 7
Large	Up to 240

PLCs were developed originally for machine tool control, and they do not see widespread use in powerplants for process control. They are, however, admirably suited for applications that are sequential or on/off. Data highways can be supplied with PLCs; in many cases, PLCs can be incorporated into networks that already include distributed control and minicomputers.

The PLC has three principal parts: a power supply, *input/output* (I/O) terminations, and a processor; the latter contains the programmed electronics and memory. In small units, representing less than a dozen relays, these three components may be combined into one housing, but more commonly each will be packaged separately. PLCs are rugged, suitable for mounting in process areas. Table 4.2 shows typical PLC specifications.

Programming. PLCs are programmed off-line, from a keyboard with a small [12-in (305-mm) diagonal screen] CRT display to support a procedure that literally builds a circuit on the screen, simulating the type of ladder diagram found in electrical wiring schematics. Plant electricians are familiar with this type of drawing. Some PLC programming languages are structured to handle boolean logic expressions, which may be more familiar to college-trained technicians and engineers who work with the instrumentation.

The ladder diagram represents a series of elements by symbols shown on successive lines corresponding to the sequence of operations in the circuit. It is formalized in a system of drafting entitled *Joint Industry Conference Electrical Standards for Industrial Equipment* or, more simply, the *JIC standards*.

The combination of a start/stop momentary-contact pushbutton and hold-in relay, for example, may be represented by a ladder diagram (Fig. 4.22), where *R* is a relay whose contacts hold in the start circuit and energize a motor control center relay *M;* 202, 203, and 22 represent programming addresses; 0-75 is an output address.

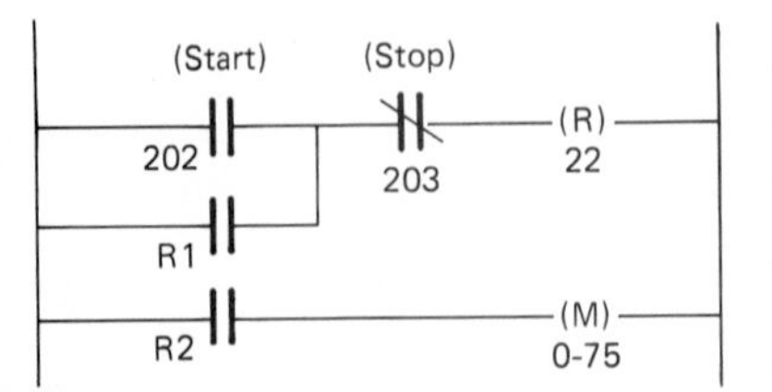

FIG. 4.22 Ladder diagram. *(Courtesy J. A. Moore.)*

Boolean logic expressions are derived from boolean algebra, a branch of mathematics in which formulas represent true statements corresponding to the flow and nonflow of signals. The boolean equivalent expressions for the start/stop circuit could be as follows: IF (202 and 203) OR (202 and 22), THEN 22; IF 22, THEN 0-75.

Scanning. During program execution, all the inputs associated with a processor are scanned, and the status (presence or absence of a voltage signal) is stored in memory. Computation is accomplished for each segment of the program (that is, each rung of the ladder) by pulling data from referenced memory addresses, processing them, and returning the values to their memory locations. The control logic dictated by the program is executed, and output values are computed and updated.

The PLC scan time is generally the time to execute the instructions in a program, one after the other. The published scan time may mean only the time it takes to scan the ladder logic, or it may mean the time for doing internal diagnostics, updating input and output values, and communicating with peripherals. Scan rates are fast, and the entire process may take only a few milliseconds; 2 to 5 ms/K of ladder logic is a commonly advertised figure. This is the basis for saying that PLCs can keep up with fast counter inputs and provide very precise control where a sequence of relay contact closures must be maintained precisely to avoid sneak circuits.

Alarms and Safety Circuits. Alarms and safety circuits triggered by relays can be handled by PLCs. An outstanding reason for using them is the reliability of the safety equipment itself. Consider the operation of standby units, for example. When a power line failure is detected, a microprocessor-based circuit starts a software-generated timing circuit. When the timer times out, the unit checks to see whether the power is still off and, if so, initiates the starting sequence of an auxiliary unit. With digital equipment, this can be done in the software by using a single device.

Solid-state design has also increased the flexibility of standard equipment, extending its application. Annunciators have been enhanced by microprocessor control in that they can be programmed to accomplish the first-out and latch-before-reset functions that required a choice of internal relays (and increased spare-parts stocking) in the hard-wired varieties.

The new safety factor introduced by digital control relates to the ability of digital equipment to announce its own faults, something that hard-wired equipment was not able to do. Digital equipment operates at such high speed that it is able to run continuous diagnostic checks on its own operations. It is common for every card in a system to incorporate some sort of timer circuit that will time out in milliseconds if conditions deviate from normal.

Computers

Modern systems technology is proposing hierarchies of digital instrumentation, putting the remote-located information gathering and processing equipment at the lowest level, the departmental central operating room at the next level, and, at an upper level, network interconnections between the central operations of all the departments of a plant and the mainframe computer that also handles payroll, inventory, purchasing, product management, and scheduling.

Minicomputer. Intermediate in this hierarchical scheme is the minicomputer as a peripheral for department operations. The minicomputer has the same basic parts as the special-purpose computer—processors, memory, mass storage of information, I/O communications—all interfaced to a high-speed, bidirectional bus. But, through programming, it can be adapted for many applications. Its service routines are powerful, its instruction set is large, and it is responsive to many external as well as internal influences through the use of interrupts. Also, the minicomputer can be *multitasking,* that is, able to perform more than one program at a time. This method of enhancing system efficiency takes advantage of the fact that a waiting time exists between the transfer of information and the processing of it.

Such a machine may be connected by a link to the central processing units of distributed control systems or PLCs, be constantly updated with process information, and simultaneously supply supervisory control settings to operating systems. Minicomputers no longer see service for direct digital control, but they can function as supervisory control devices. More important, the large memory and sophisticated bus structure of minicomputers allow them to be programmed for process modeling and optimizing, to analyze data statistically and detect developing conditions from trends in data, and to make management decisions about production and maintenance scheduling on the basis of real-time information they receive from process areas.

Personal Computer. Personal computers are becoming highly attractive as the central processing unit for networks of single-loop controllers, as peripheral machines for distributed control systems, and as programming devices for PLCs. The personal computer is an intermediately priced machine whose electronics can be packaged in a manner suitable for installation in process areas.

The personal computer has many of the attributes of the larger minicomputers, but it has less memory and slower speed (Fig. 4.23). Its operating system, preprogrammed in *read only memory* (ROM), reads and copies programs stored on floppy or hard disks into a working area of *random access memory* (RAM). It can, therefore, perform any task for which software exists on disk. Originally designed for applications at home, it has become a fixture in many offices because of the word processing, data base, and spreadsheet programs written for it.

Industrial departments are finding applications for these same types of programs. In addition, much software has been written to permit data acquisition, data manipulation, computation, and even automatic control. Personal computers are most useful for applications involving the collection and processing of large

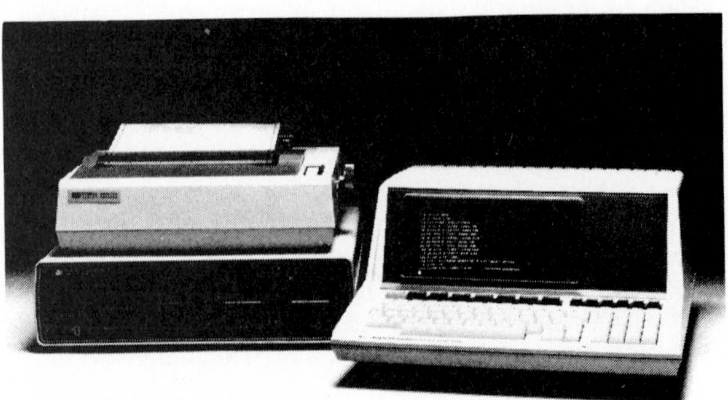

FIG. 4.23 Personal computer.

amounts of data, formatting of special reports, and storage of recipes involving a variety of parameter and instructional changes. Boiler efficiency testing is one application fitting this description.

At this time, the personal computer can do only one job at a time, although as 32-bit computers begin to appear on the market, multitasking is becoming closer to being possible. Until that happens, if a personal computer is monitoring a network of controllers and an engineer wants to use it to write a report, she or he can leave the controllers operating with their last set of operating parameters, remove the disk containing the controller program, replace it with a word processing program disk, and write the report. Control will continue, but the operator contact is lost. Innovative programming, however, has combined such tasks as controller operation, information gathering, and spreadsheet use, where the real-time values developed for the controller system can be made available to the spreadsheet computation program and be used for operation analysis.

ACKNOWLEDGMENT

Illustrations in this chapter are courtesy of *Power* magazine except where noted otherwise.

CHAPTER 6.5
THE MAN/MACHINE INTERFACE

J. A. Moore
Leeds & Northrup, A Unit of
General Signal Corp.
North Wales, PA

INTRODUCTION

The expression *man/machine interface* is a popular term that describes the relationship of an operator to the industrial process he or she is controlling. In the context of powerplant instrumentation, it refers (1) to the information about steam and power generation that the operator derives visually from instruments indicating process status and (2) to the instructions the operator is able to send manually to devices that regulate directly (regulators, motors) or indirectly (controllers) the activities of the process.

In the early days of instrumentation, visual indication came from an indicating device mounted on a process vessel or pipeline. It was observed by operators with their hands on a valve wheel that they turned to change the valve opening until process conditions suited them. The first indicators had bourdon-tube gages for pressure measurement and mercury-in-glass thermometers for temperature measurement.

As pneumatic transmitters and controllers became common, they allowed the interface to move out of the processing area and into the control room. Here, a number of indicating and controlling instruments could be brought together on a flat-surface panel so that the operator could see cause and effect for a number of variables at one time. Thermocouples and resistance thermometers were wired to indicators driven by galvanometers for temperature indication. Pressures and drafts were piped into the room directly, with manometers for indication.

Controllers, whether panel-mounted or in recorder cases, had pressure gages for indications of process-variable and control-output pressures. The operators could switch between manual and automatic control, adjust the controlled output, and tune the controller. Out-of-limit process conditions were indicated by a row of lights and a horn. A pushbutton was provided for silencing the horn when an alarm sounded.

This degree of transmission separated operators from the process by a few hundred feet. They worked at a combination panel and console. Generally, the controlling instruments were mounted on flat panels, at eye level. Rows of indicators might be mounted above the controllers or at the same elevation. Rotary switches, and pushbuttons for motor control, were mounted on a sloping apron or console, below the indicating and control devices (Fig. 5.1).

Shortly after the end of World War II, electronic instruments began to displace pneumatic devices. Because of electrical transmission, the man/machine interface moved still further from the process. Panels became more and more filled with indication and control instrumentation, and the operators no longer worked in the process area.

Control rooms built from 1945 to 1970 still contained analog-type instrumentation. A significant addition was the television screen. Screens monitoring flame conditions or boiler drum level might be mounted on a wall or panel. Plants which

FIG. 5.1 Control room circa 1950, featuring pneumatic instruments.

FIG. 5.2 Modern control room.

had computers for control, calculations, or data processing displayed process information on TV screens that might be table-mounted.

With the advent of microprocessors, large-scale integrated circuits, and data highways, the appearance of instrumentation in powerplant control rooms is changing radically (Fig. 5.2), as is the interface between the operator and the instruments. Much analog instrumentation is still in use, however, and will be for many years, so it deserves to be discussed here. Descriptions of each of the categories of instrumentation providing an interface follow, in the context of the foregoing time frame. Then the reasons for changing the interface are discussed, followed by a description of the latest types of interfaces.

INTERFACE INSTRUMENTS: 1945 TO 1970

During this quarter century, the man/machine interface blossomed. Such interface devices as recorders, indicators, annunciators, and controllers, many of them still in service, emerged as modern-day instruments capable of providing operators with the means for controlling their plants.

Recorders

Recorders provide information at the interface in the form of a graph, with the process-variable values displayed on one axis and time on the other. The graph is created by a pen trace, positioned as the process variable changes, on a chart moved at a constant rate. Real-time values are shown on a graduated scale by a pointer that moves with the pen across the chart. Pneumatic impulse clock motors are available for service in hazardous locations, but for powerplant applications 120-V, 60-Hz clock motors are the choice.

The graph on the chart provides a record of process values during past time, which is useful to the operator for making decisions in the context of recent past performance. These values are also useful to management, furnishing information about operating conditions and past performance at specific times. The operator's need for historical information is usually satisfied with a 4- to 6-h record. Management's needs may extend back a number of years, and so records are archived by chart storage.

To make the recording effective for the operator, the thickness, density, and clarity of the inked trace must be considered. They are functions not only of pen or print wheel design, but also of pen travel speed, chart speed, and printing rate. Pen development has progressed via such designs as bucket pens, ink capsules with capillary feed, ball-point pens, and felt-tip pens. The operator receives recorded process information at the interface from round-chart, strip-chart, and miniature recorders.

FIG. 5.3 Round-chart recorder.

Round-Chart Recorder. Records drawn on a 10- or 12-in (254- or 305-mm) diameter chart show a history of process-variable change as the chart moves past the pen (Fig. 5.3). The effective chart width is 4 or 5 in (102 or 127 mm). A hole in the center of the chart fits over a hub rotated by a clock motor. One revolution in 24 h is most common, although some instruments make 7-day records.

Process-variable change causes a mechanical motion of a helix or bellows, translated through a linkage to movement of an arm with a pen at the end across the chart. As the variable value changes, the pen arm pivots to move the pen across the chart. Chart ranges should be chosen so that normal operating values are between 60 and 70 percent of the distance from the center. This is a good rule for any recorder or indicator, but particularly for round-chart recorders because the space representing a time interval is much smaller at the inner edge of the chart than at the outer edge. This smaller interval compresses the pen traces for readings at the bottom of the chart to the point of illegibility.

Although round-chart recorders see less service today, they do have some popularity because their price is relatively low. Two or three records can be made on one chart, providing comparisons of related variables. High and low alarms can be obtained from contacts actuated by the pen arms.

Strip-Chart Recorders. The strip-chart record is made on a chart that unwinds from a 10-in (254-mm) wide roll, rewinding on a take-up roll (Fig. 5.4). The chart usually travels from the top of the case toward the bottom. The pen follows the horizontal axis, across the case. Thus, the trace is a rectilinear graph, and the time axis may show a continuous record of operation for several weeks of operation, depending on chart speed. The position of a pointer (moving with the pen) in relation to the gradations of a scale at the top of the chart provides a real-time indication of the measured variable. Because of the greater width of visible chart

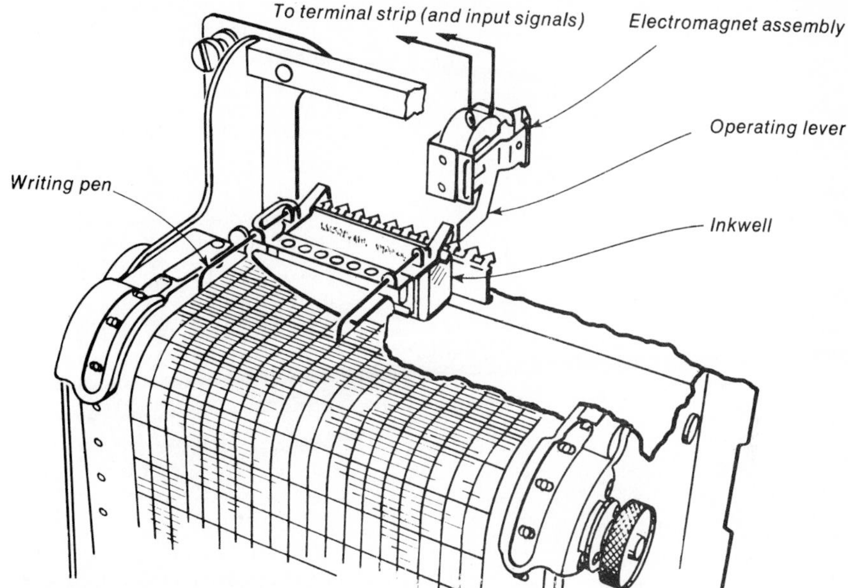

FIG. 5.4 Strip-chart recorder.

and uniform time-axis intervals, the large strip-chart record is about 10 times more readable than the round-chart record.

One, two, or three separate pens may be installed to make as many records. Strip charts may record as many as 24 points; for numbers larger than 3, a print wheel is normally selected. This impacts the chart, to imprint a dot and number. The combination of printing rate and chart travel must be compatible, so that the dots are close enough to preserve the continuity of the record. Unless the traces can be distributed across the chart so that each trace is distinct, the advantage of recording many points is reduced, since overprinting destroys legibility.

Large-case instruments usually allow a choice of pen and chart speeds. To prevent smearing, the rate of pen movement across the chart must correspond to chart speed. These are typical recommended speed relationships for pen travel time (in seconds) and chart speed in inches per hour (millimeters per hour): 10 s and 1 in/h (25 mm/h); 5 and 2 (51); 2.5 and 4 (102); 1 and 8 (204).

Extension housings added to the recorder case to hold electromechanical relays permit high and low alarms to be obtained from selected points of multipoint recorders.

Miniature Recorder. To reduce the panel space needed for recording, strip-chart and case sizes have shrunk to a point where 6-in (152-mm) high cases, with widths of 3 or 6 in (76 or 152 mm), can be panel-mounted in rows, side by side. This design is referred to, appropriately, as *miniature* (Fig. 5.5). The chart roll is only 4 in (102 mm) wide, and the pen moves across it. Pen movement is usually from top to bottom, simulating the pointer movement of a vertical-scale indicator.

There are three types of miniature recorders: (1) A 6-in (152-mm) wide case

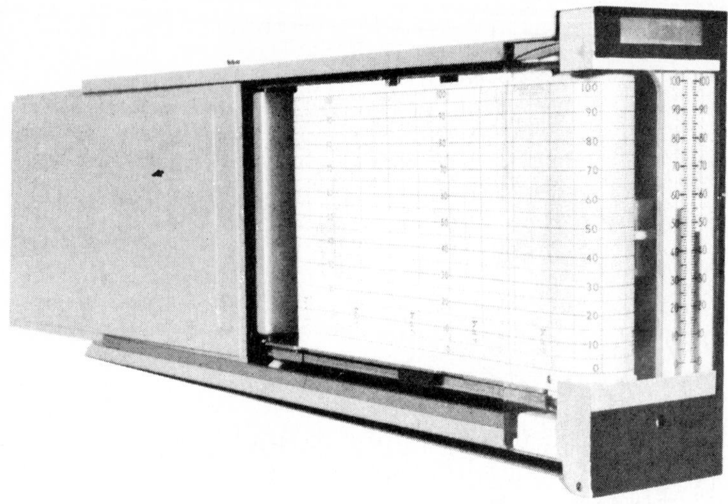

FIG. 5.5 Miniature recorder.

displays 6 h of recorded variable values, and previously recorded information can be examined by unrolling the completed portion. (2) A 3-in (76-mm) wide case exposes only the last 2 h of record and rerolls the rest in the rear of the chart mechanism for later viewing by withdrawing the chassis from the case. (3) A 6-in (152-mm) wide case uses folded paper, accordion style; 3 h of record is visible, and the previous traces can be examined by unfolding the used chart.

For all types, up to three pens can be installed. The real-time values appear on scales at the side of the chart. Gas-discharge tubes, or a pointer moving with the pen, provide the indication. Charts are long enough for a 1-month record.

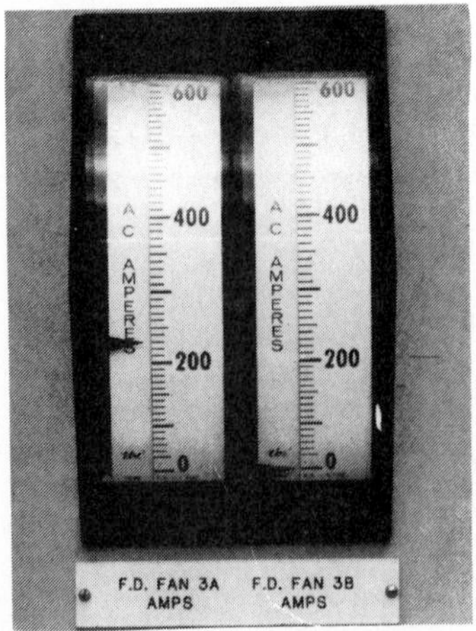

FIG. 5.6 Vertical-scale indicators mounted side by side.

Indicators

Display of present value is provided by an indicating element moved across a graduated scale as the measured process variable changes. Scales may be circular, horizontal, or vertical, but the side-by-side mounting of vertical-scale indicators is the popular panel arrangement (Fig. 5.6). The scale is usually read from left to right, bottom to top, or clockwise. The indicating element may be a pointer, the top of a column of liquid or illuminated gas, or a point of light. Very long-range indicators are constructed by moving an elongated scale past a fixed pointer. The scale may be imprinted onto the circumference of a circular drum, or on the surface of a flexible tape, with the unused portion reeled up. Most generally selected is a fixed vertical-scale, electric signal-actuated indicator. A popular size is 2¼ in (57 mm) wide by 6 in (152 mm) high, with a 4½-in (114-mm) long scale.

Indicator Internals. The movement of the pointer is provided by rotation of a coil of fine wire. The rotation is caused by passage of the signal current through the coil, which is suspended between jeweled bearings, or between thin strips of metal, taut in tension. The taut band suspension overcomes the friction that eventually builds up in bearings, so the observer does not have to tap the instrument to verify the reading. It is particularly useful for long-scale indicators and instruments subject to shock and vibration. Movements are the D'Arsonval galvanometer or moving-iron types. The latter has a fixed vane and an adjustable moving vane mounted inside a field coil. When the coil is energized, it magnetizes the vanes. The fixed vane exerts an upscale force, and the adjustable vane attracts the moving iron in the upscale direction.

Movement of the pointer is provided by rotation of a tightly stretched (taut band suspension) metal band. The accuracy of this indicator type is about 2 percent of full scale. A zero adjustment should be available so that the pointer position, when there is no input, can be set at the bottom gradation of the scale.

Draft Indicators. Draft indicators are specific to powerplant instrumentation. They are larger than the panel-mounted type described above and have the pointer at the end of a long arm. The pointer moves in an arc over the face of a curved scale about 8½ in (216 mm) long. Mounted in a case 4 in (102 mm) wide and 10 in (254 mm) high and illuminated, the indicator can be read from across the control room. It is customary to mount a bank of draft indicators above the panel, where the operator can see them from a distance by looking up. Actuating elements may be a bellows, bourdon tube, or diaphragm.

Pressure ranges are available from 0 to 5 psi (0 to 34.5 kPa) through 0 to 30 psi (0 to 207 kPa); vacuum ranges from −10 psi to 0 (−68.9 kPa to 0) through −30 psi to 0 (−207 kPa to 0); draft ranges from −5 to +5 in H_2O (−1.25 to +1.25 kPa) through −30 to +15 in H_2O (−7.47 to +3.74 kPa). Accuracy is 1 percent of range.

Annunciators

The operator interface with an industrial annunciator is a central collection point for surveillance of off-limit conditions. Panel-mounted annunciators may have as many as six rows of points, with as many as eight points in each row (Fig. 5.7). A change of condition is announced by an audible alarm and a visible backlighted nameplate identifying the specific trouble source. The abnormal condition is acknowledged when the audible alarm is silenced by the operator, but some degree of illumination of the nameplate will continue until the condition returns to its normal state.

FIG. 5.7 Control room annunciator.

When the process fault is corrected, the lighting of the annunciator window may be removed automatically, or it may have to be reset by the operator, depending on the alarm sequence. A number of alarm sequences are specified by the Instrument Society of America's *Recommended Practices*

RP-18.1. The two most common are the flashing and first-out sequences, shown in Table 5.1.

Counters

Six- or eight-digit counters, with digits the size and shape of an automobile speedometer, may be mounted on the panel to provide the operator information about accumulated fuel consumption and product generation. Counting is usually the result of a remotely generated electric impulse. Six-digit counters may be advanced mechanically, but eight-digit counters usually rely on electronic circuits and indication. Counters may total to a predetermined value, preset on a second, adjustable, row of digits. At the completion of the count, alarm contacts may close and the count reset to a starting value automatically. Alternatively, it may be reset by the operator's depressing a reset pushbutton.

Controllers

The recorder, indicator, counter, and annunciator allow the operator to observe process conditions. So does the front plate of a controller in that it displays the process variable whose value is to be controlled, the setpoint (or reference value to which it is controlled), and the value of the output signal. The controller also provides a means for the operator to change process conditions. She or he can change the reference value, select automatic or manual modes of operation, and in the manual mode change output values.

Tuning adjustments—gain, reset, and rate—are available to the operator with pneumatic controllers but not electronic controllers whose split-architecture design, common in powerplant installations, no longer has these settings at the interface. Split architecture separates the front plate from the rest of the instru-

TABLE 5.1 Alarm Sequences

Process-variable condition	Annunciator condition	Light	Horn
	Flashing		
Normal	Normal	Off	Off
Abnormal	Alert	Flashing	On
Abnormal	Acknowledged	On, steady	Off
Normal again	Normal	Off	Off
Normal	Test	On	Off
	First-out		
Normal	Normal	Off	Off
Abnormal	Alert		
	Initial	Flashing	On
	After	On	Off
Abnormal	Acknowledged		
	Initial	On	Off
	After	On	Off
Normal again	Normal	Off	Off
Normal	Test	On	Off

ment, putting the electronics and process connections in another housing or relay rack. The front plate becomes in effect a manual/automatic station, and tuning adjustments are the prerogative of the instrument technician.

Switches and Pushbuttons

The console portion of the instrument panel also contains (1) rotary switches for the selection and transfer of standby devices, meters, alternate circuits, and the establishment of power sources or permissive conditions (Fig. 5.8) and (2) pushbuttons for motor control and lights indicating operating status.

To make them easy to grasp and turn, rotary switches have knurled or pistol-grip handles. They may be locked in position, requiring pulling to turn, or may turn freely. Lateral contacts can be provided that operate when the handle is pulled out. They can be specified to make contact before breaking or to have the opposite action. There may be as many as eight switching positions with as many as 18 contacts. The switches are specified for specific contact operation as they are turned through their positions. Examples of special contact arrangements frequently found in powerplants are the power factor, wattmeter transfer, synchronizing, governor or rheostat motor control, circuit-breaker control, and voltmeter switching or transfer.

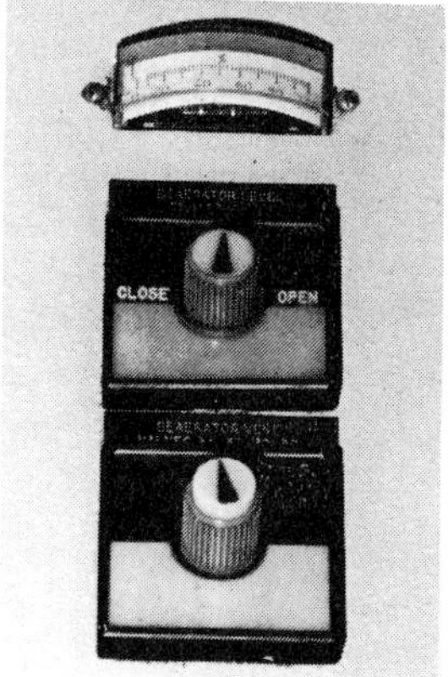

FIG. 5.8 Switches to control deaerator.

Pushbuttons and indicating lights are most often associated with motor control. Motors are seldom actuated directly. Usually they receive power through the contacts of relays located in a motor control center. The motor control relays are energized by pushbuttons on the operator's panel console. Indicating lights mounted in the pushbutton, or adjacent to it, show motor running status. By convention, a red light indicates that a motor is running, a green light that it is not.

Television Screens

Television cameras have been used for many years to create images of drum level gages and of burner-firing conditions. These have been displayed on conventional 19-in (483-mm) television screens mounted adjacent to the control panel. Information that the operator could otherwise obtain only by leaving the control room is thus made available at the operating interface. Toward the end of the period under discussion, other *cathode-ray tube* (CRT) displays appeared, bringing information from computers used to monitor (and in some cases operate) the process. Displays were alphanumeric.

This is one-way communication; the operator can ask for information but does not input commands. If a computer controls a process through conventional instrumentation such as supervisory control, an additional switch must be available on the panel-mounted *manual/automatic* (M/A) station. The operator actuates this switch to select computer control. An indicating light on the M/A station panel acknowledges that the computer is accessible and has taken over control.

OPERATOR INTERFACE: FURTHER EVOLUTION

Reduction of case size, combined with the potential for electric transmission to access large amounts of data from remote locations, has resulted in a concentration of instrumentation on large panels. Attending to them can be a full-time occupation for operators. A survey made by Honeywell Inc. in the early 1970s[1–3] reached two conclusions: (1) Conventional panel arrangements did not take advantage of the operators' natural reactions to the requirements of their duties, and (2) a completely different design offered significant opportunities for improvement.

Today's Philosophy

The basis for a new man/machine interface was introduced with Honeywell's TDC-2000 distributed control system and has become an industry standard for this type of instrumentation. It is summarized in the following list of observations and recommendations:

- Information comes to the operator in parallel, but she or he operates in a serial fashion.
- Information coming in parallel from a large number of displays is more than an operator can absorb. It would be better done on an exception basis—showing only what is changing from normal or what is out of limits.
- The operator can stand back and observe several instruments at one time, but cannot make adjustments or see exact values. He or she can stand at the panel, making adjustments and watching one value closely, but is unable to see related values on other instruments. Panel arrangements do not necessarily group related displays.
- Operators rely on pattern recognition rather than depending on explicit values.
- Analog displays show qualitative information, digital displays quantitative information. Making the best use of both by mixing them is recommended.
- Operators need trended information.
- The operator's period of greatest activity is during start-up, shutdown, and process upset. During these events she or he may need to control manually.

A strong recommendation was made that a single interface, similar to the cockpit of a large airplane, provide all associated functions at one location and within easy reach of a single operator. Microprocessor-based digital instrumentation, in conjunction with digital communications, has promoted such a man/machine interface, capitalizing on CRT displays. Keyboards allow operators to define the information they want to see, calling it over data highways from remote collecting locations, and provide the means for them to address the controlling devices with commands for manual operation.

CRT Displays

There are six types of displays in the digital-based man/machine interface. Their composition and use vary from supplier to supplier, but the following descriptions are considered to be representative of the majority commercially available.

Detail Display. Each individual function representing an instrument can appear in an individual full-screen display (Fig. 5.9). The current status of the function is shown in a manner similar to the conventional A/M station that the operator is familiar with. Digital values for process-variable input, setpoint, and output-control signal appear next to bar graph representations of the same values. Whether the function is in manual or automatic mode is displayed. The operator can change mode from a keyboard and can generate setpoint and output-signal values by keyboard manipulation.

The same display shows all the operating parameters specific to the function sources of inputs, tuning values, scaling constants, output rates and limits, and alarm limits. The interface does not extend permission to the operator to enter or change all this information. Normally, he or she will be restricted to the activities available from split-architecture instruments by a security code, that is, A/M selection, setpoint changes in automatic or manual, and output changes in manual only.

Group Display. When observing or manipulating process conditions, the operator works with the group display (Fig. 5.10). This is a collection of pictures of functions, each duplicating the bar graph representation appearing in the detail display for the same function. A common grouping of pictures, sometimes called

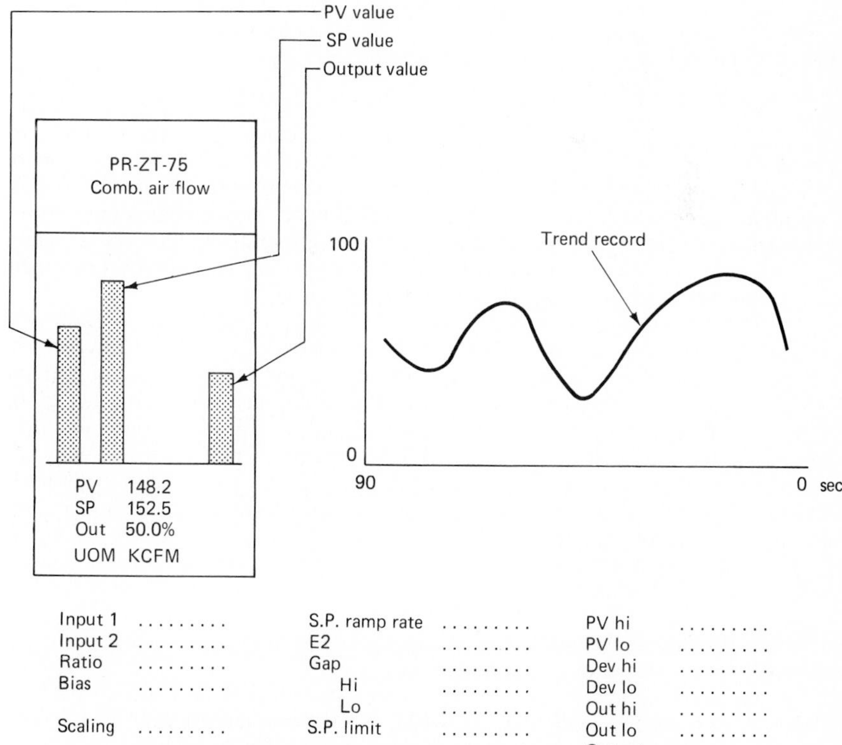

FIG. 5.9 Detail display with fill-in-the-blank programming. *(Courtesy J. A. Moore.)*

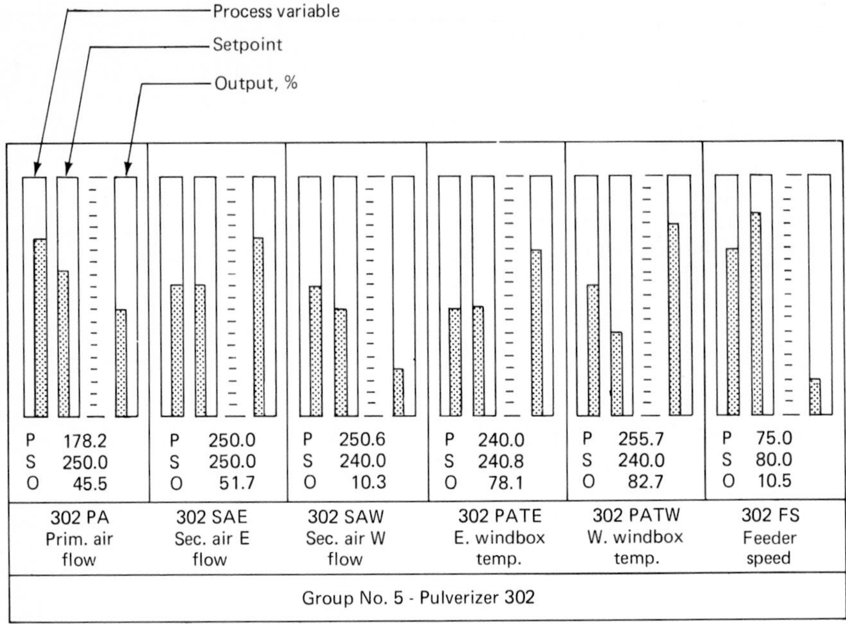

FIG. 5.10 Group display. *(Courtesy J. A. Moore.)*

templates, is in two rows of six or eight templates each. The selection of templates and their arrangement relative to each other are arbitrary and can be changed. This is in contrast to the fixed arrangement of instruments on the conventional panel, which the operator must accept even though complementary instruments are scattered throughout an array. The group display may be designed to show only those templates that function as a system, and in an arrangement that corresponds to the flow of information through them.

For example, one group in a display could include *proportional, integral, and derivative* (PID) controllers for fuel flow, air-fuel ratio, draft, and oxygen; the master controller for header pressure; comparators for cross-limiting of fuel and air; and indication of miscellaneous temperatures and pressures. Another group might be dedicated to three-element feedwater control instrumentation, and other groups to instrumentation associated with pulverizers. The same instrument representation can appear in several different groups, and an operator desiring to observe a specific group can call its display to the screen from the station keyboard. The number of groups that can be displayed is a function of the amount of memory reserved for them, but the capacity is always enough for the requirements of powerplant instrumentation.

Overview Display. Of equal importance with the group display, the overview display shows the status of the functions in a number of groups (Fig. 5.11). Each group is displayed on a portion of the screen; an overview may display 12 to 16 groups. The display is symbolic, showing a straight line representing normal conditions (process variable at setpoint, in the case of a PID controller), with deviation from setpoint indicated by a vertical line, which extends above or below the

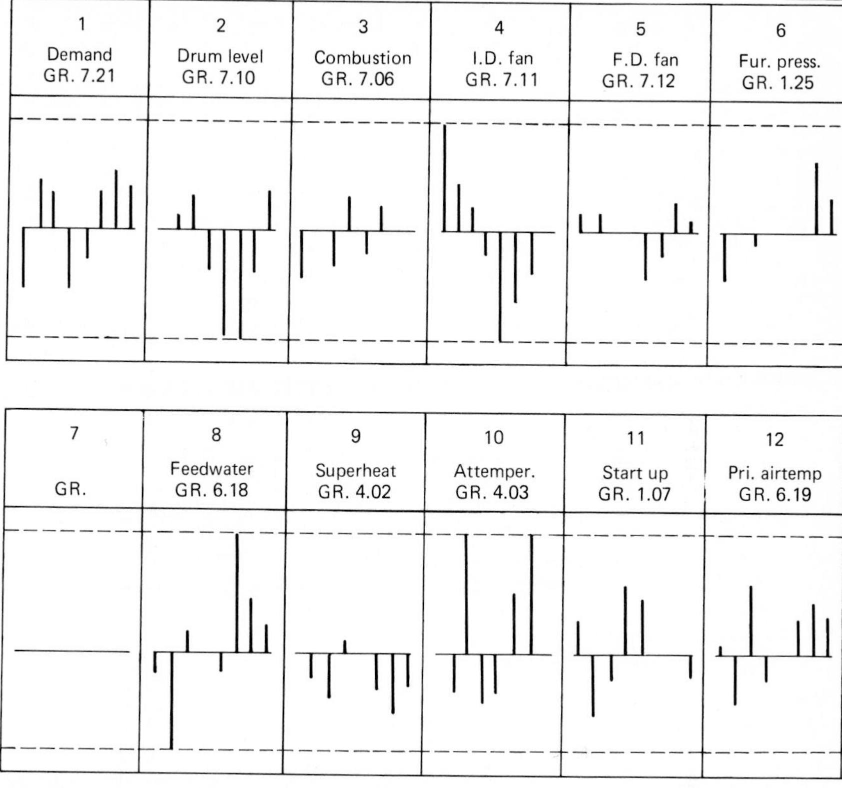

FIG. 5.11 Overview display. *(Courtesy J. A. Moore.)*

line of normality a distance proportional to the amount of deviation. Thus, the display is one of exception from normality and takes advantage of the operator's natural inclination to look for patterns.

The overview display will normally be applied in this fashion: Since the groups selected to make it up will probably represent conditions at a number of locations, the operator can scan the display to detect any abnormalities. An out-of-limit deviation will create an alarm condition, usually annunciated on the screen by a line or even an entire area on the screen changing color; also, an audible alarm may sound. Acknowledging the alarm will, in many cases, call up the display of the group in alarm, and the operator can then determine the function directly concerned with the problem, put it into manual mode and take over control, and compare the effects of changes with the status of the associated functions. If the operator wants to look at the details of the specific function to determine its configuration, she or he can do so with further keystrokes to bring up the detail display of the function in question.

Graphic Display. A graphic display is a two-dimensional picture composed on the CRT screen, usually in color (Fig. 5.12). Its most spectacular form simulates a flow sheet, showing a portion of a process—tanks, piping, motors, etc.—with

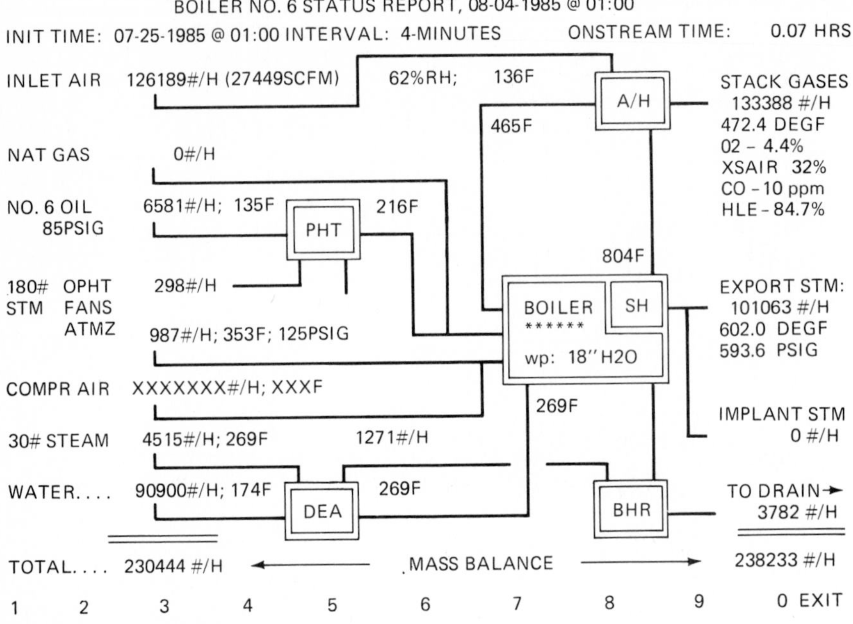

FIG. 5.12 Graphic display.

symbols for instrumentation positioned in relation to the equipment on which it is mounted. The picture can be equivalent to a group display, with the instrument symbols showing the same real-time information as the group display templates. The operator can communicate from the group display, selecting functions by moving a cursor to the desired spot on the screen and changing parameters from the keyboard. Alternatively, some systems allow selection of function by entering the tag name of the instrument from the keyboard.

Other types of graphic displays include bar graphs, annunciator windows, and trend records. However, the display of process equipment remains the most common application of this feature. It is a useful interface, not only for operating but also for training. The display can often be made dynamic, with color and shape changes activated by inputs from real-time process variables. The operator can observe valves opening and closing, pipes and tanks filling and emptying, and colors changing in accordance with temperature changes or process reaction progress (see box).

> CRT DISPLAY CRITERIA
>
> The effectiveness of the display for the operator's viewing is a function of three criteria: amount of information, color contrast, and resolution. With respect to the first, a picture must not be too "busy," or the operator will be distracted by the detail in it. A rule of thumb is that no more than 25 percent of the total area of the screen should be covered. A better rule for the group display is to restrict the number of loops in the group to four and to show equipment symbolically to establish relationships between instruments and process.

With color contrast, colors are formed on the screen by exciting a phosphor coating with electrons fired from three guns: one creates green images, the second red, and the third blue. From these three primary colors (or their absence), displays can appear in red, green, yellow, blue, magenta, cyan, black, and white.

In regard to resolution, the display is made by exciting areas of the screen called *pixels,* a contraction of the term *picture element* (see chart). A pixel is made up of a dot matrix, and the bits in a binary word describing the characteristics of each pixel define foreground color, background color, and blinking or steady display, among other parameters. The display is formed by scanning across rows, dropping down to the next row, and exciting each pixel in turn. The more pixels there are, the greater the definition that can be achieved. There is a trade-off, however, because the cost of generator picture memory, of faster scan rates, of screen size, and of response time increases.

A 1000 × 1000 matrix of rows and columns of pixels produces high resolution at high cost. Lines drawn on a screen with only a 500 × 500 matrix would be thick and jagged, and it would be impossible to create acceptable circles or sloping lines. A good resolution for process graphics is provided by a screen with 750 rows and 1000 pixels per row. The performance of the video ampliers that turn the electron beams on and off, the response time of a phosphor to acquire color and then to lose it, and the accuracy of convergence of the three primary-color electron beams are other qualities affecting resolution.

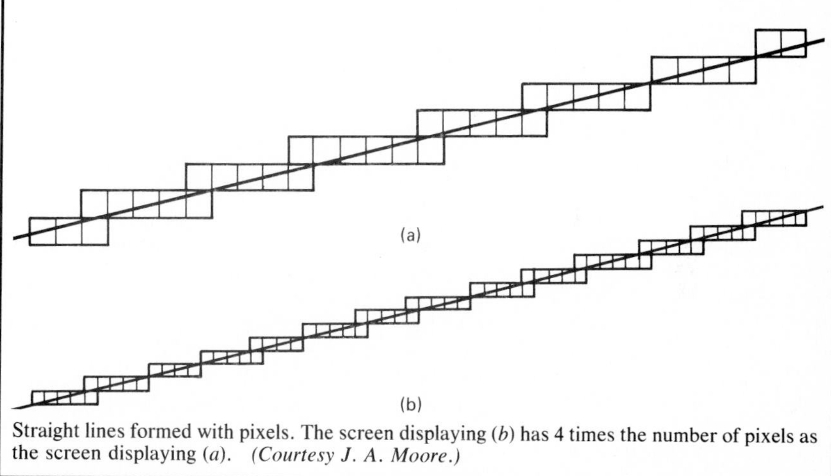

Straight lines formed with pixels. The screen displaying (*b*) has 4 times the number of pixels as the screen displaying (*a*). *(Courtesy J. A. Moore.)*

Trend Display. The continuous record of a variable, with time as the horizontal axis and the measured value as the vertical axis, is called a *trend display* (Fig. 5.13). Trend displays serve the same purpose as recorder charts because they can be saved on magnetic tape or disk and the information can be archived. It is customary to display up to four records of process variables at one time, sometimes in conjunction with an additional four records of discrete inputs. The discrete input record shows when contacts were closed, for example, or when a motor was running.

Time axes can be programmed so that a screen display represents any period from 90 s up to 7 days. Options include *zooming,* which is enlarging a portion of a record to increase the resolution. The application determines the optimum time period for display. For example, during start-up of a pressure control system

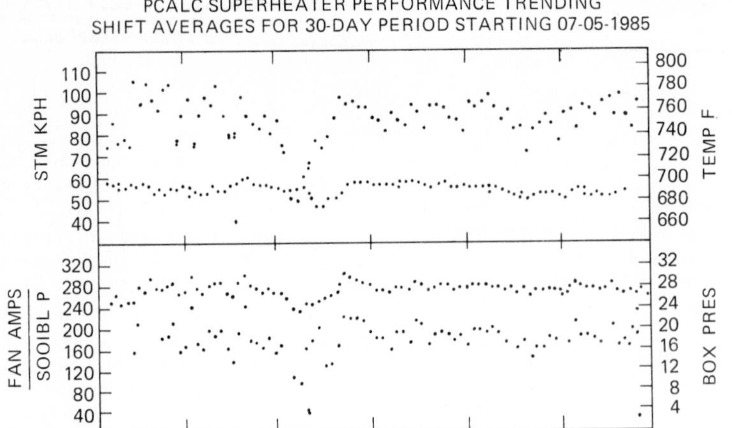

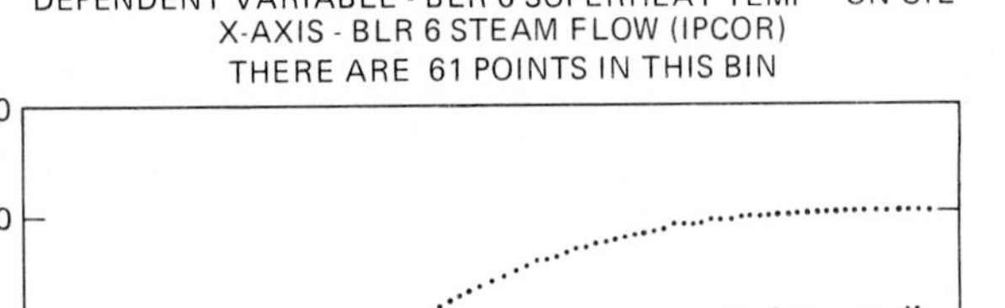

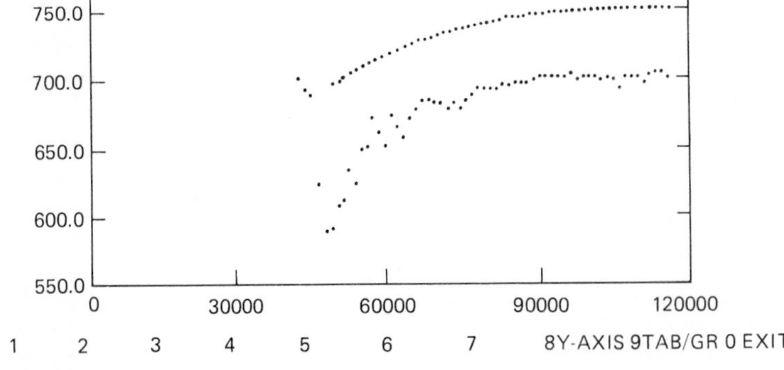

FIG. 5.13 Trend display.

when the loop is being tuned, a 90-s delay record provides a picture of pressure response to setpoint changes that is a great help in making parameter settings. If the response time is slow, as would be the case for a temperature control loop, a display time of 10 min might be more appropriate. Another useful option allows operation before and after an upset to be reviewed by recalling archival information; the interaction of several variables can be reviewed to determine primary causes of failure.

Alphanumeric Display. Lists and tables were the subject of computer-generated displays when video screens first appeared in control rooms, and displays are still viewed for that purpose. There are several important extensions to the use of alphanumeric characters in today's control rooms, however.

Menus. Here, the CRT screen interface serves one purpose at a time, but many variations are possible. One way to organize the programs that control various modes of operation is to store information in a data base, that is, in files that can be retrieved as their stored information is required. A list of the various modes of operation—for example, operating, configuring, utility access—serves as a menu of applications. In many systems, a keystroke entering the number of a menu item also programs a pointer to call the necessary files of information for the application desired.

Messages. Warnings of impending upsets, and directions for operator action to be taken in the event of specific process conditions, are displayed when limit alarms are triggered by external conditions. Shown in various colors, the screen display can incorporate a priority alerting the operator to the importance of the message.

Diagnostics. The digital system is constantly running background programs to check its own operating conditions. These checks (1) diagnose potential circuit problems or the location of actual failures and (2) indicate the condition on the interface with the operator so that repairs or replacement of printed-circuit cards can be made immediately, to minimize downtime of control equipment.

INTERFACE ENHANCEMENTS: CONVENTIONAL INSTRUMENTS

Strip-chart recorders, single-loop controllers, and other conventional control devices have not been eliminated by the invasion of digital control systems. On the contrary, they have been enhanced by such leading-edge devices as microprocessors, and their usefulness as man/machine interfaces has been increased accordingly. Enhancements to the devices previously discussed are reviewed now.

Strip-Chart Recorders

The printed scale has been replaced by a *light-emitting diode* (LED) large enough to be read by the operator from across the control room. The LED displays values for each point being printed in numbers that eliminate the necessity for interpolation of pointer position between lines on a scale. Stored memory permits the programming of each point through a small keyboard in the instrument chassis. The program accommodates the type of input source and provides input linearization, range selection, printing interval, and alarm limits.

A serial port permits real-time information, or stored values from previous times, to be transmitted over a data highway or by telephone line to a remotely located CRT display. This action has the effect of a double interface. For example, recorder values could be available at a water treatment facility for perusal by the local operator, but be available at the same time to the utility engineer at his or her office or laboratory.

Indicators

The most significant change in the appearance of indicators as an interface is that graduated scales have been replaced by digital readouts. In control rooms, this change occurred about the time digital watches first came on the market, although digital indication in meters and data acquisition readouts had been around for years.

A digital readout is created by LEDs and permits a value to be read directly instead of inferred from the position of a pointer over scale gradations. The numbers are large enough to be read from a distance. Operators prefer LED indication because of readability and credibility. A reading of 276.5, for example, will be accepted by an operator with confidence, even though the analog source on which it is based may be in error.

Integer digits from 0 to 9 are formed on a seven-segment display. Figure 5.14 shows one way that the bits of an 8-bit word can drive the LEDs in a display to form a specific number. Another binary word indexes an address to enable an individual group of segments to be activated, and the bit arrangement in the LED activating word identifies the segments that form the desired digit.

Controllers

Several changes have evolved at the interface provided by single-loop controllers. Selector switches, knobs, and pushbuttons have been replaced by a keyboard on the instrument front plate. Often, instead of depressible keys, a printed touch pad is the choice. Pressure on a printed key pushes a conductive plastic material into contact with a power source, and a circuit is completed that identifies the key function. Printing permits flexibility in keyboard design, and the typical touch pad is rated for several million operations.

Adjustments of output, setpoint, and tuning may be made by touching two key

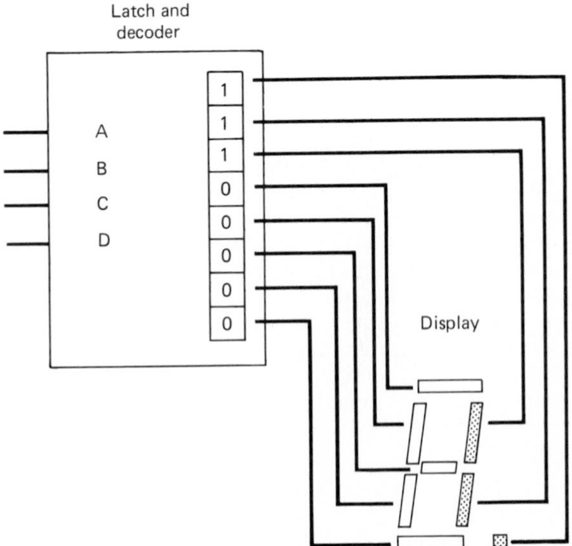

FIG. 5.14 LED activation in seven-segment display. Bits sent by a signal decoder activate LEDs to display "1." (shaded area). *(Courtesy J. A. Moore.)*

pads (Fig. 5.15). Identified by arrows or a pointing-finger symbol, one key will drive the value up, the other down. Symbols are used to identify key functions as well as printed names, taking advantage of the operator's familiarity through pattern recognition.

LED indication of process variables, control output, setpoint, and tuning values is as described above. A single indicator may be shared, and the specific category of indication desired is called up by a coded keyboard entry. Alternatively, gas-discharge tube indicators have replaced vertical- and horizontal-scale indicators.

Controllers have become microprocessor-based digital control devices, and a PID circuit is now an algorithm on a chip. Some controllers include as many as 50 functions in an algorithm library, addressable from a hand-held programming device or from a separate keyboard built into the chassis of the instrument. By means of the keyboard, a single instrument can compute functions that cannot be measured directly. This sophistication increases the operator's productivity because it interfaces him or her to the indication of values that previously had to be inferred, such as Btu consumption, efficiencies, enthalpy, and ratios of fuels.

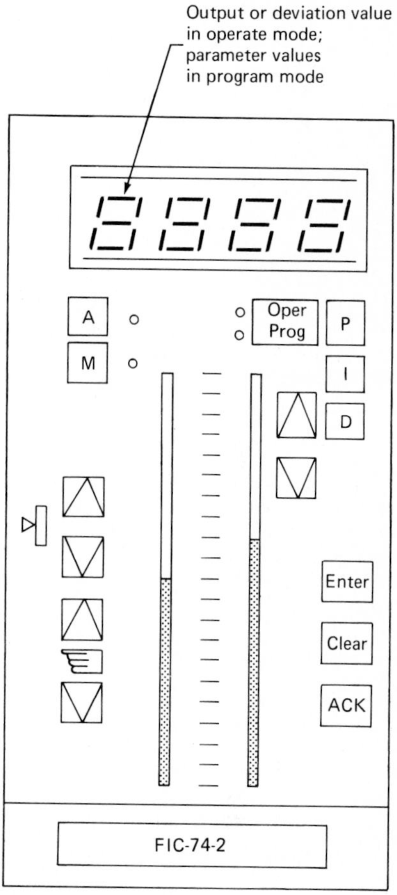

FIG. 5.15 Front plate for single-loop digital controller. *(Courtesy J. A. Moore.)*

Personal computers can also serve as an interface with the operator, as well as with management and engineering personnel, by report-formatting capabilities through word processor programs. Other personal computer programs permit combining process data with data bases for record keeping, calculations, and scheduling. CRT displays, for computers equipped with enhanced graphics adapter (EGA) cards, can simulate multipoint strip-chart recorders with a clarity and resolution that exceed the printed record of the real-time recorder.

Peripheral Equipment

Peripheral equipment communicating with microprocessor-based devices in the control room includes printers, keyboards, bulk storage in the form of disk drives, and video copiers. Of these, the keyboard must be discussed, because it is a direct interface between the operator and the process.

Keyboards. Keyboards are used by the operator in distributed control systems to call up displays, to enter setpoint and output values and make manual/automatic selection of the control mode, and to issue commands to the digital operating sys-

FIG. 5.16 Keyboard. *(Courtesy Loveland Controls Corp.)*

tem (Fig. 5.16). Since it may be desirable to restrict the operator's access to some of these activities, keyboards are often equipped with a lock that must be opened with a key to allow full keyboard operation. With operation locked out, the cursor disappears from the CRT screen and no entries may be made, although displays may be read. An added level of protection is provided by programming, which may be written to select specific actions that may not be performed, even with the keyboard unlocked.

Keyboard design simplifies operator use with special function keys which, through programming, allow the execution of a series of operations with a single keystroke. In addition to an alphanumeric selection of keys, some are provided for alarm acknowledgment, graphic display creation, and motor control. Printed touch pads are often included because they allow special design requirements to be met. When a printed touch-pad keyboard is designed, it is desirable to include (1) an audible signal indicating a key-switch closure to the operator and (2) some resistance to pressure to create a tactile feedback.

CRT Screens. CRT screen design is changing to improve the operator's interface to the process. The principal change is the touch screen, which allows the operator to point to the portion of a display that she or he needs to address, instead of moving a cursor through keyboard control. When the finger touches the screen, a matrix of infrared beams or of wires imbedded in the screen allows the vertical and horizontal axes crossing at the point of finger contact to indicate the location and initiate an appropriate action. Some designs average the points under finger contact to pinpoint locations as small as a single pixel.

ACKNOWLEDGMENT

Illustrations in this chapter are courtesy of *Power* magazine except where noted otherwise.

REFERENCES

1. Renzo Dallimonti, ''Future Operator Consoles for Improved Decision-Making and Safety,'' *Instrumentation Technology,* Aug. 1972, pp. 23–28.
2. Ibid., ''New Designs for Process Control Consoles,'' *Instrumentation Technology,* Aug. 1973, pp. 50–53.
3. Ibid., ''Human Factors in Control Center Design,'' *Instrumentation Technology,* May 1976, pp. 39–44.
4. Thomas B. Sheridan, ''Interface Needs for Coming Industrial Controls,'' *Instrumentation Technology,* Feb. 1979, pp. 37–39.
5. Renzo Dallimonti, ''Challenge for the 80s: Making Man-Machine Interfaces More Effective,'' *Control Engineering,* Jan. 1982, pp. 26–30.
6. Arthur K. McCready, ''Man-Machine Interfacing for the Process Industries,'' *Intech,* Mar. 1982, pp. 41–44.

Index

About the Editor

Thomas C. Elliott has over 30 years of experience as an editor of technical materials. Currently, he is an editor for *Power* magazine, where he has worked for 16 years. Prior to this, he was a publisher of technical books for the American Technical Society and handbook editor for the American Society of Heating, Refrigerating and Air-Conditioning Engineers. Mr. Elliott is a graduate of the Illinois Institute of Technology.